The geologic time scale

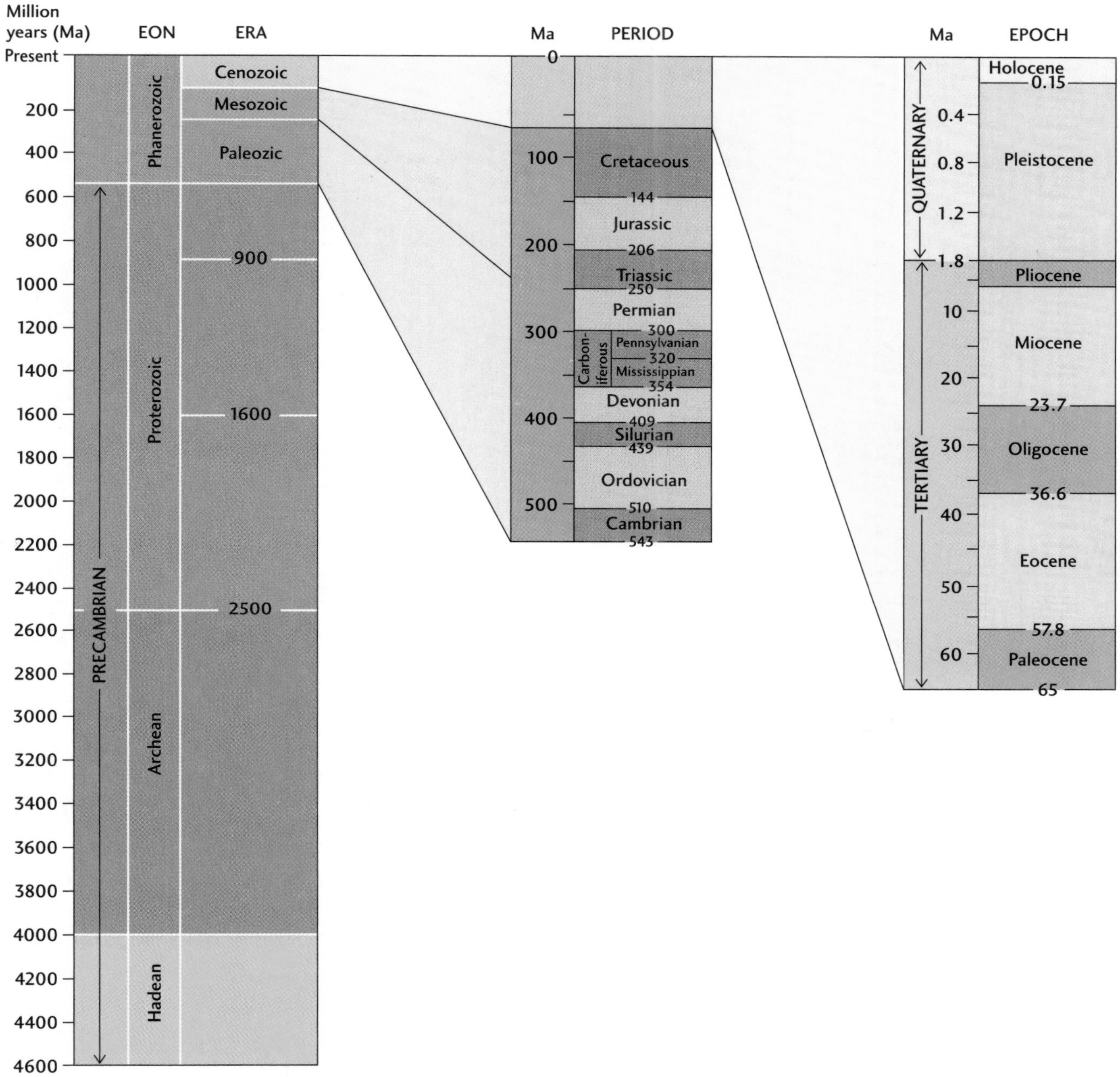

Million years (Ma): Present, 200, 400, 600, 800, 1000, 1200, 1400, 1600, 1800, 2000, 2200, 2400, 2600, 2800, 3000, 3200, 3400, 3600, 3800, 4000, 4200, 4400, 4600

EON	ERA
Phanerozoic	Cenozoic
	Mesozoic
	Paleozoic
Proterozoic	900
	1600
	2500
Archean	
Hadean	

PRECAMBRIAN

Ma: 0, 100, 200, 300, 400, 500

PERIOD	
Cretaceous	144
Jurassic	206
Triassic	250
Permian	300
Carboniferous: Pennsylvanian	320
Carboniferous: Mississippian	354
Devonian	409
Silurian	439
Ordovician	510
Cambrian	543

Ma: 0.4, 0.8, 1.2, 1.8, 10, 20, 30, 40, 50, 60

	EPOCH	
QUATERNARY	Holocene	0.15
	Pleistocene	1.8
TERTIARY	Pliocene	
	Miocene	23.7
	Oligocene	36.6
	Eocene	57.8
	Paleocene	65

The ribbon of geologic time, from the formation of the solar system to the present

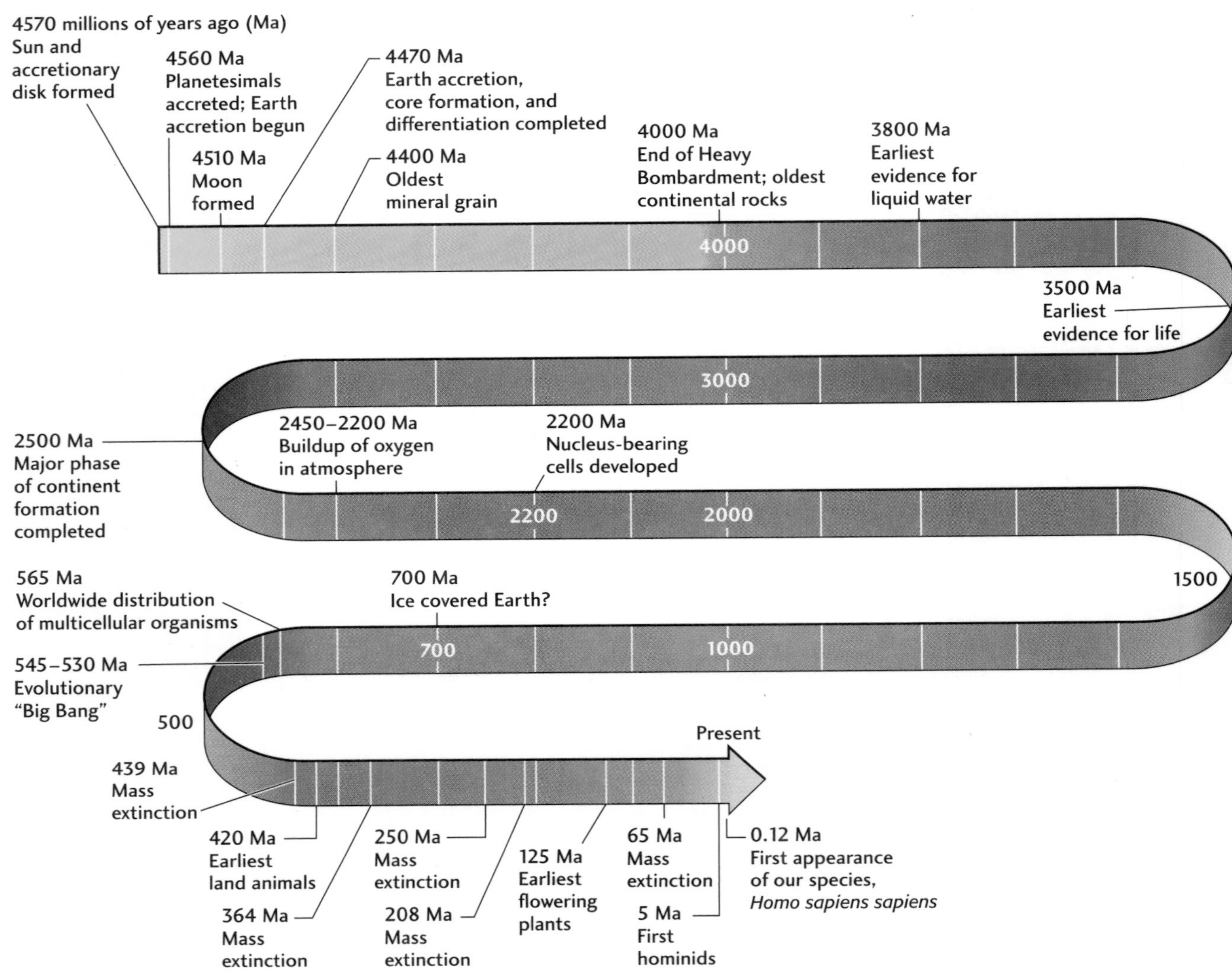

UNDERSTANDING EARTH

UNDERSTANDING EARTH

FOURTH EDITION

FRANK PRESS
The Washington Advisory Group

RAYMOND SIEVER
Harvard University

JOHN GROTZINGER
Massachusetts Institute of Technology

THOMAS H. JORDAN
University of Southern California

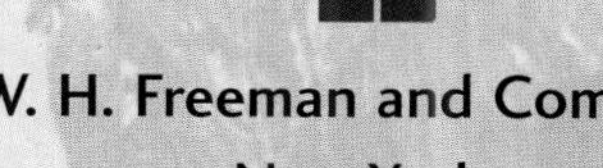

W. H. Freeman and Company
New York

PUBLISHER: **SUSAN FINNEMORE BRENNAN**

ACQUISITIONS EDITOR: **VALERIE RAYMOND**

DEVELOPMENT EDITOR: **RANDI ROSSIGNOL**

NEW MEDIA/SUPPLEMENTS EDITOR: **REBECCA PEARCE**

MARKETING MANAGER: **MARK SANTEE**

PROJECT EDITOR: **MARY LOUISE BYRD**

COVER AND TEXT DESIGNER: **VICTORIA TOMASELLI**

COVER PHOTOGRAPH: **ART WOLFE**

ILLUSTRATIONS: **JOHN WOOLSEY**

ILLUSTRATION COORDINATOR: **BILL PAGE**

PHOTO EDITOR: **MEG KUHTA**

PHOTO RESEARCHER: **ELYSE RIEDER**

PRODUCTION COORDINATOR: **JULIA DEROSA**

COMPOSITION: **SHERIDAN SELLERS, W. H. FREEMAN AND COMPANY, ELECTRONIC PUBLISHING CENTER**

PRINTING AND BINDING: **RR DONNELLEY & SONS COMPANY**

Library of Congress Cataloging-in-Publication Data
Press, Frank
Understanding earth / Frank Press . . . [et al.].—4th ed.
p. cm.
Includes bilbiographical references and index.
ISBN 0-7167-9617-1
1. Earth sciences. I. Title.

QE28.P9 2003 2003049090
550—dc21

Printed in the United States of America

First printing 2003

W. H. Freeman and Company
41 Madison Avenue
New York, NY 10010
Houndmills, Basingstoke RG21 6XS, England

www.whfreeman.com

TO OUR CHILDREN,
AND OUR CHILDREN'S CHILDREN:
MAY THEY LIVE IN HARMONY
WITH EARTH'S ENVIRONMENT.

MEET THE AUTHORS

Frank Press

Frank Press has made pioneering contributions to the fields of geophysics, oceanography, lunar and planetary sciences, and natural resource exploration. He was a member of the team that discovered the fundamental difference between oceanic and continental crust and built the instruments used in the research. Dr. Press was on the faculties of Columbia University, the California Institute of Technology (Caltech), and the Massachusetts Institute of Technology (MIT). In addition, he served as president of the U.S. National Academy of Sciences and as a senior fellow at the Department of Terrestrial Magnetism, Carnegie Institution of Washington. He is currently with The Washington Advisory Group. In 1993, Frank Press was awarded the Japan Prize by the emperor for his work in the Earth sciences.

Dr. Press has advised four presidents on scientific issues. Jimmy Carter appointed him Science Advisor to the President. Bill Clinton awarded him the National Medal of Science. Three times, *U.S. News & World Report* surveys named him the country's most influential scientist.

Raymond Siever

Raymond Siever is an internationally known expert in sedimentary petrology, geochemistry, and the evolution of oceans and the atmosphere. Dr. Siever is a long-time member of Harvard University's Department of Earth and Planetary Sciences, and he chaired the geology department for eight years. He was one of the first sedimentologists to apply the techniques of geochemistry to the study of sedimentary rocks, especially sandstones and cherts.

In addition to cowriting the popular geology text *Earth* with Frank Press, Dr. Siever wrote (with F. J. Pettijohn and Paul Potter) the classic textbook *Sand and Sandstone* (Springer-Verlag). Dr. Siever is a Fellow of the Geological Society of America and the American Academy of Arts and Sciences and has been honored with distinguished awards from the Society of Sedimentary Geology, the Geochemical Society, and the American Association of Petroleum Geologists.

John Grotzinger

John Grotzinger is a field geologist interested in the evolution of Earth's surficial environments and biosphere. His research addresses the chemical development of the early oceans and atmosphere, the environmental context of early animal evolution, and the geologic factors that regulate sedimentary basins. He has contributed to the basic geologic framework of a number of sedimentary basins and orogenic belts in northwestern Canada, northern Siberia, southern Africa, and the western United States. These field-mapping studies are the starting point for more topical laboratory-based studies involving geochemical, paleontological, and geochronological techniques. He received a B.S. in geoscience from Hobart College in 1979, an M.S. in geology from the University of Montana in 1981, and a Ph.D. in geology from Virginia Tech in 1985. He spent three years as a research scientist at the Lamont-Doherty Geological Observatory before joining the MIT faculty in 1988. From 1979 to 1990, he was engaged in regional mapping for the Geological Survey of Canada.

In 1998, Dr. Grotzinger was named the Waldemar Lindgren Distinguished Scholar at MIT and in 2000 became the Robert R. Shrock Professor of Earth and Planetary Sciences. In 1998, he was appointed director of MIT's Earth Resources Laboratory. He received the Presidential Young Investigator Award of the National Science Foundation in 1990, the Donath Medal of the Geological Society of America in 1992, and the Henno Martin Medal of the Geological Society of Namibia in 2001. He is a member of the American Academy of Arts and Sciences and the U.S. National Academy of Sciences.

Thomas H. Jordan

Tom Jordan is a geophysicist whose interests include the composition, dynamics, and evolution of the solid Earth. He has conducted research into the nature of plate tectonic return flow, the formation of a thickened tectosphere beneath the ancient continental cratons, and the question of mantle stratification. He has developed a number of seismological techniques for elucidating structural features in the Earth's interior that bear on these and other geodynamic problems. He has also worked on modeling plate motions, measuring neotectonic deformations in plate-boundary zones, quantifying various aspects of seafloor morphology, and characterizing large earthquakes. He received his Ph.D. in geophysics and applied mathematics at Caltech in 1972 and taught at Princeton University and the Scripps Institution of Oceanography before joining the MIT faculty as the Robert R. Shrock Professor of Earth and Planetary Sciences in 1984. He served as the head of MIT's Department of Earth, Atmospheric and Planetary Sciences for the decade 1988–1998. He recently moved from MIT to the University of Southern California (USC), where he is the W. M. Keck Professor of Geological Sciences and Director of the Southern California Earthquake Center.

Dr. Jordan received the James B. Macelwane Medal of the American Geophysical Union in 1983 and the George P. Woollard Award of the Geological Society of America in 1998. He is a member of the American Academy of Arts and Sciences, the U.S. National Academy of Sciences, and the American Philosophical Society.

BRIEF CONTENTS

CONTENTS

PREFACE

New Voices

It has been said that science is a history of superseded theories. New theories and innovative approaches to research and teaching are mostly the work of the next generation of scientist-authors. John Grotzinger of MIT and Tom Jordan of USC have joined the author team and will succeed Frank Press and Raymond Siever in future editions. We are lucky to partner with colleagues who share the philosophy and idealism represented in our book and bring a vision of the future to it as well. John's and Tom's influence is apparent in every chapter of the book and in its overall reorganization, most is evident in the prominent Earth systems approach and in the early coverage of plate tectonic theory.

A New Vision

When the first edition of *Earth* was published, the concept of plate tectonics was still new. For the first time, an all-encompassing theory could be used as a framework for learning about the immense forces at work in Earth's interior. Given this new paradigm, our strategy was to make the learning of Earth science as process-based as possible. This new picture of Earth as a dynamic, coherent system was central to *Earth* and to its successor, *Understanding Earth.*

Now, with *Understanding Earth,* Fourth Edition, we are taking another step forward. One might characterize it as an attempt to answer the question: what comes after plate tectonics? We present geology as a unified, process-based science with the power to convey global meaning to geologic features wherever they are found. To do so, we draw on powerful new laboratory and field tools and new theoretical approaches.

New technology such as GPS and continuous satellite monitoring of Earth from space allows us to view plates in motion, mountains being raised and eroded, crustal strain building up before an earthquake, global warming, glaciers retreating, sea level rising—all in almost real time. It is remarkable that we can now use earthquake waves to image the flow of the solid mantle hundreds and thousands of kilometers deep, revealing patterns of rising plumes and subducting plates. These new technologies also reveal startling new insights into links between climate and tectonics that have been poorly understood in the past, such as the possibility that the flow of metamorphic rocks through mountain belts may be strongly influenced by surface weathering patterns. The view of Earth as a system of interacting components subject to interference by humankind can no longer be called ideologically based opinion—it is backed by solid scientific evidence. The power of geology has never been greater. Geological science now informs the decisions of public policy leaders in government, industry, and community organizations.

Early Coverage of Plate Tectonics

Chapter 2, Plate Tectonics: The Unifying Theory, allows us to take full advantage of tectonic theory as a framework for discussing key geologic processes. Early coverage of the basic tenets of tectonic theory means that the theory can be invoked throughout the text, providing the big picture as well as the link connecting geologic phenomena. For instance, **Chapter 4** now presents metamorphism in terms of plate interactions, **Chapter 8** offers a new section on plate tectonics and sedimentary basins, and **Chapter 9** has a significantly

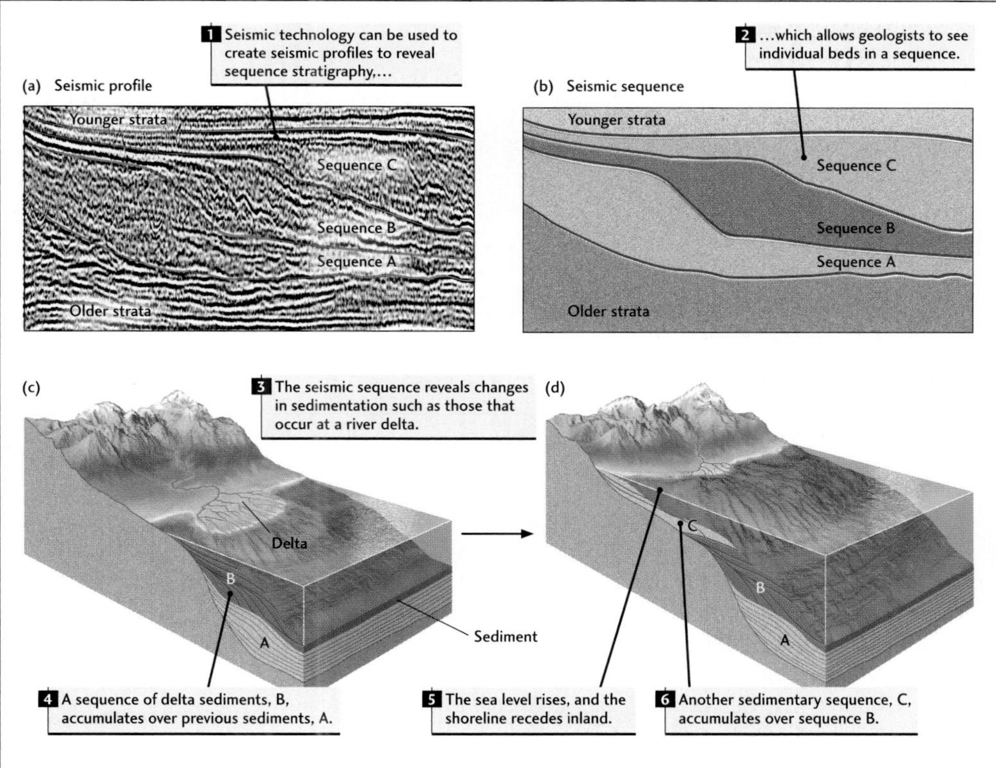

Figure 10.10 Comparison of seismic profiles (a) with seismic sequences (b), reveals the depositional process that creates bedding patterns. When tectonic subsidence or events such as global climate change have caused the sea level to rise, two deltaic sequences are found (c and d).

updated section on pressure-temperature-time paths and their significance for interpreting tectonic processes, including exhumation and uplift. The section of the book dedicated to surface processes is capped with a revised **Chapter 18**, in which landscape evolution integrates previous chapters and makes the case for significant interactions between climate and tectonics. This process-based treatment of a revitalized branch of Earth science is made possible only by having introduced plate tectonics early on.

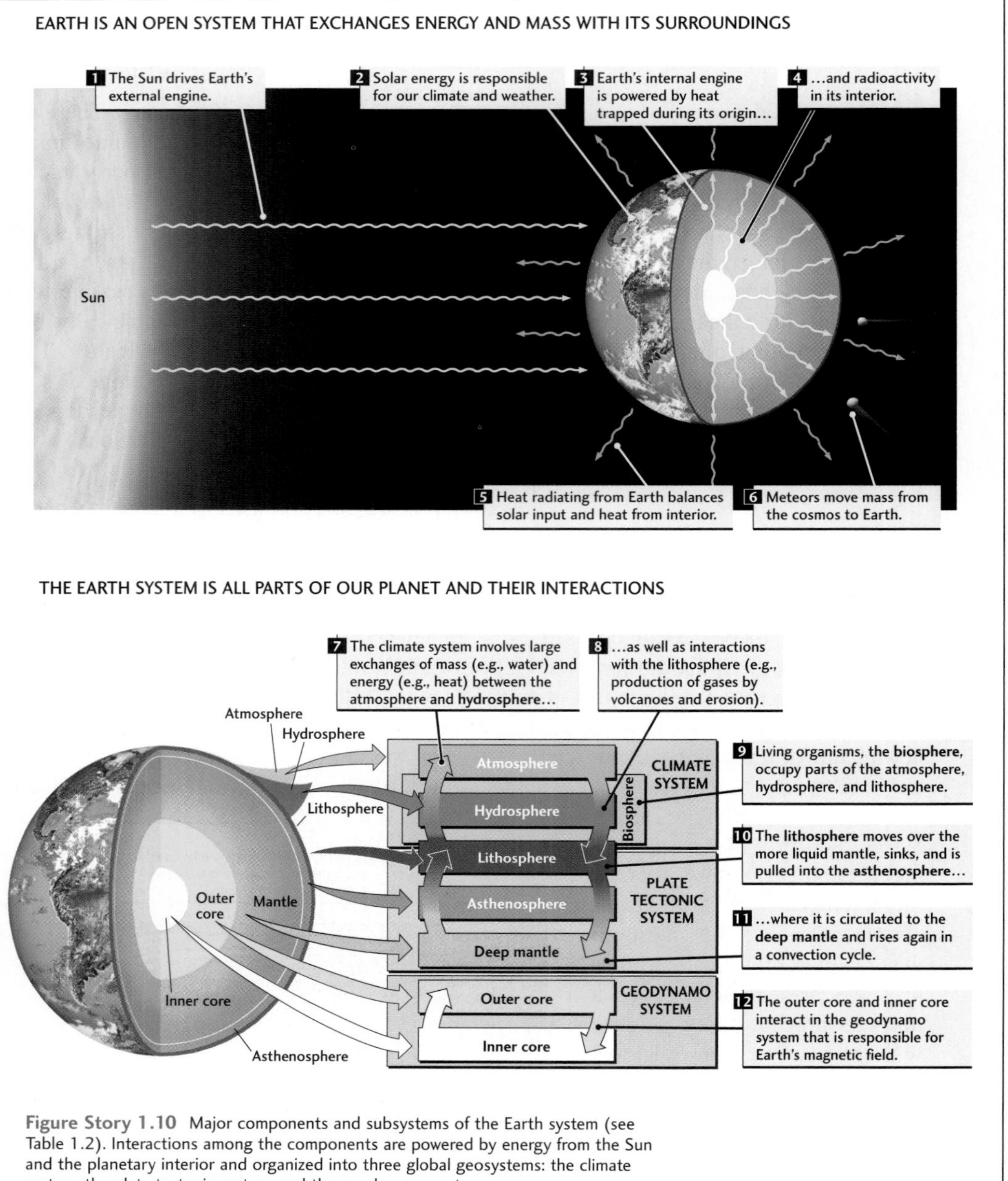

Figure Story 1.10 Major components and subsystems of the Earth system (see Table 1.2). Interactions among the components are powered by energy from the Sun and the planetary interior and organized into three global geosystems: the climate system, the plate tectonic system, and the geodynamo system.

Viewing Earth as a System

We begin with an expanded discussion of the Earth system in **Chapter 1**. The components of the Earth system are described, and the exchanges of energy and matter through the system are illustrated. This discussion serves as the springboard for the Earth systems perspective that pervades the text.

Chapter 5, Igneous Rocks: Solids from Melts, now includes a section entitled Spreading Centers as Magmatic Geosystems.

Volcanoes (**Chapter 6**) are discussed as geosystems, coupled to plate motions and interacting with the atmosphere, the oceans, and the biosphere.

The discussion of weathering in **Chapter 7** emphasizes the relationship between the climate geosystem and weathering. In **Chapter 9**, Metamorphism; **Chapter 18**, Landscape Evolution; and **Chapter 19**, Earthquakes, we stress the interactions among metamorphism, climate, plate tectonics, and the earthquake behavior of regional fault systems.

Chapter 21 delves into the convective engines of Earth's deep interior, which drive the plate tectonic and geodynamo systems.

Chapter 23 concludes with a discussion of how greenhouse gas emissions from fossil-fuel burning and other human activities may be changing Earth's climate system.

New Topics and Updates Throughout

- New material on exoplanets, early introduction of Earth system concepts, and a new section on Earth through geologic time (**Chapter 1**)
- A new section on spreading centers as magmatic geosystems (**Chapter 5**)
- A new section on volcanoes as geosystems, new material on large igneous provinces, and updated coverage of hot spots and the mantle plume hypothesis (**Chapter 6**)
- New sections on coral reefs and the evolutionary process and on plate tectonics and sedimentary basins (**Chapter 8**)

- Updated coverage of pressure-temperature paths (**Chapter 9**)
- Updated material on dome and basin formation (**Chapter 11**)
- New material on ice streaming, the instability of the West Antarctic ice sheet, and Snowball Earth (**Chapter 16**)
- Updated discussions of seafloor topography and methods for surveying the seafloor (**Chapter 17**)
- New sections on foreshocks and aftershocks, shaking intensity, plate boundaries and earthquakes, and regional fault systems (**Chapter 19**)
- An updated chapter on continental evolution, with a strong focus on North America that brings together recent insights about the history of mountain building and the formation of stable cratons (**Chapter 20**)
- An updated chapter on the deep interior, including new sections on mantle tomography, the geoid, and the geodynamo (**Chapter 21**)
- A completely revised chapter on Earth's environment and human impacts (**Chapter 23**)

Telling Stories with Words and Pictures

The most visible improvement in this new edition is the artwork. Our enduring goal to tell a story rather than provide aggregated facts is now apparent in the illustrations, particularly the new *Figure Stories*. Figure Stories bring photographs, line drawings, and text together to walk students through the major ideas behind important geologic processes.

Many more illustrations pair photographs and maps with schematics, so that students will see the context for the geologic phenomena as well as the underlying geologic features of what we can see with our eyes. Finally, much more descriptive text appears in the illustrations to help point students to their most important features.

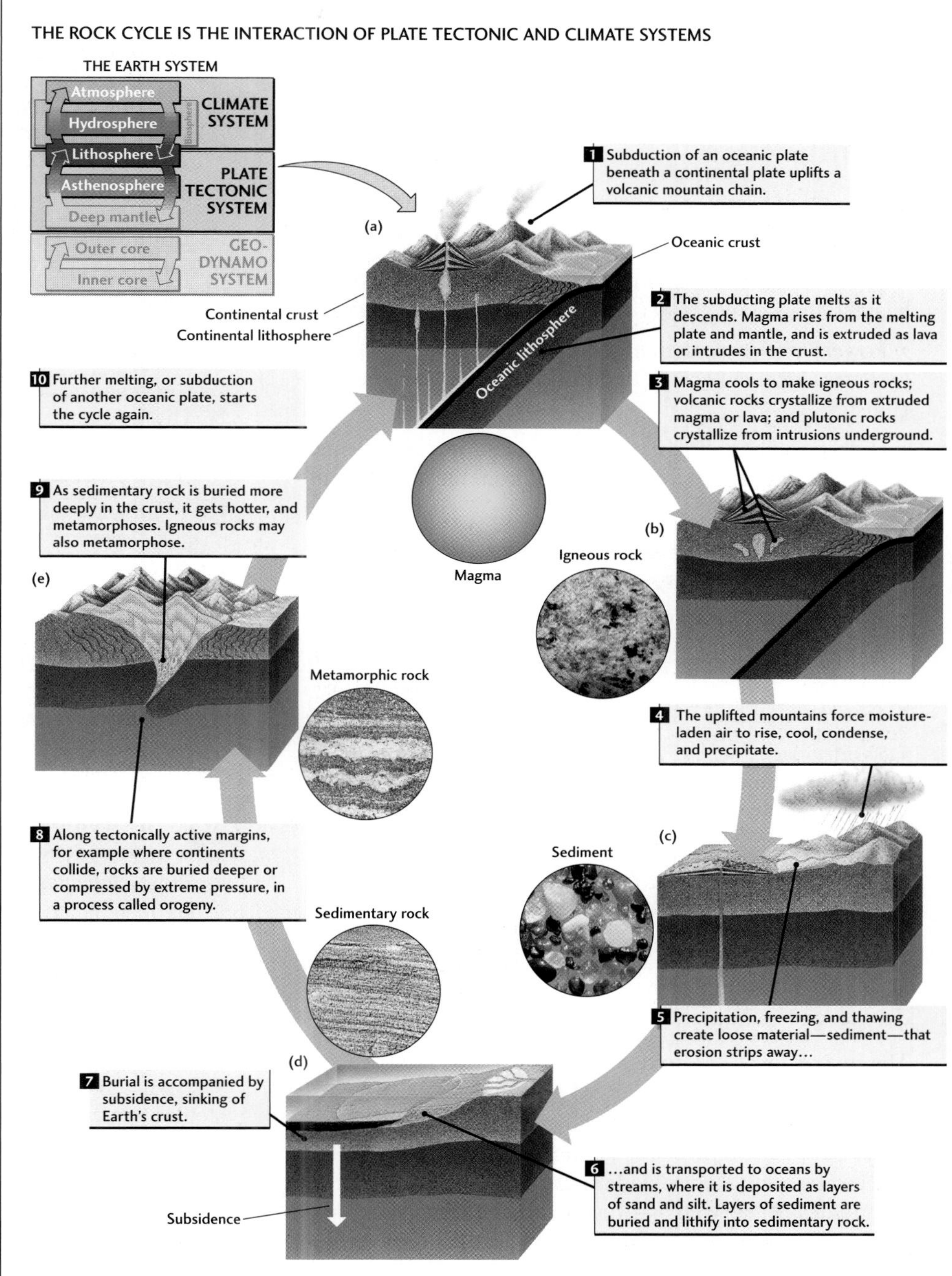

Figure Story 4.9 The rock cycle, as proposed by James Hutton more than 200 years ago. Rocks subjected to weathering and erosion form sediments, which are deposited, buried, and lithified. After deep burial, the rocks undergo metamorphism, melting, or both. Through orogeny and volcanic processes, rocks are uplifted, only to be recycled again. [*Igneous* (granite): J. Ramezani. *Metamorphic* (gneiss): Breck P. Kent. *Sedimentary* (sandstone): Breck P. Kent. *Sediment* (loose sand and gravel): Rex Elliott.]

Media and Supplements Package

A selection of electronic media and printed supplementary materials designed to support both instructors and students is available to users of this new edition of *Understanding Earth*. By focusing primarily on the importance of visualizing key concepts in geology, we are providing instructors with the presentation tools they need to help their students truly understand Earth's processes, and students with the study tools they need to study geology effectively and apply their newly acquired knowledge.

For Instructors

The **Instructor's Resource CD-ROM** (ISBN 0-7167-5782-6) contains

- High-resolution PowerPoint presentations, organized by chapter, that include *every* figure and table from the text
- PowerPoint presentations with Lecture Notes prepared by Peter Copeland and William Dupré of the University of Houston
- High-resolution JPEGs of *every* image in the text and from the Slide Set (including images from the slide sets of previous editions)
- Microsoft Word files of the *Test Bank* for easy editing and printing
- Adobe Acrobat files of the *Instructor's Manual*
- Animations, including over 40 *new* Macromedia Flash animations of the actual textbook figures that can be easily incorporated into PowerPoint presentations
- Short videos

The **Test Bank** (ISBN 0-7167-5784-2 [print] and ISBN 0-7167-5783-4 [CD-ROM]), written by Simon M. Peacock of Arizona State University and Sondra Peacock, includes approximately 50 multiple-choice questions for each chapter (over 1000 total), some of which incorporate illustrations from the text. The CD-ROM provides the *Test Bank* files in an electronic format that allows professors to edit, resequence, and add questions.

The **Instructor's Resource Manual** (ISBN 0-7167-5781-8), written by Peter Kresan and Reed Mencke of the University of Arizona, includes chapter-by-chapter sample lecture outlines, ideas for cooperative learning activities and exercises that can be easily copied and used as handouts and quizzes, and guides to the Web and Instructor's CD resources. The *Instructor's Manual* also includes an instructional design section that contains teaching tips from many instructors at the University of Arizona Learning Center. The *Instructor's Manual* is also available on both the Instructor's CD and the Companion Web Site.

The **Overhead Transparency Set** (ISBN 0-7167-5780-X) includes *every* textbook figure and table in full-color acetate transparencies.

The **Slide Set with Lecture Notes** (ISBN 0-7167-5779-6), prepared by Peter Kresan of the University of Arizona, contains approximately 100 additional images that are all fully annotated in the accompanying booklet of lecture notes.

The **Companion Web Site** at **www.whfreeman.com/understandingearth** provides access to all student materials on the Web site in addition to a password-protected

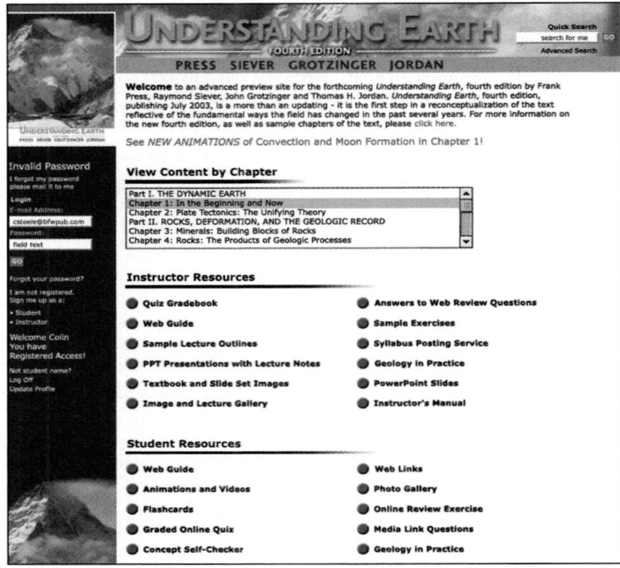

Instructor's site that contains all the PowerPoint presentations and JPEGs available on the Instructor's CD, the *Instructor's Manual,* and the Quiz Gradebook (which keeps track of students' Graded Online Quiz scores).

For Students

The **Companion Web Site*** at **www.whfreeman.com/understandingearth** includes many study tools that allow students to visualize geological processes and practice their newly acquired knowledge. The Companion Web Site contains

- Animations, including more than 40 *new* animated figures from the textbook
- Online Review Exercises, which include interactive exercises, virtual reality field trips, drag-and-drop exercises, and matching exercises
- Flashcards
- Online Quizzing
- Concept Self-Checker
- Geology in Practice exercises, inquiry-based learning activities that ask students to apply their newly acquired knowledge and think like geologists

*The Student Companion Web Site is also available on request as a CD-ROM. Please contact your W. H. Freeman Sales Representative for more details.

• Photo Gallery, additional photographs of geologic phenomena

• Current Events in Geology, an archive of geologically relevant articles from popular news sources, updated monthly

The **Student Study Guide** (ISBN 0-7167-5776-1)**,** written by Peter Kresan and Reed Mencke of the University of Arizona, includes tips on studying geology, chapter summaries, practice exams, and practice exercises that incorporate figures from the text and Web resources.

The **Lecture Notebook** (ISBN 0-7167-5778-8) is a workbook containing all the figures from the text in black and white with space for students to take notes.

The **EarthInquiry** series, developed by the **American Geological Institute** in collaboration with experienced geology instructors, is a collection of Web-based investigative activities that provides a direct way for students to explore and work with the vast amount of geological data now accessible via the Web. Covering such diverse topics as earthquakes and plate boundaries and the recurrence interval of floods, each EarthInquiry module asks students to analyze real-time data in order to develop a deeper understanding of fundamental geoscience concepts. Each module consists of a password-protected Web component and an accompanying workbook.

For more information about EarthInquiry, or to read about the various modules currently available, please visit: **www.whfreeman.com/earthinquiry.**

Acknowledgments

It is a challenge both to geology instructors and to authors of geology textbooks to compress the many important aspects of geology into a single course and to inspire interest and enthusiasm in their students. To meet this challenge, we have called on the advice of many colleagues who teach in all kinds of college and university settings. From the earliest planning stages of each edition of this book, we have relied on a consensus of views in designing an organization for the text and in choosing which topics to include. As we wrote and rewrote the chapters, we again relied on our colleagues to guide us in making the presentation pedagogically sound, accurate, and accessible and stimulating to students. To each one we are grateful.

The following instructors were involved in the planning or reviewing stages of this new edition:

Jeffrey M. Amato
New Mexico State University
Suzanne L. Baldwin
Syracuse University
Charles W. Barnes
Northern Arizona University
Carrie E. S. Bartek
University of North Carolina, Chapel Hill
Roger Bilham
University of Colorado
Michael D. Bradley
Eastern Michigan University
George R. Clark
Kansas State University
Mitchell Colgan
College of Charleston
Yildirim Dilek
Miami University
Craig Dietsch
University of Cincinnati
Grenville Draper
Florida International University
Pow-foong Fan
University of Hawaii
Mark D. Feigenson
Rutgers University
Katherine A. Giles
New Mexico State University
Michelle Goman
Rutgers University
Julian W. Green
University of South Carolina, Spartanburg
Jeff Greenberg
Wheaton College
David H. Griffing
University of North Carolina
Douglas W. Haywick
University of Southern Alabama
Michael Heaney III
Texas A&M University
Alisa Hylton
Central Piedmont Community College
James Kellogg
University of South Carolina at Columbia
David T. King, Jr.
Auburn University
Jeff Knott
California State University at Fullerton
Richard Law
Virginia Tech
Patricia D. Lee
University of Hawaii, Manoa
Laurie A. Leshin
Arizona State University
Kelly Liu
Kansas State University
J. Brian Mahoney
University of Wisconsin, Eau Claire
Bart S. Martin
Ohio Wesleyan University
Gale Martin
Community College of Southern Nevada
Robert D. Merrill
California State University, Fresno
James Mills
DePauw University
Henry Mullins
Syracuse University
John E. Mylroie
Mississippi State University
Stephen A. Nelson
Tulane University
William Parker
Florida State University
Philip Piccoli
University of Maryland
Loren A. Raymond
Appalachian State University
Mary Jo Richardson
Texas A&M University
Gary D. Rosenberg
Indiana University/Purdue University, Indianapolis
Malcolm Rutherford
University of Maryland
William E. Sanford
Colorado State University
Donald P. Schwert
North Dakota State University
Thomas Sharp
Arizona State University
Sam Swanson
University of Georgia, Athens
Jody Tinsley
Clemson University
James A. Tyburczy
Arizona State University
Michael A. Velbel
Michigan State University
Elisabeth Widom
Miami University

We remain indebted to the following instructors who helped shape earlier editions of *Understanding Earth:*

Wayne M. Ahr
Texas A & M University
Gary Allen
University of New Orleans
N. L. Archbold
Western Illinois University
Allen Archer
Kansas State University
Richard J. Arculus
University of Michigan, Ann Arbor
Philip M. Astwood
University of South Carolina
R. Scott Babcock
Western Washington University
Evelyn J. Baldwin
El Camino Community College
John M. Bartley
University of Utah
Lukas P. Baumgartner
University of Wisconsin, Madison
Richard J. Behl
California State University, Long Beach
Kathe Bertine
San Diego State University
David M. Best
Northern Arizona University
Dennis K. Bird
Stanford University
Stuart Birnbaum
University of Texas, San Antonio
David L. S. Blackwell
University of Oregon
Arthur L. Bloom
Cornell University
Phillip D. Boger
State University of New York, Geneseo
Stephen K. Boss
University of Arkansas
Robert L. Brenner
University of Iowa
David S. Brumbaugh
Northern Arizona University
Edward Buchwald
Carleton College
David Bucke
University of Vermont
Robert Burger
Smith College
Timothy Byrne
University of Connecticut
J. Allan Cain
University of Rhode Island
F. W. Cambray
Michigan State University
Ernest H. Carlson
Kent State University
Max F. Carman
University of Houston
Jams R. Carr
University of Nevada
L. Lynn Chyi
University of Akron
Allen Cichanski
Eastern Michigan University
G. S. Clark
University of Manitoba
Roger W. Cooper
Lamar University
Spencer Cotkin
University of Illinois
Peter Dahl
Kent State University
Jon Davidson
University of California, Los Angeles
Larry E. Davis
Washington State University
Robert T. Dodd
State University of New York, Stony Brook
Bruce J. Douglas
Indiana University
Carl N. Drummond
Indiana University/Purdue University, Fort Wayne
William M. Dunne
University of Tennessee, Knoxville
R. Lawrence Edwards
University of Minnesota
C. Patrick Ervin
Northern Illinois University
Stanley Fagerlin
Southwest Missouri State University
Jack D. Farmer
University of California, Los Angeles
Stanley C. Finney
California State University, Long Beach
Charlie Fitts
University of Southern Maine
Tim Flood
Saint Norbert College
Richard M. Fluegeman, Jr.
Ball State University
Michael F. Follo
Colby College
Richard L. Ford
Weaver State University
Nels F. Forsman
University of North Dakota
Charles Frank
Southern Illinois University
William J. Frazier
Columbus College
Robert B. Furlong
Wayne State University
Sharon L. Gabel
State University of New York, Oswego
Alexander E. Gates
Rutgers University
Dennis Geist
University of Idaho
Gary H. Girty
San Diego State University
William D. Gosnold
University of North Dakota
Richard H. Grant
University of New Brunswick
Jeffrey K. Greenberg
Wheaton College
Bryan Gregor
Wright State University
G. C. Grender
Virginia Polytechnic Institute and State University
Mickey E. Gunter
University of Idaho
David A. Gust
University of New Hampshire
Kermit M. Gustafson
Fresno City College
Bryce M. Hand
Syracuse University
Ronald A. Harris
West Virginia University
Richard Heimlich
Kent State University
Tom Henyey
University of Southern California
Eric Hetherington
University of Minnesota
J. Hatten Howard III
University of Georgia
Herbert J. Hudgens
Tarrant County Junior College
Warren D. Huff
University of Cincinnati
Ian Hutcheon
University of Calgary
Mohammad Z. Iqbal
University of Northern Iowa
Neil Johnson
Appalachian State University
Ruth Kalamarides
Northern Illinois University
Frank R. Karner
University of North Dakota
Alan Jay Kaufman
University of Maryland
Phillip Kehler
University of Arkansas, Little Rock
Cornelius Klein
Harvard University
Peter L. Kresan
University of Arizona
Albert M. Kudo
University of New Mexico
Robert Lawrence
Oregon State University
Don Layton
Cerritos College
Peter Leavens
University of Delaware
Barbara Leitner
University of Montevallo
John D. Longshore
Humboldt State University

Stephen J. Mackwell
Pennsylvania State University
Erwin Mantei
Southwest Missouri State University
Peter Martini
University of Guelph
G. David Mattison
Butte College
Florentin Maurrasse
Florida International University
George Maxey
University of North Texas
Joe Meert
Indiana State University
Lawrence D. Meinert
Washington State University, Pullman
Jonathan S. Miller
University of North Carolina
Kula C. Misra
University of Tennessee, Knoxville
Roger D. Morton
University of Alberta
Peter D. Muller
State University of New York, Oneonta
J. Nadeau
Rider University
Andrew Nyblade
Pennsylvania State University
Peggy A. O'Day
Arizona State University
Kieran O'Hara
University of Kentucky
William C. Parker
Florida State University
Simon M. Peacock
Arizona State University
E. Kirsten Peters
Washington State University, Pullman
Donald R. Prothero
Occidental College
Terrence M. Quinn
University of South Florida
C. Nicholas Raphael
Eastern Michigan University
J. H. Reynolds
West Carolina University
Robert W. Ridkey
University of Maryland
James Roche
Louisiana State University
William F. Ruddiman
University of Virginia
Charles K. Scharnberger
Millersville University
James Schmitt
Montana State University
Fred Schwab
Washington and Lee University
Jane Selverstone
University of New Mexico
Steven C. Semken
Navajo Community College
D. W. Shakel
Pima Community College
Charles R. Singler
Youngstown State University
David B. Slavsky
Loyola University of Chicago
Douglas L. Smith
University of Florida
Richard Smosma
West Virginia University
Donald K. Sprowl
University of Kansas
Steven M. Stanley
Johns Hopkins University
Don Steeples
University of Kansas
Randolph P. Steinen
University of Connecticut
Dorothy L. Stout
Cypress College
Bryan Tapp
University of Tulsa
John F. Taylor
Indiana University of Pennsylvania
Kenneth J. Terrell
Georgia State University
Thomas M. Tharp
Purdue University
Nicholas H. Tibbs
Southeast Missouri State University
Herbert Tischler
University of New Hampshire
Jan Tullis
Brown University
James A. Tyburczy
Arizona State University
Kenneth J. Van Dellen
Macomb Community College
J. M. Wampler
Georgia Tech
Donna Whitney
University of Minnesota
Elisabeth Widom
Miami University, Oxford
Rick Williams
University of Tennessee
Lorraine W. Wolf
Auburn University

Others have worked with us more directly in writing and preparing manuscript for publication. At our side always were the editors at W. H. Freeman and Company: Randi Rossignol and Valerie Raymond. Mary Louise Byrd supervised the process from final manuscript to printed text. Diana Siemens and Eleanor Wedge were our copyeditor and proofreader. Rebecca Pearce coordinated the media supplements. Vicki Tomaselli designed the text, and Meg Kuhta and Elyse Rieder edited and obtained the many beautiful photographs. We thank Sheridan Sellers, our compositor and layout artist, Julia DeRosa, our production manager, and Bill Page, our illustration coordinator. Special thanks to our illustrator, John Woolsey, for his outstanding ideas and drawings.

First image of the whole Earth showing the Antarctic and African continents, taken by the *Apollo 17* astronauts on December 7, 1972. [NASA.]

CHAPTER

1

Building a Planet

"I tell my wife that her fresh glass of water isn't so fresh. It has atoms in it that are 14 billion years old."

ASTRONOMER ANDY MCWILLIAM

Earth is a unique place, home to millions of organisms, including ourselves. No other planet we've yet discovered has the same delicate balance of conditions necessary to sustain life. *Geology* is the science that studies Earth: how it was born, how it evolved, how it works, and how we can help preserve its habitats for life. We have organized the discussion of geology in this book around three basic concepts that will appear in almost every chapter: (1) Earth as a system of interacting components, (2) plate tectonics as a unifying theory of geology, and (3) changes in the Earth system through geologic time. **This chapter gives a broad picture of how geologists think. It starts with the scientific method, the objective approach to the physical universe on which all scientific inquiry is based. Throughout this book, you will see the scientific method in action as you discover how geologists gather and interpret information about our planet. After this introduction, we will describe the most generally accepted scientific explanations for how Earth formed and why it continues to change.**

We will see that our planet works as a system of many interacting components beneath its solid surface and in its atmosphere and oceans. Many of these components—for example, the Los Angeles air basin, the Great Lakes, Hawaii's Mauna Loa volcano, and the Brazilian rain forest—are themselves complex subsystems. To understand the various parts of Earth, we often study its subsystems separately, as if each existed alone. To get a complete perspective on how Earth works, however, we must learn about the ways its subsystems interact with one another—how gases from volcanoes can trigger climate change, for example, or how living organisms can modify the atmosphere and, in turn, be affected by atmospheric changes.

We must also understand how the Earth system evolves through time. You may find as you read these pages that your idea of time will start to change. A geologist's view of time must accommodate spans so large that we sometimes have trouble comprehending them. Geologists estimate that Earth is about 4.5 billion years old. More than 3 billion years ago, living cells developed on Earth, but our human origins date back only a few million years—a mere few hundredths of a percent of

Earth's existence. The scales that measure individual lives in decades and mark off periods of written human history in hundreds or thousands of years are inadequate to study Earth. Geologists must explain features that evolve over tens of thousands, hundreds of thousands, or many millions of years.

The Scientific Method

The goal of all science is to explain how the universe works. The **scientific method,** on which all scientists rely, is a general research plan based on methodical observations and experiments (**Figure 1.1**). Scientists believe that physical events have physical explanations, even if they may be beyond our present capacity to understand them.

When scientists propose a *hypothesis*—a tentative explanation based on data collected through observations and experiments—they present it to the community of scientists for criticism and repeated testing against new data. A hypothesis that is confirmed by other scientists gains credibility, particularly if it predicts the outcome of new experiments.

A hypothesis that has survived repeated challenges and accumulated a substantial body of experimental support is elevated to the status of a *theory.* Although the explanatory and predictive powers of a theory have been demonstrated, it can never be considered finally proved. The essence of science is that no explanation, no matter how believable or appealing, is closed to question. If convincing new evidence indicates that a theory is wrong, scientists may modify it or discard it. The longer a theory holds up to all scientific challenges, however, the more confidently it is held.

A scientific *model* is a representation of some aspect of nature based on a set of hypotheses (usually including some established theories). Comparing a model's predictions with observations is a powerful way to test whether the hypotheses that went into the model are mutually consistent. These days, models are often formulated as computer programs that attempt to simulate the behavior of natural systems through numerical calculations. Computer simulations are important because they allow us to understand aspects of long-term system behavior that neither field observation nor laboratory experiments alone can elucidate.

To encourage discussion of their ideas, scientists share them and the data on which they are based. They present their findings at professional meetings, publish them in professional journals, and explain them in informal conversations with colleagues. Scientists learn from one another's work as well as from the discoveries of the past. Most of the great concepts of science, whether they emerge as a flash of insight or in the course of painstaking analysis, result from untold numbers of such interactions. Albert Einstein put it this way: "In science . . . the work of the individual is so bound up with that of his scientific predecessors and contemporaries that it appears almost as an impersonal product of his generation."

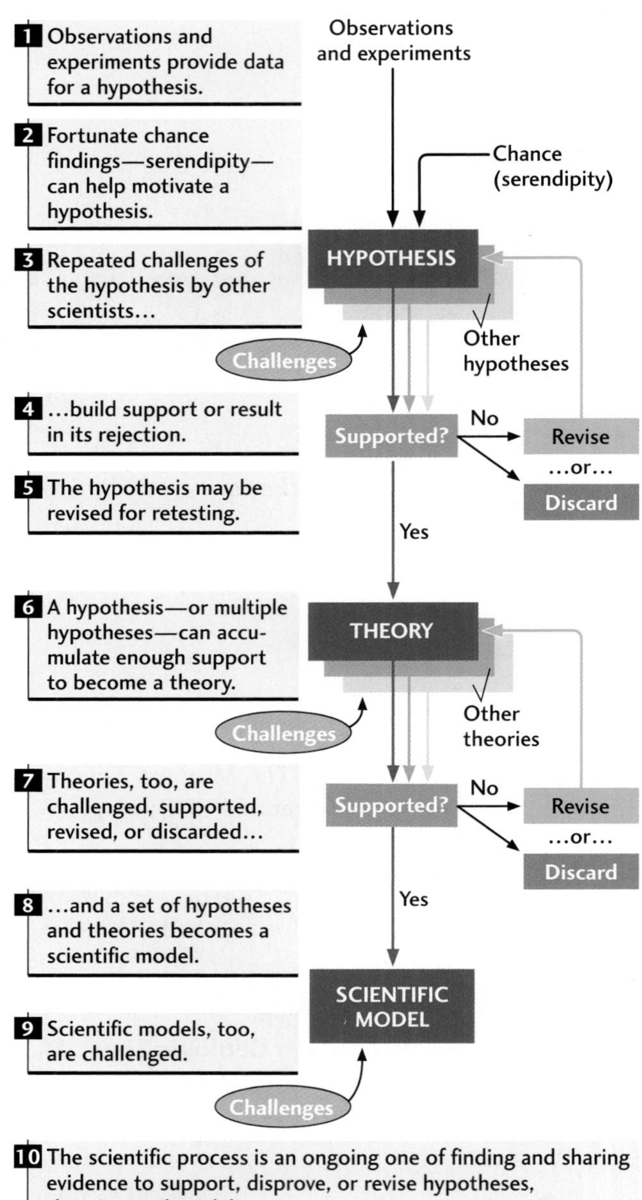

Figure 1.1 An outline of the scientific method.

Because such free intellectual exchange can be subject to abuses, a code of ethics has evolved among scientists. Scientists must acknowledge the contributions of all others on whose work they have drawn. They must not fabricate or falsify data, draw on the work of others without attribution, or otherwise be deceitful in their work. And they must accept responsibility for training the next generation of researchers and teachers. As important as any of these principles are values that are basic to science. Bruce Alberts, the president of the National Academy of Sciences, has aptly described these values as "honesty, generosity, a respect for evidence, openness to all ideas and opinions."

The Modern Theory and Practice of Geology

Like many sciences, geology depends on laboratory experiments and computer simulations to describe the physical and chemical properties of Earth's materials and to model natural processes that take place on Earth's surface and in its interior. Geology has its own particular style and outlook, however. It is an "outdoor science" in that it is grounded in observations and experiments conducted in the field and collected by remote sensing devices, such as Earth-orbiting satellites. Specifically, geologists compare direct observation of processes as they occur in the modern world with what they infer from the *geologic record.* The geologic record is the information preserved in rocks formed at various times through Earth's long history.

In the eighteenth century, the Scottish physician and geologist James Hutton advanced a historic principle of geology that can be summarized as "the present is the key to the past." Hutton's concept became known as the **principle of uniformitarianism,** and it holds that the geologic processes we see in action today have worked in much the same way throughout geologic time.

The principle of uniformitarianism does not mean that all geologic phenomena are slow. Some of the most important processes happen as sudden events. A large meteoroid that impacts Earth—a **bolide**—can gouge out a vast crater in a matter of seconds. A volcano can blow its top and a fault can rupture the ground in an earthquake just as quickly. Other processes do occur much more slowly. Millions of years are required for continents to drift apart, for mountains to be raised and eroded, and for river systems to deposit thick layers of sediments. Geologic processes take place over a tremendous range of scales in both space and time (**Figure 1.2**).

Nor does the principle of uniformitarianism imply that the only significant geologic phenomena are the ones we

Figure 1.2 Geologic phenomena can stretch over thousands of centuries or can occur with dazzling speed. *(left)* The Grand Canyon, in Arizona. [John Wang/PhotoDisc/Getty Images.] *(right)* Meteor Crater, Arizona. [John Sanford/Photo Researchers.]

see occurring today. Some processes have not been directly observed in the two and a half centuries since Hutton formulated his famous principle, yet they are no doubt important in the current Earth system. In recorded history, humans have never witnessed a large bolide impact, but we know they have occurred many times in the geologic past and will certainly happen again. The same can be said for the vast volcanic outpourings that have covered areas bigger than Texas with lava and poisoned the global atmosphere with volcanic gases. The long-term evolution of Earth is punctuated by many extreme, though infrequent, events involving rapid changes in the Earth system.

From Hutton's day onward, geologists have observed nature at work and used the principle of uniformitarianism to interpret features found in old rock formations. Despite the success of this approach, however, Hutton's principle is too confining for geologic science as it is now practiced. Modern geology must deal with the entire range of Earth's history. As we will see, the violent processes that shaped Earth's early history were distinctly different from those that operate today.

The Origin of Our System of Planets

The search for the origins of the universe and our own small part of it goes back to the earliest recorded mythologies. Today the generally accepted scientific explanation is the Big Bang theory, which holds that our universe began about 13 to 14 billion years ago with a cosmic "explosion." Before that moment, all matter and energy were compacted into a single, inconceivably dense point. Although we know little of what happened in the first fraction of a second after time began, astronomers have acquired a general understanding of the billions of years that followed. During that time, in a process that still continues, the universe has expanded and thinned out to form the galaxies and stars. Geologists focus on the past 4.5 billion years of this vast expanse, a time during which our solar system—the star that we call the Sun and the planets that orbit it—formed and evolved. Specifically, geologists look to the formation of the solar system to understand the formation of Earth.

The Nebular Hypothesis

In 1755, the German philosopher Immanuel Kant suggested that the origin of the solar system could be traced to a rotating cloud of gas and fine dust. Discoveries made in the past few decades have led astronomers back to this old idea, now called the **nebular hypothesis.** Equipped with modern telescopes, they have found that outer space beyond our solar

Figure 1.3 Evolution of the solar system.

system is not as empty as we once thought. Astronomers have recorded many clouds of the type that Kant surmised, and they have named them *nebulae.* They have also identified the materials that form these clouds. The gases are mostly hydrogen and helium, the two elements that make up all but a small fraction of our Sun. The dust-sized particles are chemically similar to materials found on Earth.

How could our solar system take shape from such a cloud? This diffuse, slowly rotating cloud contracted under the force of gravity, which is the attraction between pieces of matter because of their mass (**Figure 1.3**). Contraction, in turn, accelerated the rotation of the particles (just as ice skaters spin more rapidly when they pull in their arms), and the faster rotation flattened the cloud into a disk.

The Sun Forms Under the pull of gravity, matter began to drift toward the center, accumulating into a protostar, the precursor of our present Sun. Compressed under its own weight, the material in the proto-Sun became dense and hot. The internal temperature of the proto-Sun rose to millions of degrees, at which point nuclear fusion began. The Sun's nuclear fusion, which continues today, is the same nuclear reaction that occurs in a hydrogen bomb. In both cases, hydrogen atoms, under intense pressure and at high temperature, combine (fuse) to form helium. Some mass is converted into energy in the process. This conversion is represented by Albert Einstein's well-known equation $E = mc^2$, where E is the amount of energy released by conversion of mass (m) and c is the speed of light. Because c is a very large number (about 300,000 km/s) and c^2 is huge, a small amount of mass can yield an enormous amount of energy. The Sun releases some of that energy as sunshine; an H-bomb releases it as a great explosion.

The Planets Form Although most of the matter in the original nebula was concentrated in the proto-Sun, a disk of gas and dust, called the *solar nebula,* remained to envelop it. The solar nebula grew hot as it flattened into a disk. It became hotter in the inner region, where more of the matter accumulated, than in the less dense outer regions. Once formed, the disk began to cool, and many of the gases condensed. That is, they changed into their liquid or solid form, just as water vapor condenses into droplets on the outside of a cold glass and water solidifies into ice when it cools below the freezing point. Gravitational attraction caused the dust and condensing material to clump together in "sticky" collisions as small, kilometer-sized chunks, or *planetesimals.* In turn, these planetesimals collided and stuck together, forming larger, Moon-sized bodies. In a final stage of cataclysmic impacts, a few of these larger bodies—with their larger gravitational attraction—swept up the others to form our nine planets in their present orbits.

As the planets formed, those in orbits close to the Sun and those in orbits farther from the Sun developed in markedly different ways. The composition of the inner planets is quite different from that of the outer planets.

• *The Inner Planets* The four inner planets, in order of closeness to the Sun, are Mercury, Venus, Earth, and Mars (**Figure 1.4**). They are also known as the terrestrial ("Earthlike") planets. In contrast with the outer planets, the four inner planets are small and made up of rocks and metals. They grew close to the Sun, where conditions were so hot that most of the volatile materials—those that become gases and boil away at relatively low temperatures—could not be retained. Radiation and matter streaming from the Sun blew away most of the hydrogen, helium, water, and other light gases and liquids on these planets. Dense metals such as iron and the other heavy, rock-forming substances from which the inner planets formed were left behind. From the age of meteorites that occasionally strike Earth and that are believed to be remnants of preplanetary time, we deduce that the inner planets began to accrete about 4.56 billion years ago. Theoretical calculations indicate that they would have grown to planetary size in the remarkably short time of less than 100 million years.

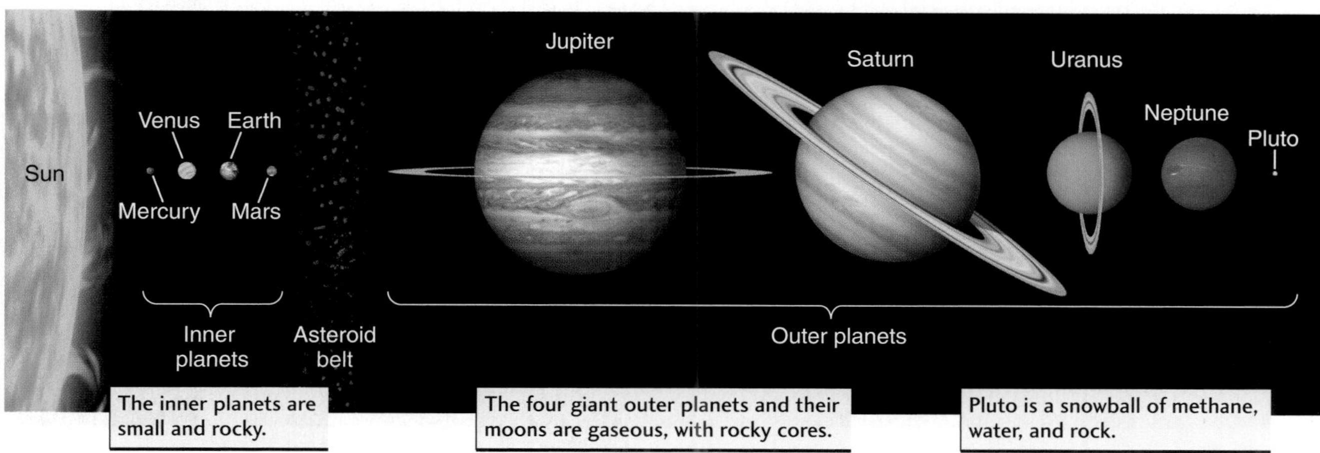

Figure 1.4 The solar system. The figure shows the relative sizes of the planets and the asteroid belt separating the inner and outer planets.

• *The Giant Outer Planets* Most of the volatile materials swept from the region of the terrestrial planets were carried to the cold outer reaches of the solar system to form the giant outer planets made up of ices and gases—Jupiter, Saturn, Uranus, and Neptune—and their satellites. The giant planets were big enough and their gravitational attraction strong enough to enable them to hold onto the lighter nebular constituents. Thus, although they have rocky cores, they (like the Sun) are composed mostly of hydrogen and helium and the other light constituents of the original nebula.

This standard model of solar system formation should be taken only for what it is: a tentative explanation that many scientists think best fits the known facts. Perhaps the model comes close to what actually happened. More important, however, this model gives us a way to think about the origin of the solar system that can be tested by observing the planets of our solar system and by studying other stars. American and Russian spacecraft carrying planetary probes have returned data about the nature and composition of the atmospheres and surfaces of Mercury, Venus, Mars, Jupiter, Saturn, Uranus, Neptune, and the Moon. A startling finding is that, in our solar system consisting of 9 planets and at least 60 satellites, no two bodies are the same!

Other Solar Systems

For ages, scientists and philosophers have speculated that there may be planets around stars other than our Sun. In the 1990s, using large telescopes, astronomers discovered planets orbiting nearby Sun-like stars. In 1999, the first family of *exoplanets*—the solar systems of another star—was found. These planets are too dim to be seen directly by telescopes, but their existence can be inferred from their slight gravitational pull on the stars that they orbit, causing to-and-fro movements of the star that can be measured. Currently, more than 90 exoplanets have been identified. Most planets found in this way are Jupiter-sized or larger and are close to their parent stars—many within scorching distance. Earth-sized planets are too small to be detected by this technique, but astronomers might be able to find such planets using other methods. For example, in about 10 years or so, spacecraft above Earth's atmosphere should be able to search for the dimming of the parent star's light as an orbiting planet passes in front of it along the line of sight to Earth.

We are fascinated by planetary systems around other stars because of what they might teach us about our own origins. Our overriding interest, though, is in the profound scientific and philosophical implications posed by the question: "Is anyone else out there?" In about 20 years, a spacecraft named *Life Finder* could be carrying instruments to analyze the atmospheres of exoplanets in our galaxy for signs of the presence of some kind of life. Based on what we know about biological processes, life on an exoplanet would probably be carbon-based and require liquid water. The benign temperatures we enjoy on Earth—not too far outside the range between the freezing and boiling points of water—appear to be essential. An atmosphere is needed to filter harmful radiation from the parent star, and so the planet must be large enough for its gravitational field to keep the atmosphere from escaping into space. For a habitable planet with advanced life *as we know it* to exist would require conditions even more limiting. For example, if the planet were too massive, delicate organisms such as humans would be too weak to withstand its larger gravitational force. Are these requirements too restrictive for life to exist elsewhere? Many scientists think not, considering the billions of Sun-like stars in our own galaxy.

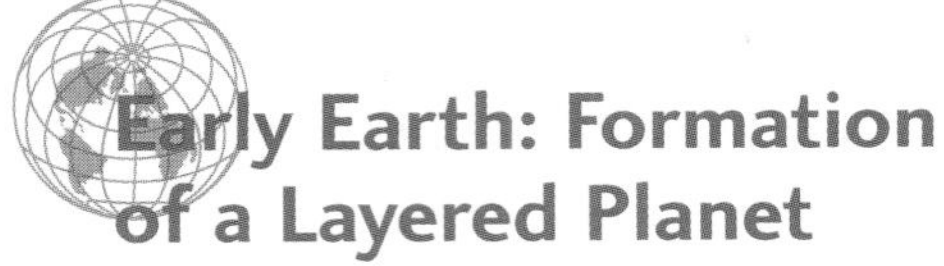

Early Earth: Formation of a Layered Planet

How did Earth evolve from a rocky mass to a living planet with continents, oceans, and an atmosphere? The answer lies in **differentiation:** the transformation of random chunks of primordial matter into a body whose interior is divided into concentric layers that differ from one another both physi-

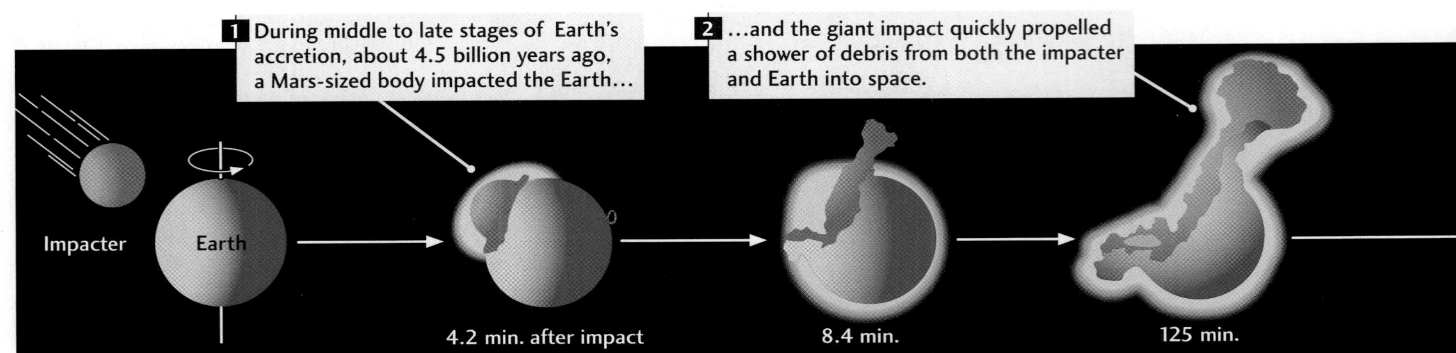

Figure 1.5 Computer simulation of the origin of the Moon by the impact of a Mars-sized body on Earth. [*Solid Earth Sciences and Society*, National Research Council, 1993.]

cally and chemically. Differentiation occurred early in Earth's history, when the planet got hot enough to melt.

Earth Heats Up and Melts

To understand Earth's present layered structure, we must return to the time when Earth was still subject to violent impacts by planetesimals and larger bodies. A moving object carries kinetic energy, or energy of motion. (Think of how the energy of motion crushes a car in a collision.) A planetesimal colliding with Earth at a typical velocity of 15–20 km/s would deliver as much energy as 100 times its weight in TNT. When planetesimals and larger bodies crashed into the primitive Earth, most of this energy of motion was converted into heat, another form of energy. The impact energy of a body perhaps twice the size of Mars colliding with Earth would be equivalent to exploding several trillion 1-megaton nuclear bombs (a single one of these terrible weapons would destroy a large city), enough to eject a vast amount of debris into space and to generate enough heat to melt most of what remained of Earth.

Many scientists now think that such a cataclysm did indeed occur during the middle to late stages of Earth's accretion. The giant impact created a shower of debris from both Earth and the impacting body and propelled it into space. The Moon aggregated from this debris (**Figure 1.5**). Earth would have re-formed as a largely molten body. This huge impact sped up Earth's rotation and changed its spin axis, knocking it from vertical with respect to Earth's orbital plane to its present 23° inclination. All this occurred about 4.5 billion years ago, between the beginning of Earth's accretion (4.56 billion years ago) and the age of the oldest Moon rocks brought back by the *Apollo* astronauts (4.47 billion years).

In addition to the giant impact, another source of heat would have caused melting early in Earth's history. Several elements (uranium, for example) are *radioactive,* which means that they disintegrate spontaneously by emitting subatomic particles. As these particles are absorbed by the surrounding matter, their energy of motion is transformed into heat. Radioactive heating would have contributed to heating and melting in the young Earth. Radioactive elements, though present only in small amounts, have had an enormous effect on Earth's evolution and continue to keep the interior hot.

Differentiation Begins

Although Earth probably began as an unsegregated mixture of planetesimals and other remnants of the solar nebula, it did not retain this form for long. Large-scale melting occurred as a result of the giant impact. Some workers in this field speculate that 30 to 65 percent of Earth melted, forming an outer layer hundreds of kilometers thick, which they call a "magma (molten rock) ocean." The interior, too, heated to a "soft" state in which its components could move around. Heavy material sank to the interior to become the core, and lighter material floated to the surface and formed the crust. The rising lighter matter brought interior heat to the surface, where it could radiate into space. In this way, Earth cooled and mostly solidified and was transformed into a differentiated or zoned planet with three main layers: a central core and an outer crust separated by a mantle (**Figure 1.6**). A summary of the timing of the events that describe Earth's origin and evolution into a differentiated planet is shown in Figure 1.12.

Earth's Core Iron, which is denser than most of the other elements, accounted for about a third of the primitive planet's material. The iron and other heavy elements such as nickel sank to form a central **core.** Scientists have found that the core, which begins at a depth of about 2900 km, is molten on the outside but solid in a region called the *inner core,* which extends from a depth of about 5200 km to Earth's center at about 6400 km. The inner core is solid because the pressures at the center are too high for iron to melt (the temperature at which any material melts increases with increasing pressure).

Earth's Crust Other molten materials were less dense than the parent substances from which they separated, so they floated toward the surface of the magma ocean. There

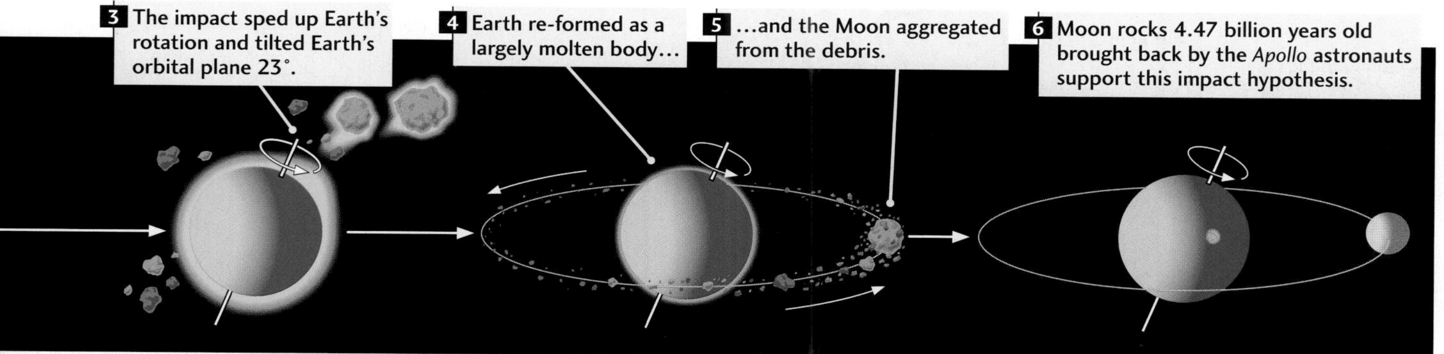

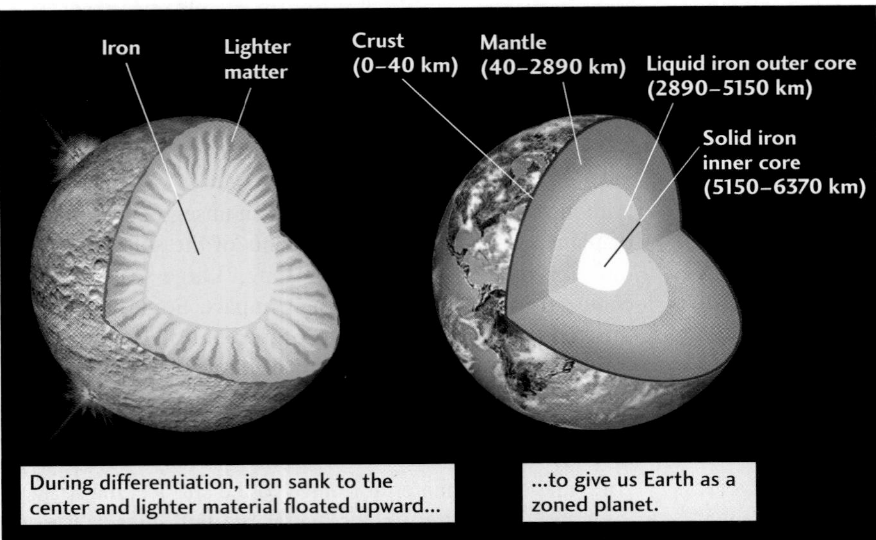

Figure 1.6 The differentiation of early Earth resulted in a zoned planet with a dense iron core, a crust of light rock, and a residual mantle between them.

they cooled to form Earth's solid **crust,** a thin outer layer about 40 km thick. The crust contains relatively light materials with low melting temperatures. Most of these materials are easily melted compounds of the elements silicon, aluminum, iron, calcium, magnesium, sodium, and potassium, combined with oxygen. All of these materials, other than iron, are among the lightest of the solid elements. (Chapter 3 discusses chemical compounds and the elements from which they form.)

Recently, in western Australia, a crystal fragment of the mineral zircon was found that is 4.3 to 4.4 billion years old, which would make it the oldest terrestrial material yet discovered. Chemical analysis indicates that the sample formed near the surface in the presence of water under relatively cool conditions. If this finding is confirmed by additional data and experiments, we can conclude that Earth may have cooled enough for a crust to form only 100 million years after the planet re-formed following the giant impact.

Earth's Mantle Between the core and the crust lies the **mantle,** a region that forms the bulk of the solid Earth. The mantle is the material left in the middle zone after most of the heavy matter sank and the light matter rose toward the surface. The mantle ranges from about 40 to 2900 km in depth. It consists of rocks of intermediate density, mostly compounds of oxygen with magnesium, iron, and silicon.

There are more than 100 elements, but chemical analysis of rocks indicates that only 8 make up 99 percent of Earth's mass (**Figure 1.7**). In fact, about 90 percent of Earth consists of only 4 elements: iron, oxygen, silicon, and magnesium. When we compare the relative abundance of elements in the crust with their abundance in the whole Earth, we find that iron accounts for a full 35 percent of Earth's mass. Because of differentiation, however, there is little iron in the crust, where the light elements predominate. As you can see in Figure 1.7, the crustal rocks on which we stand are almost 50 percent oxygen.

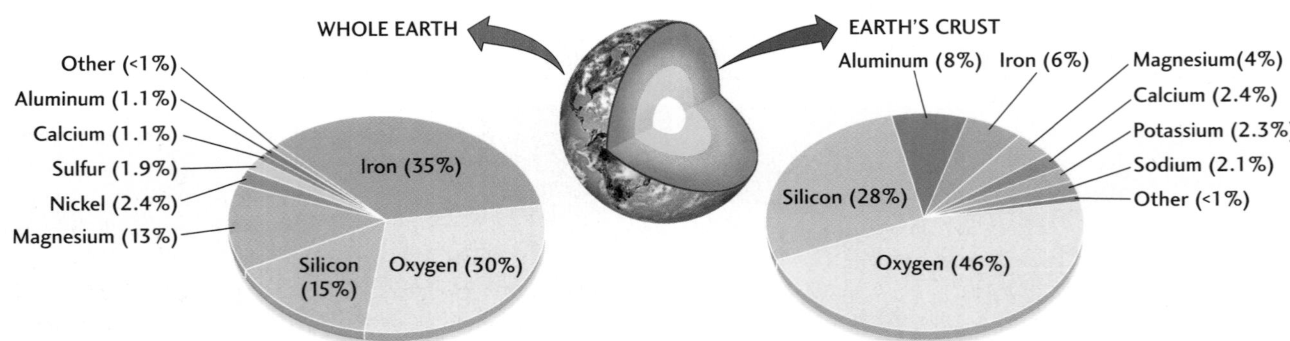

Figure 1.7 The relative abundance of elements in the whole Earth compared with that of elements in Earth's crust, given as percentages by weight. Differentiation created a light crust, depleted of iron and rich in oxygen, silicon, aluminum, calcium, potassium, and sodium. About 90 percent of Earth consists of only four elements: iron, oxygen, silicon, and magnesium. Note also that oxygen, silicon, and aluminum alone account for more than 80 percent of the crust.

Earth's Continents, Oceans, and Atmosphere Form

Early melting led to the formation of Earth's crust and eventually the continents. It brought lighter materials to Earth's outer layers and allowed even lighter gases to escape from the interior. These gases formed most of the atmosphere and oceans. Even today, trapped remnants of the original solar nebula continue to be emitted as primitive gases in volcanic eruptions.

Continents The most visible features of Earth's crust are the continents. Continental growth began soon after differentiation, and it has continued throughout geologic time. We have only the most general notion of what led to the formation of continents. We think that magma floated up from Earth's molten interior to the surface, where it cooled and solidified to form a crust of rock. This primeval crust melted and solidified repeatedly, causing the lighter materials to separate from the heavier ones and float to the top to form the primitive nucleus of the continents. Rainwater and other components of the atmosphere eroded the rocks, causing them to decompose and disintegrate. Water, wind, and ice then loosened rocky debris and moved it to low-lying places. There it accumulated in thick layers, forming beaches, deltas, and the floors of adjacent seas. Repetition of this process through many cycles built up the continents.

Oceans and Atmosphere Some geologists think that most of the air and water on Earth today came from volatile-rich matter of the outer solar system that impacted the planet after it was formed. For example, the comets we see are composed largely of water ice plus frozen carbon dioxide and other gases. Countless comets may have bombarded Earth early in its history, bringing water and gases that subsequently gave rise to the early oceans and atmosphere.

Many other geologists believe that the oceans and atmosphere can be traced back to the "wet birth" of Earth itself. According to this hypothesis, the planetesimals that aggregated into our planet contained ice, water, and other volatiles. Originally, the water was locked up (chemically bound as oxygen and hydrogen) in certain minerals carried by the aggregating planetesimals. Similarly, nitrogen and carbon were chemically bound in minerals. As Earth heated and its materials partially melted, water vapor and other gases were freed, carried to the surface by magmas, and released through volcanic activity.

The gases released from volcanoes some 4 billion years ago probably consisted of the same substances that are expelled from present-day volcanoes (though not necessarily in the same relative abundances): primarily hydrogen, carbon dioxide, nitrogen, water vapor, and a few other gases (**Figure 1.8**). Almost all of the hydrogen escaped to outer space, while the heavier gases enveloped the planet. This early atmosphere lacked the oxygen that makes up 21 percent of the atmosphere today. Oxygen did not enter the atmosphere until photosynthetic organisms evolved, as described later in this chapter.

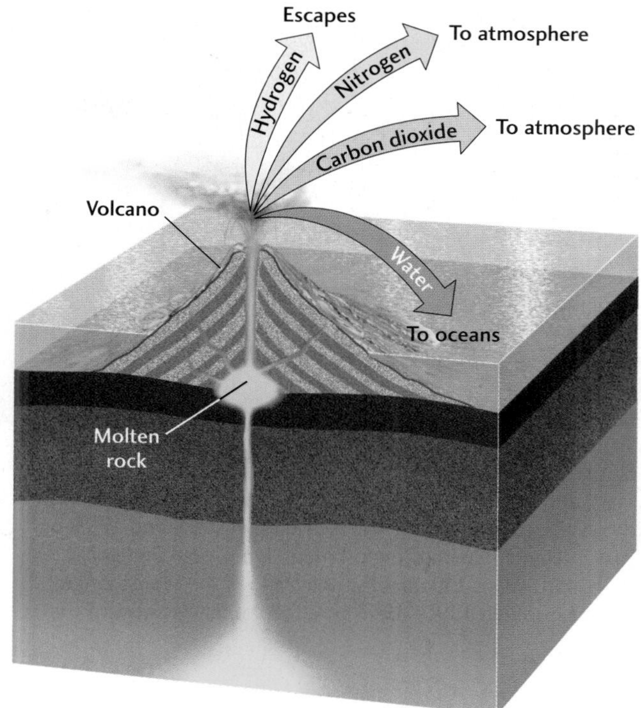

Figure 1.8 Early volcanic activity contributed enormous amounts of water vapor, carbon dioxide, and other gases to the atmosphere and oceans and solid materials to the continents. Photosynthesis by microorganisms removed carbon dioxide and added oxygen to the primitive atmosphere. Hydrogen, because it is light, escaped into space.

Diversity of the Planets

By about 4 billion years ago, Earth had become a fully differentiated planet. The core was still hot and mostly molten, but the mantle was fairly well solidified, and a primitive crust and continents had developed. The oceans and atmosphere had formed, probably from substances released from Earth's interior, and the geologic processes that we observe today were set in motion.

But what of the other planets? Did they go through the same early history? Information transmitted from our planetary spacecraft indicates that all the terrestrial planets have undergone differentiation, but their evolutionary paths have varied.

Mercury has a faint atmosphere, mostly helium. Atmospheric pressure at the surface is less than a trillionth that of Earth. There is no surface wind or water to erode and smooth its ancient surface. It looks like the Moon: intensely cratered and covered by a layer of rock debris, the fractured remnants of billions of years of meteorite impacts. Because it has essentially no atmosphere and is located close to the Sun, the planet warms to a surface temperature of 467°C during the day and cools to –173°C at night. This is the largest temperature range known in the solar system (other

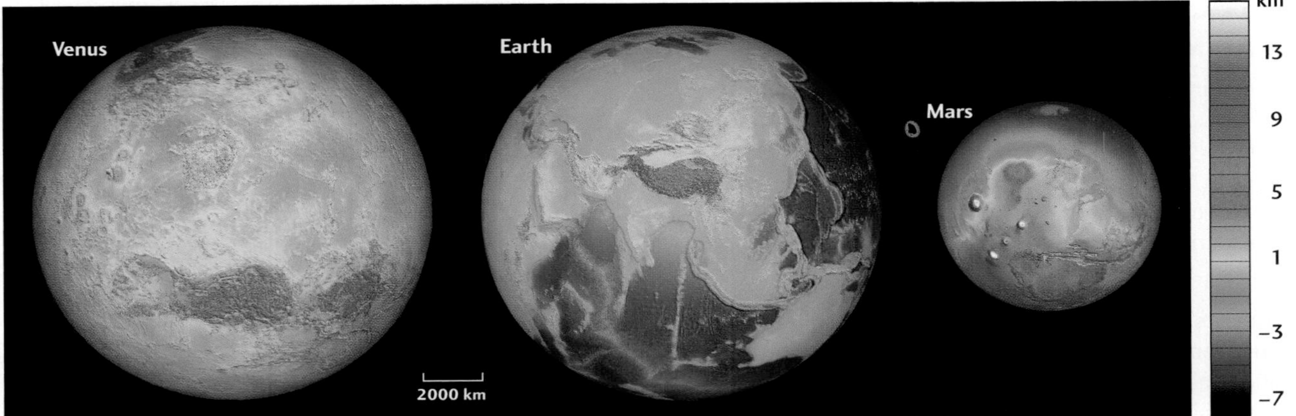

Figure 1.9 A comparison of the solid surfaces of Earth, Mars, and Venus, all at the same scale. Mars topography, which shows the greatest range, was measured in 1998 and 1999 by a laser altimeter aboard the orbiting *Mars Global Surveyor* spacecraft. Venus topography, which shows the lowest range, was measured in 1990–1993 by a radar altimeter aboard the orbiting *Magellan* spacecraft. Earth topography, which is intermediate in range and dominated by continents and oceans, has been synthesized from altimeter measurements of the land surface, ship-based measurements of ocean bathymetry, and gravity-field measurements of the seafloor surface from Earth-orbiting spacecraft. [Courtesy of Greg Neumann/MIT/GSFC/NASA.]

than that found on the Sun, which has a much larger range on its surface). Scientists are puzzled by the origin of Mercury's huge iron core. It makes up 70 percent of its mass, also a record for solar system planets.

Venus evolved into a planet with surface conditions surpassing most descriptions of Hell. It is wrapped in a heavy, poisonous, incredibly hot (475°C) atmosphere composed mostly of carbon dioxide and clouds of corrosive sulfuric acid droplets. A human standing on its surface would be crushed by the pressure, boiled by the heat, and eaten away by the sulfuric acid. Radar images that see through the thick cloud cover show that at least 85 percent of Venus is covered by lava flows. The remaining surface is mostly mountainous—evidence that the planet has been geologically active (**Figure 1.9**). Venus is Earth's twin in mass and size. How it could evolve into a planet so different from Earth is a question that intrigues planetary geologists.

Mars has undergone many of the same geologic processes that have shaped Earth (Figure 1.9). But Mars has a thin atmosphere composed almost entirely of carbon dioxide. No liquid water is present on the surface today—the planet is too cold and its atmosphere is too thin, so water would either freeze or evaporate. Networks of valleys and dry river channels, however, indicate that liquid water was abundant on the surface of Mars before 3.5 billion years ago. Some of the rocks observed by *Sojourner,* the robot rover of the 1997 Mars *Pathfinder* mission, show evidence of being worn by flowing water. Spacecraft circling Mars have recently found evidence that large amounts of water ice may be stored below the surface and sequestered in polar ice caps. Life might have formed on the wet Mars of billions of years ago and may exist today as microbes below the surface. NASA is designing spacecraft that could answer the question of life on Mars in a few years!

Most of the surface of Mars is older than 3 billion years. On Earth, in contrast, most surfaces older than about 500 million years have been obliterated by geologic activity. The following chapters will describe how these active processes have shaped the face of our planet throughout its long history.

Other than Earth, the *Moon* is the best-known body in the solar system because of its proximity and the programs of manned and unmanned exploration. As stated earlier, the favored theory of the Moon's origins proposes that it coalesced as a largely molten body after a giant impact ejected its matter from Earth. In bulk, the Moon's materials are lighter than Earth's because the heavier matter of the giant impactor and its primeval target remained embedded in Earth. The Moon has no atmosphere and, like Venus, is mostly bone dry, having lost its water in the heat generated by the giant impact. There is some new evidence from spacecraft observations that water ice may be present in small amounts deep within sunless craters at the Moon's north and south poles.

The surface that we see today is that of a very old, geologically dead body. Two terrains dominate the lunar surface. The oldest is the light-colored highlands. These rough and heavily cratered regions cover about 80 percent of the surface. The highlands are the ejected impact debris from early in lunar history when the Moon was battered by large asteroids. The remaining 20 percent of the surface is made up of the younger dark plains called *maria* (Latin for "seas" because that's the way they look from Earth). The maria were formed later when the large impact basins were subsequently flooded by lava.

The giant gaseous outer planets—*Jupiter, Saturn, Uranus,* and *Neptune*—will remain a puzzle for a long time. These huge gas balls are so chemically distinct and so large that they must have followed an evolutionary course entirely different from that of the much smaller terrestrial planets. We understand even less about the outermost planet, tiny *Pluto,* a strange frozen mixture of gas, ice, and rock and the only major planet not yet visited by our spacecraft.

Bombardment from Space

The crater-pocked surfaces of the Moon, Mars, Mercury, and other bodies are evidence of an important piece of the solar system's early history: the *Heavy Bombardment period* (see Figure 1.9). During this period, which may have lasted for 600 million years after the planets formed, the planets swept up and collided with the residual matter left behind when they were assembled. Geologic activity on Earth has obliterated the effects of this bombardment.

Space is littered with asteroids, meteoroids, comets, and other debris left over from the beginnings of our solar system. Smaller pieces of debris heat up and vaporize in Earth's atmosphere before they reach the ground, whereas larger pieces make it through. At present, some 40,000 tons of extraterrestrial material fall on Earth each year, mostly as dust and unnoticed small objects. Although the rate of impacts is now orders of magnitude smaller than it was in the Heavy Bombardment period, a large chunk of matter 1–2 km in size still collides with Earth every few million years or so. Although such collisions have become rare, telescopes are being assigned to search space and warn us in advance of sizable bodies that might slam into Earth. NASA astronomers recently predicted "with non-negligible probability" (1 chance in 300) that an asteroid 1 km in diameter will collide with Earth in March 2880. Such an event would threaten civilization.

A major impact did occur 65 million years ago. The bolide, not much larger than about 10 km, caused the extinction of half of Earth's species, including all dinosaurs. This event may have made it possible for mammals to become the dominant species and paved the way for humankind's emergence. Table 1.1 describes the effects of

Table 1.1 Bolide Impacts and Their Effects on Life on Earth

	Example(s)	Most Recent	Planetary Effects	Effects on Life
Supercolossal radius (R) > 2000 km	Moon-forming event	4.45×10^9 years ago	Melts planet	Drives off volatiles; wipes out life on Earth
Colossal $R > 700$ km	Pluto	More than 4.3×10^9 years ago	Melts crust	Wipes out life on Earth
Huge $R > 200$ km	4 Vesta (large asteroid)	About 4.0×10^9 years ago	Vaporizes oceans	Life may survive below surface
Extralarge $R > 70$ km	Chiron (largest active comet)	3.8×10^9 years ago	Vaporizes upper 100 m of oceans	Pressure-cooks photo zone;[a] may wipe out photosynthesis
Large $R > 30$ km	Comet Hale-Bopp	About 2×10^9 years ago	Heats atmosphere and surface to about 1000 K	Cauterizes continents
Medium $R > 10$ km	K/T impactor; 433 Eros (largest near-Earth asteroid)	65×10^6 years ago	Causes fires, dust, darkness; chemical changes in ocean and atmosphere; large temperature swings	Causes extinction of half of species; K/T event made dinosaurs extinct
Small $R > 1$ km	About size of 500 near-Earth asteroids	About 300,000 years ago	Causes global dusty atmosphere for months	Interrupts photosynthesis; individuals die but few species extinct; threatens civilization
Very small $R > 100$ m	Tunguska event (Siberia)	1908	Knocked over trees tens of kilometers away; caused minor hemispheric effects; dusty atmosphere	Newspaper headlines; romantic sunsets; increased birth rate

[a]Regions of Earth reached by sunlight; i.e., the atmosphere and top 100 m of ocean.

SOURCE: Modified from J. D. Lissauer, *Nature* 402: C11–C14.

impacts of various sizes on our planet and its life. The poet Robert Frost may have had in mind the vulnerability of life on Earth when he wrote

Some say the world will end in fire,
Some say in ice.
From what I've tasted of desire
I hold with those who favor fire.
But if I had to perish twice,
I think I know enough of hate
To say that for destruction ice
Is also great
And would suffice.

Earth as a System of Interacting Components

Although Earth has cooled down from its fiery beginnings, it remains a restless planet, continually changing through such geologic activity as earthquakes, volcanoes, and glaciation. This activity is powered by two heat engines: one internal, the other external. A heat engine—for example, the gasoline engine of an automobile—transforms heat into mechanical motion or work. Earth's internal engine is powered by the heat energy trapped during the planet's violent origin and generated by radioactivity in its deep interior. The internal heat drives motions in the mantle and core, supplying the energy to melt rock, move continents, and lift up mountains. Earth's external engine is driven by solar energy—heat supplied to Earth's surface by the Sun. Heat from the Sun energizes the atmosphere and oceans and is responsible for our climate and weather. Rain, wind, and ice erode mountains and shape the landscape, and the shape of the landscape, in turn, changes the climate.

All the parts of our planet and all their interactions, taken together, constitute the **Earth system.** Although Earth scientists have long thought in terms of natural systems, it was not until the latter part of the twentieth century that they were equipped with the tools needed to investigate how the Earth system actually works. Principal among these were networks of instruments and Earth-orbiting satellites to collect information about the Earth system on a global scale and electronic computers powerful enough to calculate the mass and energy transfers within the system. The major components of the Earth system are described in Table 1.2 and depicted in **Figure Story 1.10**. We have discussed some of them already; we will define the others below.

We will talk about many facets of the Earth system in later chapters. Let's get started by thinking about some of its basic features. Earth is an *open system* in the sense that it exchanges mass and energy with the rest of the cosmos. Radiant energy from the Sun energizes the weathering and erosion of Earth's surface, as well as the growth of plants, which feed almost all living things. Our climate is controlled by the balance between the solar energy coming into the Earth system and the energy Earth radiates back into

Table 1.2 Major Components of the Earth System

Solar Energy Energizes These Components	
Atmosphere	Gaseous envelope extending from Earth's surface to an altitude of about 100 km
Hydrosphere	Surface waters comprising all oceans, lakes, rivers, and groundwaters
Biosphere	All organic matter related to life near Earth's surface
Earth's Internal Heat Energizes These Components	
Lithosphere	Strong, rocky outer shell of the solid Earth that comprises the crust and uppermost mantle down to an average depth of about 100 km; forms the tectonic plates
Asthenosphere	Weak, ductile layer of mantle beneath the lithosphere that deforms to accommodate the horizontal and vertical motions of plate tectonics
Deep mantle	Mantle beneath the asthenosphere, extending from about 400 km deep to the core-mantle boundary (about 2900 km deep)
Outer core	Liquid shell composed primarily of molten iron, extending from about 2900 km to 5150 km in depth
Inner core	Inner sphere composed primarily of solid iron, extending from about 5150 km deep to Earth's center (about 6400 km deep)

EARTH IS AN OPEN SYSTEM THAT EXCHANGES ENERGY AND MASS WITH ITS SURROUNDINGS

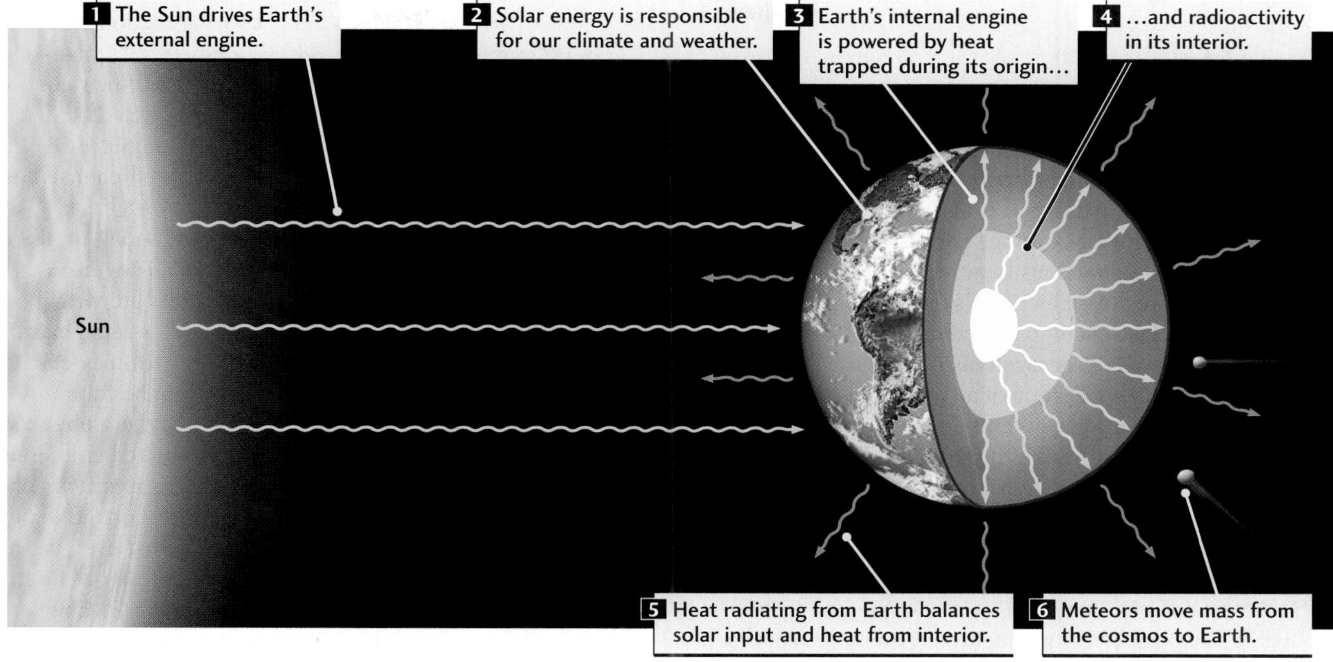

THE EARTH SYSTEM IS ALL PARTS OF OUR PLANET AND THEIR INTERACTIONS

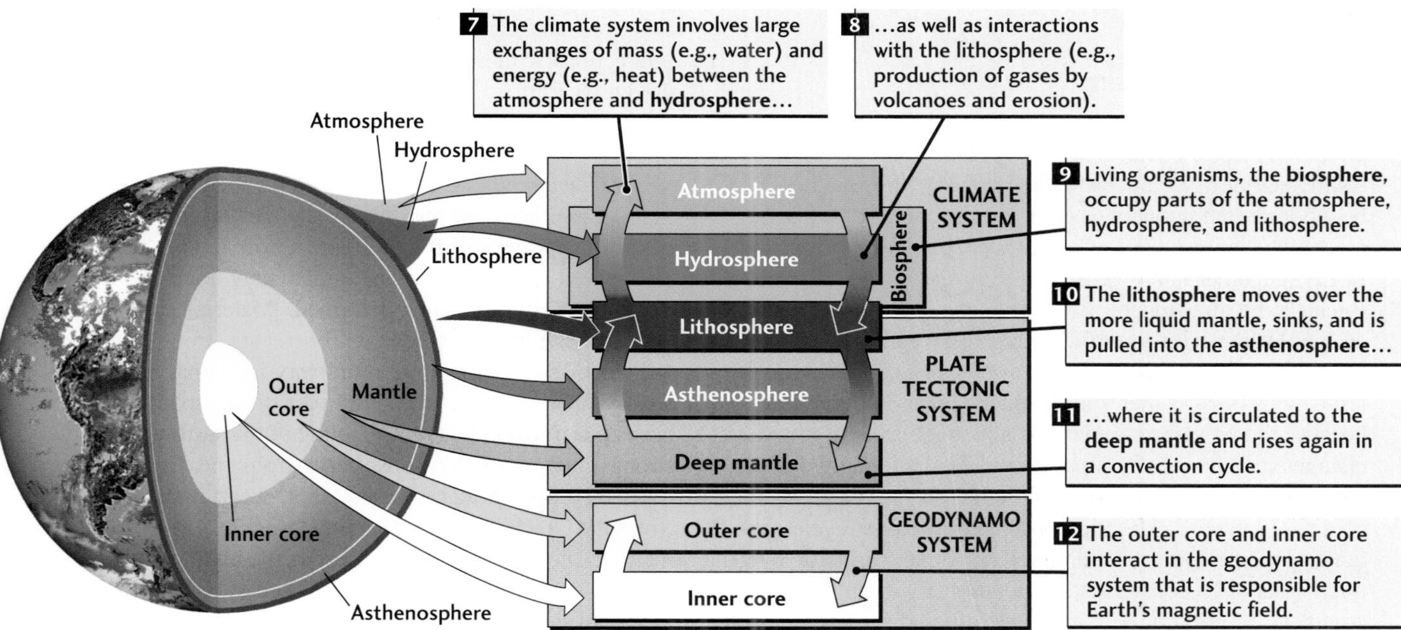

Figure Story 1.10 Major components and subsystems of the Earth system (see Table 1.2). Interactions among the components are powered by energy from the Sun and the planetary interior and organized into three global geosystems: the climate system, the plate tectonic system, and the geodynamo system.

space. The mass transfers between Earth and space decreased markedly after the Heavy Bombardment period, but they still play an active role in the Earth system—just ask the dinosaurs!

Although we think of Earth as a single system, it is a challenge to study the whole thing all at once. Instead, we will focus our attention on parts of the system we are trying to understand. For instance, in the discussion of recent climate changes, we will primarily consider interactions among the atmosphere, hydrosphere, and biosphere that are driven by solar energy. Our coverage of how the continents formed will focus on interactions between the crust and the deeper mantle that are driven by Earth's internal energy. Specialized subsystems that encompass interesting types of terrestrial behavior are called **geosystems**. The Earth system can be thought of as the collection of all these open, interacting (and often overlapping) geosystems.

In this section, we will introduce two important geosystems that operate on a global scale: the climate system and the plate tectonic system. A third global system is the geodynamo, which is responsible for Earth's magnetic field. It is an important part of how Earth works as a planet and a key tool for exploring the interior. The geodynamo is discussed in Chapter 21. Its importance to understanding plate tectonics is discussed in Chapter 2. Later in the book, we will have occasion to discuss a number of smaller geosystems. Here are three examples: volcanoes that erupt hot lava (Chapter 6), hydrologic systems that give us our drinking water (Chapter 13), and petroleum reservoirs that produce oil and gas (Chapter 22).

The Climate System

Weather is the term we use to describe the temperature, precipitation, cloud cover, and winds observed at a point on Earth's surface. We all know how variable the weather can be—hot and rainy one day, cool and dry the next—depending on the movements of storm systems, warm and cold fronts, and other rapidly changing atmospheric disturbances. Because the atmosphere is so complex, even the best forecasters have a hard time predicting the weather more than four or five days in advance. However, we can guess in rough terms what our weather will be much farther into the future, because the prevailing weather is governed primarily by the changes in solar energy input on seasonal and daily cycles: summers are hot, winters cold; days are warmer, nights cooler. *Climate* is a description of these weather cycles obtained by averaging temperature and other variables over many years of observation. In addition to mean values, a complete description of climate also includes measures of how variable the weather has been, such as the highest and lowest temperatures ever recorded on a given day.

The **climate system** includes all the properties and interactions of components within the Earth system needed to determine the climate on a global scale and discover how the climate changes with time. The problem is incredibly complicated because climate is not a behavior of the atmosphere alone. It is sensitive to many processes involving the hydrosphere, biosphere, and solid Earth (see Figure Story 1.10). To understand these interactions, scientists build numerical models—virtual climate systems—on big computers, and they compare the results of their computer simulations with observed data. (In March 2002, Japan unveiled the world's largest and fastest computer, the Earth Simulator, dedicated to modeling Earth's climate and other geosystems.)

Scientists gain confidence in their models if there is good agreement with the observed data. They use the disagreements to figure out where the models are wrong or incomplete. They hope to improve the models enough through testing with many types of observations that they can accurately predict how climate will change in the future. A particularly urgent problem is to understand the global warming that might be caused by human-generated emissions of carbon dioxide and other "greenhouse" gases. Part of the public debate about global warming centers on the accuracy of computer predictions. Skeptics argue that even the most sophisticated computer models are unreliable because they lack many features of the real Earth system. In Chapter 23, we will discuss some aspects of how the climate system works and the practical problems of climate change caused by human activities.

The Plate Tectonic System

Some of Earth's more dramatic geologic events—volcanic eruptions and earthquakes, for example—also result from interactions within the Earth system. These phenomena are driven by Earth's internal heat, which escapes through the circulation of material in Earth's solid mantle, a process known as *convection.*

We have seen that Earth is zoned by chemistry: its crust, mantle, and core are chemically distinct layers that segregated during early differentiation. Earth is also zoned by strength, a property that measures how much an Earth material can resist being deformed. Material strength depends on chemical composition (bricks are strong, soap bars are weak) and temperature (cold wax is strong, hot wax is weak). In some ways, the outer part of the solid Earth behaves like a ball of hot wax. Cooling of the surface forms the strong outer shell or **lithosphere** (from the Greek *lithos,* meaning "stone") that encases a hot, weak **asthenosphere** (from the Greek *asthenes,* meaning "weak"). The lithosphere includes the crust and the top part of the mantle down to an average depth of about 100 km. When subjected to force, the lithosphere tends to behave as a rigid and brittle shell, whereas the underlying asthenosphere flows as a moldable, or ductile, solid.

According to the remarkable theory of **plate tectonics,** the lithosphere is not a continuous shell; it is broken into about a dozen large "plates" that move over Earth's surface at rates of a few centimeters per year. Each plate acts as a

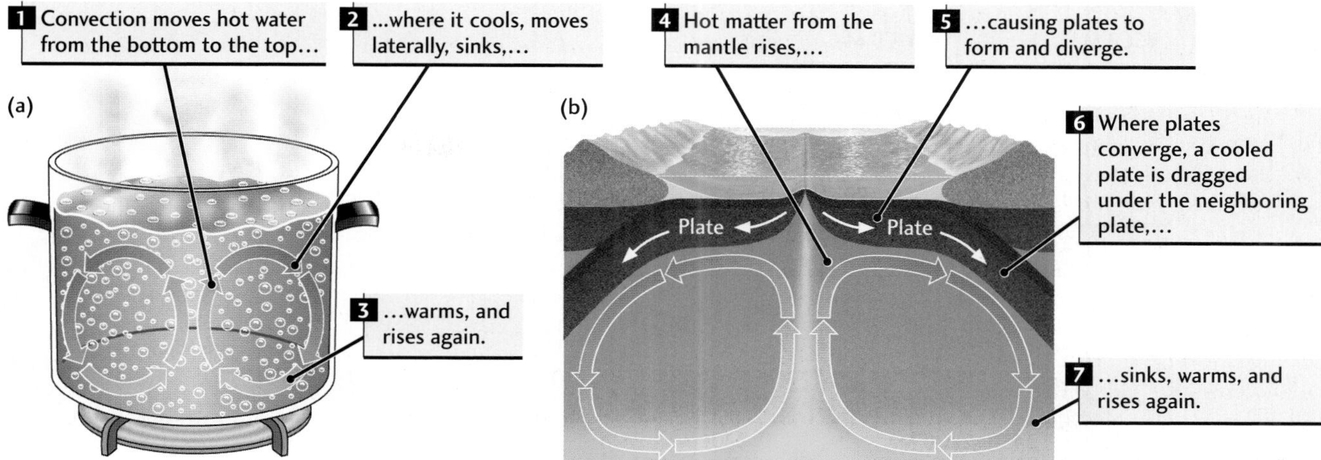

Figure 1.11 (a) Boiling water is a familiar instance of convection. (b) A simplified view of convection currents in Earth's interior.

distinct rigid unit that rides on the asthenosphere, which also is in motion. The lithosphere that forms a plate may be just a few kilometers thick in volcanically active areas and perhaps 200 km thick or more beneath the older, colder parts of the continents. The discovery of plate tectonics in the 1960s furnished scientists with the first unified theory to explain the worldwide distribution of earthquakes and volcanoes, continental drift, mountain building, and many other geologic phenomena. Chapter 2 will be devoted to a detailed description of plate tectonics.

Why do the plates move across Earth's surface instead of locking up into a completely rigid shell? The forces that push and pull the plates around the surface come from the heat engine in Earth's solid mantle, which causes convection. In general terms, convection is a mechanism of energy and mass transfer in which hotter material rises and cooler material sinks. We tend to think of convection as a process involving fluids and gases—circulating currents of water boiling in a pot, smoke rising from a chimney, or heated air floating up to the ceiling as cooled air sinks to the floor—but it can also occur in solids that are at high enough temperatures to be weak and ductile. We note that the flow in ductile solids is usually slower than fluid flow, because even "weak" solids (say, wax or taffy) are more resistant to deformation than ordinary fluids (say, water or mercury).

Convection can occur in a flowing material, either a liquid or a ductile solid, that is heated from below and cooled from above. The heated matter rises under the force of buoyancy because it has become less dense than the matter above it. When it reaches the surface, it gives up heat and cools as it moves sideways, becoming denser. When it gets heavier than the underlying material, it sinks under the pull of gravity, as depicted in **Figure 1.11**. The circulation continues as long as enough heat remains to be transferred from the hot interior to the cool surface.

The movement of the plates is the surface manifestation of convection in the mantle, and we refer to this entire system as the **plate tectonic system.** Driven by Earth's internal heat, hot mantle material rises where plates separate and begins to gel the lithosphere. The lithosphere cools and becomes more rigid as it moves away from this divergent boundary. Eventually, it sinks into the asthenosphere, dragging material back into the mantle at boundaries where plates converge (Figure 1.11b). As with the climate system (which involves a wide range of convective processes in the atmosphere and oceans), scientists study plate tectonics using computer simulations to represent what they think are the most important components and interactions. They revise the models when their implications disagree with actual data.

Earth Through Geologic Time

So far, we have discussed two major topics: how Earth formed from the early solar system, and how two global geosystems work today. What happened during the intervening 4.5 billion years? To answer this question, we begin with an overview of geologic time from the birth of the planet to the present. Later chapters will fill in the details.

An Overview of Geologic Time

Comprehending the immensity of geologic time can be a challenge for the uninitiated. The popular writer John McPhee has eloquently noted that geologists look into the

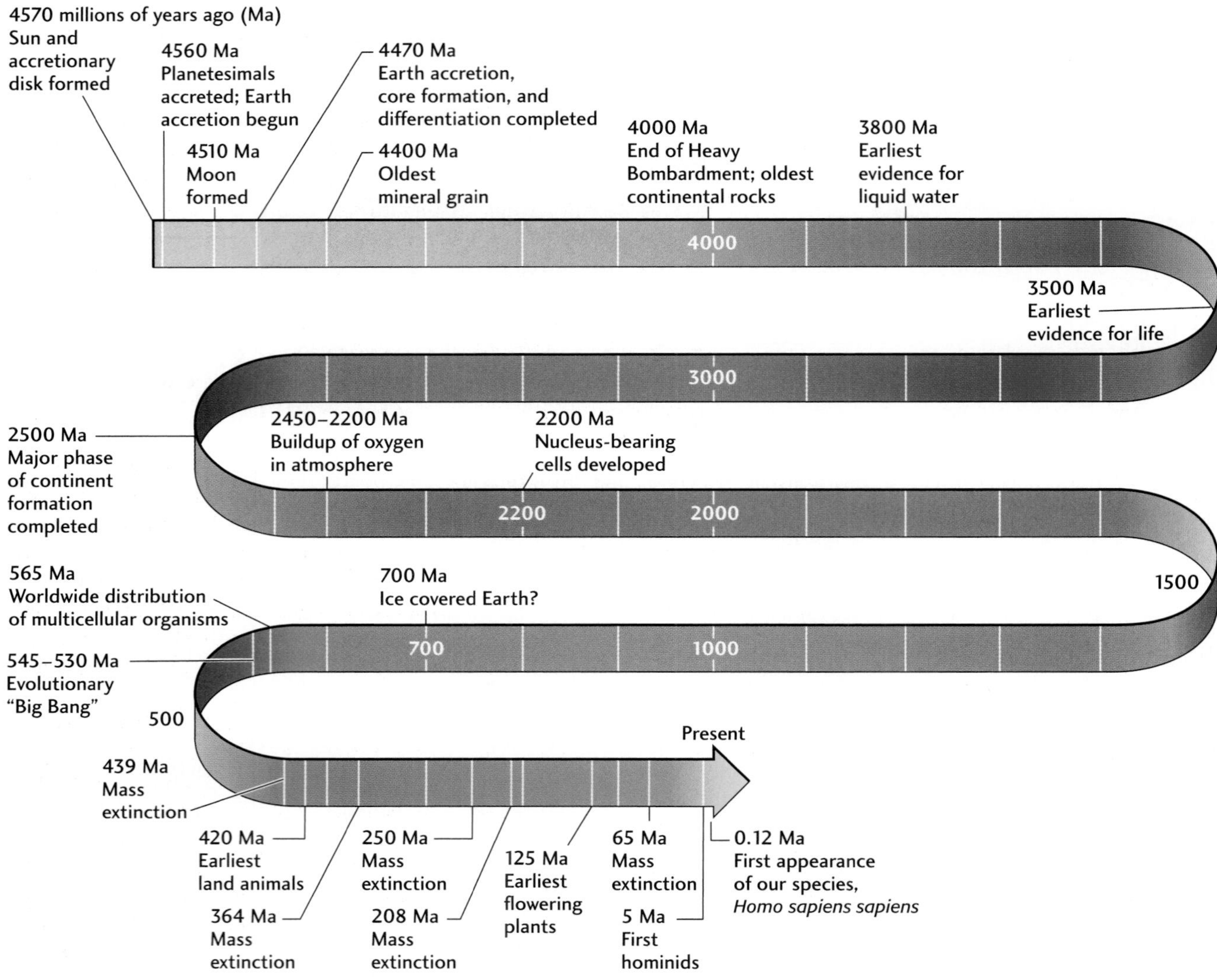

Figure 1.12 The ribbon of geologic time from the formation of the solar system to the present, measured in billions of years and marked by some major events and transitions in Earth's history. Our hominid ancestors become evident in the geologic record about 5 million years ago, only a tenth of one percent of Earth's total age. The span of human existence (about 120,000 years) is less than the thickness of the line at the end of the ribbon! Some of the events shown here remain speculative, and the timings of many are imprecise.

"deep time" of Earth's early history (measured in billions of years), just as astronomers look into the "deep space" of the outer universe (measured in billions of light-years). **Figure 1.12** presents geologic time as a ribbon marked with some major events and transitions.

We have already described the current theories of planetary accretion and differentiation during the first 500 million years of Earth's history. These can be appropriately called the geologic "dark ages," because very little of the rock record was able to survive the Heavy Bombardment period. The oldest rocks now found on Earth's surface are about 4 billion years old. Rocks as old as 3.8 billion years show evidence of erosion by water, indicating the existence of a hydrosphere. By 2.5 billion years ago, enough low-density crust had collected at Earth's surface to form large continental masses. The geologic processes that subsequently modified these continents were very similar to those we see operating today in plate tectonics.

Beginning about 2.5 billion years ago, the fossil record of early life on Earth became progressively richer, revealing a diverse set of adaptive behaviors by Earth's early pioneers. Some of these behaviors had global influence, resulting in progressive oxygenation of the atmosphere and ocean over the following 2 billion years. By deciphering this geologic record, we can reconstruct the history of biological evolution.

The Evolution of Life

All living organisms and the organic matter they produce, taken together, constitute Earth's **biosphere** (from the Greek *bios,* "life"). The evolution of life involved complex interactions among the biosphere, atmosphere, hydrosphere, and lithosphere.

Life Begins A little more than 4 billion years ago, Earth's early atmosphere and hydrosphere had already formed. Light gases such as hydrogen had escaped to space, leaving behind heavier gases such as water vapor, carbon dioxide, and sulfur dioxide. This early atmosphere allowed almost all the components of sunlight to reach Earth's surface—including ultraviolet (UV) rays, which are damaging to life. At the same time, there was enough carbon dioxide and water vapor in the atmosphere to keep the Earth warm by trapping heat radiating from Earth's surface. This phenomenon is known as the *greenhouse effect,* by analogy with the warming of a greenhouse in which glass lets in visible light but lets little heat escape.

Life somehow began in Earth's greenhouse, despite the intense UV radiation and the harsh, oxygen-poor atmosphere. Direct evidence, though currently disputed, lies in the preservation of early **fossils** (traces of organisms of past geologic ages that have been preserved in the crust). Fossils of primitive bacteria have been found in rocks dated at 3.5 billion years. A more compelling, though indirect, line of evidence is provided by the chemical composition of organic matter preserved in rocks of this age. These chemical remains of ancient organisms are rapidly surpassing the fossil evidence as the principal basis for understanding the early evolution of life on Earth.

There is a strong likelihood, however, that life had originated sometime before, perhaps 4 billion years ago or even earlier. The first step toward the evolution of primitive bacteria is thought to have been the assembly of large organic molecules from gases such as methane and ammonia. The energy for these transformations was supplied by strong UV radiation. This step has been explored in many chemical experiments that show how various building blocks of life could have formed. Somehow these organic molecules aggregated and formed systems capable of growth and metabolism. These systems were not really life, because they were not self-reproducing, so they are called *protolife.* Some scientists argue that protolife was concentrated in hot springs fed by volcanoes on the floor of the ocean.

The next critical step was the development of the first truly self-replicating molecule: ribonucleic acid (RNA). This single-stranded molecule—like its double-stranded cousin, deoxyribonucleic acid (DNA)—is intimately involved in the process of self-replication. The "RNA world" was transitory and soon evolved into a more complex "DNA world," which has characterized the biosphere for the rest of geologic history.

Not all scientists subscribe to these hypotheses. A small number believe that cometary impacts brought to Earth not only the atmosphere's gases and the oceans but also life itself. According to this view, life on Earth began when infalling comets—snowballs of frozen gases—"colonized" the planet. One scientist has proposed that the constant bombardment of Earth in these early times might well have destroyed life soon after it was synthesized. If this were the case, life would have started anew several times.

These early stages in the origin of life probably did not significantly affect the early atmosphere, which remained dominated by nitrogen and carbon dioxide.

Oxygen Becomes a Major Gas in the Atmosphere Early organisms must have fed on the relatively small amounts of organic matter produced by inorganic chemical processes or recycled from other organisms. A major change occurred when life that could make its own food by *photosynthesis* evolved. Photosynthesis is the process by which plants and other green organisms use chlorophyll (which colors them green) and the energy from sunlight to produce carbohydrates from carbon dioxide and water.

The evolution of photosynthesis early in Earth's history had immense consequences. A by-product of photosynthesis is oxygen gas (O_2). As organic matter from photosynthetic life was buried, carbon was removed from the atmosphere and oxygen accumulated. From fossil evidence, it seems likely that this process was under way by 2.5 billion years ago. Geologists have found iron-bearing rocks as old as 2.5 billion years that were oxidized ("rusted") during their formation, indicating that there was more oxygen in the atmosphere by that time. The increase to modern levels of atmospheric oxygen is now thought to have occurred in a series of stepwise increases over a period perhaps as long as 2 billion years.

As atmospheric oxygen molecules diffused upward into the stratosphere (upper atmosphere), they were transformed by solar radiation into ozone (O_3), creating a *stratospheric ozone layer.* The ozone layer absorbs certain portions of the solar UV radiation before it reaches the surface, where it can damage and cause mutations in animal and plant cells. Without this protective shield, life is unlikely to have flourished on land.

Biology's "Big Bang" Compared with life today, life on early Earth was a primitive affair, consisting mostly of small, single-celled organisms that floated near the surface of the oceans or lived on the seafloor. Between 1 billion and 2 billion years ago, life became multicellular when algae and seaweeds originated. Then, for reasons not well understood, the first animals appeared on the scene about 600 million years ago, evolving in a set of waves. The first

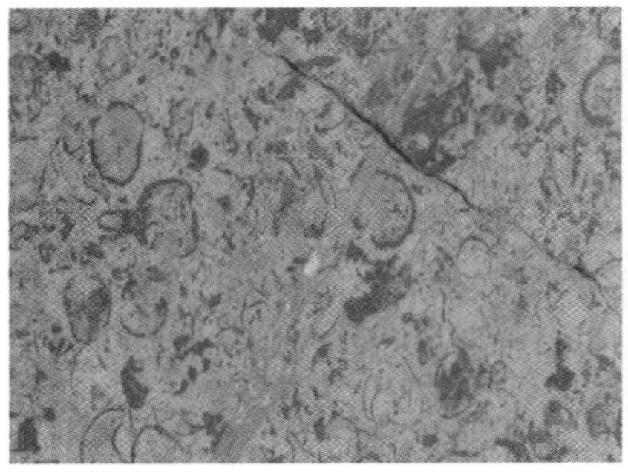

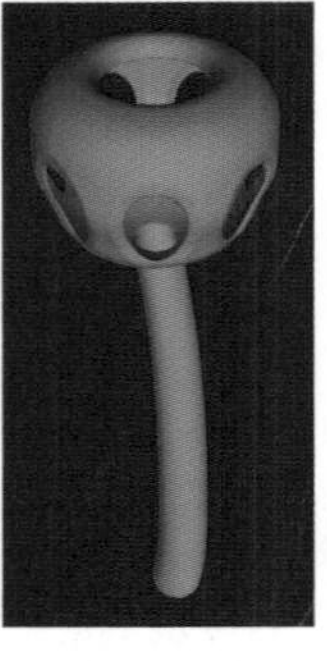
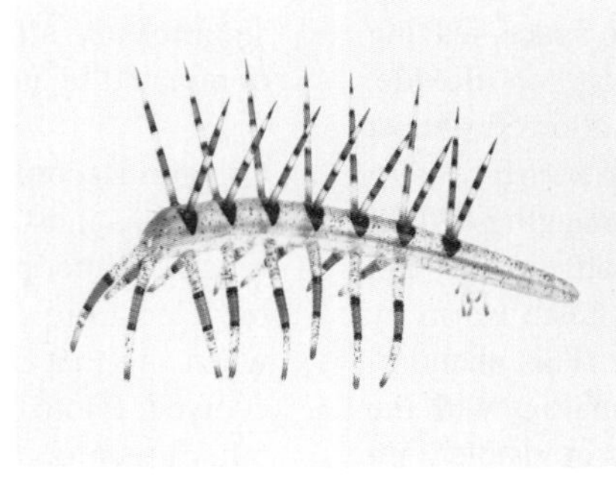
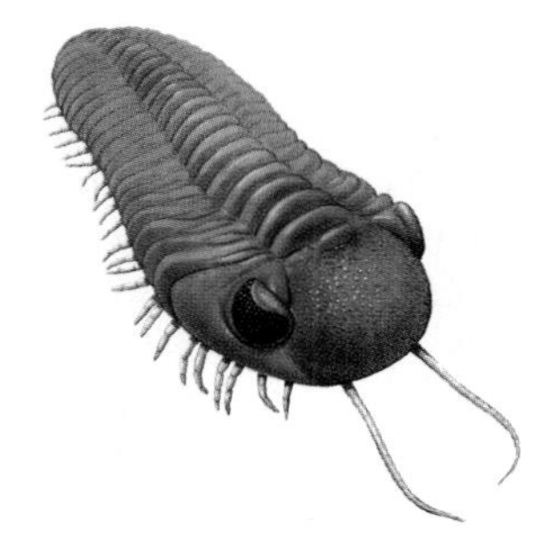

Namacalathus ***Hallucigenia*** Trilobites

Figure 1.13 Fossils that record the Cambrian explosion include Precambrian calcified fossils (*left*) that were the first organisms to use calcite in making a shell. These became extinct at the Precambrian-Cambrian boundary, along with other organisms, and helped pave the way for a strange new group of organisms, including *Hallucigenia* (*center*) and the more familiar trilobites (*right*). Both of these organisms formed weak shells made of organic material similar to fingernails. [(*left, top*) John Grotzinger; *(bottom)* W. A. Watters. (*center, top*) National Museum of Natural History/Smithsonian Institution; (*bottom*) Chase Studio/Photo Researchers. *(right, top)* Courtesy of Musée cantonal de géologie, Lausanne. Photo by Stefan Ansermet; (*bottom*) Chase Studio/Photo Researchers.]

wave featured simple jellyfish- and fernlike forms with soft bodies, as well as hard-bodied forms with shapes resembling wineglasses with holes (**Figure 1.13**). Most quickly went extinct, although a few may have served as prototypes for the second wave, which constituted the greatest diversification of new life-forms in Earth's history.

In a brief period starting 543 million years ago and probably lasting less than 10 million years, eight entirely new branches (phyla) of the animal kingdom were established, including ancestors to essentially all animals inhabiting Earth today. These organisms were a menagerie of strange beasts, described by the science essayist Steven J. Gould as the stuff of "a science fiction film." One particularly bizarre creature has been called *Hallucigenia* (see Figure 1.13). Forms more familiar to us include earthworms and their marine relatives, sea stars (starfish) and sand dollars, mollusks, insects and crustaceans, and the chordates that eventually evolved into the higher animals (including us). Other animal types, now extinct, such as the primitive-looking trilobites (see Figure 1.13), also sprang into existence. It was during this evolutionary explosion, sometimes referred to as biology's "Big Bang," that animals with hard, calcium-rich body parts first left shelly fossils in the geologic record.

Mass Extinctions from Extreme Events Although biological evolution is often viewed as a very slow process, broad evolutionary trends were frequently punctuated by brief periods of rapid change. A prime example is the evolutionary explosion we have just described. Equally spectacular were the mass extinctions, during which many types of animals and plants suddenly disappeared from the geologic record. Five of these huge turnovers are indicated in Figure 1.12. The last, discussed earlier in this chapter, was caused by a major bolide impact 65 million years ago. This extreme event ended the Age of Dinosaurs. Benthic reef communities were also extinguished and reorganized. Large pelagic forms in the deep blue sea were wiped out, along with most other primary producers. These events led to what some people have called the "Strangelove Ocean."

The causes of the other mass extinction are still being debated. In addition to bolide impacts, scientists have proposed other types of extreme events, such as rapid climate changes brought on by glaciations and massive eruptions of volcanic material. The evidence is often ambiguous or inconsistent. For example, the largest extinction event of all time took place about 250 million years ago, wiping out 95 percent of all species. A bolide impact has been proposed by some investigators, but the geologic record shows that the ice sheets expanded at this time and seawater chemistry changed, consistent with a major climate crisis. Simultaneously, an enormous volcanic eruption covered an area in Siberia almost half the size of the United States with 2 or 3 million cubic kilometers of lava. This mass extinction has been dubbed "Murder on the Orient Express," because there are so many suspects!

In this introductory chapter, we have covered many topics that will be developed more fully in succeeding chapters. We began by discussing how scientists think and work and then described how they have unraveled the history of our planet and its interacting systems. We have followed this story from Earth's fiery beginnings through its changes over geologic time and from the origin of life to the appearance of humans. Not every topic in this preview will survive unchanged over the decades or even until the next edition of this book. Geology is undergoing intensive research by your teachers and other scientists who will fill in details, challenge hypotheses, and present new ones in their pursuit of new knowledge.

SUMMARY

What is geology? Geology is the science that deals with Earth—its history, its composition and internal structure, and its surface features.

How do geologists study Earth? Geologists, like other scientists, use the scientific method. They share the data that they develop and check one another's work. A hypothesis is a tentative explanation of a body of data. If it is confirmed repeatedly by other scientists' experiments, it may be elevated to a theory. Many theories are abandoned when subsequent experimental work shows them to be false. Confidence grows in those theories that withstand repeated tests and are able to predict the results of new experiments.

How did our solar system originate? The Sun and its family of planets probably formed when a primeval cloud of gas and dust condensed about 4.5 billion years ago. The planets vary in chemical composition in accordance with their distance from the Sun and with their size.

How did Earth form and evolve over time? Earth probably grew by accretion of colliding chunks of matter. Soon after it formed, it was struck by a giant bolide. Matter ejected into space from both Earth and the impactor reassembled to form the Moon. The impact melted much of Earth. Radioactivity also contributed to early heating and melting. Heavy matter, rich in iron, sank toward Earth's center, and lighter matter floated up to form the outer layers that became the crust and continents. Outgassing contributed to the growth of the oceans and a primitive atmosphere. In this way, Earth was transformed into a differentiated planet with chemically distinct zones: an iron core; a mantle that is mostly magnesium, iron, silicon, and oxygen; and a crust rich in the light elements oxygen, silicon, aluminum, calcium, potassium, and sodium and in radioactive elements.

What are the basic elements of plate tectonics? The lithosphere is not a continuous shell; it is broken into about a dozen large "plates." Driven by convection in the mantle, plates move over Earth's surface at rates of a few centimeters per year. Each plate acts as a distinct rigid unit, riding on the asthenosphere, which also is in motion. The lithosphere begins to gel from rising hot mantle material where plates separate, cooling and becoming more rigid as it moves away from this divergent boundary. Eventually, it sinks into the asthenosphere, dragging material back into the mantle at boundaries where plates converge.

How do we study Earth as a system of interacting components? When trying to understand a complex system such as Earth, we find that it is often easier to break the system down into subsystems (geosystems) to see how they work and interact with one another. Two major global systems are the climate system, which mainly involves interactions among the atmosphere, hydrosphere, and biosphere driven by heat from the Sun, and the plate tectonic system, which mainly involves interactions among Earth's solid components (lithosphere, asthenosphere, and deep mantle) driven by the planet's internal heat.

What are some major events in Earth's history? Earth formed as a planet about 4.5 billion years ago. Rocks as old as 4 billion years have survived in Earth's crust. The earliest evidence of life has been found in rocks about 3.5 billion years old. By 2.5 billion years ago, the oxygen content of the atmosphere was rising due to photosynthesis by early plant life. Animals appeared suddenly about 600 million years ago, diversifying rapidly in a great evolutionary explosion. The subsequent evolution of life was marked by a series of mass extinctions, the last caused by a large bolide impact 65 million years ago, which killed off the dinosaurs. Our species first appeared about 40,000 years ago.

Key Terms and Concepts

asthenosphere (p. 14)
biosphere (p.16)
bolide (p. 3)
climate system (p.14)
core (p. 7)
crust (p. 8)
differentiation (p. 6)
Earth system (p. 12)
fossil (p. 17)
geosystems (p. 14)
hydrosphere (p. 14)
lithosphere (p. 14)
mantle (p. 8)
nebular hypothesis (p. 4)
plate tectonics (p. 14)
plate tectonic system (p. 15)
principle of uniformitarianism (p. 3)
scientific method (p. 2)

Exercises

This icon indicates that there is an animation available on the Web site that may assist you in answering a question.

1. What is the difference between an experiment, a hypothesis, a theory, and a fact?

2. What factors made Earth a particularly congenial place for life to develop?

3. How and why do the inner planets differ from the giant outer planets?

4. What caused Earth to differentiate, and what was the result?

5. How does the chemical composition of Earth's crust differ from that of its deeper interior? From that of its core?

6. How does viewing Earth as a system of interacting components help us to understand our planet? Give an example of an interaction between two or more geosystems.

7. Describe the central idea of the theory of plate tectonics.

Thought Questions

This icon indicates that there is an animation available on the Web site that may assist you in answering a question.

1. How does the discovery of planets orbiting other stars contribute to the debate about the possibility of life elsewhere in the cosmos? What are the scientific and philosophical implications of the existence of life on the planets of other stars? In addition to those listed on page 6, what other conditions would be required for advanced life on another planet? (For example: A magnetic field? A rotating planet?)

2. What are the advantages and disadvantages of living on a differentiated planet? On a geologically active planet?

3. If a giant impact such as the one that formed the Moon had occurred after life had formed on Earth, what would have been the consequences?

4. If you were an astronaut landing on an unexplored planet, how would you decide whether the planet was differentiated and whether it was still geologically active?

5. How do the terms *weather* and *climate* differ? Express the relationship between climate and weather as a simple equation in words: Climate equals *x* of weather over *y* years.

6. Not every planet has a geodynamo. Why? If Earth did not have a magnetic field, what would be different about our planet from the one we know?

7. Knowing how the Moon was created, what might you expect as a result if you are told that a large meteor has collided with a planet twice its size? What could be the effect of this collision on the interior composition of this planet? How would the result of the impact differ if the meteor were significantly smaller than the planet?

Suggested Readings

Ahrens, T. J. 1994. The origin of the Earth. *Physics Today* 47: 38–45.

Allegre, C. 1992. *From Stone to Star.* Cambridge, Mass.: Harvard University Press.

Alley, Richard B. 2001. The key to the past. *Nature* 409: 289.

Becker, L. 2002. Repeated blows. *Scientific American* (March): 77–83.

Cole, G. H. A. 2001. Exoplanets. *Astronomy and Geophysics* (February): 1.13–1.17.

Doyle, L. R., H.-J. Deeg, and T. M. Brown. 2000. Searching for shadows of other Earths. *Scientific American* (July): 60–65

Exploring Space. 1990. Special issue of *Scientific American.*

Golombeck, M. P. 1998. The Mars *Pathfinder* mission. *Scientific American* (July): 40–49.

Hallam, A. 1973. *A Revolution in the Earth Sciences: From Continental Drift to Plate Tectonics.* Oxford: Clarendon Press.

Halliday, A. N., and M. J. Drake. 1999. Colliding theories (origin of Earth and Moon). *Science* 283: 1861–1864.

Kerr, R. A. 2001. Rethinking water on Mars and the origin of life. *Science* 292: 39–40.

Lissauer, J. J. 1999. How common are habitable planets? *Nature* 402: C11–C13.

Merritts, D., A. de Wet, and K. Menking. 1998. *Environmental Geology.* New York: W. H. Freeman.

NASA. 2002. *Living on a Restless Planet.* Pasadena, CA: Jet Propulsion Laboratory. See also http:\\solidearth.jpl.nasa.gov.

National Academy of Sciences. 1999. *Science and Creationism.* Washington, D.C.: National Academy Press.

National Academy of Sciences. 1998. *Teaching About Evolution and the Nature of Science.* Washington, D.C.: National Academy Press.

National Research Council. 1993. *Solid-Earth Sciences and Society.* Washington, D.C.: National Academy Press.

Roberts, F. 2001. The origin of water on Earth. *Science* 293: 1056–1058.

Scientific American. 2000. *The Frontiers of Space.* New York: Scientific American Press.

Smith, D. E., et al. 1999. The global topography of Mars and implications for surface evolution. *Science* 284: 1495–1503.

Stanley, S. M. 1999. *Earth System History.* New York: W. H. Freeman.

Westbroek, P. 1991. *Life as a Geologic Force.* New York: W. W. Norton.

Wetherill, G. W. 1990. Formation of the Earth. *Annual Review of Earth Planetary Sciences* 8: 205–256.

Woolfson, M. 2000. The origin and evolution of the solar system. *Astronomy and Geophysics* 41: 1.2–1.18.

Mount Everest, Nepal, the highest mountain in the world, as viewed from Kala Pattar.
[Michael C. Klesius/National Geographic/Getty Images.]

CHAPTER 2

Plate Tectonics: The Unifying Theory

"What is now proved was once only imagin'd."
WILLIAM BLAKE

The lithosphere—Earth's strong, rigid outer shell of rock—is broken into about a dozen plates, which slide by, converge with, or separate from each other as they move over the weaker, ductile asthenosphere. Plates are created where they separate and are recycled where they converge, in a continuous process of creation and destruction. Continents, embedded in the lithosphere, drift along with the moving plates. The theory of **plate tectonics** describes the movement of plates and the forces acting between them. It also explains the distribution of many large-scale geologic features that result from movements at plate boundaries: mountain chains, rock assemblages, structures on the seafloor, volcanoes, and earthquakes. Plate tectonics provides a conceptual framework for a large part of this book and, indeed, for much of geology. **This chapter lays out the plate tectonics theory and examines how the forces that drive plate motions are related to the mantle convection system.**

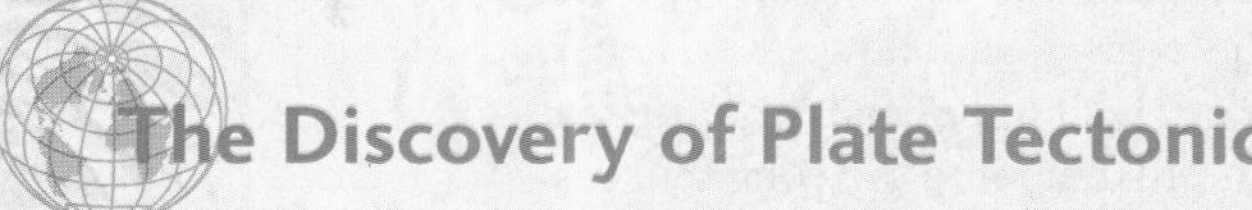

The Discovery of Plate Tectonics

In the 1960s, a great revolution in thinking shook the world of geology. For almost 200 years, geologists had developed various theories of *tectonics* (from the Greek *tekton,* meaning "builder")—the general term they used to describe mountain building, volcanism, and other processes that construct geologic features on Earth's surface. It was not until the discovery of plate tectonics, however, that a single theory could satisfactorily explain the whole range of geologic processes. Not only is plate tectonics all-encompassing, it is also elegant: many observations can be explained by a few simple principles. In the history of science, simple theories that explain many observations usually turn out to be the most enduring. Physics had a comparable revolution at the beginning of the twentieth century, when the theory of relativity unified the physical laws that govern space, time, mass, and motion. Biology had a comparable revolution in the middle of the twentieth century, when the discovery of DNA allowed biologists to explain how organisms transmit the information that controls their growth, development, and functioning from generation to generation.

The basic ideas of plate tectonics were put together as a unified theory of geology less than 40 years ago. The scientific synthesis that led to plate tectonics, however, really began much earlier in the twentieth century, with the recognition of evidence for continental drift.

Continental Drift

Such changes in the superficial parts of the globe seemed to me unlikely to happen if the earth were solid to the center. I therefore imagined that the internal parts might be a fluid more dense, and of greater specific gravity than any of the solids we are acquainted with, which therefore might swim in or upon that fluid. Thus the surface of the earth would be a shell, capable of being broken and disordered by the violent movements of the fluid on which it rested.

(Benjamin Franklin, 1782, in a letter to French geologist Abbé J. L. Giraud-Soulavie)

The concept of **continental drift**—large-scale movements of continents over the globe—has been around for a long time. In the late sixteenth century and in the seventeenth century, European scientists noticed the jigsaw-puzzle fit of the coasts on both sides of the Atlantic, as if the Americas, Europe, and Africa were at one time assembled together and had subsequently drifted apart. By the close of the nineteenth century, the Austrian geologist Eduard Suess put some of the pieces of the puzzle together and postulated that the combined present-day southern continents had once formed a single giant continent called *Gondwanaland* (or *Gondwana*). In 1915, Alfred Wegener, a German meteorologist who was recovering from wounds suffered in World War I, wrote a book on the breakup and drift of continents. In it, he laid out the remarkable similarity of rocks, geologic structures, and fossils on opposite sides of the Atlantic. In the years that followed, Wegener postulated a supercontinent, which he called **Pangaea** (Greek for "all lands"), that broke up into the continents as we know them today (**Figure 2.1**).

Although Wegener was correct in asserting that the continents had drifted apart, his hypotheses about how fast the continents were moving and what forces were pushing them across Earth's surface turned out to be wrong, which reduced his credibility among other scientists. After about a decade of spirited debate, physicists convinced geologists that Earth's outer layers were too rigid for continental drift to occur, and Wegener's ideas fell into disrepute except among a few geologists in Europe, South Africa, and Australia.

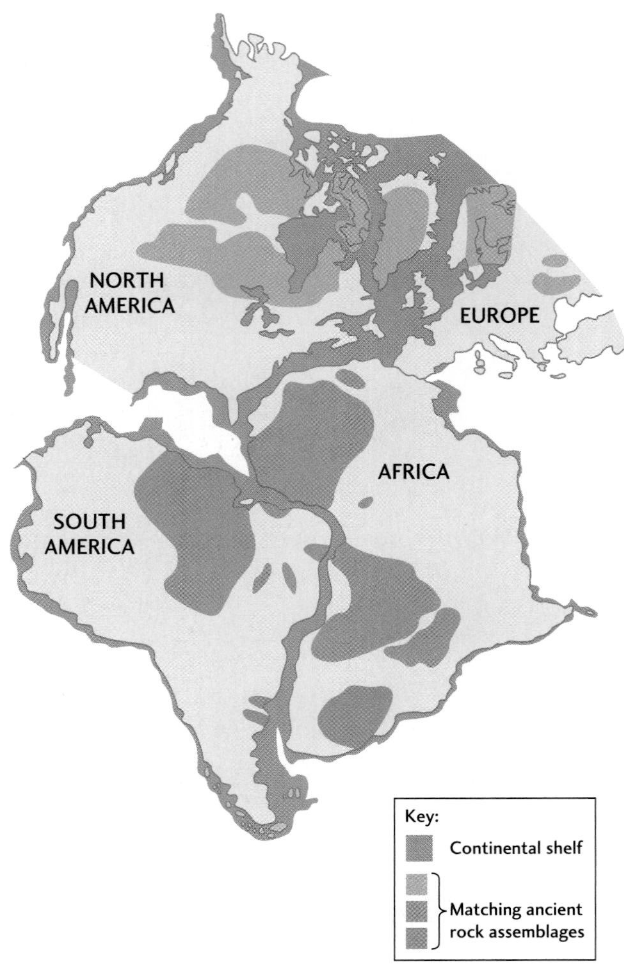

Figure 2.1 The jigsaw-puzzle fit of continents bordering the Atlantic Ocean formed the basis of Alfred Wegener's theory of continental drift. In his book *The Origin of Continents and Oceans,* Wegener cited as additional evidence the similarity of geologic features on opposite sides of the Atlantic. The matchup of ancient crystalline rocks is shown in adjacent regions of South America and Africa and of North America and Europe. [Geographic fit from data of E. C. Bullard; geological data from P. M. Hurley.]

The advocates of the drift hypothesis pointed not only to geographic matching but also to geologic similarities in rock ages and trends in geologic structures on opposite sides of the Atlantic (see Figure 2.1). They also offered arguments, accepted now as good evidence of drift, based on fossil and climatological data. Identical fossils of a reptile 300 million years old, for example, are found only in Africa and South America, suggesting that the two continents were joined at the time (**Figure 2.2**). The animals and plants on different continents showed similarities in evolution until the postulated breakup time. After that, they followed divergent evo-

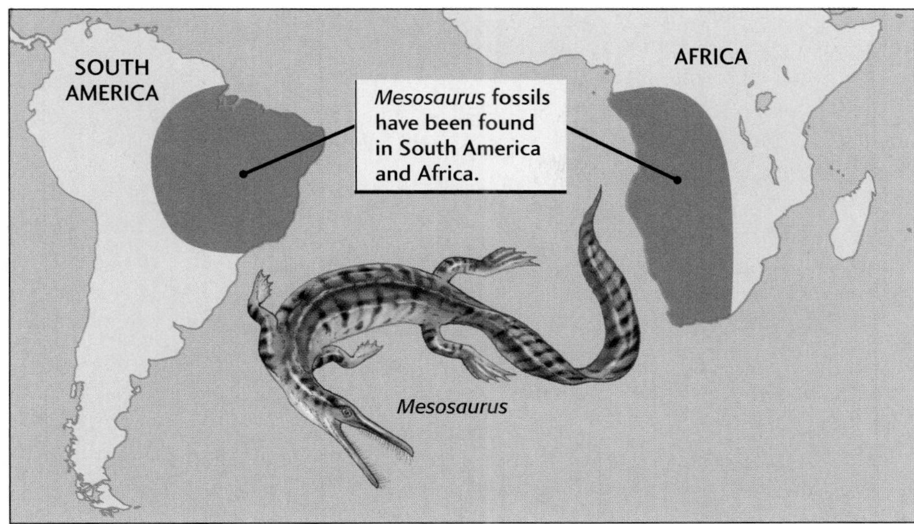

Figure 2.2 Fossils of the reptile *Mesosaurus*, 300 million years old, are found in South America and Africa and nowhere else in the world. If *Mesosaurus* could swim across the South Atlantic Ocean, it could have crossed other oceans and should have spread more widely. The observation that it did not suggests that South America and Africa must have been joined at that time. [After A. Hallam, "Continental Drift and the Fossil Record," *Scientific American* (November 1972): 57–66.]

lutionary paths, presumably because of the isolation and changing environments of the separating continental masses. In addition, rock deposits associated with glaciers that existed some 300 million years ago are now distributed across South America, Africa, India, and Australia. If the southern continents are reassembled into Gondwanaland near the South Pole, a single continental glacier could account for all the glacial deposits.

Seafloor Spreading

The geologic evidence did not convince the skeptics who maintained that continental drift was physically impossible. No one had yet come up with a plausible driving force that could have split Pangaea and moved the continents apart. Wegener, for example, thought the continents floated like boats across the solid oceanic crust, dragged along by the tidal forces of the Sun and Moon!

The breakthrough came when scientists realized that convection in Earth's mantle (discussed in Chapter 1) could push and pull the continents apart, creating new oceanic crust through the process of **seafloor spreading.** In 1928, the British geologist Arthur Holmes came close to expressing the modern notions of continental drift and seafloor spreading when he proposed that convection currents "dragged the two halves of the original continent apart, with consequent mountain building in the front where the currents are descending, and the ocean floor development on the site of the gap, where the currents are ascending." Given the physicists' arguments that Earth's crust and mantle are rigid and immobile, Holmes conceded that "purely speculative ideas of this kind, specially invented to match the requirements, can have no scientific value until they acquire support from independent evidence."

Convincing evidence began to emerge as a result of extensive exploration of the seafloor after World War II. The mapping of the undersea Mid-Atlantic Ridge and the discovery of the deep, cracklike valley, or rift, running down its center sparked much speculation (**Figure 2.3**). Geologists found that almost all of the earthquakes in the Atlantic Ocean occurred near this rift valley. Because most earthquakes are generated by tectonic faulting, these results indicated that the rift was a tectonically active feature. Other mid-ocean ridges with similar shapes and earthquake activity were found in the Pacific and Indian oceans.

In the early 1960s, Harry Hess of Princeton University and Robert Dietz of the Scripps Institution of Oceanography proposed that the crust separates along the rifts in mid-ocean ridges and that new seafloor forms by upwelling of hot new crust into these cracks. The new seafloor—actually the top of newly created lithosphere—spreads laterally away from the rift and is replaced by even newer crust in a continuing process of plate creation.

The Great Synthesis: 1963–1968

The seafloor spreading hypothesis put forward by Hess and Dietz in 1962 explained how the continents could drift apart

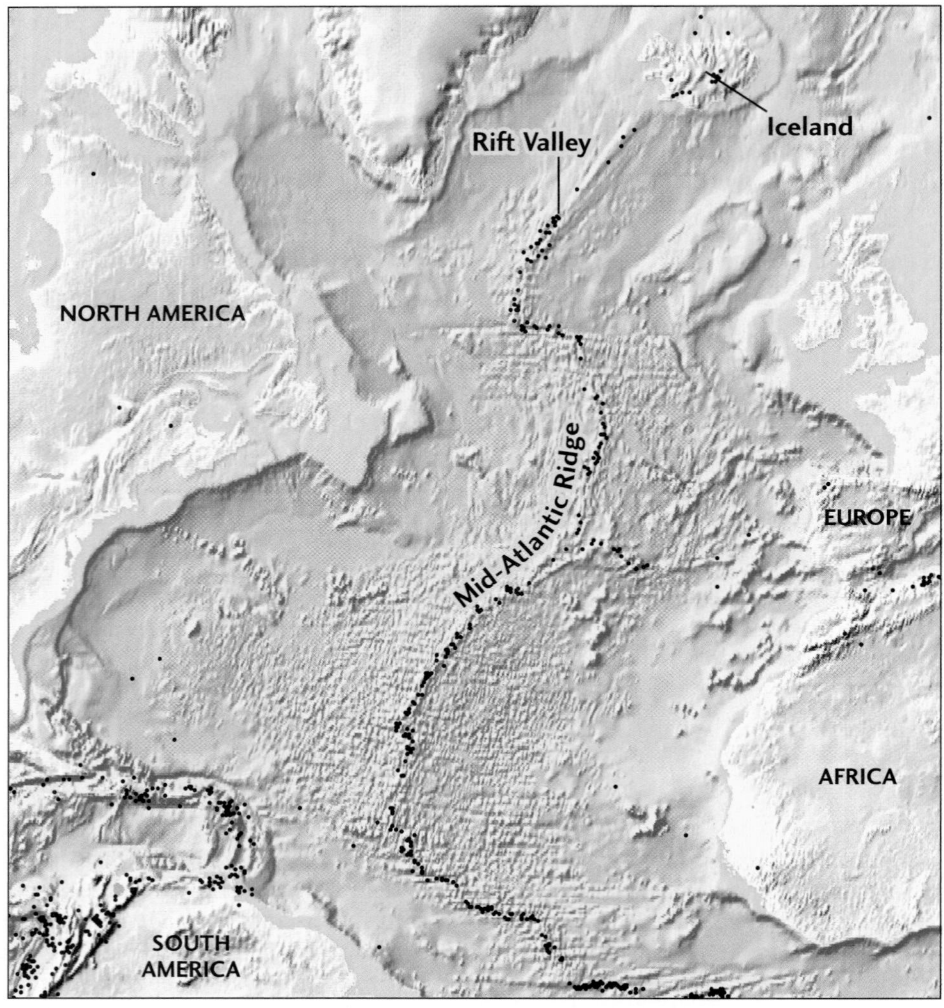

Figure 2.3 The North Atlantic Ocean floor, showing the cracklike rift valley running down the center of the Mid-Atlantic Ridge and associated earthquakes (black dots).

through the creation of new lithosphere at mid-ocean rifts. Could the seafloor and its underlying lithosphere be destroyed by recycling back into Earth's interior? If not, Earth's surface area would have to increase over time, so our planet would have to get bigger and bigger. For a period in the early 1960s, some physicists and geologists actually believed in this idea of an expanding Earth, based on a now-discredited modification of Einstein's theory of gravitation. Other geologists recognized that the seafloor was indeed being recycled in the regions of intense volcanic and earthquake activity around the margins of the Pacific Ocean basin, known collectively as the Ring of Fire (**Figure 2.4**). The details of this process, however, remained obscure.

In 1965, the Canadian geologist J. Tuzo Wilson first described tectonics around the globe in terms of rigid "plates" moving over Earth's surface. He characterized the three basic types of boundaries where plates move apart, come together, or slide past each other. In a quick succession of discoveries and theoretical advances, other scientists showed that almost all current tectonic deformations are concentrated at these boundaries. They measured the rates and directions of the tectonic motions, and they demonstrated that these motions are mathematically consistent with a system of rigid plates moving on the planet's spherical surface. The basic elements of the plate tectonics theory were established by the end of 1968. By 1970, the evidence for plate tectonics had become so persuasive in its abundance that almost all Earth scientists embraced the theory. Textbooks were revised, and specialists began to consider the implications of the new concept for their own fields. For a timeline of the landmark events leading to the theory of plate tectonics, see Appendix 3.

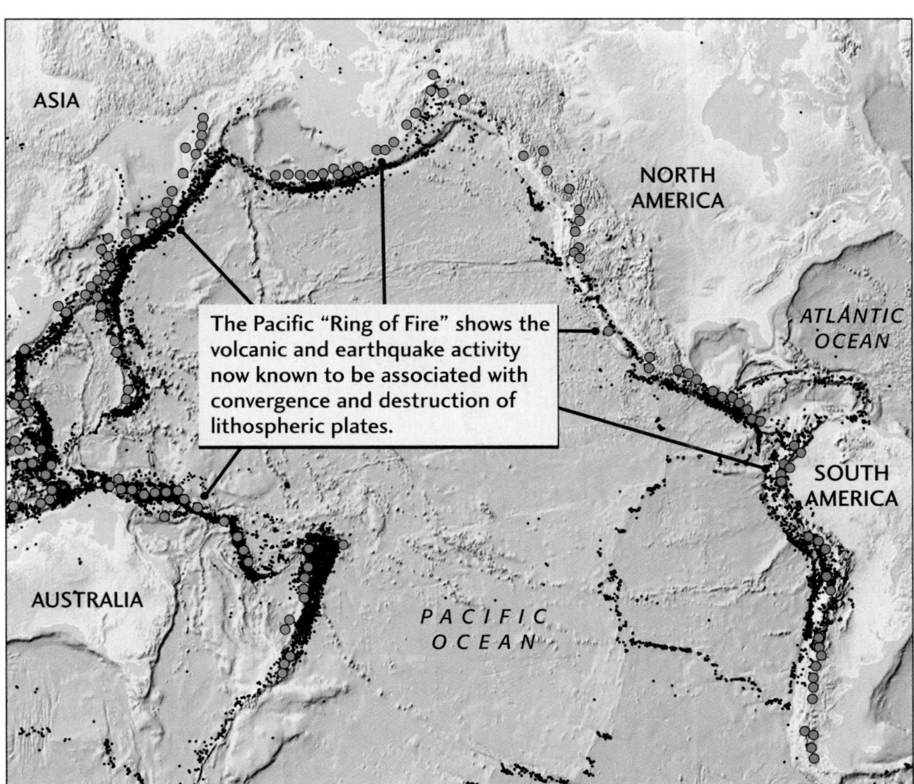

Figure 2.4 The Pacific Ring of Fire, showing active volcanoes (large red circles) and earthquakes (small black dots).

The Mosaic of Plates

According to the theory of plate tectonics, the rigid lithosphere is not a continuous shell but is broken into a mosaic of about a dozen large, rigid plates that are in motion over Earth's surface. Each plate moves as a distinct rigid unit, riding on the asthenosphere, which is also in motion. The major plates and their present-day motions are represented in **Figure 2.5**. The largest is the Pacific Plate, which comprises much (though not all) of the Pacific Ocean basin. Some of the plates are named after the continents they contain, but in no case is a plate identical with a continent. The North American Plate, for instance, extends from the Pacific coast of the continent of North America to the middle of the Atlantic Ocean, where it meets the Eurasian and African plates.

In addition to the major plates, there are a number of smaller ones. An example is the tiny Juan de Fuca Plate, a piece of oceanic lithosphere trapped between the giant Pacific and North American plates just offshore of the northwestern United States. Others are continental fragments, such as the small Anatolian Plate, which includes much of Turkey. (Not all of the smaller plates are shown in Figure 2.5.)

If you want to see geology in action, go to a plate boundary. Depending on which boundary you visit, you will find earthquakes; volcanoes; mountains; long, narrow rifts; and more. Many geologic features develop through the interactions of plates at their boundaries. The three basic types of plate boundaries are depicted in Figure 2.5 and discussed in the following pages.

- At *divergent boundaries,* plates move apart and new lithosphere is created (plate area increases).
- At *convergent boundaries,* plates come together and one is recycled back into the mantle (plate area decreases).
- At *transform-fault boundaries,* plates slide horizontally past each other (plate area remains constant).

Like many models of nature, the three types of plates shown in Figure 2.5 are idealized. Besides these basic types, there are "oblique boundaries" that combine divergence or convergence with some amount of transform faulting. Moreover, what actually goes on at a plate boundary depends on the type of lithosphere involved, because continental and oceanic lithosphere behave rather differently. The continental crust is made of rocks that are both lighter and weaker than either the oceanic crust or the mantle beneath the crust. Later chapters will examine this compositional difference in more detail, but for now you need only to keep in mind two consequences: (1) because it is lighter, continental crust is not as

EARTH'S LITHOSPHERE IS MADE OF MOVING PLATES

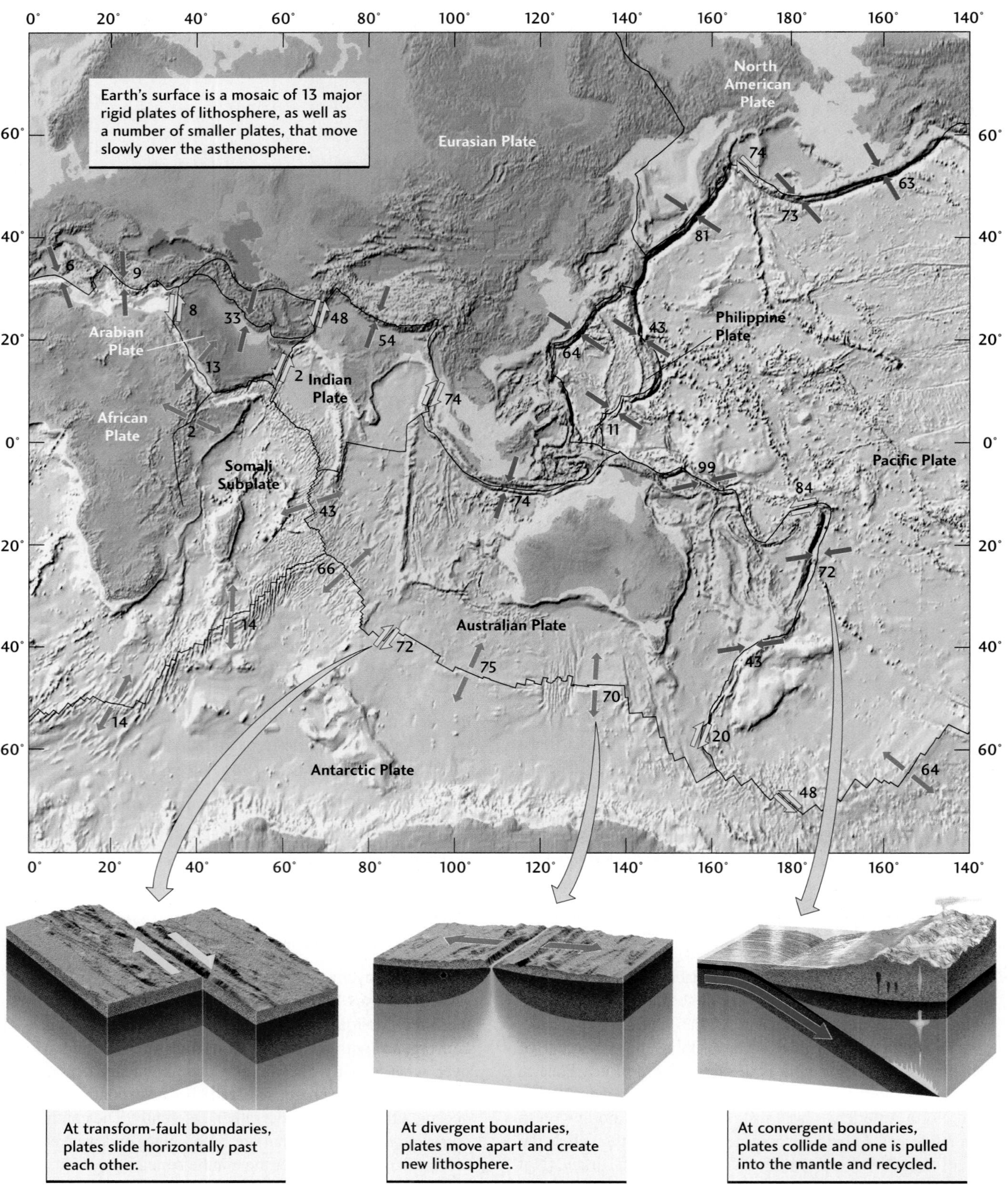

At transform-fault boundaries, plates slide horizontally past each other.

At divergent boundaries, plates move apart and create new lithosphere.

At convergent boundaries, plates collide and one is pulled into the mantle and recycled.

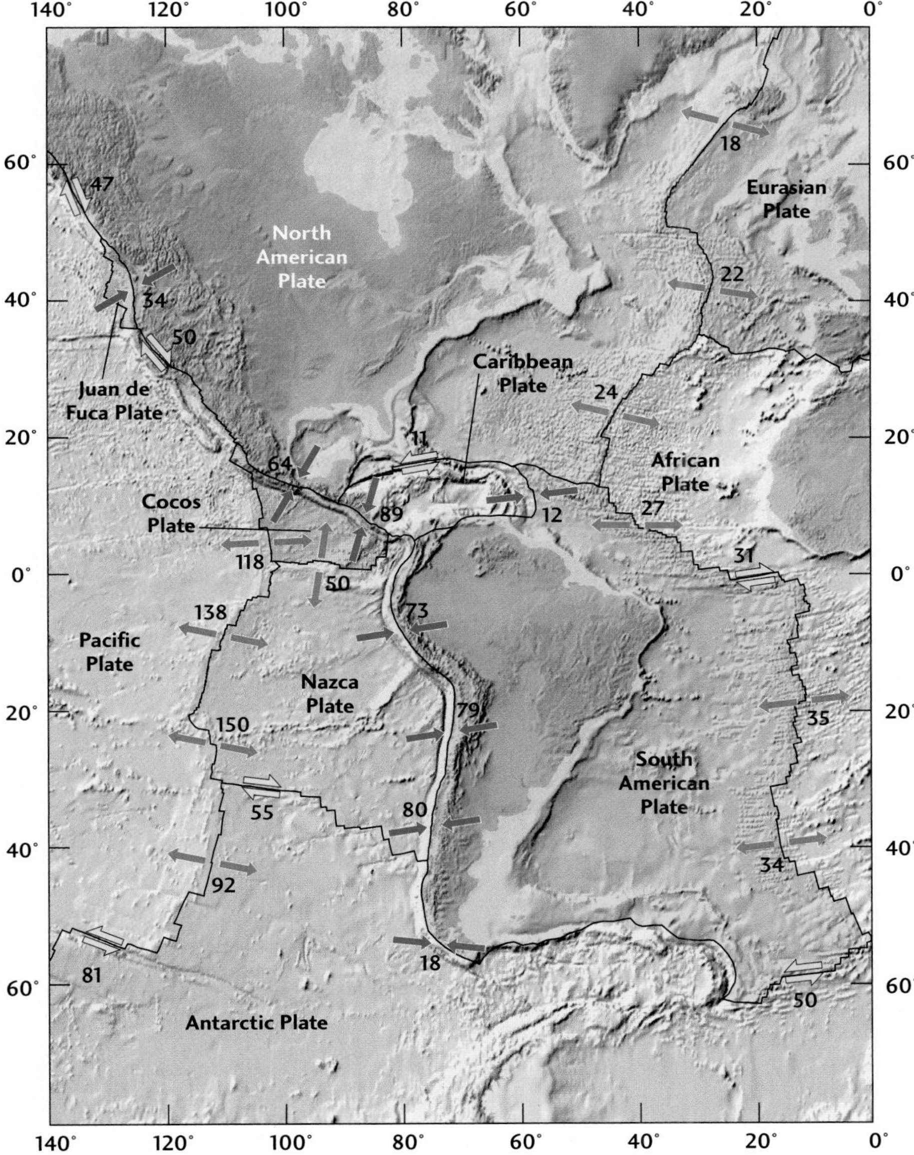

Figure 2.5 Earth's plates today, showing the plate boundaries. This flattened view of Earth's land and undersea topography shows the three basic types of plate boundaries: divergent boundaries, where plates separate (←→); convergent boundaries, where plates come together (→←); and transform-fault boundaries, where plates slide past each other (⇄). These arrows show in what directions the plates are moving with respect to each other at their common boundaries. The numbers next to the arrows indicate the relative plate speeds in millimeters per year. [Plate boundaries by Peter Bird, UCLA.]

easily recycled as oceanic crust, and (2) because continental crust is weaker, plate boundaries that involve continental crust tend to be more spread out and more complicated than oceanic plate boundaries.

Divergent Boundaries

Divergent boundaries within the ocean basins are narrow rifts that approximate the idealization of plate tectonics. Divergence within the continents is usually more complicated and distributed over a wider area. This difference is illustrated in **Figure 2.6**.

Oceanic Plate Separation On the seafloor, the boundary between separating plates is marked by a mid-ocean ridge that exhibits active volcanism, earthquakes, and rifting caused by tensional (stretching) forces that are pulling the two plates apart. Figure 2.6a shows what happens in one example, the Mid-Atlantic Ridge. Here seafloor spreading is at work as the North American and Eurasian plates sepa-

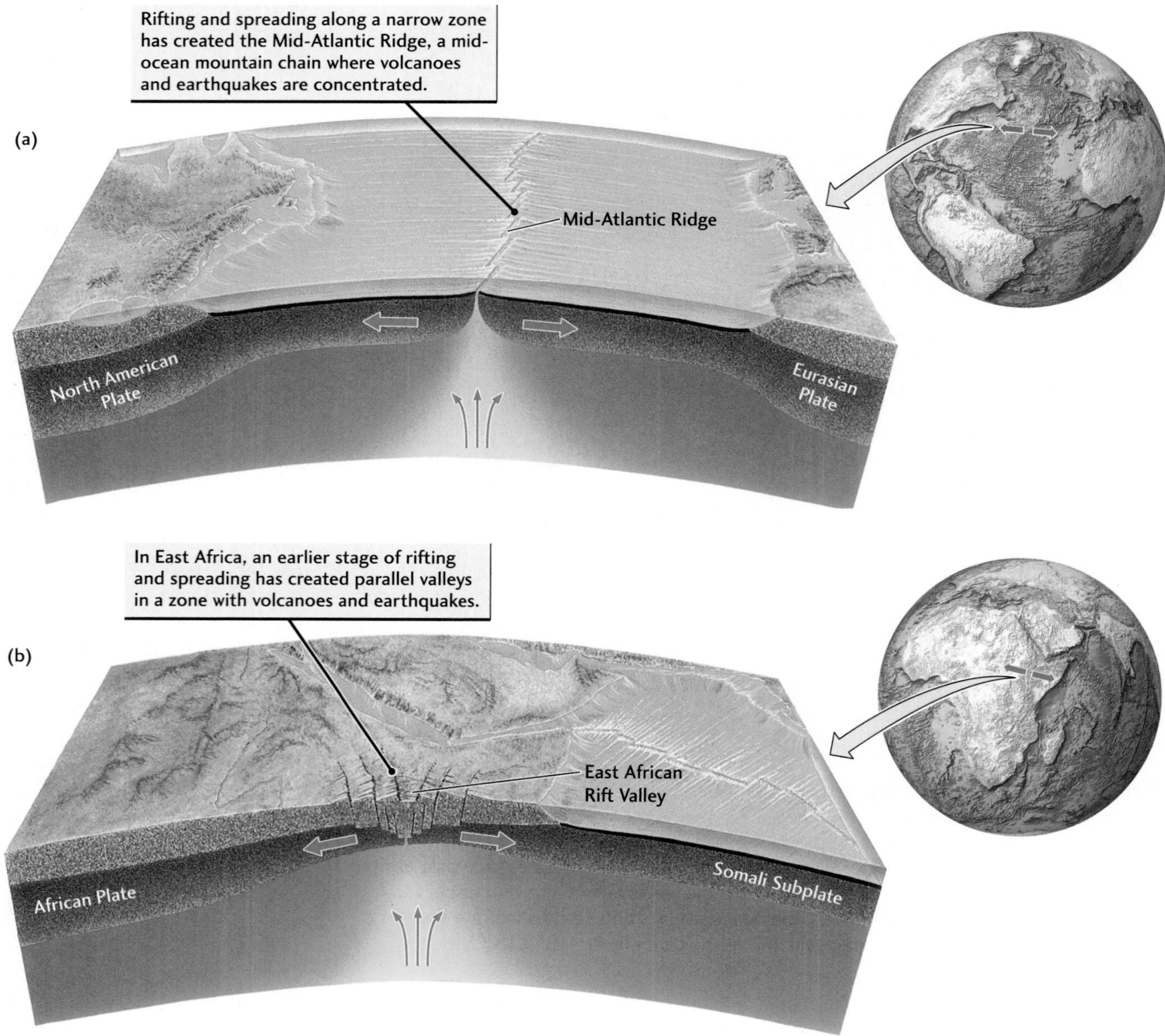

Figure 2.6 (a) Rifting and seafloor spreading along the Mid-Atlantic Ridge create a mid-ocean volcanic mountain chain where volcanism, faulting, and earthquakes are concentrated along a narrow spreading center. (b) Early stages of rifting and plate separation now occurring within eastern Africa, where multiple rift valleys with their associated faulting, volcanism, and earthquakes are distributed over a wider zone.

Figure 2.7 The Mid-Atlantic Ridge, a divergent plate boundary, surfaces above sea level in Iceland. The cracklike rift valley filled with new volcanic rocks indicates that plates are being pulled apart. [Gudmundur E. Sigvaldason, Nordic Volcanological Institute.]

rate and new Atlantic seafloor is created by mantle upwelling. (A more detailed portrait of the Mid-Atlantic Ridge was shown in Figure 2.3.) The island of Iceland exposes a segment of the otherwise submerged Mid-Atlantic Ridge, providing geologists with an opportunity to view the process of plate separation and seafloor spreading directly (Figure 2.7). The Mid-Atlantic Ridge is discernible in the Arctic Ocean north of Iceland and connects to a nearly globe-encircling system of mid-ocean ridges that winds through the Indian and Pacific oceans, ending along the western coast of North America. These **spreading centers** have created the millions of square kilometers of oceanic crust that now floor the world's oceans.

Continental Plate Separation Early stages of plate separation, such as the Great Rift Valley of East Africa (Figure 2.6b), can be found on some continents. These divergent boundaries are characterized by rift valleys, volcanic activity, and earthquakes distributed over a wider zone than oceanic spreading centers. The Red Sea and the Gulf of California are rifts that are further along in the spreading process (Figure 2.8). In these cases, the continents have separated enough for new seafloor to form along the spreading axis, and the rift valleys have been flooded by the ocean. Sometimes continental rifting may slow down or stop before the continent splits apart and a new ocean basin opens. The Rhine Valley along the border of Germany and France is a weakly active continental rift that may be this type of "failed spreading center." Will the East African Rift continue to open up, causing the Somali Subplate to split away from Africa completely and form a new ocean basin, as happened between Africa and the island of Madagascar? Or will the spreading slow down and eventually stop, as appears to be happening in western Europe? Geologists don't know the answers.

Convergent Boundaries

Plates cover the globe, so if they separate in one place, they must converge somewhere else, to conserve Earth's surface area. (As far as we can tell, our planet is not expanding!) Where plates collide head-on, they form convergent boundaries. The profusion of geologic events resulting from plate collisions makes convergent boundaries the most complex type observed in plate tectonics.

Ocean-Ocean Convergence If the two plates involved are oceanic, one descends beneath the other in a process known as **subduction** (Figure 2.9a). The oceanic lithosphere of the subducting plate sinks into the asthenosphere and is eventually recycled by the mantle convection system. This downbuckling produces a long, narrow deep-sea trench. In the Marianas Trench of the western Pacific, the ocean reaches its greatest depth, about 10 km—deeper than the height of Mount Everest. As the cold lithospheric slab descends, the pressure increases; the water trapped in the rocks of subducted oceanic crust is "squeezed out" and rises into the asthenosphere above the slab. This fluid causes the mantle to melt, producing a chain of volcanoes, called an **island arc,** on the seafloor behind the trench. The subduction of the Pacific Plate has formed the volcanically active Aleutian Islands west of Alaska, as well as the abundant island arcs of the western Pacific. Earthquakes that occur as deep as 600 km beneath these island arcs delineate the cold slabs of lithosphere as they descend into the mantle.

Ocean-Continent Convergence If one plate has a continental edge, it overrides the oceanic plate, because continental crust is lighter and much less easily subducted than oceanic crust (Figure 2.9b). The continental ledge crumples and is uplifted into a mountain chain roughly parallel to the deep-sea trench. The enormous forces of collision and subduction produce great earthquakes along the subduction interface. Over time, materials are scraped off the descending slab and incorporated into the adjacent mountains, leaving geologists

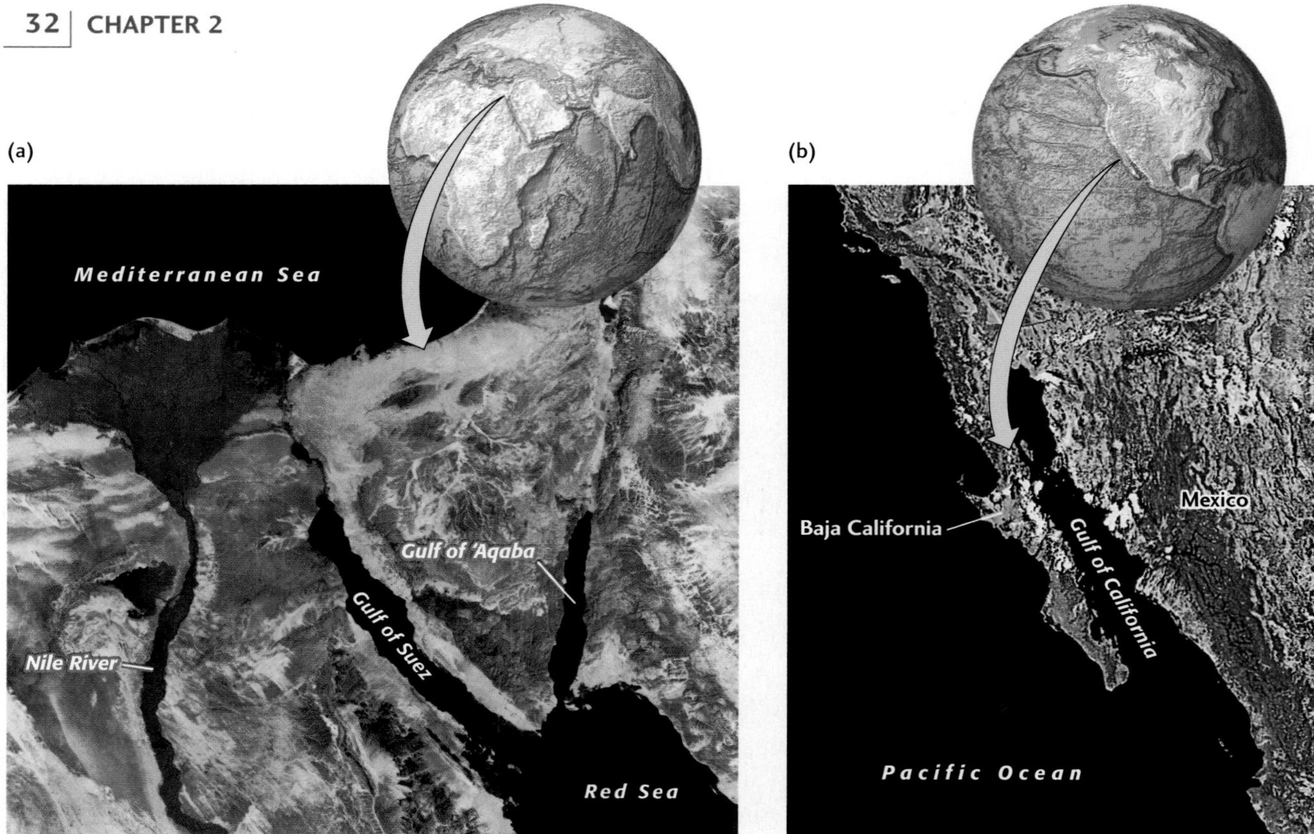

Figure 2.8 (a) The Red Sea (*lower right*) divides to form the Gulf of Suez on the left and the Gulf of 'Aqaba on the right. The Arabian Peninsula, on the right, splitting away from Africa on the left, has opened these great rifts, which are now flooded by the sea. The Nile River (*far left*) flows north into the Mediterranean Sea (*top*). [Earth Satellite Corporation.] (b) The Gulf of California, an opening ocean resulting from plate motions, marks a widening rift between Baja California and the Mexican mainland. [Worldsat International/Photo Researchers.]

with a complex (and often confusing) record of the subduction process. As in the case of ocean-ocean convergence, the water carried down by the subducting oceanic plate causes melting in the mantle wedge and the formation of volcanoes in the mountain belts behind the trench.

The west coast of South America, where the South American Plate collides with the oceanic Nazca Plate, is a subduction zone of this type. A great chain of high mountains, the Andes, rises on the continental side of the collision boundary, and a deep-sea trench lies just off the coast. The volcanoes here are active and deadly. One of them, Nevado del Ruiz in Colombia, killed 25,000 people when it erupted in 1985. Some of the world's greatest earthquakes also have been recorded along this boundary. Another example occurs where the small Juan de Fuca Plate subducts beneath the North American Plate off the coast of western North America. This convergent boundary gives rise to the dangerous volcanoes of the Cascade Range, which produced the Mount St. Helens eruption of 1980. As our understanding of the Cascadia subduction zone grows, scientists are increasingly worried that a great earthquake could occur there and cause considerable damage along the coasts of Oregon, Washington, and British Columbia.

Continent-Continent Convergence Where plate convergence involves two continents (Figure 2.9c), oceanic-type subduction cannot occur. The geologic consequences of such a collision are considerable. The collision of the Indian and Eurasian plates, both with continents at their leading edges, provides the best example. The Eurasian Plate overrides the Indian Plate, but India and Asia remain afloat, creating a double thickness of crust and forming the highest mountain range in the world, the Himalaya, as well as the vast, high plateau of Tibet. Severe earthquakes occur in the crumpling crust of this and other continent-continent collision zones.

Transform-Fault Boundaries

At boundaries where plates slide past each other, lithosphere is neither created nor destroyed. Such boundaries are **transform faults:** fractures along which relative displacement occurs as horizontal slip between the adjacent blocks. Transform-fault boundaries are typically found along mid-ocean ridges where the continuity of a divergent boundary is broken and the boundary is offset in a steplike pattern. The San Andreas fault in California, where the Pacific Plate slides by the North American Plate, is a

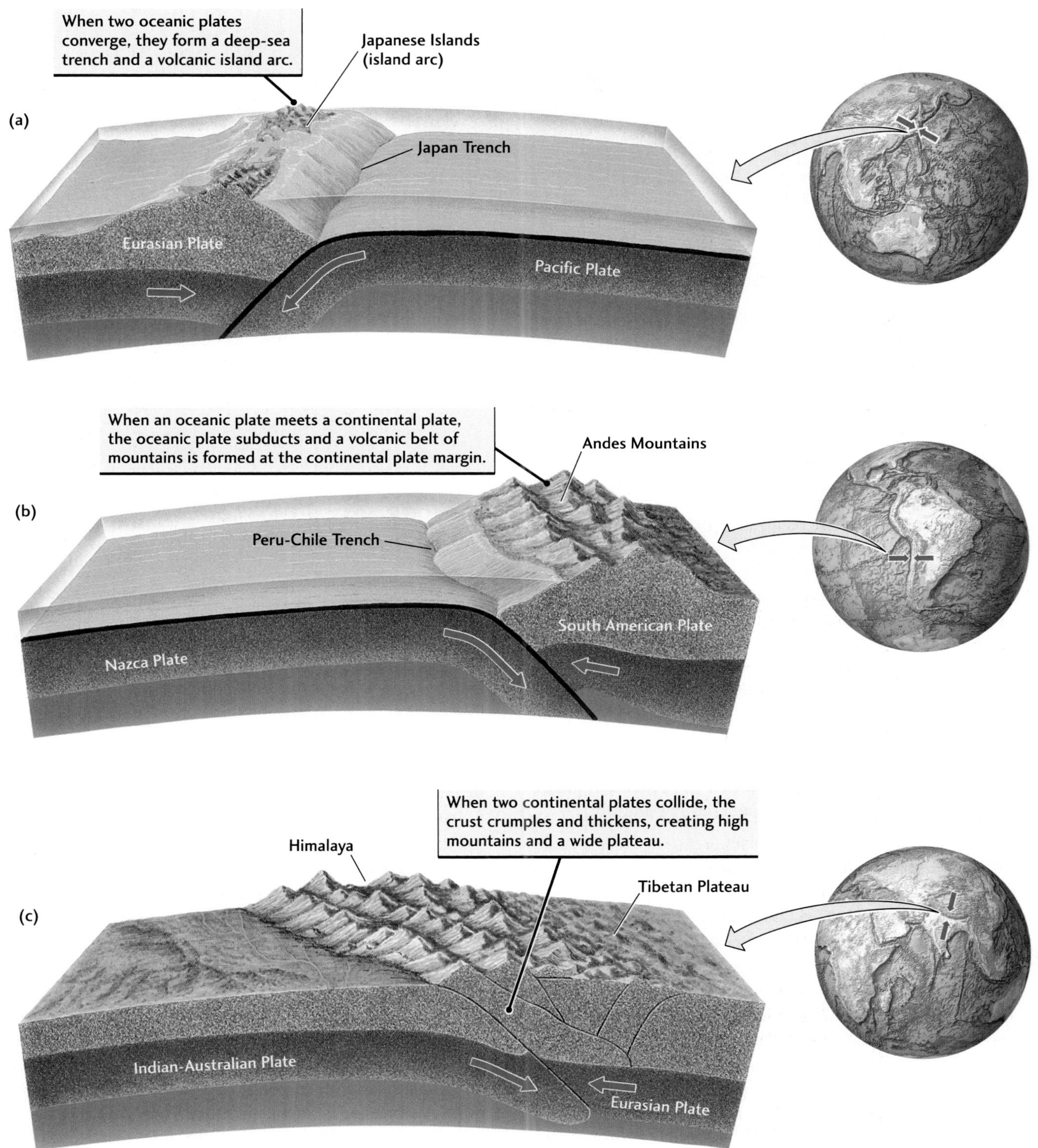

Figure 2.9 Three types of convergent boundaries. (a) Subduction of an oceanic plate beneath another oceanic plate, which forms a deep-sea trench and a volcanic island arc. (b) Subduction of an oceanic plate at a continental margin, which forms a mountainous volcanic belt within the deforming margin of the continent instead of an island arc. (c) A continent-continent plate collision, which crumples and thickens the continental crust, creating high mountains and a wide plateau.

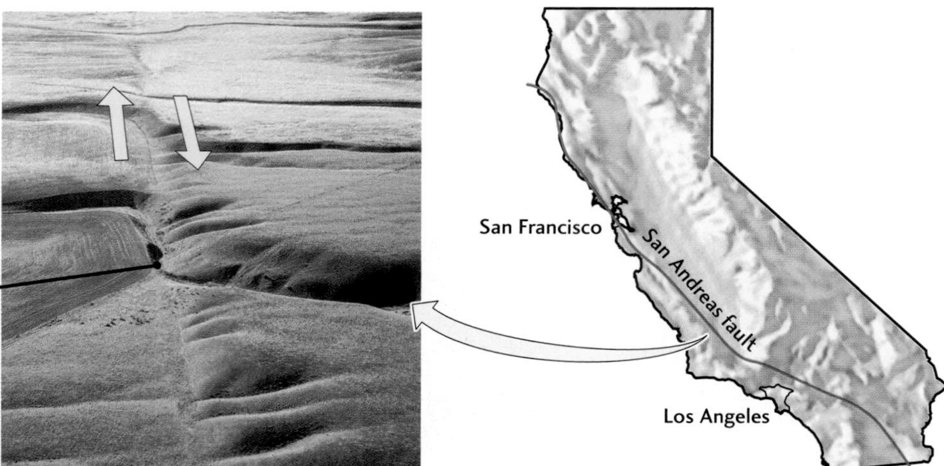

Figure 2.10 The view northwest along the San Andreas fault in the Carrizo Plain of central California. The San Andreas is a transform fault, forming a portion of the sliding boundary between the Pacific Plate on the left and the North American Plate on the right. Notice how the movement on the fault has offset the streams flowing across it. [John Shelton.]

prime example of a transform fault on land, as shown in **Figure 2.10**. Because the plates have been sliding past each other for millions of years, rocks facing each other on the two sides of the fault are of different types and ages. Large earthquakes, such as the one that destroyed San Francisco in 1906, can occur on transform-fault boundaries. There is much concern that a sudden slip may occur along the San Andreas fault or related faults near Los Angeles and San Francisco within the next 25 years or so, resulting in an extremely destructive earthquake.

Transform faults can also connect divergent plate boundaries with convergent boundaries and convergent boundaries with other convergent boundaries. Can you find examples of these types of transform-fault boundaries in Figure 2.5?

Combinations of Plate Boundaries

Each plate is bounded by some combination of divergent, convergent, and transform boundaries. As we can see in Figure 2.5, the Nazca Plate in the Pacific is bounded on three sides by zones of divergence, where new lithosphere is generated along mid-ocean ridge segments offset in a stepwise pattern by transform faults, and on one side by the Peru-Chile subduction zone, where lithosphere is consumed at a deep-sea trench. The North American Plate is bounded on the east by the Mid-Atlantic Ridge, a divergence zone; on the west by the San Andreas fault and other transform boundaries; and on the northwest by subduction zones and transform boundaries that run from Oregon to the Aleutians.

Rates and History of Plate Motions

How fast do plates move? Do some plates move faster than others, and if so, why? Is the velocity of plate movements today the same as it was in the geologic past? Geologists have developed ingenious methods to answer these questions and thereby gain a better understanding of plate tectonics. In this section, we will examine three of these methods.

The Seafloor as a Magnetic Tape Recorder

In World War II, extremely sensitive instruments were developed to detect submarines by the magnetic fields emanating from their steel hulls. Geologists modified these instruments slightly and towed them behind research ships to measure the local magnetic field created by magnetized rocks beneath the sea. Steaming back and forth across the ocean, seagoing scientists discovered regular patterns in the strength of the local magnetic field that completely surprised them. In many areas, the magnetic field alternated between high and low values in long, narrow bands, called **magnetic anomalies,** that were parallel and almost perfectly symmetrical with respect to the crest of the mid-ocean ridge. An example is shown in **Figure Story 2.11**. The detection of these patterns was one of the great discoveries that confirmed seafloor spreading and led to the plate tectonic

Figure Story 2.11 An oceanographic survey over the Reykjanes Ridge, part of the Mid-Atlantic Ridge southwest of Iceland, showed a pattern of magnetic field strength (upper right). This figure illustrates how scientists worked out the explanation of this pattern in terms of two theories: (1) that Earth's magnetic field reverses its direction in intervals of tens of thousands of years, and (2) that seafloor spreading gradually moves newly magnetized crust away from the ridge crest.

MAGNETIC MAPPING CAN MEASURE THE RATE OF SEAFLOOR SPREADING

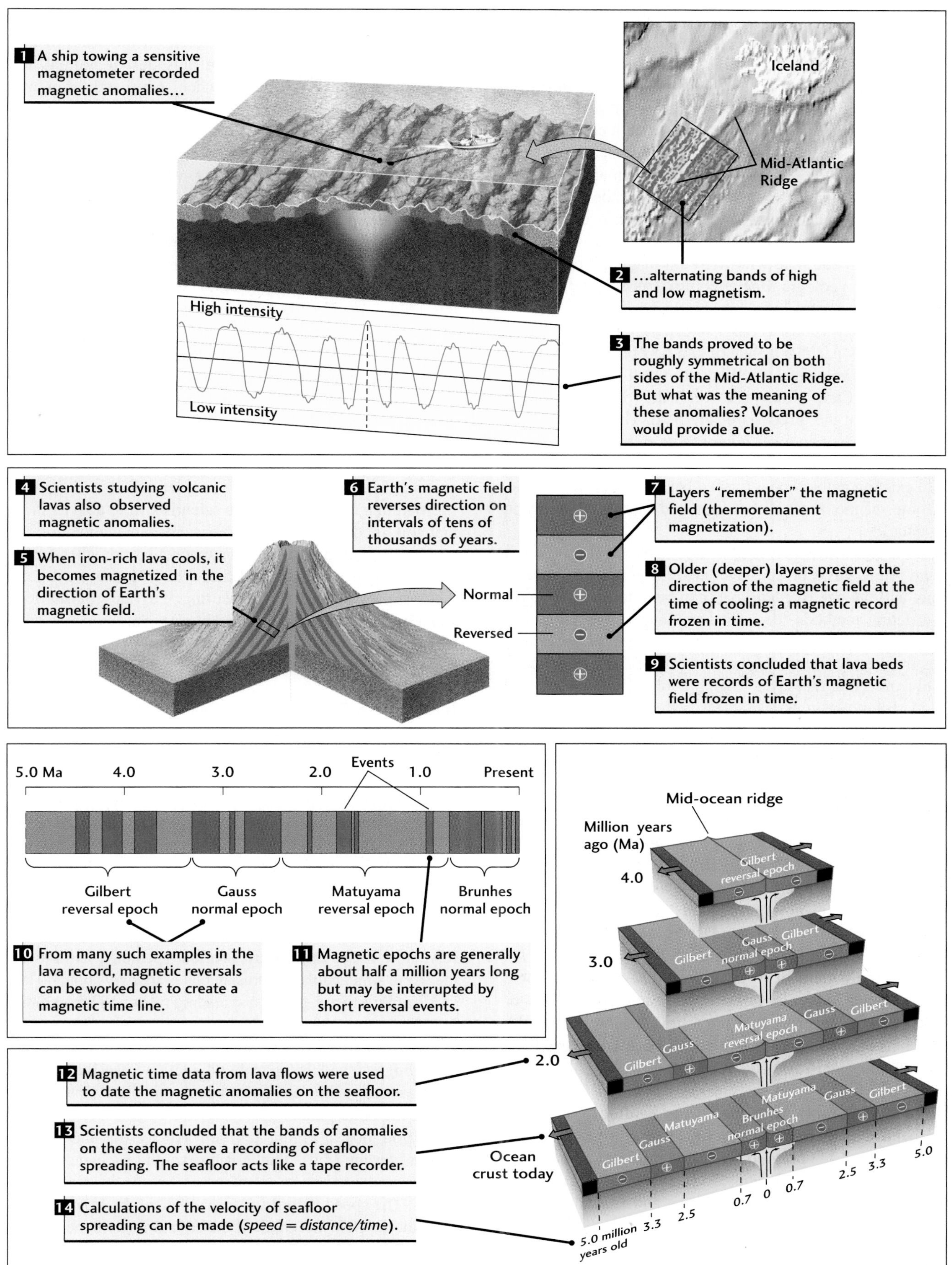

theory. It also allowed geologists to measure plate motions far back into geologic time. To understand these advances, we need to look more closely at how rocks become magnetized.

The Rock Record of Magnetic Reversals on Land For some 400 years, scientists have known that a compass needle points to the north magnetic pole (close to the geographic north pole) because of Earth's magnetic field. Imagine how stunned they were a few decades ago when they found evidence in the rock record that, over geologic time, the magnetic field often reverses itself—that is, it flips its north magnetic pole with its south magnetic pole. Over about half of geologic time, a compass needle would have pointed to the south!

In the early 1960s, geologists discovered that a precise record of this peculiar behavior can be obtained from layered flows of volcanic lava. When iron-rich lavas cool in the presence of Earth's magnetic field, they become slightly magnetized in the direction of the field. This phenomenon is called *thermoremanent magnetization,* because the rock "remembers" the magnetization long after the magnetizing field existing at the time it formed has changed. (How rocks become magnetized will be discussed more thoroughly in Chapter 21.)

In layered lava flows, each layer of rock from the top down represents a progressively earlier period of geologic time, and the age of each layer can be determined by precise dating methods (described in Chapter 10). Measurements of the thermoremanent magnetization of rock samples from each layer yield the direction of Earth's magnetic field frozen into that layer when it cooled (see Figure Story 2.11). By repeating these measurements at hundreds of places around the world, geologists have worked out the detailed history of reversals going back into geologic time. The **magnetic time scale** of the past 5 million years is given in Figure Story 2.11.

About half of all rocks studied are found to be magnetized in a direction opposite that of Earth's present magnetic field. Apparently, the field has flipped frequently over geologic time, and normal fields (same as now) and reversed fields (opposite to now) are equally likely. Major periods when the field is normal or reversed are called *magnetic epochs;* they seem to last about half a million years or so, although the pattern of reversals going back into geologic time becomes highly irregular. Superimposed on the major epochs are transient, short-lived reversals of the field, known as *magnetic events,* which may last anywhere from several thousand to 200,000 years.

Magnetic Anomaly Patterns on the Seafloor The peculiar banded magnetic patterns found on the seafloor (see Figure Story 2.11) puzzled scientists until 1963, when two Englishmen, F. J. Vine and D. H. Mathews—and, independently, two Canadians, L. Morley and A. Larochelle—made a startling proposal. Based on the new evidence for magnetic reversals that land geologists had collected from lava flows, they reasoned that the high and low magnetic bands corresponded to bands of rock on the seafloor that were magnetized during ancient episodes of normal and reversed magnetism. That is, when a research ship was above rocks magnetized in the normal direction, it would record a locally stronger field, or a *positive magnetic anomaly,* and when it was above rocks magnetized in the reversed direction, it would record a locally weaker field, or a *negative magnetic anomaly.*

This idea provided a powerful test of the seafloor spreading hypothesis, which stated that new seafloor is created along the rift at the crest of a mid-ocean ridge as the plates move apart (see Figure Story 2.11). Magma flowing up from the interior solidifies in the crack and becomes magnetized in the direction of Earth's field at the time. As the seafloor splits and moves away from the ridge, approximately half of the newly magnetized material moves to one side and half to the other, forming two symmetrical magnetized bands. Newer material fills the crack, continuing the process. In this way, the seafloor acts like a tape recorder that encodes the history of the opening of the oceans by magnetically imprinting the reversals of Earth's magnetic field.

Within a few years, marine scientists were able to show that this model provided a consistent explanation for the symmetric patterns of seafloor magnetic anomalies found on mid-ocean ridges around the world. Moreover, it gave them a precise tool for measuring the rates of seafloor spreading now and in the geologic past. This evidence contributed substantially to the discovery and confirmation of plate tectonics.

Inferring Seafloor Ages and Relative Plate Velocity By using the ages of reversals that had been worked out from magnetized lavas on land, geologists could assign ages to the bands of magnetized rocks on the seafloor. They could then calculate how fast the ocean opened up by using the formula *speed = distance/time,* where distance was measured from the ridge axis and time equaled seafloor age. For instance, the magnetic anomaly pattern in Figure Story 2.11 showed that the boundary between the Gauss normal epoch and the Gilbert reversal epoch, which had been dated from lava flows at 3.3 million years, was located about 30 km away from the Reykjanes Ridge crest. Here, seafloor spreading had moved the North American and Eurasian plates apart by about 60 km in 3.3 million years, giving a spreading rate of 18 km per million years or, equivalently, 18 mm/year.

On a divergent plate boundary, the combination of the spreading rate and the spreading direction gives the **relative plate velocity:** the velocity at which one plate moves relative to the other plate.

If you look at Figure 2.5, you will see that the spreading rate for the Mid-Atlantic Ridge south of Iceland is fairly low compared to many other places on the mid-ocean ridges. The speed record for spreading can be found on the East Pacific Rise just south of the equator, where the Pacific and Nazca plates are separating at a rate of about 150 mm/

year—an order of magnitude faster than the rate in the North Atlantic. A rough average for the mid-ocean ridges around the world is 50 mm/year. This is approximately the rate at which your fingernails grow, and it shows that, geologically speaking, the plates move very fast indeed. These spreading rates provide important data for the study of the mantle convection system, a topic we will return to later in this chapter.

We can follow the magnetic time scale through many reversals of Earth's magnetic field. The corresponding magnetic bands on the seafloor, which can be thought of as age bands, have been mapped in detail from the ridge crests across the ocean basins over a time span exceeding 100 million years.

The power and convenience of using seafloor magnetization to work out the history of ocean basins cannot be overemphasized. Simply by steaming back and forth over the ocean, measuring the magnetic fields of the rocks of the seafloor and correlating the pattern of reversals with the time sequence worked out by the methods just described, geologists determined the ages of various regions of the seafloor without even examining rock samples. In effect, they learned how to "replay the tape."

The simplicity and elegance of seafloor magnetization made it a very effective tool. But it was an indirect or remote sensing method, in that rocks were not recovered from the seafloor and their ages were not directly determined in the laboratory. Direct evidence of seafloor spreading and plate movement was still needed to convince the few remaining skeptics. Deep-sea drilling supplied it.

Deep-Sea Drilling

In 1968, a program of drilling into the seafloor was launched as a joint project of major oceanographic institutions and the National Science Foundation (Feature 2.1). Later, many nations joined the effort. This global experiment aimed to drill through, retrieve, and study seafloor rocks from many places in the world's oceans. Using hollow drills, scientists brought up cores containing sections of seafloor rocks. In some cases, the drilling penetrated thousands of meters below the seafloor surface. Geologists thus had an opportunity to work out the history of the ocean basins from direct evidence.

One of the most important facts geologists sought was the age of each sample. Small particles falling through the

2.1 Drilling in the Deep Sea

The deep-sea drilling vessel *JOIDES Resolution* is 143 m long, and amidships it carries a drilling derrick 61 m high. With size and capacity to drill in the deepest ocean, it is the only ship of its kind in the world. It can lower drill pipe thousands of meters to the seafloor and drill thousands of meters into the sediments and underlying basaltic crust.

Before the ship could accomplish such a feat, a technological breakthrough was required. A way had to be found to hold the ship stationary during the drilling process, regardless of currents and winds; otherwise, the drill pipe would break off. The problem was solved by development of a positioning device that uses sound waves transmitted by acoustic beacons planted on the seafloor. Any change in the ship's position is sensed by a computer that monitors changes in the time of arrival of the sound pulses from each beacon. The same computer controls the speed and steering to keep the vessel stationary.

Deep-sea drilling was the answer to those who said, when lunar exploration started, "Better to explore the ocean's bottom than the back side of the Moon." We ended up doing both. The deep-sea drilling program, now known as the Ocean Drilling Program, is more than 35 years old and has become international in scope.

Scientists aboard a drilling ship take samples from cores of sediment recovered from the seafloor. These samples can be analyzed to reveal the history of ocean basins and ancient climatic conditions. [Courtesy of Ocean Drilling Program/TAMA.]

ocean water—dust from the atmosphere, organic material from marine plants and animals—begin to accumulate as seafloor *sediments* as soon as new oceanic crust forms. Therefore, the age of the oldest sediments in the core, those immediately on top of the crust, tells the geologist how old the ocean floor is at that spot. The age of sediments is obtained primarily from the fossil skeletons of tiny, single-celled animals that live in the ocean and sink to the bottom when they die (see Chapter 10). It was found that the sediments in the cores become older with increasing distance from mid-ocean ridges and that the age of the seafloor at any one place agrees almost perfectly with the age determined from magnetic reversal data. This agreement validated magnetic dating of the seafloor and clinched the concept of seafloor spreading.

Measurements of Plate Motions by Geodesy

In his publications advocating continental drift, Alfred Wegener made a big mistake: he proposed that North America and Europe were drifting apart at a rate of nearly 30 m/year—a thousand times faster than the Atlantic seafloor is actually spreading! This unbelievably high speed was one of the reasons that many scientists roundly rejected his notions of continental drift. Wegener made this estimate by incorrectly assuming that the continents were joined together as Pangaea as recently as the last ice age (which occurred only about 20,000 years ago). His belief in a rapid rate also involved some wishful thinking. In particular, he hoped that the drift hypothesis could be confirmed by repeated, accurate measurements of the distance across the Atlantic Ocean using astronomical positioning.

Figure 2.12 One of the 250 GPS stations in a network that collects satellite observations along faults in southern California. These instruments use signals from GPS satellites orbiting Earth to detect small displacements of the surface, from which plate motions and plate-boundary deformations can be computed. Observations of these motions may help scientists to evaluate future occurrences of earthquakes. [Southern California Earthquake Center.]

Astronomical Positioning Astronomical positioning—measuring the positions of stars in the night sky to determine where you are—is a technique of *geodesy,* the ancient science of measuring the shape of the Earth and locating points on its surface. Surveyors have used astronomical positioning for centuries to determine geographic boundaries on land, and sailors have used it to locate their ships at sea. Four thousand years ago, Egyptian builders used astronomical positioning to aim the Great Pyramid due north.

Wegener imagined that geodesy could be used to measure continental drift in the following way. Two observers, one in Europe and the other in North America, would simultaneously determine their positions relative to the fixed stars. From these positions, they would calculate the distance between their two observing posts at that instant. They would then repeat this distance measurement from the same observing posts sometime later, say, after 1 year. If the continents are drifting apart, then the distance should have increased, and the value of the increase would determine the speed of the drift.

For this technique to work, however, one must determine the relative positions of the observing posts accurately enough to measure the motion. In Wegener's day, the accuracy of astronomical positioning was poor; uncertainties in fixing intercontinental distances exceeded 100 m. Therefore, even at the high rates of motion he was proposing, it would take a number of years to observe drift. He claimed that two astronomical surveys of the distance between Europe and Greenland (where he worked as a meteorologist), taken 6 years apart, supported his high rate, but he was wrong again. We now know that the spreading of the Mid-Atlantic Ridge from one survey to the next was only about a tenth of a meter—a thousand times too small to be observed by the techniques that were then available.

Owing to the high accuracy required to observe plate motions directly, geodetic techniques did not play a significant role in the discovery of plate tectonics. Geologists had to rely on the evidence for seafloor spreading from the geologic record—the magnetic stripes and ages from fossils described earlier. Beginning in the late 1970s, however, an astronomical positioning method was developed that used signals from distant "quasi-stellar radio sources" (quasars) recorded by huge dish antennas. This method can measure intercontinental distances to an amazing accuracy of 1 mm. In 1986, a team of scientists published a set of measurements based on this technique that showed the distance between antennas in Europe (Sweden) and North America (Massachusetts) had increased 19 mm over a period of 5 years, very close to the amount predicted by geologic models of plate tectonics. Wegener's dream of directly measuring continental drift by astronomical positioning was realized at last!

Postscript: Today, the Great Pyramid of Egypt is not aimed directly north, as stated previously, but slightly east of north. Did the ancient Egyptian astronomers make a mistake in orienting the pyramid 40 centuries ago? Archaeologists think probably not. Over this period, Africa drifted enough to rotate the pyramid out of alignment with true north.

Global Positioning System Doing geodesy with big radio telescopes is an expensive operation and is not a practical tool for detailed investigations of plate tectonic motions in remote areas of the world. Since the mid-1980s, geologists have been able to take advantage of a new constellation of 24 Earth-orbiting satellites, called the Global Positioning System (GPS), to make the same types of measurements with the same astounding accuracy using inexpensive, portable radio receivers not much bigger than this book (**Figure 2.12**). GPS receivers record high-frequency radio waves keyed to precise atomic clocks aboard the satellites. The satellite constellation serves as an outside frame of reference, just as the fixed stars and quasars do in astronomical positioning.

The changes in distance between land-based GPS receivers placed on different plates, recorded over several years, agree in both magnitude and direction with those found from magnetic anomalies on the seafloor. These experiments indicate that plate motions are remarkably steady over periods of time ranging from a few years to millions of years. Geologists are now using GPS to measure plate motions on a yearly basis at many locations around the globe (**Figure 2.13**).

Besides determining plate velocities, GPS observations have shown that the convergence between the Nazca and

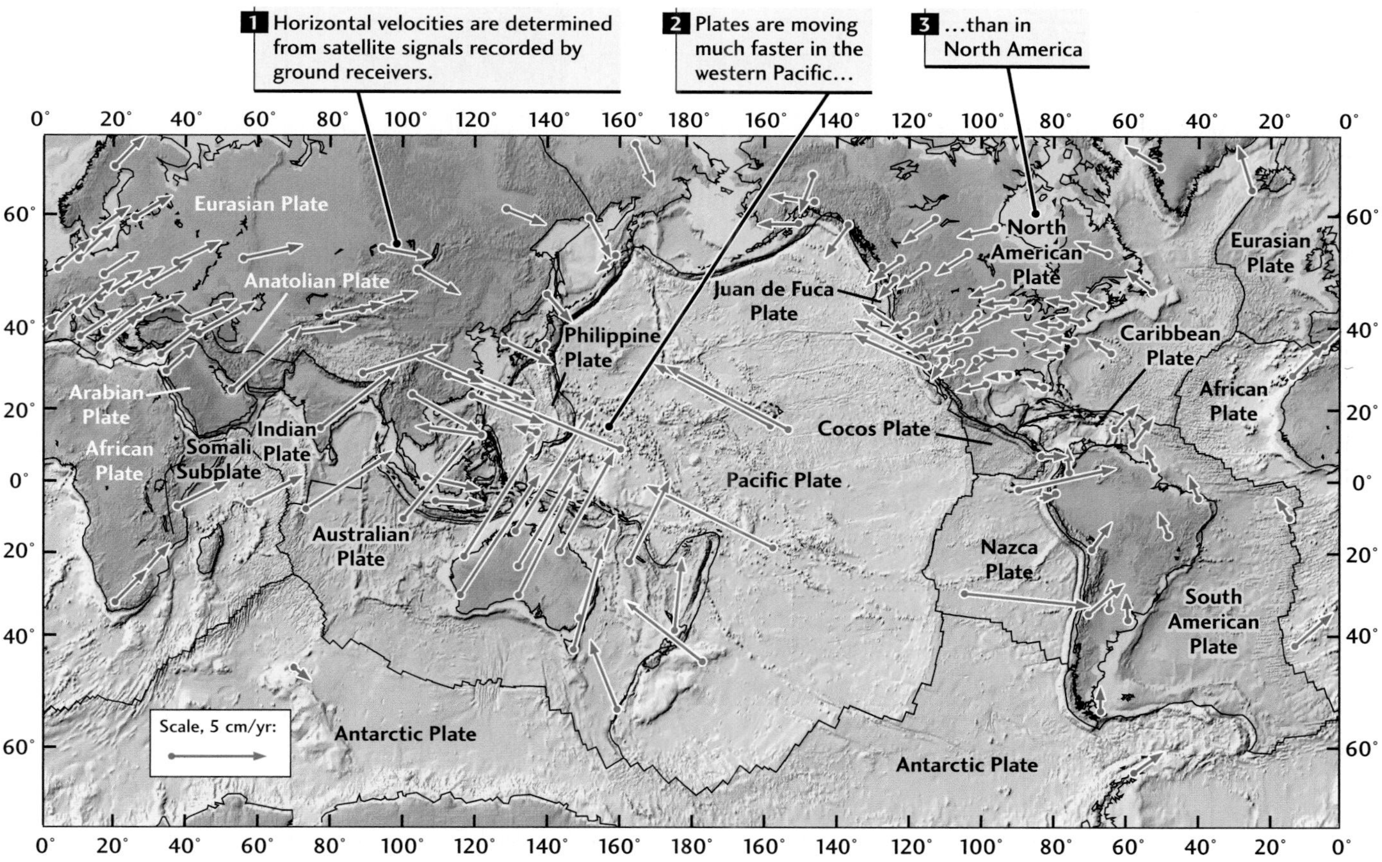

Figure 2.13 The Global Positioning System (GPS) is used to measure plate motions at many locations on Earth. The velocities shown here are determined for stations that continuously record GPS data. [Michael Heflin, JPL/CalTech.]

South American plates can be divided into three parts. About 40 percent is smooth, continuous slip between the two plates. Some 20 percent occurs as deformation at this plate boundary, which causes the uplift of the Andes Mountains. About 40 percent occurs in great earthquakes when the interface between the two plates ruptures and slips suddenly. By definition, a rigid plate should not deform. What is happening here? We will learn more about the process of plate deformation when we study earthquakes in Chapter 19.

Postscript: GPS receivers are now used in automobiles, as part of a navigating system that will lead the driver to a specific street address. It is interesting that the scientists who developed the atomic clocks used in GPS did so for research in fundamental physics and had no idea that they would be creating a multibillion-dollar industry. Along with the transistor, laser, and many other technologies, GPS demonstrates the serendipitous manner in which basic research repays the society that supports it.

The Grand Reconstruction

The supercontinent of Pangaea was the only major landmass that existed 250 million years ago. One of the great triumphs of modern geology is the reconstruction of events that led to the assembly of Pangaea and to its later fragmentation into the continents that we know today. Let's use what we have learned about plate tectonics to see how this feat was accomplished.

Seafloor Isochrons

The colorful map in **Figure 2.14** shows the ages of the world's ocean floors as determined by magnetic reversal data and fossils from deep-sea drilling. Each colored band represents a span of time corresponding to the age of the crust within that band. The boundaries between bands, called **isochrons,** are contours that connect rocks of equal age. Isochrons tell us the time that has elapsed since the

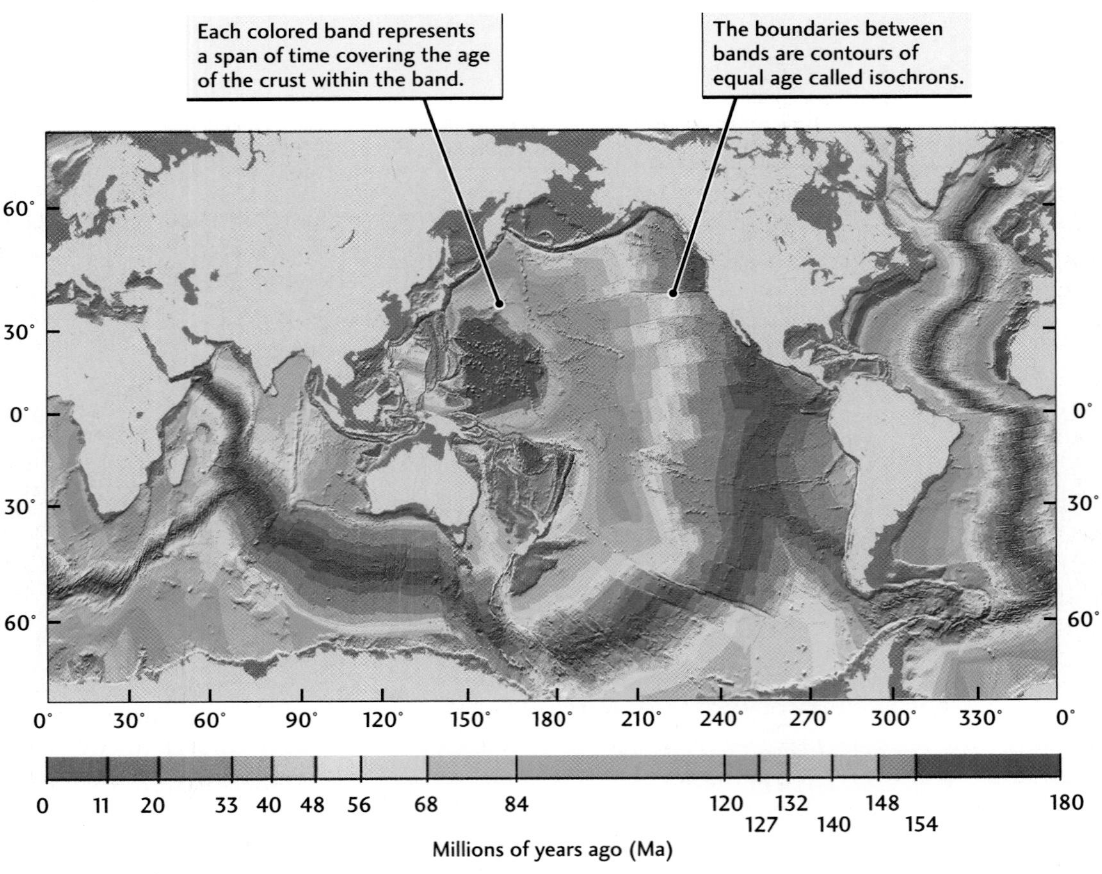

Figure 2.14 Age of seafloor crust. Each colored band represents a span of time covering the age of the crust within the band. The boundaries between bands are contours of equal age called isochrons. Isochrons give the age of the seafloor in millions of years since its creation at mid-ocean ridges. Light gray indicates land. Dark gray indicates shallow water over continental shelves. Mid-ocean ridges, along which new seafloor is extruded, coincide with the youngest seafloor (red). [*Journal of Geophysical Research* 102 (1997): 3211–3214. Courtesy of R. Dietmar Müller.]

crustal rocks were injected as magma into a mid-ocean rift and, therefore, the amount of spreading that has occurred since they formed. Notice how the seafloor becomes progressively older on both sides of the mid-ocean rifts. For example, the distance from a ridge axis to a 140-million-year isochron (boundary between green and blue bands) indicates the extent of new ocean floor created over that time span. The more widely spaced isochrons (the wider colored bands) of the eastern Pacific signify faster spreading rates than those in the Atlantic.

In 1990, after a 20-year search, geologists found the oldest oceanic rocks by drilling into the seafloor of the western Pacific. These rocks turned out to be about 200 million years old, only about 4 percent of Earth's age. This date indicates how geologically young the seafloor is compared with the continents. Over a period of 100 million to 200 million years in some places and only tens of millions of years in others, the ocean lithosphere forms, spreads, cools, and subducts back into the underlying mantle. In contrast, the oldest continental rocks are about 4 billion years old.

Reconstructing the History of Plate Motions

Earth's plates behave as rigid bodies. That is, the distances between three points on the same rigid plate—say, New York, Miami, and Bermuda on the North American Plate—do not change very much, no matter how far the plate moves. But the distance between, say, New York and Lisbon increases because the two cities are on different plates that are being separated along a narrow zone of spreading on the Mid-Atlantic Ridge. The direction of the movement of one plate in relation to another depends on geometric principles that govern the behavior of rigid plates on a sphere. Two primary principles are

1. *Transform boundaries indicate the directions of relative plate movement.* With few exceptions, no overlap, buckling, or separation occurs along typical transform boundaries in the oceans. The two plates merely slide past each other without creating or destroying plate material. Look for a transform boundary if you want to deduce the direction of relative plate motion, because the orientation of the fault is the direction in which one plate slides with respect to the other, as Figure 2.10 shows.

2. *Seafloor isochrons reveal the positions of divergent boundaries in earlier times.* Isochrons on the seafloor are roughly parallel and symmetrical with the ridge axis along which they were created. Figure 2.14 illustrates this observation. Because each isochron was at the boundary of plate separation at an earlier time, isochrons that are of the same age but on opposite sides of an ocean ridge can be brought together to show the positions of the plates and the configuration of the continents embedded in them as they were in that earlier time.

The Breakup of Pangaea

Using these principles, geologists have reconstructed the opening of the Atlantic Ocean and the breakup of Pangaea. The supercontinent of Pangaea is shown as it existed 240 million years ago in **Figure 2.15**a. It began to break apart with the rifting of North America away from Europe about 200 million years ago (Figure 2.15b). The opening of the North Atlantic was accompanied by the separation of the northern continents (Laurasia) from the southern continents (Gondwanaland, or Gondwana) and the rifting of Gondwanaland along what is now the eastern coast of Africa (Figure 2.15c). The breakup of Gondwanaland separated South America, Africa, India, and Antarctica, creating the South Atlantic and Southern oceans and narrowing the Tethys Ocean (Figure 2.15d). The separation of Australia from Antarctica and the ramming of India into Eurasia closed the Tethys Ocean, giving us the world we see today (Figure 2.15e).

The plate motions have not ceased, of course, so the configuration of the continents will continue to evolve. A plausible scenario for the distribution of continents and plate boundaries 50 million years into the future is displayed in Figure 2.15f.

The Assembly of Pangaea by Continental Drift

The isochron map in Figure 2.14 tells us that all of the seafloor extant on Earth's surface has been created since the breakup of Pangaea. We know from the geologic record in older continental mountain belts, however, that plate tectonics was operating for billions of years before this breakup. Evidently, seafloor spreading took place as it does today, and there were previous episodes of continental drift and collision. The seafloor created in these earlier times has been destroyed by subduction back into the mantle, so we must rely on the older evidence preserved on continents to identify and chart the movements of these "paleocontinents."

Old mountain belts such as the Appalachians of North America and the Urals, which separate Europe from Asia, help us locate ancient collisions of the paleocontinents. In many places, the rocks reveal ancient episodes of rifting and subduction. Rock types and fossils also indicate the distribution of ancient seas, glaciers, lowlands, mountains, and climates. The knowledge of ancient climates enables geologists to locate the latitudes at which the continental rocks formed, which in turn helps them to assemble the jigsaw puzzle of ancient continents. When volcanism or mountain building produces new continental rocks, they also record the direction of Earth's magnetic field, just as oceanic rocks do when they are created by seafloor spreading. Like a compass frozen in time, the fossil magnetism of a continental fragment records its ancient orientation and position.

Figure 2.15 shows one of the latest efforts to depict the pre-Pangaean configuration of continents. It is truly

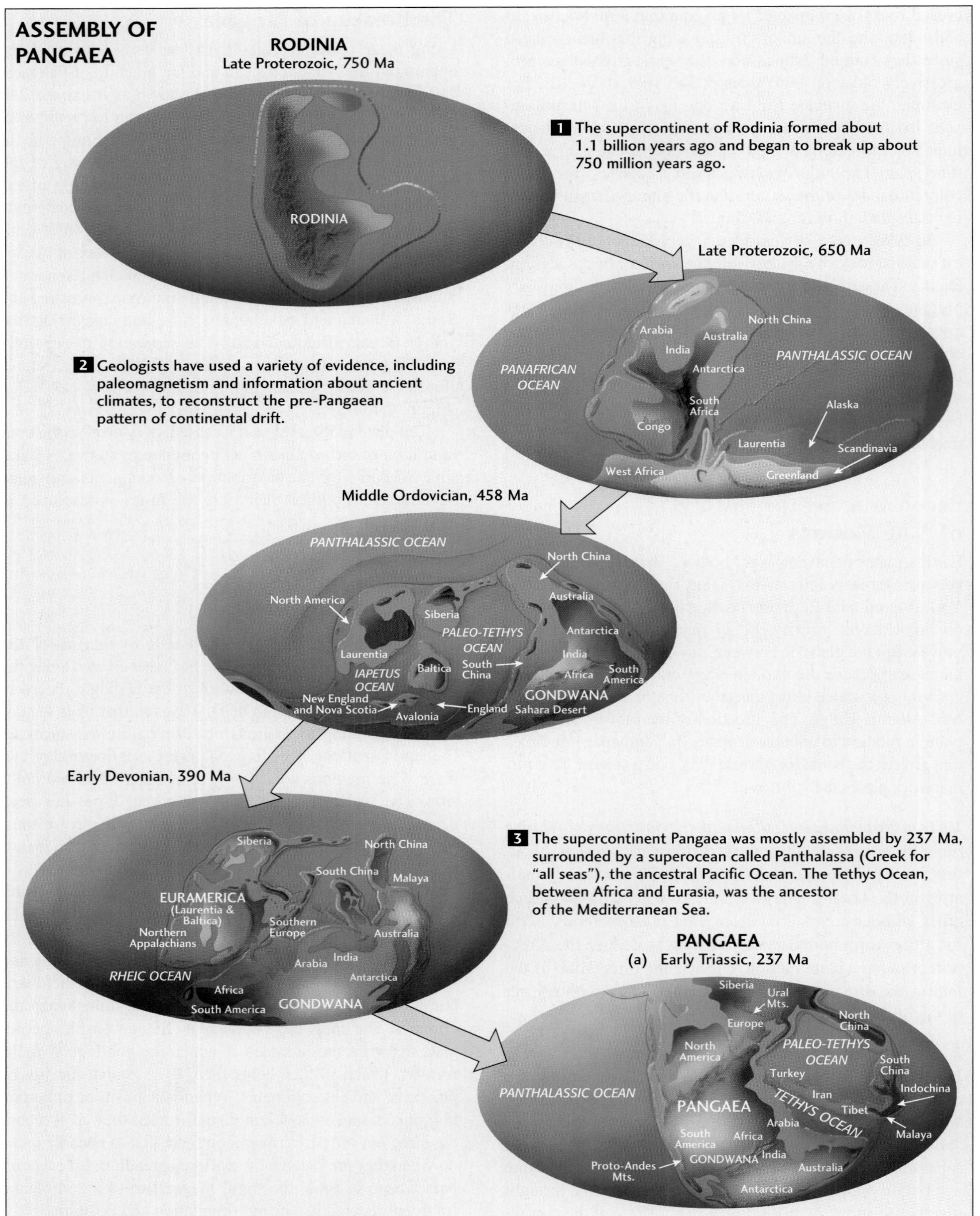

Figure 2.15 The assembly and breakup of Pangaea, from 750 million years ago to 50 million years into the future. [Paleogeographic maps by Christopher R. Scotese, 2003 PALEOMAP Project (www.scotese.com).]

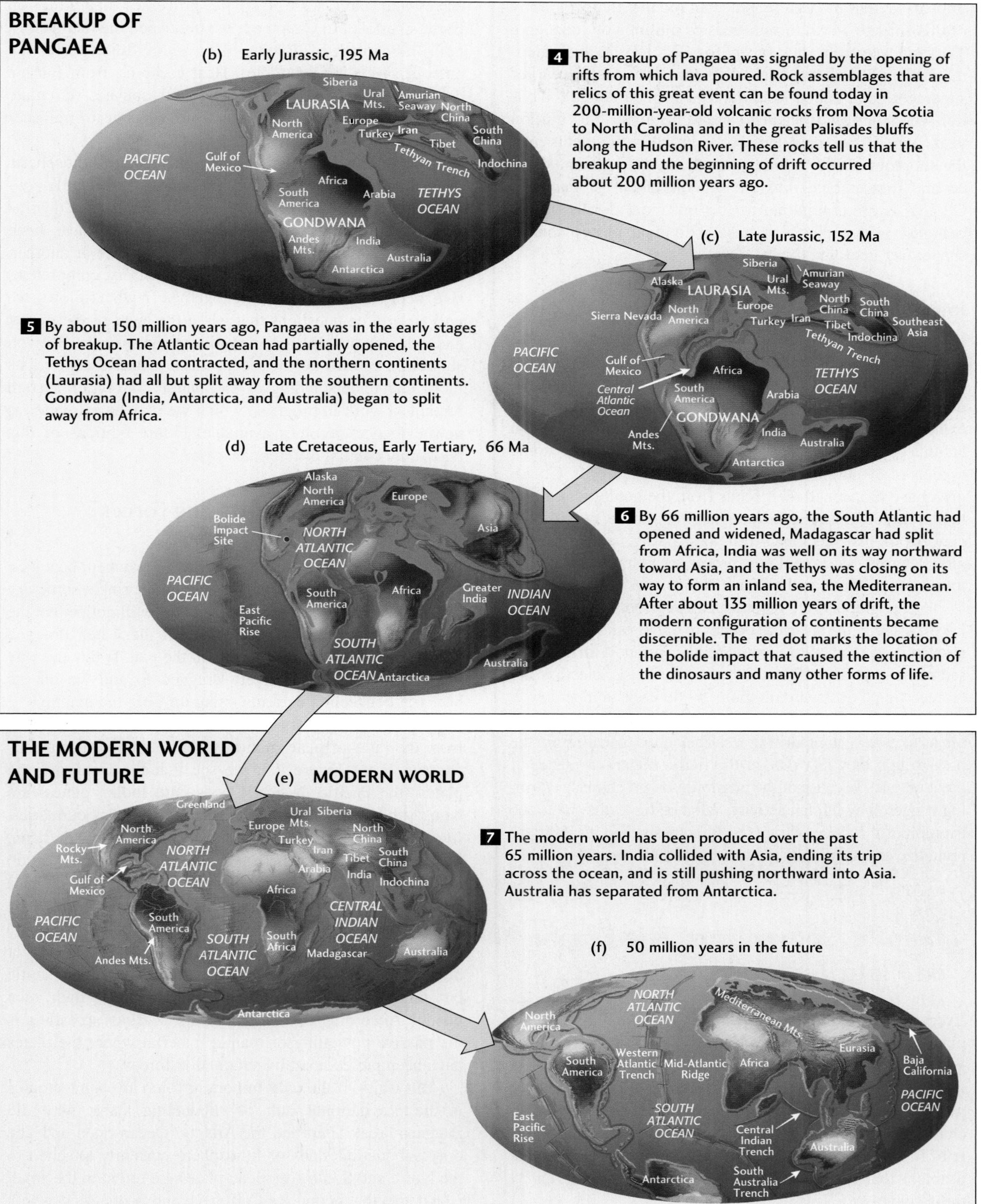

4 The breakup of Pangaea was signaled by the opening of rifts from which lava poured. Rock assemblages that are relics of this great event can be found today in 200-million-year-old volcanic rocks from Nova Scotia to North Carolina and in the great Palisades bluffs along the Hudson River. These rocks tell us that the breakup and the beginning of drift occurred about 200 million years ago.

5 By about 150 million years ago, Pangaea was in the early stages of breakup. The Atlantic Ocean had partially opened, the Tethys Ocean had contracted, and the northern continents (Laurasia) had all but split away from the southern continents. Gondwana (India, Antarctica, and Australia) began to split away from Africa.

6 By 66 million years ago, the South Atlantic had opened and widened, Madagascar had split from Africa, India was well on its way northward toward Asia, and the Tethys was closing on its way to form an inland sea, the Mediterranean. After about 135 million years of drift, the modern configuration of continents became discernible. The red dot marks the location of the bolide impact that caused the extinction of the dinosaurs and many other forms of life.

7 The modern world has been produced over the past 65 million years. India collided with Asia, ending its trip across the ocean, and is still pushing northward into Asia. Australia has separated from Antarctica.

impressive that modern science can recover the geography of this strange world of hundreds of millions of years ago. The evidence from rock types, fossils, climate, and paleomagnetism has allowed scientists to reconstruct an earlier supercontinent, called *Rodinia,* that formed about 1.1 billion years ago and began to break up about 750 million years ago. They have been able to chart its fragments over the subsequent 500 million years as these fragments drifted and reassembled into the supercontinent of Pangaea. Geologists are continuing to sort out more details of this complex jigsaw puzzle, whose individual pieces change shape over geologic time.

Implications of the Grand Reconstruction

Hardly any branch of geology remains untouched by this grand reconstruction of the continents. Economic geologists have used the fit of the continents to find mineral and oil deposits by correlating the rock formations in which they exist on one continent with their predrift continuations on another continent. Paleontologists have rethought some aspects of evolution in the light of continental drift. Geologists have broadened their focus from the geology of a particular region to a world-encompassing picture, because the concept of plate tectonics provides a way to interpret, in global terms, such geologic processes as rock formation, mountain building, and climate change.

Oceanographers are reconstructing currents as they might have existed in the ancestral oceans to understand the modern circulation better and to account for the variations in deep-sea sediments that are affected by such currents. Scientists are "forecasting" backward in time to describe temperatures, winds, the extent of continental glaciers, and the level of the sea as they were in predrift times. They hope to learn from the past so that they can predict the future better—a matter of great urgency because of the possibility of greenhouse warming triggered by human activity. What better testimony to the triumph of this once outrageous hypothesis than its ability to revitalize and shed light on so many diverse topics?

Mantle Convection: The Engine of Plate Tectonics

Everything discussed so far might be called descriptive plate tectonics. But a description is hardly an explanation. We will not fully understand plate tectonics until we have a more comprehensive theory that can explain why plates move. Finding such a theory remains one of the outstanding challenges confronting scientists who study the Earth system. In this section, we will discuss several aspects of the problem that have been central to the recent research by these scientists.

As Arthur Holmes and other early advocates of continental drift realized, mantle convection is the "engine" that drives the large-scale tectonic processes operating on Earth's surface. In Chapter 1, we described the mantle as a hot solid capable of flowing like a sticky fluid (warm wax or cold syrup, for example). Heat escaping from Earth's deep interior causes this material to convect (circulate upward and downward) at speeds of a few tens of millimeters per year.

Almost all scientists now accept that the lithospheric plates somehow participate in the flow of this mantle convection system. As is often the case, however, "the devil is in the details." Many different hypotheses have been advanced on the basis of one piece of evidence or another, but no one has yet come up with a satisfactory, comprehensive theory that ties everything together. In what follows, we will pose three questions that get at the heart of the matter and give you our opinions regarding their answers. But you should be careful not to accept these tentative answers as facts. Our understanding of the mantle convection system remains a work in progress, which we may have to alter as new evidence becomes available. Future editions of this book may contain different answers!

Where Do the Plate-Driving Forces Originate?

Here's an experiment you can do in your kitchen: heat up a pan of water until it is about to boil and sprinkle some dry tea leaves into the center of the pan. You will notice that the tea leaves move across the surface of the water, dragged along by the convection currents in the pan. Is this the way plates move about, passively dragged to and fro on the backs of convection currents rising up from the mantle?

The answer appears to be no. The main evidence comes from the rates of plate motion we discussed earlier in this chapter. From Figure 2.5, we see that the faster-moving plates (the Pacific, Nazca, Cocos, and Indian plates) are being subducted along a large fraction of their boundaries. In contrast, the slower-moving plates (the North American, South American, African, Eurasian, and Antarctic plates) do not have significant attachments of downgoing slabs. These observations suggest that rapid plate motions are caused by the gravitational pull exerted by the cold (and thus heavy) slabs of old lithosphere. In other words, the plates are not dragged along by convection currents from the deep mantle but rather "fall back" into the mantle under their own weight. According to this hypothesis, seafloor spreading is the passive upwelling of mantle material where the plates have been pulled apart by subduction forces.

But wait—if the only important force in plate tectonics is the gravitational pull of subducting slabs, why did Pangaea break apart and the Atlantic Ocean open up? The only subducting slabs of lithosphere currently attached to the North and South American plates are found in the small island arcs that bound the Caribbean and Scotia seas, which are thought to be too wimpy to drag the Atlantic apart. One possibility is that the overriding plates, as well as the sub-

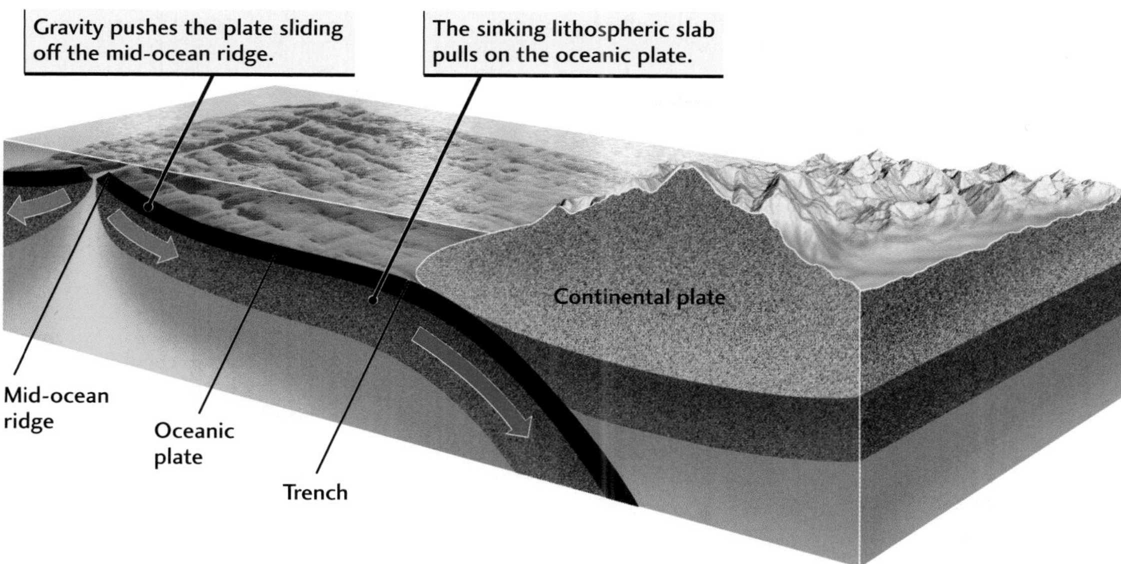

Figure 2.16 A schematic cross section through the outer part of Earth, illustrating two of the forces thought to be important in driving plate tectonics: the pulling force of a sinking lithospheric slab and the pushing force of plates sliding off a mid-ocean ridge. [Adapted from Figure 1 in an article by D. Forsyth and S. Uyeda, *Geophys. J. Roy. Astr. Soc.* 43 (1975): 163–200.]

ducting plates, are pulled toward their convergent boundaries. For example, as the Nazca Plate subducts beneath South America, it may cause the plate boundary at the Peru-Chile Trench to retreat toward the Pacific, "sucking" the South American Plate to the west.

Another possibility is that Pangaea acted like an insulating blanket, preventing heat from getting out of Earth's mantle (as it usually does through the process of seafloor spreading). This heat presumably built up over time, causing hot bulges to form in the mantle beneath the supercontinent. These bulges raised Pangaea (slightly) and caused it to rift apart in a kind of "landslide" off the top of the bulges. These gravitational forces continued to drive subsequent seafloor spreading as the plates "slid downhill" off the crest of the Mid-Atlantic Ridge. Earthquakes that sometimes occur in plate interiors show direct evidence of the compression of plates by these "ridge push" forces.

As you can tell from this brief discussion, the driving forces of plate tectonics probably involve several types of interactions. All of them are manifestations of convection in the mantle, in the sense that they involve hot matter rising in one place and cold matter sinking in another (**Figure 2.16**). Although many questions remain, we can be reasonably sure that (1) the plates themselves play an active role in this system, and (2) the forces associated with the sinking slabs and elevated ridges are probably the most important in governing the rates of plate motion. Scientists are attempting to resolve other issues raised in this discussion by comparing observations with detailed computer models of the mantle convection system. Some results will be discussed in Chapter 21.

How Deep Does Plate Recycling Occur?

For plate tectonics to work, the lithospheric material that goes down in subduction zones must be recycled through the mantle and eventually come back up as new lithosphere created along the spreading centers of the mid-ocean ridges. How deep into the mantle does this recycling process extend? That is, where is the lower boundary of the mantle convection system?

The deepest it can reach is about 2900 km below Earth's outer surface, where a sharp boundary separates the mantle from the core. The iron-rich liquid below this *core-mantle boundary* is much denser than the solid rocks of the mantle, preventing any significant interchange of material between the two layers. We can thus imagine a system of "whole-mantle" convection in which the material from the plates circulates all the way through the mantle, down as far as the core-mantle boundary (**Figure 2.17**a).

In the early days of plate tectonics theory, however, many scientists were convinced that plate recycling takes place at much shallower depths in the mantle. The evidence came from deep earthquakes that mark the descent of lithospheric slabs in subduction zones. The greatest depth of these earthquakes varies among subduction zones, depending on how cold the descending slabs are, but geologists found that no earthquakes were occurring below about 700 km. Moreover,

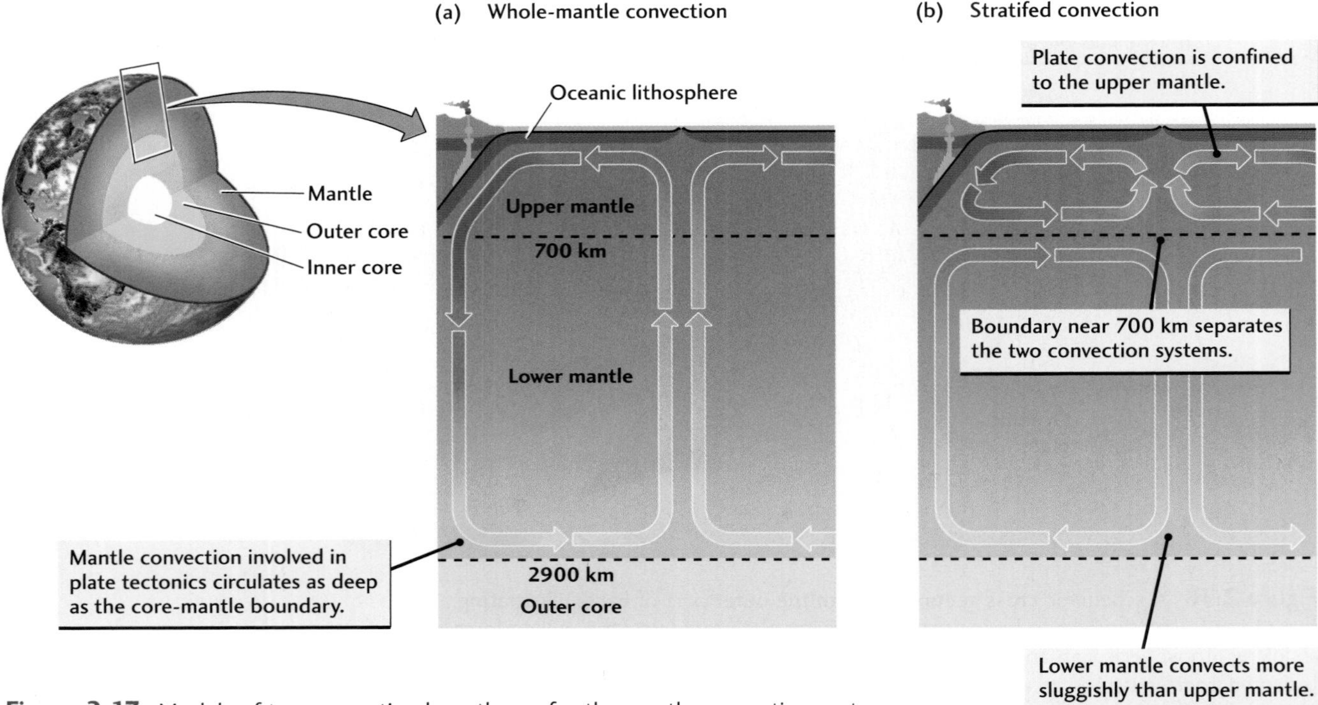

Figure 2.17 Models of two competing hypotheses for the mantle convection system.

the properties of earthquakes at these great depths indicated that the slabs were encountering more rigid material that slowed and perhaps blocked their downward progress.

Based on this and other evidence, scientists concluded that the mantle might be divided into two layers: an upper mantle system in the outer 700 km, where the recycling of lithosphere takes place, and a lower mantle system, from 700 km deep to the core-mantle boundary, where convection is much more sluggish. According to this hypothesis, called "stratified convection," the separation of the two systems is maintained because the upper system consists of lighter rocks than the lower system and thus floats on top, in the same way the mantle floats on the core (Figure 2.17b).

The way to test these two competing hypotheses is to look for "lithospheric graveyards" below the convergent zones where old plates have been subducted. Old subducted lithosphere is colder than the surrounding mantle and can therefore be "seen" using earthquake waves (much as doctors use ultrasound waves to look into your body). Moreover, there should be lots of it down there. From our knowledge of past plate motions, we can estimate that, just since the breakup of Pangaea, lithosphere equivalent to the surface area of Earth has been recycled back into the mantle. Sure enough, scientists have found regions of colder material in the deep mantle under North and South America, eastern Asia, and other sites adjacent to plate collision boundaries. These zones occur as extensions of descending lithospheric slabs, and some appear to go down as far as the core-mantle boundary. From this evidence, most scientists have concluded that plate recycling takes place through whole-mantle convection rather than stratified convection.

What Is the Nature of Rising Convection Currents?

Mantle convection implies that what goes down must come up. Scientists have learned a lot about downgoing convection currents because they are marked by narrow zones of cold subducted lithosphere that can be detected by earthquake waves. What about the rising currents of mantle material needed to balance subduction? Are there concentrated, sheetlike upwellings directly beneath the mid-ocean ridges? Most scientists who study the problem think not. Instead, they believe that the rising currents are slower and spread out over broader regions. This view is consistent with the idea, discussed above, that seafloor spreading is a rather passive process: pull the plates apart almost anywhere, and you will generate a spreading center.

There is one big exception, however: a type of narrow, jetlike upwelling called a *mantle plume* (**Figure 2.18**). The best evidence for mantle plumes comes from regions of intense, localized volcanism (called *hot spots*), such as Hawaii, where huge volcanoes are being formed in the middle of plates, far away from any spreading center. The plumes are thought to be slender cylinders of fast-rising material, less than 100 km across, that come from the deep mantle, perhaps forming in very hot regions near the core-mantle boundary. Mantle plumes are so intense that they

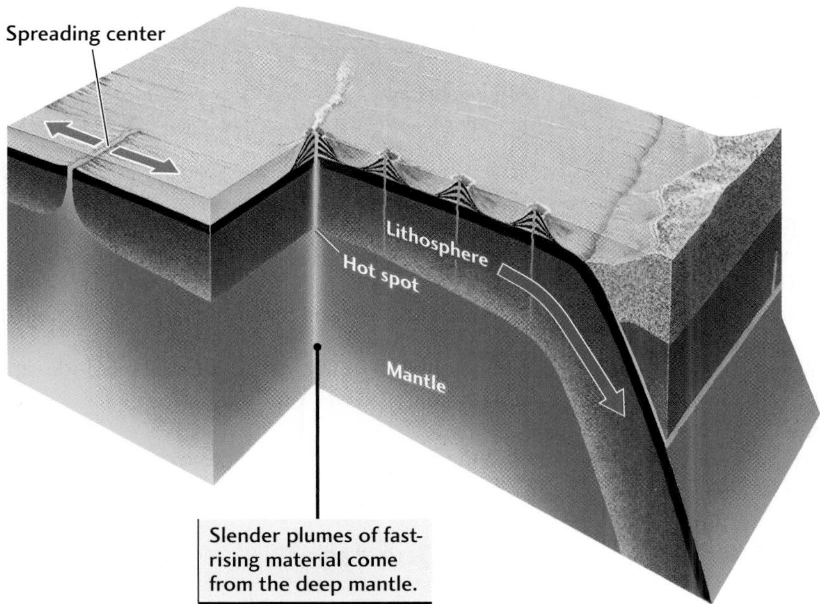

Figure 2.18 A model of the mantle plume hypothesis.

can literally burn holes in the plates and erupt tremendous volumes of lava. Plumes may be responsible for the massive outpourings of lava—millions of cubic kilometers—found in such places as Siberia and the Columbia Plateau of eastern Washington and Oregon. Some of these lava "floods" were so large and occurred so quickly that they may have changed Earth's climate and killed off many life-forms in mass extinction events (see Chapter 1). We will describe plume volcanism in more detail in Chapter 6.

The plume hypothesis was first put forward by one of the founders of plate tectonics, W. Jason Morgan of Princeton University, in 1970, soon after the plate theory had been established. Like other aspects of the mantle convection system, however, the observations that bear on rising convection currents are indirect, and the plume hypothesis remains very controversial.

The Theory of Plate Tectonics and the Scientific Method

Earlier, we considered the scientific method and the ways in which it guides the work of geologists. In the context of the scientific method, plate tectonics is not a dogma but a confirmed theory whose strength lies in its simplicity, its generality, and its consistency with many types of observations. Theories can always be overturned or modified. As we have seen, competing hypotheses have been advanced about how convection generates plate tectonics. But the theory of plate tectonics—like the theories of the age of Earth, the evolution of life, and genetics—explains so much so well and has survived so many efforts to prove it false that geologists treat it as fact.

The question remains, why wasn't plate tectonics discovered earlier? Why did it take the scientific establishment so long to move from skepticism about continental drift to acceptance of plate tectonics? Scientists work in different styles. Scientists with particularly inquiring, uninhibited, and synthesizing minds are often the first to perceive great truths. Although their perceptions frequently turn out to be false (think of the mistakes Wegener made in proposing continental drift), these visionary people are often the first to see the great generalizations of science. Deservedly, they are the ones that history remembers.

Most scientists, however, proceed more cautiously and wait out the slow process of gathering supporting evidence. Continental drift and seafloor spreading were slow to be accepted largely because the audacious ideas came far ahead of the firm evidence. The oceans had to be explored, new instruments had to be developed and used, and the deep sea had to be drilled to see what was there before the majority could be convinced. Today, many scientists are still waiting to be convinced of ideas about how the mantle convection system really works.

SUMMARY

What is the theory of plate tectonics? According to the theory of plate tectonics, the lithosphere is broken into about a dozen rigid, moving plates. Three types of plate boundaries are defined by the relative motion between plates: divergent, convergent, and transform fault.

What are some of the geologic characteristics of plate boundaries? In addition to earthquake belts, many large-scale geologic features, such as narrow mountain belts and chains of volcanoes, are associated with plate boundaries. Convergent boundaries are marked by deep-sea trenches, earthquake belts, mountains, and volcanoes. The Andes and the trenches of the west coast of South America are modern examples. Ancient convergences may show as old mountain belts, such as the Appalachians and the Urals. Divergent boundaries are typically marked by volcanic activity and earthquakes at the crest of a mid-ocean ridge, such as the Mid-Atlantic Ridge. Transform-fault boundaries, along which plates slide past each other, can be recognized by their linear topography, earthquake activity, and offsets in magnetic anomaly bands.

How can the age of the seafloor be determined? We can measure the age of the ocean's floor by comparing magnetic anomaly bands mapped on the seafloor with the sequence of magnetic reversals worked out on land. The procedure has been verified and extended by deep-sea drilling. Geologists can now draw isochrons for most of the world's oceans, enabling them to reconstruct the history of seafloor spreading over the past 200 million years. Using this method and other geologic data, geologists have constructed a detailed model of how Pangaea broke apart and the continents drifted into their present configuration.

What is the engine that drives plate tectonics? The plate tectonic system is driven by mantle convection, and the energy comes from Earth's internal heat. The plates themselves play an active role in this system. For example, the most important forces in plate tectonics come from the cooling lithosphere as it slides away from spreading centers and sinks back into the mantle in subduction zones. Lithospheric slabs extend as deep as the core-mantle boundary, indicating that the whole mantle is involved in the convection system that recycles the plates. Rising convection currents may include mantle plumes, which are intense upwellings from the deep mantle that cause localized volcanism in hot spots such as Hawaii.

Key Terms and Concepts

continental drift (p. 24)
island arc (p. 31)
isochron (p. 40)
magnetic anomaly (p. 34)
magnetic time scale (p. 36)
Pangaea (p. 24)
plate tectonics (p. 23)
relative plate velocity (p. 36)
seafloor spreading (p. 25)
spreading center (p. 31)
subduction (p. 31)
transform fault (p. 32)

Exercises

This icon indicates that there is an animation available on the Web site that may assist you in answering a question.

1. Give a modern geographic example of each of the three types of plate boundaries.

2. What evidence suggests that Pangaea ever existed?

3. How can the rate of motion between plates be calculated?

4. What happens when two continents collide at a convergent boundary? Give some examples.

5. What are the driving forces of plate tectonics?

Thought Questions

This icon indicates that there is an animation available on the Web site that may assist you in answering a question.

1. What would Earth be like if plate tectonics did not exist?

2. If plate tectonics explains so much of geology, why was it not until the 1960s that most geologists accepted the concept?

3. Would you characterize plate tectonics as a hypothesis, a theory, or a fact? Why?

4. In Figure 2.14, the isochrons are symmetrically distributed in the Atlantic Ocean but not in the Pacific. For example, the oldest seafloor (in darkest blue) is found in the western Pacific Ocean, but not in the eastern Pacific. Why?

5. Can you design an experiment that would demonstrate what forces are most important in driving plate tectonics?

6. What evidence would you look for to see if plate tectonics is occurring on other terrestrial planets?

7. Positive and negative magnetic anomaly bands reminiscent of those found on Earth were recently discovered on Mars. Surface features associated with plate tectonics, however, are absent. Speculate about the possible meaning of these observations.

8. The theory of plate tectonics was not widely accepted until the discovery of the magnetic striping of the ocean floor. In light of earlier observations—the jigsaw-puzzle fit of the continents, the occurrence of fossils of the same life-forms on both sides of the Atlantic, and paleoclimatic conditions—why is the magnetic striping such a key piece of evidence?

9. Earth is the only planet in our solar system that displays both plate tectonics and a broad diversity of life-forms. Is there any reason to believe that plate tectonics has effects on the presence and nature of life-forms? If so, what are these effects?

Suggested Readings

Anderson, R. N. 1986. *Marine Geology.* New York: Wiley.

Beloussov, V. V. 1979. Why I do not accept plate tectonics. *Eos* 60: 207–210. (See also comments on this paper by A. M. S. Sengor and K. Burke on the same pages.)

Carlson, R. W. 1997. Do continents part passively or do they need a shove? *Science* 278: 240–241.

Conrad, C. P., and Caroline Lithgow-Bertelloni. 2002. How mantle slabs drive plate tectonics. *Science* 298 (October 4): 207–209.

Cox, A. (ed.). 1973. *Plate Tectonics and Geomagnetic Reversals.* San Francisco: W. H. Freeman.

Dalziel, I. W. D. 1995. Earth before Pangaea. *Scientific American* (January): 58–63.

Hallam, A. 1973. *A Revolution in the Earth Sciences: From Continental Drift to Plate Tectonics.* New York: Oxford University Press (Clarendon Press).

Kearey, P., and F. J. Vine. 1990. *Global Tectonics.* Oxford: Blackwell Scientific.

Macdonald, K. C. 1998. Exploring the mid-ocean ridge. *Oceanus* 41(1): 2–8.

McKenzie, D. 1999. Plate tectonics on Mars? *Nature* 399: 307–308.

Muller, R. D., et al. 1997. Digital isochrons of the world's ocean floor. *Journal of Geophysical Research* 102: 3211–3214.

National Academy of Sciences. 1999. When the Earth moves: Seafloor spreading and plate tectonics. *Beyond Discovery: The Path from Research to Human Benefit.* www.beyonddiscovery.org.

Norabuena, E., et al. 1998. Space geodetic observations of Nazca–South America convergence across the central Andes. *Science* 279: 358–362.

Oreskes, N. 1999. *The Rejection of Continental Drift.* New York: Oxford University Press.

Oreskes, N., ed., with Homer Le Grand. 2001. *Plate Tectonics: An Insider's History of the Modern Theory of Earth.* Boulder, CO: Westview Press.

Parsons, T., et al. 1998. A new view into the Cascadia subduction zone and volcanic arc: Implications for earthquake hazards along the Washington margin. *Geology* 26: 199–202.

Phillips, J. D., and D. Forsyth. 1972. Plate tectonics, paleomagnetism, and the opening of the Atlantic. *Bulletin of the Geological Society of America* 83(6): 1579–1600.

Richards, M. A. 1999. Prospecting for Jurassic slabs. *Nature* 397: 203–204.

Scotese, C. R. 2001. *Atlas of Earth History,* Vol. 1, *Paleogeography.* PALEOMAP Project, Arlington, TX.

Smith, Deborah K., and J. R. Cann. 1993. Building the crust at the Mid-Atlantic Ridge. *Nature* 365: 707–714.

Van der Hilst, R. D., S. Widiyantoro, and E. R. Engdahl. 1997. Evidence for deep mantle circulation from global tomography. *Nature* 386: 578–584.

Wessel, G. R. 1986. *The Geology of Plate Margins.* Geological Society of America, Map and Chart Series MC-59.

Crystals of amethyst, a variety of quartz. The planar surfaces are crystal faces, whose geometries are determined by the underlying arrangement of the atoms that make up the crystals. [Breck P. Kent.]

CHAPTER

3

Minerals: Building Blocks of Rocks

"Everything should be made as simple as possible, but no simpler."

ALBERT EINSTEIN

In Chapter 2, we saw how plate tectonics describes Earth's large-scale structure and dynamics, but we touched only briefly on the wide variety of materials that appear in plate tectonic settings. **In this chapter and the next, we focus on rocks, the records of geologic processes, and on minerals, the building blocks of rocks.**

Rocks and minerals help determine various parts of the Earth system, much as concrete, steel, and plastic determine the structure, design, and architecture of large buildings. To tell Earth's story accurately, geologists often adopt a "Sherlock Holmes" approach: they use current evidence to deduce the processes and events that occurred in the past at some particular place. The kinds of minerals found in volcanic rocks, for example, give evidence of eruptions that brought molten rock, at temperatures perhaps as high as 1000°C, to Earth's surface. The minerals of a granite reveal that it crystallized deep in the crust under the very high temperatures and pressures that occur when two continental plates collide and form mountains like the Himalayas. Understanding the geology of a region allows us to make informed guesses about where undiscovered deposits of economically important mineral resources might lie.

We turn now to the focus of this chapter: **mineralogy**—the branch of geology that studies the composition, structure, appearance, stability, occurrence, and associations of minerals.

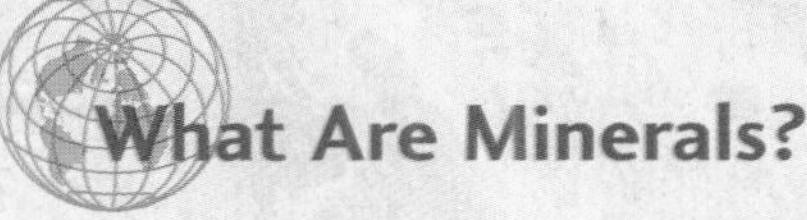

What Are Minerals?

Minerals are the building blocks of rocks: with the proper tools, most rocks can be separated into their constituent minerals. A few kinds of rocks, such as limestone,

Figure 3.1 The mineral calcite is found in the shells of many organisms, such as foraminifera. [*left:* Lester V. Bergman/Corbis; *right:* Cushman Foundation for Foraminiferal Research, 1987.]

contain only a single mineral (in this case, calcite). Other rocks, such as granite, are made of several different minerals. To identify and classify the many kinds of rocks that compose the Earth and understand how they formed, we must know about minerals.

Geologists define a **mineral** as a *naturally occurring, solid crystalline substance, generally inorganic, with a specific chemical composition.* Minerals are homogeneous: they cannot be divided by mechanical means into smaller components.

Let us examine each part of our definition of a mineral in a little more detail.

Naturally Occurring . . . To qualify as a mineral, a substance must be found in nature. Diamonds mined from the Earth in South Africa are minerals. Synthetic versions produced in industrial laboratories are not considered to be minerals. Nor are the thousands of laboratory products invented by chemists.

Solid Crystalline Substance . . . Minerals are solid substances—they are neither liquids nor gases. When we say that a mineral is *crystalline,* we mean that the tiny particles of matter, or atoms, that compose it are arranged in an orderly, repeating, three-dimensional array. Solid materials that have no such orderly arrangement are referred to as *glassy* or *amorphous* (without form) and are not conventionally referred to as minerals. Windowpane glass is amorphous, as are some natural glasses formed during volcanic eruptions. Later in this chapter, we will explore in detail the process by which crystalline materials form.

Generally Inorganic . . . Minerals are defined as inorganic substances and so exclude the organic materials that make up plant and animal bodies. Organic matter is composed of organic carbon, the form of carbon found in all organisms, living or dead. Decaying vegetation in a swamp may be geologically transformed into coal, which also is made of organic carbon; but although it is found as a natural deposit, coal is not traditionally considered a mineral. Many minerals are, however, secreted by organisms. One such mineral, calcite (**Figure 3.1**), forms the shells of oysters and many other organisms, and it contains inorganic carbon. The calcite of these shells, which constitute the bulk of many limestones, fits the definition of a mineral because it is inorganic and crystalline.

. . . With a Specific Chemical Composition The key to understanding the composition of Earth's materials lies in knowing how the chemical elements are organized into minerals. What makes each mineral unique is the combination of its chemical composition and the arrangement of its atoms in an internal structure. A mineral's chemical composition either is fixed or varies within defined limits. The mineral quartz, for example, has a fixed ratio of two atoms of oxygen to one atom of silicon. This ratio never varies, although quartz is found in many different kinds of rock. The components of the mineral olivine—iron, magnesium, and silicon—always have a fixed ratio. Although the ratio of iron to magnesium atoms may vary, the sum of those atoms in relation to the number of silicon atoms always forms a fixed ratio.

The Atomic Structure of Matter

A modern dictionary lists many meanings for the word *atom* and its derivatives. One of the first is "anything considered the smallest possible unit of any material." To the ancient

Greeks, *atomos* meant "indivisible." John Dalton (1766–1844), an English chemist and the father of modern atomic theory, proposed that atoms are particles of matter of several kinds that are so small that they cannot be seen with any microscope and so universal that they compose all substances. In 1805, Dalton hypothesized that each of the various chemical elements consists of a different kind of atom, that all atoms of any given element are identical, and that chemical compounds are formed by various combinations of atoms of different elements in definite proportions.

By the early twentieth century, physicists, chemists, and mineralogists, building on Dalton's ideas, had come to understand the structure of matter much as we do today. We now know that an **atom** is the smallest unit of an element that retains the physical and chemical properties of that element. We also know that atoms are the small units of matter that combine in chemical reactions and that atoms themselves are divisible into even smaller units.

The Structure of Atoms

Understanding the structure of atoms allows us to predict how chemical elements will react with one another and form new crystal structures. For more detailed information about the structure of atoms, see Appendix 4.

The Nucleus: Protons and Neutrons At the center of every atom is a dense **nucleus** containing virtually all the mass of the atom in two kinds of particles: protons and neutrons (**Figure 3.2**). A **proton** has a positive electrical charge of +1. A **neutron** is electrically neutral—that is, uncharged. Atoms of the same chemical element may have different numbers of neutrons, but the number of their protons does not vary. For instance, all carbon atoms have six protons.

Electrons Surrounding the nucleus is a cloud of moving particles called **electrons,** each with a mass so small that it is conventionally taken to be zero. Each electron carries an electrical charge of –1. The number of protons in the nucleus of any atom is balanced by the same number of electrons in the cloud surrounding the nucleus, so an atom is electrically neutral. Thus the nucleus of the carbon atom is surrounded by six electrons (see Figure 3.2).

Atomic Number and Atomic Mass

The number of protons in the nucleus of an atom is called its **atomic number.** Because all atoms of the same element have the same number of protons, they also have the same atomic number. All atoms with six protons, for example, are carbon atoms (atomic number 6). In fact, the atomic number of an element can tell us so much about an element's behavior that the periodic table organizes elements according to their atomic number (**Figure 3.3**). Elements in the same vertical group, such as carbon and silicon, tend to react similarly. For more detail about the periodic table, see Appendix 4.

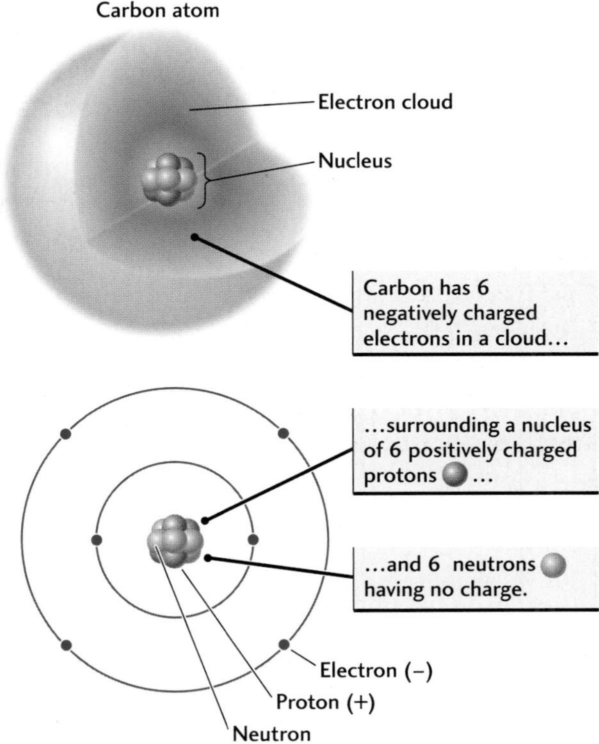

Figure 3.2 Electron structure of the carbon atom (carbon-12). The electrons, each with a charge of –1, are represented as a negatively charged cloud surrounding the nucleus, which contains six protons, each with a charge of +1, and six neutrons, each with zero charge. The size of the nucleus is greatly exaggerated; it is much too small to show at a true scale.

The **atomic mass** of an element is the sum of the masses of its protons and neutrons. (Electrons, because they have so little mass, are not included in this sum.) Although the number of protons is constant, atoms of the same chemical element may have different numbers of neutrons and therefore different atomic masses. These various kinds of atoms are called **isotopes.** Isotopes of the element carbon, for example, all with six protons, exist with six, seven, and eight neutrons, giving atomic masses of 12, 13, and 14.

In nature, the chemical elements exist as mixtures of isotopes, so their atomic masses are never whole numbers. Carbon's atomic mass, for example, is 12.011. It is close to 12 because the isotope carbon-12 is overwhelmingly abundant. The relative abundance of the different isotopes of an element on Earth is determined by processes that enhance the abundance of some isotopes over others. Carbon-12, for example, is favored by some reactions, such as photosynthesis, in which organic carbon compounds are produced from inorganic carbon compounds.

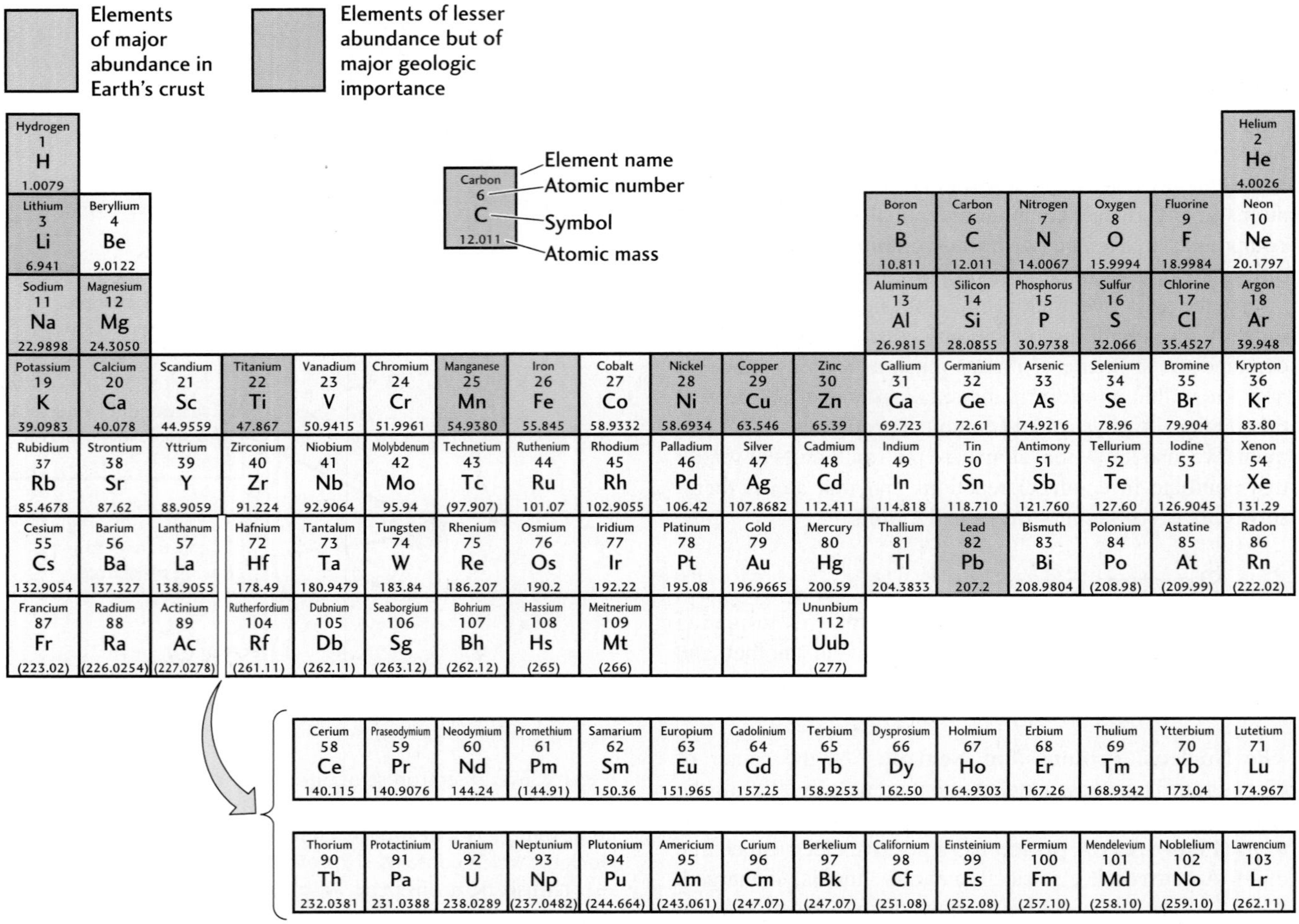

Elements of major abundance in Earth's crust

Elements of lesser abundance but of major geologic importance

Carbon / 6 / C / 12.011 — Element name / Atomic number / Symbol / Atomic mass

Hydrogen 1 H 1.0079																	Helium 2 He 4.0026
Lithium 3 Li 6.941	Beryllium 4 Be 9.0122											Boron 5 B 10.811	Carbon 6 C 12.011	Nitrogen 7 N 14.0067	Oxygen 8 O 15.9994	Fluorine 9 F 18.9984	Neon 10 Ne 20.1797
Sodium 11 Na 22.9898	Magnesium 12 Mg 24.3050											Aluminum 13 Al 26.9815	Silicon 14 Si 28.0855	Phosphorus 15 P 30.9738	Sulfur 16 S 32.066	Chlorine 17 Cl 35.4527	Argon 18 Ar 39.948
Potassium 19 K 39.0983	Calcium 20 Ca 40.078	Scandium 21 Sc 44.9559	Titanium 22 Ti 47.867	Vanadium 23 V 50.9415	Chromium 24 Cr 51.9961	Manganese 25 Mn 54.9380	Iron 26 Fe 55.845	Cobalt 27 Co 58.9332	Nickel 28 Ni 58.6934	Copper 29 Cu 63.546	Zinc 30 Zn 65.39	Gallium 31 Ga 69.723	Germanium 32 Ge 72.61	Arsenic 33 As 74.9216	Selenium 34 Se 78.96	Bromine 35 Br 79.904	Krypton 36 Kr 83.80
Rubidium 37 Rb 85.4678	Strontium 38 Sr 87.62	Yttrium 39 Y 88.9059	Zirconium 40 Zr 91.224	Niobium 41 Nb 92.9064	Molybdenum 42 Mo 95.94	Technetium 43 Tc (97.907)	Ruthenium 44 Ru 101.07	Rhodium 45 Rh 102.9055	Palladium 46 Pd 106.42	Silver 47 Ag 107.8682	Cadmium 48 Cd 112.411	Indium 49 In 114.818	Tin 50 Sn 118.710	Antimony 51 Sb 121.760	Tellurium 52 Te 127.60	Iodine 53 I 126.9045	Xenon 54 Xe 131.29
Cesium 55 Cs 132.9054	Barium 56 Ba 137.327	Lanthanum 57 La 138.9055	Hafnium 72 Hf 178.49	Tantalum 73 Ta 180.9479	Tungsten 74 W 183.84	Rhenium 75 Re 186.207	Osmium 76 Os 190.2	Iridium 77 Ir 192.22	Platinum 78 Pt 195.08	Gold 79 Au 196.9665	Mercury 80 Hg 200.59	Thallium 81 Tl 204.3833	Lead 82 Pb 207.2	Bismuth 83 Bi 208.9804	Polonium 84 Po (208.98)	Astatine 85 At (209.99)	Radon 86 Rn (222.02)
Francium 87 Fr (223.02)	Radium 88 Ra (226.0254)	Actinium 89 Ac (227.0278)	Rutherfordium 104 Rf (261.11)	Dubnium 105 Db (262.11)	Seaborgium 106 Sg (263.12)	Bohrium 107 Bh (262.12)	Hassium 108 Hs (265)	Meitnerium 109 Mt (266)			Ununbium 112 Uub (277)						

Cerium 58 Ce 140.115	Praseodymium 59 Pr 140.9076	Neodymium 60 Nd 144.24	Promethium 61 Pm (144.91)	Samarium 62 Sm 150.36	Europium 63 Eu 151.965	Gadolinium 64 Gd 157.25	Terbium 65 Tb 158.9253	Dysprosium 66 Dy 162.50	Holmium 67 Ho 164.9303	Erbium 68 Er 167.26	Thulium 69 Tm 168.9342	Ytterbium 70 Yb 173.04	Lutetium 71 Lu 174.967
Thorium 90 Th 232.0381	Protactinium 91 Pa 231.0388	Uranium 92 U 238.0289	Neptunium 93 Np (237.0482)	Plutonium 94 Pu (244.664)	Americium 95 Am (243.061)	Curium 96 Cm (247.07)	Berkelium 97 Bk (247.07)	Californium 98 Cf (251.08)	Einsteinium 99 Es (252.08)	Fermium 100 Fm (257.10)	Mendelevium 101 Md (258.10)	Noblelium 102 No (259.10)	Lawrencium 103 Lr (262.11)

Figure 3.3 The periodic table of elements, which organizes the elements (from left to right in a row) in order of atomic number. The table highlights the elements of particular geologic importance.

Chemical Reactions

The structure of an atom determines its chemical reactions with other atoms. **Chemical reactions** are interactions of the atoms of two or more chemical elements in certain fixed proportions that produce new chemical substances—chemical compounds. For example, when two hydrogen atoms combine with one oxygen atom, they form a new chemical compound that we call water (H_2O). The properties of a chemical compound formed in the course of a reaction may be entirely different from those of its constituent elements. For example, when an atom of sodium, a metal, combines with an atom of chlorine, a noxious gas, they form the chemical compound sodium chloride, better known as table salt. We represent this compound by the chemical formula *NaCl,* the symbol *Na* standing for the element sodium and the symbol *Cl* for the element chlorine. (Every chemical element has been assigned its own symbol, which we use as a kind of shorthand for writing chemical formulas and equations.)

Chemical compounds, such as minerals, are formed either by **electron transfer** between the reacting atoms or by **electron sharing** between the reacting atoms. In the reaction between sodium (Na) and chlorine (Cl) atoms to form sodium chloride (NaCl), the sodium atom loses one electron, which the chlorine atom gains (**Figure 3.4**). Because the chlorine atom has gained a negatively charged electron, it is now negatively charged, Cl^-. Likewise, the loss of an electron gives sodium a positive charge, Na^+. The compound NaCl itself remains electrically neutral because the positive charge on Na^+ is exactly balanced by the negative charge on Cl^-. A positively charged ion is called a **cation,** and a negatively charged ion is called an **anion.**

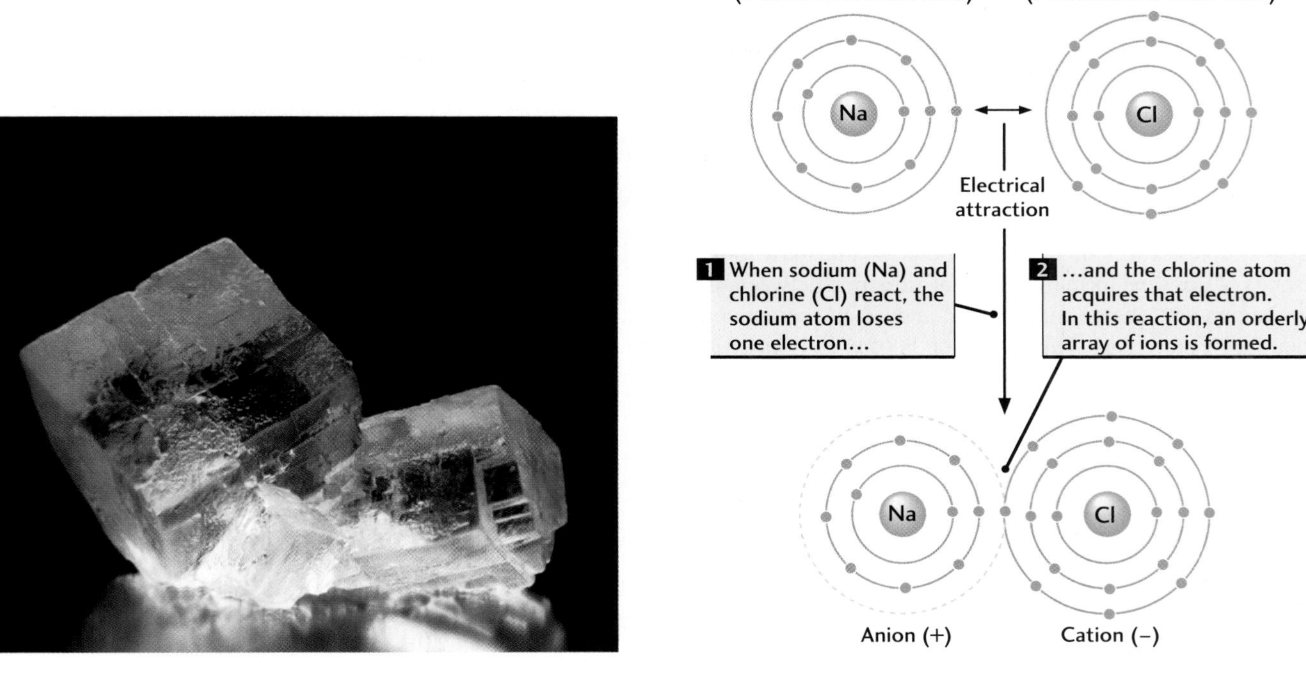

Figure 3.4 Table salt, NaCl, is formed by the reaction of chlorine and sodium atoms. [C. D. Winter/Photo Researchers.]

Atoms that do not react by gaining or losing electrons combine chemically by sharing electrons. Carbon and silicon, two of the most abundant elements in Earth's crust, tend to form compounds by electron sharing. Diamond is a compound composed entirely of carbon atoms sharing electrons (**Figure 3.5**).

Chemical Bonds

The ions or atoms of elements that make up compounds are held together by electrical forces of attraction between electrons and protons, which we call chemical bonds. The electrical attractions of shared electrons or of gained or lost

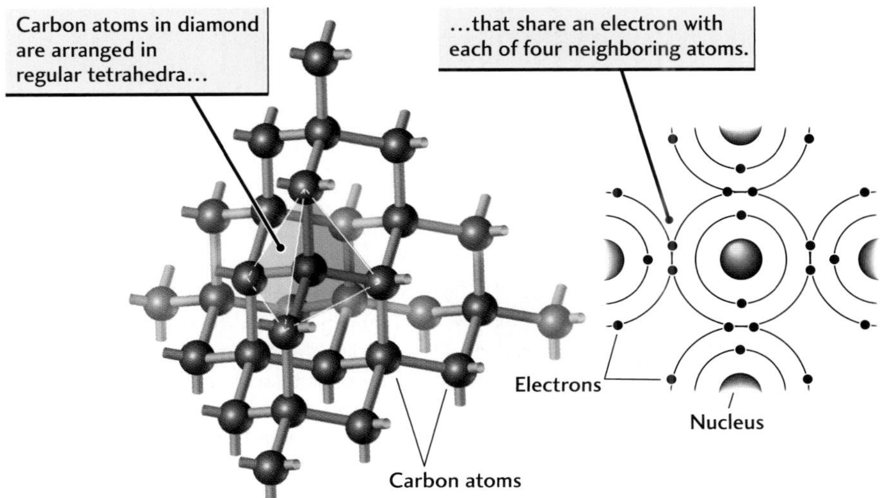

Figure 3.5 Electron sharing in diamond. The mineral diamond is composed of the single element carbon. Each carbon atom shares its four electrons with four adjacent carbon atoms.

electrons may be strong or weak, and the bonds created by these attractions are correspondingly strong or weak. Strong bonds keep a substance from chemically decomposing into its elements or into other compounds. They also make minerals hard and keep them from cracking or splitting. Two major types of bonds are found in most rock-forming minerals: ionic bonds and covalent bonds.

Ionic Bonds

The simplest form of chemical bond is the **ionic bond.** Bonds of this type form by electrical attraction between ions of the opposite charge, such as Na^+ and Cl^- in sodium chloride (see Figure 3.4). This attraction is of exactly the same nature as the static electricity that can make clothing of nylon or silk cling to the body. The strength of an ionic bond decreases greatly as the distance between ions increases. Bond strength increases as the electrical charges of the ions increase. Ionic bonds are the dominant type of chemical bonds in mineral structures; *about 90 percent of all minerals are essentially ionic compounds.*

Covalent Bonds

Elements that do not readily gain or lose electrons to form ions and instead form compounds by sharing electrons are held together by **covalent bonds,** which are generally stronger than ionic bonds. One mineral with a covalently bonded crystal structure is diamond, consisting of the single element carbon. Carbon atoms have four electrons and acquire four more by electron sharing. In diamond, every carbon atom (not an ion) is surrounded by four others arranged in a regular *tetrahedron,* a four-sided pyramidal form, each side a triangle (see Figure 3.5). In this configuration, each carbon atom shares an electron with each of its four neighbors, resulting in a very stable configuration. Figure 3.5 shows a network of carbon tetrahedra linked together.

Atoms of metallic elements, which have strong tendencies to lose electrons, pack together as cations, and the freely mobile electrons are shared and dispersed among the ions. This free electron sharing results in a kind of covalent bond that we call a **metallic bond.** It is found in a small number of minerals, among them the metal copper and some sulfides.

The chemical bonds of some minerals are intermediate between pure ionic and pure covalent bonds because some electrons are exchanged and others are shared.

The Atomic Structure of Minerals

Minerals can be viewed in two complementary ways: as crystals (or grains) that we can see with the naked eye and as assemblages of submicroscopic atoms organized in an ordered three-dimensional array. We are now prepared for a closer look at the orderly forms that characterize mineral structure and at the conditions under which minerals form. Later in this chapter, we will see that the crystal structures of minerals are manifested in their physical properties. First, however, we turn to the question of how minerals form.

How Do Minerals Form?

Minerals form by the process of **crystallization,** the growth of a solid from a gas or liquid whose constituent atoms come together in the proper chemical proportions and crystalline arrangement. (Remember that the atoms in a mineral are arranged in an ordered three-dimensional array.) The bonding of carbon atoms in diamond, a covalently bonded mineral, is one example of crystallization and crystal structure. Carbon atoms bind together in tetrahedra, each tetrahedron attaching to another and building up a regular three-dimensional structure from a great many atoms (see Figure 3.5). As a diamond crystal grows, it extends its tetrahedral structure in all directions, always adding new atoms in the proper geometric arrangement. Diamonds can be synthesized under very high pressures and temperatures that mimic conditions in Earth's mantle.

The sodium and chloride ions that make up sodium chloride, an ionically bonded mineral, also crystallize in an orderly three-dimensional array. In **Figure 3.6**a, we can see

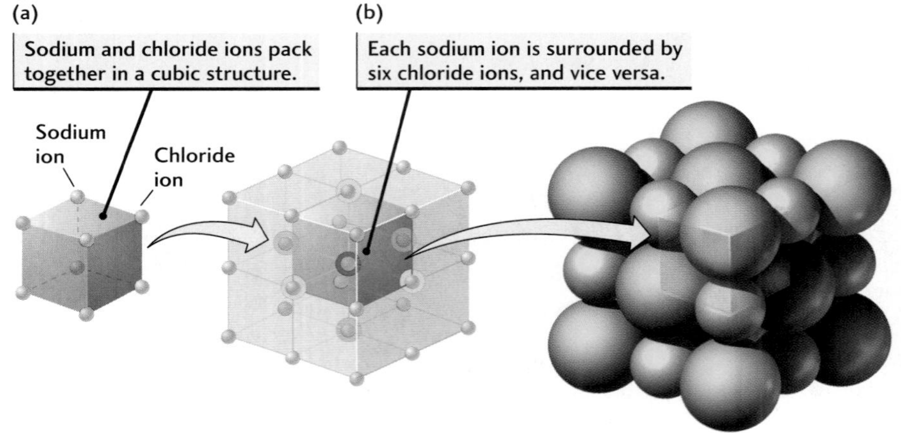

Figure 3.6 Structure of sodium chloride. (a) The dashed lines between the ions show the cubic geometry of this mineral; they do not represent bonds. Note that each sodium ion is surrounded by six chloride ions. (Ions are not drawn to scale.) (b) The relative sizes of sodium and chloride ions allow them to pack together in a cubic structure. Ions here are shown in their correct relative sizes.

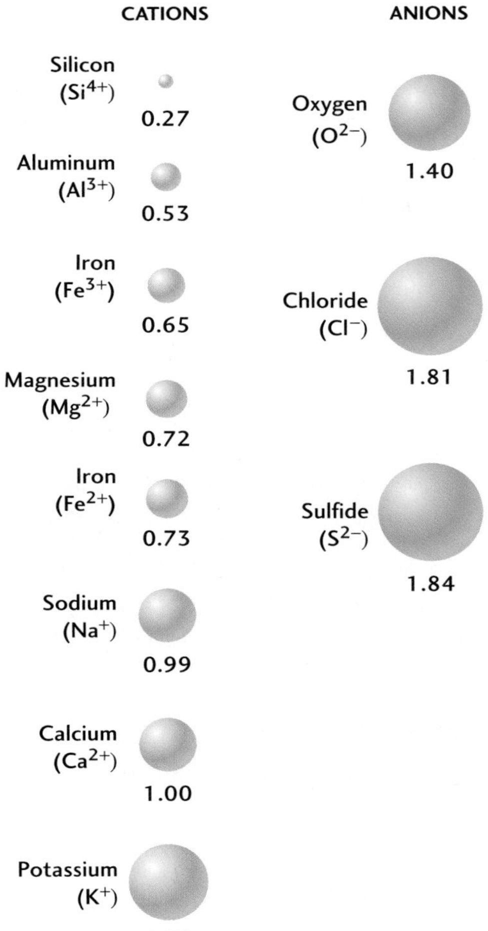

Figure 3.7 Sizes of ions as they are commonly found in rock-forming minerals. Ionic radii are given in 10^{-8} cm. [After L. G. Berry, B. Mason, and R. V. Dietrich, *Mineralogy.* San Francisco: W. H. Freeman, 1983.]

the geometry of their arrangement, with each ion of one kind surrounded by six ions of the other in a series of *cubic* structures extending in three directions. We can think of ions as if they were solid spheres, packed together in close-fitting structural units. Figure 3.6b shows the relative sizes of the ions in NaCl. There are six neighboring ions in NaCl's basic structural unit. The relative sizes of the sodium and chloride ions allow them to fit together in a closely packed arrangement.

Many of the cations of abundant minerals are relatively small; most anions are large (**Figure 3.7**). This is the case with the most common Earth anion, oxygen. Because anions tend to be larger than cations, it is apparent that most of the space of a crystal is occupied by the anions and that cations fit into the spaces between them. As a result, crystal structures are determined largely by how the anions are arranged and how the cations fit between them.

Cations of similar sizes and charges tend to substitute for one another and to form compounds having the same crystal structure but differing chemical composition. *Cation substitution* is common in minerals containing the silicate ion (SiO_4^{4-}). This process is illustrated by olivine, a silicate mineral abundant in many volcanic rocks.

Iron (Fe) and magnesium (Mg) ions are similar in size, and both have two positive charges, so they easily substitute for each other in the structure of olivine. The composition of pure magnesium olivine is Mg_2SiO_4; the pure iron olivine is Fe_2SiO_4. The composition of olivine with both iron and magnesium is given by the formula $(Mg,Fe)_2SiO_4$, which simply means that the number of iron and magnesium cations may vary, but their combined total (expressed as a subscript 2) does not vary in relation to each SiO_4^{4-} ion. The proportion of iron to magnesium is determined by the relative abundance of the two elements in the molten material from which the olivine crystallized. In many silicate minerals, aluminum (Al) substitutes for silicon (Si). Aluminum and silicon ions are so similar in size that aluminum can take the place of silicon in many crystal structures. The difference in charge between aluminum (3+) and silicon (4+) ions is balanced by an increase in the number of other cations, such as sodium (1+).

Crystallization starts with the formation of microscopic single **crystals,** ordered three-dimensional arrays of atoms in which the basic arrangement is repeated in all directions. The boundaries of crystals are natural flat (plane) surfaces called *crystal faces.* The crystal faces of a mineral are the external expression of the mineral's internal atomic structure. **Figure 3.8** pairs drawings of perfect crystals (which

Figure 3.8 Perfect crystals. A perfect crystal is rare, but no matter how irregular the shapes of the faces may be, the angles are always exactly the same. [(a) Andrew Romeo–coolrox.com; (b) Breck P. Kent.]

are very rare in nature) with photographs of two actual minerals. The simple geometric cubes of garnet crystals correspond to the cubic arrangement of its ions. The six-sided (hexagonal) shape of the quartz crystal corresponds to its hexagonal internal atomic structure.

During crystallization, the initially microscopic crystals grow larger, maintaining their crystal faces as long as they are free to grow. Large crystals with well-defined faces form when growth is slow and steady and space is adequate to allow growth without interference from other crystals nearby. For this reason, most large mineral crystals form in open spaces in rocks, such as fractures or cavities (**Figure 3.9**).

Often, however, the spaces between growing crystals fill in, or crystallization proceeds too rapidly. Crystals then grow over one another and coalesce to become a solid mass of crystalline particles, or *grains*. In this case, few or no grains show crystal faces (see Figure 3.9). Large crystals that can be seen with the naked eye are relatively unusual, but many microscopic minerals in rocks display crystal faces.

Unlike crystalline minerals, glassy materials—which solidify from liquids so quickly that they lack any internal atomic order—do not form crystals with plane faces. Instead they are found as masses with curved, irregular surfaces. The most common glass is volcanic glass.

Figure 3.10 A halite deposit from the Sultanate of Oman. This deposit, which is more than 500 million years old, is preserved in a drill core from the deep subsurface. The black band is concentrated organic matter derived from microorganisms that lived in extreme hypersaline environments that formed when the ancient ocean dried up. [John Grotzinger.]

When Do Minerals Form?

Lowering the temperature of a liquid below its freezing point is one way to start the process of crystallization. In water, for example, 0°C is the temperature below which crystals of ice, a mineral, start to form. Similarly, a magma—a hot, molten liquid rock—crystallizes solid minerals when it cools. As a magma falls below its melting point, which may be higher than 1000°C, crystals of silicate minerals such as olivine or feldspar begin to form. (Geologists usually refer to melting points of magmas rather than freezing points, because freezing normally implies cold.)

Figure 3.9 This sample of amethyst crystals displays large, well-formed crystals. [José Manuel Sanchis Calvete/Corbis.]

Another set of conditions that can produce crystallization occurs as liquids evaporate from a solution. A solution is formed when one chemical substance is dissolved in another, such as salt in water. As the water evaporates from a salt solution, the concentration of salt eventually gets so high that the solution is said to be saturated—it can hold no more salt. If evaporation continues, the salt starts to **precipitate,** or drop out of solution as crystals. Deposits of halite or table salt form under just these conditions when seawater evaporates to the point of saturation in some hot, arid bays or arms of the ocean (**Figure 3.10**).

Diamond and graphite exemplify the dramatic effects that temperature and pressure can have on mineral formation. Diamond and graphite (the material that is used as the "lead" in pencils) are **polymorphs,** alternative structures for a single chemical compound (**Figure Story 3.11**). These

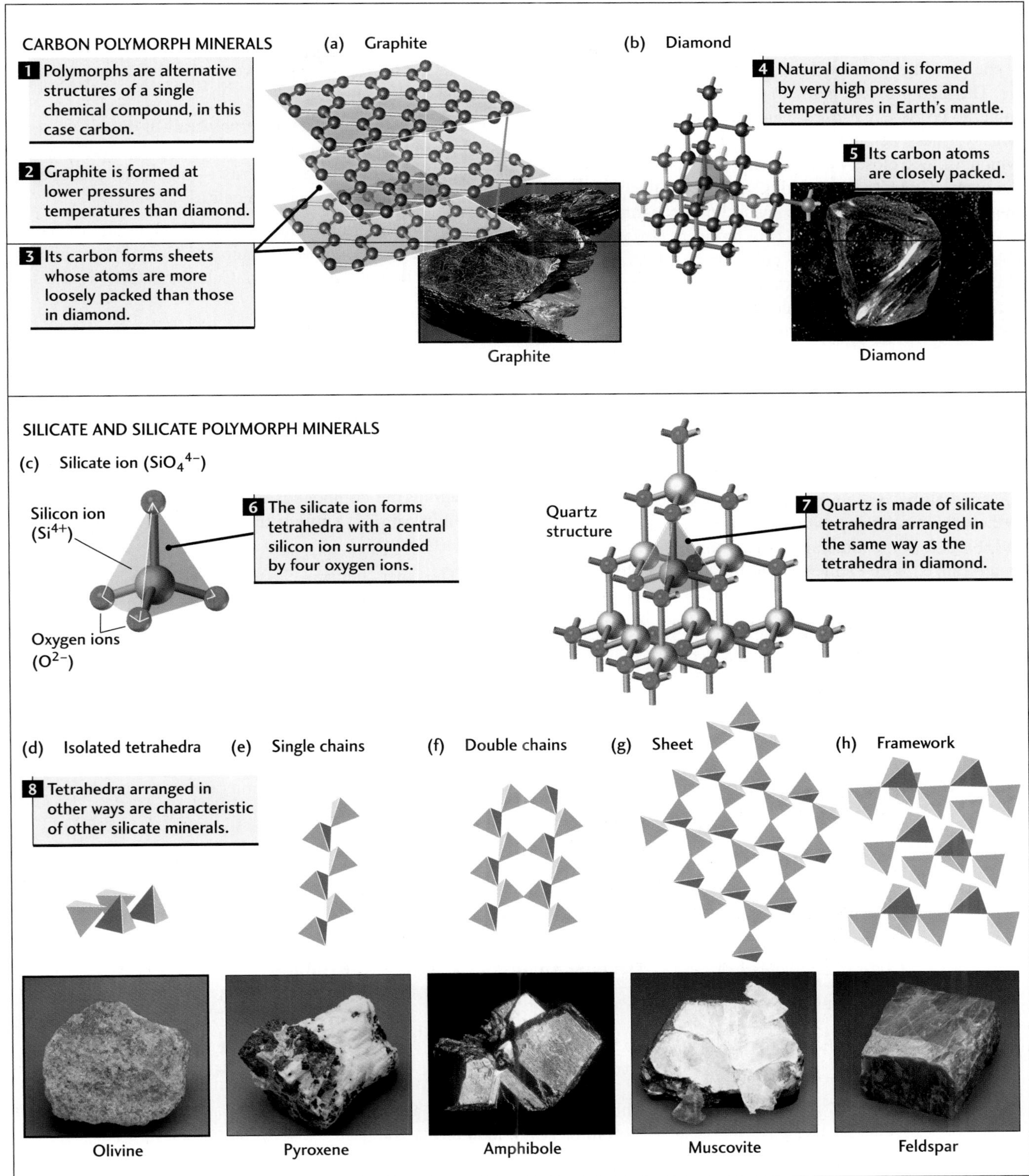

Figure Story 3.11 Polymorphs of carbon and silicate. [*Graphite:* John A. Jaszczak, Michigan Technological University. *Diamond:* Charles O'Rear/Corbis. *Olivine:* Chip Clark. *Pyroxene:* Chip Clark. *Amphibole (hornblende):* George Whiteley/Photo Researchers. *Muscovite (mica):* Chip Clark. *Feldspar:* Chip Clark.]

two minerals, both formed from carbon, have different crystal structures and very different appearances (Figure Story 3.11a and 11b). From experimentation and geological observation, we know that diamond forms and remains stable at the very high pressures and temperatures of Earth's mantle. The high pressure in the mantle forces the atoms in diamond to be closely packed. Diamond therefore has a higher density (mass per unit volume), 3.5 g/cm^3, than graphite, which is less closely packed and has a density of only 2.1 g/cm^3. Graphite forms and is stable at relatively moderate pressures and temperatures, such as those in Earth's crust.

Low temperatures also can produce closer packing. Quartz and cristobalite are polymorphs of silica (SiO_2). Quartz forms at low temperatures and is relatively dense (2.7 g/cm^3). Cristobalite, which forms at a higher temperature, has a more open structure and is therefore less dense (2.3 g/cm^3).

Rock-Forming Minerals

All minerals have been grouped into eight classes according to their chemical composition; six of those classes are listed in Table 3.1. Some minerals, such as copper, occur naturally as un-ionized pure elements, and they are classified as *native elements.* Most others are classified by their anions. Olivine, for example, is classed as a silicate by its silicate anion, SiO_4^{4-}. Halite (NaCl) is classed as a halide by its chloride anion, Cl^-. So is its close relative sylvite, potassium chloride (KCl).

Although many thousands of minerals are known, geologists commonly encounter only about 30 of them. These minerals are the building blocks of most crustal rocks, and so they are called *rock-forming minerals.* Their relatively small number corresponds to the small number of elements that are found in major abundance in Earth's crust. As we learned in Chapter 1, 99 percent of the crust is made up of only nine elements.

In the following pages, we consider the most common rock-forming minerals:

- *Silicates,* the most abundant minerals in Earth's crust, are composed of oxygen (O) and silicon (Si)—the two most abundant elements in the crust—mostly in combination with the cations of other elements.
- *Carbonates* are minerals made of carbon and oxygen in the form of the carbonate anion (CO_3^{2-}) in combination with calcium and magnesium. Calcite ($CaCO_3$) is one such mineral.
- *Oxides* are compounds of the oxygen anion (O^{2-}) and metallic cations; an example is the mineral hematite (Fe_2O_3).
- *Sulfides* are compounds of the sulfide anion (S^{2-}) and metallic cations, a group that includes the mineral pyrite (FeS_2).
- *Sulfates* are compounds of the sulfate anion (SO_4^{2-}) and metallic cations, a group that includes the mineral anhydrite ($CaSO_4$).

The other chemical classes of minerals, including native elements and halides, are not as common as the rock-forming minerals.

Silicates

The basic building block of all silicate mineral structures is the *silicate ion.* It is a tetrahedron—a pyramidal structure with four sides—composed of a central silicon ion (Si^{4+}) surrounded by four oxygen ions (O^{2-}), giving the formula

Table 3.1 Some Chemical Classes of Minerals

Class	Defining Anions	Example
Native elements	None: no charged ions	Copper metal (Cu)
Oxides and hydroxides	Oxygen ion (O^{2-}) Hydroxyl ion (OH^-)	Hematite (Fe_2O_3) Brucite ($Mg[OH]_2$)
Halides	Chloride (Cl^-), fluoride (F^-), bromide (Br^-), iodide (I^-)	Halite (NaCl)
Carbonates	Carbonate ion (CO_3^{2-})	Calcite ($CaCO_3$)
Sulfates	Sulfate ion (SO_4^{-2})	Anhydrite ($CaSO_4$)
Silicates	Silicate ion (SiO_4^{-4})	Olivine (Mg_2SiO_4)

SiO_4^{4-} (Figure Story 3.11c). Because the silicate ion has a negative charge, it often bonds to cations to form electrically neutral minerals. The silicate ion typically bonds with cations such as sodium (Na^+), potassium (K^+), calcium (Ca^+), magnesium (Mg^{2+}), and iron (Fe^{2+}). Alternatively, it can share oxygen ions with other silicon-oxygen tetrahedra. Tetrahedra may be isolated (linked only to cations), or they may be linked to other silica tetrahedra in rings, single chains, double chains, sheets, or frameworks, some of which are shown in Figure Story 3.11.

Isolated Tetrahedra Isolated tetrahedra are linked by the bonding of each oxygen ion of the tetrahedron to a cation (Figure Story 3.11d). The cations, in turn, bond to the oxygen ions of other tetrahedra. The tetrahedra are thus isolated from one another by cations on all sides. Olivine is a rock-forming mineral with this structure.

Single-Chain Linkages Single chains also form by sharing oxygen ions. Two oxygen ions of each tetrahedron bond to adjacent tetrahedra in an open-ended chain (Figure Story 3.11e). Single chains are linked to other chains by cations. Minerals of the pyroxene group are single-chain silicate minerals. Enstatite, a pyroxene, is composed of iron or magnesium ions, or both, and is limited to a chain of tetrahedra in which the two cations may substitute for each other, as in olivine. The formula $(Mg,Fe)SiO_3$ represents this structure.

Double-Chain Linkages Two single chains may combine to form double chains linked to each other by shared oxygen ions (Figure Story 3.11f). Adjacent double chains linked by cations form the structure of the amphibole group of minerals. Hornblende, a member of this group, is an extremely common mineral in both igneous and metamorphic rocks. It has a complex composition that includes calcium (Ca^{2+}), sodium (Na^+), magnesium (Mg^{2+}), iron (Fe^{2+}), and aluminum (Al^{3+}).

Sheet Linkages In sheet structures, each tetrahedron shares three of its oxygen ions with adjacent tetrahedra to build stacked sheets of tetrahedra (Figure Story 3.11g). Cations may be interlayered with tetrahedral sheets. The micas and clay minerals are the most abundant sheet silicates. Muscovite, $KAl_3Si_3O_{10}(OH)_2$, is one of the most common sheet silicates and is found in many types of rocks. It can be separated into extremely thin, transparent sheets. Kaolinite, $Al_2Si_2O_5(OH)_4$, which also has this structure, is a common clay mineral found in sediments and is the basic raw material for pottery.

Frameworks Three-dimensional frameworks form as each tetrahedron shares all its oxygen ions with other tetrahedra. Feldspars, the most abundant minerals in Earth's crust, are framework silicates (Figure Story 3.11h), as is another of the most common minerals, quartz (SiO_2).

Silicate Compositions Chemically, the simplest silicate is silicon dioxide, also called silica (SiO_2), which is found most often as the mineral quartz. When the silicate tetrahedra of quartz are linked, sharing two oxygen ions for each silicon ion, the total formula adds up to SiO_2.

In other silicate minerals, the basic units—rings, chains, sheets, and frameworks—are bonded to such cations as sodium (Na^+), potassium (K^+), calcium (Ca^{2+}), magnesium (Mg^{2+}), and iron (Fe^{2+}). As noted in the discussion of cation substitution, aluminum (Al^{3+}) substitutes for silicon in many silicate minerals.

Carbonates

The nonsilicate mineral calcite (calcium carbonate, $CaCO_3$) is one of the abundant minerals in Earth's crust and is the chief constituent of a group of rocks called limestones (**Figure 3.12**). Its basic building block, the carbonate ion (CO_3^{2-}), consists of a carbon ion surrounded by three oxygen ions in a triangle, as shown in Figure 3.12b. The carbon atom shares electrons with the oxygen atoms. Groups of carbonate ions are arranged in sheets somewhat like the sheet silicates and are bonded by layers of cations (Figure 3.12c). The sheets of carbonate ions in calcite are separated by layers of calcium ions. The mineral dolomite,

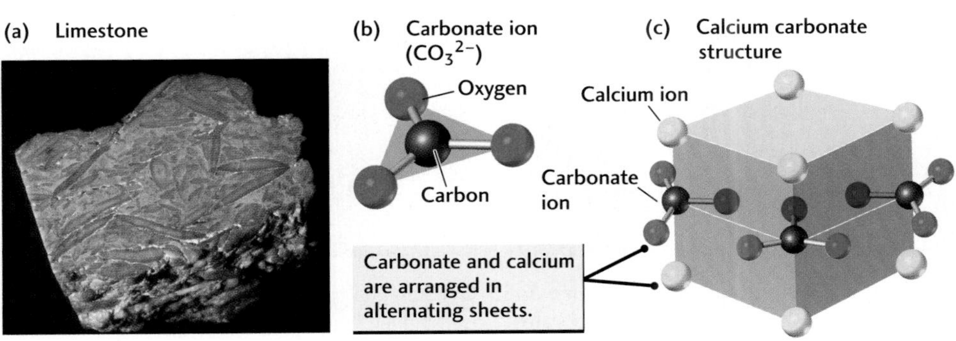

Figure 3.12 Carbonate minerals, such as calcite (calcium carbonate, $CaCO_3$), have a layered structure. (a) Limestone. [Leonard Lessin/Peter Arnold, Inc.] (b) Top view of the carbonate building block, a carbon ion surrounded in a triangle by three oxygen ions, with a net charge of −2. (c) View of the alternating layers of calcium and carbonate ions.

Figure 3.13 Nonsilicate minerals: (*left*) hematite, (*right*) spinel. [Chip Clark.]

$CaMg(CO_3)_2$, another major mineral of crustal rocks, is made up of the same carbonate sheets separated by alternating layers of calcium ions and magnesium ions.

Oxides

Oxide minerals are compounds in which oxygen is bonded to atoms or cations of other elements, usually metallic ions such as iron (Fe^{2+} or Fe^{3+}). Most oxide minerals are ionically bonded, their structures varying with the size of the metallic cations. This group is of great economic importance because it includes the ores of most of the metals, such as chromium and titanium, used in the industrial and technological manufacture of metallic materials and devices. Hematite (Fe_2O_3), shown in **Figure 3.13**, is a chief ore of iron.

Another of the abundant minerals in this group, spinel, is an oxide of two metals, magnesium and aluminum ($MgAl_2O_4$). Spinel (Figure 3.13) has a closely packed cubic structure and a high density (3.6 g/cm^3), reflecting the conditions of high pressure and temperature under which it forms. Transparent, gem-quality spinel resembles ruby and sapphire and is found in the crown jewels of England and Russia.

Sulfides

The chief ores of many valuable minerals—such as copper, zinc, and nickel—are members of the sulfide group. This group includes compounds of the sulfide ion (S^{2-}) with metallic cations. In the sulfide ion, a sulfur atom has gained two electrons in its outer shell. Most sulfide minerals look like metals, and almost all are opaque. The most common sulfide mineral is pyrite (FeS_2), often called "fool's gold" because of its goldish metallic appearance (**Figure 3.14**).

Sulfates

The basic building block of all sulfates is the sulfate ion (SO_4^{2-}). It is a tetrahedron made up of a central sulfur atom surrounded by four oxygen ions (O^{2-}). One of the most abundant minerals of this group is gypsum, the primary component of plaster (**Figure 3.15**). Gypsum forms when seawater evaporates. During evaporation, Ca^{2+} and SO_4^{2-}, two ions that are abundant in seawater, combine and precipitate as layers of sediment, forming calcium sulfate ($CaSO_4 \cdot 2H_2O$). (The dot in this formula signifies that two water molecules are bonded to the calcium and sulfate ions.)

Another calcium sulfate, anhydrite ($CaSO_4$), differs from gypsum in that it contains no water. Its name is derived from the word *anhydrous,* meaning "free from water." Gypsum is stable at the low temperatures and pressures found at Earth's surface, whereas anhydrite is stable at the higher temperatures and pressures where sedimentary rocks are buried.

Figure 3.14 A sample of pyrite, also known as "fool's gold." [Lester V. Bergman/Corbis.]

Figure 3.15 Gypsum is a sulfate formed when seawater evaporates. [José Manuel Sanchis Calvete/Corbis.]

Physical Properties of Minerals

Geologists use their knowledge of mineral composition and structure to understand the origins of rocks. First, they must identify the minerals that compose a rock. To do so, they rely greatly on chemical and physical properties that can be observed relatively easily. In the nineteenth and early twentieth centuries, geologists carried field kits for the rough chemical analysis of minerals that would help in identification. One such test is the origin of the phrase "the acid test." It consists of dropping diluted hydrochloric acid (HCl) on a mineral to see if it "fizzes" (**Figure 3.16**). The fizzing indicates that carbon dioxide (CO_2) is escaping, which means that the mineral is likely to be calcite, a carbonate mineral.

In the rest of this chapter, we will review the physical properties of minerals, many of which contribute to their practical and decorative value (see Feature 3.1).

Table 3.2 Mohs Scale of Hardness

Mineral	Scale Number	Common Objects
Talc	1	
Gypsum	2	——— Fingernail
Calcite	3	——— Copper coin
Fluorite	4	
Apatite	5	——— Knife blade
Orthoclase	6	——— Window glass
Quartz	7	——— Steel file
Topaz	8	
Corundum	9	
Diamond	10	

Hardness

Hardness is a measure of the ease with which the surface of a mineral can be scratched. Just as a diamond, the hardest mineral known, scratches glass, so a quartz crystal, which is harder than feldspar, scratches a feldspar crystal. In 1822, Friedrich Mohs, an Austrian mineralogist, devised a scale (now known as the **Mohs scale of hardness**) based on the ability of one mineral to scratch another. At one extreme is the softest mineral (talc); at the other, the hardest (diamond) (Table 3.2). The Mohs scale is still one of the best practical tools for identifying an unknown mineral. With a knife blade and a few of the minerals on the hardness scale, a field geologist can gauge an unknown mineral's position on the scale. If the unknown mineral is scratched by a piece of quartz but not by the knife, for example, it lies between 5 and 7 on the scale.

Recall that covalent bonds are generally stronger than ionic bonds. The hardness of any mineral depends on the strength of its chemical bonds: the stronger the bonds, the harder the mineral. Crystal structure varies in the silicate group of minerals, and so does hardness. For example, hardness varies from 1 in talc, a sheet silicate, to 8 in topaz, a silicate with isolated tetrahedra. Most silicates fall in the 5 to 7 range on the Mohs scale. Only sheet silicates are relatively soft, with hardnesses between 1 and 3.

Figure 3.16 The acid test. One easy but effective way to identify certain minerals is to drop diluted hydrochloric acid (HCl) on the substance. If it fizzes, indicating the escape of carbon dioxide, the mineral is likely to be calcite. [Chip Clark.]

3.1 What Makes Gems So Special?

No one can be sure when the first human picked up a mineral crystal and kept it for its rare beauty, but we do know that gems were being worn as necklaces and other adornments at the dawn of civilization in Egypt, at least 4000 years ago. These early Egyptians were undoubtedly attracted to the color and play of light on the polished surfaces of such minerals as carnelian, lapis lazuli, and turquoise. Color and luster (the ability to reflect light) are two qualities that still serve to define gemstones. Although the value placed on a gemstone varies in different cultures and historical periods, other required qualities seem to be beauty, transparency, brilliance, durability, and rarity. Most minerals have some of these remarkable qualities, but the stones considered most precious are ruby; sapphire; emerald; and, of course, diamond.

A diamond—geologically speaking, at least—may not be forever, as the advertisers claim, but it is special. Its glitter is unique, as are the play of colors and the sparkle that it emits. The source of these qualities is the way in which diamond refracts, or bends, light. These characteristics are enhanced by diamond's remarkable ability to split perfectly along certain directions of the crystal, which diamond cutters use to advantage in carefully cutting gem-quality stones. Diamond's multiple facets (faces superficially similar to crystal faces) can be polished to enhance this sparkle. These facets can be ground only by other diamonds, because diamond is the hardest mineral known—so hard that it can scratch any other mineral and remain undamaged. This mineral's tightly packed crystal structure and strong covalent bonds between carbon atoms give it these characteristics, which allow it to be identified with certainty by mineralogists and jewelers.

Rubies and sapphires are gem-quality varieties of the common mineral corundum (aluminum oxide), which is widespread and abundant in a number of rock types. Although not as hard as diamond, corundum is extremely hard. Small amounts of impurities produce the intense colors that we value. Ruby, for example, is red because of small amounts of chromium, the same substance that gives emeralds their green color.

Less valuable, sometimes called semiprecious, gemstones are topaz, garnet, tourmaline, jade, turquoise, and zircon. Most, like garnet, are common constituents of rocks, occurring mostly as small imperfect crystals with many impurities and poor transparency. But, under special conditions, gem-quality garnets form. From time to time, some minerals that are not ordinarily considered gems may enjoy sudden—perhaps temporary—popularity. Hematite (iron oxide) currently enjoys this status, appearing in necklaces and bracelets.

Sapphire (blue) and diamond (colorless) brooch by Fortunato Pio Castellani, Smithsonian Institution, nineteenth century. [Aldo Tutino/Art Resource.]

Within groups of minerals having similar crystal structures, increasing hardness is related to other factors that also increase bond strength:

- *Size:* The smaller the atoms or ions, the smaller the distance between them, the greater the electrical attraction, and thus the stronger the bond.
- *Charge:* The larger the charge of ions, the greater the attraction between ions and thus the stronger the bond.
- *Packing of atoms or ions:* The closer the packing of atoms or ions, the smaller the distance between them and thus the stronger the bond.

Size is an especially important factor for most metallic oxides and for most sulfides of metals with high atomic numbers—such as gold, silver, copper, and lead. Minerals of these groups are soft, with hardnesses of less than 3, because their metallic cations are so large. Carbonates and sulfates, groups

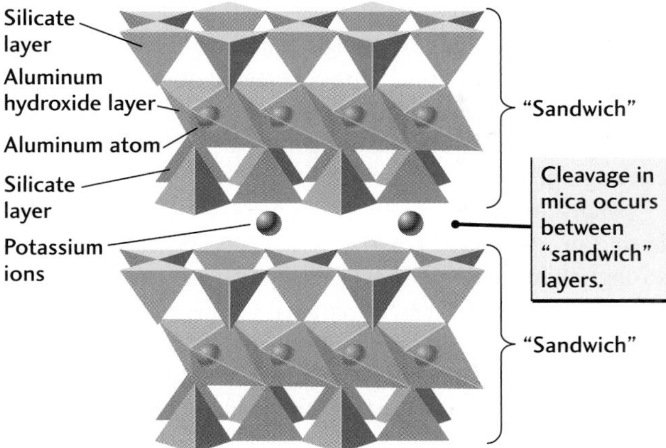

Figure 3.17 Cleavage of mica. The diagram shows the cleavage planes in the mineral structure, oriented perpendicular to the plane of the page. Horizontal lines mark the interfaces of silica-oxygen tetrahedral sheets and sheets of aluminum hydroxide bonding the two tetrahedral layers into a "sandwich." Cleavage takes place between composite tetrahedral–aluminum hydroxide sandwiches. The photograph shows thin sheets separating along the cleavage planes. [Chip Clark.]

in which the structures are packed less densely, also are soft, with hardnesses of less than 5. In all these groups, the hardness reflects the strength of the chemical bonds.

Cleavage

Cleavage is the tendency of a crystal to break along flat planar surfaces. The term is also used to describe the geometric pattern produced by such breakage. Cleavage varies inversely with bond strength: high bond strength produces poor cleavage; low bond strength produces good cleavage. Because of their strength, covalent bonds generally give poor or no cleavage. Ionic bonds are relatively weak, so they give excellent cleavage.

If the bonds between some of the planes of atoms or ions in a crystal are weak, the mineral can be made to split along those planes. Muscovite, a mica sheet silicate, breaks along smooth, lustrous, flat, parallel surfaces, forming transparent sheets less than a millimeter thick. Mica's excellent cleavage results from weakness of the bonds between the sandwiched layers of cations and tetrahedral silica sheets (**Figure 3.17**).

Cleavage is classified according to two primary sets of characteristics: (1) the number of planes and pattern of cleavage, and (2) the quality of surfaces and ease of cleaving.

Number of Planes; Pattern of Cleavage The number of planes and patterns of cleavage are identifying hallmarks of many rock-forming minerals. Muscovite, for example, has only one plane of cleavage, whereas calcite and dolomite have three excellent cleavage directions that give them a rhomboidal shape (**Figure 3.18**).

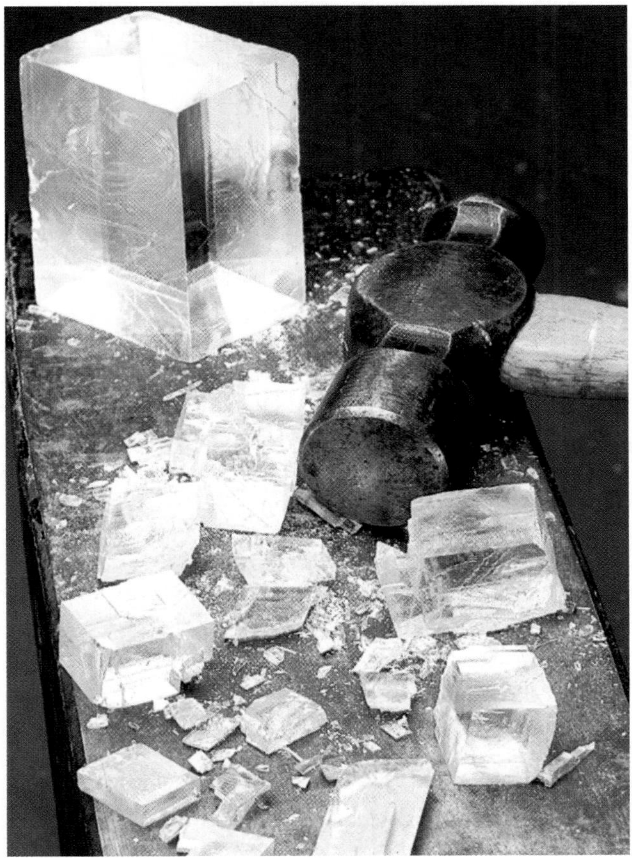

Figure 3.18 Example of rhomboidal cleavage in calcite. Calcite can be cleaved by a light hammer blow on a chisel oriented parallel to one of its planes. [Chip Clark.]

A crystal's structure determines its cleavage planes and its crystal faces. Crystals have fewer cleavage planes than possible crystal faces. Faces may be formed along any of numerous planes defined by rows of atoms or ions. Cleavage occurs along any of those planes across which the bonding is weak. All crystals of a mineral exhibit its characteristic cleavage, whereas only some crystals display particular faces.

Galena (lead sulfide, PbS) and halite (sodium chloride, NaCl) cleave along three planes, forming perfect cubes. Distinctive angles of cleavage help identify two important groups of silicates, the pyroxenes and amphiboles, that otherwise often look alike (**Figure 3.19**). Pyroxenes have a single-chain linkage and are bonded so that their cleavage planes are almost at right angles (about 90°) to each other. In cross section, the cleavage pattern of pyroxene is nearly a square. In contrast, amphiboles, the double chains, bond to give two cleavage planes, at about 60° and 120° to each other. They produce a diamond-shaped cross section.

Quality of Surfaces; Ease of Cleaving A mineral's cleavage is assessed as perfect, good, or fair, according to the quality of surfaces produced and the ease of cleaving. Muscovite can be cleaved easily, producing extremely high quality, smooth surfaces; its cleavage is *perfect.* The single- and double-chain silicates (pyroxenes and amphiboles, respectively) show *good* cleavage. Although these minerals break easily along the cleavage plane, they also break across it, producing cleavage surfaces that are not as smooth as those of mica. *Fair* cleavage is shown by the ring silicate beryl. Beryl's cleavage is less regular, and the mineral breaks relatively easily along directions other than cleavage planes.

Many minerals are so strongly bonded that they lack even fair cleavage. Quartz, a framework silicate, is so strongly bonded in all directions that it breaks only along irregular surfaces. Garnet, an isolated tetrahedral silicate, also is bonded strongly in all directions and so has no cleavage. This absence of a tendency to cleave is found in most framework silicates and in silicates with isolated tetrahedra.

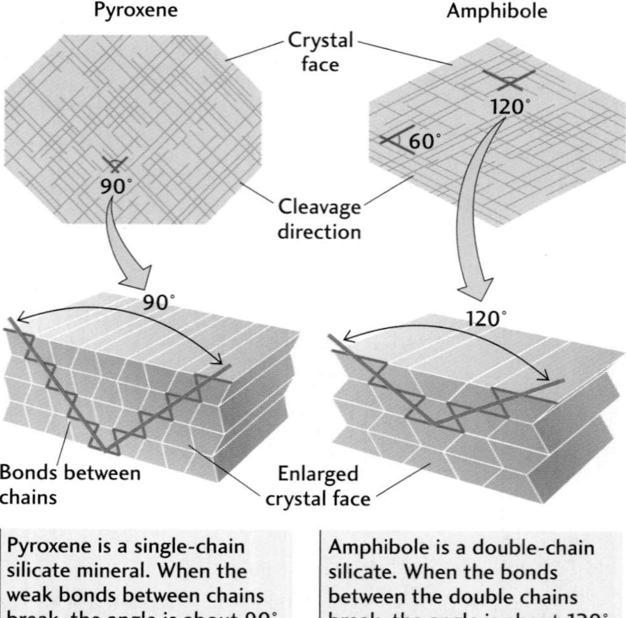

Figure 3.19 Comparison of cleavage directions and typical crystal faces in pyroxene and amphibole. These two minerals often look very much alike, but their angles of cleavage differ. These angles are frequently used to identify and classify them.

Fracture

Fracture is the tendency of a crystal to break along irregular surfaces other than cleavage planes. All minerals show fracture, either across cleavage planes or—in minerals such as quartz, with no cleavage—in any direction. Fracture is related to how bond strengths are distributed in directions that cut across crystal planes. Breakage of these bonds results in irregular fractures. Fractures may be *conchoidal,* showing smooth, curved surfaces like those of a thick piece of broken glass. A common fracture surface with an appearance like split wood is described as *fibrous* or *splintery.* The shape and appearance of many kinds of irregular fractures depend on the particular structure and composition of the mineral.

Luster

The way in which the surface of a mineral reflects light gives it a characteristic **luster.** Mineral lusters are described by the terms listed in Table 3.3. Luster is controlled by the kinds of atoms present and their bonding, both of which affect the way light passes through or is reflected by the mineral. Ionically bonded crystals tend to be glassy, or vitreous, but covalently bonded materials are more variable. Many tend to have an adamantine luster, like that of diamond. Metallic luster is shown by pure metals, such as gold, and by many sulfides, such as galena (lead sulfide, PbS). Pearly luster results from multiple reflections of light from planes beneath the surfaces of translucent minerals, such as the mother-of-pearl inner surfaces of many clam shells, which are made of the mineral aragonite. Luster, although an important criterion for field classification, depends heavily on the visual perception of reflected light. Textbook descriptions fall short of the actual experience of holding the mineral in your hand.

Color

The **color** of a mineral is imparted by light—either transmitted through or reflected by crystals, irregular masses, or a streak. **Streak** refers to the color of the fine deposit of mineral dust left on an abrasive surface, such as a tile of

Table 3.3 Mineral Luster

Luster	Characteristics
Metallic	Strong reflections produced by opaque substances
Vitreous	Bright, as in glass
Resinous	Characteristic of resins, such as amber
Greasy	The appearance of being coated with an oily substance
Pearly	The whitish iridescence of such materials as pearl
Silky	The sheen of fibrous materials such as silk
Adamantine	The brilliant luster of diamond and similar minerals

unglazed porcelain, when a mineral is scraped across it. Such tiles are called *streak plates* (**Figure 3.20**). A streak plate is a good diagnostic tool because the uniform small grains of mineral that are present in the powder on the ceramic tile permit a better analysis of color than does a mass of the mineral. A mass formed of hematite (Fe_2O_3), for example, may be black, red, or brown, but this mineral will always leave a trail of reddish brown dust on a streak plate.

Figure 3.20 Hematite may be black, red, or brown, but it always leaves a reddish brown streak when scratched along a ceramic plate. [Breck P. Kent.]

Color is a complex and not yet fully understood property of minerals. It is determined both by the kinds of ions found in the pure mineral and by trace impurities.

Ions and Mineral Color The color of pure substances depends on the presence of certain ions, such as iron or chromium, that strongly absorb portions of the light spectrum. Olivine that contains iron, for example, absorbs all colors except green, which it reflects, so we see this type of olivine as green. We see pure magnesium olivine as white (transparent and colorless).

Trace Impurities and Mineral Color All minerals contain impurities. Instruments can now measure even very small quantities of some elements—as little as a billionth of a gram in some cases. Elements that make up much less than 0.1 percent of a mineral are reported as "traces," and many of them are called trace elements.

Some trace elements can be used to interpret the origins of the minerals in which they are found. Others, such as the trace amounts of uranium in some granites, contribute to local natural radioactivity. Still others, such as small dispersed flakes of hematite that color a feldspar crystal brownish or reddish, are notable because they give a general color to an otherwise colorless mineral. Many of the gem varieties of minerals, such as emerald (green beryl) and sapphire (blue corundum), get their color from trace impurities dissolved in the solid crystal (see Feature 3.1). Emerald derives its color from chromium; the source of sapphire's blue color is iron and titanium.

The color of a mineral may be distinctive, but it is not the most reliable clue to its identity. Some minerals always show the same color; others may have a range of colors. Many minerals show a characteristic color only on freshly broken surfaces or only on weathered surfaces. Some—precious opals, for example—show a stunning display of colors on reflecting surfaces. Others change color slightly with a change in the angle of the light shining on their surfaces.

Specific Gravity and Density

One can easily feel the difference in weight between a piece of hematite iron ore and a piece of sulfur of the same size by hefting the two pieces. A great many common rock-forming minerals, however, are too similar in **density**—mass per unit volume (usually expressed in grams per cubic centimeter, g/cm^3)—for such a simple test. Scientists therefore need some easy method for measuring this property of minerals. A standard measure of density is **specific gravity,** which is the weight of a mineral in air divided by the weight of an equal volume of pure water at 4°C.

Density depends on the atomic mass of a mineral's ions and the closeness with which they are packed in its crystal structure. Consider the iron oxide magnetite, with a density of 5.2 g/cm^3. This high density results partly from the high atomic mass of iron and partly from the closely packed structure that magnetite has in common with the other members of the spinel group of minerals (see page 62). The density of the iron silicate olivine, 4.4 g/cm^3, is lower than that of magnetite for two reasons. First, the atomic mass of silicon, one of the elements from which olivine is formed, is lower than that of iron. Second, this olivine has a more openly packed structure than do minerals of the spinel group. The density of magnesium olivine is even lower, 3.32 g/cm^3, because magnesium's atomic mass is much lower than that of iron. Increases in density caused by increases in pressure affect the way that minerals transmit light, heat, and earthquake waves. Experiments at extremely high pressures have shown that olivine converts into the denser structure of the spinel group at pressures corresponding to a depth of 400 km. At a greater depth, 670 km, mantle materials are further transformed into silicate minerals with the even more densely packed structure of the mineral perovskite (calcium titanate, $CaTiO_3$). Because of the huge volume of the lower mantle, perovskite is probably the most abundant mineral in the Earth as a whole. Some perovskite minerals have been synthesized to be high-temperature semiconductors, which conduct electricity without loss of current and may have large commercial potential. Mineralogists experienced with natural perovskites helped unravel the structure of these newly created materials. Temperature also affects density: the higher the temperature, the more open and expanded the structure and thus the lower the density.

Crystal Habit

A mineral's **crystal habit** is the shape in which its individual crystals or aggregates of crystals grow. Crystal habits are often named after common geometric shapes, such as blades, plates, and needles. Some minerals have such a distinctive crystal habit that they are easily recognizable. An example is quartz, with its six-sided column topped by a pyramid-like set of faces. These shapes indicate not only the planes of atoms or ions in the mineral's crystal structure but also the typical speed and direction of crystal growth. Thus, a needlelike crystal is one that grows very quickly in one direction and very slowly in all other directions. In contrast, a plate-shaped crystal (often referred to as *platy*) grows fast in all directions that are perpendicular to its single direction of slow growth. Fibrous crystals take shape as multiple long, narrow fibers, essentially aggregates of long needles. *Asbestos* is a generic name for a group of silicates with a more or less fibrous habit that allows the crystals to become embedded in the lungs after having been inhaled (see Feature 3.2).

Some varieties of asbestos (see Feature 3.2) are examples of minerals with deleterious properties. Other minerals, such as arsenic-containing pyrites, are poisonous when ingested, and still others release toxic fumes when heated. Mineral dust diseases are found in many miners, who may face large occupational exposures. An example is silicosis, a disease of the lungs caused by inhaling quartz dust.

Table 3.4 summarizes the mineral physical properties that we discussed in this section.

Table 3.4 Physical Properties of Minerals

Property	Relation to Composition and Crystal Structure
Hardness	Strong chemical bonds give high hardness. Covalently bonded minerals are generally harder than ionically bonded minerals.
Cleavage	Cleavage is poor if bonds in crystal structure are strong, good if bonds are weak. Covalent bonds generally give poor or no cleavage; ionic bonds are weak and so give excellent cleavage.
Fracture	Type is related to distribution of bond strengths across irregular surfaces other than cleavage planes.
Luster	Tends to be glassy for ionically bonded crystals, more variable for covalently bonded crystals.
Color	Determined by kinds of atoms or ions and trace impurities. Many ionically bonded crystals are colorless. Iron tends to color strongly.
Streak	Color of fine powder is more characteristic than that of massive mineral because of uniformly small size of grains.
Density	Depends on atomic weight of atoms or ions and their closeness of packing in crystal structure. Iron minerals and metals have high density; covalently bonded minerals have more open packing and so have lower density.
Crystal habit	Depends on planes of atoms or ions in a mineral's crystal structure and the typical speed and direction of crystal growth.

3.2 Asbestos: Health Hazard, Overreaction, or Both?

In the past two decades, mention of asbestos—once used extensively as a fireproof insulator and flame retardant in plaster, ceiling and floor tile, and automobile insulation—has come to provoke fear. Asbestos has been linked to several fatal lung diseases—such as asbestosis (characterized by progressive lung stiffening and difficulty in breathing) and mesothelioma, a cancer that attacks the lining of the lung. The specific link seems to be heavy exposure to certain minerals lumped under the commercial name *asbestos*. The exact way in which these substances cause the diseases is still not well understood, but the sharp fibrous crystal habit of some of the minerals has been implicated.

The health problems associated with exposure to asbestos came to public attention when lawyers representing workers and their families brought class-action lawsuits against some of the major companies that fabricated asbestos products. The lawsuits accused the companies of responsibility for the deaths and disabilities of large numbers of people formerly employed in asbestos factories. In the wake of settlements awarded as a result of these lawsuits, some of the companies went bankrupt. The public became intensely concerned about asbestos-containing materials in schools, hospitals, and other public and private buildings. Many states now require disclosure of such materials during negotiations for the sale of private homes.

Most responsible scientists think that the frenzied concern about all forms of asbestos is an overreaction. At the center of this debate lies mineralogy.

Six distinct minerals are lumped under the commercial term *asbestos:* chrysotile, a sheet silicate member of the serpentine group; crocidolite, a double-chain silicate member of the amphibole group; and four other double-chain silicates in the serpentine group. Although crocidolite does form sharp fibers and has been heavily implicated in lung diseases, many of the other minerals do not form such fibers and have not been associated with lung disease. U.S. government regulations nevertheless apply to all of these minerals.

Physicians and mineralogists hold widely varying opinions about whether all forms of asbestos should be eliminated from buildings and factories. Any decision must consider several facts. First, heavy exposure to crocidolite is dangerous—especially for smokers, who are much more prone to asbestos-related lung diseases than are nonsmokers. Prolonged occupational exposure, such as of workers in asbestos factories, to some other forms of asbestos also may create a danger.

Abundant evidence exists, however, to indicate that people exposed for long periods of time to moderate amounts of chrysotile, the most commonly used asbestos in North America, show no asbestos-related lung disease. The lack of a correlation between exposure to specific asbestos minerals, verified by laboratory analysis, and the occurrence of specific lung diseases also contributes to confusion about the danger of exposure to asbestos. The situation is particularly ambiguous with respect to the danger posed to the general population by asbestos in large public buildings.

Many medical scientists and mineralogists doubt that we need to spend the $50 billion to $150 billion that it would cost to clean up relatively harmless chrysotile and the four forms of amphibole that do not form sharp fibers. Before any decision can be made with certainty, we need more mineralogical evaluation and additional medical studies to determine the nature of the problem confronting us.

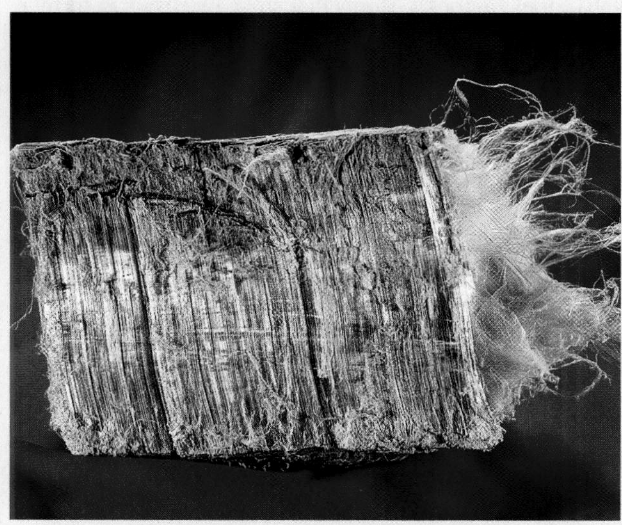

Asbestos (chrysotile). Fibers are readily combed from the solid mineral. [Runk/Schoenberger/Grant Heilman Photography.]

Minerals and the Biological World

The crystal habit and chemical composition of some minerals have made them important in the biological world. The simplest examples are the calcium carbonate minerals calcite and aragonite, used as shell materials by a wide range of invertebrate animals such as clams and oysters. We have only to feel our own bones to recognize the importance of apatite, a calcium phosphate mineral that makes up vertebrate bone.

SUMMARY

What is a mineral? Minerals, the building blocks of rocks, are naturally occurring, inorganic solids with specific crystal structures and chemical compositions that either are fixed or vary within a defined range. A mineral is constructed of atoms, the small units of matter that combine in chemical reactions. An atom is composed of a nucleus of protons and neutrons, surrounded by electrons. The atomic number of an element is the number of protons in its nucleus, and its atomic mass is the sum of the masses of its protons and neutrons.

How do atoms combine to form the crystal structures of minerals? Chemical substances react with one another to form compounds either by gaining or losing electrons to become ions or by sharing electrons. The ions in a chemical compound are held together by ionic bonds, which form by electrostatic attraction between positive ions (cations) and negative ions (anions). Atoms that share electrons to form a compound are held together by covalent bonds. When a mineral crystallizes, atoms or ions come together in the proper proportions to form a crystal structure, which is an orderly three-dimensional geometric array in which the basic arrangement is repeated in all directions.

What are the major rock-forming minerals? Silicates, the most abundant minerals in Earth's crust, are crystal structures built of silicate tetrahedra linked in various ways. Tetrahedra may be isolated (olivines) or in single chains (pyroxenes), double chains (amphiboles), sheets (micas), or frameworks (feldspars). Carbonate minerals are made of carbonate ions bonded to calcium or magnesium or both. Oxide minerals are compounds of oxygen and metallic elements. Sulfide and sulfate minerals are composed of sulfur atoms in combination with metallic elements.

What are the physical properties of minerals? A mineral's physical properties, which indicate its composition and structure, include hardness—the ease with which its surface is scratched; cleavage—its ability to split or break along flat surfaces; fracture—the way in which it breaks along irregular surfaces; luster—the nature of its reflection of light; color—imparted by transmitted or reflected light to crystals, irregular masses, or a streak (the color of a fine powder); density—the mass per unit volume; and crystal habit—the shapes of individual crystals or aggregates.

Minerals You Should Know

[*Row 1:* George Whiteley/Photo Researchers; Chip Clark; Charles O'Rear/Corbis. *Row 2:* Charles D. Winters/Photo Researchers; Chip Clark; Andrew Romeo–coolrox.com. *Row 3:* All by Chip Clark. *Row 4:* Vincent Cronin; Chip Clark; Chip Clark. *Row 5:* All by Chip Clark.]

Key Terms and Concepts

anion (p. 54)
atom (p. 53)
atomic mass (p. 53)
atomic number (p. 53)
cation (p. 54)
chemical reaction (p. 54)
cleavage (p. 65)
color (p. 66)
covalent bond (p. 56)
crystal (p. 57)
crystal habit (p. 68)
crystallization (p. 56)
density (p. 67)
electron (p. 53)
electron sharing (p. 54)
electron transfer (p. 54)
fracture (p. 66)
hardness (p. 63)
ionic bond (p. 56)
isotope (p. 53)
luster (p. 66)
metallic bond (p. 56)
mineral (p. 52)
mineralogy (p. 51)
Mohs scale of hardness (p. 63)
neutron (p. 53)
nucleus (p. 53)
polymorph (p. 58)
precipitate (p. 58)
proton (p. 53)
specific gravity (p. 67)
streak (p. 66)

Exercises

This icon indicates that there is an animation available on the Web site that may assist you in answering a question.

1. Define a mineral.

2. What is the difference between an atom and an ion?

3. Draw the atomic structure of sodium chloride.

4. What are two types of chemical bonds?

5. What are the two polymorphs of carbon?

6. List the basic structures of silicate minerals.

7. How does the cleavage of mica relate to its atomic structure?

8. Name three groups of minerals, other than silicates, based on their chemical composition.

9. What two factors account for the densities of mantle minerals?

10. How would a field geologist measure hardness?

11. What mineral tests would you conduct to distinguish hematite from magnetite?

12. What is the difference between the carbonate minerals calcite and dolomite?

13. Which of the asbestos minerals poses a health hazard?

Thought Questions

This icon indicates that there is an animation available on the Web site that may assist you in answering a question.

1. Joan and Alex are comparing rubies. Joan's is natural and Alex's is synthetic. Are both of them looking at minerals? Why or why not?

2. Hydrogen (H), the lightest element, has an atomic number of 1 and an atomic mass of 1.008 in nature. What does this information tell you about possible isotopes of hydrogen?

3. Draw a simple diagram to show how silicon and oxygen in silicate minerals share electrons. Model your diagram on Figure 3.5.

4. Use iron and magnesium in silicate minerals to illustrate cation substitution.

5. Diopside, a pyroxene, has the formula $(Ca,Mg)_2Si_2O_6$. What does this formula tell you about its crystal structure and cation substitution?

6. Oxygen exists as three isotopes with atomic masses of exactly 16, 17, and 18. The atomic mass of oxygen found in nature is approximately 16. What does this information tell you about the relative abundance of the three isotopes in nature?

7. In some bodies of granite, we can find very large crystals, some as much as a meter across, yet these crystals tend to have few crystal faces. What can you deduce about the conditions under which these large crystals grew?

8. What physical properties of sheet silicates are related to their crystal structure and bond strength?

9. How might you identify and differentiate between a single- and a double-chain silicate?

10. What physical properties would make calcite a poor choice for a gemstone?

11. Choose two minerals from Appendix 5 that you think might make good abrasive or grinding stones for sharpening steel, and describe the physical property that causes you to believe they would be suitable for this purpose.

12. Aragonite, with a density of 2.9 g/cm^3, has exactly the same chemical composition as calcite, with a density of 2.7 g/cm^3. Other things being equal, which of these two minerals is more likely to have formed under high pressure?

13. What properties of talc make it suitable for face and body powder?

14. There are at least eight physical properties one can use to identify an unknown mineral. Which ones are most useful in discriminating between minerals that look similar? Describe a strategy that would allow you to determine that an unknown clear calcite crystal is indeed not the same as a known clear crystal of quartz.

15. There is much controversy over the removal of asbestos from buildings in the United States and Canada. It is clear that some forms of asbestos do pose a health hazard. Based on what you have learned about mineral structures, what aspect of the structure of asbestos causes the hazard? Given the double-chain silicate structure of asbestos, are other double-chain silicates potentially dangerous?

16. Coal, which forms from decaying vegetation and is, therefore, a natural substance, is not considered to be a mineral. However, when coal is heated to high temperatures and buried in high-pressure areas, it transforms into the mineral graphite. Why is it, then, that coal is not considered a mineral but graphite is? Explain your reasoning.

17. Back in the late 1800s, gold miners used to "pan" for gold by placing sediment from rivers into a pan and filtering water through the pan while swirling the pan's content. Specifically, the miners wanted to be certain that they had found real gold and not pyrite ("fool's gold"). Why did this method work? What mineral property does the process of panning for gold utilize? What is another possible method for distinguishing between gold and pyrite?

Short-Term Project

Why Is the Hope Diamond Blue?

In late 1955, Robert H. Wentorf, Jr., achieved something close to alchemy. He bought a jar of peanut butter at a local food co-op, took it to his lab, and then turned a glob of the spread into a few tiny (green) diamonds.* After all, peanuts are rich in proteins, which are rich in nitrogen. Synthetic diamonds are often black due to inclusions of graphite, and traces of nitrogen within the crystal structure can turn diamonds brown, yellow, or green.

Diamonds form in the upper mantle at depths below about 200 km where temperatures and pressures cause graphite, diamond's sister mineral, to collapse into a more tightly packaged crystal structure (see Figure Story 3.11). Most diamonds weigh less than 1 carat (0.2 gram) and are colorless, pale yellow, or brown. Larger diamonds, especially colored ones, are rare. The largest faceted diamond is the yellow 545.67-carat Golden Jubilee, unveiled in 1995. The Hope Diamond, on display at the Smithsonian Institute, may be the best-known gemstone in the world. It is a very rare 45-carat blue diamond.

What produces the color of gems and minerals? Color is produced by the interaction of light with matter. You know how a prism or even water drops break white light into a spectrum or rainbow of colored light. When light hits the surface of or penetrates a crystal, it interacts with atoms; some components of the light may be absorbed while others are transmitted. As white light passes through the Hope diamond, the crystal absorbs red light and transmits blue.

Why are many diamonds colorless and others blue, red, yellow, green, or brown? Trace amounts of impurities and imperfections in the crystal structure of a mineral can change how the mineral interacts with light and cause it to be colored. Although diamonds are pure carbon, it takes only one atom of another element among the million atoms to turn a diamond blue. Rubies and sapphires are both crystals of aluminum oxide; yet rubies are red and sapphires are blue. Go to http://www.whfreeman.com/understandingearth to figure out what causes the Hope diamond to be blue, rubies to be red, and sapphires to be blue.

Short-Term Team Projects

Asbestos

Children were in danger. News reports of asbestos in the New York City schools touched off a wave of public outrage, and the Board of Education postponed the start of the 1993 school year for three weeks to complete an asbestos-removal program.

Reports of asbestos in public buildings almost always focus on the risk of lung disease. Seldom, however, do these reports include interviews with mineralogists. As a result, the public knows too little about asbestos to ask the right questions and make an informed judgment.

You now have the chance to educate the public. Working in a team of four students over the next two weeks, prepare a half-hour radio program representing a variety of perspectives on the asbestos issue. The program will show how useful a basic knowledge of mineralogy can be in deciding important policy questions. Assemble a cast of experts with opposing positions on the issue. From what

*Ivan Amato, Diamond fever. *Science News,* August 4, 1990, p. 72.

disciplines would they come? With what arguments and facts would they support their positions? Write a script for the program and deliver an oral summary of your list of experts and their positions.

Suggested Readings

Atkins P., and L. Jones. 2002. *Chemical Principles,* 2d ed. New York: W. H. Freeman.

Berry, L. G., B. Mason, and R. V. Dietrich. 1983. *Mineralogy,* 2d ed. San Francisco: W. H. Freeman.

Dietrich, R. V., and B. J. Skinner. 1990. *Gems, Granites, and Gravels.* Cambridge: Cambridge University Press.

Hazen, Robert M. 2001. Life's rocky start. *Scientific American* 284 (4): 76-85.

Keller, P. C. 1990. *Gemstones and Their Origins.* New York: Chapman & Hall.

Klein, C., and C. S. Hurlbut, Jr. 1999. *Manual of Mineralogy,* 21st ed., revised. New York: Wiley.

Patch, Suzanne Steinem. 1999. *Blue Mystery: The Story of the Hope Diamond,* 2d ed. New York: Harry N. Abrams.

Perkins, D. *Mineralogy.* 1998. Englewood Cliffs, N.J.: Prentice Hall.

Prinz, M., G. Harlow, and J. Peters. 1978. *Simon & Schuster's Guide to Rocks and Minerals.* New York: Simon & Schuster.

Varon, Lynn. 2001. Internet gems: Web sites for rockhounds and lapidaries. *Rock and Gem* 31 (6): 34–36, 90.

Uplifted sediments at Moab, Utah. These sandstones and shales represent a former cycle of uplift, erosion, and sedimentation. Erosion of the ridges by streams that are tributary to the main river is part of the current cycle. [Breck P. Kent.]

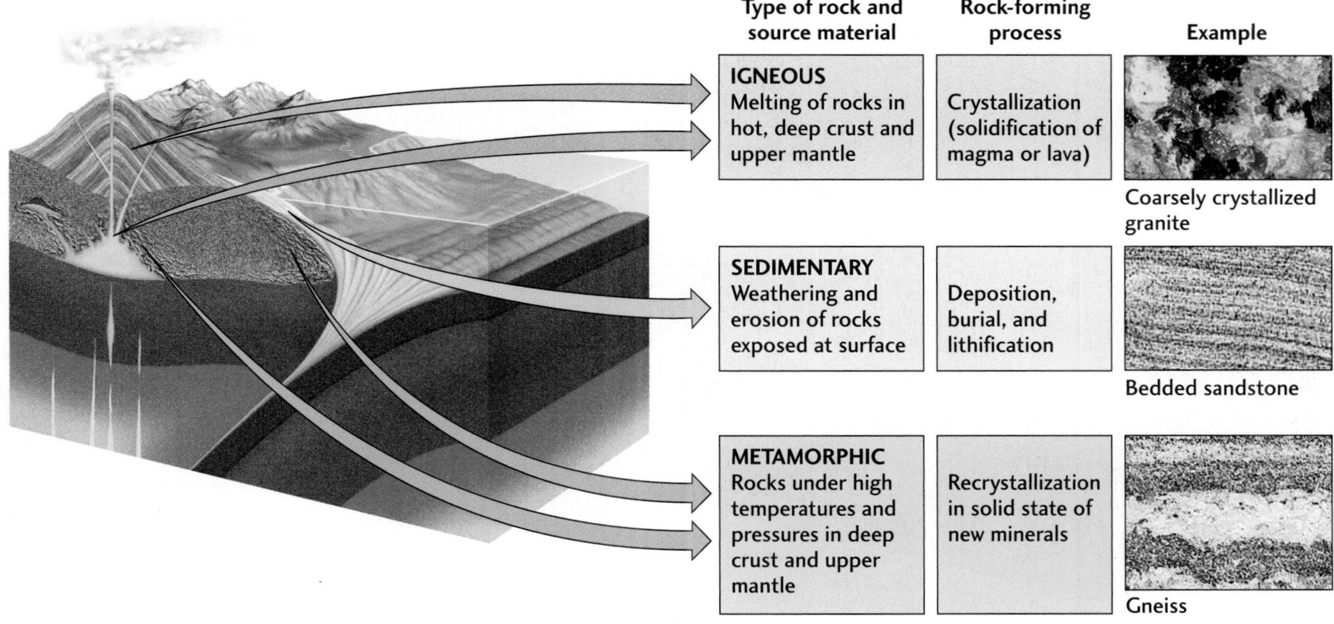

Figure 4.2 The minerals and textures of the three great rock groups are formed in different places in and on Earth by different geologic processes. As a result, geologists use mineralogical and chemical analyses to determine the origins of rocks and the processes that formed them. Granite, composed of quartz, feldspar, and mica crystals. [J. Ramezani.] Bedded sedimentary rock, made up of sandstones. [Breck P. Kent.] This crumpled and deformed metamorphic rock is a gneiss. [Breck P. Kent.]

the other two types—and see how these processes are all driven by plate tectonics and climate.

Igneous Rocks

Igneous rocks (from the Latin *ignis,* meaning "fire") form by crystallization from a magma, a mass of melted rock that originates deep in the crust or upper mantle. Here temperatures reach the 700°C or more needed to melt most rocks. When a magma cools slowly in the interior, microscopic crystals start to form. As the magma cools below the melting point, some of these crystals have time to grow to several millimeters or larger before the whole mass is crystallized as a coarse-grained igneous rock. But when a magma erupts from a volcano onto Earth's surface, it cools and solidifies so rapidly that individual crystals have no time to grow gradually. In that case, many tiny crystals form simultaneously, and the result is a fine-grained igneous rock. Geologists distinguish two major types of igneous rocks—intrusive and extrusive—on the basis of the sizes of their crystals.

Intrusive Igneous Rocks

Intrusive igneous rocks crystallize when magma intrudes into unmelted rock masses deep in Earth's crust. Large crystals grow as the magma cools, producing coarse-grained rocks. Intrusive igneous rocks can be recognized by their interlocking large crystals, which grew slowly as the magma gradually cooled (**Figure 4.3**). *Granite* is an intrusive igneous rock.

Extrusive Igneous Rocks

Extrusive igneous rocks form from rapidly cooled magmas that erupt at the surface through volcanoes. Extrusive igneous rocks, such as *basalt,* are easily recognized by their glassy or fine-grained texture (see Figure 4.3).

Common Minerals

Most of the minerals of igneous rocks are silicates, partly because silicon is so abundant and partly because many silicate minerals melt at the high temperatures and pressures

Figure 4.3 The formation of *extrusive igneous rocks* (basalt is shown here) [Chip Clark] and *intrusive igneous rocks* (granite is shown here) [J. Ramezani].

reached in deeper parts of the crust and in the mantle. The common silicate minerals found in igneous rocks include quartz, feldspar, mica, pyroxene, amphibole, and olivine (Table 4.1).

Table 4.1 Some Common Minerals of Igneous, Sedimentary, and Metamorphic Rocks

Igneous Rocks	Sedimentary Rocks	Metamorphic Rocks
Quartz*	Quartz*	Quartz*
Feldspar*	Clay minerals*	Feldspar*
Mica*	Feldspar*	Mica*
Pyroxene*	Calcite	Garnet*
Amphibole*	Dolomite	Pyroxene*
Olivine*	Gypsum	Staurolite*
	Halite	Kyanite*

An asterisk indicates that a mineral is a silicate.

Sedimentary Rocks

Sediments, the precursors of sedimentary rocks, are found on Earth's surface as layers of loose particles, such as sand, silt, and the shells of organisms. These particles form at the surface as rocks undergo weathering and erosion. **Weathering** is all of the chemical and physical processes that break up and decay rocks into fragments of various sizes. The fragmented rock particles are then transported by **erosion,** the set of processes that loosen soil and rock and move them to the spot where they are deposited as layers of sediment (**Figure 4.4**). Weathering and erosion produce two types of sediments:

- **Clastic sediments** are physically deposited particles, such as grains of quartz and feldspar derived from a weathered granite. (*Clastic* is derived from the Greek word *klastos,* meaning "broken.") These sediments are laid down by running water, wind, and ice and form layers of sand, silt, and gravel.

- **Chemical and biochemical sediments** are new chemical substances that form by precipitation when some of a rock's components dissolve during weathering and are carried in river waters to the sea. These sediments include layers of such minerals as halite (sodium chloride) and calcite (calcium carbonate, most often found in the form of reefs and shells).

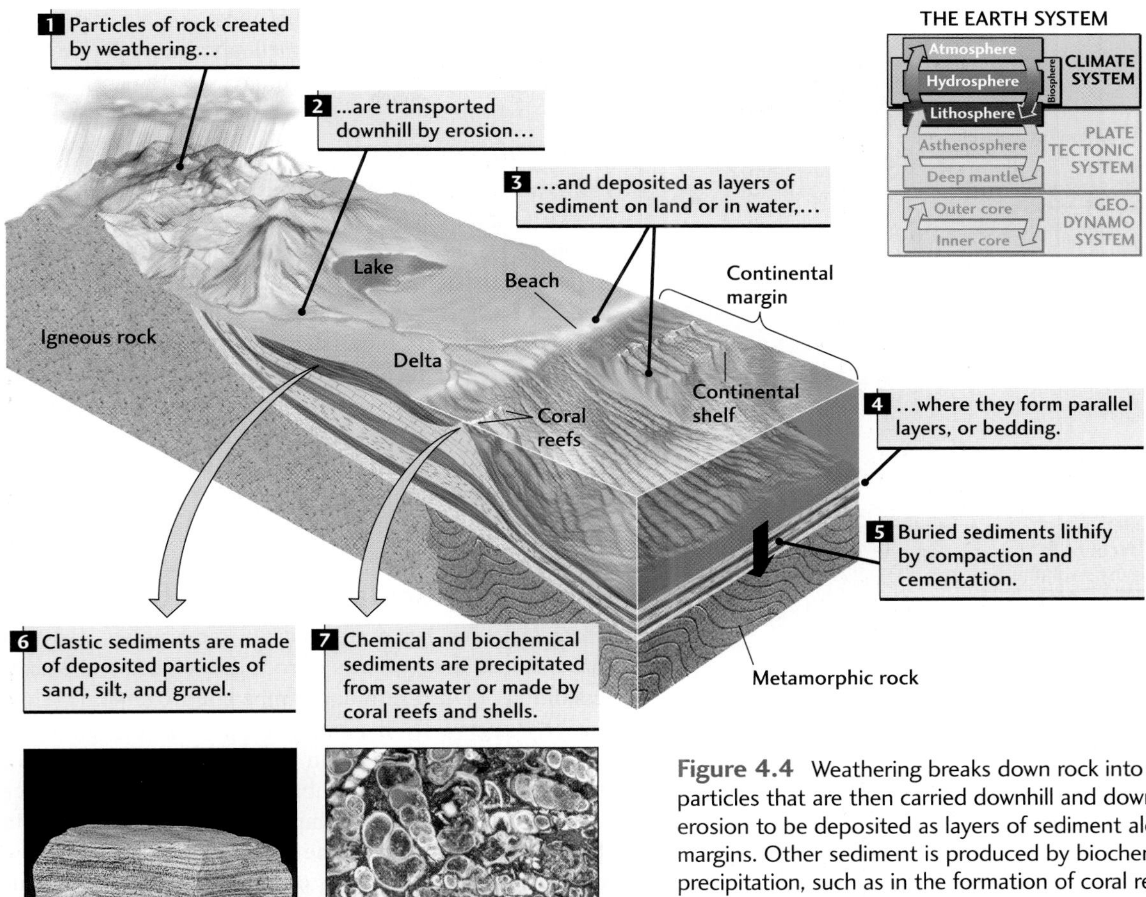

Figure 4.4 Weathering breaks down rock into smaller particles that are then carried downhill and downstream by erosion to be deposited as layers of sediment along continental margins. Other sediment is produced by biochemical precipitation, such as in the formation of coral reefs. As layers accumulate and are buried deeper and deeper, they lithify, hardening into sedimentary rock. (*left*) Laminated sandstone [Breck P. Kent]; (*right*) fossiliferous limestone [Peter Kresan].

From Sediment to Solid Rock

Lithification is the process that converts sediments into solid rock, and it occurs in one of two ways:

- By *compaction,* as grains are squeezed together by the weight of overlying sediment into a mass denser than the original.

- By *cementation,* as minerals precipitate around deposited particles and bind them together.

Sediments are compacted and cemented after burial under additional layers of sediment. Thus sandstone forms by the lithification of sand particles, and limestone forms by the lithification of shells and other particles of calcium carbonate.

Layers of Sediment

Sediments and sedimentary rocks are characterized by **bedding,** the formation of parallel layers of sediment as particles settle to the bottom of the sea, a river, or a land surface. Because sedimentary rocks are formed by surface processes, they cover much of Earth's land surface and seafloor. Although most rocks found at Earth's surface are sedimentary, their volume is small compared to the igneous and metamorphic rocks that make up the main volume of the crust because they are difficult to preserve (**Figure 4.5**).

Common Minerals

The common minerals of clastic sediments are silicates, because silicate minerals predominate in rocks that weather to form sedimentary particles (see Table 4.1). The most abundant minerals in clastic sedimentary rocks are quartz, feldspar, and clay minerals.

The most abundant minerals of chemically or biochemically precipitated sediments are carbonates, such as calcite, the main constituent of limestone. Dolomite, also found in limestone, is a calcium-magnesium carbonate formed by precipitation during lithification. Two other chemical sediments—gypsum and halite—form by precipitation as seawater evaporates.

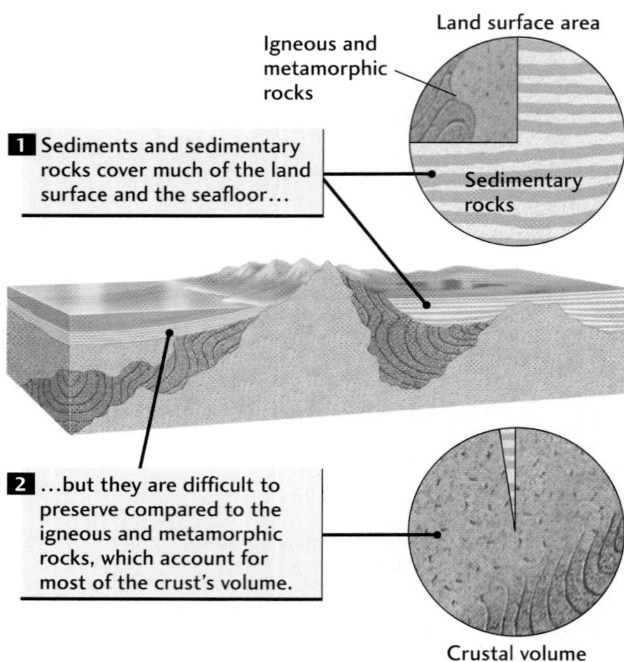

Figure 4.5 Sediments and sedimentary rocks cover much of Earth's land surface and the seafloor.

Metamorphic Rocks

Metamorphic rocks take their name from the Greek words for "change" (*meta*) and "form" (*morphe*). These rocks are produced when high temperatures and pressures deep in the Earth cause any kind of rock—igneous, sedimentary, or other metamorphic rock—to change its mineralogy, texture, or chemical composition while maintaining its solid form. The temperatures of metamorphism are below the melting points of the rocks (about 700°C) but high enough (above 250°C) for the rocks to change by recrystallization and chemical reactions.

Regional and Contact Metamorphism

Metamorphism may take place over a widespread area or a limited one (**Figure 4.6**). **Regional metamorphism** occurs where high pressures and temperatures extend over large regions, as happens where plates collide. Regional metamorphism accompanies plate collisions that result in mountain building and the folding and breaking of sedimentary layers that were once horizontal. Where high temperatures are restricted to smaller areas, such as the rocks near and in contact with an intrusion, rocks are transformed by **contact metamorphism.**

Many regionally metamorphosed rocks, such as schists, have characteristic **foliation,** wavy or flat planes produced when the rock was structurally deformed into folds. Granular textures are more typical of most contact metamorphic rocks and of some regional metamorphic rocks formed by very high pressure and temperature.

Common Minerals

Silicates are the most abundant minerals of metamorphic rocks because the parent rocks are also rich in silicates (see Table 4.1). Typical minerals of metamorphic rocks are quartz, feldspar, mica, pyroxene, and amphibole—the same kinds of silicates characteristic of igneous rocks. Several other silicates—kyanite, staurolite, and some varieties of garnet—are characteristic of metamorphic rocks alone. These minerals form under conditions of high pressure and temperature in the crust and are not characteristic of igneous rocks. They are therefore good indicators of metamorphism. Calcite is the main mineral of marbles, which are metamorphosed limestones.

Pressure-Temperature-Time Paths

Plate tectonics makes regional metamorphism a dynamic process in which volumes of rock are subjected to changing conditions of pressure and temperature over time. Consequently, regionally metamorphosed rocks contain distinctive mineral assemblages in which earlier assemblages are overprinted by later assemblages. Such rocks thus record regimes of pressure and temperature which change in time. **Pressure-temperature-time paths** are recorded not only by changes in the assemblages of minerals but also by changes in the chemical compositions of the minerals themselves. Metamorphic pressure-temperature paths are discussed in detail in Chapter 9.

Where We See Rocks

Rocks are not found in nature conveniently divided into separate bodies—igneous here, sedimentary there, metamorphic in another place. Instead, we find them arranged in patterns determined by the geologic history of a region. Geologists map these patterns both at the surface and as projected into the interior, and they try to deduce the geologic past from the present variety and distribution of the rocks.

If we were to drill a hole into any spot on Earth, we would find rocks that represent the geologic history of that region. In the top few kilometers of most regions, we would probably find sedimentary rock. Drilling deeper, perhaps 6 to 10 km down, we would eventually penetrate an underlying area of older igneous and metamorphic rock.

In fact, thousands of relatively shallow holes have been drilled on the continents in the search for oil, water, and mineral resources. These holes are major sources of information, mainly about sedimentary rocks and their history. In the quest for more data on the deep continental crust, the governments of several countries—including the United States,

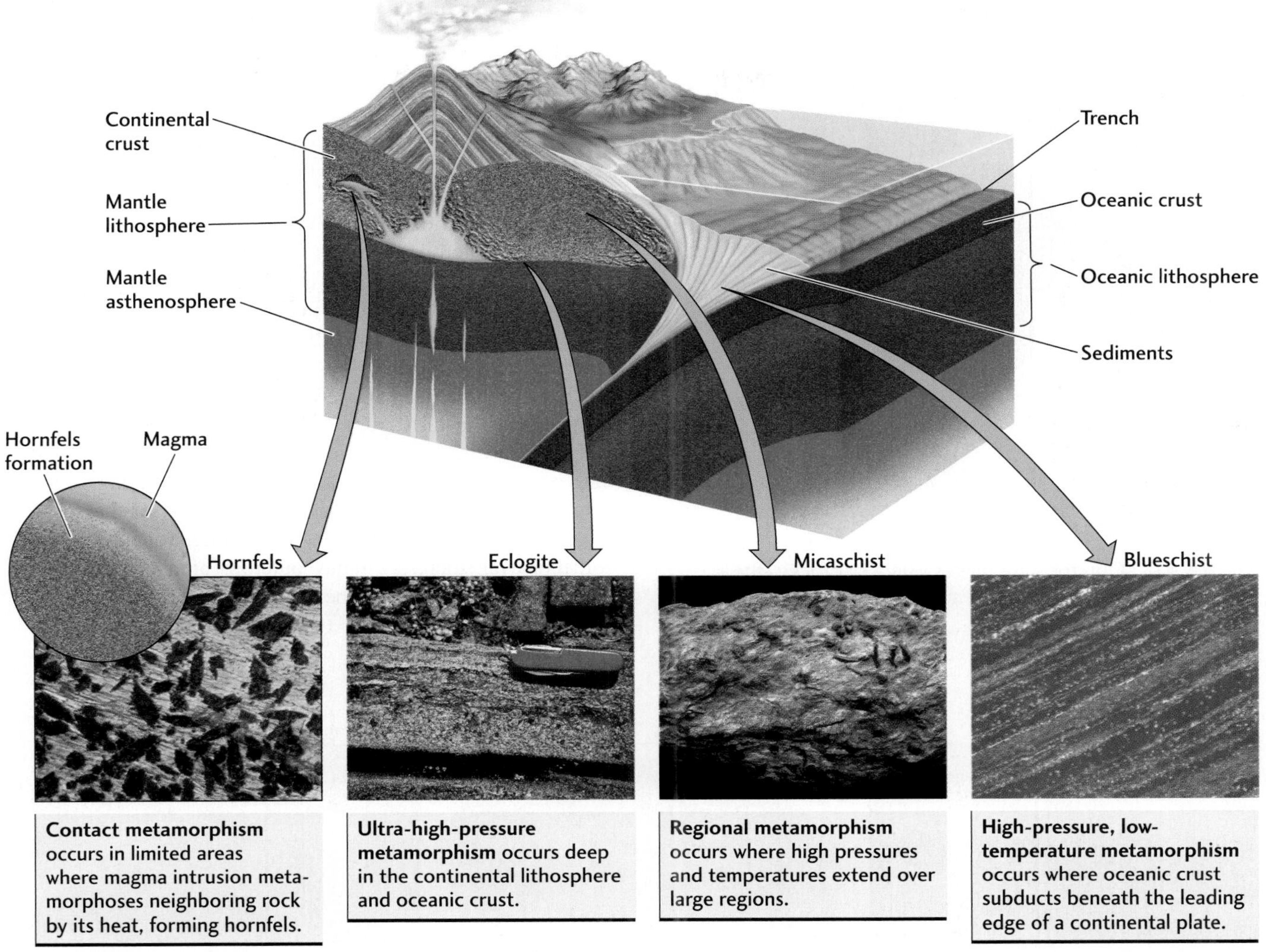

Figure 4.6 Metamorphic rocks form under four main conditions. Example of rocks shown here are (from left to right) *hornefels* [Biophoto Associates/Photo Researchers], *eclogite* [Julie Baldwin], *micaschist* [John Grotzinger], and *blueschist* [Mark Cloos].

Germany, and Russia—have drilled to great depths on the continents. The deepest hole, in Russia, is more than 12 km deep, exceeding the depth of any commercial drilling.

A large part of our knowledge about the rocks of the ocean floor comes from the hundreds of holes punched down by the Deep Sea Drilling Project, an ongoing program to drill the world's seafloor for geologic information. Started by the United States in the late 1960s, at the same time that plate tectonics swept the geological community, it is now an international venture (the Ocean Drilling Program) carried on with the cooperation of the major maritime countries of the world.

Even with all these sources of information about what lies beneath Earth's surface, geologists continue to rely on the rocks exposed in **outcrops,** places where **bedrock**—the underlying rock beneath the loose surface materials—is laid bare (**Figure 4.7**). Outcrops vary from region to region because they exemplify the geologic structure of the Earth at a particular spot. On a trip across North America, we might run across many kinds of outcrops (**Figure 4.8**). Starting at the Pacific, we would encounter sea cliffs from Mexico to Canada (Figure 4.8a). From the West Coast to the Rocky Mountain front, which stretches from New Mexico in the south to Alberta, Canada (Figure 4.8b), outcrops of all kinds of rock are abundant in the canyons, mountainsides, and cliffs of the relatively dry mountainous regions of the western third of the continent.

From the Rockies eastward to the Appalachian Mountains, the landscape is dominated by the plains and prairies of the American Midwest and the Prairie Provinces of

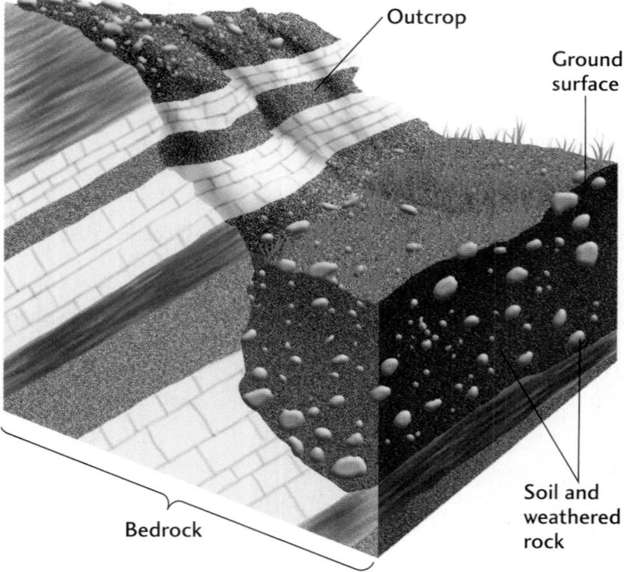

Figure 4.7 Outcrops are places where bedrock—the underlying rock beneath loose surface materials such as soil and boulders—is laid bare.

Canada. In this region, outcrops are scarce because most of the sedimentary bedrock is covered by soil and the sediments deposited by such rivers as the Missouri and their tributaries. Here outcrops occur in the low hills and gentle valleys (Figure 4.8c).

Low coastal plains cover the region from southeastern New Jersey to the Carolinas and Georgia in the east and Texas, Louisiana, Mississippi, and Alabama in the south. Here barely lithified, relatively soft sedimentary rocks are exposed in outcrops similar to those of the Great Plains. Good exposures can be found in the occasional bluffs along the shoreline. Farther south, in Florida, outcrops of limestone can be found in low hills and along the chain of islands called the Florida Keys (Figure 4.8d). To the north, outcrops become more numerous once we reach the Appalachians. In the hilly, rugged landscape of New England and the Maritime Provinces of Canada, we can find good outcrops, with the best exposures displayed along rocky coastlines. In this more humid climate, most of the rocks of the low ridges are covered by abundant vegetation and soil; however, there are many outcrops along rocky cliffs and ledges, especially on the higher ridges and mountains (Figure 4.8e).

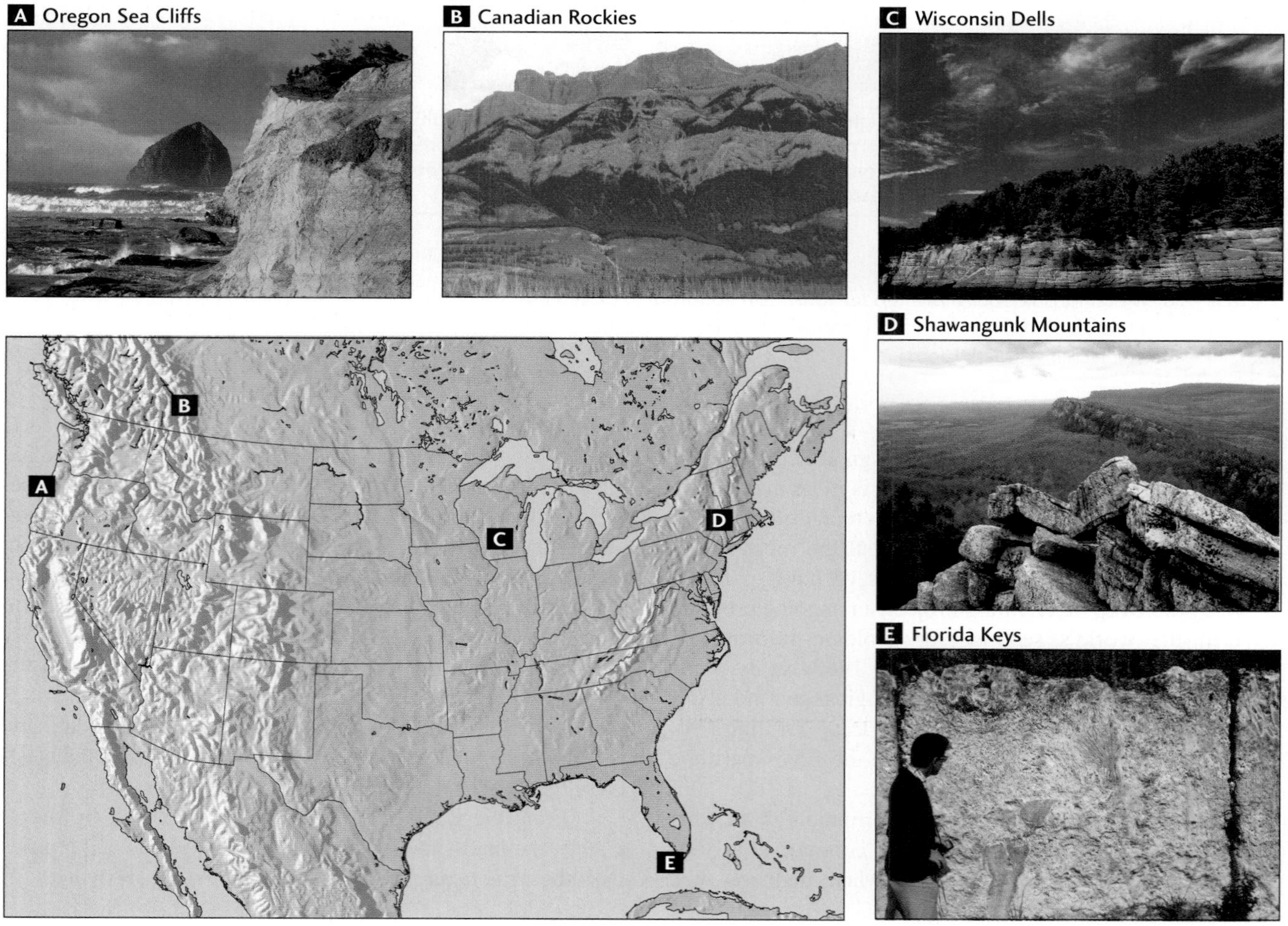

A Oregon Sea Cliffs

B Canadian Rockies

C Wisconsin Dells

D Shawangunk Mountains

E Florida Keys

As this travelogue indicates, the presence and types of outcrops depend on the nature of the landscape, which in turn depends on the geologic structure of the region, its history, and its present climate. In later chapters, we will explore in more detail how rock types relate to geologic structures (Chapter 10) and to landscapes (Chapter 19). Now, however, we turn to the rock cycle, which—in combination with plate tectonics—reveals how the three groups of rocks interrelate and how they reflect geologic structure and history.

The Rock Cycle: Interactions Between the Plate Tectonic and Climate Systems

The rock cycle is the result of interactions between two of the three fundamental Earth systems: plate tectonics and climate. Driven by interactions between these two systems, material and energy are transferred among the Earth's interior, the land surface, the oceans, and the atmosphere. For example, the melting of subducting lithospheric slabs and the formation of magma result from processes operating within the plate tectonic system. When these molten rocks erupt, matter and energy are transferred to the land surface, where the material (newly formed rocks) is subject to weathering by the climate system. The same process injects volcanic ash and carbon dioxide gas high into the atmosphere, where they may affect global climate. As global climate changes, perhaps becoming warmer or cooler, the rate of rock weathering changes, which in turn influences the rate at which material (sediment) is returned to Earth's interior.

The idea of Earth as a system had not yet been proposed when the Scotsman James Hutton described the rock cycle in an oral presentation in 1785 before the Royal Society of Edinburgh. Ten years later, he presented the cycle in more detail in his book *Theory of the Earth with Proof and Illustrations.* As is often the case in the history of science, other scientists—both in England and on the European continent—also had recognized elements of the cyclic nature of geologic change. Hutton's role was that of synthesizer: he presented the larger picture that has enabled us to understand the process.

We give an account here of one particular cycle, recognizing that such cycles vary with time and place. We begin with a magma deep in the Earth, where temperatures and pressures are high enough to melt any kind of rock: igneous, metamorphic, or sedimentary (**Figure Story 4.9**). Hutton called the melting of rocks deep in Earth's crust the plutonic episode, after Pluto, the Roman god of the underworld. We now refer to all igneous intrusives as **plutonic rocks,** whereas the extrusives are known as **volcanic rocks.** When preexisting rocks melt, all their component minerals are destroyed and their chemical elements are homogenized in the resulting hot liquid. As the magma cools, crystals of new minerals grow and form new igneous rock. Melting and the formation of igneous rock take place mainly along the boundaries of colliding or diverging tectonic plates, as well as in mantle plumes, as you will see in later chapters.

The cycle begins with subduction of an oceanic plate beneath a continental plate. The igneous rocks that form at the boundaries where plates collide, together with associated sedimentary and metamorphic rocks, are then uplifted into a high mountain chain as a section of Earth's crust crumples and deforms. Geologists call this process, which begins with plate collision and ends in mountain building, **orogeny.** After uplift, the rocks of the crust overlying the uplifted igneous rock slowly weather. Weathering creates loose material that erosion then strips away, exposing the igneous rock at the surface.

The exposed igneous rock now weathers, and chemical changes occur in some of its minerals. Iron minerals, for example, may "rust" to form iron oxides. High-temperature minerals such as feldspars may become low-temperature clay minerals. Minerals such as pyroxene may dissolve completely as rain pours over them. The weathering of the igneous rock again produces various sizes and kinds of rock debris and dissolved material, which are carried away by erosion. Some of these materials are transported over land by water and wind. Much of the debris is transported by streams to rivers and ultimately to the ocean. In the ocean, debris is deposited as layers of sand, silt, and other sediments formed from dissolved material, such as the calcium carbonate from shells.

Figure 4.8 Outcrops found in North America. (a) Rocky cliffs on the Pacific coast at Cape Kiwanda, Oregon. Shoreline cliffs such as these provide ready accessibility to bedrock for the geologist. [Fred Hirschmann.] (b) The spectacular mountains of the Canadian Rockies afford both geologists and mountain climbers the opportunity to study rocks. [D. Robert Franz and Lorri Franz/Corbis.] (c) The Wisconsin Dells is a local favorite of Upper Midwest geologists. [David Dvorak, Jr.] (d) Shawangunk Mountains, New York. Even though these mountains are ancient (part of the Appalachian Mountain chain), excellent outcrops of bedrock are formed along steep slopes. [Carr Clifton.] (e) In Florida, some of the best outcrops occur in the Keys, where ancient reefs are exposed in the cores of the islands. [Robert N. Ginsburg.]

THE ROCK CYCLE IS THE INTERACTION OF PLATE TECTONIC AND CLIMATE SYSTEMS

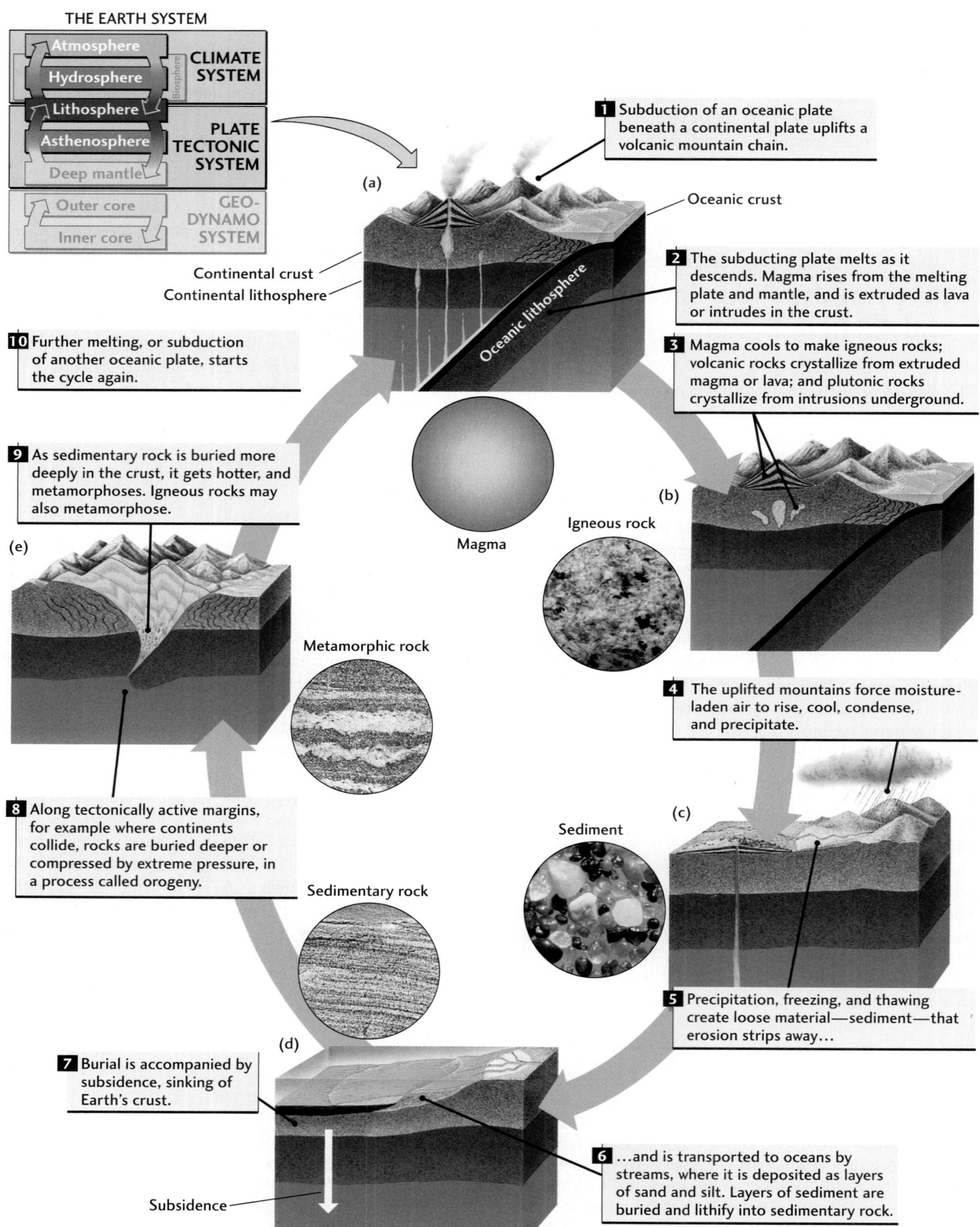

The sediments laid down in the sea, as well as those deposited on land by water and wind, are buried under successive layers of sediment, where they slowly lithify into sedimentary rock. Burial is accompanied by **subsidence**—a depression or sinking of the Earth's crust. As subsidence continues, additional layers of sediment can accumulate.

In some cases—for example, along tectonically active plate margins—subduction forces sedimentary rocks to descend to progressively greater depths (see Figure 4.6). As the lithified sedimentary rock is buried more deeply in the crust, it gets hotter. When the depth of burial exceeds 10 km and temperatures climb to more than 300°C, the minerals in the still-solid rock start to change into new minerals that are more stable at the higher temperatures and pressures of the deeper parts of the crust. The process that transforms the sedimentary rocks into metamorphic rocks is metamorphism. With further heating, the rocks may melt and form a new magma from which igneous rocks will crystallize, starting the cycle all over again.

As noted earlier, this series of processes is only one variation among the many that may take place in the rock cycle. Any type of rock—metamorphic, sedimentary, or igneous—can be uplifted during an orogeny and weathered and eroded to form new sediments. Some stages may be omitted: as a sedimentary rock is uplifted and eroded, for example, metamorphism and melting are skipped. And stages may take place out of sequence, as when an igneous rock formed in the interior is metamorphosed before it is uplifted. Also, as we know from deep drilling, some igneous rocks many kilometers deep in the crust may never have been uplifted or exposed to weathering and erosion.

The rock cycle never ends. It is always operating at different stages in various parts of the world, forming and eroding mountains in one place and laying down and burying sediments in another. The rocks that make up the solid Earth are recycled continuously, but we can see only the surface parts of the cycle. We must deduce the recycling of the deep crust and the mantle from indirect evidence.

One process that geologists were unaware of in Hutton's time is seafloor weathering or metasomatism, which was only recognized following the discovery of plate tectonics. The process involves chemical exchanges between seawater and the seafloor at mid-ocean ridges. It significantly supplements ordinary surface weathering in the return of important elements to Earth's interior. In the absence of seafloor metasomatism, the chemical composition of the ocean and atmosphere would be quite different.

Earth's Unique Systems and Rock Cycle

The rock cycle we have just described is unique to our planet because Earth's plate tectonic and climate systems differ from those on the other terrestrial planets. There are no sedimentary rocks on the Moon and Venus, for example, because they lack a hydrosphere and atmosphere and their climate is profoundly different from ours. All the rocks found on the surface of Venus have been affected and modified in various ways by the very high temperatures and the sulfuric-acid-rich atmosphere that characterize its present climate. The absence of water on the surface of Mars and the thin Martian atmosphere enable us to say that weathering and erosion on today's Mars follow a different path than they do on Earth. These examples show how the basic systems and the interactions among them that characterize a planet control how that planet works.

With this introduction to the rock world, we are ready to begin the study of rocks. In Chapters 5 and 6, we look at the geologic origin of magmas, the types of igneous rocks that form when magmas crystallize, the larger picture of plate tectonic control of igneous processes, and the dynamics of volcanoes and their eruptions. In Chapters 7 and 8, we explore weathering, the characteristics of sedimentary particles, and the ways in which various sediments and sedimentary rocks are produced. We complete our consideration of rocks in Chapter 9 by examining how high heat and pressure affect preexisting rocks, transforming them into metamorphic rocks, and how metamorphism relates to plate tectonics and orogeny.

SUMMARY

What determines the properties of the various kinds of rocks that form in and on Earth's surface? Mineralogy (the kinds and proportions of minerals that make up a rock) and texture (the sizes, shapes, and spatial arrangement of its crystals or grains) define a rock. The mineralogy and texture of a rock are determined by the geologic conditions, including chemical composition, under which it formed, either in the interior under various conditions of high temperature and pressure or at the surface, where temperatures and pressures are low.

Figure Story 4.9 The rock cycle, as proposed by James Hutton more than 200 years ago. Rocks subjected to weathering and erosion form sediments, which are deposited, buried, and lithified. After deep burial, the rocks undergo metamorphism, melting, or both. Through orogeny and volcanic processes, rocks are uplifted, only to be recycled again. [*Igneous* (granite): J. Ramezani. *Metamorphic* (gneiss): Breck P. Kent. *Sedimentary* (sandstone): Breck P. Kent. *Sediment* (loose sand and gravel): Rex Elliott.]

What are the three types of rock and how do they form? Igneous rocks form by the crystallization of magmas as they cool. Intrusive igneous rocks form in Earth's interior and have large crystals. Extrusive igneous rocks, which form at the surface where lavas and ash erupt from volcanoes, have a glassy or fine-grained texture. Sedimentary rocks form by the lithification of sediments after burial. Sediments are derived from the weathering and erosion of rocks exposed at Earth's surface. Metamorphic rocks form by alteration in the solid state of igneous, sedimentary, or other metamorphic rocks as they are subjected to high temperatures and pressures in the interior.

How does the rock cycle describe the formation of rocks as the products of geologic processes? The rock cycle relates geologic processes to the formation of the three types of rocks from one another. We can view the processes by starting at any point in the cycle. We began with the formation of igneous rocks by crystallization of a magma in the interior of the Earth. Igneous rocks then are uplifted to the surface in the mountain-building process. There, they are exposed to weathering and erosion, which produce sediment. The sediment is cycled back to the interior by burial and lithification into sedimentary rock. Deep burial leads to metamorphism or melting, at which point the cycle begins again. Plate tectonics is the mechanism by which the cycle operates.

Key Terms and Concepts

bedding (p. 79)
bedrock (p. 81)
chemical and biochemical sediments (p. 79)
clastic sediments (p. 79)
contact metamorphism (p. 80)
erosion (p. 79)
extrusive igneous rocks (p. 78)
foliation (p. 80)
igneous rocks (p. 75)
intrusive igneous rocks (p. 77)
lithification (p. 79)
metamorphic rocks (p. 77)
mineralogy (p. 75)
orogeny (p. 83)
outcrop (p. 81)
plutonic rocks (p. 83)
pressure-temperature-time path (p. 81)
regional metamorphism (p. 79)
rock (p. 75)
rock cycle (p. 77)
sedimentary rocks (p. 76)
sediments (p. 78)
subsidence (p. 83)
texture (p. 75)
volcanic rocks (p. 83)
weathering (p. 78)

Exercises

1. What are the differences between extrusive and intrusive igneous rocks?

2. What are the differences between regional and contact metamorphism?

3. What are the differences between clastic and chemical or biochemical sedimentary rocks?

4. List three common silicate minerals found in each group of rocks: igneous, sedimentary, and metamorphic.

5. Of the three groups of rocks, which form at Earth's surface and which in the interior of the crust?

6. Where on the continents can you see bare rock?

Thought Questions

This icon indicates that there is an animation available on the Web site that may assist you in answering a question.

1. What geologic processes transform a sedimentary rock into an igneous rock?

2. Name a mineral found only in sedimentary rocks that you might use to distinguish between a fine-grained sedimentary rock formed from lithified mud and an extrusive igneous rock.

3. As a magma cools, what might cause differences in the sizes of the crystals of two intrusive igneous rocks, one with crystals about 1 cm in diameter, the other with crystals about 2 mm in diameter?

4. Which igneous intrusion would you expect to have a wider contact metamorphic zone: one intruded by a very hot magma or one intruded by a cooler magma?

5. Describe the geologic processes by which an igneous rock is transformed into a metamorphic rock and then exposed to erosion.

6. Describe the kinds of outcrop that are found in various places in your hometown. If none, explain how you would determine the nature of the buried bedrock.

7. How does plate tectonics explain plutonism?

8. Using the rock cycle, trace the path from a magma to a granite intrusion to a metamorphic gneiss to a sandstone. Be sure to include the role of tectonics and the specific processes that create the rocks.

9. In the early history of Earth, there were no oceans and only a limited atmosphere. Plate tectonics either was not active or was far less developed during this period than it is now. How would these conditions, compared to those that exist today, affect the rock cycle and the incidence of the major rock types?

10. Although igneous, metamorphic, and sedimentary rocks differ markedly, they are all classified in much the same manner. What characteristics of rocks are common to the classification of all three rock types?

11. Where are igneous rocks most likely to be found? How could you be certain that the rock is igneous and not sedimentary or metamorphic?

12. On a field trip in Arizona you find a rock sample near a large outcrop and bring this sample back to the lab for identification. A thin section of the rock shows that the rock contains traces of quartz, feldspars, mica, and hornblende. Identify the rock and explain how you arrived at this conclusion.

Short-Term Team Projects

Identifying Building Stones

Building stones often are a clue to local geology. Local stone is both cost-effective and a source of community pride. With a partner, examine the building stones that you find on campus or in your community. Choose four to six types of stone that look different and note their locations on a local map. Draw each stone on a separate piece of paper and note such features as color, grain size, the presence or absence of layering, and whether the stone appears to contain one mineral or more than one. Also describe any evidence of chemical or physical weathering and judge how good the stone is for building. Then decide whether the stone is most likely to be igneous, metamorphic, or sedimentary, and explain why. Finally, compare a geologic map of the area with your findings in the field and explain why the stones that you described do or do not record the local geology. Submit an organized folder containing your drawings, observations, and inferences.

Suggested Readings

Blatt, H., and R. J. Tracy. 1996. *Petrology: Igneous, Sedimentary, and Metamorphic,* 2d ed. New York: W. H. Freeman.

Dietrich, R. V., and B. J. Skinner. 1980. *Rocks and Rock Minerals.* New York: Wiley.

Ernst, W. G. 1969. *Earth Materials.* Englewood Cliffs, N.J.: Prentice Hall.

Prinz, M., G. Harlow, and J. Peters. 1978. *Simon & Schuster's Guide to Rocks and Minerals.* New York: Simon & Schuster.

Spear, Frank. S. 1993. Metamorphic phase equilibria and pressure-temperature-time paths. Mineralogical Society of America, Monograph 22.

Columnar basalts, Devil's Post Pile National Monument, eastern Sierra Nevada. Masses of this kind of extrusive igneous rock fracture along columnar joints when they cool. [Jerry L. Ferrara/Photo Researchers.]

CHAPTER

5

Igneous Rocks: Solids from Melts

"A month in the laboratory can often save an hour in the library."

F. H. WESTHEIMER

More than 2000 years ago, the Greek scientist and geographer Strabo traveled to Sicily to view the volcanic eruptions of Mount Etna. He observed that the hot liquid lava spilling down from the volcano onto Earth's surface cooled and hardened into solid rock within a few hours. Two millennia later, eighteenth-century geologists began to understand that some sheets of rock that cut across other rock formations also had formed by the cooling and solidifying of magma. In this case, the magma had cooled slowly because it had remained buried in Earth's crust. Today we know that deep in Earth's hot crust and mantle, rocks melt and rise toward the surface. Some magmas solidify before they reach the surface, and some break through and solidify on the surface. Both processes produce **igneous rocks.**

As we saw in Chapter 4, much of Earth's crust is composed of igneous rock, some metamorphosed and some not. It follows that understanding the processes that melt and resolidify rocks is a key to understanding how Earth's crust forms. **In this chapter, we examine the wide range of igneous rocks, both intrusive and extrusive, and the processes by which they form.**

We also learned in Chapter 4 that a wide variety of igneous rocks are created by plate tectonics. Specifically, igneous rocks form at spreading centers where plates move apart and at convergent boundaries where one plate descends beneath another. Although we still have much to learn about the exact *mechanisms* of melting and solidification, we do have good answers to some fundamental questions: How do igneous rocks differ from one another? Where do igneous rocks form? How do rocks solidify from a melt? Where do melts form?

In answering these questions, we will focus on the central role of igneous processes in the Earth system. When melted rock is transported from magma chambers in Earth's interior to volcanoes, for example, a variety of gases are also carried along. These gases, especially carbon dioxide and sulfur gases, affect the atmosphere and oceans. In this way, magmas may alter climate—an unexpected relationship drawn from analysis of the Earth system.

How Do Igneous Rocks Differ from One Another?

Today we classify rock samples in the same way that some geologists in the late nineteenth century did:

- By texture
- By mineral and chemical composition

Texture

Two hundred years ago, the first division of igneous rocks was made on the basis of texture, an aspect of the rock that largely reflects differences in mineral *crystal size*. Geologists classified rocks as either coarsely or finely crystalline (see Chapter 4). Crystal size is a simple characteristic that geologists can easily distinguish in the field. A coarse-grained rock such as granite has separate crystals that are easily visible to the naked eye. In contrast, the crystals of fine-grained rocks such as basalt are too small to be seen, even with the aid of a magnifying lens. **Figure 5.1** shows samples of granite and basalt, accompanied by photomicrographs of very thin, transparent slices of each rock. Photomicrographs, which are simply photographs taken through a microscope, give an enlarged view of minerals and their textures. Textural differences were clear to early geologists, but more work was needed to unravel the meaning of those differences.

First Clue: Volcanic Rocks Early geologists observed volcanic rocks forming from lava during volcanic eruptions. (*Lava*, you may recall from Chapter 4, is the term that we apply to magma flowing out on the surface.) Geologists noted that when lava cooled rapidly, it formed either a finely crystalline rock or a glassy one in which no crystals could be distinguished. Where lava cooled more slowly, as in the middle of a thick flow many meters high, somewhat larger crystals were present.

Second Clue: Laboratory Studies of Crystallization The second clue to the meaning of texture came when experimental scientists in the nineteenth century began to understand the nature of crystallization. Anyone who has frozen an ice cube knows that water solidifies to ice in a few hours as its temperature drops below the freezing point. If you have ever attempted to retrieve your ice cubes before they were completely solid, you may have seen thin ice crystals forming at the surface and along the sides of the tray. During crystallization, the water molecules take up fixed positions in the solidifying crystal structure, and they are no longer able to move freely, as they did when the water was liquid. All other liquids, including magmas, crystallize in this way.

The first tiny crystals form a pattern. Other atoms or ions in the crystallizing liquid then attach themselves in such a way that the tiny crystals grow larger. It takes some time for the atoms or ions to "find" their correct places on a growing crystal, and large crystals form only if they have time to grow slowly. If a liquid solidifies very quickly, as a magma does when it erupts onto the cool surface of the Earth, the crystals have no time to grow larger. Instead, a large number of tiny crystals form simultaneously as the liquid cools and solidifies.

Third Clue: Granite—Evidence of Slow Cooling The study of volcanoes allowed early geologists to link finely crystalline textures with quick cooling at Earth's surface and to see finely crystalline igneous rocks as evidence of former volcanism. But in the absence of direct observation, how could geologists deduce that coarse-grained rocks form by slow cooling deep in the interior? Granite—one of the commonest rocks of the continents—turned out to be the crucial clue (**Figure 5.2**). James Hutton, one of geology's

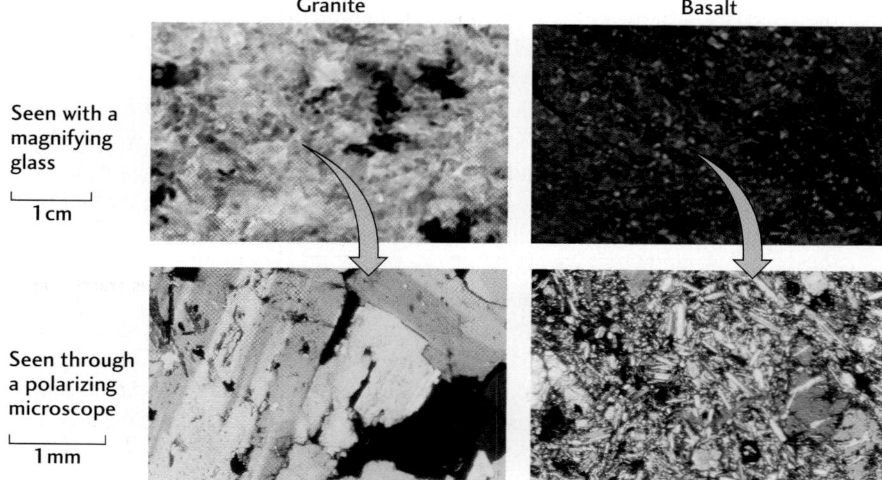

Figure 5.1 Igneous rocks were first classified by texture. Early geologists assessed texture with a small hand-held magnifying lens. Modern geologists have access to high-powered polarizing microscopes, which produce photomicrographs of thin, transparent rock slices like those shown here. [Hand-sample photographs by Chip Clark. Photomicrographs by Raymond Siever.]

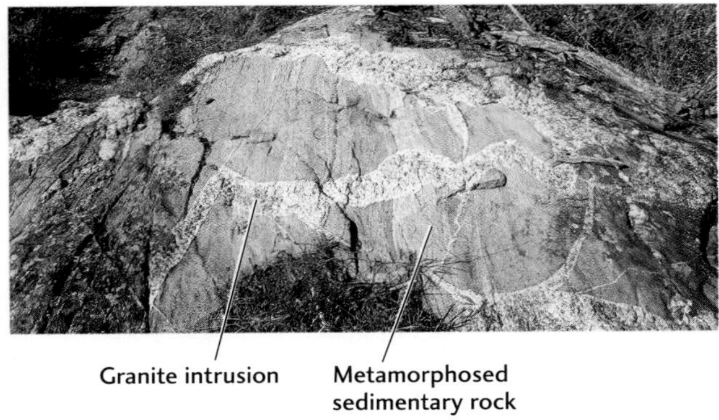

Figure 5.2 Granite intrusion (light colored) cutting across metamorphosed sedimentary rock. [Tom Bean/DRK.]

founding fathers, saw granites cutting across and disrupting layers of sedimentary rocks as he worked in the field in Scotland. He noticed that the granite had somehow fractured and invaded the sedimentary rocks, as though the granite had been forced into the fractures as a liquid.

As Hutton looked at more and more granites, he began to focus on the sedimentary rocks bordering them. He observed that the minerals of the sedimentary rocks in contact with the granite were different from those found in sedimentary rocks at some distance from the granite. He concluded that the changes in the sedimentary rocks must have resulted from great heat and that the heat must have come from the granite. Hutton also noted that granite was composed of crystals interlocked like the pieces of a jigsaw puzzle (see Figure 5.1). By this time, chemists had established that a slow crystallization process produces this pattern.

Hutton assessed these three lines of evidence and proposed that granite forms from a hot molten material that solidifies deep in the Earth. The evidence was conclusive, because no other explanation could accommodate all the facts. Other geologists, who saw the same characteristics of granites in widely separated places in the world, came to recognize that granite and many similar coarsely crystalline rocks were the products of magma that had crystallized slowly in the interior of the Earth.

Intrusive Igneous Rocks The full significance of textural distinction in igneous rocks is now clear. As we have seen, texture is linked to the rapidity and therefore the place of cooling. Slow cooling of magma in Earth's interior allows adequate time for the growth of the interlocking large coarse crystals that characterize intrusive igneous rocks (**Figure 5.3**). An **intrusive igneous rock** is one that has forced its way into surrounding rock. This surrounding rock

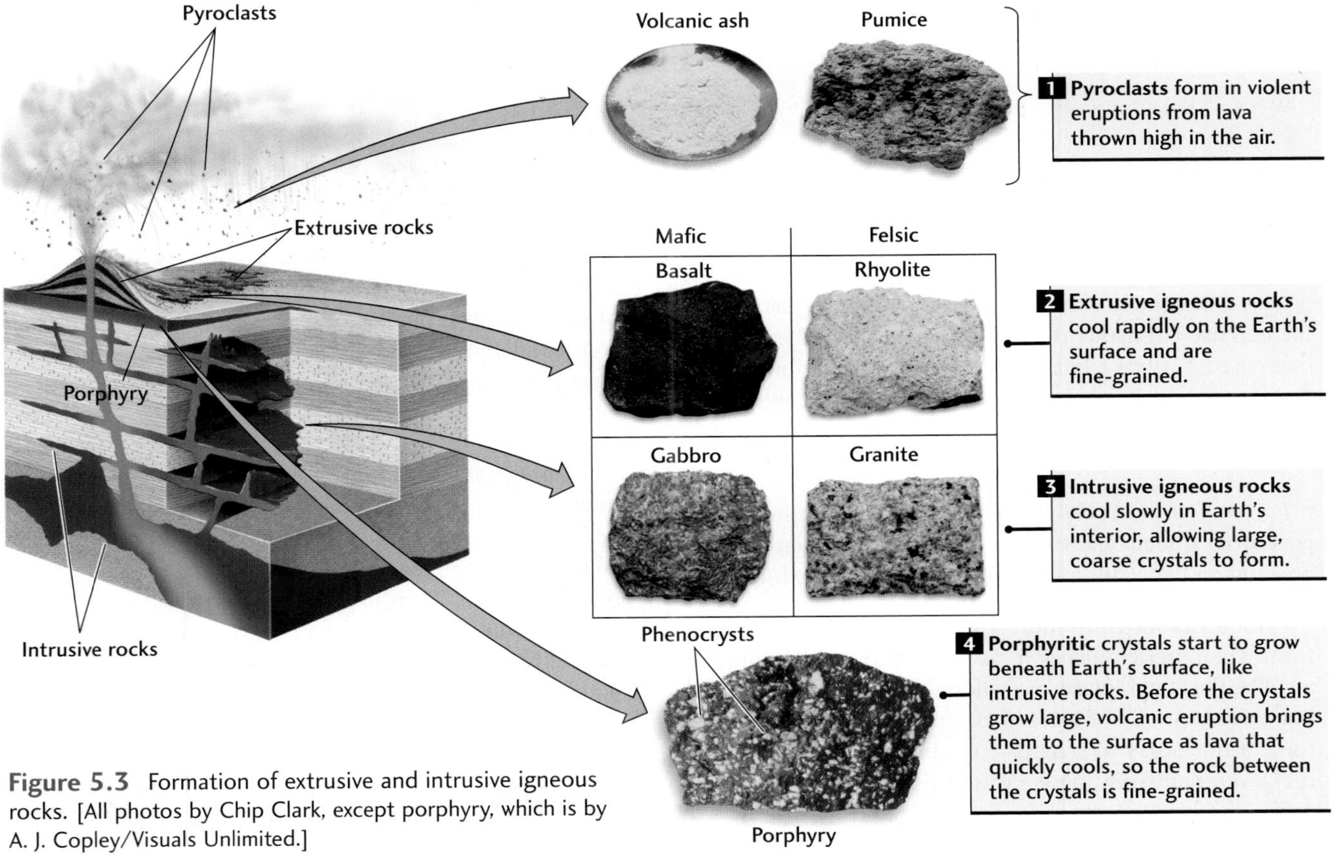

Pyroclasts
Extrusive rocks
Porphyry
Intrusive rocks
Volcanic ash
Pumice
1 **Pyroclasts** form in violent eruptions from lava thrown high in the air.
Mafic
Felsic
Basalt
Rhyolite
2 **Extrusive igneous rocks** cool rapidly on the Earth's surface and are fine-grained.
Gabbro
Granite
3 **Intrusive igneous rocks** cool slowly in Earth's interior, allowing large, coarse crystals to form.
Phenocrysts
Porphyry
4 **Porphyritic** crystals start to grow beneath Earth's surface, like intrusive rocks. Before the crystals grow large, volcanic eruption brings them to the surface as lava that quickly cools, so the rock between the crystals is fine-grained.

Figure 5.3 Formation of extrusive and intrusive igneous rocks. [All photos by Chip Clark, except porphyry, which is by A. J. Copley/Visuals Unlimited.]

Table 5.1 Common Minerals of Igneous Rocks

Compositional Group	Mineral	Chemical Composition	Silicate Structure
FELSIC	Quartz	SiO_2	Frameworks
	Potassium feldspar	$KAlSi_3O_8$	
	Plagioclase feldspar	$\begin{cases} NaAlSi_3O_8 \\ CaAl_2Si_2O_8 \end{cases}$	
	Muscovite (mica)	$KAl_3Si_3O_{10}(OH)_2$	Sheets
MAFIC	Biotite (mica)	$\left.\begin{matrix} K \\ Mg \\ Fe \\ Al \end{matrix}\right\} Si_3O_{10}(OH)_2$	
	Amphibole group	$\left.\begin{matrix} Mg \\ Fe \\ Ca \\ Na \end{matrix}\right\} Si_8O_{22}(OH)_2$	Double chains
	Pyroxene group	$\left.\begin{matrix} Mg \\ Fe \\ Ca \\ Al \end{matrix}\right\} SiO_3$	Single chains
	Olivine	$(Mg,Fe)_2SiO_4$	Isolated tetrahedra

is called **country rock.** (Later in this chapter, we will examine some special forms of intrusive igneous rocks.)

Extrusive Igneous Rocks Rapid cooling at Earth's surface produces the finely grained texture or glassy appearance of the **extrusive igneous rocks** (see Figure 5.3). These rocks, composed partly or largely of volcanic glass, form when lava or other volcanic material erupts from volcanoes. For this reason, they are also known as volcanic rocks. They fall into two major categories:

• *Lavas* Volcanic rocks formed from lavas range in appearance from smooth and ropy to sharp, spiky, and jagged, depending on the conditions under which the rocks formed.

• *Pyroclastic rocks* In more violent eruptions, **pyroclasts** form when broken pieces of lava are thrown high into the air. The finest pyroclasts are **volcanic ash,** extremely small fragments, usually of glass, that form when escaping gases force a fine spray of magma from a volcano. All volcanic rocks lithified from these pyroclastic materials are called **tuff** (see Chapter 6 for more details).

One pyroclastic rock is **pumice,** a frothy mass of volcanic glass with a great number of *vesicles.* Vesicles are holes that remain after trapped gas has escaped from the solidifying melt. Another wholly glassy volcanic rock is **obsidian;** unlike pumice, it contains only tiny vesicles and so is solid and dense. Chipped or fragmented obsidian produces very sharp edges, and Native Americans and many other hunting groups used it for arrowheads and a variety of cutting tools.

A **porphyry** is an igneous rock that has a mixed texture in which large crystals "float" in a predominantly fine crystalline matrix (see Figure 5.3). The large crystals, called *phenocrysts,* formed while the magma was still below Earth's surface. Then, before other crystals could grow, a volcanic eruption brought the magma to the surface, where it quickly cooled to a finely crystalline mass. In some cases, porphyries are developed as intrusive igneous rocks, for example, where magmas are emplaced and then cool quickly at very shallow levels in the crust. Porphyry textures are very important to geologists because they indicate that different minerals grow at different rates, a point that will be emphasized later in this chapter. In Chapter 6, we will look more closely at how these volcanic rocks and others form during volcanism. Now, however, we turn to the second way in which the family of igneous rocks is subdivided.

Chemical and Mineral Composition

We have seen that igneous rocks can be subdivided according to their texture. They can also be subdivided on the basis of their chemical and mineral compositions. Volcanic glass, which is formless even under the microscope, is often classified by chemical analysis. One of the earliest classifications of igneous rocks was based on a simple chemical analysis of their silica (SiO_2) content. Silica, as noted in Chapter 4, is abundant in most igneous rocks and accounts for 40 to 70 percent of their total weight. We still refer to rocks rich in silica, such as granite, as silicic.

Modern classifications now group igneous rocks according to their relative proportions of silicate minerals (Table 5.1). These minerals are described in Appendix 5. The silicate minerals—quartz, feldspar (both orthoclase and plagioclase), muscovite and biotite micas, the amphibole and pyroxene groups, and olivine—form a systematic series. Felsic minerals are high in silica; mafic minerals are low in silica. The adjectives *felsic* (from *fel*dspar and *si*lica) and *mafic* (from *ma*gnesium and *f*erric, from Latin *ferrum,* "iron") are applied to both the minerals and the rocks that have high contents of these minerals. Mafic minerals crystallize at higher temperatures—that is, earlier in the cooling of a magma—than do felsic minerals.

As the mineral and chemical compositions of igneous rocks became known, geologists soon noticed that some extrusive and intrusive rocks were identical in composition and differed only in texture. **Basalt,** for example, is an extrusive rock formed from lava. Gabbro has exactly the same mineral composition as basalt but forms deep in Earth's crust (see Figure 5.3). Similarly, rhyolite and granite are identical in composition but differ in texture. Thus extrusive and intrusive rocks form two chemically and mineralogically parallel sets of igneous rocks. Conversely, most chemical and mineral compositions can appear in either extrusive or intrusive rocks. The sole exceptions are very highly mafic rocks that rarely or never appear as extrusive igneous rocks.

Figure 5.4 is a model that portrays these relationships. Notice that the horizontal axis plots silica content as a

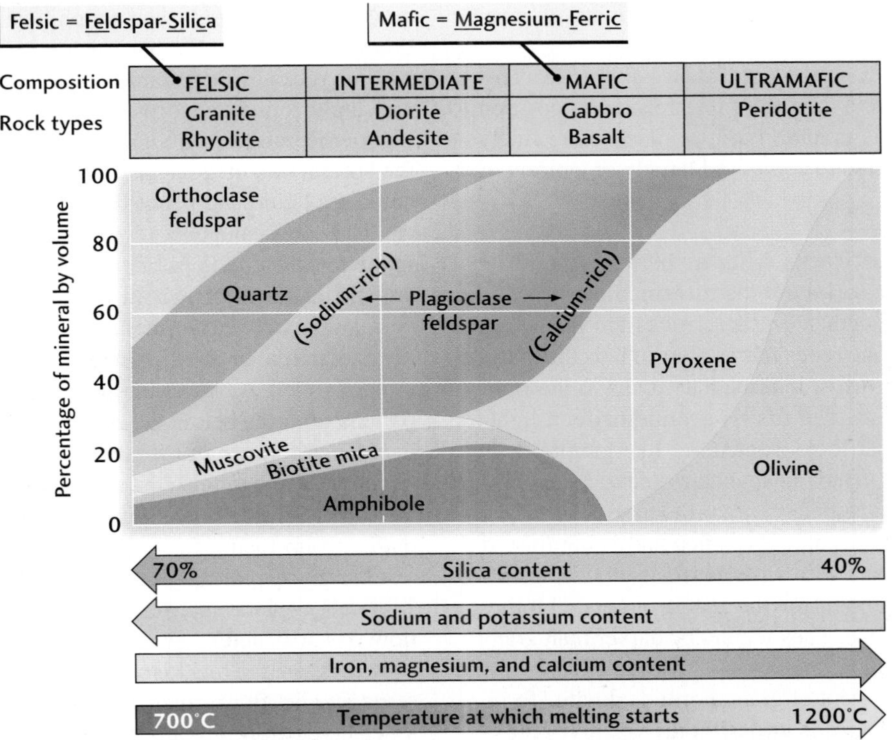

Figure 5.4 Classification model of igneous rocks. The vertical axis measures the mineral composition of a given rock as a percentage of its volume. The horizontal axis is a scale of silica content by weight. Thus, if you know by chemical analysis that a coarsely textured rock sample is about 70 percent silica, you could determine that its composition is about 6 percent amphibole, 3 percent biotite, 5 percent muscovite, 14 percent plagioclase feldspar, 22 percent quartz, and 50 percent orthoclase feldspar. Your rock would be granite. Although rhyolite has the same mineral profile, its fine texture would eliminate it.

percentage of a given rock's weight. The percentages given—from high silica content at 70 percent to low silica content at 40 percent—cover the range found in igneous rocks. The vertical axis displays a scale measuring the mineral content of a given rock as a percentage of its volume. If you know the silica content of a rock sample, you can determine its mineral composition and, from that, the type of rock.

We can use Figure 5.4 to help with the discussion of the intrusive and extrusive igneous rocks. We begin with the felsic rocks, on the extreme left of the model.

Felsic Rocks **Felsic rocks** are poor in iron and magnesium and rich in minerals that are high in silica. Such minerals include quartz, orthoclase feldspar, and plagioclase feldspar. Plagioclase feldspars contain both calcium and sodium. As Figure 5.4 indicates, they are richer in sodium near the felsic end and richer in calcium near the mafic end. Thus, just as mafic minerals crystallize at higher temperatures than felsic minerals, calcium-rich plagioclases crystallize at higher temperatures than the sodium-rich plagioclases.

Felsic minerals and rocks tend to be light in color. **Granite,** one of the most abundant intrusive igneous rocks, contains about 70 percent silica. Its composition includes abundant quartz and orthoclase feldspar and a lesser amount of plagioclase feldspar (see the left side of Figure 5.4). These light-colored felsic minerals give granite its pink or gray color. Granite also contains small amounts of muscovite and biotite micas and amphibole.

Rhyolite is the extrusive equivalent of granite. This light-brown to gray rock has the same felsic composition and light coloration as granite, but it is much more finely grained. Many rhyolites are composed largely or entirely of volcanic glass.

Intermediate Igneous Rocks Midway between the felsic and mafic ends of the series are the **intermediate igneous rocks.** As their name indicates, these rocks are neither as rich in silica as the felsic rocks nor as poor in it as the mafic rocks. We find the intrusive intermediate rocks to the right of granite in Figure 5.4. The first is **granodiorite,** a light-colored felsic rock that looks something like granite. It is also similar to granite in having abundant quartz, but its predominant feldspar is plagioclase, not orthoclase. To its right is **diorite,** which contains still less silica and is dominated by plagioclase feldspar, with little or no quartz. Diorites contain a moderate amount of the mafic minerals biotite, amphibole, and pyroxene. They tend to be darker than granite or granodiorite.

The volcanic equivalent of granodiorite is **dacite.** To its right in the extrusive series is **andesite,** the volcanic equivalent of diorite. Andesite derives its name from the Andes, the volcanic mountain chain in South America.

Mafic Rocks **Mafic rocks** are high in pyroxenes and olivines. These minerals are relatively poor in silica but rich in magnesium and iron, from which they get their characteristic dark colors. **Gabbro,** with even less silica than is found in the intermediate igneous rocks, is a coarsely grained, dark gray intrusive igneous rock. Gabbro has an abundance of mafic minerals, especially pyroxene. It contains no quartz and only moderate amounts of calcium-rich plagioclase feldspar.

Basalt, as we have seen, is dark gray to black and is the fine-grained extrusive equivalent of gabbro. Basalt is the most abundant igneous rock of the crust, and it underlies virtually the entire seafloor. On the continents, extensive thick sheets of basalt make up large plateaus in some places. The Columbia Plateau of Washington State and the remarkable formation known as the Giant's Causeway in Northern Ireland are two examples. The Deccan basalts of India and the Siberian basalts of northern Russia represent enormous outpourings of basalt that appear to coincide closely with two of the greatest periods of mass extinction in the fossil record. These great episodes of basalt formation, and the mechanisms responsible for them, are discussed further in Chapter 6.

Ultramafic Rocks **Ultramafic rocks** consist primarily of mafic minerals and contain less than 10 percent feldspar. Here, at the very low silica composition of only about 45 percent, we find **peridotite.** It is a coarsely grained, dark greenish-gray rock made up primarily of olivine with small amounts of pyroxene and amphibole. Peridotites are the dominant rock in Earth's mantle and are the source of basaltic rocks formed at mid-ocean ridges. Ultramafic rocks are rarely found as extrusives. Because they form at such high temperatures, through the accumulation of crystals at the bottom of a magma chamber, they have only rarely constituted a liquid and hence do not form typical lavas.

The names and exact compositions of the various rocks in the felsic-to-mafic series are less important than the systematic changes shown in Table 5.2. There is a strong correlation between mineralogy and the temperatures of crystallization or melting. As Table 5.2 indicates, mafic minerals melt at higher temperatures than felsic minerals. At temperatures below the melting point, minerals crystallize. Therefore, mafic minerals also crystallize at higher temperatures than felsic minerals. We can also see from the table that silica content increases as we move from the mafic group to the felsic group. Increasing silica content results in increasingly complex silicate structures (see Table 5.1), which interfere with a melted rock's ability to flow. As a structure grows more complex, the ability to flow decreases. Thus **viscosity**—the measure of a liquid's *resistance* to flow—increases as silica content increases. Viscosity is an important factor in the behavior of lavas, as we will see in Chapter 6.

It is clear that knowing a rock's minerals can yield important information about the conditions under which the rock's parent magma formed and crystallized. To interpret this information accurately, however, we must understand more about igneous processes. We turn to that topic next.

Table 5.2 Changes in Some Major Chemical Elements from Felsic to Mafic Rocks

	Felsic	Intermediate		Mafic
Coarse-Grained (intrusive)	Granite	Granodiorite	Diorite	Gabbro
Fine-Grained (extrusive)	Rhyolite	Dacite	Andesite	Basalt

← Silica increasing
← Sodium increasing
← Potassium increasing
Calcium increasing →
Magnesium increasing →
Iron increasing →

← (Viscosity increasing)

(Melting temperature increasing) →

How Do Magmas Form?

We know from the way Earth transmits earthquake waves that the bulk of the planet is solid for thousands of kilometers down to the core-mantle boundary (see Chapter 21). The evidence of volcanic eruptions, however, tells us that there must also be liquid regions where magmas originate. How do we resolve this apparent contradiction? The answer lies in the processes that melt rocks and create magmas.

How Do Rocks Melt?

Although we do not yet understand the exact mechanisms of melting and solidification, geologists have learned a great deal from laboratory experiments designed to determine how rocks melt. From these experiments, we know that a rock's melting point depends on its composition and on conditions of temperature and pressure (Table 5.3).

Temperature and Melting As geologists ran experiments on rocks in the early twentieth century, they discovered that a rock does not melt completely at any given temperature. The **partial melting** that these early geologists discovered occurs because the minerals that compose a rock melt at different temperatures. As temperatures rise, some minerals melt and others remain solid. If the same conditions are maintained at any given temperature, the same mixture of solid rock and melt is maintained. The fraction of rock that has melted at a given temperature is called a *partial melt.* To visualize a partial melt, think of how a chocolate chip cookie would look if you heated it to the point at which the chocolate chips melted while the main part of the cookie stayed solid.

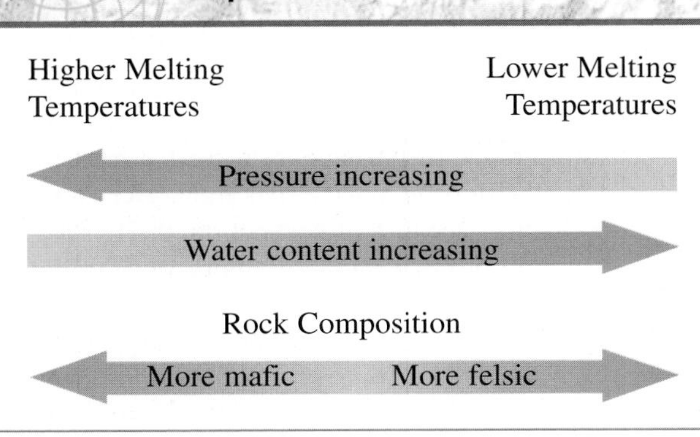

The ratio of liquid to solid in a partial melt depends on the composition and melting temperatures of the minerals that make up the original rock. It also depends on the temperature at the depth in the crust or mantle where melting takes place. At the lower end of its melting range, a partial melt might be less than 1 percent of the volume of the original rock. Much of the hot rock would still be solid, but appreciable amounts of liquid would be present as small droplets in the tiny spaces between crystals throughout the mass. In the upper mantle, for example, some basaltic partial melts can be produced by only 1 to 2 percent melting of peridotite. However, 15 to 20 percent partial melting of mantle peridotite to form basaltic magmas is common beneath mid-ocean ridges. At the high end of the melting temperature range, much of the rock would be liquid, with lesser amounts of unmelted crystals in it. An example would be a reservoir of a basalt magma and crystals just beneath a volcano such as the island of Hawaii.

Geologists in the early twentieth century seized on the new knowledge of partial melts to help them determine how different kinds of magma form at different temperatures and in different regions of Earth's interior. As you can imagine, the composition of a partial melt in which only the minerals with the lowest melting points have melted may be significantly different from the composition of a completely melted rock. Thus, basaltic magmas that form in different regions of the mantle may have somewhat different compositions. From this observation, geologists could deduce that the different magmas come from different proportions of partial melt.

Pressure and Melting Did you know that if you put water in a strong container and squeezed it very hard, it would freeze? It turns out that rocks do exactly the same thing in response to changes in pressure. To get the whole story on melting, we must consider pressure, which increases with depth in the Earth as a result of the increased weight of overlying rock. Geologists found that as they melted rocks under various pressures, higher pressures led to higher melting temperatures. Thus rocks that would melt at Earth's surface would remain solid at the same temperature in the interior. For example, a rock that melts at 1000°C at Earth's surface might have a much higher melting temperature, perhaps 1300°C, at depths in the interior. There, pressures are many thousands of times greater than the pressure at Earth's surface. It is the effect of pressure that explains why rocks in most of the crust and mantle do not melt. Rock can melt only where both its mineral composition and the temperature and pressure conditions are right. Just as an increase in pressure can keep a rock solid, a decrease in pressure can make a rock melt, given a suitably high temperature. Because of convection, the Earth's mantle rises at mid-ocean ridges—at more or less constant temperature. As the mantle material rises and the pressure decreases below a critical point, solid rocks melt spontaneously, without the introduction of any additional heat. This process, known as **decompression melting,** produces the greatest volume of molten rock anywhere on Earth. It is the process by which most basalts on the seafloor form. You will learn more about pressure and its effects on rocks in Earth's interior in Chapter 21.

Water and Melting The many experiments on melting temperatures and partial melting paid other dividends as well. One of them was a better understanding of the role of water in rock melting. Geologists knew from analyses of natural lavas that there was water in some magmas, so they added small amounts of water to the rocks they were melting. They discovered that the compositions of partial and complete melts vary not only with temperature and pressure but also with the amount of water present.

Consider, for example, the effect of dissolved water on pure albite, the high-sodium plagioclase feldspar, at the low pressures of the Earth's surface. If only a small amount of water is present, pure albite will remain solid at temperatures just over 1000°C, hundreds of degrees above the boiling point of water. At these temperatures, the water in the albite is present as a vapor (gas). If large amounts of water are present, the melting temperature of the albite will decrease, dropping to as low as 800°C. This behavior follows the general rule that dissolving some of one substance (in this case, water) in another (in this case, albite) lowers the melting point of the solution. If you live in a cold climate, you are probably familiar with this principle because you know that towns and municipalities sprinkle salt on icy roads to lower the melting point of the ice.

By the same principle, the melting temperature of the albite—and of all the feldspars and other silicate minerals—drops considerably in the presence of large amounts of water. In this case, the melting points of various silicates decrease in proportion to the amount of water dissolved in the molten silicate. This is an important point in our knowledge of how rocks melt. Water content is a significant factor in determining the melting temperatures of mixtures of sedimentary and other rocks. Sedimentary rocks contain an especially large volume of water in their pore spaces, more than is found in igneous or metamorphic rocks. As we will discuss later in this chapter, the water in sedimentary rocks plays an important role in melting in Earth's interior.

The Formation of Magma Chambers

Most substances are less dense in the liquid form than in the solid form. The density of a melted rock is lower than the density of a solid rock of the same composition. In other words, a given volume of melt would weigh less than the same volume of solid rock. Geologists reasoned that large bodies of magma could form in the following way. If the less dense melt were given a chance to move, it would move upward—just as oil, which is less dense than water, rises to the surface of a mixture of oil and water. Being liquid, the partial melt could move slowly upward through pores and along the boundaries between crystals of the overlying rocks. As the hot drops of melted rock moved upward, they would coalesce with other drops, gradually forming larger pools of molten rock within Earth's solid interior.

The ascent of magmas through the mantle and crust may be slow or rapid. Geologists have estimated that ascent rates range from 0.3 m/year to almost 50 m/year. Ascent times may be tens of thousands or even hundreds of thousands of years. As they ascend, magmas may mix with other melts; they also may affect the melting of lithospheric crust. We now know that the large pools of molten rock envisioned by early geologists form **magma chambers**—magma-filled cavities in the lithosphere that form as ascending drops of melted rock push aside surrounding solid rock. A magma chamber may encompass a volume as large as several cubic kilometers. Geologists are still studying the exact manner in which magma chambers form, and we cannot yet say exactly what they look like in three dimensions. We think of them as large, liquid-filled cavities in solid rock, which expand as more of the surrounding rock melts or as liquid migrates in through cracks and other small openings between crystals. Magma chambers contract as they expel magma to the surface in eruptions. We know for sure that magma chambers exist because earthquake waves can show us the depth, size, and general outlines of the chambers underlying some active volcanoes.

With this knowledge of how rocks melt to form magmas, we can now consider where in Earth's interior various kinds of magmas form.

Where Do Magmas Form?

Our understanding of igneous processes stems from geological inferences as well as laboratory experimentation. Our inferences are based mainly on data from two sources. The first is volcanoes on land and under the sea—everywhere that molten rock erupts. Volcanoes give us information about where magmas are located. The second source of data is the record of temperatures measured in deep drill holes and mine shafts. This record show that the temperature of Earth's interior increases with depth. Using these measurements, scientists have been able to estimate the rate at which temperature rises as depth increases.

In some locations, the temperatures recorded at a given depth are much higher than the temperatures recorded at the same depth in other locations. These results indicate that some parts of Earth's mantle and crust are hotter than others. For example, the Great Basin of the western United States is an area where the North American continent is being stretched and thinned, with the result that temperature increases at an exceptionally rapid rate, reaching 1000°C at a depth of 40 km, not far below the base of the crust. This temperature is almost high enough to melt basalt. By contrast, in tectonically stable regions, such as the interior parts of continents, temperature increases much more slowly, reaching only 500°C at the same depth.

We now know that various kinds of rock can solidify from magmas formed by partial melting. And we know that increasing temperatures in Earth's interior could create magmas. Let's turn now to the question of why there are so many different types of igneous rocks.

Magmatic Differentiation

The processes we've discussed so far demonstrate how rocks melt to form magmas. But what accounts for the variety of igneous rocks? Does this variety arise from magmas of different chemical compositions made by the melting of different kinds of rocks? Or do some processes produce variety from an originally uniform parent material?

Again, the answers to these questions came from laboratory experiments. Geologists mixed chemical elements in proportions that simulated the compositions of natural igneous rocks, then melted these mixtures in high-temperature furnaces. As the melts cooled and solidified, the experimenters carefully observed the temperatures at which crystals formed and recorded the chemical compositions of those crystals. This research gave rise to the theory of **magmatic differentiation,** a process by which rocks of varying composition can arise from a uniform parent magma. Magmatic differentiation occurs because different minerals crystallize at different temperatures. During crystallization, the composition of the magma changes as it is depleted of the chemical elements withdrawn to make the crystallized minerals.

In a kind of mirror image of partial melting, the first minerals to crystallize from a cooling magma are the ones that were the last to melt in partial-melting experiments. This initial crystallization withdraws chemical elements from the melt, changing the magma's composition. Continued cooling crystallizes the minerals that had melted at the next lower temperature range in the melting experiments. Again, the magma's chemical composition changes as various elements are withdrawn. Finally, as the magma solidifies completely, the last minerals to crystallize are the ones that melted first. This is how the same parent magma, because of its changing chemical composition throughout the crystallization process, can give rise to different igneous rocks.

Fractional Crystallization: Laboratory and Field Observations

Fractional crystallization is the process by which the crystals formed in a cooling magma are segregated from the remaining liquid. This segregation happens in several ways (**Figure Story 5.5**). In the simplest scenario, crystals formed in a magma chamber settle to the chamber's floor and are thus removed from further reaction with the remaining liquid. The magma then migrates to new locations, forming new chambers. Crystals that had formed early would be

FRACTIONAL CRYSTALLIZATION EXPLAINS THE COMPOSITION OF A BASALTIC INTRUSION

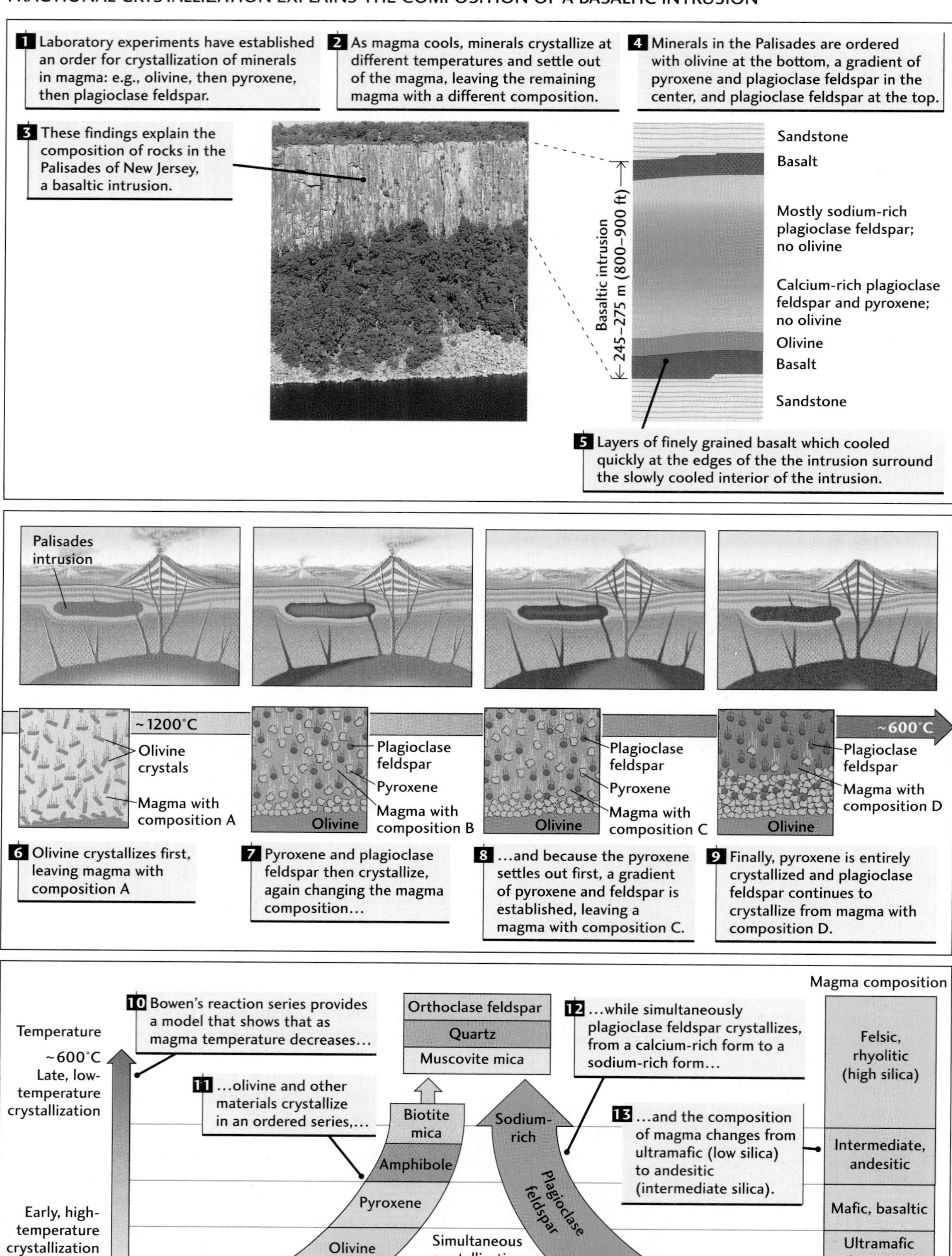

Figure Story 5.5 Magma differentiated by fractional crystallization. The Palisades intrusion is the result of fractional crystallization. [Photo by Breck P. Kent.]

segregated from the remaining magma, which would continue to crystallize as it cooled.

A good test case for the theory of fractional crystallization is provided by the Palisades, a line of imposing massive cliffs that faces the city of New York on the west bank of the Hudson River. This igneous formation is about 80 km long and, in places, more than 300 m high. It was intruded as a melt of basaltic composition into almost horizontal sedimentary rocks. It contains abundant olivine near the bottom, pyroxene and plagioclase feldspar in the middle, and mostly plagioclase feldspar near the top. These variations in mineral composition from top to bottom made the Palisades a perfect site for testing the theory of fractional crystallization. Such testing showed how laboratory experiments could help explain field observations.

From experiments on the melting of rocks with about the same proportions of the various minerals found in the Palisades intrusion, geologists knew that the temperature of the melt had to have been about 1200°C. The parts of the magma within a few meters of the relatively cold upper and lower contacts with the surrounding sedimentary rocks cooled quickly, forming a fine-grained basalt and preserving the chemical composition of the original melt. But the hot interior of the intrusion cooled more slowly, as evidenced by the slightly larger crystals found in the intrusion's interior.

The ideas of fractional crystallization lead us to think that the first mineral to crystallize from the slowly cooling interior would have been olivine. This heavy mineral would sink through the melt to the bottom of the intrusion. It can be found today in the Palisades intrusion as a coarse-grained, olivine-rich layer just above the chilled, fine-grained basaltic layer along the bottom contact. Continued cooling would have produced pyroxene crystals, followed almost immediately by calcium-rich plagioclase feldspar. These minerals, too, settled out through the magma and accumulated in the lower third of the Palisades intrusion. The abundance of plagioclase feldspar in the upper parts of the intrusion is evidence that the melt continued to change composition until successive layers of settled crystals were topped off by a layer of mostly sodium-rich plagioclase feldspar crystals.

Being able to explain the layering of the Palisades intrusion as the result of fractional crystallization was an early success of the first version of the theory of magmatic differentiation. It firmly tied field observations to laboratory experiments and was solidly based on chemical knowledge. More than two-thirds of a century of geological research has passed since the Palisades was first seen as a test case, and we now know that the Palisades actually has a more complex history. This history includes several injections of magma and a more complicated process of olivine settling. Nevertheless, the Palisades intrusion remains a valid example of fractional crystallization.

Granite and Basalt: Magmatic Differentiation

Studies of the lavas of volcanoes showed that basaltic magmas are common—far more common than the rhyolitic magmas that correspond in composition to granites. How could the granites that are so abundant in the crust have been derived from basaltic magmas?

The original idea of magmatic differentiation was that a basaltic magma would gradually cool and differentiate into a cooler, more silicic melt by fractional crystallization. The early stages of this differentiation would produce andesitic magma, which might erupt to form andesitic lavas or solidify by slow crystallization to form diorite intrusives. Intermediate stages would make magmas of granodiorite composition. If this process were carried far enough, its late stages would form rhyolitic lavas and granite intrusions (see Figure Story 5.5).

Field and laboratory work in the latter part of the twentieth century revealed the need to change earlier theories of magmatic differentiation. One line of research showed that so much time would be needed for small crystals of olivine to settle through a dense, viscous magma that they might never reach the bottom of a magma chamber. Other researchers demonstrated that many layered intrusions—similar to but much larger than the Palisades—do not show the simple progression of layers predicted by simple magmatic theory.

The biggest problem, however, was the source of granite. The first sticking point is that the great volume of granite found on Earth could not have formed from basaltic magmas by magmatic differentiation, because large quantities of liquid volume are lost by crystallization during successive stages of differentiation. To produce the existing amount of granite, an initial volume of basaltic magma 10 times the size of a granitic intrusion would be required. That abundance would imply the crystallization of huge quantities of basalt underlying granite intrusions. But geologists could not find anything like that amount of basalt. Even where great volumes of basalts are found—at mid-ocean ridges—there is no wholesale conversion into granite through magmatic differentiation.

Most in question is the original idea that all granitic rocks evolve from the differentiation of a single type of magma, a basaltic melt. Instead, geologists discovered that the melting of varied source rocks of the upper mantle and crust is responsible for much variation in magma composition:

1. Rocks in the upper mantle might partially melt to produce basaltic magma.

2. A mixture of sedimentary rocks and basaltic oceanic rocks such as those found in subduction zones might melt to form andesitic magma.

3. A melt of sedimentary, igneous, and metamorphic continental crustal rocks might produce granitic magma.

Magmatic differentiation does operate, but its mechanisms are much more complex than first recognized.

- Partial melting is of great importance in the production of magmas of varying composition. Magmatic differentiation can be achieved by the partial melting of mantle and crustal rocks over a range of temperatures and water contents.

• Magmas do not cool uniformly; they may exist transiently at a range of temperatures within a magma chamber.

• The differences in temperature in and among magma chambers may cause the chemical composition of the magma to vary from one region to another.

• A few magmas are immiscible—they do not mix with one another, just as oil and water do not mix. When such magmas coexist in one magma chamber, each forms its own crystallization products.

• Some magmas that are miscible—that *do* mix—may give rise to a crystallization path different from that followed by any one magma alone.

We now know more about the physical processes that interact with crystallization within magma chambers (**Figure 5.6**). Magma at various temperatures in different parts of a magma chamber may flow turbulently, crystallizing as it circulates. Crystals may settle, then be caught up in currents again, and eventually be deposited on the chamber's walls. The margins of such a magma chamber may be a "mushy" zone of crystals and melt lying between the solid rock border of the chamber and the completely liquid magma within the main part of the chamber. And, at some mid-ocean ridges, such as the East Pacific Rise, a mushroom-shaped magma chamber may be surrounded by hot basaltic rock with only small amounts (1 to 3 percent) of partial melt.

Forms of Magmatic Intrusions

As noted earlier, geologists cannot directly observe the shapes of intrusive igneous rocks formed when magmas intrude the crust. We can only deduce their shapes and distributions from evidence gained in geological fieldwork done millions of years after the rocks formed, long after the magma cooled and the rocks were uplifted and exposed to erosion.

We do have indirect evidence of current magmatic activity. Earthquake waves, for example, show us the general outlines of magma chambers that underlie some active volcanoes. But they cannot reveal the detailed shapes or sizes of intrusions arising from those magma chambers. In some nonvolcanic but tectonically active regions, such as an area near the Salton Sea in southern California, measurements of temperatures in deep drill holes reveal a crust much hotter than normal, which may be evidence of an intrusion at depth.

But in the end, most of what we know about intrusive igneous rock is based on the work of field geologists who have examined and compared a wide variety of outcrops and have reconstructed their history. Their studies have resulted in the description and classification of the many irregular and variable forms of intrusive bodies. In the following pages, we consider some of these bodies: plutons, sills and dikes, and veins. **Figure 5.7** illustrates a variety of extrusive and intrusive structures.

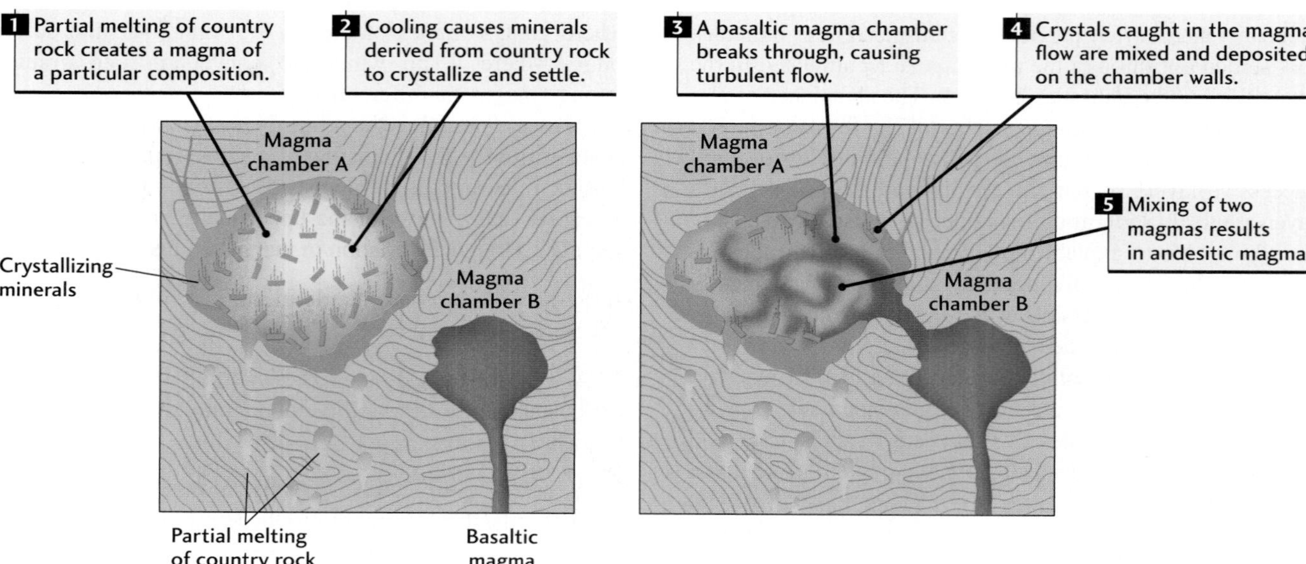

Figure 5.6 Modern ideas of magmatic differentiation. Geologists now recognize that magmatic differentiation does operate, but its mechanisms are more complex than first recognized. Melting is usually partial. Some magmas derived from rocks of varying compositions may mix, whereas other magmas are immiscible. Crystals may be transported to various parts of the magma chamber by turbulent currents in the liquid.

Plutons

Plutons are large igneous bodies formed at depth in Earth's crust. They range in size from a cubic kilometer to hundreds of cubic kilometers. We can study these large bodies when uplift and erosion uncover them or when mines or drill holes cut into them. Plutons are highly variable, not only in size but also in shape and in their relationship to the surrounding country rock.

This wide variability is due in part to the different ways in which magma makes space for itself as it rises through the crust. Most magmas intrude at depths greater than 8 to 10 km. At these depths, few holes or openings exist, because the great pressure of the overlying rock would close them. But the upwelling magma overcomes even that high pressure.

Magma rising through the crust makes space for itself in three ways (**Figure 5.8**) that collectively may be referred to as *magmatic stoping:*

1. *Wedging open the overlying rock.* As the magma lifts that great weight, it fractures the rock, penetrates the cracks, wedges them open, and so flows into the rock. Overlying rocks may bow up during this process.

2. *Breaking off large blocks of rock.* Magma can push its way upward by breaking off blocks of the invaded crust. These blocks, known as *xenoliths,* sink into the magma, melt, and blend into the liquid, in some places changing the composition of the magma.

3. *Melting surrounding rock.* Magma also makes its way by melting walls of country rock.

Most plutons show sharp contacts with country rock and other evidence of the intrusion of a liquid magma into solid rock. Other plutons grade into country rock and have structures vaguely resembling those of sedimentary rocks. The features of these plutons suggest that they formed by partial or complete melting of preexisting sedimentary rocks.

Batholiths, the largest plutons, are great irregular masses of coarse-grained igneous rock that by definition cover at least 100 km^2 (see Figure 5.7). The rest of the plutons, similar but smaller, are called **stocks.** Both batholiths and stocks are **discordant intrusions;** that is, they cut across the layers of the country rock that they intrude.

Batholiths are found in the cores of tectonically deformed mountain belts. Accumulating geological field evidence shows that batholiths are thick, horizontal, sheetlike or lobate bodies extending from a funnel-shaped central region. Their bottoms may extend 10 to 15 km deep, and a few are estimated to go even deeper. The coarse grains of batholiths result from slow cooling at great depths.

Sills and Dikes

Sills and dikes are similar to plutons in many ways, but they are smaller and have a different relationship to the layering of the surrounding intruded rock. A **sill** is a tabular, sheetlike body formed by the injection of magma between

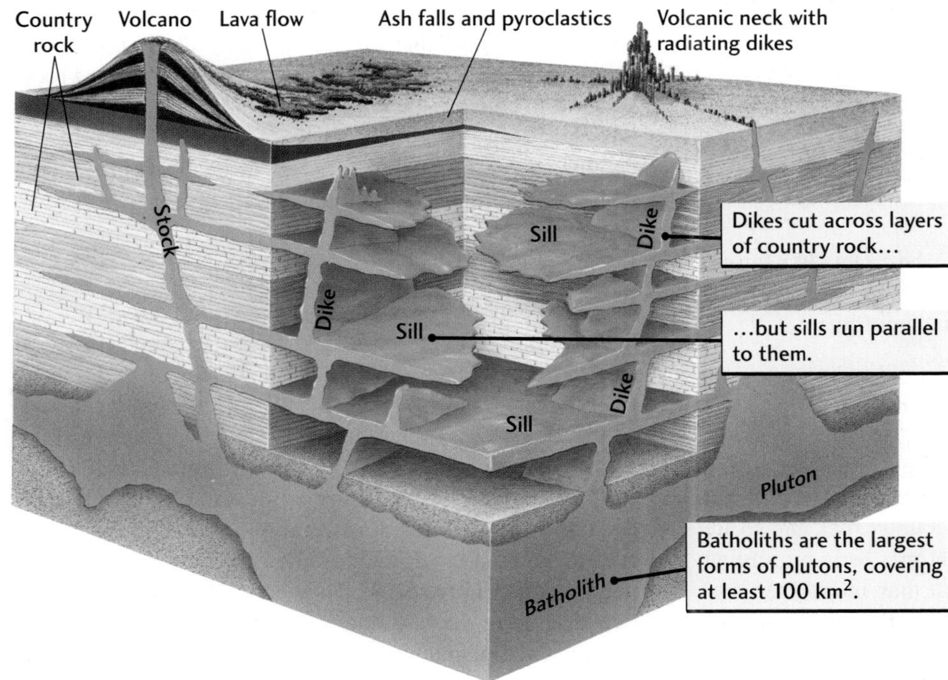

Figure 5.7 Basic extrusive and intrusive igneous structures. Notice that dikes cut across layers of country rock, but sills run parallel to them. Batholiths are the largest forms of plutons.

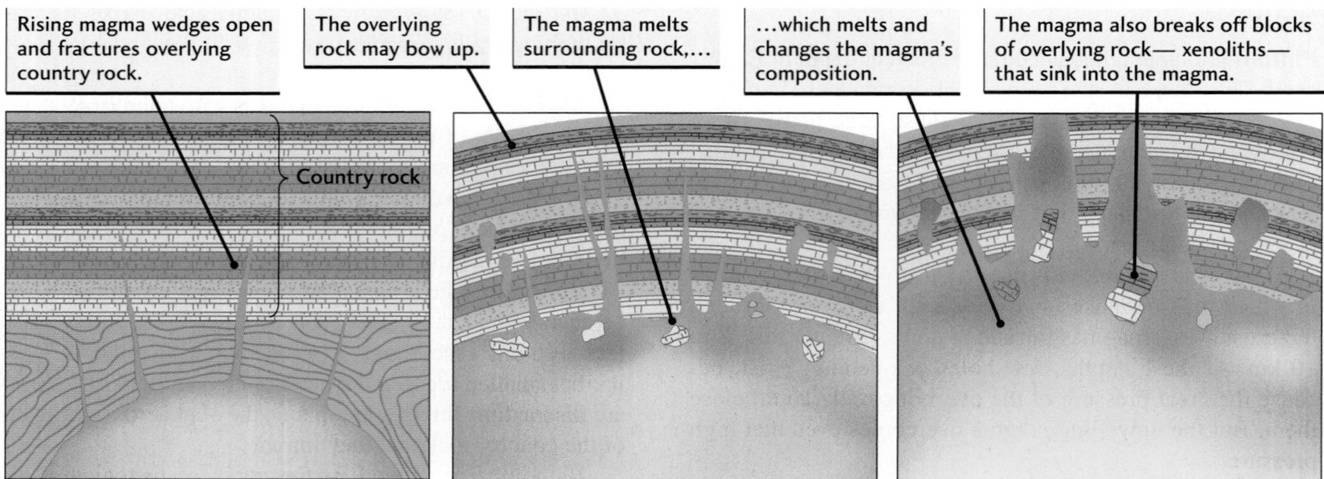

Figure 5.8 Magmas make their way into country rock in three basic ways: by invading cracks and wedging open overlying rock, by breaking off rock, and by melting surrounding rock. Pieces of broken-off country rock, called xenoliths, can become completely dissolved in the magma. If many xenoliths are dissolved and the country rock differs in composition from the magma, the composition of the magma will change.

parallel layers of preexisting bedded rock (**Figure 5.9**). Sills are **concordant intrusions;** that is, their boundaries lie parallel to these layers, whether or not the layers are horizontal. Sills range in thickness from a single centimeter to hundreds of meters, and they can extend over considerable areas. Figure 5.9a shows a large sill at Big Bend National Park in Texas. The 300-m-thick Palisades intrusion (see Figure Story 5.5) is another large sill.

Sills may superficially resemble layers of lava flows and pyroclastic material, but they differ from these layers in four ways:

1. They lack the ropy, blocky, and vesicle-filled structures that characterize many volcanic rocks (see Chapter 6).

2. They are more coarsely grained than volcanics because the sills have cooled more slowly.

3. Rocks above and below sills show the effects of heating: their color may have been changed or they may have been mineralogically altered by contact metamorphism.

4. Many lava flows overlie weathered older flows or soils formed between successive flows; sills do not.

Dikes are the major route of magma transport in the crust. They are like sills in being tabular igneous bodies, but dikes cut across layers of bedding in country rock and so are discordant (Figure 5.9b). Dikes sometimes form by forcing open fractures that formed earlier, but they more often create channels through new cracks opened by the pressure of magmatic injection. Some individual dikes can be followed in the field for tens of kilometers. Their widths vary from many meters to a few centimeters. In some dikes, xenoliths provide good evidence of disruption of the surrounding rock during the intrusion process. Dikes rarely exist alone; more typically, swarms of hundreds or thousands of dikes are found in a region that has been deformed by a large igneous intrusion.

The texture of dikes and sills varies. Many are coarsely crystalline, with an appearance typical of intrusive rocks. Many others are finely grained and look much more like volcanic rocks. Because texture corresponds to the rate of cooling, we know that the fine-grained dikes and sills invaded country rock nearer the Earth's surface, where rocks are cold in comparison with intrusions. Their fine texture is the result of fast cooling. The coarse-grained ones formed at depths of many kilometers and invaded warmer rocks whose temperatures were much closer to their own.

Veins

Veins are deposits of minerals found within a rock fracture that are foreign to the host rock. Irregular pencil- or sheet-shaped veins branch off the tops and sides of many intrusive bodies. They may be a few millimeters to several meters across, and they tend to be tens of meters to kilometers long or wide. The well-known Mother Lode of the Gold Rush of 1849 in California is a vein of quartz bearing crystals of gold. Veins of extremely coarse-grained granite cutting across a much finer grained country rock are called **pegmatites** (**Figure 5.10**). They crystallized from a water-rich magma in the late stages of solidification. Pegmatites provide ores of many rare elements, such as lithium and beryllium.

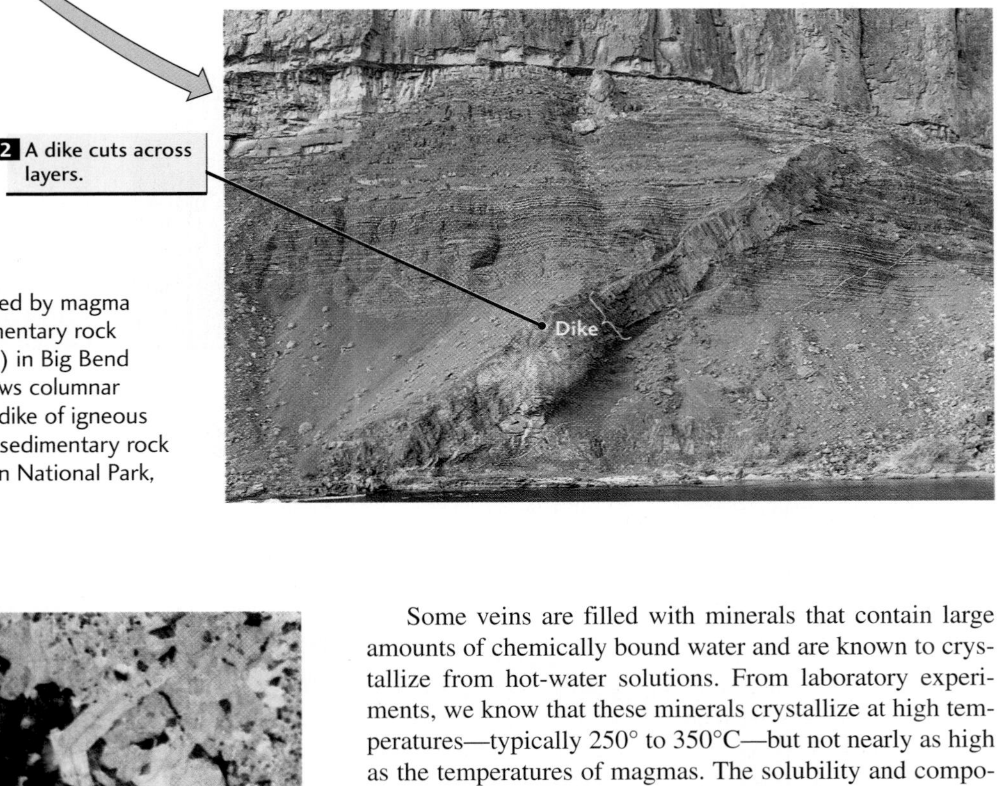

1 A sill runs parallel to country rock layers.

Sill

2 A dike cuts across layers.

Dike

(a)

(b)

Figure 5.9 (a) A dark sill formed by magma intruded between layers of sedimentary rock (bedded rock at top and bottom) in Big Bend National Park, Texas. The sill shows columnar jointing. [Tom Bean/DRK.] (b) A dike of igneous rock (dark) intruded into shaley sedimentary rock (reddish brown) in Grand Canyon National Park, Arizona. [Tom Bean/DRK.]

Figure 5.10 A granite pegmatite dike. The center of the dike displays the coarse crystallinity associated with slow cooling. The finer crystals along the boundaries of the dike cooled more rapidly. [Martin Miller.]

Some veins are filled with minerals that contain large amounts of chemically bound water and are known to crystallize from hot-water solutions. From laboratory experiments, we know that these minerals crystallize at high temperatures—typically 250° to 350°C—but not nearly as high as the temperatures of magmas. The solubility and composition of the minerals in these **hydrothermal veins** indicate that abundant water was present as the veins formed. Some of the water may have come from the magma itself, but some may be from underground water in the cracks and pore spaces of the intruded rocks. Groundwaters originate as rainwater seeps into the soil and surface rocks. Hydrothermal veins are abundant along mid-ocean ridges. In these areas, seawater infiltrates cracks in basalt and circulates down into hotter regions of the basalt ridge, emerging at hot vents on the seafloor in the rift valley between the spreading plates. Hydrothermal processes at mid-ocean ridges are examined in more detail in Chapter 6.

Igneous Activity and Plate Tectonics

Since the advent of the theory of plate tectonics in the 1960s, geologists have been trying to fit the facts and theories of igneous rock formation into its framework. We noted that batholiths, for example, are found in the cores of many mountain ranges. These mountain ranges were formed by the convergence of two plates. This observation implies a connection between plutonism and the mountain-building process and between both of them and plate tectonics—the forces responsible for plate movements.

Laboratory experiments have established the temperatures and pressures at which different kinds of rock melt, and this information gives us some idea of where melting may take place. For example, mixtures of sedimentary rocks, because of their composition and water content, melt at temperatures several hundred degrees lower than the melting point of basalt. This information leads us to expect that basalt starts to melt near the base of the crust in tectonically active regions of the upper mantle and that sedimentary rocks melt at shallower depths than basalt. The geometry of plate motions is the link that we need to tie tectonic activity and rock composition to melting (**Figure 5.11**). Two types of plate boundaries are associated with magma

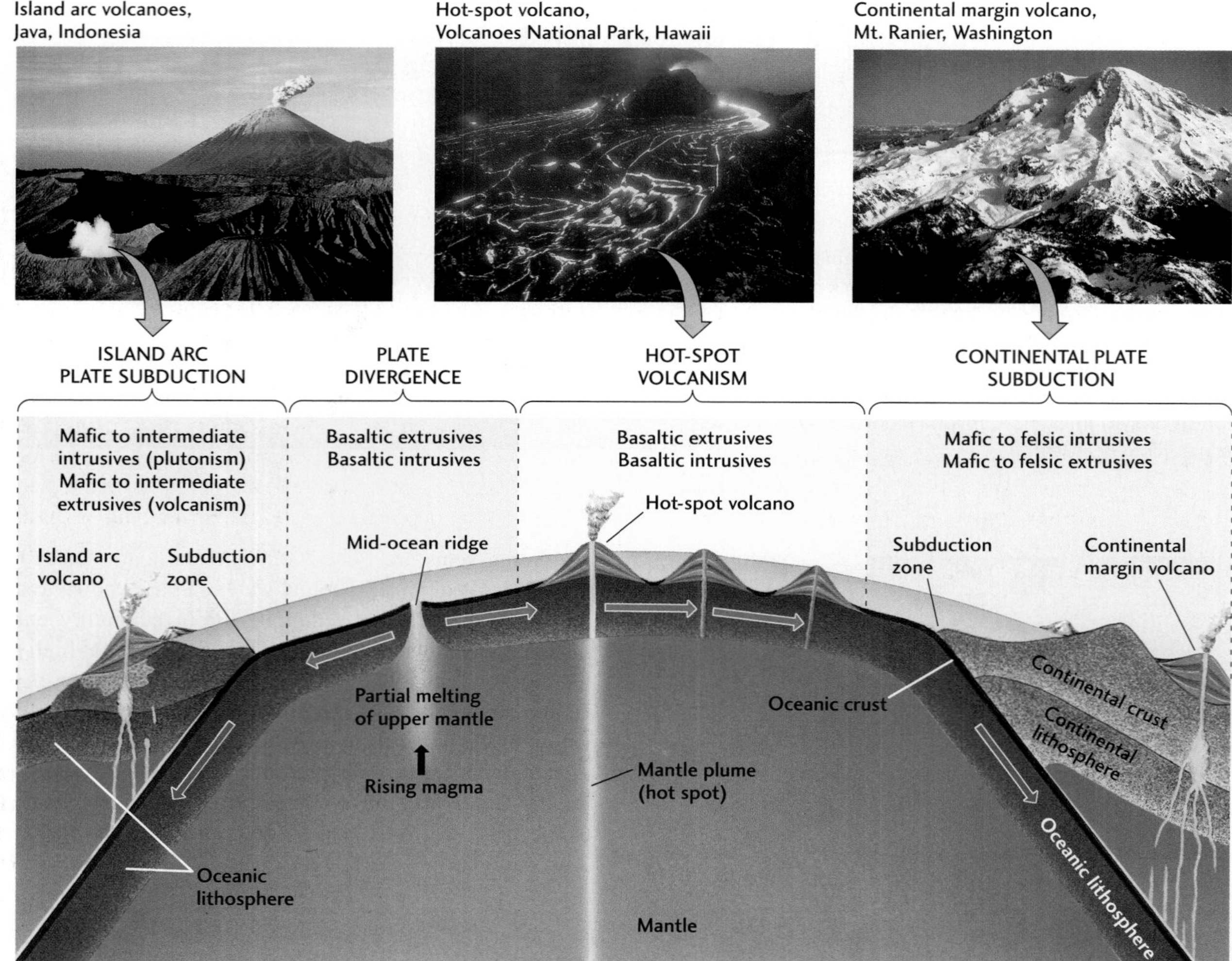

Figure 5.11 Magma forms under conditions that are strongly connected to movements of lithospheric plates. These movements control where rocks of the crust and upper mantle melt and whether they will be intruded or extruded. [*left to right:* Mark Lewis/Stone/Getty Images; G. Brad Lewis/Stone/Getty Images; Pat O'Hara/DRK.]

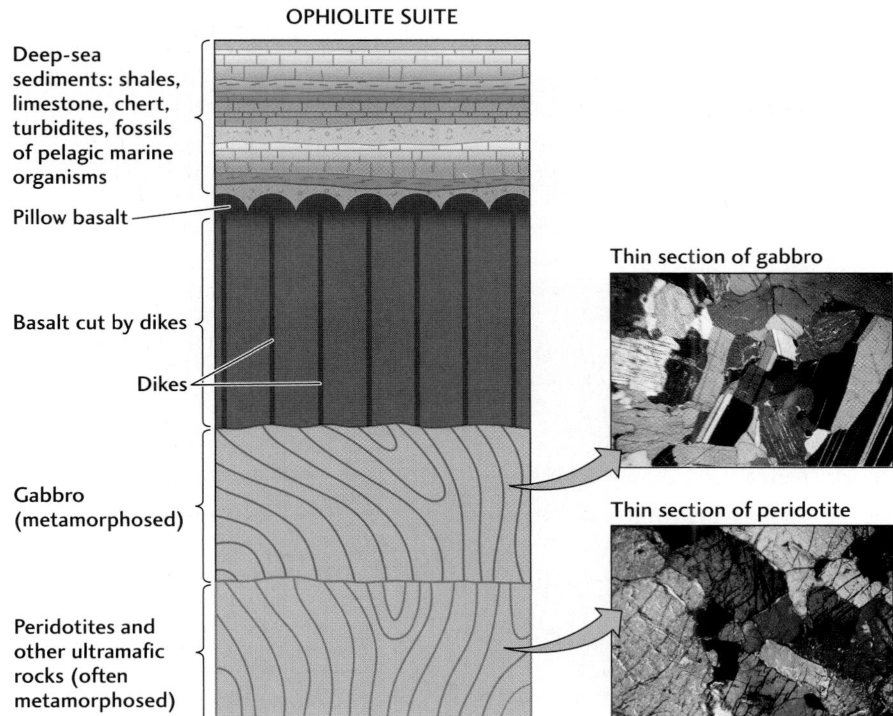

Figure 5.12 Idealized section of an ophiolite suite. The combination of deep-sea sediments, submarine pillow lavas, sheeted basaltic dikes, and mafic igneous intrusions indicates a deep-sea origin. Ophiolites are fragments of ocean lithosphere emplaced on a continent as a result of plate collisions. Peridotite—a dominant rock in the mantle—undergoes decompression melting, to form gabbro, which is then erupted to form volcanic pillow basalt (see Figure 5.13). [Photos courtesy of T. L. Grove.]

formation: mid-ocean ridges, where the divergence of two plates causes the seafloor to spread, and subduction zones, where one plate dives beneath another. Mantle plumes, though not associated with plate boundaries, are also the result of partial melting and form at or near the core-mantle boundary deep within Earth's interior (see Chapter 6).

The significant site of igneous rock formation is the globally extensive mid-ocean ridge network. Throughout the great length of this network, basaltic magmas derived from decompression melting of the mantle well up along rising convection currents. Magma is extruded as lavas, fed by magma chambers below the ridge axis. At the same time, gabbroic intrusions are emplaced at depth.

Spreading Centers as Magmatic Geosystems

Before the advent of plate tectonics, geologists were puzzled by unusual assemblages of rocks that were characteristic of the seafloor but were found on land. Known as **ophiolite suites,** these assemblages consist of deep-sea sediments, submarine basaltic lavas, and mafic igneous intrusions (**Figure 5.12**). Using data gathered from deep-diving submarines, dredging, deep-sea drilling, and seismic exploration, geologists can now explain these rocks as fragments of oceanic crust that were transported by seafloor spreading and then raised above sea level and thrust onto a continent in a later episode of plate collision. On some of the more complete ophiolite sequences preserved on land, we can literally walk across rocks that used to lie along the boundary between Earth's oceanic crust and mantle. Ocean drilling has penetrated to the gabbro layer of the seafloor but not to the crust-mantle boundary below. And sound-wave profiles have found several small magma chambers similar to the one shown in **Figure 5.13**.

How does seafloor spreading work as a magmatic geosystem? We can think of this system as a huge machine that processes mantle material to produce oceanic crust. Figure 5.13 is a highly schematic and simplified representation of what may be happening, based in part on studies of ophiolites found on land and information gleaned from ocean drilling and sound-wave profiling of the kind shown in Feature 2.1 (page 37).

Input Material: Peridotite in the Mantle The raw material fed into the machinery of seafloor spreading comes from the asthenosphere of the convecting mantle. The dominant rock type in the asthenosphere is peridotite. Geologists estimate that the mineral composition of the average peridotite in the mantle is chiefly olivine, with smaller amounts of pyroxene and garnet. The temperatures in the asthenosphere are sufficiently hot to melt a small fraction of this peridotite (less than 1 percent), but not high enough to generate substantial volumes of magma.

Process: Decompression Melting The seafloor-spreading machine generates magma from peridotite in the mantle by the process of decompression melting. Recall from earlier in this chapter that the melting temperature of a mineral

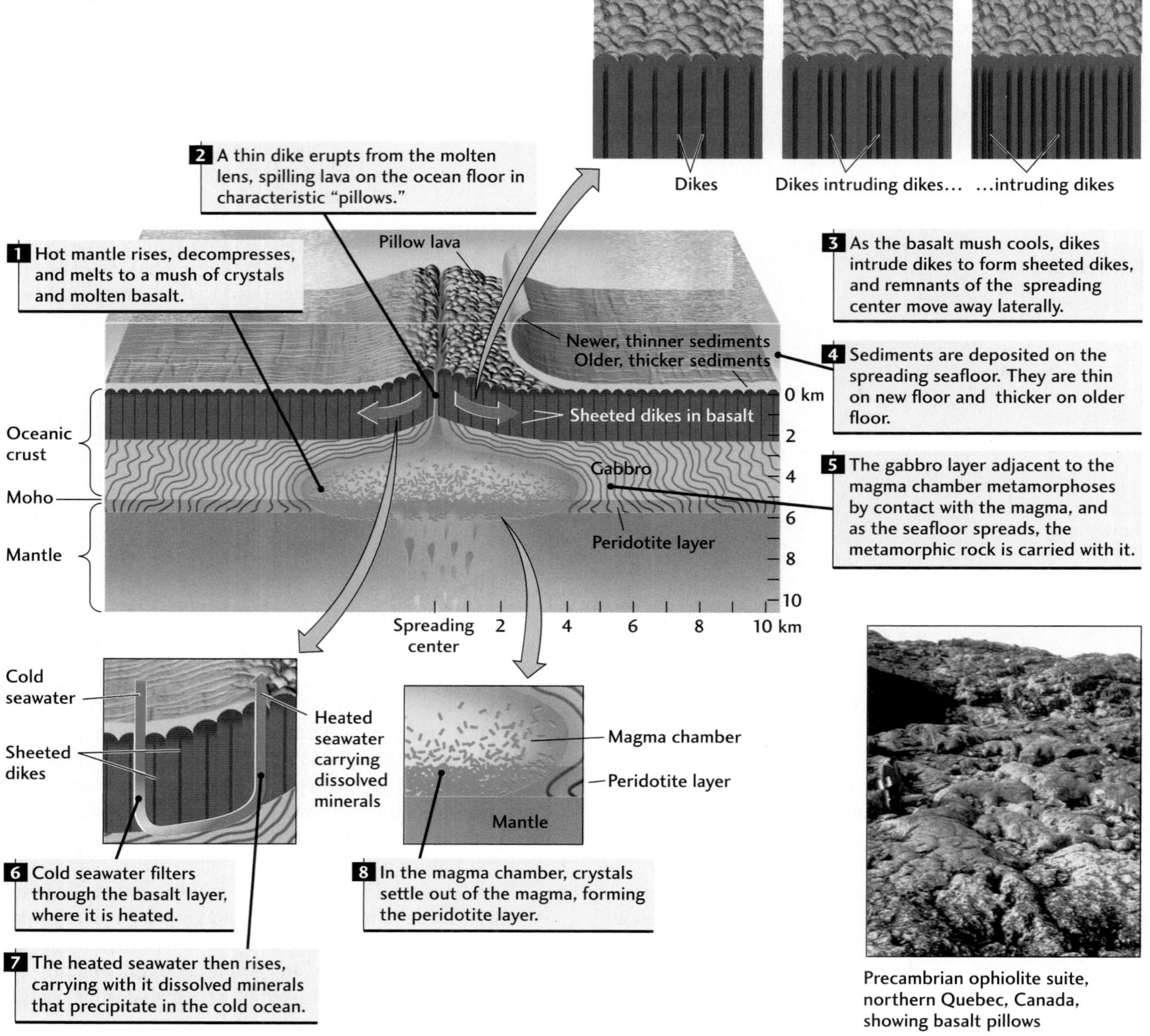

Figure 5.13 A schematic representation of the magmatic geosystem at seafloor spreading centers. Inset: Precambrian ophiolite suite. The basalt pillows shown here are from the volcanic layer of the suite. [M. St. Onge/Geological Survey of Canada.]

depends on its pressure as well as its composition: decreasing the pressure will generally decrease a mineral's melting temperature. Consequently, if a mineral is near its melting point and the pressure on it is lowered while its temperature is kept constant, the mineral will melt.

As the plates pull apart, the partially molten peridotites are sucked inward and upward toward the spreading centers. Because the peridotites rise too fast to cool off, the decrease in pressure melts a large fraction of the rock (up to 15 percent). The buoyancy of the melt causes it to rise faster than the denser surrounding rocks, separating the liquid from the remaining crystal mush to produce large volumes of magma.

Output Material: Oceanic Crust Plus Mantle Lithosphere The peridotites subjected to this process do not melt evenly; the garnet and pyroxene minerals melt more than the olivine. For this reason, the magma generated by decompression melting is not peridotitic in composition; rather, it is enriched in silica and iron and has the same composition as basalt (see Figure 5.13).

This basaltic melt accumulates in a magma chamber below the mid-ocean ridge crest, from which it separates into three layers:

1. Some magma rises through the narrow, tensional cracks that open where the plates separate and erupts into the ocean, forming the basaltic pillow lavas that cover the seafloor (see Figure 5.13).

2. Some magma freezes in the tensional cracks as vertical, sheeted dikes of gabbro (the intrusive form of basalt).

3. The remaining magma freezes as "massive gabbros" as the underlying magma chamber is pulled apart by seafloor spreading.

These igneous units—pillow lavas, sheeted dikes, and massive gabbros—are the basic layers of the crust that geologists have found throughout the world's oceans.

The seafloor-spreading machine also emplaces another layer beneath this oceanic crust: the residual peridotite from which the basaltic magma was originally derived. Geologists consider this layer to be part of the mantle, but its composition is different from that of the convecting asthenosphere. In particular, the extraction of basaltic melt makes the residual peridotite richer in the mineral olivine and stronger than ordinary mantle material. Geologists now believe it is this olivine-rich layer at the top of the mantle that gives the oceanic plates their great rigidity.

A thin blanket of deep-sea sediments begins to cover the newly formed ocean crust. As the seafloor spreads, the layers of sediments, lavas, dikes, and gabbros are transported away from the mid-ocean ridge, where this characteristic sequence of rocks that make up the oceanic crust is assembled—almost like a production line.

Rocks Formed at Subduction Zones

Other kinds of magmas underlie regions in which volcanoes are highly concentrated, such as the Andes Mountains of South America and the Aleutian Islands of Alaska. Both of these regions resulted from the sinking of one plate under another. Subduction zones are major sites of rock melting (**Figure 5.14**), and magmas of varying composition are

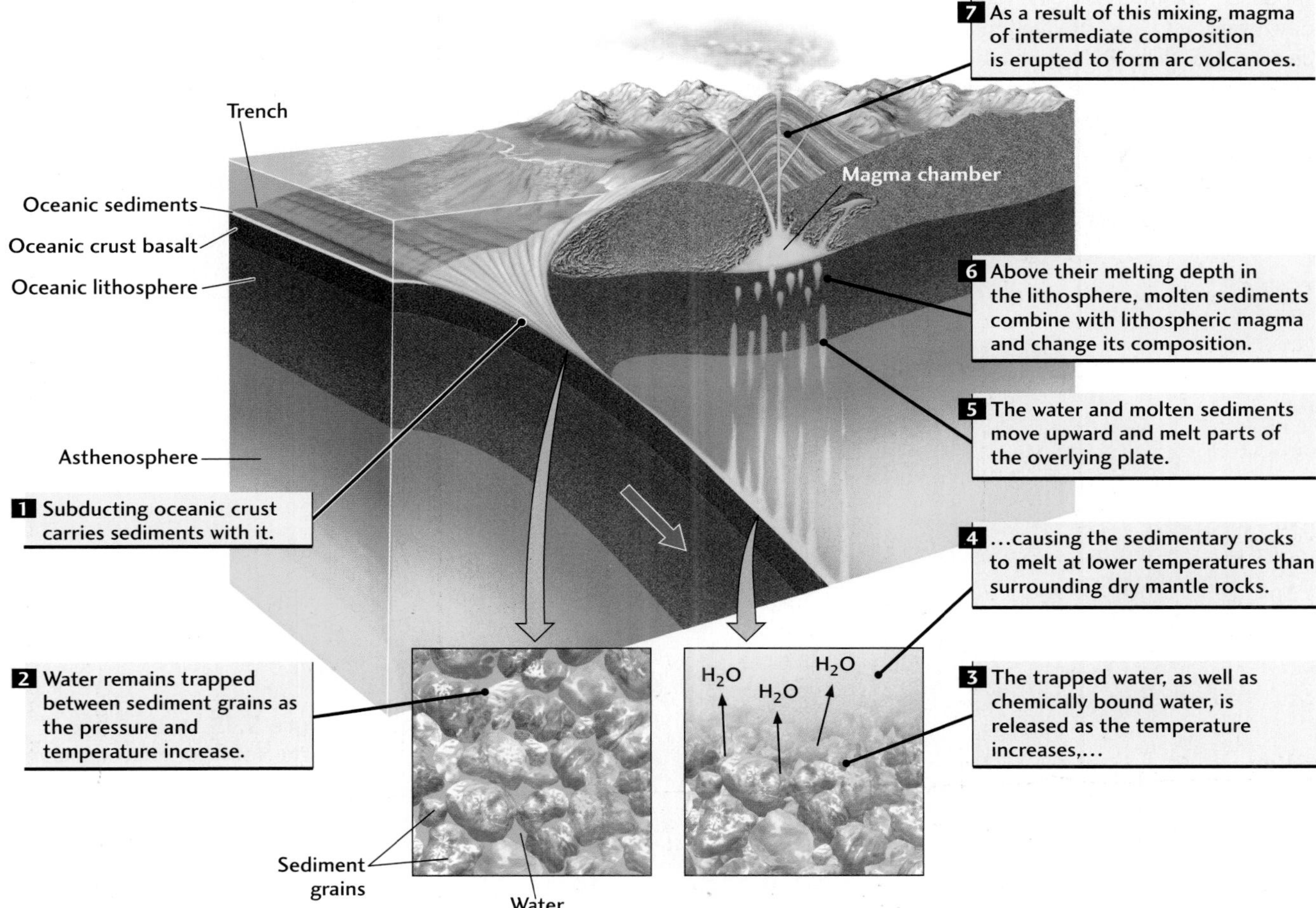

Figure 5.14 This diagrammatic cross section of a subduction zone shows an irregular magma chamber formed by the ascent of a partial melt.

formed, depending on how much and what kind of materials from the seafloor are subducted. These materials include water, mixtures of seafloor sediments, and mixtures of basaltic and felsic crust. These chemical variations are clues that the volcanic geosystems of convergent plate boundaries operate differently from those of divergent boundaries. When an ocean plate collides with and overrides another ocean plate, several complex processes are set in motion (**Figure 5.15**).

The basic mechanism is **fluid-induced melting,** rather than decompression melting. The fluid is primarily water. Before the oceanic lithosphere is subducted at a convergent boundary, a lot of water has been incorporated into its outer layers. We have already discussed one process responsible for this—hydrothermal activity during the formation of the lithosphere. Some of the seawater circulating through the crust near the spreading center reacts with the basalt to form new minerals with water bound into their structures. In addition, shales, the most abundant sedimentary rocks, are very high in clay minerals, which contain much water chemically bound in their crystal structure. As the lithosphere ages and moves across the ocean basin, sediments containing water are deposited on its surface. Some of the sediments get scraped off at the deep-ocean trench where the plate subducts, but much of this water-laden material is carried downward into the subduction zone. As the pressure increases, the water is "squeezed out" of the minerals in the outer layers of the descending slab and rises buoyantly into the mantle wedge above the slab. At moderate depths of about 5 km, some of this water is released by metamorphic chemical reactions as the temperature increases to about 150°C and basalt is converted to *amphibolite*, which is composed of ampibole and plagioclase feldspar (see Chapter 9). As other chemical reactions take place, additional water is released at greater depths, ranging from 10 to 20 km. Finally, at depths of greater than 100 km, the subducted slab undergoes an additional metamorphic transition, induced by the increased pressure, in which amphibolite is converted to *eclogite*, which is composed of pyroxene and garnet (see Chapter 9). At the same time, the slab encounters temperatures in the range of 1200° to 1500°C. The increase in both pressure and temperature in the subducting slab releases all the remaining water in addition to other materials. The released water induces peridotite, the major constituent in the wedge of mantle above the subducted plate, to melt.

Because water lowers the melting temperature of rock, as discussed earlier in this chapter, it induces melting in the mantle wedge, generating magmas that rise to form magma chambers in the overlying crust. Most of the mafic magma accumulates at the base of the crust of the overriding plate, and some of it intrudes into the crust. The magmas produced by this type of fluid-induced melting are essentially basaltic in composition, although their chemistry is different from (more variable than) that of mid-ocean ridge basalts. The composition of the magmas is further altered during their residence in the crust. Within the magma chambers, the process of fractional crystallization increases their silica content, producing eruptions of andesitic lavas. In cases where the overlying plate is continental, the heat from the magmas can melt the felsic rocks in the crust, forming magmas with even higher silica contents, such as dacitic and rhyolitic composi-

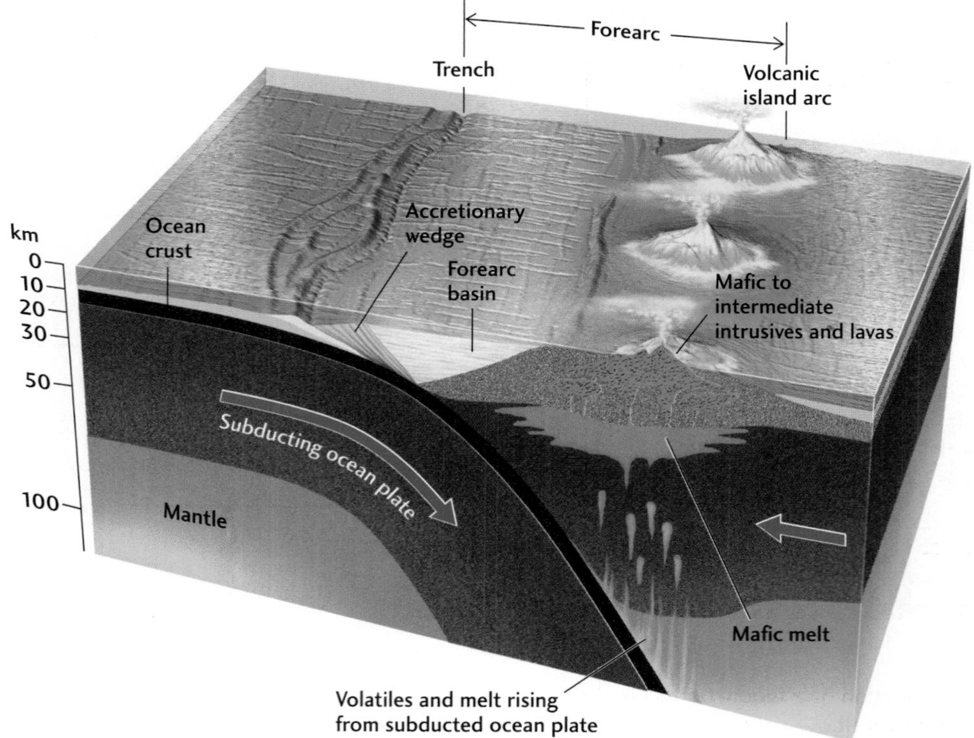

Figure 5.15 A schematic diagram of fluid-induced melting. Subducting oceanic lithosphere contains water trapped within sediments. As these sediments descend, they heat up, expelling their water as vapor. This vapor helps induce melting of the mantle, in addition to the descending slab.

tions (see Table 5.2). The contribution of slab fluids to the magma is inferred because trace elements known to be present in ocean crust and sediments are found in the magma.

Mantle Plumes

Basalts similar to those produced at mid-ocean ridges are found in thick accumulations over some parts of continents distant from plate boundaries. In the states of Washington, Oregon, and Idaho, the Columbia and Snake rivers flow over a great area covered by this kind of basalt, which solidified from lavas that flowed out millions of years ago. Large quantities of basalt also are erupted in isolated volcanic islands far from plate boundaries, such as the Hawaiian Islands. In such places, slender, pencil-like plumes of hot mantle rise from deep in the Earth, perhaps as deep as the core-mantle boundary. Mantle plumes that reach the surface, most of them far from plate boundaries, form the "hot spots" of the Earth and are responsible for the outpouring of huge quantities of basalt. The basalt is produced during decompression melting of the mantle. Mantle plumes and hot spots are discussed in more detail in Chapter 6.

To sum up, basaltic magmas form in the upper mantle beneath mid-ocean ridges and from plumes of deep origin that give rise to interplate and intraplate hot spots. Magmas of varying composition form over subduction zones, depending on how much felsic material and water are incorporated in the mantle wedge rocks overlying the subduction zone.

SUMMARY

How are igneous rocks classified? All igneous rocks can be divided into two broad textural classes: (1) the coarsely crystalline rocks, which are intrusive and therefore cooled slowly, and (2) the finely crystalline ones, which are extrusive and cooled rapidly. Within each of these broad categories, the rocks are classified as felsic, mafic, or intermediate on the chemical basis of their silica content or by the mineralogical equivalent, the proportions of lighter-colored, felsic minerals and darker, mafic minerals.

How and where do magmas form? Magmas form at places in the lower crust and mantle where temperatures and pressures are high enough for at least partial melting of water-containing rock. Basalt can partially melt in the upper mantle, where convection currents bring hot rock upward at mid-ocean ridges. Mixtures of basalt and other igneous rocks with sedimentary rocks, which contain significant quantities of water, have lower melting points than dry igneous rocks. Thus different source rocks may be melted at different temperatures and thereby affect magma compositions.

How does magmatic differentiation account for the variety of igneous rocks? If a melt underwent fractional crystallization because the crystals were separated and therefore did not react with the melt, the final rocks may be more silicic than the earlier, more mafic crystals. Fractional crystallization can produce mafic igneous rocks from earlier stages of crystallization and differentiation, and felsic rocks from later stages, but it does not adequately explain the abundance of granite. Magmatic differentiation of basalt does not explain the composition and abundance of igneous rocks. Different kinds of igneous rocks may be produced by variations in the compositions of magmas caused by the melting of different mixtures of sedimentary and other rocks and by mixing of magmas.

What are the forms of intrusive igneous rocks? Large igneous bodies are plutons. The largest plutons are batholiths, which are thick tabular masses with a central funnel. Stocks are smaller plutons. Less massive than plutons are sills, which are concordant with the intruded rock, lying parallel to its layering, and dikes, which are discordant with the layering, cutting across it. Hydrothermal veins form where water is abundant, either in the magma or in surrounding country rock.

How are igneous rocks related to plate tectonics? The two major sites of magmatic activity are mid-ocean ridges, where basalt wells up from the upper mantle, and subduction zones, where a series of differentiated magmas produces both extrusives and intrusives in island or continental volcanic arcs as the subducting oceanic lithosphere moves down into the deep crust and upper mantle.

Key Terms and Concepts

andesite (p. 94)
basalt (p. 93)
batholith (p. 101)
concordant intrusion (p. 102)
country rock (p. 92)
dacite (p. 94)
decompression melting (p. 96)
dike (p. 102)
diorite (p. 94)
discordant intrusion (p. 101)
extrusive igneous rock (p. 92)
felsic rock (p. 94)
fluid-induced melting (p. 108)
fractional crystallization (p. 97)
gabbro (p. 94)
granite (p. 94)
granodiorite (p. 94)
hydrothermal vein (p. 103)
igneous rock (p. 89)
intermediate igneous rock (p. 94)
intrusive igneous rock (p. 91)
mafic rock (p. 94)
magma chamber (p. 97)
magmatic differentiation (p. 97)

obsidian (p. 92)
ophiolite suite (p. 105)
partial melting (p. 95)
pegmatite (p. 102)
peridotite (p. 94)
pluton (p. 101)
porphyry (p. 92)
pumice (p. 92)
pyroclast (p. 92)
rhyolite (p. 94)
sill (p. 101)
stock (p. 101)
tuff (p. 92)
ultramafic rock (p. 94)
vein (p. 102)
viscosity (p. 94)
volcanic ash (p. 92)

Exercises

This icon indicates that there is an animation available on the Web site that may assist you in answering a question.

1. Why are intrusive igneous rocks coarsely crystalline and extrusive rocks finely crystalline?

2. What kinds of minerals would you find in a mafic igneous rock?

3. What kinds of igneous rock contain quartz?

4. Name two intrusive igneous rocks with a higher silica content than that of gabbro.

5. What is the difference between a magma formed by fractional crystallization and one formed by ordinary cooling?

6. How does fractional crystallization lead to magmatic differentiation?

7. Where in the crust, mantle, or core would you find a partial melt of basaltic composition?

8. In which plate tectonic settings would you expect magmas to form?

9. Why do melts migrate upward?

10. Where on the ocean floor would you find basaltic magmas being extruded?

11. Much of Earth's crustal area, and nearly all of its mantle, are composed of basaltic or ultramafic rocks. Why are granitic and andesitic rocks as plentiful as they are on Earth? Where do the materials that constitute these rocks come from?

Thought Questions

This icon indicates that there is an animation available on the Web site that may assist you in answering a question.

1. How would you classify a coarse-grained igneous rock that contains about 50 percent pyroxene and 50 percent olivine?

2. What kind of rock would contain some plagioclase feldspar crystals about 5 mm long "floating" in a dark gray matrix of crystals of less than 1 mm?

3. What differences in crystal size might you expect to find between two sills, one intruded at a depth of about 12 km, where the country rock was very hot, and the other at a depth of 0.5 km, where the country rock was moderately warm?

4. If you were to drill a hole through the crust of a mid-ocean ridge, what intrusive or extrusive igneous rocks might you expect to encounter at or near the surface? What intrusive or extrusive igneous rocks might you expect at the base of the crust?

5. Assume that a magma with a certain ratio of calcium to sodium starts to crystallize. If fractional crystallization occurs during the solidification process, will the plagioclase feldspars formed after complete crystallization have the same ratio of calcium to sodium that characterized the magma?

6. In what way is a zoned pluton evidence of fractional crystallization?

7. Why are plutons more likely than dikes to show the effects of fractional crystallization?

8. What might be the origin of a rock composed almost entirely of olivine?

9. What processes create the unequal sizes of crystals in porphyries?

10. Water is abundant in the sedimentary rocks and oceanic crust of subduction zones. How would the water affect melting in these zones?

11. Much of Earth's crustal area, and nearly all of its mantle, are composed of basaltic or ultramafic rocks. Why are granitic and andesitic rocks as plentiful as they are on Earth? Where do the materials that constitute these rocks come from?

Short-Term Project

Reading the Geologic Story of Igneous Rocks: Solids from Melts

All rocks tell a story. The story is deciphered from various clues: texture, mineral and chemical composition, association with other rocks, and geologic setting. With careful analysis and interpretation of the rock record, the geologic history of a region can be deciphered—we can read the rock record like words on a page.

Since igneous rocks are formed from a melt, called *magma,* crystals of minerals can grow in the melt like ice crystals in freezing water. As the melt cools, it becomes a crystal slush and finally transforms into a solid with interlocking crystals. The size of the crystals largely depends on the cooling rate. Typically, volcanic rocks cool relatively quickly on Earth's surface and therefore contain smaller crystals. Plutonic rocks solidify slowly within the crust and contain larger crystals. Igneous rocks may also contain gas bubbles, inclusions of other rock fragments, glass, and a fine-grained matrix of ash and pumice.

Magma generation on Earth is largely restricted to active tectonic plate boundaries and hot spots. Processes and physical conditions associated with plate boundaries cause rocks in the mantle and crust to melt. The composition of the melt depends on the rocks from which it was generated and what happens in the magma chamber before it totally solidifies. Therefore, magma composition is strongly linked to where within Earth the melt formed and, in turn, to the type of active plate boundary. For example, decompression of rising hot bodies of soft, plastic ultramafic rock within the mantle is thought to produce partial melts of basaltic composition beneath hot spots and divergent plate boundaries. Essentially, no two igneous rock bodies have exactly the same texture and composition. In fact, the characteristics of igneous rock often vary within one rock body—hence, a pluton because of all the variables that can affect rock composition and texture.

Information on and images of various types of igneous rock are available on the text's Web site: www.whfreeman.com/understandingearth. The information provided is essentially the same information that a professional geologist would gather from field and laboratory work. *Given these observations, decipher the story each rock has to tell.*

Suggested Readings

Barker, D. S. 1983. *Igneous Rocks.* Englewood Cliffs, N.J.: Prentice Hall.

Blatt, H., and R. J. Tracy. 1996. *Petrology: Igneous, Sedimentary, and Metamorphic,* 2d ed. New York: W. H. Freeman.

Coffin, M. F., and O. Eldholm. 1993. Large igneous provinces. *Scientific American* (June): 42–49.

Philpotts, A. R. 1990. *Igneous and Metamorphic Petrology.* Englewood Cliffs, N.J.: Prentice Hall.

Raymond, L. A. 1995. *Petrology.* Dubuque, Iowa: Wm. C. Brown.

Mount Rainier looming over Tacoma, Washington. Our most dangerous volcano!
[John McAnulty/Corbis.]

CHAPTER

6

Volcanism

"What a book a devil's chaplain might write on the clumsy, wasteful, blundering, low, and horribly cruel works of nature."

CHARLES DARWIN

Imagine a volcanic eruption in which an area of ground the size of New York City collapses, a region larger than Vermont is buried under hot ash that snuffs out all life, and fields 1000 or 2000 km away are blanketed by 20 cm of ash and rendered infertile. Imagine that the volcanic dust thrown high into the stratosphere dims the Sun for a year or two, so that there are no summers. Unbelievable? Yet it has happened at least twice in what is now the United States: at Yellowstone in Wyoming 600,000 years ago and at Long Valley, California, 760,000 years ago. This was long before humans first reached North America 30,000 years ago, but not very long ago on the 4.5-billion-year geologic time scale. We know that these events took place because volcanic rocks formed by the eruptions have been identified and dated.

A large portion of Earth's crust, oceanic and continental, is made up of volcanic rock. Beginning as magma deep inside the Earth, volcanic rock serves as a kind of window through which we can dimly perceive Earth's interior.

In this chapter, we examine volcanism, the process by which magma from the interior of the Earth rises through the crust, emerges onto the surface as lava, and cools into hard volcanic rock. We will look at the major types of lava, eruptive styles and the landforms they create, and the kinds of environmental disruption that volcanoes can cause. We will see how plate tectonics and mantle convection can explain why the majority of volcanoes occur at plate boundaries and a few occur at "hot spots" within plates. We will give examples of how volcanoes interact with other components of the Earth system, particularly the atmosphere, the oceans, and the biosphere. Finally, we will consider ways to mitigate the destructive potential of volcanoes and benefit from their chemical riches and heat energy.

Ancient philosophers were awed by volcanoes and their fearsome eruptions of molten rock. In their efforts to explain volcanoes, they spun myths about a hot, hellish underworld below Earth's surface. Basically, they had the right idea. Modern researchers, using science rather than mythology, also see in volcanoes evidence of Earth's internal heat.

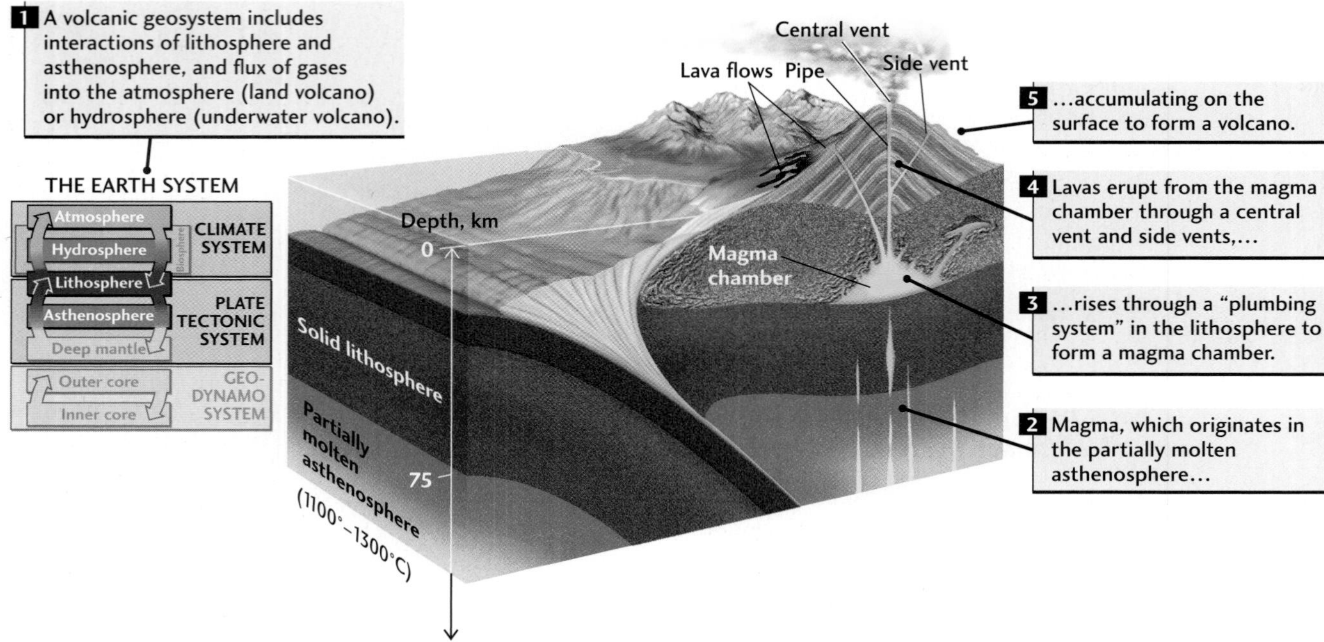

Figure 6.1 Simplified diagram of a volcanic geosystem.

Volcanoes as Geosystems

Temperature readings of rocks as far down as humans have drilled (about 10 km) show that the Earth does get hotter with depth. Geologists now believe that temperatures at depths of 100 km and more—within the asthenosphere—reach at least 1300°C, high enough for the rocks there to begin to melt. For this reason, they identify the asthenosphere as a main source of *magma,* the molten rock below Earth's surface that we call *lava* after it erupts. Sections of the solid lithosphere that ride above the asthenosphere may also melt to form magma.

Because magmas are liquids, they are less dense than the rocks that produced them. Therefore, as magma accumulates, it begins to rise buoyantly (float upward) through the lithosphere. In some places, the melt may find a path to the surface by fracturing the lithosphere along zones of weakness. In other places, the rising magma melts a path to the surface. Some of the magma eventually reaches the surface and erupts as lava. A **volcano** is a hill or mountain constructed from the accumulation of lava and other erupted materials.

Rocks, magmas, and the interactions needed to describe the entire sequence of events from melting to eruption, taken together, form a **volcanic geosystem. Figure 6.1** is a simplified diagram of a volcanic geosystem. Magmas rising through the lithosphere pool in a magma chamber, usually at shallow depths in the crust. This reservoir periodically empties to the surface through a pipelike conduit in repeated cycles of eruptions. Lava can also erupt from vertical cracks and other vents on the flanks of a volcano.

Volcanoes are important geosystems for three reasons: (1) volcanism is a fundamental tectonic process in the construction of Earth's crust; (2) volcanic eruptions are a major natural hazard for human society; and (3) lavas from volcanoes provide geologists with samples from which they can infer properties of Earth's interior. The complexity of volcanic geosystems is reflected in how these samples are chemically modified as they are generated and transported to the surface. As we saw in Chapter 5, only a very small fraction of the asthenosphere melts in the first place. A magma gains chemical components as it melts the surrounding rocks while rising through the lithosphere. It loses other components when crystals settle out in magma chambers. And its gaseous constituents escape to the atmosphere or ocean as it erupts at the surface. By accounting for these modifications, geologists can extract important information from lavas—clues to the chemical composition and physical state of the upper mantle. From preserved volcanic rock, they can also learn much about eruptions that occurred millions or even billions of years ago.

Volcanic Deposits

The chemical and mineralogical compositions of a lava have much to do with the way that it erupts and the kind of landforms it creates when it solidifies. The major types of lavas and the rocks that they form vary according to the magmas from which they were derived. In Chapter 5, we saw that igneous rocks and their precursor magmas can be divided into three major groups—felsic, intermediate, and mafic—on the basis of their chemical composition (see Table 5.2). Igneous rocks are further classified as intrusive (they cooled slowly below the surface and are coarse-grained as a result) or extrusive (they cooled quickly on the surface and are fine-grained). The major intrusive igneous rocks are granite (felsic), diorite (intermediate), and gabbro (mafic). The major extrusive counterparts are rhyolite (felsic), andesite (intermediate), and—most common of all—basalt (mafic). These classifications are summarized in Figure 5.4. With this framework in mind, let us examine the types of lava and how they flow and solidify.

Types of Lava

The several types of lava leave behind different landforms: volcanic mountains that vary in shape and solidified lava flows that vary in character. These variations result from differences in the chemical composition, gas content, and temperature of lavas. The higher the silica content and the lower the temperature, for example, the more viscous (more resistant to flow) the lava is and the more slowly it moves. The more gas a lava contains, the more violent its eruption is likely to be.

Basaltic Lavas Basaltic lava, dark in color, erupts at 1000° to 1200°C—close to the temperature of the upper mantle. Because of its high temperature and low silica content, basaltic lava is extremely fluid and can flow downhill fast and far. Streams flowing as fast as 100 km/hour have been observed, although velocities of a few kilometers per hour are more common. In 1938, two daring Russian volcanologists measured temperatures and collected gas samples while floating down a river of molten basalt on a raft of colder solidified lava. The surface temperature of the raft was 300°C, and the river temperature was 870°C. Lava streams that traveled more than 50 km from their source have been witnessed in historical times.

Basaltic lava flows vary according to the conditions under which they erupt. Important examples include

• *Flood Basalts* Highly fluid basaltic lava that erupts on flat terrain can spread out in thin sheets as a flood of lava. Successive flows often pile up into immense basaltic lava plateaus, called **flood basalts,** such as those at the great Columbia Plateau of Oregon and Washington (**Figure 6.2**).

• *Pahoehoe and Aa* Cooling basaltic lava flowing downhill falls into one of two categories, according to its surface form: pahoehoe (pronounced pa-ho´-ee-ho´-ee) or aa (ah-ah). **Figure 6.3** shows examples of both.

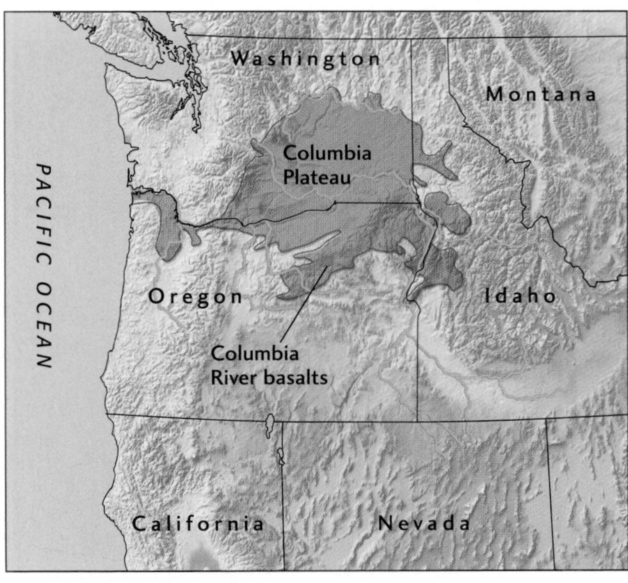

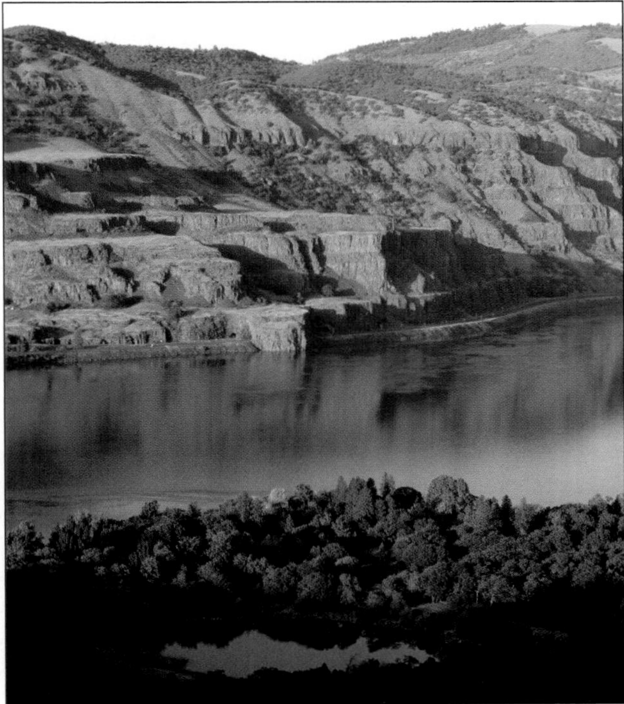

Figure 6.2 View of Columbia Plateau, Washington. Successive flows of flood basalts piled up to build this immense plateau, which covers a large area of Washington and Oregon. [Dave Schiefelbein.]

Pahoehoe (the word is Hawaiian for "ropy") forms when a highly fluid lava spreads in sheets and a thin, glassy elastic skin congeals on its surface as it cools. As the molten liquid continues to flow below the surface, the skin is dragged and twisted into coiled folds that resemble rope.

"*Aa*" is what the unwary exclaim after venturing barefoot onto lava that looks like clumps of moist,

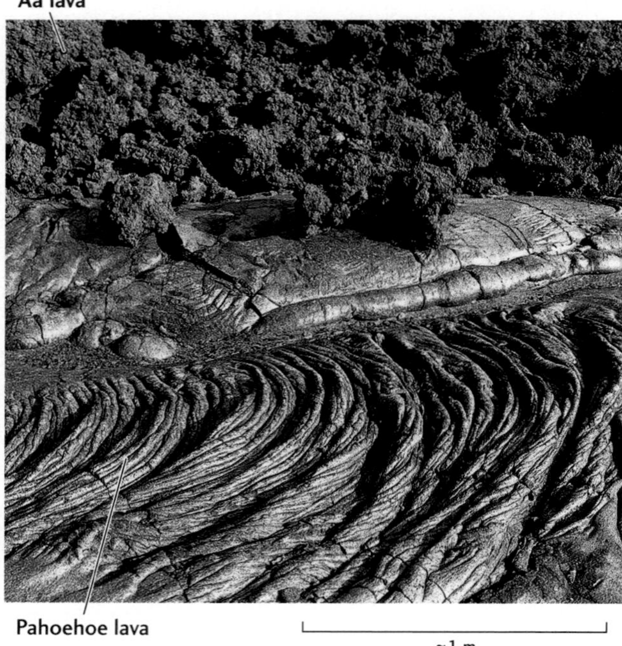

Figure 6.3 Two types of lava, ropy pahoehoe (*bottom*) and jagged blocks of aa (*top*). Mauna Loa volcano, Hawaii. [Kim Heacox/DRK.]

freshly plowed earth. Aa is lava that has lost its gases and consequently is more viscous than pahoehoe. Thus, it moves more slowly, which allows a thick skin to form. As the flow continues to move, the thick skin breaks into rough, jagged blocks. The blocks ride on the viscous, massive interior, piling up a steep front of angular boulders that advances like a tractor tread. Aa is truly treacherous to cross. A good pair of boots may last about a week on it, and the traveler or geologist can count on cut knees and elbows.

A single downhill basaltic flow commonly has the features of pahoehoe near its source, where the lava is still fluid and hot, and of aa farther downstream, where the flow's surface—having been exposed to cool air longer—has developed a thicker outer layer.

• *Pillow Lavas* A geologist who comes across **pillow lavas**—piles of ellipsoidal, pillowlike blocks of basalt about a meter wide—knows that they formed in an underwater eruption (see Figure 5.13) even if they are now on dry land. In fact, pillow lavas are an important indicator that a region was once under water. Geologist-divers have actually observed pillow lavas forming on the ocean floor off Hawaii. Tongues of molten basaltic lava develop a tough, plastic skin on contact with the cold ocean water. Because lava inside the skin cools more slowly, the pillow's interior develops a crystalline texture, whereas the quickly chilled skin solidifies to a crystal-less glass.

Rhyolitic Lavas Rhyolite, the most felsic lava, is light in color. It has a lower melting point than basalt and erupts at temperatures of 800° to 1000°C. It is much more viscous than basalt because of its lower temperature and higher silica content. Rhyolite moves 10 or more times more slowly than basalt, and, because it resists flow, it tends to pile up in thick, bulbous deposits.

Andesitic Lavas Andesite, with an intermediate silica content, has properties that fall between those of basalts and rhyolites.

Textures of Lavas

Lavas have other features that reflect the temperatures and pressures under which they formed. They can have a glassy or fine-grained texture, such as that of obsidian, if they cool quickly. Coarse-grained textures, such as those of volcanic tuffs, result if they cool slowly beneath the surface. They can also contain little bubbles, created when pressure falls suddenly as the lava rises and cools. Lava is typically charged with gas, like soda in an unopened bottle. When lava rises, the pressure on it decreases, just as the pressure on a soda drops when its bottle cap is removed. And just as the carbon dioxide in the soda creates bubbles when it is released, the water vapor and other dissolved gases escaping from lava create gas cavities, or *vesicles* (**Figure 6.4**). A frothy texture in solidified lava provides geologists with details of the rock's volcanic origins. An extremely vesicular volcanic rock, usually rhyolitic in composition, is called *pumice.* Some pumice has so much void space that it is light enough to float on water.

Pyroclastic Deposits

Water and gases in magmas can have even more dramatic effects on eruptive styles. Before a magma erupts, the con-

Figure 6.4 Vesicular basalt sample. [Glenn Oliver/Visuals Unlimited.]

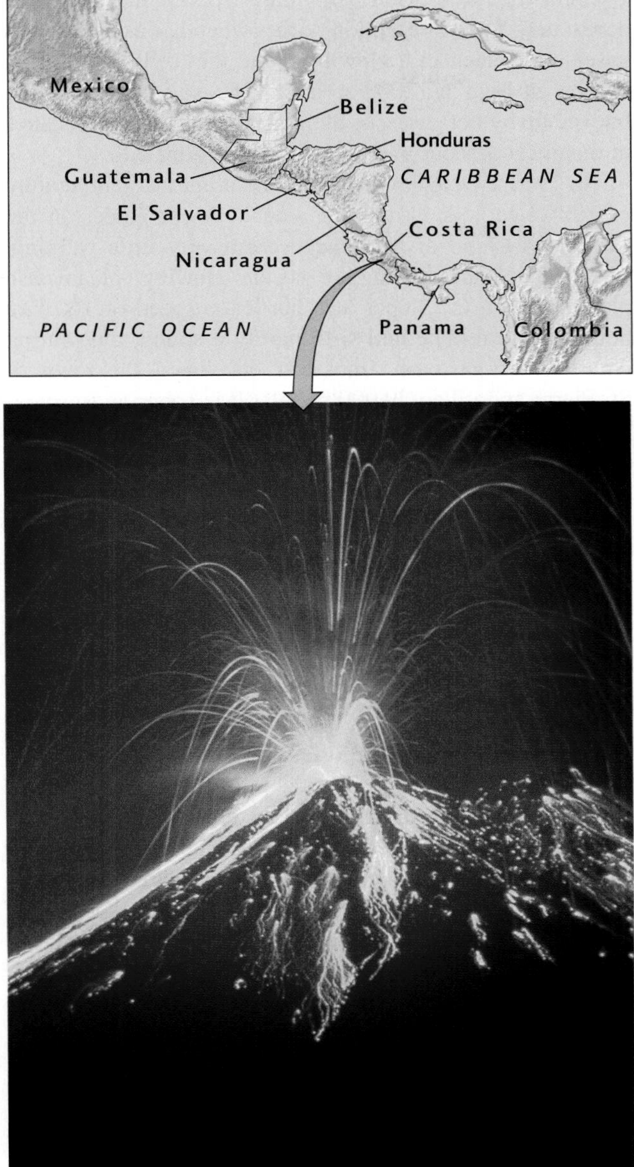

Figure 6.5 Pyroclastic eruption at Arenal volcano, Costa Rica. [Gregory G. Dimijian/Photo Researchers.]

fining pressure of the overlying rock keeps these volatiles from escaping. When the magma rises close to the surface and the pressure drops, the volatiles may be released with explosive force, shattering the lava and any overlying solidified rock into fragments of various sizes, shapes, and textures (**Figure 6.5**). Explosive eruptions are particularly likely with gas-rich, viscous rhyolitic and andesitic lavas.

Volcanic Ejecta Fragmentary volcanic rocks ejected into the air are called **pyroclasts.** These rocks, minerals, and glasses are classified according to size. The finest fragments, less than 2 mm in diameter, are called *volcanic ash.*

Figure 6.6 Volcanologist Katia Krafft examines a volcanic bomb ejected from Asama volcano, Japan. [Science Source/Photo Researchers.]

Fragments ejected as blobs of lava that become rounded and cooled in flight or chunks torn loose from previously solidified volcanic rock can be much larger. These are called *volcanic bombs* (**Figure 6.6**). Volcanic ejecta as big as houses have been thrown more than 10 km in violent eruptions. Volcanic ash fine enough to stay aloft can be carried great distances. Within two weeks of the 1991 eruption of Mount Pinatubo in the Philippines, its volcanic dust was traced all the way around the world by Earth-orbiting satellites.

Sooner or later pyroclasts fall to Earth, usually building deposits near their source. As they cool, hot sticky fragments become welded together (lithified). The rocks created from smaller fragments are called *volcanic tuffs;* those formed from larger fragments are called *volcanic breccias* (**Figure 6.7**).

Figure 6.7 Volcanic breccia. [Doug Sokell/Visuals Unlimited.]

Figure 6.8 A pyroclastic flow plunged down the slopes of Mount Unzen, in Japan, in June 1991. Note the fireman and fire engine in the foreground, trying to outrun the hot ash cloud descending on them. Three scientists who were studying this volcano died when they were engulfed by a similar flow. [AP/Wide World Photos.]

Pyroclastic Flows One particularly spectacular and often devastating form of eruption occurs when hot ash, dust, and gases are ejected in a glowing cloud that rolls downhill at speeds as high as 200 km/hour. The solid particles are buoyed up by hot gases, so there is little frictional resistance in this incandescent *pyroclastic flow* (**Figure 6.8**).

In 1902, a pyroclastic flow with an internal temperature of 800°C exploded from the side of Mont Pelée, on the Caribbean island of Martinique, with very little warning. The avalanche of choking hot gas and glowing volcanic ash plunged down the slopes at a hurricane speed of 160 km/hour. In one minute and with hardly a sound, the searing emulsion of gas, ash, and dust enveloped the town of St. Pierre and killed 29,000 people. It is sobering to scientists who render advice to others to recall the statement of one Professor Landes, issued the day before the cataclysm: "The Montagne Pelée presents no more danger to the inhabitants of St. Pierre than does Vesuvius to those of Naples." Professor Landes perished with the others. In 1991, French volcanologists Maurice and Katia Krafft (Maurice took the photograph of Katia in Figure 6.15) were killed by a pyroclastic flow at Mount Unzen, Japan.

Eruptive Styles and Landforms

Now that we have examined various types of volcanic materials that pour or blow out of Earth's interior, we can look more closely at styles of eruptions and the characteristic formations they leave behind (**Figure 6.9**). Eruptions do not always create a majestically symmetrical cone. Volcanic landforms vary in shape with the properties of the lava and the conditions under which it erupts.

Central Eruptions

Central eruptions create the most familiar of all volcanic features—the volcanic mountain shaped like a cone. These eruptions discharge lava or pyroclastic materials from a *central vent,* an opening atop a pipelike feeder channel rising from the magma chamber through which the material ascends to erupt at the Earth's surface.

Figure 6.9 Eruptive styles and landforms. (a) A shield volcano: Mauna Loa, Hawaii. [U.S. Geological Survey.] (b) A volcanic dome: Mount St. Helens after the 1980 eruption. [Lyn Topinka/USGS Cascades Volcano Observatory.] (c) A cinder cone: Cerro Negro, Nicaragua, in 1968. [Mark Hurd Aerial Surveys.] (d) A composite volcano: Fujiyama, Japan. [Corbis.] (e) A crater: Mt. Etna, Sicily. [Fabrizio Villa/AP/Wide World Photos.] (f) A caldera: Crater Lake, Oregon, which fills a caldera 8 km in diameter. [Greg Vaughn/Tom Stack & Associates.]

ERUPTIVE STYLES AND LANDFORMS

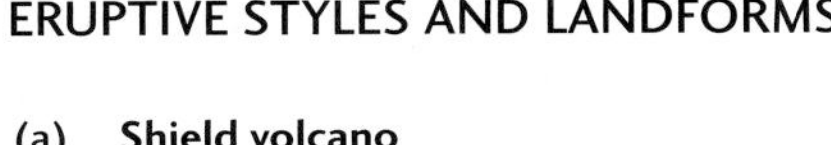

(a) Shield volcano

A shield volcano is built up by the accumulation of thousands of thin basaltic flows that spread as gently sloping sheets. Each layer in the diagram represents many hundreds of thin flows. Magma can erupt on the flanks of a volcano as well as from the central vent.

(b) Volcanic dome

Volcanic domes are bulbous masses of felsic lava, which are so viscous that instead of flowing, they pile up over the vent. The photo shows a growing dome within the crater of Mount St. Helens after its 1980 eruption.

(c) Cinder-cone volcano

In a cinder-cone volcano, ejected material is deposited as layers that dip away from the crater at the summit. The vent beneath the crater is filled with fragmental debris. The photo is of Cerro Negro, shown erupting in 1968, a cinder cone built on an older terrain of lava flows. Cerro Negro erupted again in 1995 and in 1999.

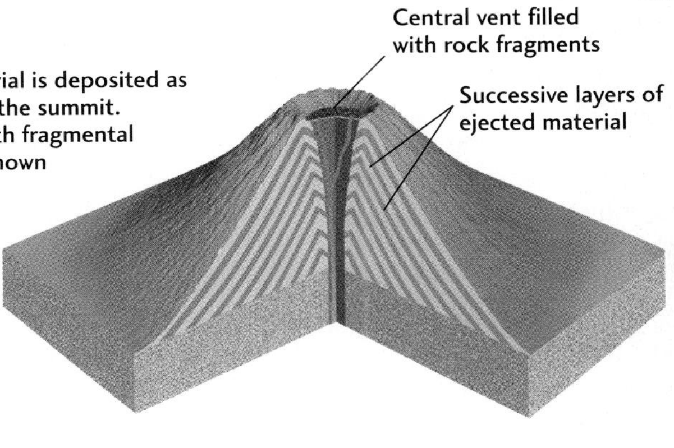

(d) Composite volcano

A composite volcano is built up of alternating layers of pyroclastic material and lava flows. Lava that has solidified in fissures forms riblike dikes that strengthen the cone.

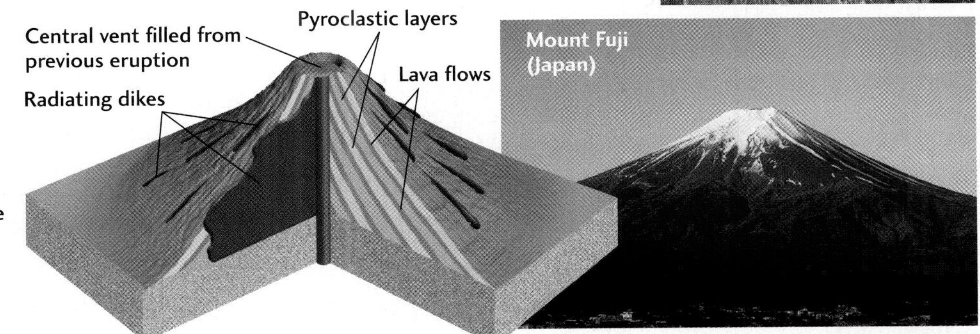

(e) Crater

Craters are found at the summits of most volcanoes. After an eruption, lava often sinks back into the vent and solidifies, to be blasted out by a later pyroclastic explosion.

(f) Caldera

A violent eruption can empty a volcano's magma chamber, which then cannot support the overlying rock. It collapses, leaving a large, steep-walled basin called a caldera.

Shield Volcanoes A lava cone is built by successive flows of lava from a central vent. If the lava is basaltic, it flows easily and spreads widely. If flows are copious and frequent, they create a broad, shield-shaped volcano many tens of kilometers in circumference and more than 2 km high. The slopes are relatively gentle. Mauna Loa, on Hawaii, is the classic example of a **shield volcano** (see Figure 6.9a). Although it rises only 4 km above sea level, it is actually the world's tallest structure: measured from its base on the seafloor, Mauna Loa is 10 km high. It has a base diameter of 120 km—an area roughly three times the size of Rhode Island. It grew to this enormous magnitude by the accumulation of thousands of lava flows, each only a few meters thick, over a period of about 1 million years. In fact, the island of Hawaii actually consists of the tops of a series of overlapping active shield volcanoes emerging through the ocean surface.

Volcanic Domes In contrast to basaltic lavas, felsic lavas are so viscous that they can just barely flow. They usually produce a *volcanic dome,* a rounded, steep-sided mass of rock. Domes look as though lava had been squeezed out of a vent like toothpaste, with very little lateral spreading. Domes often plug vents, trapping gases. Pressure increases until an explosion occurs, blasting the dome into fragments. Mount St. Helens exhibited such an explosion when it erupted in 1980 (see Figure 6.9b).

Cinder-Cone Volcanoes When volcanic vents discharge pyroclasts, the solid fragments build up and form *cinder cones.* The profile of a cone is determined by the maximum angle at which the debris remains stable instead of sliding downhill. The larger fragments, which fall near the summit, form very steep but stable slopes. Finer particles are carried farther from the vent and form gentle slopes at the base of the cone. The classic concave-shaped volcanic cone with its summit vent is formed in this way (see Figure 6.9c).

Composite Volcanoes When a volcano emits lava as well as pyroclasts, alternating lava flows and beds of pyroclasts build a concave-shaped *composite volcano,* or **stratovolcano.** The composite volcano is the most common form of such large volcanoes as Fujiyama in Japan (see Figure 6.9d), Mount Vesuvius and Mount Etna in Italy, and Mount Rainier in Washington State.

Craters A bowl-shaped pit, or *crater,* is found at the summit of most volcanoes, centered on the vent. During an eruption, the upwelling lava overflows the crater walls. When eruption ceases, the lava that remains in the crater often sinks back into the vent and solidifies. When the next eruption occurs, the material is literally blasted out of the crater in a pyroclastic explosion. The crater later becomes partly filled by the debris that falls back into it. Because a crater's walls are steep, they may cave in or become eroded over time. In this way, a crater can grow in diameter to several times that of the vent and hundreds of meters deep. The crater of Mount Etna in Sicily, for example, is currently 300 m (more than three football fields) in diameter (see Figure 6.9e).

Calderas After a violent eruption in which large volumes of magma are discharged from a magma chamber a few kilometers below the vent, the empty chamber may no longer be able to support its roof. In such cases, the overlying volcanic structure can collapse catastrophically, leaving a large, steep-walled, basin-shaped depression, much larger than the crater, called a **caldera** (see Figure 6.9f). Calderas are impressive features, ranging in size from a few kilometers to as much as 50 km or more in diameter, or about the area of greater New York City. The evolution of a typical caldera is shown in **Figure 6.10**.

After some hundreds of thousands of years, fresh magma can reenter the collapsed magma chamber and reinflate it, forcing the caldera floor to dome upward again and creating a *resurgent caldera.* The cycle of eruption, collapse, and resurgence may occur repeatedly over geologic time. The collapse of a large resurgent caldera is one of the most destructive natural phenomena on Earth. Yellowstone Caldera in Wyoming, marked today by a leftover relic, Old Faithful geyser, ejected some 1000 km^3 of pyroclastic debris during its eruptive stages about 600,000 years ago, more than a thousand times the amount of material ejected by Mount St. Helens in 1980. Ash deposits fell over much of what is now the United States. Other resurgent calderas are Valles Caldera in New Mexico, Long Valley Caldera located in California 300 km east of San Francisco, the still-active Kilauea (Hawaii) and Rabaul (Papua New Guinea) calderas, and the dormant Crater Lake in Oregon.

Monitoring caldera unrest is very important today because of the potential for widespread destruction. Fortunately, no catastrophic collapses with global consequences have occurred during recorded history, but geologists are wary of an increasing occurrence of small earthquakes in Yellowstone and Long Valley calderas and other indications of activity in the magma chambers in the crust below. For example, carbon dioxide leaking into soil from magma deeper in the crust has been killing trees since 1992 on Mammoth Mountain, a volcano on the boundary of Long Valley Caldera. A forest service ranger was almost asphyxiated by volcanic carbon dioxide leaking into a cabin on the flanks of Mammoth Mountain. Other indications of resurgence of the Long Valley Caldera are uplift of the center of the caldera by more than half a meter in the past 20 years and the occurrence of nearly continuous swarms of small earthquakes. More than a thousand occurred in a single day in 1997.

Phreatic Explosions When hot, gas-charged magma encounters groundwater or seawater, the vast quantities of superheated steam generated cause *phreatic,* or steam, explosions (**Figure 6.11**). One of the most destructive volcanic eruptions in history, that of Krakatoa in Indonesia, was a phreatic explosion.

Diatremes When hot matter from the deep interior escapes explosively, the vent and the feeder channel below it are often left filled with breccia as the eruption wanes. The resulting structure is called a **diatreme.** Shiprock, which towers over the surrounding plain in New Mexico, is a dia-

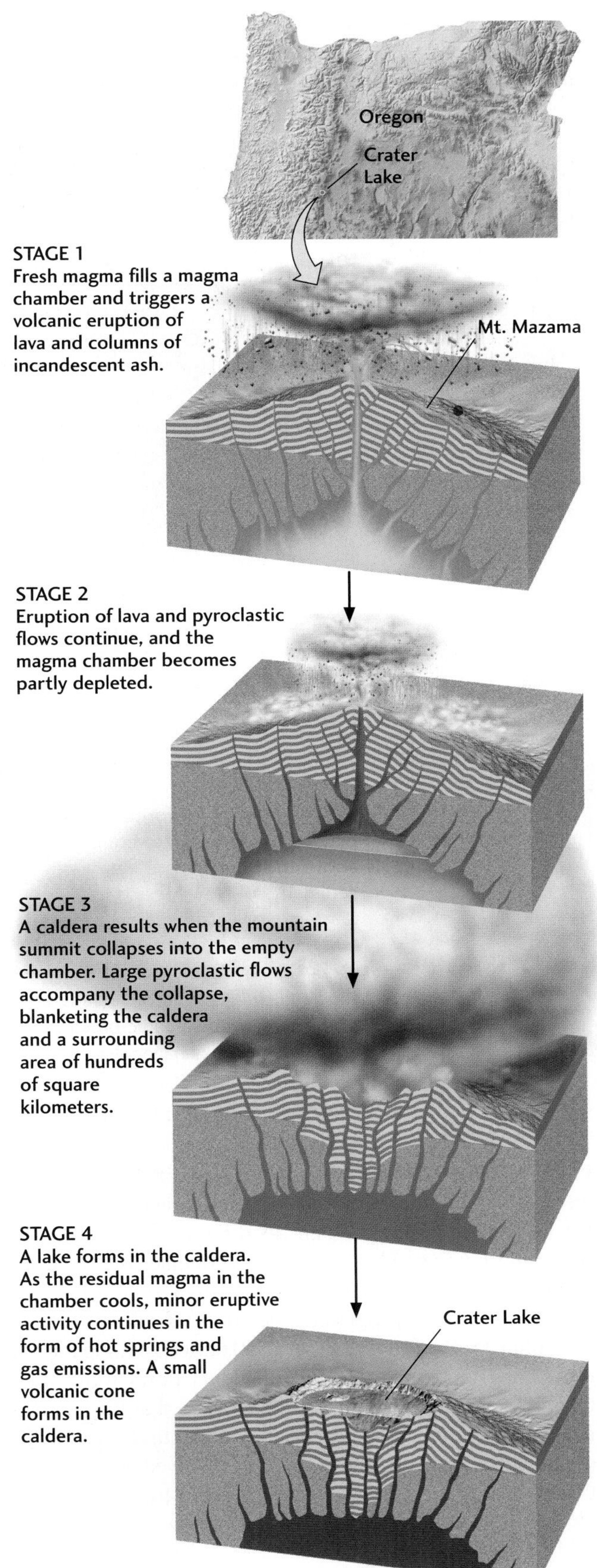

Figure 6.10 Stages in the evolution of Crater Lake Caldera.

treme exposed by the erosion of the sedimentary rocks through which it originally burst. To transcontinental air travelers, Shiprock looks like a gigantic black skyscraper in the red desert (**Figure 6.12**).

The eruptive mechanism that produces diatremes has been pieced together from the geologic record. The kinds of minerals and rocks found in some diatremes could have been formed only at great depths—100 km or so, well within the upper mantle. Gas-charged magmas force their way upward from these depths by fracturing the lithosphere and exploding into the atmosphere, ejecting gases and solid fragments from the deep crust and mantle, sometimes at supersonic speed. Such an eruption would probably look like the exhaust jet of a giant rocket upside down in the ground blowing rocks and gases into the air.

Perhaps the most exotic diatremes are *kimberlite pipes,* named after the fabled Kimberley diamond mines of South Africa. Kimberlite is a volcanic type of peridotite—a rock containing mostly the mineral olivine. Kimberlite pipes also contain a variety of mantle fragments, including diamonds, that were pulled into the magma as it exploded toward the surface. The extremely high pressures needed to squeeze carbon into the mineral diamond are reached only at depths in the Earth greater than 150 km. From careful studies of diamonds and other mantle fragments found in kimberlite pipes, geologists have been able to reconstruct sections of the mantle as if they had a drill core down to more than 200 km. These studies provide strong support for the theory that the upper mantle is made primarily of peridotite.

Figure 6.11 Phreatic explosion on Nisino-sima, a new volcano that rose above the sea in 1973 after a submarine eruption in the Pacific Ocean about 900 km south of Tokyo. [Hydrographic Department, Maritime Safety Agency, Japan.]

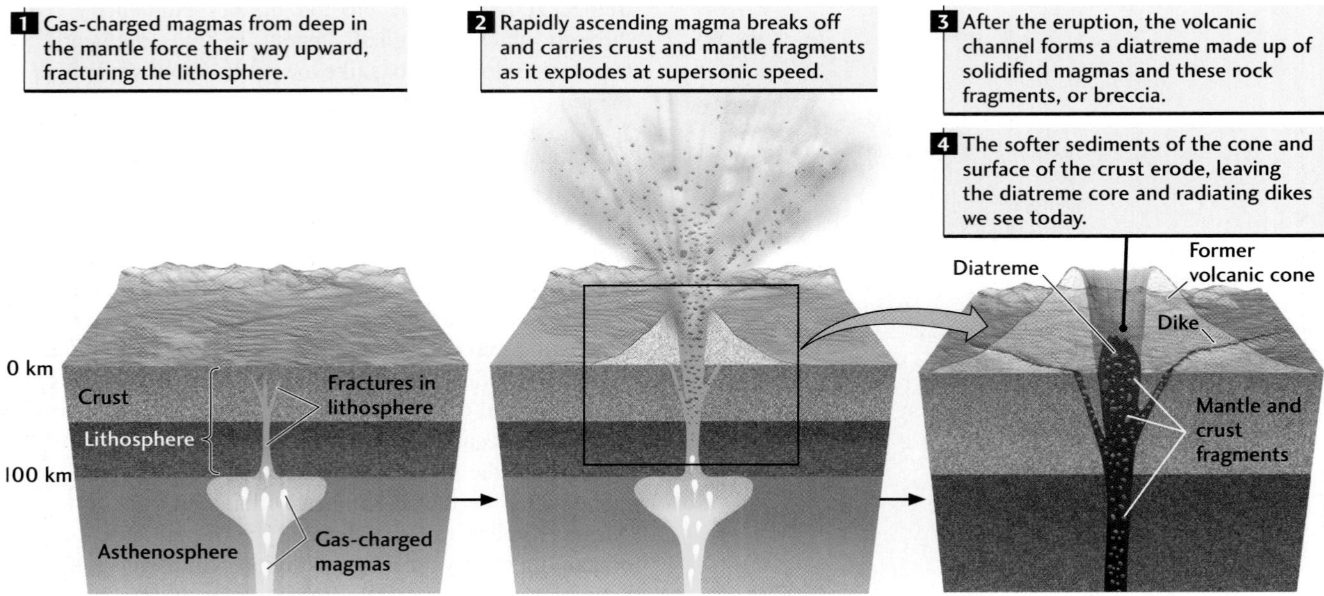

Figure 6.12 The formation of a diatreme. Shiprock, towering 515 m above the surrounding flat-lying sediments of New Mexico, is a diatreme, or volcanic pipe, exposed by erosion of the softer sedimentary rocks that once enclosed it. Note the vertical dike, one of six radiating from the central volcanic pipe. [Jim Wark, Index Stock Imagery.]

Fissure Eruptions

Imagine basaltic lava flowing out of a crack in the Earth's surface tens of kilometers long, flooding vast areas. Such **fissure eruptions** have occurred on Earth countless times in the past 4 billion years (**Figure 6.13**). Among the geologically important fissure eruptions are those that occur along mid-ocean ridges. Historically, humans have witnessed such an eruption only once, in 1783 on Iceland, which is an exposed segment of the Mid-Atlantic Ridge. One-fifth of the Icelandic population perished from starvation as a result. A fissure 32 km long opened and spewed out some 12 km^3 of basalt (see Figure 6.13), enough to cover Manhattan to a height about halfway up the Empire State Building. Fissure eruptions continue on Iceland, although on a smaller scale than the 1783 catastrophe.

Flood Basalts (Basaltic Lava Plateaus) The geologic record contains ample evidence of prehistoric basaltic flooding from great fissures. When flood basalts erupt from fissures, the lavas build a plain or accumulate as a plateau, rather than piling up as a volcanic mountain as they do when they erupt from a central vent. The flood basalts that made the Columbia Plateau (see Figure 6.2) buried 200,000 km^2 of preexisting topography. Some individual flows were more than 100 m thick, and some were so fluid that they spread more than 60 km from their source. An entirely new landscape with new river valleys has since evolved atop the lava that buried the old surface. Plateaus made by flood basalts are found on every continent.

Ash-Flow Deposits Fissure eruptions of pyroclastic materials have produced extensive sheets of hard volcanic tuffs

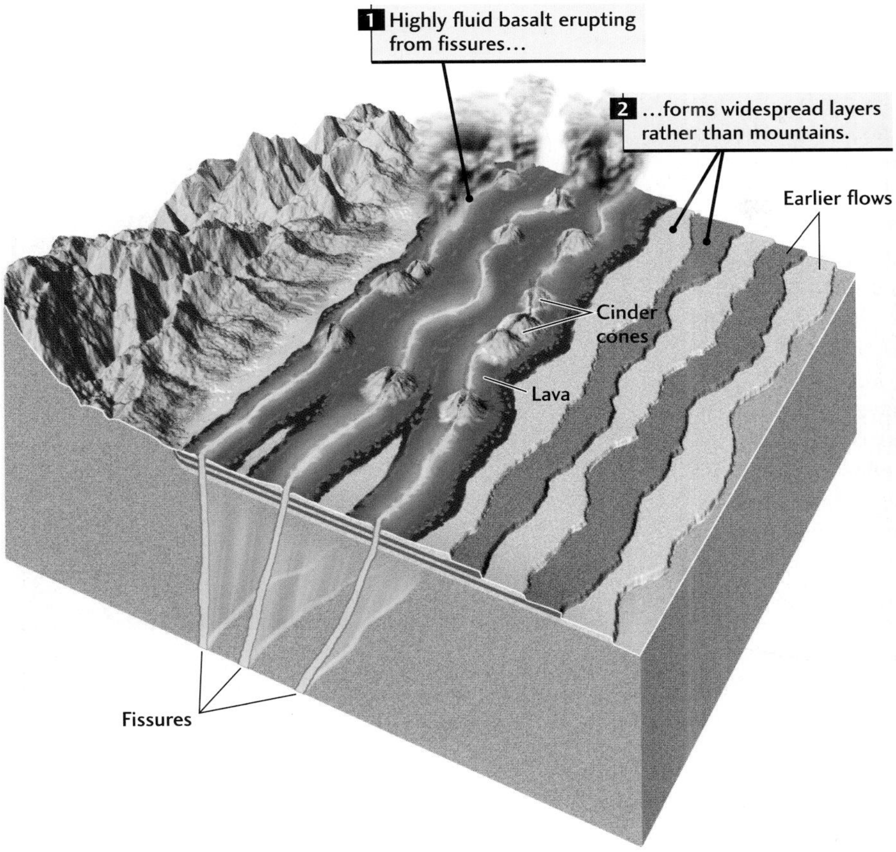

1 Highly fluid basalt erupting from fissures...
2 ...forms widespread layers rather than mountains.
Earlier flows
Cinder cones
Lava
Fissures

Figure 6.13 In a fissure eruption of highly fluid basalt, lava rapidly flows away from fissures and forms widespread layers, rather than building up into a volcanic mountain. [After R. S. Fiske, U.S. Geological Survey.] Volcanic cones along the Laki fissure (Iceland) that opened in 1783 and erupted the largest flow of lava on land in the course of recorded history. [Tony Waltham.]

called *ash-flow deposits* (**Figure 6.14**). As far we know, humans have never witnessed one of these spectacular events. The early Tertiary ash-flow deposits of the Great Basin in Nevada and adjacent states, formed in this way, cover an area of about 200,000 km^2 and are as much as 2500 m thick in some places. Yellowstone National Park, in Wyoming, has been covered by a number of ash-flow sheets. A succession of forests there were buried by these deposits.

Other Volcanic Phenomena

Lahars Among the most dangerous volcanic events are the torrential mudflows of wet volcanic debris called **lahars.** They can occur when a pyroclastic flow meets a river or a snowbank; when the wall of a crater lake breaks, suddenly releasing water; when a lava flow melts glacial ice; or when heavy rainfall transforms new ash deposits into mud. One extensive layer of volcanic debris in the Sierra Nevada of California contains 8000 km^3 of material of lahar origin, enough to cover all of Delaware with a deposit more than a kilometer thick. Lahars have been known to carry huge boulders for tens of kilometers. When Nevado del Ruiz in the Colombian Andes erupted in 1985, lahars triggered by the melting of glacial ice near the summit plunged down the slopes and buried the town of Armero 50 km away, killing more than 25,000 people.

Flank Collapse A volcano is constructed from thousands of deposits of lava or ash or both, which is not the best way

Figure 6.14 Welded tuff from an ash-flow deposit in the Great Basin of northern Nevada. [John Grotzinger.]

to build a stable structure. A volcano's sides can become too steep and break or slip off. In recent years, volcanologists have discovered many prehistoric examples of catastrophic structural failures in which a big piece of a volcano broke off, perhaps triggered by an earthquake, and slid downhill in a massive, destructive landslide. On a worldwide basis, such *flank collapses* occur at an average rate of about four times per century. The collapse of one side of Mount St. Helens was the most damaging part of its 1980 eruption (see the photograph in Feature 6.1).

Surveys of the seafloor off Hawaii have revealed many giant landslides on the underwater flanks of the Hawaiian ridge. When they occurred, these massive movements most likely triggered huge tsunamis (sea waves). In fact, coral-bearing marine sediments were found on one of the Hawaiian Islands, some 300 m above sea level. These sediments were probably deposited by giant tsunamis that were excited by a prehistoric flank collapse.

The southern flank of Kilauea volcano is advancing toward the sea at a rate of 250 mm/year, which is relatively fast, geologically speaking. This advance became more worrisome, however, when it suddenly accelerated by a factor of several hundred on November 8, 2000, probably triggered by heavy rainfall a few days earlier. A network of motion sensors detected an ominous surge in velocity of about 50 mm/day lasting for 36 hours, after which the normal slow motion was reestablished. Someday, maybe thousands of years from now but perhaps sooner, the flank probably will break off and slide into the ocean. This catastrophic event would trigger a tsunami that could prove disastrous for Hawaii, California, and other Pacific coastal areas.

Figure 6.15 Volcanologist Katia Krafft in a heatproof suit examining a lava flow on Kilauea volcano, Hawaii. [Maurice Krafft/Photo Researchers.]

Volcanic Gases The nature and origin of volcanic gases are of considerable interest and importance. It is thought that, over geologic time, these gases created the oceans and the atmosphere and may even affect today's climate. Volcanic gases have been collected by courageous volcanologists and analyzed to determine their composition (**Figure 6.15**). Water vapor is the main constituent of volcanic gas (70–95 percent), followed by carbon dioxide, sulfur dioxide, and traces of nitrogen, hydrogen, carbon monoxide, sulfur, and chlorine. Eruptions can release enormous amounts of these gases. Some volcanic gas may come from deep within the Earth, making its way to the surface for the first time. Some may be recycled groundwater and ocean water, recycled atmospheric gas, or gas that had been trapped in earlier generations of rocks.

Geosystem Interactions: Volcanism and Climatic Change The relationship between volcanic eruptions and changes in weather and climate provides a good example of Earth system interactions. For instance, the 1982 eruption of El Chichón in southern Mexico and the 1991 eruption of Mount Pinatubo injected sulfurous gases into the atmosphere, 10 km above the Earth. Through various chemical reactions, the gases formed an aerosol (a fine airborne mist) representing tens of millions of metric tons of sulfuric acid. This aerosol blocked enough of the Sun's radiation from reaching Earth's surface to lower global temperatures for a year or two. The eruption of Mount Pinatubo, one of the largest explosive eruptions of the century, led to a global cooling of at least 0.5°C in 1992. (Chlorine emissions from Pinatubo have also hastened the loss of ozone in the atmosphere, nature's shield that protects the biosphere from the Sun's ultraviolet radiation.) The debris lofted into the atmosphere during the 1815 eruption of Mount Tambora in Indonesia caused even greater cooling. The Northern Hemisphere suffered a very cold summer with snowstorms in 1816. The drop in temperature and the ash fallout caused widespread crop failures. As a result, more than 90,000 people perished in that "year without a summer." This terrible year inspired Byron's gloomy poem "Darkness":

> I had a dream, which was not all a dream.
> The bright sun was extinguish'd, and the stars
> Did wander darkling in the eternal space,
> Rayless, and pathless, and the icy earth
> Swung blind and blackening in the moonless air;
> Morn came and went—and came, and brought no day.
> And men forgot their passions in the dread
> Of this their desolation; and all hearts
> Were chill'd into a selfish prayer for light.

A surprising example of the interplay between geosystems is an unexpected interaction between climate and vol-

Figure 6.16 Fumarole becoming encrusted with sulfur deposits on the Merapi volcano, in Indonesia. [Roger Ressmeyer/Corbis.]

canism. As we have just seen, volcanoes can affect climate, but it appears that climate can also affect volcanoes. Geologists recently found statistical evidence that heavy rainfall can cause dome collapse, one of the most dangerous types of volcanic eruptions because it can trigger pyroclastic flows (see Figure 6.8). They reported that several eruptions on the island of Montserrat followed heavy rains. This is too small a sample to support a hypothesis. However, other scientists found statistical connections between seasons of heavy rains and eruptions of Mount Etna and Mount St. Helens. A U.S. Geological Survey expert believes that taking rainfall into account could improve the prediction of some eruptions. The mechanism connecting rainfall and eruptions is still unclear. We speculate that rain seeping down into hot magma becomes superheated steam and causes an explosion followed by a collapse. Or perhaps the weight of accumulated seasonal rainfall causes a collapse.

Hydrothermal Activity Volcanic activity does not stop when lava or pyroclastic materials cease to flow. For decades, sometimes centuries, after a major eruption volcanoes continue to emit gas fumes and steam through small vents called *fumaroles.* These emanations contain dissolved materials that precipitate onto surrounding surfaces as the water evaporates or cools. Various sorts of encrusting deposits (such as travertine) form, including some that contain valuable minerals (**Figure 6.16**).

Fumaroles are a surface manifestation of **hydrothermal activity**—the circulation of water through hot volcanic rocks and magmas. Interaction between Earth's hydrosphere and lithosphere is an important process in many volcanic geosystems. Circulating groundwater that reaches buried magma (which retains heat for hundreds of thousands of years) is heated and returned to the surface as hot springs and geysers. A *geyser* is a hot-water fountain that spouts intermittently with great force, frequently accompanied by a thunderous roar (**Figure 6.17**). The best-known geyser in the United States is Old Faithful in Yellowstone National Park, which erupts about every 65 minutes, sending a jet of hot water as high as 60 m into the air.

The steam and hot water formed by hydrothermal activity can be tapped as a source of geothermal energy. Hydrothermal activity is also responsible for the deposition of unusual minerals that concentrate relatively rare elements, particularly metals, into ore bodies of great economic value (see Chapter 22). As we will see, hydrothermal interactions are especially intense at the spreading centers of mid-ocean ridges, where huge volumes of water and magma come into contact.

Figure 6.17 Old Faithful geyser, in Yellowstone National Park, which erupts 60 m high about every 65 minutes. [Simon Fraser/SPL/Photo Researchers.]

6.1 Monitoring Volcanoes

Kilauea: Tracking the Flow of Magma

The giant shield volcano Mauna Loa and the smaller Kilauea on its eastern flank make up the southern half of the island of Hawaii. Because the U.S. Geological Survey operates a volcano observatory on the rim of Kilauea Caldera, this volcano, which is actively erupting, is perhaps the best studied in the world.

A modern network of instruments and laboratory facilities is used to track the movement of magma within the volcano and the changing chemistry of the erupting lavas and gases. *Seismographs,* which measure movements within the Earth, detect and locate the small earthquakes that are often correlated with movements of magma. Seismographs locate earthquakes as deep as 55 km beneath Kilauea. Such quakes often mark the entrance of magma into the channels leading from the asthenosphere through the lithosphere to Earth's surface. The upward migration of the magma can be followed over a period of months because seismic disturbances occur progressively nearer the surface as the magma rises. *Tiltmeters,* which measure tilting of the ground, indicate when the volcano begins to swell as the rising magma fills a magma chamber not far below the summit.

The first sign that an outbreak of lava is imminent is a swarm of small earthquakes, thousands of them, signifying that the magma is splitting rock as it forces its way to the surface. Very often, geologists know where the eruption will occur from the location of the earthquakes and changes in their pattern. In January 1960, for example, U.S. Geological Survey scientists detected a swarm of earthquakes not far from the village of Kapoho, on the flank of Kilauea. As they expected, an eruption broke out, destroying Kapoho but causing no casualties, because the village had been evacuated. A new landscape was created as the lava flowed to the sea. Six-meter-high walls were built in a futile attempt to divert the lava and save a seashore community. When it was all over, the tiltmeters showed that the volcano had deflated, signifying that the magma chamber below had been drained in the Kapoho eruption. The cycle is repeated every few years.

Mount St. Helens: Dangerous but Predictable

In the contiguous United States, Mount St. Helens is the most active and explosive volcano. It has a documented 4500-year history of destructive lava flows, hot pyroclastic flows, lahars, and distant ash falls. Beginning on March 20, 1980, a series of small to moderate earthquakes under the volcano signaled the start of a new eruptive phase after 123 years of dormancy. The earthquakes moved the U.S. Geological Survey to issue a formal hazard alert. The first outburst of ash and steam erupted from a newly opened crater on the summit one week later. In April, the seismic tremors increased, indi-

Destruction of property engulfed by a lava flow in the May 1990 eruption of Kilauea volcano, Hawaii. [James Cachero/Corbis.]

cating that magma was moving beneath the summit, and an ominous swelling of the northeastern flank was noticed. The U.S. Geological Survey issued a more serious warning, and people were ordered out of the vicinity. On May 18, the climactic eruption began abruptly. A large earthquake apparently triggered the collapse of the north side of the mountain, loosening a massive landslide, the largest ever recorded anywhere. As a huge flow of debris plummeted down the mountain, gas and steam under high pressure were released in a tremendous lateral blast that blew out the northern flank of the mountain. U.S. Geological Survey geologist David A. Johnston was monitoring the volcano from his observation post 8 km to the north. He must have seen the advancing blast wave before he radioed his last message: "Vancouver, Vancouver, this is it!" A northward-directed jet of superheated (500°C) ash, gas, and steam roared out of the breach with hurricane force, devastating a zone 20 km outward from the volcano and 30 km wide. A vertical eruption sent an ash plume 25 km into the sky, twice as high as a commercial jet flies. The ash cloud drifted to the east and northeast with the prevailing winds, bringing darkness at noon to an area 250 km to the east and depositing ash as deep as 10 cm on much of Washington, northern Idaho, and western Montana. The energy of the blast was equivalent to about 25 million tons of TNT. The volcano's summit was destroyed, its elevation reduced by 400 m; and its northern flank disappeared. In effect, the mountain was "hollowed out."

Local devastation was spectacular. Within an inner blast zone extending 10 km from the volcano, the thick forest was denuded and buried under several meters of pyroclastic debris. Beyond this zone, out to 20 km, trees were stripped of their branches and blown over like broken matchsticks aligned radially away from the volcano. As far as 26 km away, the hot blast was so intense that it overturned a truck and melted its plastic parts. Some fishermen were severely burned and survived only by jumping into a river. More than 60 other people were killed by the blast and its effects. A lahar formed when the landslide and pyroclastic debris—fluidized by groundwater, melted snow, and glacial ice—flowed 28 km down the valley of the Toutle River. The valley bottom was filled to a depth of 60 m. Beyond this debris pile, muddy water flowed into the Columbia River, where sediments clogged the ship channel and stranded many vessels in Portland. Mount St. Helens may continue erupting for 20 years or more until its present episode of activity comes to an end. Although much of the devastated area is still barren, after a decade almost 20 percent of the surface showed evidence of revegetation: native species of grass, legumes, and young trees began to come back.

Mount St. Helens before and after the cataclysmic eruptions of May 1980. [*Before:* Emil Muench/Photo Researchers. *Erupting:* U.S. Geological Survey. *After:* David Weintraub/Photo Researchers.]

The Global Pattern of Volcanism

Before the advent of the plate tectonics theory, geologists noted a concentration of volcanoes around the rim of the Pacific Ocean and nicknamed it the Ring of Fire. The explanation of the Ring of Fire in terms of plate subduction was one of the great successes of the new theory. We have already examined how lava compositions vary with plate tectonic setting (see Figure 5.11). In this section, we show how plate tectonics can explain essentially all major features in the global pattern of volcanism.

Figure 6.18 displays the locations of the world's active volcanoes that occur on land or above the ocean's surface. About 80 percent are found at boundaries where plates converge, 15 percent where plates separate, and the remaining few within the plate interiors. There are many more active volcanoes than shown on this figure, however. Most of the lava erupted on Earth's surface comes from vents beneath the sea, located on the spreading centers of the mid-ocean ridges.

Basalt Production at Spreading Centers

Each year, approximately 3 km^3 of basaltic lava is erupted along the mid-ocean ridges in the process of seafloor spreading. This is a truly enormous volume. In comparison, all of the active volcanoes along the convergent boundaries (about 400) generate volcanic rock at a rate of less than 1 km^3/year. Enough magma has been erupted in seafloor spreading over the past 200 million years to create all of the present-day seafloor, which covers nearly two-thirds of Earth's surface. The scale of this "factory" where ocean crust is created is a few kilometers wide and deep, and it extends intermittently along the thousands of kilometers of mid-ocean ridge (**Figure 6.19**). The magma and volcanic rocks are formed during decompression melting of mantle peridotite, as discussed in Chapter 5 (see Figure 5.13).

Hydrothermal Activity at Spreading Centers

Tensional fissures at the mid-ocean ridge crests allow seawater to circulate throughout the new oceanic crust. Heat from the hot volcanic rocks and deeper magmas drives a vigorous convective process, which pulls cold seawater into the crust, heats it through contact with magma, and expels the hot water back into the overlying ocean (see Figure 5.15).

Given the common occurrence of hot springs and geysers in volcanic geosystems on land, the evidence for pervasive hydrothermal activity at spreading centers (which are usually immersed in water) should come as no surprise. Nevertheless, geologists were amazed once they recognized the intensity of the convection and discovered some of its chemical and biological consequences. The most spectacu-

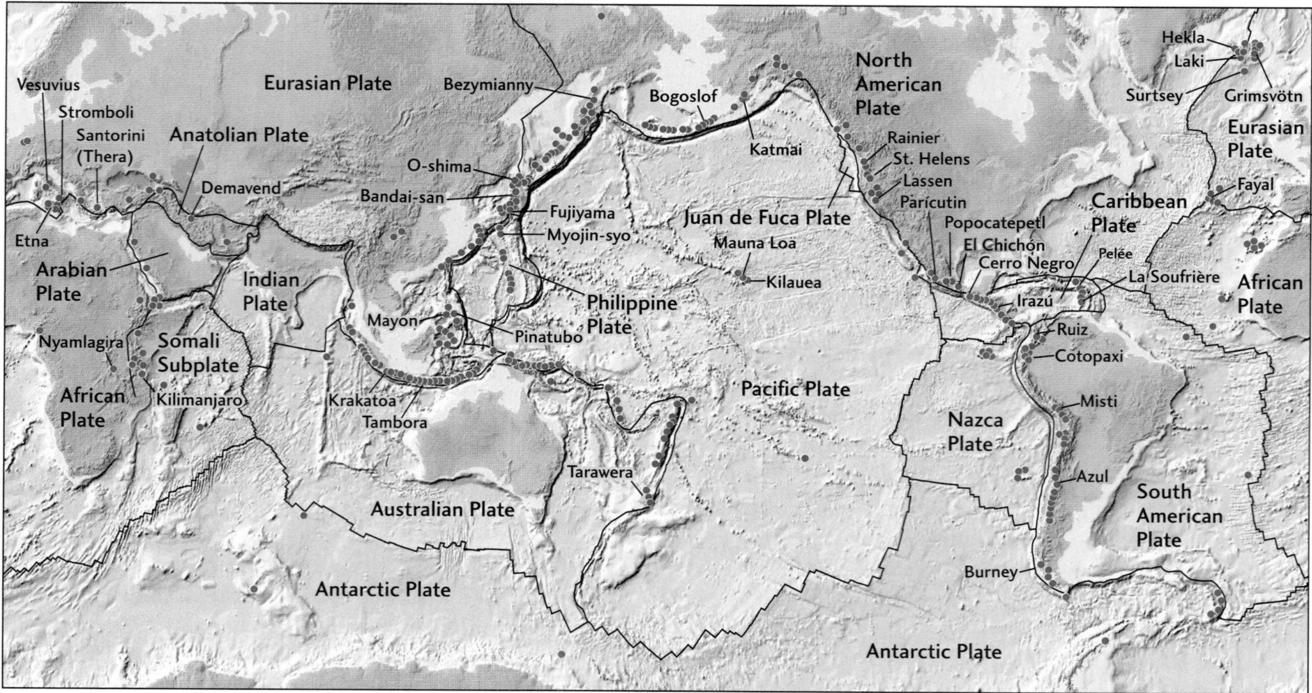

Figure 6.18 The active volcanoes of the world with vents on land or above the ocean surface (red dots). About 80 percent are found at boundaries where plates collide, 15 percent where plates separate, and the remaining few at intraplate hot spots. Black lines are plate boundaries. Not shown on this map are the numerous axial volcanoes of the mid-ocean ridge system.

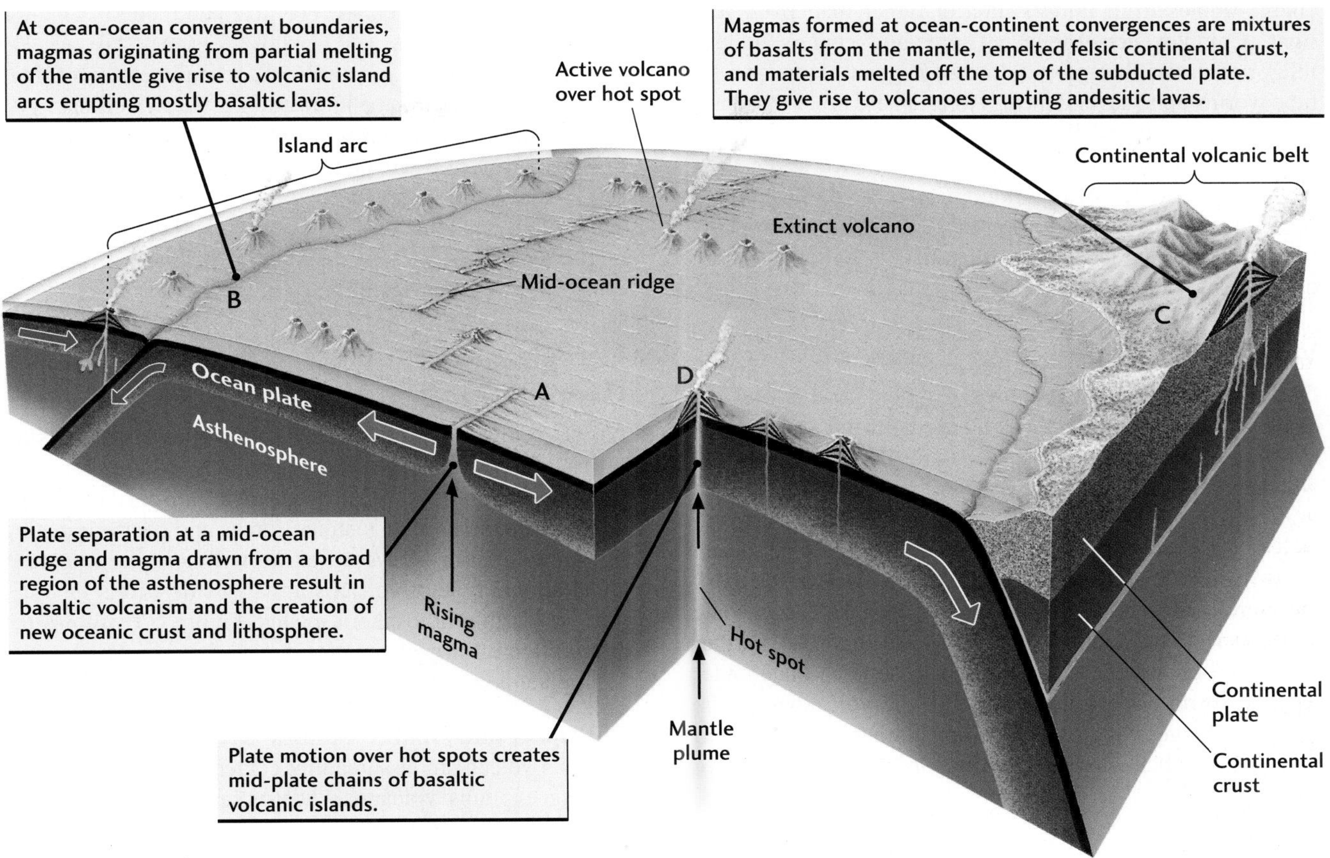

Figure 6.19 The global pattern of volcanism can be explained by plate tectonics. (A) Plate separation at a mid-ocean ridge and magma drawn from a broad region of the asthenosphere result in basaltic volcanism and the creation of the suboceanic crust and lithosphere. (B) At ocean-ocean convergent boundaries, magmas originating from partial melting of the mantle give rise to volcanic island arcs erupting mostly basaltic lavas. (C) Magmas formed at ocean-continent convergences are mixtures of basalts from the mantle, remelted felsic continental crust, and materials melted off the top of the subducted plate. They give rise to volcanoes erupting andesitic and rhyolitic lavas. (D) Plate motion over hot spots accounts for the creation of mid-plate chains of basaltic volcanic islands. Iceland (not shown) is a hot spot located at a mid-ocean ridge.

lar manifestations of this process were first found in the eastern Pacific Ocean in 1977. Plumes of hot, mineral-laden water with temperatures as high as 350°C spout as "black smokers" through hydrothermal vents on the mid-ocean ridge crest. The rates of fluid flow turned out to be very high. Marine geologists have estimated that the entire volume of the ocean is circulated through the cracks and vents of the spreading centers in only 10 million years.

Scientists have come to realize that the hydrosphere-lithosphere interactions at spreading centers profoundly affect the geology, chemistry, and biology of the oceans.

• The creation of new lithosphere accounts for almost 60 percent of the energy flowing out of Earth's interior. Circulating seawater cools the new lithosphere very efficiently and therefore plays a major role in the outward transport of Earth's internal heat.

• Hydrothermal activity leaches metals and other elements from the new crust, injecting them into the oceans. These elements contribute as much to seawater chemistry as the mineral components of all the world's rivers bring into the oceans from the continents.

• Metal-rich minerals precipitate out of the circulating seawater and form ores of zinc, copper, and iron in shallow parts of the oceanic crust. The ores form when seawater sinks through porous volcanic rocks; is heated; and leaches these elements from the lavas, dikes, gabbros, and perhaps even the underlying mush. When the heated seawater, enriched with dissolved minerals, rises and reenters the cold ocean, the ore-forming minerals precipitate (see Figure 17.7).

The hydrothermal energy and nutrients from circulating seawater feed unusual colonies of strange organisms whose energy comes from Earth's interior rather than from sunlight. Complex ecosystems at spreading centers include microorganisms that provide food for giant clams and for tube worms up to several meters long. These types of microorganisms also populate thermal springs with temperatures above the boiling point of water. The presence at the roots of the "tree of life" of microorganisms that do not require sunlight and that can survive in these energetic, chemically rich environments leads some scientists to believe that life on Earth may have begun at ocean spreading centers.

Volcanism in Subduction Zones

One of the most striking features of subduction zones is the chain of volcanoes that parallels the convergent boundary above the sinking slab of oceanic lithosphere, regardless of whether the overriding plate is oceanic or continental (see Figure 6.19). The Aleutian and Mariana island arcs are typical of the volcanic chains that form during ocean-ocean convergence. The volcanoes of the Andes Mountains mark the convergence boundary where the oceanic Nazca Plate slides beneath the continental South American Plate. Farther north, the subduction of the small Juan de Fuca Plate under western North America gives rise to the volcanoes of the Cascade Range, which stretch from northern California to British Columbia. Mount St. Helens and Mount Rainier are among them.

Magmas that feed subduction-zone volcanoes are produced by fluid-induced melting (see Chapter 5) and are more varied than the basalts of mid-ocean ridge volcanism. The lavas range from mafic to felsic—that is, from basalts to andesites to rhyolites. Volcanoes over the deeper parts of the subduction zone, where melting is going on, extrude basaltic, andesitic, and rhyolitic lavas and pyroclasts, forming a wide range of volcanic rocks. These volcanoes and the volcanic rocks that they expel form the islands of ocean volcanic arcs, such as the Aleutian Islands of Alaska.

Where subduction takes place beneath a continent, the many volcanoes and the volcanic rocks coalesce to form a mountainous arc on land. Subduction of an offshore oceanic plate has generated one such arc—the Cascade Range with its active volcanoes (see Figure 5.11), such as Mount St. Helens—in northern California, Oregon, and Washington. Between the island arc and the deep-sea trench some 200 km seaward is an area called the forearc (see Figure 5.15). This area includes a forearc basin—a depressed zone that fills with sediments derived from the arc—and an accretionary wedge formed from the pile of sediments and ocean crust scraped off the descending plate.

The terrain of the Japanese Islands is a prime example of the complex of intrusives and extrusives that evolves over many millions of years at a subduction zone. Everywhere in this small country are all kinds of extrusive igneous rocks of various ages, mixed in with mafic and intermediate intrusives, metamorphosed volcanic rocks, and sedimentary rocks derived from erosion of the igneous rocks. The erosion of these various rocks has contributed to the distinctive landscapes portrayed in so many classical and modern Japanese paintings.

In all these ways, igneous rocks reveal the major forces shaping the Earth and the interaction of Earth's climate systems with the plate tectonic system. Each plate tectonic setting produces its own pattern of igneous rocks: the lava flows and pyroclastics extruded from volcanoes; the batholiths, dikes, and sills intruded at depth; and the wide variety of rocks that come from magmas of distinctive compositions following their own routes of differentiation.

Intraplate Volcanism: The Mantle Plume Hypothesis

As we have just learned, Earth's plate boundaries are sites of prolific volcanism. At these locations, where plates are first created and then destroyed, melts are produced as a direct response to plate tectonic forces and heat originating in Earth's interior. However, as we will now learn, there is an additional type of volcanism that occurs within the interiors of Earth's lithospheric plates. Figure 6.18 shows the distribution of *intraplate volcanism* —that is, volcanoes far from plate boundaries.

Hot Spots and Mantle Plumes Decompression melting explains volcanism at spreading centers, and fluid-induced melting explains volcanism above subduction zones, but what mechanism produces intraplate volcanism? Consider the Hawaiian Islands, in the middle of the Pacific Plate (Figure 6.18). This island chain begins with the active volcanoes on the Big Island of Hawaii and continues to the northwest as a string of progressively older, extinct, eroded, and submerged volcanic ridges and mountains. In contrast to the seismically active mid-ocean ridges, the Hawaiian chain is not marked by frequent large earthquakes (except near its volcanic center). It is essentially aseismic (without earthquakes) and is therefore called an *aseismic ridge.* Volcanically active **hot spots** at the beginnings of progressively older aseismic ridges can be found elsewhere in the Pacific and in other large ocean basins. The active volcanoes of Tahiti, at the southeastern end of the Society Islands, and the Galápagos Islands, at the western end of the aseismic Nazca Ridge, are two examples.

Once the general pattern of plate motions had been worked out, geologists were able to show that these aseismic ridges approximated the volcanic tracks that the plates would make over a set of hot spots fixed relative to one another, as if they were blowtorches anchored in Earth's mantle (**Figure 6.20**). Based on this evidence, they hypothesized that hot spots were the volcanic manifestations of hot, solid material rising in narrow, cylindrical jets from deep within the mantle (perhaps as deep as the core-mantle boundary), called **mantle plumes**. When the peridotites brought up in a mantle plume reach lower pressures at shallow depths, they begin to melt, producing basaltic magma. The magma penetrates the lithosphere and erupts at the surface. The current position of the plate over the hot spot is marked

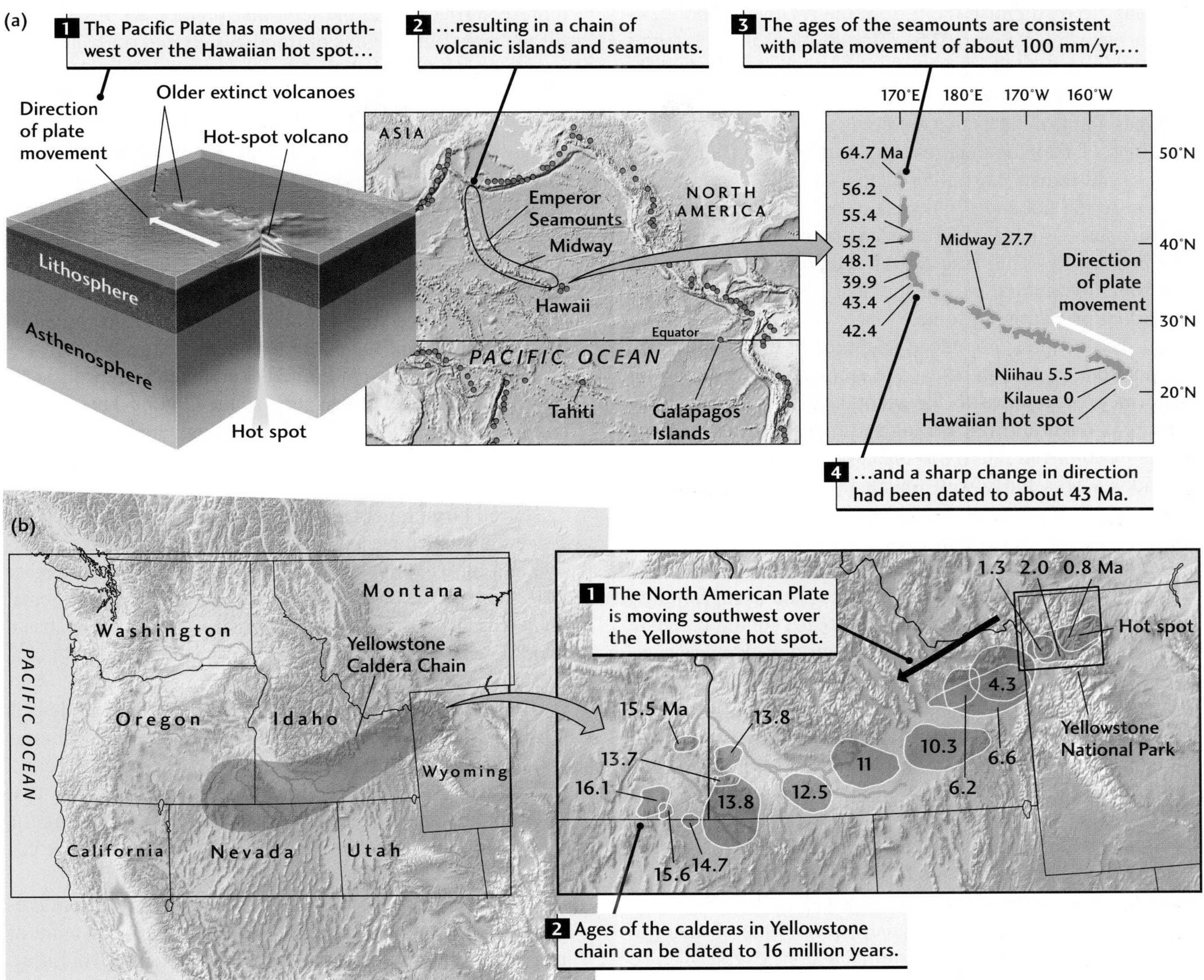

Figure 6.20 Plate motion generates a trail of progressively older volcanoes. (a) The Hawaiian Island chain and its extension into the northwestern Pacific, showing the northwestward trend toward progressively older ages. (b) The Yellowstone volcanic track marks the movement of the North American Plate over a hot spot during the past 16 million years. [Wheeling Jesuit University/NASA Classroom of the Future.]

by an active volcano, which becomes inactive as the plate moves it away from the hot spot. The plate motion thus generates a trail of extinct, progressively older volcanoes. As shown in Figure 6.20, the Hawaiian Islands fit this pattern rather nicely, yielding a rate of movement of the Pacific Plate over the Hawaiian hot spot of about 100 mm/year.

Some aspects of intraplate volcanism within continents have also been explained using the plume hypothesis. Yellowstone is an example. As described earlier in this chapter, the Yellowstone Caldera in northwestern Wyoming, only 600,000 years old, is still volcanically active with geysers, boiling springs, uplift, and earthquakes. It is the youngest member of a chain of sequentially older and now extinct calderas that supposedly mark the movement of the North American Plate over the Yellowstone hot spot (see Figure 6.20). The oldest member of the chain, a volcanic area in Oregon, erupted about 16 million years ago. If the Yellowstone hot spot is indeed fixed, a simple calculation indicates that the North American Plate moved over it to the southwest at a rate of about 25 mm/year during the past 16 million years. Accounting for the relative motion of the Pacific and North American plates, this rate and direction are consistent with the plate motions inferred from Hawaii.

Assuming that hot spots are anchored by plumes coming from the deep mantle, geologists can use the worldwide distribution of their volcanic tracks to compute how the global system of plates is moving with respect to the deep mantle. The results are sometimes called "absolute plate motions" to distinguish them from the relative motions between plates (see Figure 2.5). The absolute plate motions derived from the hot-spot tracks have helped geologists to understand what forces are driving the plates (see Chapter 2, pages 44–45). Plates that are being subducted along large fractions of their boundaries—such as the Pacific, Nazca, Cocos, and Indian-Australian plates—are moving fast with respect to the hot spots, whereas plates without much subducting slab—such as the Eurasian and African plates—are moving slowly. This observation supports the hypothesis that the gravitational pull of the dense, sinking slabs is an important plate-driving force.

If the hot spots are indeed anchored in the mantle, then the hot-spot tracks can be used to reconstruct the history of plate motions relative to the mantle, just as the magnetic isochrons have allowed geologists to reconstruct how the plates have moved relative to one another. As we have seen, this idea works fairly well for recent plate motions. Over longer periods of time, however, a number of problems arise. For instance, according to the fixed-hot-spot hypothesis, the sharp bend in the Hawaiian aseismic ridge 43 million years ago (where it becomes the north-trending Emperor seamount chain; see Figure 6.20) should coincide with an abrupt shift in Pacific Plate motion. However, no sign of a shift is evident in the magnetic isochrons, leading some geologists to question the fixed-hot-spot hypothesis. Others have pointed out that, in a convecting mantle, plumes would not necessarily remain fixed relative to one another but might be "blowin' in the wind" of the shifting convection currents.

Almost all geologists accept the notion that hot-spot volcanism is caused by some type of upwelling in the mantle beneath the plates. However, the mantle plume hypothesis—that these upwellings are narrow conduits of material rising from the deep mantle—remains controversial. Even more controversial is the idea that plumes are responsible for the great outpourings of flood basalts and other large igneous provinces.

Large Igneous Provinces The origin of fissure eruptions of basalt on continents—such as those that formed the Columbia River plateau and even larger lava plateaus in Brazil-Paraguay, India, and Siberia—is a major puzzle. The geologic record shows that immense amounts of lava, up to several million cubic kilometers, can be released in a period as short as a million years. During these events, the lava eruption rates at a single spot on Earth's surface can equal the eruption rate of the entire mid-ocean ridge system!

Flood basalts are not limited to continents; they also create large oceanic plateaus, such as the Ontong-Java Plateau on the northern side of the island of New Guinea and major parts of the Kerguelen Plateau in the southern Indian Ocean (**Figure 6.21**). These features are all examples of what geologists call **large igneous provinces** (LIPs). LIPs are defined as voluminous emplacements of predominantly mafic extrusive and intrusive rock whose origins lie in processes other than "normal" seafloor spreading. LIPs include continental flood basalts and associated intrusive rocks, ocean basin flood basalts, and the aseismic ridges of hot spots. A global map of these provinces is given in Figure 6.21.

The volcanism that covered much of Siberia with lava is of special interest because it occurred at the same time as the greatest extinction of species in the geologic record, about 250 million years ago. Some geologists think that the eruption caused the extinction, perhaps by polluting the atmosphere with volcanic gases that triggered a major climate change.

Many geologists believe that almost all LIPs were created at hot spots by mantle plumes. However, the amount of lava erupting from the most active hot spot on Earth today,

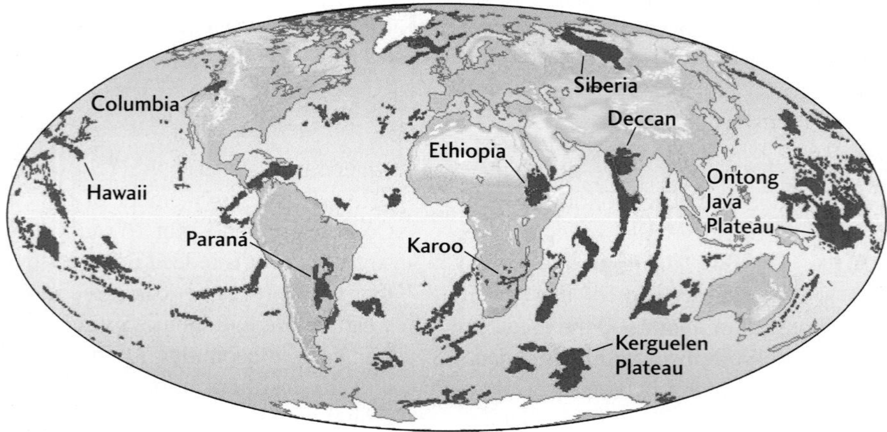

Figure 6.21 The global distribution of large igneous provinces on continents and in the ocean basins. These provinces are marked by unusually large outpourings of basaltic magmas through hot spots and fissure eruptions. They are hypothesized to be the result of major melting events caused by the arrival of plume heads at Earth's surface. [After M. Coffin and O. Eldholm, Figure 1, *Rev. Geophys.* 32:1–36, 1994.]

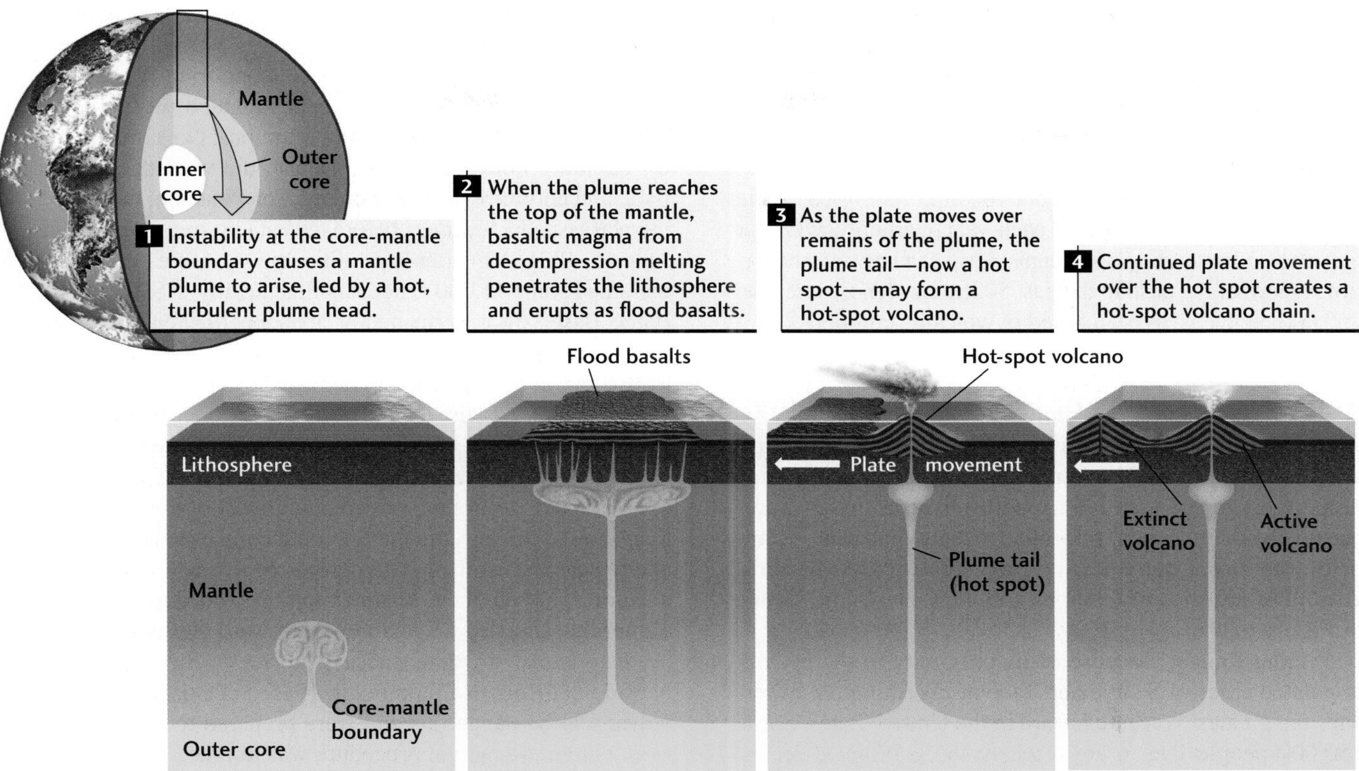

Figure 6.22 A speculative model for the formation of flood basalts and other large igneous provinces. A new plume rises from the core-mantle boundary, led by a hot, turbulent plume head. When the plume head reaches the top of the mantle, it flattens, generating a huge volume of basaltic magma, which erupts as flood basalts.

Hawaii, is paltry compared to the enormous outpourings during flood basalt episodes. What explains these unusual bursts of basaltic magma from the mantle? Some geologists have speculated that they are caused when a new plume rises from the core-mantle boundary. According to this hypothesis, a large, turbulent blob of hot material—a "plume head"—leads the way. When this plume head reaches the top of the mantle, it generates a huge quantity of magma by decompression melting, which erupts in massive flood basalts (**Figure 6.22**). Others dispute this hypothesis, pointing out that continental flood basalts often seem to be associated with preexisting zones of weakness in the continental plates—suggesting that the magmas are generated by convective processes localized in the upper mantle. Sorting out the origins of LIPs is one of the most exciting areas of current geological research.

Volcanism and Human Affairs

Large volcanic eruptions are not just of academic interest to geologists; they are also a significant natural hazard to human society. Of the many volcanoes that have affected communities worldwide, one had a particularly powerful effect on Western civilization. Teams of archeologists and marine geologists have pieced together the story of the demise of Thera (formerly Santorini), a volcanic island in the Aegean Sea. The eruption of Thera in 1623 B.C. appears to have been far more violent than that of Krakatoa. The center of the island collapsed, forming a caldera 7 km by 10 km in diameter. It is visible today as a lagoon as much as 500 m deep, with two small active volcanoes in the center. The lagoon is rimmed by two crescent-shaped islands known for their wine exports and scenic beauty and still subject to destructive earthquakes. The volcanic debris and tsunami resulting from this ancient catastrophe destroyed dozens of coastal settlements over a large part of the eastern Mediterranean. Some scientists have attributed the mysterious disappearance of the Minoan civilization to this cataclysm, and it may be the original source of the "lost continent of Atlantis" myth. The course of history was probably changed by this one volcanic event. It could happen again.

The date of eruption is known from a layer of volcanic ash transported to Greenland from Thera. The ash layer was found in 1994 in an ice core extracted from deep within the Greenland ice sheet. Annual cycles of snow deposition can be counted in a core, just as rings in a tree are counted to determine its age. The ice core goes back 7000 years and provides evidence of some 400 ancient volcanic eruptions.

Can volcanic eruptions be predicted? To some extent they can. Important premonitory signals are uplift and tilting of the ground and the occurrence of heightened earthquake activity (see Feature 6.1). Improving our ability to predict eruptions is important because there are about 100 high-risk volcanoes in the world, and some 50 erupt each year. Certainly, with our growing understanding of volcanism, we can improve the terrible record of the past. In the past 500 years alone, more than 250,000 people have been killed by volcanic eruptions.

Reducing the Risks of Hazardous Volcanoes

Of Earth's 500 to 600 active volcanoes, one of six has claimed human lives. Volcanoes can kill people and damage property by tephra (ash fall), mudflows, lavas, gases, pyroclastic flows, tsunamis, and other events that can accompany eruptions. **Figure 6.23** portrays some of these hazards. Historical statistics of fatalities and their causes are shown in **Figure 6.24**.

Mount Rainier, because of its proximity to the heavily populated cities of Seattle and Tacoma, Washington, is perhaps the greatest volcanic hazard in the United States. Some 150,000 people live in areas where the geological record shows evidence of floods and lahars that swept down from the volcano over the past 6000 years. An eruption that was not predicted could kill thousands of people and cripple the economy of the Pacific Northwest, according to the National Academy of Sciences.

On the positive side, scientists monitoring Mount St. Helens (see Feature 6.1) and Mount Pinatubo were able to issue warnings of imminent major eruptions. Government infrastructures were in place to evaluate the warnings and to issue and enforce evacuation orders. For Pinatubo, the warning was issued a few days before the cataclysmic eruption on June 15, 1991. A quarter of a million people were evacuated, including some 16,000 residents of the nearby U.S. Clark Air Force Base (since permanently abandoned). Tens of thousands of lives were saved from the lahars that destroyed everything in their paths. Casualties were limited to the few who disregarded the order. In 1994, 30,000 residents of Rabaul, Papua New Guinea, were successfully evacuated by land and sea hours before two volcanoes on either side of the town erupted, destroying or damaging most of it. Many owe their lives to the government for conducting evacuation drills and to scientists at the local volcano observatory who issued a warning when their seismographs recorded the ground tremor that signaled magma moving toward the surface.

In contrast to these success stories is the tragedy at Nevado del Ruiz in Colombia in 1985. Scientists knew this volcano to be dangerous and were prepared to issue warnings, but no evacuation procedure was in place. As a result, 25,000 lives were lost in the lahars triggered by a minor eruption. Similarly, an international team of scientists provided several days' advance warning of the 2002 eruption of

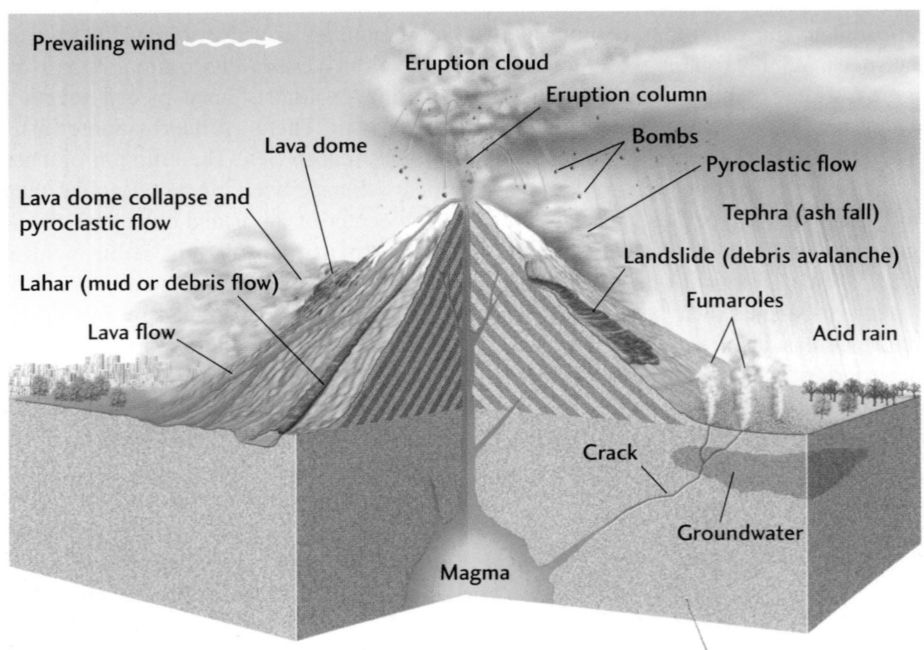

Figure 6.23 Some of the volcanic hazards that can kill people and destroy property. [B. Meyers et al., U.S. Geological Survey.]

(a)

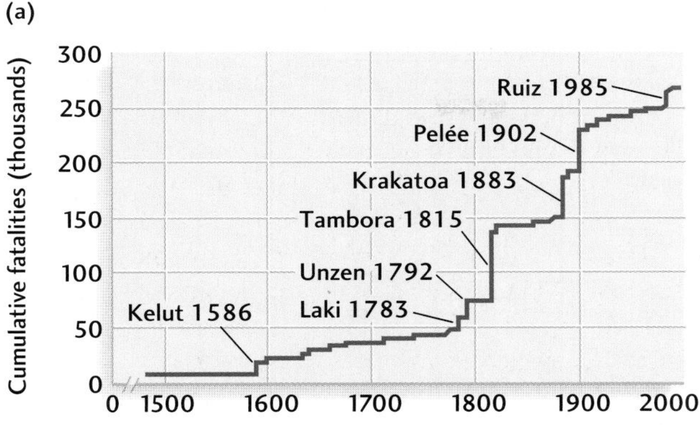

(b)

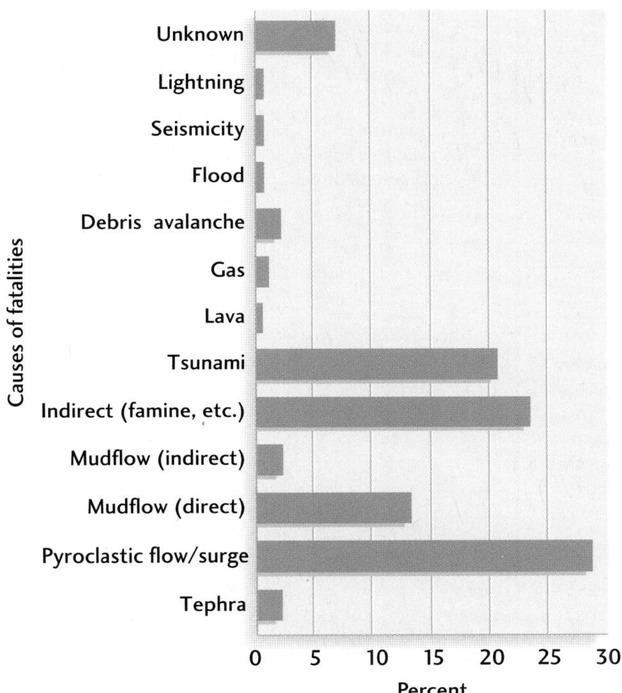

Figure 6.24 (a) Cumulative volcano statistics since A.D. 1500. The seven eruptions that dominate the record, each claiming 10,000 or more victims, are named. These account for two-thirds of the total deaths. (b) Cause of volcano fatalities since A.D. 1500. [After T. Simkin, L. Siebert, and R. Blong. *Science* 291: 255, 2001.]

Mount Nyiragongo, a volcano in the East African Rift Valley near Goma in Congo (Kinshasa). Unfortunately, a functioning government was not in place and the warning went unheeded. The suffering was great for the 400,000 people in this poverty-stricken country who fled their homes without guidance or preparation.

Volcanology has progressed to the point that we can identify the world's dangerous volcanoes and characterize their potential hazards from deposits laid down in earlier eruptions. These hazard assessments can be used to guide zoning regulations to restrict land use—the most effective measure to reduce casualties. Instrumented monitoring (see Feature 6.1) can detect signals such as earthquakes, swelling of the volcano, and gas emissions that warn of impending eruptions. People at risk can be evacuated if the authorities are organized and prepared. Volcanic eruptions cannot be prevented, but their catastrophic effects can be significantly reduced by a combination of science and enlightened public policy.

With the growth in air travel, a volcanic hazard that is attracting increased attention is the encounter by commercial jet passenger planes of volcanic ash lofted into the air traffic lanes from erupting volcanoes. According to the U.S. Geological Survey, over a period of 25 years more than 60 airplanes have been damaged by such accidents. One Boeing 747 temporarily lost all four engines when ash from an erupting volcano in Alaska was sucked into the engine and caused it to flame out. Fortunately, the pilot was able to make an emergency landing. Warnings of eruptions ejecting volcanic ash near air traffic lanes are now being issued by several countries.

The work of volcanologists is hazardous—fourteen have lost their lives since 1991. In one case, a team of volcanologists gathering data in the crater of Galeras volcano in the Colombian Andes was caught in a pyroclastic eruption. Several members lost their lives in the ash and incandescent boulders that exploded from the crater. Professor Stanley Williams of Arizona State University was a member of that team. He is now at work developing an instrument that can analyze volcanic gases in the crater of an active volcano and transmit this information to scientists at a safe distance. An impending eruption might be predicted from the kinds of gases that are emitted. The International Association of Volcanology has recommended "that scientists approach active craters only when absolutely necessary, that large groups avoid hazardous areas, and that everyone wear hard hats and protective clothing." Could these geologists have avoided this tragic loss of life by following this advice?

Can volcanic eruptions be controlled? Not likely, although in special circumstances and on a small scale, the damage can be reduced. Perhaps the most successful attempt to control volcanic activity was made on the Icelandic island of Heimaey in January 1973. By spraying advancing lava with seawater, Icelanders cooled and slowed the flow, preventing the lava from blocking the port entrance and saving some homes from destruction.

In the years ahead, the best policy for protecting the public will be the establishment of more warning and evacuation systems and more rigorous restriction of settlements in potentially dangerous locations. But even these precautions may not help. Dormant or long-extinct volcanoes can come to life suddenly—as Vesuvius and Mount St. Helens did after hundreds of years. (Some potentially dangerous volcanoes in

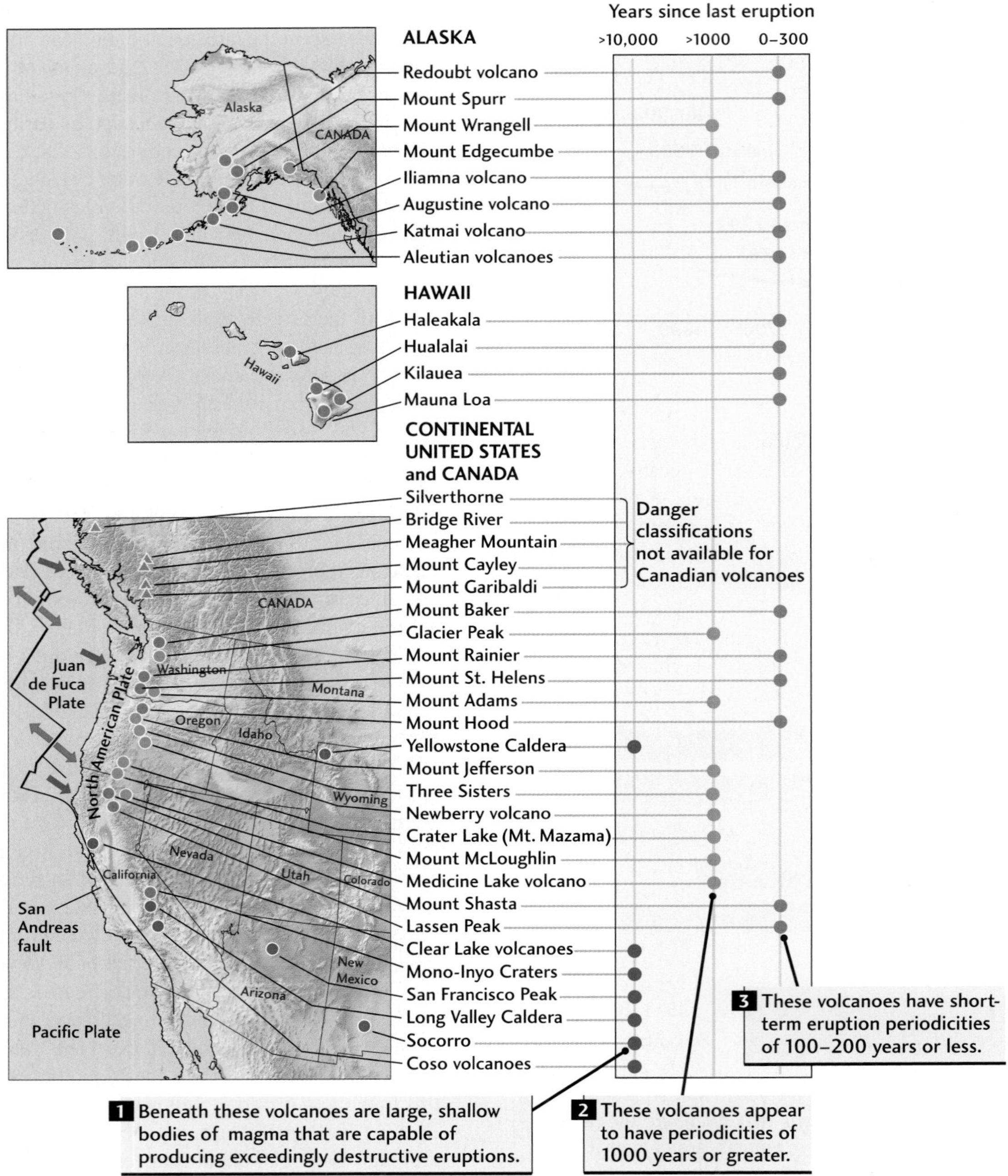

Figure 6.25 Locations of potentially hazardous volcanoes in the United States and Canada. Volcanoes within each U.S. group are color-coded blue, green, and red in the order of declining probable cause for concern, subject to revision as studies progress (danger classifications are not available for Canadian volcanoes). Note the relationship between the volcanoes, which extend from northern California to British Columbia, and the subduction plate boundary between the North American Plate and the Juan de Fuca Plate. [After R. A. Bailey, P. R. Beauchemin, F. P. Kapinos, and D. W. Klick, U.S. Geological Survey.]

the United States and Canada are identified in **Figure 6.25**.) An even more difficult problem of prediction is posed by eruptions like that of Paricutín, which rose up with little warning from a small hole in a Mexican cornfield in 1943. Whole towns were quickly buried by ash and lava as the new volcano grew by repeated eruptions. Learning how to sense the movements of deep lava in relation to possible new outlets to the surface is a real challenge for geologists.

Reaping the Benefits of Volcanoes

We have seen something of the beauty of volcanoes and something of their destructiveness. Volcanoes contribute to our well-being in many ways. In Chapter 1, we mentioned that the atmosphere and the oceans may have originated in volcanic episodes of the distant past. Soils derived from volcanic materials are exceptionally fertile because of the mineral nutrients they contain. Volcanic rock, gases, and steam are also sources of important industrial materials and chemicals, such as pumice, boric acid, ammonia, sulfur, carbon dioxide, and some metals. Seawater circulating through fissures in the ocean-ridge volcanic system is a major factor in the formation of ores and in maintaining the chemical balance in the oceans. Thermal energy from volcanism is being harnessed in more and more places. Most of the houses in Reykjavík, Iceland, are heated by hot water tapped from volcanic springs. Geothermal steam, originating in water heated by contact with hot volcanic rocks below the surface, is exploited as a source of energy for the production of electricity in Italy, New Zealand, the United States, Mexico, Japan, and the former Soviet Union.

SUMMARY

Why does volcanism occur? Volcanism occurs when molten rock inside the Earth rises buoyantly to the surface (because it is less dense than surrounding rock). The basic cause is Earth's internal heat, a vestige of its violent origin.

What are the three major categories of lava? Lavas are classified as felsic (rhyolite), intermediate (andesite), or mafic (basalt) on the basis of the decreasing amounts of silica and the increasing amounts of magnesium and iron they contain. The chemical composition and gas content of lava are important factors in the form an eruption takes.

How are the structure and terrain of a volcano related to the kind of lava it emits and the style of its eruption? Basalt can be highly fluid. On continents, it can erupt from fissures and flow out in thin sheets to build a lava plateau. A shield volcano grows from repeated eruptions of basalt from vents. Silicic magma is more viscous and, when charged with gas, tends to erupt explosively. The resultant pyroclastic debris may pile up into a cinder cone or cover an extensive area with ash-flow sheets. A stratovolcano is built of alternating layers of lava flows and pyroclastic deposits. The rapid ejection of magma from a magma chamber a few kilometers below the surface, followed by collapse of the chamber's roof, results in a large surface depression, or caldera. Giant resurgent calderas are among the most destructive natural cataclysms.

How is volcanism related to plate tectonics? The ocean crust forms from basaltic magma that rises from the asthenosphere into fissures of the ocean ridge-rift system where plates separate. All three major types of lava—basaltic (mafic), andesitic (intermediate), and rhyolitic (felsic)—can erupt in convergent zones. The basalts derive from partial melting of the mantle above the subducted plate, induced by water streaming off subducted oceanic crust. Basalts are typical of volcanic islands found at ocean-ocean plate convergences. Andesites and rhyolites are more commonly found in the volcanic belts of ocean-continent convergent plate boundaries. The addition to basaltic magma of silica and other elements derived from the remelting of felsic continental crust or from the melting of seafloor sediments and crust atop the downgoing slab can produce andesites and rhyolites. Within plates, basaltic volcanism occurs above hot spots, which are manifestations of plumes of hot material that rise from deep in the mantle.

What are some beneficial and hazardous effects of volcanism? In Earth's evolution, volcanic eruptions released much of the water and gases that formed the oceans and atmosphere. Geothermal heat drawn from areas of recent volcanism is of growing importance as a source of energy. An important ore-forming process takes place when groundwater circulates around buried magma or seawater circulates through ocean-floor rifts. Volcanic eruptions have killed some 250,000 people in the past 500 years. For a summary of how this happens, see Figure 6.24.

Key Terms and Concepts

andesitic lava (p. 116)
ash-flow lava (p. 122)
basaltic lava (p. 115)
caldera (p. 120)
cinder-core volcano (p. 120)
composite volcano (p. 120)
crater (p. 120)
diatreme (p. 121)
fissure eruption (p. 122)
flood basalt (p. 115)
hot spot (p. 130)
hydrothermal activity (p. 127)

lahar (p. 122)

large igneous province (p. 132)

mantle plume (p. 130)

pillow lava (p. 114)

pyroclast (p. 117)

rhyolitic lava (p. 116)

shield volcano (p. 120)

stratovolcano (p. 120)

volcanic dome (p. 120)

volcanic geosystem (p. 114)

volcano (p. 114)

Exercises

This icon indicates that there is an animation available on the Web site that may assist you in answering a question.

1. The asthenosphere has been identified as a major source of magma. Why? What forces magma to rise to the surface?

2. What is the difference between magma and lava? Give examples of types of volcanic rocks and their coarse-grained, intrusive counterparts.

3. Describe the principal styles of eruptions and the deposits and landforms that each style produces.

4. What is the association between plate boundaries and volcanism? Can the eruptive style and the composition of volcanic deposits be correlated with plate boundaries?

5. Under what circumstances do lahars occur? Hot springs? Ash-flow deposits?

6. Name the most dangerous features of volcanoes.

7. What signals an impending eruption?

Thought Questions

This icon indicates that there is an animation available on the Web site that may assist you in answering a question.

1. What public policy initiatives do the eruptions of Mount St. Helens, Mount Pinatubo, and Nevado del Ruiz suggest should be undertaken in such areas as zoning, land use, insurance, warning systems, and public education?

2. Do a risk-benefit analysis of volcanoes—that is, tabulate both their dangers and their contributions to humankind—and decide whether you would prefer an Earth with or without them.

3. What have we learned about the Earth's interior from volcanoes?

4. The viscosity of lava depends to a great degree upon its mineralogy. What is the relationship between lava mineralogy and volcano morphology?

5. While on a field trip, you come across a volcanic formation that resembles a field of sandbags. The individual ellipsoidal-shaped forms have a smooth, glassy surface texture. What type of lava is this, and what information does this give you about its history?

6. Explain why some of the volcanoes present in the Hawaiian Islands are dormant while others remain active. Would the situation be different if the Hawaiian Islands were in a subduction zone? Explain your answer.

Suggested Readings

A.G.U. Special Report. 1992. *Volcanism and Climatic Change.* Washington, D.C.: American Geophysical Union.

Bruce, Victoria. 2001. *No Apparent Danger: The True Story of Volcanic Disaster at Galeras and Nevado del Ruiz.* New York: HarperCollins.

Decker, R. W., and B. Decker. 1989. *Volcanoes,* revised and updated. New York: W. H. Freeman.

Dvorak, J. J., and D. Dzurisin. 1997. Volcano geodesy: The search for magma reservoirs and the formation of eruptive vents. *Reviews of Geophysics* 35: 343–384.

Dvorak, J. J., C. Johnson, and R. I. Tilling. 1982. Dynamics of Kilauea volcano. *Scientific American* (August): 46–53.

Edmond, J. M., and K. L. Von Damm. 1992. Hydrothermal activity in the deep sea. *Oceanus* (Spring): 74–81.

Fisher, R. V., G. Heiken, and J. B. Hulen. 1997. *Volcanoes: Crucibles of Change.* Princeton, N.J.: Princeton University Press.

Francis, P. 1983. Giant volcanic calderas. *Scientific American* (June): 60–70.

Heiken, G. 1979. Pyroclastic flow deposits. *American Scientist* 67: 564–571.

Humphreys, E. D., et al. 2000. Beneath Yellowstone: Evaluating plume and nonplume models using teleseismic images of the upper mantle. *GSA Today* 10: 1–7.

Krakaner, J. 1996. Geologists worry about dangers of living "under the volcano." *Smithsonian* (July): 33–125.

McPhee, J. 1990. Cooling the lava. In *The Control of Nature.* New York: Farrar, Straus & Giroux.

MELT Seismic Team. 1998. Imaging the deep seismic structure beneath a mid-ocean ridge: The MELT experiment. *Science* 280: 1215–1218.

National Research Council. 1994. *Mount Rainier: Active Cascade Volcano.* Washington, D.C.: National Academy Press.

Sigurdsson, H., ed. 2000. *The Encyclopedia of Volcanoes.* New York: Academic Press.

Simkin, T., L. Siebert, and R. Blong. 2001. Volcano fatalities: Lessons from the historical record. *Science* 291: 255.

Tilling, R. I. 1989. Volcanic hazards and their mitigation: Progress and problems. *Reviews of Geophysics* 27(2): 237–269.

Vink, G. E., and W. J. Morgan. 1985. The Earth's hot spots. *Scientific American* (April): 50–57.

White, R. S., and D. P. McKenzie. 1989. Volcanism at rifts. *Scientific American* (July): 62–72.

Williams, Stanley, and F. Montaigne. 2001. *Surviving Galeras.* Boston: Houghton Mifflin.

Winchester, S. 2003. *Krakatoa: The Day the World Exploded, August 27, 1883.* New York: HarperCollins.

Wright, T. L., and T. C. Pierson. 1992. Living with volcanoes. *U.S. Geological Survey Circular 1073.* Washington, D.C.: U.S. Government Printing Office.

A view of the southern escarpment of Oman along the Arabian Sea.
[Petroleum Development Oman.]

CHAPTER

7

Weathering and Erosion

"The heights of our land are thus leveled with the shores; our fertile plains are formed from the ruins of mountains."

James Hutton (1788)

Solid as the hardest rocks may seem, all rocks—like rusting old automobiles and yellowed old newspapers— eventually weaken and crumble when exposed to water and the gases of the atmosphere. Unlike those cars and newspapers, however, rocks may take thousands of years to disintegrate. **In this chapter, we look closely at two geologic processes that break down and fragment rocks: weathering and erosion.**

Weathering is the general process by which rocks are broken down at Earth's surface. Weathering produces all the clays of the world, all soils, and the dissolved substances carried by rivers to the ocean. Rocks weather in two ways:

- **Chemical weathering** occurs when the minerals in a rock are chemically altered or dissolved. The blurring or disappearance of lettering on old gravestones and monuments is caused mainly by chemical weathering.
- **Physical weathering** takes place when solid rock is fragmented by mechanical processes that do not change its chemical composition. The rubble of broken stone blocks and columns that were once stately temples in ancient Greece is primarily the result of physical weathering. Physical weathering also caused the cracks and breaks in the ancient tombs and monuments of Egypt.

Chemical and physical weathering reinforce each other. Chemical decay weakens fragments of rocks and makes them more susceptible to breakage. The smaller the pieces produced by physical weathering, the greater the surface area available for chemical weathering.

Weathering, Erosion, and the Rock Cycle

After tectonics and volcanism have made mountains, chemical decay and physical breakup join with rainfall, wind, ice, and snow to wear away those mountains. **Erosion** is the set of processes that loosen and transport soil and rock downhill or downwind. These processes carry away weathered material on Earth's surface and deposit it elsewhere. As erosion moves weathered solid material, it exposes fresh, unaltered rock to weathering.

Weathering and erosion are major geologic processes in the rock cycle and in Earth systems, as described in Chapter 4. Along with tectonics and volcanism (two other elements of the rock cycle), weathering and erosion shape Earth's surface and alter rock materials, converting igneous and other rocks into sediment and forming soil. In some instances, weathering and erosion are inseparable. When a rock such as pure limestone or rock salt is weathered by rainwater, for example, all the material is completely dissolved and carried away in the water as ions in solution. Chemically weathered material contributes most of the dissolved matter in the oceans.

The early sections of this chapter emphasize chemical weathering, because it is in some ways the fundamental driving force of the whole process. The effects of physical weathering, always important, depend largely on chemical decay. First, however, we examine the factors that control weathering.

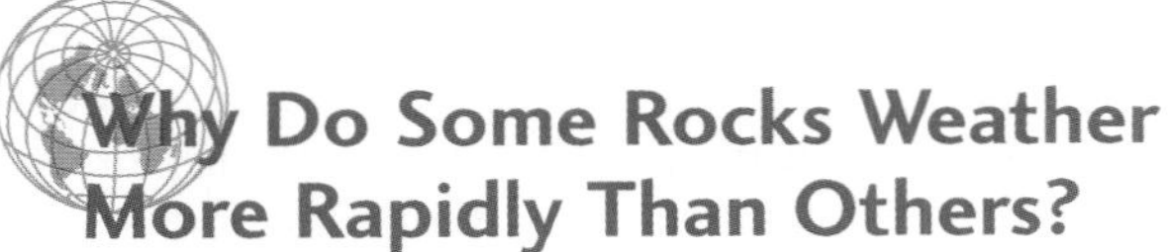

Why Do Some Rocks Weather More Rapidly Than Others?

All rocks weather, but the manner and rate of their weathering vary. The four key factors that control the fragmentation and decay of rocks are the properties of the parent rock, the climate, the presence or absence of soil, and the length of time the rocks are exposed to the atmosphere. These four factors are summarized in Table 7.1.

The Properties of the Parent Rock

The nature of a parent rock affects weathering because (1) various minerals weather at different rates and (2) a rock's structure affects its susceptibility to cracking and fragmentation. Old inscriptions on gravestones offer clear evidence of the varying rates at which rocks weather. The carved letters on a recently erected gravestone stand out in sharp relief from the stone's polished surface. After a hundred years in a moderately rainy climate, however, the surface of a limestone will be dull and the letters inscribed on it will have almost melted away, much as the name on a bar of soap disappears after a few washes (**Figure 7.1**). Granite, on the other hand, will show only minor changes. The differences in the weathering of granite and limestone result from their different mineral compositions. Given enough time, however, even a resistant rock will ultimately decay. After several hundred years, the granite monument will also have weathered appreciably, and its surface and letters will be somewhat dulled and blurred. If we used a hand lens to look more closely at the weathered granite, we would see different patterns of weathering in its constituent mineral grains. The feldspar crystals would show signs of corrosion, and their surfaces would be chalky and covered with a thin layer of soft clay. The chemical composition of the outer layers of the grains of feldspar would have changed, creating a new mineral. The quartz crystals would appear fresh—clear and unaltered.

Table 7.1 Major Factors Controlling Rates of Weathering

	Weathering Rate: Slow →		→ Fast
PROPERTIES OF PARENT ROCK			
Mineral solubility in water	Low (e.g., quartz)	Moderate (e.g., pyroxene, feldspar)	High (e.g., calcite)
Rock structure	Massive	Some zones of weakness	Very fractured or thinly bedded
CLIMATE			
Rainfall	Low	Moderate	High
Temperature	Cold	Temperate	Hot
PRESENCE OR ABSENCE OF SOIL AND VEGETATION			
Thickness of soil layer	None—bare rock	Thin to moderate	Thick
Organic content	Low	Moderate	High
LENGTH OF EXPOSURE			
Short	Moderate	Long	

Figure 7.1 Early-nineteenth-century gravestones at Wellfleet, Massachusetts. The stone on the right is limestone and is so weathered that it is unreadable. The stone on the left is slate, which retains its legibility under the same conditions. [Courtesy of Raymond Siever.]

A rock's structure also affects physical weathering. Granite monuments may remain unbroken and uncracked even after centuries of exposure, though they may show evidence of some chemical weathering. Intrusive igneous rocks, including many granites, may be massive—that is, large masses that show no changes in rock type or structure. Massive rocks have no planes of weakness that contribute to cracking or fragmentation. In contrast, shale, a sedimentary rock that splits easily along thin bedding planes, breaks into small pieces so quickly that only a few years after a new road is cut through a shale, the rock will become rubble.

Climate: Rainfall and Temperature

A tour of graveyards across the North American continent, from the southern United States to northern Canada and Alaska, would reveal that the rate of both chemical and physical weathering varies not only with the properties of the rock but also with the climate—the amount of rainfall and the temperature. High temperatures and heavy rainfall increase the growth rate of organisms and thus promote chemical weathering. Cold and dryness impede the process. Old gravestones in hot, humid Florida are badly chemically weathered, but those of the same age in the equally hot but arid Southwest are hardly affected. And gravestones in cold, dry arctic regions show even less chemical weathering than those found in the Southwest. In cold climates, water can't dissolve minerals because it is frozen. In arid regions, water is relatively unavailable. In both cases, populations of organisms are at a minimum and chemical weathering proceeds slowly.

On the other hand, climates that minimize chemical weathering may enhance physical weathering. For example, freezing water may act as a wedge, widening cracks and pushing a rock apart.

The Presence or Absence of Soil

Soil, one of our most valuable natural resources, is composed of fragments of bedrock, clay minerals formed by the chemical alteration of bedrock minerals, and organic matter produced by organisms that live in the soil. Although soil is itself a product of weathering, its presence or absence affects the chemical and physical weathering of other materials. An old nail that has been buried in soil usually will be so badly rusted that you can snap it like a matchstick. Yet a nail pried from the wood of a centuries-old house may still be strong, covered with only a thin layer of rust. Similarly, a mineral in the soil of a lowland valley may be badly altered and corroded, whereas the same mineral exposed in a nearby cliff of bedrock will be much less weathered. Although the cliff is exposed to occasional rain, the bare rock is usually dry, and weathering proceeds very slowly. No soil forms on the cliff because rain quickly carries loosened particles down to lower areas, where they can accumulate.

Soil production is a *positive-feedback process*—that is, the product of the process advances the process itself. Once soil starts to form, it works as a geological agent to weather rock more rapidly. The soil retains rainwater, and it hosts a variety of vegetation, bacteria, and other organisms. These life-forms create an acidic environment that, in combination with moisture, promotes chemical weathering, which alters or dissolves minerals. Plant roots and organisms tunneling through the soil promote physical weathering by helping to create fractures in a rock. Chemical and physical weathering, in turn, lead to the production of more soil.

The Length of Exposure

The longer a rock weathers, the greater its chemical alteration, dissolution, and physical breakup. Rocks that have been exposed at Earth's surface for many thousands of years form a rind—an external layer of weathered material ranging from several millimeters to several centimeters thick—that surrounds the fresh, unaltered rock. In dry climates, some rinds have grown as slowly as 0.006 mm per 1000 years.

Recently erupted volcanic lavas and ash deposits have had a very short period of exposure at Earth's surface and so are relatively unweathered. Because we know the dates of modern eruptions, we can measure the times required for various degrees of weathering to occur. In the years since the eruptions of Mount St. Helens in 1980, for example, the volcanic ash deposits have weathered appreciably and have altered to form other minerals. After the same length of time, masses of solidified lava are still relatively fresh. The difference in the extent of weathering is due mainly to the fact that the ash is made up of very small particles, which weather faster than the more massive volcanic rocks.

We now consider the two types of weathering—chemical and physical—in more detail.

Chemical Weathering

Rocks chemically weather when their constituent minerals react with air and water. In these chemical reactions, some minerals dissolve. Others combine with water and such components of the atmosphere as oxygen and carbon dioxide to form new minerals. We can deduce some chemical reactions from observations in the field. We can get a better picture of the mechanisms of chemical weathering if we combine field observations with laboratory experiments that simulate natural processes. We begin our investigation by examining the chemical weathering of feldspar, the most abundant mineral in Earth's crust.

The Role of Water in Weathering: Feldspar and Other Silicates

Feldspar is a key mineral in a great many igneous, sedimentary, and metamorphic rocks. Many other kinds of rock-forming silicate minerals also weather much as feldspar does. Feldspar is one of many silicates that are altered by chemical reactions to form the water-containing minerals known as clay minerals. Feldspar's behavior during weathering helps us understand the weathering process in general, for two reasons:

1. There is an overwhelming abundance of silicate minerals in the Earth.

2. The chemical processes of dissolution and alteration that characterize feldspar weathering also characterize weathering in other kinds of minerals.

Earlier in this chapter, when we described the minerals one would find in a weathered granite gravestone, we noted that the feldspar crystals would be corroded and altered. A more extreme example of feldspar weathering can be found in granite boulders in soils of the humid tropics. There, many of the factors that promote weathering—heavy rainfall, high temperature, the presence of soil, and abundant organic activity—are present. Granite boulders found in the tropics are so weakened that they can be easily kicked or pounded into a heap of loose mineral grains. Most of the feldspar particles in these boulders have been altered to clay. Greatly magnified under an electron microscope, any remaining feldspar grains would display corrosion and be coated with a clay rind (**Figure 7.2**). Quartz crystals, in contrast, would be relatively intact and unaltered.

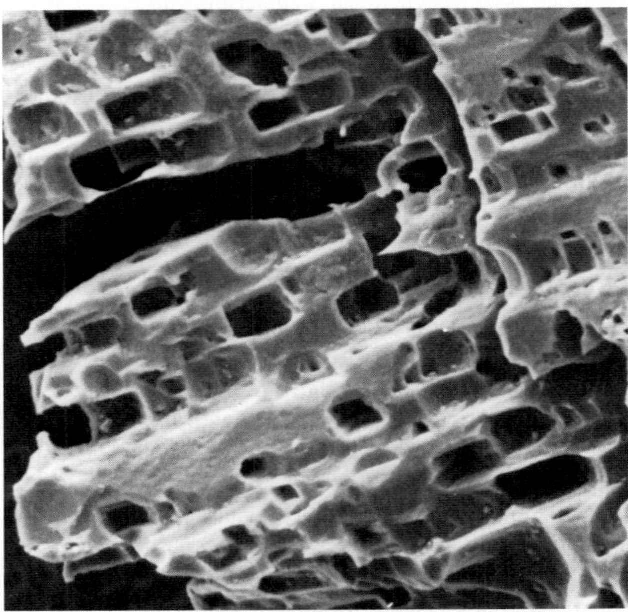

Figure 7.2 A scanning electron micrograph of feldspar etched and corroded by chemical weathering in soil. [From R. A. Berner and G. R. Holden, Jr., "Mechanism of Feldspar Weathering: Some Observational Evidence," *Geology* 5(1977): 369.]

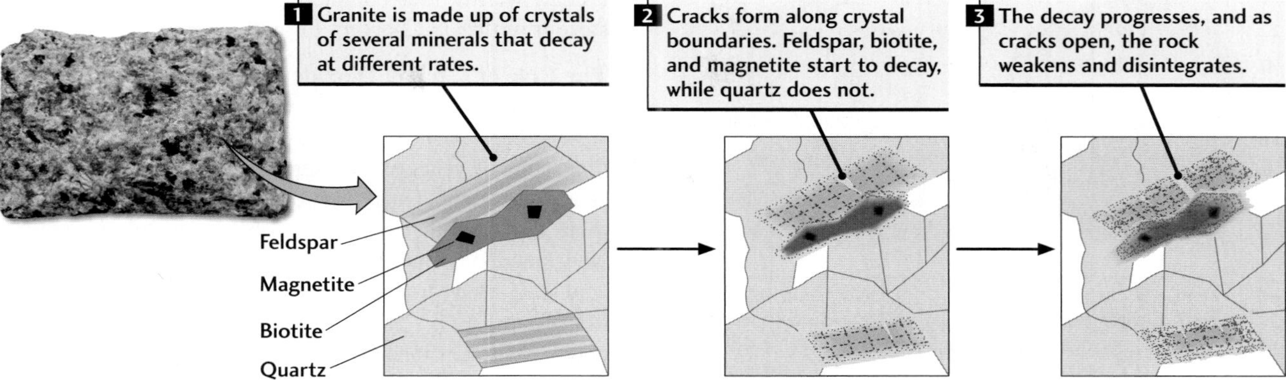

Figure 7.3 Diagrammatic microscopic views of stages in the disintegration of granite. [Chip Clark.]

In a sample of unweathered granite, the rock is hard and solid because the interlocking network of quartz, feldspar, and other crystals holds it tightly together. When the feldspar is altered to a loosely adhering clay, however, the network is weakened and the mineral grains are separated (**Figure 7.3**; see also Figure 5.1). In this instance, chemical weathering, by producing the clay, also promotes physical weathering because the rock now fragments easily along widening cracks at mineral boundaries.

The white to cream-colored clay produced by the weathering of feldspar is **kaolinite,** named for Gaoling, a hill in southwestern China where it was first obtained. Chinese artisans had used pure kaolinite as the raw material of pottery and china for centuries before Europeans borrowed the idea in the eighteenth century.

Only in the severely arid climates of some deserts and polar regions does feldspar remain relatively unweathered. This observation points to water as an essential component of the chemical reaction by which feldspar becomes kaolinite. Kaolinite is a hydrous (that is, containing water in the crystal structure) aluminum silicate. In the reaction that produces kaolinite, the solid feldspar undergoes **hydrolysis** (*hydro* means "water" and *lysis* means "to loosen"). The feldspar is broken down and also loses several chemical components. The alteration is analogous to the chemical reaction that takes place when we make coffee. Solid coffee reacts chemically with hot water to make a solution—the liquid coffee. The reaction extracts caffeine and other components of the solid bean, leaving behind spent coffee grounds. Similarly, rainwater filters into the ground, altering feldspar in rock particles and leaving behind the kaolinite as a residue (**Figure 7.4**).

The only part of a solid that reacts with a fluid is the solid's surface, so as we increase the surface area of the solid, we speed up the reaction. For example, as we grind coffee beans into finer and finer particles, we increase the ratio of their surface area to their volume. The finer the coffee beans are ground, the faster their reaction with water, and the stronger the brew becomes. Similarly, the smaller the fragments of minerals and rocks, the greater the surface area. The ratio of surface area to volume increases greatly as the average particle size decreases, as shown in **Figure 7.5**.

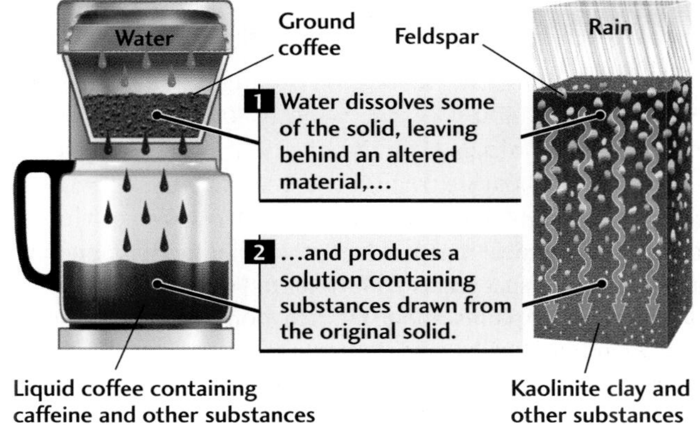

Figure 7.4 The process by which feldspar decays is analogous to the brewing of coffee.

Dissolving Feldspar in Pure Water To learn more about the weathering of feldspar, we can perform a simple laboratory experiment: immerse feldspar in pure water and analyze the solution for the kinds of material that have dissolved. First we grind the common feldspar of granite, orthoclase ($KAlSi_3O_8$), to a powder. This speeds the reaction by exposing more surface area to the water. When we analyze samples taken from the solution after some time has passed, we find small amounts of potassium and silica (SiO_2) dissolved in the water. The reaction of feldspar with water releases dissolved silica and dissolved potassium ions (K^+) and leaves behind

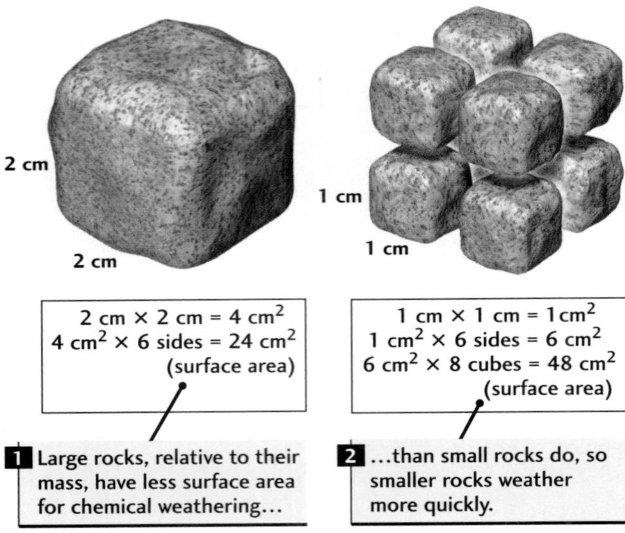

Figure 7.5 As a rock mass breaks into smaller pieces, more surface becomes available for the chemical reactions of weathering.

the new mineral kaolinite, $Al_2Si_2O_5(OH)_4$. We can now describe a chemical reaction for the weathering of granite by saying that feldspar reacts with water to form kaolinite.

Two major points about this reaction give us information about the gains and losses of material as feldspar weathers:

1. Potassium and silica are "lost" by the feldspar and appear as dissolved materials in the water solution.

2. Water is absorbed into the kaolinite crystal structure. This absorption of water is called *hydration* and is one of the major processes of weathering.

Carbon Dioxide, Weathering, and the Climate Geosystem

Variability in the atmosphere's concentration of carbon dioxide leads to corresponding variability in the rate of weathering (**Figure Story 7.6**). Higher levels of carbon dioxide in the atmosphere lead to higher levels in the soil, which increases the rate of weathering (Figure Story 7.6a). In addition, carbon dioxide, a greenhouse gas, makes Earth's climate warmer and thus promotes weathering. Weathering, in turn, converts carbon dioxide into bicarbonate ions (Figure Story 7.6b) and so decreases the amount of carbon dioxide in the atmosphere. This decrease in carbon dioxide eventually results in a cooler climate. In this way, the weathering at the surface of a feldspar grain is linked to the causes of global climate change. As more and more carbon dioxide is used up through weathering and the climate cools, weathering decreases again. As weathering decreases, the amount of carbon dioxide in the atmosphere builds up again, and the climate warms, thus completing the cycle. We consider this system further in Chapters 17 and 23.

The Role of Carbon Dioxide in Weathering The reaction of feldspar with pure water in a laboratory is an extremely slow process. Under laboratory conditions, it would take thousands of years to weather even a small amount of feldspar completely. Thus, this reaction cannot account for the more rapid weathering that we observe widely in nature. If we wanted to, we could speed weathering by adding a strong acid (such as hydrochloric acid) to our solution and dissolve the feldspar in a few days. An acid is a substance that releases hydrogen ions (H^+) to a solution. A strong acid produces abundant hydrogen ions; a weak one, relatively few. The strong tendency of hydrogen ions to combine chemically with other substances makes acids excellent solvents.

On Earth's surface, the most common natural acid—and the one responsible for increasing weathering rates—is carbonic acid (H_2CO_3). This weak acid forms when carbon dioxide (CO_2) gas from the atmosphere dissolves in rainwater:

$$\begin{array}{ccccc} \text{carbon dioxide} & + & \text{water} & \rightarrow & \text{carbonic acid} \\ CO_2 & & H_2O & & H_2CO_3 \end{array}$$

We are familiar with everyday solutions of carbon dioxide in water in the form of carbonated soft drinks. The bottler carbonates the liquid by pumping carbon dioxide into it under pressure. A large quantity of carbon dioxide becomes dissolved in the beverage, making it acidic. As you open the bottle, the pressure drops and the dissolved gas bubbles out of solution, making the solution less acidic. The amount of carbon dioxide dissolved in the liquid decreases as the amount of gaseous carbon dioxide in contact with the water decreases. When the amount of carbon dioxide in the beverage reaches the amount in the atmosphere, no more carbon dioxide bubbles out and the beverage is "flat" and only mildly acidic, like rainwater.

The amount of carbon dioxide dissolved in rainwater is small because the amount of carbon dioxide gas in the atmosphere is small. About 0.03 percent of the molecules in Earth's atmosphere are carbon dioxide. Small as that number is, it makes carbon dioxide the fourth most abundant gas, just behind argon (0.9 percent), oxygen (21 percent), and nitrogen (78 percent). The amount of carbonic acid formed in rainwater is very small, only about 0.0006 g/l. As the burning of oil, gas, and coal increases the amount of carbon dioxide in the atmosphere, the amount of carbonic acid in rain increases slightly.

Most of the acidity of acid rain, however, comes not from carbon dioxide but from sulfur dioxide and nitrogen gases, which react with water to form strong sulfuric and nitric acids, respectively. These acids promote weathering to a greater degree than carbonic acid does. Volcanoes and coastal marshes emit gases of carbon, sulfur, and nitrogen into the atmosphere, but by far the largest source is industrial pollution. (See Chapter 23 for more information about acid rain.)

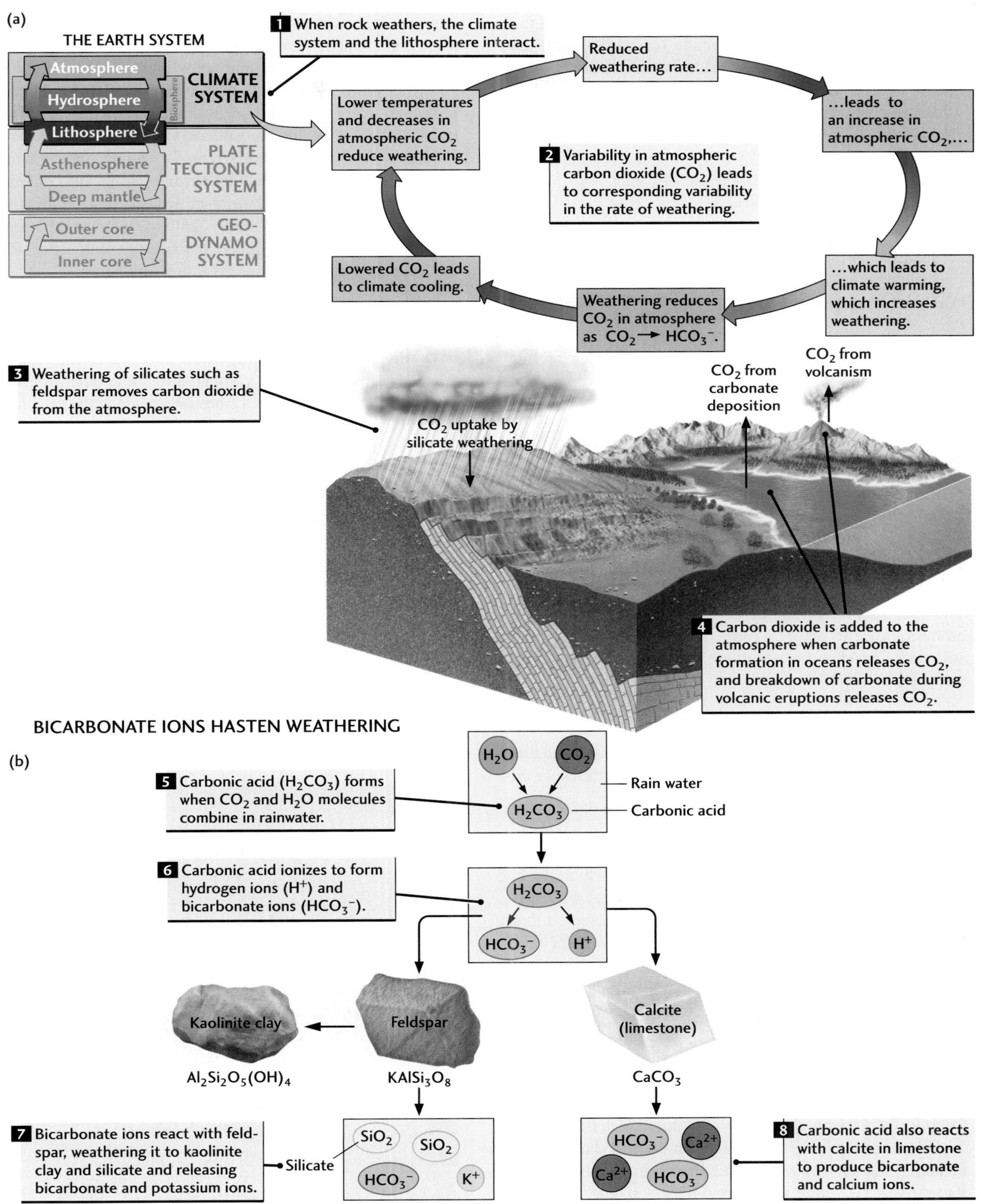

Figure Story 7.6 Atmospheric carbon dioxide concentration influences climate and weathering. (a) The cyclic nature of atmospheric carbon dioxide concentrations, weathering, and climate. (b) Carbonic acid from rainwater weathers feldspar and calcite.

Although rainwater contains only a relatively small amount of dissolved carbon dioxide (carbonic acid), that amount is enough to weather feldspars and dissolve great quantities of rock over long periods of time. We can now write the chemical reaction for the weathering of feldspar with water:

feldspar	+	carbonic acid	+	water →
$2KAlSi_3O_8$		$2H_2CO_3$		H_2O

dissolved kaolinite	+	dissolved silica	+	dissolved potassium ions	+	dissolved bicarbonate ions
$Al_2Si_2O_5(OH)_4$		$4SiO_2$		$2K^+$		$2HCO_3^-$

This simple weathering reaction illustrates the three main chemical effects of chemical weathering on silicates. It leaches, or dissolves away, cations and silica. It *hydrates,* or adds water to, the minerals. And it makes the solutions less acidic. Specifically, the carbonic acid in rainwater helps to weather feldspar in the following way (Figure 7.6b):

- A small proportion of carbonic acid molecules ionize, forming hydrogen ions (H^+) and bicarbonate ions (HCO_3^-) and thus making the water droplets slightly acidic.
- The slightly acidic water dissolves potassium ions and silica from feldspar, leaving a residue of kaolinite, a solid clay. The hydrogen ions from the acid combine with the oxygen atoms of the feldspar to form the water in the kaolinite structure. The kaolinite becomes part of the soil or is carried away as sediment.
- The solution becomes less acidic as the reaction goes on.
- The dissolved silica, potassium ions, and bicarbonate ions are carried away by rain and river waters and ultimately are transported to the ocean.

In Nature: Feldspar in Outcrops and in Moist Soil Now that we understand the chemical reaction by which acidic water weathers feldspar, we can better understand why feldspars on bare rock surfaces are much better preserved than those buried in damp soils. The chemical reaction for feldspar weathering gives us two separate but related clues: the amount of water and the amount of acid available for the chemical reaction. The feldspar on a bare rock weathers only while the rock is moist with rainwater. During all the dry periods, the only moisture that touches the bare rock is dew. The feldspar in moist soil is constantly in contact with the small amounts of water retained in spaces between grains in the soil. Thus feldspar weathers continuously in moist soil.

There is more acid in the water in the soil than there is in falling rain. Rainwater carries its original carbonic acid into the soil. As water filters through the soil, it picks up additional carbonic acid and other acids produced by the roots of plants, by the many insects and other animals that live in the soil, and by the bacteria that degrade plant and animal remains. Recently, it was discovered that some bacteria release organic acids, even in waters hundreds of meters deep in the ground. These organic acids then weather feldspar and other minerals in rocks below the surface. Bacterial respiration in soils may increase the soils' carbon dioxide to as much as 100 times the atmospheric value!

Rock weathers more rapidly in the tropics than in temperate and cold climates mainly because plants and bacteria grow quickly in warm, humid climates, contributing the carbonic acid and other acids that promote weathering. Additionally, most chemical reactions, weathering included, speed up with an increase in temperature.

Fast Weathering of Carbonates

Even olivine, the most rapidly weathered silicate mineral, is relatively slow to dissolve compared to some other nonsilicate rock-forming minerals. Limestone, made of the calcium-magnesium carbonate minerals calcite and dolomite, weathers very quickly in humid regions. Old limestone buildings show the effects of dissolution by rainwater (**Figure 7.7**). Underground water dissolves away great quantities of carbonate minerals, hollowing out caves in limestone formations. Farmers and gardeners use ground limestone to counterbalance acidity in soil because of the stone's ability to dissolve rapidly over the growing season. When limestone dissolves, no clay minerals are formed. The solid dissolves completely, and its components are carried off in solution.

Carbonic acid promotes the weathering of limestone just as it does the weathering of silicates (see Figure Story 7.6b). The overall reaction by which calcite, the major min-

Figure 7.7 Weathered limestone blocks and columns of 2500-year-old Greek ruins at Segesta, Italy, show pitted, etched surfaces caused by chemical dissolution. [Ric Ergenbright.]

eral of limestones, dissolves in rain or other water containing carbon dioxide is

calcite	+	carbonic acid	→	calcium ions	+	bicarbonate ions
$CaCO_3$		H_2CO_3		Ca^{2+}		$2HCO_3^-$

The reaction proceeds only in the presence of water, which contains the carbonic acid and dissolved ions. When calcite dissolves, the calcium ions and bicarbonate ions are carried away in solution. Dolomite, $CaMg(CO_3)_2$, another abundant carbonate mineral, dissolves in the same way, producing equal amounts of magnesium ions and calcium ions.

Chemical Stability: A Speed Control for Weathering

Much more of Earth's surface is covered by silicate rocks than by carbonate rocks such as limestone. But because carbonate minerals dissolve faster and in greater amounts than silicates do, the weathering of limestone accounts for more of the total chemical weathering of Earth's surface each year than does any other rock.

Why do weathering rates vary so widely among different minerals? Minerals weather at different rates because there are differences in their chemical stability in the presence of water at given surface temperatures.

Chemical stability is a measure of a substance's tendency to remain in a given chemical form rather than to react spontaneously to become a different chemical substance. Chemical substances are stable or unstable in relation to a specific environment or set of conditions. Feldspar, for example, is stable under the conditions found deep in Earth's crust (high temperatures and small amounts of water) but unstable under the conditions at Earth's surface (lower temperatures and abundant water). Iron metal in a meteoroid in outer space is chemically stable. It may remain unaltered for billions of years because it is not exposed to oxygen or water. If the meteoroid were to fall to Earth, where it would be exposed to oxygen and water, the iron metal would be chemically unstable and would react spontaneously to form iron oxide. As we saw earlier, limestone weathers faster than granite—a result of limestone's lower stability in Earth's surface environments. Two characteristics of a mineral—its solubility and its rate of dissolution—help determine its chemical stability.

Solubility The solubility of a specific mineral is measured by the amount of the mineral dissolved in water when the solution is saturated. Saturation is the point at which the water cannot hold any more of the dissolved substance. The higher a mineral's solubility, the lower its stability under weathering. Rock salt, for example, is unstable under weathering. It is highly soluble in water (about 350 g per liter of water) and is leached from a soil by even small amounts of water. Quartz, in contrast, is stable under most weathering conditions. Its solubility in water is very low (only about 0.008 g per liter of water), and it is not easily leached from a soil.

Table 7.2 Relative Stabilities of Common Minerals Under Weathering

Stability of Minerals	Rate of Weathering
MOST STABLE	Slowest
Iron oxides (hematite)	↓
Aluminum hydroxides (gibbsite)	
Quartz	
Clay minerals	
Muscovite mica	
Potassium feldspar (orthoclase)	
Biotite mica	
Sodium-rich feldspar (albite)	
Amphiboles	
Pyroxene	
Calcium-rich feldspar (anorthite)	
Olivine	
Calcite	
Halite	
LEAST STABLE	Fastest

Rate of Dissolution A mineral's rate of dissolution is measured by the amount of the mineral that dissolves in an unsaturated solution in a given length of time. The faster the mineral dissolves, the less stable it is. Feldspar dissolves at a much faster rate than quartz, and, primarily for that reason, it is less stable than quartz under weathering.

Relative Stability of Common Rock-Forming Minerals Knowing the relative chemical stabilities of various minerals enables us to ascertain the intensity of weathering in a given area. In a tropical rain forest, only the most stable minerals will be left on an outcrop or in the soil, and so we know that the weathering there is intense. In an arid region such as the desert of North Africa, where weathering is minimal, alabaster (gypsum) monuments remain intact, as do many other unstable minerals. Table 7.2 shows the relative stabilities of all the common rock-forming minerals. Salt and carbonate minerals are the least stable, iron oxides the most stable.

Other Silicates Forming Other Clays

Clay minerals are a principal component of soils and sediments all over Earth's surface. One would expect this to be the case because the rock-forming silicates constitute a large fraction of Earth's crust and because weathering at the surface is widespread. Deposits of clay pure enough to be used as the raw material for pottery, chinaware, and industrial ceramics are found in some uncommon soils and sedimentary rocks. Brick, clay tiles, and other structural ceramics require material that is less pure.

Clays form through the weathering of a variety of silicate minerals, not just feldspar. Amphibole, mica, and other silicates of granite weather to form clays in much the same way as feldspar does. As the mineral weathers, it absorbs water and loses silica and ions such as sodium, potassium, calcium, and magnesium to the solution. The kinds of clays formed depend on two factors:

1. The composition of the parent silicates
2. The climate

The clay mineral montmorillonite, for example, which swells when it absorbs large quantities of water, typically forms from the weathering of volcanic ash. Montmorillonite is a common weathering product of other silicates, too, especially in semiarid environments such as the high plains of the southwestern United States. As a result of the great variety of parent silicates, climates, and clay minerals, predicting the course of silicate weathering for specific clay minerals is a task for specialized geologists and soil scientists.

Not all silicates weather to form clay minerals. Rapidly weathering silicates, such as some pyroxenes and olivines, may dissolve completely in humid climates, leaving no clay residue. Quartz, one of the slowest to weather of the abundant silicate minerals, also dissolves without forming any clay mineral.

Silicate weathering can form materials other than clay minerals. One such product is **bauxite,** an ore composed of aluminum hydroxide that is the major source of aluminum metal. Bauxite forms when clay minerals derived from weathered silicates continue to weather until they have lost all their silica and ions other than aluminum. Bauxite is found in tropical regions, where rainfall is heavy and weathering intense.

The Role of Oxygen in Weathering: From Iron Silicates to Iron Oxides

Iron is one of the eight most abundant elements in Earth's crust, but iron metal, the chemical element in its pure form, is rarely found in nature. It is present only in certain kinds of meteorites that fall to Earth from other places in the solar system. Most of the iron ores used for the production of iron and steel are formed by weathering. These ores are composed of iron oxide minerals originally produced during the weathering of iron-rich silicate minerals, such as pyroxene and olivine. The iron released by dissolution of these minerals combines with oxygen from the atmosphere to form iron oxide minerals. This chemical reaction is called **oxidation** because it is the chemical combination of an element with oxygen. Like hydrolysis, oxidation is one of the important chemical weathering processes.

The iron in minerals may be present in one of three forms: metallic iron, ferrous iron, or ferric iron. In the metallic iron found only in meteorites, the iron atoms are uncharged: they have neither gained nor lost electrons by reaction with another element. In the **ferrous iron** (Fe^{2+}) found in silicate minerals such as pyroxene, the iron atoms have lost two of the electrons they have in the metallic form and thus have become ions. The iron in the most abundant iron oxide at Earth's surface, **hematite** (Fe_2O_3), is **ferric iron** (Fe^{3+}). The iron atoms in ferric iron have lost three electrons. Ferrous iron ions oxidize by losing an additional electron, going from 2^+ (ferrous) to 3^+ (ferric). All the iron oxides formed at Earth's surface are ferric. The electrons lost by the iron are gained by oxygen atoms as they become oxygen ions (O^{2-}). Thus, oxygen atoms from the atmosphere oxidize ferrous iron to ferric iron.

When an iron-rich mineral such as pyroxene is exposed to water, its silicate structure dissolves, releasing silica and ferrous iron to solution, where the ferrous iron is oxidized to the ferric form (**Figure 7.8**). The strength of the chemical bonds between ferric iron and oxygen make ferric iron insoluble in most natural surface waters. It therefore precipi-

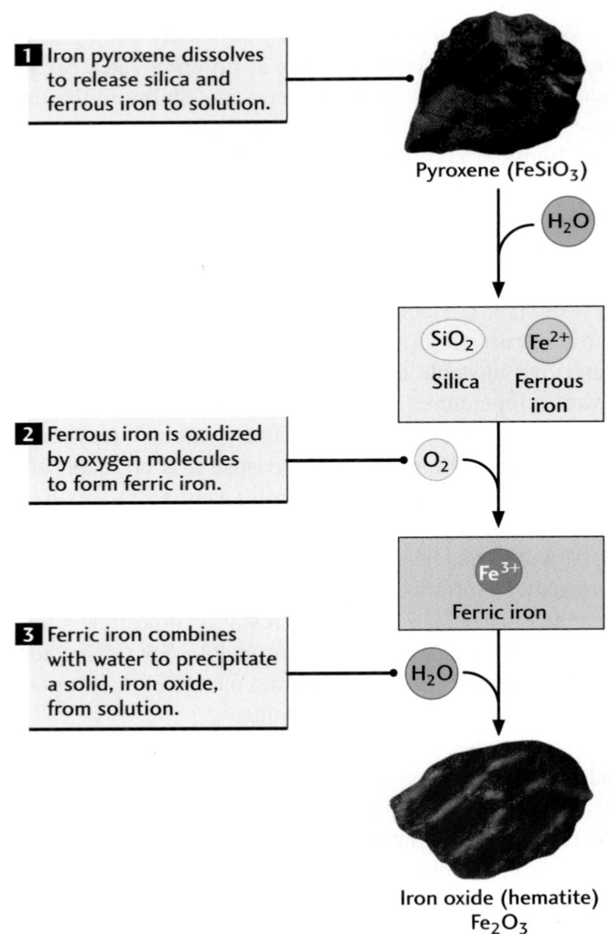

Figure 7.8 The general course of chemical reactions by which an iron-rich mineral, such as pyroxene, weathers in the presence of oxygen and water.

Figure 7.9 Red and brown iron oxides color weathering rocks in Monument Valley, Arizona. [Betty Crowell.]

tates from the solution, forming a solid ferric iron oxide. We are familiar with ferric iron oxide in another form—rusting iron, which is produced when iron metal is exposed to the atmosphere.

We can show this overall weathering reaction of iron-rich minerals by the following equation:

$$\begin{array}{ccccccc} \text{iron pyroxene} & + & \text{oxygen} & \rightarrow & \text{hematite} & + & \text{dissolved silica} \\ 4FeSiO_3 & & O_2 & & 2Fe_2O_3 & & 4SiO_2 \end{array}$$

Although the equation does not show it explicitly, water is required for this reaction to proceed.

Iron minerals, which are widespread, weather to the characteristic red and brown colors of oxidized iron (**Figure 7.9**). Iron oxides are found as coatings and encrustations that color soils and weathered surfaces of iron-containing rocks. The red soils of Georgia and other warm, humid regions are colored by iron oxides. Iron minerals weather so slowly in frigid regions that iron meteorites frozen in the ice of Antarctica are almost entirely unweathered.

Physical Weathering

Now that we have surveyed chemical weathering alone, we can turn to its partner, physical weathering. We can see the workings of physical weathering most clearly by examining its role in arid regions, where chemical weathering is minimal.

Physical Weathering in Arid Regions

Weathered outcrops in arid regions are covered by a rubble of various-sized fragments, from individual mineral grains only a few millimeters in diameter to boulders more than a meter across. The differences in size result from varying degrees of physical weathering and from patterns of breakage of the parent rock. As physical weathering proceeds, the larger particles are cracked and broken into smaller ones. Some of these fragments are broken along planes of weakness in the parent rock (**Figure 7.10**). Grains of sand form when individual crystals of various minerals, such as quartz, break apart from one another or when fine-grained rocks such as basalt are fragmented.

Although physical weathering is the most common form of weathering in dry regions, even there chemical weathering has prepared the way. Slight chemical alterations of feldspar and other minerals weaken the cohesive forces that hold the crystals in a rock together. As small cracks form and widen, individual quartz or feldspar crystals are freed by a combination of physical and chemical weathering and fall to the ground. Fractures enlarge, and large blocks of rock are separated from an outcrop.

Physical Weathering in All Regions

The chemical weathering that promotes physical weathering is itself promoted by fragmentation, which opens channels through which water and air can penetrate and react with minerals inside a rock. The breaking up of a rock into

Figure 7.10 Weathered, enlarged joint patterns developed in two directions in rocks at Point Lobos State Reserve, California. [Jeff Foott/DRK.]

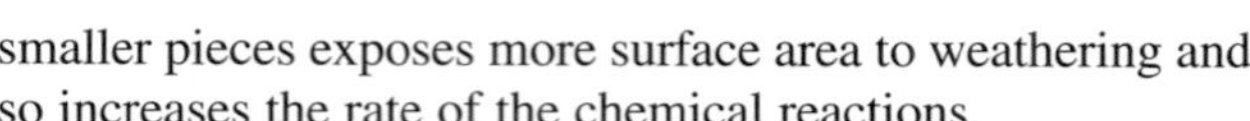

Figure 7.11 Organisms such as these tree roots invade fractured rock, widening cracks and promoting further chemical and physical weathering. [Peter Kresan.]

smaller pieces exposes more surface area to weathering and so increases the rate of the chemical reactions.

Physical weathering does not always depend on chemical weathering. Some processes, such as the freezing of water in cracks, break up unweathered rock masses. Some rocks are made particularly susceptible to physical weathering by fracturing that occurs when tectonic forces bend and break rocks in the course of mountain building. On the Moon, physical fragmentation works alone, because there is no water to make chemical weathering possible. On that lifeless terrain, rocks are broken into boulders and fine dust by the impacts of large and small meteorites.

What Determines How Rocks Break?

Rocks can break for a variety of reasons, including stress along natural zones of weakness and biological and chemical activity.

Natural Zones of Weakness Rocks have natural zones of weakness along which they tend to crack. In sedimentary rocks such as sandstone and shale, these zones are the bedding planes formed by the successive layers of solidified sediment. Metamorphic rocks such as slate form parallel planes of fractures that enable them to be split easily to form roofing tiles. Granites and other rocks are massive. Massive rocks tend to crack along regular fractures spaced one to several meters apart, called **joints** (see Chapter 11 for further discussion of joints). These and less regular fractures form while rocks are still deeply buried in Earth's crust. Through uplift and erosion, the rocks rise slowly to Earth's surface. There, freed from the weight of overlying rock, the fractures open slightly. Once the fractures open a little, both chemical and physical weathering work to widen the cracks.

Activity by Organisms The activity of organisms affects both chemical and physical weathering. Bacteria and algae invade cracks, producing microfractures. These organisms, both those in cracks and those that may encrust the rock, produce acid, which then promotes chemical weathering. In some regions, acid-producing fungi are active in soils, contributing to chemical weathering. Animals burrowing or moving through cracks can break rock. Many of us have seen a crack in a rock that has been widened by a tree root. Physical weathering takes place when the force of the growing root system helps to pry cracks apart (**Figure 7.11**).

Frost Wedging One of the most efficient mechanisms for widening cracks is **frost wedging**—breakage resulting from the expansion of freezing water. As water freezes, it exerts an outward force strong enough to wedge open a crack and split a rock (**Figure 7.12**). This is the same process that can crack open the engine block of a car that is not protected by antifreeze. Frost wedging is most important where water episodically freezes and thaws, such as in temperate climates and in mountainous regions.

Mineral Crystallization Other expansive forces that can split rocks are generated when minerals crystallize from solutions in rock fractures. As minerals crystallize in a tiny crack, they force the crack open further because the minerals grow very inefficiently, building large porous networks that prop the crack open. This phenomenon is most common in arid regions, where dissolved substances derived from chemical weathering may crystallize as a solution evaporates. The minerals produced in this way include calcium carbonate (commonly), gypsum (occasionally), and rock salt (rarely).

Alternating Heat and Cold A recurring idea among geologists who study weathering is that rocks can break as a re-

Figure 7.12 Gneiss boulder, 3 m high, fractured by frost wedging. Taylor Valley, Victoria Land, Antarctica. [Michael Hambrey.]

sult of the daily cycle of hot days and cold nights in a desert, where temperatures at twilight may drop from 43° to 15°C in an hour. Rock might be weakened by its expansion in the heat and contraction in the cold. Campfires and forest fires have shown us that fires built over a rock can crack its surface. Various laboratory simulations of natural breakage caused by temperature extremes have failed to confirm the hypothesis. Supporters of the idea, however, point out that a few months of laboratory experiments are no match for the thousands of years of expansion in the heat and contraction in the cold that might weaken a rock and cause it to fracture. The matter remains unsettled.

Exfoliation and Spheroidal Weathering There are two forms of rock breakage not directly related to earlier fractures or joints caused by previous weathering. **Exfoliation** is a physical weathering process in which large flat or curved sheets of rock fracture and are detached from an outcrop. These sheets may look like the layers peeled from a large onion (**Figure 7.13**). **Spheroidal weathering** is also a cracking and splitting off of curved layers from a generally spherical boulder, but usually on a much smaller scale (**Figure 7.14**). Even though exfoliation and spheroidal weathering are common, no generally accepted explanation of their origin has yet emerged. It is apparent that these two processes are other instances in which chemical weathering induces cracks that follow the shape of the surface in some way, particularly in strongly jointed rock. Some geologists have suggested that both exfoliation and spheroidal weathering result from an uneven distribution of expansion and contraction caused by chemical weathering and temperature changes.

Other Forces Rivers excavate bedrock valleys by beating on the bedrock of the channel with transported rocks and by hurling their own force against the bedrock in waterfalls and rapids. The scouring and plucking action of glaciers can also break up rock masses, as we will see in Chapter 16. And in Chapter 18, we will see how waves, pounding on rocky shores with a force equal to hundreds of tons per square meter, fracture exposed bedrock.

Physical Weathering and Erosion

As we've emphasized, weathering and erosion are closely related, interacting processes. Physical weathering and erosion

Figure 7.13 Exfoliation on Half Dome, Yosemite National Park, California. [Tony Waltham.]

Figure 7.14 This remarkable structure is a product of spheroidal weathering. As cabbage-leaf-like layers fall away, a rock core remains, retaining the original shape of the rock on a smaller scale. [Michael Follo.]

are closely tied to how wind, water, and ice work to transport weathered material.

Physical weathering fractures a large rock into smaller pieces, which are more easily transported and therefore more easily eroded than the larger mass. The first steps in the erosion process are the downhill movements of masses of weathered rock, such as landslides, and the transportation of individual particles by flows of rainwater down slopes. The steepness of the slopes affects both physical and chemical weathering, which in turn affect erosion. Weathering and erosion are more intense on steep slopes, and their action makes slopes gentler. Wind may blow away the finer particles, and glacial ice can carry away large blocks torn from bedrock.

Chemical weathering rates are low at high altitudes, where temperatures are generally low, soil is thin or absent, and vegetation is sparse. Physical weathering is greater at high altitudes and in glacial terrains, where the ice tears apart the rock. We can see that the sizes of the materials formed by physical weathering are closely related to various erosional processes. As weathered material is transported, it may again change in size and shape, and its composition may change as a result of chemical weathering. When transportation stops, deposition of the sediment formed by weathering begins.

Figure 7.15 is a chart summarizing the processes of physical and chemical weathering. All these processes,

WEATHERING FACTORS			
1. Duration of weathering		Less weathering, erosion, and soil formation over short periods of time	More weathering, erosion, and soil formation over long periods of time
2. Bedrock type		More stable minerals, (e.g., quartz), result in lower weathering	Less stable minerals, (e.g., feldspar), result in higher weathering
3. Climate	Lower temperatures	Less chemical weathering (dissolution, alteration to aid physical weathering, production of clay materials)	More physical weathering (thermal expansion and contraction, frost wedging, breakage of bedrock, fragmentation to smaller sizes)
	Higher temperatures	Less physical weathering	More chemical weathering
	Rainfall amount	Little rainfall (less dissolution of minerals, physical weathering, fragmentation, erosion)	Heavy rainfall (more dissolution of minerals, production of clay materials, production of small size particles, erosion)
	Rainfall acidity	Low acidity (less dissolution of minerals, less physical weathering)	High acidity (more dissolution of minerals, more production of clay materials)
4. Topography	Steep slopes	Less chemical weathering	More physical weathering, more erosion
	Gentle slopes	Less physical weathering, less erosion	More chemical weathering

Figure 7.15 Summary chart of weathering. Factors considered are the proportion of stable and unstable minerals in a rock, climate (including temperature and rainfall), and topography (including steep and gentle slopes).

which are detailed in Chapters 12 through 16, contribute to the formation of different kinds of landscape (Chapter 17).

Soil: The Residue of Weathering

Not all weathering products are eroded and immediately carried away by streams or other transport agents. On moderate and gentle slopes, plains, and lowlands, a layer of loose, heterogeneous weathered material remains overlying the bedrock. It may include particles of weathered and unweathered parent rock, clay minerals, iron and other metal oxides, and other products of weathering. Engineers and construction workers refer to this entire layer as "soil." Geologists, however, prefer to call this material **regolith,** reserving the term *soil* for the topmost layers, which contain organic matter and can support life. We can easily see the difference between regolith and soil if we consider the regolith found on the Moon. The Moon's regolith is a loose layer of fragmented rocks and dust, but it is quite sterile. It contains little or no organic matter and cannot support life. The organic matter in Earth's soil is **humus,** the remains and waste products of the many plants, animals, and bacteria living in it. Leaf litter contributes significantly to the soil of forests.

Soils vary in color, from the brilliant reds and browns of iron-rich soils to the black of soils rich in organic matter. Soils also vary in texture. Some are full of pebbles and sand; others are composed entirely of clay. Soils are easily eroded, and so they do not form on very steep slopes or where high altitude or frigid climate prevents plant growth.

Soil is such an essential part of our environment and economy that a separate field of study, soil science, developed in the twentieth century. Soil scientists, agronomists, geologists, and engineers study the composition and origin of soils, their suitability for agriculture and construction, and their value as a guide to climate conditions in the past. Many scientists are focusing special attention on ways to combat the serious threat of soil erosion (see Feature 7.1).

Soil Profiles

A road or trench cut through a soil reveals its vertical structure, the soil profile (**Figure 7.16**). The soil's topmost layer, typically not much more than a meter or two thick, is usually the darkest, containing the highest concentration of organic matter. This topmost layer of soil is called the **A-horizon** (a particular level in a rock section is commonly called a "horizon"). In a thick soil that has formed over a long period of time, the inorganic components of this top layer are mostly clay and insoluble minerals such as quartz. Soluble minerals have been leached from this layer. Beneath the topmost section is the **B-horizon,** where organic matter is sparse. In this layer, soluble minerals and iron oxides have accumulated in small pods, lenses, and coatings. The lowest layer, the **C-horizon,** is slightly altered bedrock, broken and decayed, mixed with clay from chemical weathering. The transition from one horizon to another is usually indistinct.

Soils are also described as either residual or transported. Residual soils evolve in one place from bedrock to well-developed soil horizons. Most soils are residual, and they form faster and become thicker where weathering is intense. Even under intense weathering, the A-horizon may take thousands of years to develop to the point at which it can support crops. Soil forms slowly because the most active chemical weathering occurs only during short intervals of rain. During dry periods, the reactions continue, but very slowly and only when some moisture remains in the soil. When soil dries out fully between rains, chemical weathering stops almost completely.

Transported soils may accumulate in some limited areas of lowlands after being eroded from surrounding slopes and carried downhill. They can often be recognized because they are more similar in texture and composition to soils than to ordinary sediments. However, they may be confused with ordinary sediment laid down by rivers, wind, and ice. In some cases, parts of the original upland soil profile are preserved. These soils owe their thickness to deposition rather than to weathering in place. Transported soils are common.

Climate, Time, and Soil Groups

Climate strongly affects weathering and therefore has a great influence on the characteristics of the soil formed on any given parent rock. The characteristics of a soil in a warm and humid region, for instance, differ from those of soils in arid and temperate regions. Soil scientists have mapped soil characteristics over much of the world, with the hope of preventing soil erosion and fostering efficient agricultural practices. For our purposes, we can distinguish three major soil groups—pedalfers, laterites, and pedocals—on the basis of their mineralogy and chemical composition, which also correlate with climate (see Figure 7.16).

Temperate Climates: Pedalfers The characteristics of soils in regions with moderate rainfall and temperatures depend on the climate, the type of parent rock, and the length of time the soil has had to develop and thicken. Intense weathering decreases the influence of the parent rock, as does the length of exposure to weathering. Soil developed after a relatively short time on a granite bedrock in a climate with moderate temperatures and humidity may differ greatly from soil formed on limestone under the same conditions, because the influence of the parent rock remains strong. The soil on the granite may contain remnants of the silicate minerals and be dominated by the clay minerals forming from feldspar, a chief constituent of the parent rock. The soil on the limestone may have a few remnants of calcium carbonate, but most of the limestone fragments will have dissolved. The clay minerals will be mainly those

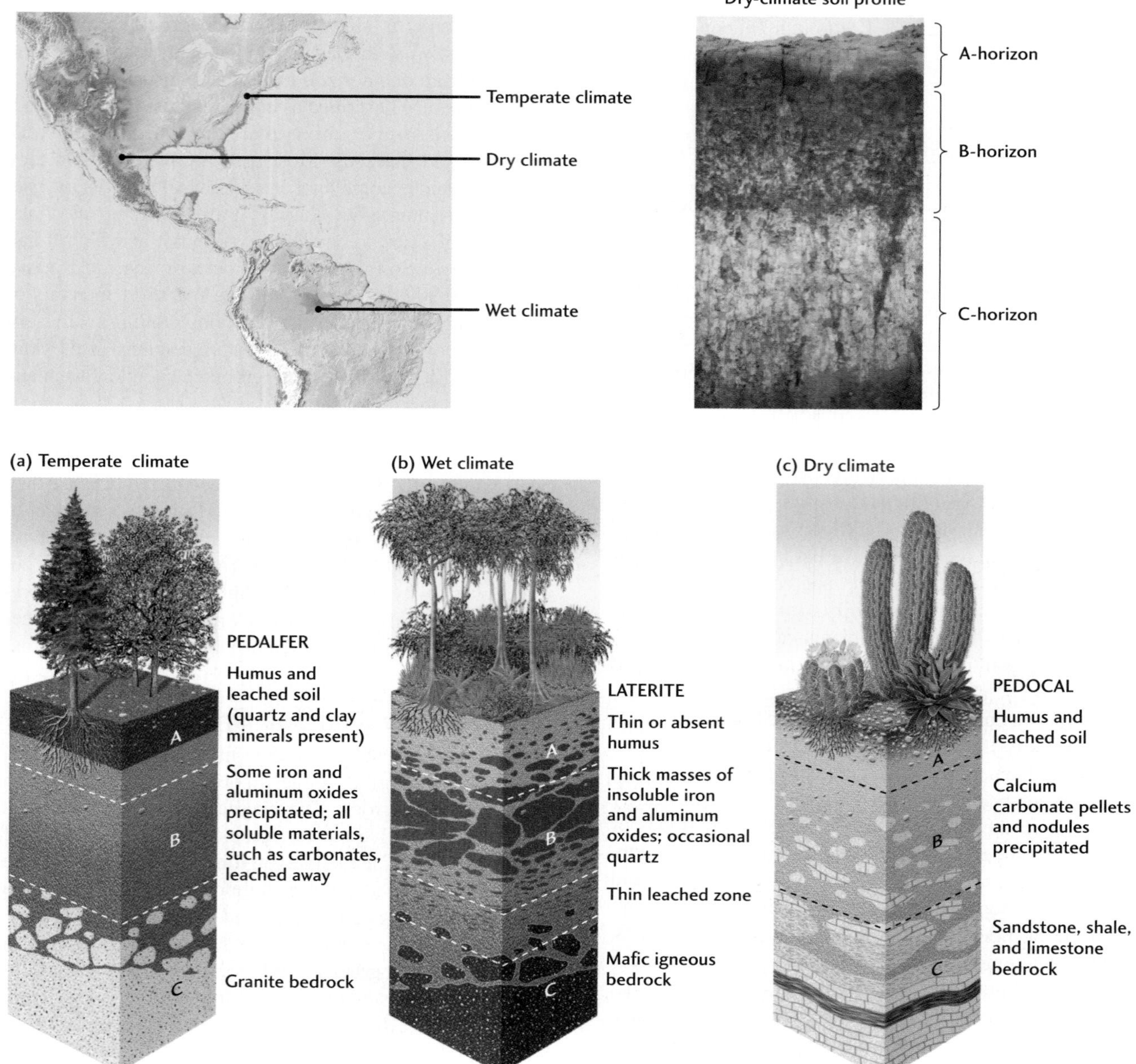

Figure 7.16 Soil profile. The thickness of the soil profile depends on the climate, the length of time during which the soil has been forming, and the composition of the parent rock. The transition from one horizon to another is usually indistinct. (a) Pedalfer soil profile developed on granite in a region of high rainfall. The only mineral materials in the upper parts of the soil profile are iron and aluminum oxides and silicates such as quartz and clay minerals, all of which are very insoluble. Calcium carbonate is absent. (b) Laterite soil profile developed on a mafic igneous rock in a tropical region. In the upper zone, only the most insoluble precipitated iron and similar oxides remain, plus occasional quartz. All soluble materials, including even relatively insoluble silica, are leached; thus the whole soil profile may be considered to be an A-horizon directly overlying a C-horizon. (c) Pedocal soil profile developed on sedimentary bedrock in a region of low rainfall. The A-horizon is leached; the B-horizon is enriched in calcium carbonate precipitated by evaporating soil waters. [Photo courtesy of Department of Primary Industries, Victoria, Australia.]

EARTH POLICY

7.1 Soil Erosion

Before settlers began to farm the prairies of the United States and Canada in the nineteenth century, the soils were moderately thick. Soil formed relatively quickly in this area because the glaciation that ended about 10,000 years ago had deposited an abundance of easily weathered, ground-up bedrock material. But soils form very slowly; even in rapid weathering environments, they may gain as little as 2 mm per year.

Because soils take such a long time to form, they cannot be renewed quickly after they have been eroded. There is some balance between the moderate natural erosion of soils by streams and the wind and the slow formation of new soil. If soil forms and erodes at roughly the same rate, its thickness remains constant. If soil erodes more slowly than it forms, it grows thick. If it erodes much more rapidly than it forms, new soil has no opportunity to form and the existing soil is quickly lost.

As a general rule, it takes 30 years to form an inch of topsoil. That inch can be lost in less than a decade as a result of poor farming practices and excessive grazing. Agriculture accelerates erosion because plowing breaks up the soil and eliminates the erosion-resistant natural plant cover. Soil erosion has been especially severe in many places in the world. One such place is the North American prairies, where plows have bitten deeply and conservation practices were long ignored. The loosened soils have thinned and been carried away by the region's rivers. The prairies are now vulnerable to further erosion as dust storms strike in long periods of drought (see Chapter 15).

Contour plowing can lessen the damage caused by agriculture. This practice covers a field with curved furrows that follow the contours of the natural slopes, rather than running parallel to property lines that intersect slopes. Contour plowing inhibits erosion because much of the rainwater is channeled along the contoured furrows instead of running off downhill.

Despite widespread promotion of such agricultural techniques, however, soils are still eroding much more rapidly than they can be replaced. In parts of the United States and Canada, more than 10 tons of topsoil per acre of cropland are lost annually. In the United States alone, 2 billion tons of topsoil are lost to erosion each year—twice the amount of soil formed in the same period. A loss on that scale is equivalent to destroying 780,000 acres of cropland. Comparable losses are hitting other agricultural regions of the world, including the steppes of Russia, parts of Africa, and the decimated rain forests in Madagascar and Brazil. The direct and indirect costs of soil erosion are estimated to be $44 billion in the United States and $400 billion throughout the entire world. If such losses persist in this new century, agricultural yields from thinned soils will decrease, with the inevitable abandonment of agriculture in the regions most seriously affected. And once soil is gone, it takes thousands of years to form again. The estimation and prediction of soil erosion are a backdrop to policy decisions on optimal agricultural methods and ways of maintaining agriculture at a sustainable level while preserving Earth's soil cover.

Soil erosion in Madagascar. [Peter Johnson/Corbis.]

found as impurities in the parent limestone. After many thousands of years, however, the differences between the two soils may dwindle or even disappear. Both soils may develop the same clay minerals, depending on the exact nature of the climate, and both will have lost all soluble minerals in the upper layers. Thus, young soils are affected by the composition of the parent rock, but older soils are affected primarily by climate.

Areas of moderate to high rainfall in the eastern United States, most of Canada, and much of Europe are home to the **pedalfer** group of soils (see Figure 7.16a). Pedalfers take their name from *pedon,* the Greek word for "ground" or "soil," and *al* and *fe,* from the chemical symbols for aluminum and iron. The upper and middle layers of pedalfers contain abundant insoluble minerals such as quartz, clay minerals, and iron alteration products. Carbonates and other more soluble minerals are absent. Pedalfers make good agricultural soils.

Wet Climates: Laterites In warm, humid climates, weathering is fast and intense and soils become thick. The higher the temperature and humidity, the lusher the vegetation. Abundant vegetation and moisture and warm temperatures speed up chemical weathering so greatly that the uppermost layer of the soil is leached of all soluble and easily weathered minerals. The residue of this fast weathering is **laterite,** a deep-red soil in which feldspar and other silicates have been completely altered, leaving behind mostly aluminum and iron oxides and hydroxides (see Figure 7.16b). Silica and calcium carbonate have been leached from laterite. Although these soils may support lush vegetation in equatorial jungles, they are not very productive for crop plants. Most of the organic matter is constantly recycled from the surface to the vegetation, with at best a very thin layer of humus at the top of the soil. The clearing of trees and tilling of the soil allow its superficial layer of rich humus to oxidize quickly and disappear, uncovering the infertile layer below.

For this reason, many laterites can be farmed intensively for only a few years before they become barren and have to be abandoned. Large regions of India are now in this condition. As sections of the Amazon rain forest of Brazil are cleared, they too become barren after only a few years. To restore a forest on laterite soil under natural conditions would take many thousands of years.

Dry Climates: Pedocals Lack of water and absence of vegetation impede weathering, so soils are thin in arid regions. In cold arid regions, where chemical weathering is very slow, the parent rock's influence is dominant, even where soils have been forming over long periods of time. As a result, A-horizons contain many unweathered minerals and fragments of the parent rock. When rainfall is too sparse to dissolve significant amounts of soluble minerals, these minerals may remain in the A-horizon.

Pedocals are the dominant soils in arid regions (see Figure 7.16c). These soils are rich in calcium from the calcium carbonate and other soluble minerals they contain and are low in organic matter. Pedocals are found in the southwestern United States and similar climates. Much of the soil water is drawn up near the surface and evaporates between rainfalls, leaving behind precipitated nodules and pellets of calcium carbonate, mostly in the middle layer of the soil. The combination of mineralogy and dryness is less favorable for a large population of organisms in the soil, making pedocals less fertile than pedalfers.

Paleosols: Working Backward from Soil to Climate Recently, there has been much interest in ancient soils that have been preserved as rock in the geologic record. Some of these soils are more than a billion years old. These *paleosols,* as they are called, are being studied as guides to ancient climates and even to the amounts of carbon dioxide and oxygen in the atmosphere in former times. The mineralogy of paleosols billions of years old provides evidence that there was no oxidation of soils at that early stage of Earth's history and therefore that oxygen had not yet evolved to become a major fraction of the atmosphere.

Humans as Weathering Agents

People tend not to think of themselves as part of the natural landscape, but they are. Humans are responsible for acid rain, which enhances chemical weathering. Chemical weathering is promoted by physical weathering, and humans speed up both processes by their innumerable activities that break up rocks, from digging foundations and constructing highways to conducting massive mining operations. It has been estimated that road building alone moves 3000 trillion tons of rock and soil throughout the world each year.

Soils may be long-term reservoirs of pollution after they are contaminated by spills of toxic materials. Contaminants leak slowly from the soil into ground and surface waters. Humans have also loaded the ground with salt, pesticides, and petroleum products, all of which may significantly interfere with plant growth and thus accelerate soil erosion.

Weathering Makes the Raw Material of Sediment

Soil is only one of many products of weathering. Processes that weather rocks and break them apart produce fragments that vary greatly in size and shape, from huge boulders 5 m across to small particles too fine to see without a microscope. Pieces larger than a large grain of sand (2 mm in diameter) tend to be rock fragments containing mineral grains of the parent rock. Sand and silt grains are usually individual crystalline grains of any of the various minerals that make up the rock (**Figure 7.17**). The smallest pieces of weathered material are the fine clay minerals produced by the chemical alteration of silicate minerals. A certain mass of granite, for example, is broken down by weathering into the following classes:

- Mineral and rock fragments of the parent rock that are still identifiable as such
- Solid products of chemical alteration, such as clay minerals and iron oxides
- Ions dissolved in rainwater and soil water

The mass of all these products, minus the water and carbon dioxide derived from the atmosphere, equals the original mass of the granite that was weathered.

Figure 7.17 Sand is a sediment composed of grains of various minerals broken down from various parent rocks in the course of chemical and physical weathering. [Rex Elliott.]

In this way, a granite rock is transformed into the raw material of sediment and the salts of the sea. The weathered products are eventually carried away from the weathering site by wind, water, and ice. Ultimately they are deposited as various kinds of sediment, such as the sand, silt, and mud found in river valleys and the limestones and evaporites of the world's oceans. As the rock cycle proceeds, this sediment is buried by additional deposits and slowly turns into sedimentary rock, the subject of Chapter 8.

SUMMARY

What is weathering and how is it geologically controlled? Rocks are broken down at Earth's surface by chemical weathering—the chemical alteration or dissolution of a mineral—and by physical weathering—the fragmentation of rocks by mechanical processes. Erosion wears away the land and transports the products of weathering, the raw material of sediment. The nature of the parent rock affects weathering because various minerals weather at different rates. Climate strongly affects weathering: warmth and heavy rainfall speed weathering; cold and dryness slow it down. The presence of soil accelerates weathering by providing moisture and an acidic environment, which promote chemical weathering and the growth of plant roots that aid physical weathering. All else being equal, the longer a rock weathers, the more completely it breaks down.

How does chemical weathering work? Potassium feldspar ($KAlSi_3O_8$), or orthoclase, weathers in much the same way as do many other kinds of silicate minerals. Feldspar weathers in the presence of water by hydrolysis. In the chemical reaction of orthoclase with water, potassium (K) and silica (SiO_2) are lost to the water solution and the solid feldspar changes into the clay mineral kaolinite, $Al_2Si_2O_5(OH)_4$. Carbon dioxide (CO_2) dissolved in water promotes chemical weathering by providing acid in the form of carbonic acid (H_2CO_3). Water in the ground and streams on the surface carry away dissolved ions and silica. Iron (Fe), which is found in ferrous form in many silicates, weathers by oxidation, producing ferric iron oxides in the process. Carbonates weather by dissolving completely, leaving no residue. These processes operate at varying rates, depending on the chemical stability of minerals under weathering conditions.

What are the processes of physical weathering? Physical weathering breaks rocks into fragments, either along crystal boundaries or along joints in rock masses. Physical weathering is promoted by chemical weathering, which weakens grain boundaries; by frost wedging, crystallization of minerals, and burrowing and tunneling by organisms and tree roots, all of which expand cracks; by fires; and perhaps by alternating extremes of heat and cold. Patterns of breakage such as exfoliation and spheroidal weathering result from interactions between chemical and physical weathering processes.

How do soils form as products of weathering? Soil is a mixture of clay minerals, weathered rock particles, and organic matter that forms as organisms interact with weathering rock and water. Weathering is controlled by climate and the activity of organisms, so soils form faster in warm, humid climates than in cold, dry ones. Young soils are affected by the composition of the parent rock, but older soils are affected primarily by climate. The three major soil groups are pedalfers, found in temperate climates; laterites, found in the humid tropics; and pedocals, which form in warm, dry climates.

Key Terms and Concepts

A-horizon (p. 155)
bauxite (p. 150)
B-horizon (p. 155)
chemical stability (p. 149)
chemical weathering (p. 141)
C-horizon (p. 155)
erosion (p. 141–142)
exfoliation (p. 153)
ferric iron (p. 150)
ferrous iron (p. 150)
frost wedging (p. 152)
hematite (p. 150)
humus (p. 155)
hydrolysis (p. 145)
joint (p. 152)
kaolinite (p. 145)
laterite (p. 155)
oxidation (p. 150)
pedalfer (p. 155)
pedocal (p. 158)
physical weathering (p. 141)
regolith (p. 155)
soil (p. 143)
spheroidal weathering (p. 153)
weathering (p. 141)

Exercises

This icon indicates that there is an animation available on the Web site that may assist you in answering a question.

1. What do the various kinds of rocks used for monuments tell us about weathering?

2. What rock-forming minerals found in igneous rocks weather to clay minerals?

3. How does abundant rainfall affect weathering?

4. Which weathers faster, a granite or a limestone?

5. How does physical weathering affect chemical weathering?

6. How do climates affect chemical weathering?

7. What are the main factors that control the development of different soil types?

8. What accelerates soil erosion?

Thought Questions

This icon indicates that there is an animation available on the Web site that may assist you in answering a question.

1. You are planning to use polished decorative limestone facings for monuments in Tucson, Arizona (a hot, arid region), and Seattle, Washington (a cool, rainy region). How do you think each of these monuments might look after a hundred years?

2. In northern Illinois, you can find two soils developed on the same kind of bedrock: one is 10,000 years old and the other is 40,000 years old. What differences would you expect to find in their compositions or profiles?

3. Which igneous rock would you expect to weather faster, a granite or a basalt? What factors influenced your choice?

4. Assume that a granite with crystals about 4 mm across and a rectangular system of joints spaced about 0.5 to 1 m apart is weathering at Earth's surface. What size would you ordinarily expect the largest weathered particle to be?

5. You have been comparing weathering in two rocks: (a) a basalt rich in volcanic glass, with very small crystals of pyroxene and calcium-rich feldspar that are about 0.5 mm across; and (b) a gabbro of exactly the same mineral composition but with large crystals about 3 mm across. What can you say about the speed of weathering in these two rocks?

6. Why do you think a road built of concrete, an artificial rock, tends to crack and develop a rough, uneven surface in a cold, wet region even when it is not subjected to heavy traffic?

7. Pyrite is a mineral in which ferrous iron is combined with sulfide ion. What major chemical process weathers pyrite?

8. Rank the following rocks in the order of the rapidity with which they weather in a warm, humid climate: a sandstone made of pure quartz, a limestone made of pure calcite, a granite, a deposit of rock salt (halite, NaCl).

9. What differences would you expect to find between the weathering of a pure magnesium olivine (Mg_2SiO_4) and the weathering of a pure iron olivine (Fe_2SiO_4)?

10. What would a world look like if there were no weathering at the surface?

11. Go to a local graveyard and compile a one-page description of the visible evidence of weathering of different kinds of rock used for gravestones. If you find no such evidence, explain why.

Short-Term Project

How Do Hoodoos Form?

As explained in this chapter, chemical and physical weathering processes reinforce each other. *Hoodoos* (a technical geologic term) are whimsically shaped pinnacles of rock and balanced boulders that are products of both physical and chemical weathering, commonly in dry climates. Bryce Canyon in Utah is a national park in large part because of its spectacular display of hoodoos (see the photo at the text's Web site). Hoodoos typically form in rock types such as sediments (as in Bryce), volcanic tuff, and granites. A set of vertical fractures (joints) is also a prerequisite for well-developed hoodoos.

At Bryce Canyon, soft siltstones and limestones from an ancient lake bed are exposed along the rim of the Paunsaugunt Plateau. Headward erosion into the rim of the plateau has created a maze of hoodoos. Snowmelt and heavy summer thunderstorms generate the runoff that readily erodes the soft sediments. Given the high elevation of the plateau, many days of freeze and thaw conditions also enhance physical weathering. The plateau rim is receding at a rate of 9 to 48 inches per 100 years, based on the study of growth rings of trees impacted by the erosion of the rim.

How do the hoodoos at Bryce Canyon form? Why are hoodoos and most weathered rocks typically rounded? To what degree does physical weathering enhance chemical weathering? Explore the answers to these questions at the text's Web site, http://www.whfreeman.com/understandingearth.

Suggested Readings

Blatt, H. 1992. *Sedimentary Petrology,* 2d ed. New York: W. H. Freeman.

Carroll, D. 1970. *Rock Weathering.* New York: Plenum.

Colman, S. M., and D. P. Dethier. 1986. *Rates of Chemical Weathering of Rocks and Minerals.* New York: Academic Press.

Gauri, K. L. 1978. The preservation of stone. *Scientific American* (June): 126.

Loughnan, F. C. 1969. *Chemical Weathering of the Silicate Minerals.* New York: Elsevier.

Martini, I. P., and W. Chesworth (eds.). 1992. *Weathering, Soils, and Paleosols.* New York: Elsevier.

Nahon, D. B. 1991. *Introduction to the Petrology of Soils and Chemical Weathering.* New York: Wiley.

Retallack, G. J. 1990. *Soils of the Past.* Boston: Unwin Hyman.

Sedimentary rocks exposed at Eagle Mountain, Nevada, were formed in an ancient ocean, about 500 million years ago. These rocks are mostly limestones and dolostones, which formed in shallow oceans when calcifying animals and plants died, leaving their shells as sediment. [John Grotzinger.]

CHAPTER

8

Sediments and Sedimentary Rocks

"A man should examine for himself the great piles of superimposed strata, and watch the rivulets bringing down mud, and the waves wearing away the sea-cliffs, in order to comprehend something about the duration of past time, the monuments of which we see all around us."

CHARLES DARWIN

Much of Earth's surface, including its seafloor, is covered with sediments. These layers of loose particles have diverse origins. Most sediments are created by weathering of the continents. Some are the remains of organisms that secreted mineral shells. Yet others consist of inorganic crystals that precipitated when dissolved chemicals in oceans and lakes combined to form new minerals.

Sedimentary rocks were once sediments, and so they are records of the conditions at Earth's surface when and where the sediments were deposited. Geologists can work backward to infer the sources of the sediments from which the rocks were formed and the kinds of places in which the sediments were originally deposited. For example, the top of Mount Everest is composed of fossiliferous limestones. This evidence indicates that long before Mount Everest was uplifted, this point—now the highest in the world—was part of the floor of an ocean.

The kind of analysis used to make inferences about rock formations at the top of Mount Everest applies just as well to ancient shorelines, mountains, plains, deserts, and swamps in other regions. In one area, for example, sandstone may record an earlier time when beach sands accumulated along a shoreline that no longer exists. In a bordering area, carbonate reefs may have been laid down along the perimeter of a tropical island. Beyond, there may have been a nearshore area in which the sediments were shallow marine carbonate muds that later became thin-bedded limestone. In reconstructing such environments, we can map the continents and oceans of long ago.

Further inferences are also possible. We can work backward to understand former plate tectonic settings and movements by studying sedimentary rocks, which reveal their origin in volcanic arcs, rift valleys, or collisional mountains. In some cases, where the components of sediments and sedimentary rocks are derived from the weathering of preexisting rocks, we can form hypotheses about the ancient climate and weathering regime. We can also use sedimentary rocks formed by precipitation from seawater to read the history of change of Earth's climate and seawater chemistry.

The study of sediments and sedimentary rocks has great practical value as well. Oil and gas, our most valuable sources of energy, are found in these rocks. So is much of the uranium used for nuclear power. Coal, which is a distinct type of sedimentary rock, is also used to generate energy. Phosphate rock used for fertilizer is sedimentary, as is much of the world's iron ore. Knowing how these kinds of sediments form helps us to find and use these limited resources.

Finally, because virtually all sedimentary processes take place at or near Earth's surface where we humans live, they provide a background for our understanding of environmental problems. We once studied sedimentary rocks primarily to understand how to exploit the natural resources just mentioned. Increasingly, however, we study these rocks to improve our understanding of Earth's environment.

In this chapter, we will see how geologic processes such as weathering, transportation, sedimentation, and diagenesis produce sediments and sedimentary rocks. We will describe the compositions, textures, and structures of sediments and sedimentary rocks and examine how they correlate with the kinds of environments in which the sediments and rocks are laid down. Throughout the chapter, we will apply our understanding of sediment origins to the study of human environmental problems and to the exploration for energy and mineral resources.

Sedimentary Rocks and the Rock Cycle

Sediments, and the sedimentary rocks formed from them, are produced during the surface stage of the rock cycle (discussed in Chapter 4). In other words, they form after rocks have been brought up from the interior by tectonics and before they are returned to the interior by burial. The processes that make up the sedimentary stages of the rock cycle are reviewed in **Figure 8.1**.

- *Weathering* Physical weathering fractures rocks; chemical weathering converts minerals and rocks into altered solids, solutions, and precipitates.

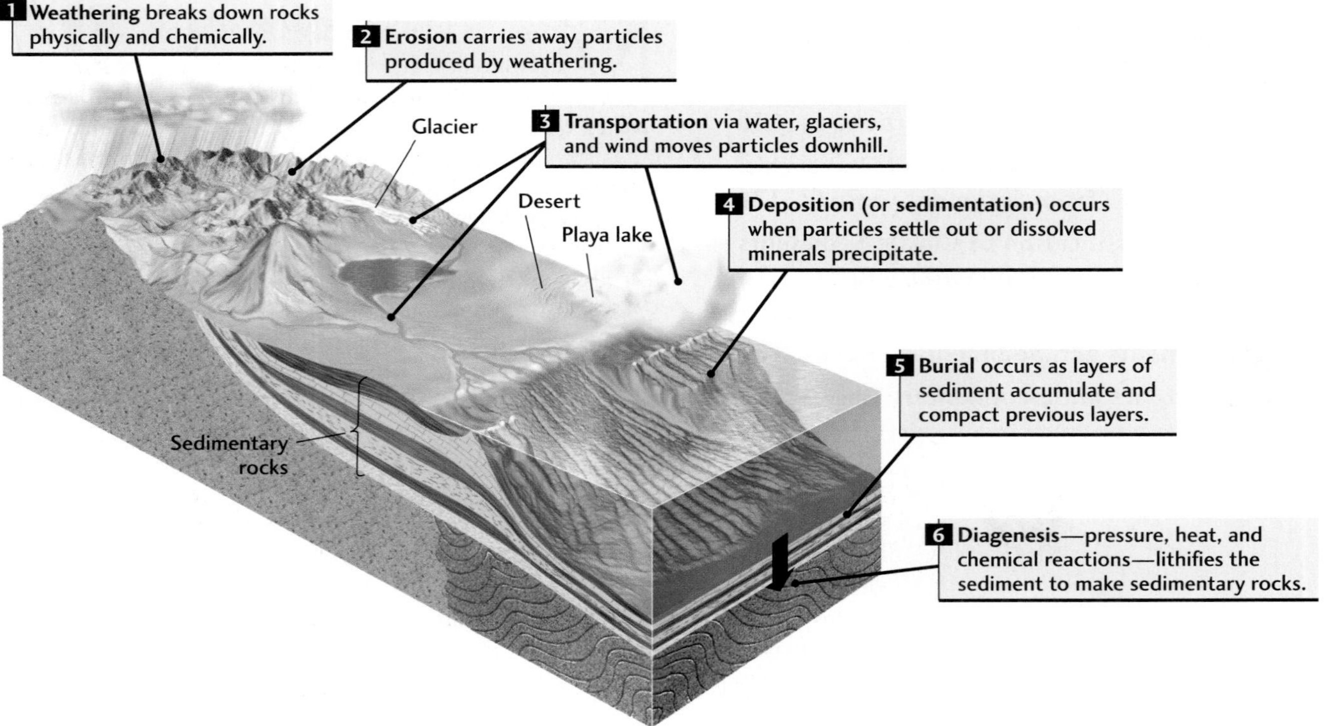

Figure 8.1 The sedimentary stages of the rock cycle comprise several overlapping processes: physical and chemical weathering; erosion; transportation; deposition; burial; and diagenesis.

- *Erosion* Erosion mobilizes the particles produced by weathering.
- *Transportation* Currents of wind and water and the moving ice of glaciers transport particles to new locations downhill or downstream.
- *Deposition* (also called *sedimentation*) Sedimentary particles settle out as winds die down, water currents slow, or glacier edges melt. These particles form layers of sediment on land or under the sea. In the ocean or in land aquatic environments, chemical precipitates form and are deposited, and the shells of dead organisms are broken up and deposited.
- *Burial* As layers of sediment accumulate, older deposited material is compacted and eventually buried in Earth's crust.
- *Diagenesis* Diagenesis refers to the physical and chemical changes—including pressure, heat, and chemical reactions—by which buried sediments are lithified and acquire a new identity as sedimentary rocks.

Sediments in the Earth System

We can view the sedimentary stages of the rock cycle as interactions between the plate tectonic and climate geosystems that govern Earth's surface and shallow crust. For example, weathering and erosion result from interactions of the solid crust with the atmosphere and oceans. In a typical interaction, weathering and transportation might increase if rainfall increases. Transportation brings the materials to deposition sites. As deposits accumulate, they are buried by later arrivals of sediment. As sediments are buried, they begin to undergo diagenesis. In this chapter and succeeding ones, we will see how specific processes, such as stream flow, play roles in these system interactions.

Weathering and Erosion Yield the Raw Materials: Particles and Dissolved Substances

As we saw in Chapter 7, chemical weathering and mechanical fragmentation of rock at the surface make both solid and dissolved products, and erosion carries away these materials. The end products are grouped either as clastic sediments or as chemical and biochemical sediments.

Clastic Sediments **Clastic particles** are physically transported rock fragments produced by the weathering of preexisting rocks. **Clastic sediments** are accumulations of clastic particles. The particles range in size from boulders to pebbles to particles of sand, silt, and clay. They also vary widely in shape. Natural breakage along joints, bedding planes, and other fractures in the parent rock determines the shapes of boulders, cobbles, and pebbles. Sand grains tend to inherit their shapes from the individual crystals formerly interlocked in the parent rock.

Clastic sediments are sometimes called *siliciclastic* because they are produced by the weathering of rocks composed largely of silicate minerals. The mixture of minerals in clastic sediments varies. Minerals such as quartz are resistant to weathering and thus are found unaltered in clastic sediments. There may be partly altered fragments of minerals, such as feldspar, that are less resistant to weathering and so less stable. Still other minerals in clastic sediments, such as clay minerals, may be newly formed. Varying intensities of weathering can produce different sets of minerals in sediments derived from the same parent rock. Where weathering is intense, the sediment will contain only clastic particles made of chemically stable minerals, mixed with clay minerals. Where weathering is slight, many minerals that are unstable under surface conditions will survive as clastic particles. Table 8.1 shows three possible sets of minerals in a typical granite outcrop.

Chemical and Biochemical Sediments The dissolved products of weathering are ions or molecules in the waters of soils, rivers, lakes, and oceans. These dissolved substances are precipitated from water by chemical and biochemical reactions. **Chemical sediments** form at or near their place of deposition, usually from seawater. **Biochemical sediments** comprise the undissolved mineral remains of organisms as well as minerals precipitated by biological processes. We distinguish these two sediment types here for

Table 8.1 Minerals Remaining in Clastic Sediments Derived from an Average Granite Outcrop Under Varying Intensities of Weathering

INTENSITY OF WEATHERING		
Low	**Medium**	**High**
Quartz	Quartz	Quartz
Feldspar	Feldspar	Clay minerals
Mica	Mica	
Pyroxene	Clay minerals	
Amphibole		

convenience only; in practice, many chemical and biochemical sediments overlap. In most of the world, much more rock is fragmented by physical weathering than is dissolved by chemical weathering. Thus clastic sediments are about 10 times more abundant than chemical and biochemical sediments in the Earth's crust.

Biochemical sediments in shallow marine environments consist of layers of biologically precipitated sedimentary particles, such as whole or fragmented shells. Sometimes shells can be transported, further broken up, and deposited as **bioclastic sediments.** These shallow-water sediments consist predominantly of two calcium carbonate minerals—calcite and aragonite—in variable proportions. Other minerals such as phosphates and sulfates are only locally abundant.

In the deep ocean, biochemical sediments are made of the shells of only a few kinds of organisms They are composed predominantly of the calcium carbonate mineral calcite, but silica may be precipitated broadly over some parts of the deep ocean. Because these biochemical particles accumulate in very deep water where agitation by sediment-transporting currents is rare, the shells rarely form bioclastic sediments.

Other chemical sediments form inorganically. For example, the evaporation of seawater often leads to the precipitation of layers composed of gypsum or halite. These sediments form in arid climates and in places where an arm of the sea becomes isolated enough that evaporation concentrates the dissolved chemicals in seawater to the point of precipitation.

Transportation and Deposition: The Downhill Journey to Sedimentation Sites

After clastic particles and dissolved ions have formed by weathering and erosion, they start a journey to a sedimentation area. This journey may be very long; for example, it might span thousands of kilometers from the tributaries of the Mississippi in the highlands of the Rocky Mountains to the swamps of Louisiana.

Most agents of transportation carry material downhill. Rocks falling from a cliff, sand carried by a river flowing to the sea, and glacial ice slowly creeping downhill are all responses to gravity. Although winds may blow material from a low elevation to a higher one, in the long run gravity is relentless, and windblown sand and dust settle in response to its pull. When a windblown particle drops to the ocean and settles through the water, it is "trapped." It can be picked up again only by an ocean current, which can transport it only to another depositional site on the seafloor.

In marine environments, chemically or biochemically produced particles may be transported from the area where they formed to a nearby site of deposition. However, marine currents that transport sediment, such as tidal currents (see Chapter 18), act over a shorter distance than big rivers on land. The short transport distance for chemical or biochemical sediments contrasts with the much greater distances over which siliciclastic sediments are transported.

Currents as Transport Agents for Clastic Particles Most sediments are transported by air and water currents. The enormous quantities of all kinds of sediment found in the oceans result primarily from the transporting capabilities of rivers, which annually carry a solid and dissolved sediment load of about 25 billion tons (250×10^{14} g).

Air currents move material, too, but in far smaller quantities than do rivers or ocean currents. As particles are lifted into the fluid air or water, the current carries them downwind or downriver. The stronger the current—that is, the faster it flows—the larger the particles it transports.

Current Strength, Particle Size, and Sorting Sedimentation starts where transportation stops. For clastic particles, the driving force of sedimentation is largely the effect of gravity. Particles tend to settle under the pull of gravity. This tendency works against the ability of a current to carry a particle. Although it is a basic law of physics that, in a vacuum, particles of all sizes and densities fall to Earth at the same speed, this law does not apply to particles in a fluid. In a fluid, large grains settle faster than small ones. The settling velocity is proportional to the density of the particle and to its size. The most common minerals in sediments have roughly the same density (about 2.6 to 2.9 g/cm^3). We therefore use size, which is more conveniently measured than density, as an indicator of the settling velocity of minerals in sediments.

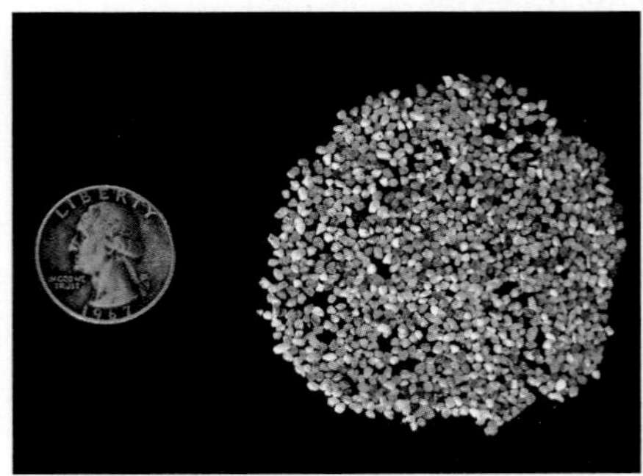

Well-sorted sand

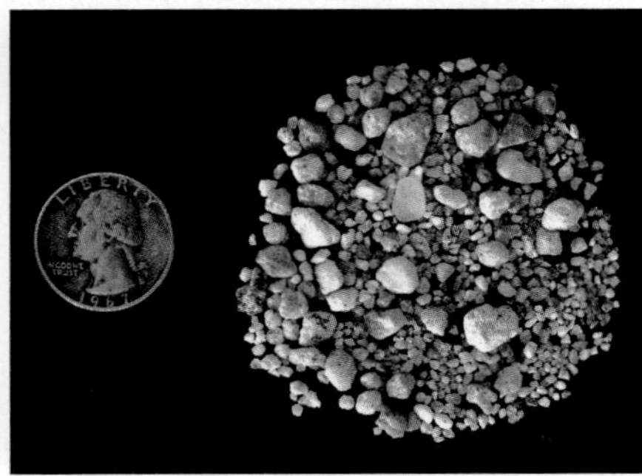

Poorly sorted sand

Figure 8.2 As currents decrease in velocity, sediment is segregated according to particle size. The relatively homogeneous group of sand grains on the left is well sorted; the group on the right is poorly sorted. [Bill Lyons.]

As a current that is carrying particles of various sizes begins to slow, it can no longer keep the largest particles suspended, and they settle. As the current slows even more, smaller particles settle. When the current stops completely, even the smallest particles settle. More specifically, currents segregate particles in the following ways:

- *Strong currents* (faster than 50 cm/s) carry gravels, along with an abundant supply of coarse and fine detritus. Such currents are common in swiftly flowing streams in mountainous terrains, where erosion is rapid. Beach gravels are deposited where ocean waves erode rocky shores. Glaciers produce and deposit clastic debris of all sizes as they move forward.

- *Moderately strong currents* (20–50 cm/s) lay down sand beds. Currents of moderate strength are common in most rivers, which carry and deposit sand in their channels. Rapidly flowing floodwaters may spread sands over the width of a river valley. Winds also blow and deposit sand, especially in deserts, and waves and currents deposit sand on beaches and in the ocean.

- *Weak currents* (slower than 20 cm/s) carry muds composed of the finest clastic particles. Weak currents are found on the floor of a river valley when floodwaters recede slowly or stop flowing entirely. Generally, muds are deposited in the ocean some distance from shore, where currents are too slow to keep even fine particles in suspension. Much of the floor of the open ocean is covered with mud particles originally transported by surface waves and currents or by the wind. These particles slowly settle to depths where currents and waves are stilled and, ultimately, all the way to the bottom of the ocean.

As you can see, currents may begin by carrying particles of widely varying size, which then become separated as the velocity of the current varies. A strong, fast current may lay down a bed of gravel, while keeping sands and muds in suspension. If the current weakens and slows, it will lay down a bed of sand on top of the gravel. If the current then stops altogether, it will deposit a layer of mud on top of the sand bed. This tendency for variations in current velocity to segregate sediments according to size is called **sorting.** A well-sorted sediment consists mostly of particles of a uniform size. A poorly sorted sediment contains particles of many sizes (**Figure 8.2**).

Particles are generally transported intermittently rather than steadily. Fast currents give way to weak flows or stop entirely. A river may transport large quantities of sand and gravel when it floods, but it will drop them as the flood recedes, only to pick them up again and carry them even farther in the next flood. Strong winds may carry large amounts of dust for a few days and then die down and deposit the dust as a layer of sediment. Similarly, strong tidal or other shallow-water currents along some ocean margins may transport eroded particles of previously deposited calcium carbonate sediments to places farther offshore and drop them there. Both chemical and physical weathering processes continue during transportation. Like transportation agents, the slower weathering processes tend to operate intermittently.

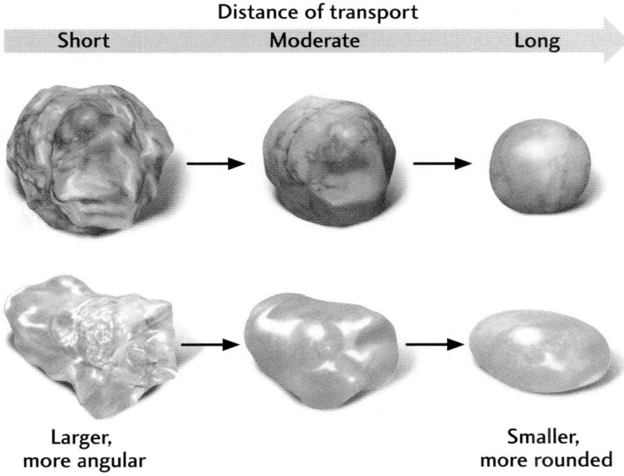

Figure 8.3 Transportation reduces the size and angularity of clastic particles. Grains become rounded and slightly smaller as they are transported, although the general shape of the grain may not change significantly.

While water and wind currents are transporting particles, physical weathering continues. Physical weathering processes affect particles in two ways: they reduce particle size, and they round the originally angular fragments (**Figure 8.3**). As particles are transported, they tumble and strike one another or rub against bedrock. Pebbles or large grains that collide forcefully may break into two or more smaller pieces. Weaker hits may chip small pieces from edges and corners. Abrasion by bedrock, together with grain impacts, also rounds the particles, wearing off and smoothing sharp edges and corners. These effects apply mostly to the larger particles; there is little abrasion of sand and silt by impact.

Chemical Weathering Is Intermittent Although clastic material is still in contact with the chief agents of chemical weathering—water and the oxygen and carbon dioxide of the atmosphere—slow weathering reactions do not have much effect during the brief periods when material is actually being transported by a current. Most chemical weathering occurs during the long intermittent periods when the sediment is temporarily deposited before being picked up again by the current. For example, when a river floods its valley for a few days, it deposits sand, silt, and clay. After the flood recedes, chemical weathering of the deposits resumes and continues until the next flood. That flood may pick up the sediment of the earlier flood and redeposit it farther downstream, where it begins to undergo chemical weathering again.

In this way, episodes of transportation and deposition can alternate with episodes of chemical weathering. Deposition may be intermittent, and the total time between the formation of clastic debris and its final deposition may be many hundreds or thousands of years, depending on the distance to the final depositional area and the number of stop-offs along the way. Clastic particles eroded by the head-waters of the Missouri River in the mountains of western Montana, for example, take hundreds of years to travel the 3200 km down the Missouri and Mississippi rivers to the Gulf of Mexico. During the long journey, physical weathering and intermittent chemical weathering may affect the particles these rivers carry.

Oceans and Lakes: Chemical Mixing Vats The driving force of chemical and biochemical sedimentation is chemistry rather than gravity. Chemical substances dissolved in water during weathering are carried along with the water as a homogeneous solution. Materials such as dissolved calcium ions are part of the water solution itself, so gravity cannot cause them to settle out of it. As dissolved materials flow down rivers, they ultimately enter lake waters or the ocean.

Oceans may be thought of as huge chemical mixing tanks. Rivers, rain, wind, and glaciers constantly bring in dissolved materials. Smaller quantities of dissolved materials enter the ocean by hydrothermal chemical reactions between seawater and hot basalt at mid-ocean ridges. The ocean continuously loses water by evaporation at the surface. The inflow and outflow of water to and from the oceans are so exactly balanced that the amount of water in the oceans remains constant over such geologically short times as years, decades, or even centuries. Over a time scale of thousands to millions of years, however, the balance may shift. During the Ice Ages, for example, significant quantities of seawater were converted into glacial ice, and sea level was drawn down by more than 100 m.

The entry and exit of dissolved materials, too, are balanced. Each of the many dissolved components of seawater participates in some chemical or biochemical reaction that eventually precipitates it out of the water and onto the seafloor. As a result, the ocean's **salinity**—the total amount of dissolved substances in a given volume of seawater—remains constant. Totaled over all the oceans of the world, precipitation balances the total inflow of dissolved material from continental weathering and from hydrothermal activity at mid-ocean ridges—yet another way in which the Earth system maintains equilibrium.

We can see some of the mechanisms that sustain this chemical balance by considering the element calcium. Calcium is an important component of the most abundant biochemical precipitate formed in the oceans, calcium carbonate ($CaCO_3$). Calcium is dissolved when limestone and silicates containing calcium, such as some feldspars and pyroxenes, are weathered on land and brought as ions (Ca^{2+}) to the oceans. There, a wide variety of marine organisms biochemically combine the calcium ions with bicarbonate ions (HCO_3^-), also present in seawater, to form their calcium carbonate shells. The calcium that entered the ocean as dissolved ions leaves it as a solid sediment when the organisms die and their shells settle and accumulate as calcium carbonate sediment on the seafloor. Ultimately, postdepositional processes transform the calcium carbonate sediment into limestone. The chemical balance that keeps the levels of calcium dissolved in the ocean constant is thus regulated in part by the activities of organisms.

Nonbiological mechanisms also maintain chemical balance in the oceans. For example, sodium ions (Na^+) brought into the oceans react chemically with chloride ions (Cl^-) to form the precipitate sodium chloride (NaCl). This happens when evaporation raises the amounts of sodium and chloride ions past the point of saturation. As we saw in Chapter 3, solutions crystallize minerals when they become so saturated with dissolved materials that they can hold no more. The intense evaporation required to crystallize salt takes place in warm, shallow arms of the sea.

Organic sedimentation is yet another type of biochemical precipitation. Vegetation may be preserved from decay in swamps and accumulate as a rich organic material, **peat,** which contains more than 50 percent carbon. Peat is ultimately buried and transformed by diagenesis into coal. In both lake and ocean waters, the remains of algae, bacteria, and other microscopic organisms may accumulate in sediments as organic matter that can be transformed into oil and gas.

Sedimentary Environments

Of the many ways in which sedimentation can be classified, geologists have found the concept of sedimentary environments most useful. A **sedimentary environment** is a geographic location characterized by a particular combination of geologic processes and environmental conditions (**Figure Story 8.4**). Sedimentary environments are often grouped by their locations on the continents, near shorelines, or in the ocean. Environmental conditions include

- The kind and amount of water (ocean, lake, river, arid land)
- The topography (lowland, mountain, coastal plain, shallow ocean, deep ocean)
- Biological activity

Geologic processes include the currents that transport and deposit sediments (water, wind, ice), the plate tectonic settings that may affect sedimentation and burial of sediments, and volcanic activity. Thus, a beach environment brings together the dynamics of waves approaching and breaking on the shore, the resultant currents, and the distribution of sediments on the beach.

Sedimentary environments are related to plate tectonic settings. For example, the environment of a deep-ocean trench is found at a subduction zone, whereas thick alluvial (river) deposits are typically associated with mountains formed by the collision of continents. Alluvial deposits are also found along the borders of continental rift valleys. Sedimentary environments may be affected or determined by climate as well

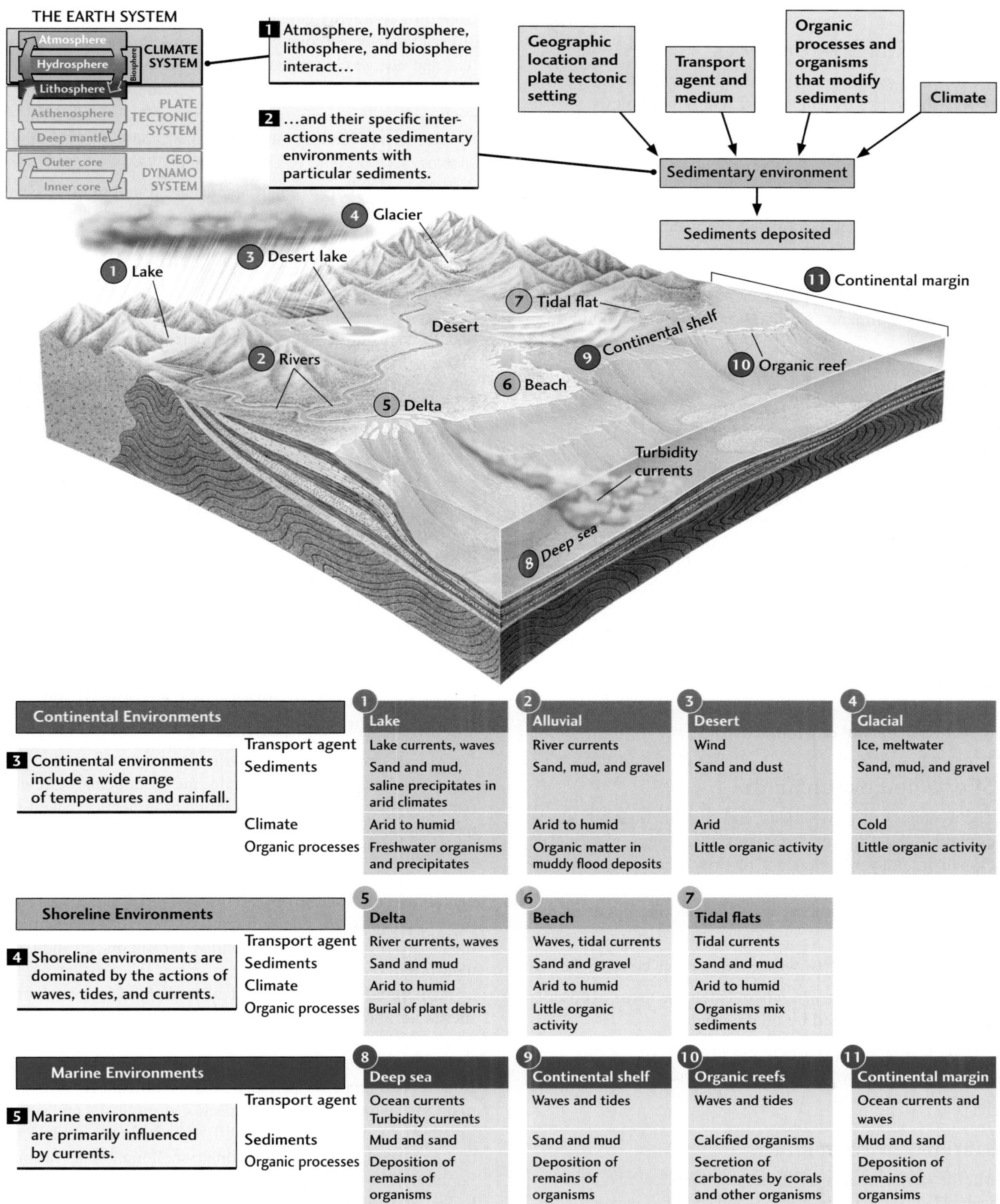

Continental Environments		1 Lake	2 Alluvial	3 Desert	4 Glacial
3 Continental environments include a wide range of temperatures and rainfall.	Transport agent	Lake currents, waves	River currents	Wind	Ice, meltwater
	Sediments	Sand and mud, saline precipitates in arid climates	Sand, mud, and gravel	Sand and dust	Sand, mud, and gravel
	Climate	Arid to humid	Arid to humid	Arid	Cold
	Organic processes	Freshwater organisms and precipitates	Organic matter in muddy flood deposits	Little organic activity	Little organic activity

Shoreline Environments		5 Delta	6 Beach	7 Tidal flats
4 Shoreline environments are dominated by the actions of waves, tides, and currents.	Transport agent	River currents, waves	Waves, tidal currents	Tidal currents
	Sediments	Sand and mud	Sand and gravel	Sand and mud
	Climate	Arid to humid	Arid to humid	Arid to humid
	Organic processes	Burial of plant debris	Little organic activity	Organisms mix sediments

Marine Environments		8 Deep sea	9 Continental shelf	10 Organic reefs	11 Continental margin
5 Marine environments are primarily influenced by currents.	Transport agent	Ocean currents Turbidity currents	Waves and tides	Waves and tides	Ocean currents and waves
	Sediments	Mud and sand	Sand and mud	Calcified organisms	Mud and sand
	Organic processes	Deposition of remains of organisms	Deposition of remains of organisms	Secretion of carbonates by corals and other organisms	Deposition of remains of organsims

Figure Story 8.4 A sedimentary environment is characterized by a particular set of environmental conditions and geologic processes.

as by tectonics. For example, a desert environment has an arid climate; a glacial environment has a cold one.

Continental Environments

Sedimentary environments on continents are diverse, owing to the wide range of temperature and rainfall on the surface of the land. These environments are built around rivers, deserts, lakes, and glaciers (see Figure Story 8.4).

• An *alluvial environment* includes a river channel, the borders of the channel, and the flat valley floor on either side of the channel that is covered by water when the river floods. Rivers are present on all the continents but Antarctica, and so alluvial deposits are widespread. Organisms are abundant in the muddy flood deposits and are responsible for organic sediments. Climates vary from arid to humid.

• A *desert environment* is arid. Sediment in a desert forms by a combination of wind action and the work of rivers that flow (mostly intermittently) through it. The aridity inhibits organic growth, so organisms have little effect on the sediment. Desert sand dunes provide a special sandy environment.

• A *lake environment* is controlled by the relatively small waves and moderate currents of inland bodies of fresh or saline water. Chemical sedimentation of organic matter and carbonates may occur in freshwater lakes. Saline lakes such as those found in deserts evaporate and precipitate a variety of evaporite minerals, such as halite. The Great Salt Lake is an example.

• A *glacial environment* is dominated by the dynamics of moving masses of ice and is characterized by a cold climate. Vegetation is present but has small effects on sediment. At the melting border of a glacier, meltwater streams form a transitional alluvial environment.

Shoreline Environments

The dynamics of waves, tides, and currents on sandy shores dominate shoreline environments (see Figure Story 8.4). Organisms may be abundant in these shallow waters, but they do not influence clastic sedimentation much, except where carbonate sediments are abundant. Shoreline environments include

• *Deltaic environments,* where rivers enter lakes or the sea

• *Tidal flat environments,* where extensive areas exposed at low tide are dominated by tidal currents

• *Beach environments,* where the strong waves approaching and breaking on the shore distribute sediments on the beach, laying down strips of sand or gravel

Marine Environments

Marine environments are usually subdivided on the basis of water depth, which determines the kinds of currents found in ocean settings (see Figure Story 8.4). Alternatively, they can be classified on the basis of distance from land.

• *Continental-shelf environments* are located in the shallow waters off continental shores, where sedimentation is controlled by relatively gentle currents. Sedimentation may be either clastic or chemical, depending on the source of clastics and the extent of carbonate-producing organisms or evaporite conditions.

• *Organic reefs* are composed of carbonate structures formed by carbonate-secreting organisms built up on continental shelves or on oceanic volcanic islands.

• *Continental-margin environments* are found in the deeper waters at the edges of the continents, where sediment is deposited by turbidity currents. A turbidity current is a turbulent submarine avalanche of sediment and water that moves downslope.

• *Deep-sea environments* include all the floors of the deep ocean, far from the continents, where the quiet waters are disturbed only occasionally by ocean currents. These environments include the continental slope, which is built up by turbidity currents traveling far from continental margins; the abyssal plains, which accumulate sediments provided mostly by the skeletons of plankton that settle from above; and the mid-ocean ridges.

We have seen that sedimentary environments can be defined by location. They can also be categorized according to the kinds of sediments found in them or according to the dominant type of sedimentation. Grouping in this manner produces two broad classes: clastic sedimentary environments and chemical and biochemical sedimentary environments.

Clastic versus Chemical and Biochemical Sedimentary Environments

Clastic sedimentary environments are those dominated by clastic sediments. They include the continental alluvial (stream), desert, lake, and glacial environments, as well as the shoreline environments transitional between continental and marine: deltas, beaches, and tidal flats. They also include oceanic environments of the continental shelf, continental margin, and deep-ocean floor where clastic sands and muds are deposited. The sediments of these clastic environments are often called **terrigenous sediments,** to indicate their origin on land.

Chemical and biochemical sedimentary environments are those characterized principally by chemical and biochemical precipitation (Table 8.2). By far the most abundant are *carbonate environments*—marine settings where calcium carbonate, principally of biochemical origin, is the main sediment. Hundreds of species of mollusks and other invertebrate organisms, as well as calcareous (calcium-containing) algae, secrete carbonate shell materials. Various populations of these organisms live at different depths of water, both in quiet areas and in places where waves and currents are strong. As they die, their shells accumulate to form sediment.

Table 8.2 Major Chemical and Biochemical Sedimentary Environments

Environment	Agent of Precipitation	Sediments
SHORELINE AND MARINE		
Carbonate (includes reef, bank, deep sea, etc.)	Shelled organisms, some algae; inorganic precipitation from seawater	Carbonate sands and muds, reefs
Evaporite	Evaporation of seawater	Gypsum, halite, other salts
Siliceous: deep sea	Shelled organisms	Silica
CONTINENTAL		
Evaporite	Evaporation of lake water	Halite, borates, nitrates, carbonates, other salts
Swamp	Vegetation	Peat

Except for those of the deep sea, carbonate environments are found mostly in the warmer tropical or subtropical regions of the oceans, where chemical conditions favor the precipitation of calcium carbonate. These regions include organic reefs, carbonate sand beaches, tidal flats, and shallow carbonate banks. In a few places, carbonate sediments may form in cooler waters that are supersaturated with carbonate—waters that are generally below 20°C, such as some regions of the Antarctic Ocean south of Australia. Carbonate sediments in cool waters are mainly calcite shell materials.

An *evaporite environment* is created when the warm seawater of an arid inlet or arm of the sea evaporates more rapidly than it can mix with the connected open marine seawater. The degree of evaporation and the length of time it has proceeded control the salinity of the evaporating seawater and thus the kinds of sediment formed. Evaporite environments also form in lakes with no outlet rivers. Such lakes may produce sediments of halite, borate, nitrates, and other salts.

Siliceous environments are special deep-sea environments named for the remains of silica shells deposited in them. The organisms that secrete silica grow in surface waters where nutrients are abundant. Their shells settle to the ocean floor and accumulate as layers of siliceous sediment.

Sedimentary Structures

All kinds of bedding and a variety of other surfaces formed at the time of deposition are called **sedimentary structures. Bedding,** or *stratification,* is a common feature of sediments and sedimentary rocks. The parallel layers of different grain sizes or compositions indicate successive depositional surfaces. Bedding may be thin, on the order of centimeters or even millimeters thick. At the opposite extreme, bedding may be meters or even many meters thick. Most bedding is horizontal, or nearly so, at the time of deposition. Some types of bedding, however, form at a high angle relative to horizontal.

Cross-Bedding

Cross-bedding consists of sets of bedded material deposited by wind or water and inclined at angles as large as 35° from the horizontal (**Figure 8.5**). Cross-beds form when

Figure 8.5 Cross-bedding in a desert environment. The varying directions of cross-bedding in this sandstone are due to changes in wind direction at the time the sand dunes were deposited. Navajo sandstone, Zion National Park, southwestern Utah. [Peter Kresan.]

grains are deposited on the steeper, downcurrent (lee) slopes of sand dunes on land or of sandbars in rivers and under the sea. Cross-bedding of wind-deposited sand dunes may be complex, a result of rapidly changing wind directions. Cross-bedding is common in sandstones (**Figure 8.6**) and is also found in gravels and some carbonate sediments. Cross-bedding is easier to see in sandstones than in sands, which must be trenched or otherwise excavated to see a cross section.

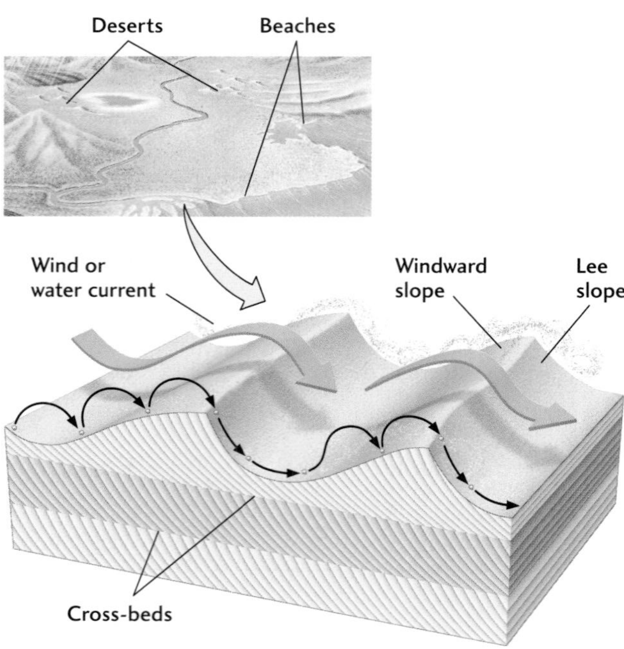

Figure 8.6 Cross-beds form when grains are deposited on the steeper, downcurrent (lee) slope of a dune or ripple.

Graded Bedding

Graded bedding is most abundant in continental slope and deep-sea sediments deposited by a special variety of bottom current called a turbidity current (see Chapter 17). Each layer in a graded bed progresses from coarse grains at the base to fine grains at the top. The grading indicates a waning of the current that deposited the grains. A graded bed comprises one set of coarse-to-fine beds, normally ranging from a few centimeters to several meters thick, that formed horizontal or nearly horizontal layers at the time of deposition. Accumulations of many individual graded beds can reach a total thickness of hundreds of meters. A bed formed as a result of deposition from a turbidity current is called a *turbidite.*

Ripples

Ripples are very small dunes of sand or silt whose long dimension is at right angles to the current. They form low, narrow ridges, or corrugations, most only a centimeter or two high, separated by wider troughs. These sedimentary structures are common in both modern sands and ancient sandstones (**Figure 8.7**). Ripples can be seen on the surfaces of windswept dunes, on underwater sandbars in shallow streams, and under the waves at beaches. Geologists

Figure 8.7 *(left)* Ripples in modern sand on a beach [Raymond Siever]. *(right)* Ancient ripple-marked sandstone [Reg Morrison/Auscape].

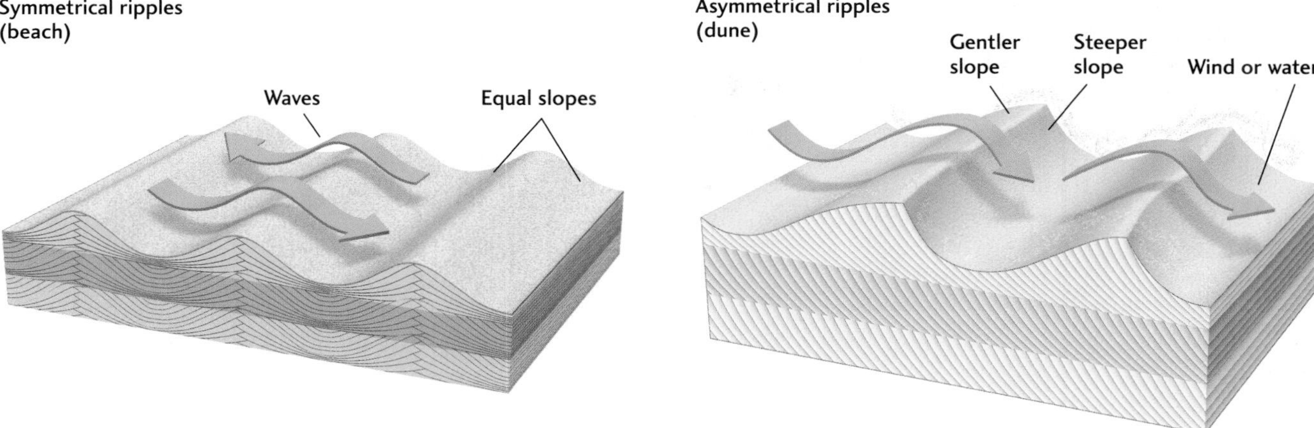

Figure 8.8 The shapes of ripples on beach sand, produced by the back-and-forth movements of waves, are symmetrical. Ripples on dunes and river sandbars, produced by the movement of a current in one direction, are asymmetrical.

can distinguish the symmetrical ripples made by waves moving back and forth on a beach from the asymmetrical ripples formed by currents moving in a single direction over river sandbars or windswept dunes (**Figure 8.8**).

Bioturbation Structures

Bedding in many sedimentary rocks is broken or disrupted by roughly cylindrical tubes a few centimeters in diameter that extend vertically through several beds. These sedimentary structures are remnants of burrows and tunnels excavated by clams, worms, and many other marine organisms that live on the bottom of the sea. These organisms rework existing sediments by burrowing through muds and sands—a process called **bioturbation.** They ingest sediment for the bits of organic matter it contains and leave behind the reworked sediment, which fills the burrow (**Figure 8.9**). From bioturbation structures, geologists can deduce the behaviors of organisms that burrowed the sediment and thus reconstruct the sedimentary environment.

Figure 8.9 Bioturbation structures. These tracks are thought to have been made by trilobites living in mid-Cambrian muddy sediment in Montana about 500 million years ago. This rock is crisscrossed with fossilized tracks and tunnels originally made as the organisms burrowed through the mud. [Chip Clark.]

Bedding Sequences

Bedding sequences are built of interbedded and vertically stacked layers of sandstone, shale, and other sedimentary rock types. A bedding sequence may consist of cross-bedded sandstone, overlain by bioturbated siltstone, overlain in turn by rippled sandstone—in any combination of thicknesses for each rock type in the sequence. Bedding sequences give geologists insight into the history of events that occurred at Earth's surface long ago.

Bedding sequences help geologists to reconstruct how all the sediments were deposited. **Figure 8.10** shows a bedding sequence typically formed by rivers. A river lays down repetitive sequences that form as its channel migrates back and forth across the valley floor. The lower part of each sequence represents the sediments deposited in the deepest part of the channel, where the current was strongest. The upper part represents the sediments deposited in the shallow parts of the channel, where the current was weakest. Typically, a bedding sequence formed in this manner will consist of sediments that grade upward from coarse to fine. Large-scale cross-bedded layers will be found at the base, then small-scale cross-bedding, and then horizontal bedding at the top. The top layer of horizontally bedded, fine-grained

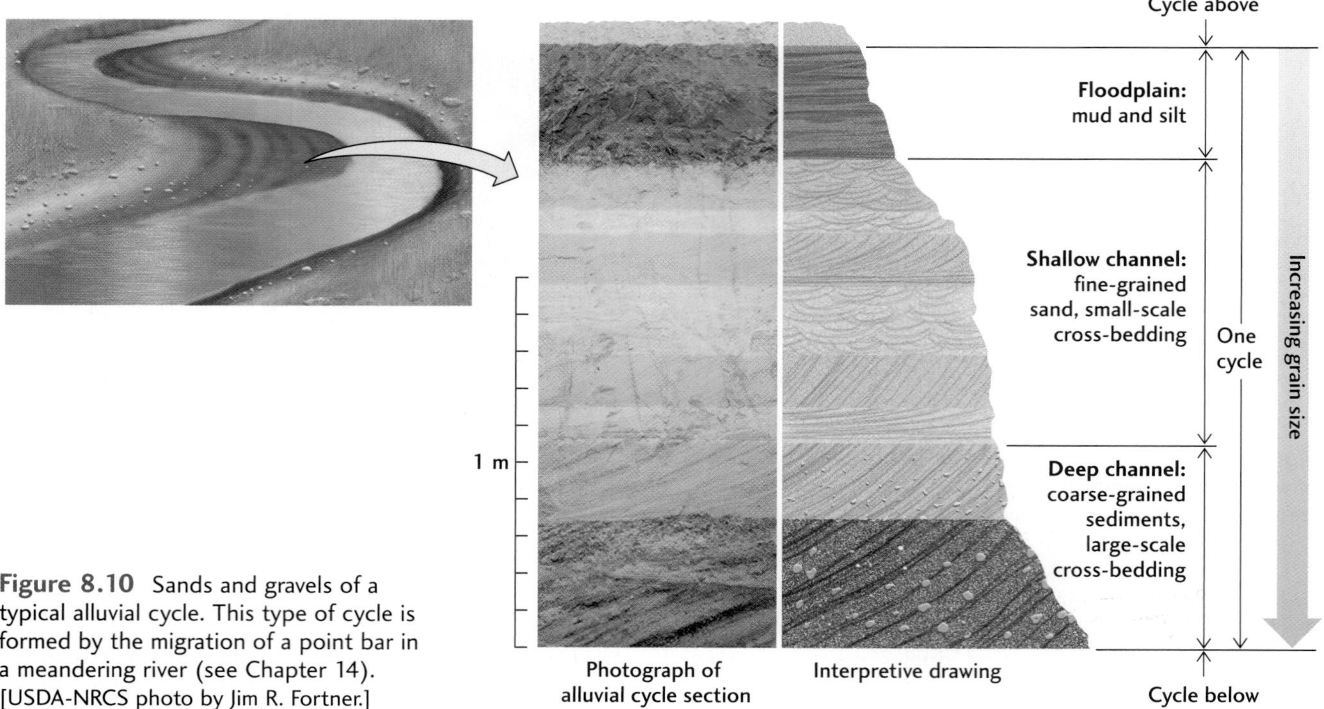

Figure 8.10 Sands and gravels of a typical alluvial cycle. This type of cycle is formed by the migration of a point bar in a meandering river (see Chapter 14). [USDA-NRCS photo by Jim R. Fortner.]

sediment accumulates in the shallowest part of the channel, where the weakest current allows the finest-grained sediment to accumulate in horizontal layers by settling out of suspension. Today advanced computer models relate bedding sequences of sands deposited in alluvial environments to their causative factors. Other characteristic bedding sequences can be used to recognize deposition at the shoreline and in the deep sea. (These sequences are discussed further in Chapter 17.)

Burial and Diagenesis: From Sediment to Rock

Most of the clastic particles produced by weathering and erosion of the land end up as marine sediments deposited in various parts of the world's oceans, brought there by rivers, wind, and glaciers. A smaller amount of clastic sediment is laid down on land.

Similarly, most chemical and biochemical sediments are deposited on the floor of the ocean, although some are deposited in lakes and swamps. Once clastic, chemical, or biochemical sediments reach the ocean's floor, they are trapped there. An ocean current may pick them up again and transport them to a different depositional site on the seafloor. In most parts of the deep ocean, however, bottom currents are not strong enough to erode the sediment after it has settled. Therefore, compared to sediments laid down on land, a larger fraction of sediment deposited on the ocean floor is buried and preserved for a long time.

Diagenesis: Heat, Pressure, and Chemistry Transform Sediment into Rock

After sediments are deposited and buried, they are subject to **diagenesis**—the many physical and chemical changes that continue until the sediment or sedimentary rock is either exposed to weathering or metamorphosed by heat and pressure (**Figure 8.11**). Burial promotes these changes because buried sediments are subjected to increasingly high temperatures and pressures in Earth's interior.

Temperature increases with depth in the Earth's crust at an average rate of 30ºC for each kilometer of depth. At a depth of 4 km, buried sediments may reach 120°C or more, the temperature at which certain types of organic matter buried in sediments may be converted to oil and gas. Pressure also increases with depth—on average, about 1 atmosphere for each 4.4 m of depth. This increased pressure is responsible for the compaction of buried sediments. Both cementation and compaction result in **lithification,** the hardening of soft sediment into rock.

Cementation **Cementation** is a major chemical diagenetic change in which minerals are precipitated in the pores of

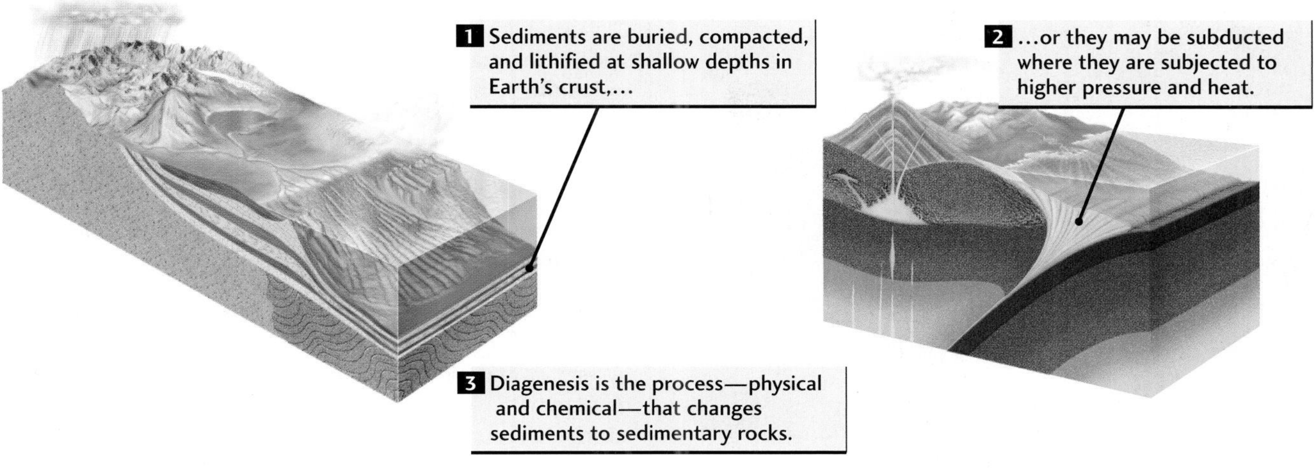

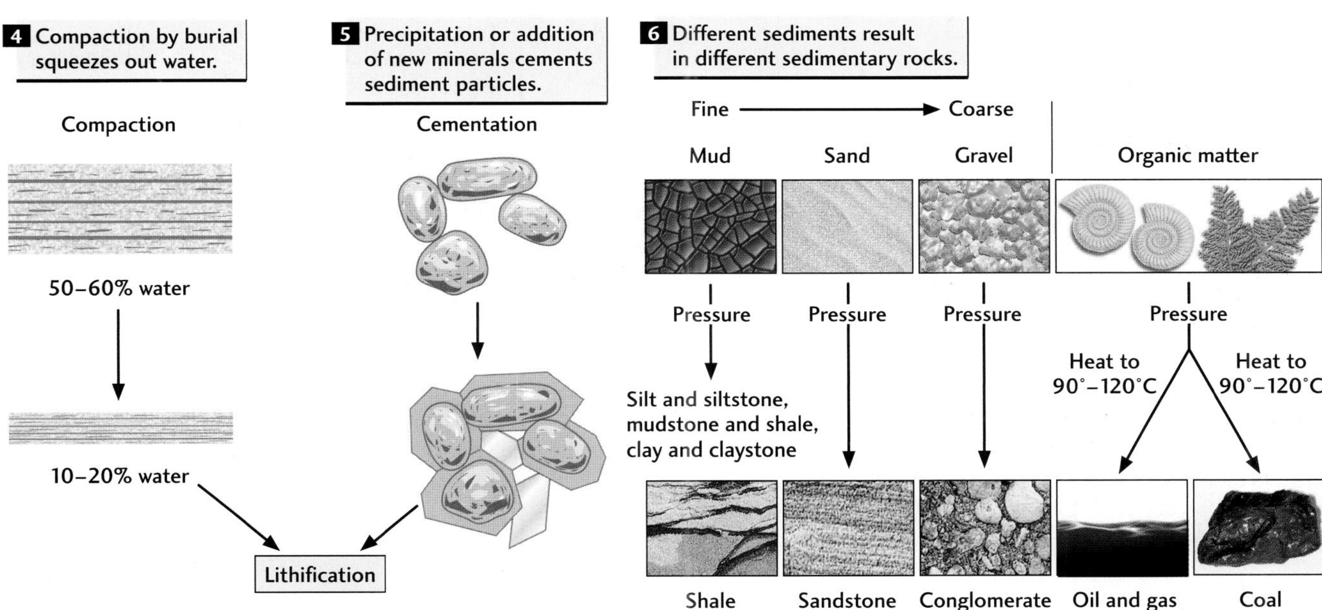

Figure 8.11 Diagenetic processes produce changes in composition and texture. Most of the changes tend to transform a loose, soft sediment into a hard, lithified sedimentary rock. [*Shale:* D. Cavagnaro/Visuals Unlimited. *Sandstone and conglomerate:* Breck P. Kent. *Oil and gas and coal:* John Woolsey.]

sediments, forming cements that bind clastic sediments and rocks. Cementation decreases **porosity,** the percentage of a rock's volume consisting of open pores between grains. Cementation also results in lithification, the hardening of soft sediment into rock. In some sands, for example, calcium carbonate is precipitated as calcite, which acts as a cement that binds the grains and hardens the resulting mass into sandstone (**Figure 8.12**). Other minerals such as quartz may cement sands, muds, and gravels into sandstone, mudstone, and conglomerate.

Compaction The major physical diagenetic change is **compaction,** a decrease in the volume and porosity of a sediment. Compaction occurs when the grains are squeezed closer together by the weight of overlying sediment. Sands are fairly well packed during deposition, so they do not compact much. However, newly deposited muds, including carbonate muds, are highly porous. Often, more than 60 percent of the sediment is water in pore spaces. As a result, muds compact greatly after burial, losing more than half their water.

Figure 8.12 This photomicrograph of sandstone shows quartz grains (white and gray) cemented by calcite (brightly colored and variegated) introduced after deposition. [Peter Kresan.]

Classification of Clastic Sediments and Sedimentary Rocks

We can now use our knowledge of sedimentation to classify sediments and their lithified counterparts, sedimentary rocks. The major divisions are, again, the clastic and the chemical and biochemical. Clastic sediments and sedimentary rocks constitute more than three-quarters of the total mass of all types of sediments and sedimentary rocks in the Earth's crust. We therefore begin with them.

Classification by Particle Size

Clastic sediments and rocks are classified primarily by the sizes of the grains, yielding the following three broad categories (Table 8.3):

- *Coarse:* gravel and conglomerate
- *Medium:* sand and sandstone
- *Fine:* silt and siltstone; mud, mudstone, and shale; clay and claystone

Classifying the various clastic sediments and rocks on the basis of their particle size in effect distinguishes them by one of the important conditions of sedimentation—current strength. As we have seen, the larger the particle, the stronger the current needed to move and deposit it. This relationship between current strength and particle size is the reason like-sized particles tend to accumulate in sorted beds. That is, most sand beds do not contain pebbles or mud, and most muds consist only of particles finer than sand.

Of the various types of clastic sediments and sedimentary rocks, the fine-grained clastics are by far the most abundant—about three times more common than the coarser clastics (**Figure 8.13**). The abundance of the fine-grained clastics, which contain larger amounts of clay minerals, is due to the chemical weathering into clay minerals of the large quantities of feldspar and other silicate minerals in Earth's crust. We turn now to a consideration of each of the three groups of clastic sediments and sedimentary rocks in more detail.

Coarse-Grained Clastics: Gravel and Conglomerate

Gravel is the coarsest clastic sediment, consisting of particles larger than 2 mm in diameter and including pebbles,

Table 8.3 Major Classes of Clastic Sediments and Sedimentary Rocks

Particle Size	Sediment	Rock
COARSE	GRAVEL	
Larger than 256 mm	Boulder	Conglomerate
256–64 mm	Cobble	
64–2 mm	Pebble	
MEDIUM		
2–0.062 mm	SAND	Sandstone
FINE	MUD	
0.062–0.0039 mm	Silt	Siltstone
Finer than 0.0039 mm	Clay	Mudstone (blocky fracture) Shale (breaks along bedding) Claystone

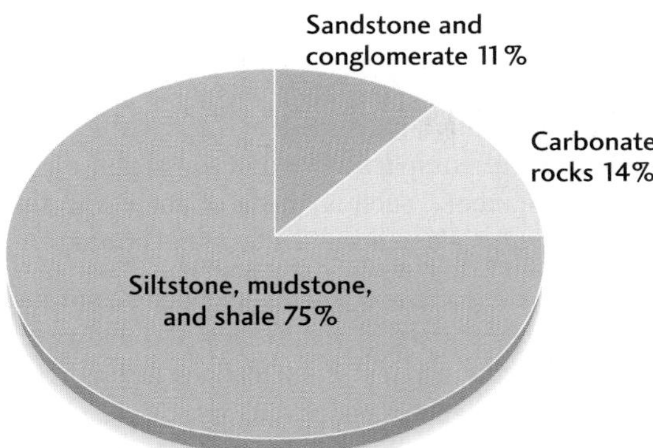

Figure 8.13 The relative abundance of the major sedimentary rock types. In comparison with these three types, all other sedimentary rock types—including evaporites, cherts, and other chemical sedimentary rocks—exist in only minor amounts.

cobbles, and boulders (see Table 8.3). **Conglomerates** are the lithified equivalents of gravel (**Figure 8.14**). Pebbles, cobbles, and boulders are easy to study and identify because of their large size, This is important because their size can tell us the speed of the currents that transported them. In addition, their composition can tells us about the nature of the distant terrain where they were produced.

There are relatively few environments—mountain streams, rocky beaches with high waves, and glacier meltwaters—in which currents are strong enough to transport pebbles. Strong currents also carry sand, and we almost always find sand between the pebbles. Some of it was deposited with the gravel, and some infiltrated into the spaces between fragments after the gravel was deposited. Pebbles and cobbles become rounded very quickly by abrasion in the course of transport on land or in water.

Medium-Grained Clastics: Sand and Sandstone

Sand consists of medium-sized particles, ranging from 0.062 to 2 mm in diameter (see Table 8.3). These sediments are moved even by moderate currents, such as those of rivers, waves at shorelines, and the winds that blow sand into dunes. Sand particles are large enough to be seen with the naked eye, and many of their features are easily discerned with a low-power magnifying glass. The lithified equivalent of sand is **sandstone** (see Figure 8.14).

Sizes and Shapes of Sand Grains The medium-sized clastics, sand particles, are subdivided into fine, medium, and coarse. The average size of the grains in any one sandstone can be an important clue to both the strength of the current that carried them and the sizes of the crystals eroded from the parent rock. The range and relative abundance of the various sizes are also significant. If all the grains are close to the average size, the sand is well sorted. If many grains are much larger or smaller than the average, the sand is poorly sorted. The degree of sorting can help distinguish, for example, between sands of beaches (well sorted) and muddy sands deposited by glaciers (poorly sorted). The shapes of sand grains also can be important clues to their origin. Sand grains, like pebbles, are rounded during transport. Angular grains imply short distances of transportation; rounded ones indicate long journeys down a large river system.

Mineralogy of Sands and Sandstones Within each textural category, clastics can be further subdivided by mineralogy, which can help identify the parent rocks. Thus, there are quartz-rich and feldspar-rich sandstones. Some sands are bioclastic; they formed when material such as carbonate originally precipitated as a shell but then was broken and transported by currents Thus, the mineral composition of sands and sandstones indicates the source areas that were eroded to produce the sand grains. Sodium- and potassium-rich feldspars with abundant quartz, for example, might

Conglomerate

Sandstone

Shale

Figure 8.14 Clastic sedimentary rocks. [*Conglomerate and sandstone:* Breck P. Kent. *Shale:* D. Cavagnaro/Visuals Unlimited.]

indicate that the sediments were eroded from a granitic terrain. Other minerals, as we will see in Chapter 9, would be indicative of metamorphic parent rocks.

The mineralogical composition of parent rocks can also be correlated with plate tectonic settings. Sandstones containing abundant fragments of mafic volcanic rocks, for example, are derived from the volcanic arcs of subduction zones.

Major Kinds of Sandstone Sandstones fall into several major groups on the basis of their mineralogy and texture (**Figure 8.15**):

- **Quartz arenites** are made up almost entirely of quartz grains, usually well sorted and rounded (see Figure 8.15). These pure quartz sands result from extensive weathering that occurred before and during transport and removed everything but quartz, the most stable mineral.

- **Arkoses** are more than 25 percent feldspar; the grains tend to be poorly rounded and less well sorted than those of pure quartz sandstones. These feldspar-rich sandstones come from rapidly eroding granitic and metamorphic terrains where chemical weathering is subordinate to physical weathering.

- **Lithic sandstones** contain many fragments derived from fine-grained rocks, mostly shales, volcanic rocks, and fine-grained metamorphic rocks.

- **Graywacke** is a heterogeneous mixture of rock fragments and angular grains of quartz and feldspar, the sand grains being surrounded by a fine-grained clay matrix. Much of this matrix is formed by the chemical alteration and mechanical compaction and deformation of relatively soft rock fragments, such as those of shale and some volcanic rocks, after deep burial of the sandstone formation.

Both groundwater geologists and petroleum geologists have a special interest in sandstones. Groundwater geologists examine the origins of sandstones to predict possible supplies of water in areas of porous sandstone, such as those found in the western plains of North America. Petroleum geologists must know about the porosity and cementation of sandstones because much of the oil and gas discovered in the past 150 years has been found in buried sandstones. In addition, much of the uranium used for nuclear power plants and weapons has come from diagenetic uranium in sandstones.

Fine-Grained Clastics: Silt and Siltstone; Mud, Mudstone, and Shale; Clay and Claystone

The finest-grained clastic sediments and sedimentary rocks are the silts and siltstones; the muds, mudstones, and shales; and the clays and claystones. These sediments consist of particles that are all less than 0.062 mm in diameter, but the sediments vary widely in their range of grain sizes and min-

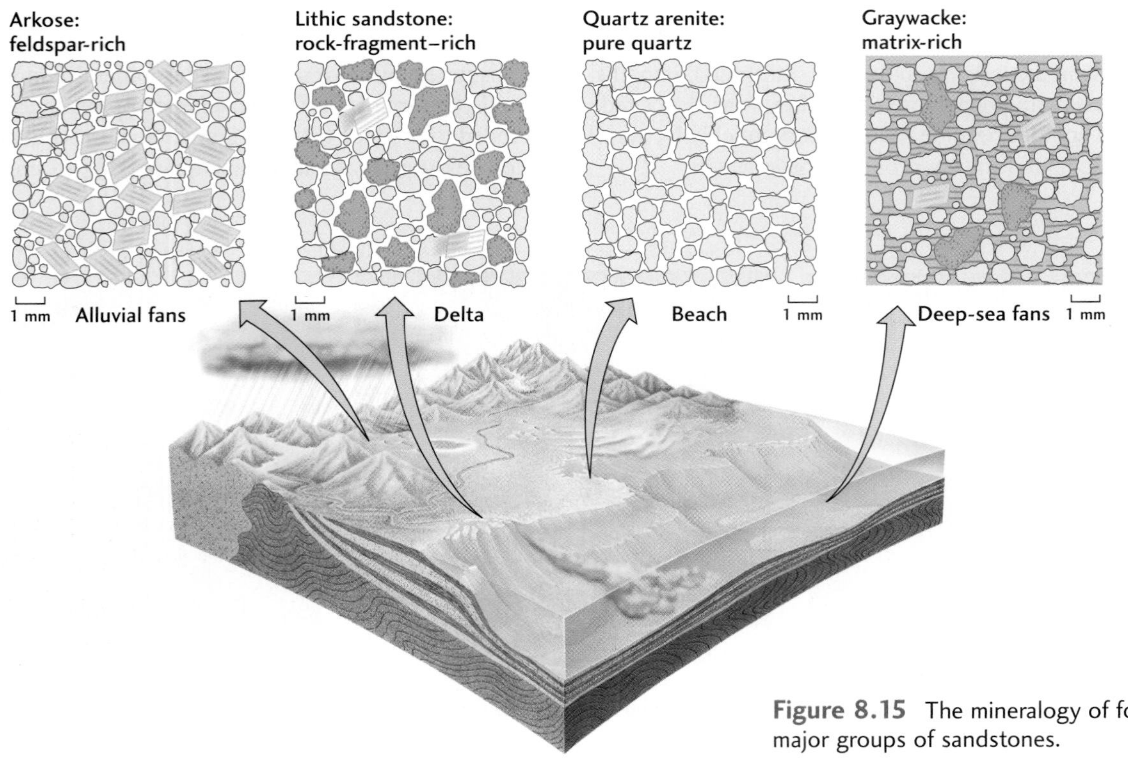

Figure 8.15 The mineralogy of four major groups of sandstones.

eral compositions. Fine-grained sediments are deposited by the gentlest currents, which allow the finest particles to settle slowly to the bottom in quiet waves.

Silt and Siltstone **Siltstone** is the lithified equivalent of **silt,** a clastic sediment in which most of the grains are between 0.0039 and 0.062 mm in diameter. Siltstone looks similar to mudstone or very fine grained sandstone.

Mud, Mudstone, and Shale **Mud** is a clastic sediment, mixed with water, in which most of the particles are less than 0.062 mm in diameter (see Table 8.3). Thus, mud can be made of silt- or clay-sized sediments, or varying quantities of both. This general term is very useful in fieldwork because it is often difficult to distinguish between silt- and clay-sized sediment without detailed study using a microscope. Muds are deposited by rivers and tides. After a river has flooded its lowlands and the flood recedes, the current slows and mud settles, some of it containing abundant organic matter. This mud contributes to the fertility of river bottomlands. Muds are left behind by ebbing tides along many tidal flats where wave action is mild. Much of the deep-ocean floor, where currents are weak or absent, is blanketed by mud.

The fine-grained rock equivalents of muds are mudstones and shales. **Mudstones** are blocky and show poor or no bedding. Bedding may have been well marked when the sediments were first deposited but was then lost by bioturbation (see Figure 8.9). **Shales** (see Figure 8.14) are composed of silt plus a significant component of clay, which causes them to break readily along bedding planes. Many muds, mudstones, and shales are more than 10 percent carbonate, forming deposits of calcareous shales. Black, or organic, shales contain abundant organic matter. Some, called oil shales, contain large quantities of oily organic material, which makes them a potentially important source of oil. (We consider the oil shales in more detail in Chapter 22.)

Clay and Claystone **Clay** is the most abundant component of fine-grained sediments and sedimentary rocks and consists largely of clay minerals. Clay-sized particles are less than 0.0039 mm in diameter (see Table 8.3). Rocks made up exclusively of clay-sized particles are called **claystones.**

Classification of Chemical and Biochemical Sediments and Sedimentary Rocks

We have seen that clastic sediments and sedimentary rocks give us information about continental parent rocks and weathering. Chemical and biochemical sediments tell us about chemical conditions in the ocean, the predominant environment of sedimentation (Table 8.4). **Carbonate**

Table 8.4 Classification of Biochemical and Chemical Sediments and Sedimentary Rocks

Sediment	Rock	Chemical Composition	Minerals
BIOCHEMICAL			
Sand and mud (primarily bioclastic)	Limestone	Calcium carbonate ($CaCO_3$)	Calcite (aragonite)
Siliceous sediment	Chert	Silica (SiO_2)	Opal, chalcedony, quartz
Peat, organic matter	Organics	Carbon compounds Carbon compounded with oxygen and hydrogen	(coal), (oil), (gas)
CHEMICAL			
No primary sediment (formed by diagenesis)	Dolostone	Calcium-magnesium carbonate ($CaMg[CO_3]_2$)	Dolomite
Iron oxide sediment	Iron formation	Iron silicate; oxide (Fe_2O_3); carbonate	Hematite, limonite, siderite
Evaporite sediment	Evaporite	Sodium chloride (NaCl); calcium sulfate ($CaSO_4$)	Gypsum, anhydrite, halite, other salts
No primary sediment (formed by diagenesis)	Phosphorite	Calcium phosphate ($Ca_3[PO_4]_2$)	Apatite

environments, by far the most abundant chemical and biochemical sedimentation environments, occur in marine settings where calcium carbonate is the main sediment. Chemical sedimentation takes place in some lakes, particularly those of arid regions where evaporation is intense, such as the Great Salt Lake of Utah. Such sediments account for only a very small fraction relative to the amounts deposited along the ocean's shorelines, on continental shelves, and in the deep ocean.

Classification by Chemical Composition

Chemical and biochemical sediments and sedimentary rocks are classified by their chemical composition (see Table 8.4). For marine sediments, this classification is based on the chemical elements dissolved in seawater.

We divide nonclastic sediments into chemical and biochemical sediments to emphasize the importance of organisms as the chief mediators of this kind of sedimentation. The shells of organisms, biochemically precipitated, account for much of the carbonate sediment of the world, and carbonate is by far the most abundant nonclastic sediment. The chemical sediments are precipitated by inorganic processes alone and are less abundant.

Carbonate Sediments and Sedimentary Rocks: Limestone and Dolostone

Carbonate sediments and **carbonate rocks** form from the accumulation of carbonate minerals precipitated organically or inorganically. The precipitation characteristically takes place as part of the growth process of carbonate-secreting organisms. However, precipitation may also occur during sedimentation or diagenesis. The minerals precipitated are either calcium carbonates or calcium-magnesium carbonates. Carbonate rocks are abundant because of the large amounts of calcium and carbonate present in seawater. Carbonate is derived from the carbon dioxide in the atmosphere. Calcium and carbonate come from the easily weathered limestone on the continents. Calcium is also supplied by weathering of feldspars and other minerals in igneous and metamorphic rocks.

Most carbonate sediments of shallow marine environments are bioclastic. They were originally secreted biochemically as shells by organisms living near the surface or on the bottom of the oceans. Their skeletons constitute individual pieces or clasts of carbonate. Some carbonate deposits are formed of bioclastic debris that was broken and transported by currents, but most carbonates simply represent the in situ accumulation of various calcium carbonate skeletons. They are found from the coral reefs of the Pacific and Caribbean to the shallow banks of the Bahama Islands. Although the oceanic abyssal plain is less accessible for study than these spectacular vacation spots, it is where most carbonate is deposited today.

Oceanic Sources of Carbonate Sediments Most of the carbonate sediments deposited on the ocean's abyssal plains are derived from the calcite shells and skeletons of **foraminifera,** tiny single-celled organisms that live in surface waters, and from other organisms that secrete calcium carbonate. When the organisms die, their shells and skeletons settle to the seafloor and accumulate there as sediment (**Figure Story 8.16**). In addition to calcite, most carbonate sediments contain aragonite, a less stable form of calcium carbonate. As noted earlier in this chapter, some organisms precipitate calcite, some precipitate aragonite, and some precipitate both.

Reefs are moundlike or ridgelike organic structures constructed of the carbonate skeletons of millions of organisms. In the warm seas of the present, most reefs are built by corals and other less well known calcifying organisms such as algae. In contrast with the soft, loose sediment produced in other environments, the calcium carbonate of the corals and other organisms forms a rigid, wave-resistant structure—a cemented buttress of solid limestone—that is built up to and slightly above sea level (see Figure Story 8.16; also see Feature 8.1). The solid limestone of the reef is produced directly by the action of organisms; there is no soft sediment stage. On and around these reefs live hundreds of species of other carbonate-precipitating organisms, many of them similar to the familiar clams, algae, and snails of our shorelines. Marine algae also precipitate carbonate. These single-celled organisms are similar to primitive plants and grow on reef tracts and in other carbonate environments.

Inorganic Precipitation of Carbonate Sediments A significant fraction of the carbonate mud in lagoons and on shallow banks such as those of the Bahama Islands is inorganically precipitated directly from seawater. Microorganisms may help facilitate this process, but their role is still uncertain. The chemical basis for inorganic carbonate precipitation is the relative abundance of calcium (Ca^{2+}) and bicarbonate (HCO_3^-) ions in seawater. The warm, tropical parts of the ocean are supersaturated with calcium carbonate and precipitate carbonate by the following chemical reaction:

$$\underset{\substack{\text{calcium ions}\\\text{(dissolved)}}}{Ca^{2+}} + \underset{\substack{\text{bicarbonate ions}\\\text{(dissolved)}}}{2HCO_3^-} \rightarrow \underset{\substack{\text{calcium carbonate}\\\text{(precipitated)}}}{CaCO_3} + \underset{\substack{\text{carbonic acid}\\\text{(dissolved)}}}{H_2CO_3}$$

When multicellular organisms secrete their carbonate shells, they are effectively producing the same chemical reaction, but in this case by biochemical means. The chemical reaction that precipitates calcium carbonate is the reverse of the reaction that takes place in the chemical weathering of limestones (see Chapter 7).

ORGANISMS CREATE CARBONATE PLATFORM SYSTEMS

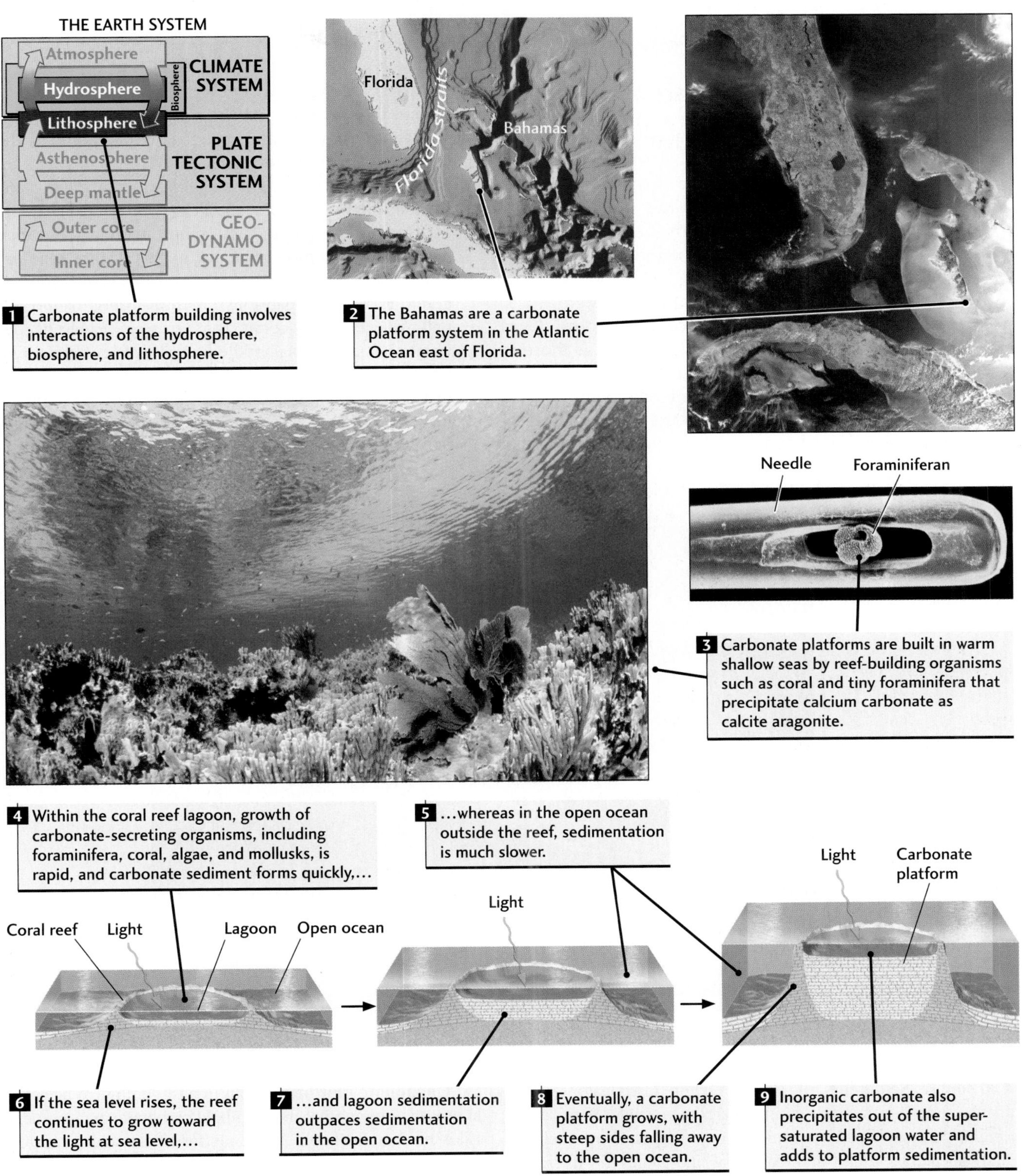

Figure Story 8.16 The growth of a carbonate platform. (*top right*) Carbonate platforms in the Bahamas [NASA]. (*left center*) Coral reef, Key Largo, Florida [Stephen Frink/Index Stock Imagery]. (*right center*) Scanning electron micrograph of a foraminiferan in the eye of a needle [Chevron Corporation].

8.1 Darwin's Coral Reefs and Atolls

For more than 200 years, coral reefs have attracted explorers and travel writers. Ever since Charles Darwin sailed the oceans on the *Beagle* from 1831 to 1836, these reefs have been a matter of scientific discussion, too. Darwin was one of the first to analyze the geology of coral reefs, and his theory of their origin is still accepted today.

The coral reefs that Darwin studied were atolls, islands in the open ocean with circular lagoons. Coral reefs also form at continental margins, such as those of the Florida Keys. The outermost part of a reef is a slightly submerged, wave-resistant reef front, a steep slope facing the ocean. The reef front is composed of the interlaced skeletons of actively growing coral and calcareous algae, forming a tough, hard limestone. Behind the reef front is a flat platform extending into a shallow lagoon. An island may lie at the center of the lagoon. Parts of the reef, as well as a central island, are above water and may become forested. A great many plant and animal species inhabit the reef and the lagoon.

Coral reefs are generally limited to waters less than about 20 m deep because, below that depth, seawater does not transmit enough light to enable reef-building corals to grow. (Exceptions are some kinds of individual—noncolonial—corals that grow in much deeper waters.) Darwin explained how coral reefs could be built up from the bottom of the dark, deep ocean. The process starts with a volcano building up to the surface from the seafloor. As the volcano temporarily or permanently becomes dormant, coral and algae colonize the

Bora Bora atoll, South Pacific Ocean. Reefal organisms have built a barrier around the volcanic island, forming a protected lagoon. [Jean-Marc Truchet/Stone/Getty Images.]

Carbonate Sediments of Mixed Origin Carbonate sediments contain some fine-grained carbonate muds of mixed origin, consisting partly of microscopic fragments of shells and calcareous algae and partly of inorganic (or microbial) precipitates. These muds form in the lagoons behind reefs and on extensive shallow banks such as those of the Bahamas. Such deposits are found on passive continental margins and fringing volcanic island arcs in the warmer parts of the oceans, where they are not flooded with clastic sediment.

Carbonate platforms, both in past geological ages and at present, are another major mixed carbonate environment. Like the Bahama banks, these platforms are shallow, extensive flat areas where both biological and nonbiological carbonates are deposited. Below the level of the platform are carbonate ramps, gentle slopes to deeper waters that also accumulate carbonate sediment, much of it fine-grained. In other times and places, platforms may be rimmed carbonate shelves in which there is a clear demarcation of the shelf

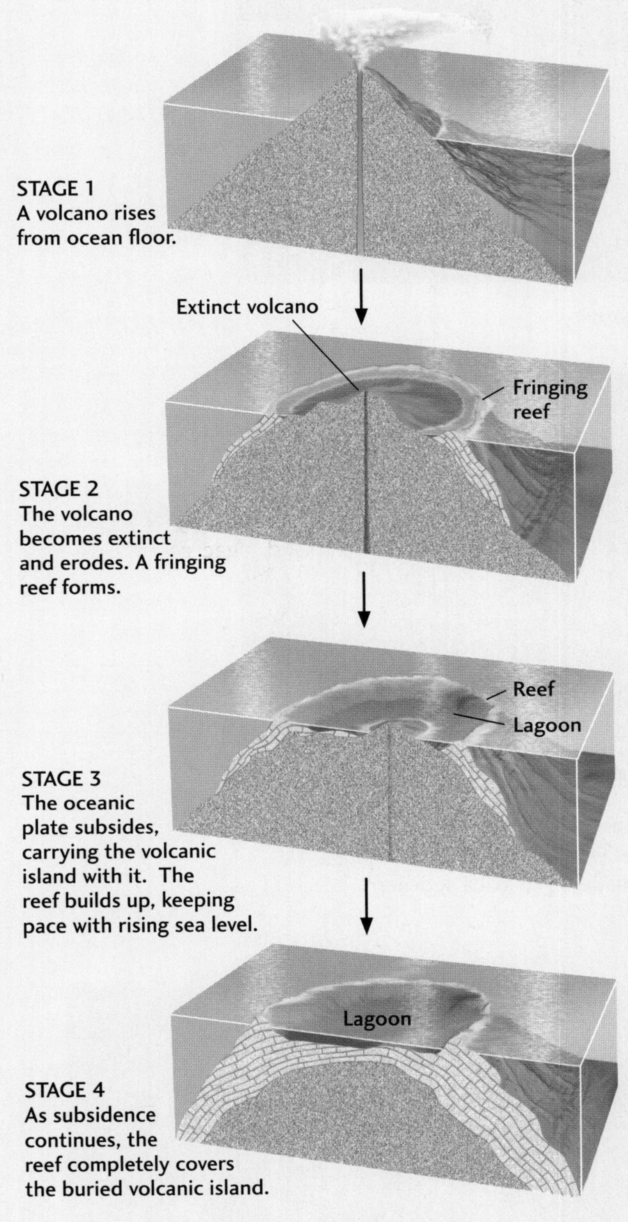

shore and build fringing reefs—coral reefs similar to atolls that grow around the edges of a central volcanic island. Erosion may then lower the volcanic island almost to sea level.

Darwin reasoned that if such a volcanic island were to subside slowly beneath the waves, actively growing coral and algae might keep pace with the subsidence, continuously building up the reef so that the island remained. In this way, the volcanic island would disappear and we would be left with an atoll. More than 100 years after Darwin proposed his theory, deep drilling on several atolls penetrated volcanic rock below the coralline limestone and confirmed the theory. And, some decades later, the theory of plate tectonics explained both volcanism and the subsidence that resulted from plate cooling and contraction.

Evolution of a coral reef from a subsiding volcanic island, first proposed by Charles Darwin in the nineteenth century.

margin by reefs of various organisms and by buildups of shoals of bioclastic and other materials. Below the rims are steep slopes covered with detritus derived from the rim materials. The role of now-extinct organisms in building ancient reefs will be discussed shortly.

Limestone The dominant biochemical sedimentary rock lithified from carbonate sediments is **limestone,** which is composed mainly of calcium carbonate ($CaCO_3$) in the form of the mineral calcite (**Figure 8.17**a; see Table 8.4). Limestone is formed from carbonate sand and mud and, in some cases, ancient reefs.

Dolostone: Formed by Diagenesis Another abundant carbonate rock is **dolostone,** made up of the mineral dolomite, which is composed of calcium-magnesium carbonate, $CaMg(CO_3)_2$ (see Table 8.4). Dolostones are diagenetically altered carbonate sediments and limestones. The mineral

Figure 8.17 Chemical and biochemical sedimentary rocks. [Breck P. Kent.] (a) Limestone, lithified from carbonate sediments; (b) gypsum and (c) halite, marine evaporites that crystallize out of shallow seawater basins; and (d) chert, made up of silica sediment.

dolomite does not form as a primary precipitate from ordinary seawater, and no organisms secrete shells of dolomite. Instead, the original calcite or aragonite of a carbonate sediment is converted into dolomite after deposition. Some calcium ions in the calcite or aragonite are exchanged for magnesium ions from seawater (or magnesium-rich groundwater) slowly passing through the pores of the sediment. This converts the calcium carbonate mineral, $CaCO_3$, into dolomite, $CaMg(CO_3)_2$.

Reefs and Evolutionary Processes Today reefs are constructed mainly by corals; but at earlier times in Earth's history, other organisms—such as a now-extinct variety of mollusk (**Figure 8.18**)—built wave-resistant structures, some of them cemented buttresses of solid limestone. The successive pulses of organism diversification and extinction displayed by reef-building organisms over geologic time illustrate how ecology and environmental change help regulate the process of evolution. Today, natural and human-generated effects threaten the growth of coral reefs, which are very sensitive to environmental change. In 1998, an El Niño event raised sea surface temperatures to the point where many reefs in the

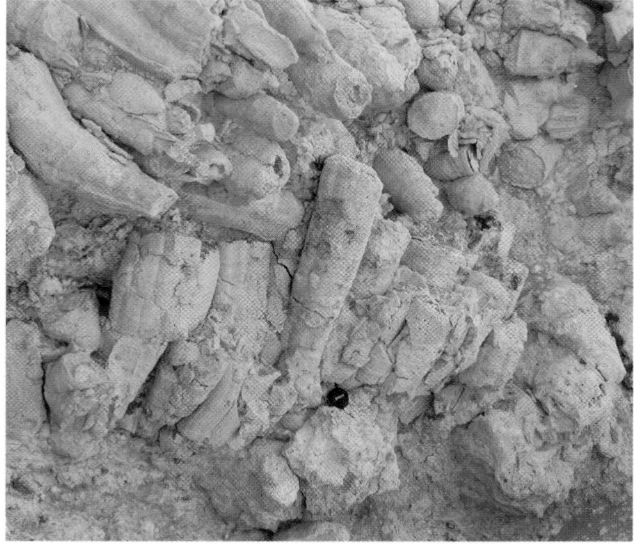

Figure 8.18 Reefal limestone made of extinct clams (rudists) in the Cretaceous Shuiba formation, Sultanate of Oman. [John Grotzinger.]

western Indian Ocean were killed. The Florida Keys reefs are dying off for a completely different reason: they're getting too much of a good thing. It turns out that groundwaters originating in the farmlands of the Florida peninsula are seeping out near the reefs and exposing them to lethal concentrations of nutrients.

The marine environments where carbonate sedimentation produces rigid limestone structures—including reefs, carbonate banks, and deep-water deposits in the open ocean—are considered further in Chapter 17.

Evaporite Sediments: Sources of Halite, Gypsum, and Other Salts

Evaporite sediments and **evaporite rocks** are precipitated inorganically from evaporating seawater and from water in arid-region lakes that have no river outlets.

Marine Evaporites Marine evaporites are the chemical sediments and sedimentary rocks formed by the evaporation of seawater. This evaporite environment is created when the warm seawater of an arid inlet or arm of the sea evaporates more rapidly than it can mix with the connected open marine seawater. The degree of evaporation controls the salinity of the evaporating seawater and thus the kinds of sediments formed. The sediments and rocks produced in these environments contain minerals formed by the crystallization of sodium chloride (halite), calcium sulfate (gypsum and anhydrite), and other combinations of the ions commonly found in seawater. As evaporation proceeds, seawater becomes more concentrated and minerals crystallize in a set sequence. As dissolved ions precipitate to form each mineral, the evaporating seawater changes composition.

Seawater has the same composition in all the oceans, which explains why marine evaporites are so similar the world over. No matter where seawater evaporates, the same sequence of minerals always forms. The history of evaporite minerals shows that the composition of the world's oceans has stayed more or less constant over the past 1.8 billion years. Before that time, however, the precipitation sequence may have been different, indicating that seawater composition changed.

The great volume of many marine evaporites, some hundreds of meters thick, shows that they could not have formed from the small amount of water that could be held in a shallow bay or pond. A huge amount of seawater must have evaporated. The way in which such large quantities of seawater evaporate is very clear in bays or arms of the sea that meet the following conditions (**Figure 8.19**):

- The freshwater supply from rivers is small.
- Connections to the open sea are constricted.
- The climate is arid.

In such locations, water evaporates steadily, but the openings allow seawater to flow in to replenish the evaporating

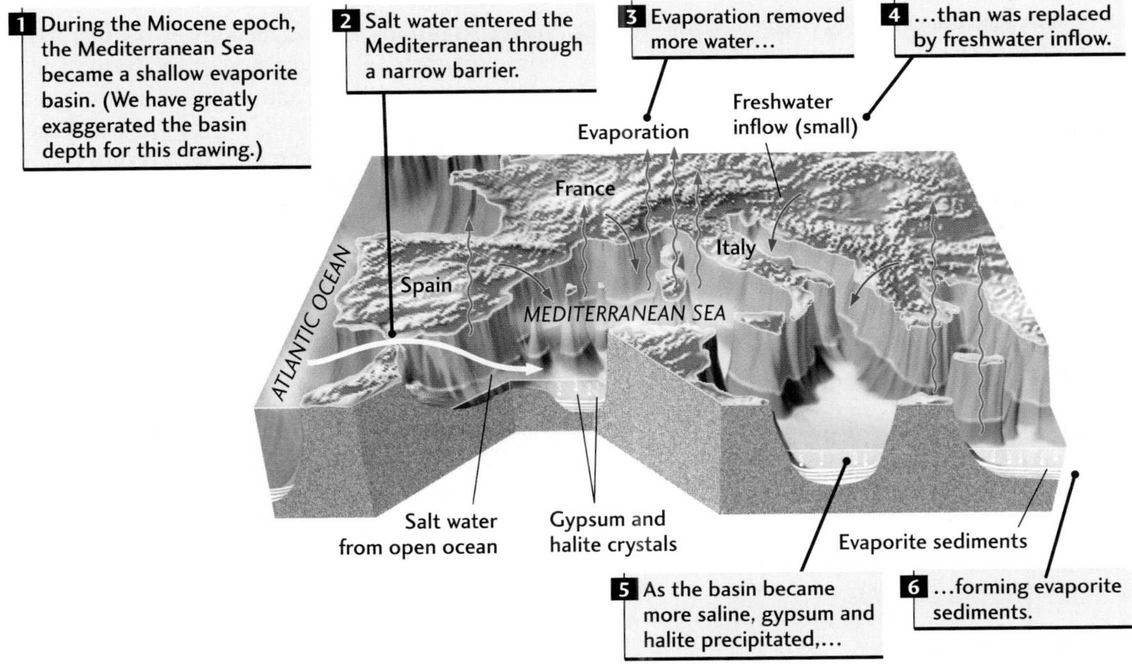

Figure 8.19 A marine evaporite environment. When seawater evaporated in a shallow basin, such as in the Mediterranean Sea, with a restricted connection to the open ocean, gypsum formed as an evaporite sediment. A further increase in salinity led to the crystallization of halite.

waters of the bay. As a result, those waters stay at constant volume but become more saline than the open ocean. The evaporating bay waters remain more or less constantly supersaturated and steadily deposit evaporite minerals on the floor of the evaporite basin.

As seawater evaporates, the first precipitates to form are the carbonates. Continued evaporation leads to the precipitation of gypsum, calcium sulfate ($CaSO_4 \times 2H_2O$) (see Figure 8.17b). By the time gypsum precipitates, almost no carbonate ions are left in the water. Gypsum is the principal component of plaster of Paris and is used in the manufacture of wallboard, which lines the walls of most new houses built today.

After still further evaporation, the mineral halite (NaCl)—one of the most common chemical sediments precipitated from evaporating seawater—starts to form (see Figure 8.17c). Halite, you may remember from Chapter 3, is table salt. Deep under the city of Detroit, Michigan, beds of salt laid down by an evaporating arm of an ancient ocean are commercially mined.

In the final stages of evaporation, after the sodium chloride is gone, magnesium and potassium chlorides and sulfates precipitate. The salt mines near Carlsbad, New Mexico, contain commercial quantities of potassium chloride. Potassium chloride is often used as a substitute for table salt (sodium chloride) by people with certain dietary restrictions.

This sequence of precipitation has been studied in the laboratory and is matched by the bedding sequences found in certain natural salt formations. Most of the world's evaporites consist of thick sequences of dolomite, gypsum, and halite and do not contain the final-stage precipitates. Many do not even go as far as halite. The absence of the final stages indicates that the water did not evaporate completely but was replenished by normal seawater as evaporation continued.

Nonmarine Evaporites Evaporite sediments also form in arid-region lakes that typically have few or no river outlets. In such lakes, evaporation controls the lake level, and incoming salts derived from chemical weathering accumulate. The Great Salt Lake of Utah is one of the best known of these lakes. River waters enter the lake, bringing salts dissolved in the course of weathering. In the dry climate of Utah, evaporation has more than balanced the inflow of fresh water from rivers and rain. As a result, concentrated dissolved ions in the lake make it one of the saltiest bodies of water in the world—eight times saltier than seawater.

In arid regions, small lakes may collect unusual salts, such as borates (compounds of the element boron), and some become alkaline. The water in this kind of lake is poisonous. Economically valuable resources of borates and nitrates (minerals containing the element nitrogen) are found in the sediments beneath some of these lakes.

Silica Sediment: Source of Chert

One of the first sedimentary rocks to be used for practical purposes by our prehistoric ancestors was **chert,** which is made up of chemically or biochemically precipitated silica (SiO_2) (see Figure 8.17d). Early hunters used it for arrowheads and other tools because it could be chipped and shaped to form hard, sharp implements. A common name for chert is *flint,* and the terms are virtually interchangeable. The silica in most cherts is in the form of extremely fine crystalline quartz. Some geologically young cherts consist of opal, a less well crystallized form of silica.

Like calcium carbonate, much silica sediment is precipitated biochemically, secreted by ocean-dwelling organisms. These organisms grow in surface waters where nutrients are abundant. When they die, they sink to the deep-ocean floor, where their shells accumulate as layers of silica sediment. After these silica sediments are buried by later sediments, they are diagenetically cemented into chert. Chert may also form as diagenetic nodules and irregular masses replacing carbonate in limestones and dolomites.

Phosphorite Sediment

Among the many other kinds of chemical and biochemical sediments deposited in the sea are phosphorites. Sometimes called phosphate rock, **phosphorite** is composed of calcium phosphate precipitated from phosphate-rich seawater in places where currents of deep, cold water containing phosphate and other nutrients rise along continental margins. The phosphorite forms diagenetically by the interaction between muddy or carbonate sediments and the phosphate-rich water.

Iron Oxide Sediment: Source of Iron Formations

Iron formations are sedimentary rocks that usually contain more than 15 percent iron in the form of iron oxides and some iron silicates and iron carbonates. Most of these rocks formed early in Earth's history, when there was less oxygen in the atmosphere and, as a result, iron dissolved more easily. In soluble form, iron was transported to the sea and precipitated where oxygen was being produced by microorganisms.

Organic Particles: Source of Coal, Oil, and Gas

Coal is a biochemically produced sedimentary rock composed almost entirely of organic carbon formed by the diagenesis of swamp vegetation. Coal is classified as an **organic sedimentary rock,** a group of sedimentary rocks that consist entirely or partly of organic carbon–rich deposits formed by the decay of once-living material that has been buried.

Oil and **gas** are fluids that are not normally classed with sedimentary rocks. They can be considered organic sediments, however, because they form by the diagenesis of organic material in the pores of sedimentary rocks. Deep burial changes organic matter originally deposited along with inorganic sediments into a fluid that then escapes to other, porous formations and becomes trapped there. As noted earlier in this chapter, oil and gas are found mainly in sandstones and limestones (see Chapter 22).

Plate Tectonics and Sedimentary Basins

Sedimentary environments, the composition and texture of sediments, and the geometry of the basins in which sediments accumulate are all related to plate tectonic settings. For example, the environment of a deep-ocean trench is found at a subduction zone, whereas thick alluvial (river) deposits are typically associated with mountains formed by the collision of continents. Alluvial deposits are also found along the borders of continental rift valleys.

Sandstones that build deltas adjacent to a stable continent will tend to be quartz-rich, well rounded, and well sorted. Sandstones deposited in deep-ocean trenches in arc-continent collision zones, in contrast, will consist of abundant igneous or metamorphic rock fragments and will be less well rounded and sorted.

Sediments accumulate in depressions formed by **subsidence** of the Earth's crust, where they are buried and converted into thick piles of sedimentary rock. During subsidence, a broad area of the crust sinks (subsides) relative to the elevation of the crust of surrounding areas. Subsidence is induced partly by the additional weight of sediments on the crust but is driven mostly by tectonic mechanisms, such as regional downfaulting.

Tectonic Mechanisms of Basin Subsidence

Sedimentary basins are regions of considerable extent (at least 10,000 km^2) where the combination of deposition and subsidence has formed thick accumulations of sediment and sedimentary rock. Studies have been stimulated primarily by exploration for oil and gas, which are abundant in sedimentary basins. Our growing knowledge has enabled us to infer the deep structure of basins and thus to learn more about the continental lithosphere.

Rift Basins and Thermal Sag Basins When plate separation begins within a continent, the mechanism of basin subsidence involves stretching, thinning, and heating of the underlying lithosphere by the forces of plate separation (**Figure 8.20**). A long, narrow rift develops, with great down-dropped crustal blocks. Hot ductile mantle rises and fills the space created by the thinned lithosphere and crust, initiating the volcanic eruption of basaltic rocks in the rift zone. **Rift basins** are deep, narrow, and elongate, with thick successions of sedimentary rocks and also extrusive and intrusive igneous rocks. The rift valleys of East Africa, the rifted valley of the Rio Grande, and the Jordan Valley in the Middle East are current examples of rift basins (**Figure 8.21**).

At later stages, when rifting has led to seafloor spreading and the newly formed continental plates are drifting away from each other, the basin subsidence mechanism principally involves cooling of the lithosphere that was thinned and heated during the earlier rifting stage (Figure 8.20). In this case, cooling leads to an increase in the density of lithosphere, which in turn leads to its subsidence and the development of offshore **thermal sag basins.** Sediments are supplied from erosion of the adjacent land to form **continental shelf deposits.**

The sedimentary basins off the Atlantic coasts of North and South America, Europe, and Africa are products of this process. These basins began to form when the supercontinent Pangaea split about 200 million years ago and the American plates separated from the European and African plates. Figure 8.20 shows the wedge-shaped deposit of sediments underlying the Atlantic continental shelf and margin of the United States, which were formed in this way. The offshore basins continue to receive sediments for a long time because the trailing edge of the continent subsides slowly and because the continents provide a tremendous area from which sediments can be derived. The load of the growing mass of sediment further depresses the crust, so the basins can receive still more material from the land. As a result of these two effects, the deposits can accumulate in an orderly fashion to thicknesses of 10 km or more.

The Illinois Basin is an example of a thermal sag basin that began with a failed continental rift. About a billion years ago, the proto–North American continent was split by a rift valley running north-northeast from an area south of the present confluence of the Ohio and Mississippi rivers to the northeastern part of the lower peninsula of Michigan. This valley subsided as it rifted apart and received river-borne sediments that accumulated to at least 3 km in thickness. Over millions of years, as the rift became inactive, the original tectonic valley was gradually buried beneath sediment. Subsidence continued as the formerly hot crust of the rift zone slowly cooled and began to sag. By about 500 million years ago, the crust subsided to the point where it was below sea level, and shallow oceans spread over this region of the continent. Sediments of the subsiding basin became marine sands and then shallow marine carbonate sediments, including reefs at times. Today these reefs form prolific oil and gas reservoirs and are the targets of continuing exploration.

This carbonate sedimentation was brought to a close about 330 million years ago as large rivers transported sand, gravel, and mud (eroded from the rising Appalachian Mountains) into the now spoon-shaped sag basin. This alluvial sedimentation accumulated another kilometer of sediment in the basin, this time with abundant organic sediment that gradually converted into the coal beds that are so prevalent in the region. The basin finally stopped subsiding about 250 million years ago, yet the underlying rift still moves along the buried major faults and causes earthquakes, such as the severe earthquake at New Madrid, Missouri, in 1812.

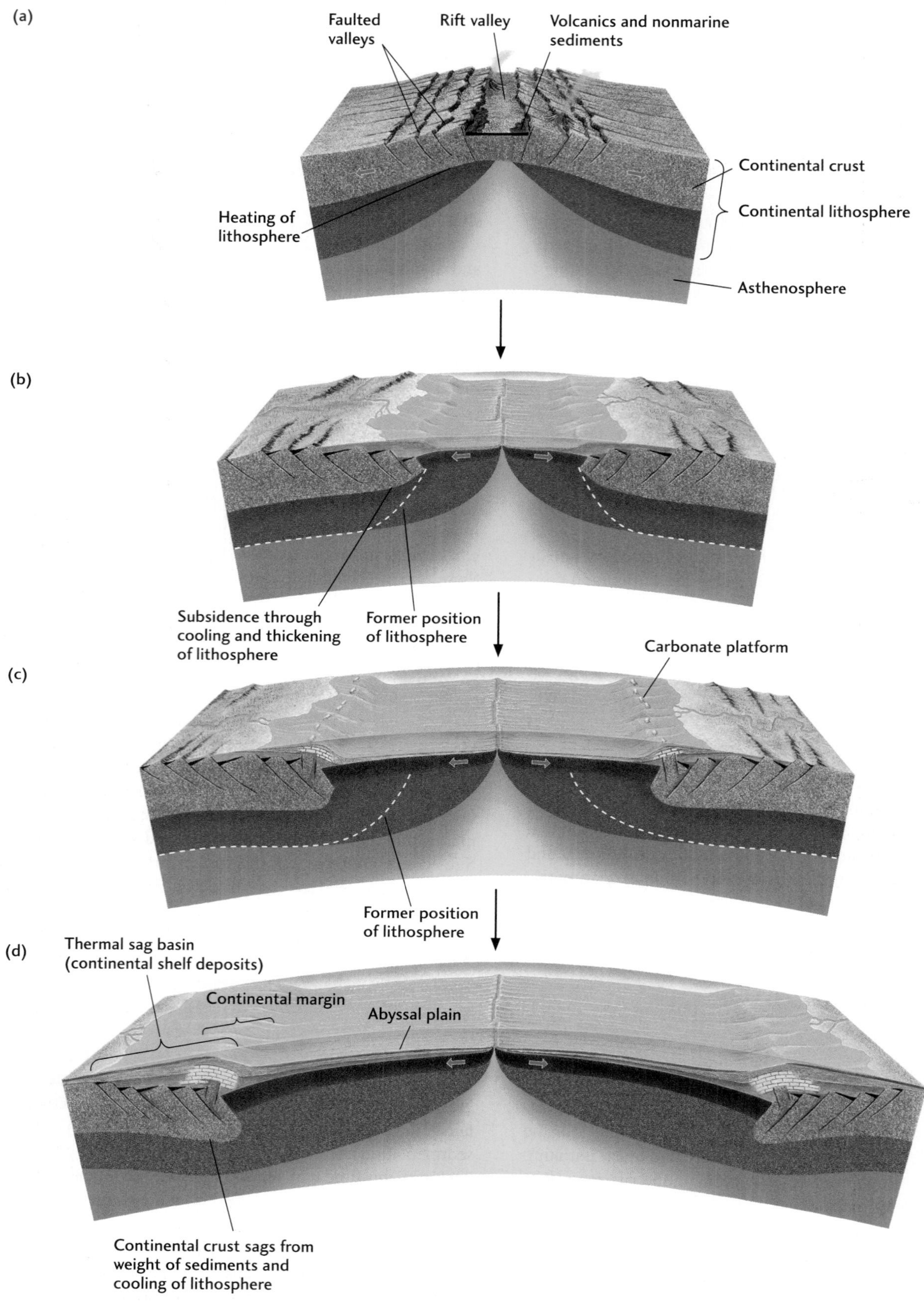
(a)
Faulted valleys
Rift valley
Volcanics and nonmarine sediments
Continental crust
Continental lithosphere
Heating of lithosphere
Asthenosphere
(b)
Subsidence through cooling and thickening of lithosphere
Former position of lithosphere
(c)
Carbonate platform
Former position of lithosphere
(d)
Thermal sag basin (continental shelf deposits)
Continental margin
Abyssal plain
Continental crust sags from weight of sediments and cooling of lithosphere

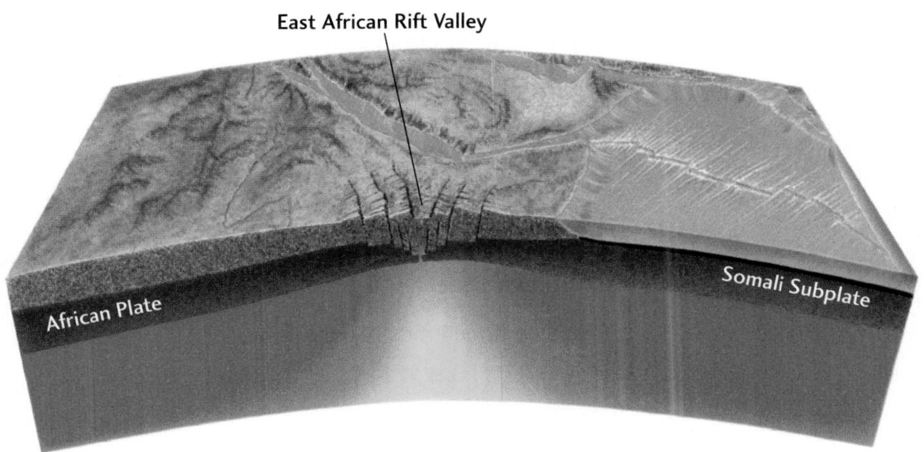

Figure 8.21 Rifting and plate separation within a continent.

Flexural Basins A third type of basin develops within zones of tectonic convergence, where one lithospheric plate pushes up over the other. When this occurs, the weight of the overriding plate causes the overridden plate to bend or flex down, producing a **flexural basin.** The Indo-Gangetic Basin, named after the Indus and Ganges rivers that flow through it, is subsiding in response to the flexural down-bending of the Indian continent as it collides with, is pushed under, and bends beneath Asia.

SUMMARY

What are the major processes that form sedimentary rock? Weathering and erosion produce the clastic particles that compose sediment and the dissolved ions that precipitate to form biochemical and chemical sediments. Currents of water and wind, and the flow of ice, transport the sediment to its ultimate resting place, the site of sedimentation. Sedimentation (also called deposition), a settling of particles from the transporting agent, produces bedded sediments in river channels and valleys, on sand dunes, and at the edges and floors of the oceans. Lithification and diagenesis harden the sediment into sedimentary rock.

What are the two major divisions of sediments and sedimentary rocks? Sediments and sedimentary rocks are classified as clastic or chemical and biochemical. Clastic sediments form from the fragments of parent rock produced by physical weathering and the clay minerals produced by chemical weathering. Water and wind currents and ice carry these solid products to the oceans and sometimes deposit them along the way. Chemical and biochemical sediments originate from the ions dissolved in water during chemical weathering. These ions are transported in solution to the oceans, where they are mixed into seawater. Through chemical and biochemical reactions, the ions are precipitated from solution, and the precipitated particles settle to the ocean floor.

How do we classify the major kinds of clastic sediments and chemical and biochemical sediments? Clastic sediments and sedimentary rocks are classified by the sizes of their particles: as gravels and conglomerates; sands and sandstones; silts and siltstones; muds, mudstones, and shales; clays and claystones. This method of classifying sediments emphasizes the importance of the strength of the current as it transports and deposits solid materials. The chemical and biochemical sediments and sedimentary rocks are classified on the basis of their chemical compositions. The most abundant of these rocks are the carbonate rocks—limestone and dolostone. Limestone is made up largely of biochemically precipitated shell materials. Dolostone is formed by the diagenetic alteration of limestones. Other chemical and biochemical sediments include evaporites; siliceous sediments such as chert; phosphorites; iron formations; and peat and other organic matter that is transformed into coal, oil, and gas.

Figure 8.20 The development of sedimentary basins on a rifted continental margin. (a) A rift develops in Pangaea as hot mantle materials well up and the ancient continent stretches and thins. Volcanics and Triassic nonmarine sediments are deposited in the faulted valleys. (b) Seafloor spreading begins. The lithosphere cools and contracts under the receding continental margins, which subside below sea level. (c) Evaporites, deltaic sediments, and carbonates are deposited. (d) These deposits are then covered by Jurassic and Cretaceous sediments derived from continental erosion. The Atlantic margins of Europe, Africa, and North and South America have histories similar to this one.

Key Terms and Concepts

bedding (p. 171)
bedding sequence (p. 173)
biochemical sediment (p. 165)
bioclastic sediment (p. 166)
bioturbation (p. 173)
carbonate environment (pp. 179–180)
carbonate platform (p. 182)
carbonate rock (p. 180)
carbonate sediment (p. 180)
cementation (p. 174)
chemical sediment (p. 165)
chert (p. 186)
clastic particle (p. 165)
clastic sediment (p. 165)
clay (p. 179)
claystone (p. 179)
coal (p. 186)
compaction (p. 175)
conglomerate (p. 177)
continental shelf deposit (p. 187)
cross-bedding (p. 171)
diagenesis (p. 174)
dolostone (p. 183)
evaporite rock (p. 184)
evaporite sediment (p. 184)
flexural basin (p. 189)
foraminifera (p. 180)
gas (p. 186)
graded bedding (p. 172)
gravel (p. 177)
iron formation (p. 186)
limestone (p. 183)
lithification (p. 174)
mud (p. 179)
mudstone (p. 179)
oil (p. 186)
organic sedimentary rock (p. 186)
peat (p. 168)
phosphorite (p. 186)
porosity (p. 175)
reef (p. 180)
rift basin (p. 187)
ripple (p. 172)
salinity (p. 168)
sand (p. 177)
sandstone (p. 177)
sedimentary basin (p. 187)
sedimentary environment (p. 168)
sedimentary structure (p. 171)
shale (p. 179)
siliceous environment (p. 171)
silt (p. 179)
siltstone (p. 179)
sorting (p. 167)
subsidence (p. 187)
terrigenous sediment (p. 170)
thermal sag basin (p. 187)

Exercises

This icon indicates that there is an animation available on the Web site that may assist you in answering a question.

1. What processes change sediment into sedimentary rock?

2. How do clastic sedimentary rocks differ from chemical and biochemical sedimentary rocks?

3. How and on what basis are the clastic sedimentary rocks subdivided?

4. What kind of sedimentary rocks were originally formed by the evaporation of seawater?

5. Define sedimentary environment, and name three clastic environments.

6. Name three kinds of sandstone.

7. Name two kinds of carbonate rocks and explain how they differ.

8. How do organisms produce or modify sediments?

9. Name two ions that take part in the precipitation of calcium carbonate in a sedimentary environment.

10. In what two kinds of sedimentary rocks are oil and gas found?

Thought Questions

This icon indicates that there is an animation available on the Web site that may assist you in answering a question.

1. Weathering of the continents has been much more widespread and intense in the past 10 million years than it was in earlier times. How might this observation be borne out in the sediments that now cover Earth's surface?

2. In what respects might you consider a volcanic ash fall a sediment?

3. A geologist is heard to say that a particular sandstone was derived from a granite. What information could she have gleaned from the sandstone to lead her to that conclusion?

4. You are looking at a cross section of a rippled sandstone. What sedimentary structure would tell you the direction of the current that deposited the sand?

5. You discover a bedding sequence that has a conglomerate at the base; grades upward to a sandstone and then to a shale; and finally, at the top, grades to a limestone of cemented carbonate sand. What changes in the area of the sediment's source or in the sedimentary environment would have been responsible for this sequence?

6. From the base upward, a bedding sequence begins with a bioclastic limestone, passes upward into a dense carbonate rock made of carbonate-cementing organisms (including algae normally found with coral), and ends with beds of dolomite. Deduce the possible sedimentary environments represented by this sequence.

7. In what sedimentary environments would you expect to find carbonate muds?

8. How can you use the sizes and sorting of sediments to distinguish between sediments deposited in a glacial environment and those deposited on a desert?

9. Describe the beach sands that you would expect to be produced by the beating of waves on a coastal mountain range consisting largely of basalt.

10. What role do transporting currents play in the origin of some kinds of limestone?

11. Name a sedimentary rock that is essentially the product of diagenesis and that has no exact equivalent as a sediment.

12. Where are reefs likely to be found?

13. An ocean bay is separated from the open ocean by a narrow, shallow inlet. What kind of sediment would you expect to find on the floor of the bay if the climate were warm and arid? What kind of sediment would you find if the climate were cool and humid?

14. How are chert and limestone similar in origin?

Short-Term Team Projects

Carbonate Sediments

The three main minerals of carbonate rocks are calcite, $CaCO_3$; aragonite, also $CaCO_3$; and dolomite, $CaMg(CO_3)_2$. Calcite and aragonite are polymorphs; they have the same chemical composition but different crystal structures. Some organisms make their shells and skeletons of aragonite, whereas others make them of calcite. Interestingly, fossils of very old organisms are never made of aragonite. The explanation is that aragonite is less stable than calcite and eventually breaks down to form calcite.

The origin of dolomite, in contrast, is unresolved. Dolomite is a mineral found in many sedimentary sequences, but it is not a constituent of shells or newly deposited carbonate sediment. Many sedimentologists regard it as a secondary mineral that forms when calcite or aragonite (primary minerals) combine with magnesium. Others think that dolomite does sometimes form as a primary mineral. Maybe you can help to resolve this question. For this short-term project, you and a partner should propose and explain a hypothesis for the formation of dolomite. Describe how you could test your hypothesis, perhaps using standard equipment in a college laboratory. Ask your professor to help you set up such an experiment in the laboratory and follow through with the test for a few weeks.

Inferring the Origin of Interesting Formations

The Shawangunk conglomerate is a well-known sedimentary unit that runs through New Jersey and southern New York. Parts of it are exposed in outcrops. The Shawangunk conglomerate consists mostly of quartz pebbles, so it is very resistant to erosion and forms spectacular cliffs at the eastern margins of the Catskill Mountains. Only in a few locations does it display pebbles of different rock types.

What type of source area would supply the coarse-grained sediments for the Shawangunk conglomerate? How could you determine the direction of the source area at the time the conglomerate was deposited? What might the few diverse pebble types tell you about the source? Using a geologic map of your own region, locate a coarse conglomerate unit and determine the source of its constituent particles. If there are no coarse conglomerates in your area, determine why this is the case.

Suggested Readings

Blatt, H., and R. J. Tracy. 1996. Sedimentary rocks. In *Petrology,* 2d ed., pp. 215–350. New York: W. H. Freeman.

Goreau, T. F., N. I. Goreau, and T. J. Goreau. 1979. Corals and coral reefs. *Scientific American* (August): 124–136.

Leeder, M. R. 1982. *Sedimentology.* London: Allen and Unwin.

Mack, W. N., and E. A. Leistikow. 1996. Sands of the world. *Scientific American* (August): 62–67.

McLane, M. 1995. *Sedimentology.* New York: Oxford University Press.

Prothero, D. R., and F. Schwab. 1996. *Sedimentary Geology.* New York: W. H. Freeman.

Siever, R. 1988. *Sand.* New York: Scientific American Library.

These rocks show both the banding and the deformation into folds characteristic of sedimentary rocks metamorphosed into marble, schist, and gneiss. Sequoia National Forest, California. [Gregory G. Dimijian/Photo Researchers.]

CHAPTER 9

Metamorphic Rocks

"Speak to the earth, and it should teach thee."

JOB 11:8

We are all familiar with some ways in which heat and pressure can transform materials. Frying raw ground meat changes it into a hamburger composed of chemical compounds very different from those in the raw meat. Cooking batter in a waffle iron not only heats up the batter but also puts pressure on it, transforming it into a rigid solid. In similar ways, rocks change as they encounter high temperatures and pressures. Deep in Earth's crust, tens of kilometers below the surface, temperatures and pressures are high enough to metamorphose rock without being high enough to melt it. Increases in heat and pressure and changes in the chemical environment can alter the mineral compositions and crystalline textures of sedimentary and igneous rocks, *even though they remain solid all the while.* The result is the third large class of rocks: the **metamorphic,** or "changed form," **rocks,** which have undergone changes in mineralogy, texture, chemical composition, or all three.

Metamorphic changes bring a preexisting rock into equilibrium with new surroundings. Given enough time—short by geologic standards but usually a million years or more—the rock changes mineralogically and texturally until it is in equilibrium with the new temperatures and pressures. A limestone filled with fossils, for example, might be transformed into a white marble in which no trace of fossils remains. The mineral and chemical composition of the rock may be unaltered, but its texture may have changed drastically from small calcite crystals to large, intergrown crystals. Shale, a well-bedded rock so finely grained that no individual mineral crystal can be seen with the naked eye, might become a schist in which the original bedding is obscured and the texture is dominated by large crystals of mica. In these metamorphic transformations, both mineral composition and texture changed, but the overall chemical composition of the rocks remained the same.

Clay minerals are silicates but differ from micas in that they contain lots of water molecules trapped between silicate sheets in the crystal structure. During metamorphism, most of this water is lost as the clay minerals are transformed to mica. Other rocks, such as those altered by heat or by fluids derived from igneous activity, change in mineralogy, texture, and chemical composition. Some silicate minerals are so diagnostic of metamorphism that their mere presence indicates a rock is metamorphic. These minerals include kyanite, andalusite, and sillimanite; staurolite; garnet; and epidote. Other minerals are common in metamorphic rocks but are also found in some igneous rocks. These include garnet, quartz, muscovite, amphibole, and feldspar.

There are many reasons that geologists study metamorphic rocks, but all relate to one common objective: *to understand how Earth's crust has evolved over geologic history.*

This chapter examines the causes of metamorphism, the types of metamorphism that take place under certain sets of conditions, and the origins of the various textures that characterize metamorphic rocks.

Metamorphism and the Earth System

Metamorphism, like all other geologic processes, is part of the Earth system. A major component of metamorphism is

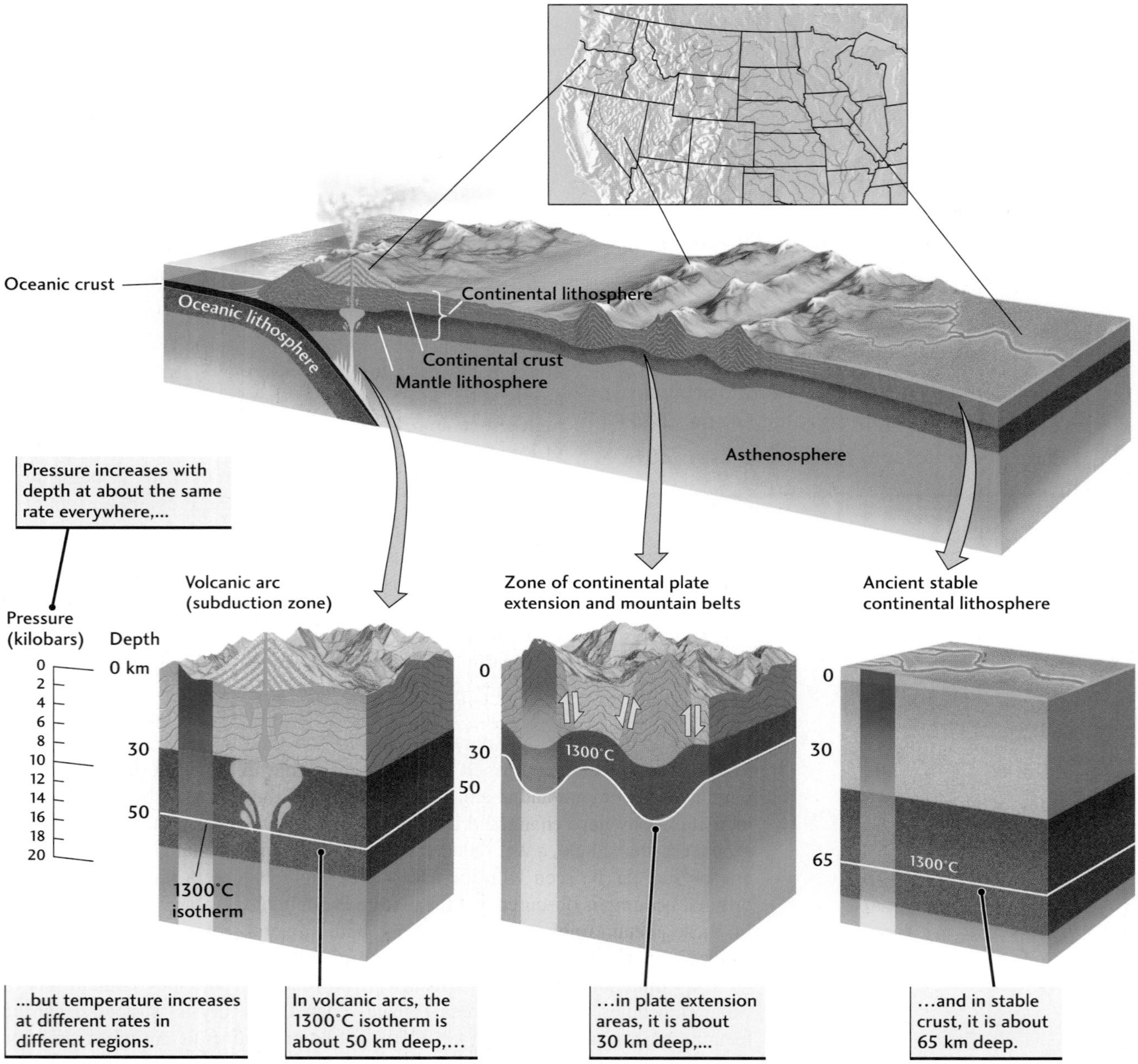

Oceanic crust
Oceanic lithosphere
Continental lithosphere
Continental crust
Mantle lithosphere
Asthenosphere

Pressure increases with depth at about the same rate everywhere,...

Volcanic arc (subduction zone)
Zone of continental plate extension and mountain belts
Ancient stable continental lithosphere

Pressure (kilobars): 0, 2, 4, 6, 8, 10, 12, 14, 16, 18, 20
Depth: 0 km, 30, 50

1300°C isotherm

...but temperature increases at different rates in different regions.

In volcanic arcs, the 1300°C isotherm is about 50 km deep,...

...in plate extension areas, it is about 30 km deep,...

...and in stable crust, it is about 65 km deep.

Figure 9.1 Pressure and temperature increase with depth in all regions, as shown in this cross section of a volcanic region, a continental region, and an ancient stable continental lithosphere region. Pressure increases with depth at more or less the same rate everywhere, but temperature increases at different rates in different regions. (Pressure is measured in kilobars; 1 kilobar is approximately equal to 1000 times the atmospheric pressure at Earth's surface.)

Earth's internal heat, which drives metamorphism that is controlled by temperature. Thus, Earth's interior heat powers the parts of the Earth system that govern metamorphic—and igneous—processes. As we will see later in this chapter, metamorphism results in the release of water vapor, carbon dioxide, and other gases. These gases leak to the surface and contribute to the atmosphere, thus affecting processes that depend on atmospheric composition, such as weathering.

Causes of Metamorphism

Sediments and sedimentary rocks belong to Earth's surface environments, whereas igneous rocks belong to the melts of the lower crust and mantle. Metamorphic rocks exposed at the surface are mainly the products of processes acting on rocks at depths ranging from the upper to the lower crust. Most have formed at depths of 10 to 30 km, the middle to lower half of the crust. Although most metamorphism takes place at depth, it can also occur at Earth's surface. We can see metamorphic changes in the baked surfaces of soils and sediments just beneath volcanic lava flows.

The internal heat of the Earth, its pressure, and its fluid composition are the three principal factors that drive metamorphism. The contribution of pressure is the result of vertically oriented forces exerted by the weight of overlying rocks and horizontally oriented forces developed as the rocks are deformed.

Temperature increases with depth at different rates in different regions of the Earth, ranging from 20° to 60°C per kilometer of depth (Figure 9.1). In much of Earth's crust, temperature increases at a rate of 30°C per kilometer of depth. At that rate, the temperature will be about 450°C at a depth of 15 km—much higher than the average temperature of the surface, which ranges from 10° to 20°C in most regions. (Refer to the discussion of geotherms in Chapter 21.) The pressure at a depth of 15 km comes from the weight of all the overlying rock and amounts to about 4000 times the pressure at the surface.

High as these temperatures and pressures may seem, they are only in the middle range of metamorphism, as Figure 9.2 shows. We refer to the metamorphic rocks formed under the lower temperatures and pressures of shallower crustal regions as *low-grade rocks* and the ones formed at deeper zones of higher temperatures and pressures as *high-grade rocks.* As the grade of metamorphism changes, the mineral assemblages within metamorphic rocks also change, allowing geologists to define a set of *metamorphic facies,* which we will describe later.

The Role of Temperature

Heat greatly affects a rock's mineralogy and texture. In Chapter 5, we learned how important the influence of heat can be in breaking chemical bonds and altering the existing crystal structures of igneous rocks. Heat has an equally important role in the formation of metamorphic rocks. For example,

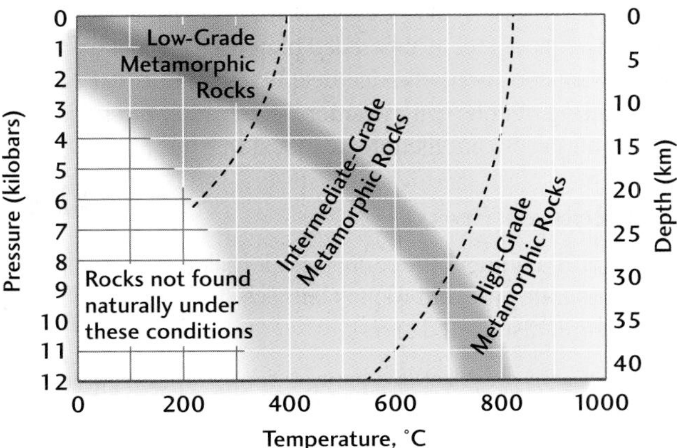

Figure 9.2 Temperatures, pressures, and depths at which low- and high-grade metamorphic rocks form. The dark band shows common rates at which temperature and pressure increase with depth over much of the continents.

plate tectonic processes may move sediments and rocks from the surface of the Earth to the interior, where temperatures are higher. As the rock adjusts to the new temperature, its atoms and ions recrystallize, linking up in new arrangements and creating new mineral assemblages. Many new crystals will grow larger than they were in the original rock. If deformation occurs at the same time, the rock may become banded as minerals of different compositions are segregated into separate planes. (See the chapter opening photograph.)

The increase of temperature with increasing depth is called a *geothermal gradient.* The geothermal gradient varies depending on plate tectonic setting, but on average it is about 30°C per kilometer of depth. In areas where the continental lithosphere has been thinned by plate extension, such as in Nevada's Great Basin, the geothermal gradient is *steep* (for example, 50°C per kilometer of depth). In areas where the continental lithosphere is old and thick, such as beneath central North America, the geothermal gradient is *shallow* (for example, 20°C per kilometer of depth) (see Figure 9.1). As sedimentary rocks containing clay minerals are buried deeper and deeper, the clay minerals begin to recrystallize and form new minerals such as mica. With additional burial to greater depths—and temperatures—the micas become unstable and begin to recrystallize into new minerals such as garnet. We see, then, that because different minerals crystallize and remain stable at different temperatures, the metamorphic geologist, like the igneous geologist, can use a rock's composition as a kind of *geothermometer* to gauge the temperature at which the rock formed. Given a specific assemblage of minerals in a metamorphic rock, the geologist can infer the temperature at which the rock formed.

Plate tectonic processes such as subduction and continental collision, which transport rocks and sediments into the hot depths of the crust, are the primary mechanisms that form most metamorphic rocks. In addition, limited metamorphism may occur where rocks are subjected to elevated temperatures adjacent to recently intruded plutons. The heat is locally intense but does not penetrate deeply. Heat pulses produced by intruding plutons can metamorphically alter the surrounding country rock, but the effect is local in extent.

The Role of Pressure

Pressure, like temperature, changes a rock's texture as well as its mineralogy. Solid rock is subjected to two basic kinds of pressure, also called **stress.**

1. *Confining pressure* is a general force applied equally in all directions, such as the pressure a swimmer feels when submerged in a pool. Just like a swimmer moving to greater depths in the pool, a rock descending to greater depths in the Earth will be subjected to progressively increased confining pressure.

2. *Directed pressure* is force exerted in a particular direction, as when a ball of clay is squeezed between thumb and forefinger. Directed pressure, or *differential stress,* is usually concentrated within zones or along discrete planes. The compressive force that occurs where plates converge is a form of directed pressure, and it results in deformation of the rocks near the plate boundaries. Heat reduces the strength of a rock, so directed pressure is likely to cause severe folding and deformation of metamorphic rocks in mountain belts where temperatures are high. Rocks subjected to differential stress may be severely distorted, becoming flattened in the direction the force is applied and elongated in the direction perpendicular to the force.

Metamorphic minerals may be compressed, elongated, or rotated to line up in a particular direction, depending on the kind of stress applied to the rocks. Thus, directed pressure guides the shape and orientation of the new metamorphic crystals formed as the minerals recrystallize under the influence of both heat and pressure. During the recrystallization of micas, for example, the crystals grow with the planes of their sheet-silicate structures aligned perpendicular to the directed stress.

Pressure, like temperature, increases with depth in the Earth. Pressure is usually recorded in *kilobars* (1000 bars, abbreviated as kbar) and increases at a rate of 0.3 to 0.4 kbar per kilometer of depth (see Figure 9.1). One bar is approximately equivalent to the pressure of air at the surface of the Earth. A diver who is touring the deeper part of a coral reef at a depth of 10 m would experience another 1 bar of pressure. The pressure to which a rock is subjected deep in the Earth is related to both the *thickness* of the overlying rocks and the *density* of those rocks.

Minerals that are stable at the lower pressure near Earth's surface become unstable and recrystallize to new minerals under the increased pressure at depth in the crust. Laboratory studies have yielded data on the pressures required for these changes. Using these data, we can examine the mineralogy and texture of metamorphic rock samples and infer what the pressures were in the area where they formed. Thus, metamorphic mineral assemblages can be used as pressure gauges, or *geobarometers.* Given a specific assemblage of minerals in a metamorphic rock, the geologist can place constraints on the range of pressures, and therefore depths, at which the rock formed.

The Role of Fluids

Metamorphism can significantly alter a rock's mineralogy by introducing or removing chemical components that dissolve in water. Hydrothermal fluids produced during metamorphism carry dissolved carbon dioxide as well as chemical substances—such as sodium, potassium, silica, copper, and zinc—that are soluble in hot water under pressure. As hydrothermal solutions percolate up to the shallower parts of the crust, they react with the rocks they penetrate, changing their chemical and mineral compositions and sometimes completely replacing one mineral with another without changing the rock's texture. This kind of change in a rock's bulk composition by fluid transport of chemical substances into or out of the rock is called **metasomatism.** Many valuable deposits of copper, zinc, lead, and other metal ores are formed by this kind of chemical substitution (see Chapter 22).

Hydrothermal fluids accelerate metamorphic chemical reactions. Atoms and ions dissolved in the fluid can migrate through a rock and react with the solids to form new minerals. As metamorphism proceeds, the water itself reacts with the rock when chemical bonds between minerals and water molecules form or break.

Where do these chemically reactive fluids originate? Although most rocks appear to be completely dry and of extremely low porosity, they characteristically contain fluid in minute pores (the spaces between grains). This water is derived from chemically bound water in clay, not from sedimentary pore waters, which are largely expelled during diagenesis. In other, hydrous minerals, such as mica and amphibole, water forms part of their crystalline structures. The carbon dioxide dissolved in hydrothermal fluids is derived largely from sedimentary carbonates—limestones and dolostones.

Types of Metamorphism

Geologists can duplicate metamorphic conditions in the laboratory and determine the precise combinations of pressure, temperature, and chemical composition under which transformations might take place. But to understand how any particular combination relates to the geology of metamorphism—that is, when, where, and how these conditions came about in the Earth—geologists categorize metamorphic rocks on the basis of the geological circumstances of their origins. We describe these categories next; **Figure 9.3** locates them in relation to major plate tectonic settings.

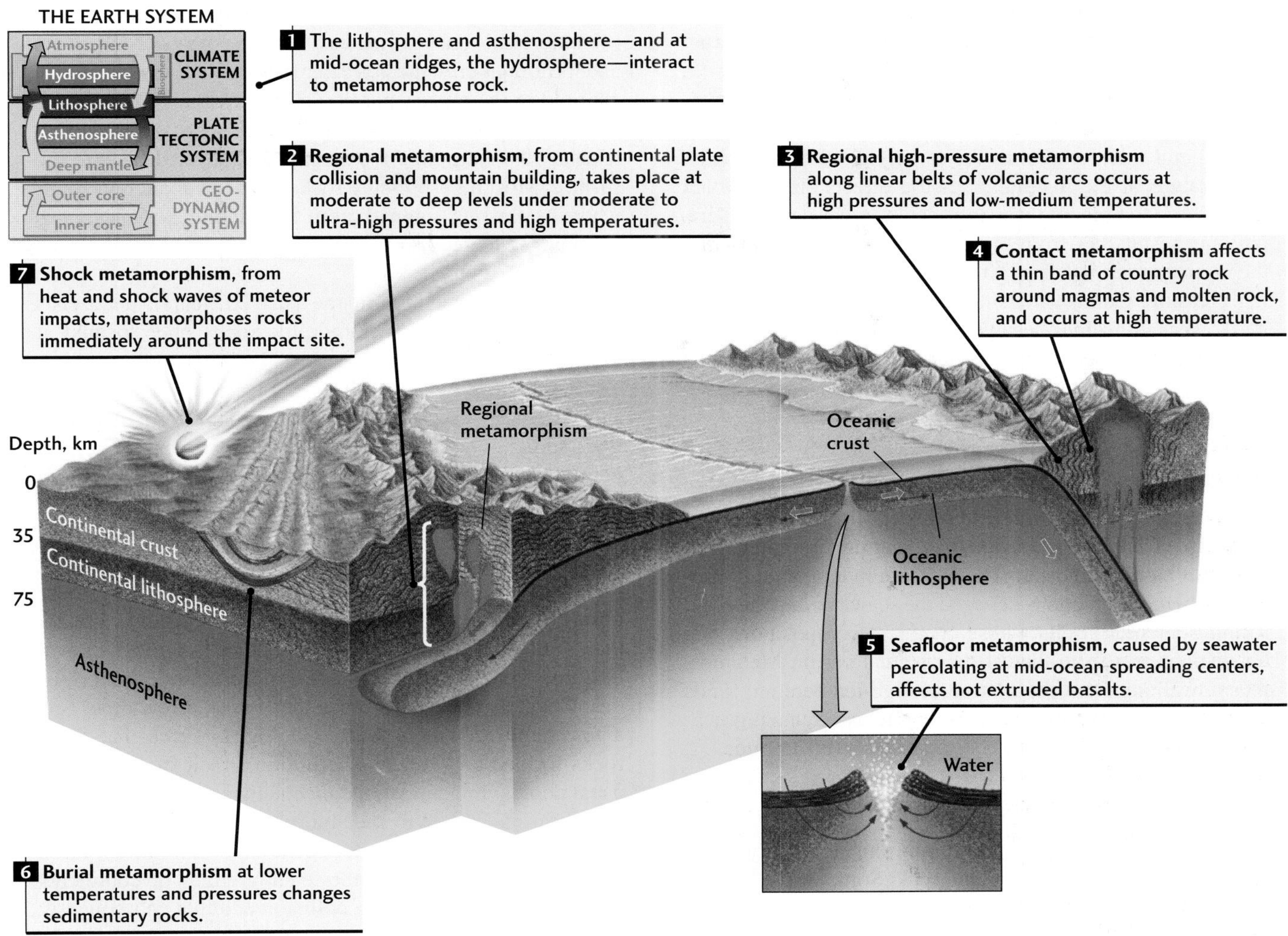

Figure 9.3 The main types of metamorphism and where they occur.

Regional Metamorphism

Regional metamorphism, the most widespread type of metamorphism, takes place where both high temperature and high pressure are imposed over large parts of the crust. We use this term to distinguish this type of metamorphism from more localized changes near igneous intrusions or faults. Regional metamorphism is a characteristic feature of convergent plate tectonic settings. It occurs within volcanic island arcs, such as the Andes Mountains of South America, and in the cores of mountain chains produced during the collision of continents, such as the Himalayan Mountains of central Asia. These mountain belts are often linear features, so ancient (and modern) zones of regional metamorphism are often linear in their distribution. In fact, geologists usually interpret belts of regionally extensive metamorphic rocks as representing sites of former mountain chains that were eroded over millions of years, exposing the rocks at Earth's surface.

Some regional metamorphic belts are created by high temperatures and moderate to high pressures near the volcanic arcs formed where subducted plates sink deep into the mantle. Regional metamorphism under very high pressures and temperatures takes place at deeper levels of the crust along boundaries where colliding continents deform rock and raise high mountain belts. During regional metamorphism, rocks typically will be transported to significant depths in the Earth's crust, only to be exhumed by subsequent uplift and erosion at Earth's surface. However, a full understanding of the patterns of regional metamorphism, including how rocks respond to systematic changes in temperature and pressure over time, depends on the specific tectonic setting. We will discuss this topic later in the chapter.

Contact Metamorphism

Igneous intrusions metamorphose the immediately surrounding rock by propagating their heat outward, which subjects the minerals of the preexisting rock to new conditions. This type of localized transformation, called **contact metamorphism,** normally affects only a thin region of country rock along the contact. In many contact metamorphic rocks, especially at the margins of shallow intrusions, the mineral transformations are largely related to the high temperature of the magma. Pressure effects are important only where the magma was intruded at great depths. Here, pressure results not from the intrusion forcing its way into the country rock but from the presence of regional confining pressure. Contact metamorphism by extrusives is limited to very thin zones because lavas cool quickly at the surface and their heat has little time to penetrate deep into the surrounding rocks and cause metamorphic changes.

Seafloor Metamorphism

Another type of metamorphism, called **seafloor metamorphism** or *metasomatism,* is often associated with mid-ocean ridges (see Chapter 5). Seawater percolating through the hot, fractured basalts is heated. The increase in temperature promotes chemical reactions between the seawater and the rock, forming altered basalts whose chemical compositions differ distinctively from that of the original basalt. Metamorphism resulting from percolation of high-temperature fluids also takes place on continents when fluids circulating near igneous intrusions metamorphose the rocks they intrude.

Other Types of Metamorphism

There are other types of metamorphism that produce smaller amounts of metamorphic rock. Some of these, such as ultra-high-pressure metamorphism, are extremely important in helping geologists to understand conditions deep within the Earth.

Low-Grade (Burial) Metamorphism Recall from Chapter 8 that when sedimentary rocks are gradually buried during subsidence of the crust, they slowly heat up as they come into equilibrium with the temperature of the crust around them. In this process, diagenesis alters their mineralogy and texture. Diagenesis grades into **low-grade,** or **burial, metamorphism,** which is caused by the progressive increase in pressure exerted by the growing pile of overlying sediments and sedimentary rocks and by the increase in heat associated with increased depth of burial in the Earth.

Depending on the local geothermal gradient, low-grade metamorphism typically begins at depths of 6 to 10 km, where temperatures range between 100° and 200°C and pressures are less than 3 kbar. This knowledge is of great importance to the oil and gas industry, which defines "economic basement" as the depth where low-grade metamorphism begins. Oil and gas wells are rarely drilled below this depth because temperatures above 130°C convert organic matter trapped in sedimentary rocks into methane or carbon dioxide rather than crude oil and natural gas.

High-Pressure and Ultra-High-Pressure Metamorphism Metamorphic rocks formed at high pressures (8–12 kbar) and ultra-high pressures (greater than 28 kilobars) are rarely exposed at the surface for geologists to study. These rocks are unusual because they form at such great depths that it takes a very long time for them to be recycled back to the surface. Most high-pressure rocks form in subduction zones where sediments scraped from subducting oceanic plates are plunged to depths of over 30 km, where they experience pressures of up to 12 kbar.

Relatively recently (in the past 20 years), geologists have recognized that metamorphic rocks once located at the base of the crust can sometimes be found at the surface. These rocks—called **eclogites**—typically contain minerals such as *coesite* (a very dense, high-pressure form of quartz) that indicate pressures of greater than 28 kbar, suggesting depths of over 80 km. Such rocks formed under moderate to high temperatures, ranging up to 800°–1000°C. In a few cases, these rocks contain *microscopic diamonds,* indicative of pressures greater than 40 kbar and depths greater than 120 km! Surprisingly, outcrop exposures of these ultra-high-pressure metamorphic rocks cover areas greater than 400 km by 200 km. The only other two rocks known to come from these depths are diatremes and kimberlites (see Chapter 6), igneous rocks that form narrow pipes just a few hundred meters wide. Geologists agree that these rocks formed by "volcanic" eruption, albeit from very unusual depths. In contrast, the mechanisms required to bring the ultra-high-pressure metamorphic rocks to the surface are hotly debated. In the most general interpretation, these rocks represent pieces of the leading edges of continents that were subducted during collision and that subsequently rebounded (via some unknown mechanism) back to the surface before they had to time to recrystallize at lower pressures.

Shock Metamorphism **Shock metamorphism** occurs when a meteorite collides with Earth. Meteorites are fragments of comets or asteroids that have been brought to Earth by its gravitational field. Upon impact, the energy represented by the meteorite's mass and velocity is transformed to heat and shock waves that pass through the impacted country rock. The country rock can be shattered and even partially melted to produce *tektites,* which look like droplets of glass. In some cases, quartz is transformed into coesite and *stishovite,* two of its high-pressure forms.

Most large impacts on Earth have left no trace of a meteorite because these bodies are usually destroyed in the collision with Earth. The occurrence of coesite and craters with distinctive fringing fracture textures, however, often provides key evidence of these collisions. Earth's dense atmosphere causes most meteoroids to burn up before they strike its surface, so shock metamorphism is rare on Earth. On the surface of the Moon, however, shock metamorphism is pervasive. It is characterized by extremely high pressures of many tens to hundreds of kilobars.

Plate Tectonics and Types of Metamorphism

Plate tectonics provides a framework for understanding metamorphic rocks. Different types of metamorphism are likely to occur in different tectonic settings (see Figure 9.3):

- *Plate interiors.* Contact metamorphism, burial metamorphism, and perhaps regional metamorphism occur at the base of the crust. Shock metamorphism is likely to be best preserved in this setting because of the large exposed area of plate interiors.
- *Divergent plate margins.* Seafloor metamorphism and contact metamorphism around intruding plutons in the ocean crust are found at divergent plate margins.
- *Convergent plate margins.* Regional metamorphism, high-pressure and ultra-high-pressure metamorphism, and contact metamorphism around intruding plutons are found at convergent plate boundaries.
- *Transform plate margins.* In oceanic settings, seafloor metamorphism may occur. In both oceanic and continental settings, one will find extensive shearing along the plate boundary, producing *cataclastic* deformation textures at shallow levels and *mylonitic* deformation textures at deep levels in the crust. (We will discuss deformation in Chapter 11.)

Metamorphic Textures

Metamorphism imprints new textures on the rocks that it alters (**Figure Story 9.4**). The texture of a metamorphic rock is determined by the sizes, shapes, and arrangement of its constituent crystals. Some metamorphic textures depend on the particular kinds of minerals formed, such as the micas, which are platy. Variation in grain size is also important. In general, geologists find that the grain size of crystals increases as metamorphic grade increases. Each textural variety tells us something about the metamorphic process that created it.

Foliation and Cleavage

The most prominent textural feature of regionally metamorphosed rocks is **foliation,** a set of flat or wavy parallel planes produced by deformation. These foliation planes may cut the bedding at any angle or be parallel to the bedding (Figure Story 9.4a).

A major cause of foliation is the presence of platy minerals, chiefly the micas and chlorite. Platy minerals tend to crystallize as thin platelike crystals. The planes of all the platy crystals are aligned parallel to the foliation, an alignment called the *preferred orientation* of the minerals (Figure Story 9.4b). As platy minerals crystallize, the preferred orientation is usually perpendicular to the main direction of the forces squeezing the rock as it is deformed during metamorphism. Preexisting minerals may acquire a preferred orientation and thus produce foliation by rotating until they lie parallel to the developing plane.

Minerals whose crystals have an elongate, pencil-like shape also tend to assume a preferred orientation during metamorphism, the crystals normally lining up parallel to the foliation plane. Rocks that contain abundant amphiboles, typically metamorphosed mafic volcanics, have this kind of texture.

The most familiar form of foliation is seen in slate, a common metamorphic rock, which is easily split along smooth, parallel surfaces into thin sheets. This *slaty cleavage* (not to be confused with the cleavage of a mineral such as muscovite) develops along moderately thin, regular intervals in the rock.

Classification of Foliated Rocks

The **foliated rocks** are classified according to four main criteria (Figure Story 9.4c):

1. The size of their crystals
2. The nature of their foliation
3. The degree to which their minerals are segregated into lighter and darker bands
4. Their metamorphic grade

Figure Story 9.4d shows examples of the major types of foliated rocks. In general, foliation progresses from one texture to another, reflecting the increase in temperature and pressure. In this progression, a shale may metamorphose first to a slate, then to a phyllite, then to a schist, then to a gneiss, and finally to a migmatite.

Slate **Slates** are the lowest grade of foliated rocks. These rocks are so fine-grained that their individual minerals cannot be seen easily without a microscope. They are commonly produced by the metamorphism of shales or, less frequently, of volcanic ash deposits. Slates usually range from dark gray to black, colored by small amounts of organic material originally present in the parent shale. Slate splitters learned long ago to recognize this foliation and use it to make thick or thin slates for roofing tiles and blackboards. We still use flat slabs of slate for flagstone walks in parts of the country where slate is abundant.

Phyllite The **phyllites** are of a slightly higher grade than the slates but are similar in character and origin. They tend to have a more or less glossy sheen resulting from crystals of mica and chlorite that have grown a little larger than those of slates. Phyllites, like slates, tend to split into thin sheets, but less perfectly than slates.

Schist At low grades of metamorphism, crystals of platy minerals are generally too small to be seen, foliation is closely spaced, and layers are very thin. As metamorphic rocks are more intensely metamorphosed into higher grades, the platy crystals grow large enough to be visible to

REGIONAL METAMORPHISM CHANGES ROCK TEXTURE

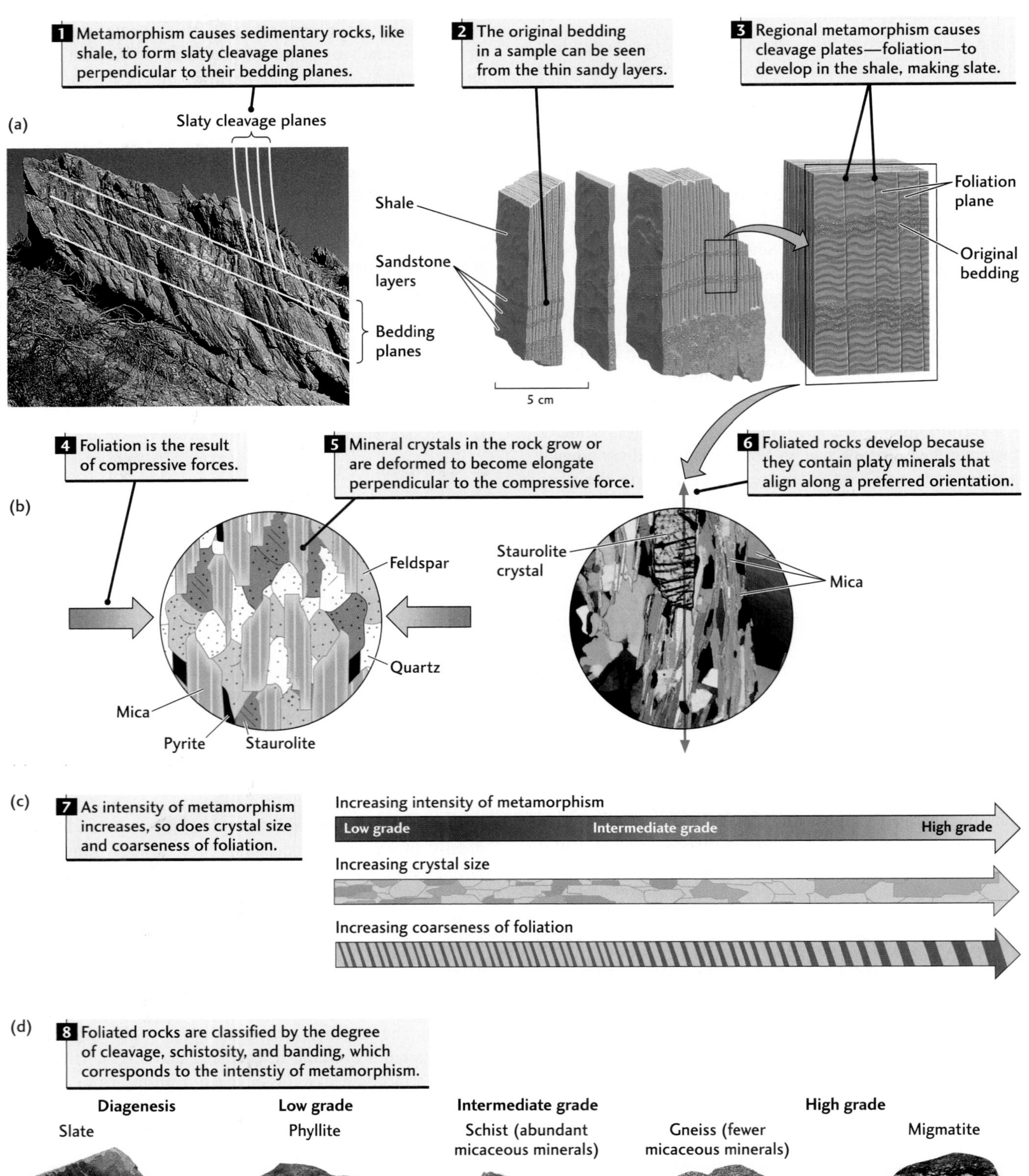

the naked eye, and the minerals may tend to segregate in lighter and darker bands. This parallel arrangement of sheet minerals produces the pervasive coarse, wavy foliation known as schistosity, which characterizes **schists.** Schists are among the most abundant metamorphic rock types. They contain more than 50 percent platy minerals, mainly the micas muscovite and biotite. Schists may contain thin layers of quartz, feldspar, or both, depending on the quartz content of the original shale.

Gneiss Even coarser foliation is shown by high-grade **gneisses,** light-colored rocks with coarse bands of segregated light and dark minerals throughout the rock. The banding of gneisses into light and dark layers results from the segregation of lighter-colored quartz and feldspar and darker amphiboles and other mafic minerals. Gneisses are coarse-grained, and the ratio of granular to platy minerals is higher than it is in slate or schist. The result is poor foliation and thus little tendency to split. Under conditions of high pressure and temperature, mineral assemblages of the lower-grade rocks containing micas and chlorite change into new assemblages dominated by quartz and feldspars, with lesser amounts of micas and amphiboles.

Migmatite Temperatures higher than those necessary to produce gneiss may begin to melt the country rock. In this case, as with igneous rocks (see Chapter 5), the first minerals to melt will be those with the lowest melting temperatures. Therefore only part of the country rock melts, and the melt may migrate only a short distance before freezing again. Rocks produced in this way are badly deformed and contorted, and they are penetrated by many veins, small pods, and lenses of melted rock. The result is a mixture of igneous and metamorphic rock called **migmatite.** Some migmatites are mainly metamorphic, with only a small proportion of igneous material. Others have been so affected by melting that they are considered almost entirely igneous.

Granoblastic Rocks

Granoblastic rocks are composed mainly of crystals that grow in equant (equidimensional) shapes, such as cubes and spheres, rather than in platy or elongate shapes (**Figure 9.5**). These rocks may result from metamorphism in which deformation is absent, such as contact metamorphism. Granoblastic (nonfoliated) rocks include hornfels, quartzite, marble, greenstone, amphibolite, and granulite. All granoblastic rocks, except hornfels, are defined by their mineral composition rather than their texture because all of them are massive in appearance.

Quartzite

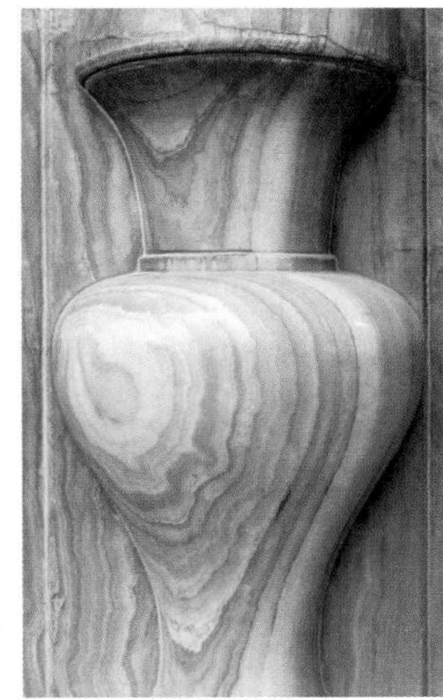

Marble

Figure 9.5 Granoblastic (nonfoliated) metamorphic rocks. (a) Quartzite [Breck P. Kent]; (b) marble. [Diego Lezama Orezzoli/Corbis].

Hornfels is a high-temperature contact metamorphic rock of uniform grain size that has undergone little or no deformation. Its platy or elongate crystals are oriented randomly, and foliated texture is absent. Hornfels has a granular texture overall, even though it commonly contains pyroxene, which makes elongate crystals, and some micas.

Quartzites are very hard, nonfoliated white rocks derived from quartz-rich sandstones. Some quartzites are massive, unbroken by preserved bedding or foliation. Others contain thin bands of slate or schist, relics of former interbedded layers of clay or shale.

Figure Story 9.4 Slaty cleavage, the most familiar form of foliation, develops along thin, regular intervals. (a) This outcrop, in southwestern Montana, shows slaty cleavage [Martin Miller]. (b) The photomicrograph of schist shows the preferred orientation of mica and staurolite crystals [S. Dobos.]. (c) and (d) Classification of foliated rocks. Cleavage, schistosity, and banding do not generally correspond to the original bedding direction of the sedimentary rock. [*Slate:* Andrew J. Martinez/Photo Researchers. *Phyllite:* Courtesy of Kurt Hollocher, Union College. *Schist:* Biophoto Associates/Photo Researchers. *Gneiss:* Breck P. Kent. *Migmatite:* Kip Hodges.]

Marbles are the metamorphic products of heat and pressure acting on limestones and dolomites. Some white, pure marbles, such as the famous Italian Carrara marbles prized by sculptors, show a smooth, even texture of intergrown calcite crystals of uniform size. Other marbles show irregular banding or mottling from silicate and other mineral impurities in the original limestone (see Figure 9.5b).

Greenstones are metamorphosed mafic volcanic rocks. Many of these low-grade rocks form when mafic lavas and ash deposits react with percolating seawater or other solutions. Large areas of the seafloor are covered with basalts slightly or extensively altered in this way at mid-ocean ridges. An abundance of chlorite gives these rocks their greenish cast.

Amphibolite can be a nonfoliated rock made up of amphibole and plagioclase feldspar. It is typically the product of medium- to high-grade metamorphism of mafic volcanics. Foliated amphibolites are produced when deformation occurs.

The high-grade metamorphic rock **granulite** has a granoblastic texture. Lower-grade granulite is often referred to as *granofels*. Granofels are medium- to coarse-grained rocks in which the crystals are equant and show only faint foliation at most. They are formed by the metamorphism of shale, impure sandstone, and many kinds of igneous rock.

Large-Crystal Textures

New metamorphic minerals may grow into large crystals surrounded by a much finer grained matrix of other minerals. These large crystals are **porphyroblasts** and are found in both contact and regionally metamorphosed rocks (**Figure 9.6**). They grow as the chemical components of the matrix are reorganized and thus replace parts of the matrix. Porphyroblasts form when there is a strong contrast between the

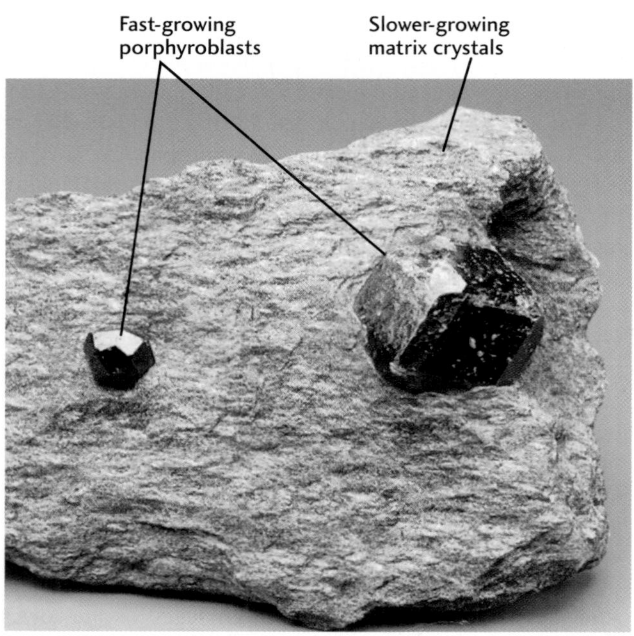

Figure 9.6 Garnet porphyroblasts in a schist matrix. [Chip Clark.]

Table 9.1 Classification of Metamorphic Rocks on Texture

Classification	Characteristics	Rock Name	Typical Parent Rock
Foliated	Distinguished by slaty cleavage, schistosity, or gneissic foliation; mineral grains show preferred orientation	Slate Phyllite Schist Gneiss	Shale, sandstone
Granoblastic (nonfoliated)	Granular, characterized by coarse or fine interlocking grains; little or no preferred orientation	Hornfels	Shale, volcanics
		Quartzite	Quartz-rich sandstone
		Marble	Limestone, dolomite
		Argillite	Shale
		Greenstone	Basalt
		Amphibolite[a]	Shale, basalt
		Granulite[b]	Shale, basalt
Porphyroblastic	Large crystals set in fine matrix	Slate to gneiss	Shale

[a]Typically contains much amphibole, which may show alignment of long, narrow crystals.
[b]High-temperature, high-pressure rock.

chemical and crystallographic properties of the matrix and those of the porphyroblast minerals. This contrast causes the porphyroblast crystals to grow faster than the slow-growing minerals of the matrix, at the expense of the matrix. Porphyroblasts vary in size, ranging from a few millimeters to several centimeters in diameter. Their composition also varies. Garnet and staurolite are two common minerals that form porphyroblasts, but many others are also found. The precise composition and distribution of porphyroblasts of these two minerals can be used to infer the paths of pressure and temperature that occurred during metamorphism.

Table 9.1 summarizes the textural classes of metamorphic rocks and their main characteristics.

Regional Metamorphism and Metamorphic Grade

Metamorphic rocks form under a wide range of conditions, and their minerals and textures are clues to the pressures and temperatures in the crust where and when they formed. Geologists who study the formation of metamorphic rocks constantly seek to determine the intensity and character of metamorphism more precisely than is indicated by a designation of "low grade" or "high grade." To make these finer distinctions, geologists read minerals as though they were pressure gauges and thermometers. The techniques are best illustrated by their application to regional metamorphism.

Mineral Isograds: Mapping Zones of Change

When geologists study broad belts of regionally metamorphosed rocks, they can see many outcrops, some showing one set of minerals, some showing others. Different parts of these belts may be distinguished by their *index minerals,* the characteristic minerals that define metamorphic zones representing a restricted range of pressures and temperatures (**Figure Story 9.7**). For example, one may cross from a region of unmetamorphosed shales to a belt of weakly metamorphosed slates and then to a belt of high-grade schists (Figure Story 9.7a). At the slate belt, a new mineral—chlorite—appears. Moving in the direction of increasing metamorphism, the geologist may successively encounter other metamorphic mineral zones, the schists becoming progressively more foliated (Figure Story 9.7b).

We can make a map of these zones where one metamorphic grade changes to another. To do so, geologists define the zones by drawing lines called *isograds* that connect the places where index minerals first appear. Isograds are used in Figure Story 9.7a to show a series of rocks produced by the regional metamorphism of a shale. A pattern of isograds tends to follow the structural grain of a region as folds and faults reveal it. An isograd based on a single index mineral, such as the biotite isograd, is a good approximate measure of metamorphic pressure and temperature.

To determine pressure and temperature more precisely, geologists examine a group of two or three minerals whose textures indicate that they crystallized together. For example, a sillimanite isograd would be represented by the chemical reaction of muscovite and quartz to produce potassium feldspar (K-feldspar) and sillimanite, liberating water (as water vapor) in the process:

$$\underset{KAl_3Si_3O_{10}(OH)_2}{\text{muscovite}} + \underset{SiO_2}{\text{quartz}} \rightarrow \underset{KAlSi_3O_8}{\text{K-feldspar}} + \underset{Al_2SiO_5}{\text{sillimanite}} + \underset{H_2O}{\text{water}}$$

Many groups of minerals have been carefully studied in the laboratory to determine more exactly the pressures and temperatures at which they formed. The results are used to calibrate the field mapping of isograds.

Isograds reveal the pressures and temperatures at which minerals form, so the isograd sequence in one metamorphic belt may differ from that in another belt. The reason for this difference is that pressure and temperature do not increase at the same rate in all geologic settings. As we discussed earlier in this chapter, pressure increases more rapidly than temperature in some places and more slowly in others (see Figures 9.1 and 9.2).

Metamorphic Grade and Parent-Rock Composition

The kind of metamorphic rock that results from a given grade of metamorphism depends partly on the mineral composition of the parent rock. The metamorphism of slate shown in Figure Story 9.7b and c reveals the effects of metamorphic conditions on rocks rich in clay minerals, quartz, and perhaps some carbonate minerals. The metamorphism of mafic volcanic rocks, composed predominantly of feldspars and pyroxene, follows a different course (**Figure 9.8**a).

In the regional metamorphism of a basalt, for example, the lowest-grade rocks characteristically contain various **zeolite** minerals. The silicate minerals in this class contain water in cavities within the crystal structure. Zeolite minerals form by alteration at very low temperatures and pressures. Rocks that include this group of minerals are thus identified as zeolite grade.

Overlapping with the zeolite grade is a higher grade of metamorphosed mafic volcanic rocks, the **greenschists,** whose abundant minerals include chlorite. Next are the *amphibolites,* which contain large amounts of amphiboles. The highest grade of metamorphosed mafic volcanics comprises the **pyroxene granulites,** coarse-grained rocks containing pyroxene and calcium plagioclase.

INDEX MINERALS, GRADE, AND FACIES DESCRIBE METAMORPHISM

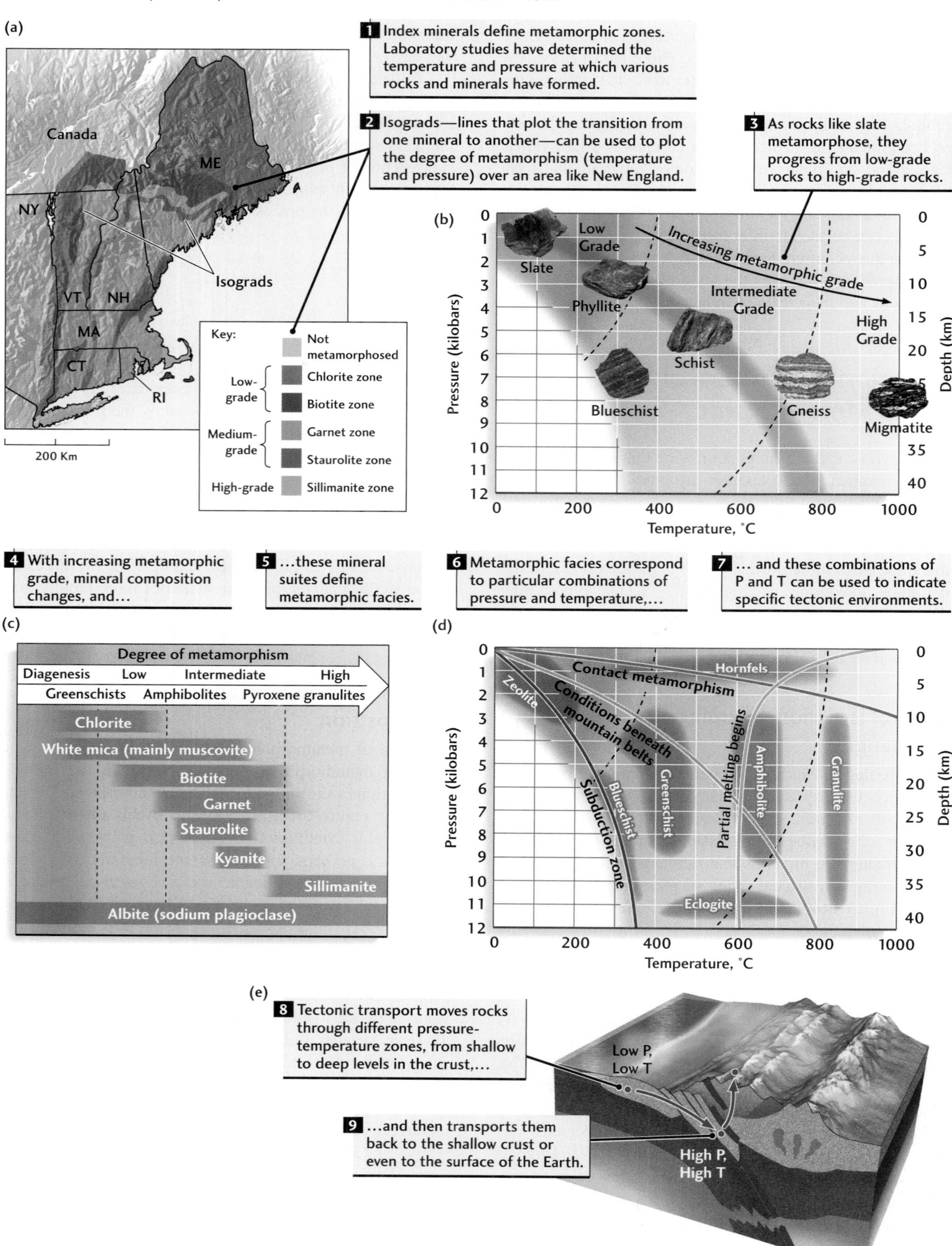

(a)

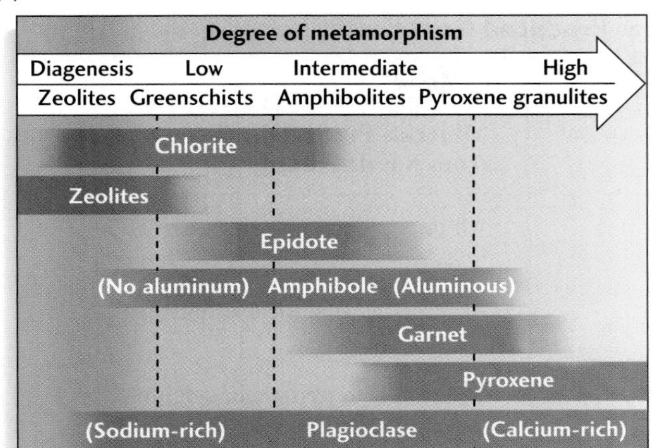

(b)

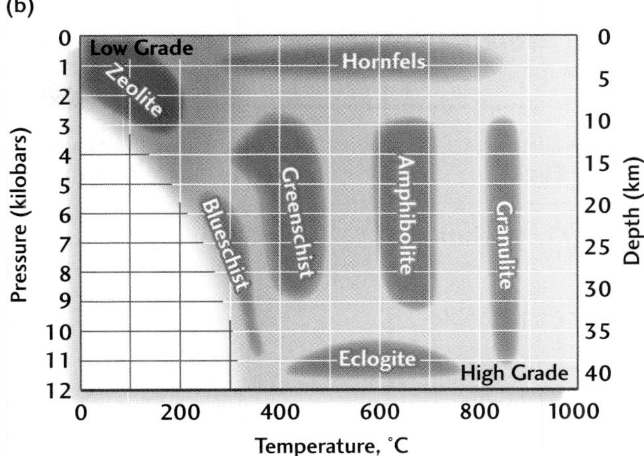

Figure 9.8 (a) Changes in the mineral composition of mafic rocks, metamorphosed under conditions ranging from low grade to high grade. (b) The metamorphic facies of mafic rock types.

Pyroxene granulites are the products of high-grade metamorphism in which the temperature is high and the pressure is moderate. The opposite situation, where the pressure is high and the temperature moderate, produces rocks of **blueschist** grade with various starting compositions, from mafic volcanic rocks to shaley sedimentary rocks. The name comes from the abundance in these rocks of glaucophane, a blue amphibole. Still another metamorphic rock, formed at extremely high pressures and moderate to high temperatures, is *eclogite,* which is rich in garnet and pyroxene.

Metamorphic Facies

We can put all this information about metamorphic grades—derived from parent rocks of many different chemical compositions—on a graph of temperature and pressure (see Figure Story 9.7d and Figure 9.8b). **Metamorphic facies** are groupings of rocks of various mineral compositions formed under different grades of metamorphism from different parent rocks. The terms *high grade* and *low grade* are used to convey a general sense of the degree of metamorphism. By designating particular metamorphic facies, we can be more specific about the degree of metamorphism preserved in rocks. Two essential points characterize the concept of metamorphic facies:

1. Different kinds of metamorphic rocks are formed from parent rocks of different composition at the same grade of metamorphism.

2. Different kinds of metamorphic rocks are formed under different grades of metamorphism from parent rocks of the same composition.

Table 9.2 lists the major minerals of the metamorphic facies produced from shale and basalt. Because parent rocks vary so greatly in composition, there are no sharp boundaries between metamorphic facies (see Figure Story 9.7d and Figure 9.8b).

The analysis of metamorphic facies allows us to interpret the tectonic processes responsible for metamorphism (see Figure Story 9.7e).

Figure Story 9.7 (a) Index minerals are used to determine isograds in this New England terrain. (b) The parent rocks, shales, have been metamorphosed as a result of progressive increases in temperature and pressure. (c) Changes in the mineral compositions of shales metamorphosed under conditions ranging from low grade to high grade. (d) The various types of metamorphic rocks may be grouped in accordance with the pressure and temperature conditions under which they are formed. There are no sharp boundaries between any of these facies. (e) Metamorphism occurs when rocks are tectonically transported to deeper levels in the crust and then back to the surface again. [*Slate:* Andrew J. Martinez/Photo Researchers. *Phyllite:* Kurt Hollocher. *Blueschist:* Courtesy of Mark Cloos. *Schist:* Biophoto Associates/Photo Researchers. *Gneiss:* Breck P. Kent. *Migmatite:* Kip Hodges.]

Table 9.2 Major Minerals of Metamorphic Facies Produced from Parent Rocks of Different Composition

Facies	Minerals Produced from Shale Parent	Minerals Produced from Basalt Parent
Greenschist	Muscovite, chlorite, quartz, sodium-rich plagioclase feldspar	Albite, epidote, chlorite
Amphibolite	Muscovite, biotite, garnet, quartz, plagioclase feldspar	Amphibole, plagioclase feldspar
Granulite	Garnet, sillimanite, plagioclase feldspar, quartz	Calcium-rich pyroxene, calcium-rich plagioclase feldspar
Eclogite	Garnet, sodium-rich pyroxene, quartz	Sodium-rich pyroxene, garnet

Plate Tectonics and Metamorphism

Soon after the theory of plate tectonics was proposed, geologists started to see how patterns of metamorphism fit into the larger framework of plate tectonic movements that cause volcanism and **orogeny.** Orogeny means "mountain making," particularly by the folding and thrusting of rock layers, often with accompanying magmatic activity. Regional metamorphic belts are frequently associated with continental collisions that build mountains. In the cores of the major mountain belts of the world, from the Appalachians to the Alps, we find long belts of regionally metamorphosed and deformed sedimentary and volcanic rocks that parallel the lines of folds and faults in the mountains.

Metamorphic Pressure-Temperature Paths

The concept of metamorphic grade, introduced above, is completely *static*. This means that the grade of metamorphism can inform us of the maximum pressure or temperature to which a rock was subjected, but it says nothing about where the rock encountered these conditions or how it was transported back to Earth's surface. It is important to understand that most metamorphism is a *dynamic* process, not a static event. Metamorphism generally is characterized by changing conditions of pressure and temperature, and the history of these changes is called a **metamorphic P-T path.** The P-T path can be a sensitive recorder of many important factors that influence metamorphism—such as the sources of heat, which change temperatures, and the rates of tectonic transport, which change pressures. Therefore, the analysis of P-T paths in metamorphic rocks may provide considerable insight into the plate tectonic settings responsible for metamorphism (see Figure Story 9.7e).

To obtain a P-T path, geologists must analyze specific metamorphic minerals in the laboratory. One of the most widely used minerals is garnet, which serves as a sort of recording device (**Figure 9.9**). During metamorphism, garnet grows steadily, and as the pressure and temperature of the surrounding environment change, the composition of the garnet changes. The oldest part of the garnet is its core and the youngest is its outer edge, so the variation in composition from core to edge will yield the history of metamorphic conditions. From a measured value of garnet composition in the lab, the corresponding values for pressure and temperature can be obtained and then plotted as a P-T path. P-T paths have two segments. The *prograde* segment indicates increasing pressure and temperature, and the *retrograde* segment indicates decreasing pressure and temperature.

Continent-Ocean Convergence

The rock assemblages that form when a plate carrying a continent on its leading edge converges with a subducting oceanic plate are shown in **Figure 9.10**. Thick sediments eroded from the continent rapidly fill the adjacent depressions in the seafloor around the subduction zone. As it descends, the cold oceanic slab stuffs the region below the inner wall of the trench (the wall closer to land) with these sediments and with deep-sea sediments and ophiolite shreds scraped off the descending plate. Regions of this sort, located between the magmatic arc on the continent and the trench offshore, are enormously complex and variable. The deposits are all highly folded, intricately sliced, and metamorphosed. They are difficult to map in detail but are recognizable by their distinctive mix of materials and structural features. Such a chaotic mixture is called a **mélange** (French for "mixture"). The metamorphism is the kind characteristic of high pressure and low temperature, because the material

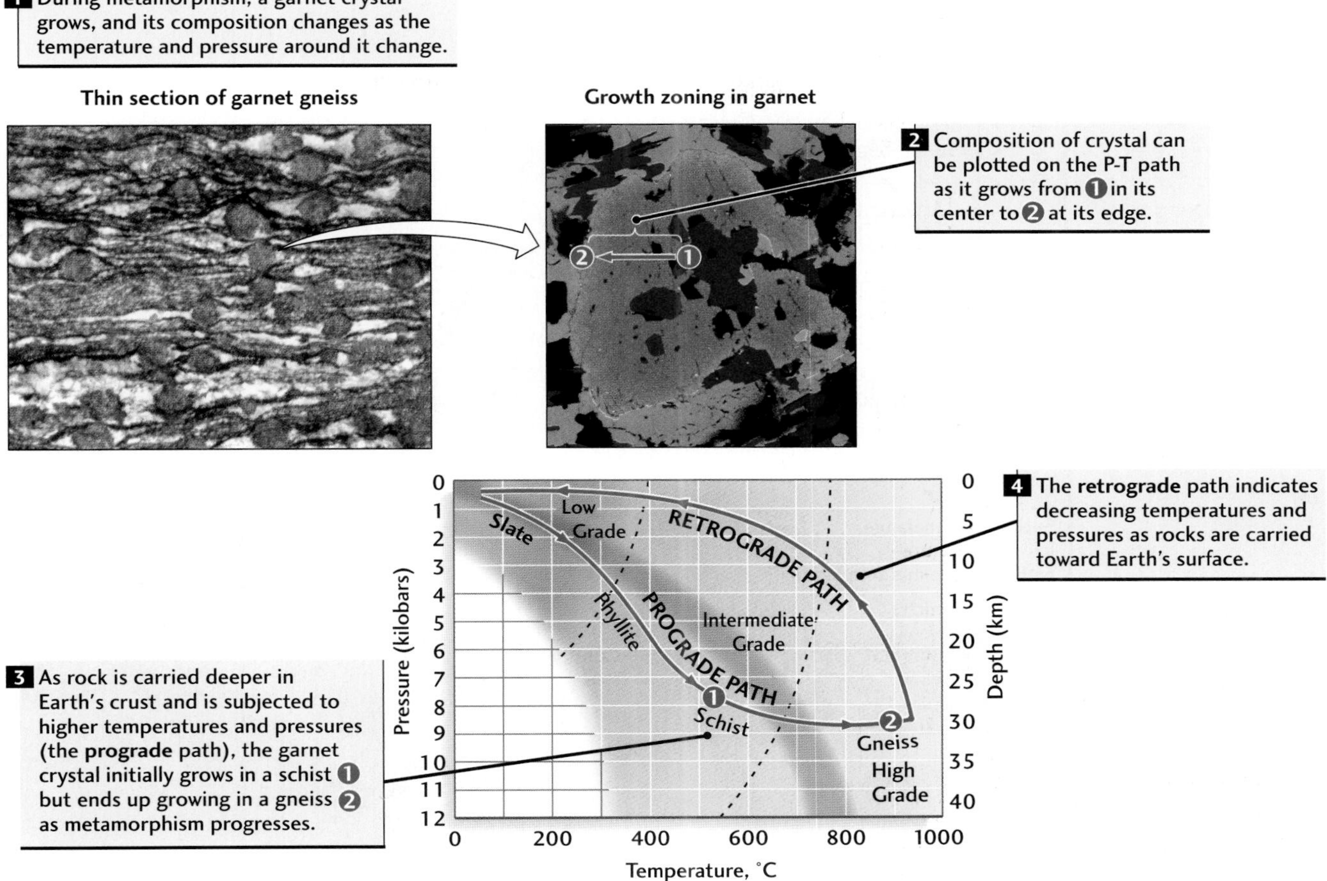

Figure 9.9 Metamorphic pressure-temperature paths. The path that a metamorphic rock typically follows begins with an increase in pressure and temperature, the prograde path, followed by a decrease in pressure and temperature, the retrograde path. [Photos courtesy of Kip Hodges.]

may be carried relatively rapidly to depths as great as 30 km, where recrystallization occurs in the environment of the still-cold subducting slab.

Subduction-Related Metamorphism *Blueschists*—the metamorphosed volcanic and sedimentary rocks whose minerals (Figure 9.10) indicate that they were produced under very high pressures but at relatively low temperatures—form in the forearc region of a subduction zone, the area between the seafloor trench and the volcanic arc. Here sediments are carried down the subduction zone along the surface of a cool subducting lithospheric slab. The subducted plate moves down so quickly that there is little time for it to heat up, whereas the pressure increases rapidly.

Eventually, as part of the subduction process, the material rises back to the surface. This **exhumation** occurs because of two effects: buoyancy and circulation. Imagine trying to push a basketball below the surface of a swimming pool. The air-filled basketball has a lower density than the surrounding water, so it tends to rise back to the surface. In a similar way, the subducted metamorphic rocks are driven upward by their inherent buoyancy relative to the surrounding crust. But what "pushes" the material down to begin with? A natural circulation sets up in the subduction zone. You can think of a subduction zone as an eggbeater. As the eggbeater rotates, it moves the froth in a circular direction. What moves in one direction eventually moves in the opposite direction because of the circular motion. In an analogous way, the sinking slab in a subduction zone sets up a circular motion of material above the slab, first pulling material down to great depths, then returning it to the surface.

Figure 9.10 shows the typical P-T path for rocks subjected to blueschist-grade metamorphism during subduction and exhumation. The P-T path is superimposed on the metamorphic facies diagram. Note that the P-T path forms a loop on this diagram. The prograde part of the path represents subduction, as shown by a rapid increase in pressure for only a relatively small increase in temperature. During

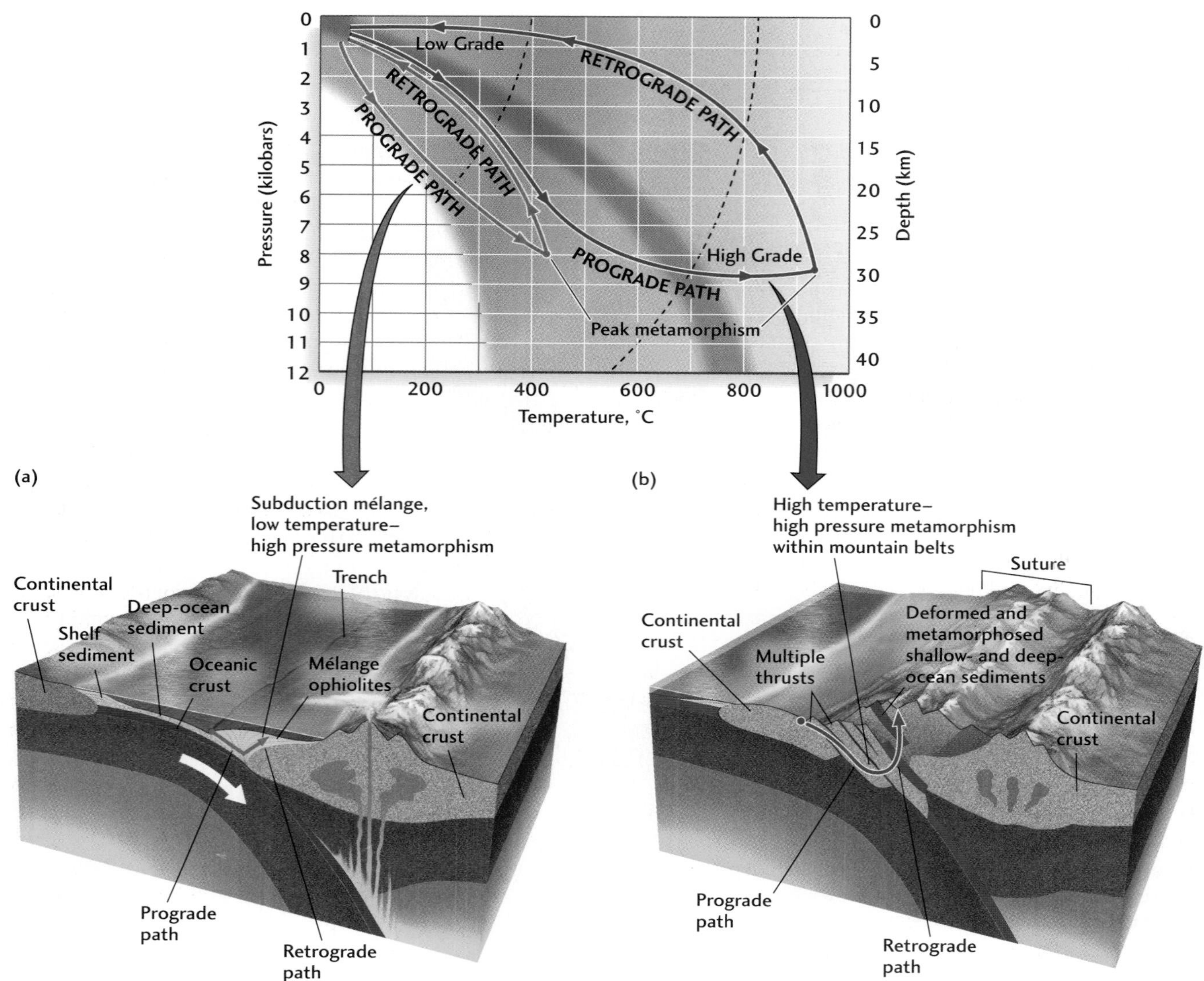

Figure 9.10 P-T paths and rock assemblages associated with (a) ocean-continent plate convergence and (b) continent-continent plate convergence. The P-T paths differ by illustrating the lower geothermal gradients present in subduction zones. Rocks transported to similar depths—and pressures—beneath mountain belts become much hotter at an equivalent depth.

exhumation, the path loops back around because temperature is still slowly increasing, but now pressure is rapidly decreasing. The retrograde part of the P-T path represents the exhumation process, described above.

Evidence of Ancient Ocean-Continent Convergence The essential elements of these collisional rock assemblages have been found at many places in the geologic record. One can see mélange in the Franciscan formation of the California Coast Ranges and in the parallel belt of arc magmatism in the Sierra Nevada to the east. These rocks mark the Mesozoic collision between the North American Plate and the Farallon Plate, which has disappeared by subduction (see Figure 20.6). The location of mélange on the west and magmatism on the east shows that the now-absent Farallon Plate was the subducted one, overridden by the North American Plate on the east. Analysis of the P-T paths for metamorphic minerals in the blueschist-grade Franciscan mélange reveals a loop similar to that shown in Figure 9.10, indicating rapid descent to high pressures, which is a diagnostic attribute of subduction.

Other examples of mélange-arc pairs can be found along the continental margins framing the Pacific Basin—in Japan, for instance. The central Alps were uplifted by the convergence of a Mediterranean plate with the European continent. The Andes Mountains (from which the name of the volcanic rock andesite is derived), near the west coast of South America, are products of a collision between ocean and continental plates. Here the Nazca Plate collides with and is subducted under the South American Plate.

Continent-Continent Collision

Plates may have continents embedded in them, and a continent can collide with another continent, as shown in Figure 9.10b. Because continental crust is buoyant, both continents may resist subduction and stay afloat. As a result, they collide, and a wide zone of intense deformation develops at the boundary where the continents grind together. The remnant of such a boundary left behind in the geologic record is called a *suture.* The intense deformation that occurs during orogeny results in a much-thickened continental crust in the collision zone, often producing high mountains such as the Himalayas. Belts of magmatism characteristically form at depth within the core of the mountain range adjacent to the suture. Ophiolites are often found near the suture; they are relics of an ancient ocean that disappeared in the convergence of two plates (see Chapter 5).

As continents collide and the lithosphere thickens, the deep parts of the continental crust heat up and metamorphose to different grades. In deeper zones, melting may begin at the same time. In this way, a complex mixture of metamorphic and igneous rocks forms the cores of orogenic belts that evolve during mountain building. Millions of years afterward, when erosion has stripped off the surface layers, the cores are exposed at the surface, providing the geologist with a rock record of the metamorphic processes that formed the schists, gneisses, and other metamorphic rocks.

P-T paths for metamorphic rocks produced by continental collision have a different shape from those produced by subduction alone. Continental collision generates higher temperatures than subduction. Therefore, as a rock is pushed to greater depths during collision, the temperature that corresponds to a given pressure will be higher (see Figure 9.10b). The P-T path begins at the same place as the path for subduction but shows a more rapid increase in temperature as greater pressures and depths are reached. Geologists generally interpret the prograde segment of a collisional P-T path as indicating burial of rocks beneath high mountains during orogeny. The retrograde segment then represents uplift and exhumation of the buried rocks during the collapse of mountains, either by erosion or by postcollision stretching and thinning of the continental crust.

The prime example of a collision of continents is the Himalayas, which began to form some 50 million years ago when the Indian continent collided with the Asian continent. The collision continues today: India is moving into Asia at a rate of a few centimeters per year, and the uplift is still going on, together with faulting and very rapid erosion rates caused by rivers and glaciers.

Exhumation: Links Between the Plate Tectonic and Climate Geosystems

Forty years ago, plate tectonics theory provided a ready explanation for how metamorphic rocks could be produced by seafloor spreading, plate subduction, and continental collision. By the mid-1980s, the study of P-T paths provided a more highly resolved picture of the specific tectonic mechanisms involved in the deep burial of rocks. At the same time, however, it surprised geologists by providing an equally well resolved image of the subsequent, and often very rapid, uplift and exhumation of these deeply buried rocks. Since the time of this discovery, geologists have been searching for exclusively tectonic mechanisms that could bring these rocks back to Earth's surface so quickly. One popular idea is that mountains, having been built to great elevations during collision-induced crustal thickening, suddenly fail by gravitational collapse. The old saying "what goes up must go down" applies here, but with surprisingly fast results. So fast, in fact, that some geologists don't believe this is the only important effect—other forces must also be at work.

As we will learn in Chapter 18, geologists who study landscapes have discovered that extremely high erosion rates can be produced by glaciers and rivers in tectonically active mountainous regions. Over the past decade, these geologists have presented a new hypothesis that links rapid rates of uplift and exhumation to rapid erosion rates. The idea here is that *climate,* not just tectonics alone, drives the flow of rocks from the deep crust to the shallow crust through the process of rapid erosion. Thus, tectonics—which acts through orogeny and mountain building—and climate—which acts through weathering and erosion—*interact* to control the flow of metamorphic rocks to Earth's surface. After decades of emphasis on solely tectonic explanations for regional and global Earth processes, it now seems that two apparently unrelated disciplines of geology—metamorphism and surface processes—may be linked in a very elegant way. As one geologist exclaimed: "Savory the irony should the metamorphic muscles that push mountains to the sky be driven by the pitter patter of tiny raindrops."

SUMMARY

What factors cause metamorphism? Metamorphism—alteration in the solid state of preexisting rocks—is caused by increases in pressure and temperature and by reactions with chemical components introduced by migrating fluids. As pressures and temperatures deep within the crust increase as a result of tectonic or igneous activity, the chemical components of the parent rock rearrange themselves into a new set of minerals that are stable under the new conditions. Rocks metamorphosed at relatively low pressures and temperatures are referred to as low-grade rocks. Those metamorphosed at high temperatures and pressures are called high-grade rocks.

Chemical components of a rock may be added or removed during metamorphism, usually by the influence of fluids migrating from nearby intrusions.

What are the various types of metamorphism? The three major types of metamorphism are (1) regional metamorphism, during which large areas are metamorphosed by high pressures and temperatures generated during orogenies; (2) contact metamorphism, during which rocks surrounding magmas are metamorphosed primarily by the heat of the igneous body; and (3) seafloor metamorphism, during which hot fluids percolate through and metamorphose various crustal rocks. Three additional types are (1) low-grade, or burial, metamorphism, during which deeply buried sedimentary rocks are altered by the more or less normal increases in pressure and temperature with depth in the crust; (2) high-pressure and ultra-high-pressure metamorphism, in which rocks may be subjected to pressures as great as 40 kbar, equivalent to depths greater than 120 km; and (3) shock metamorphism, which results from the impact of meteorites. Country rock is shattered by propagating shock waves and, in the process, quartz can be transformed to its denser, high-pressure forms, coesite and stishovite.

What are the chief kinds of metamorphic rocks? Metamorphic rocks fall into two major textural classes: the foliated (displaying fracture cleavage, schistosity, or other forms of preferred orientation of minerals) and the granoblastic, or nonfoliated. The kinds of rocks produced by metamorphism depend on the composition of the parent rock and the grade of metamorphism. The regional metamorphism of a shale leads to zones of foliated rocks of progressively higher grade, from slate to phyllite, schist, gneiss, and migmatite. These zones are marked by isograds defined by the first appearance of an index mineral. Regional metamorphism of mafic volcanic rocks progresses from zeolite grade to greenschist and then to amphibolite and pyroxene granulite. Among granoblastic rocks, marble is derived from the metamorphism of limestone, quartzite from quartz-rich sandstone, and greenstone from basalt. Hornfels is the product of contact metamorphism of fine-grained sedimentary rocks and other types of rock containing an abundance of silicate minerals. According to the concept of metamorphic facies, rocks of the same grade may differ because of variations in the chemical composition of the parent rocks, whereas rocks of the same composition may vary because of different grades of metamorphism.

How do metamorphic rocks relate to plate tectonic processes? During both subduction and continental collision, preexisting rocks and sediments are pushed to great depths in the Earth, where they are subjected to increasing pressure and temperature that result in metamorphic mineral reactions. The shape of metamorphic P-T paths provides insight into the manner in which these rocks are metamorphosed. In convergent margin settings, P-T paths indicate rapid subduction of rocks and sediments to sites with high pressures and relatively low temperatures. In settings where subduction leads to continental collision, rocks are pushed down to depths where pressure and temperature are both high. In both settings, the P-T paths form loops. The loops show that after the rocks experienced the maximum pressures and temperatures, they were pushed back up to shallow depths. This process of exhumation may be driven by the collapse of mountain belts either through enhanced weathering and erosion at Earth's surface or through tectonic stretching and thinning of the continental crust.

Key Terms and Concepts

amphibolite (p. 202)
blueschist (p. 203)
contact metamorphism (p. 198)
eclogite (p. 198)
exhumation (p. 208)
foliated rock (p. 199)
foliation (p. 199)
gneiss (p. 201)
granoblastic rock (p. 201)
granulite (p. 202)
greenschist (p. 203)
greenstone (p. 201)
high- (and ultra-high) pressure metamorphism (p. 198)
hornfels (p. 201)
low-grade (burial) metamorphism (p. 198)
marble (p. 201)
mélange (p. 207)
metamorphic facies (p. 203, 205)
metamorphic P-T path (p. 206)
metamorphic rock (p. 193)
metasomatism (p. 196)
migmatite (p. 201)
orogeny (p. 206)
phyllite (p. 199)
porphyroblast (p. 202)
pyroxene granulite (p. 203)
quartzite (p. 201)
regional metamorphism (p. 197)
schist (p. 201)
seafloor metamorphism (p. 198)
shock metamorphism (p. 198)
slate (p. 199)
stress (p. 196)
zeolite (p. 203)

Exercises

MEDIA LINK *This icon indicates that there is an animation available on the Web site that may assist you in answering a question.*

 **1.** What types of metamorphism are related to igneous intrusions?

2. What does preferred orientation refer to in a metamorphic rock?

3. Name a mineral that is commonly found in a schist and shows preferred orientation.

4. Name two granoblastic metamorphic rocks.

5. What is a porphyroblast?

6. Contrast a schist and a gneiss.

7. What is an isograd?

8. What is the difference between a granite and a slate?

9. How are metamorphic facies related to temperatures and pressures?

10. In which plate tectonic settings would you expect to find regional metamorphism?

Thought Questions

1. What kinds of preferred orientation of minerals would you expect to find in an amphibolite?

2. Why are there no metamorphic rocks formed under natural conditions of very low pressure and temperature, as shown in Figure 9.2?

3. How is slaty cleavage related to deformation?

4. Are cataclastic rocks more likely to be found in a continental rift valley or in a volcanic arc?

5. Would you choose to rely on chemical composition or type of foliation to determine metamorphic grade? Why?

6. You have mapped an area of metamorphic rocks and have observed north-south isograd lines running from kyanite in the east to chlorite in the west. Were metamorphic temperatures higher in the east or in the west?

7. Contrast the minerals found in the contact metamorphism of a pure limestone and of a limestone containing appreciable layers of shale.

8. Which kind of pluton would produce the highest grade metamorphism, a granite intrusion 20 km deep or a gabbro intrusion at a depth of 5 km?

9. Why would you not expect to find rocks produced by burial metamorphism at a mid-ocean ridge?

10. Subduction zones are generally characterized by high pressure—low temperature metamorphism. In contrast, continental collision zones are marked by moderate pressure—high temperature metamorphism. Which region has a higher geothermal gradient? Explain.

Suggested Readings

Blatt, H., and R. Tracy. 1996. *Petrology: Igneous, Sedimentary, and Metamorphic,* 2d ed. New York: W. H. Freeman.

Hyndman, D. W. 1985. *Petrology of Igneous and Metamorphic Rocks.* New York: Wiley.

Liou, J. G., S. Maruyama, and W. G. Ernst. 1997. Seeing a mountain in a grain of garnet. *Science* 276:48–49.

Raymond, L. A. 1995. *Petrology.* Dubuque, Iowa: Wm. C. Brown.

Spear, Frank S. 1993. Metamorphic phase equilibria and pressure-temperature-time paths. Mineralogical Society of America Monograph 22.

Wynn, Jeffrey C., and Eugene Shoemaker. 1998. The day the sands caught fire. *Scientific American* 279:64–21.

Trilobites preserved as fossils in rocks about 365 million years old. Ontario, Canada.
[William E. Ferguson.]

CHAPTER

10

The Rock Record and the Geologic Time Scale

"If you can look into the seeds of time and say which grain will grow and which will not, speak then to me."

William Shakespeare

Although Earth appears to be solid and stable on our human time scale, there is hardly a place on Earth that is not moving both vertically and horizontally, however slowly. The geological processes that shape Earth's surface and give structure to its interior take place over millions of years. On that time scale, continents, oceans, and mountains have moved great distances. A key job of the geologist is to learn the patterns and rates of these movements.

In this chapter, we examine some of the ways in which geologists deal with these extraordinarily long intervals, both to understand the processes and to reconstruct the geologic history of the planet.

One of the most important reasons for placing geological processes in their proper sequence is to learn about the evolution of the planet that we see today. When were the Rocky Mountains formed? How often has Earth experienced ice ages, with their continent-sized glaciers? What was happening in East Africa when early humans were evolving? To answer these kinds of questions, we need a tool for organizing and dating the rock record. We need a geologic calendar to determine both the sequence in which rock layers formed and their ages, and we need a way to compare the ages of rocks situated continents apart. Two centuries of modern geological research have resulted in such a tool: the geologic time scale. The geologic time scale enables scientists to determine the age of the Earth, to unravel complex geologic histories, and even to study the origin and evolution of life.

Timing the Earth

Geologists differ from most other scientists in their attitude toward time. Physicists and chemists study processes that last only fractions of a second: the splitting of an

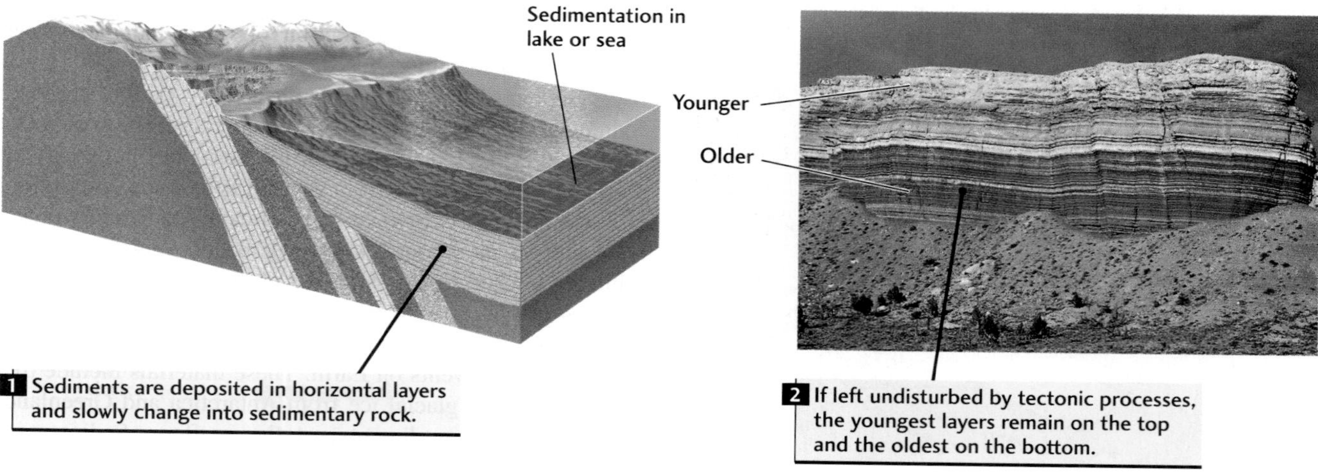

Figure 10.3 Layers of sedimentary rock in Marble Canyon, part of the Grand Canyon. The Grand Canyon was cut by the Colorado River through what is now northern Arizona. These layers record millions of years of geologic history. [Fletcher & Baylis/Photo Researchers.]

stratigraphic succession is the chronology of the constituent beds and the sedimentary conditions that they imply.

With a geologic timekeeper, or "stratigraphic clock," geologists can tell whether one rock layer is older than another, although they cannot necessarily tell how much older in years. One might expect that a stratigraphic succession would provide a measure of time in actual years if sediments had accumulated continuously at a steady rate, had compacted a constant amount as they lithified, and had not eroded. If we knew that muddy sediments accumulated at a rate of 10 m per million years, for instance, then 100 m of mudstone would represent 10 million years of deposition.

In practice, however, we cannot accurately gauge time in years from stratigraphy, for several reasons. First, as we saw in Chapter 8, sediments do not accumulate at a constant rate in any sedimentary environment. During a flood, a river may deposit a layer of sand several meters thick in its channel over just a few days, whereas in the years between floods, it will deposit a sand layer only a few centimeters thick. Even on the deep ocean floor, where it may take 1000 years for a layer of mud 1 mm thick to accumulate, sedimentation is unsteady and the thickness of sediment cannot be used for precise timekeeping. In addition, the rate at which sediment is deposited varies widely in different sedimentary environments.

The second reason that stratigraphy is an imprecise timekeeper is that the rock record does not tell us how many years have passed between periods of deposition. Many places on the floor of a river valley receive sediment only during times of flood. The times between floods are not represented by any sediment. Over the course of Earth's history, in various places there have been long intervals, some lasting *millions* of years, in which no sediments were deposited at all. In other places and at other times, sedimentary rocks may have been removed by erosion. Although we often can tell where a gap in the record exists, we rarely can say exactly how long an interval it represents.

The final reason—and the one most important to geologists who wish to compare the geologic histories of different parts of the Earth—is that stratigraphy alone cannot be used to determine the relative ages of two widely separated beds. A geologist might be able to establish the relative age of one bed or a series of beds by following an outcrop a limited distance, but there is no way of knowing whether a rock layer in Arizona, say, is older or younger than one in northern Canada.

Those early geologists found that fossils were the key to detecting missing time intervals and correlating the relative ages of rocks at different geographic locations. Fossils became the single most important tool for constructing an accurate geologic time scale for the entire planet.

Fossils as Timepieces

To a great many students today, it must be obvious that fossils are the remains of ancient organisms. Some look very much like animals now living, although others—such as the trilobites in the photograph at the beginning of this chapter—are the remains of extinct life-forms. Fossils can be shells, teeth, bones, impressions of plants, or tracks of animals (**Figure 10.4**). The most common fossils in rocks of the past half-billion years are the shells of invertebrate animals, such as clams, oysters, and the ammonite group shown in Figure 10.4a. Much less common are the bones of vertebrates, such as mammals, reptiles, and dinosaurs. Yet fossil dinosaur bones do exist, and they have told us much

(a)

(b)

Figure 10.4 Animal and plant fossils. (a) Ammonite fossils, ancient examples of a large group of invertebrate organisms that are now largely extinct. Their sole representative in the modern world is the chambered nautilus. [Chip Clark.] (b) Petrified Forest, Arizona. These ancient logs are millions of years old. Their substance was completely replaced by silica, which preserved all the original details of form. [Tom Bean.]

about the nature of these long-extinct beasts. Plant fossils are abundant in some rocks, particularly those associated with coal beds, where leaves, twigs, branches, and even whole tree trunks can be recognized (Figure 10.4b). Fossils are not found in intrusive igneous rocks, because the original biological material is lost in a hot melt. Fossils are rarely found in metamorphic rocks, because any remains of organisms are usually too changed and distorted to be recognized.

The ancient Greeks were probably the first to surmise that fossils are records of ancient life, but it was not until modern times that the concept took hold and its consequences were explored. One of the first modern thinkers to make the connection between fossils and once-living organisms was Leonardo da Vinci, in the fifteenth century. In the seventeenth century, Nicolaus Steno compared what he called "tongue-stones" found in the Mediterranean region with similarly shaped teeth of modern sharks and concluded that the stones were the remains of ancient life.

By the end of the eighteenth century, after hundreds of fossils and their relationships to modern organisms had been described and catalogued, the evidence that fossils are the remains of formerly living creatures was overwhelming. Thus **paleontology,** the study of the history of ancient life from the fossil record, took its place beside geology, the study of Earth's history from the rock record.

The dividends from the study of fossils did not accrue to geology alone. Young Charles Darwin's famous voyage as naturalist on the *Beagle* (1831–1836) vastly extended his knowledge of the great variety of fossil organisms and of what their presence in rocks portended. He also had an opportunity to see a host of unfamiliar animal and plant species in their native habitats. In 1859, Darwin proposed the theory of evolution. It revolutionized scientific thinking about the origins of the millions of species of animal and plant life and provided a sound theoretical framework for paleontology.

Well before Darwin, in 1793, William Smith, a surveyor working in southern England, had recognized that fossils could be used to date the relative ages of sedimentary rocks. Smith was fascinated by the variety of fossils, and he collected them from the sequences of rock strata that he saw exposed along canals and outcrops. He observed that different layers contained different kinds of fossils, and he was able to tell one layer from another by the characteristic fossils in each. He established a general order for the sequence of fossils and strata, from lowermost (oldest) to uppermost (youngest) rock layers. Regardless of the location, Smith could predict the stratigraphic position of any particular layer or set of layers in any outcrop in southern England, basing his prediction on its fossil assemblages. This stratigraphic ordering of fossils is known as a *faunal succession.*

Smith was the first person to use faunal succession to correlate rocks from different outcrops. In each outcrop, he identified distinct formations. A **formation** is a series of rock layers in a region that has about the same physical properties and may contain the same assemblage of fossils. Some formations consist of a single rock type, such as limestone. Others are made up of thin, interlayered beds of different kinds of rocks, such as sandstone and shale. However they vary, each formation comprises a distinctive set of rock layers that can be recognized and mapped as a unit.

Using his knowledge of faunal successions, Smith matched up the similarly aged formations found in different outcrops. By noting the vertical order in which the formations were found in each place, he compiled a composite stratigraphic succession for the entire region. His composite series showed how the complete succession would have looked if the formations at different levels in all the various outcrops could have been brought together in a single spot. **Figure 10.5** shows such a composite for a series of three formations.

Relying on this approach of combining faunal successions with stratigraphic successions, geologists of the past two centuries have carefully correlated formations throughout the world. The result, as we will see, is a geologic time scale for the entire Earth.

Unconformities: Markers of Missing Time

In putting together sequences of formations, geologists often find places in which a formation is missing. Either it was never deposited, or it was eroded away before the next strata were laid down. The boundary along which the two existing formations meet is called an **unconformity**—a surface between two layers that were not laid down in an unbroken sequence (**Figure 10.6**). An unconformity represents time, just as sedimentary rocks do.

An unconformity not only represents time but also implies tectonic forces that raised the land above sea level, where it became eroded. Alternatively, unconformities may

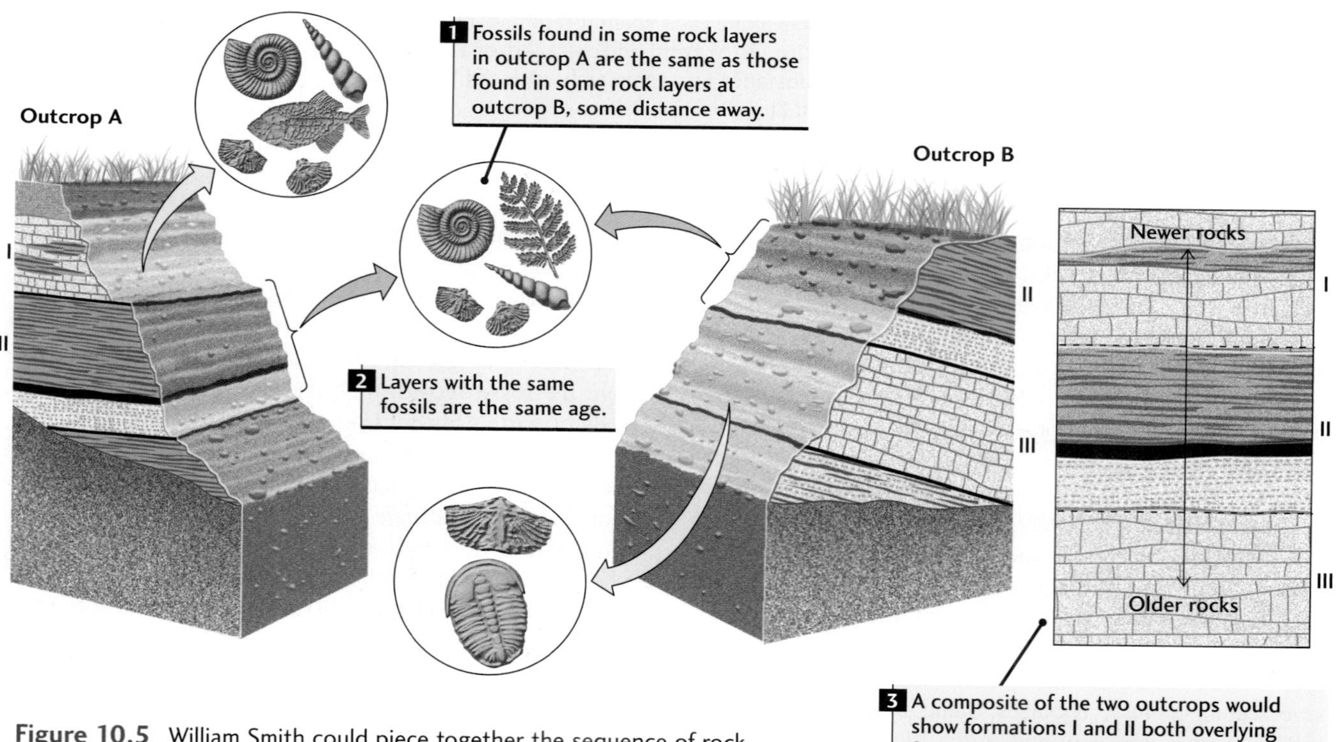

Figure 10.5 William Smith could piece together the sequence of rock layers of different ages containing different fossils by correlating outcrops found in southern England. In this example, formations I and II were exposed at outcrop A and formations II and III at outcrop B.

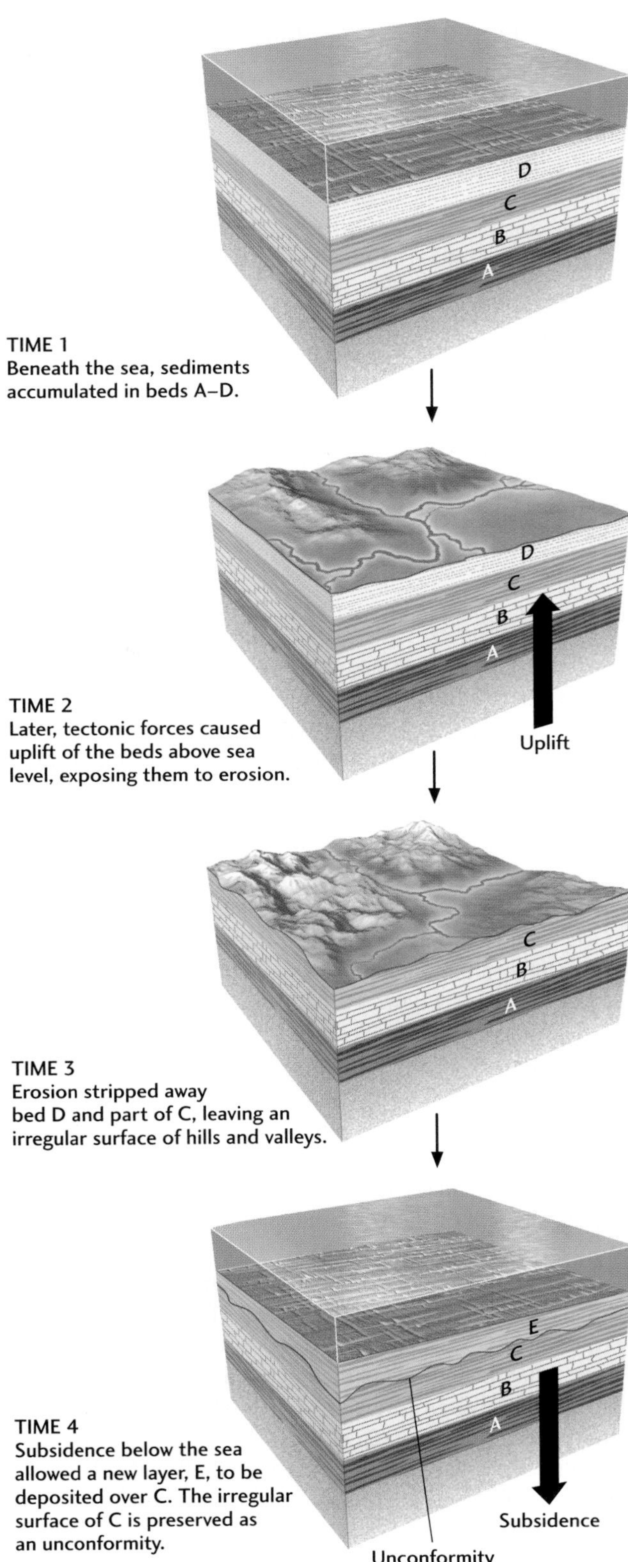

Figure 10.6 An unconformity is a surface between two layers that were not laid down in an unbroken sequence. In the series of events outlined here, an unconformity is created through uplift and erosion, followed by subsidence and another round of sedimentation.

represent times in which the land was eroded as sea level fell globally. Sea level could be lowered by, for example, the withdrawal of water from the oceans to form glacial ice caps at the poles.

Geologists classify different kinds of unconformities by the relationships between the upper and lower sets of layers. An unconformity in which the upper set of layers overlies an erosional surface developed on an undeformed, still-horizontal lower set of beds is a **disconformity.** An unconformity in which the upper beds overlie metamorphic or igneous rock is a **nonconformity.** An unconformity in which the upper layers overlie lower beds that have been folded by tectonic processes and then eroded to a more or less even plane is an **angular unconformity.** In an angular unconformity, the two sets of layers have bedding planes that are not parallel. **Figure 10.7** depicts a dramatic angular unconformity found in the Grand Canyon. **Figure 10.8** illustrates the process by which angular unconformities may form.

Figure 10.7 The Great Unconformity in the Grand Canyon, Colorado, is an angular unconformity between the horizontal Tapeats sandstones above and the steeply folded Precambrian Wapatai shales below. [GeoScience Features Picture Library.]

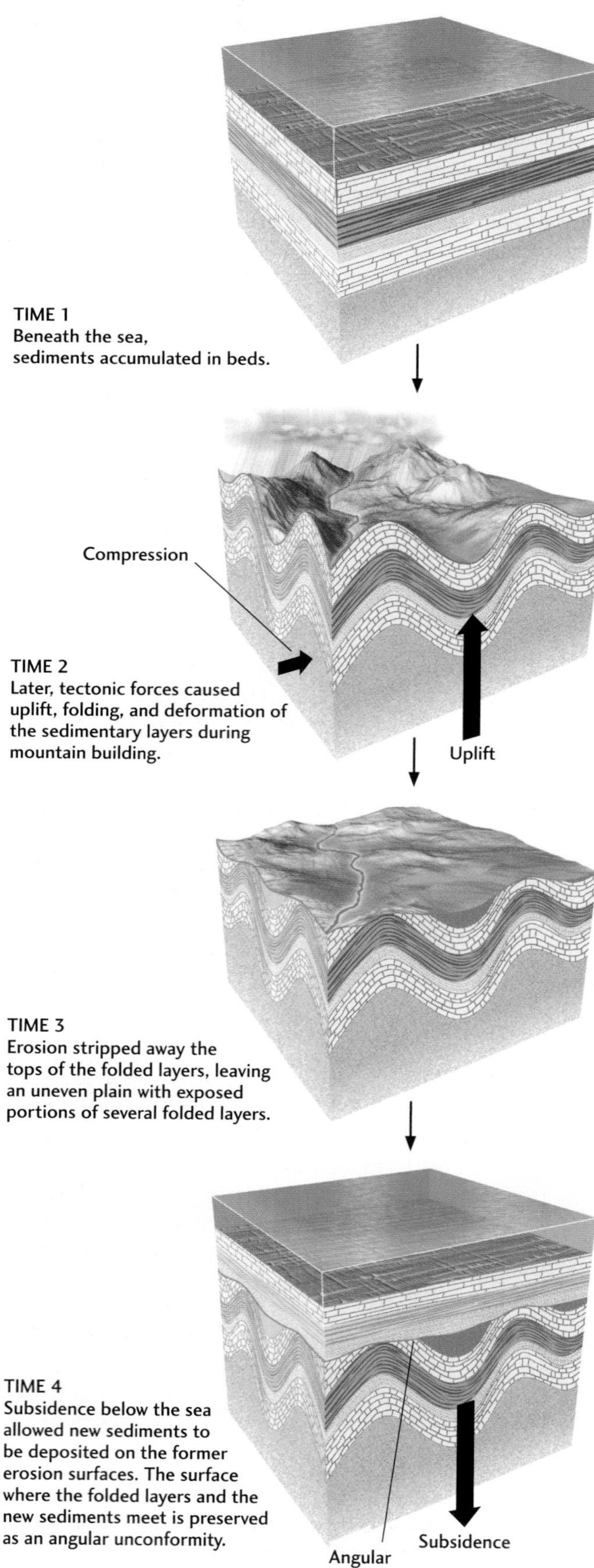

Cross-Cutting Relationships

Other disturbances of the layering of sedimentary rocks also provide clues for dating. Recall from Chapter 5 that discordant dikes or other magmatic intrusions can cut through and interrupt sedimentary layers. Faults displace bedding planes as they break apart blocks of rock (as we will see in Chapter 11). Faults can also displace dikes and sills. Intrusions and faults can be fitted into stratigraphic successions, so they help us place geological events within a relative time frame. We know that deformational or intrusive events must have taken place after the affected sedimentary layers were deposited, and so such deformations or intrusions must be younger than the rocks that they cut (**Figure 10.9**). If the intrusions or fault displacements are eroded and planed off at an unconformity and then overlain by a younger series of formations, we know that the intrusions or faults are older than the younger series of beds.

Sequence Stratigraphy

In the past three decades, a new form of stratigraphy—**sequence stratigraphy**—has been developed. This kind of stratigraphy was originally known as *seismic stratigraphy* because it took advantage of great improvements in exploration seismology (see Chapter 21) that allowed geologists to see individual beds and packages of beds on a seismic profile, or cross section (**Figure 10.10**). Subtle geometric features not discernible in outcrop exposures often are strikingly revealed in seismic profiles (Figure 10.10a). The basic unit used for sequence stratigraphy is the *sequence,* a series of sedimentary beds bounded above and below by unconformities (Figure 10.10b). An example is the alluvial sedimentary sequence discussed in Chapter 8, representing just one cycle of stream deposits. The sequences used for sequence stratigraphy are generally larger packets of beds that might extend over many such cycles. The unconformities that define sequences represent fluctuations in sea level that allowed land erosion to take place. The beds that make up the sequences show internal patterns that are diagnostic of changes in sedimentation.

For example, in a large river delta, sediment is laid down as the river enters the sea. This sediment slowly builds up the seafloor to sea level, thereby creating new land. Alluvial sediments then accumulate on this new surface, and—given millions of years—the delta may advance many miles into the sea (Figure 10.10c). If the sea level were to rise as a result of either global climatic changes or tectonic subsidence, the shoreline would be shifted many

Figure 10.8 An angular unconformity is a surface of erosion that separates two sets of layers having bedding planes that are not parallel. This sequence shows how such a surface can form.

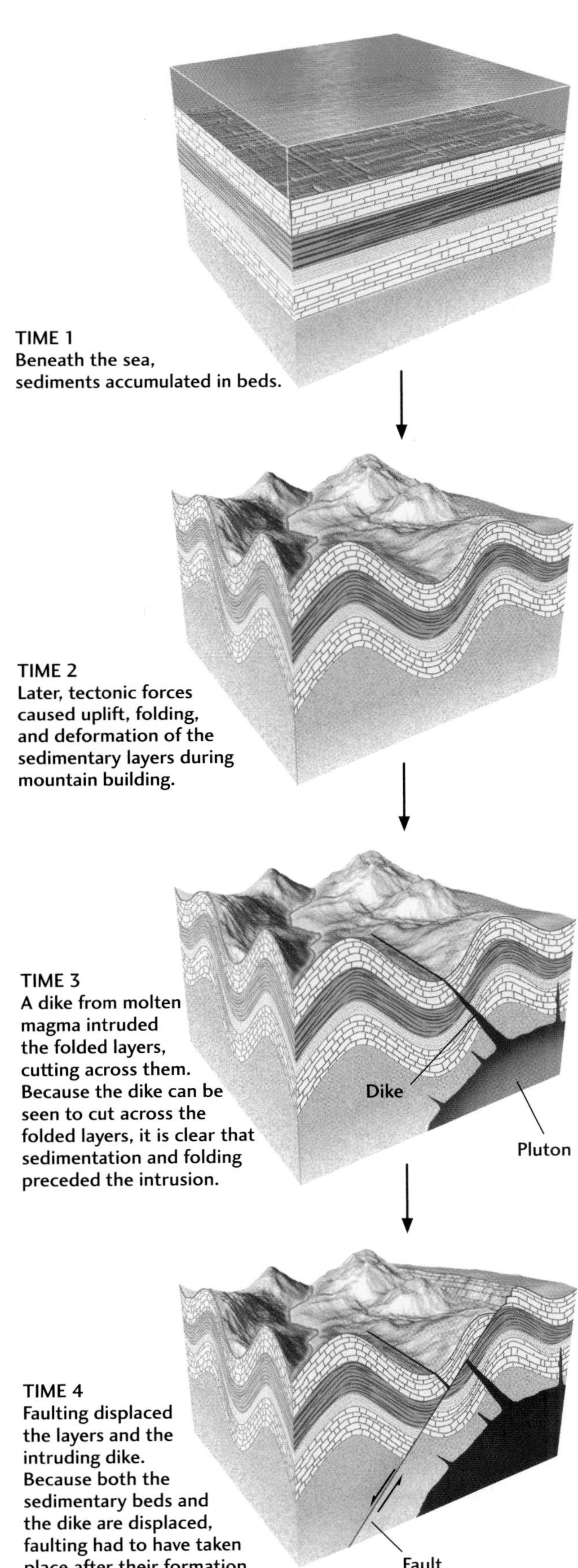

miles inland. A new delta would start to be constructed overlying the former deltaic sequence, as illustrated in Figure 10.10d.

Using their knowledge of bedding patterns, sequence stratigraphers can match sequences of the same geologic age over wide areas. From such information, we can reconstruct a region's geologic history, including changes in sea level, in relation to successions of sequences.

The Geologic Time Scale

We have now seen several ways of ordering rock strata and correlating them with a time sequence of geologic events:

- We can determine the relative ages of sedimentary rocks both by the simple rule of superposition and by the local and global fossil record.
- We can use deformation and angular unconformities to date tectonic episodes in relation to the stratigraphic sequence.
- We can use cross-cutting relationships to establish the relative ages of igneous bodies or faults cutting through sedimentary rocks.

Combining all three methods, we can decipher the history of geologically complicated regions (**Figure Story 10.11**).

In the nineteenth and twentieth centuries, geologists used these relative dating principles and pieced together information from outcrops all over the world to work out the entire **geologic time scale,** a relative-age calendar of Earth's geologic history. Each time interval on this scale is correlated with a corresponding set of rocks and fossils. Although the geologic time scale is still being refined, its main divisions have remained constant for the past century.

As **Figure 10.12** illustrates, the geologic time scale is divided into four major time units, in order of decreasing length: eons, eras, periods, and epochs. An **eon** is the largest division of history.

Archean Eon The oldest eon is the Archean (from the Greek *archaios,* "ancient"). Archean rocks range from the earliest rocks known, about 4 billion years old, to rocks 2.5 billion years old. During the Archean eon, the geodynamo, plate tectonics, and the climate system were established. The cores of most continents formed at this early time in Earth's history, when the plate tectonic system operated somewhat differently from the way that it did later on in Proterozoic and more recent times. Fossils of primitive unicellular microorganisms are found in some sedimentary rocks of this age.

Figure 10.9 Cross-cutting relationships allow us to place geological events within the relative time frames given by the stratigraphic succession.

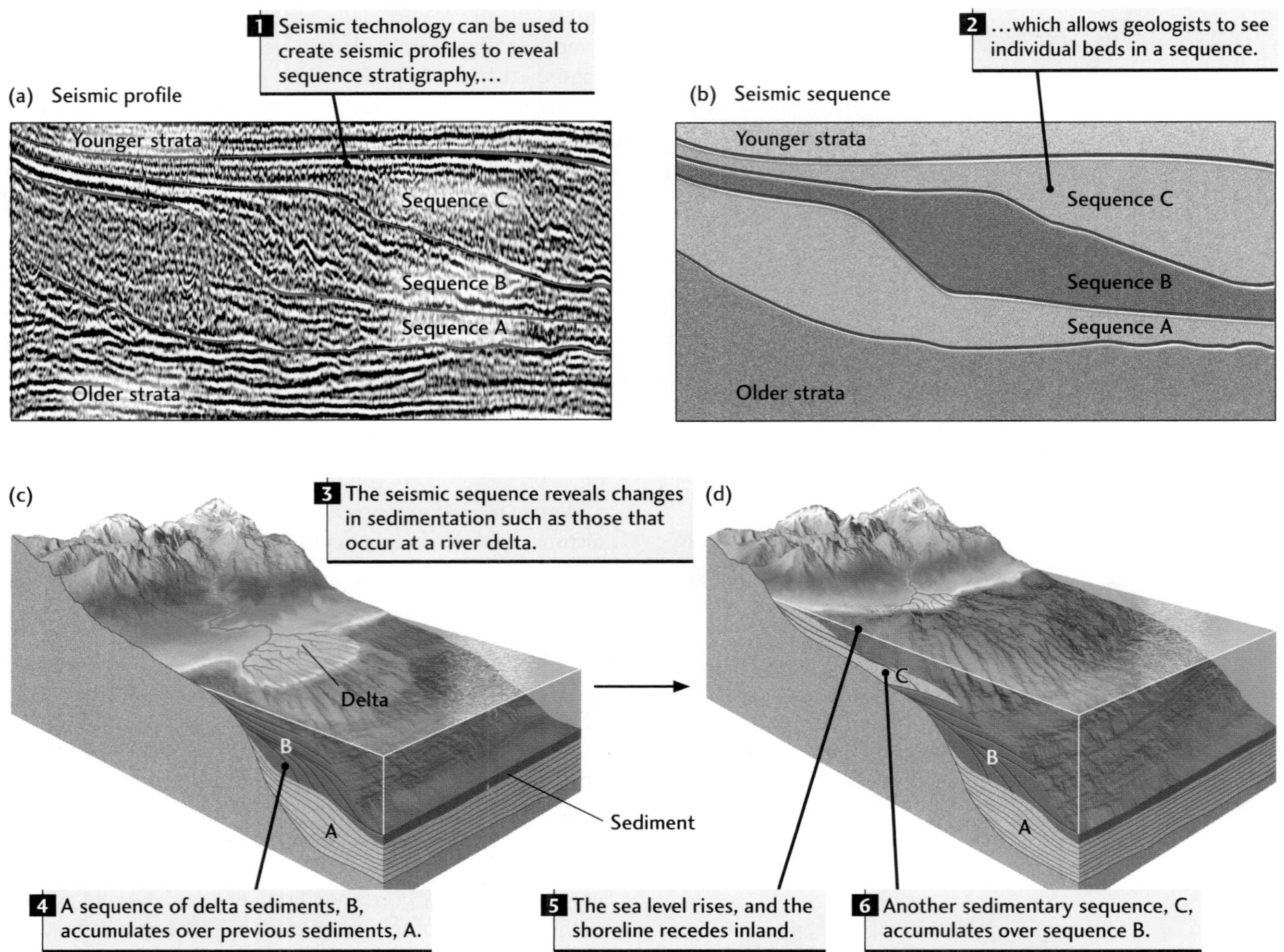

Figure 10.10 Comparison of seismic profiles (a) with seismic sequences (b), reveals the depositional process that creates bedding patterns. When tectonic subsidence or events such as global climate change have caused the sea level to rise, two deltaic sequences are found (c and d).

Proterozoic Eon The next set of rocks formed during the Proterozoic (from the Greek *proteros,* "earlier," and *zoi,* "life") eon (from 2.5 billion to 543 million years ago). During Proterozoic times, interactions between the plate tectonic and climate geosystems were close to the state that they attained in later geologic times, with some significant exceptions. One exception was the precipitation from seawater of tremendous quantities of iron oxide. As oxygen formed on the early Earth, it combined with the reduced (unoxidized) iron present in the oceans to form iron oxide, which then precipitated and was deposited on the seafloor. Precipitation of iron oxide kept the level of oxygen in the atmosphere very low until the iron was all used up. Oxygen did not approach present levels until later in the Proterozoic

Figure Story 10.11 A cross section of four formations enables geologists to reconstruct the stages of an area's geologic history. From field mapping, a geologist makes a cross section of four formations: A, deformed metamorphic rocks; B, a granite pluton; D, sandstones, limestones, and shales containing marine fossils; F, sandstones containing land fossils. Formations A and D are separated by an unconformity (C). Formations D and F are separated by an angular unconformity (E).

GEOLOGISTS USE CROSS-CUTTING RELATIONSHIPS TO ESTABLISH A RELATIVE CHRONOLOGY

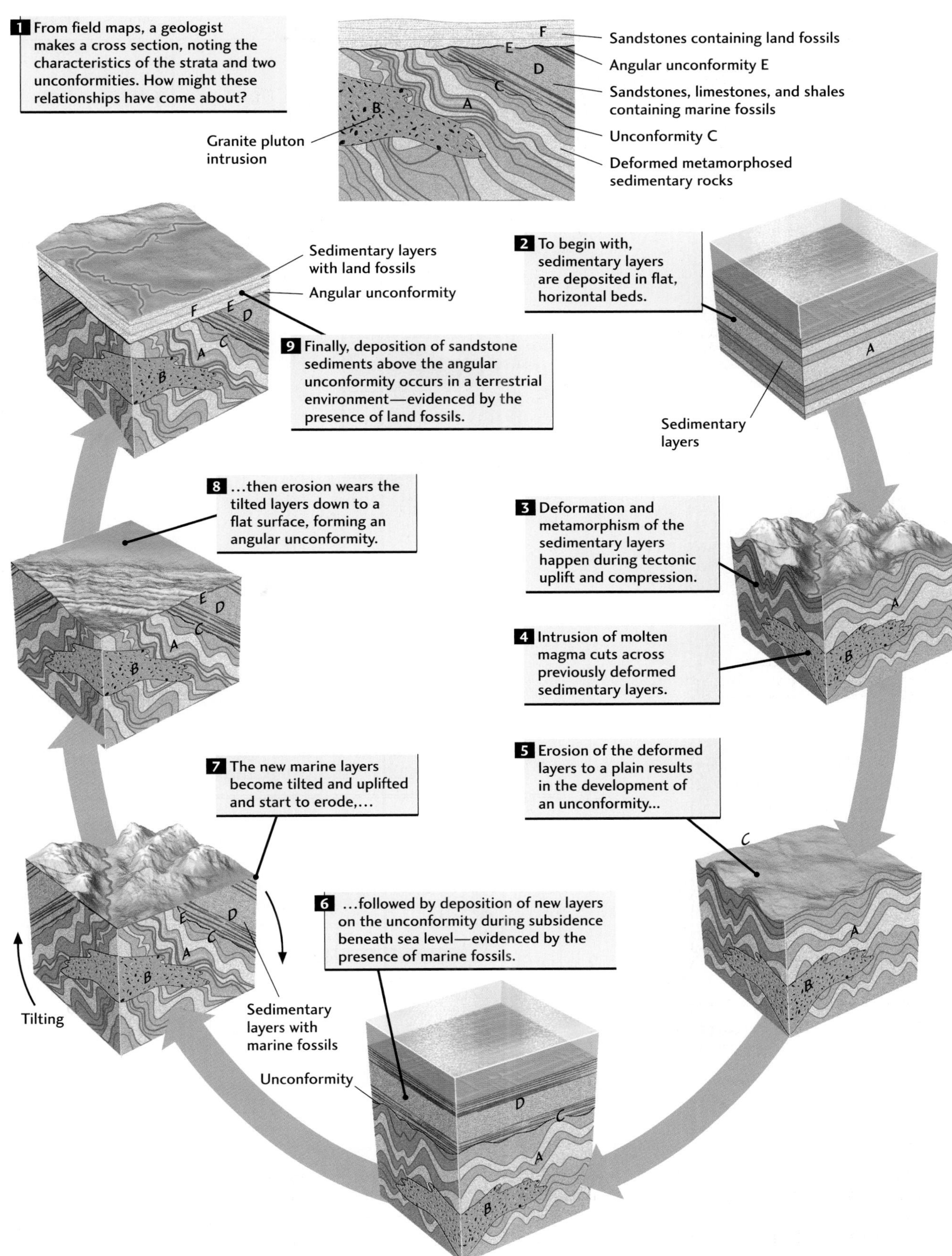

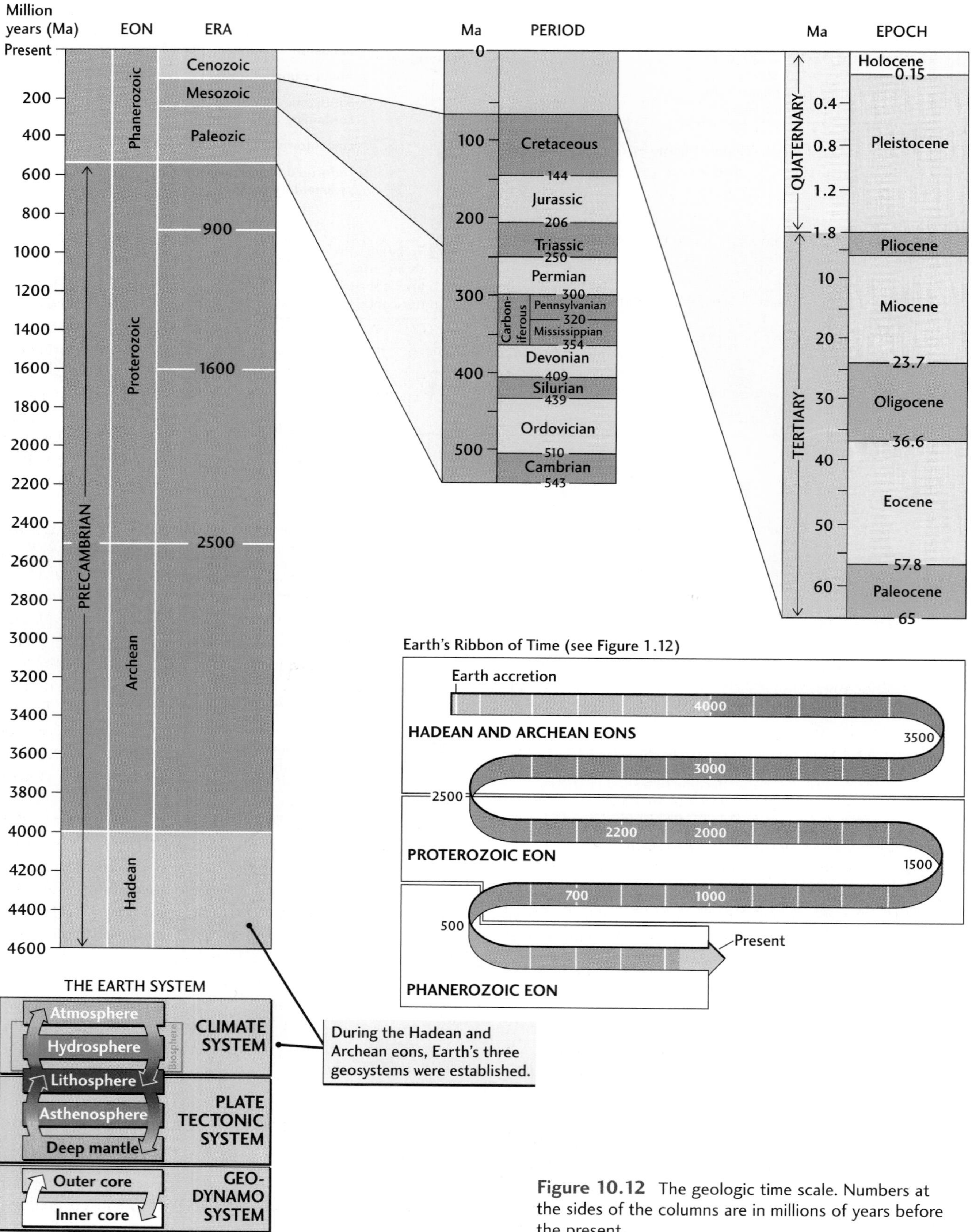

Figure 10.12 The geologic time scale. Numbers at the sides of the columns are in millions of years before the present.

and may have encouraged single-celled life-forms to evolve into multicellular algae and animals, which are preserved in the later Proterozoic fossil record.

Phanerozoic Eon The most recent and best-understood eon, covering the past 543 million years, is the Phanerozoic (from the Greek *phaneros,* "visible," and *zoi,* "life"). Many rock formations of this age contain an abundance of shells and other fossils, such as vertebrate bones. With rare exceptions, the world's oil and gas reserves formed during this time in Earth's history. The Phanerozoic is subdivided into three **eras:**

- Paleozoic ("old life") era: from 543 million to 251 million years ago
- Mesozoic ("middle life") era: from 251 million to 65 million years ago
- Cenozoic ("recent life" era: from 65 million years ago to the present

The eras are then subdivided into **periods,** most of which are named either for the geographic locality in which the formations are best displayed or were first described or for some distinguishing characteristic of the formations. The Jurassic period, for example, is named for the Jura Mountains of France and Switzerland and the Carboniferous period for the coal-bearing sedimentary rocks of Europe and North America.

Periods are further subdivided into **epochs;** the geologically best known are the subdivisions of the Tertiary period, such as the Pliocene epoch (spanning the interval from 6 million to 1.8 million years ago). In Feature 10.1, the geologic time scale is used to interpret one of the world's most spectacular outcrops, the Grand Canyon.

In working out this geologic time scale, geologists had to change their way of thinking about the Earth. James Hutton, known as the father of modern geology, and Charles Lyell, author of one of the earliest and most influential geology textbooks (*Principles of Geology,* first volume published in 1829), led geologists to understand that the planet was not shaped by a series of catastrophic events over a mere few thousand years, as many people believed. Rather, Earth was the product of ordinary geological processes operating steadily over much longer time intervals. As noted in Chapter 4, Hutton was among the first to grasp the cyclical nature of geologic change: the rock cycle of erosion, weathering, sedimentation, burial, igneous and tectonic activity, and mountain building.

Also inherent in Hutton's and Lyell's thinking was the principle of uniformitarianism—which, you may recall from Chapter 1, states that the processes we see shaping the Earth today are the same as those that have been operating during all of Earth's history. Although different kinds of sediments may have been deposited at different rates in different places throughout Earth's history, we can be reasonably sure that the depositional processes that laid down sediments millions and billions of years ago were working then in the same way as they do today.

Isotopic Time: Adding Dates to the Time Scale

The geologic time scale based on studies of stratigraphy and fossils is a relative scale. With it, geologists can say whether one formation is older than another, but they cannot determine precisely when a rock formed. It's like knowing that World War I preceded World War II but not knowing the specific years in which each conflict began and ended. Nineteenth-century geologists could only estimate that it might take millions of years for one fossil assemblage to change to another. Although they had estimated times for various geological processes, such as the laying down of sediments on lake bottoms or the erosion of a river valley, they could not devise a way to measure the duration of those processes accurately.

Some nineteenth-century physicists had estimated the age of the Earth and the solar system from astronomical and physical principles and calculated it to be many millions of years old. But these educated estimates were based on the physical principles of the day (some now outmoded), and they varied a good deal, from 25 million to 75 million years. Then, in 1896, an advance in modern physics paved the way for reliable and accurate measurements of geologic time in actual years. Henri Becquerel, a French physicist, discovered radioactivity in uranium. Within less than a year of Becquerel's find, the French chemist Marie Sklodowska-Curie discovered and isolated a different and highly radioactive element, radium.

In 1905, the physicist Ernest Rutherford suggested that radioactivity could be used to measure the exact age of a rock. He was able to tell the age in years of a uranium mineral from measurements in his laboratory. In the next few years, the ages of many more rocks were determined as dating methods were refined and more radioactive elements were found. This was the start of **isotopic dating,** the use of naturally occurring radioactive elements to determine the ages of rocks. When Rutherford announced the results of his first measurements, it became clear that the Earth is billions of years old and that the Phanerozoic eon alone was a little more than half a billion years long.

Radioactive Atoms: The Clocks in Rocks

How do geologists use radioactivity to determine the age of a rock? What the pioneers of nuclear physics discovered was that atoms of uranium, radium, and several other radioactive

10.1 The Grand Canyon Sequence and Regional Correlation of Strata

The rocks of the Grand Canyon and other parts of the Colorado Plateau area have many stories to tell. They record a long history of sedimentation in a variety of environments, sometimes on land and sometimes under the sea. Several unconformities mark erosion intervals. The rocks contain a succession of fossils, which reveals the evolution of new organisms and the extinction of old ones. From correlation of the rock sequences exposed at different localities, geologists can reconstruct a geologic history over a billion years long.

The lowermost—and therefore oldest—rocks exposed at the Grand Canyon are dark igneous and metamorphic rocks forming the Vishnu group, known to be about 1.6 billion years old based on isotopic dating techniques.

Above the Vishnu are the younger Precambrian Grand Canyon beds. These beds contain fossils of millimeter-scale single-cell microorganisms. A nonconformity separates the Vishnu and Grand Canyon beds, signifying a period of structural deformation accompanying metamorphism of the Vishnu and then erosion before the deposition of the Grand Canyon beds. The tilting of the Grand Canyon beds at an angle from their originally horizontal position shows that they, too, were folded after deposition and burial.

An angular unconformity divides the Grand Canyon beds from the overlying horizontal Tapeats Sandstone. This unconformity indicates a long period of erosion after the lower rocks had been tilted. The Tapeats Sandstone and Bright Angel Shale can be dated as Cambrian by their fossils, many of which are trilobites.

Above the Bright Angel Shale is a group of horizontal limestone and shale formations (Muav Limestone, Temple Butte Limestone, Redwall Limestone) that represent about 200 million years from the late Cambrian period to the end of the Mississippian period. There are so many time gaps represented by disconformities in these rocks that less than 40 percent of the Paleozoic is actually represented by rock strata.

The next set of strata, high up on the canyon wall, is the Supai group of formations (Pennsylvanian and Permian), which contains fossils of land plants like those found in coal beds of North America and other continents. Overlying the Supai is the Hermit, a sandy red shale.

Continuing up the canyon wall, we find another continental deposit, the Coconino Sandstone, which contains vertebrate animal tracks. The animal tracks suggest that the Coconino was formed in a terrestrial environment during Permian times. At the top of the cliffs at the canyon rim are two more formations of Permian age: the Toroweap, made mostly of limestone, overlain by the Kaibab, a massive layer of sandy and cherty limestone. These two formations record subsidence below sea level and the deposition of marine sediments.

The succession of strata at the Grand Canyon, though picturesque and informative, represents an incomplete picture of Earth history. Younger periods of geologic time are not preserved, and we must travel to locations in Utah such as Zion and Bryce Canyon National Parks to fill in this younger history. At Zion, we find equivalents of the Kaibab and Moenkopi, which allow us to correlate back to the Grand Canyon area and establish a link. In contrast to the Grand Canyon area, however, the Zion rocks extend upward in age to Jurassic time, including ancient sand dunes represented by sandstones of the Navajo formation. If we travel a bit further, we find that the Navajo formation also occurs in Bryce Canyon but that at this location the strata stack upwards to the Tertiary age Wasatch formation.

The correlation of strata among these three areas of the Colorado Plateau shows how widely separated localities—each with an incomplete record of geologic time—can be pieced together to build a composite record of Earth history.

Generalized stratigraphic section of the rock units in the Grand Canyon, Zion Canyon, and Bryce Canyon sequences. [John Wang/Photo Disc/Getty Images; David Muench/Corbis; and Tim Davis/Photo Researchers, respectively.]

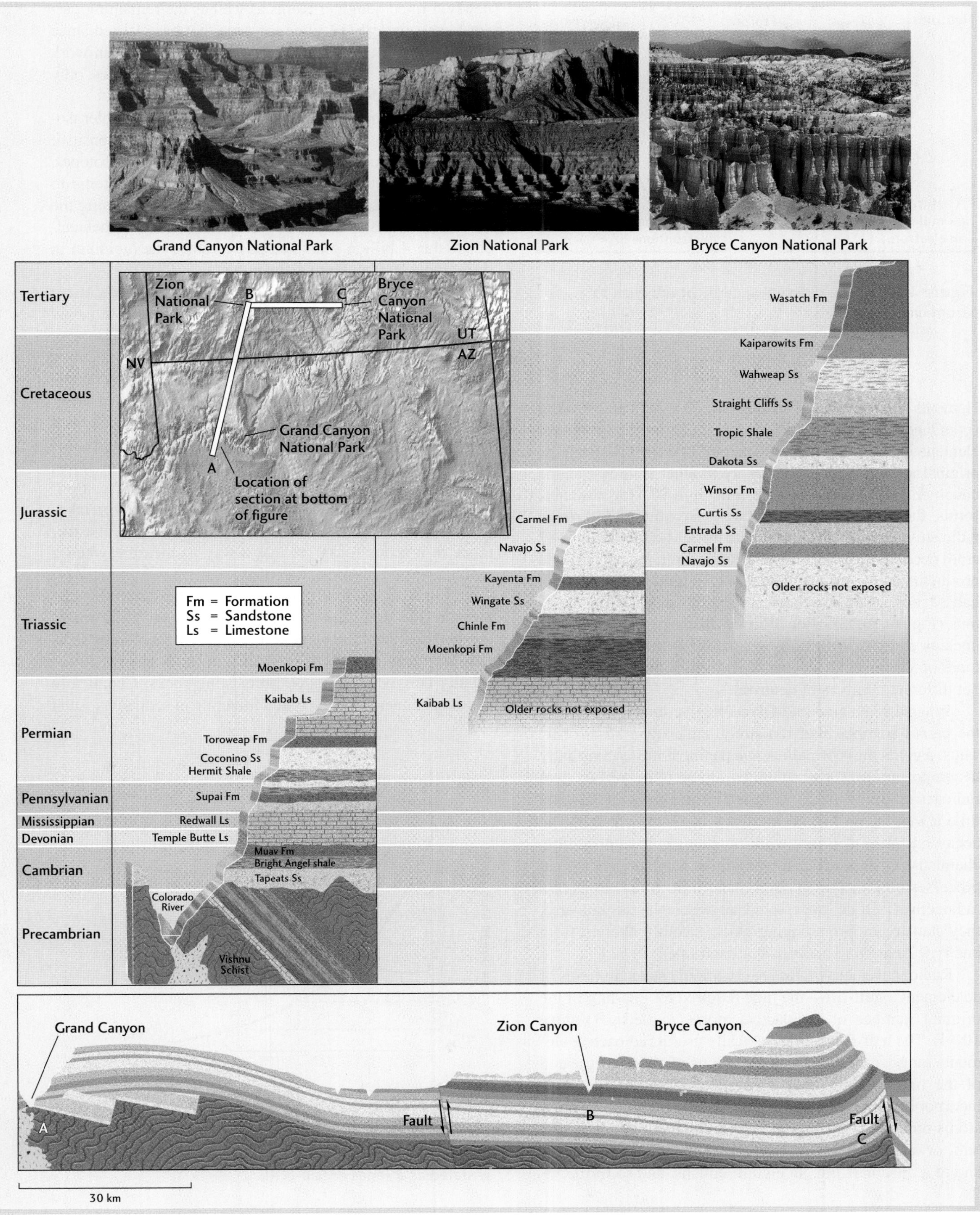
Grand Canyon National Park
Zion National Park
Bryce Canyon National Park
Tertiary
Cretaceous
Jurassic
Triassic
Permian
Pennsylvanian
Mississippian
Devonian
Cambrian
Precambrian
Zion National Park
B
C
Bryce Canyon National Park
UT
AZ
NV
Grand Canyon National Park
A
Location of section at bottom of figure
Fm = Formation
Ss = Sandstone
Ls = Limestone
Moenkopi Fm
Kaibab Ls
Toroweap Fm
Coconino Ss
Hermit Shale
Supai Fm
Redwall Ls
Temple Butte Ls
Muav Fm
Bright Angel shale
Tapeats Ss
Colorado River
Vishnu Schist
Carmel Fm
Navajo Ss
Kayenta Fm
Wingate Ss
Chinle Fm
Moenkopi Fm
Kaibab Ls
Older rocks not exposed
Wasatch Fm
Kaiparowits Fm
Wahweap Ss
Straight Cliffs Ss
Tropic Shale
Dakota Ss
Winsor Fm
Curtis Ss
Entrada Ss
Carmel Fm
Navajo Ss
Older rocks not exposed
Grand Canyon
Zion Canyon
Bryce Canyon
A
Fault
B
Fault
C
30 km

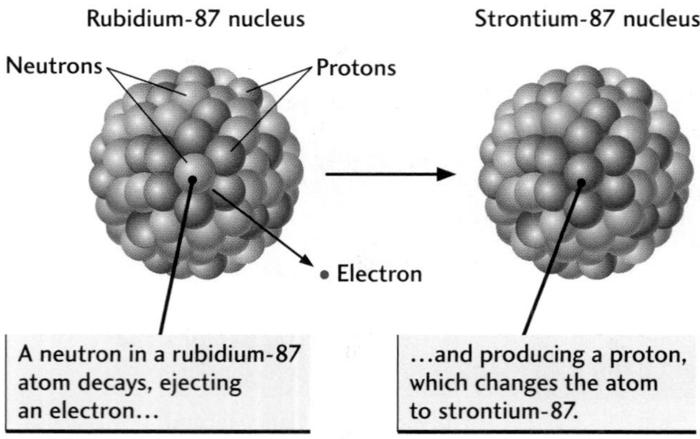

Figure 10.13 The radioactive decay of rubidium to strontium.

elements are unstable. The nucleus of a radioactive atom spontaneously disintegrates, forming an atom of a different element and emitting radiation, a form of energy. We call the original atom the parent; its decay product is known as the daughter. The parent isotope rubidium-87, for instance, forms the stable daughter isotope strontium-87 through radioactive decay. A neutron in the nucleus of a rubidium-87 atom decomposes, ejecting an electron from the nucleus and producing a new proton. The former rubidium atom, which had 37 protons, thus becomes a strontium atom, with 38 protons (**Figure 10.13**). (Recall from Chapter 3 that an atomic nucleus consists of protons and neutrons and that the isotopes of a given element contain the same number of protons but different numbers of neutrons.)

When a given amount of a radioactive substance decays, the parent isotopes alter randomly, rather than all at once. Thus, a given mass of radioactive parent atoms is constantly decomposing to form daughter atoms. The reason that radioactive decay offers a dependable means of keeping time is that the probability of decay is a fixed number. The decay rate does not vary with the changes in temperature, chemistry, or pressure that typically accompany geological processes on Earth or on other planets. So when atoms of a radioactive isotope are created anywhere in the universe, they start to act like a ticking clock, steadily altering from one type of atom to another at a fixed rate.

Radioactive decay rates are commonly stated in terms of an element's **half-life**—the time required for one-half of the original number of radioactive atoms to decay (**Figure 10.14**). The half-lives of geologically useful radioactive elements range from thousands to billions of years. At the end of the first half-life after a radioactive isotope has been incorporated into a new mineral, half the number of parent atoms remain. At the end of a second half-life, half of that half, or one-quarter of the original number, are left. At the end of a third half-life, an eighth remains, and so forth.

If we know the decay rate and can count the number of newly formed daughter atoms as well as the remaining parent atoms, we can calculate the time that has elapsed since the radioactive clock began to tick. In effect, we can work back to the time when there were no daughter isotopes, only those of the undecayed parent element.

Geologists measure the ratio of parent and daughter isotopes with a mass spectrometer, a very precise and sensitive instrument that can detect even minute quantities of isotopes. Suppose we determine the ratio of rubidium-87 atoms to strontium-87 atoms in a rock sample to be 19:1. Using the known rubidium-to-strontium decay rate, we could then calculate that 4 billion years had passed since the rubidium in our sample had begun to disintegrate.

To geologists, this is the age of the rock—or, more exactly, the time since rubidium was first trapped in newly formed minerals of the rock. Rubidium and strontium, like other elements, are incorporated into minerals as they crystallize from a magma or recrystallize during metamorphism. During crystallization, the ratio of rubidium to strontium is homogenized, which sets the radioactive clock back to zero. Inside the newly formed mineral, the radioactive decay of rubidium-87 continues, and new strontium atoms start to accumulate, which changes their ratio. These strontium atoms cannot escape unless recrystallization takes place again. Thus, rubidium-87 and other radioactive isotopes in igneous rocks provide a way to measure when a magma was emplaced and cooled.

Radioactive isotopes in metamorphic rocks enable us to measure the time that has elapsed since they were metamorphosed. Rubidium decay is not used for dating sedimentary rocks, because the minerals in clastic sediments are usually derived from older, preexisting rocks. Chemically and biochemically precipitated minerals in sediments, such

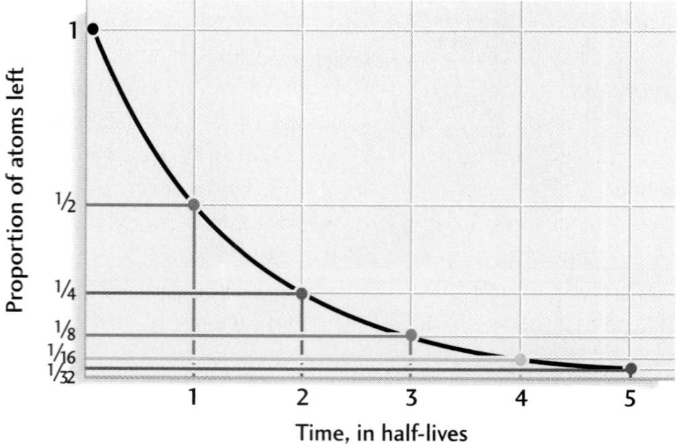

Figure 10.14 The number of radioactive atoms in any mineral declines at a fixed rate over time. This rate of decay is stated as a series of half-lives.

Table 10.1 Major Radioactive Elements Used in Radiometric Dating

Isotopes		Half-Life of Parent (years)	Effective Dating Range (years)	Minerals and Materials That Can Be Dated
Parent	Daughter			
Uranium-238	Lead-206	0.7 billion	10 million–4.6 billion	Zircon Apatite
Uranium-235	Lead-207	4.5 billion	10 million–4.6 billion	Zircon Apatite
Potassium-40	Argon-40	1.3 billion	50,000–4.6 billion	Muscovite Biotite Hornblende
Rubidium-87	Strontium-87	47 billion	10 million–4.6 billion	Muscovite Biotite Potassium feldspar
Carbon-14	Nitrogen-14	5730	100–70,000	Wood, charcoal, peat Bone and tissue Shell and other calcium carbonate Groundwater, ocean water, and glacier ice containing dissolved carbon dioxide

as carbonates, usually contain too little newly precipitated rubidium for accurate isotopic analysis. Sometimes, however, enough carbon from seawater is incorporated into a sedimentary rock at the time of deposition to enable geologists to determine the rock's geologic age.

Geologists use a number of naturally occurring radioactive elements to determine the ages of rocks (Table 10.1). Each radioactive element has its own decay rate. Those that decay slowly over billions of years, such as rubidium-87, are used to measure the ages of old rocks. Radioactive elements that decay rapidly over only a few tens of thousands of years, such as carbon-14, are useful for determining the ages of very young rocks. Isotopic dating is possible only if a measurable number of parent and daughter atoms remain in the rock. For example, if the rock is very old and the decay rate is fast, almost all the parent atoms will already have been transformed. In that case, we could determine that the istotopic clock has run down, but we would have no way of knowing how long ago it stopped.

Carbon-14: Clocking "Recent" Activity

Carbon-14, which decays to nitrogen-14, has a half-life of 5730 years. In a rock that is 30,000 years old, then, more than five half-lives have passed, and only a little less than 1/32 of the original amount of carbon-14 remains. By the time 70,000 years have passed, too few carbon-14 atoms are left to count accurately. Carbon-14 is thus most suitable for measuring times in the relatively recent geologic past.

Carbon-14 is an especially important tool for dating fossil bone, shell, wood, and other organic materials in very young sediments because all these materials contain carbon, including a small amount of carbon-14. Carbon is an essential element in the living cells of all organisms. For example, as green plants grow, they continuously incorporate into their tissues a small amount of carbon-14, along with stable carbon isotopes, from carbon dioxide in the atmosphere. When a plant dies, it stops absorbing carbon dioxide, and no new carbon of any kind is added. At that moment, the amount of carbon-14 in relation to the stable carbon isotopes is identical with that in the atmosphere. However, the amount of carbon-14 in the dead tissue, which is incorporated as fossil organic material in a sediment, steadily decreases as radioactive atoms decay. The nitrogen-14 daughter atoms of carbon-14 are gaseous and thus leak from the sediment, so we cannot measure the daughters accurately. We can, however, compare the amount of carbon-14 left in the plant material with the original amount that would have been in equilibrium with the atmosphere, which is assumed to be approximately constant over the relatively short periods of time being measured. This comparison yields the time that has elapsed since the plant died.

Limits and Uses of Isotopic Dating

Isotopic dating cannot give an accurate reading for every rock that a geologist samples. If a rock containing uranium were to lose some of its lead through weathering, for example, we might get an erroneously young age. Or if an igneous rock were metamorphosed, the daughter isotopes that had accumulated since the crystallization of the magma might be lost, resetting the clock to the time of metamorphism rather than the time of initial formation. In addition to these factors, the accuracy and precision of isotopic dating depend on accurate measurement of the often minute amounts of daughter atoms found in rocks. Techniques have now advanced to the point where lead from a single zircon crystal can be used to date a rock. In fact, accuracy has improved so much in recent years that Paleozoic and Precambrian rocks can be dated with an error of no more than a few hundred thousand years—a remarkable improvement over the errors of more than 50 million years just a few decades ago.

One of the boundaries of the geologic time scale that has shifted a good deal as a result of better isotopic dating and the acquisition of new samples from previously unexplored territory is the boundary between the Cambrian period and Precambrian time (the Archean and Proterozoic eons). In the past few years, this boundary has changed several times, from the former date of 570 million years ago to the current, more firmly placed value of 543 million years ago. This date is particularly important because it relates to some of the most significant developments in the evolution of multicellular organisms.

The geologic time scale has many uses, aside from its value to geologists. Anthropologists who reconstruct the path of human evolution and archeologists who date various human settlements use the most recent parts of the time scale, covering the past 5 million to 10 million years. Seismologists who study earthquake-prone regions use the time scale and stratigraphy of the recent past to map movements on faults that gave rise to earthquake shocks.

Another Geologic Clock

Time is so central to the study of the Earth that geologists continue to search for additional ways to gauge geologic time. Paleomagnetic stratigraphy, for example, is under steady development as an adjunct to isotopic dating. As we learned in Chapter 2, Earth's magnetic field reverses about every half-million years. These periodic reversals are recorded in the orientation of magnetic minerals in rocks, especially those on the seafloor. The magnetic time scale has been calibrated both by radioactive age determinations and by the stratigraphic ages of overlying and underlying fossiliferous formations.

From Three Lines of Evidence: A Reliable Dating Tool

Once geologists had determined isotopic ages and linked them to their earlier studies of fossils and stratigraphy, they could add absolute dates to the geologic time scale (see Figure 10.11). With these three sources of information, geologists can deduce the approximate ages of rock formations, even those that do not contain materials amenable to radiometric analysis. For example, if we know from isotopic dating that an igneous intrusion is 500 million years old, then sedimentary layers cross-cut by that intrusion must be older than 500 million years. And if these same layers overlie metamorphic rocks radiometrically dated at 550 million years old, then we know that the sedimentary layers were formed 500 million to 550 million years ago. If these sedimentary rocks contain fossils indicating Cambrian and late Ordovician stratigraphic ages, we know the absolute ages of parts of these geological periods. With this kind of bracketing of ages, geologists worked out the entire geologic time scale. After almost a century of isotopic dating and continued work on the stratigraphy of the world, this time scale remains undisputed in all major respects.

Estimating the Rates of Very Slow Earth Processes

Now that we have a way of dating rocks, let us see what the time scale can tell us about the rates of some slow geological processes. Consider the opening of an ocean, in which seafloor plates spread away from each other at mid-ocean ridges. The southern Atlantic Ocean from South America to Africa is a little more than 5000 km wide; that is how far the two continents have separated from each other. The seafloor at the edges of these continents is the oldest, having formed at the inception of spreading. From fossils, we know that these sediments are about 100 million years old (middle Cretaceous age). Thus, the average rate of spreading of this part of the ocean floor is 5000 km every 100 million years, or about 5 cm/year. In other parts of the oceans, spreading rates may be as low as 2 to 4 cm/year (Mid-Atlantic Ridge) or as high as 10 to 17 cm/year (East Pacific Rise).

This kind of rough calculation gives a surprisingly good estimate of the rate of seafloor spreading. It was put to the test in 1987, when scientists were first able to measure the spreading rate of the Atlantic seafloor directly, using the technology of long-ranging laser beams and satellites. Their results agreed with the spreading rates that marine geologists had calculated from the age and position of the seafloor.

The dating methods described in this chapter also help us understand another slow process, one of direct concern to the people of California: the movement of crustal blocks along the San Andreas and adjacent faults, which has been responsible for many major earthquakes. **Figure 10.15**

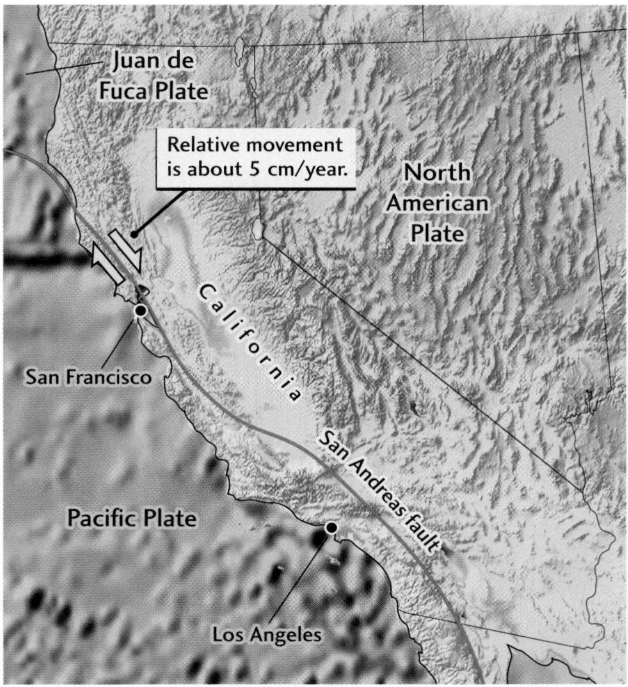

Figure 10.15 We can calculate a relative movement of about 5 cm/year between the North American and Pacific plates by the offset of the geological formations split by the fault, by satellite measurements, and by seafloor spreading rates.

shows how the Pacific Plate slides past the North American Plate along this transform boundary. One way to determine the rate of this movement is to measure the offsets of distinctive geologic formations of various ages split by the plate boundary faults. Dividing the distance that now separates these formations by the time since they formed and split gives the rate of movement. Seafloor spreading rates and satellite measurements provide other estimates of plate motions along the boundary. Using these observations, we can estimate that the average movement over the past several million years was about 5 to 6 cm/year. If that rate continues, Los Angeles, on the Pacific Plate, will be contiguous with San Francisco, on the North American Plate, in about 10 million years.

We can also measure rates of vertical movement by dating marine deposits that are now above sea level. Parts of the Alpine range, for instance, contain marine fossils known to be about 15 million years old. These fossils, now elevated 3000 m above sea level, were originally deposited on the shallow seafloor close to sea level. The sedimentary rocks containing the fossils must have been uplifted an average of about 2 mm/decade, although the rates may have been higher or lower for shorter time intervals or in different parts of the mountain range.

The final case we will consider here is erosion. Erosional processes are continuously wearing down the land surface. These processes are so slow that two photographs of a river valley taken 96 years apart show little difference (**Figure 10.16**). We can estimate erosion rates by adding up all the disintegrated and dissolved products of erosion that rivers and the wind are carrying away from a land area. The rate at which the North American continent erodes has been estimated to be 3 mm each century. At this rate, it would take 100 million years to reduce a 3000-m-high mountain to sea level.

Thus, in these particular cases, it took about 100 million years to open an ocean, 15 million years to raise a mountain range, and 100 million years to erode it. As we will see, however, these time intervals are relatively short compared with the entire history of the planet. During that history, Earth has undergone many cycles of mountain building and erosion.

Figure 10.16 Two photographs of Bowknot Bend on the Green River in Utah taken nearly 100 years apart show that little has changed in the configuration of rocks and formations in that time interval.

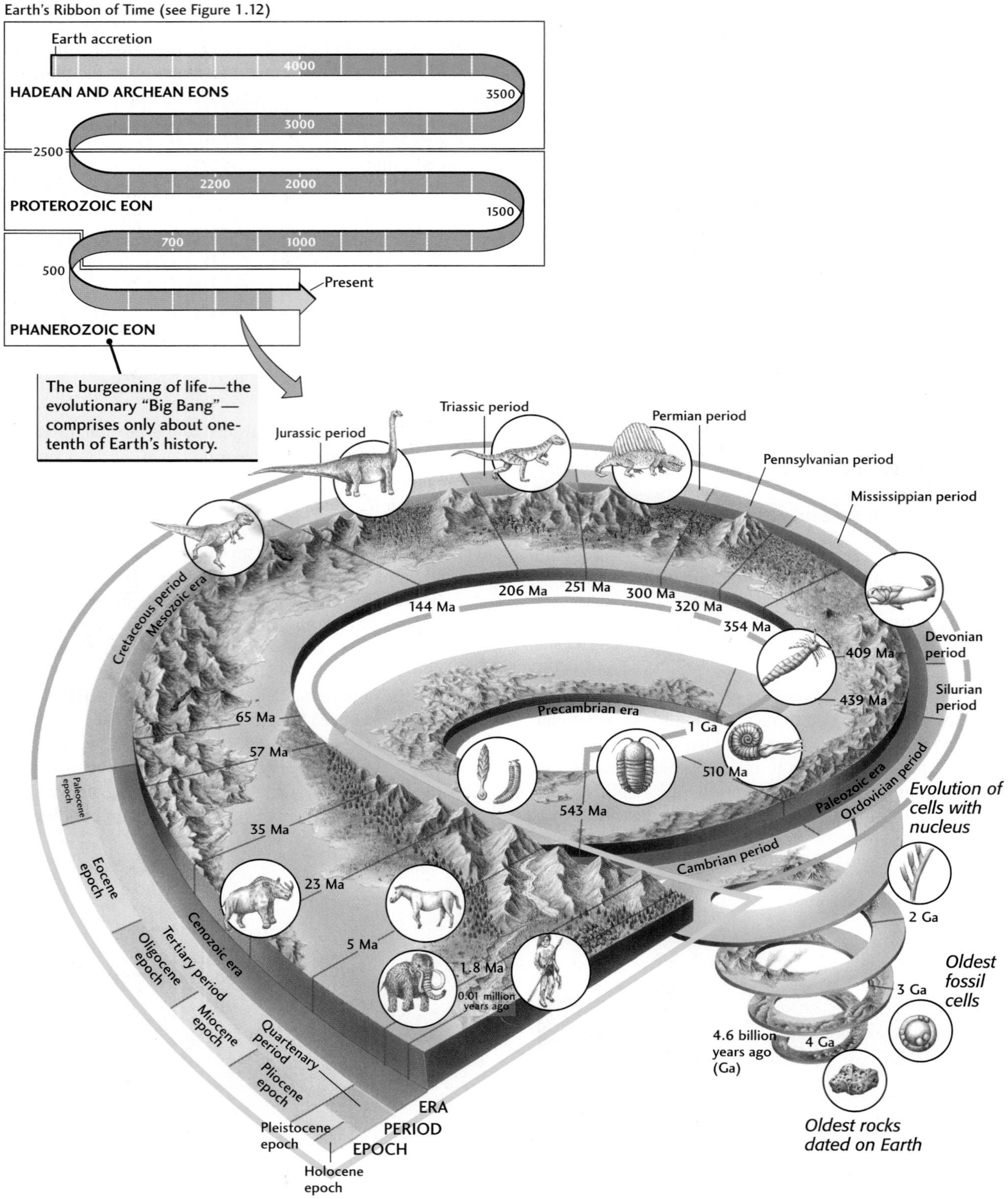

Earth's Ribbon of Time (see Figure 1.12)

Earth accretion
4000
HADEAN AND ARCHEAN EONS
3500
3000
2500
2200
2000
PROTEROZOIC EON
1500
700
1000
500
Present
PHANEROZOIC EON

The burgeoning of life—the evolutionary "Big Bang"—comprises only about one-tenth of Earth's history.

Triassic period
Permian period
Jurassic period
Pennsylvanian period
Mississippian period
Cretaceous period
Mesozoic era
144 Ma
206 Ma
251 Ma
300 Ma
320 Ma
354 Ma
Devonian period
409 Ma
439 Ma
Silurian period
Precambrian era
1 Ga
65 Ma
57 Ma
510 Ma
Paleocene epoch
543 Ma
Paleozoic era
Ordovician period
Evolution of cells with nucleus
35 Ma
Cambrian period
Eocene epoch
23 Ma
2 Ga
5 Ma
Cenozoic era
Tertiary period
Oligocene epoch
1.8 Ma
0.01 million years ago
Oldest fossil cells
3 Ga
Miocene epoch
Quartenary period
Pliocene epoch
4.6 billion years ago (Ga)
4 Ga
ERA
PERIOD
EPOCH
Pleistocene epoch
Holocene epoch
Oldest rocks dated on Earth

Figure 10.17 Geologic timeline of Earth's history. Abbreviations: Ma, million years ago; Ga, billion years ago.

An Overview of Geologic Time

We can now combine the geologic time scale with absolute dates from an analysis of radioactive decay and the evolution of organisms to construct a timeline for the whole history of the Earth, beginning at its birth 4.6 billion years ago (see the front endpapers). **Figure 10.17** shows the whole sweep of geologic time as a spiral pathway, each revolution of the spiral corresponding to 1 billion years. From this illustration, we can see how small a proportion of Earth's history is taken up by the eras of the Phanerozoic eon and what a tiny amount of time has elapsed since humans evolved.

As another way of comprehending this extraordinarily long period of time, think of Earth's age as being only one calendar year. On January 1, Earth was formed. During January and part of early February, Earth became organized into core, mantle, and crust. About February 21, life evolved. During all of spring, summer, and early fall, Earth evolved continents and ocean basins something like those of today, and plate tectonics became active. On October 25, at the beginning of the Cambrian period, complex organisms, including those with shells, arrived. On December 7 reptiles evolved, and on Christmas Day the dinosaurs became extinct. Modern humans, *Homo sapiens,* appeared on the scene at 11:00 P.M. on New Year's Eve, and the last glacial age ended at 11:58:45 P.M. Three-hundredths of a second before midnight, Columbus landed on a West Indian island. And a few thousandths of a second ago, you were born.

SUMMARY

How do geologists know how old a rock is and whether one rock is older than another? Geologists determine the order in which rocks formed by studying their stratigraphy, fossils, and arrangement in the field. An undeformed sequence of sedimentary rock layers will be horizontal, with each layer younger than the layers beneath it and older than the ones above it. In addition, because animals and plants have evolved progressively through time, their fossil remains change in known ways in the stratigraphic sequence. Knowing the faunal succession makes it easier for geologists to spot missing sedimentary layers, known as unconformities. Most important, fossils enable geologists to correlate rocks located throughout the world.

How did geologists create a geologic time scale that is applicable throughout the world? Using fossils to correlate rocks of the same geological age and piecing together the sequences exposed in hundreds of thousands of outcrops throughout the world, geologists compiled a stratigraphic sequence applicable everywhere on Earth. The composite sequence represents the geologic time scale. The use of isotopic dating allowed scientists to assign absolute dates to the units of the time scale. Isotopic dating is based on the behavior of radioactive elements, in which unstable parent atoms are transformed into daughter isotopes at a constant rate. When radioactive elements are locked into minerals as rocks form, the number of daughters increases as the number of parents decreases. By measuring parents and daughters, we can calculate absolute ages.

Why is the geologic time scale important to geology? The geologic time scale enables geologists to reconstruct the chronology of events that have shaped the planet. The time scale has been instrumental in validating and studying plate tectonics and in estimating the rates of geological processes too slow to be monitored directly, such as the opening of an ocean over millions to hundreds of millions of years. The development of the time scale revealed that Earth is much older than early geologists and others had imagined and that it has undergone almost constant change as a result of slow processes working throughout the planet's history. The creation of the geologic time scale paralleled the development of paleontology and the theory of evolution, one of the most revolutionary and powerful ideas in science.

Key Terms and Concepts

angular unconformity (p. 219)
disconformity (p. 219)
eon (p. 221)
epoch (p. 225)
era (p. 225)
formation (p. 218)
geologic time (p. 215)
geologic time scale (p. 221)
half-life (p. 228)
isotopic age (p. 214)
isotopic dating (p. 225)
nonconformity (p. 219)
paleontology (p. 217)
period (p. 225)
principle of original horizontality (p. 215)
principle of superposition (p. 215)
relative age (p. 214)
sequence stratigraphy (p. 220)
stratification (p. 215)
stratigraphic succession (p. 215)
stratigraphy (p. 215)
unconformity (p. 218)

Exercises

MEDIA LINK *This icon indicates that there is an animation available on the Web site that may assist you in answering a question.*

1. Name the geological periods, from youngest to oldest.

2. Give the absolute times of the beginnings of the Paleozoic, the Mesozoic, and the Cenozoic eras.

3. To what element does rubidium-87 decay radioactively?

4. Give the age of a sediment that might be dated by carbon-14.

5. What geologic events are implied by an angular unconformity?

6. What is the principle of superposition?

7. What is the principle of original horizontality?

8. How do angular unconformities differ from disconformities and nonconformities?

9. What property of fossils do geologists use to date the formations in which they are found?

10. How does determining the ages of igneous rocks help to date fossils?

Thought Questions

This icon indicates that there is an animation available on the Web site that may assist you in answering a question.

1. As you pass by an excavation in the street, you see a cross section showing paving at the top, soil below the paving, and bedrock at the base. You also notice that a vertical water pipe extends from a drain in the street into a sewer in the soil. What can you say about the relative ages of the various layers and the water pipe?

2. How would you be able to ascertain the relative ages of several volcanic ash falls exposed in an outcrop?

3. What evidence could you give to a friend to support the idea that a particular formation was many millions of years old?

4. Construct a diagram similar to the one in Figure Story 10.11 to show the following series of geologic events: (a) sedimentation of a limestone formation; (b) uplift and folding of the limestone; (c) erosion of the folded terrain; (d) subsidence of the terrain and sedimentation of a sandstone formation.

5. Many fine-grained muds are deposited at a rate of about 1 cm per 1000 years. At this rate, how long would it take to accumulate a stratigraphic sequence half a kilometer thick?

6. What radioactive elements might you use to date a schist that is approximately 1 billion years old?

7. What geologic event is dated by radioactive decay of a mineral in a schist?

8. What geologic event is dated by radioactive decay of a mineral in a basalt?

9. Name a geologic event that can be dated by both stratigraphic methods and isotopic dating.

10. From the Grand Canyon section shown in Feature 10.1, give one example each of an angular unconformity, a disconformity, and a nonconformity.

11. Do you think it would be possible to use isotopic dating to determine the age of a basalt on the Moon, which has a composition very much like that of the basalts on Earth? What radioactive decay scheme might you use?

12. You would like to know the date when a dormant volcano in South America was last active. What methods could you use to determine this date?

Short-Term Team Projects

Geologic Time

John McPhee, a well-known geology writer, popularized the calendar year as a way to express geologic time in human terms. For this project, you and a classmate will team up to create your own metaphor for geologic time.

The calendar-year metaphor entails a direct time-for-time conversion. You may compare geologic time to another time scale, such as a life span, or to other measures, such as distance, volume, or weight. A personal metaphor, such as comparing geologic time to the distance between your home states, is easily remembered.

Your metaphor should include the following events: origin of Earth; oldest rocks; first life on the planet; transition to an oxygen-rich atmosphere; origin of multicellular life; first land plants; first land animals; appearance and extinction of dinosaurs; first hominids; first anatomically modern humans; three events of your choice from human history; your birthdays.

You will need to know the dates of the events, and you will need to calculate the percentage of geologic time between them. For example, if the Earth formed 4.6 billion years ago and the oldest rocks on Earth are about 3.8 billion years old, you can calculate that 0.8 billion years, or about 17 percent of geologic time, elapsed between the planet's formation and the solidification of its first known rocks: $4.6 - 3.8 = 0.8$; $0.8/4.6 = 0.17$. Calculate the percentage differences between the other geologic events and scale your metaphor accordingly.

Present your team project as a visual image. A distance-for-time metaphor, for example, might be represented by a map and markers along the way. Write a two-page description, including why you chose the metaphor and how to interpret the comparative time scales.

Suggested Readings

Berry, W. B. N. 1987. *Growth of a Prehistoric Time Scale.* Palo Alto, Calif.: Blackwell Scientific.

Faure, G. 1986. *Principles of Isotope Geology,* 2d ed. New York: Wiley.

Palmer, A. R. 1984. *Decade of North American Geologic Time Scale.* Map and Chart Series MC-50. Boulder, Colo.: Geological Society of America.

Simpson, G. G. 1983. *Fossils.* New York: Scientific American Books.

Stanley, S. M. 1999. *Earth System History.* New York: W. H. Freeman.

Winchester, S. 2002. *The Map That Changed the World: William Smith and the Birth of Modern Geology.* New York: Perennial.

Folded sedimentary rocks, northwest Canada. The folds have a wavelength of about 1 km. [John Grotzinger.]

CHAPTER

11

Folds, Faults, and Other Records of Rock Deformation

"Inequality is the cause of all local movements."

Leonardo da Vinci

The eighteenth- and nineteenth-century scientists who laid the foundations of modern geology concluded that most sedimentary rocks were originally deposited as soft horizontal layers at the bottom of the sea and hardened over time. But they were puzzled by the many hardened rocks that were tilted, bent, or fractured. They wondered: What forces could have deformed these hard rocks in this way? Can we reconstruct the history of the rocks from the patterns of deformation found in the field? Today's geologists would add: How does the deformation relate to plate tectonics? We answer these questions in this chapter and in later ones.

In this chapter, we look at how geologists collect and interpret field observations to reconstruct the geologic history of a region.

They might ask: What kinds of rocks were originally formed? What happened to them then? If they were deformed by tectonic forces, what was the nature of the deformation and what kinds of forces caused it? In their search for answers, geologists found similar patterns of deformation throughout the world. A few concepts can explain in general why and how rocks were deformed. We must learn the features of deformed rocks, however, before we can see and then explain the patterns of deformation.

Folding and **faulting** are the most common forms of deformation in the sedimentary, metamorphic, and igneous rocks that make up Earth's crust. Folds in rocks are like folds in clothing. Just as cloth pushed together from opposite sides bunches up in folds, layers of rock slowly compressed by forces in the crust are pushed into

Figure 11.1 An outcrop of originally horizontal rock layers bent into folds by compressive tectonic forces. [Phil Dombrowski.]

folds (**Figure 11.1**). Tectonic forces can also cause a rock formation to break and slip on both sides of a fracture, parallel to the fracture (**Figure 11.2**). This is a fault. Geologic folds and faults can range in size from centimeters to tens of kilometers or more. Many mountain ranges are actually a series of large folds or faults, or both, that have been weathered and eroded. Geologists now believe the forces that move the large plates are ultimately responsible for the deformations found in most local areas.

Interpreting Field Data

Scientists are trained to be keen observers so that they can gather accurate information on which to base their theories. To figure out how rock formations are deformed, geologists need precise information about the geometry of the beds that they can see. A basic source of information is the outcrop, where the bedrock that underlies the surface everywhere is exposed—that is, not obscured by soil or loose boulders. Figure 11.1 shows an example: a sedimentary bed that has been bent into a fold. Often, however, folded rocks are only partly exposed in an outcrop and can be seen only as an inclined layer (**Figure 11.3**). The orientation of the layer is an important clue the geologist can use to piece together a picture of the overall deformed structure. It takes only two measurements to describe the orientation of a layer of rock exposed at a given location: the strike and the dip.

Measuring Strike and Dip

Figure 11.4 shows how the strike and the dip are observed and measured in the field. The **strike** is the compass direc-

Figure 11.2 Small-scale faults. Once-continuous rock layers have been permanently displaced along fractures by extensional tectonic forces. Artificial cut near Mount Carmel Junction, southern Utah. [Tom Bean.]

Figure 11.3 Dipping limestone and shale beds on the coast of Somerset, England. Children are walking along the strike of beds that dip to the left. Dip is the angle of steepest descent of the bed from the horizontal; strike is at right angles to the dip direction. [Chris Pellant.]

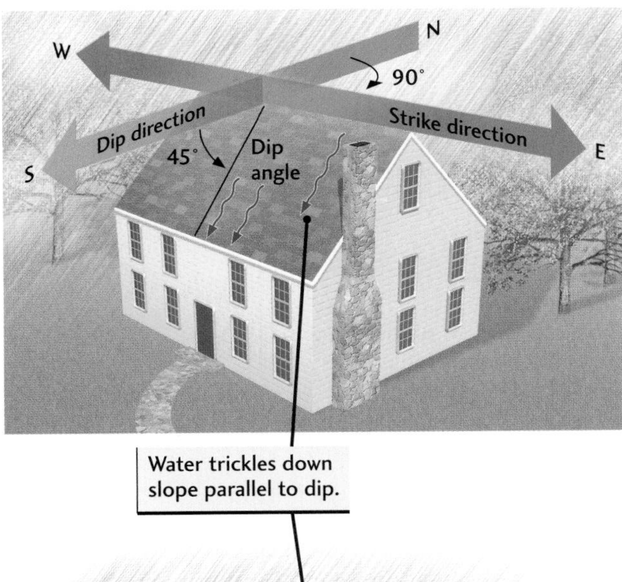

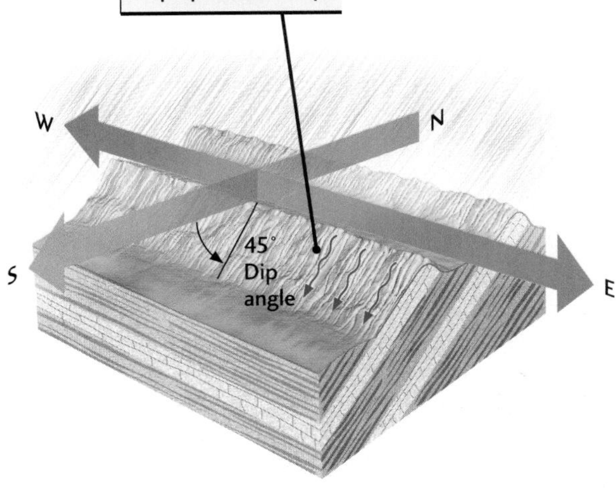

Figure 11.4 Geologists use the strike and dip of a formation to define its orientation at a particular place. [After A. Maltman, *Geological Maps: An Introduction* (New York: Van Nostrand Reinhold, 1990), p. 37.]

tion of a rock layer as it intersects with a horizontal surface. The **dip,** which is measured at right angles to the strike, is simply the amount of tilting—that is, the angle at which the bed inclines from the horizontal. For example, a geologist might describe the particular outcrop on the right side of Figure 11.4 as "a bed of coarse-grained sandstone striking west and dipping 45 degrees south."

Constructing a Geological Map and Cross Section

A geological map is a convenient tool for recording and organizing such information as the locations of outcrops, the nature of their rocks, and the dips and strikes of inclined layers. (Appendix 6 gives more information on geological maps.) Also helpful in piecing together a geological story is a geological cross section—a diagram showing the features that would be visible if a vertical slice were made through part of the crust. A natural cross section can often be observed in the vertical face of a cliff, a quarry, or a road cut. A cross section can also be constructed from the information on a geological map. **Figure 11.5** shows both a simple geological map of an area where sedimentary rocks, originally horizontal, were bent into a fold and the cross section derived from it.

Notice that the map and cross section in Figure 11.5 include portions of the fold that eroded away long ago. How does the geologist reconstruct the deformed shapes of the rock layers, even when erosion has removed parts of a formation? The process is like putting together a three-dimensional jigsaw puzzle with missing pieces. Common sense and intuition play important roles. The geologist may first notice such features as sedimentary layers and surmise that they were originally deposited as horizontal beds at the bottom of the sea. If the layers are now found to be inclined, as they are in Figure 11.5, the implication is that later events

Geological map

3 Arrows and angles show dips of the formations.

Key

5 Conglomerate (youngest formation)

4 Limestone

3 Shale

2 Red sandstone

1 Brown sandstone (oldest formation)

Geological cross section

1 The arrangement of the layers indicates that formation 1 is at the bottom and therefore is the oldest exposed at the surface.

2 Flanking formations are successively younger.

Figure 11.5 A geological map and a cross section derived from it. The arrangement of the layers indicates that formation 1 is at the bottom and therefore is the oldest; flanking formations are successively younger. The dashed lines reconstruct the eroded portion of the fold by connecting identical formations (shown by the same colors and numbers) and matching the observed dip.

tilted or bent the rocks. The law of superposition (younger beds are laid down over older beds, as we saw in Chapter 10) tells the geologist that the bed numbered 1 is the oldest and that the flanking, overlying beds are successively younger. Using the dip of the layers, the geologist constructs a cross section—a vertical cut as it would exist along the traverse marked A–B on the map. The geologist finds that the beds on both sides of formation 1 are identical. By linking them up with dashed lines that match the observed dips, the geologist reconstructs the boundaries of the portions of the beds removed by erosion. To complete the cross section, the geologist projects the trend of the beds below the ground, even though the rocks cannot be seen.

Figure 11.5 tells the story of an ancient ocean, now gone, in which a succession of sedimentary rocks was deposited on the seafloor. These beds, once horizontal, were subjected to crustal forces of compression, bent into folds, and raised above sea level. Erosion could then remove a major portion of the section, leaving the present-day remnant portrayed by the map and cross section. The outcropping beds in Figure 11.3 probably have a similar history.

How Rocks Become Deformed

The tectonic forces that deform rocks can be of three types: **compressive forces,** which squeeze and shorten a body; **tensional forces,** which stretch a body and tend to pull it apart; and **shearing forces,** which push two sides of a body in opposite directions. (To visualize shearing, think of what happens to a deck of cards held between your palms when you move your hands parallel to each other but in opposite directions. The cards slide past one another, deforming the deck.) **Figure Story 11.6** shows how rock bodies become deformed when subjected to horizontally directed compression, tension, and shear. These same kinds of forces are

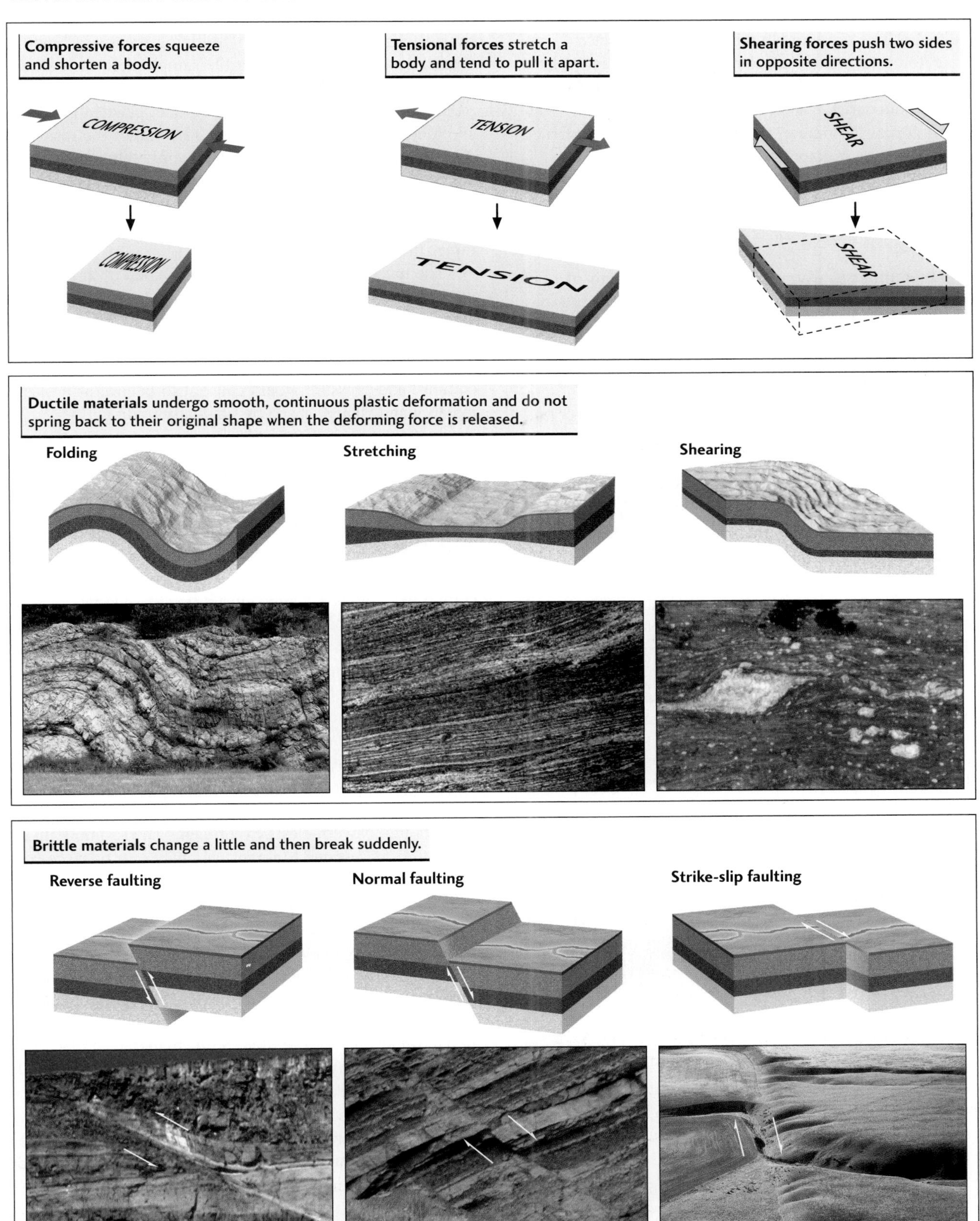

Figure Story 11.6 Rocks are deformed by folding or by faulting when they are subjected to different kinds of tectonic forces. Geologists see the pattern of deformation in the field and infer the nature of the forces that caused it. [(*top row*) Breck Kent, Kip Hodges, John Grotzinger. (*bottom row*) Gerald and Buff Corsi/Visuals Unlimited, Peter Kaufman, John S. Shelton.]

active at the three types of boundaries between plates. Compressive forces dominate at convergent boundaries, where plates are pushed together. Tensional forces dominate at divergent boundaries, where plates are pulled apart. And shearing forces dominate at transform-fault boundaries, where plates slide horizontally past each other.

What Determines Whether a Rock Bends or Breaks?

For years geologists were baffled by the problem of how rocks, which seem strong and rigid, could be distorted into folds or broken along faults by tectonic forces. Although geology relies heavily on field observations, geologists conducted laboratory experiments to resolve the puzzle of why rock formations fold in one place and fracture in another. Different types of rocks were squeezed at the low pressures and temperatures characteristic of conditions near the surface, as well as at temperatures and pressures high enough to simulate conditions at a depth of about 30 km in the Earth.

In one such experiment, the investigators applied compressive force by pushing down with a piston on one end of a small cylinder of marble. At the same time, they applied pressure to all sides of the sample to simulate the confining pressure to which materials deep in the crust are subjected by the weight of the overlying rock. (Confining pressure is like the pressure that you feel all over your body in a deep underwater dive.) Under the low confining pressures found at shallow depths in the crust, the marble sample deformed by fracturing. Under the high confining pressures found at greater crustal depths, the marble sample deformed slowly and steadily without fracturing. It behaved as a pliable, or *ductile,* material—that is, it deformed plastically into a shortened, bulging shape. **Figure 11.7** shows the results of this experiment. In other experiments, geologists have shown that if rocks are hotter when forces are applied, this kind of smooth, plastic deformation occurs more readily (just as chocolate softens and bends when it is warmed, rather than breaking).

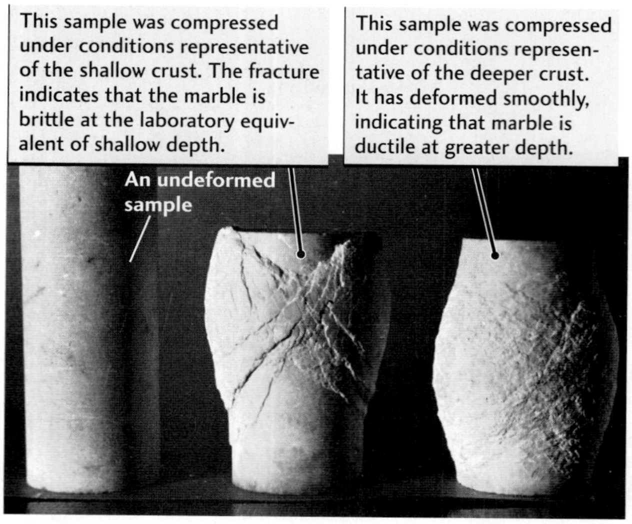

Figure 11.7 Results of laboratory experiments conducted to discover how a rock—in this case, marble—is deformed by compressive forces. [M. S. Patterson, Australian National University.]

The investigators concluded that if a bed of this particular marble were subjected to tectonic forces near the surface, where temperature and confining pressure are relatively low, it would tend to deform by fracturing and faulting. If it were deeper than a few kilometers, however, the increased temperature and pressure would cause it to behave as a plastic material and change shape by gradually folding.

Brittleness and Ductility Under Natural Conditions

The experiments just described are examples of the brittle or ductile response of rocks when they are subjected to deformational forces. As force is increased, a **brittle material** undergoes little change until it breaks suddenly, A **ductile material,** by contrast, undergoes smooth and continuous plastic deformation and does not spring back to its original shape when the deforming force is released. Glass that is near room temperature is a familiar brittle material, modeling clay a ductile one. Marble is brittle at shallow depths but ductile deeper in the crust (see Figure 11.7).

Natural conditions are more complex than laboratory conditions. Tectonic forces are applied over millions of years, whereas a laboratory experiment may be performed in a few hours. Nevertheless, experiments shed some light on the way rocks respond to forces, and they give us more confidence in our interpretations of field evidence. When we see folds and fractures in the field, we can remember that some rocks are brittle, others are ductile, and the same rock can be brittle at shallow depths and ductile deep in the crust. The type of rock and the rate of deformation affect how rocks behave under applied forces. For example, we should expect the crystalline *basement rocks* (old, underlying igneous or metamorphic rocks) to be more brittle than the ductile young sediments that may cover them. We also expect more brittle faulting in rocks that have been deformed rapidly and more ductile folding in those that have been deformed slowly. (Think of Silly Putty®, which deforms as a ductile clay when you squeeze it slowly but breaks into pieces when you smash it quickly onto a hard surface.)

Deformation Textures

Tectonic movements cause the brittle parts of Earth's crust to crack and slip. As the rocks along a fault plane shear past

(a)

(b)

Figure 11.8 (a) Fault breccia. (b) Mylonite developed in the Great Slave Lake shear zone, Northwest Territories, Canada. The rock was originally a granite. As a result of intense shearing, the large, originally angular potassium feldspar crystals have been rolled and transformed into smooth balls. [(a) Courtesy of Peter Kaufman; (b) John Grotzinger.]

each other, they grind and mechanically fragment solid rock and promote recrystallization under the high pressures. Where rocks are near the surface and brittle deformation occurs, shearing produces rocks with *cataclastic textures,* such as *fault breccias* (**Figure 11.8**a).

In contrast, if shearing occurs at a depth where temperature and pressure are high enough for ductile deformation to occur, metamorphic rocks called *mylonites* are formed (Figure 11.8b). The movement of rock surfaces against each other recrystallizes and granulates minerals and strings them out in bands or streaks. The textural effects of deformation are most obvious in mylonites, but they are also prominent in cataclastic rocks. Development of mylonites typically occurs at greenschist to amphibolite grade metamorphism (see Chapter 9).

The San Andreas fault of southern California makes a good case study of how deformation textures might relate to changes in temperature and pressure with depth. This fault marks the boundary between the Pacific Plate and the North American Plate and extends right through the crust and down into the mantle. Up to a depth of a few kilometers, the fault zone is thought to be characterized by cataclastic textures, indicating deformation during low-grade metamorphism. When the rocks fracture, they generate earthquakes. At depths greater than about 15 km, however, the fault is thought to be characterized by a broad zone of deformation that produces mylonites, indicating deformation during high-grade metamorphism. Because the deformation that generates mylonites occurs by sliding and not by fracturing, no earthquakes are generated at this depth (see Chapter 19).

How Rocks Fracture: Joints and Faults

We have seen that the way rocks deform depends on the kinds of forces to which they are subjected and the conditions under which the forces are applied. Some layers crumple into folds, and some fracture. There are two kinds of fractures, joints and faults. A **joint** is a crack along which there has been no appreciable movement. A **fault** is a fracture with relative movement of the rocks on both sides of it, parallel to the fracture. Joints and faults tell geologists something about the forces a region has experienced in the past.

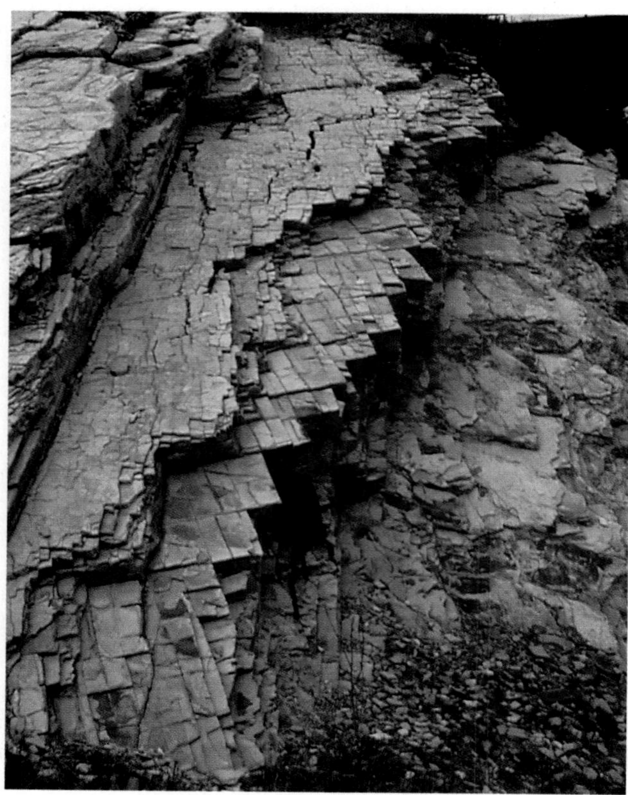

Figure 11.9 Intersecting sets of joints (backpack for scale). [Courtesy of Peter Kaufman.]

Joints

Joints are found in almost every outcrop. Some are caused by tectonic forces. Like any other easily broken material, brittle rocks fracture more easily at flaws or weak spots when they are subjected to pressure. These flaws can be tiny cracks, fragments of other materials, or even fossils. Regional forces—compressive, tensional, or shearing—that have long since vanished may leave a set of joints as their imprint.

Joints can also form as a result of nontectonic expansion and contraction of rocks. Regular patterns of joints are often found in plutons and lavas that have cooled, contracted, and cracked. Erosion can strip away surface layers, releasing the confining pressure on underlying formations and allowing the rocks to expand and split at flaws.

When a formation fractures at many places and develops joints, the joints are usually only the beginning of a series of changes that will significantly alter the formation. For example, joints provide channels through which water and air can reach deep into the formation and speed the weathering and weakening of the structure internally. If two or more sets of joints intersect, weathering may cause a formation to break into large columns or blocks (**Figure 11.9**). The circulation of hydrothermal fluids through joints can deposit minerals such as quartz and dolomite, forming *veins.* Quartz veins that fill the joints in cooling granitic intrusions sometimes contain significant amounts of gold, silver, and other valuable ore minerals. Much of the gold mined in the great California gold rush can be traced to such deposits.

Faults

The tectonic forces that cause faulting are particularly intense near plate boundaries. Faults are common features of mountain belts, which are associated with plate collisions, and of rift valleys, where plates are being pulled apart. In some transform faults, such as the San Andreas fault of California, the horizontal *offset,* which measures the relative displacement of the two plates, can amount to hundreds of kilometers over tens of millions of years (**Figure 11.10**).

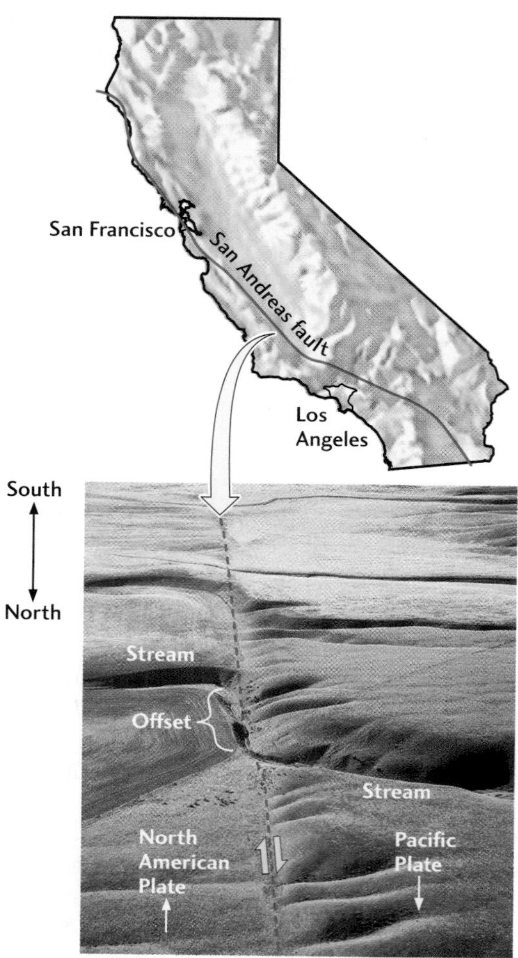

Figure 11.10 View of the San Andreas fault looking southeastward. The fault runs from top to bottom (dashed line) near the middle of the photograph. Note the offset of the stream (Wallace Creek) as it crosses the fault, which is caused by the northward movement of the Pacific Plate with respect to the North American Plate. [John S. Shelton.]

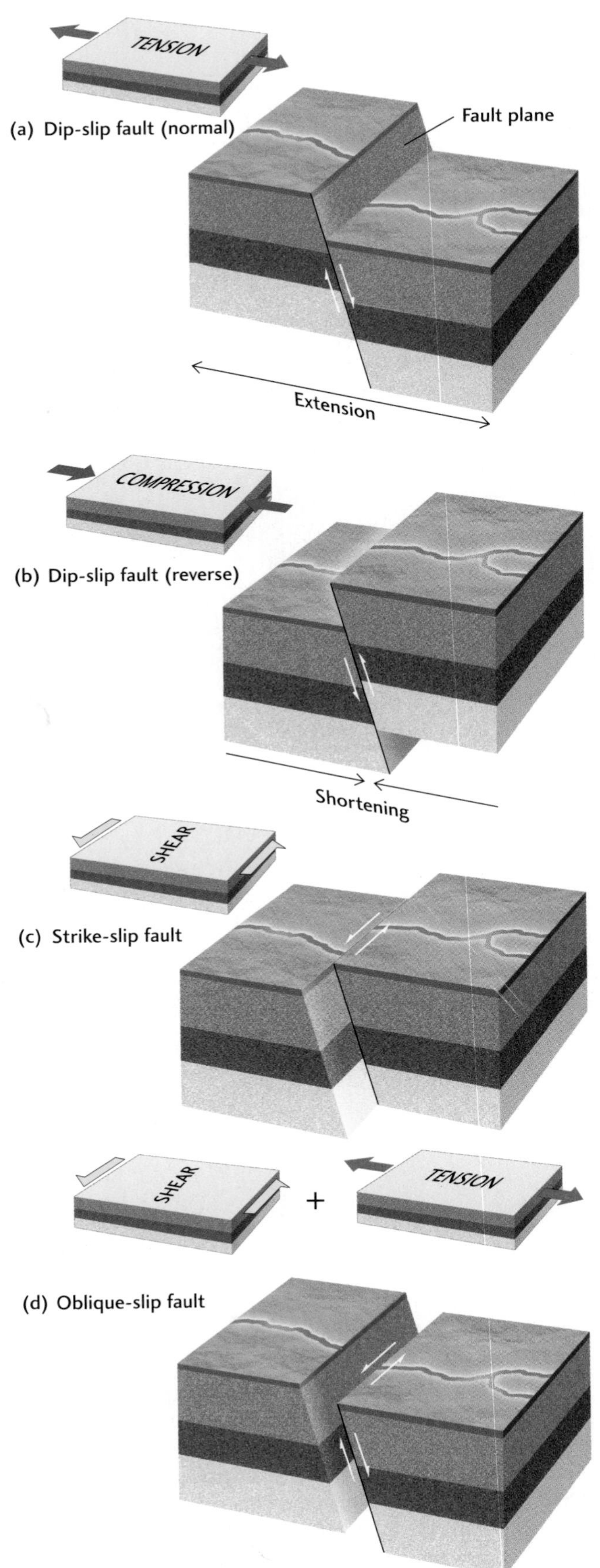

Crustal forces in the interiors of plates can also cause faulting in rocks far from plate boundaries.

Faults can be very large, like the San Andreas, or very small, like the faults shown in Figure 11.2. All faults, regardless of size, can be classified by the direction of relative movement, or slip, at the fracture (**Figure 11.11**). The surface along which the formation fractures and slips is the *fault plane.* Two terms defined earlier, *dip* and *strike,* describe the orientation of the fault plane. A *dip-slip fault* is one in which there is relative movement of the formations up (reverse) or down (normal) the dip of the fault plane (Figure 11.11a and b). A *strike-slip fault* is one in which the movement is horizontal, parallel to the strike of the fault plane (Figure 11.11c). (A strike-slip fault that forms a plate boundary is called a *transform fault.*) A movement along the strike and simultaneously up or down the dip is described as an *oblique-slip fault* (Figure 11.11d). Dip-slip faults are associated with compression or tension, and strike-slip faults indicate that horizontal shearing forces were at work, as illustrated in Figure Story 11.6. An oblique-slip fault results from shear in combination with either compression or tension.

Faults need to be characterized further, because the movement can be up or down or right or left, as Figure 11.11 indicates. In a *normal fault,* the rocks above the fault plane move down in relation to the rocks below the fault plane, causing an extension of the section. (Figure 11.2 shows a small-scale normal fault in an outcrop.) A *reverse fault,* then, is one in which the rocks above the fault plane move upward in relation to the rocks below, causing a shortening of the section. Reverse faulting results from compression. As we face a strike-slip fault, if the block on the other side is displaced to the right, the fault is a *right-lateral fault;* if the block on the other side of the fault is displaced to the left, it is a *left-lateral fault.* These movements result from shearing forces.

Finally, a reverse fault in which the dip of the fault plane is less than 45°, so that the overlying block is pushed mainly horizontally, is a *thrust fault* (**Figure 11.12**). Thrust faults in which one block has been pushed horizontally over the other are often found in intensely deformed mountain belts. These *overthrusts* occur where the crust accommodates compressional forces by breaking and shortening as one sheet overrides the other. Often the shortening may cover many tens of kilometers and involve multiple thrust faults.

Figure 11.11 Types of faults. (a) Normal faults, caused by tensional forces, result in extension. (b) Reverse faults, caused by compressive forces, result in shortening. (c) Strike-slip faults are associated with shearing forces. (d) Oblique slip suggests a combination of shear and compression or tension.

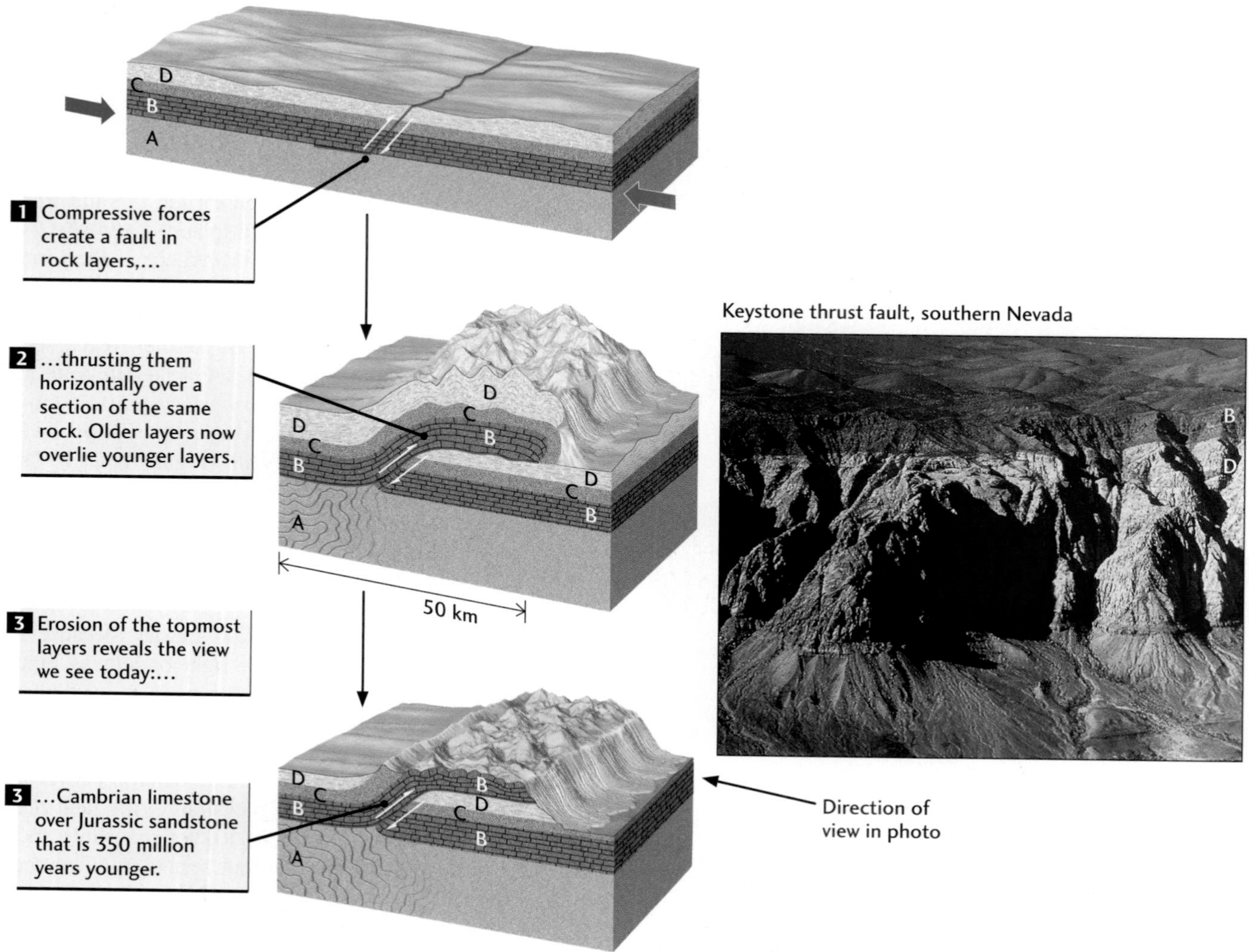

Figure 11.12 The Keystone thrust fault of southern Nevada is a large-scale overthrust sheet of a kind found in California and southern Nevada. Compressive forces have detached a sheet of rock layers—D, C, B—and thrust it a great distance horizontally over the section D, C, B, A. [Photo by John S. Shelton.]

Tensional forces, which leave behind normal faults as evidence of their action, may split a plate apart. This splitting can result in the development of a **rift valley**—a depression where blocks are dropped down relative to two flanking blocks (**Figure 11.13**). The tensional forces create a long, narrow trough bounded on each side by one or more parallel normal faults. The rift valleys of East Africa, the rifts of mid-ocean ridges, the Rhine River valley, and the Red Sea Rift are well-known rift valleys.

Geologists recognize faults in the field in several ways. A fault may form a *scarp* (small cliff) that marks the trace of the fault across the ground surface (**Figure 11.14**). If the relative movement has been large, as it is for transform faults like the San Andreas, the formations currently facing each other across the fault will probably differ in lithology and age (see Figure 11.10). The offset section of what was once a single formation is often so far away that it cannot be found. When movements are smaller, offset features can be observed and measured. (For example, look again at the small-scale fault in Figure 11.2. See if you can match up the offset beds.) In establishing the time of faulting, geologists use a simple rule: a fault must be younger than the youngest rocks that it cuts (the rocks had to be there before they could break) and older than the oldest undisrupted formation that covers it.

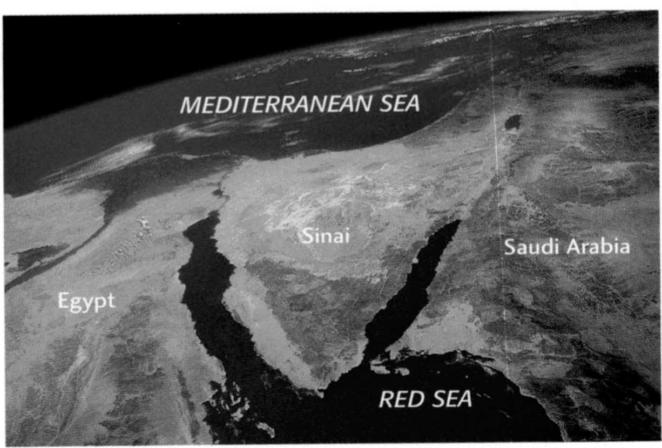

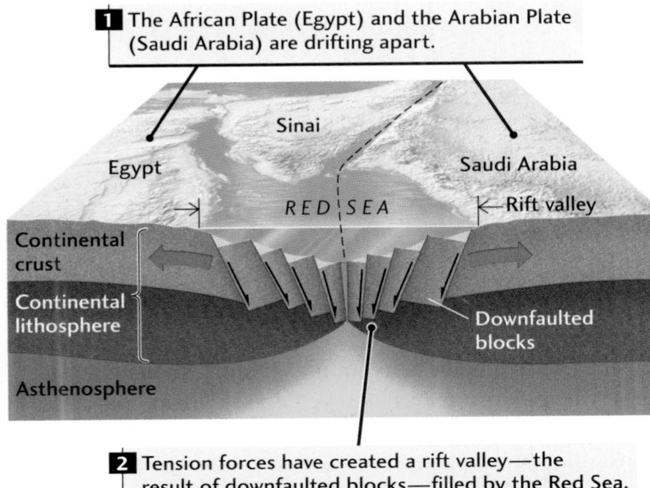

Figure 11.13 A rift valley results from tensional forces and normal faulting. The African Plate, on which Egypt rides, and the Arabian Plate, bearing Saudi Arabia, are drifting apart. The tensional forces have created a rift valley, filled by the Red Sea. [NASA/TSADO/Tom Stack.] The diagram shows parallel normal faults bounding the rift valley in the crust beneath the sea.

Figure 11.14 This scarp is a fresh surface feature that formed when a new reverse fault broke and caused a devastating earthquake in Armenia in 1988. In a few decades, the near-vertical scarp will erode into a gentler slope. [Armando Cisternas, Université Louis Pasteur.]

How Rocks Fold

Folds, like faults, are the signatures of deformation forces that result from plate tectonics. The term *fold* implies that an originally planar structure, such as a sedimentary bed, has been bent. The deformation can be produced by either horizontal or vertical forces in the crust, just as pushing in on opposite edges of a piece of paper or pushing up or down on one side or the other can fold it. Folding is a common form of deformation observed in layered rocks, most typically in mountain belts. In many young mountain systems where erosion has not yet erased them, majestic, sweeping folds can be traced, some of them with dimensions of many kilometers (as in the photograph at the beginning of this chapter). On a much smaller scale, very thin beds can be crumpled into folds a few centimeters long (**Figure 11.15**). The bending can be gentle or severe, depending on the magnitude of the applied forces, the length of time that they were applied, and the ability of the beds to resist deformation.

Figure 11.15 Small-scale folds in a Precambrian iron formation, Beresford Lake area, Manitoba. Very thin layers have been crumpled into folds a few centimeters long. [Geological Survey of Canada.]

Types of Folds

Layered rocks can fold in several basic ways in response to compressive forces, depending on the properties of the rocks and the magnitude and direction of the applied force. Geologists have developed a vocabulary to specify different types of folds and their parts. Upfolds, or arches, of layered rocks are called **anticlines;** downfolds, or troughs, are called **synclines** (**Figure Story 11.16**). The two sides of a fold are its *limbs.* The **axial plane** is an imaginary surface that divides a fold as symmetrically as possible, with one limb on either side of the plane. The line made by the lengthwise intersection of the axial plane with the beds is the *fold axis.* A fold with a *horizontal* axis is shown in Figure Story 11.16b. If the axis is not horizontal, the fold is called a *plunging fold.*

Not every fold has a vertical axial plane with limbs dipping symmetrically away from the axis, as in Figure Story 11.16c. With increasing amounts of deformation, the folds can be thrown into asymmetrical shapes, with one limb dipping more steeply than the other (Figure Story 11.16d). This can also occur if the direction of the deformational force is oblique to the layering of the beds. Such *asymmetrical folds* are common. When the deformation is intense and one limb has been tilted beyond the vertical, the fold is called an **overturned fold.** Both limbs of an overturned fold dip in the same direction, as shown in Figure Story 11.16e; but the order of the layers in the bottom limb is precisely the reverse of their original sequence—that is, older rocks are on top of younger rocks. The folds in Figure Story 11.16e have been overturned to such an extent that the axial plane is approaching the horizontal. One limb has been rotated into a completely upside-down sequence, with older beds on top of younger beds. (See Figure 11.1 for another example.)

Folds change along their axes. Follow the axis of any fold in the field and sooner or later the fold dies out or appears to plunge into the ground. **Figure 11.17** diagrams the geometry of *plunging anticlines* and *plunging synclines.* In eroded mountain belts, a zigzag pattern of outcrops may appear in the field after erosion has removed much of the surface rock. The eroded Valley and Ridge belt

Figure Story 11.16 (a) Anticlines fold upward; synclines fold downward, (b) horizontal and plunging folds, (c) symmetrical, (d) asymmetrical, and (e) overturned folds. Increasing horizontal force can deform symmetrical folds into asymmetrical and overturned folds. [(*symmetrical fold*) Courtesy of Mark McNaught. (*overturned fold*) John Grotzinger.]

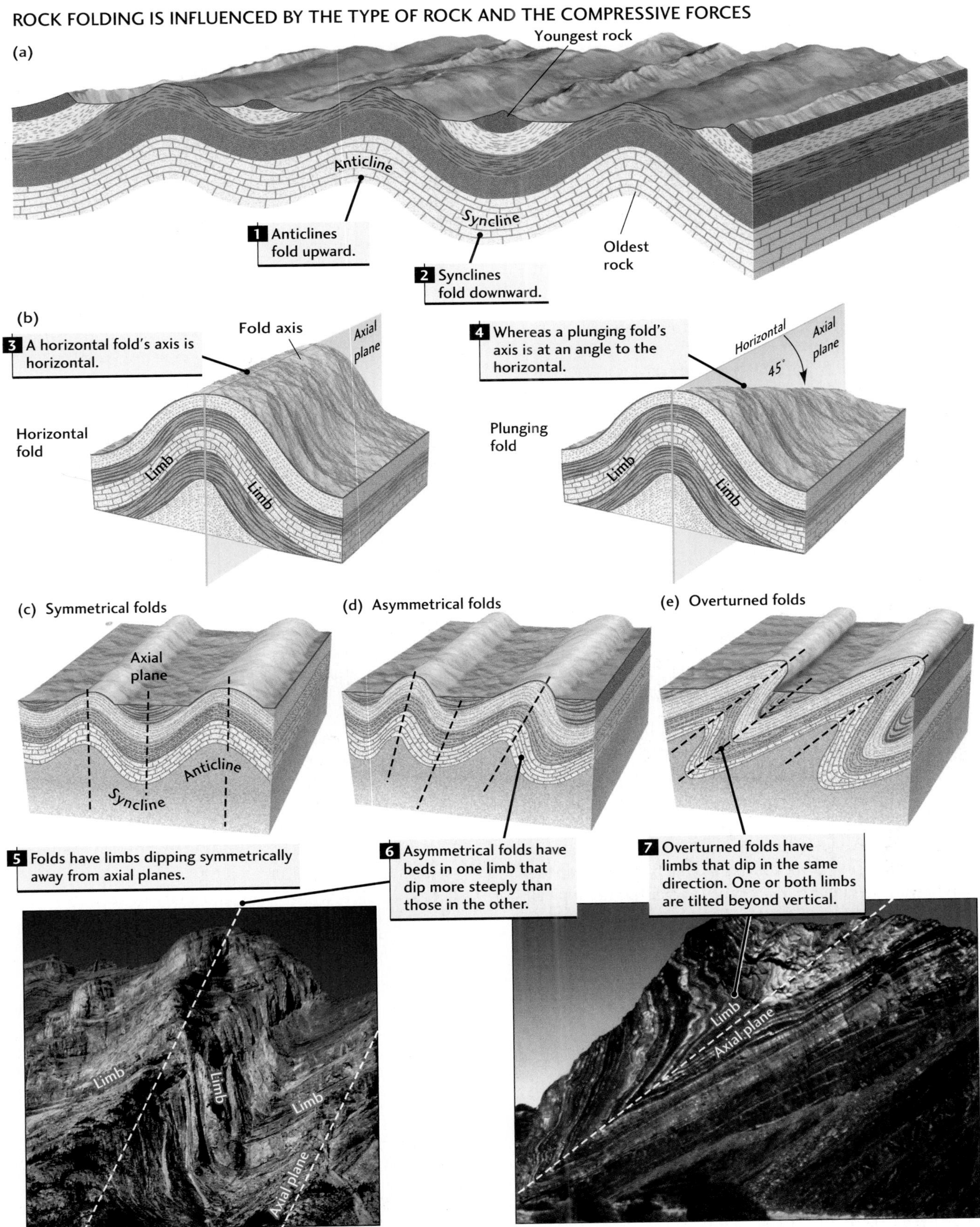
ROCK FOLDING IS INFLUENCED BY THE TYPE OF ROCK AND THE COMPRESSIVE FORCES
(a)
Youngest rock
Anticline
Syncline
1 Anticlines fold upward.
2 Synclines fold downward.
Oldest rock
(b)
Fold axis
Axial plane
3 A horizontal fold's axis is horizontal.
Horizontal fold
Limb
Limb
4 Whereas a plunging fold's axis is at an angle to the horizontal.
Horizontal
45°
Axial plane
Plunging fold
Limb
Limb
(c) Symmetrical folds
Axial plane
Anticline
Syncline
(d) Asymmetrical folds
(e) Overturned folds
5 Folds have limbs dipping symmetrically away from axial planes.
6 Asymmetrical folds have beds in one limb that dip more steeply than those in the other.
7 Overturned folds have limbs that dip in the same direction. One or both limbs are tilted beyond vertical.
Limb
Limb
Limb
Axial plane
Limb
Axial plane

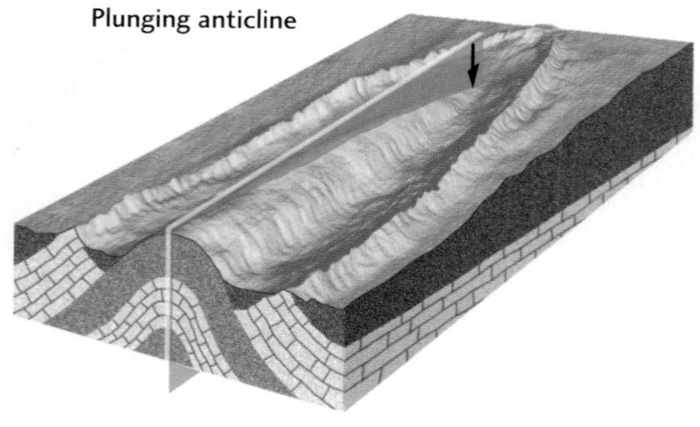

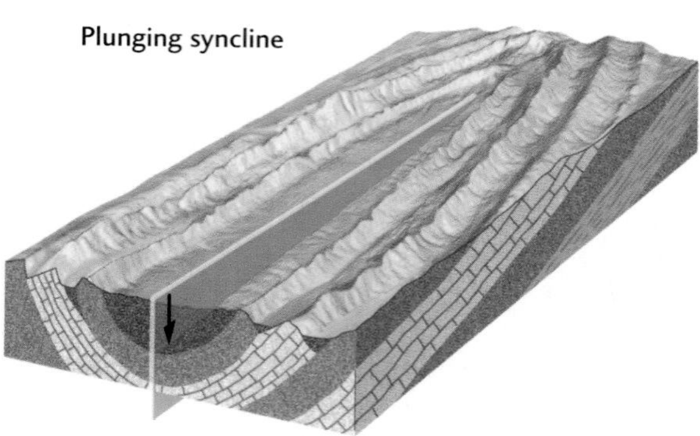

Figure 11.17 The geometry of plunging folds. Note the converging pattern of the layers where they intersect the surface.

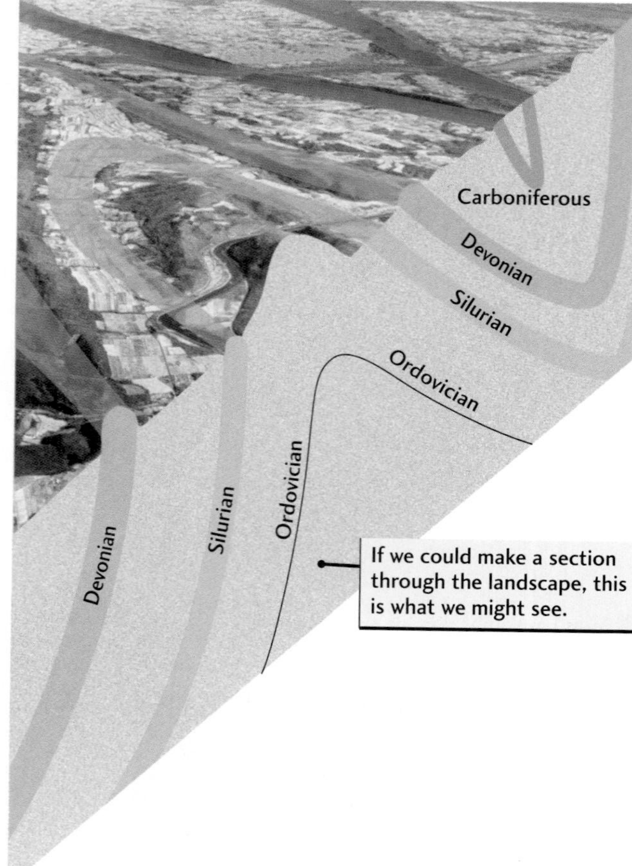

Figure 11.18 *(top)* The erosional remnants of plunging folds show a characteristic zigzag pattern in this view of the Valley and Ridge belt of the Appalachian Mountains, 48 km northwest of Harrisburg, Pennsylvania. *(bottom)* In this drawing, the imaginary trench reveals the subsurface structure. [From J. S. Shelton, *Geology Illustrated* (San Francisco: W. H. Freeman, 1966).]

of the Appalachians shows this characteristic pattern (**Figure 11.18**).

A **dome** is an anticlinal structure, a broad circular or oval upward bulge of rock layers. The flanking beds of a dome encircle a central point and dip radially away from it (**Figure 11.19**). A **basin** is a synclinal structure, a bowl-shaped depression of rock layers in which the beds dip radially toward a central point. Domes and basins are typically many kilometers in diameter. They are recognized in the field by outcrops with the characteristic circular or oval shapes seen in Figure 11.19. Domes are very important in oil geology because oil is buoyant and tends to migrate upward through permeable rocks. If the rocks at the high point of a dome are not easily penetrated, the oil becomes trapped against them.

Domes and basins can result from several types of deformation. Some domes can be attributed to rising bodies of buoyant material—such as magma, hot igneous rock, or salt—that push the overlying sediments upward. Others are caused by multiple episodes of deformation, for instance, when rocks are compressed in one direction and then again in a direction nearly perpendicular to the original direction. Some basins form when a heated portion of the crust cools

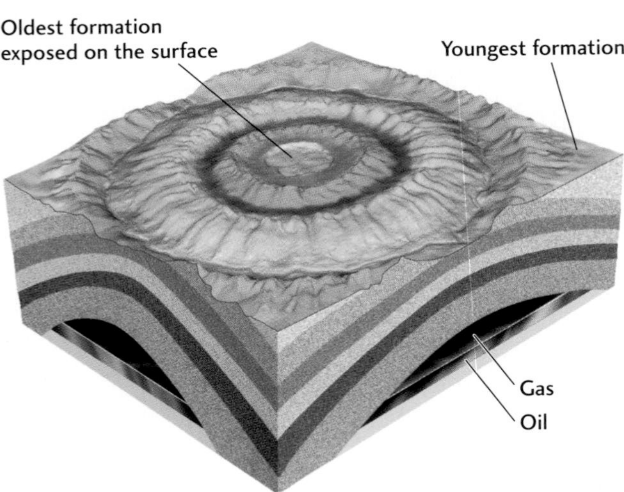

Figure 11.19 Sinclair dome, an eroded dome in strata 10 km east of Rawlings, Wyoming. The highway and railroad at the lower right suggest its dimensions. The characteristic circular or elliptical outcrop pattern of a dome and the extension to the subsurface are shown in the diagram. The oldest bed is in the core; the flanking formations are successively younger and dip away from the core. [Photo by John S. Shelton.]

and contracts, causing the overlying sediments to subside. Others result when tectonic forces stretch the crust. The weight of sediments deposited in a shallow sea can depress the crust, forming a basin. There are many domes and basins in the central portion of the United States. The Black Hills of South Dakota are an eroded dome; much of the lower peninsula of Michigan is a sedimentary basin.

What Geologists Infer from Folds

Observations in the field seldom provide geologists with complete information. Either bedrock is obscured by overlying soils, or erosion has removed much of the evidence of former structures. So geologists search for clues they can use to work out the relationship of one bed to another. For example, in the field or on a map, an eroded anticline would be recognized by a strip of older rocks forming a core bordered on both sides by younger rocks dipping away, as illustrated in Figure 11.5. An eroded syncline would show as a core of younger rocks bordered on both sides by older rocks dipping toward the core (**Figure 11.20**).

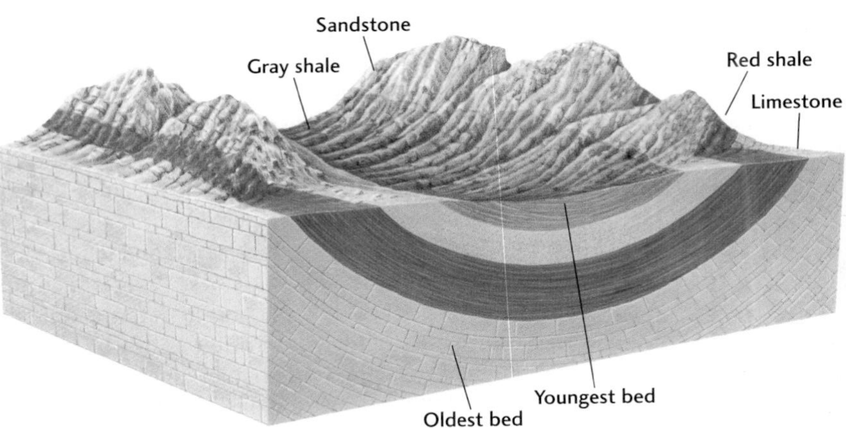

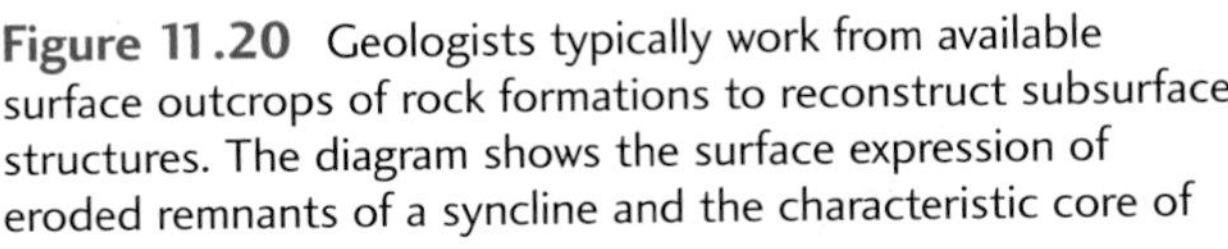

Figure 11.20 Geologists typically work from available surface outcrops of rock formations to reconstruct subsurface structures. The diagram shows the surface expression of eroded remnants of a syncline and the characteristic core of younger rocks flanked on both sides by older rocks dipping toward the core. The photo shows a syncline exposed in quarry near Middletown, Virginia. [Courtesy of Christopher M. Bailey, College of William and Mary.]

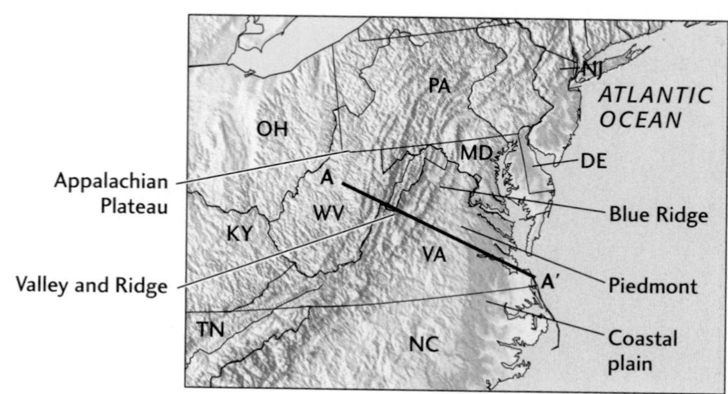

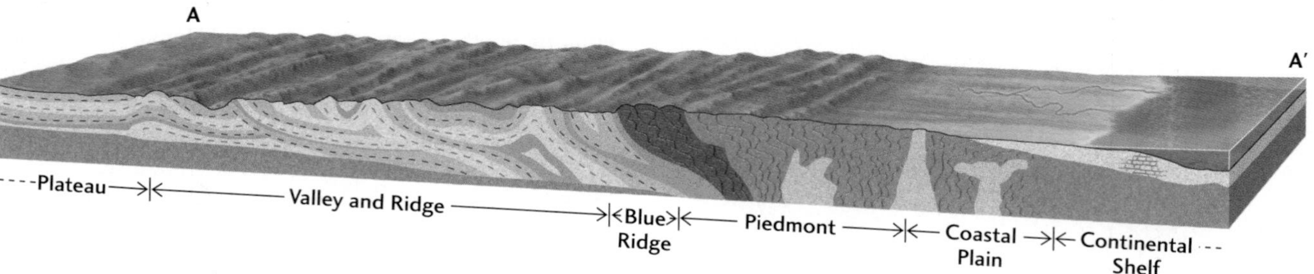

Figure 11.21 Satellite photo (above) and cross section (below) of the Valley and Ridge province of the Appalachian Mountains, the eroded remnant of a folded mountain belt. [Photo by NASA; cross section after D. Johnson, *Stream Sculpture on the Atlantic Slope* (New York: Columbia University Press, 1931).]

Folds are typically found in elongated groups. A strip of land in which the rock layers are folded—that is, a *fold belt*—suggests to a geologist that the region was compressed at one time by horizontal tectonic forces. The Valley and Ridge province of the Appalachians is a folded mountain belt (**Figure 11.21**). A satellite photograph taken from 320 km above the Valley and Ridge province is shown in Figure 11.21. We will see in Chapter 20 that an ancient plate collision accounts for the wrinkling of the once-flat layers of sedimentary rocks in this region.

Unraveling Geologic History

Usually the geologic history of a region is a succession of episodes of deformation and other geologic processes. Let us take what appears to be a complicated example and see how some of the concepts introduced in this chapter lead to a simple interpretation. The cross sections in **Figure 11.22** represent a few tens of kilometers of a geologic province that underwent a succession of events. First, horizontal layers of sediment were deposited, then they were tilted and folded by horizontal forces of compression. After that, they were uplifted above sea level. There, erosion gave them a new horizontal surface, which was covered unconformably by lava when forces deep in Earth's interior caused a volcanic eruption. In the final stage, horizontal stretching (tension forces) resulted in normal faulting, which broke the crust into blocks. The geologist sees only the last stage but visualizes the entire sequence. When the sedimentary beds have been identified, the geologist starts with the knowledge that the beds must originally have been horizontal and undeformed at the bottom of an ancient ocean. The succeeding events can then be reconstructed.

Present-day surface relief—such as we find in the Alps, the Rocky Mountains, the Pacific Coast Ranges, and the Himalaya—can be traced in large part to deformation that occurred over the past few tens of millions of years. These younger mountain systems still contain much of the information that the geologist needs to piece together the history of deformation. However, deformation that occurred hundreds of millions of years ago no longer shows as the rugged mountains that once existed. Erosion has left behind only the remnants of folds and faults expressed as low ridges and shallow valleys (see Figure 11.18). As we

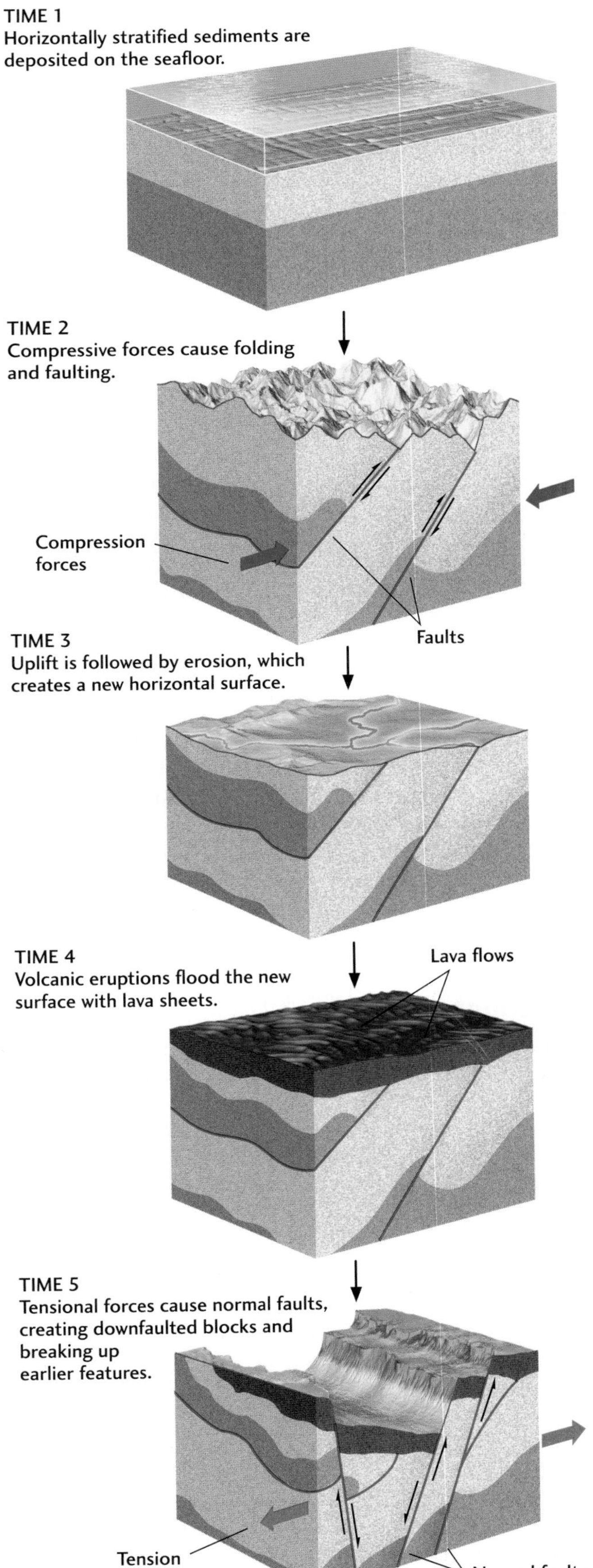

will see in Chapter 21, even older episodes of mountain building are evident from the twisted, highly metamorphosed formations that constitute the basement rocks of the interiors of continents.

Deformation—in the form of mountain belts with their structures of folds and faults, rift valleys, and strike-slip faults—leaves its unmistakable mark on the landscape. These topographic expressions are often guides to the deformation structures that shaped them. Even such relatively small-scale features as the shapes of hills and valleys and the courses of streams may be controlled by the complex interaction between underlying structures and erosion.

It is important to remember that topography is not determined by structure alone. A valley commonly forms in the trough of a syncline and a ridge at the crest of an anticline. However, we should not expect the crests of anticlines always to form ridges and the troughs of synclines always to become valleys. An important factor in the shaping of landforms on stratified rocks is the amount of resistance that the individual beds offer to weathering and erosion, as well as whether the layers are tilted, folded, or faulted.

We have seen that there are patterns in the way rocks deform that relate to the forces acting in Earth's crust. Plate movements play an important role in the generation of these forces. Geologists have learned to decipher these patterns, beginning with the formation of rocks and then reconstructing their subsequent deformation and erosion.

SUMMARY

What do laboratory experiments tell us about the way rocks deform when they are subjected to crustal forces? Laboratory studies show that some rocks deform as brittle materials, others as ductile materials. These qualities depend on the kind of rock, the temperature, the surrounding pressure, the magnitude of the force, and the speed with which it is applied.

What are some of the deformation structures that show up in rocks in the field? Among the geologic structures in rock formations that result from deformation are joints, faults, folds, domes, and basins.

Figure 11.22 Stages in the development of a fictitious geologic province. A geologist sees only the last stage and attempts to reconstruct from the structural features all the earlier stages in the history of a region.

What kinds of forces take part in the formation of these structures? Joints are caused by regional stresses or by the cooling and contraction of the rocks. Normal faults can be caused by tensional or stretching forces, which occur at boundaries where plates diverge. Reverse faults and thrusts can be produced by compressive forces, such as those found at boundaries where plates converge. Shearing forces can produce strike-slip faults. Folds are usually formed by compressive forces, such as occur at boundaries where plates collide. Some domes are produced by the intrusion of magma deep in the crust. Basins can form when tensional forces stretch the crust or when a heated portion of the crust cools and contracts. The weight of sediments deposited in a basin can contribute to its deepening.

How do geologists reconstruct the history of a region? Geologists see the end results of a succession of events: deposition, deformation, erosion, volcanism, and so forth. They deduce the deformational history of a region by identifying and fixing the ages of the rock layers, recording the geometric orientation of the beds on maps, mapping folds and faults, and reconstructing cross sections of the subsurface consistent with the surface observations. They can ascertain the relative age of a deformation by finding a younger undeformed formation lying unconformably on an older deformed bed.

Key Terms and Concepts

anticline (p. 248)	faulting (p. 237)
axial plane (p. 248)	folding (p. 237)
basin (p. 250)	joint (p. 243)
brittle material (p. 242)	overturned fold (p. 248)
compressive force (p. 240)	rift valley (p. 246)
dip (p. 239)	shearing force (p. 240)
dome (p. 250)	strike (p. 238)
ductile material (p. 242)	syncline (p. 248)
fault (p. 243)	tensional force (p. 240)

Exercises

This icon indicates that there is an animation available on the Web site that may assist you in answering a question.

1. Why do some rock layers fold and others break into faults when they are subjected to crustal forces?

2. Which types of deformation structures would be expected at the three types of plate boundaries?

3. How would you identify a fault in the field? How would you tell whether it was a normal, reverse, or strike-slip fault?

4. If you found tilted beds in the field, how would you tell if they were part of an anticline or a syncline?

5. Draw a cross section of a rift valley and indicate by arrows the directions of the forces that produced it. Do the same for a thrust fault.

6. What was the direction of the crustal forces that deformed the Appalachian block depicted in Figure 11.21?

7. Draw a geological cross section that tells the following story. A series of marine sediments are deposited and subsequently deformed into folds and thrust faults. These events are followed by erosion. Volcanic activity ensues, and lava flows over the eroded surface. A final stage of high-angle faulting breaks the crust into several upheaved and downdropped rocks.

Thought Questions

This icon indicates that there is an animation available on the Web site that may assist you in answering a question.

1. If you were asked to describe the geologic history of a region that had not yet been explored, how would you proceed?

2. Other than crustal forces, what might cause rocks to deform?

3. Although anticlines are upfolds and synclines are downfolds, we often find synclinal ridges and anticlinal valleys, as in Figure Story 11.16 and Figure 11.21. Explain why. Try to devise a sequence of rock layers and a series of events that could lead to each of these outcomes.

Short-Term Team Projects

The Formation of Folds and Faults

Devise a laboratory experiment to illustrate the formation of the several types of folds and faults. *Hint:* Use layers of differently colored clays to represent a sequence of undisturbed beds.

Suggested Readings

Hatcher, R. D., Jr. 1995. *Structural Geology: Principles, Concepts, and Problems.* New York: Macmillan.

Ramsay, J. F. 1987. *Techniques of Modern Structural Geology: Folds and Fractures.* Orlando, Fla.: Academic Press.

Twiss, R. J., and E. M. Moores. 1992. *Structural Geology.* New York: W. H. Freeman.

A giant landslide has ripped houses apart, leaving them unsupported.
[Tom McHugh/Photo Researchers.]

CHAPTER

12

Mass Wasting

"The snout of the debris flow was twenty feet high, tapering behind. Debris flows sometimes ooze along, and sometimes move as fast as the fastest river rapids. The huge dark snout was moving nearly five hundred feet a minute and the rest of the flow behind was coming twice as fast, making roll waves as it piled forward against itself—this great slug, as geologists would describe it, this discrete slug, this heaving violence of wet cement. Already included in the debris were propane tanks, outbuildings, picnic tables, canyon live oaks, alders, sycamores, cottonwoods, a Lincoln Continental, an Oldsmobile, and countless boulders five feet thick."

JOHN MCPHEE, *THE CONTROL OF NATURE*

John McPhee is a well-known author of popular books on various aspects of geology. He was writing about the rain-soaked, stormy night of February 9, 1978, in the San Gabriel Mountains, which tower over the northern part of the Los Angeles area. The Shields Canyon debris flow he describes was a **mass movement**—one of many kinds of downhill movements of masses of soil, rock, mud, or other unconsolidated (loose and uncemented) materials under the force of gravity. The masses are not pulled down primarily by the action of an erosional agent, such as wind, running water, or glacial ice. Instead, mass movements occur when the force of gravity exceeds the strength (resistance to deformation) of the slope materials. Earthquakes, floods, and other geological events trigger such movements. The materials then move down the slope, either at a slow or very slow rate or as a sudden, sometimes catastrophic, large movement. In various combinations of falling, sliding, and flowing, mass movements can displace small, almost imperceptible amounts of soil down a gentle hillside or they can be huge landslides that dump tons of earth and rock on valley floors below steep mountain slopes.

Every year, mass movements take their toll of lives and property throughout the world. In late October and early November of 1998, one of the most catastrophic hurricanes of the century, Hurricane Mitch, dropped torrential rains on Central America, saturating the ground and causing raging floods and landslides. At least 9000 people were killed, and billions of dollars in damage was done as the floods and slides laid waste to once-fertile land and crops of corn, beans, coffee, and peanuts. One of the hardest-hit places was near the Nicaragua-Honduras border, where a series of landslides and mudflows buried at least 1500 people. Dozens of villages were simply obliterated, engulfed by a sea of mud. The flanks of a crater on Casita volcano collapsed and started a series of slides and flows that were described as a moving wall of mud more than 7 m high. Those in the direct path of the avalanche could not escape, and many were buried alive as they tried to outrun the fast-moving mud.

Because mass movements are responsible for so much destruction, we want to be able to predict them, and we certainly want to refrain from provoking them by unwise interference with natural processes. We cannot prevent most natural mass movements, but we can control construction and land development to minimize losses.

Mass wasting includes all the processes by which masses of rock and soil move downhill under the influence of gravity, eventually to be carried away by other transport agents. Mass wasting is one of the consequences of weathering and rock fragmentation. It is an important part of the general erosion of the land, especially in hilly and mountainous regions. Mass movements change the landscape by scarring mountainsides as great masses of material fall or slide away from the slopes. The material that moves ends up as tongues or wedges of debris on the valley floor, sometimes piling up and damming a stream running through the valley. The scars and debris deposits, mapped in the field or from aerial photographs, are clues to past mass movements. By reading these clues, geologists may be able to predict and issue timely warnings about new movements likely to occur in the future.

In mass movements, as in many other kinds of geological processes, human interference can have severe effects. Although human engineering works seem small compared with the natural world, they are significant. In the United States alone, just one activity—excavation for houses and other buildings—breaks up and transports more than 700 million metric tons of surface materials each year, according to some calculations. This amount far exceeds the 550 million metric tons moved annually in the United States by natural processes.

This chapter examines why masses move; what characteristics differentiate mass movements; and how weather, plate tectonic setting, and human activity contribute to mass movements.

What Makes Masses Move?

Field observations have led geologists to identify three primary factors that influence mass movements (Table 12.1):

1. *The nature of the slope materials.* They may be solid masses of bedrock, the regolith and soil formed by weathering, or sediment. Slopes may be made up of **unconsolidated materials**—loose and uncemented—or **consolidated materials**—compacted and bound together by mineral cements.

2. *The steepness and stability of slopes.* This factor contributes to the tendency of materials to fall, slide, or flow under various conditions.

3. *The amount of water in the materials.* This characteristic depends on how porous the materials are and on how much rain or other water they have been exposed to.

Table 12.1 Factors That Influence Mass Movements

Nature of Slope Material	Steepness of Slope	Water Content	Stability of Slope
	UNCONSOLIDATED		
Loose sand or sandy silt	Angle of repose	Dry Wet	High Moderate
Unconsolidated mixture of sand, silt, soil, and rock fragments	Moderate	Dry Wet	High Low
	Steep	Dry Wet	High Low
	CONSOLIDATED		
Rock, jointed and deformed	Moderate to steep	Dry or wet	Moderate
Rock, massive	Moderate	Dry or wet	High
	Steep	Dry or wet	Moderate

All three factors operate in nature, but slope stability and water content are most strongly influenced by human activity, such as excavation for building and highway construction. All three produce the same result: they lower resistance to movement, and then the force of gravity takes over and the slope materials begin to fall, slide, or flow.

The Nature of Slope Materials

Slope materials vary greatly in different kinds of terrain because they are so dependent on the details of the local geology. Thus, the metamorphic bedrock of one hillside may be badly fractured by foliation, whereas another slope only a few hundred meters away is composed of massive granite. Slopes of unconsolidated material are the least stable of all.

Unconsolidated Sand and Silt The behavior of loose, dry sand and silt illustrates how the steepness and stability of slopes influence mass movements. Children's sandboxes have made nearly everyone familiar with the characteristic slope of a pile of dry sand. The angle between the slope of any pile of sand or silt and the horizontal is the same, whether the pile is a few centimeters or several meters high. For most sands and silts, the angle is about 35°. If you scoop some sand from the base of the pile very slowly and carefully, you can increase the slope angle a little, and it will hold temporarily. If you then jump on the ground near the sandpile, however, the sand will cascade down the side of the pile, which will again assume its original slope of 35°.

The original and resumed slope angle of the sandpile is its **angle of repose,** the maximum angle at which a slope of loose material will lie without cascading down. A slope that is steeper than the angle of repose is unstable and will tend to collapse to the stable angle. Sand or silt grains form piles with slopes at and below the angle of repose because of frictional forces between the individual sand grains. However, as more and more sand grains are placed on the pile, and the slope steepens, the ability of the frictional forces to prevent sliding decreases and the pile suddenly collapses.

The angle of repose varies significantly with a number of factors, one of which is the size and shape of the particles (**Figure Story 12.1**). Larger, flatter, and more angular pieces of loose material remain stable on steeper slopes. The angle of repose also varies with the amount of moisture between particles. The angle of repose of wet sand is higher than that of dry sand because the small amount of moisture between the grains tends to bind them together so that they resist movement. The source of this binding tendency is **surface tension**—the attractive force between molecules at a surface (Figure Story 12.1b). Surface tension makes waterdrops spherical and allows a razor blade or paper clip to float on a smooth water surface. Too much water, on the other hand, separates the particles and allows them to move freely over one another. Saturated sand, in which all the pore space is occupied by water, runs like a fluid and collapses to a flat pancake shape (Figure Story 12.1c). The surface tension that binds moistened sand allows beach sculptors to create elaborate sand castles (**Figure 12.2**). When the tide comes in and saturates the sand, the structures collapse.

Unconsolidated Mixtures Slope materials composed of mixtures of unconsolidated sand, silt, clay, soil, and fragments of rock will form slopes with moderate to steep angles (see Table 12.1). The platy shape of clay minerals, the organic content of soils, and the rigidity of rock fragments are all key factors that change the ability of the material to form slopes with a specific angle.

Consolidated Materials Slopes of consolidated dry materials—such as rock, compacted and cemented sediments, and vegetated soils—may be steeper and less regular than slopes consisting of loose material. They can become unstable when they are oversteepened or denuded of vegetation. See Figures 15.15 and 16.25 for examples of steep slopes in consolidated materials. The particles of consolidated sediments such as dense clays are bound together by cohesive forces associated with tightly packed particles. **Cohesion** is an attractive force between particles of a solid material that are close together.

The resistance to movement resulting from cohesion, cementation, and the binding action of plant roots is sometimes called internal friction because it is like the friction that resists the movement of any piece of matter against another. In a material with high internal friction, the particles are not as free to move as are loose particles such as sand.

Water Content

Mass movements of consolidated materials usually can be traced to the effects of moisture, often in combination with such other factors as loss of vegetation or oversteepening of the slope. When the ground becomes saturated with water, the solid material is lubricated, the internal friction is lowered, and the particles or larger aggregates can move past one another more easily. Water may seep into the bedding planes of muddy or sandy sediments, for example, and promote the slippage of beds past one another. When consolidated materials absorb large amounts of water, the pressure of the water in the pores of the material may be great enough to separate the grains and distend the mass. Then the material may start to flow like a fluid. This process is called **liquefaction.**

Soils become more susceptible to erosion and mass movement when they are stripped of vegetation, usually by burning or deforestation. When root systems no longer bind the soil, water can penetrate it more easily and the soil becomes less stable. This is exactly what happens when forest fires are followed by rains (**Figure 12.3**). When a slope of consolidated material is oversteepened, such as by a stream that undercuts part of a valley wall, the slope becomes unstable—just as steepening the angle of a sandpile makes it unstable. Sooner or later, the unstable valley wall will move down to assume a more stable angle.

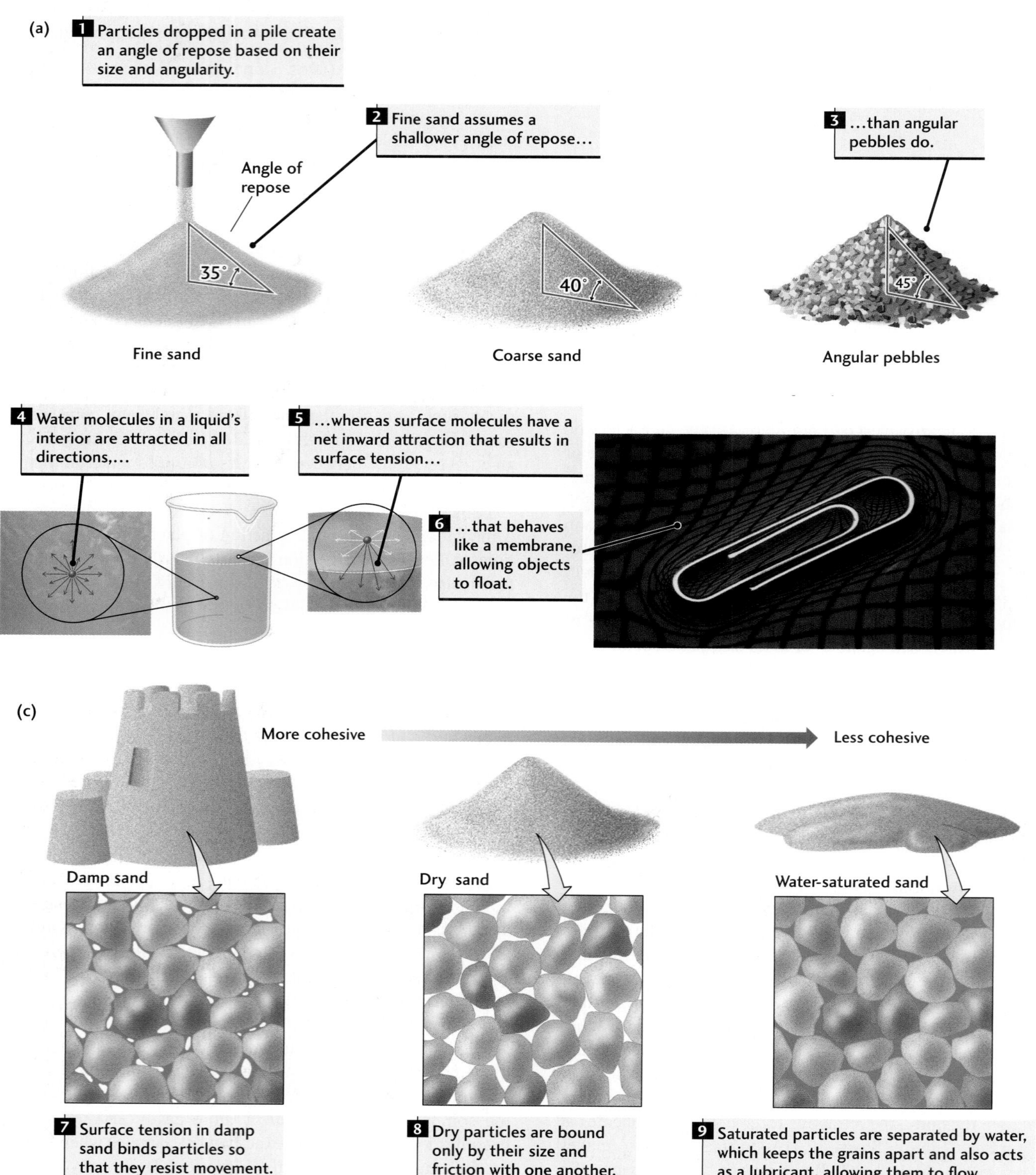

Figure Story 12.1 Slope material, slope steepness, and water content influence mass movement.

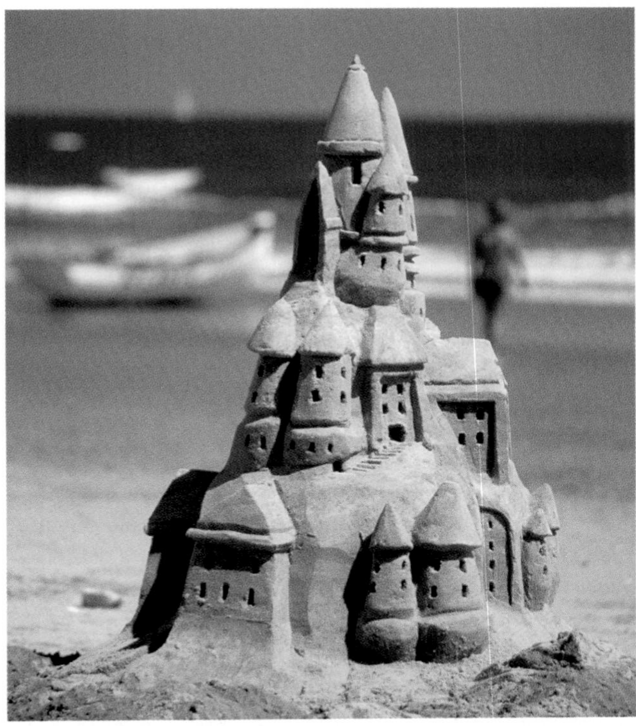

Figure 12.2 Sand castles hold their shape because they are made of damp sand. The steepness of slope is maintained by moisture between grains. [Kelly Mooney Photography/Corbis.]

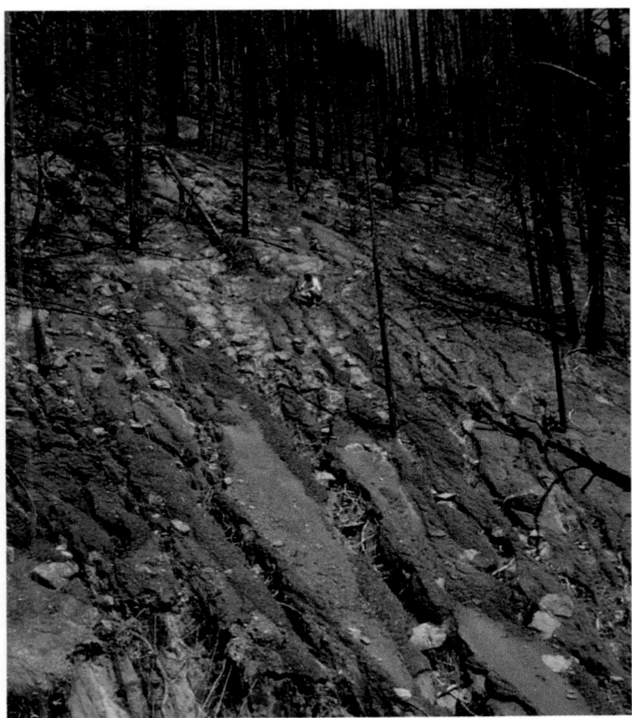

Figure 12.3 A fire that stripped soil of vegetation and root systems in Yellowstone National Park caused a weakening of the soil that made it susceptible to erosion and mass movement. [Grant Meyer.]

Steepness and Stability of Slopes

Rock slopes range from relatively gentle inclines of easily weathered shales and volcanic ash beds to vertical cliffs of hard rocks such as granite. The stability of rock slopes depends on the weathering and fragmentation of the rock. Shales, for example, tend to weather and fragment into small pieces that form a thin layer of loose rubble covering the bedrock (**Figure 12.4**). The resulting slope angle of the bedrock is similar to the angle of repose of loose, coarse sand. The weathered rubble gradually builds up to an unstable slope, and eventually some of the loose material will slide down. In many parts of the United States and Canada, highway builders have allowed unstable slopes to accumulate, leading to slides that close roads or restrict travel.

Limestones and hard, cemented sandstones in arid environments resist erosion and break into large blocks, forming steep, bare bedrock slopes above and gentler slopes covered with broken rock below. The bedrock cliffs are fairly stable, except for the occasional mass of rock that falls and rolls down to the rock-covered slope below. As rocks accumulate on this slope, it gradually steepens until it exceeds the angle of repose, becomes unstable, and slides down. Where such limestones or sandstones are interbedded with shale, slopes may be stepped. (The rocks in the Grand Canyon are an example; see Feature 10.1.) As shale slides from under the sandstone beds, the harder beds are undercut, become less stable, and eventually fall as large blocks.

The structure of the beds influences their stability, especially when the dip of the beds (the angle that the bedding makes with the horizontal; see Chapter 11) parallels the

Figure 12.4 Weathered shale slope fragmented into a thin layer of rubble, Grand Canyon. [Martin Miller.]

12.1 Reducing Loss from Landslides and Preventing Landslides

In December 1999, torrential rains triggered huge slides of mud, rocks, and trees in Vargas, Venezuela, at the foot of the Avila Mountains, resulting in the loss of as many as 30,000 lives and the evacuation of 140,000 people. About 23,000 homes were destroyed in what may be the worst natural disaster in Venezuela's history.

In some countries, losses from landslides exceed losses from all other natural hazards combined. In March 1987, for example, the vibrations of an earthquake in the mountains of Ecuador after a month of heavy rains triggered a landslide that killed at least 1000 people and ruptured the trans-Ecuadorian oil pipeline—the nation's most important source of income—at a cost of approximately $1.5 billion.

In the United States, landslides cause 25 to 50 deaths and $1 billion to $2 billion in property damage each year. Losses of life and property will continue to increase throughout the world as newly constructed homes, roads, commercial buildings, and public works push into the hilly and unstable terrains that are most susceptible to landslides.

Japan's landslide-mitigation program is one model for other countries to follow. Much of the terrain of the Japanese Islands is a combination of hilly and mountainous topography, volcanic deposits, and deformed metamorphic rocks—a mixture that makes many slopes potential landslide hazards. After several decades of severe losses of its citizens' lives and property, the Japanese government in 1958 instituted land-use management controls that continue today. The government controls how land can be used and what types of building construction are permitted. The government also conducts research and implements advanced engineering techniques, such as soil-drainage networks. It forces adherence to strict building and grading regulations and sponsors a strong program of public education and broadcasts warnings of landslide danger to local populations.

The Japanese investment in these landslide-control programs has been successful. In 1938, major landslide disasters resulted in 500 deaths and the destruction of or damage to 130,000 homes. In 1976, the worst year for landslides since the control program began, fewer than 125 lives were lost and only 2000 homes were destroyed or damaged.

As Japan and other countries have proved, we can prevent or minimize the effects of smaller mass movements and those provoked by human activity. We can take three main steps to help prevent or decrease loss of life and damage to property:

1. *Avoid construction in areas that are prone to mass movements.* This measure requires good zoning regulations based on adequate geological assessment of the terrain. A geological survey will quickly reveal the conditions in an area that make construction unsafe, such as poorly vegetated slopes covered with unconsolidated material that absorbs water easily. In areas where slopes and materials are unstable, it may be either too expensive to engineer safe structures or impossible altogether.

2. *Build in a way that does not make naturally stable slopes unstable.* Architectural and landscape design can help ensure that relatively stable areas are not made unstable by stripping the vegetation that binds the soil or by artificially oversteepening the slopes. Such designs must adapt construction plans to fit the natural situation. In Venezuela, for example, the well-forested and largely untouched southern slopes of the Avila Mountains helped mitigate damage to Caracas in the south during

angle of the slope. Bedding planes may be zones of potential weakness because the adjacent beds differ in their ability to absorb water. Such beds may become unstable, allowing masses of rock to slide along the weak bedding planes.

Triggers of Mass Movements

When the right combination of materials, moisture, and steepened slope angle makes a hillside unstable, a slide or flow is inevitable. All that is needed is a trigger. Sometimes a slide or debris flow, like the one in the San Gabriel Mountains described by John McPhee at the beginning of this chapter, is provoked by a heavy rainstorm. Many slides are set off by vibrations, such as those that occur in earthquakes. Others may be precipitated just by gradual steepening that eventually results in the sudden collapse of the slope. In some cases, we can take action to avert slides or debris flows (see Feature 12.1).

the torrential rains of December 1999. The heavily developed northern slopes, on the other hand, allowed the devastation of Vargas to the north.

3. *Engineer water drainage so that slope materials will not become waterlogged and likely to slide or flow.* Slopes that are otherwise stable may creep or slump after heavy rains if retaining walls or the construction itself prevents drainage of the water. In areas where heavy rains tend to continue over long periods, storm drainage of potentially unstable slopes is crucial. Good sewer systems are also part of slide prevention. A frequent culprit when soil becomes waterlogged is poorly designed drainage from septic tanks in a hillside home development.

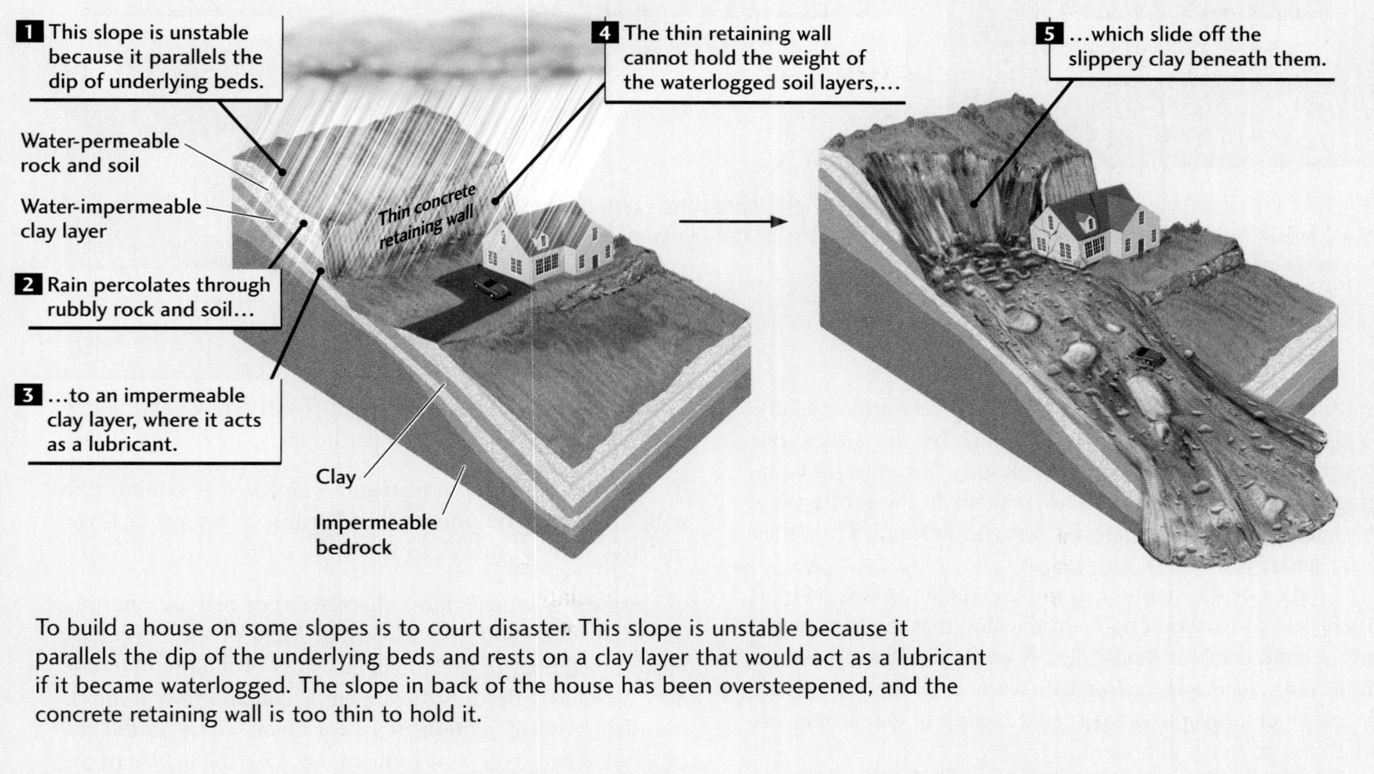

To build a house on some slopes is to court disaster. This slope is unstable because it parallels the dip of the underlying beds and rests on a clay layer that would act as a lubricant if it became waterlogged. The slope in back of the house has been oversteepened, and the concrete retaining wall is too thin to hold it.

Geological reports can help to minimize the human costs of mass movements, but only if city planners and individual home buyers heed the reports and avoid building or buying in unstable areas. Most of the damage in the great Alaskan earthquake of March 27, 1964, was caused by the slides it triggered. Mass movements of rock, earth, and snow wreaked havoc in residential areas of Anchorage, and there were major submarine slides along lakeshores and the seacoast. Huge landslides took place along the flat plains below the 30- to 35-m-high bluffs along the coast. The bluffs were composed of interbedded clays and silts. During the earthquake, the ground shook so hard that unstable, water-saturated sandy layers in the clay were transformed into fluid slurries. This is the process of liquefaction, which we discussed in the section on water content. Enormous blocks of clay and silt were shaken down from the bluffs and slid along the flat ground with the liquefied sediments, leaving a completely disrupted terrain of

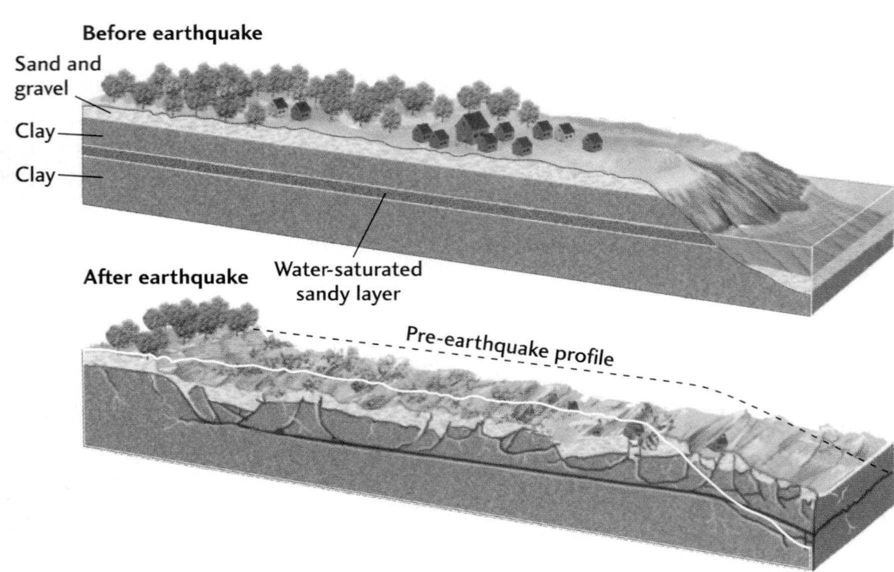

Figure 12.5 Landslide at Turnagain Heights, Alaska, triggered by the earthquake of 1964. [Steve McCutcheon/ Alaska Pictorial Service.] Cross sections of the bluffs at Anchorage, Alaska, before and after the earthquake.

jumbled blocks and broken buildings (**Figure 12.5**). Houses and roads were carried along by the slides and destroyed. The whole process took only 5 minutes, beginning about 2 minutes after the first shock of the earthquake. At one locality, three people were killed and 75 homes were destroyed.

Studies of the stability of those Alaskan slopes and the likelihood of earthquakes had indicated that the area was a prime candidate for landslides. A geological report issued more than a decade earlier had warned of the hazards of development, but the great scenic beauty of the area overwhelmed people's judgment.

Classification of Mass Movements

Although the popular press often refers to any mass movement as a landslide, there are many different kinds of mass movements, each with its own characteristics. We use the term *landslide* only in its popular sense, to refer to mass movements in general.

Geologists classify mass movements in accordance with three characteristics, as summarized in **Figure 12.6**:

1. The nature of the material (for example, whether it is rock or unconsolidated debris)

2. The velocity of the movement (from a few centimeters per year to many kilometers per hour)

3. The nature of the movement: whether it is sliding (the bulk of the material moves more or less as a unit) or flowing (the material moves as if it were a fluid)

Some movements have characteristics that are intermediate between sliding and flowing. Most of the mass may move by sliding, for example, but parts of it along the base may move as a fluid. A movement is called a flow if that is the main type of motion. It is not always easy to tell the exact mechanism of a movement, because the nature of the movement must be reconstructed from the debris deposited after the event is over.

Rock Mass Movements

Rock movements include rockfalls, rockslides, and rock avalanches of small blocks or larger masses of bedrock. During a **rockfall,** newly detached individual blocks plummet suddenly in free fall from a cliff or steep mountainside (**Figure 12.7**). The velocities of the free-falling rock are the fastest of all rock movements, but the travel distances are the shortest, generally only meters to hundreds of meters. Weathering weakens the bedrock along joints until the slightest pressure, often exerted by the expansion of water as it freezes in a crack, is enough to trigger the rockfall.

MASS MOVEMENTS ARE CLASSIFIED ACCORDING TO THE DOMINANT MATERIAL, WATER OR AIR CONTENT, AND VELOCITY OF THE MOVEMENT

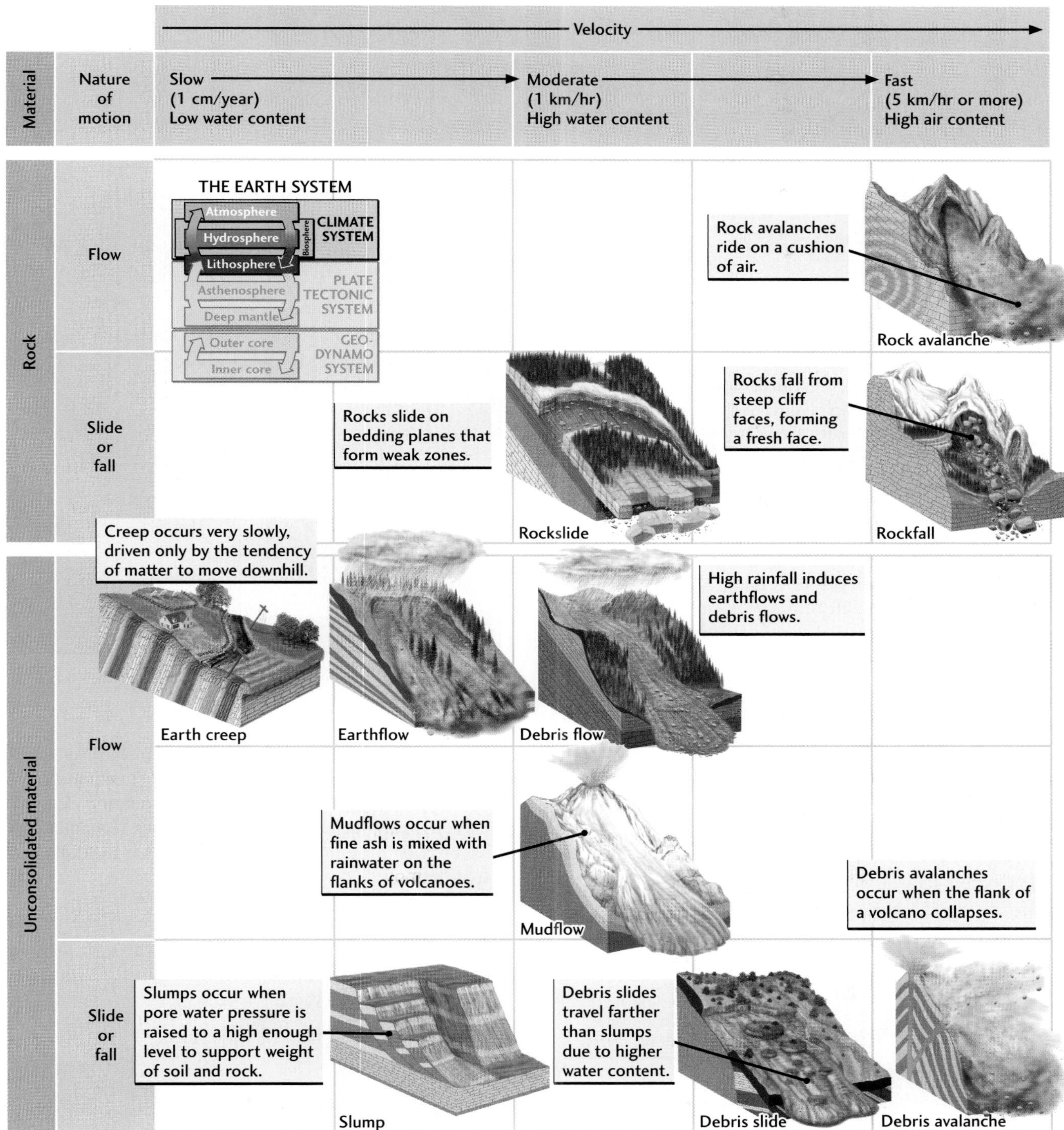

Figure 12.6 Mass movements are classified according to three factors: nature of material (consolidated or unconsolidated), velocity, and the nature of the movement.

The evidence for the origin of rockfalls is clear from the blocks in a rocky accumulation at the foot of a steep bedrock cliff. Referred to collectively as **talus,** these blocks of fallen rock can be matched with rock outcrops on the cliff. Talus accumulates slowly, building up blocky slopes along the base of a cliff over long periods of time.

In many places, rocks do not fall freely but slide down a slope, forming **rockslides.** Although these movements are

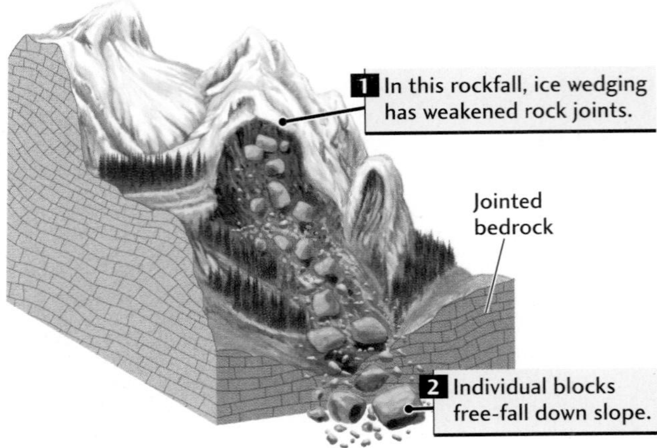

Figure 12.7 In a rockfall, individual blocks plummet in a free fall from a cliff or steep mountainside. Rockfall, Zion National Park, Utah. [Sylvester Allred/Visuals Unlimited.]

fast, they are slower than rockfalls because masses of bedrock slide more or less as a unit, often along downward-sloping bedding or joint planes (**Figure 12.8**).

Rock avalanches differ from rockslides in their much greater velocities and travel distances (**Figure 12.9**). They are composed of large masses of rocky materials that broke up into smaller pieces when they fell or slid. The pieces then flow farther downhill at velocities of tens to hundreds of kilometers per hour. Rock avalanches are typically triggered by earthquakes. They are some of the most destructive mass movements because of their large volume (many are more than a half-million cubic meters) and their fast transport of materials for thousands of meters at high velocities.

Most rock mass movements occur in high mountainous regions; they are rare in low hilly areas. Rock masses tend to move where weathering and fragmentation have attacked rocks already predisposed to breakage by structural deformation, relatively weak bedding, or metamorphic cleavage planes. In many such regions, extensive talus accumulations have been built by infrequent but large-scale rockfalls and rockslides.

Figure 12.8 In a rockslide, large masses of bedrock move more or less as a unit in a fast, downward slide. Rockslide, Elephant Rock, Yosemite National Park. [Jeff Foott/DRK.]

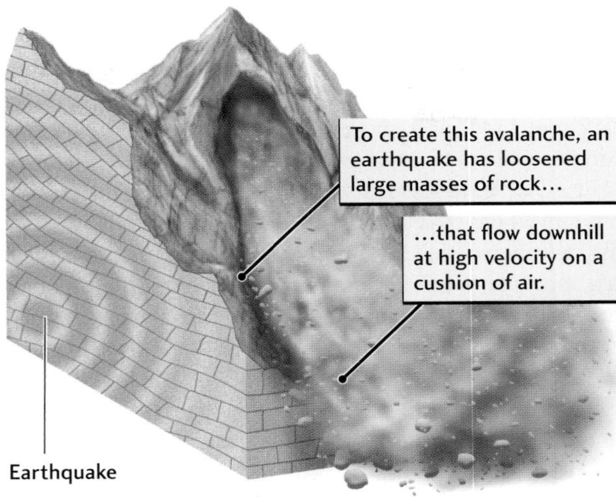

Figure 12.9 In a rock avalanche, large masses of broken rocky material flow, rather than slide, downward at high velocity. Two rock avalanches can be seen that were triggered by the November 3, 2002, earthquake along the Denali fault, Alaska. The rock avalanches traveled down south-facing mountains, across the one-and-a-half-mile-wide Black Rapids Glacier, and flowed part way up the opposite slope. [Photo by Dennis Trabant, USGS; mosaic by Rod March, USGS.]

Unconsolidated Mass Movements

Unconsolidated material, often called *debris,* includes various mixtures of soil; broken-up bedrock; trees and shrubs; and materials of human construction, from fences to cars and houses. Most unconsolidated mass movements are slower than most rock movements, largely because of the lower slope angles at which these materials move. Although some unconsolidated materials move as coherent units, many flow like very viscous fluids, such as honey or syrup. (Viscosity, you will recall, is a measure of a fluid's resistance to flow.)

The slowest unconsolidated mass movement is **creep**—the downhill movement of soil or other debris at a rate ranging from about 1 to 10 mm/year, depending on the kind of soil, the climate, the steepness of the slope, and the density of the vegetation (**Figure 12.10**). The movement is a very slow deformation of the regolith, with the upper layers of

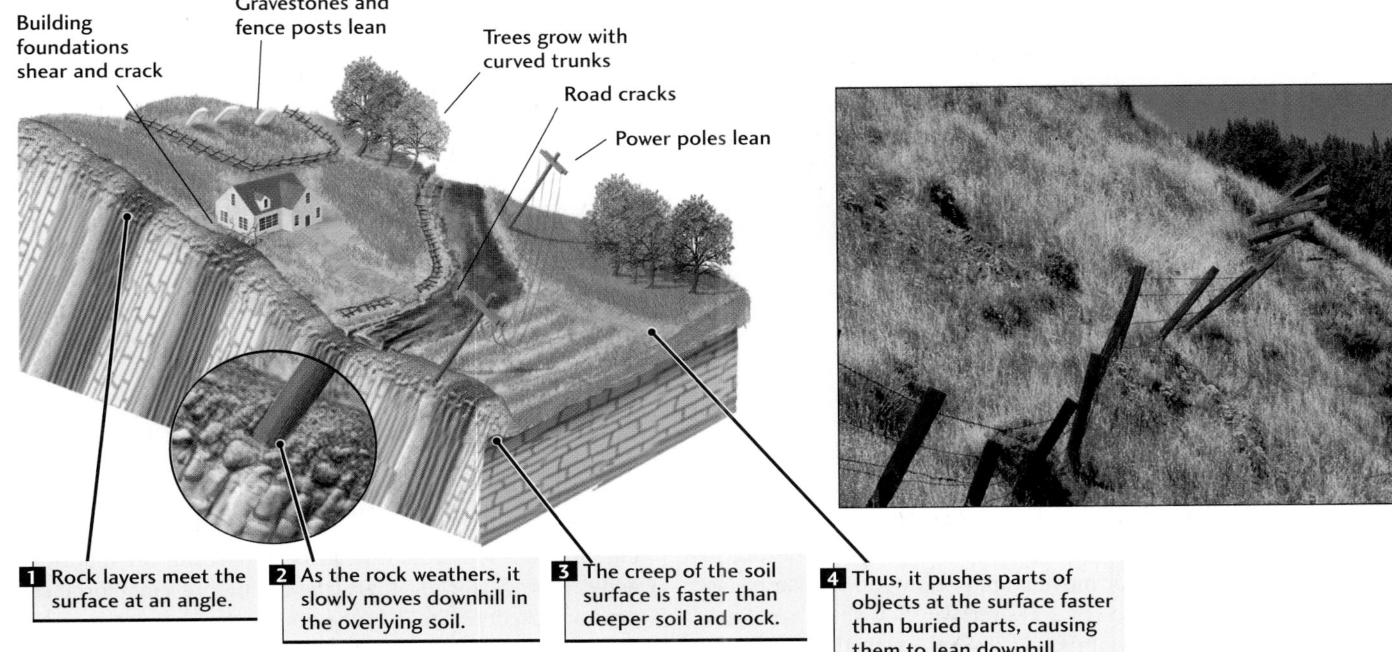

Figure 12.10 Creep is the downhill movement of soil or other debris at a rate of about 1 to 10 mm/year. A fence offset by creep in Marin County, California. [Travis Amos.]

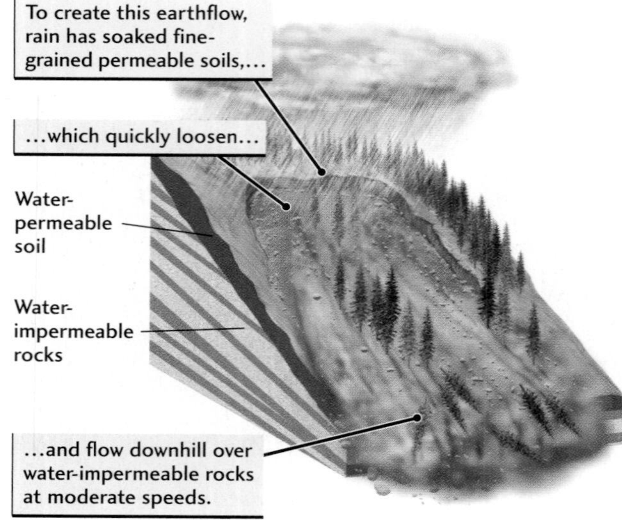

Figure 12.11 An earthflow is a movement of relatively fine grained materials that travel as fast as a few kilometers per hour. Earthflow, Hogan Creek, Denali National Park, Alaska. [Steve McCutcheon/Visuals Unlimited.]

regolith moving down the slope faster than the lower layers. Such slow movements may cause trees, telephone poles, and fences to lean or move slightly downslope. The heavy weight of the masses of soil creeping downhill can break poorly supported retaining walls and crack the walls and foundations of buildings. In icy regions where the ground is permanently frozen, *solifluction* is a type of movement that occurs when water in the surface layers of the soil alternately freezes and thaws, causing the soil to ooze downhill and carrying broken rocks and other debris with it.

Earthflows (**Figure 12.11**) and debris flows (**Figure 12.12**) are fluid mass movements that usually travel faster than creep, as much as a few kilometers per hour, primarily because they have less resistance to flow. An **earthflow** is a fluid movement of relatively fine grained materials, such as soils, weathered shales, and clay. A **debris flow** is a fluid mass movements of rock fragments supported by a muddy matrix. Debris flows contain much material coarser than sand and tend to move more rapidly than earthflows. In some cases, debris flows may reach velocities of 100 km/hour.

Mudflows are flowing masses of material mostly finer than sand, along with some rock debris, containing large amounts of water (Figure 12.12). The mud offers little resistance to flow because of the high water content and thus tends to move faster than earth or debris. Many mudflows move at several kilometers per hour. Most common in hilly and semiarid regions, mudflows start after infrequent, sometimes prolonged, rains. The previously dry, cracked mud keeps absorbing water as the rain continues. In the process, its physical properties change; the internal friction decreases, and the mass becomes much less resistant to movement. The slopes, which are stable when dry, become unstable, and any disturbance triggers movement of waterlogged masses of mud. Mudflows travel down upper valley slopes and merge on the valley floor. Where mudflows exit from confined valleys into broader, lower valley slopes and flats, they may splay out to cover large areas with wet debris. Mudflows can carry huge boulders, trees, and even houses.

Debris avalanches (**Figure 12.13**) are fast downhill movements of soil and rock that usually occur in humid mountainous regions. Their speed results from the combination of high water content and steep slopes. Water-saturated debris may move as fast as 70 km/hour, a speed comparable to that of water flowing down a moderately steep slope. A debris avalanche carries with it everything in its path. In 1962, a debris avalanche in the high Peruvian Andes traveled almost 15 km in about 7 minutes, engulfing most of eight towns and killing 3500 people. Eight years later, on May 31, 1970, an earthquake in the same active subduction zone toppled a large mass of glacial ice at the top of one of the highest mountains, Nevado de Huascarán. As the ice broke up, it mixed with the debris of the high slopes and became an ice-debris avalanche. The avalanche picked up debris as it raced downhill, increasing its speed to an almost unbelievable 200 to 435 km/hour. More than 50 million cubic meters of muddy debris roared down into the valleys, killing 17,000 people and wiping out scores of villages.

On May 30, 1990, an earthquake shook another mountainous area in northern Peru, again setting off mudflows and debris avalanches. It was the day before a memorial ceremony scheduled to commemorate the disaster that had occurred 20 years earlier. In regions where unstable slopes build up and earthquakes are frequent because of the subduction of an oceanic plate beneath a continental plate, there can be no doubt about the necessity of learning how to

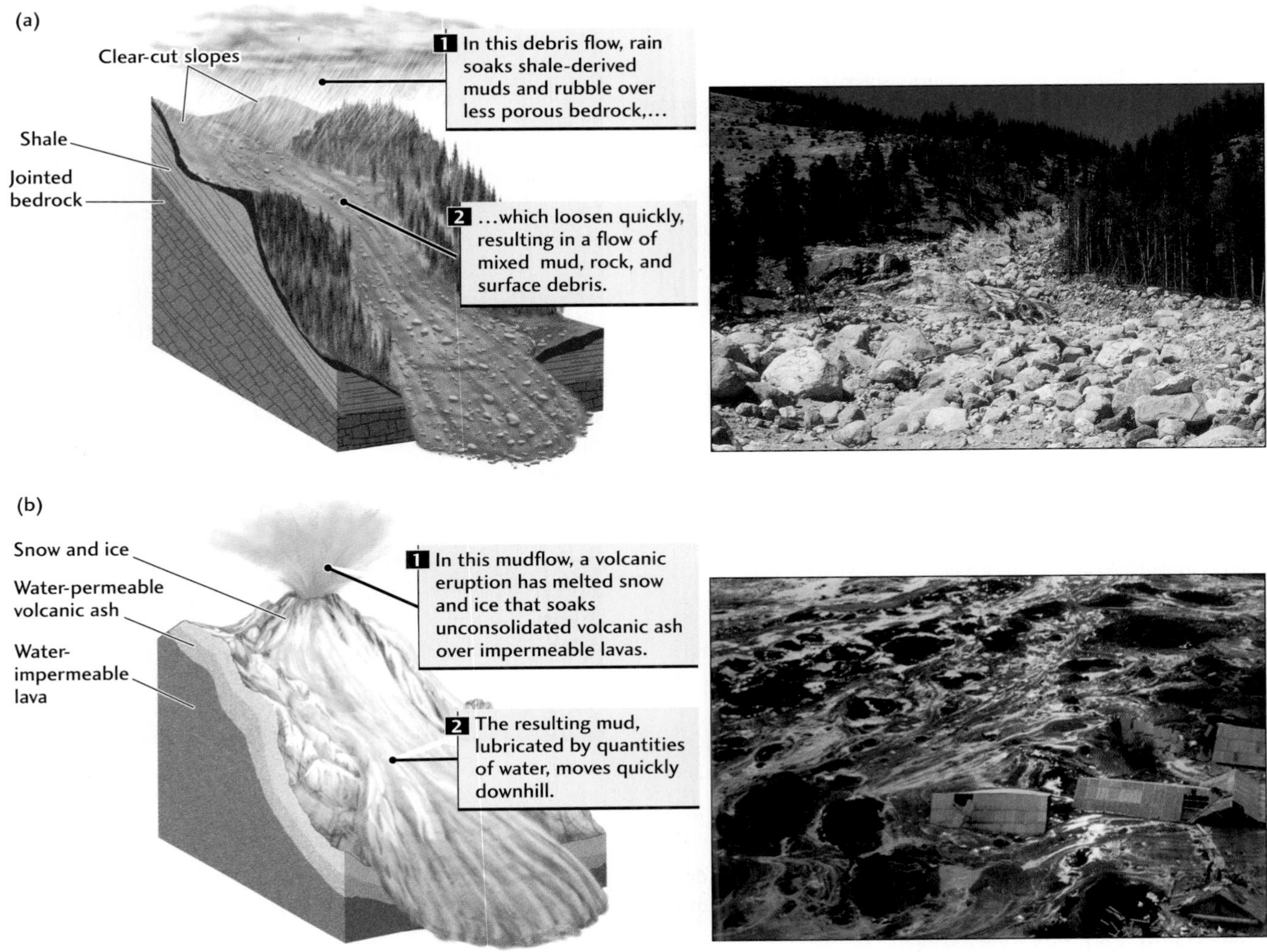

Figure 12.12 (a) A debris flow contains material that is coarser than sand and travels at rates from a few kilometers per hour to many tens of kilometers per hour. Debris flow, Rocky Mountain National Park, Colorado [E. R. Degginger]. (b) A mudflow contains large quantities of water, and many such flows move at a rate of several kilometers per hour. Mudflows tend to move faster than earthflows or debris flows. An earthquake in Tadzhikistan in January 1989 produced 15-m-high mudflows on slopes weakened by rain [Vlastimir Shone/Getty Images].

predict both earthquakes and the dangerous mass movements that follow.

Mudflows and debris avalanches are easily triggered on slopes of volcanic cinder cones when the accumulations of unconsolidated ash and other erupted materials become saturated with rainwater. Flows of volcanic debris may be set off by events associated with the volcanic eruption itself (such as earthquakes), by downpours of rain and ash from the eruption cloud, and by sudden falls of volcanic debris. More than 55 such flows, many of them connected with eruptions, have occurred on the slopes of Mount Rainier, Washington, over the past 10,000 years, according to geologists of the U.S. Geological Survey. In 1980, a giant volcanic debris avalanche triggered by the eruption of Mount St. Helens, also in Washington, roared down the northern slope of the mountain at a speed of about 200 km/hour and covered an area of more than 600 km^2 below the mountain (see Figure 12.13).

A **slump** is a slow slide of unconsolidated material that travels as a unit (**Figure 12.14**). In most places, the slump slips along a basal surface that forms a concave-upward shape, like a spoon. Faster than slumps are **debris slides** (**Figure 12.15**), in which the rock material and soil move largely as one or more units along planes of weakness, such as a waterlogged clay zone either within or at the base of the debris. During the slide, some of the debris may behave like a chaotic, jumbled flow. Such a slide may become predominantly a flow as it moves rapidly downhill and most of the material mixes as if it were a fluid.

Figure 12.13 A debris avalanche is the fastest unconsolidated flow, owing to its high water content and movement down steep slopes. In 1970, an earthquake-induced debris avalanche on Mt. Huascarán, Peru, buried the towns of Yungay and Ranrahirca. The avalanche traveled 17 km at a speed of up to 280 km/hour and is estimated to have consisted of up to 100 million cubic meters of water, mud, and rocks. The death toll from the earthquake and landslide was 66,700 persons. (*top photo*) Before the avalanche; (*bottom photo*) aftermath of the avalanche. [Lloyd Cluff/Corbis.]

Erosion of Mass-Movement Materials

Materials found on talus slopes and the debris of slides, flows, and avalanches erode easily and extensively because they have already been broken into fine grain sizes and large surface areas, which makes them more susceptible to weathering. When these weathered materials reach lower slopes, they are transferred to small streams. As a result, few mass movements are preserved in the ancient rock record, although many geologists have seen evidence of mass movements of the past few thousand years.

Submarine Mass Movements

No one has been able to witness a submarine slide, but marine geologists can reconstruct such events by identifying slide materials on the seafloor and dating the sediment layers that accumulate between these events. Submarine slides may occur at sites of steep topography, such as mid-ocean ridges and volcanic seamounts, or in places with very rapid sediment accumulation rates, such as deltas along the continental margins. For example, within the past 450,000 years, a mass measuring 4 km by 5 km slid down one wall of the

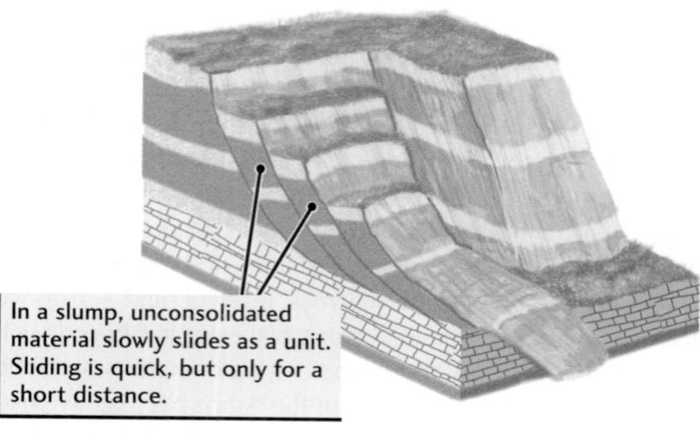

Figure 12.14 A slump is a slow slide of unconsolidated material that travels as a unit. Soil slump, Sheridan, Wyoming. [E. R. Degginger.]

Figure 12.15 A debris slide travels as one or more units and moves more quickly than a slump. A relatively recent debris slide has choked this narrow valley in the Tien Shan mountains in Kyrgystan. [Martin Miller/Visuals Unlimited.]

Mid-Atlantic Ridge rift valley. The slopes had been weakened by faulting and hydrothermal metamorphism. The debris avalanche that formed when the slope failed flowed more than 11 km into the rift valley, depositing a debris wedge of almost 20 billion cubic meters. In another submarine setting, blocky debris covers the flanks of the Hawaiian volcano Mauna Loa from the present shoreline seaward to a depth of 5 km. The debris is the remnant of a submarine slide and consists of mixed lavas and volcanic sediment. Marine volcanic slides are formed by sudden eruptions, either of submarine volcanoes or, as in this case, of a land volcano bordering the ocean.

Unlike slides on land, submarine slides may generate tsunamis, or "tidal waves" (see Chapter 19). Tsunamis can have devastating effects on coastal regions. For example, geologists estimate that coastal communities in Japan and all along the western margin of North America would suffer tremendous damage from an event as big as the ancient slide along the flanks of Mauna Loa. Submarine slides also pose a risk for the facilities of oil and gas companies that operate near the base of the continental slope. Here, failure of the slope can produce slides too small to generate damaging tsunamis but large enough to topple billion-dollar production facilities. As the search for new reserves pushes farther and farther offshore, this risk increases dramatically.

Understanding the Origins of Mass Movements

To understand how slope steepness, the nature of the slope materials, and the materials' water content interact to create mass movements, geologists study both natural mass movements and those provoked by human activities. They investigate the causes of a modern slide by combining eyewitness reports with geological studies of the source of the slide and the distribution and nature of the debris dropped into the valley below. They can infer the causes of prehistoric slides from geological evidence alone where the debris is still present and can be analyzed for size, shape, and composition.

Natural Causes of Landslides

In April 1983, about 4 million cubic meters of unconsolidated mud and debris slid down from a wall of Spanish Fork Canyon in Utah (see Figure 12.15). Near-record depths of snow had accumulated at higher elevations in the surrounding Wasatch Mountains. A warm spring melted the snow rapidly, and heavy rainstorms augmented the meltwaters, triggering debris flows and slides in unprecedented numbers in Spanish Fork and other canyons of the Wasatch Range. In this case, the nature of the materials—soil and rock that was structurally deformed, cracked, and weak—combined with the steep slopes of Spanish Fork to make the canyon susceptible to slides. The rain and melting snow saturated a mass of regolith and weathered rock on the wall of the canyon, lubricating both the mass and the bedrock surface of the slope below it. Sooner or later, the mass had to give.

The 1925 landslide in the Gros Ventre River valley of western Wyoming illustrates again how water, the nature of slope materials, and slope stability interact to produce slides (**Figure 12.16**). In the spring of that year, melting snow and heavy rains swelled streams and saturated the ground in the valley. One local rancher, out on horseback, looked up to see the whole side of the valley racing toward his ranch. From the gate to his property, he watched the slide hurtle past him at about 80 km/hour and bury everything he owned.

About 37 million cubic meters of rock and soil slid down one side of the valley that day, then surged more than 30 m up the opposite side and fell back to the valley floor. Most

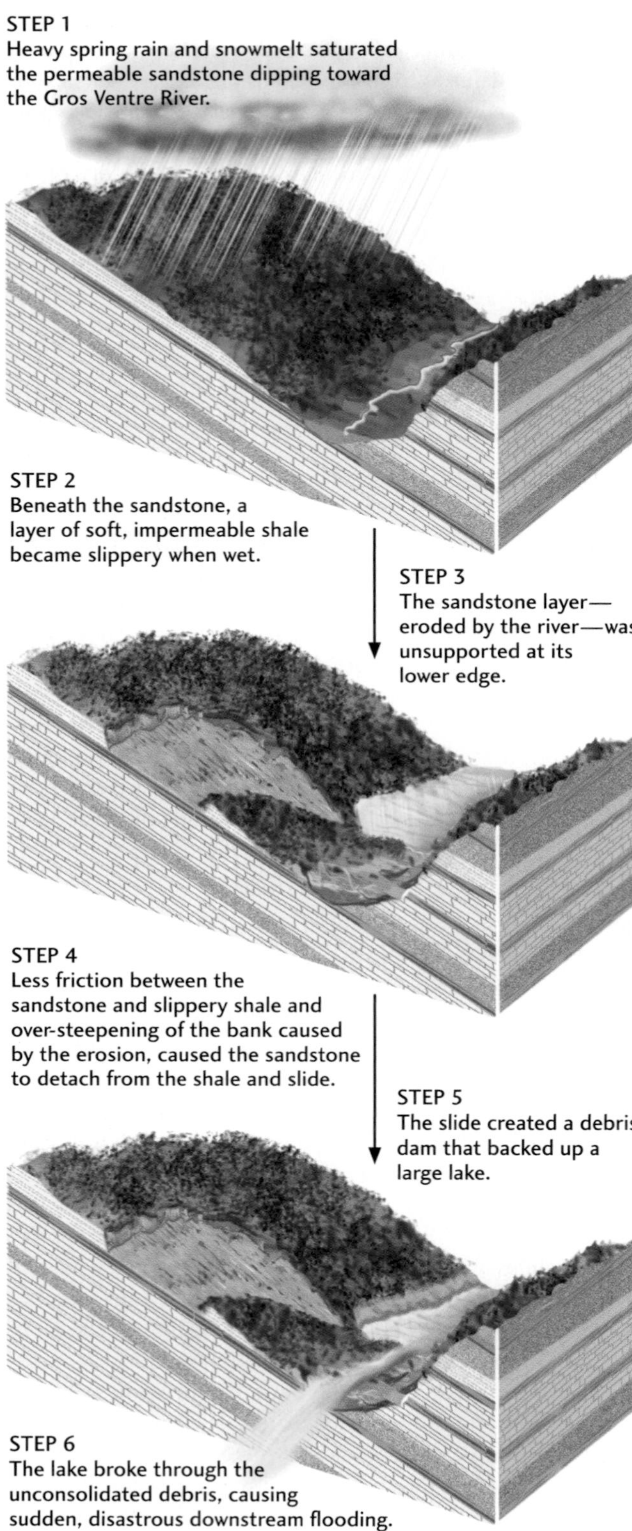

Figure 12.16 The 1925 Gros Ventre slide. [After W. C. Alden, "Landslide and Flood at Gros Ventre, Wyoming." *Transactions of the American Institute of Mining, Metallurgical, and Petroleum Engineers* (1928): 345–361.] The Gros Ventre landslide, Grand Teton National Park, Wyoming. [Photo by Stephen Trimble.]

of the slide was a confused mass of blocks of sandstone, shale, and soil, but one large section of the side of the valley, covered with soil and a forest of pine, slid down as a unit. The slide dammed the river, and a large lake grew over the next two years. Then the lake overflowed, breaking the dam and rapidly flooding the valley below.

The causes of the Gros Ventre slide were all natural. In fact, the stratigraphy and structure of the valley made a slide almost inevitable. On the side of the valley where the slide occurred, a permeable, erosion-resistant sandstone formation dipped about 20° toward the river, paralleling the slope of the valley wall. Under the sandstone were beds of soft, impermeable shale that became slippery when wet. The conditions became ideal for a slide when the river channel cut through most of the sandstone at the bottom of the valley wall and left it with virtually no support. Only friction along the bedding plane between the shale and sandstone kept the layer of sandstone from sliding. The river's removal of the sandstone's support was equivalent to scooping sand from the base of a sandpile—both cause oversteepening. The heavy rains and snow meltwater saturated the sandstone and the surface of the underlying shale, creating a slippery surface along the bedding planes at the top of the shale. No one knows what triggered the Gros Ventre slide, but at some point the force of gravity overcame friction and almost all of the sandstone slid down along the water-lubricated surface of the shale.

The formation of a dam on a river and the growth of a lake, as happened at both Spanish Fork Canyon and Gros Ventre, are common consequences of a landslide. Because most slide materials are permeable and weak, such a dam is soon breached when the lake water reaches a high level or overflows. Then the lake drains suddenly, releasing a catastrophic torrent of water (see Figure 12.16).

Plate Tectonic Settings

There is hardly an area of geology that plate tectonics has not touched, and mass wasting is no exception. We can see the connection between mass wasting and plate tectonics by considering topographic heights, steep slopes, the composition of the surface materials, and rock-weakening effects. We can outline the effects by considering the basic plate tectonic settings where mass movements are likely to occur.

Perhaps most obvious is the setting in which two plates converge, forming mountains. An ocean-continent plate convergence and its associated subduction zone, such as the Andes Mountains of South America, bring together high, steep mountain slopes and abundant volcanic eruptions, which produce ash that can easily become mudflows. Add to this the fracturing and deformation of rocks that accompany orogeny and the many earthquakes along subduction zones, and all the conditions for frequent mass movements are met. A continent-continent collision, exemplified by the Himalaya, meets all the same conditions except for volcanism.

Seafloor spreading, too, may be associated with mass movements, particularly at the steep slopes formed in continental rift valleys where divergence initiates. We have already mentioned submarine landslides in the central rift valley of the Mid-Atlantic Ridge, another setting of seafloor spreading. Transform-fault plate boundaries, such as the San Andreas fault, are also the sites of frequent mass movements, given the steep slopes that may be found along the fault and the prevalence of earthquakes. Contrast these conditions with those typical of areas far from present or former plate boundaries: low topography, relatively undisturbed rocks, gentle slopes, and absence of earthquakes.

Human Activities That Promote or Trigger Slides

Although the vast bulk of mass wasting is natural, humans may trigger landslides or make them more likely in vulnerable areas by such activities as changing natural slopes. In 1995, excavation at the base of a road-cut slope during work to widen a highway in Quebec oversteepened the side of the road cut. The road was cut through soft formations of silt and clay that were susceptible to weakening by water. After a period of heavy rain, the steepened slope of the road cut became saturated with water and suddenly gave way in a debris flow that carried away buildings, roads, and people. Three deaths can be traced to this case of poor highway engineering.

Some geological settings are so susceptible to landslides that engineers may have to forgo construction projects in these areas entirely. A landslide in one such place ultimately killed 3000 people. The place was Vaiont, an Italian alpine valley, and the time was the night of October 9, 1963. A large reservoir in the valley was impounded by a concrete dam (the second highest in the world, at 265 m) and bordered by steep walls of interbedded limestone and shale. A great debris slide of 240 million cubic meters (2 km long, 1.6 km wide, and more than 150 m thick) plunged into the deep water of the reservoir behind the dam. The debris filled the reservoir for a distance of 2 km upstream of the dam and created a giant spillover. In the violent torrent that hurtled downstream as a 70-m-high flood wave, 3000 people died.

Engineers had underestimated three warning signs at Vaiont (**Figure 12.17**):

- The weakness of the cracked and deformed layers of limestone and shale that made up the steep walls of the reservoir
- The scar of an ancient slide on the valley walls above the reservoir

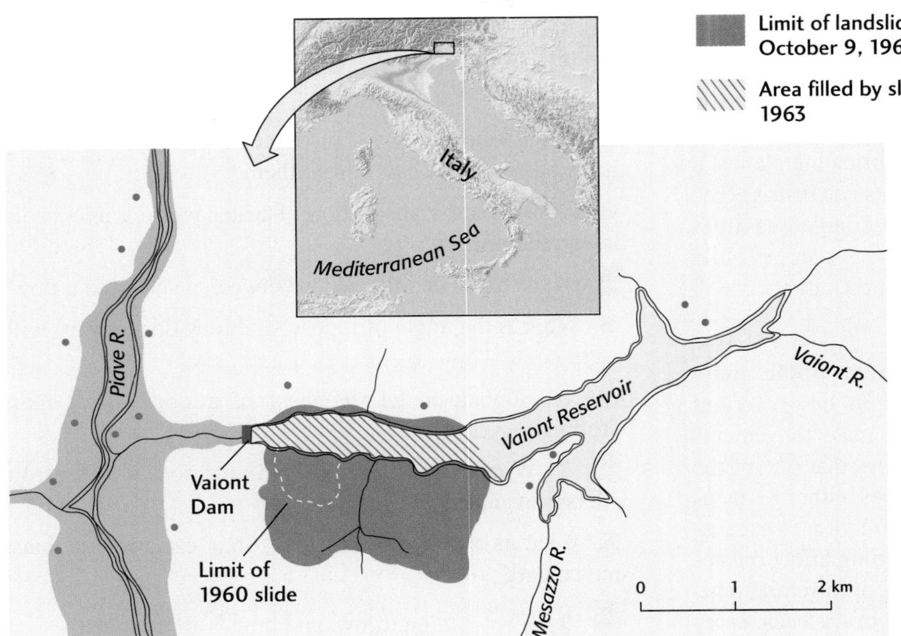

Figure 12.17 An ancient small slide on the north side of the walls of the Vaiont Dam reservoir, as well as a landslide in 1960 (shown by a dashed line in the brown area), warned of the danger of mass movement above the reservoir. In October 1963, a massive landslide that could have been predicted caused the water in the Vaiont reservoir to overflow the dam, flooding the downstream areas and killing 3000 people.

• A forewarning of danger signaled by a small rockslide in 1960, just three years earlier

Although the 1963 landslide was natural and could not have been prevented, its consequences could have been much less severe. If the reservoir had been located in a geologically safer place, where the water was less likely to spill over its walls, damage might have been limited to a lesser loss of property and far fewer deaths. We cannot prevent most natural mass movements, but we can minimize our losses through more careful control of construction and land development.

SUMMARY

What are mass movements, and what kinds of material do they involve? Mass wasting erodes and sculpts the land surface. Mass movements are slides, flows, or falls of large masses of material down slopes in response to the pull of gravity. The movements may be imperceptibly slow or far too fast to outrun. The masses consist of bedrock; consolidated material, including compacted sediment or regolith; or unconsolidated material, such as loose, uncemented sediment or regolith. Rock movements include rockfalls, rockslides, and rock avalanches. Unconsolidated material moves by creep, slump, debris slide, debris avalanche, earthflow, mudflow, and debris flow.

What factors are responsible for mass movements, and how are such movements triggered? The three factors that have the greatest bearing on the predisposition of material to move down a slope are (1) the steepness and stability of the slope, (2) the nature of the material, and (3) the water content of the material. Slopes become unstable when they are steeper than the angle of repose, the maximum slope angle that unconsolidated material will assume without cascading down. Slopes in consolidated material may also become unstable when they are oversteepened or denuded of vegetation. Water absorbed by the material contributes to instability in two ways: (1) by lowering internal friction (and thus resistance to flow) and (2) by lubricating planes of weakness in the material. Mass movements can be triggered by earthquakes or by sudden absorption of large quantities of water after a torrential rain. In many places, slopes build to a point of instability at which the slightest vibration will set off a slide, flow, or fall.

What factors are responsible for catastrophic mass movements, and how can such movements be prevented or minimized? Analysis of both natural mass movements and those induced by human activity shows that one of the main causes is the oversteepening of slopes, either by natural erosional processes or by human construction or excavation. Because water content has such a strong effect on stability, absorption of water from prolonged or torrential rains is often an important factor. The structure of the beds, especially when bedding dips parallel to the slope, can promote mass movements. Volcanic eruptions may produce a tremendous fallout of ash and other materials that build up into unstable slopes. Slides or flows of the volcanic materials are triggered by earthquakes that accompany eruptions. We can prevent or minimize loss of life and damage to property from catastrophic mass movements by refraining from steepening or undercutting slopes. Careful engineering can keep water from making materials more unstable. In some areas that are extremely prone to mass movements, development may have to be restricted.

Key Terms and Concepts

angle of repose (p. 259)
cohesion (p. 259)
consolidated materials (p. 258)
creep (p. 267)
debris avalanche (p. 268)
debris flow (p. 257)
debris slide (p. 269)
earthflow (p. 268)
liquefaction (p. 259)
mass movement (p. 257)
mass wasting (p. 258)
mudflow (p. 268)
rock avalanche (p. 264)
rockfall (p. 264)
rockslide (p. 264)
slump (p. 269)
surface tension (p. 259)
talus (p. 264)
unconsolidated materials (p. 258)

Exercises

This icon indicates that there is an animation available on the Web site that may assist you in answering a question.

1. What role do earthquakes play in the occurrence of landslides?

2. What kinds of mass movements advance so rapidly that a person could not outrun them?

3. How does the absorption of water weaken unconsolidated material?

4. What is the difference between a slide and a flow?

5. What is the angle of repose, and how does it vary with water content?

6. How does the steepness of a slope affect mass wasting?

7. What is a debris flow, and how does it differ from a debris avalanche?

8. What is the typical history of a catastrophic mass movement?

9. What is a mudflow, and how is it produced?

10. What are the differences between a rockfall, a rockslide, and a rock avalanche?

Thought Questions

This icon indicates that there is an animation available on the Web site that may assist you in answering a question.

1. You are planning a highway through hills made of unlithified sands and gravels and you want to minimize mass movements. What construction practices will you avoid?

2. Would a prolonged drought affect the potential for landslides? How?

3. What geological conditions might you want to investigate before you bought a house at the base of a steep hill of bedrock covered by a thick mantle of regolith?

4. You are excavating the base of a slope that is prone to sliding, taking care not to oversteepen it, when heavy rain begins and continues to fall for 3 days. Why might you temporarily stop heavy construction trucks from driving near the slope?

5. Would you expect a talus slope of large blocks of granite to be prone to debris flows? Explain.

6. What evidence would you look for to indicate that a mountainous area had undergone a great many prehistoric landslides?

7. What factors would make the potential for mass movements in a mountainous terrain in the rainy tropics greater or lesser than the potential in a similar terrain in a desert?

8. What kind(s) of mass movements would you expect from a steep hillside with a thick layer of soil overlying unconsolidated sands and muds after a prolonged period of heavy rain?

9. What factors weaken rock and enable gravity to start a mass movement?

10. Why would you expect that regions near a continental rift valley, where two plates are beginning to diverge, might have frequent mass movements?

Short-Term Team Projects

Environmental Changes and Land Use

While you are enrolled in your geology class, a variety of natural and human-induced environmental changes are likely to take place in your local area. There may be an earthquake, drought, flood, sinkhole collapse, or oil spill. Local officials may vote to open a new landfill, allow railroad transportation of hazardous materials through town, or develop the local waterfront.

In teams of four, prepare a series of news releases, approximately one per month, for your local media outlets in response to natural or human-induced environmental changes and to local land-use policy decisions. This project will require you to search for events pertinent to geology in local newspapers and broadcasts and to write responses from a geologist's perspective to the events that your team discovers.

Because you are writing for a nontechnical audience, you should make each news release brief (preferably only one page) and avoid the use of jargon. The title must grab the reader's attention. State the topic clearly in the first paragraph, describe the nature and significance of the event or issue, and emphasize its implications for the public.

Submit your news releases to your local newspapers and radio and television stations. Most local media outlets welcome newsworthy items about events in their community.

Suggested Readings

Bloom, A. L. 1991. *Geomorphology,* 2d ed. Englewood Cliffs, N.J.: Prentice Hall.

Chapman, D. 1995. *Natural Hazards.* Oxford: Oxford University Press.

Coch, N. K. 1995. *Geohazards: Natural and Human.* Englewood Cliffs, N.J.: Prentice Hall.

Costa, J. E., and V. R. Baker. 1981. *Surficial Geology.* New York: Wiley.

Costa, J. E., and G. F. Wieczorek (eds.). 1987. *Debris Flows/Avalanches: Process, Recognition, and Mitigation. Reviews in Engineering Geology,* vol. 7. Boulder, Colo.: Geological Society of America.

Lundgren, L. 1986. *Environmental Geology.* Englewood Cliffs, N.J.: Prentice Hall.

Porter, S. C., and G. Orombelli. 1981. Alpine rockfall hazards. *American Scientist* 69: 67–75.

Slosson, J. E., A. G. Keene, and J. A. Johnson (eds.). 1993. *Landslides/Landslide Mitigation. Reviews in Engineering Geology,* vol. 9. Boulder, Colo.: Geological Society of America.

Tyler, M. B. 1995. *Look Before You Build: Geologic Studies for Safer Land Development in the San Francisco Bay Area.* U.S. Geological Survey Circular 1130. Menlo Park, Calif.: U.S. Geological Survey.

Voight, Barry (ed.). 1978. *Rockslides and Avalanches,* vol. 1. *Natural Phenomena.* New York: Elsevier.

Angel Falls, Venezuela, is the highest waterfall on Earth. The falls plunge 3000 ft from a flat-topped mountain composed of 1.7-billion-year-old sandstones. [Michael K. Nichols/National Geographic/Getty Images.]

CHAPTER

13

The Hydrologic Cycle and Groundwater

"Water, water, every where,
Nor any drop to drink."

SAMUEL TAYLOR COLERIDGE

Geologists who specialize in the science of **hydrology** study the movements and characteristics of water on and under Earth's surface. Water is essential to a wide variety of geological processes. Rivers and glacial ice are major agents of erosion that help to shape the landscape of the continents. Water is essential to weathering, both as a solvent of minerals in rock and soil and as a transport agent that carries away dissolved and weathered materials. Water that sinks into surface materials forms large reservoirs of groundwater. Water lubricates the materials involved in landslides and other forms of mass movement. Hot water circulating over igneous bodies or through mid-ocean ridges produces hydrothermal ore deposits.

Water is vital to all life on the planet. Humans cannot survive more than a few days without it, and even the hardiest desert plants and animals need some water. The amount of water that modern civilization requires is far greater than what humans need for simple physical survival. Water is used in immense quantities for industry, agriculture, and such urban needs as sewage systems. The United States, one of the heaviest users of water in the world, has been steadily increasing its consumption since the nineteenth century. In the 35 years between 1950 and 1985 alone, water use nearly tripled, from 34 billion gallons a day to about 90 billion gallons a day. By 1990, only 5 years later, that figure almost quadrupled, to 339 billion gallons a day. Some, but not all, of this increase is a result of population growth. In the United States, water consumption, corrected for population by using figures for consumption per person, actually fell by about 20 percent from 1980 to 1995. Developed countries have started to emphasize more efficient use of finite water resources.

Hydrology is becoming more important to all of us as the demand on limited water supplies increases. To protect those supplies while we satisfy our needs, we must understand not only where to find water but also how water supplies are renewed. With that knowledge, we can use and dispose of water in ways that do not endanger future supplies.

This chapter is a survey of water in and on the Earth.

Flows and Reservoirs

We can see water moving over Earth's surface in rivers, and we can see the water in lakes and oceans. It is harder to see the massive amounts of water stored in the atmosphere and underground and the flows into and out of these storage places. As water evaporates, it vanishes into the atmosphere as vapor. As rain sinks into the ground, it becomes **groundwater**—the mass of water stored beneath Earth's surface.

Each place in which water is stored is a **reservoir.** Earth's main natural reservoirs are the oceans; glaciers and polar ice; groundwater; lakes and rivers; the atmosphere; and the biosphere. **Figure 13.1** shows the distribution of water among these reservoirs. The oceans are by far the largest repository of water on the planet. Although the total amount of water in rivers and lakes is relatively small, these reservoirs are important to human populations because they contain fresh water. The amount of groundwater is more than 100 times the amount in rivers and lakes, but much of it is unusable because it contains large quantities of dissolved material.

Reservoirs gain water from inflows, such as rain and river inflow, and lose water from outflows, such as evaporation and river outflow. If inflow equals outflow, the size of the reservoir stays the same, even though water is constantly entering and leaving. These flows mean that any given quantity of water spends a certain average time, called the residence time, in a reservoir. We will consider reservoirs and residence times more fully in Chapter 24.

How Much Water Is There?

The world's total water supply is enormous—about 1.46 billion cubic kilometers distributed among the various reservoirs. If it covered the land area of the United States, it would submerge the 50 states under a layer of water about 145 km deep. This total is constant, even though the flows from one reservoir to another may vary from day to day, year to year, and century to century. Over these geologically short time intervals, there is neither a net gain or loss of water to or from Earth's interior nor any significant loss of water from the atmosphere to outer space.

The Hydrologic Cycle: A Component of the Earth System

Water on or beneath Earth's surface cycles among the various reservoirs: the oceans, the atmosphere, and the land. The cyclical movement of water—from the ocean to the atmosphere by evaporation, to the surface through rain, to streams through runoff and groundwater, and back to the ocean—is the **hydrologic cycle. Figure 13.2** is a simplified illustration of the endless circulation of water and the amounts moved. The hydrologic cycle is a component of the Earth system and thus interacts with the atmospheric, ocean, and landscape components. (For further discussion, see Chapter 23.)

Within the range of temperatures found at Earth's surface, water shifts among the three states of matter: liquid (water), gas (water vapor), and solid (ice). These transformations power some of the main flows from one reservoir to another in the hydrologic cycle. Earth's external heat engine, powered by the Sun, drives the hydrologic cycle, mainly by evaporating water from the oceans and transporting it as water vapor in the atmosphere. Under the right conditions of temperature and humidity, water vapor condenses to the tiny droplets of water that form clouds and eventually falls as rain or snow over the oceans and continents. Some of the water that falls on land soaks into the ground by **infiltration,** the process by which water enters rock or soil through joints or small pore spaces between particles. Part of this groundwater evaporates through the soil surface. Another part is absorbed by plant roots, carried up to the leaves, and returned to the atmosphere by *transpiration*—the release of water vapor from plants. Other groundwaters may return to the surface in springs that empty into rivers and lakes.

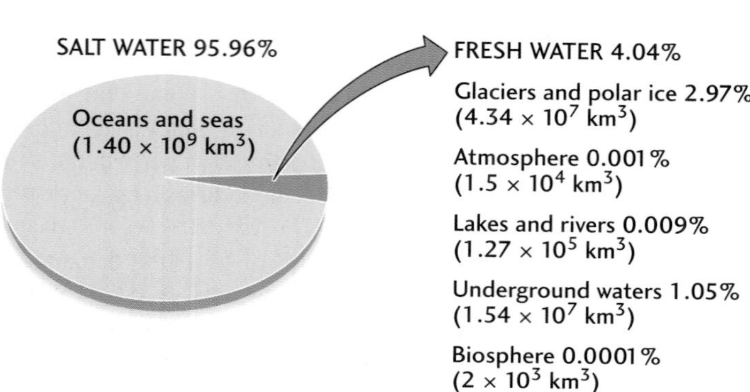

Figure 13.1 The distribution of water on Earth. [Revised from J. P. Peixoto and M. Ali Kettani, "The Control of the Water Cycle." *Scientific American* (April 1973): 46; E. K. Berner and R. A. Berner, *Global Environment* (Upper Saddle River, N.J.: Prentice Hall, 1996), pp. 2–4.]

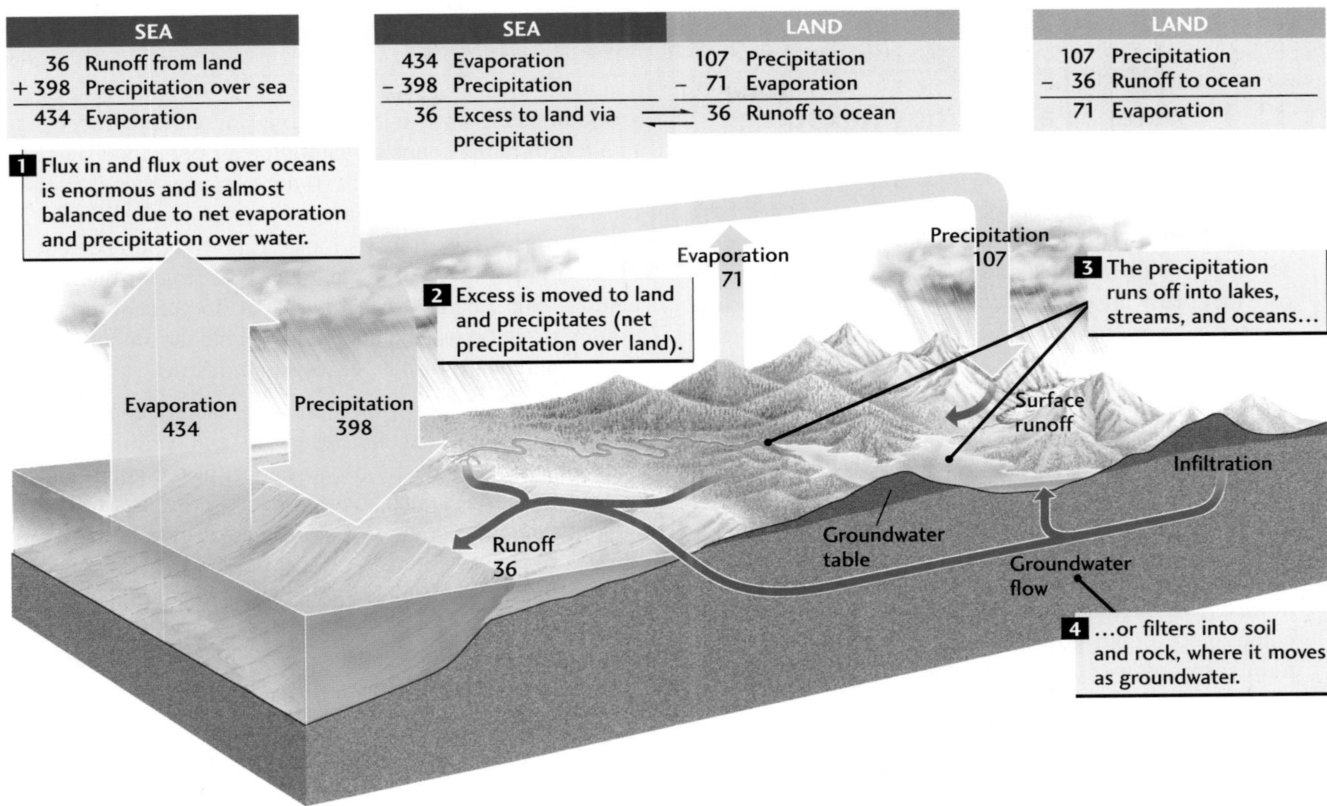

Figure 13.2 The hydrologic cycle. Water enters the atmosphere by evaporation from the oceans and continents and leaves by precipitation as rain and snow. Water lost by evaporation from the oceans is balanced by water gained from surface runoff from the continents and rainfall over the oceans. The water flow budgets are in thousands of cubic kilometers per year. [Data from E. K. Berner and R. A. Berner, *Global Environment* (Upper Saddle River, N.J.: Prentice Hall, 1996), p. 3.]

The rainwater that does not infiltrate the ground runs off the surface, gradually collecting into streams and rivers. The sum of all rainwater that flows over the surface, including the fraction that may temporarily infiltrate near-surface formations and then flow back to the surface, is called **runoff.** Some runoff may later seep into the ground or evaporate from rivers and lakes, but most of it flows into the oceans.

Snowfall may be converted into ice in glaciers, which return water to the oceans by melting and runoff and to the atmosphere by *sublimation,* the transformation from a solid (ice) directly into a gas (water vapor). Most of the water that evaporates from the oceans returns to them as rain and snow, commonly grouped together as precipitation. The remainder falls over the land and either evaporates or returns to the ocean as runoff.

Figure 13.2 shows how the total flows among reservoirs balance one another in the hydrologic cycle. The land surface, for example, gains water from precipitation and loses the same amount of water by evaporation and runoff. The ocean gains water from runoff and precipitation and loses the same amount by evaporation. As you can see from Figure 13.2, more water evaporates from the oceans than falls on them as rain. This loss is balanced by the water returned as runoff from the continents. Thus, the size of each reservoir stays constant.

How Much Water Can We Use?

As threats of water shortages loom, our use of water enters the arena of public policy debate (see Feature 13.1). The global hydrologic cycle is what ultimately controls water supplies. Almost all the water we use is fresh water—water that is not salty. Artificial desalination of seawater (the removal of salt from it) produces small but steadily growing amounts of fresh water in areas such as the arid Middle East. In the natural world, however, fresh water is supplied only by rain, rivers, lakes, some groundwaters, and water melted from snow or ice on land. All these waters are ultimately supplied by precipitation. Therefore, the practical limit to the amount of natural fresh water that we can ever

13.1 Water Is a Precious Resource: Who Should Get It?

Until recently, most people in the United States have taken our water supply for granted. Scientific analyses of available supplies and user needs, however, indicate that many areas of the country will experience water shortages more frequently. These shortages will create conflict among the several sectors of consumers—residential, industrial, agricultural, and recreational—over who has the greatest rights to the water supply.

In recent years, widely publicized droughts and mandatory restrictions on water use—such as have occurred in California, Florida, Colorado, and many other places—have alerted the public that the nation faces major water problems. Public concern waxes and wanes, however, as periods of droughts and abundant rainfall come and go, and governments are not pursuing long-term solutions with the urgency that they deserve. Here are some facts to ponder:

- A human can survive with about 2 liters of water per day. In the United States, the per capita use for all purposes is about 6000 liters per day.
- Industry uses about 38 percent and agriculture about 43 percent of the water withdrawn from our reservoirs.
- Per capita domestic water use in the United States is two to four times greater than that in western Europe, where consumers pay as much as 350 percent more for their water.
- Although the states in the western United States receive one-fourth of the country's rainfall, their per capita water use (mostly for irrigation) is 10 times greater than that of the eastern states, and at much lower prices. In California, for example, which imports most of its water, 85 percent of the water is used for irrigation, 10 percent for municipalities, and 5 percent for industry. A 15 percent reduction in irrigation use would almost double the amounts of water available for use by cities and industries.
- The traditional ways of increasing the water supply, such as building dams and reservoirs and drilling, have become extremely costly because most of the good (and therefore cheaper) sites have already been used. The building of more dams to hold larger reservoirs carries environmental costs, such as the flooding of inhabited areas, detrimental changes in river flows above and below dams, and the disturbance of fish and other wildlife habitat. Factoring in these costs has led to delays or rejection of proposals for new dams.
- The fresh water used in the United States eventually returns to the hydrologic cycle, but it may return to a reservoir that is not well located for human use, and the quality may be degraded. Recycled irrigation water is often saltier than natural fresh water and is loaded with pesticides. Polluted urban waste water ends up in the oceans.
- Global climate change may lead to reduced rainfall in western states, exacerbating the problems there and making long-term solutions even more urgent.

envision using is the amount steadily supplied to the continents by precipitation.

Hydrology and Climate

For most practical purposes, local hydrology—the amount of water there is in a region and how it flows from one reservoir to another—is more important than global hydrology. The strongest influence on local hydrology is the climate, which includes both temperature and precipitation. In warm areas where rain falls frequently throughout the year, water supplies—both at the surface and underground—are abundant. In warm, arid or semiarid regions, it rarely rains and water is a precious resource. People who live in icy climates rely on meltwaters from snow and ice. In some parts of the world, seasons of heavy rain, called monsoons, alternate with long dry seasons during which water supplies shrink, the ground dries out, and vegetation shrivels.

Wherever we live, climate and the geology of the land strongly influence the amounts of water cycled from one reservoir to another. Geologists are especially interested in how changes in precipitation and evaporation affect water supplies by altering the amounts of infiltration and runoff, which determine groundwater levels. If sea level rises as a result of global warming, groundwaters in low-lying coastal regions may become salty as seawater invades formerly fresh groundwaters.

Humidity, Rainfall, and Landscape

Many differences in climate are related to the temperature of the air and the amount of water vapor it contains. The **relative humidity** is the amount of water vapor in the air, expressed as a percentage of the total amount of water the air could hold at that temperature if saturated. When the relative humidity is 50 percent and the temperature is 15°C, for example, the amount of moisture in the air is one-half the maximum amount the air could hold at 15°C.

Warm air can hold much more water vapor than cold air. When unsaturated warm air at a given relative humidity cools enough, it becomes supersaturated and some of the vapor condenses into water droplets. The condensed water droplets form clouds. We can see clouds because they are made up of visible water droplets rather than invisible water vapor. When enough moisture has condensed into clouds and the droplets have grown too heavy to stay suspended by air currents, they fall as rain.

Most of the world's rain falls in warm, humid regions near the equator, where both the air and the surface waters of the ocean are warm. Under these conditions, a great deal of the ocean water evaporates, resulting in high humidity. When water-laden winds from these oceanic regions rise over nearby continents, the air cools and becomes supersaturated. The result is heavy rainfall over the land, even at great distances from the coast.

Landscape can alter precipitation patterns. For example, mountain ranges form **rain shadows,** areas of low rainfall on their leeward (downwind) slopes. Moisture-filled air rising over high mountains cools and precipitates rain on the windward slopes, losing much of its moisture by the time it reaches the leeward slopes (**Figure 13.3**). The air warms again as it drops to lower elevations on the other side of the mountain range. The relative humidity declines because warm air can hold more moisture before becoming saturated. This further decreases the moisture available for rain. There is a rain shadow on the eastern side of the Cascade Mountains of Oregon. Moist winds blowing over the Pacific Ocean hit the mountains' western slopes, causing heavy rainfall. The eastern slopes, on the other side of the range in the rain shadow, are dry and barren.

Unlike tropical climates, polar climates tend to be very dry. The polar oceans and the air above them are cold, so evaporation from the sea surface is minimized and the air can hold little moisture. Between the tropical and polar extremes are the temperate climates, where rainfall and temperatures are moderate.

Droughts

Droughts—periods of months or years when precipitation is much lower than normal—can occur in all climates. Arid regions are especially vulnerable to decreases in their water supplies during prolonged droughts. Lacking replenishment from precipitation, rivers may shrink and dry up, reservoirs may evaporate, and the soil may dry and crack while vegetation dies. As populations grow, demands on reservoirs increase; a drought can deplete already inadequate water supplies.

The severest droughts of the past few decades have affected lands along the southern border of the Sahara Desert, where tens of thousands of lives have been lost to famine. This long drought has expanded the desert and effectively destroyed farming and grazing in the area.

Another prolonged but less severe drought affected most of California from 1987 until February 1993, when torrential rains arrived. During the drought, groundwater and reservoirs dropped to their lowest levels in 15 years. Some control measures were instituted, but a move to

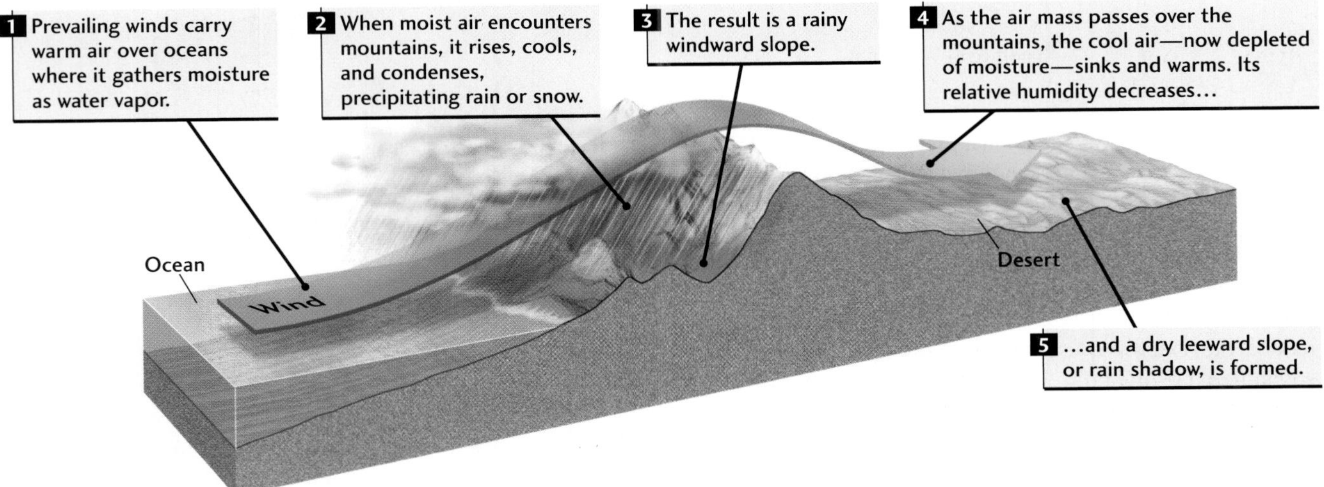

Figure 13.3 Rain shadows are areas of low rainfall on the leeward (downwind) slopes of a mountain range.

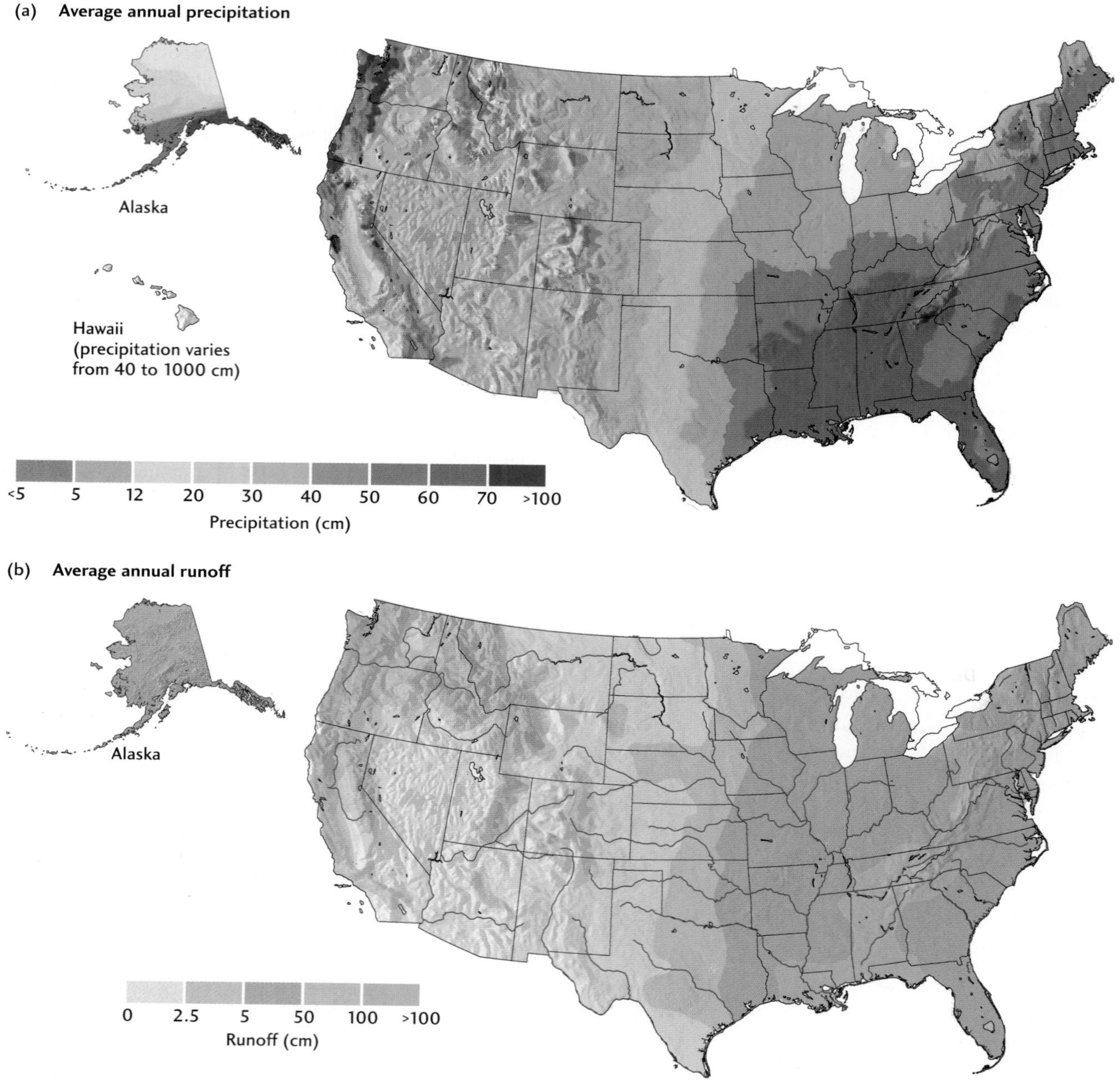

(a) **Average annual precipitation**

Alaska

Hawaii
(precipitation varies
from 40 to 1000 cm)

<5 5 12 20 30 40 50 60 70 >100
Precipitation (cm)

(b) **Average annual runoff**

Alaska

0 2.5 5 50 100 >100
Runoff (cm)

Figure 13.4 (a) Average annual precipitation in the United States. [Data from U.S. Department of Commerce, *Climatic Atlas of the United States*, 1968.] (b) Average annual runoff in the United States. [Data from USGS Professional Paper 1240-A, 1979.]

reduce the extensive use of water supplies for irrigation encountered strong political resistance from farmers and the agricultural industry (see Feature 13.1).

The midwestern United States and parts of Canada experienced a severe but short-lived drought in 1988, when surface water supplies shrank and the Mississippi River was lowered by many feet and closed to traffic. By 1989, precipitation over the region had returned to normal.

The Hydrology of Runoff

A dramatic example of how precipitation affects local stream and river runoff can be seen when weather forecasters predict flash flooding after torrential rains. When levels of precipitation and runoff are measured over a large area (such as all the states drained by a major river) and over a long period (a year, say), the relationship is less extreme but still strong. The maps of precipitation and runoff shown in **Figure 13.4** illustrate this relationship. When we compare them, we see that in areas of low precipitation—such as southern California, Arizona, and New Mexico—only a small fraction of precipitation ends up as runoff. In dry regions, much of the precipitation is lost by evaporation and infiltration. In humid areas such as the southeastern United States, a much higher proportion of the precipitation runs off in rivers. A large river may carry great amounts of water from an area with high rainfall to an area with low rainfall. The Colorado River, for example, begins in an area of moderate rainfall in Colorado and then carries its water through arid western Arizona and southern California.

Major rivers carry most of the world's surface runoff. The millions of small and medium-sized rivers carry about half the world's entire runoff; about 70 major rivers carry the other half. And the Amazon River of South America carries almost half of that. The Amazon carries about 10 times more water than the Mississippi, the largest river of North America (Table 13.1).

Table 13.1 Water Flows of Some Great Rivers

River	Water Flow (m^3/s)
Amazon, South America	175,000
La Plata, South America	79,300
Congo, Africa	39,600
Yangtze, Asia	21,800
Brahmaputra, Asia	19,800
Ganges, Asia	18,700
Mississippi, North America	17,500

Surface runoff collects and is stored in natural lakes and in artificial reservoirs created by the damming of rivers. Wetlands, such as swamps and marshlands, also act as storage depots for runoff (**Figure 13.5**). If these reservoirs are

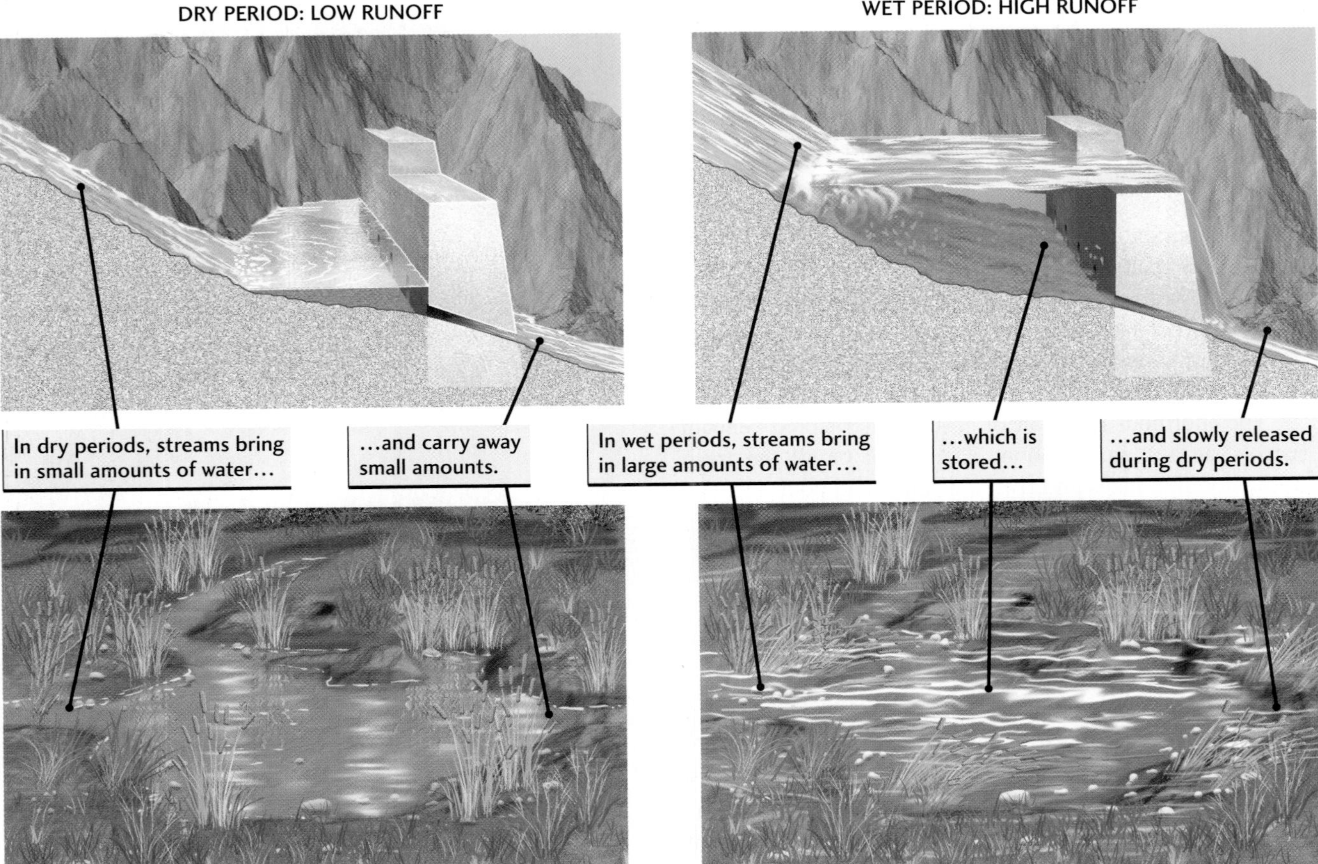

Figure 13.5 Like a natural lake or an artificial reservoir behind a dam, a wetland (such as a swamp or a marsh) stores water during times of rapid runoff and slowly releases it during periods of little runoff.

large enough, they can absorb short-term inflows from major rainfalls, holding some of the water that would otherwise spill over riverbanks. During dry seasons or droughts, the reservoirs release water to streams or to water systems built for human use. By smoothing out seasonal or yearly variations in runoff and releasing steady flows of water downstream, these reservoirs help to control flooding. For this reason, some geologists work to stop the artificial draining of wetlands for real estate development. Destruction of wetlands also threatens biological diversity, because wetlands are breeding grounds for a great many species of birds and invertebrates.

Wetlands are fast disappearing as land development continues. In the United States, more than half the original wetlands are gone. California and Ohio have kept only 10 percent of their original wetlands. The movement to protect wetlands has spawned heated controversy. The legal definition of wetland has been argued for years and has become a political football. A 1995 scientific study of the question by the National Academy of Sciences has been attacked as "political" by opponents of regulation. Some politicians who object to regulations designed to protect wetlands have asked for a 50 percent reduction in the extent of federally regulated wetlands.

Groundwater

Groundwater forms as raindrops infiltrate soil and other unconsolidated surface materials, sinking even into cracks and crevices of bedrock. We tap groundwater by drilling wells and pumping the water to the surface. Well drillers in temperate climates know that they are most likely to find a good supply of water if they drill into porous sand or sandstone beds not far below the surface. Beds that store and transmit groundwater in sufficient quantity to supply wells are called **aquifers.**

The enormous reservoir of groundwater stored beneath Earth's surface equals about 22 percent of all the fresh water stored in lakes and rivers, glaciers and polar ice, and the atmosphere. For thousands of years, people have drawn on this resource, either by digging shallow wells or by storing water that flows out onto the surface at springs. Springs are direct evidence of water moving below the surface (**Figure 13.6**).

How Water Flows Through Soil and Rock

When water moves into and through the ground, what determines where and how fast it flows? With the exception of caves, there are no large open spaces for pools or rivers of water underground. The only space available for water is the pore space between grains of sand and other particles that make up the soil and bedrock and the space in fractures. Some pores, however small and few, are found in every kind of rock and soil, but large amounts of pore space are most often found in sandstones and limestones. Recall from Chapter 8 that the amount of pore space in rock, soil, or sediment is its *porosity*—the percentage of its total volume that is taken up by pores. Porosity depends on the size and shape of the grains and how they are packed together. The more loosely packed the particles, the greater the pore space between the grains. In many sandstones, porosity is as high as 30 percent (**Figure 13.7**). Minerals that cement grains reduce porosity. The smaller the particles and the more they vary in shape, the more tightly they fit together. Porosity is

Figure 13.6 Groundwater exits from a cliff in Vasey's Paradise, Marble Canyon, Grand Canyon National Park, Arizona. This is a dramatic example of a spring formed where hilly topography allows water in the ground to flow out onto the surface. [Larry Ulrich.]

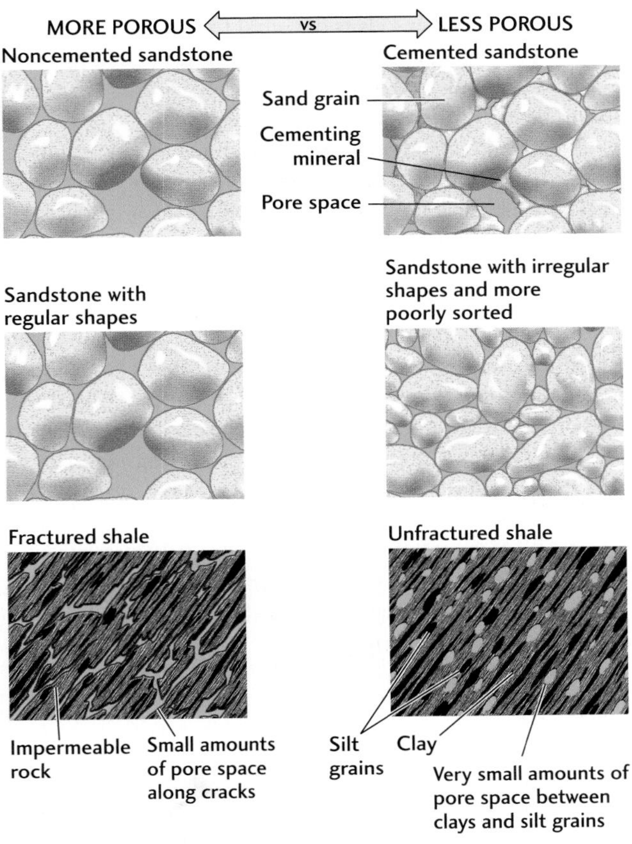

Figure 13.7 Pores in rocks are normally filled partly or entirely with water. (Pores in oil- or gas-bearing sandstones and limestones are filled with oil or gas.)

higher in sediments and sedimentary rocks (10–40 percent) than in igneous or metamorphic rocks (as low as 1–2 percent). Pore space in limestones varies, depending on how many pores were created through dissolution by groundwater or during weathering. In most unfractured shales, porosity is well under 10 percent. Fractured rocks may contain appreciable pore space—as much as 10 percent of the volume—in their many cracks. Porosity is highest of all—more than 40 percent of volume—in soils and underlying loose sand and gravel layers.

Although porosity tells us how much water a rock can hold if all its pores are filled, it gives us no information about how rapidly water can flow through the pores. Water travels through a porous material by winding between grains and through cracks. The smaller the pore spaces and the more tortuous the path, the more slowly the water travels. The ability of a solid to allow fluids to pass through it is its **permeability.** Generally, permeability increases as porosity increases. Permeability also depends on the sizes of the pores, how well they are connected, and how tortuous a path the water must travel to pass through the material.

Both porosity and permeability are important factors when one is searching for a groundwater supply. In general, a good groundwater reservoir is a body of rock, sediment, or soil with both high porosity (so that it can hold large amounts of water) and high permeability (so that the water can be pumped from it easily). A rock with high porosity but low permeability may contain a great deal of water, but because the water flows so slowly, it is hard to pump it out of the rock. Table 13.2 summarizes the porosity and permeability of various rock types.

The Groundwater Table

As well drillers bore deeper into soil and rock, the samples they bring up become wetter. At shallow depths, the material is unsaturated—the pores contain some air and are not completely filled with water. This level is called the

Table 13.2 Porosity and Permeability of Aquifer Rock Types

Rock Type	Porosity (Pore Space That May Hold Fluid)	Permeability (Ability to Allow Fluids to Pass Through)
Gravel	Very high	Very high
Coarse- to medium-grained sand	High	High
Fine-grained sand and silt	Moderate	Moderate to low
Sandstone, moderately cemented	Moderate to low	Low
Fractured shale or metamorphic rocks	Low	Very low
Unfractured shale	Very low	Very low

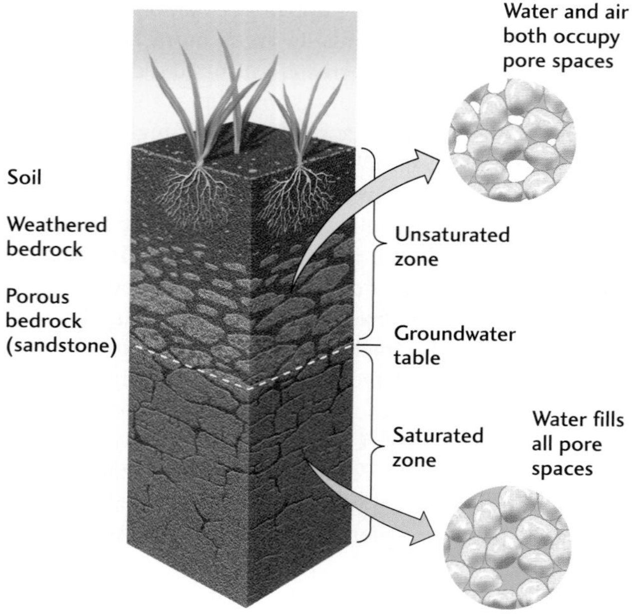

Figure 13.8 The groundwater table is the boundary between the unsaturated zone and the saturated zone. The saturated and unsaturated zones can be in either unconsolidated material or bedrock.

unsaturated zone (often termed the *vadose zone*). Below it is the **saturated zone,** the level in which the pores of the soil or rock are completely filled with water. The saturated and unsaturated zones can be in unconsolidated material or bedrock (**Figure 13.8**). The boundary between the two zones is the **groundwater table,** usually shortened to "water table." When a hole is drilled below the water table, water from the saturated zone flows into the hole and fills it to the level of the water table.

Groundwater moves under the force of gravity, and so some of the water in the unsaturated zone may be on its way down to the water table. A fraction of the water, however, will remain in the unsaturated zone, held in small pore spaces by surface tension—the attraction between the water molecules and the surfaces of the particles. Surface tension, you may recall from Chapter 12, keeps the sand on a beach moist, even though there are spaces below to which water could travel by gravity. The evaporation of water in pore spaces in the unsaturated zone is slowed both by the effect of surface tension and by the relative humidity of the air in the pore spaces, which can be close to 100 percent.

If we were to drill wells at several sites and measure the elevations of the water levels in the wells, we could construct a map of the water table. A cross section of the landscape might look like the one shown in **Figure 13.9**. The water table follows the general shape of the surface topography, but the slopes are gentler. The water table is at the surface in river and lake beds and at springs. Under the influence of gravity, groundwater moves downhill from an area where the water-table elevation is high—under a hill, for example—to places where the water-table elevation is low, such as a spring where groundwater exits to the surface.

Water enters and leaves the saturated zone through recharge and discharge. **Recharge** is the infiltration of water into any subsurface formation, often by rain or snow meltwater from the surface. Recharge may also take place through the bottom of a stream where the stream channel lies at an elevation above that of the water table (see Figure 13.9). Streams that recharge groundwater in this way are called **influent streams,** and they are most characteristic of arid regions, where the water table is deep. **Discharge** is the exit of groundwater to the surface, the opposite of recharge. When a stream channel intersects the water table, water discharges from the groundwater to the stream. Such an **effluent stream** is typical of humid areas. Effluent streams continue to flow long after runoff has stopped because they are fed by groundwater. Thus, the reservoir of groundwater may be increased by influent streams and depleted by effluent streams.

Aquifers

Groundwater may flow in unconfined or confined aquifers. In **unconfined aquifers,** the water travels through beds of more or less uniform permeability that extend to the surface in both discharge and recharge areas. The level of the reservoir in an unconfined aquifer is the same as the height of the water table.

Many permeable aquifers, typically sandstones, are bounded above and below by shale beds of low permeability. These relatively impermeable beds are **aquicludes,** and groundwater either cannot flow through them or flows through them very slowly. When aquicludes lie both over and under an aquifer, they form a **confined aquifer.**

The impermeable beds above a confined aquifer prevent rainwater from infiltrating directly into the aquifer. Instead, a confined aquifer is recharged by precipitation over the recharge area, often characterized by outcropping rocks in a topographically higher upland. Here rainwater can enter the ground because there is no aquiclude preventing infiltration. The water then travels down the aquifer underground (**Figure 13.10**). Water in a confined aquifer—known as an **artesian flow**—is under pressure. At any point in the aquifer, the pressure is equivalent to the weight of all the water in the aquifer above that point.

Figure 13.9 Dynamics of the groundwater table in permeable shallow formations in a temperate climate. The depth of the water table fluctuates in response to the balance between water added from precipitation (recharge) and water lost by evaporation and from wells, springs, and streams (discharge).

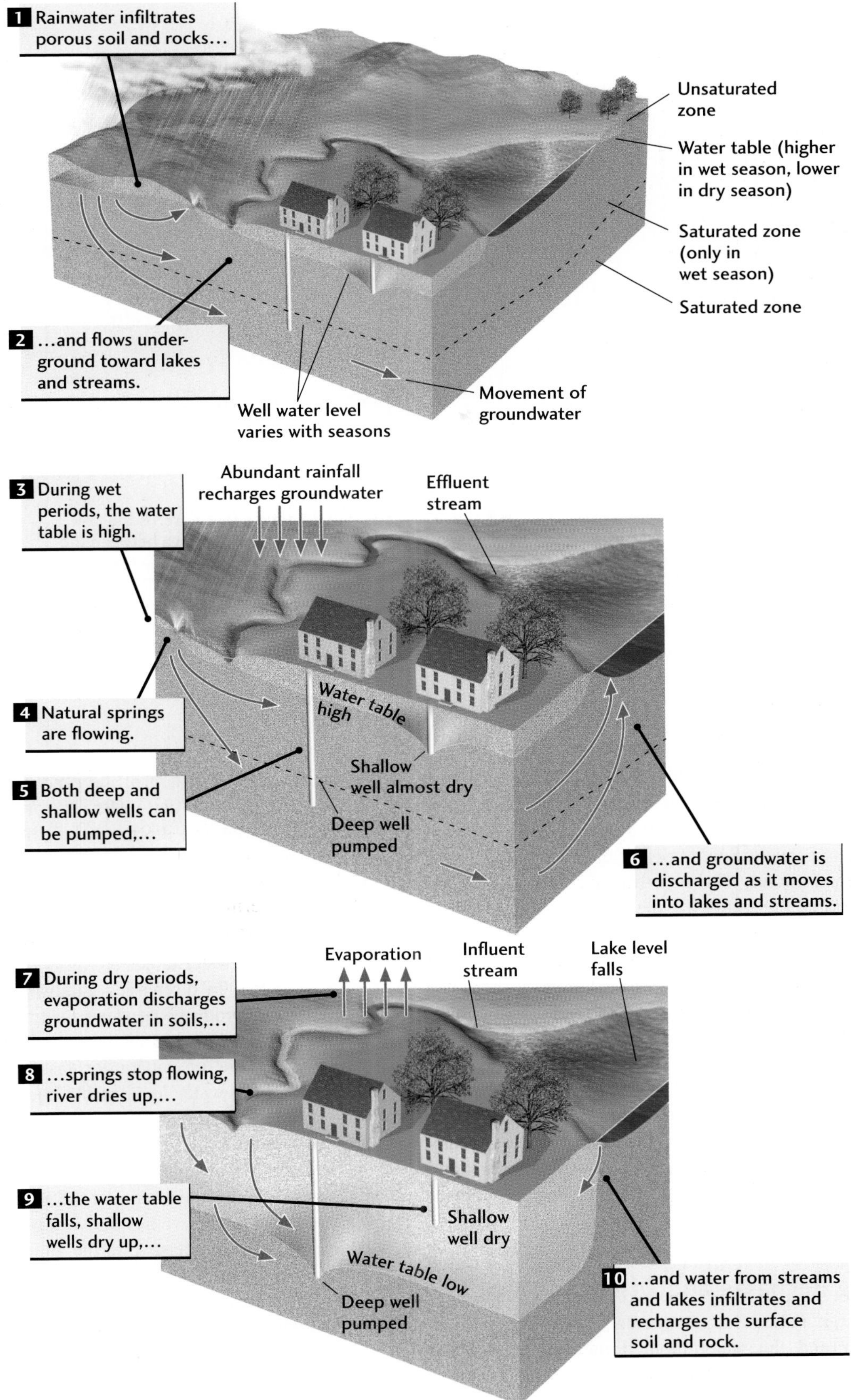
1 Rainwater infiltrates porous soil and rocks...
Unsaturated zone
Water table (higher in wet season, lower in dry season)
Saturated zone (only in wet season)
Saturated zone
2 ...and flows underground toward lakes and streams.
Well water level varies with seasons
Movement of groundwater
3 During wet periods, the water table is high.
Abundant rainfall recharges groundwater
Effluent stream
Water table high
4 Natural springs are flowing.
5 Both deep and shallow wells can be pumped,...
Shallow well almost dry
Deep well pumped
6 ...and groundwater is discharged as it moves into lakes and streams.
Evaporation
Influent stream
Lake level falls
7 During dry periods, evaporation discharges groundwater in soils,...
8 ...springs stop flowing, river dries up,...
9 ...the water table falls, shallow wells dry up,...
Shallow well dry
Water table low
Deep well pumped
10 ...and water from streams and lakes infiltrates and recharges the surface soil and rock.

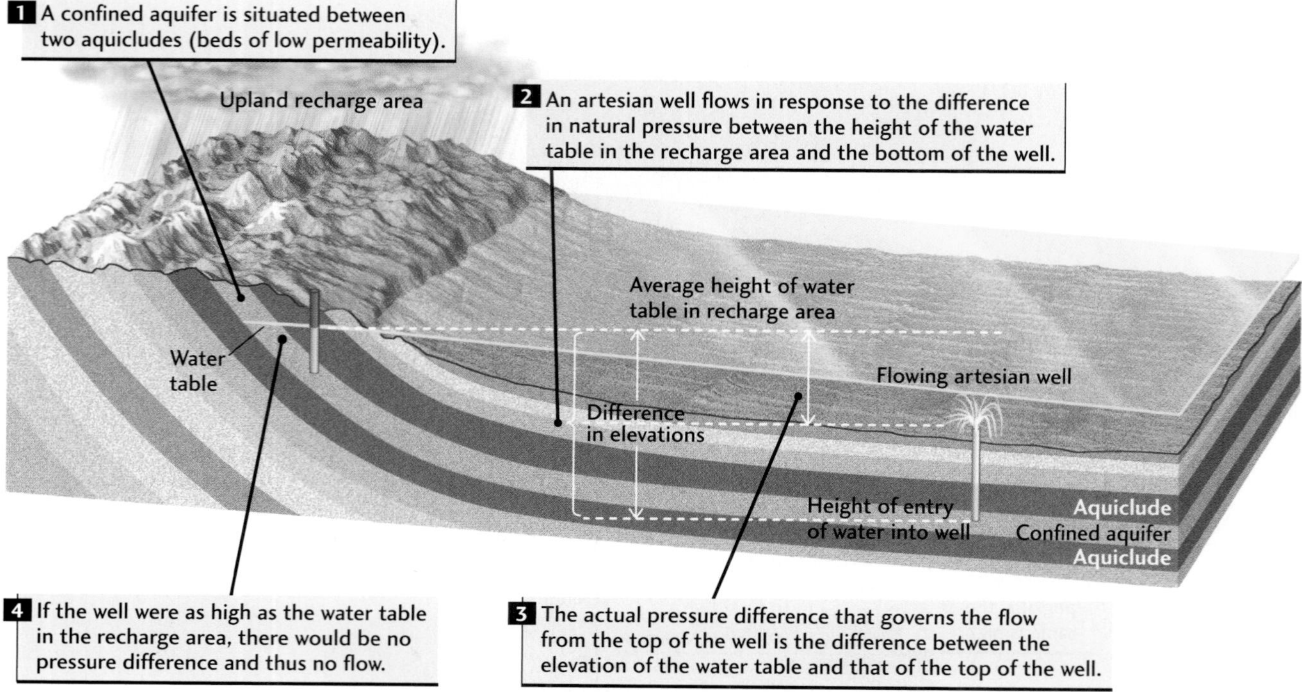

Figure 13.10 A confined aquifer is created when an aquifer is situated between two aquicludes (beds of low permeability).

If we drill a well into a confined aquifer at a point where the elevation of the ground surface is lower than that of the water table in the recharge area, the water will flow out of the well spontaneously. Such wells are called **artesian wells,** and they are extremely desirable because no energy is required to pump the water to the surface. The water is brought up by its own pressure.

In more complex geological environments, the water table may be more complicated. For example, if there is a relatively impermeable clay layer—an aquiclude—in a permeable sand formation, the aquiclude may lie below the water table in a shallow aquifer and above the water table in a deeper aquifer (**Figure 13.11**). The water table in the shallow aquifer is called a **perched water table** because it is above the main water table in the lower aquifer. Many perched water tables are small, only a few meters thick and restricted in area, but some extend for hundreds of square kilometers.

Balancing Recharge and Discharge

When recharge and discharge are balanced, the reservoir of groundwater and the water table remain constant, even though water is continually flowing through the aquifer. For recharge to balance discharge, rainfall must be frequent enough to equal the sum of the runoff from rivers and the outflow from springs and wells.

But recharge and discharge will not always be equal, because rainfall varies from season to season. Typically, the water table drops in dry seasons and rises during wet periods. A decrease in recharge, such as during a prolonged drought, will be followed by a longer-term imbalance and a lowering of the water table.

An increase in discharge, usually from increased well pumping, can produce the same imbalance. Shallow wells may end up in the unsaturated zone and go dry. When a well pumps water out of an aquifer faster than recharge can replenish it, the water level in the aquifer is lowered in a cone-shaped area around the well, called a *cone of depression* (**Figure 13.12**). The water level in the well is lowered to the depressed level of the water table. If the cone of depression extends below the bottom of the well, that well goes dry. If the bottom of the well is above the base of the aquifer, extending the well deeper into the aquifer may allow more water to be withdrawn, even at continued high pumping rates. If the rate of pumping is maintained and the well is deepened so much that the entire aquifer is tapped, however, the cone of depression can reach the bottom of the aquifer and deplete it. The aquifer will recover only if the pumping rate is reduced enough to give it time to recharge.

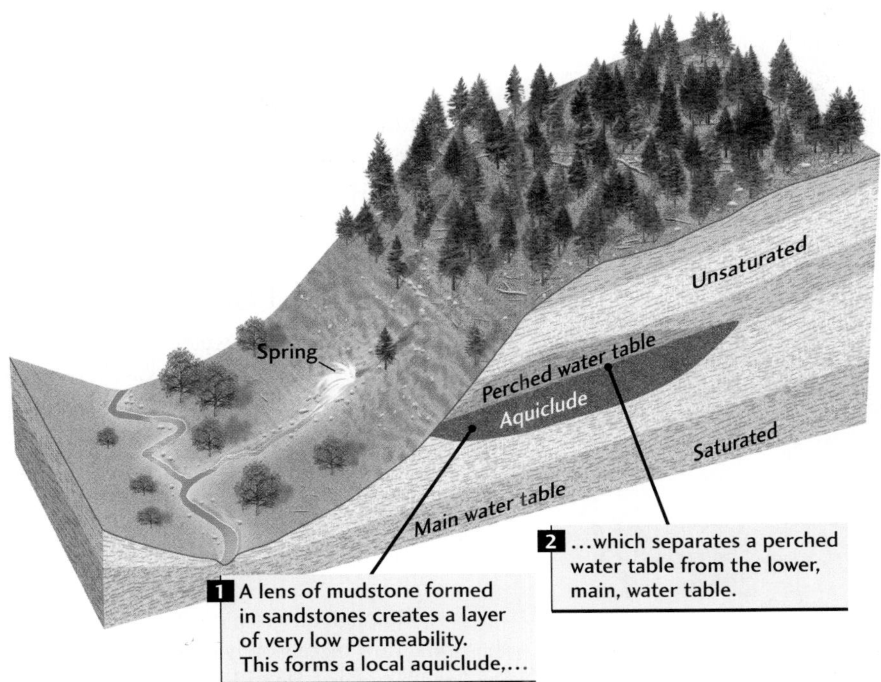

Figure 13.11 A perched water table forms in geologically complex situations—in this case, where a shale aquiclude is located above the main water table in a sandstone aquifer. The dynamics of the perched water table's recharge and discharge may be different from those of the main water table. The main water table in this example can be recharged only from its lower outcrop slopes.

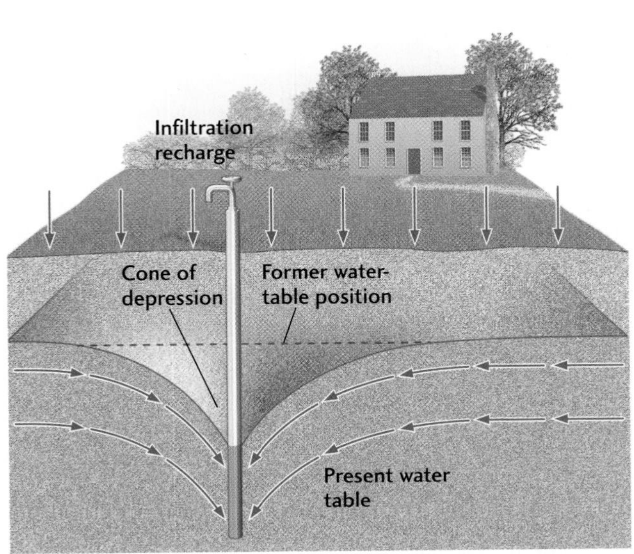

Figure 13.12 Excessive pumping in relation to recharge draws down the water table into a cone-shaped depression around a well. The water level in the well is lowered to the depressed level of the water table.

The extreme withdrawal of water not only can deplete the aquifer but also can cause another undesirable environmental effect. As the pressure of the water in pore space falls, materials formerly overlying the aquifer may subside, creating sinklike depressions (**Figure 13.13**). As water in some sediments is removed, the sediments compact, and the loss of volume is manifested in the lowering of the surface. Subsidence caused by overpumping has occurred in Mexico City and in Venice, Italy, as well as in many other regions of heavy pumping, such as the San Joaquin Valley in California. In these places, the rate of subsidence of the surface has reached almost 1 m every 3 years. Although a few experiments have attempted to reverse the subsidence by pumping water back into the groundwater system, they have not been very successful, because most compacted materials do not expand easily to their former state. The best measure is to halt further subsidence by restricting pumping.

People who live near the ocean's edge may face a different problem when pumping rates are high in relation to recharge: the incursion of salt water into the well. Near shorelines or a little offshore, an underground boundary separates salt water under the sea from fresh water under the land. This boundary slopes down and inland from the shoreline in such a way that salt water underlies the fresh

Figure 13.13 In Antelope Valley, California, overpumping of groundwater has led to fissures and sinklike depressions on Rogers Lakebed at Edwards Air Force Base. This fissure, formed in January 1991, is about 625 m long. [James W. Borchers/USGS.]

water of the aquifer (**Figure 13.14**). Under many ocean islands, a fresh-groundwater lens (shaped like a simple double-convex lens) floats on a base of seawater. The fresh water floats because it is less dense than seawater (1.00 g/cm^3 versus 1.02 g/cm^3, a small but significant difference). Normally, the pressure of fresh water keeps the saltwater margin slightly offshore.

The balance between recharge and discharge in the freshwater aquifers maintains this freshwater-seawater boundary. As long as recharge by rainwater is at least equal to discharge by pumping, the well will provide fresh water. If water is withdrawn faster than it is recharged, however, a cone of depression develops at the top of the aquifer, mirrored by an inverted cone rising from the freshwater-seawater boundary below. The cone of depression at the upper part of the aquifer makes it more difficult to pump fresh water, and the inverted cone below leads to an intake of salt water at the bottom of the well (see Figure 13.14). People living closest to the shore are the first affected. Some towns on Cape Cod in Massachusetts, on Long Island in New York, and in many other nearshore areas have had to post notices that town drinking water contains more salt than is considered healthful by environmental agencies. There is no ready solution to this problem other than to slow the pumping or, in some places, to recharge the aquifer artificially by funneling runoff into the ground.

You can see that a rise in sea level, which has been predicted as a result of global warming, would seriously alter the seawater margin. As sea level rises, the margin also rises. Seawater can then invade coastal aquifers and turn fresh groundwater into salt water.

The Speed of Groundwater Flows

The *speed* at which water moves in the ground strongly affects the balance between discharge and recharge. Most groundwaters move slowly, a fact of nature responsible for our groundwater supplies. If groundwater moved as rapidly as rivers do, aquifers would run dry after a period of time without rain, just as many small streams run dry. The slow-moving groundwater flow also makes rapid recharge impossible if groundwater levels are lowered by excessive pumping.

Although all groundwaters flow through aquifers slowly, some flow more slowly than others. In the middle of the nineteenth century, Henri Darcy, town engineer of Dijon, France, proposed an explanation for the difference in flow rates. While studying the town's water supply, Darcy measured the elevations of water in various wells and mapped the varying heights of the water table in the district. He calculated the distances that the water traveled from well to well and measured the permeability of the aquifers. Here are his findings.

- For a given aquifer and distance of travel, the rate at which water flows from one point to another is directly proportional to the drop in elevation of the water table between the two points. As the difference in elevation increases, the rate of flow increases.

- The rate of flow for a given aquifer and a given difference in elevation is inversely proportional to the flow distance that the water travels. As the distance increases, the rate decreases. The ratio between the elevation difference and the flow distance is known as the **hydraulic gradient.** Just as a ball rolls faster down a steeper slope than a gentler one, groundwater flows more quickly down a steeper hydraulic gradient. In general, groundwater does not run down the slope of the groundwater table but follows the hydraulic gradient of the flow, which may travel various paths below the water table.

- Darcy reasoned that the relationship between flow and hydraulic gradient should hold whether the water is moving through a porous sandstone aquifer or an open pipe. You might guess that the water would move more quickly through a pipe than through the tortuous turns of pore spaces in an aquifer. Darcy recognized this factor and included a measure of permeability in his final equation, and so, other things being equal, the greater the permeability and thus the greater the ease of flow, the faster the flow.

Darcy's law, which summarizes these relationships, can be expressed in a simple equation (**Figure 13.15**): the

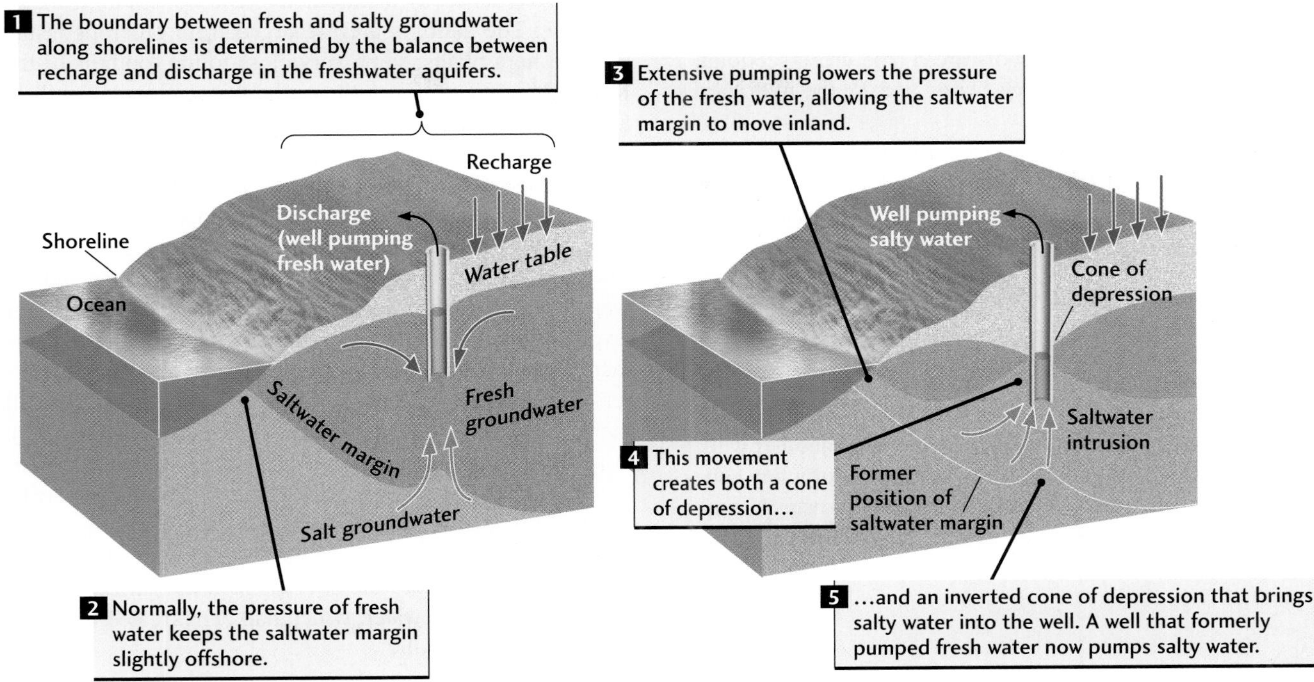

Figure 13.14 The balance between recharge and discharge maintains the freshwater-seawater boundary.

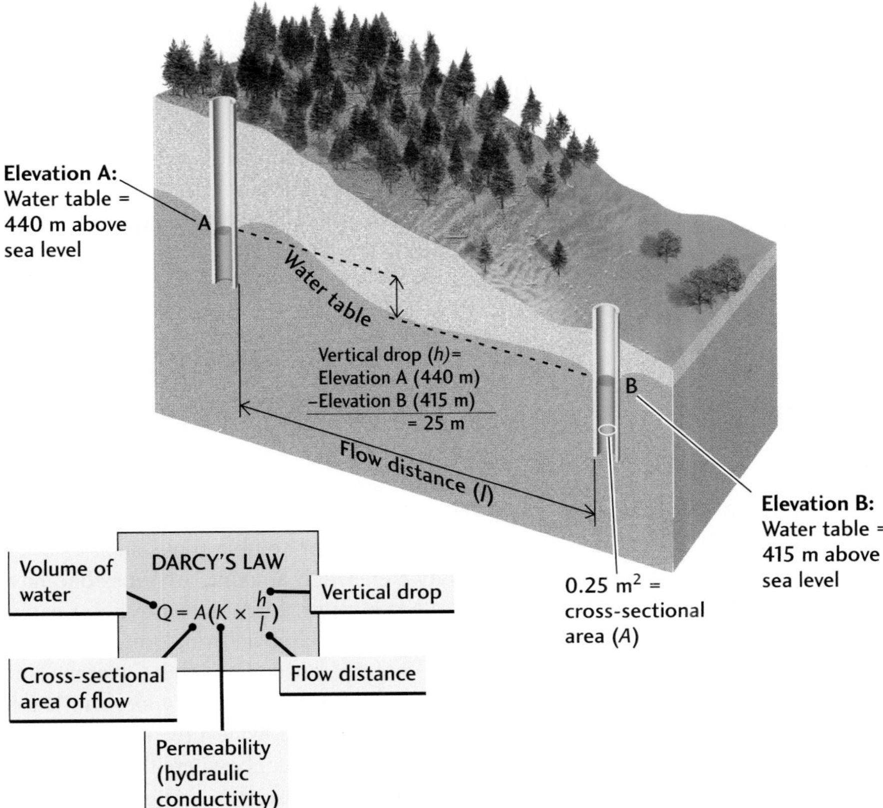

Figure 13.15 Darcy's law describes the rate of groundwater flow down a slope between two points, Elevation A and Elevation B. The volume of water flowing at a certain time (Q) is proportional to the difference in height (h) between the high and low points of the slope (here shown as the drop in the elevation of the water table between the two points), divided by the flow distance between them (the hydraulic gradient, l) and by K, a constant proportional to the permeability of the aquifer. The symbol A represents the cross-sectional area through which the water flows.

volume of water flowing in a certain time (Q) is proportional to the vertical drop (h) divided by the flow distance (l). The two remaining symbols are A, the cross-sectional area through which the water flows, and K, the hydraulic conductivity (a measure of permeability). (K also depends on the properties of the fluid, especially density and viscosity, which are important in dealing with fluids other than water.)

$$Q = A(K \times \frac{h}{l})$$

Velocities calculated by Darcy's law have been confirmed experimentally by measuring how long it takes a harmless dye introduced into one well to reach another well. In most aquifers, groundwater moves at a rate of a few centimeters per day. In very permeable gravel beds near the surface, groundwater may travel as much as 15 cm/day. (This speed is still much slower than the speeds of 20 to 50 cm/s typical of river flows.)

Water Resources from Major Aquifers

Large parts of North America rely on groundwater for all their water needs. The demand on groundwater resources has grown as populations have increased and as uses such as irrigation have expanded (**Figure 13.16**). Many areas of the Great Plains and other parts of the Midwest rest on sandstone formations, most of which are confined aquifers like the one shown in Figure 13.10. Thousands of wells have been drilled into these formations, most of which transport water over hundreds of kilometers and constitute a major resource. The aquifers are recharged from outcrops in the western high plains, some very close to the foothills of the Rocky Mountains. From there, the water runs downhill in an easterly direction.

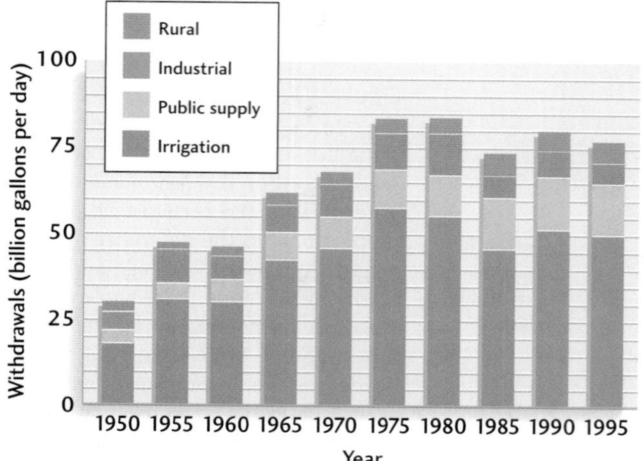

Figure 13.16 Groundwater withdrawals, United States, 1950–1995. [USGS.]

Darcy's law tells us that water flows at rates proportional to the slope of an aquifer between its recharge area and a given well. In the western plains, the slopes are gentle and water moves slowly through the aquifers, recharging them at low rates. At first, many of these wells were artesian and the water flowed freely. As more wells were drilled, the water levels dropped, and the water had to be pumped to the surface. Extensive pumping withdraws water from some aquifers faster than the slow recharge from far away can fill them, so the reservoirs are being depleted (see Feature 13.2).

Efforts to reduce excessive discharge have been supplemented by attempts to increase recharge of aquifers artificially in some areas. On Long Island, New York, for example, the water authority drilled a large system of recharge wells to put water into the aquifer from the surface. These wells pumped used water, which had been treated to purify it, back into the ground.

The water authority also constructed large, shallow basins over natural recharge areas to augment infiltration from surface waters by catching and diverting runoff, including storm and industrial waste drainage. Officials in charge of the program knew that urban development can decrease recharge by interfering with infiltration. As urbanization progresses, the impermeable materials used to pave large areas for streets, sidewalks, and parking lots prevent water from infiltrating the ground. Rainwater runoff increases, and the decrease in natural infiltration into the ground may deprive the aquifers of much of their recharge. One remedy is to catch and use the storm runoff in a systematic program of artificial recharge, as the Long Island water authority did. The multiple efforts of the water authority helped rebuild the Long Island aquifer, though not to its original levels.

Erosion by Groundwater

Every year, thousands of people visit caves, either on tours of popular attractions such as Mammoth Cave, Kentucky, or in adventurous explorations of little-known caves. These underground open spaces are produced by the dissolution of limestone—or, rarely, of other soluble rocks such as evaporites—by groundwater. Huge amounts of limestone have dissolved to make some caves. Mammoth Cave, for example, has tens of kilometers of large and small interconnected chambers, and the large room at Carlsbad Caverns, New Mexico, is more than 1200 m long, 200 m wide, and 100 m high. Limestone formations are widespread in the upper parts of the crust, but caves form only where these relatively soluble rocks are at or near the surface and enough carbon

13.2 When Do Groundwaters Become Nonrenewable Resources?

For more than 100 years, water from the Ogallala aquifer, a formation of sand and gravel, has supplied the cities, towns, ranches, and farms of western Texas and eastern New Mexico (see map). The population of the region has climbed from a few thousand late in the nineteenth century to about a million today. The Ogallala continues to provide the irrigation water needed to support agriculture, which is the area's economic base, but the water pressure in the wells has declined steadily and the water table has dropped by 30 m or more.

Natural recharge of the Ogallala aquifer of the southern plains is very slow, because rainfall is sparse, the degree of evaporation is high, and the recharge area is small. Pumping, primarily for irrigation, has been so extensive—about 6 billion cubic meters of water per year from 170,000 wells—that recharge cannot keep up. At current rates of recharge, if all pumping were to stop, it would take several thousand years for the water table to recover its original position and well pressure to be restored. Some scientists have attempted to recharge the aquifer artificially by injecting water from shallow lakes that form in wet seasons on the high plains. These experiments have managed to increase recharge, but the aquifer is still in danger over the long term.

It is estimated that the remaining supplies of water in the Ogallala will last only into the early years of this century. As this valuable underground reservoir is drained, about 5.1 million acres of irrigated land in western Texas and eastern New Mexico will dry up—and so will 12 percent of the country's supply of cotton, corn, sorghum, and wheat and a significant fraction of the feedlots for the nation's cattle.

Other aquifers in the northern plains and elsewhere in North America are in a similar condition. In three major areas of the United States—Arizona, the high plains, and California—groundwater supplies are significantly depleted. As water use increases, we all must eventually adopt sensible conservation practices.

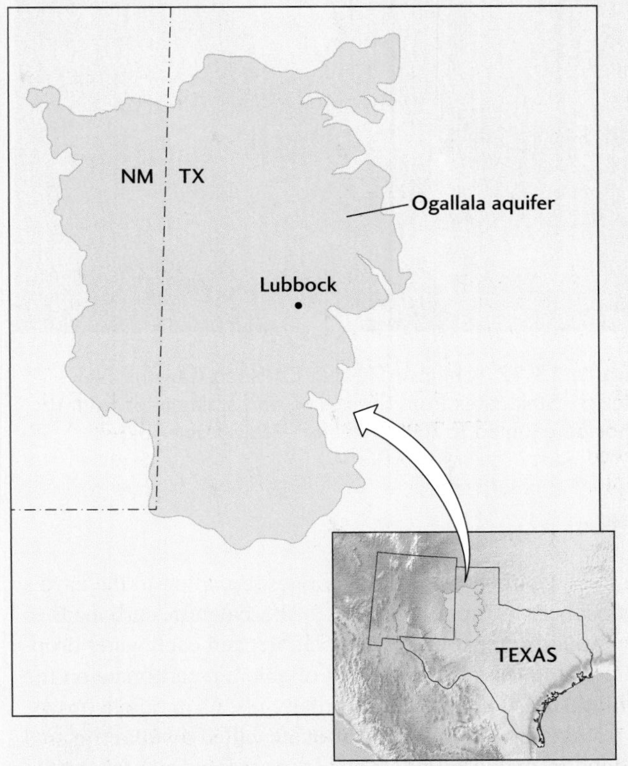

The southwestern high plains of Texas and New Mexico are underlain by the Ogallala aquifer. The blue region represents the aquifer. The general recharge area is located along the western margin of the aquifer. [USGS.]

dioxide–rich water infiltrates the surface to dissolve extensive areas of limestone.

As we saw in Chapter 7, the atmospheric carbon dioxide contained in rainwater enhances the dissolution of limestone. Water that infiltrates soil may pick up even more carbon dioxide from plant roots, bacteria, and other soil-dwelling organisms that give off this gas. As this carbon dioxide–rich water moves down to the water table, through the unsaturated zone to the saturated zone, it creates openings as it dissolves carbonate minerals. These openings enlarge as limestone dissolves along joints and fractures, forming a network of rooms and passages. Such networks form extensively in the saturated zone, where—because the caves are filled with water—dissolution takes place over all surfaces, including floors, walls, and ceilings.

We can explore caves that were once below the water table but are now in the unsaturated zone because the water table fell. In these caves, now air-filled, water saturated with calcium carbonate may seep through the ceiling. As each drop of water drips from the cave's ceiling, some of its

Figure 13.17 Chinese Theater, Carlsbad Caverns, New Mexico. Stalactites from the ceiling and stalagmites from the floor have joined to form a column. [David Muench.]

Figure 13.18 A large sinkhole formed by the collapse of a shallow underground cavern. Such collapses can occur so suddenly that moving cars are buried. Winter Park, Florida. [Leif Skoogfors/Woodfin Camp.]

dissolved carbon dioxide evaporates, escaping to the cave's atmosphere. Evaporation makes the calcium carbonate in the groundwater solution less soluble, and each water droplet precipitates a small amount of calcium carbonate on the ceiling. These deposits accumulate, just as an icicle grows, in a long, narrow spike of carbonate called a **stalactite** suspended from the ceiling. When some of the water falls to the cave floor, more carbon dioxide escapes and another small amount of calcium carbonate is precipitated on the cave floor below the stalactite. These deposits also accumulate, forming a **stalagmite.** Eventually, a stalactite and a stalagmite may grow together to form a column (**Figure 13.17**).

Unusual species of bacteria have been discovered in such caves. Some geologists think that the Carlsbad Caverns were formed partly by sulfate bacteria that produced sulfuric acid.

In some places, dissolution may thin the roof of a limestone cave so much that it collapses suddenly, producing a **sinkhole**—a small, steep depression in the land surface above a cavernous limestone formation (**Figure 13.18**). Sinkholes are characteristic of a distinctive type of topography known as *karst,* named for a region in the northern part of the former Yugoslavia. **Karst topography** is an irregular hilly terrain characterized by sinkholes, caverns, and a lack of surface streams (**Figure 13.19**). Underground drainage channels replace the normal surface drainage system of small and large rivers. Short, scarce streams often end in sinkholes, detouring underground and sometimes reappearing miles away. Karst topography is found in regions with three characteristics:

1. A high-rainfall climate, with abundant vegetation (providing carbon dioxide–rich waters)
2. Extensively jointed limestone formations
3. Appreciable hydraulic gradients

Karst terrains often have environmental problems, including surface subsidence from the collapse of underground space and potentially catastrophic cave-ins. In North and Central America, karst topography is found in limestone terrains of Indiana, Kentucky, and Florida and on the Yucatán Peninsula of Mexico. Karst is well developed on uplifted coral limestone of late Cenozoic terrains of tropical volcanic island arcs.

Water Quality

Most residents of Canada and the United States take a supply of fresh, pure water for granted. A growing number of people, however, are fearful of contaminants in their water and are buying bottled spring water or installing purifying systems in their homes. Almost all water supplies in North America are free of bacterial contamination, and the vast majority are free enough of chemicals to drink safely. Earlier, we discussed the problem of seawater incursion into the water supplies of some shoreline communities, which results in unacceptable levels of sodium in the water. A more common problem is the pollution of rivers and aquifers by toxic wastes from surface dumps.

Contamination of the Water Supply

Lead Pollution Lead is a well-known pollutant produced by industrial processes that inject contaminants into the atmosphere. When water vapor condenses in the atmosphere, lead is incorporated into raindrops, which then transport it to Earth's surface. Lead is routinely eliminated from public water supplies by chemical treatment before the water is dis-

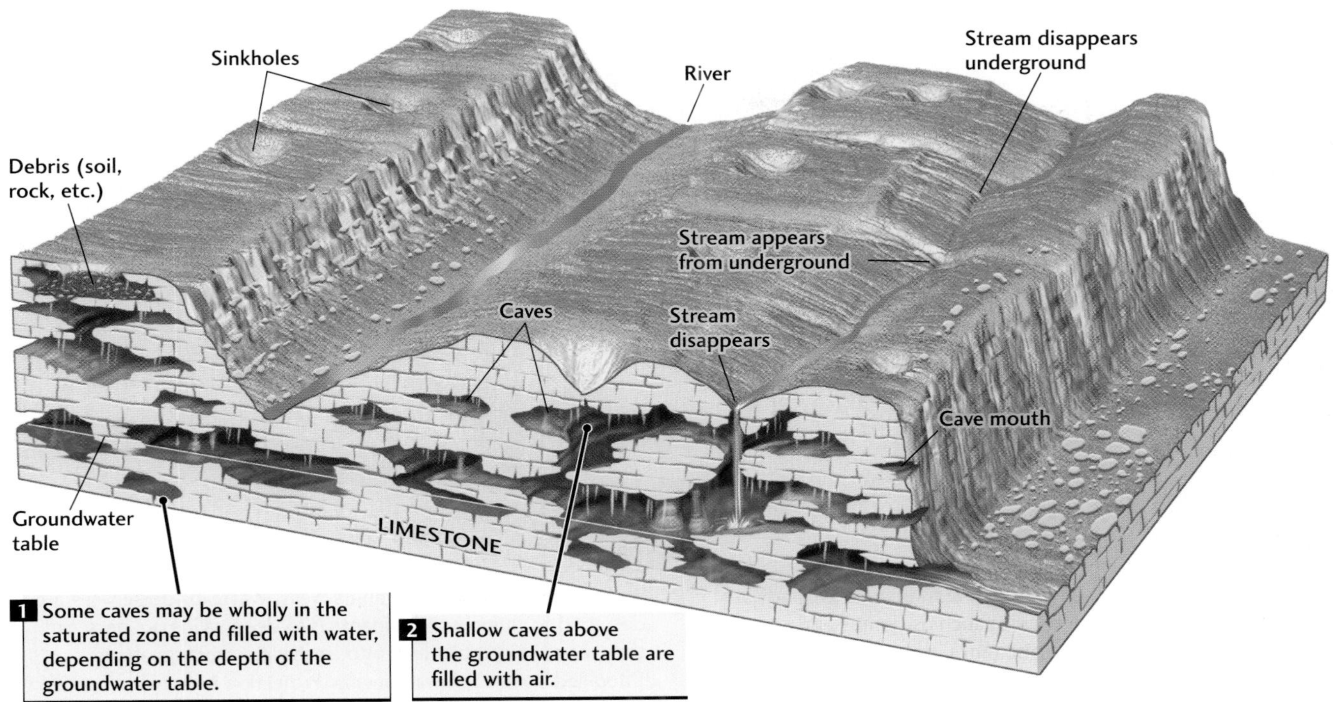

Figure 13.19 Some major features of karst topography are caves, sinkholes, and disappearing streams.

tributed through the water mains. In older homes with lead pipes, however, lead can leach into the water. In Boston, for example, an astonishing 41 percent of 174 tap-water samples taken first thing in the morning had unsafe lead levels. Even in newer construction, the lead solder used to connect copper pipes and the metals used in faucets are sources of contamination. Replacing old lead pipes with durable plastic pipes can reduce lead contamination. Even letting the water run for a few minutes to clear pipes can help.

Radioactive Wastes There is no easy solution to the problem of contamination from radioactive waste. When radioactive waste is buried underground, it may be leached by groundwaters and find its way into water-supply aquifers. Storage tanks and burial sites at the atomic weaponry plants in Oak Ridge, Tennessee, and Hanford, Washington, have already leaked radioactive wastes into shallow groundwaters (see Feature 22.1).

Microorganisms in Groundwater We have learned in the past few decades that, contrary to all expectations, bacteria can and do live in large numbers at great depths (as much as several thousand meters) in groundwaters and constitute a huge biomass. Most of these bacteria are supported by nutrients from organic matter buried with the original sediment, but geologists recently discovered bacteria that draw energy from hydrogen in rocks. The hydrogen is generated by geochemical reactions between groundwater and rocks such as basalt. These reactions, aside from serving as a source of energy for the bacteria, continue the weathering process underground.

Septic tanks, widely used in some areas that lack full sewer networks, are settling tanks buried at shallow depths in which the solid wastes from house sewage are decomposed by bacteria. To prevent contamination of potable water, septic tanks must be installed at sufficient distance from water wells in shallow aquifers.

Other Chemical Contaminants As we have seen, human activities can contaminate groundwaters (**Figure 13.20**). The disposal of chlorinated solvents—such as trichloroethylene (TCE), widely used as a cleaner in industrial processes—poses a formidable problem. These solvents persist in the environment because they are difficult to remove from contaminated waters. Buried gasoline storage tanks can leak, and road salt inevitably drains into the soil and ultimately into aquifers. Rain can wash agricultural pesticides, herbicides, and fertilizers into the soil. From the soil, they percolate downward into aquifers. In some agricultural areas where nitrate fertilizers are heavily used, groundwaters may contain high quantities of nitrate. In one recent study, 21 percent of the shallow wells sampled for drinking water exceeded the maximum amounts of nitrate (10 ppm) allowed in the United States. Such high nitrate levels pose a danger of "blue baby" syndrome (the inability to maintain healthy oxygen levels) to infants 6 months old and younger.

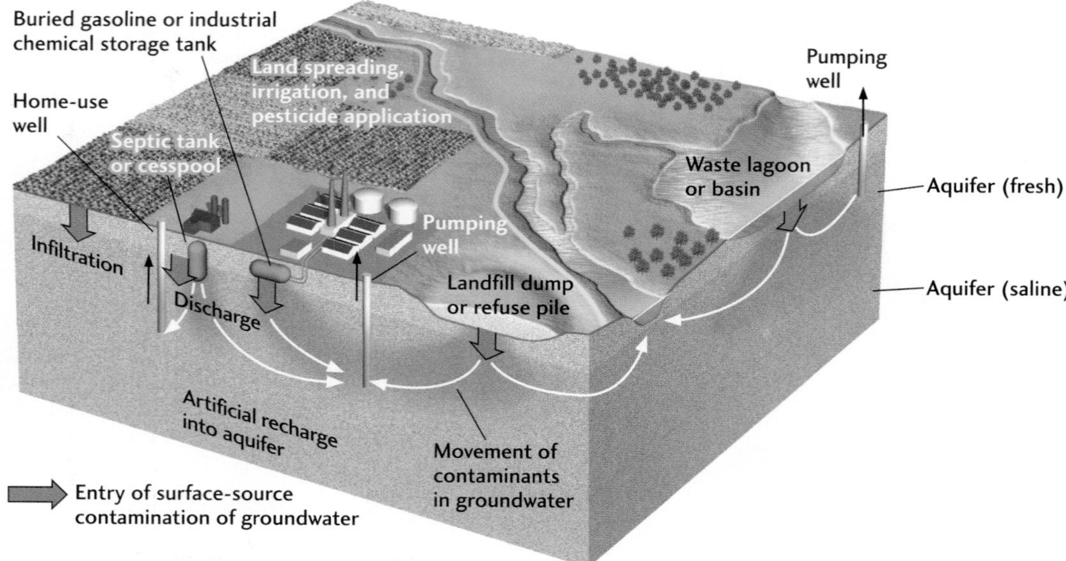

Figure 13.20 Human activities can contaminate groundwater. Contaminants from surface sources such as dumps and subsurface sources such as septic tanks enter aquifers through normal groundwater flow. Contaminants may be introduced into water supplies through pumping wells. Waste-disposal wells are designed to pump contaminants into deep saline aquifers, but they may accidentally leak into freshwater aquifers above. [Modified from U.S. Environmental Protection Agency.]

Reversing Contamination Can we reverse the contamination of water supplies? The answer is a qualified yes, but the process is costly and very slow. The faster an aquifer recharges, the easier it is to clean. If the recharge is rapid, once we close off the sources of contamination, fresh water moves into the aquifer, and in a short time the water quality is restored. Even a fast recovery, however, can take a few years.

The contamination of slowly recharging reservoirs is more difficult to reverse. The rate of groundwater movement may be so slow that contamination from a distant source may take a long time to appear. By the time it does, it is too late for rapid recovery. Even with cleaned-up recharge, some contaminated deep reservoirs hundreds of kilometers from the recharge area may not respond for many decades.

When public water supplies are polluted, we can pump the water and then treat it chemically to make it safe, but this is an expensive procedure. Alternatively, we can try to treat the water while it remains underground. In one moderately successful experimental procedure, contaminated water was funneled into a buried bunker full of iron filings that detoxified the water by reacting with contaminants. The reactions of the iron filings and contaminants produced new, nontoxic compounds that attached themselves to the iron filings.

Is the Water Drinkable?

Water that tastes agreeable and is not dangerous to health is called **potable water.** The amounts of dissolved substances in potable waters are very small, usually measured by weight in parts per million (ppm). Potable groundwaters of good quality typically contain about 150 ppm total dissolved materials, because even the purest natural waters contain some dissolved substances derived from weathering. Only distilled water contains less than 1 ppm dissolved substances.

The many cases of groundwater contamination have led to the establishment of water-quality standards based on medical studies. These studies have concentrated on the effects of ingesting average amounts of water containing various quantities of contaminant elements and compounds. For example, the Environmental Protection Agency has set the maximum allowable concentration of arsenic, whose poisonous nature is well known, at 0.05 ppm.

Groundwater is almost always free of solid particles when it seeps into a well from a sand or sandstone aquifer. The tortuous passageways of the rock or sand act as a fine filter, removing small particles of clay and other solids and even straining out bacteria and large viruses. Limestone aquifers may have larger pores and so may filter less efficiently. Any bacterial contamination found at the bottom of a well is usually introduced from the surface by the pump materials or from nearby underground sewage disposal, often when septic tanks are located too close to the well.

Some groundwaters, although perfectly safe to drink, simply taste bad. Some have a disagreeable taste of "iron" or are slightly sour. Groundwaters passing through limestone dissolve carbonate minerals and carry away calcium, magnesium, and bicarbonate ions, making the water "hard." Hard water may taste fine but does not lather readily when used with soap. Water passing through waterlogged forests or swampy soils may contain dissolved organic compounds and hydrogen sulfide.

How do these differences in taste and quality arise in safe drinking waters? Some of the highest-quality, best-tasting public water supplies come from lakes and artificial surface reservoirs, many of which are simply collecting places for rainwater. Some groundwaters taste just as good, and these tend to be waters that pass through rocks that weather only slightly. Sandstones made up largely of quartz, for example, contribute little in dissolved substances, and thus waters passing through them taste fresh.

As we have seen, the contamination of groundwaters in relatively shallow aquifers is a problem, and recovery is difficult. But are there deeper groundwaters that we can use?

Water Deep in the Crust

All rocks below the groundwater table are saturated with water. Even in the deepest wells drilled for oil, some 8 or 9 km deep, we always find water in permeable formations. At these depths, waters move so slowly—probably less than a centimeter per year—that they have plenty of time to dissolve even very insoluble minerals from the rocks through which they pass. Thus, dissolved materials become more concentrated in these waters than in near-surface waters, rendering them unpotable. For example, groundwaters that pass through salt beds, which dissolve quickly, tend to contain large concentrations of sodium chloride.

At depths greater than 12 to 15 km, deep into the basement igneous and metamorphic rocks that underlie the sedimentary formations of the upper crust, porosities and permeabilities are very low. The only pore spaces of these basement rocks are distributed along small cracks and the boundaries between crystals. Although they are saturated, they contain very little water because their porosity is so low (**Figure 13.21**). In some deeper regions of the crust, such as along subduction zones, hot waters containing dissolved carbon dioxide play an important role in the chemical reactions of metamorphism. These waters help to dissolve some minerals and precipitate others (see Chapter 9).

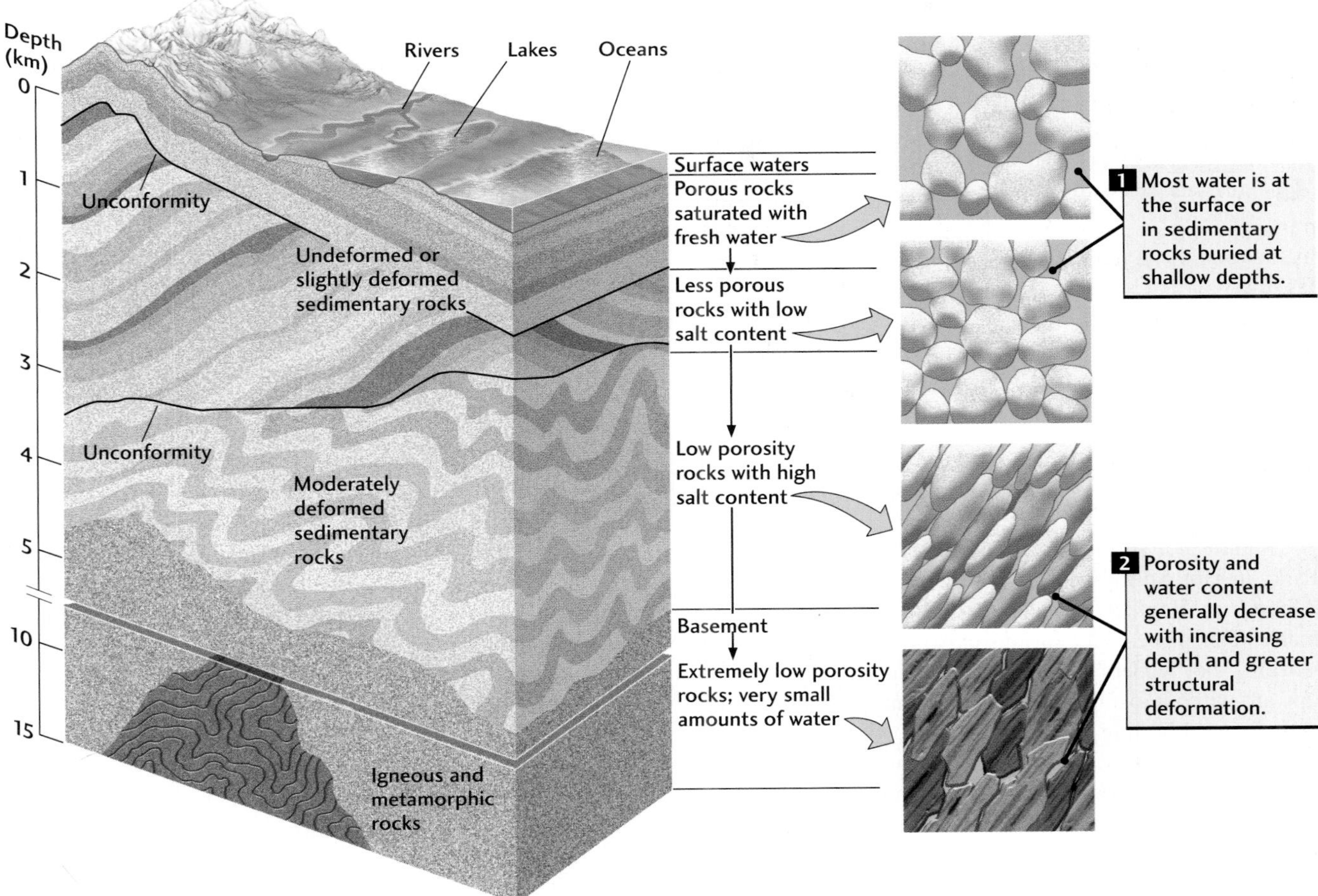

Figure 13.21 The distribution of water in a typical section of continental crust. Most water is at the surface or in sedimentary rocks buried at shallow depths. Porosity and water content generally decrease with increasing depth and greater structural deformation.

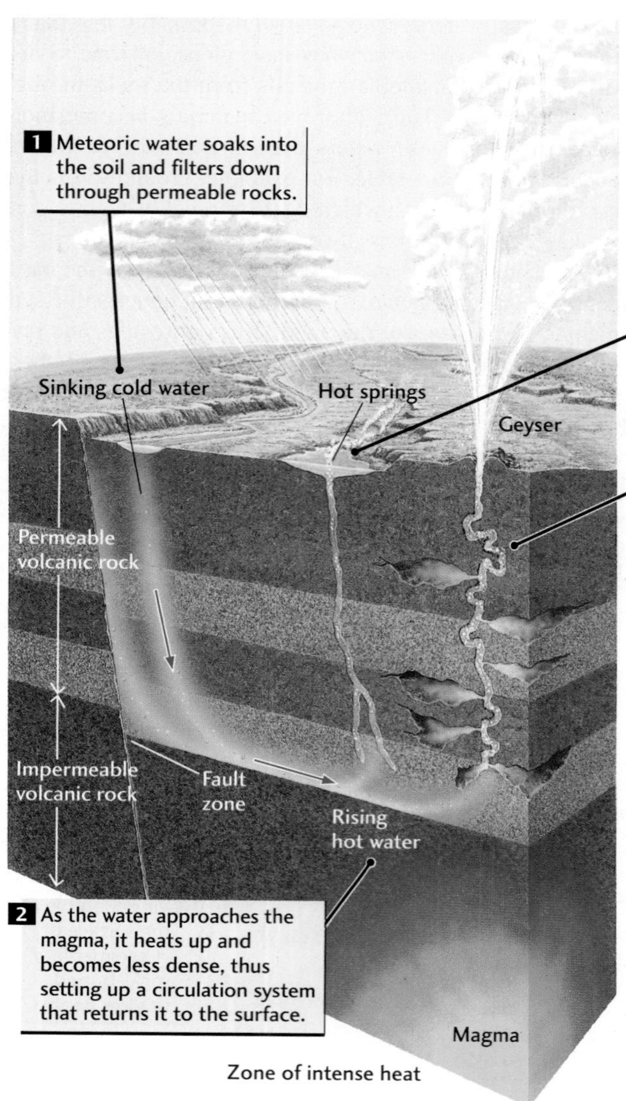

Figure 13.22 Circulation of water over a magma body produces geysers or hot springs.

Even some mantle rocks are presumed to contain water, although in very minute quantities.

Natural hot springs are found in Yellowstone National Park; in Hot Springs, Arkansas; in Banff Sulfur Springs, Alberta; in Reykjavík, Iceland; in New Zealand; and in many other places. Hot springs exist where **hydrothermal waters**—hot waters deep in the crust—migrate rapidly upward without losing much heat and emerge at the surface, sometimes at boiling temperatures.

Hydrothermal waters are loaded with chemical substances dissolved from rocks at high temperatures. As long as the water remains hot, the dissolved material stays in solution. As hydrothermal waters reach the surface and cool quickly, however, they may precipitate various minerals, such as opal (a form of silica) and calcite or aragonite (forms of calcium carbonate). Crusts of calcium carbonate produced at some hot springs build up to form the rock travertine, prized for its beauty as a polished stone. Amazingly, microorganisms that can withstand temperatures above the boiling point of water have been discovered in these environments, where they may contribute to the formation of calcium carbonate crusts. While still below the surface, hydrothermal waters also deposit some of the world's richest metallic ores as they cool, as we will see in Chapter 22.

Most hydrothermal waters of the continents come from surface waters that percolated downward to deeper regions of the crust (**Figure 13.22**). These surface waters originate primarily as **meteoric waters**—rain, snow, or other forms of water derived from the atmosphere (from the Greek *meteoron,* "phenomenon in the sky," which also gives us the word *meteorology*). Meteoric waters may be very old. It has been determined that the water at Hot Springs, Arkansas, derives from rain and snow that fell more than 4000 years ago and slowly infiltrated the ground.

Water that escapes from a magma can also contribute to hydrothermal waters. In areas of igneous activity, sinking meteoric waters are heated as they encounter hot masses of rocks. The hot meteoric waters then mix with water released from the nearby magma. Hydrothermal waters return to the surface as hot springs or geysers. Hot springs flow steadily; geysers erupt hot water and steam intermittently.

The theory explaining the intermittent eruptions of geysers is an example of geological deduction. We cannot observe the process directly because the dynamics of the underground hot-water system are hidden from sight hundreds of meters below the surface. Geysers are probably connected to the surface by a system of very irregular and crooked fractures, recesses, and openings—in contrast to the more regular and direct plumbing of hot springs. The irregular geyser plumbing sequesters some water in recesses, thus helping to prevent the bottom waters from mixing with shallower waters and cooling. The bottom waters are heated by contact with hot rock. When they reach the boiling point, steam starts to ascend and heats the shallower waters, increasing the pressure and triggering an eruption. After the pressure is released, the geyser then becomes quiet as the fractures slowly and irregularly refill with water.

In 1997, geologists reported the results of a novel technique to learn about geysers. They lowered a miniature video camera to about 7 m below the surface of a geyser. They found that the geyser shaft was constricted at that point. Farther down, the shaft widened to a large chamber containing a wildly boiling mixture of steam, water, and what appeared to be carbon dioxide bubbles. These direct observations dramatically confirmed the previous theory of how geysers work.

Geologists have turned to hydrothermal waters in the search for new and clean sources of energy. Northern California, Iceland, Italy, and New Zealand have already harnessed the steam produced by hydrothermal activity in hot springs and geysers to drive electricity-generating turbines. Hydrothermal waters may soon be put to wider use for producing power, as we will see in Chapter 22.

Although hydrothermal waters are important for power generation and ore deposits, these waters do not contribute to surface water supplies, primarily because they contain so much dissolved material.

SUMMARY

How does water move on and in the Earth in the hydrologic cycle? Water movements maintain a constant balance among the major reservoirs of water at or near Earth's surface. These reservoirs include oceans, lakes, and rivers; glaciers and polar ice; and groundwater. Evaporation from the oceans, evaporation and transpiration from the continents, and sublimation from glaciers transfer water to the atmosphere. Precipitation as rain and snow returns water from the atmosphere to the oceans and the land surface. Runoff in rivers returns part of the precipitation that falls on land to the ocean. The remainder infiltrates the ground to become groundwater. Differences in climate produce local variations in the balance among evaporation, precipitation, runoff, and infiltration.

How does water move below the ground? Groundwater forms as rain infiltrates the surface of the ground and travels through pore spaces in the soil, sediment, or rock that serve as an aquifer. Water moves through the upper, unsaturated zone through the water table into the saturated zone. Groundwater moves downhill under the influence of gravity, eventually emerging at springs, where the water table intersects the ground surface. Over the long term, recharge and discharge of a groundwater aquifer are in dynamic balance. Groundwater may flow in unconfined aquifers, which are continuous to the surface, or in confined aquifers, which are bounded by aquicludes. Confined aquifers produce artesian flows and spontaneously flowing artesian wells. Darcy's law describes the groundwater flow rate in relation to the slope of the water table and the permeability of the aquifer.

What factors govern our use of groundwater resources? As the population grows, the demands on groundwaters increase greatly, particularly where irrigation is widespread. Many aquifers, such as those of the western plains of North America, have such slow recharge rates that continued pumping in the past century has reduced the pressure in artesian wells. As pumping discharge continues to exceed recharge, such aquifers are being depleted, and there is no prospect of renewal for many years. Artificial recharge may help to renew some aquifers, but conservation will be required to preserve others. The contamination of groundwater by sewage, industrial effluents, and radioactive wastes reduces the potability of some waters and limits our resources.

What geological processes are affected by groundwater? Erosion by groundwater in humid limestone terrains produces karst topography, with caves, sinkholes, and disappearing streams. The heating of downward-percolating meteoric waters by magma bodies leads to a circulation that brings hydrothermal waters to the surface as geysers and hot springs. At great depths in the crust, more than 12 to 15 km, rocks contain extremely small quantities of water because their porosities are significantly reduced. This extreme porosity reduction results from the tremendous weight of the overlying rocks.

Key Terms and Concepts

aquiclude (p. 286)
aquifer (p. 282)
artesian flow (p. 286)
artesian well (p. 286)
confined aquifer (p. 286)
Darcy's law (pp. 290–292)
discharge (p. 286)
effluent stream (p. 286)
groundwater (p. 278)
groundwater table (p. 286)
hydraulic gradient (p. 290)
hydrologic cycle (p. 278)
hydrology (p. 277)
hydrothermal water (p. 298)
infiltration (p. 278)
influent stream (p. 286)
karst topography (p. 294)
meteoric water (p. 298)
perched water table (p. 288)
permeability (p. 285)
potable water (p. 296)
rain shadow (p. 281)
recharge (p. 286)
relative humidity (p. 280)
reservoir (p. 278)
runoff (p. 279)
saturated zone (p. 286)
sinkhole (p. 294)
stalactite (p. 294)
stalagmite (p. 294)
unconfined aquifer (p. 286)
unsaturated zone (p. 286)

Exercises

This icon indicates that there is an animation available on the Web site that may assist you in answering a question.

1. What are the main reservoirs of water at or near Earth's surface?

2. How do mountains form rain shadows?

3. What is an aquifer?

4. What is the difference between the saturated and the unsaturated zones of groundwater?

5. How do aquicludes make a confined aquifer?

6. How are recharge and discharge balanced to make a groundwater table stable?

7. How does Darcy's law relate groundwater movement to permeability?

8. How does dissolution of limestone relate to karst topography?

9. What are the sources of water in hot springs?

10. What are some common contaminants in groundwater?

11. Define the groundwater table.

Thought Questions

This icon indicates that there is an animation available on the Web site that may assist you in answering a question.

1. If the Earth warmed, causing evaporation from the oceans to greatly increase, how would the hydrologic cycle of today be altered?

2. If you lived near the seashore and started to notice that your well water had a slightly salty taste, how would you explain the change in water quality?

3. Why would you recommend against extensive development and urbanization of the recharge area of an aquifer that serves your community?

4. If it were discovered that radioactive waste had seeped into groundwater from a nuclear processing plant, what kind of information would you need to predict how long it would take for the radioactivity to appear in well water 10 km from the plant?

5. What geological processes would you infer are taking place below the surface at Yellowstone National Park, which has many hot springs and geysers?

6. Why should communities ensure that septic tanks are maintained in good condition?

7. Why are more and more communities in cold climates restricting the use of salt to melt snow and ice on highways?

8. Your new house is built on soil-covered granite bedrock. Although you think that prospects for drilling a successful water well are poor because of the granite, the well driller familiar with the area says that he has drilled many good water wells in this granite. What arguments might each of you offer to convince the other?

9. How might the hydrologic cycle have differed quantitatively during the maximum glaciation of 19,000 years ago, when a good deal of the continents was covered by ice?

10. You are exploring a cave and notice a small stream flowing on the cave floor. Where could the water be coming from?

Short-Term Team Project

Flooding

In the U.S. Senate and House of Representatives, recent college graduates work as legislative assistants. Their main job is to brief a senator or a representative about pressing issues on which the official must cast an informed vote.

You and a partner are legislative assistants to a U.S. senator from South Dakota. The senator has been contacted by constituents in Pierre, which is located on the east bank of the south-flowing Missouri River and just south of the Oahe Dam. The residents of Pierre have had repeated problems with winter flooding. The Army Corps of Engineers says that the flooding occurs because the Bad River, a tributary that flows across heavily farmed and cattle-grazed land into the Missouri River, picks up large quantities of sediment and dumps it where the two rivers join, making this area quite shallow. In the winter, thick ice forms on the slow-moving Missouri River, and liquid water must flow through a thin horizontal zone (parallel to the channel bottom) between the accumulated sediment and the thick ice. When water is released from above the dam to generate electricity, it backs up at the narrow passage and floods Pierre. The Corps's only suggestion to prevent the flooding is to build levees along the banks of the Missouri. The senator wants to know:

1. Is the sediment from the Bad River really the problem? Why or why not?

2. What might be done to curtail the sedimentation?

3. Would the construction of levees be a long-term solution to the problem? Why or why not?

4. Would simply dredging out the accumulated sediment to deepen the Missouri River channel be a long-term solution?

5. What other solutions are there besides levee construction or dredging?

6. If the Corps has been able to control the flooding along the Mississippi River, doesn't that mean that it could deal with this problem up in South Dakota?

The senator wants these questions answered in a concise two- to three-page memo. To prepare the senator to deal with the issue, your team should investigate the nature and quality of work undertaken by the Army Corps of Engineers and understand the geologic nature of the problem near Pierre.

Suggested Readings

Dolan, R., and H. G. Goodell. 1986. Sinking cities. *American Scientist* 74: 38–47.

Dunne, T., and L. B. Leopold. 1978. *Water in Environmental Planning.* San Francisco: W. H. Freeman.

Frederick, K. D. 1986. *Scarce Water and Institutional Change.* Washington, D.C.: Resources for the Future.

Freeze, R. A., and J. A. Cherry. 1979. *Groundwater.* Englewood Cliffs, N.J.: Prentice Hall.

Heath, R. C. 1983. *Basic Groundwater Hydrology.* U.S. Geological Survey Water-Supply Paper 2220.

Jennings, J. N. 1983. Karst landforms. *American Scientist* 71: 578–586.

Leopold, L. B. 1977. *Water, Rivers, and Creeks.* Sausalito, Calif.: University Science Books.

National Research Council. 1993. *Solid-Earth Sciences and Society.* Washington, D.C.: National Academy Press.

U.S. Geological Survey. 1990. *Hydrologic Events and Water Supply and Use. National Water Summary 1987.* U.S. Geological Survey Water-Supply Paper 2350.

Aerial view of the meandering Niobrara River, in Nebraska. This appearance is typical for meandering streams in plains environments. [Gary D. McMichael/Photo Researchers.]

CHAPTER

14

Streams: Transport to the Oceans

"All the rivers run into the sea; yet the sea is not full; unto the place from whence the rivers come, thither they return again. . . . The thing that hath been, it is that which shall be; and that which is done is that which shall be done: and there is no new thing under the sun."

ECCLESIASTES 1:7, 9

Streams are the major geological agents operating on the surface of the land. As they erode bedrock and transport and deposit sand, gravel, and mud, streams of all sizes—from tiny rills to major rivers—are preeminent carvers of the landscape.

In this chapter, we focus on how streams accomplish their geological work: how water flows in currents; how currents carry sediment; how streams break up and erode solid rock; and, on a larger scale, how streams carve valleys and assume a variety of forms as they channel water downstream.

Rivers are deeply embedded in the imagery of language. We speak of flowing waters, babbling brooks, raging torrents. Our everyday language uses many different words to describe waterways, but geologists give more precise meanings to some of these terms. We reserve the word **stream** for any flowing body of water, large or small, and **river** for the major branches of a large stream system.

Some body of water flows through almost every town and city in most parts of the world. These streams have served as commercial waterways for barges and steamers and as water resources for resident populations and industries. The river Nile, for example, was vital to the agricultural economy of ancient Egypt and remains important to Egypt today. Living near a river also entails risk. When rivers flood, they destroy lives and property, sometimes on a huge scale.

Streams cover most of Earth's surface and are the major shapers of the continental landscape. They erode mountains, carry the products of weathering down to the oceans, and drop billions of tons of sediment along the way in bars and flood deposits. At their mouths, on the edges of the continents, they dump even greater quantities of sediment, building new land out into the oceans.

Worldwide, streams carry about 16 billion tons of clastic sediment and an additional 2 to 4 billion tons of dissolved matter each year. Humans are responsible for much of the present stream load. According to some estimates, prehuman sediment transport was about 9 billion tons per year, less than half the present value. We increase the sediment load of streams in some places, through agriculture and accelerated erosion. In other places, we decrease the sediment load by constructing dams, which trap sediment behind their retaining walls.

Streams interact with many other parts of the Earth system. The relationship of streamflow to the precipitation of water from the atmosphere is fairly obvious. In Chapter 13, we reviewed the important role of streams in the hydrologic cycle and their connections to groundwater, evaporation, and transpiration. Less obvious are their connections to tectonics, climate, and the biological world, which we cover in this chapter.

Streams and rivers carry back to the sea the bulk of the rainwater that falls on land and much of the sediment produced by erosion of the land surface. We begin by examining how running water moves in currents and how that movement enables streams to carry various kinds of sediment.

How Stream Waters Flow

All streams, large and small, move in accordance with some basic characteristics of flowing fluids. We can illustrate two kinds of fluid flow by using lines of motion called *streamlines* (**Figure 14.1**). In **laminar flow,** the simplest kind of movement, straight or gently curved streamlines run parallel to one another without mixing or crossing between layers. The slow movement of thick syrup over a pancake, with strands of unmixed melted butter flowing in parallel but separate paths, is a laminar flow. **Turbulent flow** has a more complex pattern of movement, in which streamlines mix, cross, and form swirls and eddies. Fast-moving river waters typically show this kind of motion. Turbulence—the degree to which there are irregularities and eddies in the flow—may be low or high. Whether a flow is laminar or turbulent depends on three factors:

1. Its velocity (rate of movement).

2. Its geometry (primarily its depth).

3. Its **viscosity,** which is a measure of a fluid's resistance to flow. The more viscous (the "thicker") a fluid is, the more it resists flow. The higher the viscosity, the greater the tendency for laminar flow.

Viscosity arises from the attractive forces between the molecules of a fluid. These forces tend to impede the slipping and sliding of molecules past one another. The greater the attractive forces, the greater the resistance to mixing with

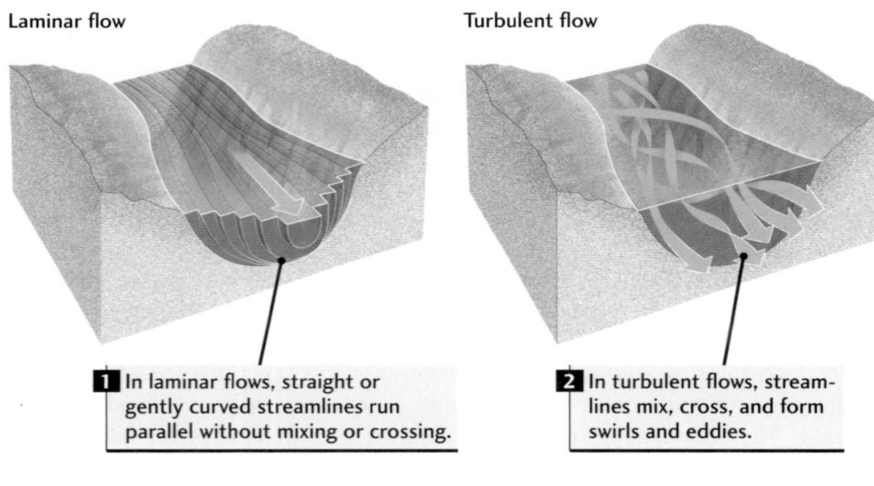

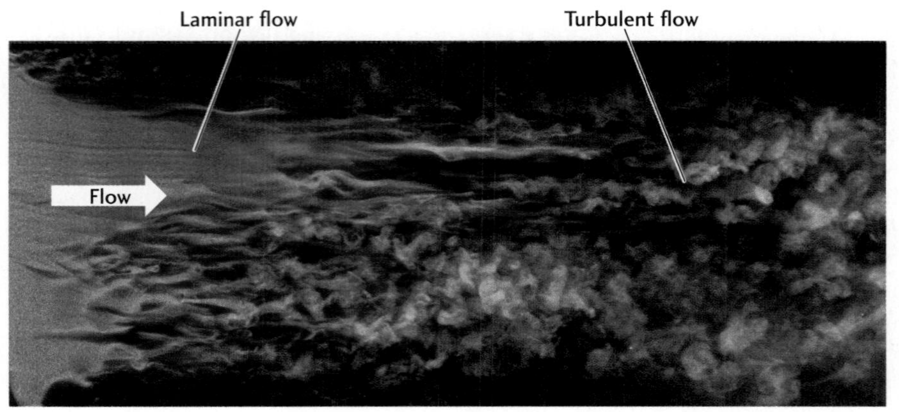

Figure 14.1 Laminar flows and turbulent flows. The photograph shows the transition from laminar to turbulent flow in water along a flat plate, revealed by the injection of a dye. Flow is from left to right. [ONERA.]

neighboring molecules and the higher the viscosity. For example, when a cold syrup or a viscous cooking oil is poured, its flow is sluggish and laminar. The viscosity of most fluids, including water, decreases as the temperature increases. Given enough heat, a fluid's viscosity may decrease sufficiently to change a laminar flow into a turbulent one.

Water has low viscosity in the common range of temperatures at Earth's surface. For this reason alone, most watercourses in nature tend to turbulent flow. In addition, the rapid movement of water in most streams makes them turbulent. In nature, we are likely to see laminar flows of water only in thin sheets of rain runoff flowing slowly down nearly level slopes. In cities, we may see small laminar flows in street gutters. Because most streams and rivers are broad and deep and flow quickly, their flows are almost always turbulent.

A stream may show turbulent flow over much of its width and laminar flow along its edge, where the water is shallow and is moving slowly. The flow velocity is highest near the center of a stream, and we commonly refer to a rapid flow as a strong current.

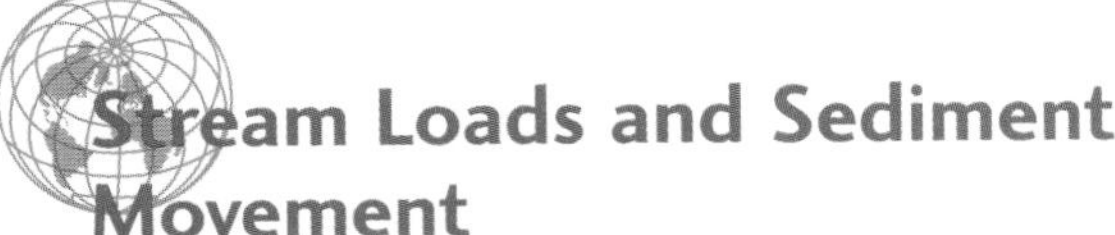

Stream Loads and Sediment Movement

Erosion and Transport

Streams vary in their ability to erode and carry sand grains and other sediment. Laminar flows of water can lift and carry only the smallest, lightest, clay-sized particles. Turbulent flows, depending on their speed, can move particles ranging from clay size to pebbles and cobbles. As turbulence lifts particles from the streambed into the flow, the flow carries them downstream. Turbulence also rolls and slides larger particles along the bottom. A stream's **suspended load** includes all the material temporarily or permanently suspended in the flow. Its **bed load** is the material the stream carries along the bed by sliding and rolling (**Figure 14.2**).

The faster the current, the larger the particles carried as suspended and bed load. A flow's ability to carry material of a given size is its **competence.** As a current increases in velocity and coarser particles are suspended, the suspended load grows. At the same time, more of the bed material is in motion, and the bed load also increases. As we would expect, the larger the volume of a flow, the more suspended load and bed load it can carry. The total sediment load carried by a flow is its **capacity.**

The velocity and the volume of a flow affect both the competence and the capacity of a stream. The Mississippi River, for example, flows at moderate speeds along most of its length and carries only fine to medium-sized particles (clay to sand), but it carries huge quantities of them. A small, steep, fast-flowing mountain stream, in contrast, may carry boulders, but only a few of them.

Settling from Suspension

A stream's ability to carry sediment depends on a balance between turbulence, which lifts particles, and the competing downward pull of gravity, which makes them settle out of

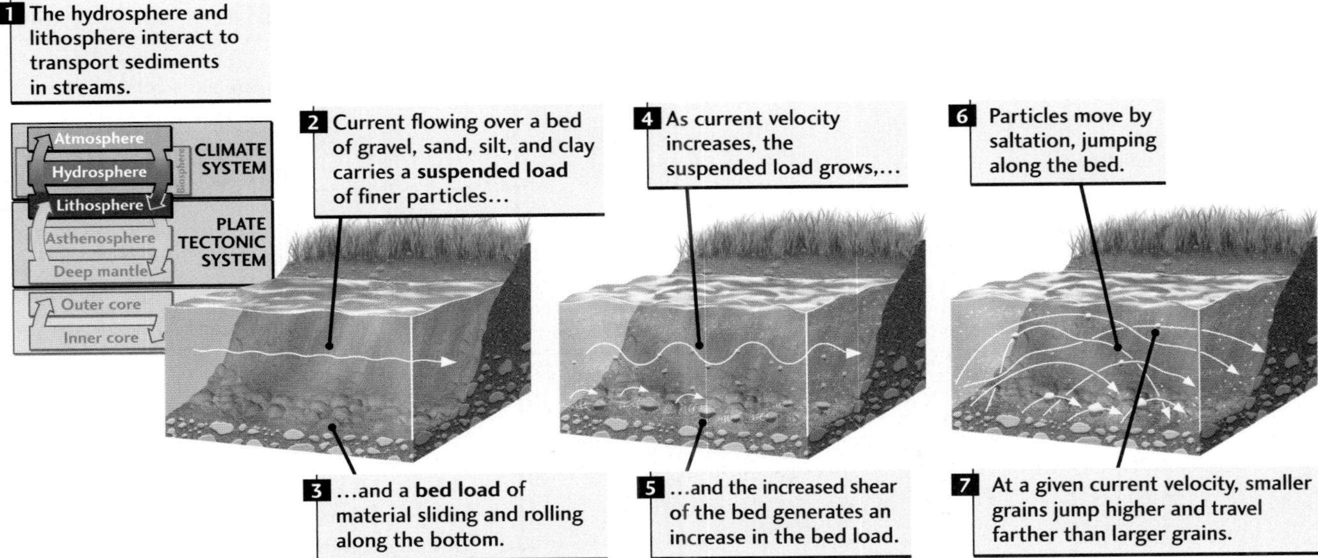

Figure 14.2 A current flowing over a bed of sand, silt, and clay transports particles in two ways: as bed load, the material sliding and rolling along the bottom; and as suspended load, the material temporarily or permanently suspended in the flow itself. Saltation is an intermittent jumping motion of grains. In general, the smaller the particle, the higher it jumps and the farther it travels.

the current and become part of the bed. The speed with which suspended particles of various weights settle to the bottom is called the **settling velocity.** Small grains of silt and clay are easily lifted into the stream and settle slowly, so they tend to stay in suspension. The settling velocity of larger particles, such as medium- and coarse-grained sand, is much faster. Most larger grains therefore stay suspended in the current only a short time before they settle.

Sand grains in a flow typically move by **saltation**—an intermittent jumping motion along the streambed. The grains are sucked up into the flow by turbulent eddies, move with the current for a short distance, and then fall back to the bottom (see Figure 14.2). If you were to stand in a rapidly flowing sandy stream, you might see a cloud of saltating sand grains moving around your ankles. The bigger the grain, the longer it will tend to remain on the bed before it is picked up. Once a large grain is in the current, it will settle quickly. The smaller the grain, the more frequently it will be picked up, the higher it will "jump," and the longer it will take to settle.

Thus, turbulent flows transport sediment by suspension (clays), saltation (sands), and rolling and sliding along the bed (sand and gravel). To study how a particular river carries sediment, geologists and hydraulic engineers measure the relationship between particle size and the force the flow exerts on the particles in the suspended and bed loads. Engineers use these data to calculate how much sediment a particular flow can move and how rapidly it can move it. This information allows them to design dams and bridges or to estimate how quickly artificial reservoirs behind dams will fill with sediment. Geologists can infer the velocities of ancient currents from the sizes of grains in sedimentary rocks.

Figure 14.3 graphs the relationship between grain size and flow velocity. This relationship, in which higher current velocities are required to transport finer grains, exists because it is easier for the flow to lift noncohesive particles (particles that do not stick together) from the bed than it is to lift cohesive particles (particles that stick together, as many clay minerals do). The finer the cohesive particles, the greater the velocity of the flow required to erode them. For these small grains, settling velocities are so slow that even a gentle current, about 20 cm/s, can keep the particles in suspension and transport sediment.

Bed Forms: Dunes and Ripples

When sand grains on a streambed are transported by saltation, they tend to form cross-bedded dunes and ripples (see Chapter 8). **Dunes** are elongated ridges of sand that can be many meters high in large rivers. **Ripples** are very small dunes—with heights ranging from less than a centimeter to several centimeters—whose long dimension is formed at right angles to the current. Although underwater ripples and dunes are harder to observe than those produced on land by air currents, they form in the same way and are just as common. As a current moves sand grains by saltation, they are eroded from the upstream side of ripples and dunes and deposited on the downstream side. The steady downstream transfer of grains across the ridges causes the ripple and dune forms to migrate downstream. The speed of this migration is much slower than the movement of individual grains and very much slower than the current. (We will look at ripple and dune migration in more detail in Chapter 15.)

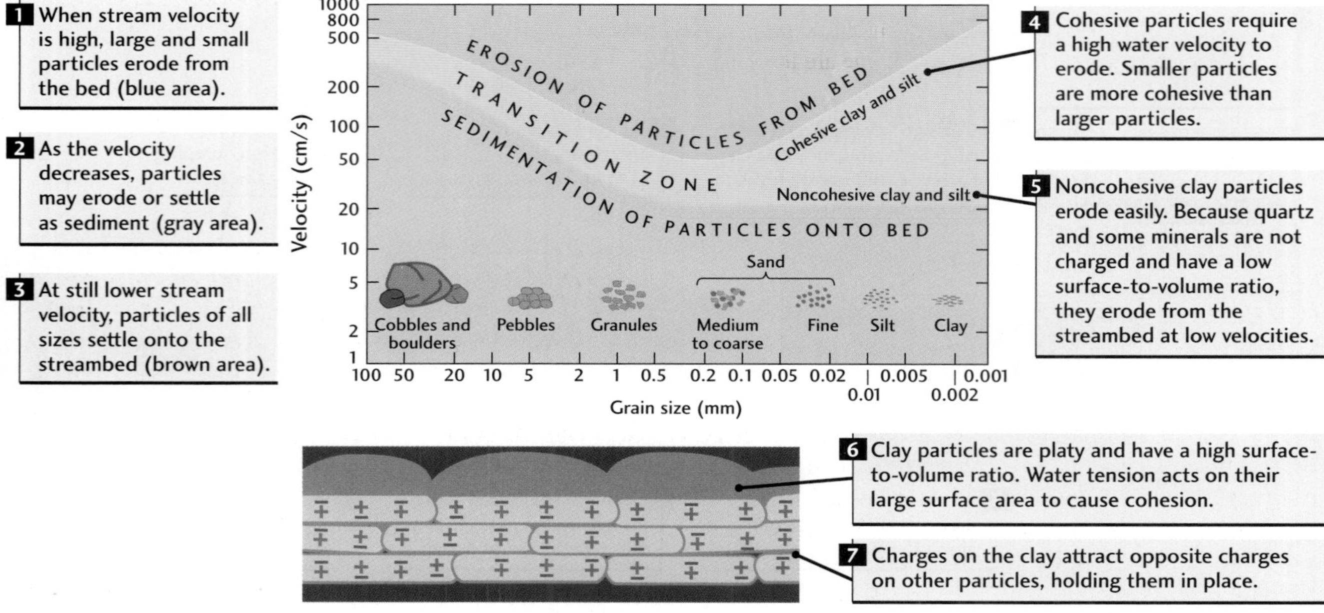

Figure 14.3 The relationship between particle size and current velocity. [After F. Hjulstrom, as modified by A. Sundborg, "The River Klarilven." *Geografisk Annaler,* 1956.]

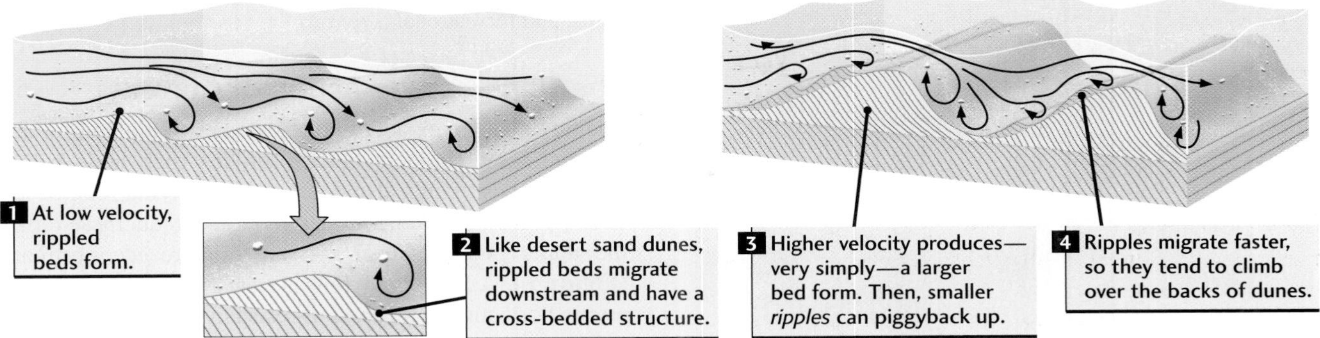

Figure 14.4 The change in the form of a sand bed with increasing flow velocity. [After D. A. Simmons and E. V. Richardson, "Forms of Bed Roughness in Alluvial Channels." *American Society of Civil Engineers Proceedings* 87: 87–105 (1961).]

The shapes of ripples and dunes and their migration speeds change as the velocity of the current increases. At the lowest current velocities, few grains are saltating and the stream's sand bed is flat. At slightly higher velocities, the number of saltating grains increases. A rippled bed forms, and ripples migrate downstream (**Figure 14.4**). As the velocity increases further, the ripples grow larger and migrate faster until, at a certain point, dunes replace the ripples. Both ripples and dunes have a cross-bedded structure, and as the current flows over their tops it can actually reverse and flow backward along their lee (downstream) side. As the dunes grow larger, small ripples form. These ripples tend to climb over the backs of the dunes because they migrate more quickly than dunes. Very high current velocities will erase the dunes and form a flat bed below a dense cloud of rapidly saltating sand grains. Most of these grains hardly settle to the bottom before they are picked up again. Some are in permanent suspension.

How Running Water Erodes Solid Rock

We can easily see a rapid current picking up loose sand from its bed and carrying it away, thus eroding the bed. At high water levels and during floods, streams can even scour and cut into banks of unconsolidated sediment, which then slump into the flow and are carried away. *Gullies*—valleys made by small streams eroding soft soils or weak rocks—cut their way upstream into higher land. The process by which rivers cut upstream, rather than downstream, is called *headward erosion.* Headward erosion commonly accompanies widening and deepening of valleys and may be extremely rapid—as much as several meters in a few years in easily erodible soils.

We cannot so easily see the much slower erosion of solid rock. Running water erodes solid rock by abrasion, by chemical and physical weathering, and by the undercutting action of currents. This topic is discussed further in Chapter 18.

Abrasion

One of the major ways a river breaks apart and erodes rock is by slow abrasion. The sand and pebbles the river carries create a sandblasting action that wears away even the hardest rock. On some riverbeds, pebbles and cobbles rotating inside swirling eddies grind deep **potholes** into the bedrock (**Figure 14.5**). At low water, we can see pebbles and sand lying quietly at the bottom of exposed potholes. The ability of a river to erode its bed depends on the river's discharge and slope. Both of these attributes are discussed later in this chapter, and in Chapter 18 we will see how they combine to shape landscapes.

Chemical and Physical Weathering

Chemical weathering, which alters a rock's minerals and weakens it along joints and cracks, helps destroy rocks in streambeds just as it does on the land surface. Physical weathering can be violent, as the crashes of boulders and the constant smaller impacts of pebbles and sand split the rock along cracks. Such impacts in a river channel break up rock much faster than slow weathering on a gently sloping hillside does. When impacts and weathering have loosened large blocks of bedrock, strong upward eddies may pull them up and out in a sudden, violent plucking action.

Rock erosion is particularly strong at rapids and waterfalls. *Rapids* are places in a stream where the flow is extremely fast because the slope of the riverbed suddenly steepens, typically at rocky ledges. The speed and turbulence of the water quickly break blocks into smaller pieces that are carried away by the strong current.

Stream-channel erosion can be a major source of sediment in extensively urbanized areas. The erosion is accelerated

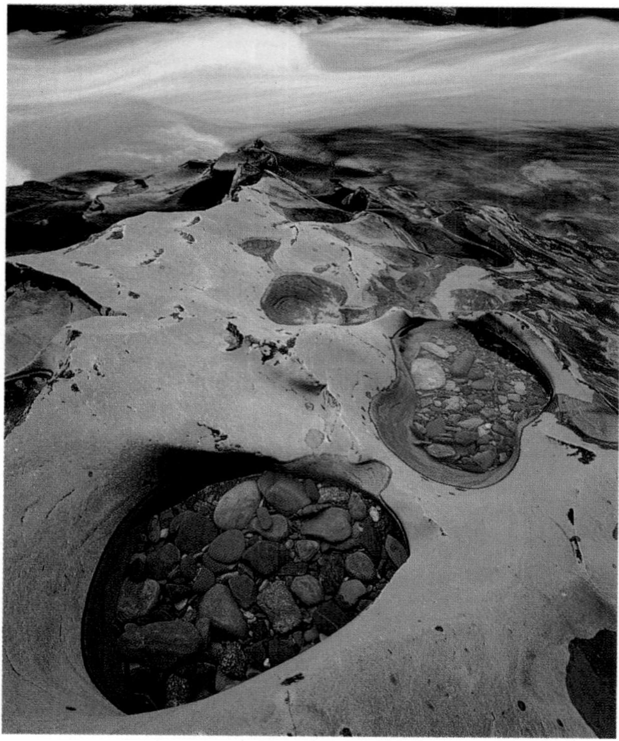

Figure 14.5 Potholes in river rock along McDonald Creek, Glacier National Park, Montana. The pebbles rotate inside the potholes, grinding deep holes in the bedrock. [Carr Clifton/Minden Pictures.]

by changing land uses, such as reservoir impoundments, dams, and recreational activities.

The Undercutting Action of Currents

The tremendous impact of huge volumes of plunging water and tumbling boulders quickly erodes rock beds below waterfalls. Waterfalls also erode the underlying rock of the cliff that forms the falls. As erosion undercuts these cliffs, the upper beds collapse and the falls recede upstream (**Figure 14.6**). Erosion by falls is fastest where the rock layers are horizontal, with erosion-resistant rocks at the top and softer rocks, such as shales, making up the lower layers. Historical records show that the main section of Niagara Falls, perhaps the best-known falls in North America, has been moving upstream at a rate of a meter per year.

Stream Valleys, Channels, and Floodplains

As streams erode Earth's surface—in some places bedrock, in others unconsolidated sediment—they create valleys. A stream **valley** encompasses the entire area between the tops of the slopes on both sides of the river. The cross-sectional profile of many river valleys is V-shaped, but many other valleys have a broad, low profile like that shown in **Figure 14.7**. At the bottom of the valley is the **channel,** the trough through which the water runs. The channel carries all the water during normal, nonflood times. At low water level, the stream may run only along the bottom of the channel. At high water levels, the stream occupies most of the channel. In broad valleys, a **floodplain**—a flat area about level with the top of the channel—lies on either side of the channel. It is the part of the valley that is flooded when the river spills over its banks, carrying with it silt and sand from the main channel.

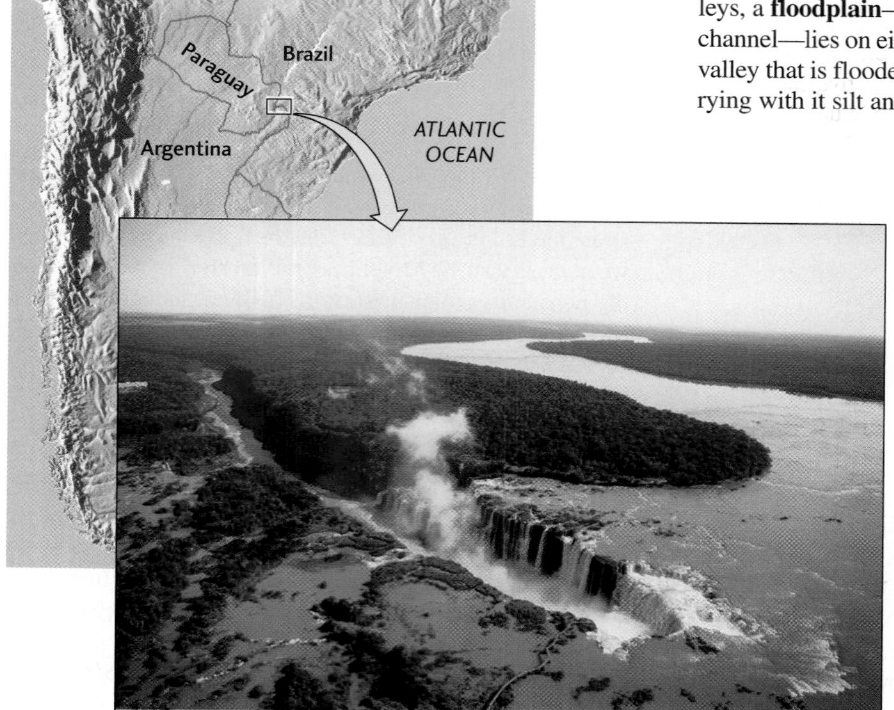

Figure 14.6 This waterfall on the Iguaçú River, Brazil, is retreating upstream as falling water and sediment pound the cliff's base and undercut it. From the center to the upper left, one can see steep walls, the remnants of the waterfall's retreat upstream. [Donald Nausbaum.]

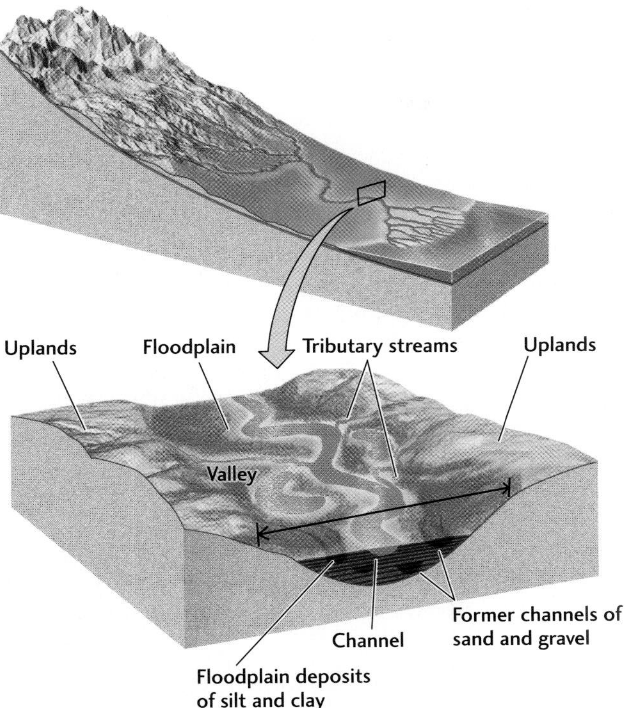

Figure 14.7 A river flows in a channel that moves over a broad, flat floodplain in a wide valley eroded from uplands. Floodplains may be narrow or absent in steep valleys.

Stream Valleys

In high mountains, stream valleys are narrow and have steep walls, and the channel may occupy most or all of the valley bottom. A small floodplain may be visible only at low water levels. In such valleys, the stream is actively cutting into the bedrock, a characteristic of tectonically active, newly uplifted highlands. In lowlands, where tectonic uplift has long since ceased, stream erosion of valley walls is helped by chemical weathering and mass wasting. With a long time to operate, these processes produce gentle slopes and floodplains many kilometers wide.

Channel Patterns

As a stream channel makes its way along the bottom of a valley, it may run straight in some stretches and take a snaking, irregular path in others, sometimes splitting into multiple channels. The channel may flow along the center of the floodplain or hug one edge of the valley. In addition to straight stretches, the two other types of channel patterns are meandering and braided.

Meanders On a great many floodplains, channels follow curves and bends called **meanders,** so named for the Maiandros (now Menderes) River in Turkey, fabled in ancient times for its winding, twisting course. Meanders are usual in streams flowing on low slopes in plains or lowlands, where channels typically cut through unconsolidated sediments—fine sand, silt, or mud—or easily eroded bedrock. Meanders are less pronounced but still common where the channel flows on higher slopes and harder bedrock. In such terrain, meandering stretches may alternate with long, relatively straight ones.

A stream that has cut deeply into the curves and bends of its channel may produce incised meanders (**Figure 14.8**). Other streams may meander on somewhat wider floodplains bounded by steep, rocky valley walls. We are not sure why these two different patterns appear. We do know that meandering is widespread not only in streams but also in a great many other kinds of flows. For example, the Gulf Stream, a powerful current in the western North Atlantic Ocean, meanders. Lava flows on Earth meander, and planetary geologists have found meanders in dry water channels and lava flows on Mars and in lava flows on Venus.

Meanders on a floodplain migrate over periods of many years, eroding the outside banks of bends, where the current is strongest (**Figure Story 14.9**). Meanders shift position from side to side, as well as downstream, in a snaking motion something like that of a long rope being snapped (Figure Story 14.9a). As the outside banks are eroded, curved sandbars called **point bars** are deposited along the inside banks, where the current is slower. Migration may be rapid: some meanders on the Mississippi shift as much as 20 m/year. As meanders move, so do the point bars, building up an accumulation of sand and silt over the part of the floodplain across which the channel migrated.

Figure 14.8 This section of the San Juan River, Utah, is a good example of an incised meander belt, a deeply eroded, meandering, V-shaped valley with virtually no floodplain. [Tom Bean.]

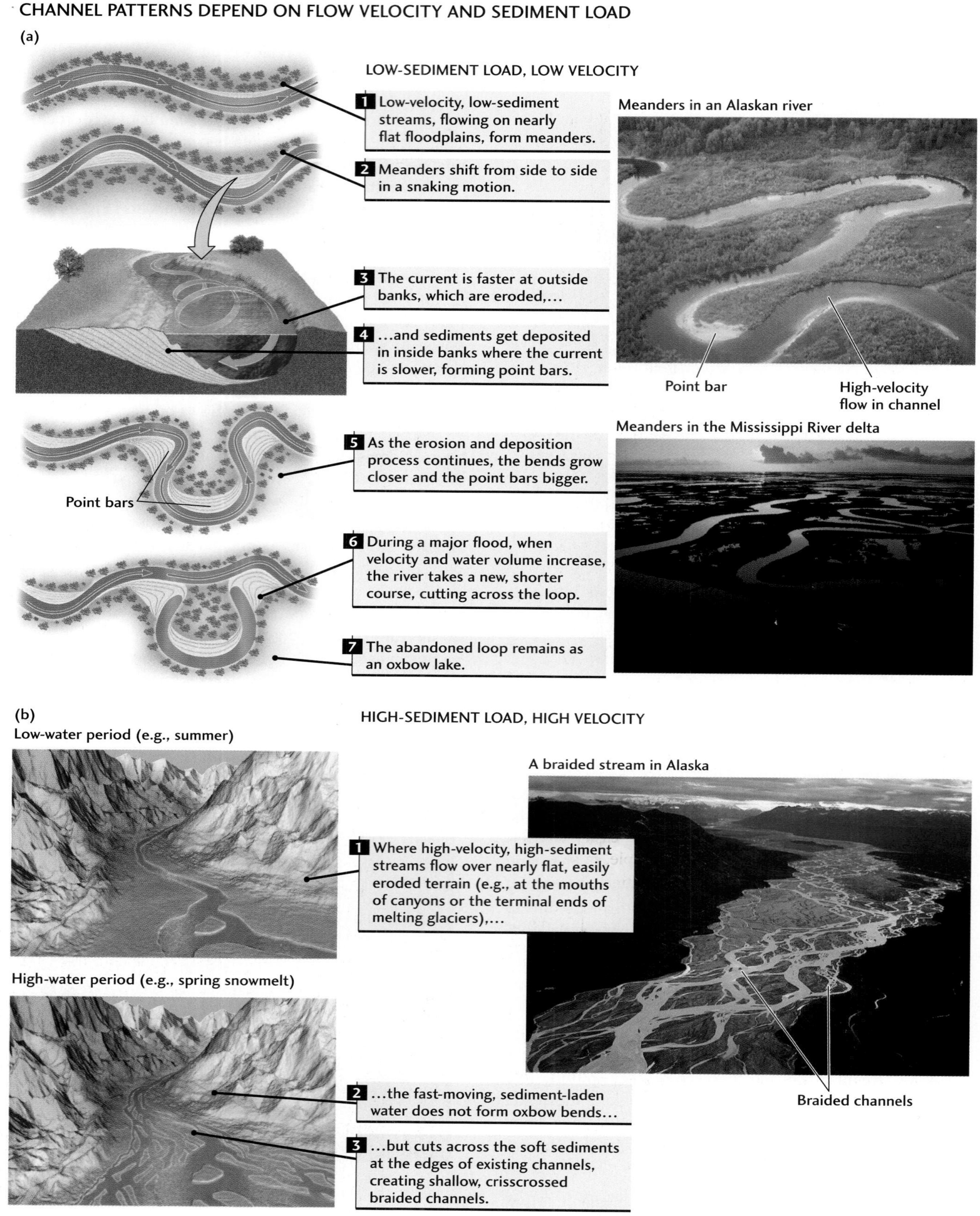

CHANNEL PATTERNS DEPEND ON FLOW VELOCITY AND SEDIMENT LOAD

(a)

LOW-SEDIMENT LOAD, LOW VELOCITY

1 Low-velocity, low-sediment streams, flowing on nearly flat floodplains, form meanders.

2 Meanders shift from side to side in a snaking motion.

3 The current is faster at outside banks, which are eroded,...

4 ...and sediments get deposited in inside banks where the current is slower, forming point bars.

5 As the erosion and deposition process continues, the bends grow closer and the point bars bigger.

Point bars

6 During a major flood, when velocity and water volume increase, the river takes a new, shorter course, cutting across the loop.

7 The abandoned loop remains as an oxbow lake.

Meanders in an Alaskan river

Point bar

High-velocity flow in channel

Meanders in the Mississippi River delta

(b)

HIGH-SEDIMENT LOAD, HIGH VELOCITY

Low-water period (e.g., summer)

1 Where high-velocity, high-sediment streams flow over nearly flat, easily eroded terrain (e.g., at the mouths of canyons or the terminal ends of melting glaciers),...

High-water period (e.g., spring snowmelt)

2 ...the fast-moving, sediment-laden water does not form oxbow bends...

3 ...but cuts across the soft sediments at the edges of existing channels, creating shallow, crisscrossed braided channels.

A braided stream in Alaska

Braided channels

Figure Story 14.9 Channel patterns. (*top photo*) A meandering river, west of Anchorage, Alaska. The white areas are point bars formed at the insides of bends. [Peter Kresan.] (*middle photo*) The lower Mississippi River and its main channel, oxbows, and oxbow lakes. [Nathan Benn/Corbis.] (*bottom photo*) Chitina River, Alaska, is a braided stream with many channels that split and rejoin. [Tom Bean.]

As meanders migrate, sometimes unevenly, the bends may grow closer and closer together, until finally the river bypasses the next loop, often during a major flood. The river takes a new, shorter course like that shown in Figure Story 14.9a. In its abandoned path, it leaves behind an **oxbow lake**—a crescent-shaped, water-filled loop.

Engineers sometimes artificially straighten and confine a meandering river, channeling it along a straight path with the aid of concrete abutments. The Army Corps of Engineers has been channeling the Mississippi River since 1878. In a period of 13 years, it decreased the length of the lower Mississippi by 243 km. Part of the severity of the Mississippi River flood of 1993 has been ascribed to channelization and the height of the artificial riverbanks designed by flood-control engineers. Without channelization, floods are more frequent but less damaging. With it, damage may be catastrophic when a flood breaches the artificially high banks, as it did in 1993. Channelization has also been criticized for destroying wetlands and much of the natural vegetation and animal life of the floodplain. Environmental concerns such as these stimulated action to restore one channelized river, the Kissimmee in central Florida, to its original meandering course. Today, restoration projects are well under way. If left to its own natural processes, the Kissimmee might have taken many decades or hundreds of years to restore itself.

Braided Streams Some streams have many channels instead of a single one. A **braided stream** is one whose channel divides into an interlacing network of channels, which then rejoin, in a pattern resembling braids of hair (Figure Story 14.9b). Braided streams are found in many settings, from broad lowland valleys to wide, downfaulted, sediment-filled valleys adjacent to mountain ranges. Braids tend to form in rivers with large variations in volume of flow combined with a high sediment load and easily erodible banks. They are well developed, for example, in sediment-choked streams formed at the edges of melting glaciers.

The Stream Floodplain

A stream channel migrating over the floor of a valley creates a floodplain. Point bars formed during migration build up the surface of the floodplain, as does sediment deposited by floodwaters when the stream overflows its banks. Erosional floodplains, covered with a thin layer of sediment, can form when a stream erodes bedrock or unconsolidated sediment as it migrates.

As floodwaters spread out over the floodplain, the velocity of the water slows and the current loses its ability to carry sediment. The floodwater velocity drops most quickly along the immediate borders of the channel. As a result, the current deposits much coarse sediment, typically sand and gravel, along a narrow strip at the edge of the channel. Successive floods build up **natural levees,** ridges of coarse material that confine the stream within its banks between floods, even when water levels are high (**Figure 14.10**).

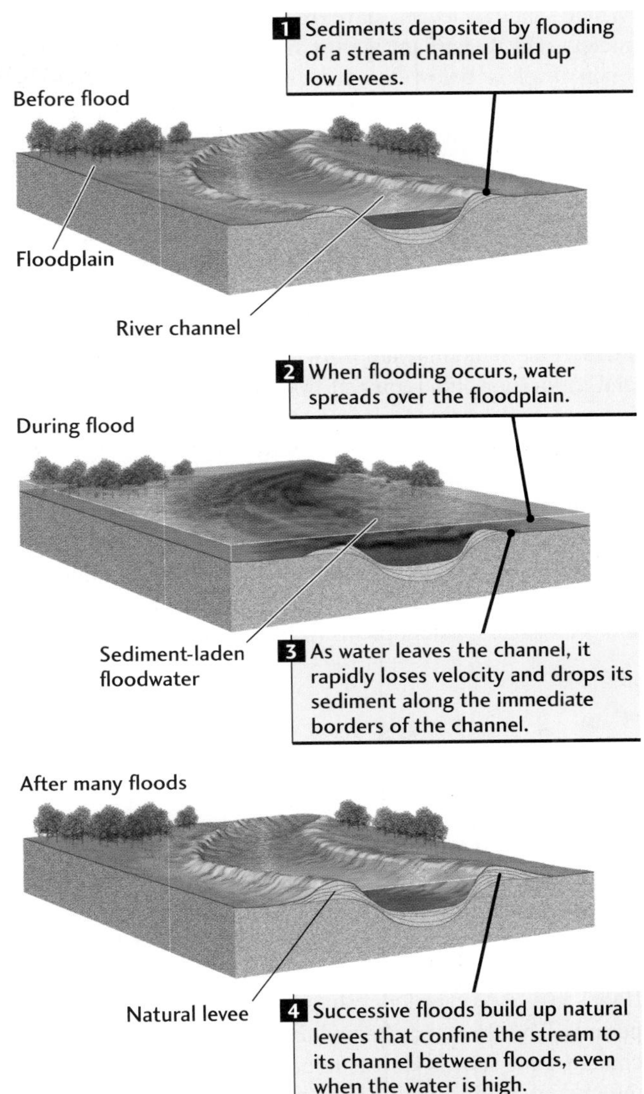

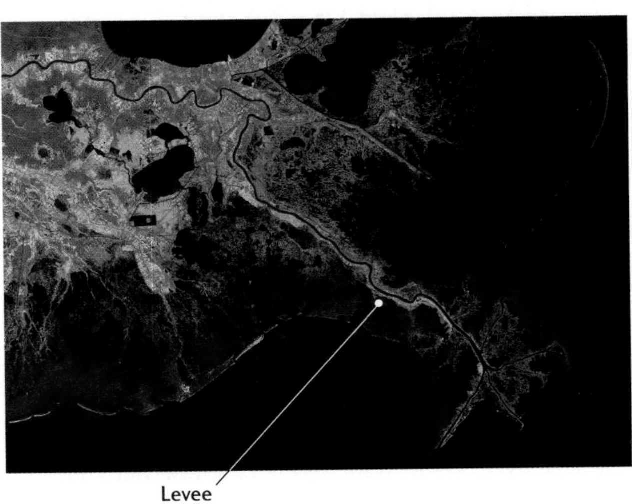

Figure 14.10 The formation of natural levees by river floods. Natural levees at Vicksburg, Mississippi. [U.S. Geological Survey National Wetlands Research Center.]

Where natural levees have reached a height of several meters and the stream almost fills the channel, the floodplain level is below the stream level. You can walk the streets of an old river town built on a floodplain, such as Vicksburg, Mississippi, and look up at the levee, knowing that the river waters are rushing by above your head.

During floods, fine sediments—silts and muds—are carried well beyond the channel banks, often over the entire floodplain, and are deposited there as floodwaters continue to lose velocity. Receding floodwaters leave behind standing ponds and pools of water. The finest clays are deposited there as the standing water slowly disappears by evaporation and infiltration. Fine-grained floodplain deposits have been a major resource for agriculture since ancient times. The fertility of the floodplains of the Nile and other rivers of the Middle East, which contributed to the evolution of the cultures that flourished there thousands of years ago, depended on frequent flooding. Today the great, broad floodplain of the Ganges in northern India continues to play an important role in India's life and agriculture. Many ancient and modern cities are sited on floodplains (see Feature 14.1).

Streams Change with Time and Distance

The flow of a stream appears steady when you view it from a bridge for a few minutes or canoe along it for a few hours, but its volume and velocity at a single place may change appreciably from month to month and season to season. Streams are dynamic systems, moving from low waters to floods in a few years and reshaping their valleys over longer periods. The flow and channel dimensions of a stream also change as it moves downstream, from narrow valleys in the upland headwaters to broader floodplains in the middle and lower courses. Most of these longer-term changes are adjustments in the normal (nonflood) volume and velocity of flow as well as the depth and width of the channel.

Discharge

We measure the size of a stream's flow by its **discharge**—the volume of water that passes a given point in a given time as it flows through a channel of a certain width and depth. (In Chapter 13, we defined discharge as the volume of water leaving an aquifer in a given time. These definitions are consistent, because they both describe volume of flow per unit of time.) Stream discharge is usually measured in cubic meters per second or cubic feet per second. Discharge in small streams may vary from about 0.25 to 300 m^3/s. The discharge of a well-studied medium-sized river in Sweden, the Klarälven, varies from 500 m^3/s at low water levels to 1320 m^3/s at high water levels. The discharge of the Mississippi River can vary from as low as 1400 m^3/s at low water levels to more than 57,000 m^3/s during floods.

To calculate discharge, we multiply the cross-sectional area (the width multiplied by the depth of the part of the channel occupied by water) by the velocity of the flow (distance traveled per second):

$$\text{discharge} = \underset{(\text{width} \times \text{depth})}{\text{cross section}} \times \underset{(\text{distance traveled per second})}{\text{velocity}}$$

Figure 14.11 illustrates this relationship. For discharge to increase, either velocity or cross-sectional area, or both, must increase. Think of increasing the discharge of a garden hose by turning the valve open more, which increases the velocity of water coming out of the end of the hose. The cross-sectional area of the hose, measured by its diameter, cannot change, so the discharge must increase. In a stream, as discharge at a particular point increases, both the velocity and the cross-sectional area tend to increase. (The velocity is

(a) **1** A river with a lower cross-sectional area and a lower velocity has a lower discharge ($3\text{ m} \times 10\text{ m} = 30\text{ m}^2 \times 1\text{ m/s} = 30\text{ m}^3\text{/s}$ discharge)...

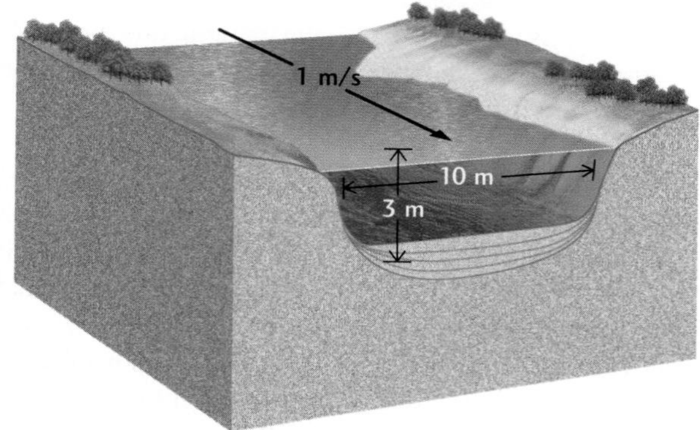

(b) **2** ...than a river with a higher cross-sectional area and a higher velocity which has a higher discharge ($9\text{ m} \times 10\text{ m} = 90\text{ m}^2 \times 2\text{ m/s} = 180\text{ m}^3\text{/s}$).

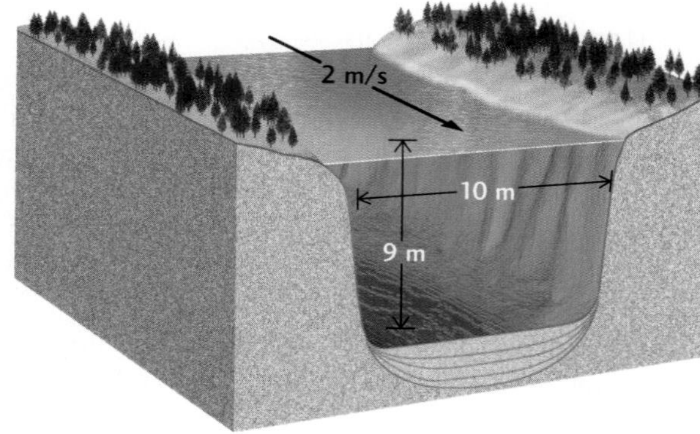

Figure 14.11 Discharge depends on velocity and cross-sectional area. A river at (a) low discharge and (b) high discharge. [Data from T. Dunne and L. B. Leopold, *Water in Environmental Planning* (San Francisco: W. H. Freeman, 1978).]

14.1 The Development of Cities on Floodplains

Floodplains are natural sites for urban settlements because they combine easy transportation along a river with access to fertile agricultural lands. Such sites, however, are subject to the floods that formed the plains.

About 4000 years ago, cities began to dot floodplains in Egypt along the Nile, in the ancient land of Mesopotamia along the Tigris and Euphrates rivers, and in Asia along the Indus River of India and the Yangtze and Huang Ho of China. Later, many of the capital cities of Europe were built on floodplains: London on the Thames, Paris on the Seine, Rome on the Tiber. Floodplain cities in North America include St. Louis on the Mississippi, Cincinnati on the Ohio, and Montreal on the St. Lawrence.

Floods periodically destroyed sections of these ancient and modern cities on the lower parts of the floodplains, but each time the inhabitants rebuilt them. Today, most large cities are protected by artificial levees that strengthen and heighten the river's natural levees. Extensive systems of dams can help control flooding that would affect these cities, but they cannot eliminate the risk entirely. In 1973, the Mississippi went on a rampage with a flood that continued for 77 consecutive days at St. Louis. It reached a record 4.03 m above flood stage (the height at which the river first overflows the channel banks). In 1993, the Mississippi and its tributaries broke loose again, shattering the old record in a flood that has been officially designated the most devastating flood in U.S. history. The flood resulted in 487 deaths and $15 billion to $20 billion in property damage. At St. Louis, the Mississippi stayed above flood stage for 144 of the 183 days between April and September. An unexpected result of this flood was widespread pollution as floodwaters leached agricultural chemicals from farmlands and then deposited them in flooded areas. Some geologists believe that the construction of levees and dikes to confine the Mississippi contributes to the record high floods. The river can no longer erode its banks and widen its channel to accommodate some of the additional water flowing during times of high discharge.

What are cities and towns in this position to do? Some have urged a halt to all construction and development on the lowest parts of the floodplains. Some have called for the elimination of federally subsidized disaster funds for rebuilding in such areas. Harrisburg, Pennsylvania, hit hard by a flood in 1972, turned some of its devastated riverfront area into a park. In a dramatic move after the 1993 Mississippi flood, the town of Valmeyer, Illinois, voted to move the entire town to high ground several miles away. The new site was chosen with the help of a team of geologists from the Illinois Geological Survey. Yet some people who have lived on floodplains all their lives want to stay and are prepared to live with the risk. The costs of protecting some river-bottom areas are prohibitive, and these places will continue to pose public policy problems.

Like many cities built on river floodplains, Liuzhou, China, is subject to flooding. This flood, in July 1996, was the largest recorded in the city's 500-year history. [Xie Jiahua/China Features/Corbis Sygma.]

also affected by the slope of the channel and the roughness of the river bottom and sides, which we can neglect for the purposes of this explanation.) The cross-sectional area increases because the flow occupies more of the channel's width and depth.

Discharge in most rivers increases downstream as more and more water flows in **tributaries,** streams that release water into larger streams. Increased discharge means that width, depth, or velocity must increase, too. Velocity does not increase downstream as much as the increase in discharge might lead us to expect, because the slope along the lower courses of a stream decreases (decreasing slope reduces velocity). Where discharge does not increase significantly downstream and slope decreases greatly, a river will flow more slowly.

Floods

In 1996, continued heavy rains swelled the Yangtze, China's longest river, to flood stage. Large parts of the countryside were inundated. More than 700 people were killed, and hundreds of thousands who fled to the nearby mountains from the lowlands of the floodplain were trapped there. Such a flood is an extreme case of increased discharge that results from a short-term imbalance between input and output. As the discharge increases, the flow velocity in the channel increases and the water gradually fills the channel. As the discharge continues to increase, the water spills over the banks. Rivers flood regularly, some at infrequent intervals, others almost every year. Some floods are large, with very high water levels lasting for days. At the other extreme are minor floods that barely break out from the channel before they recede. Small floods are more frequent, occurring every 2 or 3 years on average. Large floods are generally less frequent, usually occurring every 10, 20, or 30 years.

No one can know exactly how high—either in water height or in discharge—a flood will be in any given year, so geologists state their predictions as *probabilities,* not certainties. For a particular stream, a geologist might state that there is a 20 percent probability that a flood of a given discharge—say, 1500 m^3/sec—will occur in any one year. This probability corresponds to an average time interval—in this case, 5 years (20 percent = 1 in 5)—that we expect between two floods with a discharge of 1500 m^3/sec. We speak of a flood of this discharge as a 5-year flood. The average time interval between the occurrence of two geological events of a given magnitude is called a **recurrence interval.** A flood of greater magnitude—say, 2600 m^3/sec—on the same stream is likely to happen only once every 50 years, and it would therefore be called a 50-year flood. A graph of the annual probabilities and recurrence intervals for a range of flood discharges in one river—the Skykomish, in Washington State—is shown in **Figure 14.12**.

The recurrence interval of floods of different discharge varies from stream to stream. The interval depends on three factors:

1. The climate of the region
2. The width of the floodplain
3. The size of the channel

In a dry climate, for example, the recurrence interval of a 2600 m^3/sec flood may be much longer than the recurrence interval of a 2600 m^3/sec flood on a similar stream in an area that gets intermittent rain. For this reason, individual graphs of recurrence intervals of major rivers are necessary if towns along these rivers are to be prepared to cope with floods of various heights.

The prediction of river floods and their heights has become much more reliable as automated rainfall and river

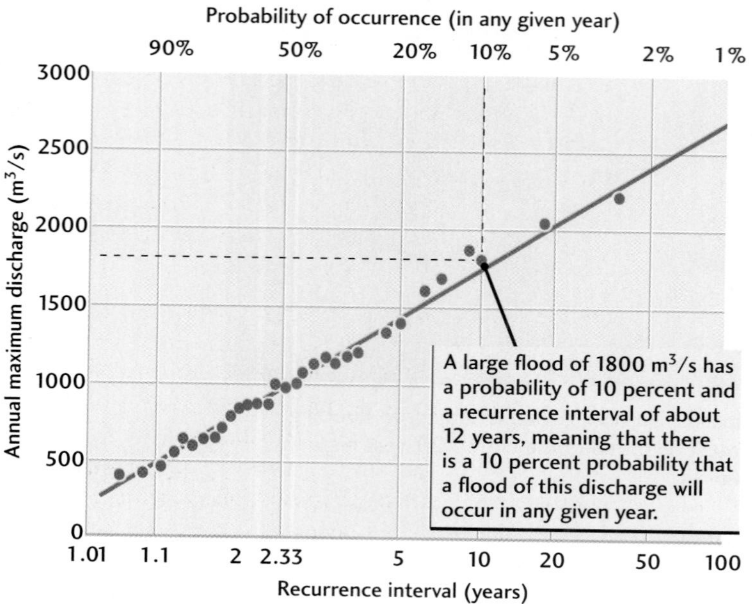

Figure 14.12 The flood-frequency curve for annual floods on the Skykomish River at Gold Bar, Washington. This curve predicts the probability that a flood of a certain discharge will occur in any given year. [After T. Dunne and L. B. Leopold, *Water in Environmental Planning* (San Francisco: W. H. Freeman, 1978).]

measurements coupled with new computer models have come into use. These methods of forecasting could allow geologists to predict rises and falls of rivers as much as several months in advance. Today, however, predictions are chiefly used for short-term forecasts, such as flood warnings of only a few days.

Longitudinal Profile and the Concept of Grade

We have seen that streamflow at any locality balances inputs and outputs, which become temporarily out of balance during floods. Studies of changes in discharge, velocity, channel dimensions, and topography (especially slope) along the entire length of a stream from its headwaters (where it begins) to its mouth (where it ends) reveal a larger-scale and longer-term balance. A stream is in dynamic equilibrium between erosion of the streambed and sedimentation in the channel and floodplain over its entire length. This equilibrium is controlled by several factors:

- Topography (including slope)
- Climate
- Streamflow (including both discharge and velocity)
- The resistance of rock to weathering and erosion

A particular combination of factors—such as high topography, humid climate, high discharge and velocity, hard rocks, and low sediment load—would make a stream erode a steep valley into bedrock and carry downstream all sediment derived from that erosion. Downstream, where the topography is lower and the stream might flow over easily erodible sediments, it would deposit bars and floodplain sediments, building up the elevation of the streambed by sedimentation.

We describe the slope of a river from headwaters to mouth by plotting the elevation of its streambed against distances from its headwaters. **Figure 14.13** plots the slope of the Platte and South Platte rivers from the headwaters of the South Platte in central Colorado to the mouth of the Platte in Nebraska. This smooth, concave-upward curve, which represents a cross-sectional view of the river, is its **longitudinal profile.** All streams, from small rills to large rivers, show this same general concave-upward profile, from notably steep near a stream's head to low, almost level, near its mouth.

Why do all streams follow this profile? The answer lies in the combination of factors that control erosion and sedimentation. All streams run downhill from their headwaters to their mouths. Erosion is greater in the higher parts of a stream's course than in the lower parts because slopes are steeper and flow velocities can be very high, which has an important influence on the erosion of bedrock (see Chapter 18). In a stream's lower courses, where it carries sediments derived from erosion of the upper courses, sedimentation becomes more significant. Differences in topography and the other factors listed above may make the longitudinal profile steeper or shallower in the upper and lower courses of a stream, but the general shape remains concave upward.

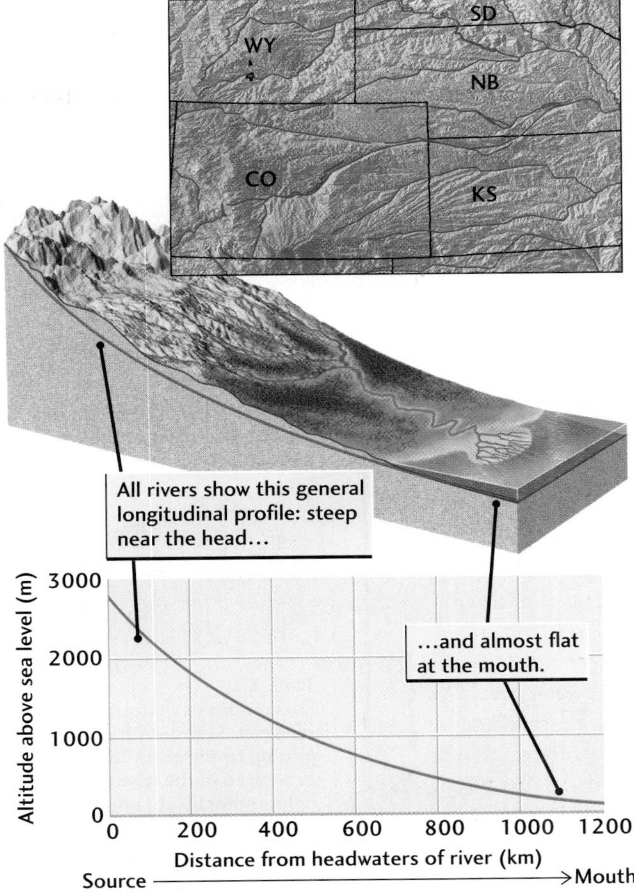

Figure 14.13 The longitudinal profile of the Platte and South Platte rivers from the headwaters of the South Platte in central Colorado to the mouth of the Platte at the Missouri River in Nebraska. [Data from H. Gannett, in *Profiles of Rivers in the United States.* USGS Water-Supply Paper 44, 1901.]

The longitudinal profile is controlled at its lower end by a stream's **base level,** the elevation at which it ends by entering a large standing body of water, such as a lake or ocean. Streams cannot cut below base level, because base level is the "bottom of the hill"—the lower limit of the longitudinal profile.

Profiles Change When Base Levels Change Changes in natural base level affect a longitudinal profile in predictable ways. **Figure 14.14** illustrates the longitudinal profile for the natural local and regional base levels of a river flowing into a lake and from the lake into the ocean. If the regional base level rises—perhaps because of faulting—the profile will show the effects of sedimentation as the river builds up channel and floodplain deposits to reach the new, higher regional base-level elevation. Damming a river artificially can create a new local base level, with similar effects on the

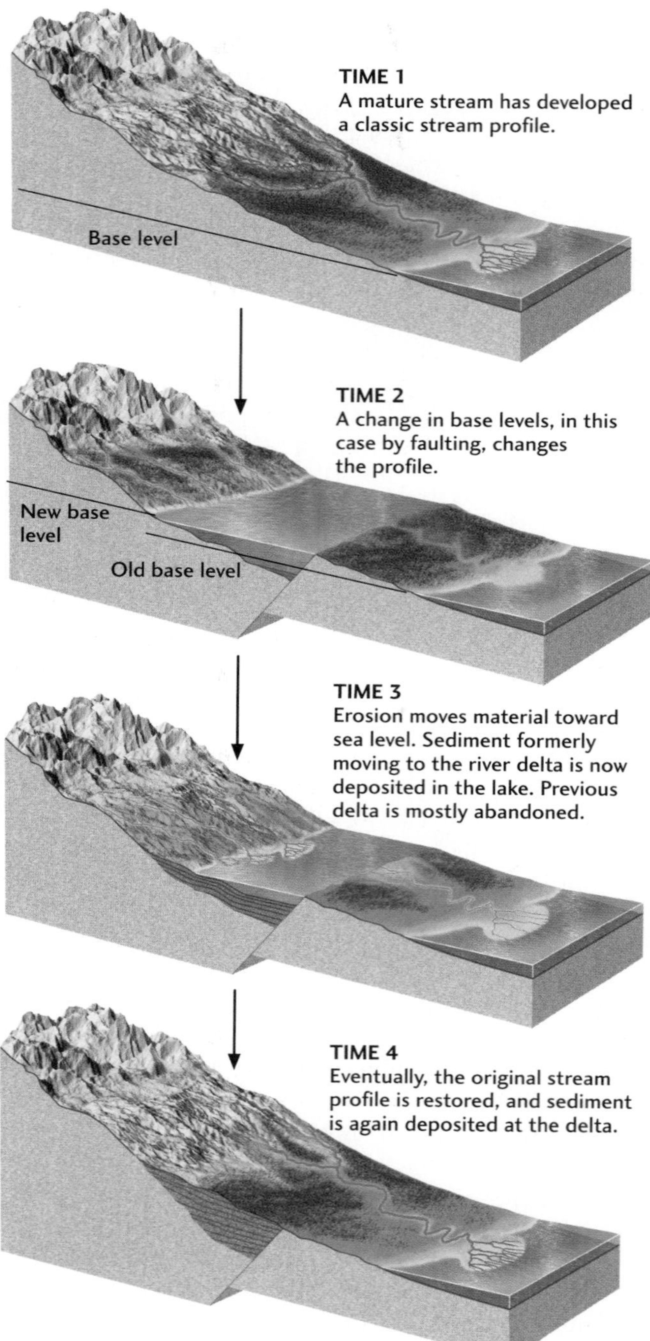

Figure 14.14 The base level of a stream controls the lower end of its longitudinal profile. The profiles illustrated here are for natural regional and local base levels of a river flowing into a lake and from the lake into an ocean. In each river segment, the profile adjusts to the lowest level that the river can reach.

longitudinal profile (**Figure 14.15**). The slope of the river upstream from the dam decreases, because the new local base level artificially flattens the river's profile at the location of the reservoir formed behind the dam. The decrease in slope lowers the river's velocity, decreasing its ability to transport sediment. The stream deposits some of the sediment on the bed, which makes the concavity somewhat shallower than it was before the dam was built. Below the dam, the river, now carrying much less sediment, adjusts its profile to the new conditions and typically erodes its channel in the section just below the dam.

This kind of erosion has severely affected sandbars and beaches in Grand Canyon National Park downstream from the Glen Canyon Dam. The erosion threatens animal habitats and archeological sites as well as beaches used for recreation. River specialists have calculated that if river discharge during floods were increased by a certain amount, enough sand would be deposited to prevent depletion by erosion. This calculation was confirmed by a huge experiment in which a controlled flood was staged at the Glen Canyon Dam in 1996. As the gates of the dam were opened, about 10 billion gallons of water spilled into the canyon at a rate fast enough to fill Chicago's 100-story Sears Tower in 17 minutes. This experiment showed that eroded areas could be brought back by sedimentation during floods.

Falling sea level also alters regional base levels and longitudinal profiles. The regional base levels of all streams flowing into the ocean are lowered, and their valleys are cut into former stream deposits. When the drop in sea level is large, as it was during the last glacial period, rivers erode steep valleys into coastal plains and continental shelves.

Graded Streams Over a period of years, a stream's profile becomes stable as the stream gradually fills in low spots and erodes high spots, thereby producing the smooth curve that represents the balance between erosion and sedimentation. That balance is governed not only by the stream's base level but also by the elevation of its headwaters and by all the other factors controlling the equilibrium of the stream profile, as discussed earlier in the chapter. At equilibrium, the stream is a **graded stream**—one in which the slope, velocity, and discharge combine to transport its sediment load, with neither sedimentation nor erosion. If the conditions that produce a particular graded-stream profile change, the stream's profile will change to reach a new equilibrium. Such changes may include depositional and erosional patterns and alterations in the shape of the channel.

In places where regional base level is constant over geologic time, the longitudinal profile represents the balance between tectonic uplift and erosion on the one hand and transport and deposition on the other. If uplift is dominant, typically in the upper courses of a stream, the profile is steep and expresses the dominance of erosion and transport. As uplift slows and the headwater region is eroded, the profile is lowered.

Alluvial Fans One place where a river must adjust suddenly to changed conditions is at a mountain front. Here, streams leave narrow mountain valleys and enter broad, rel-

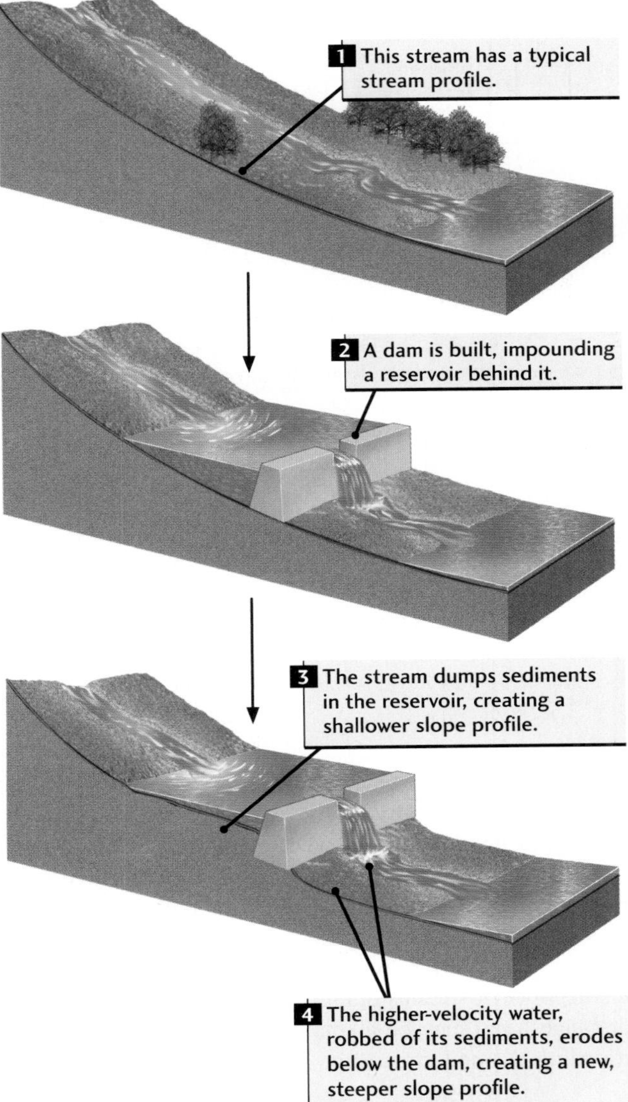

Figure 14.15 A change in the base level of a river caused by human intervention, such as the construction of a dam, alters a river's profile.

atively flat valleys at lower elevations. Along such fronts, typically at steep fault scarps, streams drop large amounts of sediment in cone- or fan-shaped accumulations called **alluvial fans** (**Figure 14.16**). This deposition results from the sudden decrease in velocity that occurs as the channel widens abruptly. To a minor extent, a lowering of the slope below the front also slows the stream velocity. The surface of the alluvial fan typically has a concave-upward shape connecting the steeper mountain profile with the gentler valley profile. Coarse materials, from boulders to sand, dominate on the steep upper slopes of the fan. Lower down, finer sands, silts, and muds are deposited. Fans from many adjacent streams along a mountain front may merge to form a long wedge of sediment whose appearance may mask the outlines of the individual fans that make it up.

Terraces Tectonic uplift can change the equilibrium of a stream valley, resulting in flat, steplike surfaces that line the stream above the floodplain. These **terraces** mark former floodplains that existed at a higher level before regional uplift or an increase in discharge caused the stream to erode into the former floodplain. Terraces are made of floodplain deposits and are often paired, one on each side of the stream, at the same level (**Figure 14.17**). The sequence of events that forms terraces starts when a stream creates a floodplain. Rapid uplift then changes the stream's equilibrium, causing it to cut down into the floodplain. In time, the stream reestablishes a new equilibrium at a lower level. It may then build another floodplain, which will also undergo uplift and be sculpted into another, lower pair of terraces.

The Effect of Climate Climate also strongly affects the longitudinal profile, primarily through the influence of temperature and precipitation on weathering and erosion (see Chapter 7). Warm temperatures and high rainfall promote weathering and erosion of soils and hillslopes and thus enhance sediment transport by streams. High rainfall also

Figure 14.16 An alluvial fan (Tucki Wash) in Death Valley, California. Alluvial fans are large cone- or fan-shaped accumulations of sediment deposited when a stream must suddenly adjust to changed conditions, as at a mountain front. [Michael Miller.]

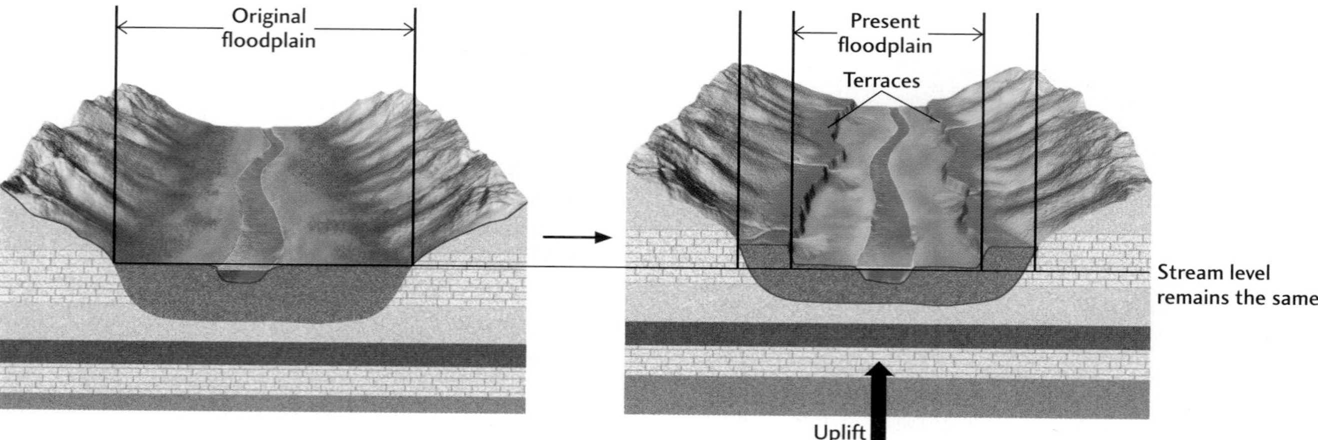

Figure 14.17 Terraces form when the land surface is uplifted and a river erodes into its floodplain and establishes a new floodplain at a lower level. The terraces are remnants of the former floodplain.

leads to greater river discharge, which results in more erosion of riverbeds. An analysis of sediment transport over the entire United States provides evidence that global climate change over the past 50 years is responsible for a general increase in streamflow. Although tectonics plays a dominant role in the formation of alluvial fans at fault scarps, short-term buildup of sediment or trenching is the result of climatic change, primarily variations in temperature.

Lakes

Lakes are accidents of the longitudinal profile, as we can see easily where a lake has formed behind a dam (see Figure 14.15). Lakes range in size from ponds only 100 m across to the world's largest and deepest lake, Lake Baikal in southwestern Siberia. This lake holds approximately 20 percent of the total amount of fresh water in the world's lakes and rivers. It is located in a continental rift zone, a typical plate tectonic setting for lakes. The damming that takes place in a rift valley results from faulting that blocks a normal exit of water. Streams can flow into a rift valley easily but cannot flow out until water builds up to a high enough level to allow them to exit. There are a great many lakes in the northern United States and Canada because glacial ice and glacial sedimentary debris have disrupted normal drainage. Sooner or later, if tectonics and climate remain stable, such lakes will drain away as new outlets form and the longitudinal profile becomes smooth. Because lakes are so much smaller than the oceans, they are more likely than the oceans to be affected by water pollution. Chemical and other industries have polluted Lake Baikal, and Lake Erie has been very polluted for many years, although there has been some improvement recently.

Drainage Networks

Every topographic rise between two streams, whether it measures a few meters or a thousand, forms a **divide**—a ridge of high ground along which all rainfall runs off down one side of the rise or the other. A **drainage basin** is an area of land, bounded by divides, that funnels all its water into the network of streams draining the area (**Figure 14.18**). A drainage basin may be a small area, such as a ravine surrounding a small stream, or a great region drained by a

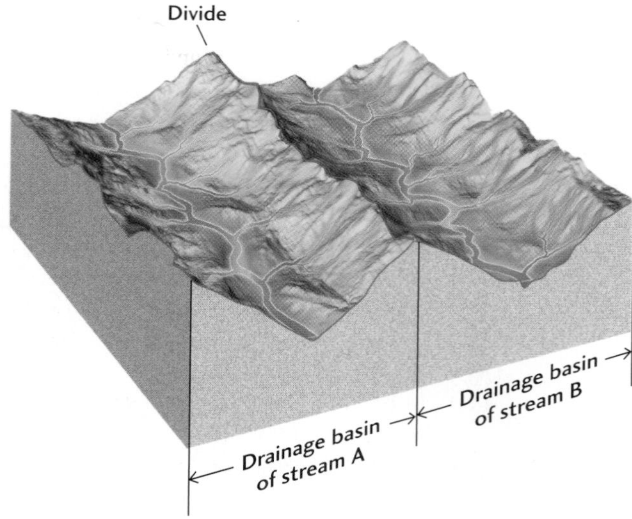

Figure 14.18 Stream valleys and drainage basins are separated by divides, which may be ridges, gentle uplands, or mountain ranges.

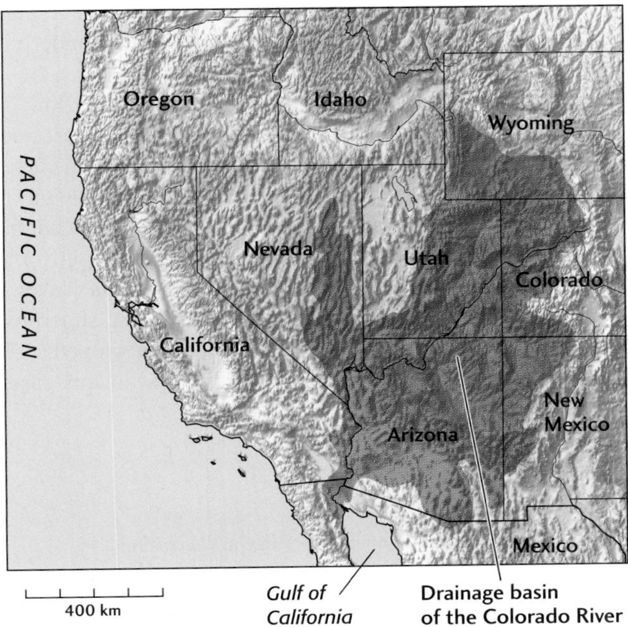

Figure 14.19 The natural drainage basin of the Colorado River covers about 630,000 km^2, a large part of the southwestern United States. This basin is surrounded by divides that separate it from the neighboring drainage basins. [After U.S. Geological Survey.]

major river and its tributaries (**Figure 14.19**). A continent has several major drainage basins separated by major divides. In North America, the continental divide along the Rocky Mountains separates all waters flowing into the Pacific Ocean from all those entering the Atlantic.

Drainage Patterns

A map showing the courses of large and small streams reveals a pattern of connections called a **drainage network.** If you followed a stream from its mouth to its head, you would see that it steadily divides into smaller and smaller tributaries, forming drainage networks that show a characteristic branching pattern (**Figure 14.20**). Branching is a general property of many kinds of networks in which material is collected and distributed. The network of the human circulatory system, for example, distributes blood to the body through a branching system of arteries and collects it through a corresponding system of veins.

Perhaps the most familiar branching networks are those of trees and roots. Most rivers follow the same kind of irregular branching pattern, called **dendritic drainage** (from the Greek *dendron,* for "tree"). This fairly random drainage pattern is typical of terrains where the bedrock is uniform, such as horizontally bedded sedimentary rocks or massive igneous or metamorphic rocks. Other patterns are *rectangular, trellis,* and *radial.*

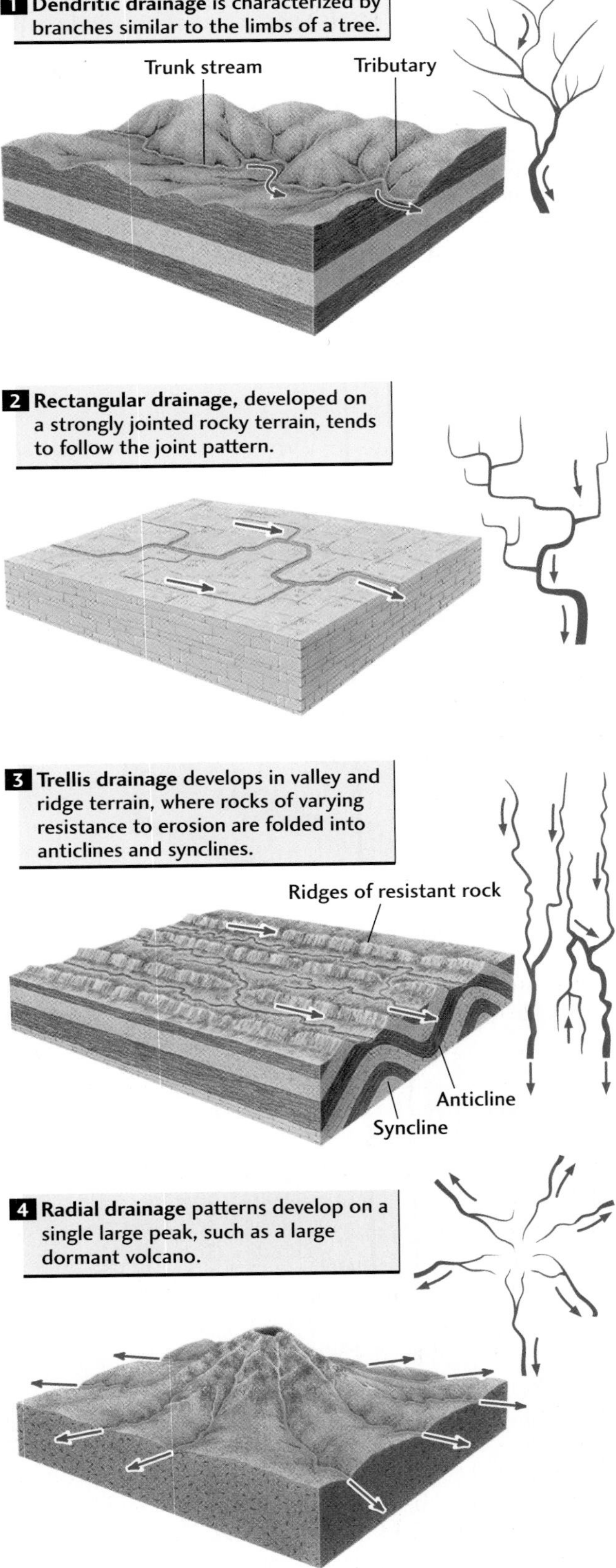

Figure 14.20 Typical drainage networks.

Drainage Patterns and Geologic History

We can observe directly or judge from historical records how most stream drainage patterns evolved. Some streams, for example, cut through erosion-resistant bedrock ridges to form steep-walled notches or gorges. What could cause a stream to cut a narrow valley directly through a ridge rather than running along the lowland on either side of it? The geologic history of the region provides the answers. If a ridge is formed by structural deformation while a preexisting stream is flowing over it, the stream may erode the rising ridge to form a steep-walled gorge, as in **Figure 14.21**. Such a stream is called an **antecedent stream** because it existed before the present topography was created, and it maintained its original course despite changes in the underlying rocks and in topography.

In another geological situation, a stream may flow in a dendritic drainage pattern over horizontal beds of sedimentary rocks that overlie folded and faulted rocks with varying resistance to erosion. Over time, the stream cuts down into the underlying rocks and erodes a gorge in the resistant bed. Such a **superposed stream** flows through resistant formations because its course was established at a higher level, on uniform rocks, before downcutting began. A superposed stream tends to continue the pattern that it developed earlier rather than adjusting to its new conditions. In this case, a

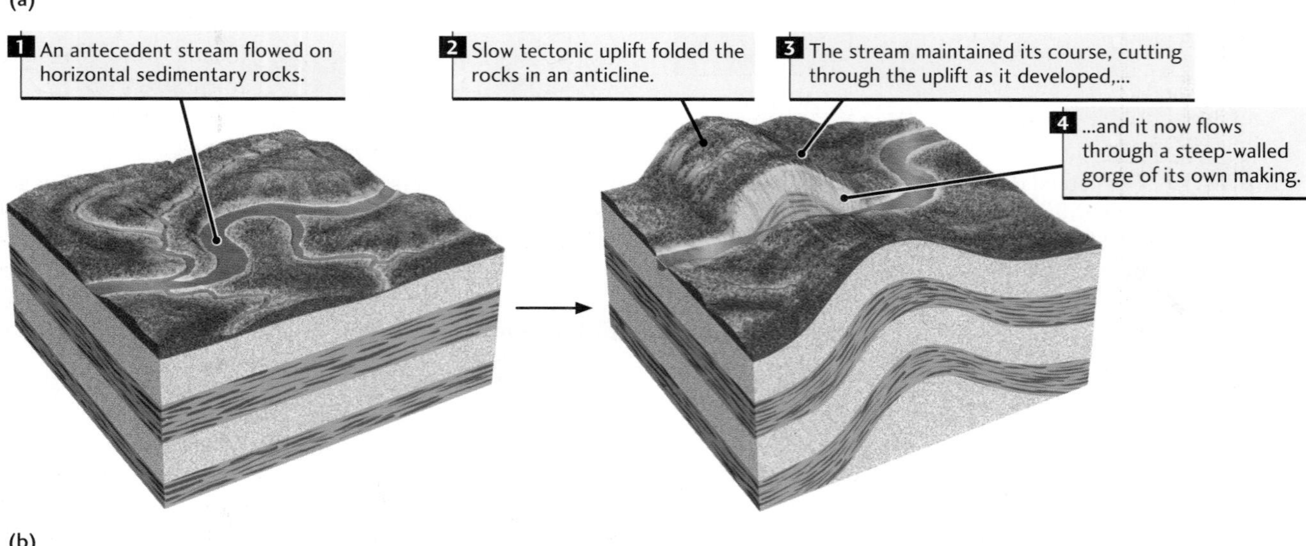

Figure 14.21 (a) An antecedent stream. (b) The Delaware Water Gap, located between Pennsylvania and New Jersey, is an antecedent stream. [Michael P. Gadomski/ Photo Researchers.]

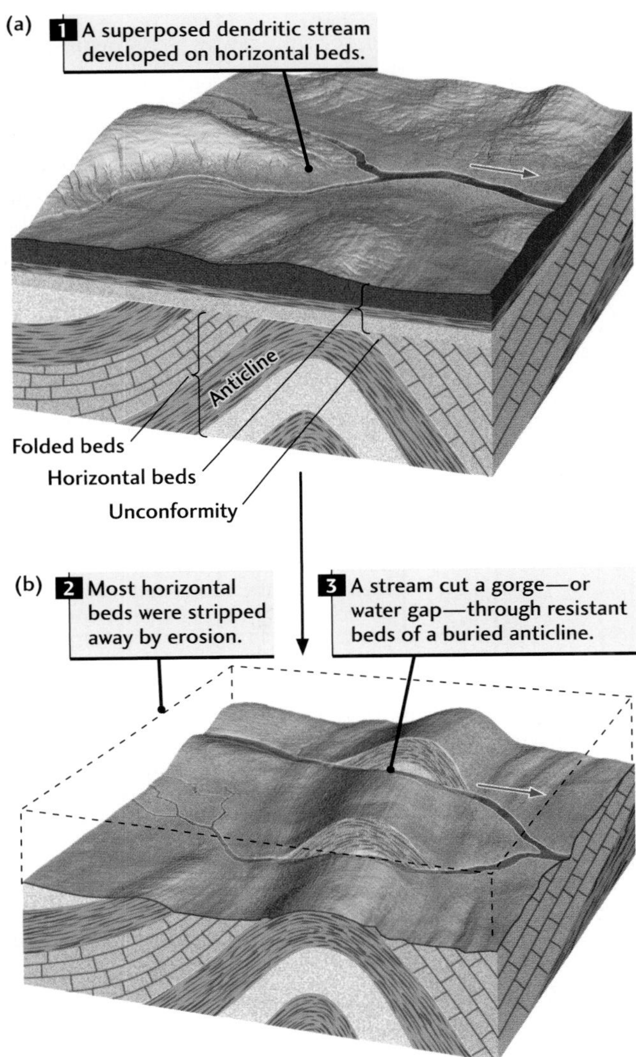

Figure 14.22 The development of a superposed stream by erosion of horizontal beds overlying folded beds of varying resistance to erosion.

dendritic pattern is forced onto a surface that would otherwise have evolved a rectangular network (**Figure 14.22**).

Deltas: The Mouths of Rivers

Sooner or later, all rivers end as they flow into a lake or an ocean, mix with the surrounding water, and—no longer able to travel downhill—gradually lose their forward momentum. The largest rivers, such as the Amazon and the Mississippi, can maintain some current many kilometers out to sea. Where smaller rivers enter a turbulent, wave-swept coast, the current disappears almost immediately beyond the river's mouth.

Delta Sedimentation

As its current gradually dies out, a river progressively loses its power to transport sediment. The coarsest material, typically sand, is dropped first, right at the mouth in most rivers. Finer sands are dropped farther out, followed by silt and, still farther out, by clay. As the floor of the lake or sea slopes to deeper water away from the shore, all the dropped materials of various sizes build up a depositional platform called a **delta.** (We owe the name *delta* to the Greek historian Herodotus, who traveled through Egypt about 450 B.C. The roughly triangular shape of the sediments deposited at the mouth of the Nile prompted him to name it after the Greek letter Δ, delta.)

As a river approaches its delta, where the slope profile is almost level with the sea, it reverses its upstream-branching drainage pattern. Instead of collecting more water from tributaries, it discharges water into **distributaries**—smaller streams that receive water and sediment from the main channel, branch off *downstream,* and thus distribute the water and sediment into many channels. Materials deposited on top of the delta, typically sand, make up horizontal **topset beds.** Downcurrent, on the outer front of the delta, fine-grained sand and silt are deposited to form gently inclined **foreset beds,** which resemble large-scale cross-beds. Spread out on the seafloor seaward of the foreset beds are thin, horizontal **bottomset beds** of mud, which are eventually buried as the delta continues to grow. **Figure 14.23** shows a typical large marine delta.

The Growth of Deltas

As a delta builds out into the sea, the mouth of its river advances seaward, leaving new land in its wake. Much of this land is a delta plain just a few meters above sea level. Such plains include large areas of wetlands—which, as noted in Chapter 13, are valuable because they store water and form the habitat of many diverse species of plants and animals. In many areas, delta wetlands have suffered a two-pronged attack. First, the extensive flood-control dams built since the 1930s have decreased the volume of sediment brought to the edge of the delta, thereby reducing the sediment supply to the wetlands. Second, massive artificial levees have prevented the small but frequent floods that nourish the delta wetlands.

A growing delta shifts the flows from some distributaries to others with shorter routes to the sea. As a result of such shifts, the delta grows in one direction for some hundreds or thousands of years and then breaks out into a new distributary and begins to grow into the sea in another direction. A major river, such as the Mississippi or the Nile, forms a large delta thousands of square kilometers in area. The delta of the Mississippi, like many other major river deltas, has been growing for millions of years. About 150 million years ago, it started out around what is now the junction of the Ohio and the Mississippi rivers, at the southern tip of Illinois. It has advanced about 1600 km since then, creating almost the

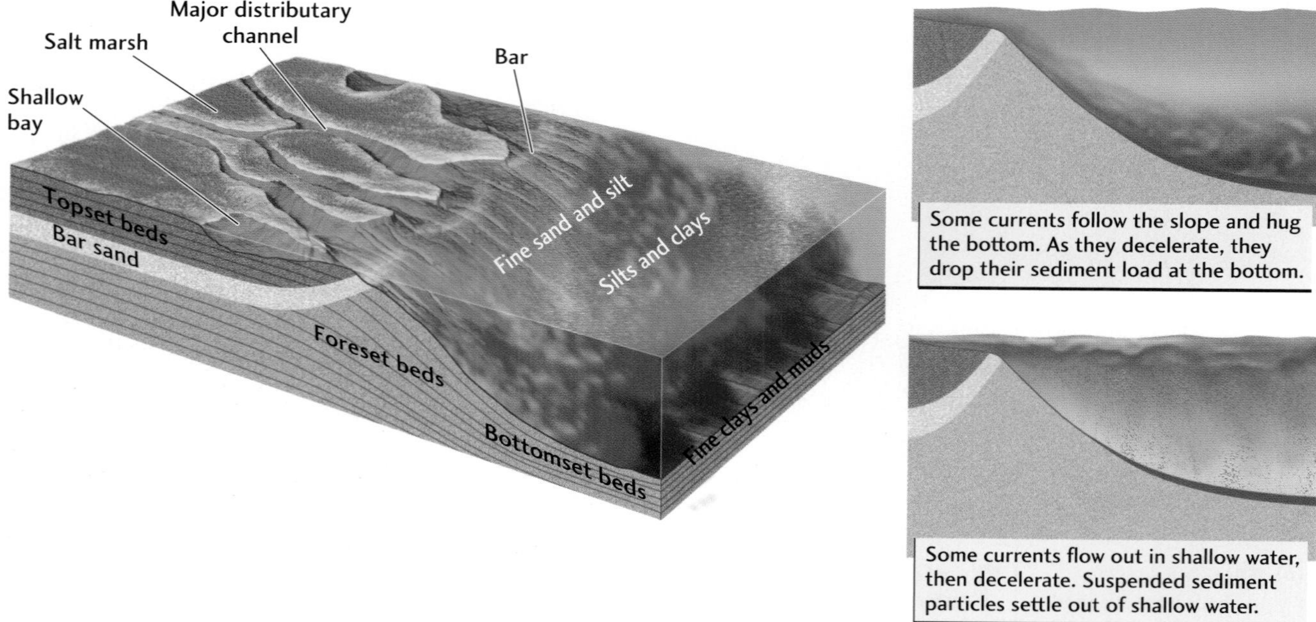

Figure 14.23 A typical large marine delta, many kilometers in extent, in which the fine-grained foreset beds are deposited at a very low angle, typically only 4° to 5° or less. Sandbars form at the mouths of the distributaries, where the currents' velocity suddenly decreases. The delta builds forward by the advance of the bar and the topset, foreset, and bottomset beds. Between distributary channels, shallow bays fill with fine-grained sediment and become salt marshes. This general structure is found on the Mississippi delta.

entire states of Louisiana and Mississippi as well as major parts of adjacent states. **Figure 14.24** shows the growth of the Mississippi delta over the past 6000 years.

Deltas grow by the addition of sediment, and they sink as the sediment becomes compacted and the crust subsides under the weight of the sediment load. Venice, built on part of the Po River delta in northern Italy, has been subsiding steadily. Both crustal subsidence and depression of the ground attributable to the pumping of water from aquifers beneath the city are responsible.

The Effects of Waves, Tides, and Tectonics

Strong waves, shoreline currents, and tides affect the growth and shape of marine deltas. Waves and shoreline currents may move sediment along the shore almost as rapidly as it is dropped there by a river. The delta front then becomes a long beach with only a slight seaward bulge at the mouth. Where tidal currents move in and out, they redistribute deltaic sediment into elongate bars parallel to the direction of the currents, which in most places is at approximately right angles to the shore.

In some places, waves and tides are strong enough to prevent a delta from forming. Instead, the sediment a river brings down to the sea is dispersed along the shoreline as beaches and bars and is transported into deeper waters offshore. The east coast of North America lacks deltas for this reason. The Mississippi has been able to build its delta because neither waves nor tides are very strong in the Gulf of Mexico.

Tectonics also exerts some control over where deltas form, because deltas require

- Uplift in the drainage basin, which provides abundant sediment
- Crustal subsidence in the delta region to accommodate the great weight and volume of sediment

Two of the world's large deltas—the Mississippi and the Rhône (in France)—derive their large sediment loads primarily from distant mountain ranges: the Rockies for the Mississippi and the Alps for the Rhône. Both are in the same type of plate tectonic setting—a passive margin originally formed from a rifted continental margin.

Few large deltas are associated with active subduction zones. The reason may be that it is unusual for a large river, such as the Columbia of Washington State and Oregon, to carry abundant sediment through a volcanic arc (the Cascade Range) to the sea. In addition, oceanic island arcs are too small in land area to provide much clastic sediment.

The continental convergence that elevated the Himalayas also formed the great deltas of the Indus and Ganges. Ultimately, plate tectonic settings influence delta location and formation.

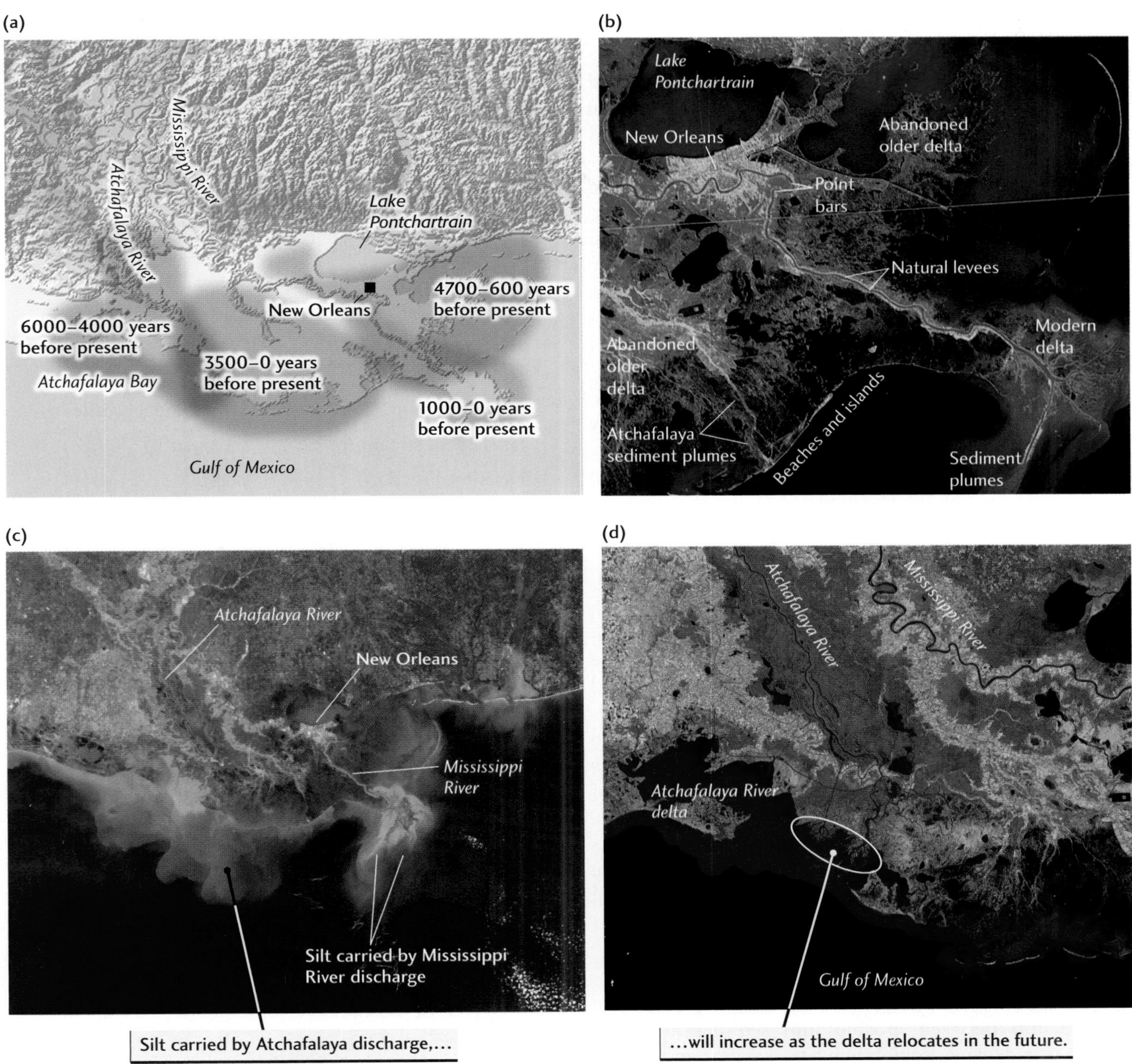

Figure 14.24 (a) The Mississippi delta. Over the past 6000 years, the river has built its delta first in one direction and then in another as water flow shifted from one major distributary to another. The modern delta was preceded by deltas deposited to the east and west. (b) The infrared-sensitive film used to record this satellite image of the Mississippi delta causes the vegetation to appear red, relatively clear water to appear dark blue, and water with suspended sediment to appear light blue. At the upper left are New Orleans and Lake Pontchartrain. Well-defined natural levees and point bars are at the center. At the lower left are beaches and islands that formed as waves and currents transported river-deposited sand from the delta. [From G. T. Moore, "Mississippi River Delta from Landsat." *Bulletin of the American Association of Petroleum Geologists* (1979).] (c) Satellite photograph of the Mississippi delta. [NASA.] (d) This image shows the discharge of sediment to the Gulf of Mexico from the Mississippi River delta and the Atchafalaya River. A major flood could divert the main flow of the Mississippi River into the Atchafalaya River, enabling a new delta to form. Construction of artificial levees by the Army Corp of Engineers has prevented this thus far. [U.S. Geological Survey National Wetlands Research Center.]

SUMMARY

How does flowing water in streams erode solid rock and transport and deposit sediment? Any fluid can move in either laminar or turbulent flow, depending on its velocity, viscosity, and flow geometry. The turbulence that characterizes most streams is responsible for transporting sediment by suspension (clays), saltation (sands), and rolling and sliding along the bed (sand and gravel). The tendency for particles to be carried in suspension is countered by the gravitational force that pulls them to settle to the bottom. The settling velocity measures the speed with which suspended particles settle to the bottom. When a streamflow slows, it loses its competence to carry sediment and deposits it, in many places as beds of rippled, cross-bedded sand. Running water erodes solid rock by abrasion; by chemical weathering that enlarges and opens cracks; by physical weathering as sand, pebbles, and boulders crash against rock; and by the plucking and undercutting actions of currents.

How do stream valleys and their channels and floodplains evolve? As a stream flows, it carves a valley with steep to gently sloping walls and a more or less broad floodplain on either side of its channel. The channel may be straight, meandering, or braided. During normal, nonflood periods, a stream's channel carries all the water and sediment within its banks. When stream discharge increases to flood stage, the sediment-laden water overflows the banks of the channel and inundates the floodplain. The velocity of the floodwater decreases as it spreads over the floodplain. The water drops sediment, which builds up natural levees and floodplain deposits. Recurrence intervals relate the probability that a flood of a given discharge or height will occur in any year to the interval of time between floods of that discharge or height.

How does a stream's longitudinal profile represent the equilibrium between erosion and sedimentation? A stream is in dynamic equilibrium between erosion and sedimentation over its entire length. Topography, discharge, velocity, and slope affect this equilibrium. A stream's longitudinal profile, always concave upward, is a cross section of the stream's elevation from its headwaters to the base level at its mouth in a lake or the ocean. Uplift at the upper end of a stream and the rise and fall of sea level at its lower end will change the profile. Alluvial fans form at mountain fronts, primarily as a result of an abrupt widening of the valley and secondarily as a result of a change in slope.

How do drainage networks work as collection systems and deltas as distribution systems for water and sediment? Rivers and their tributaries constitute an upstream-branching drainage network that collects the water and sediment running off a specific drainage basin. Each drainage basin is separated from its neighbors by a divide. Drainage networks show various branching patterns—dendritic, rectangular, trellis, or radial—depending on topography, rock type, and geologic structure. Near the mouth of a river, where it forms its delta, the river tends to branch downstream into distributaries. Rivers drop their sediment load at deltas, in topset, foreset, and bottomset beds. Deltas are modified or even absent where waves, tides, and shoreline currents are strong. Tectonics controls delta formation by uplift in the drainage basin and subsidence in the delta region.

Key Terms and Concepts

alluvial fan (p. 317)
antecedent stream (p. 320)
base level (p. 315)
bed load (p. 305)
bottomset bed (p. 321)
braided stream (p. 311)
capacity (p. 305)
channel (p. 308)
competence (p. 305)
delta (p. 321)
dendritic drainage (p. 319)
discharge (p. 312)
distributary (p. 321)
divide (p. 318)
drainage basin (p. 318)
drainage network (p. 319)
dune (p. 306)
floodplain (p. 308)
foreset bed (p. 321)
graded stream (p. 316)
laminar flow (p. 304)
longitudinal profile (p. 315)
meander (p. 309)
natural levee (p. 311)
oxbow lake (p. 311)
point bar (p. 309)
pothole (p. 307)
recurrence interval (p. 314)
ripple (p. 306)
river (p. 303)
saltation (p. 306)
settling velocity (p. 306)
stream (p. 303)
superposed stream (p. 320)
suspended load (p. 305)
terrace (p. 317)
topset bed (p. 321)
tributary (p. 314)
turbulent flow (p. 304)
valley (p. 308)
viscosity (p. 304)

Exercises

This icon indicates that there is an animation available on the Web site that may assist you in answering a question.

1. How does velocity determine whether a given flow is laminar or turbulent?

2. How does the size of a sediment grain affect the speed with which it settles to the bottom of a flow?

3. What kind of bedding characterizes a ripple or a dune?

4. How do braided and meandering river channels differ?

5. Why is a floodplain so named?

6. What is a natural levee, and how is it formed?

7. What is the discharge of a stream, and how does it vary with velocity?

8. How is a river's longitudinal profile defined?

9. What is the most common kind of drainage network developed over horizontally bedded sedimentary rocks?

10. What is a delta distributary?

Thought Questions

This icon indicates that there is an animation available on the Web site that may assist you in answering a question.

1. Why might the flow of a very small, shallow stream be laminar in winter and turbulent in summer?

2. Describe and compare the river floodplain and valley above and below the waterfall shown in Figure 14.6.

3. You live in a town on a meander bend of a major river. An engineer proposes that your town invest in new and higher artificial levees to prevent the meander from being cut off. Give the arguments, pro and con, for this investment.

4. In some places, engineers have artificially straightened a meandering stream. If such a straightened stream were then left free to adjust its course naturally, what changes would you expect?

5. If global warming produces a significant rise in sea level as polar ice melts, how will the longitudinal profiles of the world's rivers be affected?

6. In the first few years after a dam was built on it, a stream severely eroded its channel downstream of the dam. Could this erosion have been predicted?

7. Your hometown, built on a river floodplain, experienced a 50-year flood last year. What are the chances that another flood of that magnitude will occur next year?

8. Define the drainage basin where you live in terms of the divides and drainage networks.

9. What kind of drainage network do you think is being established on Mount St. Helens since its violent eruption in 1980?

10. The Delaware Water Gap is a steep, narrow valley cut through a structurally deformed high ridge in the Appalachian Mountains. How could it have formed?

11. A major river, which carries a heavy sediment load, has no delta where it enters the ocean. What conditions might be responsible for the lack of a delta?

Short-Term Team Projects

See Chapters 12 and 13.

Suggested Readings

Chorley, R. J., S. A. Schumm, and D. E. Sugden. 1984. *Geomorphology.* New York: Methuen.

Dunne, T., and L. B. Leopold. 1978. *Water in Environmental Planning.* San Francisco: W. H. Freeman.

Leopold, L. B. 1994. *A View of the River.* Cambridge, Mass.: Harvard University Press.

McPhee, J. 1989. *The Control of Nature.* New York: Farrar, Straus & Giroux.

Ritter, D. F. 1986. *Process Geomorphology,* 2d ed. Dubuque, Iowa: Wm. C. Brown.

Schumm, S. A. 1977. *The Fluvial System.* New York: Wiley Interscience.

Large sand dunes advancing over the floor of the Namib Desert, Namibia. Such dunes are formed by the wind in desert climates where sand is abundant. [Courtesy of Roger Swart.]

CHAPTER

15

Winds and Deserts

"This is my letter to the World
That never wrote to Me—
The simple News that Nature told—
With tender Majesty"

EMILY DICKINSON

At one time or another, we've all been caught in a wind strong enough to have blown us over, had we not leaned into it or held onto something solid. London, England, which rarely gets strong winds, experienced a major windstorm on January 25, 1990. Winds blowing at more than 175 km/hour ripped roofs off buildings, blew trucks over, and made it virtually impossible to walk on the streets. Some areas of the world experience hurricanes, which bring torrential rains and high winds that erode the land and drive ocean waves that erode and transport huge tonnages of sediment on beaches and in shallow waters. Many winds are strong enough to blow sand grains into the air, creating sandstorms.

In this chapter, we focus on this potent force that shapes the surface of the land, particularly in deserts, where strong winds can howl for days on end.

The wind is a major agent of erosion and deposition, moving enormous quantities of sand, silt, and dust over large regions of the continents and oceans. Wind is much like water in its ability to erode, transport, and deposit sediment. This is not surprising, because the general laws of fluid motion that govern liquids also govern gases. As we will see, however, there are differences that make the wind less powerful than water currents. Unlike a stream, whose discharge depends on rainfall, wind works most effectively in the absence of rain. We will consider Earth's deserts in particular detail because so many of the geological processes there are related to the work of the wind. The ancient Greeks called the god of winds Aeolus, and geologists today use the term **eolian** for the geological processes powered by the wind.

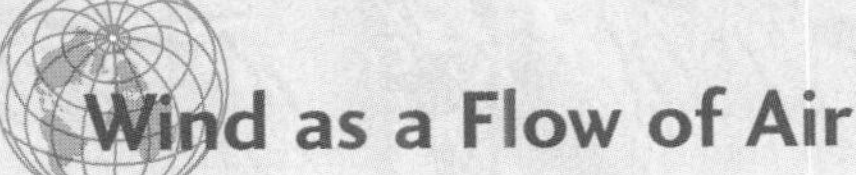

Wind as a Flow of Air

Wind is a natural flow of air that is parallel to the surface of the rotating planet. As with water flows, streamlines can be used to describe air flows. Although winds obey all the laws of fluid flow that apply to water in streams (see Chapter 14), there are some differences. Unlike the flow of water in river channels, winds are generally unconfined by solid boundaries, except for the ground surface and narrow valleys. Air flows are free to spread out in all directions, including upward into the atmosphere.

Turbulence

Like the water flowing in rivers, air flows are nearly always turbulent. As we saw in Chapter 14, turbulence depends on three characteristics of a fluid: density, viscosity, and velocity. The extremely low density and viscosity of air make it turbulent even at the velocity of a light breeze.

Wind Belts

Winds vary in speed and direction from day to day, but over the long term, they tend to come mainly from one direction. In the temperate latitudes, which include most of North America, the prevailing winds come from the west and so are referred to as the *westerlies* (**Figure 15.1**). Temperate climates are located at latitudes between 30° and 60° north and 30° and 60° south. In the tropics, which are between 30° south and 30° north of the equator, the *trade winds* (named for an archaic use of the word *trade* meaning "track" or "course") blow from the east, which is where most of the world's deserts are.

These wind belts arise because the Sun warms a given amount of land surface most intensely at the equator, where the Sun's rays are almost perpendicular to the surface. The Sun heats the Earth less quickly at high latitudes and at the poles because there the rays strike the surface at an angle (Figure 15.1). Hot air, which is less dense than cold air, rises at the equator and flows toward the poles, gradually sinking as it cools. The cold, dense air at the poles then flows back along the surface of the globe toward the equator.

The simple circulatory pattern of air flow between the equator and the poles is complicated by Earth's rotation, which deflects any current of air or water to the right in the Northern Hemisphere and to the left in the Southern Hemisphere. This effect on Earth's air flow is called the *Coriolis effect,* named after its discoverer.

The Coriolis effect on atmospheric circulation deflects warm and cold air flows moving both northward and southward in both hemispheres (see Figure 15.1). For example, as surface winds in the Northern Hemisphere blow southward into the hot equatorial belt, they are deflected to the right (westward) and hence blow from the northeast rather than from the north. These are the northeast trade winds. The Northern Hemisphere westerlies are flows that initially moved

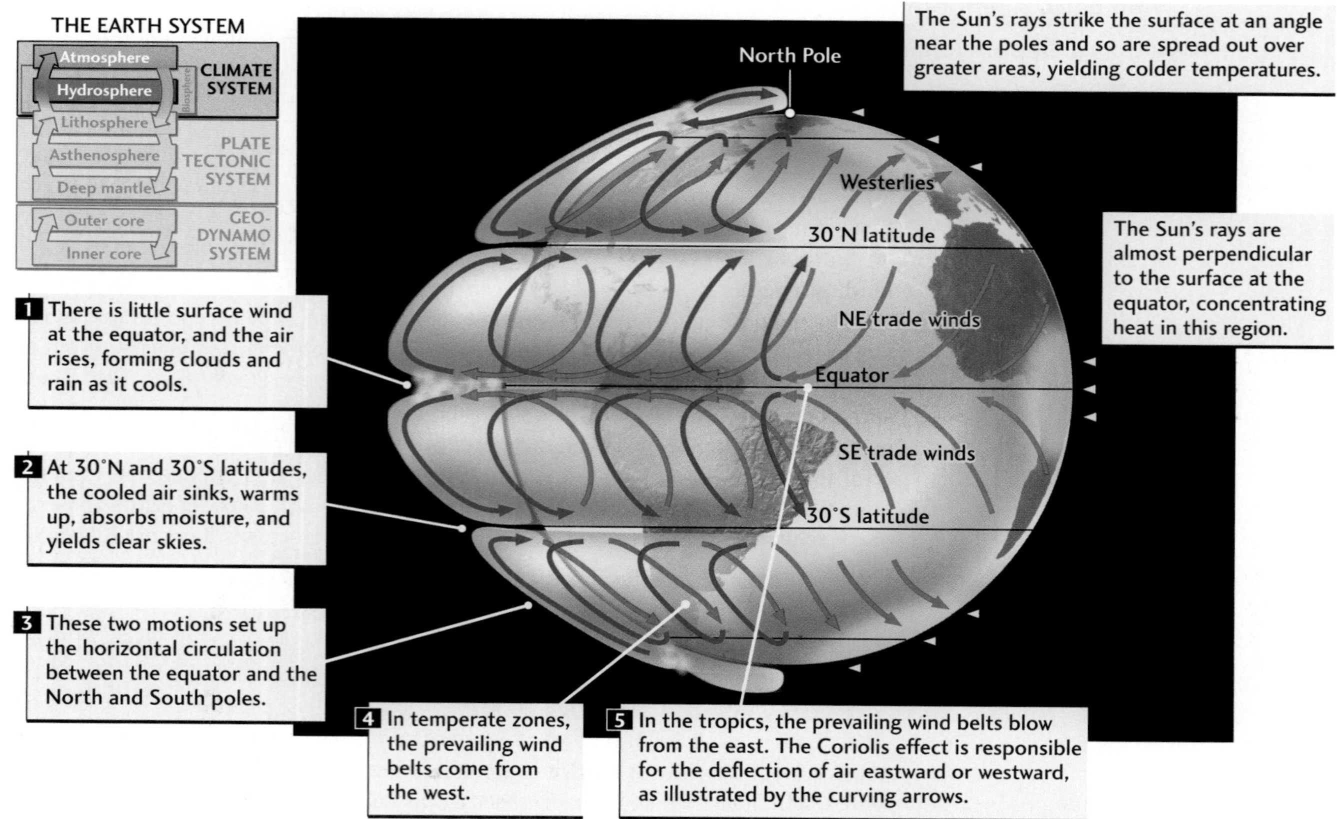

Figure 15.1 The circulation of Earth's atmosphere.

northward but were deflected to the right (eastward) and thus blow from the southwest. Near the equator, air movement is mostly upward, and so there is little wind at the surface. As the air rises, it cools, causing the cloudiness and abundant rain of the tropics. At about 30°N and 30°S latitude, some of the high-altitude poleward flow starts to sink to Earth's surface. The cold, dry air warms and absorbs moisture as it sinks, producing clear skies and arid climates. This dry, sinking air overlies many of the world's deserts, such as the Sahara.

Wind as a Transport Agent

Most of us are familiar with rainstorms or snowstorms—high winds accompanied by heavy precipitation. We may have less experience of dry storms, during which high winds blowing for days on end carry enormous tonnages of sand, silt, and dust. The amount of material the wind can carry depends on the strength of the wind, the sizes of the particles, and the surface materials of the area over which the wind blows.

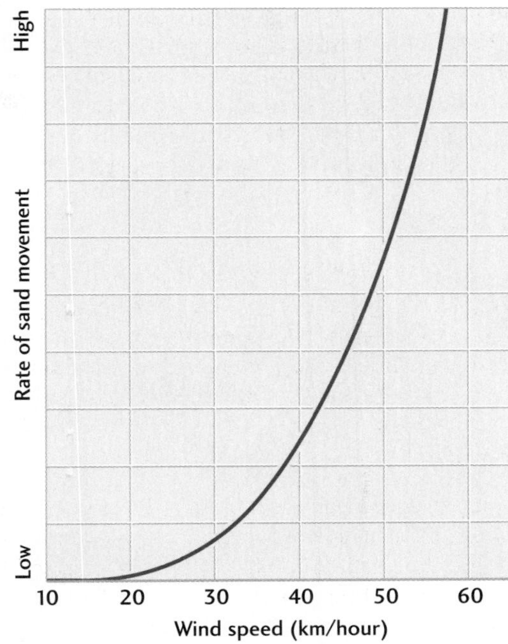

Figure 15.2 The amount of sand moved daily across each meter of width of a dune's surface in relation to wind speed. High-speed winds blowing for several days can move enormous quantities of sand. [After R. A. Bagnold, *The Physics of Blown Sand and Desert Dunes* (London: Methuen, 1941).]

Wind Strength

Figure 15.2 shows how much sand winds of various speeds can erode from a 1-m-wide strip across a sand dune's surface. A strong wind of 48 km/hour can move half a ton of sand (about equivalent in volume to two large suitcases) from this small surface area in a single day. At higher wind speeds, the amounts of sand that can be moved increase rapidly. No wonder entire houses can be buried by a sandstorm lasting several days!

Particle Size

The wind exerts the same kind of force on particles on the land surface as a river current exerts on its bed. Turbulence and forward motion combine to lift particles into the wind and carry them along, at least temporarily. Even the lightest breezes carry dust, the finest-grained material. Dust usually consists of particles less than 0.01 mm in diameter but often includes somewhat larger silt particles. Moderate winds can carry dust to heights of many kilometers, but only strong winds can carry particles larger than 0.06 mm in diameter, such as sand grains. Moderate breezes can roll and slide these grains along a sandy bed, but it takes a stronger wind to lift sand grains into the air flow. A wind usually cannot transport the largest particles, however, because air has low viscosity and density. Even though winds can be very strong, only rarely can they move large pebbles and cobbles in the way that rapidly flowing rivers do.

Surface Material

The wind can transport sand and dust only from surface materials such as dry soil, sediment, or bedrock. The wind cannot erode and transport wet soils because they are too cohesive. Wind can carry sand grains weathered from a loosely cemented sandstone, but it cannot erode grains from a granite or basalt. Most windblown sands are locally derived. Sand grains are typically buried in dunes after traveling a relatively short distance (usually no more than a few hundred kilometers), mainly by saltation near the ground. The extensive sand dunes of such major deserts as the Sahara and the wastes of Saudi Arabia are exceptions. In those great sandy regions, sand grains may have traveled more than 1000 km.

Materials Carried by the Wind

Windblown Dust Air has a staggering capacity to hold dust. Dust includes microscopic rock and mineral fragments of all kinds, especially silicates, as might be expected from their abundance as rock-forming minerals. Two of the most important sources of silicate minerals in dust are clays from soils on dry plains and volcanic dust from eruptions. Organic materials, such as pollen and bacteria, are also common components of dust. Charcoal dust is abundant downwind of forest fires. When found in buried sediments, it is evidence of forest fires in earlier geologic times. Since the beginning of the industrial revolution, we have been pumping new kinds of synthetic dust into the air—from ash produced by burning coal to the many solid chemical compounds produced by manufacturing processes, incineration of wastes, and motor vehicle exhausts.

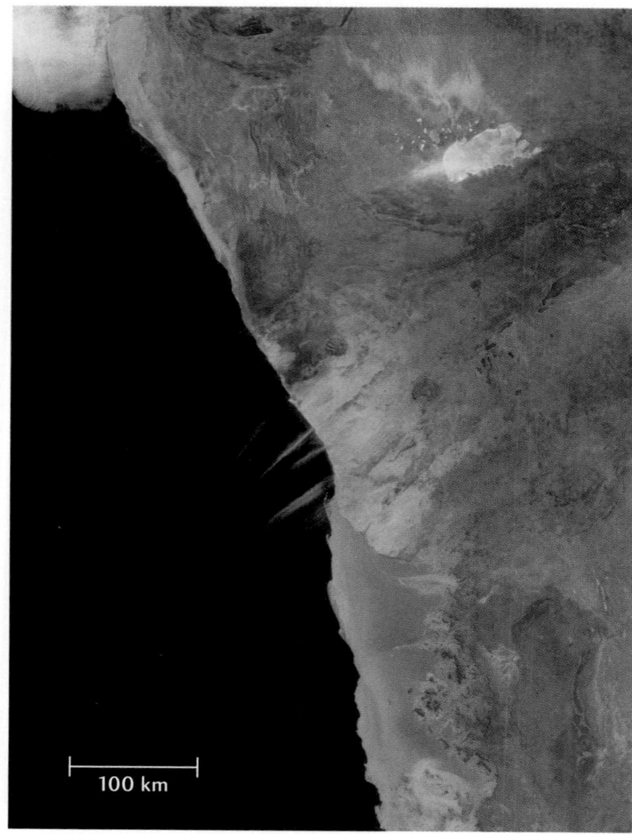

Figure 15.3 Satellite photograph of a dust storm originating in the Namib Desert in September 2002. Dust and sand are being transported from right (east) to left (west) in response to strong winds blowing out to sea. The sediment can be transported for hundreds to thousands of kilometers across the ocean. [NASA.]

In large dust storms, 1 km^3 of air may carry as much as 1000 tons of dust, equivalent to the volume of a small house. When such storms cover hundreds of square kilometers, they may carry more than 100 million tons of dust and deposit it in layers several meters thick. (See Feature 15.1 for a discussion of similar dust storms on Mars.) Fine-grained particles from the Sahara have been found as far away as England and have been traced across the Atlantic Ocean to Barbados in the Caribbean. Wind annually transports about 260 million tons of material, mostly dust, from the Sahara to the Atlantic Ocean. Scientists on oceanographic research vessels have measured airborne dust far out to sea, and today it can be observed directly from space (**Figure 15.3**). Comparison of the composition of this dust with that of deep-sea sediments in the same region indicates that windblown dust is an important contributor to oceanic sediment, supplying up to a billion tons each year. A large part of this dust comes from volcanoes, and there are individual ash beds on the seafloor marking very large eruptions.

Volcanic dust is abundant because much of it is very fine grained and it is erupted high into the atmosphere, where it can travel farther than nonvolcanic dust blown by winds closer to the surface. Volcanic explosions inject huge quantities of dust into the atmosphere. The volcanic dust from the 1991 eruption of Mount Pinatubo in the Philippine Islands circled the Earth, and most of the finest-grained particles did not settle until 1994 or 1995. By partly reflecting the Sun's rays, dust from the Pinatubo eruption cooled Earth's surface by a small amount for a few years.

Mineral dust in the atmosphere increases when agriculture, deforestation, erosion, and changing land use disrupt soils. A large amount of mineral dust may come from the Sahel region on the southern border of the Sahara, where drought and overgrazing are responsible for a heavy load of dust.

Windblown dust has complex effects on climate. Mineral dust scatters the incoming visible radiation from the Sun and absorbs the outgoing infrared (heat) radiation from the Earth. Thus, mineral dust has a net cooling effect in the visible part of the spectrum and a net warming effect in the infrared.

Scientists use the term *aerosol* for atmospheric particles of natural or synthetic origin that are small enough to be suspended in the lower levels of the atmosphere for days to weeks. Aerosols affect climate in the same ways as mineral dust does.

Windblown Sand The sand the wind transports may consist of almost any kind of mineral grain produced by weathering. Quartz grains are by far the most common because quartz is such an abundant constituent of many surface rocks, especially sandstones. Many windblown quartz grains have a frosted or matte (roughened and dull) surface (**Figure 15.4**) like the inside of a frosted light bulb. Some of the grain

Figure 15.4 Photomicrograph of frosted and rounded grains of quartz from sand dunes in Saudi Arabia. [Walter N. Mack.]

15.1 The *Pathfinder* Expedition and Martian Dust Storms

Of all the planets in the solar system, Mars is the most like Earth. Although Mars has a thinner atmosphere, it has weather that changes seasonally and an Earthlike day (called a sol) of 24 hours and 37 minutes. Mars also has a complex surface environment including ice, soil, and sediment. Meandering channels provide evidence that water once flowed on the Martian surface. Today, Mars is so cold and dry that its surface environment is considered too hostile to support life. In past times, however, the surface of Mars was probably warmer and wetter, and it may have given rise to life in ancient lakes or springs. If so, we hope to find fossil evidence of ancient Martian life on or beneath the surface.

Mars Pathfinder was the first spacecraft to set down on Mars since the two *Viking* landers in 1976. *Pathfinder* landed in a region known as Ares Vallis (Mars Valley) on July 4, 1997 (Sol 1). The mission tested important new technologies to be employed during the 2004 *Athena Rover* mission. On Sol 2, *Pathfinder*'s rover, *Sojourner*—named for Sojourner Truth, an African-American reformist of the Civil War era—rolled down a ramp and onto the Martian soil.

Over the next few sols, *Sojourner* visited several igneous rocks, affectionately named Barnacle Bill, Yogi, and Scooby Doo. In its travels, *Sojourner* drove over eolian sand and dust, which formed small dunes and strips developed in the lee of some of the bigger rocks. In the accompanying photograph, you can see gray patches of coarser sand amid the fine red sand. The coarse gray sand was concentrated as a lag deposit, a residual accumulation of coarse rock fragments on a surface after the wind has blown away finer material. Planetary geologists estimate that the winds that produced these patches blew at velocities of up to 30 m/s (108 km/hour) during great dust storms that covered the entire planet. Although these winds are not as fast as Earth's strongest winds, and Mars's atmosphere is less dense than Earth's, Martian winds are still strong enough to form an array of eolian depositional and erosional features identical to those observed on Earth. Even the pink color of the Martian atmosphere owes its origin to suspended quantities of windblown dust entrained during dust storms.

Martian dust storms directly affect our ability to study the Martian surface. The *Sojourner* and *Athena* rovers depend on solar power to move and conduct their exploratory activities. Ultimately, the duration of these activities is limited to about 90 sols—the time it takes for enough windblown dust to settle out on the rovers' solar panels to terminate power generation.

Full-color panorama of Ares Vallis (Mars Valley), taken by the *Mars Pathfinder* lander. *Pathfinder*'s rover, *Sojourner*, was photographed at many different positions to create this composite image. The ridges of sand that span the width of the image are the result of occasional high winds that transport dust and sand across the surface of Mars. A coating of windblown dust is also visible on many of the rocks. On the left-hand side of the image, *Sojourner* has just driven onto a patch of coarse, gray windblown sand. The rover is about 30 cm tall. [NASA.]

frosting is produced by wind-driven impacts, but most of it results from slow, long-continued dissolution by dew. Even the tiny amounts of dew found in arid climates are enough to etch microscopic pits and hollows into sand grains, creating the frosted appearance. Frosting is found only in eolian environments, so it is good evidence that a sand grain has been blown by the wind.

Windblown calcium carbonate grains accumulate where there are abundant fragments of shells and coral, such as in Bermuda and on many coral islands in the Pacific Ocean. The White Sands National Monument in New Mexico is a prominent example of sand dunes made of gypsum sand grains eroded from evaporite bedrock.

Wind as an Agent of Erosion

By itself, the wind can do little to erode large masses of solid rock exposed at Earth's surface. Only when rock is fragmented by chemical and physical weathering can the wind pick up particles. In addition, the particles must be dry, because wet soils and damp fragmented rock are held together by moisture. Thus, wind erodes most effectively in arid climates, where winds are strong and dry and any moisture quickly evaporates.

Sandblasting

Windblown sand is an effective natural **sandblasting** agent. The common method of cleaning buildings and monuments with compressed air and sand works on exactly the same principle: the impact of high-speed particles wears away the solid surfaces. Natural sandblasting mainly works close to the ground, where most sand grains are carried. Sandblasting rounds and erodes rock outcrops, boulders, and pebbles and frosts the occasional glass bottle.

Ventifacts are wind-faceted pebbles having several curved or almost flat surfaces that meet at sharp ridges (**Figure 15.5**). Each surface or facet is made by sandblasting of the pebble's windward side. Occasional storms roll or rotate the pebbles, exposing a new windward side to be sandblasted. Many ventifacts are found in deserts and in glacial gravel deposits, where the necessary combination of gravel, sand, and strong winds is present.

Deflation

As particles of dust, silt, and sand become loose and dry, blowing winds can lift and carry them away, gradually eroding the ground surface in a process called **deflation** (**Figure 15.6**). Deflation, which can scoop out shallow depressions or hollows, occurs on dry plains and deserts and on temporarily dried-up river floodplains and lake beds. Firmly established vegetation, even the sparse vegetation of arid and semiarid regions, retards it. Deflation occurs slowly in areas with plants because their roots bind soil and their stems and leaves disrupt the air currents and shelter the ground surface. Deflation works fast where the vegetation cover is broken, either naturally by killing drought or artificially by cultivation, construction, or motor vehicle tracks.

When deflation removes the finer-grained particles from a mixture of gravel, sand, and silt in sediments and soils, it produces a remnant surface of gravel too large for the wind to transport. Over thousands of years, as deflation removes the finer-grained particles from successive stream deposits, the gravel accumulates as a layer of **desert pavement**—a coarse, gravelly ground surface that protects the soil or sediments below from further erosion (**Figure 15.7**).

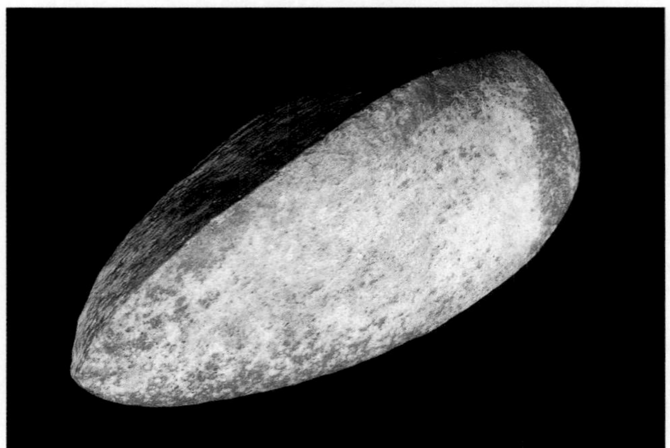

Figure 15.5 A ventifact. This wind-faceted pebble from Antarctica has been shaped by windblown sand in a frigid environment. [E. R. Degginger.]

Figure 15.6 A shallow deflation hollow in the San Luis Valley, Colorado. Wind has scoured the surface and eroded it to a slightly lower elevation. Deflation occurs in dry areas where the vegetation cover is absent or broken. [Breck P. Kent.]

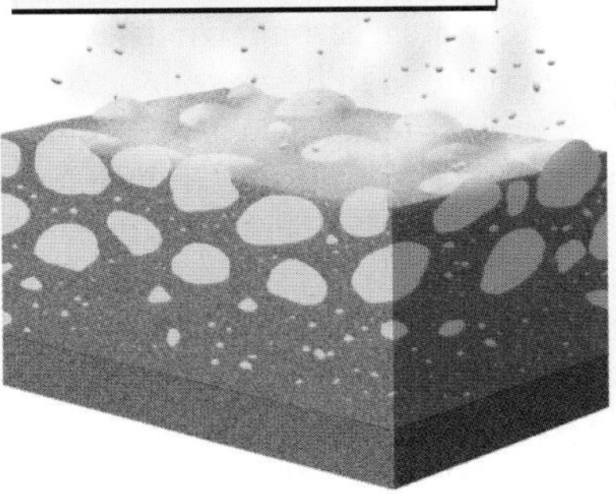

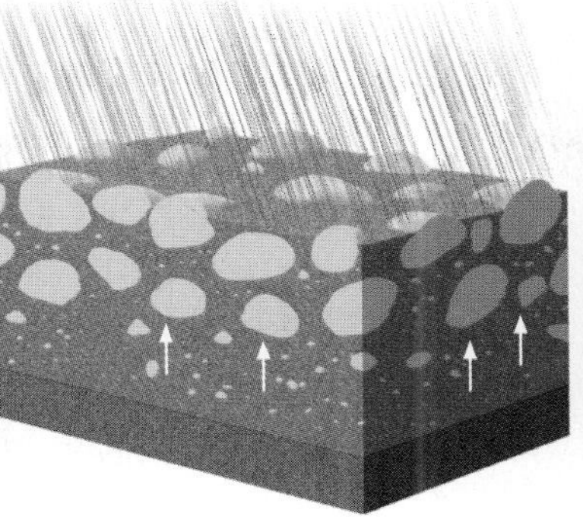

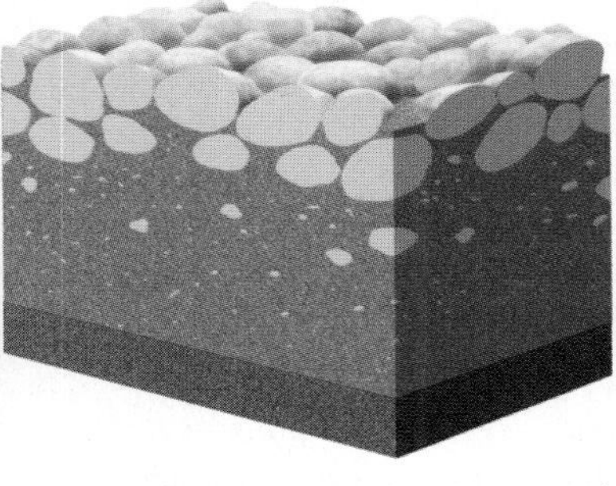

Figure 15.7 Desert pavement in the Kofa Mountains of Arizona, in the Sonoran Desert. [William E. Ferguson.] According to a new theory, the evolution of desert pavement begins when the wind blows fine-grained materials into heterogeneous soil or sediment. During rainstorms, the fine, windblown sediments infiltrate beneath the coarse layer of pebbles. Microbes living beneath the pebbles produce bubbles that help raise the pebbles and maintain their position at the surface. Over time, these processes lead to thickening of the dust accumulating beneath the pebble layer.

This theory of pavement formation is not completely accepted, because a number of pavements seem not to have formed in this way. A new theory is that some of them are formed by the deposition of windblown sediments. The coarse rock pavement stays at the surface, while windblown dust infiltrates below the surface layer of pavement, is modified by soil-forming processes, and accumulates there.

Wind as a Depositional Agent

When the wind dies down, it can no longer transport the sand, silt, and dust it carries. The coarser material is deposited in sand dunes of various shapes, ranging in size from low knolls to huge hills more than 100 m high. The finer silt and dust fall as a more or less uniform blanket of silt and clay. By observing these depositional processes working today, geologists have been able to link them with such characteristics as bedding and texture to deduce past climates and wind patterns from ancient sandstones and dust falls.

Where Sand Dunes Form

Sand dunes occur in relatively few environmental settings. Many of us have seen the dunes that form behind ocean beaches or along large lakes. Some dunes are found on the sandy floodplains of large rivers in semiarid and arid regions. Most spectacular are the fields of dunes that cover large expanses of some deserts (**Figure 15.8**). Such dunes may reach heights of several hundred meters, truly mountains of sand.

Dunes form only in settings that have a ready supply of loose sand: beach sands along coasts, sandy river bars or floodplain deposits in river valleys, and sandy bedrock formations in deserts. Another common factor in dune

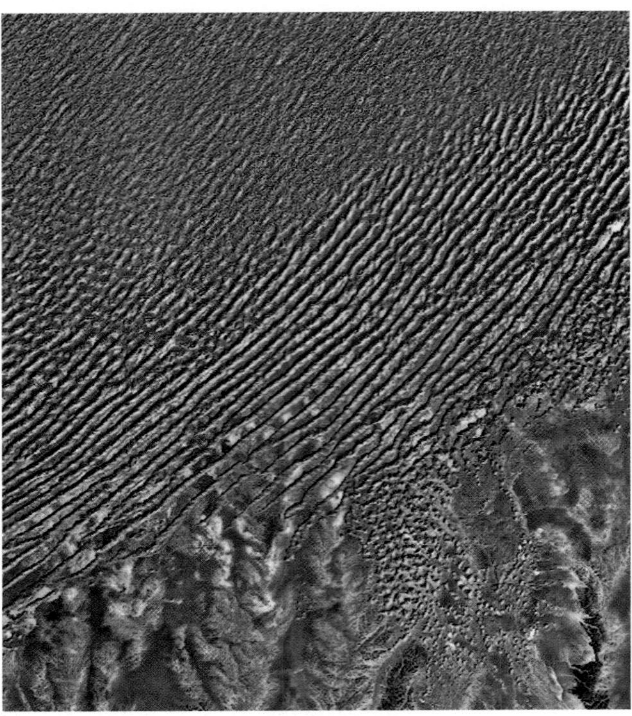

Figure 15.8 A vast expanse of sand dunes on the southern Arabian Peninsula. These linear dunes trend northeast-southwest, parallel with the prevailing northeasterly winds. At the lower right is an eroding plateau of horizontally bedded sedimentary rock. [Satellite image data processing by (ZERIM), Ann Arbor, Michigan.]

formation is wind power. On oceans and lakes, strong winds blow onshore off the water. Strong winds, sometimes of long duration, are common in deserts.

The wind cannot pick up wet materials easily, so most dunes are found in dry climates. The exception is dune belts along a coast, where sand is so abundant and dries so quickly in the wind that dunes can form even in humid climates. In such climates, soil and vegetation begin to cover the dunes only a little way inland from the beaches, and the winds no longer pick up the sand.

Dunes may stabilize and become vegetated when the climate grows more humid and then start moving again when an arid climate returns. There is geological evidence that during droughts two to three centuries ago and earlier, sand dunes in the western high plains of the United States and Canada were reactivated and migrated over the plains.

How Sand Dunes Form and Move

The wind moves sand by sliding and rolling it along the surface and by causing saltation, the jumping motion that temporarily suspends grains in a current of water or air. Saltation in air flows works the same way as it does in a river (see Figure 14.2), except that the jumps in air flows are higher and longer. Sand grains suspended in an air current often rise to heights of 50 cm over a sand bed and 2 m over a pebbly surface—much higher than grains of the same size can jump in water. The difference arises partly from the fact that air is less viscous than water and therefore does not inhibit

Figure 15.9 Wind ripples in sand at Stovepipe Wells, Death Valley, California. Although complex in form, these ripples are always transverse (at 90°) to the wind direction. [Tom Bean.]

the bouncing of the grains as much as water does. In addition, the impact of grains falling in air induces higher jumps as they hit the surface. These collisions, which the air hardly cushions, kick surface grains into the air in a sort of splashing effect. As saltating grains impact a sand bed, they can push forward grains too large to be thrown up into the air, causing the bed to creep in the direction of the wind. A sand grain striking the surface at high speed can propel another grain as far as six times its own diameter.

Almost inevitably, when the wind moves sand along a bed, it produces ripples and dunes much like those formed by water (**Figure 15.9**). Ripples in sand, like those under water, are transverse; that is, at right angles to the current. At low to moderate wind speeds, small ripples form. As the wind speed increases, the ripples become larger. Ripples migrate in the direction of the wind over the backs of larger dunes. Some wind is almost always blowing, and so a sand bed is almost always rippled to some extent.

Given enough sand and wind, any obstacle—such as a large rock or a clump of vegetation—can start a dune. Streamlines of wind, like those of water, separate around obstacles and rejoin downwind, creating a wind shadow downstream of the obstacle. Wind velocity is much lower in the wind shadow than in the main flow around the obstacle. In fact, it is low enough to allow sand grains blown into the shadow to settle there. The wind is moving so slowly that it can no longer pick up these grains, and they accumulate as a *sand drift,* a small pile of sand in the lee of the obstacle (**Figure 15.10**). As the process continues, the sand drift itself becomes an obstacle. If there is enough sand and the wind continues to blow in the same direction long enough, the drift grows into a dune. Dunes may also grow by the enlargement of ripples, just as underwater dunes do.

As a dune grows, the whole mound starts to migrate downwind by the combined movements of a host of individual grains. Sand grains constantly saltate to the top of the low-angle windward slope and then fall over into the wind shadow on the lee slope, as shown in **Figure Story 15.11**. These grains gradually build up a steep, unstable accumulation on the upper part of the lee slope. Periodically, the steepened buildup gives way and spontaneously slips or cascades down this **slip face,** as it is called, to a new slope at a lower angle. If we overlook the short-term, unstable steepenings of the slope, the slip face maintains a stable, constant slope angle—its angle of repose. As we saw in Chapter 12, this angle increases with the size and angularity of the particles.

Successive slip faces deposited at the angle of repose create the cross-bedding that is the hallmark of windblown dunes. As dunes accumulate, interfere with one another, and become buried in a sedimentary sequence, the cross-bedding is preserved even though the original shapes of the dunes are lost. Sets of sandstone cross-bedding many meters thick are evidence of high, windblown dunes. From the directions of these eolian cross-beds, geologists can reconstruct wind directions of the past (see Figure 8.6)

As more sand accumulates on the windward slope of a dune than blows off onto the slip face, the dune grows in height. Heights of 30 m are common, and huge dunes in

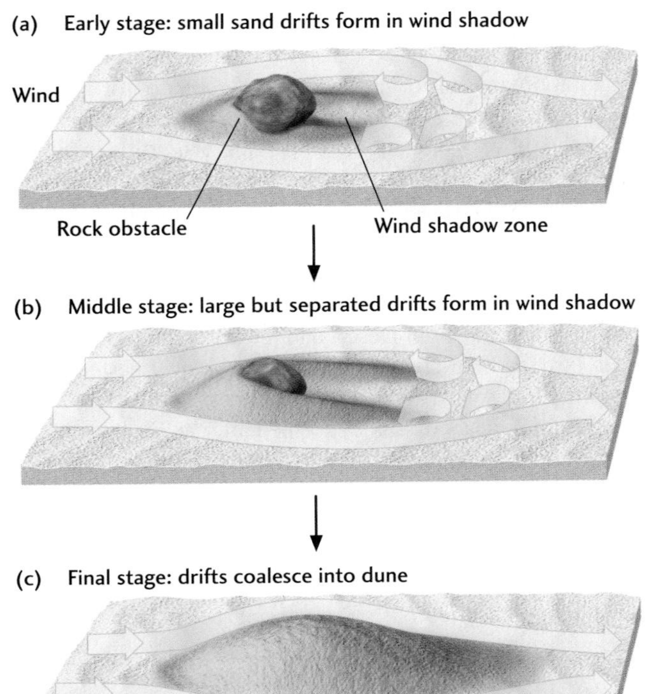

Figure 15.10 Sand dunes may form in the lee of a rock or other obstacle. By separating the wind streamlines, the rock creates a wind shadow in which the eddies are weaker than the main flow. The windborne sand grains are thus able to settle and pile up in drifts that eventually coalesce into a dune. [After R. A. Bagnold, *The Physics of Blown Sand and Desert Dunes* (London: Methuen, 1941).] Sand dunes, Owens Lake, California. [Martin Miller.]

SAND-DUNE FORMATION DEPENDS ON WIND VELOCITY AND AMOUNT OF SAND

1 A ripple or dune advances by the movements of individual grains of sand. The whole form moves forward slowly as sand erodes from the windward slope and is deposited on the leeward slope.

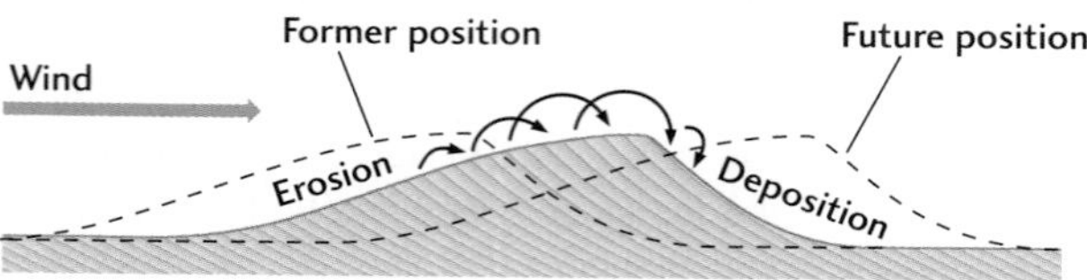

2 Particles of sand arriving on the windward slope of the dune move by saltation over the crest,...

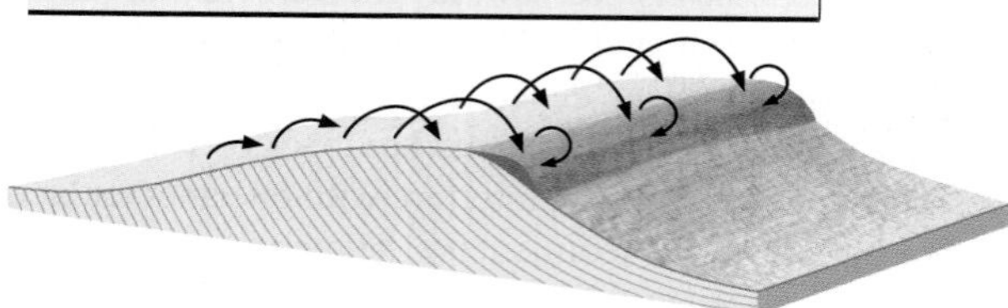

3 ...where the wind velocity decreases and the sand deposited slips down the leeward slope.

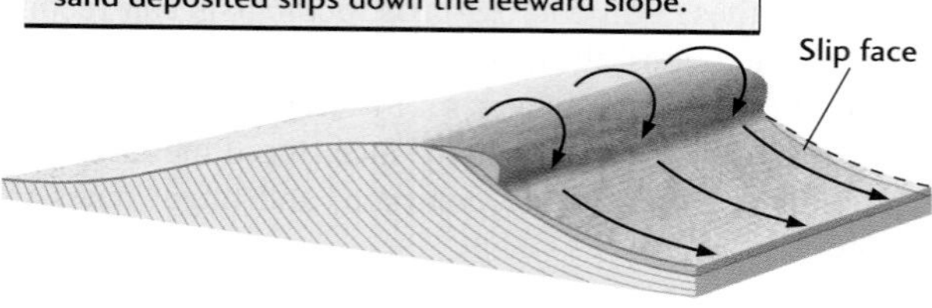

4 This process acts like a conveyor belt that moves the dune forward.

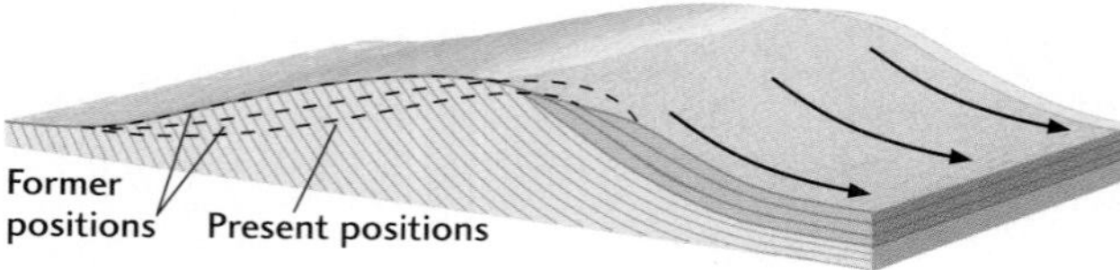

5 As the dune grows higher, the wind streamlines become compressed and their velocity increases. The dune stops growing vertically when it reaches a height at which the wind is so fast that it blows the sand grains off the dune as quickly as they are brought up.

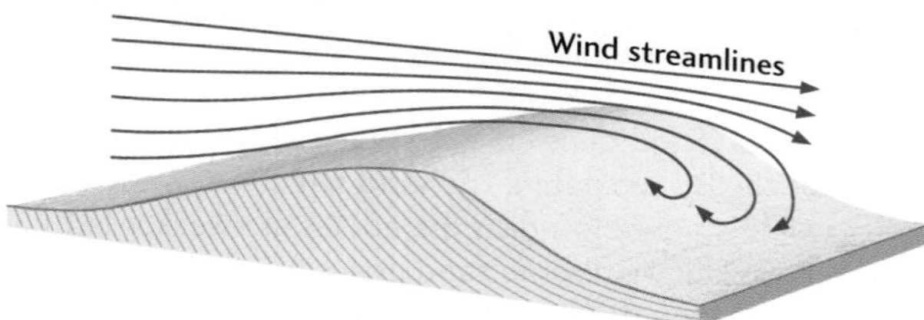

Figure Story 15.11 A ripple or dune advances by the movements of individual grains.

Saudi Arabia reach 250 m, which seems to be the limit. The explanation for the limited dune height lies in the relationship among wind streamline behavior, velocity, and topography. Wind streamlines advancing over the back of a dune become more compressed as the dune grows higher (see Figure Story 15.11). As more air rushes through a smaller space, the wind velocity increases. Ultimately, the air speed at the top of the dune becomes so great that sand grains blow off the top of the dune as quickly as they are brought up the windward slope. When this balance is reached, the height of the dune remains constant.

Dune Types

A person standing in the middle of a large expanse of dunes might be bewildered by the seemingly orderless array of undulating slopes. It takes a practiced eye to see the dominant pattern, and it may even require observation from the air. The general shapes and arrangements of sand dunes depend on the amount of sand available and the direction, duration, and strength of the wind. We cannot yet be certain of the specific mechanisms by which a particular wind regime results in one dune form or another. Geologists recognize four main types of dunes: barchans, transverse dunes, blowout dunes, and linear dunes (**Figure 15.12**).

Dust Falls and Loess

As the velocity of dust-laden wind decreases, the dust settles to form **loess,** a blanket of sediment composed of fine-grained particles. Beds of loess lack internal stratification. In compacted deposits more than a meter thick, loess tends to form vertical cracks and to break off along sheer walls during erosion (**Figure 15.13**). Geologists theorize that the vertical cracking may be caused by a combination of root penetration and uniform downward percolation of groundwater, but the exact mechanisms are still unknown.

Loess covers as much as 10 percent of Earth's land surface. The major loess deposits are found in China and North America. China has more than a million square kilometers of loess deposits (**Figure 15.14**). The great loess deposits of China extend over wide areas in the northwest; most are 30 to 100 m thick, although some exceed 300 m. The winds blowing over the Gobi Desert and the arid regions of central Asia provided the dust, which still blows over eastern Asia and the Chinese interior. Some of the loess deposits in China are 2 million years old. They formed after an increase in the elevation of the Himalaya and related mountain belts of western China introduced rain shadows and dry climates to the continental interior. The tectonic uplift of these mountain belts was responsible for the cold, dry climates of the Pleistocene epoch in much of Asia. These climates inhibited vegetation and dried out soils, causing extensive wind erosion and transportation.

The best-known loess deposit in North America is in the Upper Mississippi Valley. It originated as silt and clay deposited on extensive floodplains of rivers draining the

1 **Barchans** are crescent-shaped dunes, usually but not always found in groups. The horns of the crescent point downwind. Barchans are the products of limited sand supply and unidirectional winds.

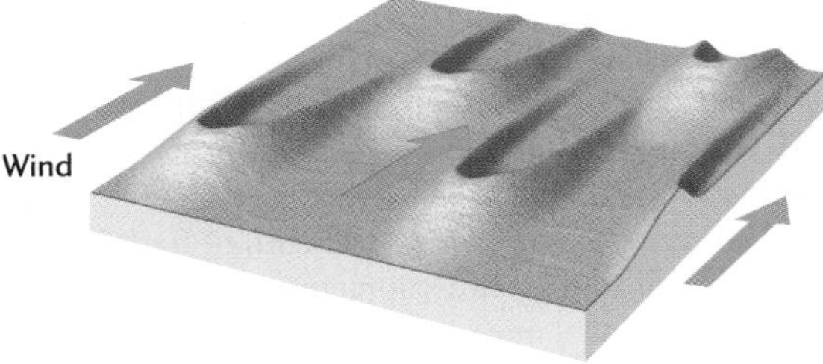

2 **Transverse dunes** are long ridges oriented at right angles to the wind direction. These dunes form in arid regions where there is abundant sand and vegetation is absent. Typically, sand-dune belts behind beaches are transverse dunes formed by strong onshore winds.

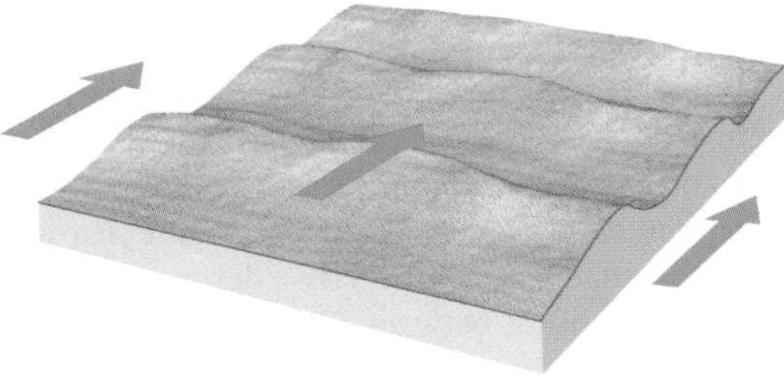

3 **Blowout dunes** are almost the reverse of barchans. The slip face of a blowout dune is convex downwind, whereas the barchan's is concave downwind.

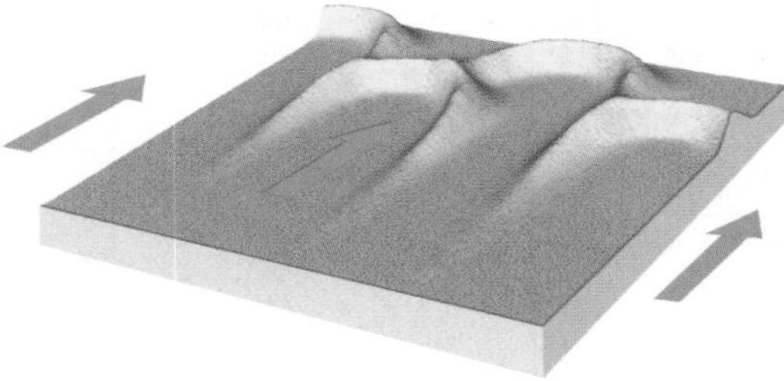

4 **Linear dunes** are long ridges of sand whose orientation is parallel to the wind direction. These dunes may reach heights of 100 m and extend many kilometers. Most areas covered by linear dunes have a moderate sand supply, a rough pavement, and winds that are always in the same general direction.

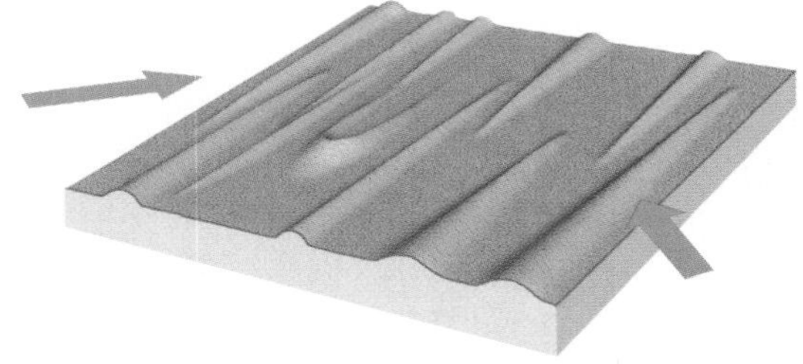

Figure 15.12 Dune types in relation to prevailing winds.

Figure 15.13 Pleistocene loess, Catalina Mountains, Arizona, showing vertical cracking. [E. R. Degginger.]

Figure 15.14 Comfortable dwelling caves hand-carved into steep cliffs of loess in central China. These deposits of windblown dust accumulated in the past 2.5 million years, reaching a thickness of as much as 400 m. [Stephen C. Porter.]

edges of melting glaciers of the Pleistocene epoch. Strong winds dried the floodplains, whose frigid climate and rapid rates of sedimentation inhibited vegetation, and blew up tremendous amounts of dust, which then settled to the east. Geologists recognized that this loess deposit is distributed as a blanket of more or less uniform thickness on both hills and valleys, all in or near formerly glaciated areas. Changes in the regional thickness of the loess in relation to the prevailing westerly winds confirm its eolian origin. The loess on the eastern sides of major river floodplains is 8 to 30 m thick, greater than on the western sides, and the thickness decreases downwind rapidly to 1 to 2 m farther east of the floodplains.

Soils formed on loess are fertile and highly productive. They also pose environmental problems, because they are easily eroded into gullies by small streams and are deflated by the wind when they are poorly cultivated.

The Desert Environment

The hot, dry deserts of the world are among the most hostile environments for humans. Yet many of us are fascinated by the strange forms of animal and plant life and the bare rocks and sand dunes found there. Of all Earth's environments, the wind is best able to do its work of erosion and sedimentation in the desert.

All told, arid regions amount to one-fifth of Earth's land area, about 27.5 million square kilometers. Semiarid plains account for an additional one-seventh. Given the reasons for the existence of large areas of deserts in the modern world—mountain building by plate tectonics, the transport of continental regions to low latitudes by continental drift, and Earth's climatic belts—we can be confident that, by the principle of uniformitarianism, extensive deserts have existed throughout geologic time.

Where Deserts Are Found

Rainfall is the major factor determining the location of the world's great deserts. The Sahara and Kalahari deserts of Africa and the Great Australian Desert get extremely low amounts of rainfall, normally less than 25 mm/year and in some places less than 5 mm/year. These deserts are found in Earth's warmest regions, between 30°N and 30°S latitudes (**Figure 15.15**). The deserts lie under virtually stationary areas of high atmospheric pressure. The Sun beats down through a cloudless sky week after week, and the relative humidity is extremely low.

Deserts also exist in the midlatitudes—between 30°N and 50°N and between 30°S and 50°S—in regions where rainfall is low because moisture-laden winds either are blocked by mountain ranges or must travel great distances from the ocean, their source of moisture. The Great Basin and Mojave deserts of the western United States, for example, lie in rain shadows created by the western coastal mountains. As we saw in Chapter 13, wind descending from the mountains warms and dries, leading to low precipitation (see Figure 13.2). The Gobi and other deserts of central Asia are so far inland that the winds reaching them have precipitated all their ocean-derived moisture long before they arrive at the interior of the continent.

Another kind of desert is found in polar regions. There is little precipitation in these cold, dry areas because the frigid air can hold only extremely small amounts of moisture. The dry valley region of southern Victoria Land in Antarctica is so dry and cold that its environment resembles that of Mars.

The Role of Plate Tectonics In a sense, deserts are a result of plate tectonics. The mountains that create rain shadows are made by collisions between converging continental and oceanic plates. The great distance separating central Asia from the oceans is a consequence of the size of the continent,

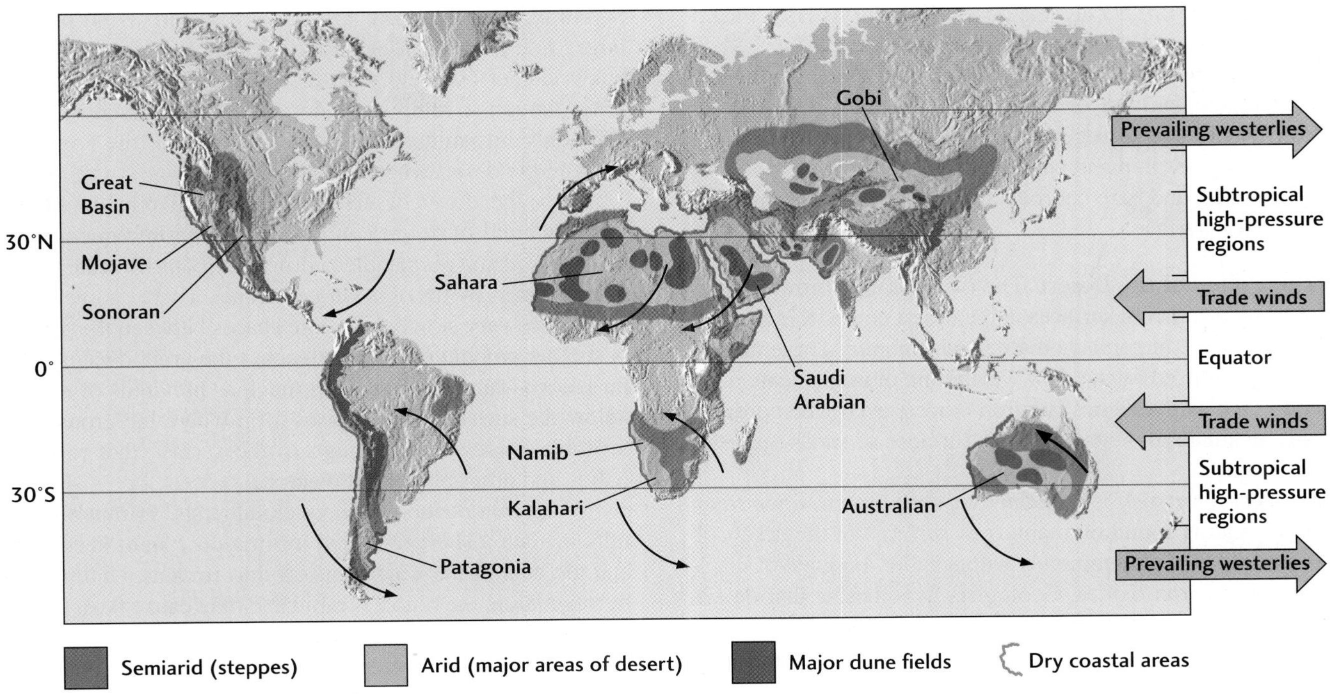

Figure 15.15 Major desert areas of the world (exclusive of polar deserts) in relation to prevailing wind directions and major mountain and plateau areas. Sand dunes make up only a small proportion of the total desert area. [After K. W. Glennie, *Desert Sedimentary Environments* (New York: Elsevier, 1970).]

a huge landmass assembled from smaller plates by continental drift. Large deserts are found at low latitudes because continental drift powered by plate tectonics moved them there from higher latitudes. If, in some future plate tectonic scenario, the North American continent were to move south by 2000 km or so, the northern Great Plains of the United States and Canada would become a hot, dry desert. Something like that happened to Australia. About 20 million years ago, Australia was far to the south of its present position, and its interior had a warm, humid climate. Since then, Australia has moved northward into an arid subtropical zone, where its interior has become a desert.

The Role of Climate Change Changes in a region's climate may transform semiarid lands into deserts, a process called **desertification.** Climatic changes that we do not fully understand may decrease precipitation for decades or even centuries. After a dry period, the region may return to a milder, wetter climate. Over the past 10,000 years, climates of the Sahara appear to have oscillated between drier and wetter conditions. We have evidence from radar imaging of the Sahara from the space shuttle *Endeavor* that an extensive system of river channels existed a few thousand years ago. Now dry and buried by more recent sand deposits, these ancient drainage systems carried abundant running water across the northern Sahara during wetter periods.

The Role of Humans These oscillating climates occurred naturally, but human activities are responsible for some desertification today. The unrestrained growth of human populations and their agriculture and increased animal grazing may result in the expansion of deserts. When population growth and periods of drought coincide, the results in semiarid regions can be disastrous.

"Making the desert bloom," the opposite of desertification, has been a slogan of some countries with desert lands. They irrigate on a massive scale to convert marginal or desert areas into productive farmlands. The Great Valley of California, where much of North America's fruits and vegetables are grown, is one example. The growth of cities such as Phoenix, Arizona, which depend on imported water, has produced islands of urbanization—complete with humid haze and air pollution—in an otherwise arid environment.

Desert Weathering

As unique as deserts are, the same geological processes operate there as elsewhere. Weathering and transportation work in the same way as they do everywhere, but with a different balance in deserts. In the desert, physical weathering predominates over chemical weathering. Chemical weathering of feldspars and other silicates proceeds slowly because the water required for the reaction to proceed is lacking. The little clay that does form is usually blown away by strong winds before it can accumulate. Slow chemical weathering and rapid wind transport combine to prevent the buildup of any significant thickness of soil, even where sparse vegetation binds some of the particles. Thus, soils are thin and patchy. Sand, gravel, rock rubble of many sizes, and bare bedrock are characteristic of much of the desert surface.

The Colors of the Desert The rusty, orange-brown colors of many weathered surfaces in the desert come from the ferric iron oxide minerals hematite and limonite. These minerals are produced by the slow weathering of iron silicate minerals such as pyroxene. The iron oxides, even when present only in small amounts, stain the surfaces of sands, gravels, and clays.

Desert varnish is a distinctive, dark brown, sometimes shiny coating found on many rock surfaces in the desert. It is a mixture of clay minerals with smaller amounts of manganese and iron oxides. Geologists hypothesize that desert varnish forms very slowly from a combination of dew, the chemical weathering that produces clay minerals and iron and manganese oxides, and the adhesion of windblown dust to exposed rock surfaces. The process is so slow that Native American inscriptions scratched in desert varnish hundreds of years ago still appear fresh, with a stark contrast between the dark varnish and the light unweathered rock beneath (**Figure 15.16**). Varnish requires thousands of years to form, and some particularly ancient varnishes in North America are of Miocene age. However, recognizing varnish as such in ancient sandstones is difficult.

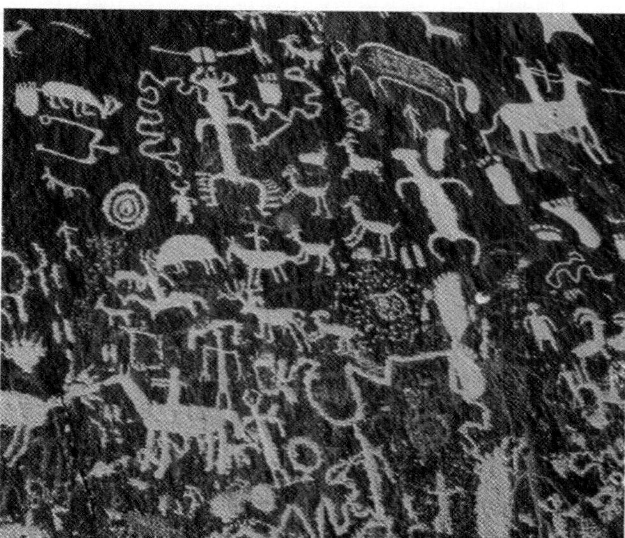

Figure 15.16 Petroglyphs scratched in desert varnish by early Native Americans; Newspaper Rock, Canyonlands, Utah. The scratches are several hundred years old and appear fresh on the varnish, which accumulated over thousands of years. [Peter Kresan.]

Streams: The Primary Agents of Erosion Wind plays a larger role in erosion in the desert than it does elsewhere, but it cannot compete with the erosive power of streams. Even though it rains so seldom that most desert streams flow only intermittently, streams do most of the erosional work in the desert when they do flow.

Even the driest desert gets occasional rain. In sandy, gravelly areas of deserts, infiltration of the infrequent rainfall into soil and permeable bedrock temporarily replenishes groundwater in the unsaturated zone. There, some of it evaporates very slowly into pore spaces between the grains. A smaller amount eventually reaches the groundwater table far below—in some places, as much as hundreds of meters below the surface. Desert oases form where the groundwater table comes close enough to the surface that roots of palms and other plants can reach it.

When rain occurs in heavy cloudbursts, so much water falls in such a short time that infiltration cannot keep pace and the bulk of the water runs off into streams. Unhindered by vegetation, the runoff is rapid and may cause flash floods along valley floors that have been dry for years. Thus, a large proportion of streamflows in the desert consist of floods. When floods occur, they have great erosive power because most loose weathering debris is not held in place by vegetation. Streams may become so choked with sediment that they look more like fast-moving mudflows than like rivers. The abrasiveness of this sediment load moving rapidly at flood velocities makes such streams efficient eroders of bedrock valleys.

Desert Sediment and Sedimentation

Alluvial Sediments As sediment-laden flash floods dry up, they leave distinctive deposits on the floors of desert valleys. A flat fill of coarse debris often covers the entire valley floor, and the ordinary differentiation of a stream into channel, levees, and floodplains is absent (**Figure 15.17**). The sediments of many other desert valleys clearly show the intermixing of stream-deposited channel and floodplain sediments with eolian sediments. The combination of stream and eolian processes occurring in the past formed extensive sheets of eolian sandstone separated by stream flood surfaces and floodplain sandstones deposited between the sheets of eolian sandstone. Large alluvial fans (see Chapter 14) are prominent features at mountain fronts in deserts because desert streams deposit much of their high sediment load on the fans. The rapid infiltration of stream water into the permeable fan material deprives the streams of the water required to carry the sediment load any farther downstream. Debris flows and mudflows make up large parts of the alluvial fans of arid, mountainous regions.

Eolian Sediments By far the most dramatic sedimentary accumulations in the desert are sand dunes. Dune fields range in size from a few square kilometers to "seas of sand" found in major deserts such as those of Namibia. These

(a)

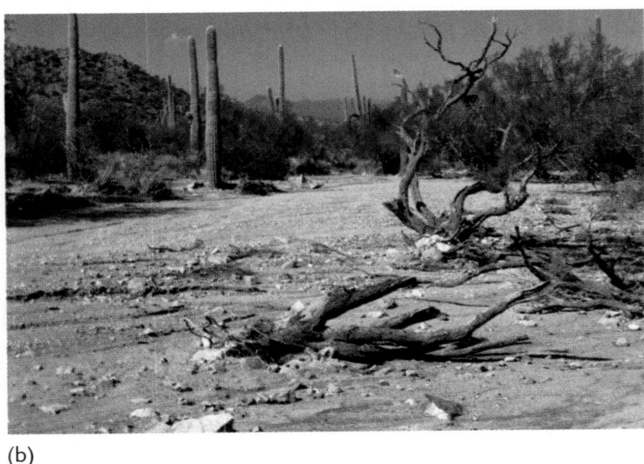

(b)

Figure 15.17 (a) Dry-wash flooding during a summer thunderstorm in Saguaro National Monument, Arizona. (b) The same dry wash a day after flooding. The coarse debris deposited by sudden desert floods may cover the entire valley floor. [Peter Kresan.]

sand seas—or *ergs*—may cover as much as 500,000 km^2, twice the size of the state of Nevada. Although film and television portrayals lead one to think that deserts are mostly sand, actually only one-fifth of the world's desert area is covered by sand. The other four-fifths are rocky or covered with desert pavement. Sand covers only a little more than one-tenth of the Sahara Desert, and sand dunes are far less common in the deserts of the southwestern United States.

Evaporite Sediments **Playa lakes** are permanent or temporary lakes that occur in arid mountain valleys or basins (**Figure 15.18**). As the lake water evaporates, dissolved weathering products are concentrated and gradually precipitated. Playa lakes are sources of evaporite minerals: sodium carbonate, borax (sodium borate), and other unusual salts. The water in playa lakes may be deadly to drink. Desert streams contribute large amounts of dissolved salts, and these salts accumulate in playa lakes when the streams redissolve evaporite minerals deposited by evaporation from earlier runoff. If evaporation is complete, the lakes become **playas,** flat beds of clay that are sometimes encrusted with precipitated salts.

Figure 15.18 A desert playa lake in Death Valley, California. Playa lakes are sources of evaporite minerals such as borax. Their waters may be deadly to drink because of high quantities of dissolved minerals. [David Muench.]

Desert Landscape

Desert landscapes are some of the most varied on Earth. Large, low, flat areas are covered by playas, desert pavements, and dune fields. Uplands are rocky, cut in many places by steep river valleys and gorges. The lack of vegetation and soil makes everything seem sharper and harsher than it would seem in landscapes of more humid climates. (We will discuss landscapes in more detail in Chapter 18.) In contrast with the rounded, soil-covered, vegetated slopes found in most humid regions, the coarse fragments of varying size produced by desert weathering form steep cliffs with masses of angular talus slopes at their bases (**Figure 15.19**).

Valleys in deserts have the same range of profiles as valleys elsewhere, but far more of them have steep walls produced by the rapid erosion caused by mass movements and

Figure 15.19 This desert landscape at Kofa Butte, Kofa National Wildlife Refuge, Arizona, shows the steep cliffs and masses of talus slopes produced by desert weathering. [Peter Kresan.]

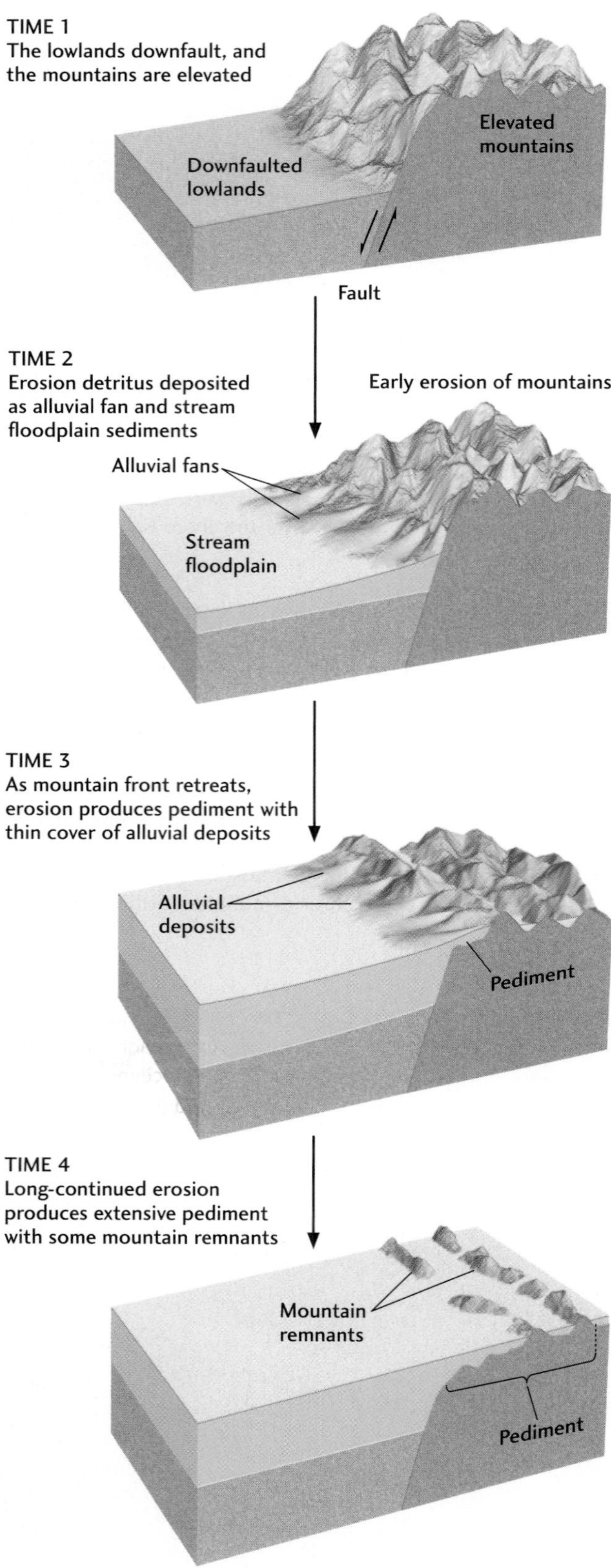

Figure 15.20 Stages in the evolution of a typical pediment, an erosional form produced in arid mountainous settings.

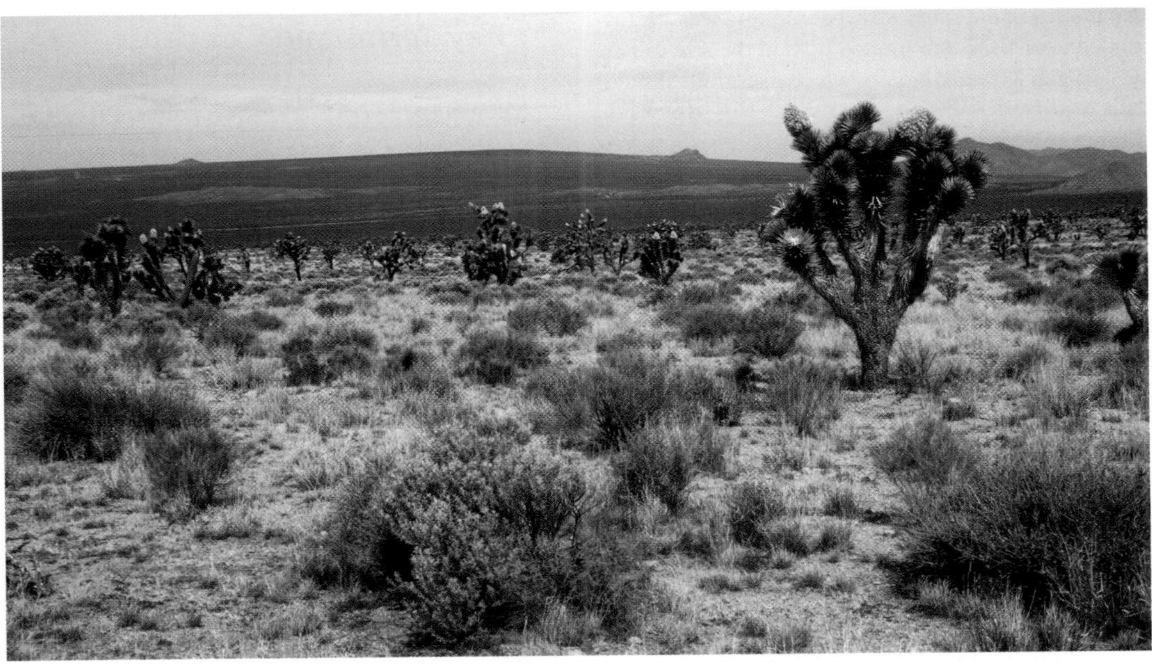

Cima Dome is a pediment in the Mojave Desert. The surface of the dome is covered by a thin veneer of alluvial sediments. The two knobs, on the left and right sides of the dome, are regarded as the final remnants of a former mountain. [Martin Miller.]

streams. Much of the landscape of deserts is shaped by rivers, but the valleys—called **dry washes** in the western United States and **wadis** in the Middle East—are usually dry.

Desert streams are widely spaced because of the relatively infrequent rainfall, but drainage patterns are generally similar to those of other terrains—with one difference. Many desert streams die out before they can reach across the desert to join larger rivers flowing to the oceans. This may happen because of infrequent rainfall and the loss of water by evaporation and infiltration or because dunes dam the streams. Damming by dunes may lead to the development of playa lakes.

A special type of eroded bedrock surface, called a **pediment,** is a characteristic landform of the desert. Pediments are broad, gently sloping platforms of bedrock left behind as a mountain front erodes and retreats from its valley (**Figure 15.20**). The pediment spreads like an apron around the base of the mountains as thin deposits of alluvial sands and gravels accumulate. Long-continued erosion eventually forms an extensive pediment below a few mountain remnants (see Figure 15.20). A cross section of a typical pediment and its mountains would reveal a fairly steep mountain slope abruptly leveling into the gentle pediment slope. Alluvial fans deposited at the lower edge of the pediment merge with the sedimentary fill of the valley below the pediment.

There is much evidence that a pediment is formed by running water that both cuts the erosional platform above and deposits an alluvial fan apron below. At the same time, the mountain slopes at the head of the pediment maintain their steepness as they retreat, instead of becoming the rounded, gentler slopes found in humid regions. We do not know how the specific rock types and erosional processes interact in an arid environment to keep slopes steep while the pediment enlarges.

SUMMARY

Where do winds form and how do they flow? Earth is encircled by belts of wind that develop in response to warming of the atmosphere at the equator, which causes the air to rise and flow toward the poles. As the air moves toward the poles, it gradually cools and begins to sink. The cold, dense air at the poles then flows back along Earth's surface to the equator. Winds vary in speed and direction from day to day, but over the long term, they tend to come mainly from one direction. The Coriolis effect, produced by Earth's rotation, deflects these moving winds to the right in the Northern Hemisphere and to the left in the Southern Hemisphere.

How do winds erode and transport sand and finer-grained sediments? Wind can pick up and transport dry sediment particles in the same way that running water transports sediment. Air flows are limited both in the size of particles they can carry (rarely larger than coarse-grained sand) and in their restricted ability to keep particles in suspension for long times. These limitations result from air's low viscosity and density. Almost all air flows are turbulent, and high winds may reach velocities of 100 km/hour or more, enhancing the wind's ability to carry sediment. Windblown materials include volcanic ash, quartz and other mineral fragments such as clay minerals, and organic materials such as pollen and bacteria. The wind can carry great amounts of sand and dust. It moves sand grains primarily by saltation and carries the finer-grained silt and clay-sized particles (dust) in suspension. Winds blow sand into cross-bedded ripples and dunes. Deflation and sandblasting are the primary ways in which winds erode Earth's surface, thereby producing desert pavement and ventifacts.

How do winds deposit sand dunes and dust? When winds die down, they deposit sand in dunes of various shapes and sizes. Dunes form in sandy desert regions, behind beaches, and along sandy floodplains, all of which are places with a ready supply of loose sand and moderate to strong winds. Dunes start as sand drifts in the lee of obstacles and grow to heights of many meters or even hundreds of meters. Dunes migrate downwind as sand grains saltate up the gentler windward slopes and fall over onto the steeper downwind slip faces. The various kinds of dunes—transverse, linear, barchan, and blowout—are determined by the speed of the wind, its constancy or variability of direction, and the abundance of sand. As the velocity of dust-laden winds decreases, the dust settles to form loess, a blanket of dust. Loess layers in many recently glaciated areas were deposited by winds blowing over the floodplains of muddy streams formed by meltwater. Loess can accumulate to great thicknesses downwind of dusty desert regions.

How do wind and water combine to shape the desert environment and its landscape? Deserts occur in the rain shadows of mountain ranges, in subtropical regions of constant high pressure, and in the interiors of some continents. In all these places, originally moisture-laden winds become dry, and rainfall is rare. Weathering mechanisms are the same in deserts as in more humid regions, but in deserts physical breakdown of rocks is predominant, with chemical weathering at a minimum because of the lack of water. Most desert soils are thin, and bare rock surfaces are common. Streams may run only intermittently, but they are responsible for much of the erosion and sedimentation in the desert, carrying away heavy loads of coarse sediment and depositing them on alluvial fans and floodplains. Naturally dammed rivers in mountainous deserts can form playa lakes, which deposit evaporite minerals as they dry up. Desert landscapes comprise dunes formed by eolian sedimentation; desert pavements; and pediments, which are broad, gently sloping platforms eroded from bedrock as mountains retreat while maintaining the steepness of their slopes.

Key Terms and Concepts

barchan (p. 337)
blowout dune (p. 337)
deflation (p. 332)
desertification (p. 339)
desert pavement (p. 332)
desert varnish (p. 340)
dry wash (p. 343)
eolian (p. 327)
linear dune (p. 337)
loess (p. 336)
pediment (p. 343)
playa (p. 341)
playa lake (p. 341)
sandblasting (p. 332)
slip face (p. 335)
transverse dune (p. 337)
ventifact (p. 332)
wadi (p. 343)

Exercises

MEDIA LINK

This icon indicates that there is an animation available on the Web site that may assist you in answering a question.

1. What types of materials and sizes of particles can the wind move?

2. What is the difference between the way wind transports dust and the way it transports sand?

3. How is the wind's ability to transport sedimentary particles linked to climate?

4. What are the main features of wind erosion?

5. Where do sand dunes form?

6. Name three types of sand dunes and show the relationship of each to wind direction.

7. What typical desert landforms are composed of sediment?

8. What are the geologic processes that form playa lakes?

9. What is desertification?

10. Where are loess deposits found?

Thought Questions

1. You have just driven a truck through a sandstorm and discover that the paint has been stripped from the lower parts of the truck but the upper parts are barely scratched. What process is responsible, and why is it restricted to the lower parts of the truck?

2. What evidence might you find in an ancient sandstone that would point to its eolian origin?

3. Compare the heights to which sand and dust are carried in the atmosphere and explain the differences or similarities.

4. Trucks continually have to haul away sand covering a coastal highway. What do you think might be the source of the sand? Could its encroachment be stopped?

5. What features of a desert landscape would lead you to believe it was formed mainly by streams, with secondary contributions from eolian processes?

6. Which of the following options would be a more reliable indication of the direction of the wind that formed a barchan: cross-bedding or the orientation of the dune's shape on a map? Why?

7. What factors determine whether sand dunes will form on a stream floodplain?

8. There are large areas of sand dunes on Mars. From this fact alone, what can you infer about conditions on the Martian surface?

9. What aspects of an ancient sandstone would you study to show that it was originally a desert sand dune?

10. What kinds of landscape features would you ascribe to the work of the wind, to the work of streams, or to both?

11. How does desert weathering differ from or resemble weathering in more humid climates?

12. What evidence would cause you to infer that dust storms and strong winds were common in glacial times?

Short-Term Team Project

See Chapter 12.

Suggested Readings

Bagnold, R. A. 1941. *The Physics of Blown Sand and Desert Dunes.* London: Methuen.

Brookfield, M. E., and S. Thomas (eds.). 1983. *Eolian Sediments and Processes.* New York: Elsevier.

Cooke, R. U., A. Warren, and A. Goudie. 1993. *Desert Geomorphology.* London: UCL Press.

Haff, P. K. 1986. Booming dunes. *American Scientist* 74: 376–381.

Idso, S. B. 1976. Dust storms. *Scientific American* (October): 108.

Mabbutt, J. A. 1977. *Desert Landforms.* Cambridge, Mass.: MIT Press.

Pye, K. 1989. *Aeolian Dust and Dust Deposits.* New York: Academic Press.

Sheridan, D. 1981. *Desertification of the United States.* Washington, D.C.: Council on Environmental Quality.

Several glaciers meet in the St. Elias Mountains, Luane National Park, Yukon, Canada. [Stephen J. Krasemann/DRK Photo.]

CHAPTER

16

Glaciers: The Work of Ice

"Great is the Earth, and the way it became what it is. Do you imagine it is stopped at this?"

WALT WHITMAN

The view from space of Earth's vast blue oceans and white clouds dramatically emphasizes the water covering our planet. But satellite pictures of polar regions and white-peaked mountain ranges also show us that much of Earth's water is frozen (**Figure 16.1**).

About 10 percent of Earth's land surface is covered by glacial ice, much of which moves slowly and steadily outward from the centers of the polar ice caps and downward from the mountain peaks. Over the short term, the ice melts at the edges of glaciers at about the same rate as it advances, and so the total ice area remains the same.

Over a longer span of time, global cooling can expand the ice sheets, as melting fails to keep pace with the accumulation and movement of glacial ice. As recently as 20,000 years ago, snow and ice covered almost three times more land surface than they do now. Only a few decades from now, global warming may contract the ice sheets, as the ice melts faster than it can accumulate and move. The worldwide effects could be immense. Melting ice would raise sea level and submerge low-lying cities. Climatic zones would migrate, changing temperate zones into semiarid ones and vice versa. There is no doubt that a knowledge of Earth's icy regions is an intensely practical subject.

In addition to being indicators of climate change, glaciers also strongly influence the shape of landscapes (see Chapter 18). Many of the landscapes of mountain belts were sculpted by glaciers that have since melted away. Glaciers erode steep-walled valleys, scrape bedrock surfaces, and pluck huge blocks from their rocky floors. In the relatively short geological time span of the recent ice ages, glaciers carved far more topography than did rivers and the wind. Glacial erosion creates enormous amounts of debris. Ice transports huge tonnages of sediment, carrying it to the edge of a glacier, where it is deposited or carried away by meltwater streams. All over Earth, glacial erosion and sedimentation affect:

- The water discharge and sediment loads of major river systems
- The quantity of sediment delivered to the oceans
- Erosion and sedimentation in coastal areas and on shallow continental shelves

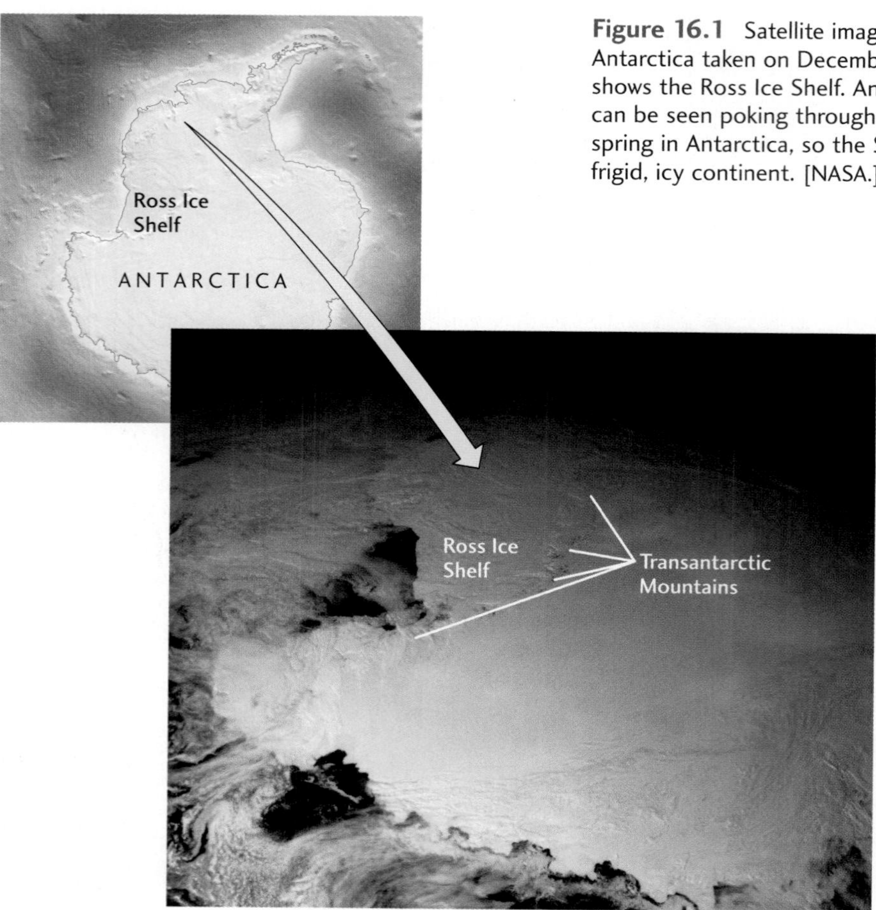

Figure 16.1 Satellite image of part of Antarctica taken on December 8, 1990. The view shows the Ross Ice Shelf. An occasional mountain can be seen poking through the ice. It is late spring in Antarctica, so the Sun never sets on the frigid, icy continent. [NASA.]

In this chapter, we take a close look at Earth's glaciers, the ways in which they change, and how Earth systems and human activities affect and are affected by these changes. We also examine the effects of glaciers as they carry and deposit loads of sediment, leaving their mark on Earth's surface as they advance and retreat.

Ice as a Rock

What is ice? To a geologist, a block of ice is a rock, a mass of crystalline grains of the mineral ice. Like most rocks, ice is hard, but it is much less dense than most rocks. It has in common with igneous rocks an origin as a frozen fluid. Like sediments, it is deposited in layers at the surface of the Earth and can accumulate to great thicknesses. Like metamorphic rocks, it is transformed by recrystallization under pressure. Glacial ice forms by the burial and metamorphism of the "sediment" snow. Loosely packed snowflakes—each a single crystal of the mineral ice—age and recrystallize into a solid mass (**Figure 16.2**). The unusual characteristic of ice as a mineral is its extremely low melting temperature—hundreds of degrees lower than the temperatures at which most minerals melt.

What Is a Glacier?

A mass of ice, like any other mass of rock or soil at Earth's surface, can move downhill. **Glaciers** are large masses of ice on land that show evidence of being in motion or of once having moved. We divide glaciers, on the basis of size and shape, into two basic types: valley glaciers and continental glaciers.

Valley Glaciers Many skiers and mountain climbers are familiar with **valley glaciers,** sometimes called *alpine glaciers* (**Figure 16.3**). These rivers of ice form in the cold heights of mountain ranges where snow accumulates, usually in preexisting valleys, and they flow down the bedrock valleys. A valley glacier usually occupies the complete width of the valley and may bury its bedrock base under hundreds of meters of ice. In warmer, low-latitude climates, valley glaciers may be found only at the heads of valleys on the highest mountain peaks. An example is the glacial ice that covers the Mountains of the Moon in east-central Africa. In

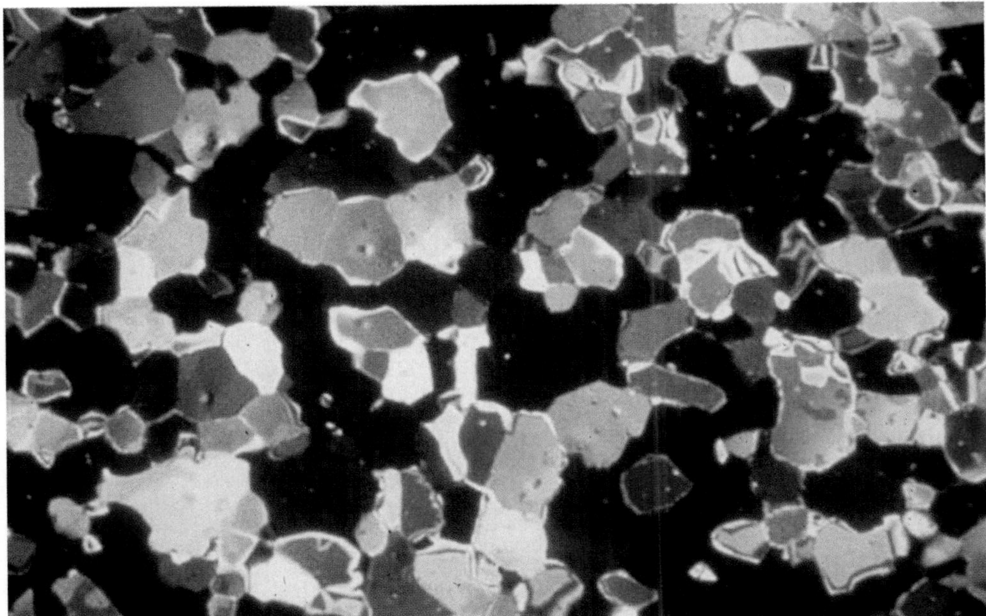

Figure 16.2 A typical mosaic of crystals of glacial ice. The tiny circular and tubular spots are bubbles of air. [Science VU/NOAA/USGS/Visuals Unlimited.]

colder, high-latitude climates, valley glaciers may descend many kilometers, down the entire length of a valley. In some places, broad lobes of ice may descend into lower lands bordering mountain fronts. Valley glaciers that flow down coastal mountain ranges may terminate at the ocean's edge, where masses of ice break off and form icebergs

Continental Glaciers A **continental glacier** is an extremely slow moving, thick sheet of ice (hence sometimes called an *ice sheet*) that covers a large part of a continent. Today, the world's largest ice sheets cover much of Greenland and Antarctica (**Figure 16.4**). The glacial ice of Greenland and Antarctica is not confined to mountain valleys but covers virtually the entire land surface. In Greenland, 2.8 million cubic kilometers of ice cover 80 percent of the island's total area of 4.5 million square kilometers. The upper surface of the ice sheet resembles an extremely wide convex lens. At its highest point, in the middle of the island,

Figure 16.3 Herbert Glacier, a valley glacier, near Juneau, Alaska. [Greg Dimijian/Photo Researchers.]

Figure 16.4 Sentinel Range, Antarctica. These mountains stick up through the thick ice of the Antarctic continental glacier. [Betty Crowell.]

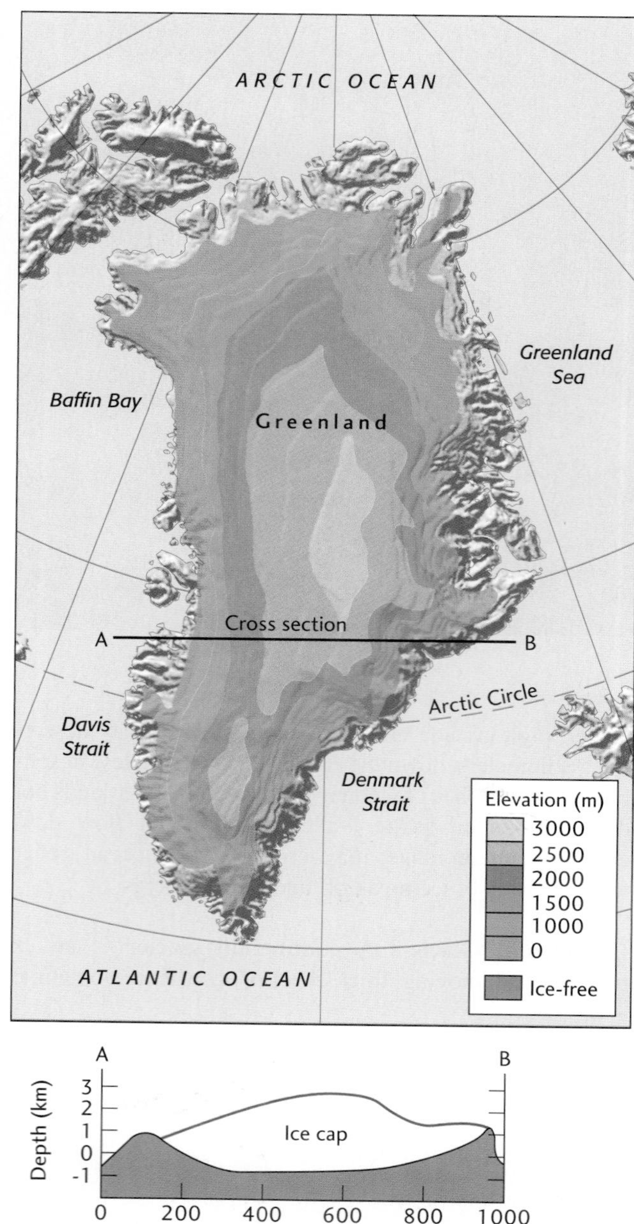

Figure 16.5 The extent of the glacial ice cap and the elevation of the ice surface on Greenland. The generalized cross section of south-central Greenland, A–B, shows the lenslike shape of the ice cap. The ice moves down and out from the thickest section. [After R. F. Flint, *Glacial and Quaternary Geology* (New York: Wiley, 1971).]

the ice is more than 3200 m thick (**Figure 16.5**). From this central area, the ice surface slopes to the sea on all sides. At the mountain-rimmed coast, the ice sheet breaks up into narrow tongues resembling valley glaciers that wind through the mountains to reach the sea. At the sea, the ice breaks off to form icebergs.

Large as the Greenland glacier is, the Antarctic ice sheet dwarfs it (**Figure 16.6**). Ice blankets 90 percent of the Antarctic continent, covering an area of about 12.5 million square kilometers and reaching thicknesses of about 3000 m. In Antarctica, as in Greenland, the ice forms a dome in the center and slopes down to the margins. In some places, thinner sheets of ice floating on the ocean are attached to the main glacier on land. The best known of these is the Ross Ice Shelf, a thick layer of ice about the size of Texas that floats on the Ross Sea.

Ice caps are the masses of ice formed at Earth's North and South Poles. Most of the Arctic ice cap, located at the lowest north latitudes, lies over water and is generally not referred to as a glacier. Almost all of the Antarctic ice cap lies over land, the continent of Antarctica, and is considered a continental glacier.

How Glaciers Form

A glacier starts with abundant winter snowfall that does not melt away in the summer. The snow is slowly converted into ice, and when the ice is thick enough, it begins to flow.

First Requirement: Low Temperatures For a glacier to form, temperatures must be low enough to keep snow on the ground year-round. These conditions occur at high latitudes (polar and subpolar regions) and high altitudes (mountains). As noted in Chapter 15, the Sun warms high latitudes less intensely because its rays strike Earth at a low angle there. High altitudes are cold, as the first high-flying airplane pilots discovered, because the lowest 10 km of the atmosphere grows steadily cooler as height above the ground surface increases. The height of the snow line—the altitude above which snow does not completely melt in summer—varies.

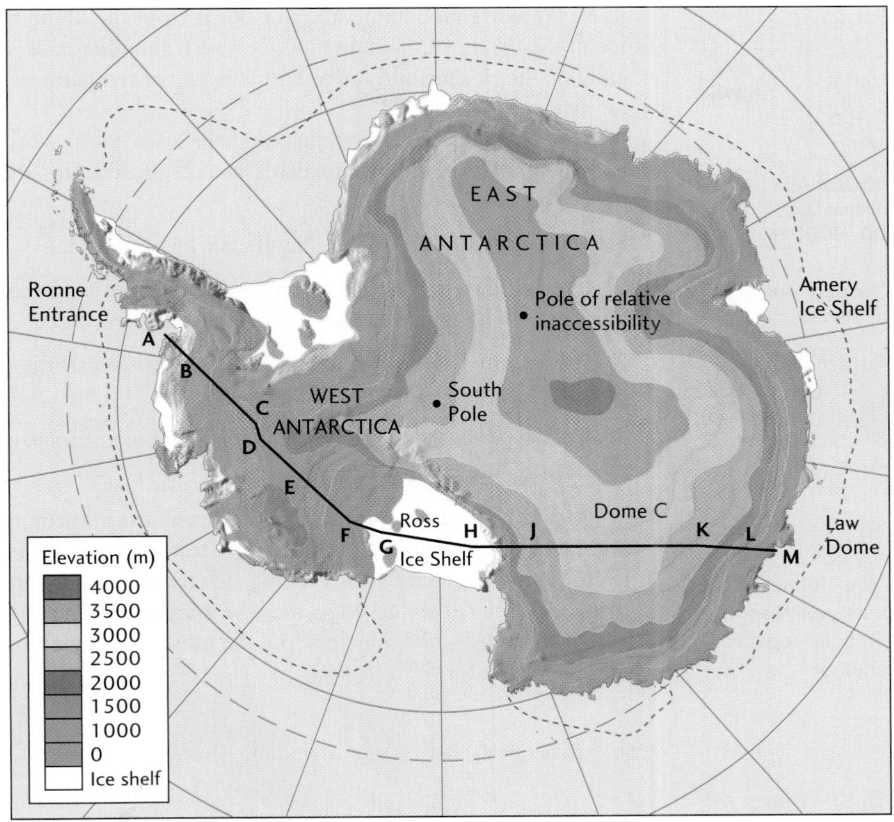

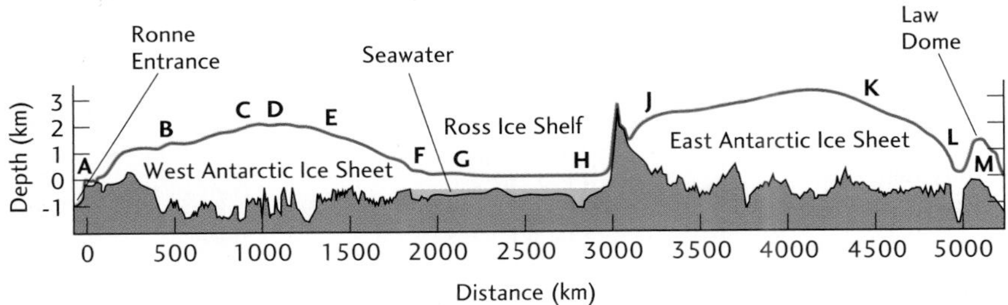

Figure 16.6 This contour map and the cross section of Antarctica show the topography of the ice sheet that covers the entire continent and the land beneath it. Ice shelves are shown in gray. [After U. Radok, "The Antarctic Ice." *Scientific American* (August 1985): 100; based on data from the International Antarctic Glaciological Project.]

Even in warm climates, glaciers will form if the mountains are high enough. Near the equator, glaciers form only on mountains that are higher than about 5500 m. This minimum altitude steadily decreases toward the poles, where snow and ice stay year-round even at sea level (**Figure 16.7**).

Second Requirement: Adequate Amounts of Snow Snow and the formation of glaciers require moisture as well as cold. Moisture-laden winds tend to drop most of their snow on the windward side of a high mountain range, so the leeward side is likely to be dry and unglaciated. Most of the high Andes of South America, for example, lie in a belt of prevailing easterly winds. Glaciers form on the moist eastern slopes, but the dry western side has little snow and ice.

Cold climates are not necessarily the snowiest. For example, Nome, Alaska, has a polar arctic climate with a yearly average maximum temperature of 9°C, but it gets only 4.4 cm of precipitation a year, virtually all of it as snow. Compare these figures with those for Caribou, Maine. Caribou has a cool climate with a yearly average maximum temperature of 25°C, and annual precipitation averages 90 cm. The average annual snowfall is a whopping 310 cm. Nevertheless, the conditions around Nome, where little of the snow melts, are better for the formation of glaciers than the conditions in Caribou, where all that snow melts in the spring. In arid climates, glaciers are unlikely to form at all, unless the temperature is so frigid all year that virtually no snow melts and all is preserved, as it is in Antarctica.

Glacial Growth: Accumulation

A fresh snowfall is a fluffy mass of loosely packed snowflakes. As the small, delicate crystals age on the ground,

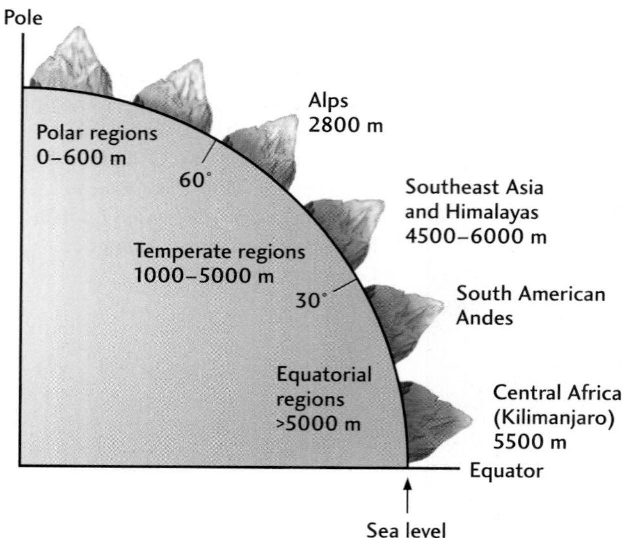

Figure 16.7 The height of the snow line—the altitude above which snow does not melt completely in summer—varies with latitude. The snow line is at or near sea level in polar regions and at heights of more than 5500 m at the equator.

they shrink and become grains (**Figure 16.8**). During this transformation, the mass of snowflakes compacts to form a dense, granular snow. As new snow falls and buries the older snow, the granular snow compacts further to an even denser form, called **firn.** Further burial and aging produce solid glacial ice as the smallest grains recrystallize, cementing all the grains together. The whole process may take only a few years, although 10 to 20 years is more likely. A typical glacier grows slightly during the winter, as snow falls on its surface. The amount of snow added to the glacier annually is its *accumulation.*

As glacial snow and ice accumulate, they entrap and preserve valuable relics of Earth's past. In 1992, Italian and Austrian scientists announced that they had discovered the body of a prehistoric human preserved in alpine ice on the border between their two countries. In northern Siberia, extinct animals such as the woolly mammoth, a great elephantlike prehistoric creature that once roamed icy terrains, have been found frozen and preserved by ancient ice. Ancient dust particles and bubbles of atmospheric gases are also preserved in glacial ice (see Figure 16.2). Chemical analysis of air bubbles found in very old, deeply buried Antarctic and Greenland ice tells us that levels of atmospheric carbon dioxide were lower during the last glaciation than they have been since the glaciers retreated.

Glacial Shrinkage: Ablation

When ice accumulates to a thickness sufficient for movement to begin, the formation of the glacier is complete. Ice, like water, flows downhill under the pull of gravity. The ice moves down a mountain valley or down from the dome of ice at the center of a continental ice sheet. In either case, it extends the glacier into lower altitudes where temperatures are warmer.

The total amount of ice that a glacier loses each year is called **ablation.** Four mechanisms are responsible for the loss of ice:

1. *Melting* As the ice melts, the glacier loses material.
2. **Iceberg calving** Pieces of ice break off and form icebergs when a glacier reaches a shoreline (**Figure 16.9**).
3. *Sublimation* In cold climates, ice can be transformed directly from the solid into the gaseous state.
4. *Wind erosion* Strong winds can erode the ice, primarily by melting and sublimation.

Most of the glacial shrinkage that results from warming and melting takes place at the glacier's leading edge. Thus, even though a glacier is advancing downward or outward from its center, the ice edge may be retreating. The two mechanisms by which glaciers lose the most ice are melting and calving.

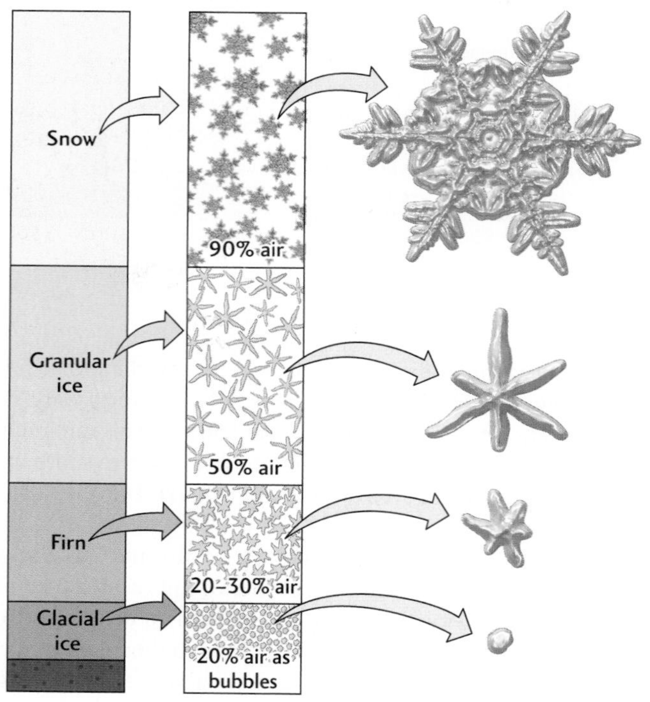

Figure 16.8 Stages in the transformation of snow crystals, first into granular ice, then into firn, and finally into glacial ice. A corresponding increase in density accompanies this transformation as air is eliminated from the crystals. [After H. Bader et al., "Der Schnee und seine Metamorphose." *Beiträge zur Geologie der Schweiz* (1939).]

Figure 16.9 Iceberg calving at Wrangell-St. Elias National Park, Alaska. Calving occurs when huge blocks of ice break off at the edge of a glacier that has moved to a shoreline. [Tom Bean.]

Glacial Budgets: Accumulation Minus Ablation

The difference between accumulation and ablation, called the glacial budget, results in the growth or shrinkage of a glacier (**Figure 16.10**). When accumulation minus ablation is zero over a long period, the glacier remains of constant size, even as it continues to flow downslope from the area where it formed. Such a glacier accumulates snow and ice in its upper reaches and ablates an equal amount in its lower parts. If accumulation exceeds ablation, the glacier grows; if accumulation is less than ablation, the glacier shrinks.

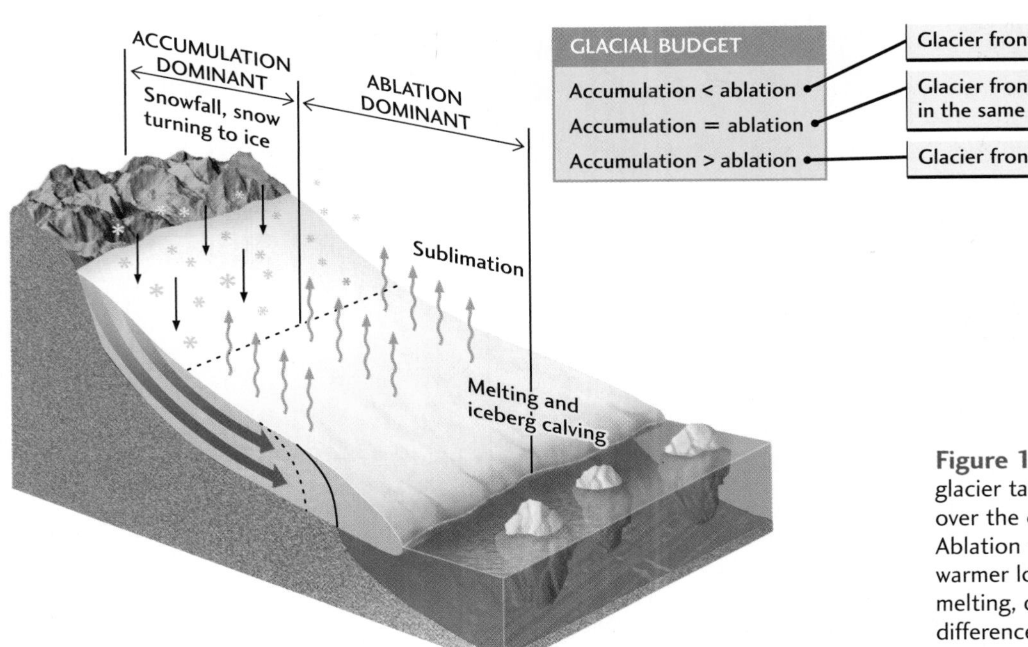

Figure 16.10 Accumulation of a glacier takes place mainly by snowfall over the colder upper regions. Ablation takes place mainly in the warmer lower regions by sublimation, melting, or iceberg calving. The difference between accumulation and ablation is the glacial budget.

Glacial budgets vary from year to year. Some show trends of growth or shrinkage in response to climatic variations over many decades. Yet, in the past several thousand years, many glaciers have remained constant, on the average. Now that many scientists are becoming concerned about the effects of global warming on Earth's climate (see Chapter 23), geologists have proposed that glacial budgets be carefully monitored. Glacial shrinkage in certain areas may be a good early warning of local or regional climate change. In 1995, for example, satellites gave us images of the ice shelves of the Antarctic Peninsula extending from West Antarctica. These images showed extensive retreats of shelf ice and the calving of an 80-km-long iceberg. This shrinkage corresponds to the warming of the west side of Antarctica by 2.5°C in the past 50 years. We can see a similar pattern of retreat, though on a smaller scale, in a small Antarctic Peninsula ice shelf between 1936 and 1992 (**Figure 16.11**). Between 1995 and 2002, the West Antarctic ice sheet remained relatively stable; only relatively small areas broke up over this period. During February and March of 2002, however, a major portion of the ice shelf collapsed and broke away from the Antarctic Peninsula. Scientists are increasingly concerned that further breakup or shrinkage of the ice sheet would raise sea level enough to flood many low-lying coastal cities periodically. We will discuss the processes leading to ice shelf collapse later in this chapter.

Glaciers: Moving Feasts for Water-Poor Regions?

As a glacier melts, great quantities of meltwater flow out from under the ice and from its edges. These meltwaters are the primary sources of cold-water streams that flow in mountain valleys below glaciers. If a glacial valley becomes dammed by glacial debris, a lake may form at the glacier's terminus.

The abundance of meltwater shows that large amounts of fresh water are tied up in glacial ice. Glaciers could be sources of fresh water for water-poor regions, if transportation from source to users can be made economical. Some geologists have even suggested that icebergs might be towed through the ocean like barges. Although this idea may not be practical very soon, it might conceivably be put to use in the future.

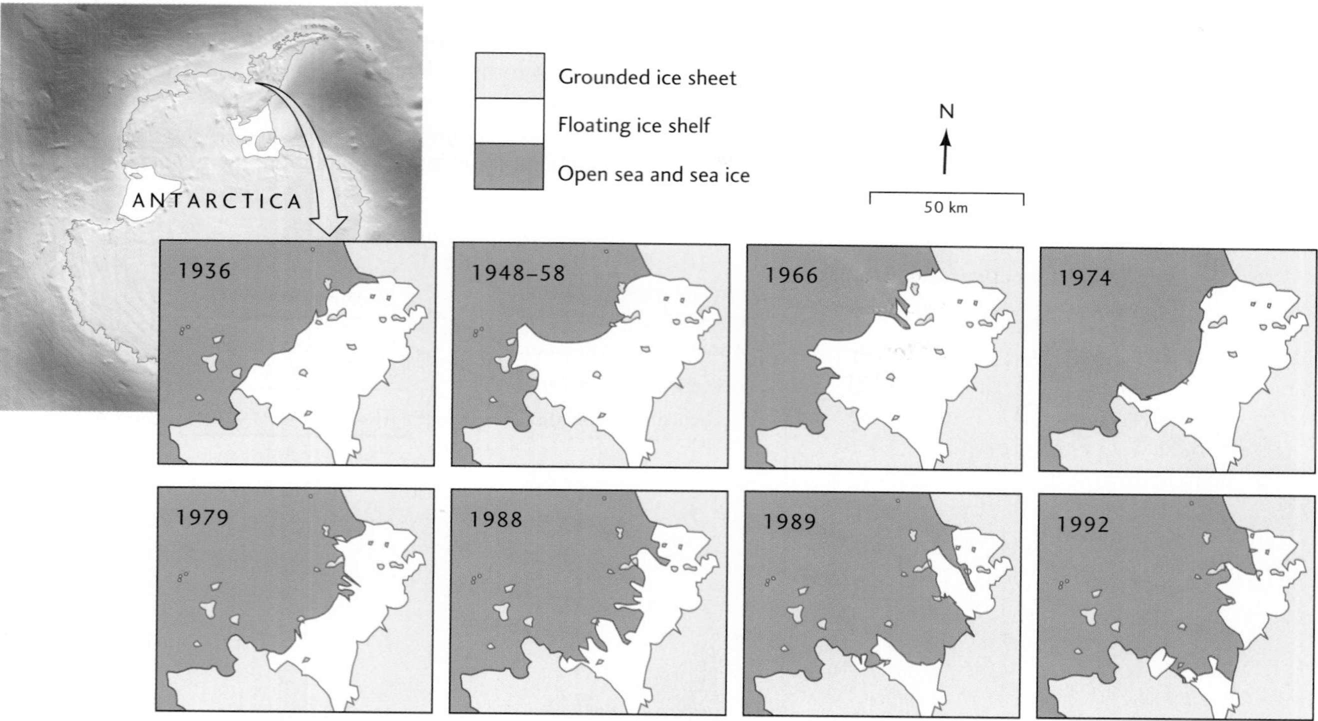

Figure 16.11 Successive stages in the retreat of a small ice shelf off the Antarctic Peninsula between 1936 and 1992. The area shown in the rectangles is about 90 km by 70 km. As the first three cells indicate, the shelf grew slightly between 1936 and 1966 but then retreated dramatically over the next 26 years, shrinking to its much smaller state by 1992. [After D. G. Vaughan and C. S. M. Doake, "Recent Atmospheric Warming and Retreat of Ice Shelves on the Antarctic Peninsula." *Nature* 379 (1996): 328–330.]

How Glaciers Move

When ice becomes thick enough—normally at least several tens of meters—for gravity to overcome the ice's resistance to movement, it starts to move and thus becomes a glacier. The ice deforms and flows slowly downhill in the same kind of laminar flow as that of thin, slow-water streams (see Chapter 14). The motion of glaciers is responsible for the immense amount of geologic work done by ice. In fact, it was seeing the results of glacial movement—erosion, transportation, and sedimentation—that led scientists to realize that ice does move. Unlike the readily observed rapid flow of a river, glacial motion is so slow that the ice seems not to move at all from day to day, giving rise to the expression "moving at a glacial pace."

The rate of glacial movement increases as the slope steepens or the ice thickens. Even on a flat surface, such as a continental lowland, ice will flow outward if it becomes thick enough. Just as a viscous fluid flows on a flat surface—honey on a slice of bread, for example—a continental glacier flows and spreads out as its thickness increases. How does ice, a solid, flow as if it were a slow-moving, viscous liquid?

Mechanisms of Glacial Flow

Glaciers flow primarily by two mechanisms, plastic flow and basal slip (**Figure Story 16.12**). In plastic flow, the movement occurs within the ice. In basal slip, the ice slides downslope as a single unit along the base of a glacier, like a brick sliding down an inclined board. The interface between the overlying ice and the underlying ground may not be a sharp boundary. Instead—especially where the ground consists of sediments or weak sedimentary rocks—the interface is a transition between ice loaded with debris and deformed ground containing appreciable ice.

Movement of Glaciers by Plastic Flow Under the great pressure within a glacier, individual crystals of ice slip tiny distances, on the order of a ten-millionth of a millimeter, over short intervals of time (Figure Story 16.12b). This movement is known as **plastic flow,** because it is like the plastic deformation of deeply buried rock (discussed in Chapter 11). The total of all the small movements of the enormous number of ice crystals that make up the glacier amounts to a large movement of the whole mass of ice. To visualize this process, think of a random pile of decks of playing cards, each deck held together by a rubber band. The whole pile can be made to shift by inducing many small slips between cards in the individual decks. As ice crystals grow under stress deeper in the glacier, their microscopic slip planes become parallel, increasing flow rates.

Plastic flow dominates in bitterly cold regions, where the ice throughout the glacier, including its base, is well below the freezing point. The basal ice is frozen to the ground, and most of the movement of these cold, dry glaciers takes place above the base by plastic deformation. What movement does occur results in detachment of pieces of bedrock or soil. Plastic flow can also occur in glaciers formed in warmer, more temperate regions. However, in these glaciers movement is often dominated by basal slip.

Movement of Glaciers by Basal Slip The other mechanism of glacial movement is **basal slip,** the sliding of a glacier along its base (Figure Story 16.12c). The amount and kind of basal slip vary, depending on the temperature at the boundary between ice and ground in relation to the melting point of ice. At the base of a glacier, the ice is under tremendous pressure from the weight of the overlying ice. The freezing point of ice decreases as pressure increases, so the ice at the base of a glacier melts sooner than ice at the surface. The melted ice becomes a lubricant at the base of the glacier. This is the same effect that makes ice skating possible. The weight of the body on the narrow skate blade provides enough pressure to melt just a little ice under the blade, which lubricates the blade so that it can slide easily along the surface. Similarly, melting of ice at the base of a glacier creates a lubricating layer of water on which the overlying ice can slide.

In addition to pressure, the temperature of the ice at the base of a glacier depends partly on the surface temperature of the ice and partly on the flow of heat from the ground below. In moderately cold, temperate regions, the air temperature at the surface is not too low so that the temperature at the base of the glacier may be high enough for some melting to occur.

In 1996, a team of Russian geologists reported finding a great lake of fresh water 4 km under the ice of central Antarctica. The lake is 200 km long and covers an area of 14,000 km^2. Lake Vostok, as it is called, is named for the Russian ice drilling station (see Feature 16.1). About 70 other bodies of fresh meltwater below the central Antarctic ice sheet have been identified, although we know relatively little about the plumbing that may connect a good many of them.

Moderately cold areas are typical of glacier-filled valleys in temperate regions. In such areas, the ice may be at the melting point not only at the base of the glacier but also higher in the glacier, particularly near its surface if the air is warmer than freezing. Plastic flow contributes a small amount of internal heating from the friction generated by microscopic slips of crystals under the great pressure of the ice. In these glaciers, water occurs in the ice as small drops between crystals and as pools in tunnels in the ice. The water throughout the glacier eases internal slip between layers of ice. In addition, ice may melt a little at the bottom and then refreeze, each time moving downhill a little bit.

Crevasses The upper parts of a glacier (shallower than about 50 m) have little pressure on them. At low pressures, the ice behaves as a rigid, brittle solid, cracking as it is dragged along by the plastic flow of the ice below. These

GLACIERS MOVE BY PLASTIC FLOW AND BASAL SLIP

(a)

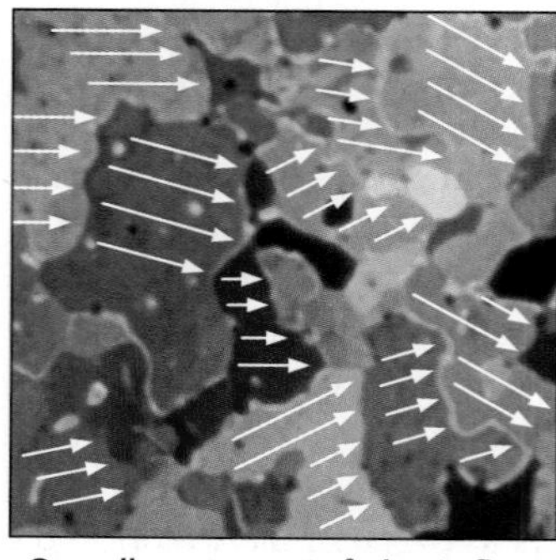

Ice crystals

The flow of a glacier is accomplished by small slips, over short intervals of time, along the microscopic planes of a great number of ice crystals. Ice crystals may stretch and rotate, or grow and recrystallize, and in some cases slide past each other.

(b) **PLASTIC FLOW**

1 Plastic flow dominates in cold regions where the ice at the base of the glacier is frozen to the bedrock or soil.

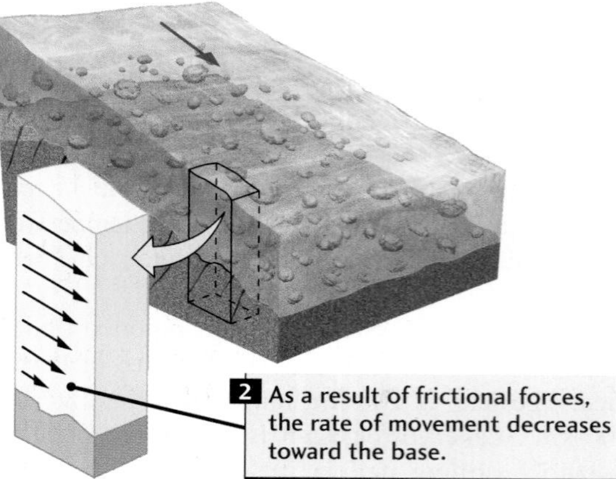

2 As a result of frictional forces, the rate of movement decreases toward the base.

3 Valley glaciers in cold regions move mostly by plastic flow. If one drives a set of stakes deep into the glacier in a line across its flow,...

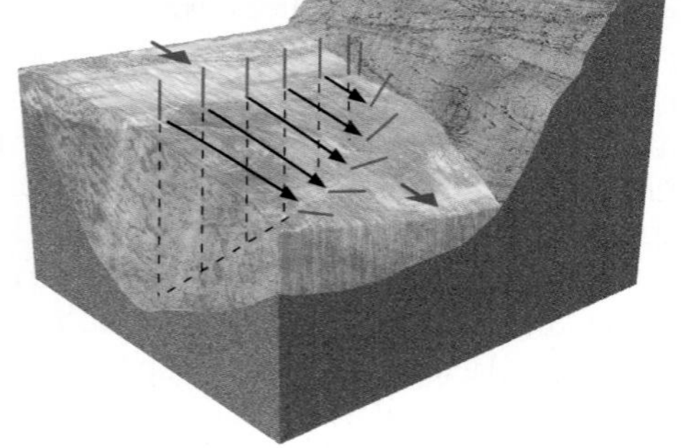

4 ...later the stakes in the center will have moved farther downhill and will slant forward. This indicates faster movement in the center and top of the glacier.

(c) **BASAL SLIP**

5 Basal slip flow dominates in temperate regions for very thick glaciers where the pressure of overlying ice melts water at the glacier's base.

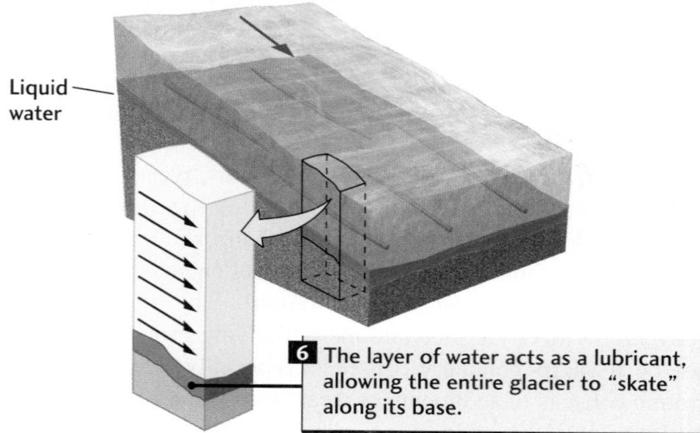

6 The layer of water acts as a lubricant, allowing the entire glacier to "skate" along its base.

7 Continental glaciers, such as those in Greenland and Antarctica, move mostly by basal slip.

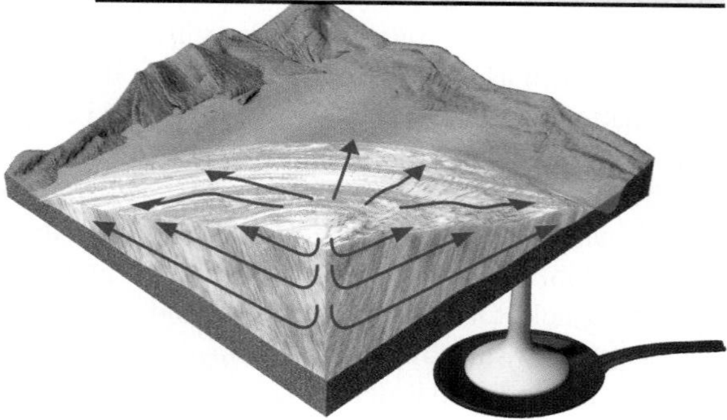

8 In continental glaciers, the ice moves down and out from the thickest section, like pancake batter poured on a griddle, as shown by the arrows.

Figure Story 16.12 Glaciers flow by two primary mechanisms.

cracks, called **crevasses,** break up the surface ice into many small and large blocks (**Figure 16.13**). Crevasses are more likely to occur in places where glacier deformation is strong—such as where the ice drags against bedrock walls, at curves in the valley, and where the slope steepens sharply. The movement of brittle surface ice at these places is a "flow" resulting from the slipping movements between these irregular blocks, similar in some ways to the microscopic slips of ice crystals but on a much larger scale.

Flow Patterns and Speeds

Flow in Valley Glaciers The speeds at which different parts of a valley glacier move vary with the depth of the ice and with the glacier's position in relation to the valley walls. Valley glaciers flow partly by basal slip and partly by plastic flow within the body of ice. Strong frictional forces at the glacier's base and sides, where it is in solid contact with the rock of the valley, inhibit the movement of the ice (see Figure Story 16.12d).

Louis Agassiz, a Swiss zoologist and geologist, was the first to measure the different speeds of ice in valley glaciers, more than a century ago. As a young professor, not yet 30, he and his students camped on a glacier to monitor its movement. They pounded stakes into the ice and measured their changes in position over a few years. The fastest movement, about 75 m in one year, was along the centerline of the glacier. The deformation of long vertical tubes pounded deep into the ice later demonstrated that ice at the base of the glacier moved more slowly than ice in the center.

The ice of some valley glaciers may move at a uniform speed. Such movement occurs only where the climate permits the glacier to slide as a single unit entirely by basal slip along a lubricating layer of meltwater next to the ground.

A sudden period of fast movement of a valley glacier, called a **surge,** sometimes occurs after a long period of little movement. Surges may last several years, and during that time the ice may speed along at more than 6 km/year—1000 times the normal velocity of a glacier. Although the mechanism of surges is not fully understood, it appears that a surge follows a buildup of water pressure in meltwater tunnels at or near the base of the glacier. This pressurized water greatly enhances basal slip.

Antarctica in Motion Antarctica may appear to be a land frozen in time, but it certainly is not still. Glaciers plow down the continent's center to the sea, icebergs snap off and crash into the ocean, and great rivers of ice snake through the ice sheet. All these movements are evidence of the dynamic relationship between this remote continent and global climate. Continental glaciers in polar climates, where basal slip is minor or absent, have the highest rates of movement in the center of the ice. Pressure there is very high, and the only frictional retarding forces are between layers of ice moving at different speeds (see Figure Story 16.12e).

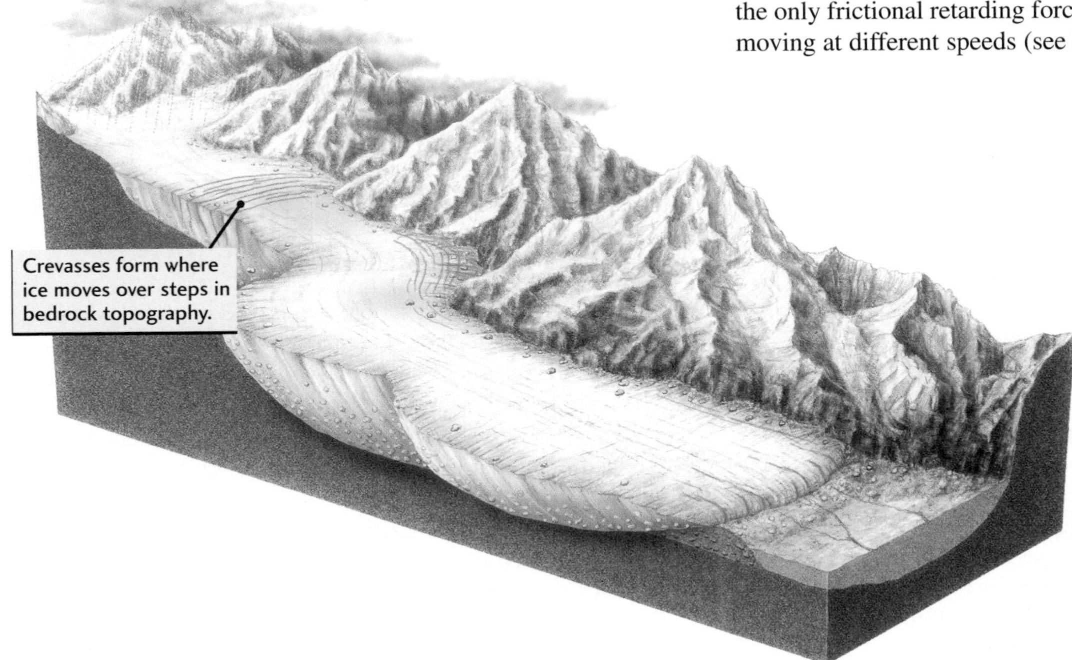

Figure 16.13 Crevasses on Mount Rainier, Washington State. [DougChurchill.com.] Crevasses tend to occur where the glacier valley curves or at sudden changes of slope.

16.1 Vostok and GRIP: Ice-Core Drilling in Antarctica and Greenland

At the Vostok science station in the frozen Antarctic, Russian scientists have been working year-round since 1960 to discover the climatological history of Earth hidden in the ice. Evidence in the ice may prove to be a source of insight into global climate change. In the 1970s, scientists at Vostok drilled boreholes 500 to 952 m deep in the eastern Antarctic ice and brought up a set of ice cores showing layers produced by annual cycles of ice formation from snow. Careful counting of the layers, working from the top down, revealed the age of the ice, much as tree rings reveal the age of a tree. This time-stratigraphic record of the ice proved to correlate with the temperatures thought to prevail when the layers formed. A high ratio of oxygen-18 to oxygen-16 in an ice layer indicates that the ice formed when atmospheric temperature was relatively high. High concentrations of carbon dioxide, methane, and other greenhouse gases in an ice layer also suggest that it formed during a period of atmospheric warming.

By the 1990s, ice borers at Vostok had drilled to a depth of 2755 m, penetrating ice not only from the last glacial period but also from the interglacial period before it. They thereby accumulated a stratigraphic record for the past 160,000 years. These cores showed that eastern Antarctica, where the ice cores were drilled, was colder and drier during the last glaciation than during the 11,000 years of the current interglacial period. Variations in the oxygen isotope ratios of ice layers confirm other evidence that variations in Earth's orbit control the cyclical alternation of glacial and interglacial epochs. The carbon dioxide content of ice layers formed during glacial intervals decreased markedly with climate cooling, but this decrease was a little later than the time of first cooling indicated by oxygen isotope ratios. We do not yet fully understand this lag. On the other hand, an increase in carbon dioxide does coincide in time with warming of the atmosphere.

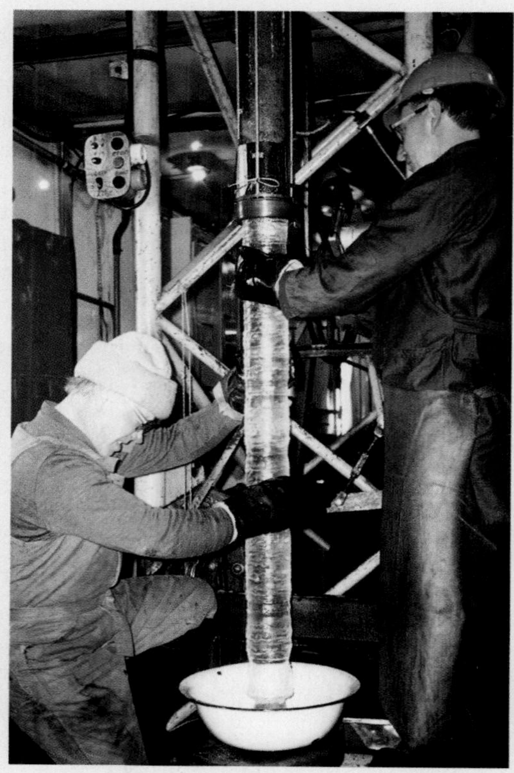

Russian scientists at Vostok science station in Antarctica carefully remove an ice core from a drill. The layers produced by annual cycles of ice formation are visible. [R. J. Delmas, Laboratoire de glaciologie et géophysique de l'environnement, Centre National de la Recherche Scientifique.]

Geologists use satellites and airborne radar to map the shapes and overall movements of glaciers. These measurements show that Antarctic glaciers flow rapidly in **ice streams** 25 to 80 km wide and 300 to 500 km long (**Figure 16.14**). These streams reach velocities of 0.3 to 2.3 m/day, compared with the flow rate of 0.02 m/day in the adjacent ice sheet. Boreholes drilled in the ice reveal that the base of the ice stream is at the melting point and that the meltwater is mixed with soft sediment. One theory is that rapid movement of the ice streams is related to deformation of the water-saturated basal sediment. Similar ice streams may form during climatic warming, which leads to ice breakup and rapid deglaciation. In a period of global warming, the ice streams would contribute to the retreat of glaciers and the instability of the West Antarctic ice sheet.

Using high-resolution radar satellite mapping, geologists have observed that several Antarctic glaciers have retreated more than 30 km in just 3 years. Over the past 20 years or so, enormous pieces of ice have snapped off Antarctic glaciers. In March 2000, an iceberg the size of Delaware calved from the Ross Ice Shelf. More recently, in February and March of 2002, a portion of the Larsen Ice Shelf the size of Rhode Island (about 3250 km^2) shattered and separated from the east side of the Antarctic Peninsula

Meanwhile, in the Arctic, a group of scientists in 1992 completed 2 years of drilling the top 3 km of ice in the Greenland ice cap. The Greenland Ice Core Project, better known as GRIP, is an outstanding example of international scientific cooperation in Europe. GRIP is sponsored by the European Science Foundation, which includes the governments of Belgium, Denmark, France, Germany, Iceland, Italy, Switzerland, and the United Kingdom. The GRIP core penetrated ice layers of the last glacial period (11,000 to 115,000 years before the present) and the last interglacial period (the Eemian, 115,000 to 135,000 years before the present). It stopped in layers going back 235,000 years, near the end of the previous interglacial period. The accompanying diagram shows the relationship between depth of the ice core and times of glacial and interglacial periods for the past 135,000 years.

A parallel core was drilled 30 km to the west of the GRIP site by the U.S. Greenland Ice-Sheet Project (known as GISP2). The GRIP and GISP2 ice cores agreed very closely over the past 110,000 years, confirmed the glacial chronology at Vostok, and gave a good picture of the changes in climate and atmospheric composition of the past 250,000 years. Analysis of the cores also corrected earlier suggestions that the climates of the Eemian interglacial period were unstable and changed appreciably on a short time scale. The GRIP and GISP2 findings confirmed the remarkably rapid climate oscillations of the last glacial period. The GRIP and Vostok cores yielded precise oxygen-18 profiles that correlate closely with those of deep-sea sediments, providing additional detail and further supporting the chronology of glacial-interglacial climate change. It has become clear that temperature changes between glacial and interglacial periods were much larger than scientists had thought.

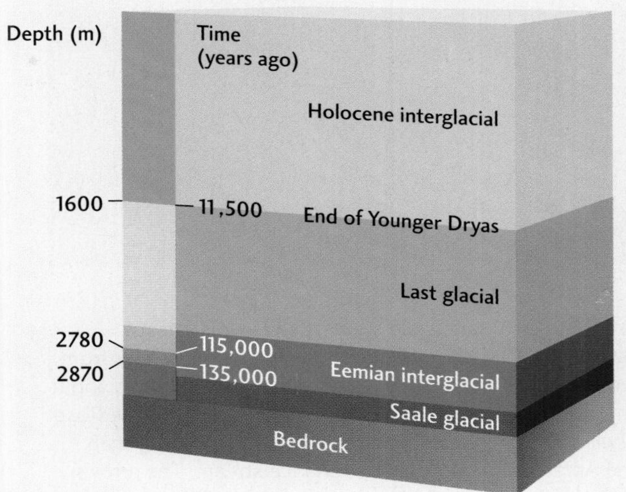

Comparison of ice depths and time before the present for recent glacial and interglacial intervals. [After J. W. C. White, "Don't Touch That Dial." *Nature* 364 (1993): 186.]

These triumphs have not been won easily. The scientists had to take extreme care not to melt and contaminate the ice cores while drilling them, transporting them to the laboratories, and storing them until the ice and the air bubbles trapped in it could be analyzed. The scientists also had to guard against misleading results—caused, for example, by the reaction of carbon dioxide with impurities in the ice. It is a tribute to the patience, ingenuity, and care of these hardy bands of researchers that the worldwide scientific community has recognized the great importance of glacial ice-core research in understanding the history and predicting the future of global climate change.

(**Figure 16.15**). The fracturing of this piece of the ice sheet produced thousands of icebergs.

Geologists who monitor the Larsen Ice Shelf were able to predict this episode of ice-sheet collapse. Field and satellite-based observations showed that the ice stream's flow rate had increased dramatically, which was interpreted as a sign of instability. If something causes the discharge rate of an ice stream to change, a stable ice shelf can suddenly become unstable. In the case of the Larsen Ice Shelf, geologists think that warming temperatures, both on the surface of the ice and in deep water at the bottom of the shelf, have conspired to accelerate ice-stream-flow velocities.

Isostasy, Ice Shelves, and Sea Level Change

Scientists who study global climate change wonder what will happen to sea level if Earth's ice shelves continue to collapse. Would it rise? It turns out that if all Earth's ice shelves were to break off into the ocean over the next few years, with no additional complications, sea level wouldn't change. The reason is that, like icebergs, ice shelves float on the ocean. When they melt, the ice is converted to water, but there is no change in sea level. For the same reason, when ice cubes in your drink melt, the level of the liquid doesn't change.

(a)

(b)

Figure 16.14 Antarctic ice streams and velocity maps. (a) Beardmore Glacier, Antarctica, showing fast flow lines from the middle background to the foreground. This glacier is about 25 km wide and flows from the Antarctic Plateau into the Ross Ice Shelf. [Wolfgang Bayer/Bruce Coleman.] (b) Velocity map for the Antarctic Lambert Glacier and ice stream. The arrows show the directions in which the ice is flowing. The areas of no motion (yellow) are either exposed land or stationary ice. The smaller, tributary glaciers generally have low velocities of 100–300 m/year (green), which gradually increase as they flow down the sloping surface of the continent and intersect the upper reaches of the Lambert Glacier. Most of the Lambert Glacier has velocities of 400–800 m/year (blue). As it extends into and across the Amery Ice Shelf and the ice sheet spreads out and thins, velocities increase to 1000–1200 m/year (pink/red). [NASA.]

Figure 16.15 Collapse of the Larsen Ice Shelf. This satellite image was acquired on March 7, 2002, toward the end of a two-month period in which the ice shelf separated from land and splintered into thousands of icebergs. The darkest colors on the right-hand side of the image are open seawater. The white parts are icebergs, the remaining parts of the ice shelf, and glaciers on land. The bright blue area is a mixture of seawater and highly fractured ice. The area of the image is about 150 km by 185 km. [NASA/GSFC/LaRC/JPL, MISR Team.]

Here's why. The idea that an ice shelf or iceberg floats on the ocean is the **principle of isostasy** (**Figure Story 16.16**). In a similar way, life jackets float because the lightweight flotation material inside them is less dense than seawater. Ice shelves and icebergs float because ice is less dense than seawater. The flotation, or buoyancy, is due to the fact that the volume of the life jacket or of the ice *below* the sea surface weighs less than the volume of water that it displaces. This buoyancy force counteracts the gravitational force that presses down on the iceberg. Bigger icebergs stand higher above the sea surface but also have a deeper root below the surface to provide greater buoyancy. In Chapters 18 and 20 we will see that the principle of isostasy can be applied to mountain belts and continents because the less dense continental crust floats on the denser mantle.

By displacing seawater as they float, icebergs and ice shelves drive sea level to a higher position. Sea level will be higher depending on how much seawater is displaced, so the bigger the iceberg or ice shelf, the higher sea level will stand. The principle of isostasy ensures that this sea level rise is proportional to the volume of seawater displaced. However, when the icebergs and ice shelves melt, there is no change in sea level—the water displaced by an iceberg is filled by the volume of water that the iceberg represents.

Melting ice can only cause sea level to rise if it is on land, not floating in water. The melting of glaciers contributes to sea level rise in two ways. In the first case, ice that is located on land will melt to produce water that fills up the ocean; the rate of sea level rise depends on the ice

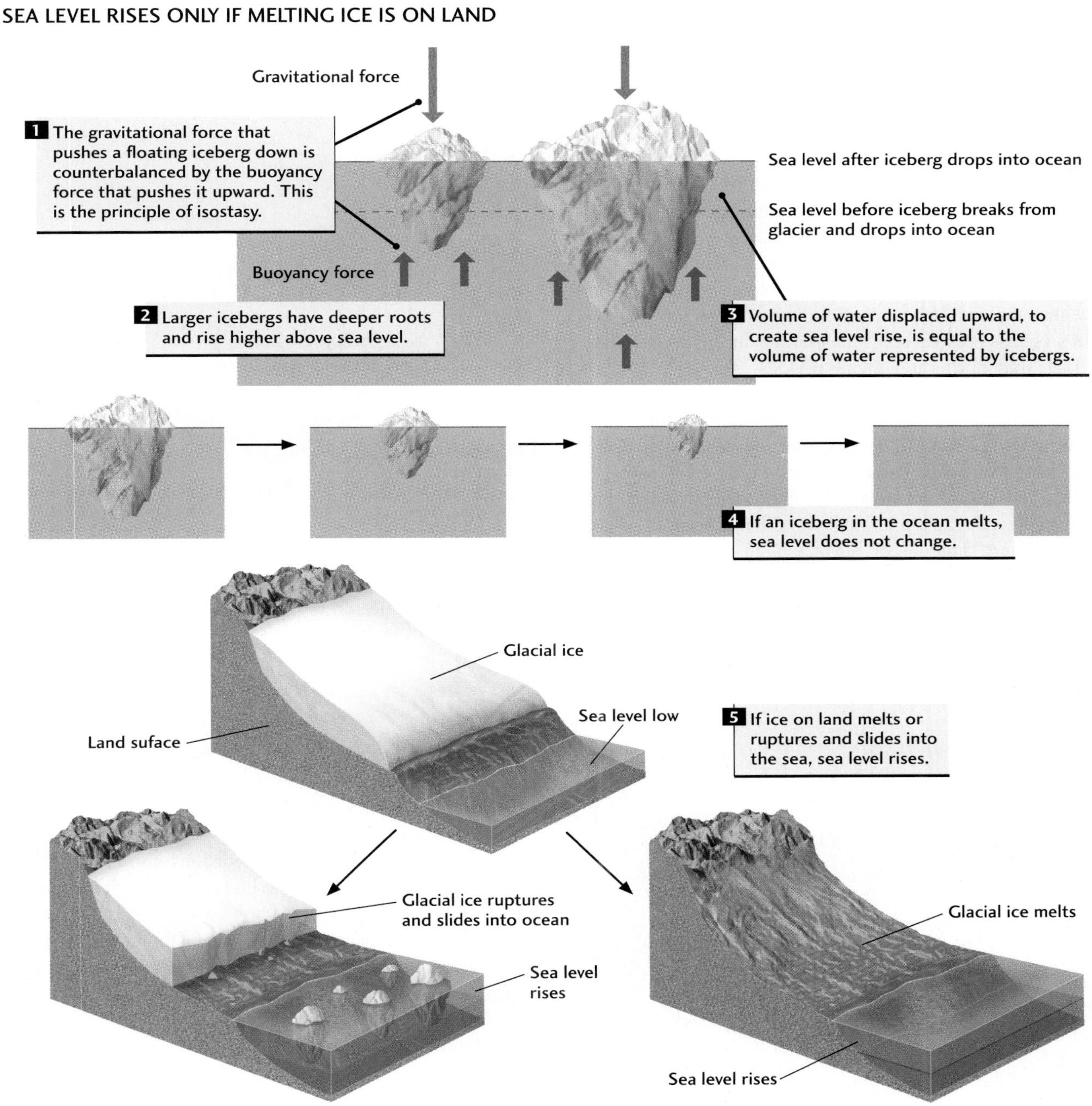

SEA LEVEL RISES ONLY IF MELTING ICE IS ON LAND

Gravitational force

1 The gravitational force that pushes a floating iceberg down is counterbalanced by the buoyancy force that pushes it upward. This is the principle of isostasy.

Buoyancy force

2 Larger icebergs have deeper roots and rise higher above sea level.

Sea level after iceberg drops into ocean

Sea level before iceberg breaks from glacier and drops into ocean

3 Volume of water displaced upward, to create sea level rise, is equal to the volume of water represented by icebergs.

4 If an iceberg in the ocean melts, sea level does not change.

Glacial ice

Sea level low

Land suface

5 If ice on land melts or ruptures and slides into the sea, sea level rises.

Glacial ice ruptures and slides into ocean

Sea level rises

Glacial ice melts

Sea level rises

Figure Story 16.16 The principle of isostasy shows that the melting of ice shelves and icebergs will not change the sea level.

melting rate. In the second case, sea level will rise if continental glaciers shed icebergs directly into the sea; sea level rise is instantaneous as blocks of ice slide off the land and into the sea. Therefore, the destruction of the ice shelves to generate sea level rise is only of concern if part of the ice shelf is grounded but then slides into the sea. In this case, the weight of the ice is no longer supported by the continent but by the buoyancy of the seawater it displaces, which rises upward to a new and higher level.

Glacial Landscapes

Just as you cannot see your own footprint in the sand while your foot is covering it, you cannot see the effects of an active glacier at its base or sides. Only when the ice melts is its geological work of erosion and sedimentation revealed. We can infer the physical processes driven by moving ice from the topography of formerly glaciated areas and the distinctive landforms left behind.

Glacial Erosion and Erosional Landforms

Glaciers have an amazing capacity to erode solid rock. A valley glacier only a few hundred meters wide can tear up and crush millions of tons of bedrock in a single year. The ice erodes this heavy load of sediment from the rock floor and sides of a glacier and carries it to the ice front, where it is deposited as the ice melts. The amount of these deposits demonstrates that ice is a far more efficient agent of erosion than water or wind. We can compare the amounts of sediment deposited in glacial and interglacial periods and the rates at which they were deposited. From such studies, we know that the total amount of sediment deposited in the world's oceans was several times larger during recent glacial times than during nonglacial periods.

At its base and sides, a glacier engulfs jointed, cracked blocks of rock and breaks and grinds them against the rock pavement below. The grinding action fragments rocks into a great range of sizes, from boulders as big as houses to fine silt- and clay-sized material called **rock flour.** Rock flour is subject to rapid chemical weathering because of its fine size—and thus larger surface area. Where glacial debris is still encased in ice and the ground is overlain by thick ice, chemical weathering is slower than on ice-free terrains. When finely pulverized rock flour is freed from the ice at the melting edge of a glacier, however, it dries into dust. As we saw in Chapter 15, the wind can blow dust for great distances, ultimately depositing it as the loess that is so common during glacial periods.

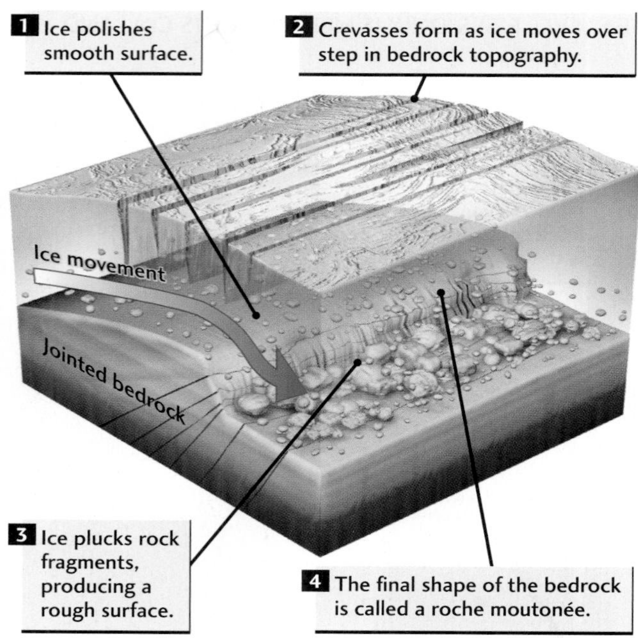

Figure 16.18 A *roche moutonée* is a small bedrock hill, smoothed by the ice on the upstream side and plucked to a rough face on the downstream side as the moving ice pulls fragments from joints and cracks.

Figure 16.17 Glacial polish, striations, and grooves on bedrock, Glacier Bay National Park, Alaska. Striations are evidence of the direction of ice movement and are especially important clues for reconstructing the movement of continental glaciers. [Carr Clifton.]

As a glacier drags rocks along its base, the rocks scratch or groove the pavement. Such abrasions are termed **striations,** and they provide more evidence of glacial movement. The orientation of striations shows us the direction of ice movement—an especially important factor in the study of continental glaciers, which lack obvious valleys. By mapping striations over wide areas formerly covered by continental glaciers, we can reconstruct the glaciers' flow patterns (**Figure 16.17**).

Advancing glacier ice smooths small hills of bedrock—known as *roches moutonées* ("sheep rocks") for their resemblance to a sheep's back—on their upstream side and plucks them to a rough, steep slope on the downstream side (**Figure 16.18**). These contrasting slopes indicate the direction of ice movement.

A valley glacier carves a series of erosional forms as it flows from its origin to its lower edge. At the head of the glacial valley, the plucking and tearing action of the ice tends to carve out an amphitheater-like hollow called a **cirque,** usually shaped like half of an inverted cone (**Figure 16.19**). With continued erosion, cirques at the heads of adjacent valleys gradually meet at the mountaintops, producing sharp, jagged crests called **arêtes** along the divide.

As a valley glacier moves down from its cirque, it excavates a new valley or deepens an existing river valley,

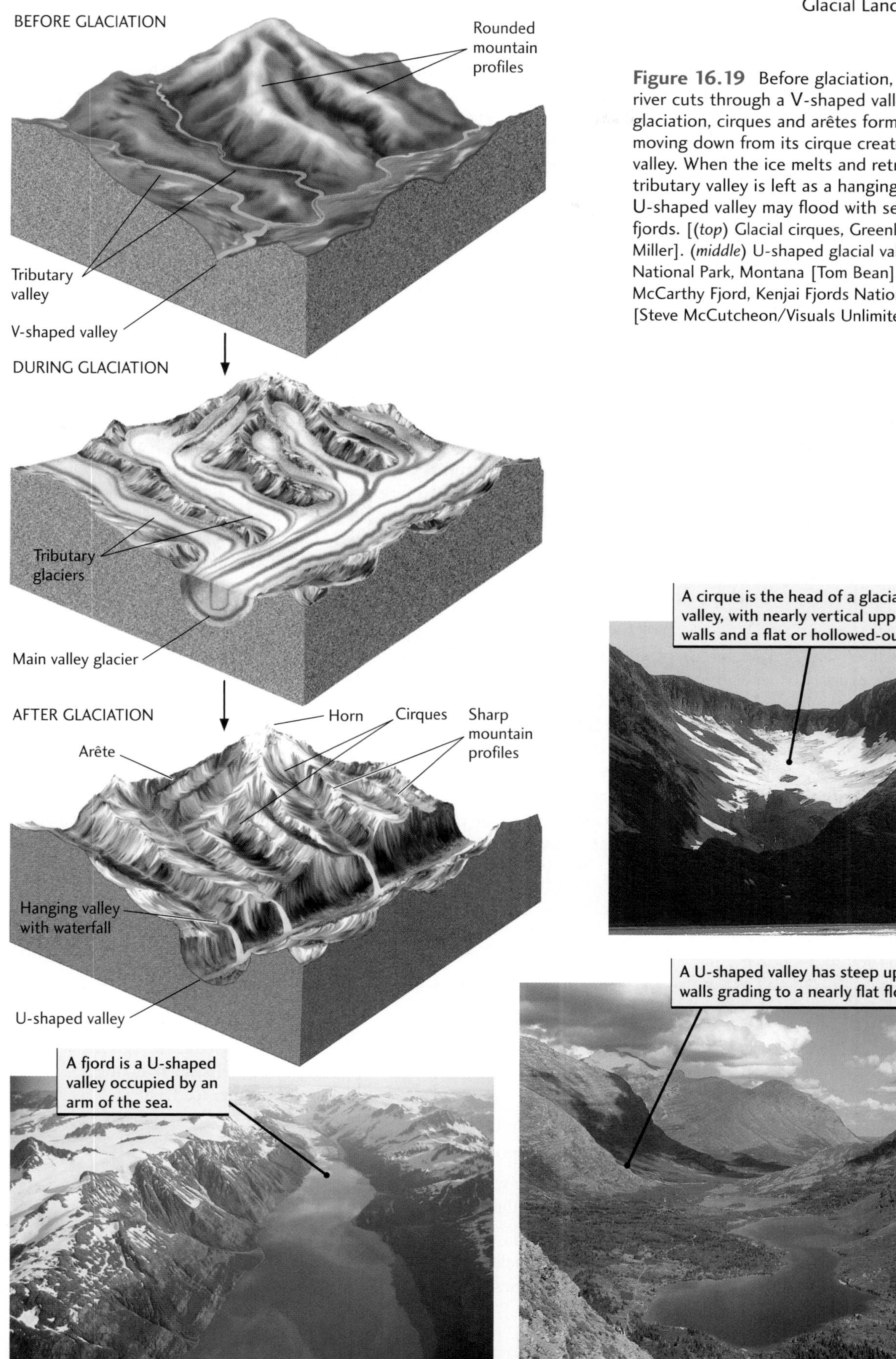

Figure 16.19 Before glaciation, a mountain river cuts through a V-shaped valley. During glaciation, cirques and arêtes form. The glacier moving down from its cirque creates a U-shaped valley. When the ice melts and retreats, the tributary valley is left as a hanging valley and the U-shaped valley may flood with seawater to form fjords. [(*top*) Glacial cirques, Greenland [Martin Miller]. (*middle*) U-shaped glacial valley, Glacier National Park, Montana [Tom Bean]. (*bottom left*) McCarthy Fjord, Kenjai Fjords National Park, Alaska [Steve McCutcheon/Visuals Unlimited].

creating a characteristic **U-shaped valley** (see Figure 16.19). Glacial valley floors are wide and flat and have steep walls, in contrast with the V-shaped valleys of many mountain rivers.

Glaciers and rivers differ not only in the shape of the valleys they create but also in how their tributaries form junctions. Although the ice surface is level where a tributary glacier joins the main valley glacier, the floor of the tributary valley may be much shallower than that of the main valley. When the ice melts, the tributary valley is left as a **hanging valley**—one whose floor lies high above the main valley floor (see Figure 16.19). After the ice is gone and rivers occupy the valleys, the junction is marked by a waterfall as the stream in the hanging valley plunges over the steep cliff separating it from the main valley below.

Unlike rivers, valley glaciers at coastlines may erode their valley floors far deeper than sea level. When the ice retreats, these steep-walled valleys—which still maintain a U-shaped profile—are flooded with seawater (see Figure 16.19). These glaciated arms of the sea, called **fjords,** create the spectacular rugged scenery for which the coasts of Alaska, British Columbia, and Norway are renowned.

Glacial Sedimentation and Sedimentary Landforms

Glaciers transport eroded rock materials of all kinds and sizes downstream, eventually depositing them where the ice melts. Ice is a most effective transporter of debris because the material it picks up does not settle out, as does the load carried by a river. Like water and wind currents, flowing ice has both a competence (an ability to carry particles of a certain size) and a capacity (a total amount of sediment that it can transport). Ice has extremely high competence: it can carry huge blocks many meters in diameter that no other transporting agent could budge. Ice also has a tremendous capacity. Some ice is so full of rock material that it is dark and looks like sediment cemented with ice.

When glacial ice melts, it deposits a poorly sorted, heterogeneous load of boulders, pebbles, sand, and clay (**Figure 16.20**). A wide range of particle sizes is the characteristic that differentiates glacial sediment from the much better sorted material deposited by streams and winds. The heterogeneous material puzzled early geologists, who were not aware of its glacial origins. They called it *drift* because it seemed to have drifted in somehow from other areas. The term **drift** is now used for *all* material of glacial origin found anywhere on land or at sea.

Some drift is deposited directly by melting ice. This unstratified and poorly sorted sediment is known as **till,** and it may contain all sizes of fragments—clay, sand, or boulders. The large boulders often contained in till are called **erratics** because of their seemingly random composition, so very different from that of local rocks.

Other deposits of drift are laid down as the ice melts and releases sediment. Meltwater streams flowing in tunnels within and beneath the ice and in streams at the ice front may pick up, transport, and deposit some of the material. Like any other waterborne sediment, this material is stratified and well sorted and may be cross-bedded. Drift that has been caught up and modified, sorted, and distributed by meltwater streams is called **outwash.** As we mentioned earlier, strong winds can blow fine-grained material from outwash floodplains and deposit it as loess.

Glacial sedimentary sequences can be identified by the distinctive textures of interbedded tills, outwash, and loess, as well as by striations and other erosional forms that may be preserved. Mapping of such sequences has allowed geologists to infer many glaciations of past geologic times.

Table 16.1 Glacial Moraines

Types of Moraine	Location with Respect to Ice Front	Comments
End moraine	At ice front	After glacier melts, seen as ridge parallel to former ice front
Terminal moraine	At ice front marking farthest advance of ice (an end moraine)	See end moraine
Lateral moraine	Along the edge of glacier where it scrapes side walls of valley	Heavy sediment load eroded from valley walls; when ice melts, seen as ridge parallel to valley walls
Medial moraine	Formed as two joined glaciers merge their lateral moraines below junction	Sediment load inherited from lateral moraines that formed it; forms ridge parallel to valley walls
Ground moraine	Beneath the ice as a layer of glacial debris	Ranges from thin and patchy to a thick blanket of till

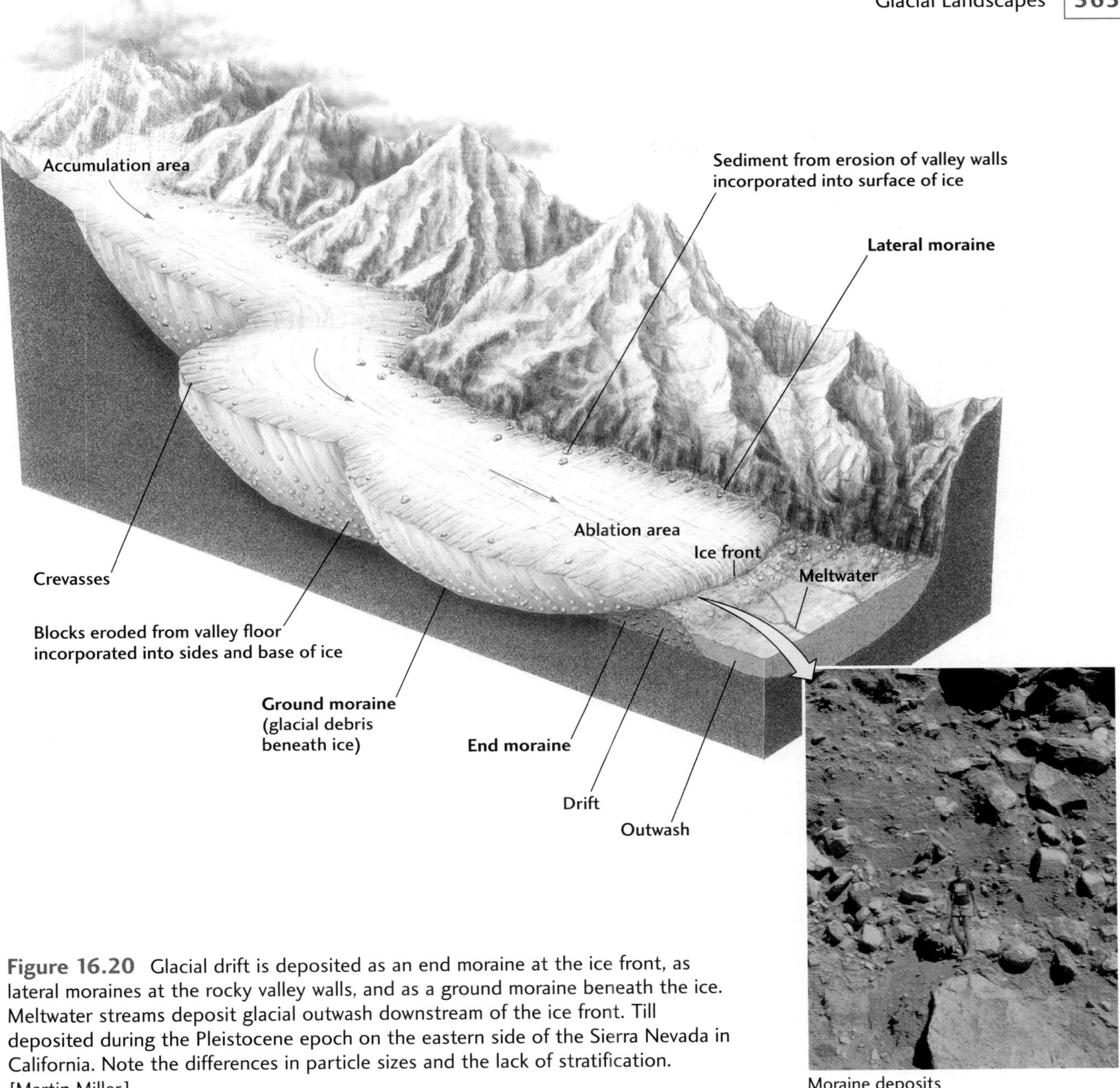

Figure 16.20 Glacial drift is deposited as an end moraine at the ice front, as lateral moraines at the rocky valley walls, and as a ground moraine beneath the ice. Meltwater streams deposit glacial outwash downstream of the ice front. Till deposited during the Pleistocene epoch on the eastern side of the Sierra Nevada in California. Note the differences in particle sizes and the lack of stratification. [Martin Miller.]

Ice-Laid Deposits A **moraine** is an accumulation of rocky, sandy, and clayey material carried by the ice or deposited as till. There are many types of moraines, each named for its position with respect to the glacier that formed it (Table 16.1 and Figure 16.20). One of the most prominent in size and appearance is an *end moraine,* formed at the ice front. As the ice flows steadily downhill, it brings more and more sediment to its melting edge. The unsorted material accumulates there as a hilly ridge of till. Regardless of the shape or location, moraines of all kinds consist of till. Figure 16.20 illustrates the processes by which various kinds of moraines form as a glacier works its way through a valley.

Some continental glacier terrains display prominent landforms called **drumlins**—large, streamlined hills of till and bedrock that parallel the direction of ice movement (**Figure 16.21**). Drumlins, usually found in clusters, are shaped like long, inverted spoons with the gentle slopes on the downstream sides. They may be 25 to 50 m high and a kilometer long. Geologists think that drumlins form under glaciers that move over a plastic mixture of subglacial sediment and water. When the plastic mass encounters a bedrock knob or other obstacle, it is subjected to increased pressure. It then loses water and solidifies, forming a streamlined mass. An alternative hypothesis proposes that drumlins are formed by ice erosion of an earlier accumulation of till.

Water-Laid Deposits Deposits of outwash from glacial meltwaters take a variety of forms. **Kames** are small hills of sand and gravel dumped at or near the edge of the ice (see Figure 16.21). Some kames are deltas built into lakes at the

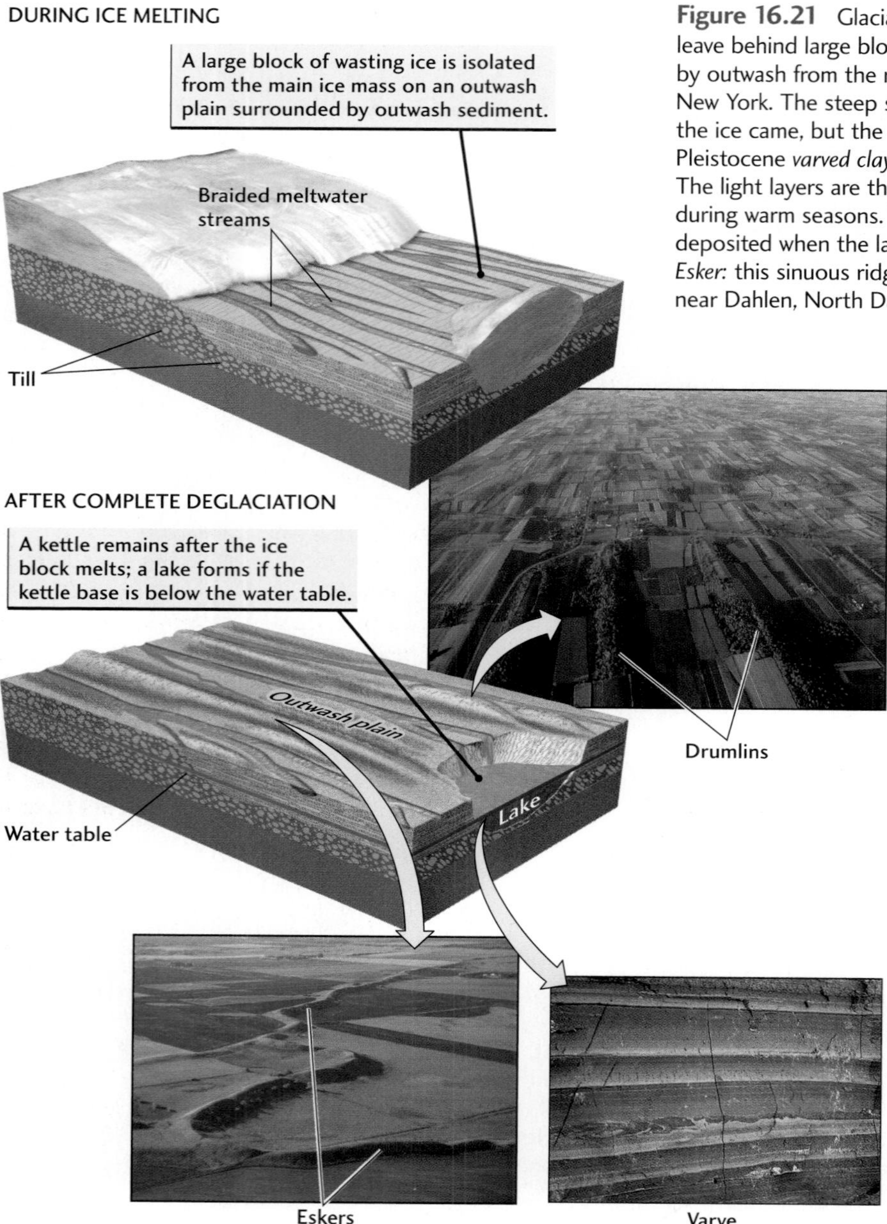

Figure 16.21 Glacial deposits. As a glacier retreats, it may leave behind large blocks of wasting ice that are gradually buried by outwash from the receding ice front. *Drumlins,* from Rochester, New York. The steep side usually faces the direction from which the ice came, but the reverse is also found [John Shelton]. Pleistocene *varved clay,* from an excavation in Stockholm, Sweden. The light layers are the coarse sediments deposited in a lake during warm seasons. The dark layers are the fine clays deposited when the lake was frozen in winter [John Shelton]. *Esker:* this sinuous ridge was formed by a glacial ice age stream near Dahlen, North Dakota [Tom Bean].

ice edge. When a lake drains, the deltas are preserved as flat-topped hills. Kames are often exploited as commercial sand and gravel pits.

Glaciated terrains are dotted with **kettles,** hollows or undrained depressions that often have steep sides and may be occupied by ponds or lakes. Modern glaciers, which may leave behind huge isolated blocks of ice in outwash plains as they melt, offer the clue to the origin of kettles. A block of ice a kilometer in diameter could take 30 years or more to melt. During that time, the melting block may be partly buried by outwash sand and gravel carried by meltwater streams, usually braided, coursing around it. By the time the block has melted completely, the margin of the glacier has retreated so far from the region that little outwash reaches the area. The sand and gravel that formerly surrounded the block of ice now surround a depression. If the bottom of the kettle lies below the groundwater table, a lake will form. Figure 16.21 shows the evolution of an outwash kettle.

Valley glaciers may deposit silts and clays on the bottom of a lake at the edge of the ice, in a series of alternating coarse and fine layers called varves (see Figure 16.21). A **varve** is a pair of layers formed in one year by seasonal freezing of the lake surface. In the summer, when the lake is free of ice, coarse silt is laid down by abundant meltwater streams flowing from the glacier into the lake. In the winter, when the surface of the lake is frozen, the finest clays settle, forming a thin layer on top of the coarse summer layer.

Some lakes formed by continental glaciers were huge, many thousands of square kilometers in extent. The till dams that created these lakes were sometimes breached and carried away at a later time, causing the lakes to drain rapidly and creating huge floods. In eastern Washington State, an area called the scablands is covered by broad, dry stream channels, relics of torrential floodwaters draining from Lake Missoula. From giant current ripples, sandbars, and the coarse gravels found there, geologists have estimated that these floodwaters flowed as quickly as 30 m/s and discharged 21 million cubic meters per second. In contrast, the velocities of ordinary river flows are measured in tens or hundreds of centimeters per second, and their discharges range from a few cubic meters per second to the 57,000 m^3/s of the Mississippi in flood.

Eskers are long, narrow, winding ridges of sand and gravel found in the middle of ground moraines (see Figure 16.21). They run for kilometers in a direction roughly parallel to the direction of ice movement. The origin of eskers is suggested by the well-sorted, water-laid character of the materials and the sinuous, channel-like course of the ridge. Eskers were deposited by meltwater streams flowing in tunnels along the bottom of a melting glacier. The tunnels themselves were opened by water seeping through crevasses and cracks in the ice.

Permafrost

The ground is always frozen in very cold regions, where the summer temperature never gets high enough to melt more than a thin surface layer. Perennially frozen soil, or **permafrost,** today covers as much as 25 percent of Earth's total land area. In addition to the soil itself, permafrost includes aggregates of ice crystals in layers, wedges, and irregular masses. The proportion of ice to soil and the thickness of the permafrost vary from region to region. Permafrost is not defined by soil moisture content, overlying snow cover, or location. Rather, it is defined solely by temperature. Any rock or soil remaining at or below 0°C for 2 or more years is permafrost.

In Alaska and northern Canada, permafrost may be as thick as 300 to 500 m. The ground below the permafrost layer, insulated from the bitterly cold temperatures at the surface, remains unfrozen. It is warmed from below by the internal heat of the Earth. Permafrost is a difficult material to handle in engineering projects—such as roads, building foundations, and the Alaska oil pipeline—because permafrost melts as it is excavated. The meltwater cannot infiltrate the still-frozen soil below the excavation, so it stays at the surface, waterlogging the soil and leading it to creep, slide, and slump. Engineers decided to build part of the Alaska pipeline above ground when an analysis showed that the pipeline would thaw the permafrost around it in some places and lead to unstable soil conditions.

Permafrost covers about 82 percent of Alaska and 50 percent of Canada, as well as great parts of Siberia (**Figure 16.22**). Outside the polar regions, it is present in high, mountainous areas, especially on the Tibetan Plateau.

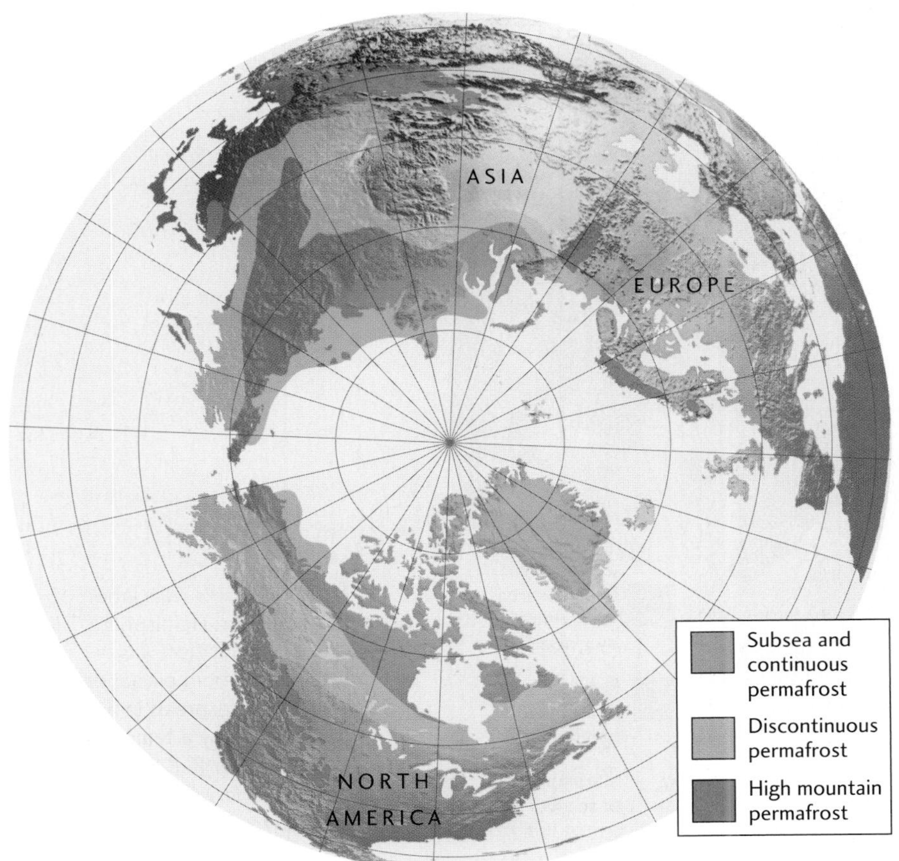

Figure 16.22 The North Pole is at the center of this map of the Northern Hemisphere showing the distribution of permafrost. The large area of high mountain permafrost at the top of the map is on the Tibetan Plateau. [After a map by T. L. Pewe, Arizona State University.]

Permafrost extends downward several hundred meters in shallow marine areas off the Arctic coasts, presenting difficult engineering problems for offshore oil drillers.

Ice Ages: The Pleistocene Glaciation

In the middle of the nineteenth century, the same Louis Agassiz who measured the speed of a Swiss alpine glacier immigrated to the United States and became a professor at Harvard University, where he continued his studies in geology and other sciences. For his research, Agassiz visited many places in the northern parts of Europe and North America, from the mountains of Scandinavia and New England to the rolling hills of the American Midwest. In all these diverse regions, Agassiz saw signs of glacial erosion and sedimentation (**Figure 16.23**). In flat plains country, he saw moraines that reminded him of the end moraines of valley glaciers. The heterogeneous material of the drift, including erratic boulders, convinced him of its glacial origin.

The areas covered by this drift were so vast that the ice that deposited them must have been a continental glacier larger than Greenland or Antarctica (**Figure 16.24**). Eventually, Agassiz and others convinced geologists and the general public that a great continental glaciation had extended the polar ice caps far into regions that now enjoy temperate climates. For the first time, people began to talk about ice

Figure 16.23 Irregular hills alternate with lakes in a terrain of glacial till, Coteau des Prairies, South Dakota. [John S. Shelton.]

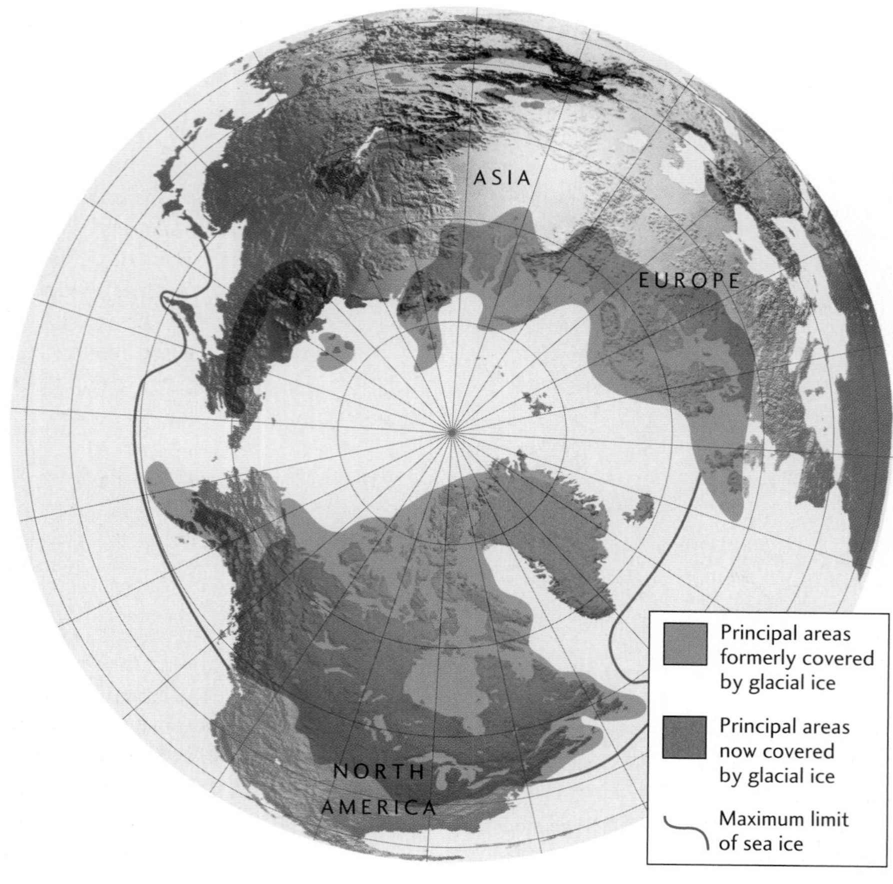

Figure 16.24 Glaciation of the Northern Hemisphere. The range of glaciation was established from glacial drift, which is widespread in areas that now enjoy a temperate climate. [After R. F. Flint, *Glacial and Quaternary Geology* (New York: Wiley, 1971).]

ages. It was also apparent that the glaciation occurred in the relatively recent past, because the drift was soft, like freshly deposited sediment. We now know the age accurately from radiometric dating of the carbon-14 in logs buried in the drift. The drift of the last glaciation was deposited during one of the most recent epochs of geologic time, the Pleistocene, which lasted from 1.8 million to 10,000 years ago. Along the east coast of the United States the southernmost advance of this ice is recorded by the enormous sand and till deposits of the terminal moraines that form Long Island and Cape Cod.

Multiple Glacial Ages: Continental and Oceanic Records

It soon became clear that there were multiple glacial ages during the Pleistocene, with warmer interglacial intervals between them. As geologists mapped glacial deposits a little more than a century ago, they became aware that there were several layers of drift, the lower ones corresponding to earlier ice ages. Between the older layers of glacial material were well-developed soils containing fossils of warm-climate plants. These soils were evidence that the glaciers retreated as the climate warmed. By the early part of the twentieth century, scientists believed that four distinct glaciations had affected North America and Europe during the Pleistocene epoch.

This idea was modified in the late twentieth century, when geologists and oceanographers examining oceanic sediment found fossil evidence of warming and cooling of the oceans. Ocean sediments presented a much more complete geologic record of the Pleistocene than continental glacial deposits. The fossils buried in Pleistocene and earlier ocean sediments were of foraminifera—small, single-celled marine organisms that secrete shells of calcium carbonate (calcite, $CaCO_3$; see Figure 3.1). These shells differ in their proportion of ordinary oxygen (oxygen-16) and the heavy oxygen isotope oxygen-18 (see Chapter 3). The ratio of oxygen-16 to oxygen-18 found in the calcite of a foraminifer's shell depends on the temperature of the water in which the organism lived. Different ratios in the shells preserved in various layers of sediment revealed the temperature changes in the oceans during the Pleistocene epoch (**Figure 16.25**). During glaciations, marine sediments are enriched in oxygen-18 because oxygen-16 is preferentially evaporated from the oceans and trapped in glacial ice.

Isotopic analysis of shells allowed geologists to measure another glacial effect. The oxygen isotope ratio of the ocean changes as a great deal of water is withdrawn from it by evaporation and is precipitated as snow to form glacial ice. The lighter oxygen-16 has a greater tendency to evaporate from the ocean surface than the heavier oxygen-18, Thus, more of the heavy isotope is left behind. Scientists could use isotopic analysis to trace the growth and shrinkage of continental glaciers, even in parts of the ocean where there may have been no great change in temperature—around the equator, for example. From this analysis of marine sediments, we have learned that there were many shorter, more regular cycles of glaciation and deglaciation than geologists had recognized from the glacial tills of the continents alone.

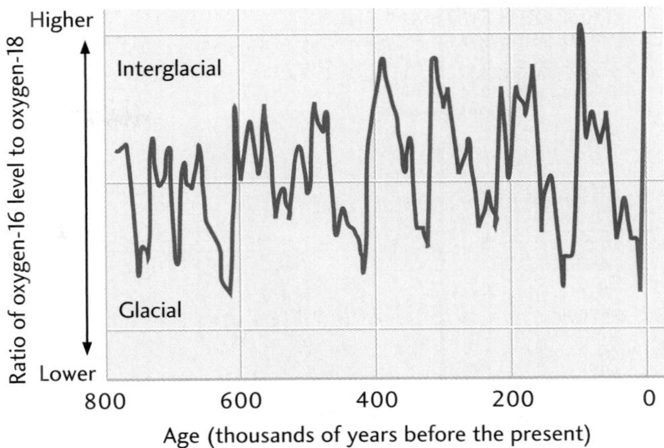

Figure 16.25 Relative changes in the ratio of oxygen-16 to oxygen-18 in the calcite of foraminifera in deep-sea sediment cores. This ratio changes in response to increases and decreases in the temperature of the water in which the foraminifera live and the extent to which evaporation and growth of glaciers enrich oxygen-18 in the ocean. Thus, this graph also indicates interglacial conditions (the peaks) and glacial conditions (the valleys). [After T. E. Graedel and P. F. Crutzen, *Atmospheric Change* (New York: W. H. Freeman, 1993).]

Climate Geosystem Interactions: Sea Level Change

The relationship between sea level change and global climate change provides an excellent illustration of interactions within the climate geosystem. As Earth warms or cools, its ice volume shrinks or grows, which directly affects the volume of the oceans. As ice volume increases, the volume of the ocean decreases and sea level falls. When ice volume decreases, sea level rises. Thus, sea level change is indirectly linked to climate change through interactions with temperature and ice volume.

You may recall from our coverage of the hydrologic cycle in Chapter 13 that much of the water that falls as snow on the continents originates in the sea. During the Pleistocene epoch, the more water there was tied up as ice in continental glaciers, the less water there was in the ocean. As continental glaciers grew, sea level dropped everywhere on the globe. The accumulation of a large amount of ice—say, 400,000 km^3—corresponds to only a small drop in sea level—about 1 m—depending on the density of the ice and the configuration of the shallow ocean and low-lying land areas. During the maximum extent of the most recent glaciation 18,000 years ago, however, sea level dropped about 130 m. This drop in sea level corresponds to an enormous

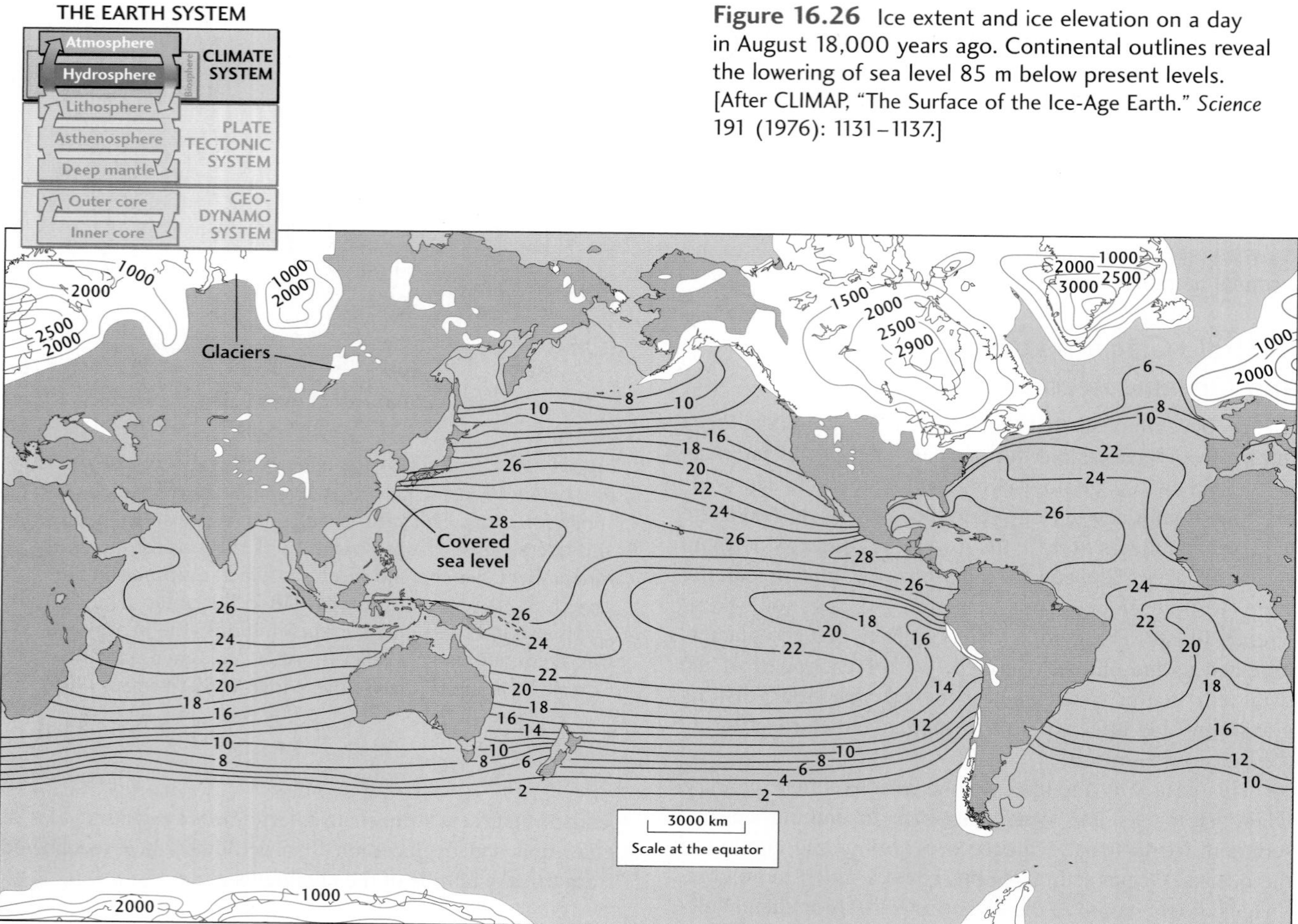

Figure 16.26 Ice extent and ice elevation on a day in August 18,000 years ago. Continental outlines reveal the lowering of sea level 85 m below present levels. [After CLIMAP, "The Surface of the Ice-Age Earth." *Science* 191 (1976): 1131–1137.]

volume of ice—about 70 million cubic kilometers, or almost three times the amount of ice on Earth today.

Figure 16.26 shows the ice distribution on the continents 18,000 years ago. Masses of ice 2 to 3.5 km thick built up over North America, Europe, and Asia. In the Southern Hemisphere, the Antarctic ice expanded, and the southern tips of South America and Africa were covered with ice. The continents were slightly larger than they are today because the continental shelves surrounding them—some more than 100 km wide—were exposed by the large drop in sea level.

Rivers extended their channels across the newly emergent continental shelves and began to erode channels in the former seafloor. Early cultures, such as those of prehistoric Egypt, were evolving in the lands beyond the ice sheets, and humans roamed these low coastal plains.

Climate Geosystem: Causes of Glacial Epochs

Geologists have long debated the causes of ice ages. Today the discussion is heightened by concern both over possible short-term warming trends that will melt part of the world's glaciers and over long-term cooling trends that might expand them (see Feature 16.2). Although geologists still dispute details, it is now generally accepted that ice ages result from interactions within the climate geosystem. Theories of ice ages center on climate change and possible factors in the atmosphere and oceans that could provoke glaciation at one time and then reverse course to cause deglaciation.

Theories of ice ages try to explain two phenomena:

1. The general cooling of polar regions, which led to the polar ice caps. Evidence from glacial sediments in Antarctica and in the ocean indicates that today's polar ice caps began to form about 10 million years ago, in the Miocene epoch, and grew slowly until they culminated in the full-blown glaciers of the Pleistocene.

2. The causes of alternating glacial and interglacial ages, which follow a more or less regular pattern throughout the Pleistocene.

A favored explanation for the general cooling of polar regions lies in the positions of the continents in relation to

16.2 Future Changes in Sea Level and the Next Glaciation

What would happen if the 25 million cubic kilometers of the planet's water now tied up as ice were to melt? The sea level would rise by about 65 m, flooding many of the world's major cities—New Orleans, London, New York, Los Angeles, Tokyo, and many others. Low-lying areas of the continents, where most of the world's population lives, would be submerged beneath shallow seas.

It is not likely that all of Earth's ice will melt for many million years to come. But it is very possible that a warming of climate in this new century will have two effects that could cause the sea level to rise:

1. The melting of some glacial ice alone would produce a small rise.
2. The extremely small expansion of water that accompanies an increase in temperature would be multiplied by the huge amount of ocean water warmed, causing the oceans to expand and the sea level to rise.

A global warming of just a few degrees might cause sea level to rise by a few meters. Although this amount may seem negligible, it is not. During storms and tidal surges, the sea level would be enough higher to cause major flooding of low-lying coastal regions. Millions of people in Bangladesh, for example, live on the deltas of the Ganges and Brahmaputra rivers, only a few meters above sea level. A 1-m rise in sea level would be devastating for these people, who have already suffered catastrophic floods in which thousands of lives were lost.

We can expect the Earth to return to a period of glaciation within the next 10,000 years, because the current interglacial age has already lasted 10,000 years and the average length of interglacial periods is about 20,000 years. We can predict the general scenario. As polar and temperate regions grow colder, perhaps because of a gradual decrease in atmospheric carbon dioxide, valley glaciers will grow larger. In parts of northern North America, Europe, and Asia, snow will stop melting in the summers. Ice will gradually accumulate in large continental glaciers, and it will eventually start to flow into the colder parts of formerly temperate regions, engulfing everything in its path. At the same time, sea level will fall, leaving the seaports of the world high and dry. A mass migration of people and animals into warmer regions will follow. Forest and plant populations will change in response to their new climates. It is hard to envision how the human population of the future will cope with such enormous changes.

The Bay of Bengal, Bangladesh. Low-lying Bangladesh is subject to disastrous flooding in the event of a rise in sea level, which could result from global warming. [James P. Blair/National Geographic.]

one another and to the poles. Continental drift driven by plate tectonics determines these positions. For most of Earth's history, there were no extensive land areas in the polar regions, and there were no ice caps. The oceans circulated through the open polar regions, transporting heat from the equatorial regions and helping the atmosphere to distribute temperatures fairly evenly over the Earth. But when large land areas drifted to positions that obstructed the efficient transport of heat by the oceans, the differences in temperature between the poles and the equator increased. As the poles cooled, the ice caps formed. We can predict that many millions of years from now, if the large polar and subpolar lands of the northern continents and Antarctica drift away from the poles, the ice caps will melt away.

So far, the alternation between glacial and interglacial ages has been explained best by astronomical cycles. These cycles cause periodic variations in the amount of heat Earth receives from the Sun. The shape of Earth's orbit around the Sun changes cyclically, being more circular at some times and more elliptical at others. The degree of ellipticity of Earth's orbit around the Sun is known as *eccentricity.* A circular orbit has low eccentricity, and a more elliptical orbit has high eccentricity (**Figure 16.27**a). The time interval of one cycle of variation from low to high eccentricity is about 100,000 years.

Another astronomical cycle involves the angle of *tilt* of Earth's rotation axis. The angle cycles between 21.5° and 24.5° over a time period of about 41,000 years. This cycle also slightly changes the heat received from the Sun (Figure 16.27b). In addition, Earth wobbles slightly on its axis of rotation, giving rise to a *precession* with a time period of about 23,000 years (Figure 16.27c). Precession, too, changes the amount of heat that Earth receives from the Sun.

Milutin Milankovitch, a Yugoslav geophysicist, first calculated these orbital cycles in the 1920s and 1930s. His work showed that glacial and interglacial stages correspond to variations in the heat Earth receives from the Sun, which result from the sum of all Earth's orbital changes. Recent work on this theory predicts regular changes in climate, with a long-term periodic glaciation every 100,000 years and shorter ones about every 40,000 and 20,000 years.

Evidence from Boreholes in the Ice

As the climatic interpretation of the marine sediment record of multiple glaciations was confirmed and grew more detailed, geologists searched for better continental records of glaciation. This search led to the development of new technologies for drilling deep boreholes in the Antarctic and Greenland ice sheets and recovering ice cores from them. In the Antarctic, Russian geologists at Vostok Station have been drilling ever deeper cores since the 1970s; in 1993, they reached a depth of 2755 m. Several groups of U.S. and European geologists have drilled a series of boreholes in the Greenland ice.

The ice cores from these boreholes have given us a detailed stratigraphic record of yearly ice layers produced by the conversion of snow into ice. At the same time, the oxygen isotope and carbon dioxide content of the ice layers has yielded a record of temperature variations In this way, geologists could determine the glacial and interglacial intervals from the ice cores (see Feature 16.1). An additional kind of information came from bubbles of air trapped at the time the ice formed. When they analyzed these tiny bubbles for various gases, Swiss and French scientists discovered that carbon dioxide in the atmosphere was lower during glacial periods and higher during interglacial periods. This finding is just what one would expect from the greenhouse effect and is a striking confirmation of the theory of global climate change that we will discuss more fully in Chapter 23. The changes in carbon dioxide and the shifts from glacial to interglacial periods do not seem to have taken place at precisely the same times, but some of the differences may be attributable to errors in measurement.

Ice cores from a mountain glacier 6 km high in the Peruvian Andes have recently caused us to revise our view of tropical climates during the last ice age. Although past work on deep-sea sediments suggested that the tropical oceans

(a) Eccentricity (100,000 years)

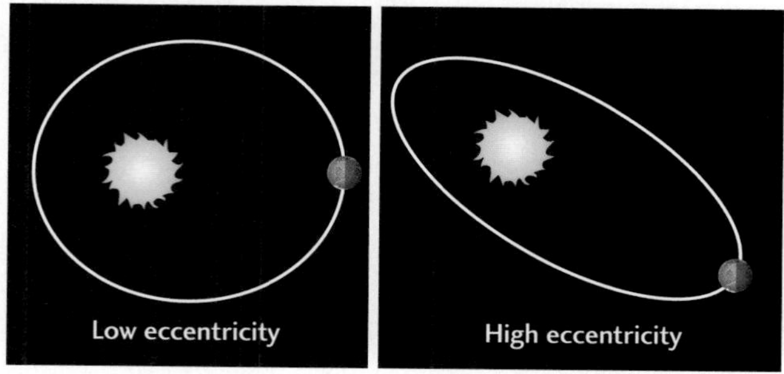

(b) Tilt (41,000 years)

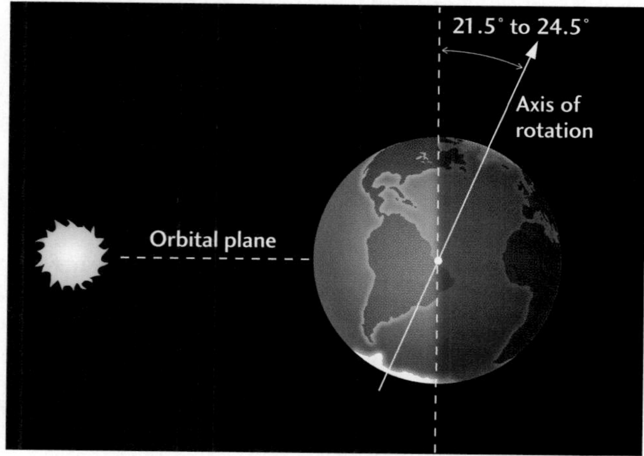

(c) Precession (23,000 years)

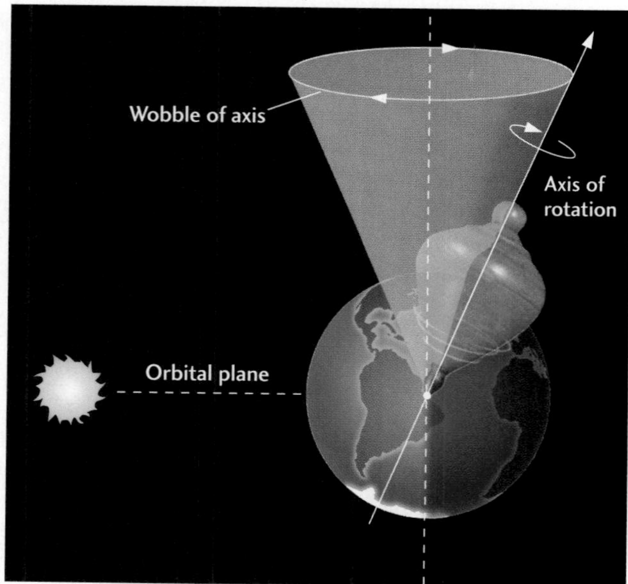

Figure 16.27 Three components of variations in Earth's orbit around the Sun and their time intervals for a single cycle. (a) *Eccentricity* is the degree of ellipticity of Earth's orbit. (b) *Tilt* is the angle between Earth's axis of rotation and the vertical to the orbital plane. (c) *Precession* is the wobble of the axis of rotation. One can imagine this motion by thinking of the wobble of a spinning top.

were warm and were hardly affected by glaciation at higher latitudes, the Peruvian core indicates that temperatures in tropical climates oscillated sharply at the end of the most recent ice age—just as the regions at higher latitudes did.

Climate Since the Last Glaciation

We have learned from ice cores in Greenland that a complicated series of events took place before, during, and after the relatively abrupt deglaciation that ended the most recent ice age. A fast warming seems to have occurred about 14,500 years ago. About 1000 years later, the climate started to cool again and started another period of glaciation about 12,500 years ago. This temporary return to glacial conditions, called the Younger Dryas period, lasted about 1000 years. Then, about 11,700 years ago, the temperature increased again, by about 6°C, and deglaciation proceeded to the present state of shrunken polar ice caps and glaciers.

The transition from the cold of the Younger Dryas to the relative warmth of the present interglacial period occurred abruptly. It started with a sharp global temperature rise of 5° to 10°C and continued to an increase of about 15°C. The initial jump was computed to have taken place in less than 10 years. The speed of the transition was a shock to geologists, who had thought the changes took place over many thousands of years. These and other fast changes in climate suggest that the global climate system operates like the rapid action of a switch, going from one state to another in a matter of a few years.

Some of the climate changes of the past 10,000 years seem to be related to cyclical changes in meltwater volume. Six oscillatory changes are correlated with continental glacier advances and retreats, which in turn correlate with the magnitude of iceberg calving. Amounts of iceberg calving are estimated from seafloor trails of very coarse grained sediment distributed from Labrador to France. These trails seem to have originated when large fleets of icebergs broke off from the ice sheets of eastern Canada and melted as they were carried to the east. Meltwater that floated on the surface accompanied the icebergs, altering the circulation of the North Atlantic and thus affecting the climate of North America and Europe.

During normal, nonglacial times, circulation somewhat like a conveyor belt delivers an enormous amount of heat to the North Atlantic Ocean and moderates Europe's climate (**Figure 16.28**). The warm surface currents flow from equatorial regions northeastward. In the North Atlantic, the water cools and becomes more saline (because less fresh water enters the sea from rivers at high latitudes) The cooler, saltier water sinks because it is has become denser. In this way, a subsurface cold current is created that flows southward as part of a general *thermohaline* circulation—so called because it is driven by differences in temperature and salinity. If the normal balance between North Atlantic cooling at high latitudes and the input of fresh water from rain, snowfall, and river runoff is disturbed by the addition of much fresh water (say, from melting glaciers), the thermohaline circulation slows, perhaps even shutting down completely. The climatic implications are poorly understood but could be dangerous. Europe might become much colder than today, for example.

Climatic events of the past thousand years have been deduced from ice cores and confirmed by human history.

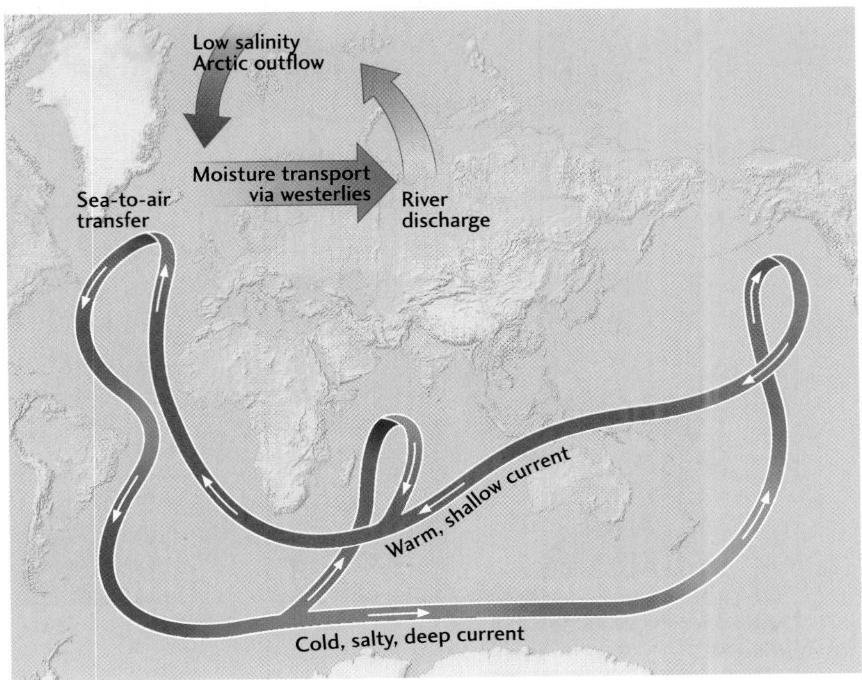

Figure 16.28 The global ocean thermohaline circulation, which involves the joint effects of temperature and salinity, is sometimes referred to as the ocean's "conveyor belt." It is important because it is responsible for a large portion of the heat transport from the tropics to higher latitudes in the present climate. For example, the Gulf Stream in the Atlantic Ocean forms part of the global thermohaline circulation, transporting warmer waters northward, thereby contributing to western Europe's relatively mild climate for its latitude.

From A.D. 1400 to 1650, Earth underwent the Little Ice Age, during which the Baltic Sea froze over and ice on the Thames River in England—which has not frozen over in modern times—reached thicknesses of several inches. There is no agreement among geologists and atmospheric scientists about the exact causes of these events. It seems certain, however, that we will learn more about climate change in relation to glaciation as the nations of the world become increasingly concerned about the effects of future climate changes.

The Record of Ancient Glaciation

The Pleistocene glaciation was not unique in Earth's history. Since the early part of the twentieth century, we have known from glacial striations and lithified ancient tills, called **tillites,** that glaciers covered parts of the continents several times in the geologic past, long before the Pleistocene. Tillites indicate major continental glaciations during Permian-Pennsylvanian time, Ordovician time, and at least twice during Precambrian time.

The Permian-Pennsylvanian glaciation was widespread over southern Gondwanaland (**Figure 16.29**), and tillites are preserved across much of the continents in the Southern Hemisphere (Figure 16.29b). The Ordovician glaciation was more limited in its distribution and is best preserved in northern Africa. The oldest Precambrian glaciation occurred about 3 billion years ago—although this interpretation is disputed. The oldest confirmed glaciation occurred about

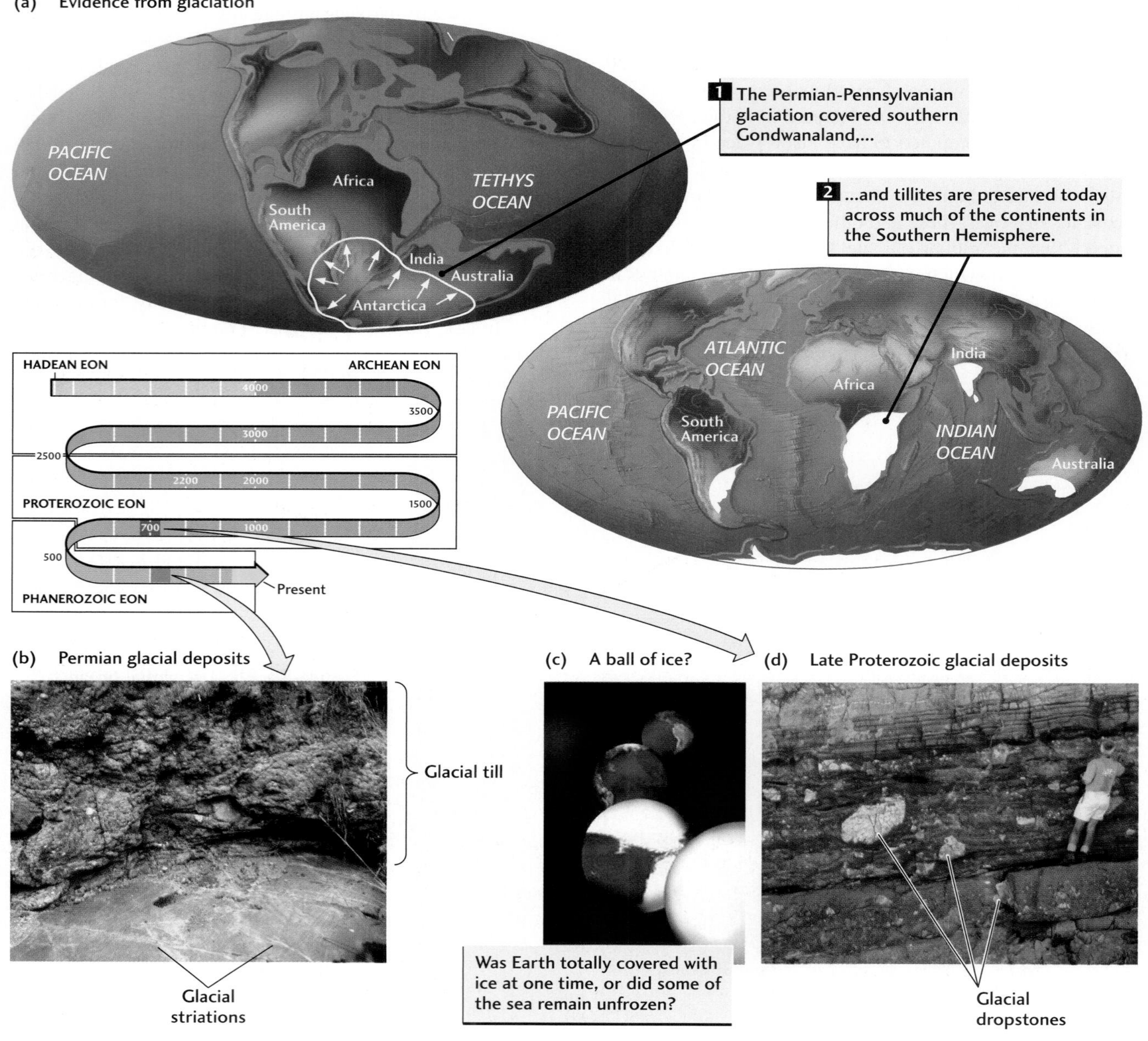

2.4 billion years ago, and its glacial deposits are preserved in Wyoming, along the Canadian portion of the Great Lakes, in northern Europe, and in South Africa. The youngest Precambrian glacial epoch, which spanned the period between 750 million and about 600 million years ago, involved several episodes of glaciation separated by warm interglacial periods. Curiously, paleomagnetic data indicate that ice may have extended all the way to the equator This evidence has provoked some geologists to speculate that Earth may have been completely covered by ice, from pole to pole. This bold hypothesis is called *Snowball Earth* (Figure 16.29c). Glacial deposits of this age have been found on every continent (Figure 16.29d).

On Snowball Earth, there was ice everywhere—even the oceans were frozen. The average global temperatures would have been Antarctic, about –40°C. Except for a few warm spots near volcanoes, no life would have survived. How could such an apocalyptic event have occurred? And how could it have ended, returning us to the Earth we know today? The answers lie in the feedbacks that occur within the climate geosystem.

Initially, as Earth cooled, ice sheets at the poles spread outward, their white surfaces reflecting more and more sunlight away from the Earth. The loss of sunlight cooled the Earth, which further expanded the ice sheets. The process would have continued until it reached the tropics, encasing the planet in a layer of ice as much as 1 km thick. This scenario is an example of positive feedback in the climate geosystem. In positive feedback, the results of a process reinforce the process itself. (We discussed positive-feedback systems in Chapter 7, in the section on soil formation.)

For millions of years, Earth remained buried in ice, while the few volcanoes that poked above the surface slowly pumped carbon dioxide into the atmosphere. The accumulating carbon dioxide trapped solar heat, eventually causing greenhouse warming. When the level of carbon dioxide in the atmosphere reached a critical level, the temperature rose abruptly as a result of greenhouse warming, the ice melted, and Earth became a hothouse. In this scenario, the increase of carbon dioxide represents a negative feedback in the climate geosystem. In negative feedback, the results of a process slow down or retard the process.

The Snowball Earth hypothesis is controversial, and some geologists disagree with the idea that the oceans may have frozen. Nevertheless, the evidence for glaciation at low latitudes is strong, and the hypothesis serves as an example of how feedbacks in Earth's climate geosystem can work to produce extreme change. Geologists have their work cut out for them as they test this hypothesis using their understanding of how Earth's climate geosystem behaves.

SUMMARY

How do glaciers form, and how do they move? Glaciers form where climates are cold enough that snow, instead of melting away in summer, is transformed into ice by recrystallization. As snow accumulates, the ice thickens, either at the tops of mountain valley glaciers or at the domed centers of continental ice sheets The thickness increases until the ice becomes so heavy that gravity starts to pull it downhill. During periods of stable climate, the size of a glacier remains constant. The ice lost by melting, sublimation, and iceberg calving in the ablation zone is balanced by new snow gained in the accumulation zone. During a warming period, a glacier shrinks as ablation exceeds accumulation. Conversely, a glacier expands in a cooling period as accumulation outpaces ablation. Glaciers move by a combination of plastic flow and basal slip. Plastic flow dominates in very cold regions, where the glacier's base is frozen to the ground. Basal slip is more important in warmer climates, where melting at the glacier's base lubricates the ice and helps it move over rock. The speed of a glacier's movement varies with the level and position of the ice.

Why do ice shelves float? Ice floats on the ocean in exactly the same way that ice cubes float in a glass of water. Ice is less dense than water, so ice cubes, icebergs, and ice shelves all float. Floating occurs because the ice is buoyed up by a force that is equal to the mass of the water that the ice displaces. This force tends to push the ice back to the surface of the water and counteracts the gravitational force that tends to make the ice sink. The condition of floating—when these forces are equivalent—is known as isostasy. Isostasy ensures that sea level does not rise when either icebergs or ice shelves melt. This is because the mass of the ice is equal to the mass of the water it displaces. When the ice melts, it simply replaces the water it displaced.

Figure 16.29 Ancient glacial epochs. (a) Maps showing distribution of Permian-Pennsylvanian glacial deposits, formed more than 350 million years ago. At this time, the continents were assembled as the supercontinent Gondwanaland and the ice was situated in the Southern Hemisphere, centered over Antarctica, as the modern-day ice fields are. The second map shows the distribution of Permian-Pennsylvanian glacial deposits today. (b) Permian glacial deposits from South Africa [John Grotzinger]. (c) The development of a late Precambrian Snowball Earth. Geologists debate the extent to which ice covered the globe, but some think even the oceans became frozen. (d) Late Proterozoic glacial deposits [John Grotzinger].

How do glaciers erode bedrock, transport and deposit sediment, and shape the landscape? Glaciers erode bedrock by scraping, plucking, and grinding bedrock into sizes ranging from boulders to fine rock flour. Valley glaciers erode cirques and arêtes at their heads; excavate U-shaped and hanging valleys in their main courses; and create fjords where they end at the ocean and erode their valleys below sea level. Glacial ice has both high competence and high capacity, which enable it to carry abundant sediments of all sizes. Glaciers transport huge quantities of sediments to the ice front, where melting releases them. The sediments may be deposited directly from the melting ice as till or picked up by meltwater streams and laid down as outwash. Moraines and drumlins are characteristic landforms deposited by ice. Kames, eskers, and kettles are laid down by meltwater. Permafrost is perennially frozen soil in very cold regions.

What are ice ages, and what causes them? Glacial drift of Pleistocene age is widespread over high-latitude regions that now enjoy temperate climates. Widespread drift is evidence that continental glaciers once expanded far beyond the polar regions. Studies of the geologic ages of glacial deposits on land and of seafloor sediments show that the Pleistocene glacial epoch consisted of multiple advances (glacial intervals) and retreats (interglacial intervals) of the continental ice sheets. Each advance corresponded to a global lowering of sea level that exposed large areas of continental shelf. During interglacial intervals, sea level rose and submerged the shelves. Although the causes of glaciation remain uncertain, the global cooling that leads to glaciation appears to result from continental drift that gradually moves continents to positions where they obstruct the transport of heat from the equator to the poles. A favored explanation for the alternation of glacial and interglacial intervals is the effect of astronomical cycles. Very small periodic changes in the eccentricity of Earth's orbit and precession of its axis of rotation alter the amount of sunlight received at Earth's surface. There is also evidence that decreased levels of carbon dioxide in the atmosphere diminish the greenhouse effect and lead to glaciation. In contrast, the present increase in atmospheric carbon dioxide resulting from the burning of fossil fuels may cause global warming.

Key Terms and Concepts

ablation (p. 352)
arête (p. 362)
basal slip (p. 355)
cirque (p. 362)
continental glacier (p. 345)
crevasse (p. 357)
drift (p. 364)
drumlin (p. 365)
erratic (p. 364)
esker (p. 367)
firn (p. 352)
fjord (p. 364)
glacier (p. 348)
hanging valley (p. 364)
ice stream (p. 358)
isostasy (p. 360)
kame (p. 365)
kettle (p. 366)
moraine (p. 365)
outwash (p. 364)
permafrost (p. 367)
plastic flow (p. 355)
rock flour (p. 362)
striation (p. 362)
surge (p. 357)
till (p. 364)
tillite (p. 374)
U-shaped valley (p. 364)
valley glacier (p. 348)
varve (p. 366)

Exercises

This icon indicates that there is an animation available on the Web site that may assist you in answering a question.

1. How are valley glaciers distinguished from continental glaciers?

2. How is snow transformed into glacial ice?

3. How does glacial growth or shrinkage result from the balance between ablation and accumulation?

4. What are the mechanisms of glacier flow?

5. How do glaciers erode bedrock?

6. Why are striations evidence of former glaciation?

7. Name three kinds of glacial sediment.

8. Name three landforms made by glaciers.

9. How does glaciation affect sea level?

10. What variations in Earth's orbit affect climate?

11. How does carbon dioxide in the atmosphere affect climate?

Thought Questions

This icon indicates that there is an animation available on the Web site that may assist you in answering a question.

1. Why is glacial ice stratified in many places?

2. Some parts of a glacier contain much sediment, others very little. What accounts for the difference?

3. Contrast the kinds of till that you would expect to find in two glaciated areas: one a terrain of granitic and metamorphic rocks; the other a terrain of soft shales and loosely cemented sands.

4. What geological evidence would you search for if you wanted to know the direction of glacial movement in a continentally glaciated region?

5. You are walking over a winding ridge of glacial drift. What evidence will you look for to discover whether you are on an esker or an end moraine?

6. One of the dangers of exploring glaciers is the possibility of falling into a crevasse. What topographic features of a valley glacier or its surroundings would you use to infer that you were on a part of the glacier that was badly crevassed?

7. You are scouting out possible commercial sand and gravel deposits in an area that was once near the edge of a continental glacier. What landforms will you look for?

8. You live in New Orleans, not far from the mouth of the Mississippi River. What might be your first indication that the world is entering a new glacial age?

9. In the century just before Columbus's voyage in 1492, the Native Americans living in the northern United States and Canada must have noticed a significant change in climate. What kinds of changes in their surroundings might they have noticed, and how long did such changes last?

10. Some geologists think that one result of continued global warming could be the shrinkage and collapse of the West Antarctic ice sheet. How might this affect the populations of North America and Europe?

11. Evidence from boreholes drilled into the ice shows that there is liquid water at the base of some glaciers. What kinds of glaciers might have liquid water at the base? What factors might be responsible for the melting of ice at the bottom of these glaciers?

12. Ice is a very unusual substance because it is less dense than its liquid form, water. However, it is only a bit less dense (0.9 gram/cm^3) than water (1.0 gram/cm^3) which accounts for why most of an iceberg is actually under water. If ice had the same density as water, how high would the top of an iceberg stand relative to sea level?

Suggested Readings

Alley, R. B., and M. L. Bender. 1998. Greenland ice cores: Frozen in time. *Scientific American* (February): 80–85.

Broecker, W. S., and G. H. Denton. 1990. What drives glacial cycles? *Scientific American* (January): 48–56.

Covey, C. 1984. The Earth's orbit and the ice ages. *Scientific American* (February): 58.

Denton, G. H., and T. J. Hughes. 1981. *The Last Great Ice Sheets.* New York: Wiley.

Hambrey, M. J., and J. Alean. 1992. *Glaciers.* Cambridge: Cambridge University Press.

Hoffman, P., and D. Schrag. 2000. Snowball Earth. *Scientific American* (January): 68.

Imbrie, J., and K. P. Imbrie. 1979. *Ice Ages: Solving the Mystery.* Short Hills, N.J.: Enslow Press.

Menzies, J. (ed.). 1995. *Modern Glacial Environments: Processes, Dynamics and Sediments.* Oxford: Butterworth-Heinemann.

Sharp, R. P. 1988. *Living Ice: Understanding Glaciers and Glaciation.* Cambridge: Cambridge University Press.

Waves pounding a rocky shoreline at Cape Arago State Park, Coos County, Oregon.
[Steve Terrill/Corbis.]

CHAPTER

17

Earth Beneath the Oceans

"O blest Atlantis! Can the legend be built on wild fancies which thy name surround? Or doth the story of thy classic ground with stern facts of Nature's face agree?"

T. VERON WOLLASTON (FIRST CRITICAL REVIEW OF DARWIN'S *ORIGIN OF SPECIES*)

For most of human history, the 71 percent of Earth's surface covered by the oceans was a mystery. The large populations who lived at the edge of the sea knew well the force of the waves and the rise and fall of the tides, but they could only guess at the nature of the seafloor deeper than the shallowest coastal waters. By the middle of the nineteenth century, occasional oceangoing ships made scientific observations of water depths, the plants and animals of the sea, and the chemistry of seawater. Then, in 1872, H.M.S. *Challenger,* a small wooden British warship converted and fitted out for the first specifically scientific study of the seas, left England for a 4-year voyage over the world's oceans. The 50 thick volumes of reports from that expedition gave the public its first knowledge of the Mid-Atlantic Ridge, one of the longest mountain ranges in the world—almost all of it underwater. The *Challenger* expedition discovered great areas of submerged hills and flat plains, extraordinarily deep trenches, and submarine volcanoes.

Today, 130 years after that pioneering voyage, hundreds of oceanographic research vessels from many countries ply the seas in search of answers to the questions first raised by the early discoveries. What tectonic forces raised the submarine mountain ranges and depressed the trenches? Why are some areas of the seafloor flat and others hilly? Although oceanographers made many important discoveries in the first half of the twentieth century, the answers to most of these questions had to await the plate tectonic revolution of the late 1960s. As we saw in Chapter 2, it was geological and geophysical observations of the ocean floors, not the continents, that led to the theory of plate tectonics.

In this chapter, we examine what early and modern researchers have discovered about the geology of Earth's ocean basins, with their submerged

mountains and valleys; underwater volcanoes; many kinds of rocks, sediments, and chemical components; and waves and tides.

The word *ocean* applies both to the five major oceans (Atlantic, Pacific, Indian, Arctic, and Antarctic) and to the single connected body of water called the *world ocean*. The term *sea* includes both the oceans and the smaller bodies of water set off somewhat from the world ocean. Thus, the Mediterranean Sea is narrowly connected with the Atlantic Ocean by the Straits of Gibraltar and with the Indian Ocean by the Suez Canal. Other seas, such as the North Sea and the Atlantic Ocean, are broadly connected. In the world ocean, seawater—the salty water of the oceans and seas—is remarkably constant in its general chemical composition from year to year and from place to place. The equilibrium maintained by the oceans is determined by the general composition of river waters entering the sea, the composition of the sediment brought into the oceans, and the formation of new sediment in the ocean.

Basic Differences in the Geology of Oceans and Continents

Plate tectonics has provided us with a basic understanding of the differences between continental and oceanic geology. The oceans have no folded and faulted mountains like those on the continents. Instead, plate tectonic deformation is largely restricted to the faulting and volcanism found at mid-ocean ridges and subduction zones. Moreover, the weathering and erosion processes discussed in previous chapters are much less important in the oceans than on the land because in the oceans there are no efficient fragmentation processes, such as freezing and thawing, or major erosive agents, such as streams and glaciers. Deep-sea currents can erode and transport sediment but cannot effectively attack the plateaus and hills of basaltic rock that form the oceanic crust.

Because tectonic deformation, weathering, and erosion are minimal over much of the seafloor, the constructive processes of volcanism and sedimentation dominate the geology of the oceans. Volcanism creates island groups in the middle of the oceans (such as the Hawaiian Islands), arcs of volcanic islands near deep oceanic trenches, and the mid-ocean ridges. Sedimentation shapes much of the rest of the ocean floor. Soft sediments of mud and calcium carbonate blanket the low hills and plains of the sea bottom and accumulate on oceanic plates as they spread from mid-ocean ridges. As the plates move farther and farther from a ridge, they accumulate sediments. Deep-sea sedimentation is more continuous than the sedimentation in most continental environments, and it therefore preserves a better record of geologic events—for example, a more detailed history of Earth's climate changes.

This record is limited in time, however, because subduction is continually swallowing the oceanic plates back into the mantle, thereby destroying oceanic sediments by metamorphism and melting. On average, it takes only a few tens of millions of years for the crust created at a mid-ocean ridge to spread across an ocean and come to a subduction zone. As we saw in Chapter 2, the oldest parts of today's ocean floor were formed in the Jurassic period, about 180 million years ago; they are currently found near the western edge of the Pacific Plate (see Figure 2.14). In the next 10 million years or so, the sedimentary record that lies atop this crust will disappear down a subduction zone.

We will begin our exploration of the oceans with a closer look at the deep ocean floor and the continental margins that bound the ocean basins.

Geology of the Deep Oceans

A topographic map of Earth's surface (**Figure 17.1**) reveals the most important geologic features submerged beneath the oceans: mid-ocean ridges, volcanic tracks of hot spots, deep-sea trenches, island arcs, and continental margins. How oceanographers have obtained this fundamental information is a fascinating story of scientific exploration.

Making a map of the deep ocean floor is no easy task. Because sunlight can penetrate only 100 m or so below the sea surface, the deep ocean is a very dark place. It is not possible to map the ocean floor using visible light, nor can we use radio waves beamed from spacecraft, as we have used to map the surface of cloud-shrouded Venus. Ironically, spacecraft photography has allowed us to map the surfaces of our planetary neighbors with much higher resolution than we have been able to map the deep ocean floor, even to this day.

Probing the Seafloor from Surface Ships

It is possible to view the seafloor directly from a deep-diving submersible. Pioneered by the French oceanographer Jacques-Yves Cousteau, these small ships can observe and photograph at great depths (**Figure 17.2**). With their mechanical arms, they can break off pieces of rock, sample soft sediment, and catch specimens of exotic deep-sea animals. Newer robotic submersibles are guided by scientists on the mother ship above. But submersibles are expensive to build and operate, and they cover small areas at best.

For most work, today's oceanographers use instrumentation to sense the seafloor topography indirectly from a ship at the surface. A shipboard echo sounder, developed in the early part of the twentieth century, sends out pulses of sound waves. When the sound waves are reflected back from the ocean bottom, they are picked up by sensitive microphones in the water. Oceanographers can compute the depth by measuring the interval between the time the pulse leaves the ship and the time it returns as a reflection. The result is an automatically traced profile of the bottom topography. Powerful echo sounders are also used to probe the stratigraphy of sedimentary layers beneath the ocean floor.

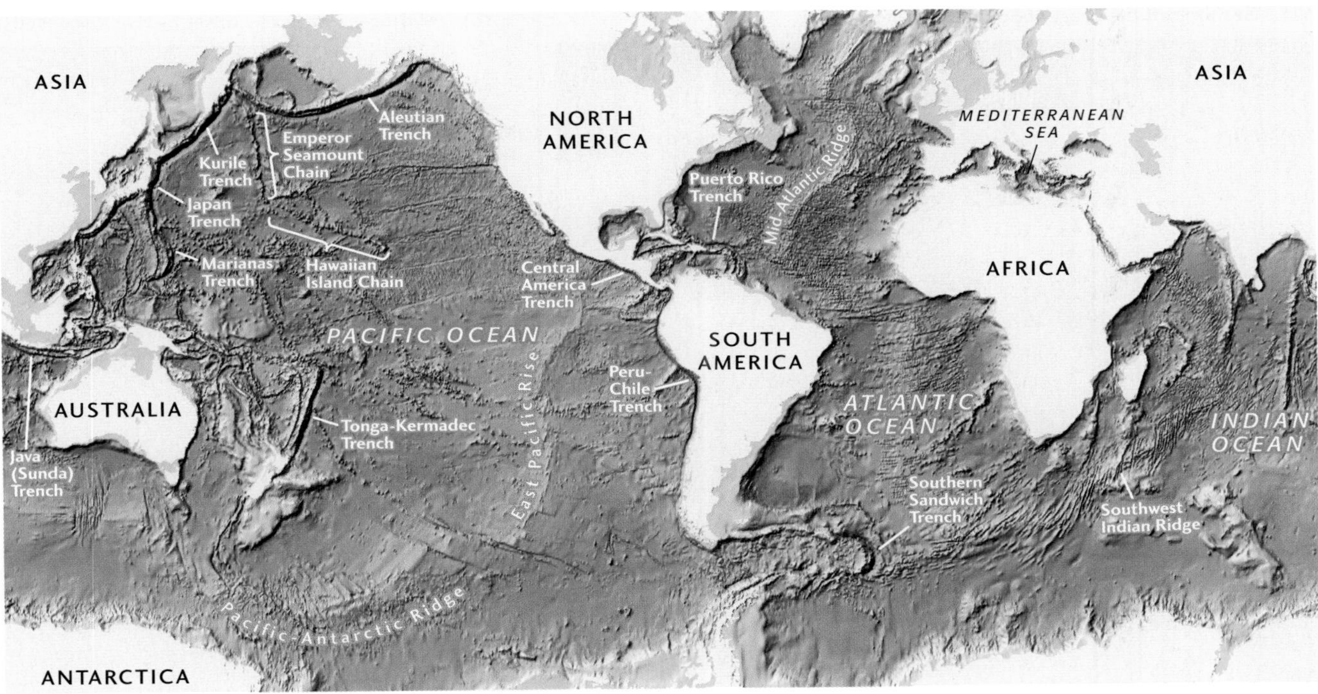

Figure 17.1 Earth's topography beneath the oceans showing the major features of the deep seafloor.

Many of today's oceanographic vessels are outfitted with hull-mounted arrays of echo sounders that can reconstruct a detailed image of seafloor topography in a swath extending as much as 10 km on either side of the ship as it steams along (see Figure 17.2). These systems can map seafloor topography over large regions with unprecedented resolution of small-scale geologic features, such as undersea volcanoes, canyons, and faults. **Figure 17.3** shows several

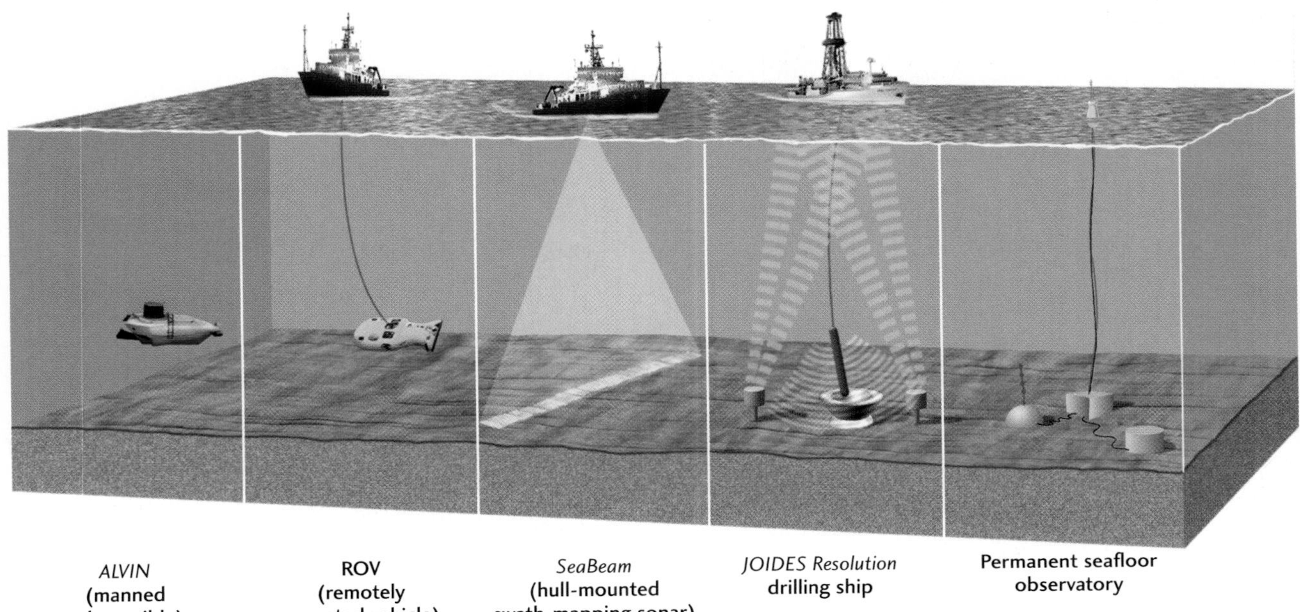

Figure 17.2 High-technology methods for exploring the deep seafloor. The manned deep submersible *ALVIN* and a remotely operated vehicle (ROV) are directed from a surface ship. *SeaBeam*, a hull-mounted multibeam echo sounder, continuously maps seafloor topography in a wide swath as a ship steams across the ocean surface. The drilling ship *JOIDES Resolution* uses bottom transponders to navigate a drill string into a reentry cone on the seafloor. Permanent unmanned seafloor observatories monitor processes in the subsurface and the overlying water column for extended periods of time.

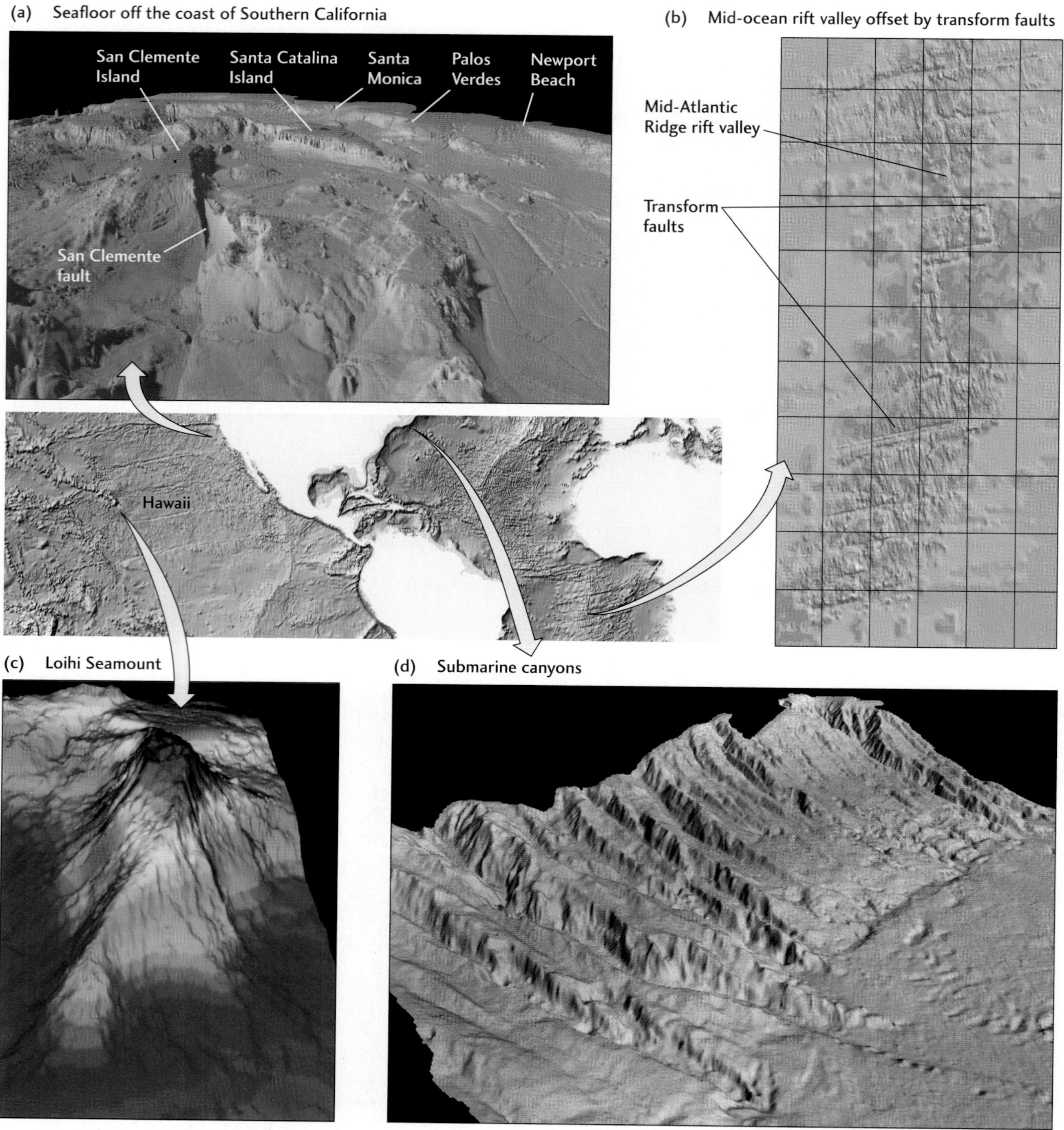

Figure 17.3 Three examples of topography of the seafloor obtained by high-resolution swath mapping by surface ships and rendered by computer processing as three-dimensional images. (a) The seafloor off the coast of southern California, showing fault-controlled structures of a geologic province known as the California Borderland. [Chris Goldfinger and Jason Chaytor, Oregon State University.] (b) Ridge between 25°S and 36°S showing the southeast-trending rift valley offset by northeast-trending transform faults. [Ridge Multibeam Database, Lamont-Doherty Earth Observatory, Columbia University.] (c) Loihi Seamount, just south and east of the Big Island of Hawaii, the newest in the string of hot-spot volcanoes that form the Hawaiian Islands chain. [Ocean Mapping Development Center, University of Rhode Island.] (d) The continental shelf (*top*), slope (*central and upper area*), and rise (*lower left*) off the coast of New England. Note the deep submarine canyons that incise the continental margin. [From L. Pratson and W. Haxby, *Geology* 24(1) (1996): 4. Courtesy of Lincoln Pratson and William Haxby, Lamont-Doherty Earth Observatory, Columbia University.]

impressive images of the seafloor derived by this type of mapping.

Other types of instruments can be towed behind a ship or lowered to the bottom to detect such properties as the magnetism of the seafloor, the shapes of undersea cliffs and mountains, and heat coming from the crust. Underwater cameras on sleds towed near the ocean bottom can photograph the details of the seafloor and the organisms that inhabit the deep. Since 1968, the U.S. Deep Sea Drilling Project and its successor, the international Ocean Drilling Program, have sunk hundreds of drill holes to depths of many hundreds of meters below the seafloor. Cores obtained from these drill holes have provided geologists with sediment and rock samples for detailed physical and chemical studies (see Figure 17.2 and Feature 2.1). Plans are now afoot to install a global network of unmanned deep-sea observatories that will send back streams of data about processes taking place on the deep seafloor and in the overlying water column. Marine geologists today work in a beehive of high technology!

Charting the Seafloor by Satellite

Despite all this fancy gear, there are still many regions of the oceans that have not been surveyed in detail by surface ships, and our knowledge of the seafloor remains fragmentary. Recently, however, scientists have developed a tool that enables a satellite to "see through" the ocean and chart the topography of the seafloor on a global scale.

The new method makes use of an altimeter mounted on a satellite. The altimeter sends pulses of radar beams that are reflected back from the ocean below, giving measurements of the distance between the satellite and the sea surface with a precision of a few centimeters. The height of the sea surface depends not only on waves and ocean currents but also on changes in gravity caused by the topography and composition of the underlying seafloor. The gravitational attraction of a seamount, for example, can cause water to "pile up" above it, producing a bulge in the sea surface of as much as 2 m above average sea level. Similarly, the diminished gravity over a deep-sea trench is evident as a depression of the sea surface of as much as 60 m below average sea level.

This method has allowed us to infer features of the ocean floor from satellite data and display them as if the seas were drained away. Marine geologists have used this technique to map new features of the seafloor not revealed by ship surveys, especially in the poorly surveyed southern oceans. The satellite data have also revealed deeper structures below the oceanic crust, including gravity anomalies associated with convection currents in the mantle, which we will discuss in Chapter 21.

Profiles Across Two Oceans

To gain an appreciation of the geologic features that lie beneath the oceans, we will take a brief tour across two of Earth's major ocean basins, the Atlantic and the Pacific, as if we were driving a deep-diving submarine along the ocean floor.

An Atlantic Profile The Atlantic profile shown in **Figure 17.4** extends from North America to Gibraltar. Starting from

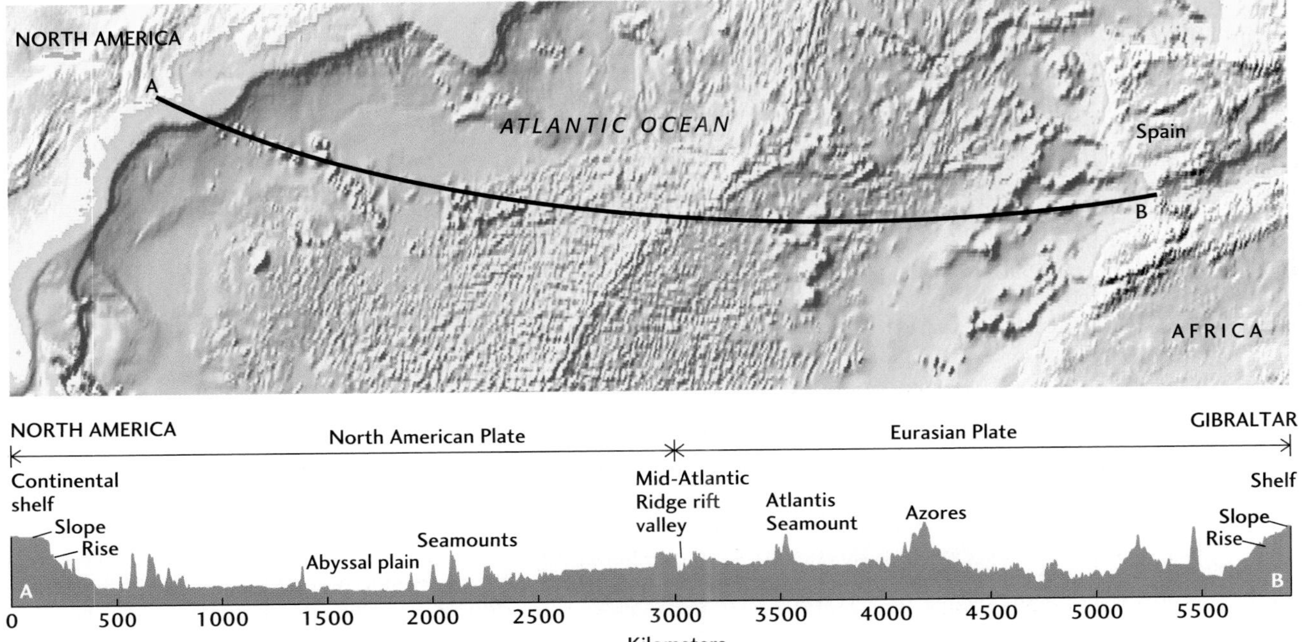

Figure 17.4 A topographic profile across the Atlantic Ocean from New England (*left*) to Gibraltar (*right*).

the coast of New England, we descend from the shoreline to depths of 50 to 200 m and travel along the **continental shelf.** This broad, flat, sand- and mud-covered platform is a slightly submerged part of the continent. After traveling about 50 to 100 km across the shelf, down a very gently inclined surface, we find ourselves at the edge of the shelf. There, we start down a steeper incline, the **continental slope.** This mud-covered slope descends at an angle of about 4°, a drop of 70 m over a horizontal distance of 1 km, which would appear as a noticeable grade if we were driving on land.

The continental slope is irregular and is marked by gullies and **submarine canyons,** deep valleys eroded into the slope and the shelf behind it (see Figure 17.3). On the lower parts of the slope, at depths of about 2000 to 3000 m, the downward incline becomes gentler. Here it merges into a more gradual downward incline called the **continental rise,** an apron of muddy and sandy sediment extending into the main ocean basin.

The continental rise is tens to hundreds of kilometers wide, and it grades imperceptibly into a wide, flat **abyssal plain** that covers large areas of the ocean floor at depths of about 4000 to 6000 m. These plains are broken by occasional submerged volcanoes, mostly extinct, called **seamounts.** As we travel along the abyssal plain, we gradually climb into a province of low abyssal hills whose slopes are covered with fine sediment. Continuing up the hills, the sediment layer becomes thinner, and outcrops of basalt appear beneath it. As we rise along this steep, irregular topography to depths of about 3000 m, we are climbing the flanks and then the mountains of the Mid-Atlantic Ridge.

Abruptly, we come to the edge of a deep, narrow valley a few kilometers wide at the top of the ridge (**Figure 17.5**). This narrow cleft, marked by active volcanism, is the rift valley where two plates separate. As we cross the valley and climb the east side, we move from the North American Plate to the Eurasian Plate. Continuing eastward, we encounter topography similar to that on the west side of the ridge, only in reverse order, because the ocean floor is more or less symmetrical on either side of the ridge. On the path we have taken, this symmetry is disturbed by some large seamounts and the volcanic Azores Islands, which mark an active hot spot, perhaps caused by the heat from an upwelling mantle plume (see Figure 2.18). Passing over the rough topography of the abyssal hills on the flank of the Mid-Atlantic Ridge, we descend to an abyssal plain, and then ascend up the continental rise, slope, and shelf off the coast of Europe.

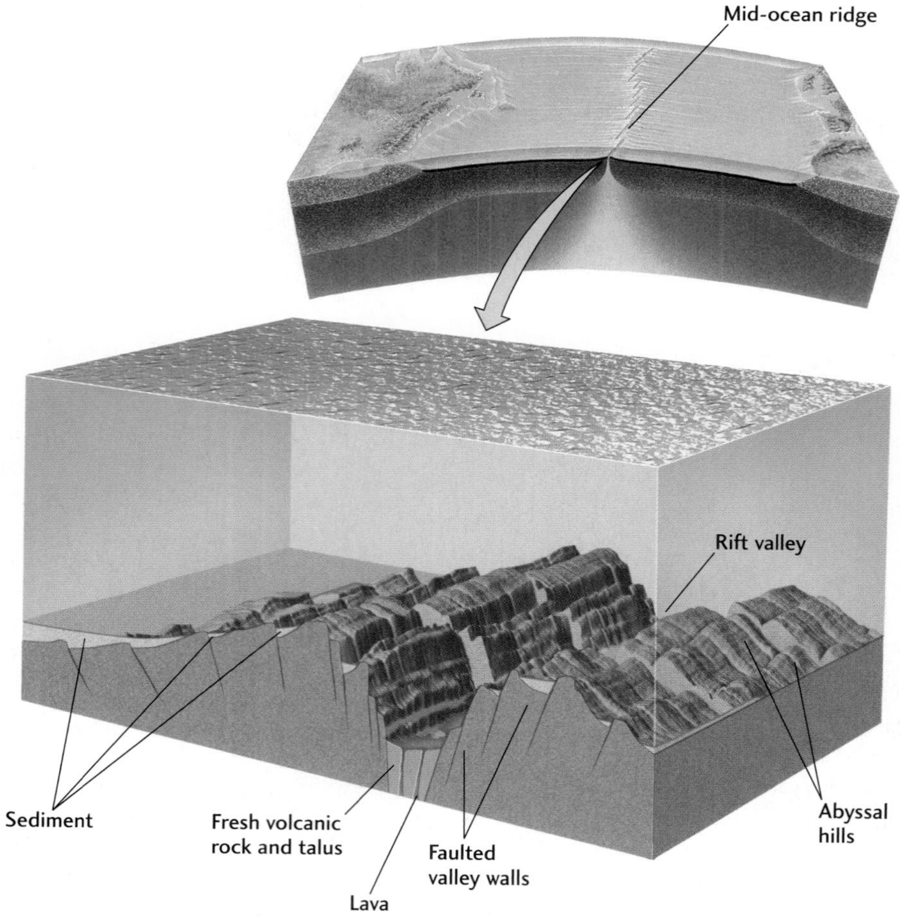

Figure 17.5 A profile of the central rift valley of the Mid-Atlantic Ridge in the FAMOUS (French-American Mid-Ocean Undersea Study) area southwest of the Azores Islands. The deep valley, where most of the basalt is extruded, is faulted. [After ARCYANA, "Transform Fault and Rift Valley from Bathyscaph and Diving Saucer." *Science* 190 (1975): 108.]

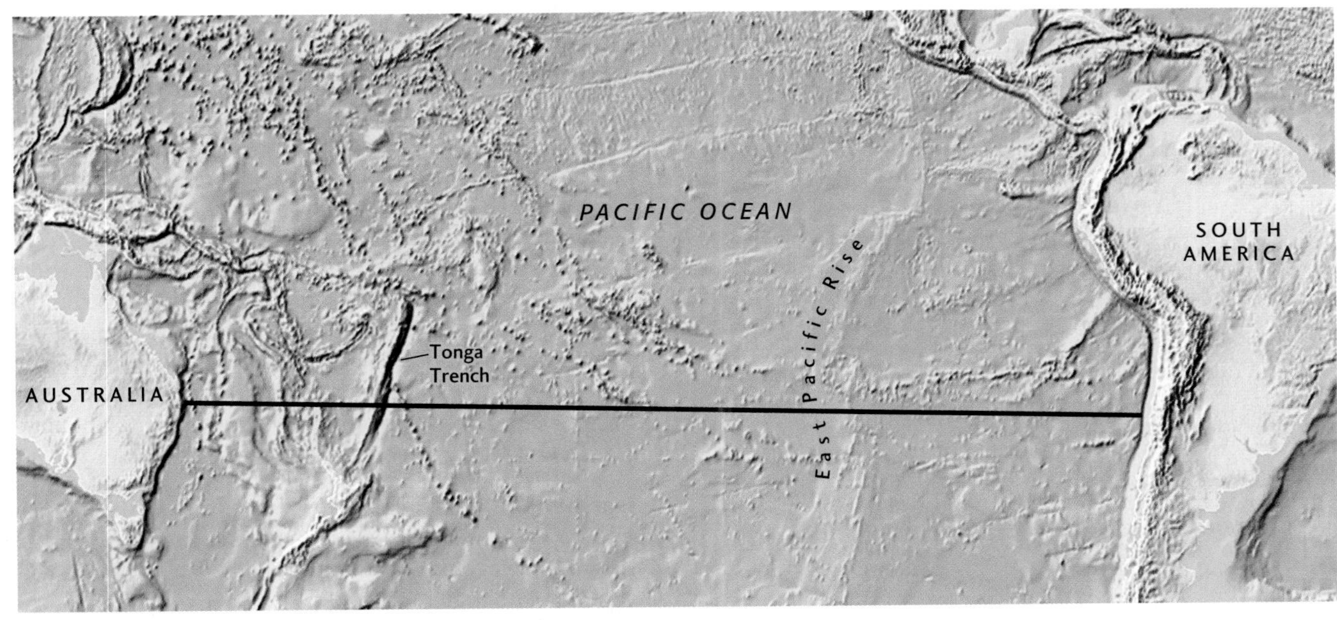

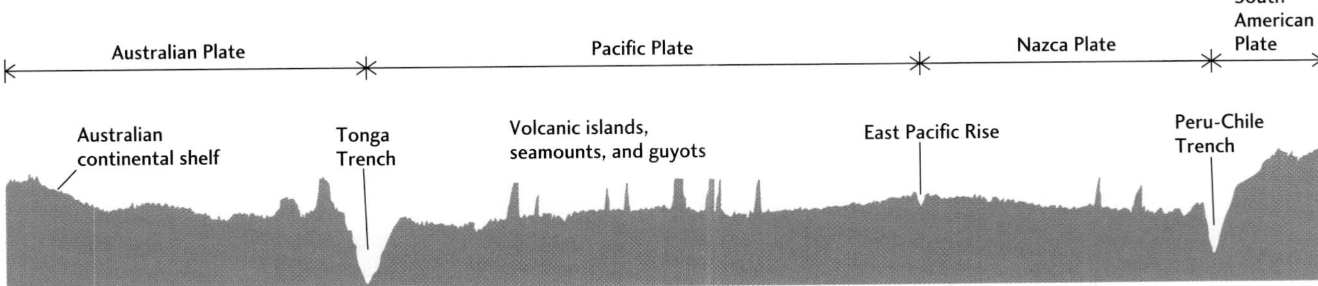

Figure 17.6 Topographic profile of the floor of the Pacific Ocean traveling from South America (*right*) to Australia (*left*).

A Pacific Profile Our second virtual tour moves westward across the Pacific, from South America to Australia (**Figure 17.6**). Beginning on the west coast of Chile, we cross a narrow continental shelf only a few tens of kilometers wide. The edge of the shelf plunges off a continental slope that is much steeper than we found in the Atlantic, extending down to 8000 m as we enter the Peru-Chile Trench. This long, deep, narrow depression of the seafloor is the surface expression of the subduction of the Nazca Plate under the South American Plate.

Continuing across the trench and up onto the abyssal hills of the Nazca Plate, we come to the East Pacific Rise, an active mid-ocean ridge. The East Pacific Rise is lower than the Mid-Atlantic Ridge, and its rate of seafloor spreading is the world's highest—at about 150 mm/year, more than six times faster than the spreading rate of the Mid-Atlantic Ridge—but it has the characteristic central rift valley and outcrops of fresh basalt. On the west side of the East Pacific Rise, we cross over to the Pacific Plate and travel westward over its broad central regions, which are studded with seamounts and volcanic islands.

We eventually arrive at another subduction zone, marked by the Tonga Trench, where the Pacific Plate returns into the mantle beneath the Australian Plate. This is one of the deepest places in all the oceans, almost 11,000 m below the ocean surface. On the west side of the trench, the seafloor rises to the volcanic Tonga and Fiji islands. Beyond this island arc, we return to the deep seafloor, now on the Australian Plate, and come to the continental rise, slope, and shelf of Australia, which is similar to the east coast of North America.

Main Features of the Deep Ocean Floor

Away from the edges of continents and subduction zones, the deep seafloor is constructed primarily by volcanism related to plate tectonic motions and secondarily by sedimentation in the open sea.

Figure 17.7 A plume of hot, mineral-laden water spouts from a "black smoker" hydrothermal vent on the East Pacific Rise. [D. B. Foster/Woods Hole Oceanographic Institution.]

Mid-Ocean Ridges Mid-ocean ridges are the sites of the most intense volcanic and tectonic activity on the deep seafloor. The main rift valley is the center of the action. The valley walls are faulted and intruded with basalt sills and dikes (see Figure 17.5), and the floor of the valley is covered with flows of basalt and talus blocks from the valley walls, mixed with a little sediment settling from surface waters.

Hydrothermal springs form on the rift valley floor as seawater percolates into fractures in the basalt on the flanks of the ridges, is heated as it moves down to hotter basalt, and finally exits at the valley floor. There, it boils up at temperatures as high as 380°C (see Chapter 6). Some springs are "black smokers," full of dissolved hydrogen sulfide and metals the hot water has leached from the basalt (**Figure 17.7**). Others are cooler "white smokers" that have a different composition. Hydrothermal springs on the seafloor produce mounds of iron-rich clay minerals, iron and manganese oxides, and large deposits of iron-zinc-copper sulfides.

Mid-ocean ridges are offset at many places by transform faults that laterally displace the rift valleys (see Figure 17.3b). Large earthquakes occur on these faults as one plate slips past the other. Rocks collected from the walls of the transform faults often have the olivine-rich compositions typical of the mantle, rather than the basaltic composition of the oceanic crust. This suggests that the magmatic process that creates oceanic crust may operate less efficiently where the spreading center abuts a fault.

Abyssal Hills and Plains The floor of the deep ocean away from the mid-ocean ridges is a landscape of hills, plateaus, sediment-floored basins, and seamounts. **Abyssal hills** are ubiquitous on the slopes of mid-ocean ridges. They are typically 100 m or so high and lineated parallel to the ridge crest. They are formed primarily by normal faulting of the basaltic oceanic crust as it moves out of the rift valley. Almost all of this faulting occurs during the first million years of the plate's existence, after which the faults bounding the hills become inactive. As the new oceanic lithosphere drifts away from the spreading center, it cools and contracts, lowering the seafloor. Its hilly, subsiding surface receives a steady rain of sediment from surface waters, gradually becoming mantled with deep-sea muds and other deposits.

Near the continental margins, terrigenous sediments moving down the continental slopes add to this sediment cover and create the flat, unbroken expanses of the abyssal plains. These plains are the flattest solid surfaces on the planet. Given their tectonic stability, they have received serious consideration as possible repositories for human-generated wastes, including radioactive materials.

Seamounts, Hot-Spot Island Chains, and Plateaus The seafloor is littered with tens of thousands of seamounts. Most are submerged, but some rise to the sea surface as volcanic islands. Seamounts and volcanic islands may be isolated or found in clusters or chains. Most, though not all, seamounts are created by eruptions near active spreading centers or where a plate overrides a mantle hot spot.

Some of the larger seamounts, called **guyots,** have flat tops, the result of erosion of an island volcano when it was above sea level. These islands have since submerged as the plate on which they were riding cooled, contracted, and subsided as it passed away from the spreading center or hot spot that formed it.

Among the most enigmatic features of the deep ocean basins are the large basalt plateaus. Some appear to have formed near "triple junctions" where three spreading centers meet. Others are associated with massive eruptions at hot spots far from spreading centers. One of the largest examples of the latter type, and probably the oldest, is the Shatsky Rise in the northwestern Pacific Ocean, about 1600 km southeast of Japan. The origin of oceanic plateaus may be related to the plume volcanism that some scientists think is also responsible for other large igneous provinces, such as the continental flood basalts discussed in Chapter 6 (see Figure 6.22).

Continental Margins

The shorelines, shelves, and slopes of the continents are together called **continental margins.** On the journeys we took across the Atlantic and Pacific, we encountered the two basic types of continental margins, passive and active (**Figure 17.8**). A **passive margin** is far from a plate boundary, like those off the east coast of North America and Australia and the west coast of Europe. The name implies quiescence: volcanoes are absent and earthquakes are few and far between. In contrast, **active margins,** like the one off the west coast of South America, are associated with subduction. Occasionally, active margins are associated with

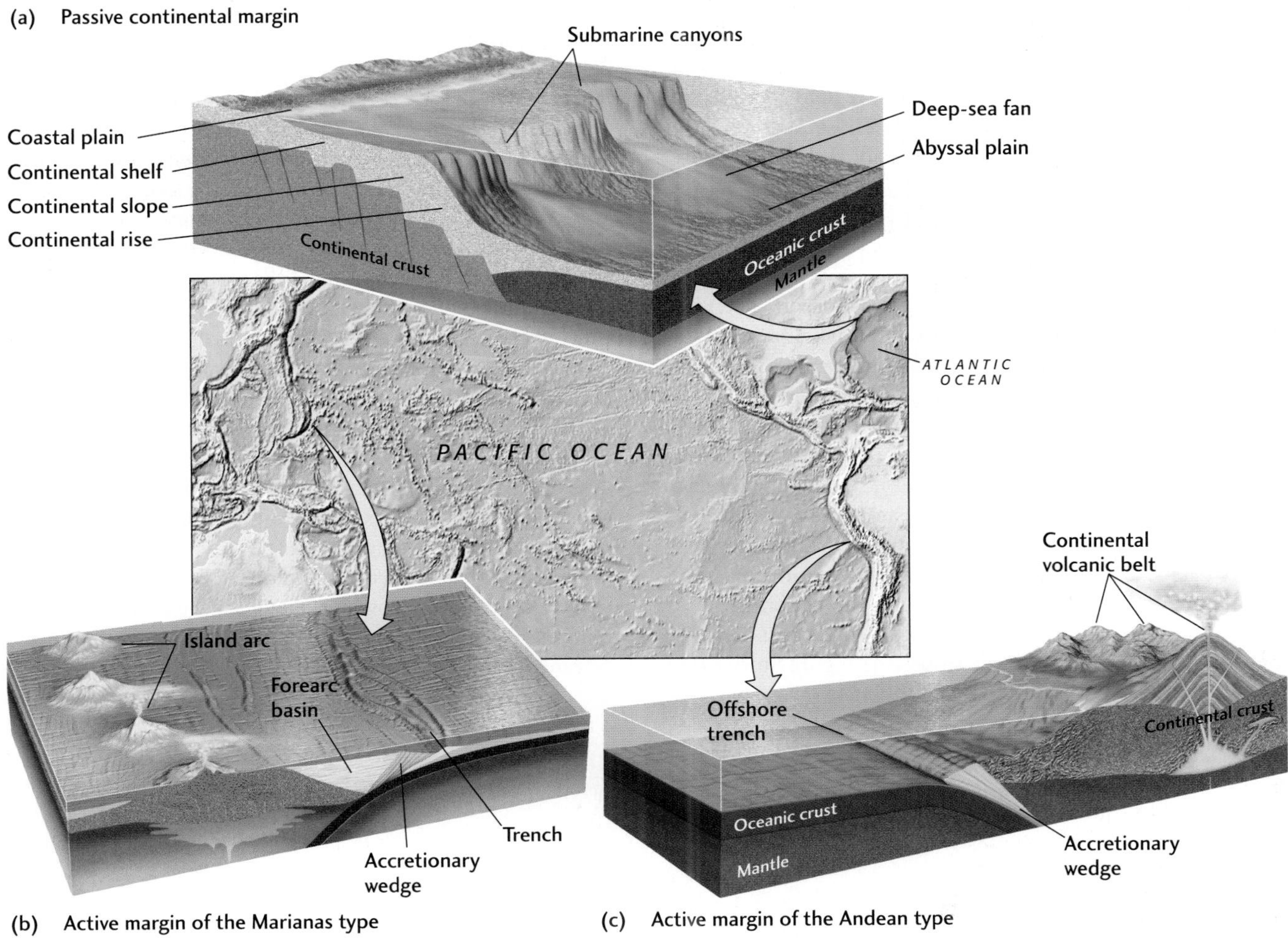

Figure 17.8 Schematic profiles of three types of continental margins. (a) Passive (continental) margin. (b) Active margin of the Marianas type. (c) Active margin of the Andean type.

transform faulting. The volcanic activity and frequent earthquakes give these continental margins their name. Active margins at subduction zones include an offshore trench and an active volcanic belt.

The continental shelves of passive margins consist of essentially flat-lying, shallow-water sediments, both terrigenous and carbonate, several kilometers thick (see Figure 17.8a). Although the same kinds of sediment can be found on active margin shelves, they are more likely to be structurally deformed and to include ash and other volcanic materials as well as deep-ocean sediments. Active margins on the eastern side of the Pacific Ocean usually feature a narrow continental shelf that falls off sharply into a deep-sea trench without much accumulation of sediments (see Figure 17.8c). Those on the western side of the Pacific (for example, off the Marianas Islands) have wider margins with substantial *forearc basins,* where thick sequences of sediment are deposited (see Figure 17.8b). These sediments come partly from erosion of the uplifted volcanic arc, but they also accumulate by being "scraped off" the subducting oceanic crust.

The Continental Shelf

The continental shelf is one of the most economically valuable parts of the ocean. Georges Bank off New England and the Grand Banks of Newfoundland, for example, have been among the world's most productive fishing grounds for 100 years or so. Recently, oil-drilling platforms have been used to extract huge quantities of oil and gas from the continental shelf, especially off the Gulf coast of Louisiana and Texas. For economic and other reasons, most of the world's nations (though not the United States) signed the international Law of the Sea treaty in 1982. The treaty governs the territorial and economic rights of nations in the world's oceans.

Because continental shelves lie at shallow depths, they are subject to exposure and submergence as a result of changes in sea level. During the Pleistocene glaciation, all of the shelves now at depths of less than 100 m were above sea level, and many of their features formed then. Shelves at high latitudes were glaciated, producing an irregular topography of shallow valleys, basins, and ridges. Those at lower latitudes are more regular, incised by occasional stream valleys.

The Continental Slope and Rise: Turbidity Currents

The waters of the continental slope and rise are too deep for the seafloor to be affected by waves and tidal currents. As a consequence, sediments that have been carried across the shallow continental shelf by waves and tides come to rest as they are draped over the slope. Continental slopes shows signs of sediment slumping and the erosional scars of gullies and submarine canyons. Deposits of sands, silts, and muds on both slope and rise indicate active sediment transport into deeper waters. For some time, geologists puzzled over what kind of current might cause both erosion and sedimentation on the slope and rise at such great depths.

The answer proved to be **turbidity currents**—flows of turbid, muddy water down a slope (see Figure Story 8.4). Because of its suspended load of mud, the turbid water is denser than the overlying clear water and flows beneath it. Turbidity currents can both erode and transport sediment. Their role in ocean processes was first understood by Philip Kuenen, a Dutch geologist and oceanographer. In 1936, Kuenen produced and filmed such currents in his laboratory by pouring muddy water into the end of a long, narrow tank with a sloping bottom. He showed that these currents could move at many kilometers per hour and that the speed was proportional to the steepness of the slope and the density of the current. Kuenen proposed the idea, revolutionary at the time, that turbidity currents operate widely in the ocean, especially on continental slopes, at depths well below any possible wave or tidal action.

Turbidity currents start when the sediment draped over the edge of the continental shelf slumps onto the continental slope (**Figure 17.9**a and b). The sudden submarine landslide, which can occur spontaneously or be triggered by an earthquake, throws mud into suspension, creating a dense, turbid layer of water near the bottom. This turbid layer starts to flow, accelerating down the slope. As the turbidity current reaches the foot of the slope and the gentler incline of the continental rise, it slows. Some of the coarser sandy sedi-

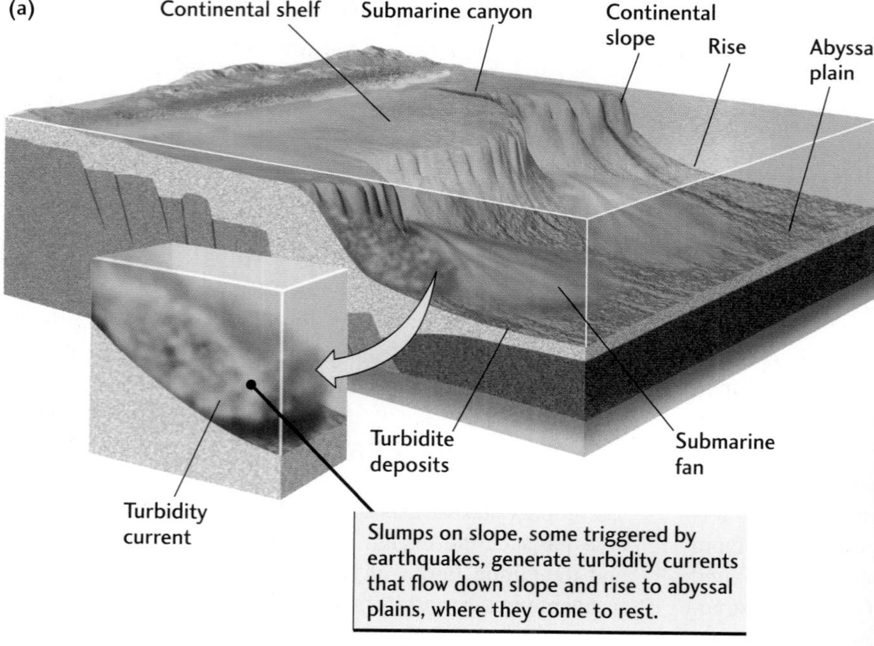

Figure 17.9 (a) How a turbidity current forms in the ocean. These currents can erode and transport large quantities of sand down the continental slope. Submarine canyons and fans are formed by turbidity currents that start on the continental shelf or slope and erode canyons in them, leading to channels on the fan. (b) Sandfall at the head of a submarine canyon at the edge of the continental shelf. These falls generate sandy flows, such as turbidity currents, that lay down fans of sandy sediment at the foot of the continental slope. [U.S. Navy.]

ment starts to settle, often forming a *submarine fan*—a deposit something like an alluvial fan on land. Some of the stronger currents continue across the rise, cutting channels in the submarine fans (see Figure 17.9a). Eventually, the currents reach the level bottom of the ocean basin, the abyssal plain, where they spread out and come to rest in graded beds of sand, silt, and mud called **turbidites.**

According to current research, submarine landslides that start turbidity currents are common. Some of them may be huge. One slide generated 8- to 10-m-thick turbidites over a large area of the western Mediterranean, accumulating a volume of 500 km^3. Submarine slides may be related to methane gas hydrates, crystalline solids composed of methane (found as a constituent of natural gas), and water. Gas hydrates are stable at the high pressures and low temperatures of many large areas of the oceans. In deeply buried sediment, the hydrate turns into gas. If sea level is lowered, as it was during the ice ages, the pressure at the bottom is reduced and the hydrate may gasify, triggering a landslide. The quantities of gas produced are enormous, and geologists have speculated about the possibility of exploiting these subsea gas hydrates.

Submarine Canyons

Submarine canyons are deep valleys eroded into the continental shelf and slope. They were discovered near the beginning of the twentieth century and were first mapped in detail in 1937. At first, some geologists thought they might have been formed by rivers. There is no question that the shallower parts of some canyons were river channels during periods of low sea level. But this hypothesis could not provide a complete explanation. Most of the canyon floors are thousands of meters deep. Even during the maximum lowering of sea level in the ice ages, rivers could have eroded only to a depth of about 100 m.

Although other types of currents have been proposed, turbidity currents are now the favored explanation for the deeper parts of submarine canyons (see Figure 17.9). Evidence supporting this conclusion comes in part from a comparison of modern canyons and their deposits with well-preserved similar deposits of the past, particularly the pattern of turbidites deposited on submarine fans.

Physical and Chemical Sedimentation in the Ocean

Almost everywhere that oceanographers search the seafloor, they find a blanket of sediment. The muds and sands mantle and cover the topography of basalt originally formed at mid-ocean ridges. The ceaseless sedimentation in the world's oceans modifies the structures formed by plate tectonics and creates its own topography at sites of rapid deposition. The sediment is mainly of two kinds: terrigenous muds and sands eroded from the continents and biochemically precipitated shells of organisms that live in the sea. In parts of the ocean near subduction zones, sediments derived from volcanic ash and lava flows are abundant. In tropical arms of the sea where evaporation is intense, evaporite sediments are deposited.

Sedimentation on Continental Margins

Terrigenous sedimentation on the continental shelf is produced by waves and tides. The waves of large storms and hurricanes move sediment over the shallow and moderate depths of the shelf, and tidal currents flow over the shelf. The waves and currents distribute the sediment brought in by rivers into long ribbons of sand and layers of silt and mud.

Biochemical sedimentation on the shelf results from the buildup of layers of the calcium carbonate shells of clams, oysters, and many other organisms living in shallow waters. Most of these organisms cannot tolerate muddy waters and are found only where terrigenous materials are minor or absent, such as along the extreme southern coast of Florida or off the coast of Yucatán in Mexico. Here, coral reefs thrive and organisms build up large thicknesses of carbonate sediment (see Chapter 8).

Deep-Sea Sedimentation

Far from the continental margins, fine-grained terrigenous and biochemically precipitated particles suspended in seawater slowly settle from the surface to the bottom. These open-ocean sediments, called **pelagic sediments,** are characterized by their great distance from continental margins, their fine particle size, and their slow-settling mode of deposition. The terrigenous materials are brownish and grayish clays, which accumulate on the seafloor at a very slow rate, a few millimeters every 1000 years. A small fraction, about 10 percent, may be blown by the wind to the open ocean.

Within the pelagic sediments, the most abundant biochemically precipitated particles are the shells of *foraminifera,* tiny single-celled animals that float in the surface waters of the sea. These calcium carbonate shells fall to the bottom after their occupants die. There they accumulate as **foraminiferal oozes,** sandy and silty sediments composed of foraminiferal shells (**Figure 17.10**). Other carbonate oozes are made up of shells of different microorganisms, called *coccoliths.*

Foraminiferal and other carbonate oozes are abundant at depths of less than about 4 km, but they are rare on the deeper parts of the ocean floor. This rarity cannot result from a lack of shells, because the surface waters are full of them everywhere and the living foraminifera are unaffected by the bottom far below. The explanation for the absence of carbonate oozes is that the shells dissolve in deep seawater below a certain depth, called the *carbonate compensation*

Figure 17.10 Scanning electron micrograph of oceanic ooze. Shown here are shells of both carbonate- and silica-secreting unicellular organisms. [Scripps Institution of Oceanography, University of California, San Diego.]

depth (CCD) (**Figure 17.11**). Because of how the oceans circulate, the deeper waters differ from shallower waters in three ways:

1. They are colder. Colder, denser polar waters sink beneath warmer tropical waters and travel toward the equator along the bottom.

2. They contain more carbon dioxide. Not only do colder waters absorb more carbon dioxide than warmer waters, but any organic matter they are carrying tends to be oxidized to carbon dioxide in the course of their long circulation.

3. They are under higher pressure. This increase in pressure results from the greater weight of overlying water.

These three factors make calcium carbonate more soluble in deep waters than in shallow ones. As the shells of dead foraminifera fall to the bottom below the CCD, they enter an environment undersaturated in calcium carbonate, and they dissolve.

Another kind of biochemically precipitated sediment, **silica ooze,** is produced by sedimentation of the silica shells of diatoms and radiolarians. *Diatoms* are green algae abundant in the surface waters of the oceans. *Radiolarians* are unicellular organisms that secrete shells of silica. After burial on the seafloor, silica oozes are cemented into the siliceous rock *chert.*

Some components of pelagic sediments form by chemical reactions of seawater with sediment on the seafloor. The most prominent examples are manganese nodules—black, lumpy accumulations ranging from a few millimeters to many centimeters across. These nodules cover large areas of the deep ocean floor, as much as 20 to 50 percent of the Pacific. Rich in nickel and other metals, they are a potential commercial resource if some economical way can be found to mine them from the seafloor.

The Edge of the Sea: Waves and Tides

Coasts, the broad regions where land meets sea, present striking contrasts of landscape (**Figure 17.12**). On the coast of North Carolina, for example, long, straight, sandy beaches stretch for miles along low coastal plains. In most of New England, by contrast, rocky cliffs bound elevated

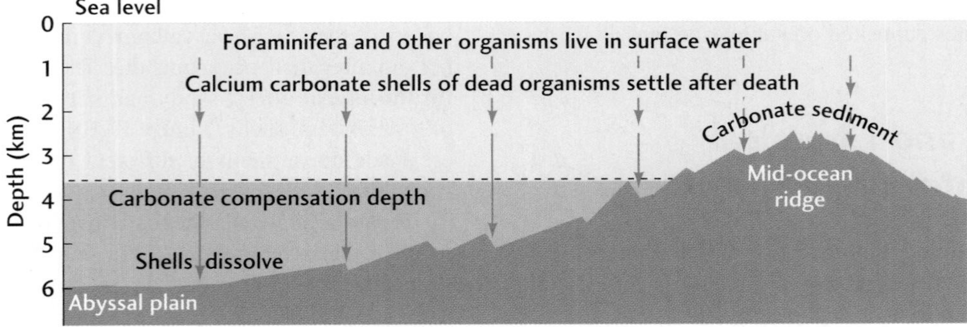

Figure 17.11 The carbonate compensation depth is the level in an ocean below which calcium carbonate dissolves. As the shells of dead foraminifera and other carbonate-shelled organisms settle into the deep waters, they enter an environment undersaturated in calcium carbonate and so dissolve.

Figure 17.12 Coastlines exhibit a variety of geologic forms. (a) Long, straight, sandy beach coastline, Pea Island, North Carolina. [Courtesy of Bill Birkemeier/U.S. Army Corp of Engineers.] (b) Rocky coastline, Mount Desert Island, Maine. This glaciated coastline is being uplifted following the end of the last ice age about 17,000 years ago. [Neil Rabinowitz/Corbis.] (c) The Twelve Apostles, a group of stacks at Port Campbell, Australia, developed in horizontal beds of sedimentary rock. These remnants of shore erosion are left as the shoreline retreats under the action of waves. [Kevin Schafer.] (d) Coral reef along the Florida coastline. [Hays Cummins, Miami University.]

shores, and the few beaches that exist are made of gravel. Many of the seaward edges of islands in the tropics, such as those in the Caribbean Sea, are coral reefs, the delight of divers. As we will see, tectonics, erosion, and sedimentation work together to create this great variety of shapes and materials.

The major geological forces operating at the *shoreline*—the line where the water surface intersects the shore—are waves and tides. Together, they erode even the most resistant rocky shores. Waves and tides create currents, which transport sediment produced by erosion of the land and deposit it on beaches and in shallow waters along the shore.

Wave Motion: The Key to Shoreline Dynamics

Centuries of observation have taught us that waves are changeable. In quiet weather, waves roll regularly into shore with calm troughs between them. In the high winds of a storm, however, waves are everywhere, moving in a confusion of shapes and sizes. They may be low and gentle far from the shore yet become high and steep as they approach land. High waves can break on the shore with fearful violence, shattering concrete seawalls and tearing apart houses built along the beach. To understand the dynamics of shorelines and to make sensible decisions about shore development, we need to understand how waves work.

The wind blowing over the surface of the water creates waves by transferring the energy of motion from air to water. As a gentle breeze of 5 to 20 km/hour starts to blow over a calm sea surface, ripples—little waves less than a centimeter high—take shape. As the speed of the wind increases to about 30 km/hour, the ripples grow to full-sized waves. Stronger winds create larger waves and blow off their tops to make whitecaps. The height of the waves increases as

- The wind speed increases
- The wind blows for longer times
- The distance over which the wind blows the water increases

Storms blow up large, irregular waves that radiate outward from the storm area, like the ripples moving outward from a pebble dropped into a pond. As the waves travel out from the storm center in ever-widening circles, they become more regular, changing into low, broad, rounded waves called *swell,* which can travel hundreds of kilometers. Several storms at different distances from a shoreline, each producing its own pattern of swell, account for the often irregular intervals between waves approaching the shore.

Wave forms travel, but the water stays in the same place. If you have seen waves in an ocean or a large lake, you have probably noticed how a piece of wood or other light material floating on the water moves a little forward as the top of the wave passes and then a little backward as the trough between waves passes. While moving back and forth, the wood stays in roughly the same place, and so does the water around it.

Small water particles at the surface or beneath the waves move in circular vertical orbits. At any given point along the path of a wave, all the water particles are at the same relative positions in their orbits regardless of their depth. The radii of the orbits are large near the water surface, but they gradually get smaller with increasing depth, as shown in **Figure Story 17.13**. Why? It is the free motion of the surface that makes the waves possible; therefore, what happens at the surface—for example, force of the wind—has less influence at greater depths and particle motions become progressively smaller. A wave form is created as many water particles move to the top of the orbit, and the wave advances as the particles continue around the orbit. The trough is created as the particles reach the bottom of the orbit.

We describe a wave form in terms of the following three characteristics (see Figure Story 17.13):

1. *Wavelength,* the distance between crests
2. *Wave height,* the vertical distance between the crest and the trough
3. *Period,* the time that it takes for successive waves to pass

We measure the velocity at which a wave moves forward by using a basic equation:

$$V = L/T$$

where V is the velocity, L is the wavelength, and T is the period. Thus, a typical wave with a length of 24 m and a period of 8 s would have a velocity of 3 m/s. The periods of most waves range from just a few seconds to as long as 15 or 20 s, with wavelengths varying from about 6 m to as much as 600 m. Consequently, wave velocities vary from 3 to 30 m/s. The orbital motion decays with depth and becomes very small below a depth of about one-half the wavelength. That is why deep divers and submarines are unaffected by the waves at the surface.

The Surf Zone

Swell becomes higher as it approaches the shoreline. There it assumes the familiar sharp-crested wave shape. These waves are called *breakers* because, as the waves come closer to shore, they break and form surf, a foamy, bubbly surface. The offshore belt along which breaking waves collapse as they approach the shore is the *surf zone.* Breaking waves pound the shore, eroding and carrying away sand, weathering and breaking up solid rock, and destroying structures built close to the shoreline.

The transformation from swell to breakers starts where the bottom shallows to less than one-half the wavelength of the swell. At that point, the small orbital motions of the water particles just above the bottom become restricted because the water can no longer move vertically. Right next to the bottom, the water can only move back and forth horizontally. Above that, the water can move vertically just a little, combining with the horizontal motion to give a flat elliptical orbit rather than a circular one (see Figure Story 17.13). The orbits become more circular the farther they are from the bottom.

The change from circular to elliptical orbits slows the whole wave, because the water particles take longer to travel around ellipses than around circles. Although the wave slows, its period remains the same because the swell keeps coming in from deeper water at the same rate. From the wave equation, we know that if the velocity decreases and the period remains constant, the wavelength must also decrease. The typical wave that we used as our example earlier might, while keeping the same period of 8 s, change to a length of 16 m and thus a velocity of 2 m/s. Thus, the waves become more closely spaced, higher, and steeper, and their wave crests become sharper.

As a wave rolls toward the shore, it becomes so steep that the water can no longer support itself, and the wave breaks with a crash in the surf zone (see Figure Story 17.13). Gently sloping bottoms cause waves to break farther from shore, and steeply sloping bottoms make waves break closer to shore. Where rocky shores are bordered by deep

WAVE MOTION IS INFLUENCED BY WATER DEPTH AND SHAPE OF THE SHORELINE

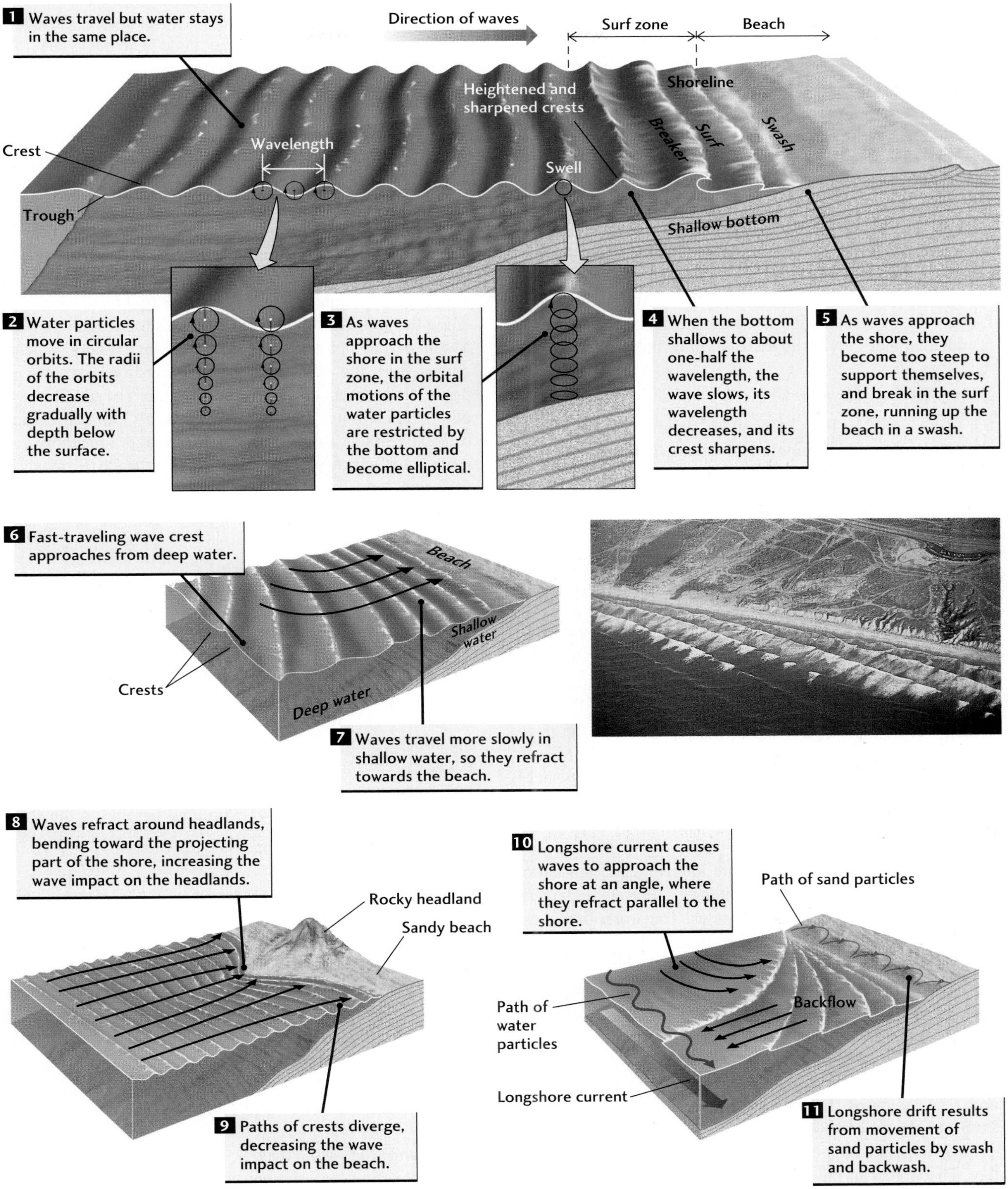

1 Waves travel but water stays in the same place.

Direction of waves

Surf zone

Beach

Shoreline

Heightened and sharpened crests

Breaker

Surf

Swash

Crest

Wavelength

Swell

Trough

Shallow bottom

2 Water particles move in circular orbits. The radii of the orbits decrease gradually with depth below the surface.

3 As waves approach the shore in the surf zone, the orbital motions of the water particles are restricted by the bottom and become elliptical.

4 When the bottom shallows to about one-half the wavelength, the wave slows, its wavelength decreases, and its crest sharpens.

5 As waves approach the shore, they become too steep to support themselves, and break in the surf zone, running up the beach in a swash.

6 Fast-traveling wave crest approaches from deep water.

Beach

Shallow water

Crests

Deep water

7 Waves travel more slowly in shallow water, so they refract towards the beach.

8 Waves refract around headlands, bending toward the projecting part of the shore, increasing the wave impact on the headlands.

Rocky headland

Sandy beach

9 Paths of crests diverge, decreasing the wave impact on the beach.

10 Longshore current causes waves to approach the shore at an angle, where they refract parallel to the shore.

Path of sand particles

Path of water particles

Backflow

Longshore current

11 Longshore drift results from movement of sand particles by swash and backwash.

Figure Story 17.13 The breaking and refraction of waves approaching a shoreline. The photo shows refracting and breaking near Oceanside, California. [John S. Shelton.]

water, the waves break directly on the rocks with a force amounting to tons per square meter, throwing water high into the air. It is not surprising that concrete seawalls built to protect buildings along the shore quickly start to crack and must be repaired constantly.

After breaking at the surf zone, the waves, now reduced in height, continue to move in, breaking again right at the shoreline. They run up onto the sloping front of the beach, forming an uprush of water called *swash.* The water then runs back down again as *backwash.* Swash can carry sand and even large pebbles and cobbles if the waves are high enough. The backwash carries the particles seaward again.

The back and forth motion of the water near the shore is strong enough to carry sand grains and even gravel. Wave action in water as deep as about 20 m can move fine sand. Large waves caused by intense storms can scour the bottom at much greater depths, down to 50 m or more.

Wave Refraction

Far from shore, the lines of swell are parallel to one another but are usually at some angle to the shoreline. As the waves approach the beach over a shallowing bottom, the rows of waves gradually bend to a direction more parallel to the shore. This bending of lines of wave crests as they approach the shore from an angle is called *wave refraction* (see Figure Story 17.13). It is similar to the bending of light rays in optical refraction, which makes a pencil half in and half out of water appear to bend at the water surface. Wave refraction begins as a wave approaches the shore at an angle. The part of the wave closest to the shore encounters the shallowing bottom first, and the orbits of the water particles in that part of the wave become more elliptical. As the orbits become more elliptical, the front of the wave slows. Then the next part of the wave meets the bottom and it, too, slows. Meanwhile, the parts closest to shore have moved into even shallower water and slowed even more. Thus, in a continuous transition along the wave crest, the line of waves bends toward the shore as it slows.

Wave refraction results in more intense wave action on projecting headlands and less intense action in indented bays, as Figure Story 17.13 illustrates. The water becomes shallow more quickly around headlands than in the surrounding deeper water on either side. Waves are refracted around headlands—that is, they are bent toward the projecting part of the shore from both sides. The waves converge around the point of land and expend proportionately more of their energy breaking there than at other places along the shore. Thus, erosion by waves is concentrated at headlands and tends to wear them away more quickly than it does straight sections of shoreline.

The opposite happens as a result of wave refraction in a bay. The waters in the center of the bay are deeper, so the waves are refracted on either side into shallower water. The energy of wave motion is diminished at the center of the bay, which makes bays good harbors for ships.

Although refraction makes waves more parallel to the shore, many waves still approach at some small angle. As the waves break on the shore, the swash moves up the beach slope perpendicular to this small angle. The backwash runs down the slope in the opposite direction at a similar small angle. The combination of the two motions results in a trajectory like a parabola that moves the water a short way down the beach (see Figure Story 17.13). Sand grains carried by swash and backwash are thus moved along the beach in a zigzag motion known as *longshore drift.*

Waves approaching the shoreline at an angle can also cause a **longshore current,** a shallow-water current that is parallel to the shore. The water movement of swash and backwash in and out from the shore at an angle creates a zigzag path of water particles that adds up to a net transport along the shore in the same direction as the longshore drift. Much of the net transport of sand along many beaches comes from this kind of current. Longshore currents are prime determiners of the shapes and extent of sandbars and other depositional shoreline features. At the same time, because of their ability to erode loose sand, longshore currents may remove much sand from a beach. Longshore drift and longshore currents working together are potent processes in the transport of large amounts of sand on beaches and in very shallow waters. In deeper but still shallow waters (less than 50 m), longshore currents—especially those running during large storms—strongly affect the bottom.

Some types of flows related to longshore currents can pose a threat to unwary swimmers. A *rip current,* for example, is a strong flow of water moving outward from the shore at right angles to the shore. It occurs when a longshore current builds up along the shore and the water piles up imperceptibly until a critical point is reached. At that point, the water breaks out to sea, flowing through the oncoming waves in a fast current. Swimmers can avoid being carried out to sea by swimming parallel to the shore to get out of the rip.

The Tides

The twice-daily rise and fall of the sea that we call **tides** have been known to mariners and shoreline dwellers for thousands of years. Many observers noticed a relationship among the position and phases of the Moon, the heights of the tides, and the times of day at which the water reaches high tide. Not until the seventeenth century, however, when Isaac Newton formulated the laws of gravitation, did we begin to understand that the tides result from the gravitational pull of the Moon and the Sun on the water of the oceans.

The Moon, the Sun, Gravity, and the Tides The gravitational attraction between any two bodies decreases as they get farther apart. Thus, the tide-producing force varies on different parts of the Earth, depending on whether they are closer to or farther from the Moon. On the side of the Earth closest to the Moon, the water, being closer to the Moon, experiences a larger gravitational attraction than the average attraction for the whole of the solid Earth. This produces a bulge in the water, seen as a tide. On the side of Earth farthest from the Moon, the solid part is pulled more towards the Moon than the water, and the water appears to be pulled away from the Earth as another bulge.

Thus, two bulges of water occur on Earth's oceans: one from the side nearest the Moon, where the net attraction is toward the Moon, and the other on the side farthest from the Moon, where the net attraction is away from the Moon (**Figure 17.14**). The net gravitational attraction between the oceans and the Moon is at a maximum on the side of Earth facing the Moon and at a minimum on the side facing away from the Moon. As Earth rotates, the bulges of water stay approximately aligned. One always faces the Moon, the other is always directly opposite. These bulges passing over the rotating Earth are the high tides.

The Sun, although much farther away, has so much mass (and thus so much gravity) that it, too, causes tides. Sun tides are a little less than half the height of Moon tides. Sun tides are not synchronous with Moon tides. Sun tides occur as the Earth rotates once every 24 hours, the length of a solar day. The rotation of the Earth with respect to the Moon is a little longer because the Moon is moving around the Earth, resulting in a lunar day of 24 hours and 50 minutes. In that lunar day, there are two high tides, with two low tides between them.

When the Moon, Earth, and Sun line up, the gravitational forces of the Sun and the Moon reinforce each other. This produces the *spring tides,* which are the highest tides; they derive their name from their height, not from the season. They appear every two weeks at full and new Moon. The lowest tides, the *neap tides,* come in between, at the first- and third-quarter Moon, when the Sun and Moon are at right angles to each other with respect to the Earth (see Figure 17.14).

Although the tides occur regularly everywhere, the difference between high and low tides varies in different parts of the ocean. As the tidal bulges of water rise and fall, they also move along the surface of the ocean, encountering obstacles, such as continents and islands, that hinder the flow of water. In the middle of the Pacific Ocean—in Hawaii, for example, where there is little constriction or obstruction of the flow of the tides—the difference between low and high tides is only 0.5 m. Near Seattle, where the shore along Puget Sound is very irregular and the tides are constricted

(a)

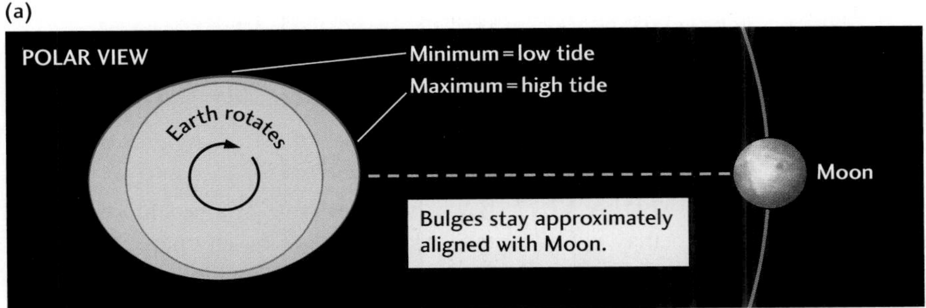

(b)

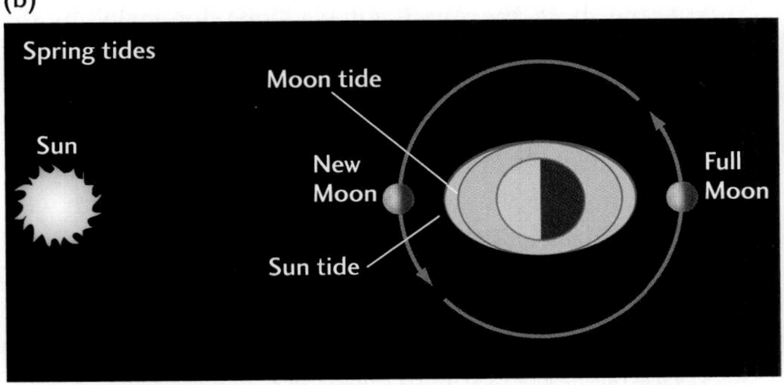

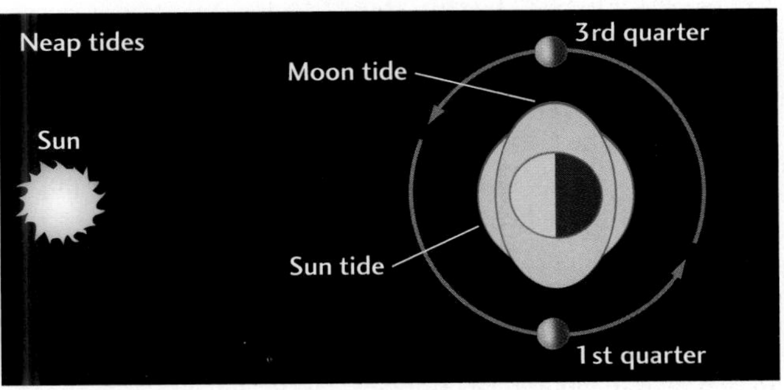

Figure 17.14 Origin of the ocean tides from gravitational attraction of the Sun and Moon. (a) The Moon's gravitational attraction causes two bulges of water on the Earth's oceans, one on the side nearest the Moon and the other on the side farthest from the Moon. As the Earth rotates, these bulges remain aligned and pass over Earth's surface, forming the high tides. The relative positions of the Earth, Moon, and Sun determine the heights of high tide during the lunar month. (b) At new and full Moon, Sun and Moon tides reinforce each other and make the highest (spring) tides. At first- and third-quarter Moon, Sun and Moon tides are in opposition, causing low (neap) tides.

through narrow passageways, the difference between the two tides is about 3 m. Extraordinary tides occur in a few places, such as the Bay of Fundy in eastern Canada, where the tidal range can be more than 12 m. Many people living along the shore need to know when tides will occur, so governments publish tide tables showing predicted tide heights and times. These tables combine local experience with knowledge of the astronomical motions of Earth and the Moon with respect to the Sun.

Tides may combine with waves to cause extensive erosion of the shore and destruction of shoreline property. Intense storms passing near the shore during a spring tide may produce *tidal surges,* waves at high tide that can overrun the entire beach and batter sea cliffs. Do not confuse tidal surges with tsunamis, commonly but incorrectly termed "tidal waves." *Tsunamis* are large ocean waves caused by undersea events such as earthquakes, landslides, and the explosion of oceanic volcanoes (see Figure 19.17 and Feature 19.1).

Tidal Currents Tides moving near shorelines cause currents that can reach speeds of a few kilometers per hour. As the tide rises, the water flows in toward the shore as a *flood tide,* moving into shallow coastal marshes and up small streams. As the tide passes the high stage and starts to fall, the *ebb tide* moves out, and low-lying coastal areas are exposed again. Such tidal currents meander across the cut channels into **tidal flats,** the muddy or sandy areas that lie above low tide but are flooded at high tide (**Figure 17.15**).

Figure 17.15 Tidal flats, such as this one at Mont-Saint-Michel, France, may be extensive areas covering many square kilometers but most often are narrow strips seaward of the beach. When a very high tide advances on a broad tidal flat, it may move so rapidly that areas are flooded faster than a person can run. The beachcomber is well advised to learn the local tides before wandering. [Thierry Prat/Corbis Sygma.]

Shorelines

We end our exploration of the oceans with shorelines, where we can observe the constant motion of ocean waters and their effects on the shore. Current environmental problems such as coastal erosion and polluted shallow waters have made the geology of shorelines and shallow seas a critical area of research. Waves, longshore currents, and tidal currents interact with the rocks and tectonics of the coast to shape shorelines into a multitude of forms (see Figure 17.12). We can see these factors at work in the most popular of shorelines—beaches.

Beaches

A beach is a shoreline made up of sand and pebbles. Beaches may change shape from day to day, week to week, season to season, and year to year. Waves and tides sometimes broaden and extend a beach by depositing sand and sometimes narrow it by carrying sand away.

Many beaches are straight stretches of sand that range from 1 km to more than 100 km long; others are smaller crescents of sand between rocky headlands. Belts of dunes border the landward edge of many beaches; bluffs or cliffs of sediment or rock border others. Beaches may have tide terraces—flat, shallow areas between the upper beach and an outer bar of sand—on their seaward sides (**Figure 17.16**).

The Structure of Beaches

Figure 17.17 shows the major parts of a beach. These parts may not all be present at all times on any particular beach. Farthest out is the *offshore,* bounded by the surf zone, where the bottom begins to become shallow enough for waves to break. The *foreshore* includes the surf zone; the tidal flat; and, right at the shore, the swash zone, a slope dominated by the swash and backwash of the waves. The *backshore* extends from the swash zone up to the highest level of the beach.

The Sand Budget of a Beach A beach is a scene of incessant movement. Each wave moves sand back and forth with swash and backwash. Both longshore drift and longshore currents move sand down the beach. At the end of a beach and to some extent along it, sand is removed and deposited in deep water. In the backshore or along sea cliffs, sand and pebbles are freed by erosion and replenish the beach. The wind that blows over the beach transports sand, sometimes offshore into the water and sometimes onshore onto the land.

All these processes together maintain a balance between adding and removing sand, resulting in a beach that may appear to be stable but is actually exchanging its material on all sides. **Figure 17.18** illustrates the **sand budget** of a beach—the inputs and outputs caused by erosion and sedi-

Figure 17.16 Tide terrace. At low tide, the outer ridge (a sandbar at high tide) is exposed. Also exposed is the shallow depression between the ridge and the upper beach, which is rippled by the tidal flow in many places. [James Valentine.]

mentation. At any point along a beach, it gains sand by the inputs: from erosion of material along the backshore; from longshore drift and longshore current; and from rivers that enter the sea along the shore, bringing in sediment. The beach loses sand from the outputs: winds carry sand to backshore dunes, longshore drift and current carry it down-current, and deep-water currents and waves transport it during storms.

If the total input balances the total output, the beach is in equilibrium and keeps the same general form. If input and output are not balanced, the beach grows or shrinks. Temporary imbalances are natural over weeks, months, or years. A series of large storms, for example, might move large amounts of sand from the beach to somewhat deeper waters on the far side of the surf zone, narrowing the beach. Then, in a slow return to equilibrium over weeks of mild weather and low waves, the sand might move onto shore and rebuild a wide beach. Without this constant shifting of the sands, beaches might be unable to recover from trash, litter, and some kinds of pollution. Within a year or two, even oil from spills will be transported or buried out of sight, although the tarry residue may later be uncovered in spots. Beaches would clean up rapidly if the littering were to stop.

Some Common Forms of Beaches Long, wide, sandy beaches grow where sand inputs are abundant, often where soft sediments make up the coast. Where the backshore is low and the winds blow onshore, wide dune belts border the beach. If the shoreline is tectonically elevated and the rocks are hard, cliffs line the shore, and any small beaches that evolve are composed of material eroded from the cliffs. Where the shore is low-lying, sand is abundant, and tidal currents are strong, extensive tidal flats are laid down and are exposed at low tide.

What happens if one of the inputs is cut off—for example, by a concrete wall built at the top of the beach to prevent erosion? Because erosion supplies sand to the beach as one of the inputs, preventing it cuts the sand supply and so shrinks the beach. Attempts to save the beach may actually destroy it (see Feature 17.1).

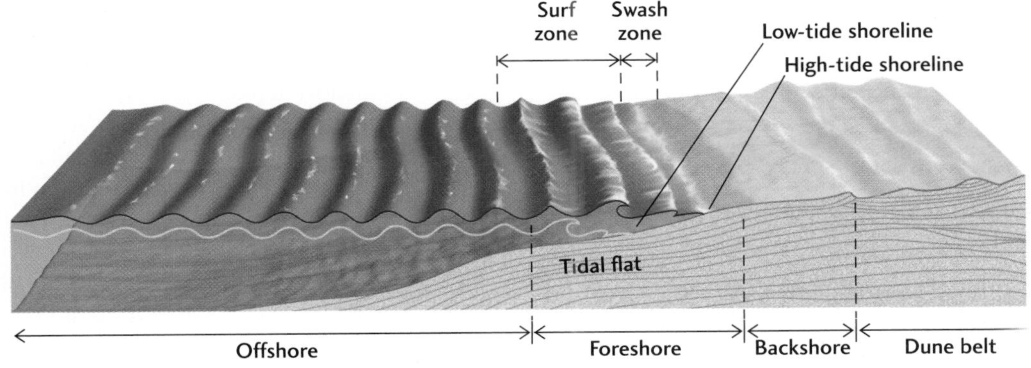

Figure 17.17 A profile of a beach, showing its major parts.

SAND BUDGET	
INPUTS	OUTPUTS
Sediments eroded from backshore cliffs by waves	Sediments transported to backshore dunes by offshore winds
Sediments eroded from upcurrent beach by longshore drift and current	Sediments transported downcurrent by longshore drift and current
Sediments brought in by rivers	Sediments transported to deep water by tidal currents and waves

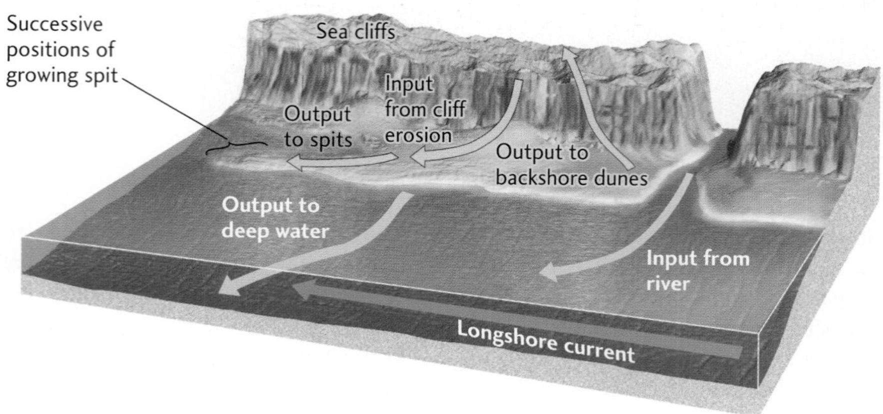

Figure 17.18 The sand budget is a balance between inputs and outputs of sand by erosion, transport, and sedimentation.

Erosion and Deposition at Shorelines

The topography of the shoreline, like that of the land interior, is a product of tectonic forces elevating or depressing Earth's crust, erosion wearing it down, and sedimentation filling in the low spots. Thus, the factors directly at work are

- Uplift of the coastal region, which leads to erosional coastal forms
- Subsidence of the coastal region, which produces depositional coastal forms
- The nature of the rocks or sediments at the shoreline
- Changes in sea level, which affect the drowning or emergence of a shoreline
- The average and storm wave heights
- The heights of the tides, which affect both erosion and sedimentation

Erosional Coastal Forms Erosion is active at tectonically uplifted rocky coasts. Along these coasts, prominent cliffs and headlands jut into the sea, alternating with narrow inlets and irregular bays with small beaches. Waves crash against rocky shorelines, undercutting cliffs and causing huge blocks to fall into the water, where they are gradually worn away. As the sea cliffs retreat by erosion, isolated remnants called *stacks* are left standing in the sea, far from the shore (see Figure 17.12c). Erosion by waves also planes the rocky surface beneath the surf zone and creates a **wave-cut terrace,** sometimes visible at low tide (**Figure 17.19**). Wave erosion continuing for long periods may straighten shorelines, as headlands retreat faster than recesses and bays.

Where relatively soft sediments or sedimentary rocks make up the coastal region, the slopes are gentler and the heights of shoreline bluffs are lower. Waves efficiently erode these softer materials; erosion of bluffs on such shores may be extraordinarily rapid. The high sea cliffs of soft glacial materials along the Cape Cod National Seashore in Massachusetts, for instance, are retreating about a meter each year. Since Henry David Thoreau walked the entire length of the beach below those cliffs in the mid-nineteenth century and wrote of his travels in *Cape Cod,* about 6 km^2 of coastal land have been eaten by the ocean, equivalent to about 150 m of beach retreat.

In recent decades, more than 70 percent of the total length of the world's sand beaches has retreated at a rate of at least 10 cm/year, and 20 percent of the total length has retreated at a rate of more than 1 m/year. Much of this loss

Figure 17.19 Multiple wave-cut terraces on the California coastline. [Photo by Dan Muhs/U.S. Geological Survey. Daniel R. Muhs, Kathleen R. Simmons, George L. Kennedy, and Thomas K. Rockwell. "The Last Interglacial Period on the Pacific Coast of North America: Timing and Paleoclimate." *Geological Society of America Bulletin* (May 2002): 569–592.]

may be traced to the damming of rivers, which decreases the sediment supply to the shoreline.

Depositional Coastal Forms Sediment builds up in areas where tectonic subsidence depresses the crust along a coast. Such coasts are characterized by long, wide beaches and wide, low-lying coastal plains of sedimentary strata. Shoreline forms include sandbars, low-lying sandy islands, and extensive tidal flats. Long beaches grow longer as longshore currents carry sand to the downcurrent end of the beach. There it builds up, first as a submerged bar, then rising above the surface and extending the beach by a narrow addition called a *spit* (**Figure 17.20**).

Long sandbars offshore may build up and become **barrier islands** that form a barricade between open ocean waves and the main shoreline. Barrier islands are common, especially along low-lying coasts of easily erodible and transportable sediments or poorly cemented sedimentary rocks where longshore currents are strong. Some of the most prominent barrier islands are found along the coast of New Jersey, at Cape Hatteras, and along the Texas coast of the Gulf of Mexico, where one, Padre Island, is 130 km long. As the bars build up above the waves, vegetation takes hold, stabilizing the islands and helping them resist wave erosion during storms. Barrier islands are separated from the coast by tidal flats or shallow lagoons. Like beaches on the main shore, barrier islands are in dynamic equilibrium with the forces shaping them. If their equilibrium is disturbed by natural changes in climate or in wave and current regimes or by real estate development, they may be disrupted or devegetated, leading to increased erosion. Under such conditions, barrier islands may even disappear beneath the sea surface. Barrier islands may also grow larger and more stable if sedimentation increases.

Over the course of hundreds of years, shorelines may undergo significant changes. Hurricanes and other intense storms, such as the "storm of the century" that hit the East Coast of the United States in March 1993, may form new inlets or elongate spits or may breach existing inlets and spits. Such changes have been documented by remapping from aerial photographs taken at various time intervals. The shoreline of Chatham, Massachusetts, at the elbow of Cape Cod, has changed enough in the past 160 years or so that a lighthouse has had to be moved. Figure 17.20 illustrates the numerous changes that have taken place in the configuration of the bars to the north and the long spit of Monomoy Island, as well as several breaches of the bars. Many homes are now at risk in Chatham, but there is little that the residents or the state can do to prevent these beach processes from taking their natural course.

Change in Sea Level as a Measure of Global Warming

Shorelines are sensitive to changes in sea level, which can alter tidal heights, change the approach of waves, and affect the path of longshore currents. The rise and fall of sea level can be local—a result of tectonic subsidence or uplift—or global—the result, for example, of glacial melting or growth (see Chapter 16). One of the primary concerns about human-induced global warming is its potential for causing sea level to rise and thereby flooding coastlines.

In periods of lowered sea level, areas that were offshore are exposed to agents of erosion. Rivers extend their courses over formerly submerged regions and cut valleys into newly exposed coastal plains. When sea level rises, flooding the lands of the backshore, river valleys are drowned, marine sediments build up along former land areas, and erosion is

17.1 Preserving Our Beaches

Orrin Pilkey of Duke University, a geologist and oceanographer, is in the forefront of scientists concerned about saving our beaches and halting massive development on fragile shorelines. Many houses built on the shoreline, battered by waves, could be saved by constructing concrete buttresses, seawalls, and other structures designed to protect shoreline property, but these structures would destroy the beach. Pilkey, a well-known researcher on coastal processes, is an advocate for the beaches of the Carolinas, which have come under heavy pressure from commercial developers. Knowing how the beach system works, he believes it is foolish to try to interfere with the natural process by which beaches remain in dynamic equilibrium with the waves and currents.

Humans are altering this equilibrium on more and more beaches by building cottages on the shore; paving beach parking lots; erecting seawalls; and constructing groins, piers, and breakwaters. The consequence of poorly thought out construction is the shrinkage of the beach in one place and its growth in another. The classic example is a narrow groin built out from shore at right angles to it. In

Beach-fill placement at the southern end of Monmouth Beach, Monmouth County, New Jersey. This erosion control project by the U.S. Army Corps of Engineers includes periodic nourishment of the restored beaches on a 6-year cycle for a period of 50 years. [U.S. Army Corps of Engineers, New York District.]

replaced by sedimentation. Today, long fingers of the sea indent many of the shorelines of the northern and central Atlantic coast. These long indentations are former river valleys that were flooded as the last glacial age ended about 10,000 years ago and the sea level rose.

Sea-level variations on geologic time scales can be measured by studies of wave-cut terraces (see Figure 17.19), but detecting global changes on short (human) time scales can be difficult. Changes can be measured locally by using a tide gauge that records sea level relative to a land-based benchmark. The major problem is that the land itself moves vertically as a result of tectonic deformation, sedimentation, and other geologic changes, and this motion gets incorporated into the tide-gauge observations. Through careful analysis, however, oceanographers have found that global sea level has risen by 10 to 25 cm over the last century.

This increase correlates with a worldwide increase in temperatures, which most scientists now believe has been caused, at least in part, by human pollution of the atmosphere with carbon dioxide and other greenhouse gases (see Chapter 23). Global warming leads to a worldwide rise in sea level in two different ways. First, it causes continental glaciers and polar ice caps to melt, which increases the amount of water in the ocean basins. Second, higher tem-

the subsequent months and years, the sand disappears from the beach on one side of the groin and greatly enlarges the beach on the other side. As landowners and developers bring suit against one another and against state governments, trial lawyers take the issue of "sand rights"—the beach's right to the sand that it naturally contains—into the courts.

The disappearance and enlargement of beaches are the predictable results of a longshore current. The waves, current, and drift bring sand toward the groin from the upcurrent direction (usually the dominant wind direction). Stopped at the groin, they dump the sand there. On the downcurrent side of the groin, the current and drift pick up again and erode the beach. On this side, however, replenishment of sand is sparse because the groin blocks the current. As a result, the beach budget is out of balance, and the beach shrinks. If the groin is removed, the beach returns to its former state.

The only way to save a beach is to leave it alone. Even if concrete walls and piers can be kept in repair with large expenditures of money, many times at public expense, the beach itself will suffer. Along some beaches, resort hotels truck in sand, but that expensive solution is temporary, too.

At Seabright, in Monmouth County, New Jersey, a beach has been built up by the U.S. Army Corps of Engineers, using half a million dump-truck loads of sand. This is the costliest beach "nourishment" project to date. As of 1999, the beach had lasted 3 years. The engineers expect to replenish the eroding beach in another 3 years.

Sooner or later, we must learn to let the beaches remain in their natural state.

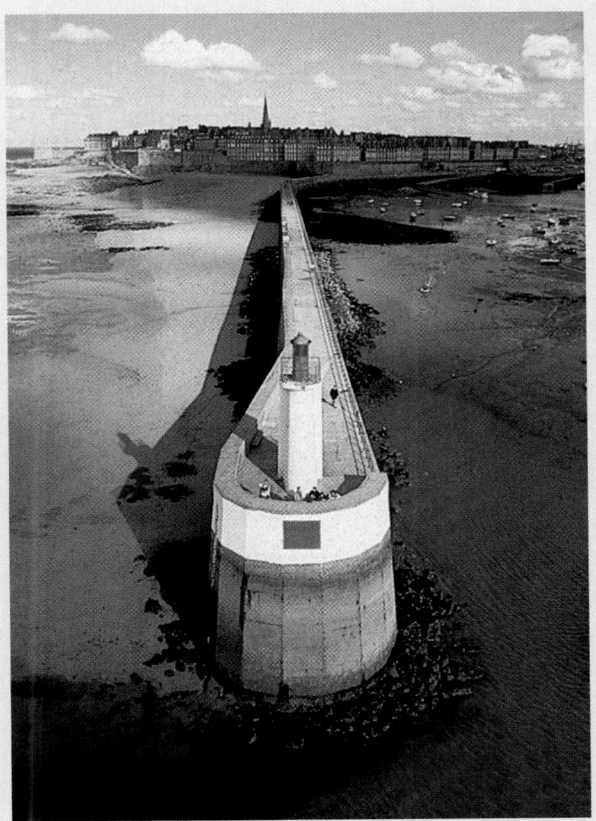

Construction of groins along a shore to control erosion of a beach may produce erosion downcurrent of the groin and loss of parts of the beach (*right of groin*) while sand piles up on the other side (*left of groin*). Longshore current flows from left to right. [Philip Plisson/Explorer.]

peratures cause the water to expand by a small fraction, increasing its volume (just as thermal expansion causes liquid to rise in a thermometer). These effects appear to be about equal in magnitude; that is, each can explain about half of the observed increase in sea level.

Satellite altimeters provide a sensitive new technique for determining the altitude of the sea surface relative to the orbit of the satellite (see the discussion on page 383). The data indicate that sea level is rising at a rate of about 4 mm/year. Some of the rise may result from short-term variations, but the magnitude of the rise is consistent with climate models that take into account greenhouse warming. These models predict that without significant worldwide efforts to reduce greenhouse gas emissions, sea level will probably rise by another 30 to 60 cm during this century.

As global warming causes the sea level to rise, we will see the effects on our beaches. Indeed, the shorelines of the world serve as barometers for impending changes caused by many types of human activities. The pollution of our inland waterways sooner or later arrives at our beaches, as sewage from city dumping and oil from ocean tankers wash up on the shore. As real estate development and construction along shorelines expands, we will see the continuing contraction and even disappearance of some of our finest beaches.

(a) Beach near Chatham Light

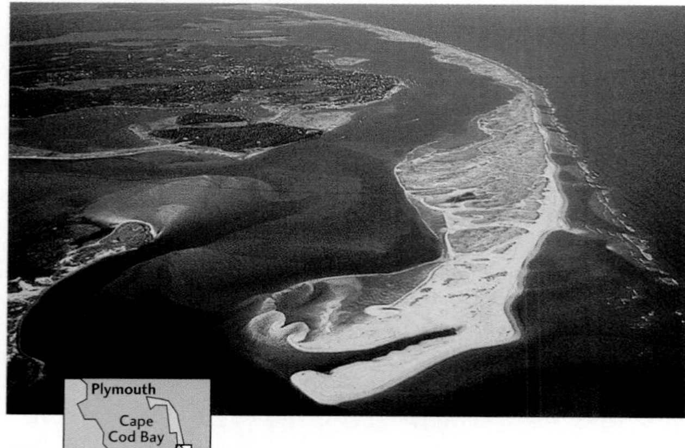

Figure 17.20 Migrating barrier islands at the southern tip of Cape Cod, Massachusetts. (a) Aerial view of Monomoy Point, Massachusetts. This spit has advanced into deep water to the south (*foreground*) from barrier islands along the main body of the Cape to the north (*background*). [Steve Dunwell/The Image Bank.] (b) Changes in the shoreline at Chatham, Massachusetts, at the elbow of Cape Cod, in the past 160 years. [After Cindy Daniels, *Boston Globe* (February 23, 1987).]

The 1987 breach in the barrier spit, shown at the right below, closed again before this photo was taken.

(b)

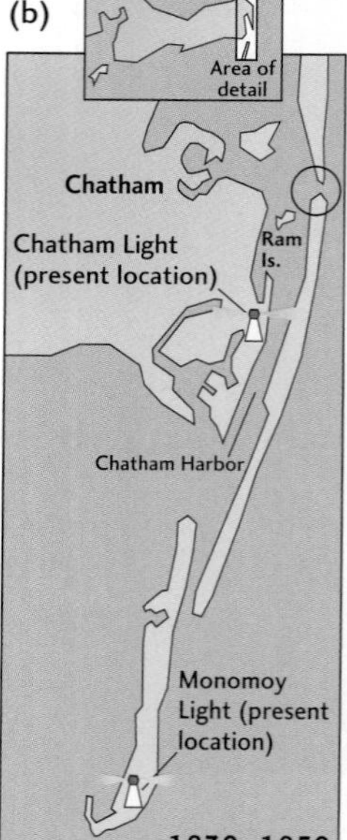

Circle shows approximate location of 1846 breach in barrier spit. Ram Island later disappears.

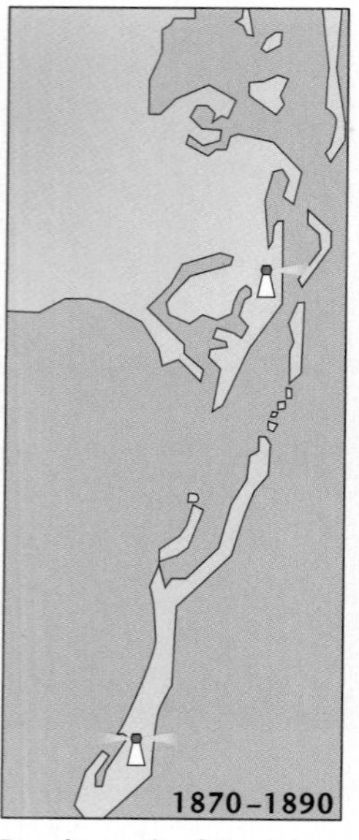

Beach south of the inlet breaks up and migrates southwest toward the mainland and Monomoy.

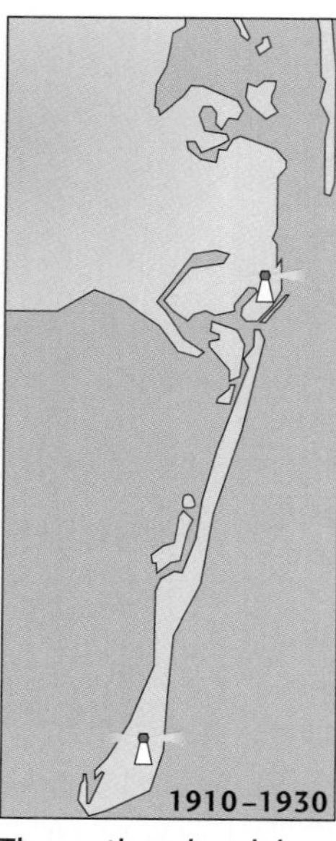

The southern beach has disappeared, and its remnants soon will connect Monomoy to the mainland.

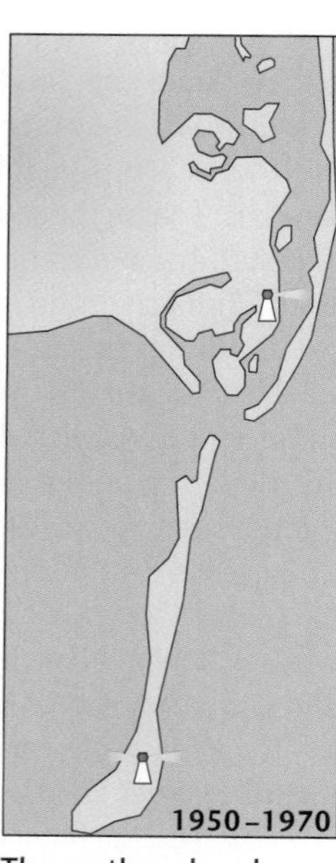

The northern beach steadily grows with cliff sediment; Monomoy breaks from the mainland.

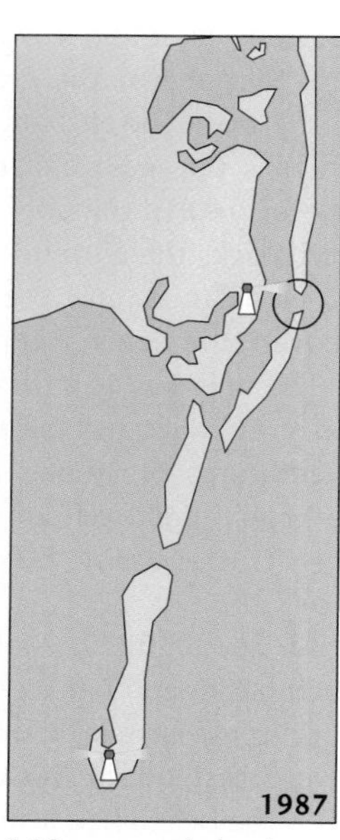

140-year cycle begins again with Jan. 2 breach in the barrier spit across from Chatham Light (circle).

SUMMARY

How does the geology of the oceans differ from that of the continents? Continents have grown and been modified over the past 4 billion years as a result of multiple plate collisions, the creation of folded and faulted mountain belts, magmatism, erosion, and sedimentation. The geology of the continents is complex, and only a portion of their history is preserved in the rock record. Ocean basins are simple in comparison. They are created at mid-ocean ridges where plates separate and are destroyed by subduction over the brief period of less than 200 million years. Deep-sea sediments provide an almost complete record of the relatively brief geologic history of ocean basins.

How is the deep seafloor formed? The deep seafloor is constructed by volcanism at mid-ocean ridges and at oceanic hot spots such as Hawaii, and by deposition of fine-grained clastic and biochemically precipitated sediments. Mid-ocean ridges are the sites of seafloor spreading and the extrusion of basalt, which produces new oceanic lithosphere. Volcanic islands, submerged seamounts and guyots, plateaus, and abyssal hills are all accumulations of volcanic rock, most of which are mantled by sediment. Deep-sea trenches form as oceanic lithosphere is pulled downward into a subduction zone.

What are the characteristics of a mid-ocean ridge? A rift valley runs down the crest, where plates separate and new oceanic lithosphere is produced. Earthquakes and active basaltic volcanoes are concentrated near the ridge crest. Seawater percolates through cracks on the flanks of such ridges and emerges in the valley floors as hydrothermal springs called "smokers". These heated waters, laden with dissolved metal oxides and sulfides, precipitate mounds of mineral-rich deposits when they mix with the cold ocean-bottom water.

What are the major components of a continental margin? A continental margin is made up of a shallow continental shelf; a continental slope that descends more or less steeply into the depths of the ocean; and a continental rise, a gently sloping apron of sediment deposited at the lower edges of the continental slope and extending to the abyssal plain farther out in the ocean. Waves and tides affect the continental shelf, but the continental slope is shaped primarily by turbidity currents. These deep-water currents form as slumps and slides on the continental slope create turbid suspensions of muddy sediment in bottom waters. Turbidity currents also produce submarine fans, submarine canyons, and canyons. Active continental margins form where oceanic lithosphere is subducted beneath a continent. Passive continental margins form where rifting and seafloor spreading carry continental margins away from plate boundaries.

What kinds of sedimentation occur in and near the oceans? The two main types are terrigenous sediments and biochemically precipitated sediments. Terrigenous sediments are primarily muds and sands eroded from the continents and deposited by wave action and tidal currents along the continental shelf, and by turbidity currents that flow down the continental margins. Biochemical sedimentation on the shelf results from the buildup of calcium carbonate from shells and coral reefs. In shallow tropical seas where evaporation is intense, evaporite sediments can also be deposited. Open-ocean (pelagic) sediments consist of reddish-brown clays and foraminiferal and silica oozes composed of the biochemically precipitated calcium carbonate and silica shells of microscopic organisms living in surface ocean waters. In parts of the ocean near subduction zones, sediments derived from volcanic ash and lava flows are abundant.

What processes shape shorelines? At the edge of the sea, waves and tides, interacting with tectonics, control the formation and dynamics of shorelines, from beaches and tidal flats to uplifted rocky coasts. Winds blowing over the sea generate waves; as the waves approach the shore, they are transformed into breakers in the surf zone. Wave refraction results in longshore currents and longshore drift, which transport sand along beaches. Tides, generated by the gravitational pull of the Moon and Sun on the water of the oceans, are agents of sedimentation on tidal flats.

Key Terms and Concepts

Exercises

MEDIA LINK *This icon indicates that there is an animation available on the Web site that may assist you in answering a question.*

1. Where and how is the deep seafloor created by volcanism?

 **2.** What plate tectonic process is responsible for deep-sea trenches?

3. What features of the seafloor are associated with plate divergences and with plate convergences?

4. Along what kinds of continental margins do we find broad continental shelves?

5. Where do turbidity currents form?

6. What are pelagic sediments?

7. What processes in the ocean are responsible for foraminiferal and silica oozes?

8. How are ocean waves formed?

9. How does wave refraction concentrate erosion at headlands?

10. How has human interference affected some beaches?

Thought Questions

1. What are the chief differences between the Atlantic and Pacific oceans in regard to topography, tectonics, volcanism, and other seafloor processes?

2. How might plate tectonics account for the contrast between the broad continental shelf off the eastern coast of North America and the narrow, almost nonexistent shelf off the western coast?

3. A major corporation hires you to determine whether New York City's garbage can be dumped at sea within 100 km offshore. What kinds of places would you explore, and what are your concerns?

4. There is very little sediment to be found on the floor of the central rift valley of the Mid-Atlantic Ridge. Why might this be so?

5. A plateau rising from the deep ocean floor to within 2000 m of the surface is mantled with foraminiferal ooze, whereas the deep seafloor below the plateau, about 5000 m deep, is covered with reddish-brown clay. How can you account for this difference?

6. What kinds of sediment might you expect to find in the Peru-Chile Trench off the coast of South America?

7. You are describing a deep-sea sediment core from the base up. In the bottom part, most of the sediment is a foraminiferal ooze; midway to the top, however, the ooze disappears and the core is brownish-gray. What could be responsible for this change?

8. You are studying a sequence of sedimentary rocks and discover that the beds near the base are shallow marine sandstones and mudstones. Above these beds is an unconformity, overlying which are nonmarine sandstones. Above this group of beds is another unconformity, and overlying it are marine beds similar to those at the base. What could account for this sequence?

9. Why would you want to know the timing of high tide if you wanted to observe a wave-cut terrace?

10. In a 100-year period, the southern tip of a long, narrow, north-south beach has become extended about 200 m to the south by natural processes. What shoreline processes could have caused this extension?

11. After a period of calm along a section of the eastern shore of North America, a severe storm with high winds passes over the shore and out to sea. Describe the state of the surf zone before the storm, during it, and after the storm has passed.

Long-Term Team Project

Coastal Protection

In 1996, the Institute of Marine and Coastal Sciences at Rutgers University published a report that calls for a major shift in policy with regard to coastal protection. Rather than recommending that coastal beaches and shore properties in high-hazard areas be defended and rebuilt, the report urges the New Jersey Department of Environmental Protection to advocate the removal of storm-damaged buildings and a policy of letting nature reclaim the beach. The report also suggests that shore towns be encouraged not to zone vulnerable areas for development. Some geologists consider this policy very wise.

Working in teams of three or four, pick a nearby or favorite vacation beach. If feasible, each group should visit its beach at the beginning of the semester and after each storm event during the semester to measure and record the width of the beach and make other observations of how the beach changes. Other sources of information might include the state's department of environmental protection, the town's chamber of commerce, and U.S. Geological Survey Web sites. Using records available from the U.S. Army Corps of Engineers under the Freedom of Information Act, investigate the approach that has been used to maintain and preserve the local shore that you have chosen to study. Has the Corps constructed seawalls or groins? Alternatively, has the Corps recommended pumping sand from offshore back onto the beach? What measures has the state taken to protect its shoreline, and what effect have they had on the physical appearance of the beach? Has the town chosen to let shore processes proceed naturally? Using your knowledge of coastal processes; your observations, if possible; and information about changes in your beach from other sources, assess the wisdom of the approach taken to preserve the beach of your study.

Suggested Readings

Anderson, R. 1986. *Marine Geology.* New York: Wiley.

Cone J. 1991. *Fire Under the Sea.* New York: William Morrow.

Davis, R. A. 1994. *The Evolving Coast.* New York: W. H. Freeman.

Dolan, R., and H. Lins. 1987. Beaches and barrier islands. *Scientific American* (July): 146.

Hardisty, J. 1990. *Beaches: Form and Process.* New York: HarperCollins Academic.

Hollister, C. D., A. R. M. Nowell, and P. A. Jumars. 1984. The dynamic abyss. *Scientific American* (March): 42.

Komar, P. 1998. *The Pacific Northwest Coast.* Durham, N.C.: Duke University Press.

Macdonald, K. C., and B. P. Luyendyk. 1981. The crest of the East Pacific Rise. *Scientific American* (May): 100.

Menard, H. W. 1986. *The Ocean of Truth: A Personal History of Global Tectonics.* Princeton, N.J.: Princeton University Press.

National Research Council. 2000. *Clean Ocean Coastal Waters.* Washington, D.C.: National Academy Press.

National Science Foundation. 2001. *Ocean Sciences at the New Millennium.* Washington, D.C. http://www.joss.ucar.edu/joss_psg/publications/decadal/Decadal.low.pdf.

Nunn, P. D. 1994. *Oceanic Islands.* Oxford: Blackwell.

Pilkey, O. H., and W. J. Neal. 1988. Coastal geologic hazards. In R. E. Sheridan and J. A. Grow (eds.), *The Geology of North America,* vol. I-2, *The Atlantic Continental Margin: U.S.* pp. 549–556. Boulder, Colo.: Geological Society of America.

Schlee, J. S., H. A. Karl, and M. E. Torresan. 1995. *Imaging the Sea Floor.* U.S. Geological Survey Bulletin 2079.

Grand Teton Mountains, Wyoming. The rugged mountains are made of granite that has been uplifted and weathered to form impressive spires. In the foreground, uplift of the plains has resulted in incision of the Snake River, which left behind a series of steplike benches in the topography. The topography of Earth records the ceaseless competition between uplift and erosion. [Point Anderson.]

CHAPTER 18

Landscapes: Tectonic and Climate Interaction

"The Colorado River made it [the Grand Canyon]. But you feel when you're there that God gave the Colorado River its instructions."

J. B. PRIESTLEY

Have you ever gazed out over the horizon and wondered why Earth's surface has the shape it does and what forces created that shape? From high, snow-capped peaks to broad, rolling plains, Earth's landscape comprises a diverse array of landforms—large and small, rough and smooth. Landscapes evolve through slow changes as uplift, weathering, erosion, transportation, and deposition combine to sculpt the land surface.

In the past, these changes were imperceptible on a human time scale, but new technologies now permit us to measure the rates of many of these processes directly. **Geomorphology**—the study of landscapes and their evolution—is a revitalized branch of the Earth sciences that has benefited tremendously from the ability to measure these slow processes. Knowing how landscapes evolve helps us to manage our land resources and appreciate the linkages between tectonics and climate. Understanding the development of landscapes represents a great challenge to geologists because it requires them to integrate many branches of Earth science.

In the most basic sense, landscapes can be viewed as the results of a competition between processes that raise Earth's surface and those that lower it. Driven by the plate tectonic geosystem, Earth's crust is uplifted into mountain ranges. Uplifted rocks are exposed to weathering and erosion, which are driven by the climate geosystem. The geomorphology of uplifted rocks is an expression of the interaction between these two geosystems.

The uplifted areas of Earth's crust can be narrow or broad, and uplift rates can be high or low. Similarly, physical and chemical weathering may operate over narrow or broad areas, and the intensity of weathering can be high or low. Thus, landscapes themselves depend on the *proportion* of these tectonic and climatic influences. Moreover, these influences interact. For example, uplift may drive regional (or even global) climate change and weathering rates, which in turn help control further uplift of mountain ranges.

In this chapter, we look closely at how tectonics and climate—and their component processes, including uplift, weathering, erosion, and sediment transport and deposition—interact in the dynamic process that sculpts the landscape.

Topography, Elevation, and Relief

We begin our study of landscape evolution with the elementary facts of any terrain that are obvious as one examines Earth's surface: the height and ruggedness, or roughness, of mountains and lowlands. **Topography** is the general configuration of varying heights that give shape to Earth's surface. We compare the heights of landscape features with sea level—the average height of the oceans throughout the world. We then express the altitude, or vertical distance above or below sea level, as **elevation.** A topographic map shows the distribution of elevations in an area and most often represents this distribution by **contours,** lines that connect points of equal elevation (**Figure 18.1**). The more closely spaced the contour lines, the steeper the slope.

Centuries ago, geologists learned to survey topography and construct maps to plot and record geological information. Although land surveying based on age-old methods is still used for some purposes, modern mapmakers rely on satellite photographs, radar imaging, airborne laser range finders, and other technologies that enable them to discern elevation and other topographic properties (**Figure 18.2**).

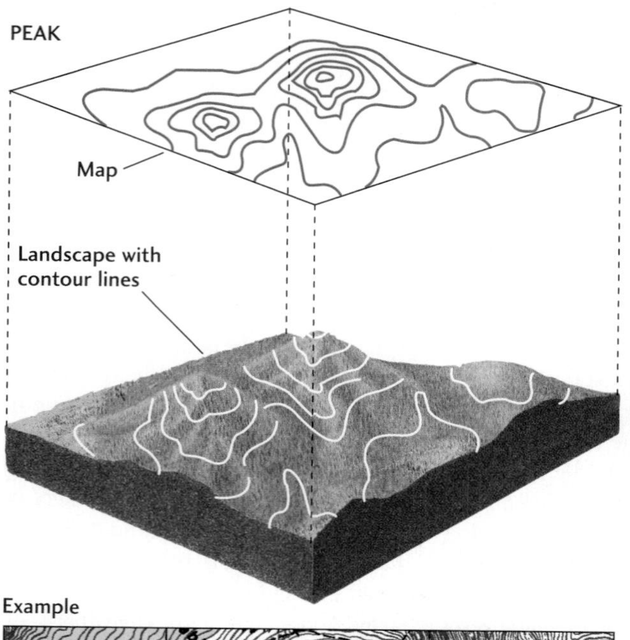

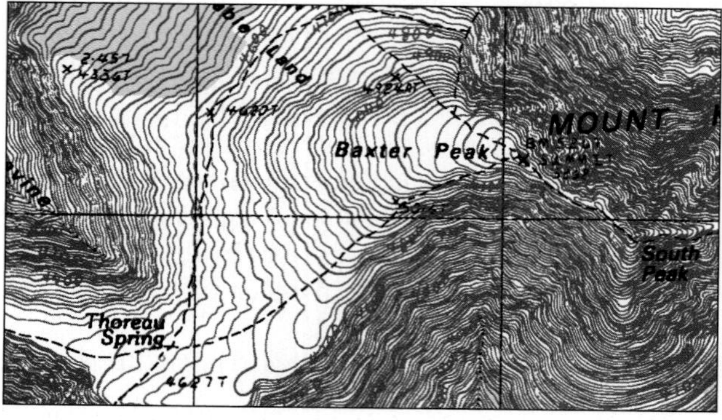

Mt. Katahdin, Maine

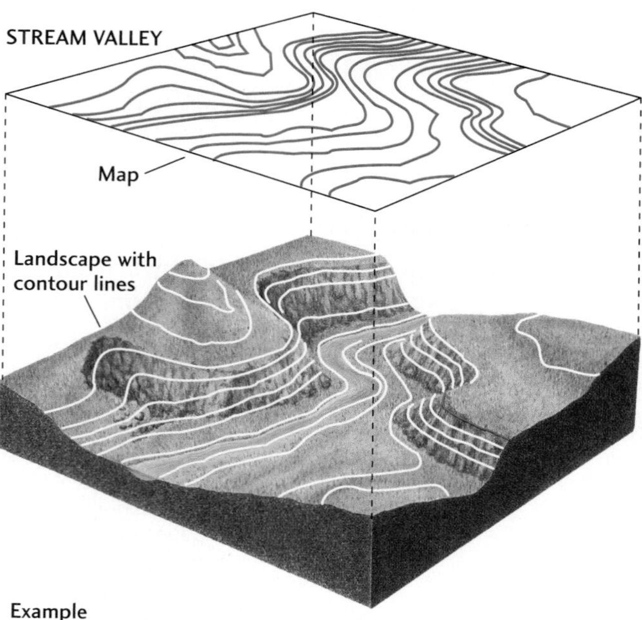

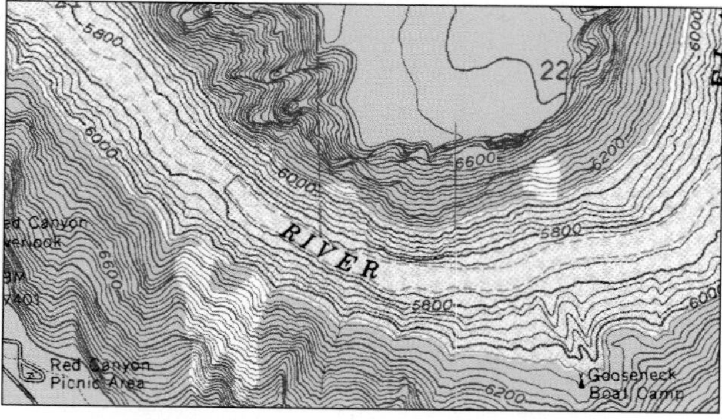

Flaming Gorge, Wyoming

Figure 18.1 The topography of a mountain peak (*left*) and a stream valley (*right*) can be accurately depicted on a flat topographic map by contours, lines that connect points of equal elevation. The more closely spaced the contour lines, the steeper the slope. [After A. Maltman, *Geological Maps: An Introduction* (New York: Van Nostrand Reinhold, 1990), p. 17. Topographical maps from USGS DRG.]

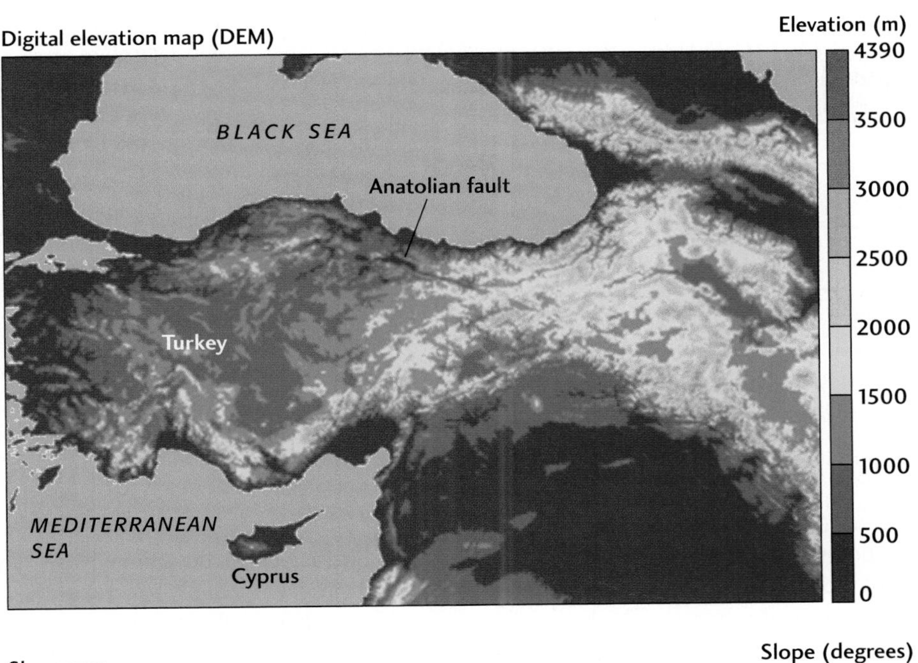

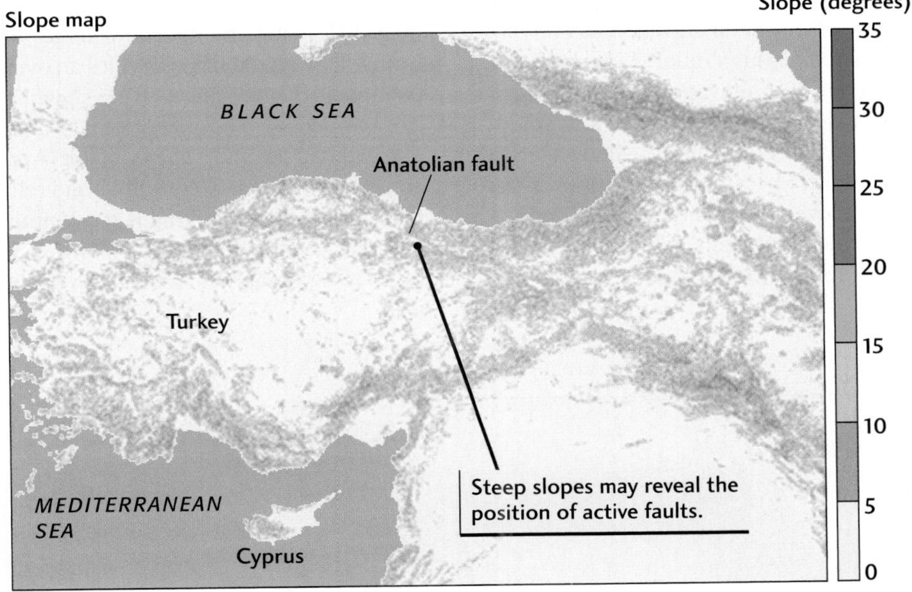

Figure 18.2 Topographic maps for Turkey and adjacent areas. Elevations can be obtained digitally from satellites to produce several types of maps. The top map shows elevation. Because the values are depicted digitally, each pixel representing one elevation value, this type of map is called a Digital Elevation Model, or DEM. In the bottom map, elevation values from the top map are used to calculate the slopes between adjacent pixels. The slopes are then represented by angles measured in degrees from horizontal. This slope map is very useful for identifying places where the changes in topography are particularly abrupt, such as along the front of mountain ranges or along active fault scarps. [Marin Clark.]

One of the properties of topography is **relief**—the difference between the highest and lowest elevations in a particular area (**Figure 18.3**). As this definition implies, relief varies with the scale of the area over it which it is measured. In studies of geomorphology, it is useful to define three fundamental components of relief: hillslope relief (the decrease in elevation between mountain summits/ridgelines and the point at which channels begin), tributary channel relief (the decrease in elevation along tributaries), and trunk channel relief (the decrease in elevation along a trunk channel).

To estimate relief in an area of interest from contours on a topographic map, we subtract the elevation of the lowest

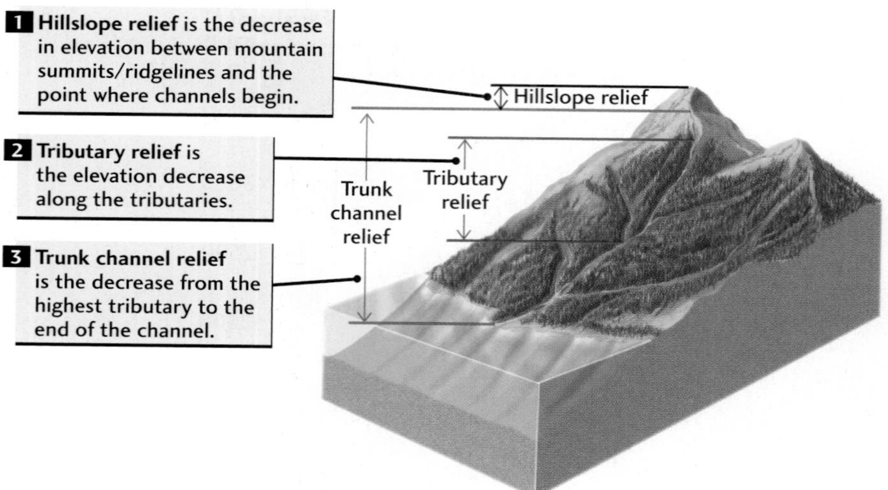

Figure 18.3 Relief is the difference between the highest and lowest elevations in a region. Three types of relief can be defined for a typical mountainous area.

contour, usually at the bottom of a river valley, from that of the highest, at the top of the highest hill or mountain. Relief is a measure of the roughness of a terrain. The higher the relief, the more rugged the topography. Mount Everest, the highest mountain in the world with an elevation of 8861 m, is located in an area of extremely high relief (**Figure 18.4**a). In general, most regions of high elevation also have high relief, and most areas of low elevation have low relief. There are exceptions, however. For example, the Dead Sea in Israel and Jordan has the lowest land elevation in the world at 392 m below sea level, but it is flanked by impressive mountains that provide significant relief in this small area of the Earth (Figure 18.4b). Other regions, such as the Tibetan Plateau of the Himalayas, may lie at high elevations but have relatively low relief.

If we fly over the United States, we can see many sorts of topography. **Figure 18.5** is a computer-processed digital map that shows the details of large- and small-scale landforms. This digital map provides an overall view of the entire contiguous United States while showing features as small as 2.5 km across. The moderate elevations and relief of the elongate ridges and valleys of the Appalachian Mountains contrast with the low elevations and relief of the midwestern plains. Even more striking is the contrast between the plains and the Rocky Mountains. As we examine these different types of topography more closely, we can characterize them

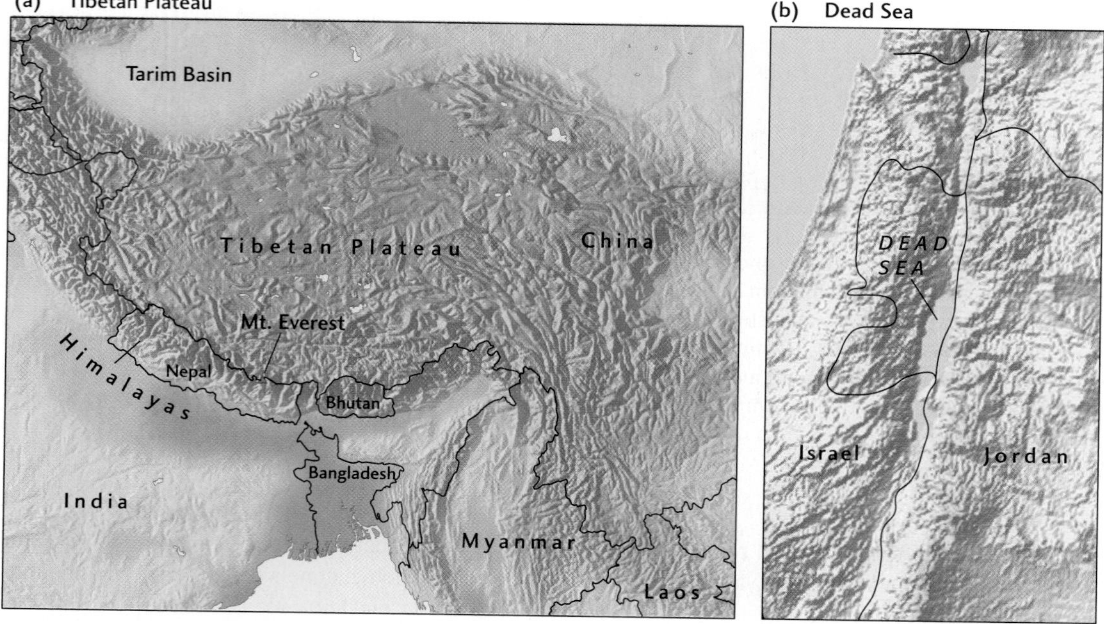

Figure 18.4 Topographic maps. (a) Mount Everest, the highest mountain in the world, and (b) the Dead Sea, the lowest land elevation in the world. [Marin Clark and Nathan Niemi.]

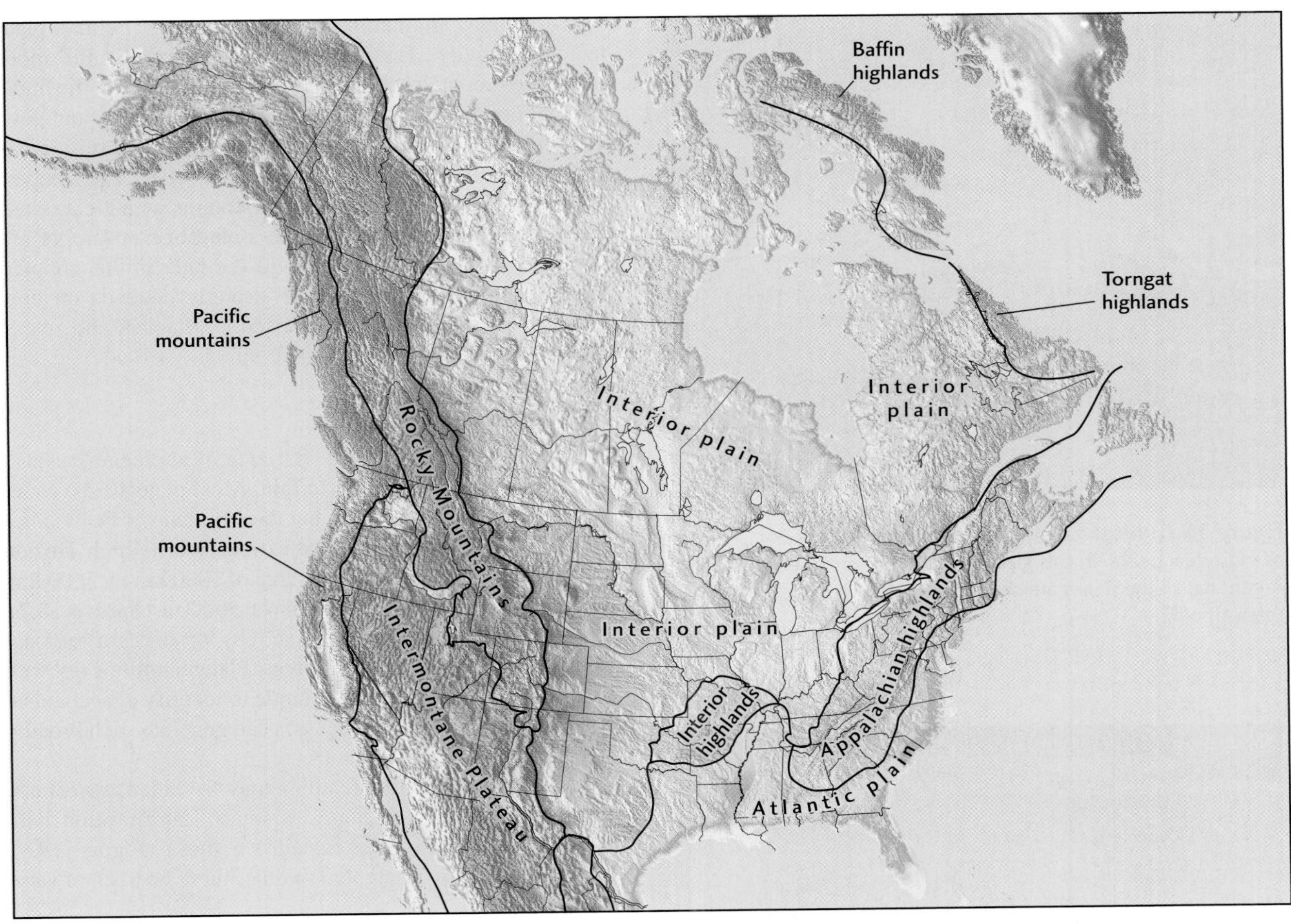

Figure 18.5 A digital shaded-relief map of landforms in the contiguous United States and Canada. [Gail P. Thelin and Richard J. Pike, U.S. Geological Survey, 1991.]

not only by elevation and relief but also by their landforms: the steepness of their slopes, the shapes of the mountains or hills, and the forms of the valleys.

Landforms: Features Sculpted by Erosion and Sedimentation

Rivers, glaciers, and the wind leave their marks on the surface of the Earth in a variety of **landforms:** rugged mountain slopes, broad valleys, floodplains, dunes, and many others. The scale of landforms ranges from regional to very local. At the largest scale (tens of thousands of kilometers), mountain belts form topographic walls along the boundaries of lithospheric plates. At the smallest scale (meters), the topography of an individual outcrop may be caused by differential weathering of the rocks of differing hardnesses that compose it. This chapter focuses mostly on the regional-scale features that define the overall topography of Earth's surface.

Mountains and Hills

We have used the word *mountain* many times in this book, yet we can define it no more precisely than to say that a mountain is a large mass of rock that projects well above its surroundings. Most mountains are found with others in ranges, where peaks of various heights are easier to distinguish than distinct separate mountains (**Figure 18.6**). Where mountains rise as single peaks above the surrounding lowlands, they are usually isolated volcanoes or erosional remnants of former mountain ranges.

We distinguish between mountains and hills only by size and custom. Elevations that would be called mountains in lower terrains are called hills in regions of high elevation. In general, however, landforms more than several hundred meters above their surroundings are called mountains.

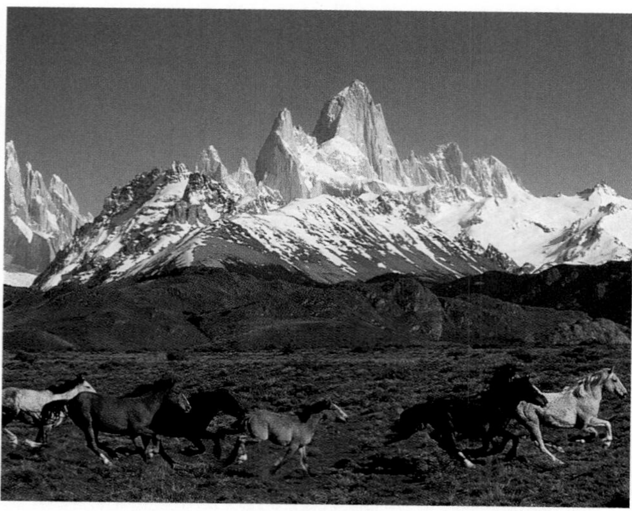

Figure 18.6 Most mountains are found in ranges, not as individual peaks. In this glacially sculpted terrain in southern Argentina, all the peaks are sharp arêtes. [Galen Rowell/ Peter Arnold.]

Mountains are manifestations of direct or indirect plate tectonic activity. The more recent the activity, the more likely the mountains are to be high. The Himalaya, the highest mountains in the world, are also among the youngest. The steepness of the slopes in mountainous and hilly areas generally correlates with elevation and relief. The steepest slopes are usually found in high mountains with great relief. The slopes of mountains with less elevation and relief are less steep and rugged. As we will see later in this chapter, the relief of a mountain range strongly depends on how much the bedrock has been incised by glaciers and rivers, relative to the amount of tectonic uplift.

Plateaus

A **plateau** is a large, broad, flat area of appreciable elevation above the neighboring terrain. Most plateaus have elevations of less than 3000 m, but the Altiplano of Bolivia lies at an elevation of 3600 m. The extraordinarily high Tibetan Plateau, which extends over an area of 1000 km by 5000 km, has an average elevation of almost 5000 m (**Figure 18.7**). Plateaus form where tectonic activity produces a regional uplift in response to vertical forces. Plateau uplift is not well understood, and geologists struggle to identify a mechanism by which plate tectonic processes can generate such broadly uplifted areas.

Smaller plateau-like features may be called *tablelands*. In the western United States, a small, flat elevation with steep slopes on all sides is called a **mesa** (**Figure 18.8**). Mesas result from differential weathering of bedrock of varying hardness.

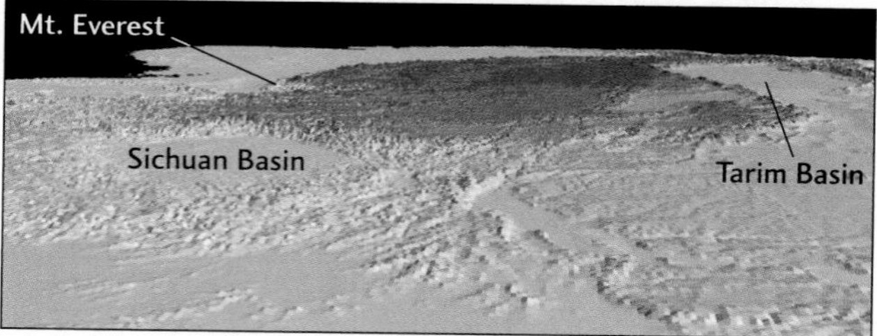

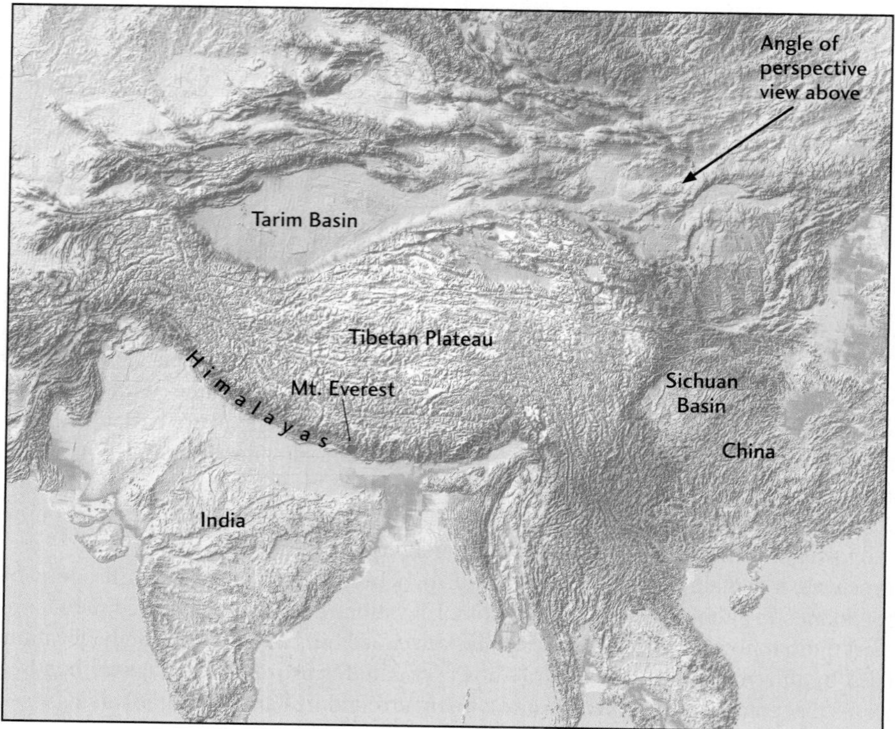

Figure 18.7 Map view of the Tibetan Plateau, the highest and most extensive on Earth. The tectonic mechanisms that produced this massive uplift are still unclear. Some geologists think that it formed when the Indian Plate was subducted beneath the Eurasian Plate, doubling the lithospheric thickness in the region where the plateau is today. Isostatic rebound of this doubly thick lithosphere would produce uplift of the plateau region. Other geologists argue that during the subduction of India beneath Asia, the Earth's lower crust may have flowed—just like the mantle—then pooled as a fluid cushion and pushed up the plateau region.

Figure 18.8 A mesa in Monument Valley, Arizona. The flat tops are held up by erosion-resistant beds. [Raymond Siever.]

Structurally Controlled Ridges and Valleys

In young mountains, during the early stages of folding and uplift, the upfolds (anticlines) form ridges and the downfolds (synclines) form valleys (**Figure 18.9**). As climate and weathering begin to predominate and erosional gullies and valleys bite deeper into the structures, the topography may become inverted, so that the anticlines form valleys and the synclines form ridges. This happens where the rocks—typically sedimentary rocks such as limestones, sandstones, and shales—exert strong control on the topography by their variable resistance to erosion. If the rocks beneath an anticline are mechanically weak, as shales are, the core of the anticline may be eroded to form an anticlinal valley (**Figure 18.10**). In a region that has been eroded for many millions of years, a pattern of linear anticlines and synclines produces a series of ridges and valleys such as those of the Appalachian Mountains in Pennsylvania and adjacent states (**Figure 18.11**).

River Valleys and Bedrock Erosion

Observations of river valleys in various regions led to one of the important early theories of geology: the idea that river valleys were created through erosion by the rivers that flowed in them. Geologists could see that sedimentary rock formations

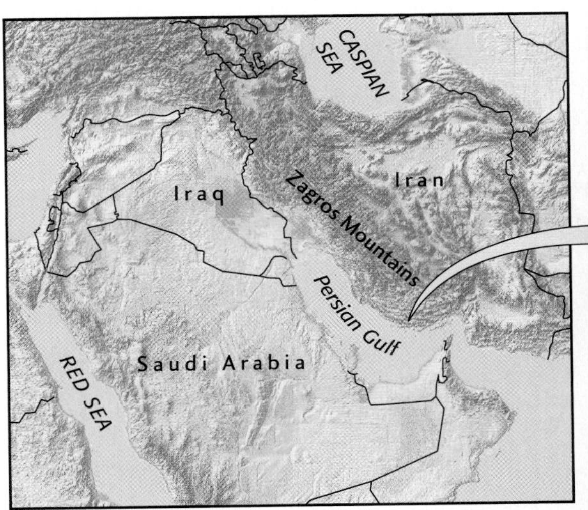

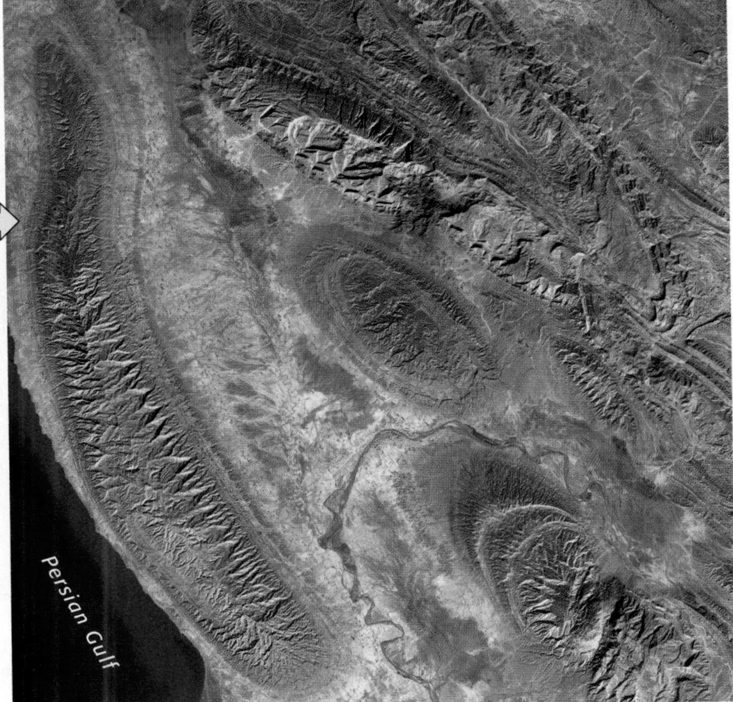

Figure 18.9 Valley and ridge topography formed on a folded terrain of sedimentary rock. The deformation is so recent (Pliocene) that erosion has not yet significantly modified the original structural forms of anticlines (ridges) and synclines (valleys). Zagros Mountains, Iran. [NASA.]

TIME 1
Harder, erosion-resistant rocks lie over softer, more erodable layers. Ridges are over anticlines and streams flow in valleys formed by synclines. Tributary streams on anticline slopes flow faster and with more power than valley streams. They erode the slopes faster than main streams erode the valleys.

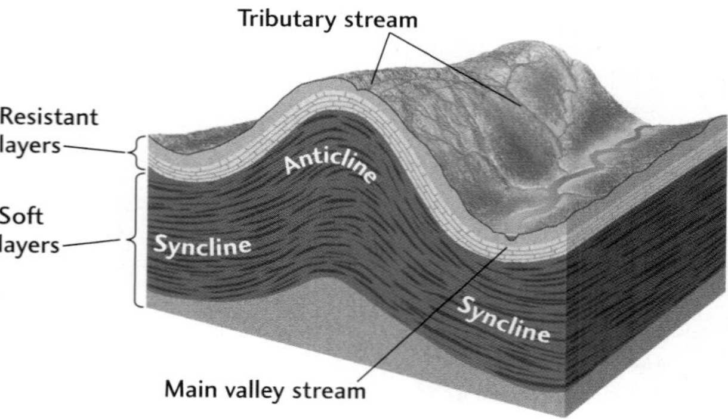

TIME 2
Tributaries over the synclines cut through resistant rock layers and start to quickly carve the softer underlying rock into steep valleys over the anticlines.

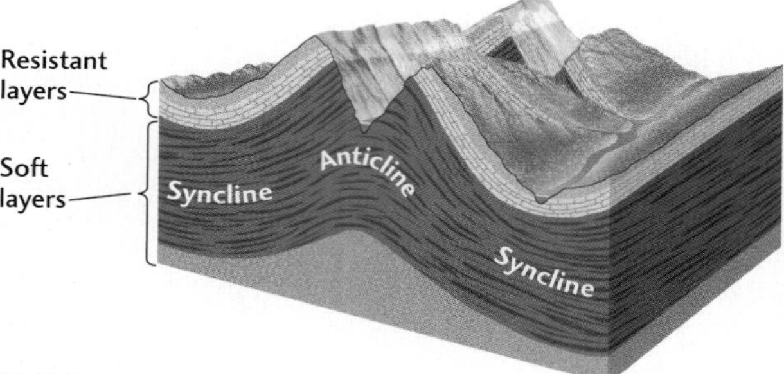

TIME 3
As the process continues, valleys form over the anticlines and ridges capped by resistant strata are left over the synclines.

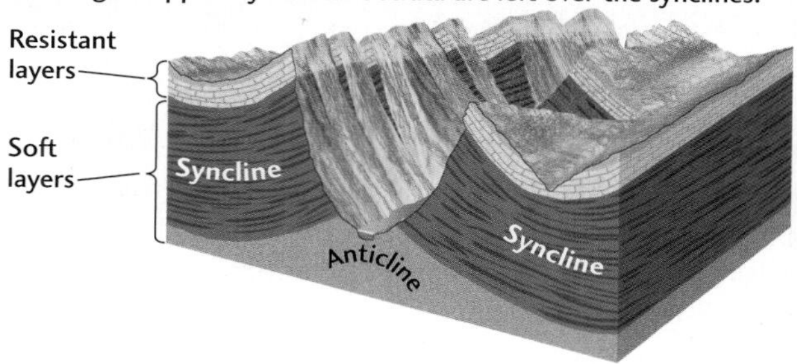

Figure 18.10 Stages in the development of ridges and valleys in folded mountains. In early stages, ridges are formed by anticlines. In later stages, the anticlines may be breached and ridges may be held up by caps of resistant rocks as erosion forms valleys in less resistant rocks.

on one side of a valley matched the same kinds of formations on the opposite side. This led them to conclude that the formations had once been deposited as a continuous sheet of sediment. The river had removed enormous quantities of the original formation by breaking up the rock and carrying it away. Today, geologists are trying to figure out which of several physical processes contribute to this erosion of bedrock.

How a river erodes bedrock depends on the river's **stream power**—which is the product of river slope and river discharge—balanced by the bedrock's ability to resist erosion—which is the product of the volume and grain size of sediment in the river channel (**Figure Story 18.12**). If stream power is high enough to wash away the sediment load, resistance to erosion is a function mostly of bedrock

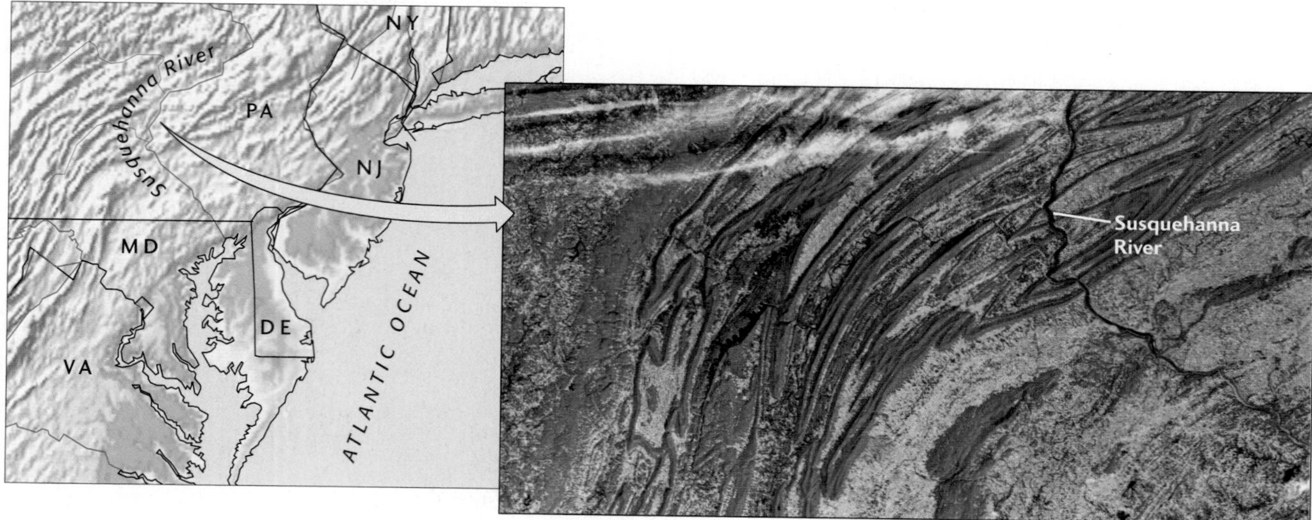

Figure 18.11 The Appalachian Valley and Ridge province has the tectonically controlled topography of linear anticlines and synclines produced by millions of years of erosion. The prominent ridges, shown in reddish orange, are held up by erosion-resistant sedimentary rocks that have been folded into an intricate series of anticlines and synclines. [Earth Satellite Corp.]

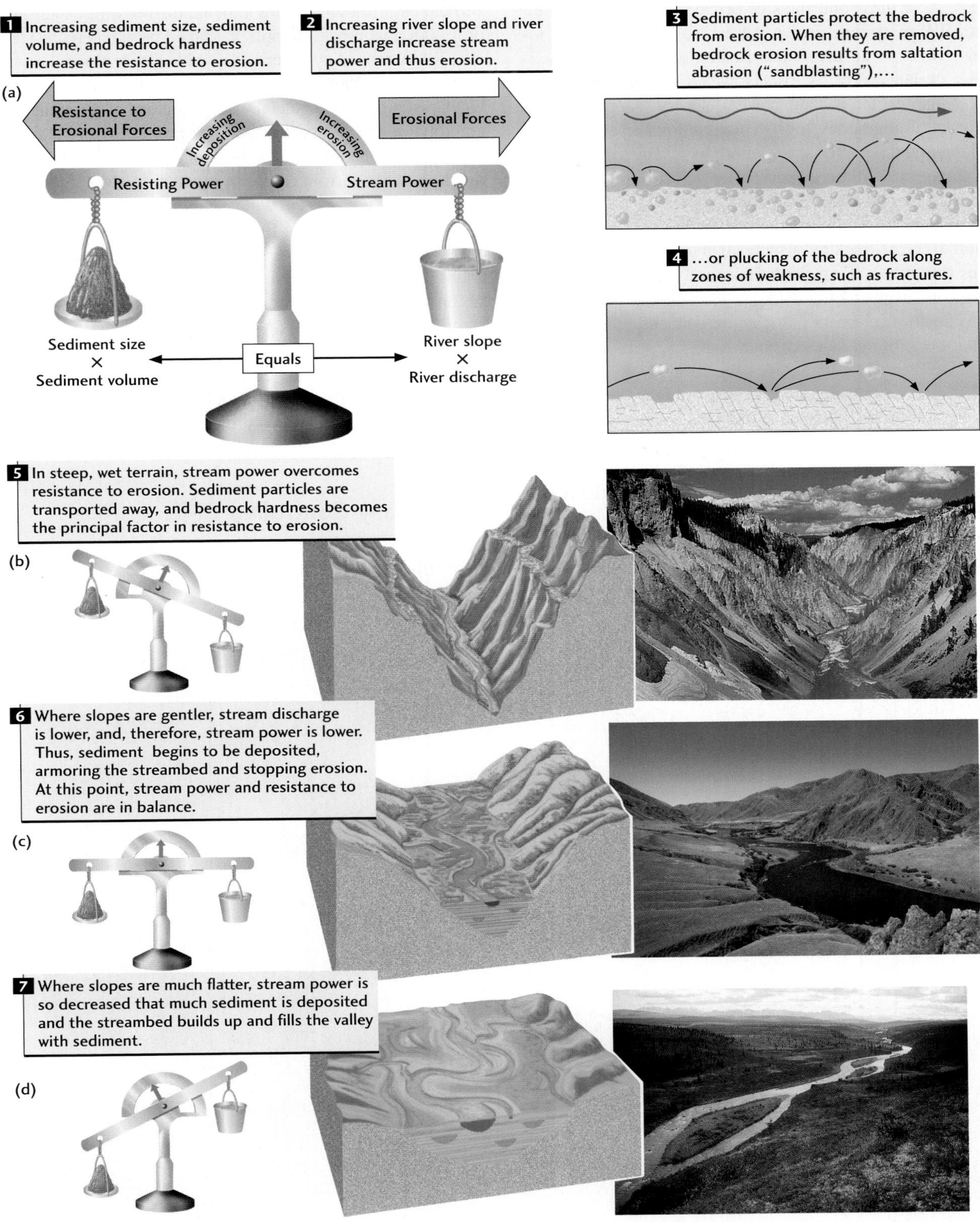

Figure Story 18.12 (a) An increase in stream slope or discharge or a decrease in sediment volume or sediment grain size will increase a river's erosion of its bed. [After D. W. Burbank and R. S. Anderson, *Tectonic Geomorphology* (Oxford: Blackwell, 2001), p. 24.] (b) Steep V-shaped walls of a canyon eroded by a stream in a mountainous region. Yellowstone River, Yellowstone National Park. [Jeff Henry/RocheJaune Pictures.] (c) Wide, open valley in a low mountainous region. Snake River, Suicide Point, Idaho. [Dave G. Houser/Corbis.] (d) River flowing through a broad, flat, lowland valley. Mulchatna River, Alaska. [Glenn Oliver/Visuals Unlimited.]

hardness. As it turns out, rates of bedrock erosion increase dramatically as stream power increases. On most days, a flowing river accomplishes little erosion because discharge, and thus stream power, is low. However, on the rare days when discharge (and stream power) is very large, erosion rates can be dramatically high. This relationship illustrates a fundamental characteristic shared by many of Earth's geosystems: rare, large events often create much more change than frequent but small ones.

Three principal processes erode bedrock. The first is **abrasion** of the bedrock by suspended and saltating sediment particles that move along the bottom and sides of the river channel. (In Chapters 14 and 15, we learned that saltation is the movement of sediment particles by bounces and impacts along the bottom and sides of a channel.) Second, the drag force of the current itself abrades the bedrock as it plucks rock fragments from the channel. Third, at higher elevations, glacial erosion forms valleys that can be subsequently occupied by rivers. Determining the relative importance of these three processes in mountainous terrains is one way geologists can distinguish between the climatic and tectonic influences on landscape evolution.

River valleys have many names—canyons, gulches, arroyos, gullies—but all have the same general geometry. A vertical cut through a young mountain river valley with little or no floodplain has a simple V-shaped profile (Figure Story 18.12b). A broad, low river valley with a wide floodplain has a cross section that is more open but is still distinct from the U-shaped profile of a glacial valley. Regions of different general topography and type of bedrock produce river valleys of varying shape and width (Figure Story 18.12c and d). Valleys range from the narrow gorges of mountain belts and erosion-resistant rock types to the wide, shallow valleys of plains and easily eroded rock types. Between these extremes, the width of a valley generally corresponds to the erosion state of the region. Valleys are somewhat broader in mountains that have begun to be lowered and rounded from erosion and are much broader in low-lying hilly topography.

A **badland** is a deeply gullied topography resulting from the fast erosion of easily erodible shales and clays, such as those of the South Dakota Badlands (**Figure 18.13**) Virtually the entire area is a proliferation of gullies and valleys, with little flat land between them.

Structurally Controlled Cliffs

The folds and faults produced by rock deformation during mountain building leave their marks on Earth's surface.

Figure 18.13 Gully erosion in the Badlands of South Dakota, formed in easily eroded sedimentary rocks. Sage Creek Wilderness, Badlands National Park. [Willard Clay.]

(a)

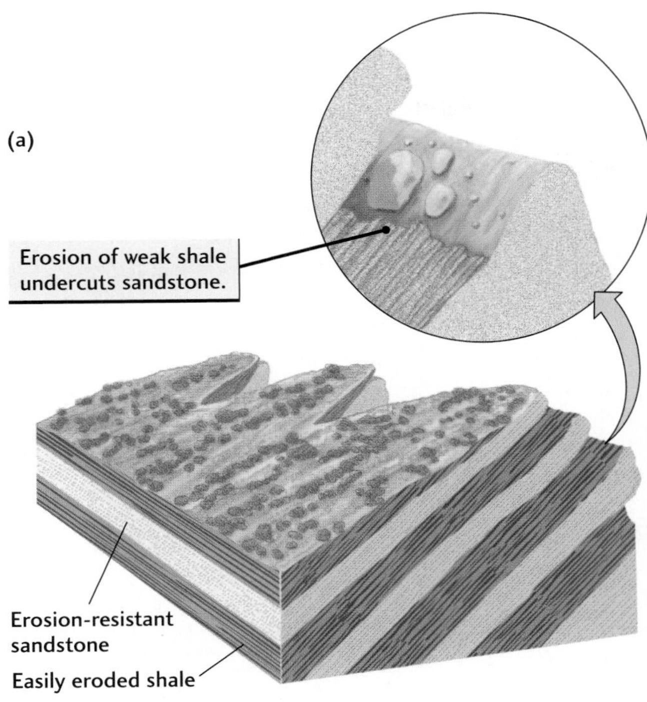

(b)

Figure 18.14 (a) Cuestas form where gently dipping beds of erosion-resistant rock, such as sandstone, are undercut by erosion of an easily eroded underlying rock, such as shale. (b) Cuestas formed on structurally tilted sedimentary rocks in Dinosaur National Monument, Colorado. [Martin Miller.]

Cuestas are asymmetrical ridges in a tilted and eroded series of beds with alternating weak and strong resistance to erosion. One side of a cuesta has a long, gentle slope determined by the dip of the erosion-resistant bed. The other side is a steep cliff formed at the edge of the resistant bed where it is undercut by erosion of a weaker bed beneath (**Figure 18.14**). Much more steeply dipping or vertical beds of hard strata erode more slowly to form **hogbacks**—steep, narrow, more or less symmetrical ridges (**Figure 18.15**). Steep cliffs are also produced by nearly vertical faults in which one side rises higher than the other.

Figure 18.15 Hogbacks are narrow ridges formed by layers of erosion-resistant sedimentary rocks that are tectonically turned up so that the beds are vertical or nearly so. These hogback ridges are in the Rocky Mountains near Denver, Colorado. [Tom Till/DRK.]

Interacting Geosystems Control Landscape

Broadly speaking, the interaction of Earth's internal and external heat engines controls landscape. The internal heat engine drives tectonics, which elevates mountains and volcanoes. The external heat engine, powered by the Sun, controls climate and weathering, which wear away the mountains and fill basins with sediment. Sunlight causes the motions of the atmosphere that produce climate, Earth's different temperature regimes, and the rainwater that runs off the continents as rivers. Thus, landscape is controlled by Earth's interacting geosystems (**Figure 18.16**).

Feedback Between Uplift and Erosion

The ceaseless competition between tectonic processes, which tend to create mountains and build topography, and surface processes, which tend to tear them down, is the subject of intensive study by geomorphologists. The interaction between the two is a **negative feedback.** In this kind of

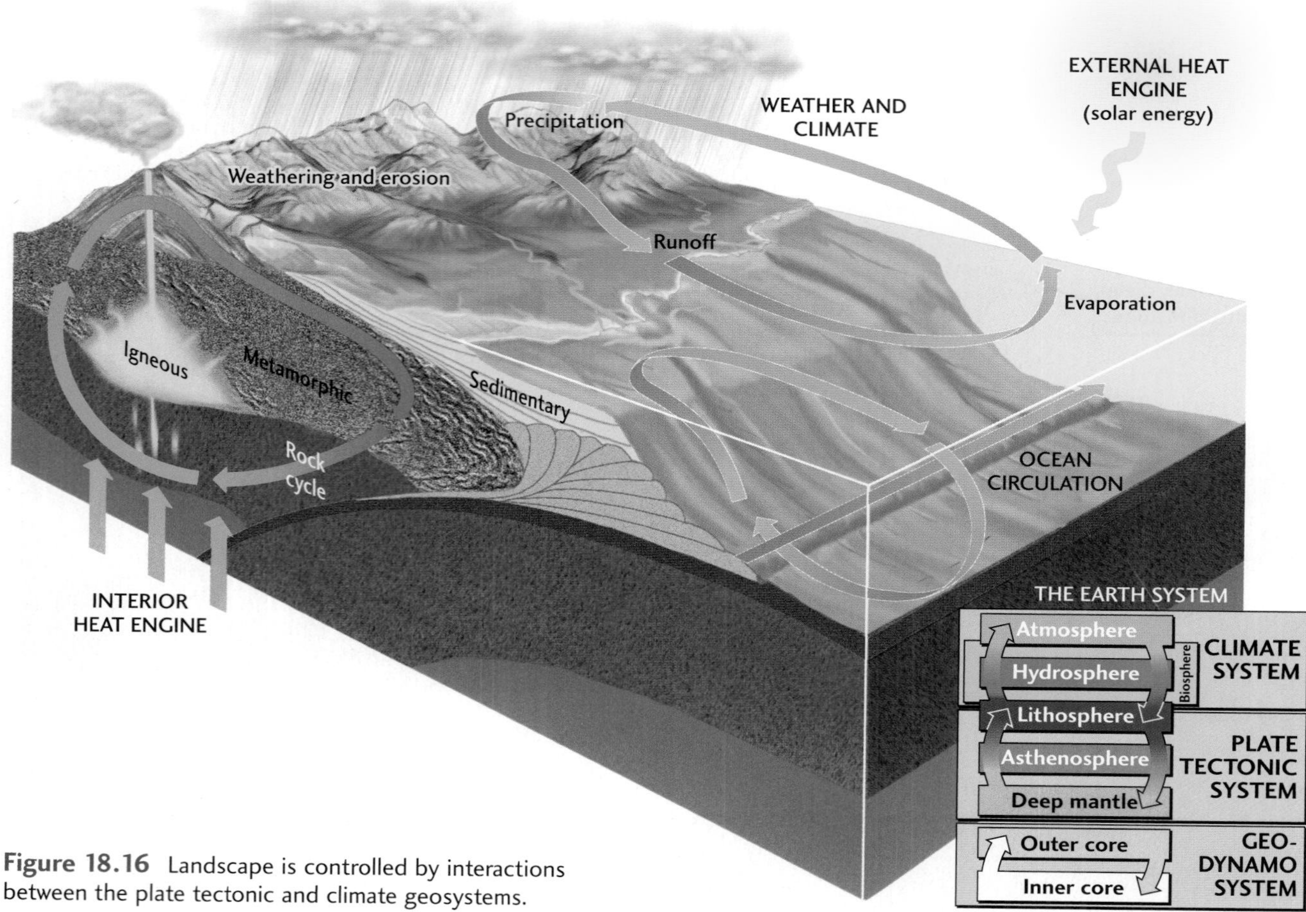

Figure 18.16 Landscape is controlled by interactions between the plate tectonic and climate geosystems.

process, one action produces an effect (the feedback) that tends to slow the original action and stabilize the process at a lower rate. For example, if you are thirsty, you will drink a glass of water quickly at first. As the drinking reduces your thirst (the feedback), you will drink more slowly, until finally your thirst is completely satisfied and you stop drinking. The process of drinking has stabilized at a rate of zero.

A similar negative-feedback process begins when strong tectonic action elevates mountains. Tectonic uplift provokes an increase in erosion rate (**Figure 18.17**). The higher the mountains grow, the faster erosion wears them down. As long as mountain building continues, elevations stay high or increase. When mountain building slows, perhaps because of a change in the rate of plate tectonic movements, the mountains rise more slowly or stop rising entirely. As growth slows or stops, erosion starts to dominate and the mountains are worn down to lower elevations. This process explains why old mountains, such as the Appalachians, are relatively low compared to the much younger Rockies. As the mountains continue to be worn down, erosion also slows, and the whole process eventually tapers off. Elevation is thus a balance between the rate of tectonic uplift and the rate of erosion.

Curiously, over the shorter time scales of thousands to millions of years, tectonics and climate can interact in a **positive feedback** (see Figure 18.17b and Feature 18.1) in which mountain summits may become *higher* as a result of erosion. This occurs because continents and mountains are buoyed up and float in the earth's mantle, much like the icebergs discussed in Chapter 16 float in seawater.

Continents and mountains float because the large volume of less dense continental crust that projects into the denser mantle provides the buoyancy. Note that the crust is thicker under a mountain because a deeper root is needed to float the additional weight of the mountain. In Chapter 20, you will see that the mantle just beneath the crust is solid rock, but that over thousands to millions of years the mantle flows very slowly like a viscous fluid when forces are applied. The *principle of isostasy* implies that, over this period of time, the mantle has little strength and behaves like a viscous liquid when it is forced to support the weight of continents and mountains.

Isostasy also implies that, as a large mountain range forms, it slowly sinks under gravity and the crust bends downward. When enough of a root bulges into the mantle, the mountain floats. If the valleys in a mountain range are deepened by erosion (see Figure 18.17b), the weight on the crust is lightened and less root is needed for flotation. As the valleys erode down, the root floats upward. This process, called *isostatic rebound,* forms the basis of *positive feedback*

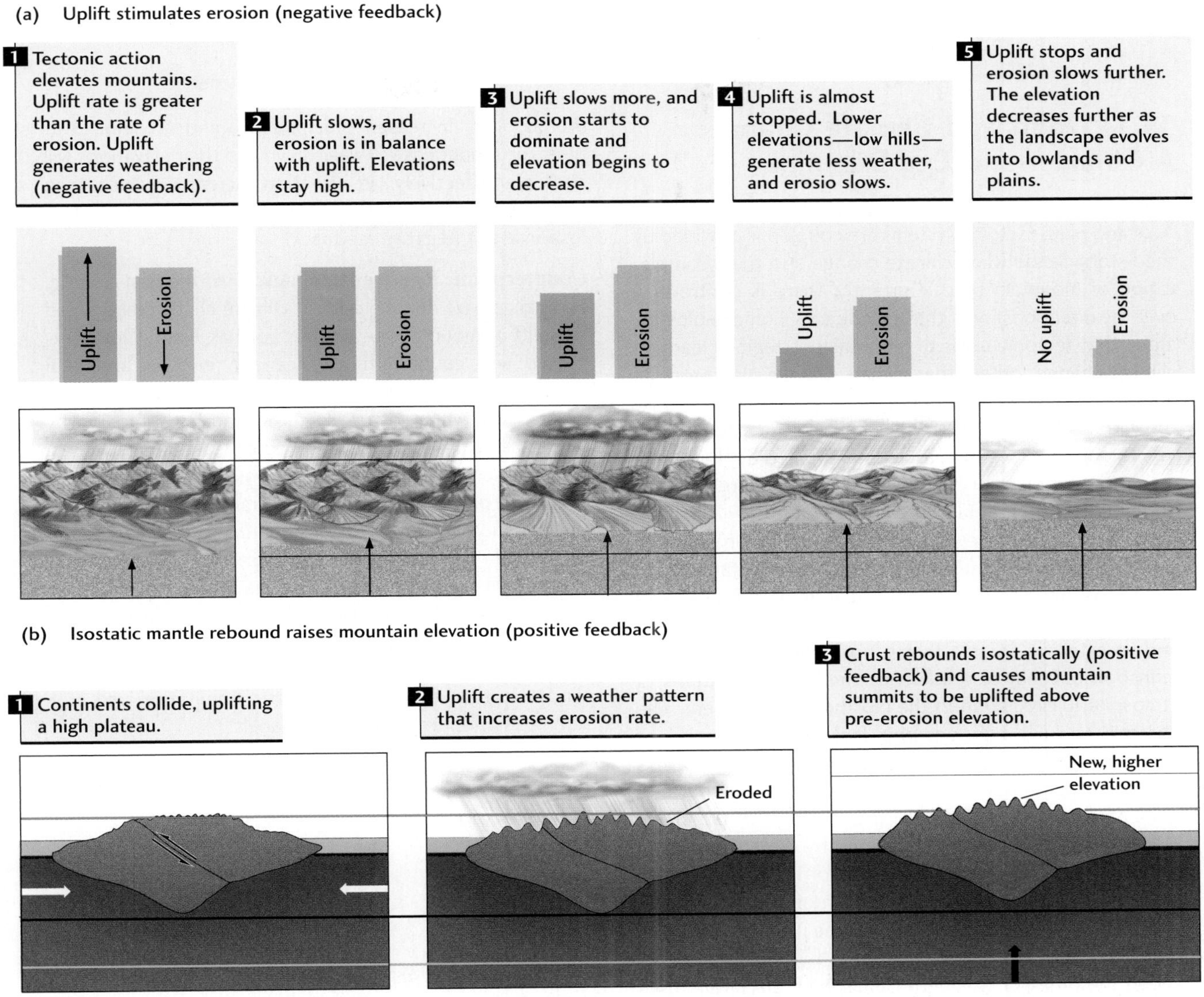

(a) Uplift stimulates erosion (negative feedback)

1 Tectonic action elevates mountains. Uplift rate is greater than the rate of erosion. Uplift generates weathering (negative feedback).

2 Uplift slows, and erosion is in balance with uplift. Elevations stay high.

3 Uplift slows more, and erosion starts to dominate and elevation begins to decrease.

4 Uplift is almost stopped. Lower elevations—low hills—generate less weather, and erosio slows.

5 Uplift stops and erosion slows further. The elevation decreases further as the landscape evolves into lowlands and plains.

Uplift | Erosion | Uplift | Erosion | Uplift | Erosion | Uplift | Erosion | No uplift | Erosion

(b) Isostatic mantle rebound raises mountain elevation (positive feedback)

1 Continents collide, uplifting a high plateau.

2 Uplift creates a weather pattern that increases erosion rate.

3 Crust rebounds isostatically (positive feedback) and causes mountain summits to be uplifted above pre-erosion elevation.

Eroded

New, higher elevation

Figure 18.17 (a) A negative-feedback loop relates uplift and erosion to surface elevation. Tectonic uplift causes an increase in erosion rate, which in turn lowers the surface elevation. The elevation is thus a balance between the rates of tectonic uplift and erosion. (b) A positive feedback response is obtained if the effects of isostasy are included. Here, continents and and mountains float because the large volume of less dense continental crust that projects into the denser mantle provides the buoyancy. If the weight of the mountain range is reduced by erosion, then the root of the mountain range will rebound isostatically, leading to an increase in the heights of mountain summits. [After D. W. Burbank and R. S. Anderson, *Tectonic Geomorphology* (Oxford: Blackwell, 2001), p. 9.]

between climate and tectonics, that results in mountain summits being elevated to new highs as a result of erosion (see Figure 18.17b). Over longer time scales, however, the mountain summits will be worn down as described above and in Figure 18.17a.

Feedback Between Climate and Topography

Glaciers, rivers, and landslides—powerful agents of erosion—operate at different rates at different altitudes. Thus climate, which changes as a function of altitude, modulates weathering and therefore also modulates erosion and uplift of mountain ranges.

In Chapter 7, we saw that the effects of climate on weathering include freezing and thawing and expansion and contraction due to heating and cooling. Climate also affects the rate at which water dissolves minerals. Rainfall and temperature, the components of climate, affect weathering and erosion through the rain that falls on bedrock and soil, the infiltration of water into the soil, mass wasting, rivers, and

18.1 Uplift and Climate Change: A Chicken-and-Egg Dilemma

One of the clearest illustrations of how Earth's climate and plate tectonic systems are coupled is provided by the feedbacks between climate change and the mean elevation of mountain belts. Currently, there is controversy over the directionality of these feedbacks. Some geologists argue that tectonic uplift of mountainous regions leads to climate change, others that climate change may promote tectonic uplift. This type of debate is well characterized by the classic chicken-and-egg dilemma: which came first?

The uplift versus climate debate is fueled by the observation that cooling of the Northern Hemisphere's climate and uplift of the Tibetan Plateau may have been synchronous. With a mean elevation of 5000 m and an area well over half the size of the United States, the Tibetan Plateau is the most imposing topographic feature on the surface of the Earth (see Figure 18.7). The Tibetan Plateau is so high and so extensive in area that it not only drives the Asian monsoon but may also influence the broader patterns of atmospheric circulation in the Northern Hemisphere. The absence of such a topographic feature would no doubt result in significant climatological differences in Earth's Northern Hemisphere.

Unfortunately, although the timing of Northern Hemisphere cooling is well calibrated by the age of glacial deposits and by the deep-sea record of temperature decrease that signals the construction of major ice sheets and glaciers, the timing of uplift of the Tibetan Plateau is not. This is where the debate begins. If Tibetan uplift preceded the onset of Northern Hemisphere glaciation, then it might be argued that tectonically induced uplift indirectly caused climate change. On the other hand, if uplift of Tibet lagged the onset of the Northern Hemisphere glaciation, then it might be argued that climate change promoted uplift as an isostatic response to enhanced erosion rates.

Point: Negative Feedback. The possibility that mountain building may have promoted glaciation of the Northern Hemisphere has been recognized for more than 100 years. Geologists who currently advocate this point of view believe that several important processes occurred in the uplift of Tibet that led to a negative feedback. In this scenario, uplift led to a change in atmospheric circulation; which led to cooling of the Northern Hemisphere; which resulted in an increase in precipitation, glaciation, and river runoff in Tibet. This in turn produced higher erosion rates; which then led to removal of CO_2—an important greenhouse gas—from the atmosphere; which in turn led to further cooling, increased precipitation, and erosion. Over time, the mountains will wear down and their elevations will decrease. Effectively, an elevation increase—which then modulates climate—results in an elevation decrease. This response is a negative feedback.

Counterpoint: Positive Feedback. Over the past decade, geologists have discovered that climate change might lead to uplift of mountainous regions such as Tibet. This unexpected and counterintuitive scenario involves an initial cooling of climate, which stimulates an increase in precipitation rates, which in turn leads to enhanced erosion by glaciers and rivers. In the absence of isostatic responses, an increase in erosion would be predicted to feed back negatively and lower the mountain ranges. However, when the influence of isostasy is considered, erosion is predicted to result in a mass decrease in the mountain range as a whole, with the result being that the mountains would be uplifted and their peaks elevated to new and higher positions (see Figure 18.17b). At these higher elevations, the mountains would feed back positively and help to modify climate further, thereby increasing precipitation and erosion rates and further enhancing uplift.

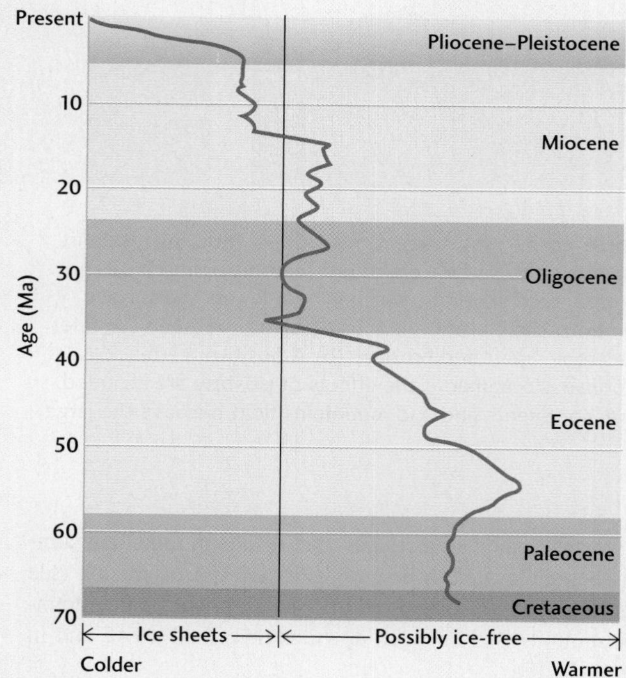

Summary of the deep-sea temperatures in the Atlantic Ocean. [After M. E. Raymo and W. F. Ruddiman, "Tectonic Forcing of Late Cenozoic Climate." *Nature* 359 (1992): 117.]

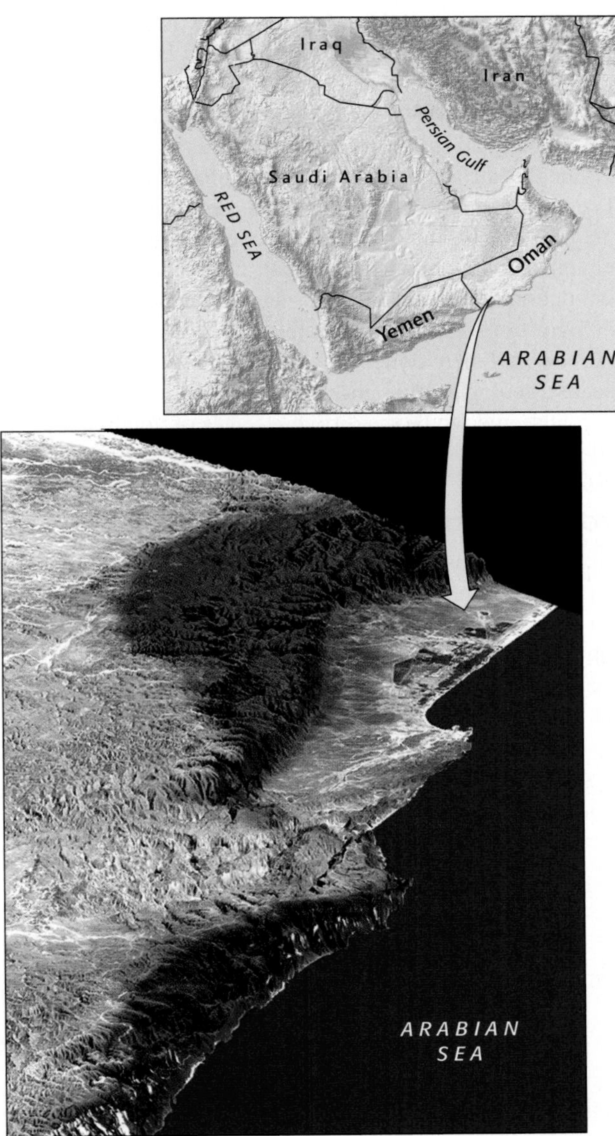

Figure 18.18 Easterly view of the escarpment along the Arabian Sea at the Yemen-Oman border. (See the photograph at the beginning of Chapter 7 for a close-up view of the escarpment.) This three-dimensional satellite image illustrates how topography determines local climate and, in turn, controls erosion and landscape development. Although the Arabian Peninsula is very arid, the steep escarpment of the Qara Mountains wrings moisture from the monsoons (seasonal rains). The moisture allows natural vegetation to grow (green along the mountain fronts and in the canyons) and soil to develop (dark brown areas). In contrast, the light-colored areas are mostly dry desert. This climate focuses erosion on the ocean side of the mountain range. The intense erosion, in turn, has caused the escarpment to retreat landward, from right to left. [NASA.]

glaciers, all of which help to break up the rock and mineral particles eroded from slopes and carry them downhill. Thus climate plays a role in topography.

High elevation and relief enhance the fragmentation and mechanical breakup of rocks, partly by promoting freezing and thawing. At high elevations, where the climate is cool, mountain glaciers scour bedrock and erode deep valleys. Fragmented debris on mountain slopes moves downhill quickly in slides and other mass movements, exposing fresh rock to attack by weathering. Rivers run faster in mountains than in lowlands and therefore erode and transport sediment more rapidly. Chemical weathering plays an important role in the erosion of high mountains, but the mechanical breakup of rocks is so rapid that most of the debris appears to be almost unweathered. The products of chemical decay—dissolved material and clay minerals—are carried down from steep mountain slopes as soon as they form. The intense erosion that occurs at high elevations produces a topography of steep slopes; deep, narrow river valleys; and narrow floodplains and drainage divides.

In lowlands, by contrast, weathering and erosion are slow and the clay mineral products of chemical weathering accumulate as thick soils. Mechanical breakup occurs, but its effects are small compared with those of chemical weathering. Most rivers run over broad floodplains and do little mechanical cutting of bedrock. Glaciers are absent, except in cold polar regions. Even in lowland deserts, strong winds merely facet and round rock fragments and outcrops rather than breaking them up. A lowland thus tends to have a gentle topography with rounded slopes, rolling hills, and flat plains.

Just as climate affects topography, topography can affect climate. For example, mountains cause rain shadows, dry areas on the leeward slopes of mountain ranges (see Figure 13.3). Rain shadows result in preferential erosion of one side of a mountain belt (**Figure 18.18**). In places such as New Guinea, where the difference between rainfall on the windward and leeward sides of the mountain belt is extreme, geologists have estimated that the uplift history of metamorphic rocks buried deep in the crust is influenced by the history of rainfall at Earth's surface!

Models of Landscape Evolution

The strong contrasts in the morphology of landscapes stimulated early geologists to speculate about their causes. Three prominent and influential geologists were William Morris Davis, Walther Penck, and John Hack. Davis believed that an initial burst of tectonic uplift is followed by a long period of erosion, with landscape morphology depending mostly on geologic age. Davis's view was so dominant during the early 1900s that it overshadowed the contemporaneous hypothesis of Walther Penck, who argued that tectonic uplift competes with erosion to control landscape morphology. Penck's ideas were not given the attention they deserved until the 1950s, more than two decades after Davis's death. In the 1960s, another conceptual breakthrough occurred when John Hack recognized that uplift could not raise elevation above some critical limit, even if sustained for long periods of time. Mountains, in the absence of erosion, will collapse under their own weight because of the finite strength of rocks.

Modern views of landscape evolution incorporate parts of all these early ideas and acknowledge that there is a natural, time-dependent progression of landscape form. Geologists now understand that the evolutionary path of landscapes depends strongly on the *time scale* over which geomorphic change occurs. The importance of the different landscape-modifying processes varies as a function of the interval of time over which the landscape is observed to change. For example, variations in climate are a very important factor in landscape evolution over the past 100,000 years but only a minor factor over time scales of 100 million years. Over these longer intervals of geologic time, the history of tectonic uplift is probably most important.

Davis's Cycle of Uplift and Erosion

William Morris Davis, a Harvard geologist in the early 1900s, studied mountains and plains throughout the world. Davis proposed a *cycle of erosion* that progresses from the high, rugged, tectonically formed mountains of youth to the rounded forms of maturity and the worn-down plains of old age and tectonic stability (**Figure 18.19**a). Davis believed that a strong, rapid pulse of tectonic uplift begins the cycle. All the topography is built during this first stage. Erosion eventually wears down the landscape to a relatively flat surface, leveling all structures and differences in bedrock. Davis saw the flat surfaces of extensive unconformities as evidence of such plains in past geologic times. Here and there, an isolated hill might stand as an uneroded remnant of former heights. Most geologists at that time accepted Davis's assumption that mountains are elevated suddenly over short geologic times and then stay tectonically fixed as erosion slowly wears them down. Davis's cycle was accepted partly because geologists thought they could find many examples of what seemed to be the different stages of youth, maturity, and old age.

Erosion Competes with Uplift

Davis's view was challenged by his contemporary, Walther Penck, who argued that the magnitude of tectonic deformation and uplift gradually increases to a climax and slowly wanes (Figure 18.19b). Unfortunately, Davis, with his greater professional stature and prolific publication style, was able to promote his own ideas more effectively. Penck's ideas did not really become broadly noticed until about 50 years after their publication.

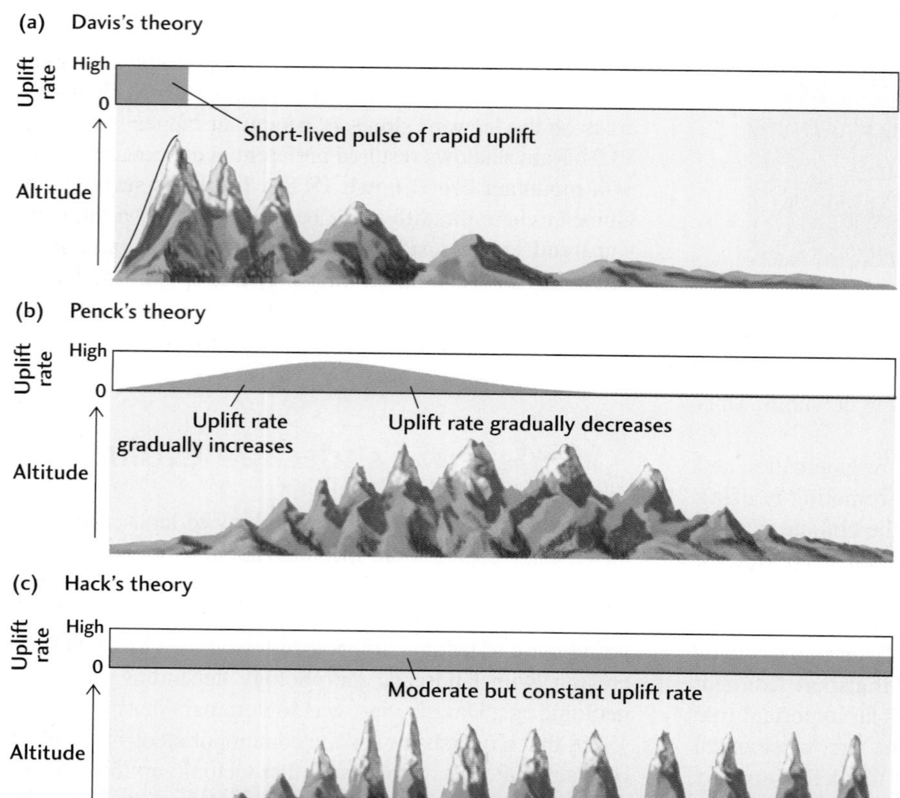

Figure 18.19 Classic models of landscape evolution resulting from tectonic uplift and erosion as proposed by William Morris Davis (a), Walther Penck (b), and John Hack (c). [After D. W. Burbank and R. S. Anderson, *Tectonic Geomorphology* (Oxford: Blackwell, 2001), p. 5.]

Table 18.1 Methods for Absolute Dating of Landscape

Method	Useful Range (Years Ago)	Materials Needed
RADIOISOTOPIC		
Carbon-14	35,000	Wood, shell
Uranium/thorium	10,000–350,000	Carbonate (corals)
Thermoluminescence (TL)	30,000–300,000	Quartz silt
Optically stimulated luminescence	0–300,000	Quartz silt
COSMOGENIC		
In situ beryllium-10, aluminum-26	3–4 million	Quartz
Helium, neon	Unlimited	Olivine, quartz
Chlorine-36	0–4 million	Quartz
CHEMICAL		
Tephrochronology	0–several million	Volcanic ash
PALEOMAGNETIC		
Identification of reversals	>700,000	Fine sediments, volcanic lava flows
Secular variations	0–700,000	Fine sediments
BIOLOGICAL		
Dendrochronology	10,000	Wood

Source: Modified from D. W. Burbank and R. S. Anderson, *Tectonic Geomorphology* (Oxford: Blackwell, 2001), p. 39.

Penck proposed that geomorphic surface processes attack the uplifting mountains throughout the interval of uplift. Eventually, as the rate of deformation wanes, rates of erosion predominate over rates of uplift, resulting in a gradual decrease in both relief and mean elevation. This model was a conceptual breakthrough because it recognized that landscape evolution may result from a competition between uplift and erosion. Davis's model, in contrast, emphasized the temporal distinction between these two processes In his model, landscape age was the primary determinant of form.

Measuring Rates of Uplift and Erosion

Over the past decade, geologists have shown that Penck's view of the competition between uplift and erosion is correct, especially for tectonically active landscapes. Choosing between alternative theories of landscape evolution requires that we determine rates of uplift and erosion in mountain building. New technologies such as the Global Positioning System (GPS) (see Figure 19.5) and radar interferometry have produced spectacular maps of crustal deformation and uplift rates. New dating methods have helped us determine the age of geomorphically important surfaces, such as river terraces (Table 18.1) that date back 1 million years.

A promising new dating scheme is based on the fact that cosmic rays penetrating the uppermost meter of exposed rock or soil lead to the production of very small quantities of certain radioactive isotopes. One of them is beryllium-10, which accumulates to a greater degree the longer the rock or soil is exposed and to a lesser degree the deeper it is buried. Geologists used beryllium-10 to compare river-terrace ages along the course of the Indus River in the Himalayas Mountains. They plotted height changes against time to find average erosion and uplift rates. Stream erosion rates in the Himalayas were found to vary between 2 mm and 12 mm/year. Elsewhere, high mountain tectonic uplift rates have been measured in the same general range, from 0.8 mm to 12 mm/year.

Landscapes Achieve Dynamic Equilibrium

John Hack elaborated on the idea that erosion competes with uplift. He believed that when uplift and erosion rates are sustained for a long period of time, landscape evolution will achieve a balance, or dynamic equilibrium (see Figure 18.19c). During the period of equilibrium, landforms may undergo minor adjustments, but the overall landscape will look more or less the same.

Hack recognized that the height of mountains could not increase forever, even if uplift rates were extremely high. We saw in Chapter 12 that rocks break if large enough stresses are imposed, so it stands to reason that if mountains become too high and steep, they will collapse under their own weight simply because of gravitational stresses. Thus, with continued uplift beyond some critical limit, slope failures and mass wasting alone will prevent further increases in elevation. Consequently, rates of uplift and rates of erosion come into a long-term balance. Unlike the models of Davis and Penck, Hack's model does not require uplift rates to decrease.

A fascinating implication of Hack's model is that a landscape does not have to evolve at all as long as uplift and erosion rates are balanced. Nevertheless, Earth's history teaches us that whatever goes up eventually does come down. Over very long time scales, the models of Davis and Penck are more accurate descriptions of how landscapes will ultimately change form (see Figure 18.17a). When erosion exceeds uplift, slopes become lower and more rounded. Because few areas of the world remain tectonically quiescent for as long as 100 million years, the perfectly flat erosion plain that Davis proposed could form only rarely in Earth's history. The model of dynamic equilibrium is perhaps most appropriate for landscapes in tectonically active areas where a given uplift rate can be sustained for more than a million years or so.

SUMMARY

What are the principal components of landscape? Landscape is described in terms of topography, which includes elevation, the altitude of the Earth's surface above or below sea level, and relief, the difference between the highest and the lowest spots in a region. Landscape also comprises the varied landforms produced through erosion and sedimentation by rivers, glaciers, mass wasting, and the wind. The most common landforms are mountains and hills, plateaus, and structurally controlled cliffs and ridges—all of which are produced by tectonic activity modified by erosion.

How do the climate and plate tectonic systems interact to control landscape? Landscape is determined by tectonics, erosion, climate, and the hardness of bedrock. Tectonics, driven by plate motions, elevates mountains and lowers tectonic valleys and basins. Erosion carves bedrock into valleys and slopes. Climate affects weathering and erosion and makes glacial and desert landscapes possible. The varying erosion resistance of bedrock types partly accounts for differences in slope and valley profiles, steep slopes being found in rocks with greater resistance.

How do landscapes evolve? The evolution of landscapes depends strongly on the competition between the forces of uplift and the forces of erosion. Landscapes begin their evolution with tectonic uplift, which in turn stimulates erosion. While tectonic uplift rates are high, erosion rates may also be high, but mountains will be high and steep. As uplift rates wane, erosion rates will still be high and thus become relatively more important; the land surface is lowered and slopes are rounded. As uplift rates decrease to nil, erosion becomes dominant and wears down the former mountains to gentle hills and broad plains. In the unusual event of long-continued tectonic inactivity, the land surface may be worn to a level plain. Climate and bedrock type strongly modify the evolutionary path in various surface environments, making desert and glacial landscapes very different.

Why don't mountains sink? Mountains, like icebergs, also float. But they float on the mantle instead of the ocean. Over long time scales (thousands to millions of years), the mantle behaves like a liquid and exerts a buoyancy force on the base of mountains, which counteracts the gravitational force. Thus, many mountain ranges are in isostatic equilibrium with the mantle. During rapid erosion of mountain ranges, summits may be uplifed to new highs because the mass of mountains is reduced during erosion, resulting in isostatic uplift.

Key Terms and Concepts

abrasion (p. 414)
badland (p. 416)
contour (p. 408)
cuesta (p. 416)
elevation (p. 408)
geomorphology (p. 407)
hogback (p. 417)
landform (p. 411)
mesa (p. 412)
negative feedback (p. 417)
plateau (p. 412)
positive feedback (p. 418)
relief (p. 409)
stream power (p. 414)
topography (p. 408)

Exercises

This icon indicates that there is an animation available on the Web site that may assist you in answering a question.

1. Give three examples of landforms.

2. What is topographic relief, and how is it related to altitude?

3. Why does relief vary with the scale of the area over it which it is measured?

4. How do faulting and uplift control topography?

5. Compare the different erosional processes in topographically high and low areas.

6. How do river slope and discharge affect stream power?

7. How does climate affect topography, and vice versa?

8. How does the balance between tectonics and erosion affect mountain heights?

9. In what regions of North America do active plate tectonic movements currently affect landscape?

Thought Questions

MEDIA LINK *This icon indicates that there is an animation available on the Web site that may assist you in answering a question.*

1. The summits of two mountain ranges lie at different elevations: range A at about 8 km and range B at about 2 km. Without knowing anything else about these mountain ranges, could you make an intelligent guess about the relative ages of the mountain-building processes that formed them?

2. If you were to climb 1 km, from a river valley to a mountaintop 2 km high in a tectonically active area versus a tectonically inactive area, which would probably be the more rugged climb?

3. A young mountain range of uniform age, rock type, and structure extends from a far northern frigid climate through a temperate zone to a southern tropical rainy climate. How would the topography of the mountain range differ in each of the three climates?

4. Describe the main landforms in a low-lying humid region where the bedrock is limestone.

5. In what landscapes would you expect to find lakes?

6. What changes could you predict for the landscape of the Himalayan Mountains and Tibetan Plateau in the next 10 million years and the next 100 million years?

7. What changes in the landscape of the Rocky Mountains of Colorado might result from a change in the present temperate but somewhat dry climate to a warmer climate with a large increase in rainfall?

8. Over short time scales (thousands of years), isostatic uplift may temporarily raise mountain summits to higher elevations. However, over the longer term (millions of years), continued erosion will reduce these summits to progressively lower elevation. As this happens, what does the concept of isostatic equilibrium predict for the depth of the base of continental lithosphere beneath mountains? Remember that mountains float on the mantle at the base of the continental lithosphere. Should this depth increase or decrease as mountains are worn down?

Suggested Readings

Burbank, D. W., and R. S. Anderson. 2001. *Tectonic Geomorphology.* Oxford: Blackwell.

Goudie, A. 1995. *The Changing Earth: Rates of Geomorphological Processes.* London: Blackwell.

Merritts, D., and M. Ellis. 1994. Introduction to special section on tectonics and topography. *Journal of Geophysical Research* 99: 12135–12141.

Pinter, N., and M. T. Brandon. 1997. How erosion builds mountains. *Scientific American* (April): 74–79.

Strain, P., and F. Engle. 1992. *Looking at Earth.* Atlanta: Turner Publishing.

Sullivan, W. 1984. *Landprints: On the Magnificent American Landscape.* New York: New York Times Book Co.

Summerfield, M. A., 1991. *Global Geomorphology.* Essex: Longman.

Workers of a rescue operation stand before a building completely demolished in the September 1985 earthquake in Mexico City, Mexico. About 10,000 people were killed in Mexico City by this earthquake. [Owen Franken/Corbis.]

CHAPTER

19

Earthquakes

"[T]hey feel the earth tremble under their feet; the sea rises boiling in the port and shatters the vessels that are at anchor. Whirlwinds of flame and ashes cover the streets and public squares; the houses crumble, the roofs are tumbled down upon the foundations, and the foundations disintegrate; thirty thousand inhabitants of every age and of either sex are crushed beneath the ruins. . . .

After the earthquake, which had destroyed three-quarters of Lisbon, the country's wise men ... decided that the spectacle of a few persons burned by a slow fire in a great ceremony is an infallible secret for keeping the earth from quaking."

VOLTAIRE, *CANDIDE*

When Voltaire set out to describe the destruction of the great earthquake of 1755 in Lisbon, Portugal, he was able to draw on the detailed reports of contemporary observers. The Lisbon earthquake is remembered not only for its tragic proportions but also for its role in helping to usher in the Enlightenment, the period in the eighteenth century during which observation and rational explanation of natural events began to gain favor over irrationality and superstition.

No description, however vivid, can replace the personal experience of the violent ground movement in a major earthquake. Outdoors, the quake would knock you down, and the noise of toppling buildings would be deafening. Bridges would sway and collapse. Indoors, the motion would shake you out of bed or throw you wall to wall in a hallway. Furniture would slide all over a room, chandeliers would sway and fall, window glass would break and spray you with shards, dishes and groceries would crash to the floor from shelves. A poorly constructed building would collapse, the floors above falling on you to crush or entrap you.

Building collapse is the major cause of earthquake casualties. Fire, landslides, and dam failures contribute to the havoc. Video pictures of the death and destruction caused by earthquakes in the past two decades in Mexico, Armenia, Iran, Japan, California, and the Philippines have brought the dimensions of such cataclysms to hundreds of millions of viewers worldwide for the first time.

This chapter will examine how seismologists (scientists who study earthquakes) locate, measure, and try to predict earthquakes.

In Chapter 2, we discussed how the concentration of earthquakes in narrow zones helped lead to the development of the plate tectonics theory. Here we will examine earthquake patterns in more detail. On a global scale, most earthquakes do indeed occur at plate boundaries, the intensely strained zones where plates converge, split apart, or slide past each other. However, in many earthquake-prone areas, such as California and Japan, the pattern of tectonic faulting that causes earthquakes is much

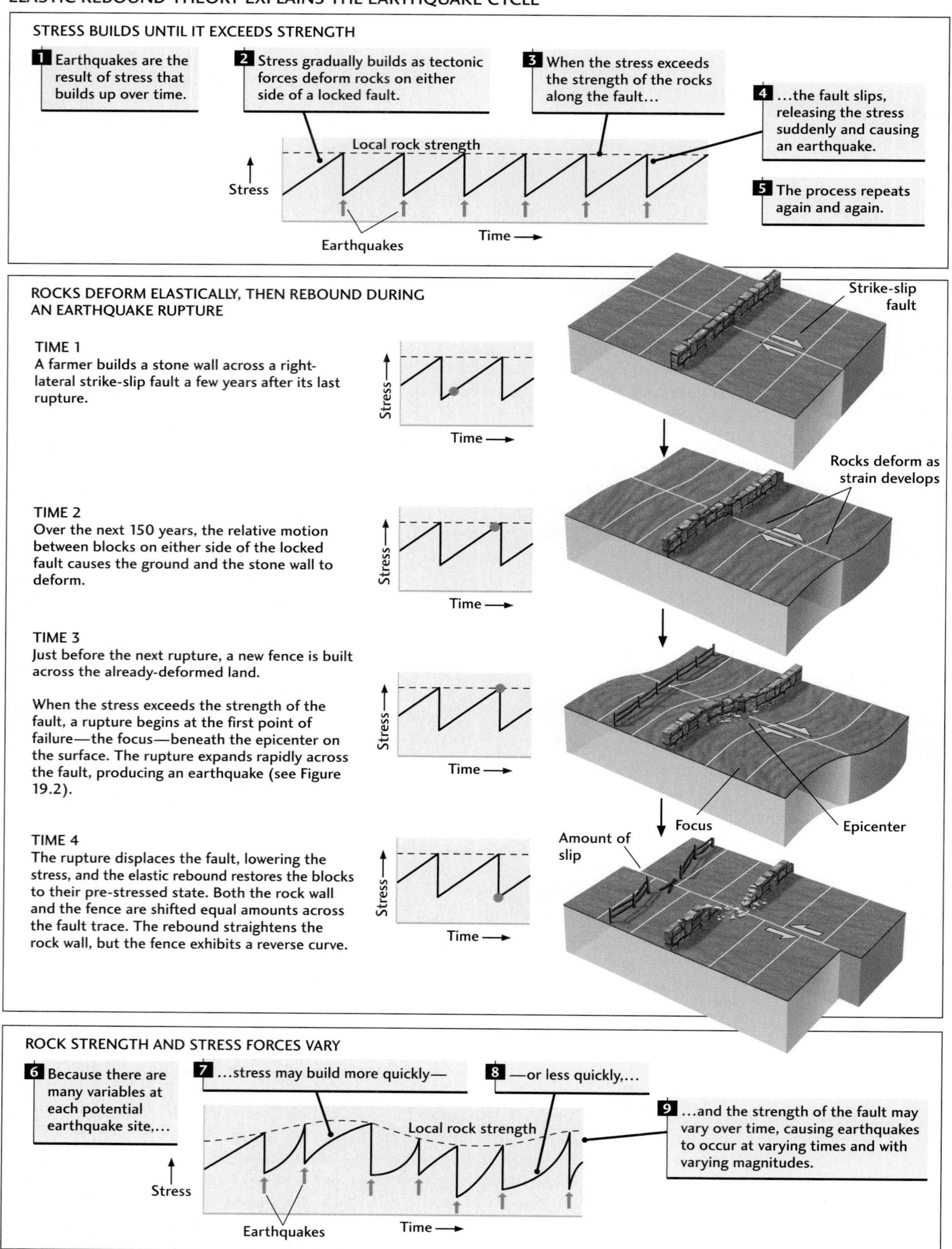

Figure Story 19.1 The elastic rebound theory of earthquake occurrence by repeated cycles of strain accumulation and release. [Lower panel after H. Kanamori and E. E. Brosky, *Physics Today* (June 2001): pp. 34–39.]

more complex than the simplest models of plate tectonics might suggest. These complexities contribute to the difficulty of predicting earthquakes. We will find that earthquakes cannot yet be predicted reliably, although their destructiveness can be reduced considerably. To do the latter, we must use our geological knowledge of where large earthquakes are likely to occur in designing buildings, dams, bridges, and other structures to withstand earthquake shaking.

What Is an Earthquake?

We have seen that plate movements generate forces in narrow zones at the boundaries between tectonic plates. These global forces leave their imprint locally in ways that can be described by the concepts of stress, strain, and strength. *Stress* is the local force per unit area that causes rocks to deform. *Strain* is a measure of the amount of deformation. Rocks fail—that is, they lose cohesion and break into two or more parts—when they are stressed beyond a critical value called the *strength.* Brittle rock formations commonly fail on faults when they are stressed beyond their strength. An **earthquake** occurs when rocks being stressed suddenly break along a new or preexisting fault. The two blocks of rock on either side of the fault slip suddenly, setting off ground vibrations, or **seismic waves** (from the Greek *seismos,* meaning "shock" or "earthquake"), that are often destructive. When the fault slips, the stress is reduced, dropping to a level below the rock strength. After the earthquake, the stress begins to increase again, and the cycle is repeated (**Figure Story 19.1**).

0 Second
Rupture expands circularly on fault plane, sending out seismic waves in all directions.

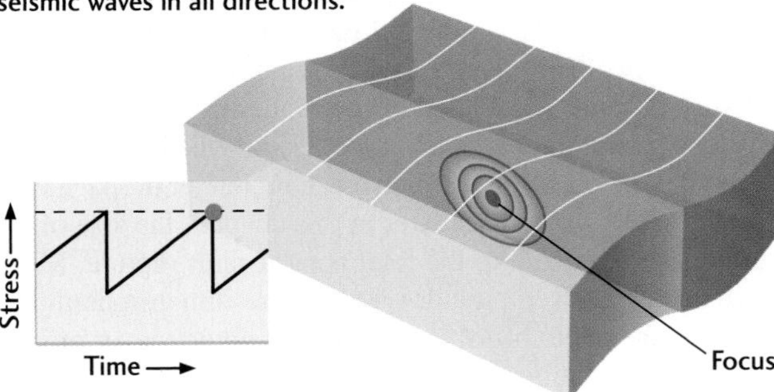

5 Seconds
Rupture continues to expand as a crack along the fault plane. When rupture front reaches the surface, displacements occur along the fault trace and rocks at the surface begin to rebound from their deformed state.

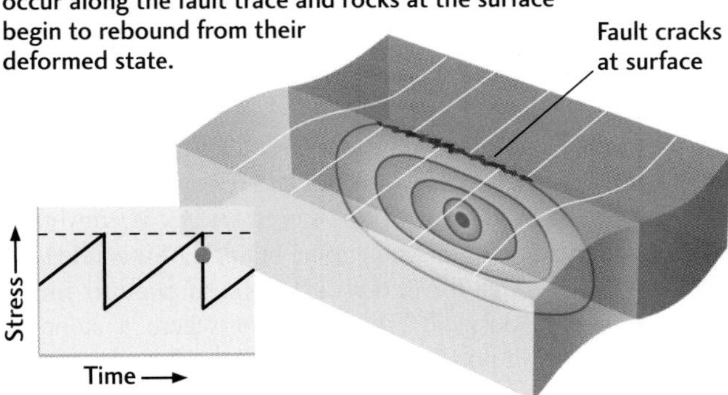

10 Seconds
Rupture front progresses down the fault plane, reducing the stress and allowing rocks on either side to rebound. Seismic waves continue to be emitted in all directions as the fault propagates.

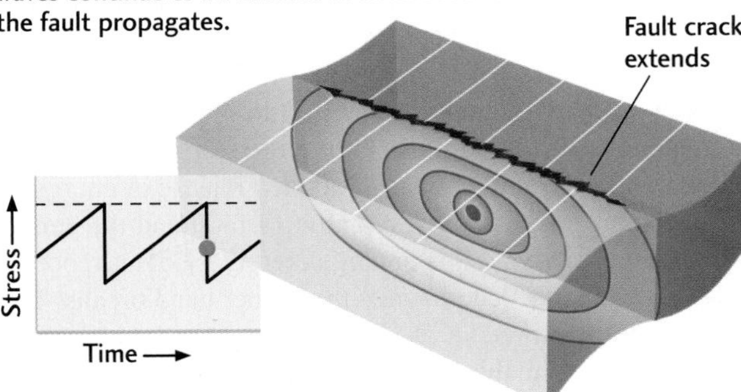

20 Seconds
Rupture has progressed along the entire length of the fault. The fault has reached its maximum displacement, and the earthquake stops.

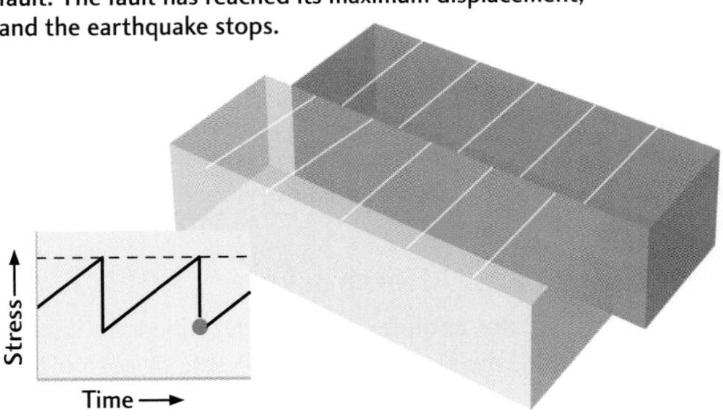

Figure 19.2 Fault rupture during an earthquake begins at the focus and spreads out over the fault plane, radiating seismic waves wherever the fault is slipping. The sequence shown here occurs in a few tens of seconds between Time 3 and Time 4 in Figure Story 19.1. The photo shows a 3-m right-lateral offset of the fence, near Bolinas, California, following the 1906 San Francisco earthquake. Displacement of the fence was observed to be similar to the fence at Time 4 in elastic rebound sequence. [G. K. Gilbert.]

Earthquakes occur most often at plate boundaries, where stresses are concentrated and strain is intense.

Elastic Rebound Explains Why Earthquakes Occur

The earthquake on the San Andreas fault that devastated San Francisco in 1906 received the most detailed study of any earthquake up to that time. A geologist who investigated that catastrophe, Henry Fielding Reid of Johns Hopkins University, advanced the **elastic rebound theory** to explain why earthquakes occur.

To visualize what happens in an earthquake, imagine the following experiment carried out across a fault between two hypothetical crustal blocks. Suppose that surveyors had painted straight lines on the ground, running perpendicular to a fault and extending from one block to the other, as in Figure Story 19.1. The two blocks are being pushed in opposite directions by plate motions. The weight of the overlying rock presses them together, however, so friction locks them together along the fault. They do not move, just as a car does not move when the emergency brake is engaged. Instead of slipping along the fault as stress builds up, the blocks are strained elastically near the fault, as shown by the bent lines in Figure Story 19.1. By *elastically,* we mean that the blocks would spring back and return to their undeformed, stress-free shape if the fault were suddenly to unlock.

As the slow plate movements continue to push the blocks in opposite directions, the strain in the rocks—evidenced by the bending of the survey lines—continues to build up, for decades or even centuries. At some point, the strength of the rocks is exceeded. The frictional bond that locks the fault can no longer hold somewhere along the fault, and it breaks. The blocks slip suddenly, and the rupture extends over a section of the fault. Figure Story 19.1 shows how the two blocks have rebounded—sprung back to their undeformed state—after the earthquake. The imaginary bent lines have straightened, and the two blocks have been displaced. The distance of the displacement is called the **slip.**

Fault Rupture During Earthquakes

The point at which the slip begins is the **focus** of the earthquake (see Figure Story 19.1). The **epicenter** is the geographic point on Earth's surface directly above the focus. For example, you might hear in a news report: "Seismologists at the California Institute of Technology report that the epicenter of last night's destructive earthquake in California was located 56 kilometers southeast of Los Angeles. The depth of the focus was 10 kilometers."

For most earthquakes occurring in continental crust, the focal depths range from about 2 to 20 km. Continental earthquakes are rare below 20 km, because under the high temperatures and pressures found at greater depths, the crust deforms as a ductile material and cannot support brittle fracture (just as hot wax flows when stressed, whereas cold wax breaks; see Chapter 11). In subduction zones, however, where cold oceanic lithosphere plunges back into the mantle, earthquakes can begin at depths as great as 690 km.

The fault rupture does not occur all at once. It begins at the focus and expands outward on the fault plane, typically at 2 to 3 km/s (**Figure 19.2**). The rupture stops where the stresses become insufficient to continue breaking the fault (such as where the rocks are stronger) or where the rupture enters ductile material in which it can no longer propagate as a fracture. As we will see later in this chapter, the size of an earthquake is related to the total area of fault rupture. Most earthquakes are very small, with rupture dimensions much less than the depth of focus, and so the rupture never breaks the surface. In large, destructive earthquakes, however, surface breaks are common. For example, the great 1906 San Francisco earthquake caused surface displacements averaging about 4 m along a 400-km section of the San Andreas fault (see Figure 19.2). Faulting in the largest earthquakes can extend as far as 1000 km, and the slip of the two blocks can be as large as 10 m. Generally, the longer the fault rupture, the larger the fault slip.

The sudden slipping of the blocks at the time of the earthquake reduces the stress on the fault and releases much of the stored strain energy. Most of this stored energy is converted to frictional heating of the fault zone, but part of it is released as seismic waves that travel outward from the rupture, much as waves ripple outward from the spot where a stone is dropped into a still pond. The focus of the earthquake generates the first seismic waves, although slipping parts of the fault continue to generate waves until the rupture stops. In a large event like the 1906 San Francisco earthquake, the propagating fault continues to produce waves for many tens of seconds. These waves can cause damage all along the fault break, even far from the epicenter.

The elastic strain energy that slowly builds up over decades when two blocks are pushed in opposite directions is analogous to the strain energy stored in a rubber band when it is slowly stretched. The sudden release of energy in an earthquake, signaled by slip along a fault and the generation of seismic waves, is analogous to the violent rebound or springback that occurs when the rubber band breaks. The elastic energy stored in the stretched rubber band is suddenly released in the backlash. In the same way, elastic energy accumulates and is stored for many decades in rocks under stress. The energy is released at the moment the fault ruptures, and some of it is radiated as seismic waves in the few minutes of an earthquake.

In its idealized form, the elastic rebound model implies that faults should exhibit a periodic buildup and release of strain energy. Sequences of real earthquakes rarely exhibit this simple behavior, however, which is one of the reasons earthquakes are so difficult to predict.

Foreshocks and Aftershocks

An example of the complexities not described by simple elastic rebound is the phenomenon of aftershocks. An **after-**

shock is an earthquake that occurs as a consequence of a previous earthquake of larger magnitude. Aftershocks follow the main earthquake in sequences, and their foci are distributed in and around the rupture plane of the main shock (**Figure 19.3**).

The number and sizes of aftershocks depend on the main shock's magnitude, and both quantities decrease with time. The aftershock sequence of a magnitude 5 earthquake might last for only a few weeks, whereas one for a magnitude 7 earthquake can persist for several years. The size of the largest aftershock is usually about one magnitude unit smaller than the main shock. In other words, a magnitude 7 earthquake might have an aftershock as big as magnitude 6. In populated regions, the shaking from large aftershocks can be very dangerous, compounding the damage caused by the main shock.

A **foreshock** (see Figure 19.3) is a small earthquake that occurs in the vicinity of, but before, a main shock. One or more foreshocks have preceded many large earthquakes, so seismologists have tried to use foreshocks to predict when and where large earthquakes might happen. Unfortunately, it is usually very hard to distinguish foreshocks from other small earthquakes that occur randomly on active faults, so this method has only rarely proved successful.

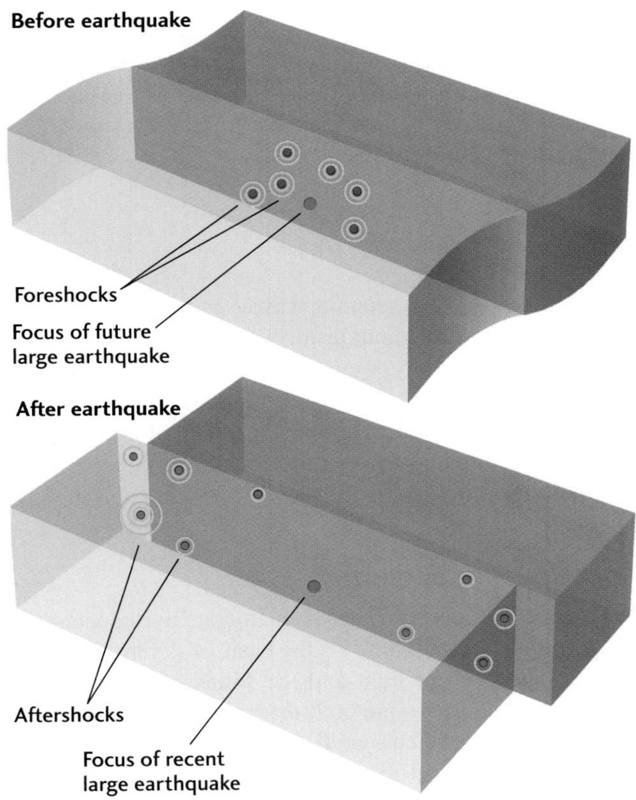

Figure 19.3 Foreshocks occur in the vicinity of, but before, the main shock. Aftershocks are smaller shocks that follow the main shock.

Studying Earthquakes

As in any experimental science, instruments and field observations provide the basic data used to study earthquakes. These data enable investigators to analyze the seismic waves that originate in earthquakes, locate earthquakes, determine their sizes and numbers, and understand their relationships to faults.

Seismographs

The modern **seismograph,** which records the seismic waves that earthquakes generate, is an important tool for studying earthquakes and probing Earth's deep interior (**Figure 19.4**). The seismograph is to the Earth scientist what the telescope is to the astronomer—a tool for peering into inaccessible regions. The ideal seismograph would be a device affixed to a stationary frame not attached to the Earth. When the ground shook, seismologists could measure the changing distance between the seismograph, which did not move, and the vibrating ground, which did. As yet, we have no way to position a seismograph that is not attached to the Earth—although modern space technology may eventually remove this limitation. So we compromise. A mass is attached to the Earth so loosely that the ground can vibrate up and down or side to side without causing much motion of the mass.

One way to achieve this loose attachment is to suspend the mass from a spring (Figure 19.4a). When seismic waves move the ground up and down, the mass tends to remain stationary because of its inertia (an object at rest tends to stay at rest), but the mass and the ground move relative to each other because the spring can compress or stretch. In this way, the vertical displacement of the Earth caused by seismic waves can be recorded by a pen on chart paper or, nowadays, digitally on a computer.

Loose attachment of the mass can also be achieved with a hinge. A seismograph that has its mass suspended on hinges like a swinging gate (Figure 19.4b) can record the horizontal motions of the ground. In modern seismographs, advanced electronic technology is used to amplify ground motion before it is recorded. These instruments can detect ground displacements as small as 10^{-8} cm—an astounding feat, considering that such small displacements are of atomic size.

Seismic Waves

Install a seismograph anywhere, and within a few hours it will record the passage of seismic waves generated by an earthquake somewhere on Earth. The waves will have traveled from the earthquake focus through the Earth and arrived at the seismograph in three distinct groups. The first waves to arrive are called primary waves or **P waves.** The secondary or **S waves** follow. Both P and S waves travel

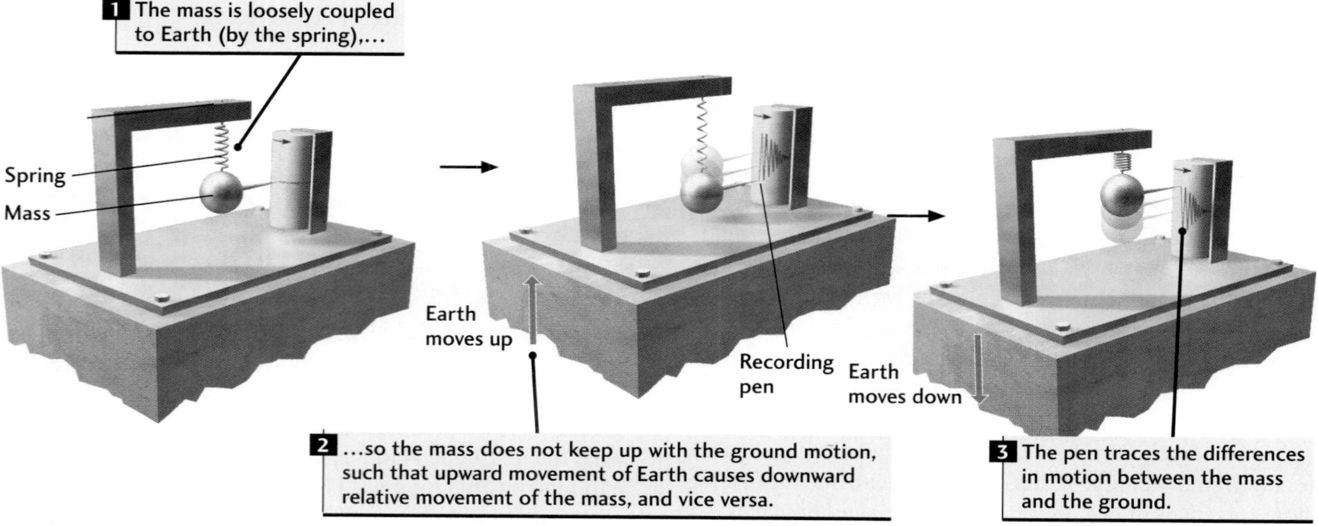

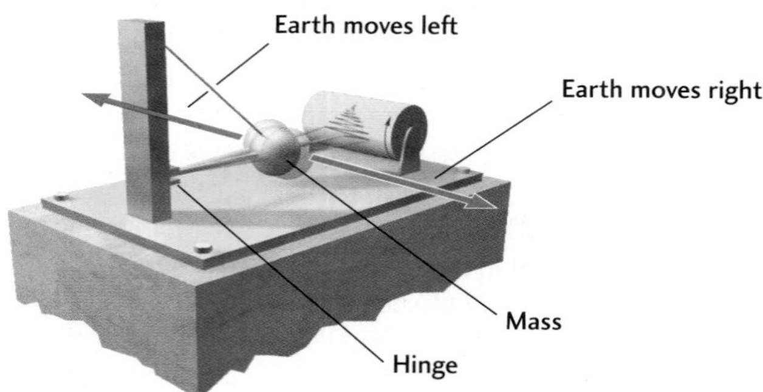

Figure 19.4 Seismographs record (a) vertical or (b) horizontal motion. Because of its loose coupling to the Earth through the spring (a) or hinge (b) and its inertia, the mass does not keep up with the motion of the ground. The pen traces the differences in motion between the mass and the ground, in this way recording the vibrations of seismic waves. A typical observatory has instruments to measure three components of ground motion: up-down, horizontal east-west, and horizontal north-south.

through Earth's interior. Finally come the **surface waves,** which travel around Earth's surface (**Figure Story 19.5**).

P waves in rock are analogous to sound waves in air, except that P waves travel through the solid rock of Earth's crust at about 6 km/s, which is about 20 times faster than sound waves travel through air. Like sound waves, P waves are *compressional waves,* so called because they travel through solid, liquid, or gaseous materials as a succession of compressions and expansions. P waves can be thought of as push-pull waves: they push or pull particles of matter in the direction of their path of travel.

S waves travel through solid rock at a little more than half the speed of P waves. They are called *shear waves* because they displace material at right angles to their path of travel. Shear waves do not exist in liquids or gases.

Surface waves are confined to Earth's surface and outer layers because, like waves on the ocean, they need a free surface to ripple. Their speed is slightly less than that of S waves. One type of surface wave sets up a rolling motion in the ground; another type shakes the ground sideways (see Figure Story 19.5).

People have felt seismic waves and witnessed their destructiveness throughout history, but not until the close of the nineteenth century were seismologists able to devise instruments to record them. Seismic waves enable us to locate earthquakes and determine the nature of faulting, and they provide our most important tool for probing Earth's deep interior.

Locating the Epicenter

Locating a quake's epicenter is analogous to deducing the distance to a lightning bolt on the basis of the time interval between the flash of light and the sound of thunder—the greater the distance to the bolt, the larger the time interval. Light travels faster than sound, so the lightning flash may be likened to the P waves of earthquakes and the thunder to the slower S waves.

The time interval between the arrival of P and S waves depends on the distance the waves have traveled from the focus. This relationship is established by recording seismic waves from an earthquake or underground nuclear explosion

THERE ARE THREE TYPES OF SEISMIC WAVES

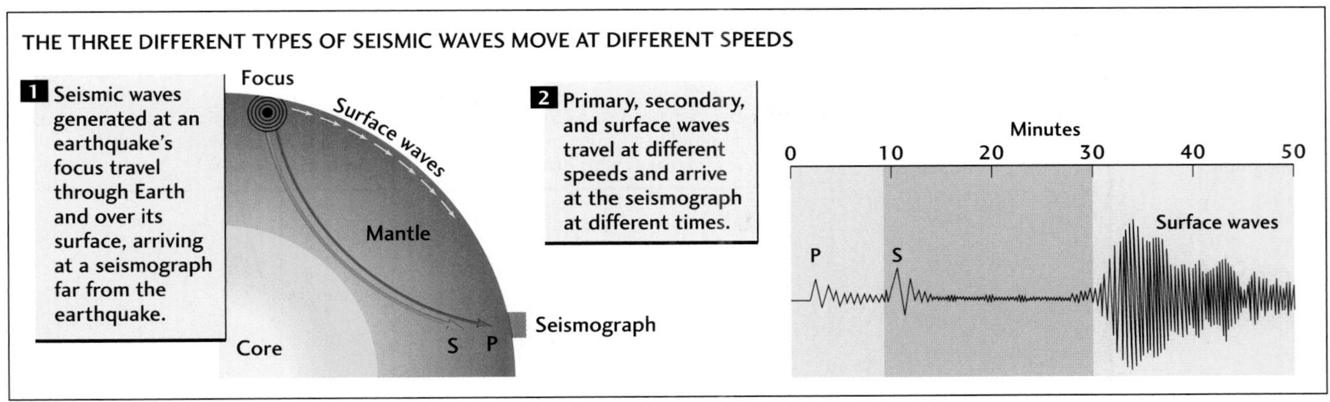

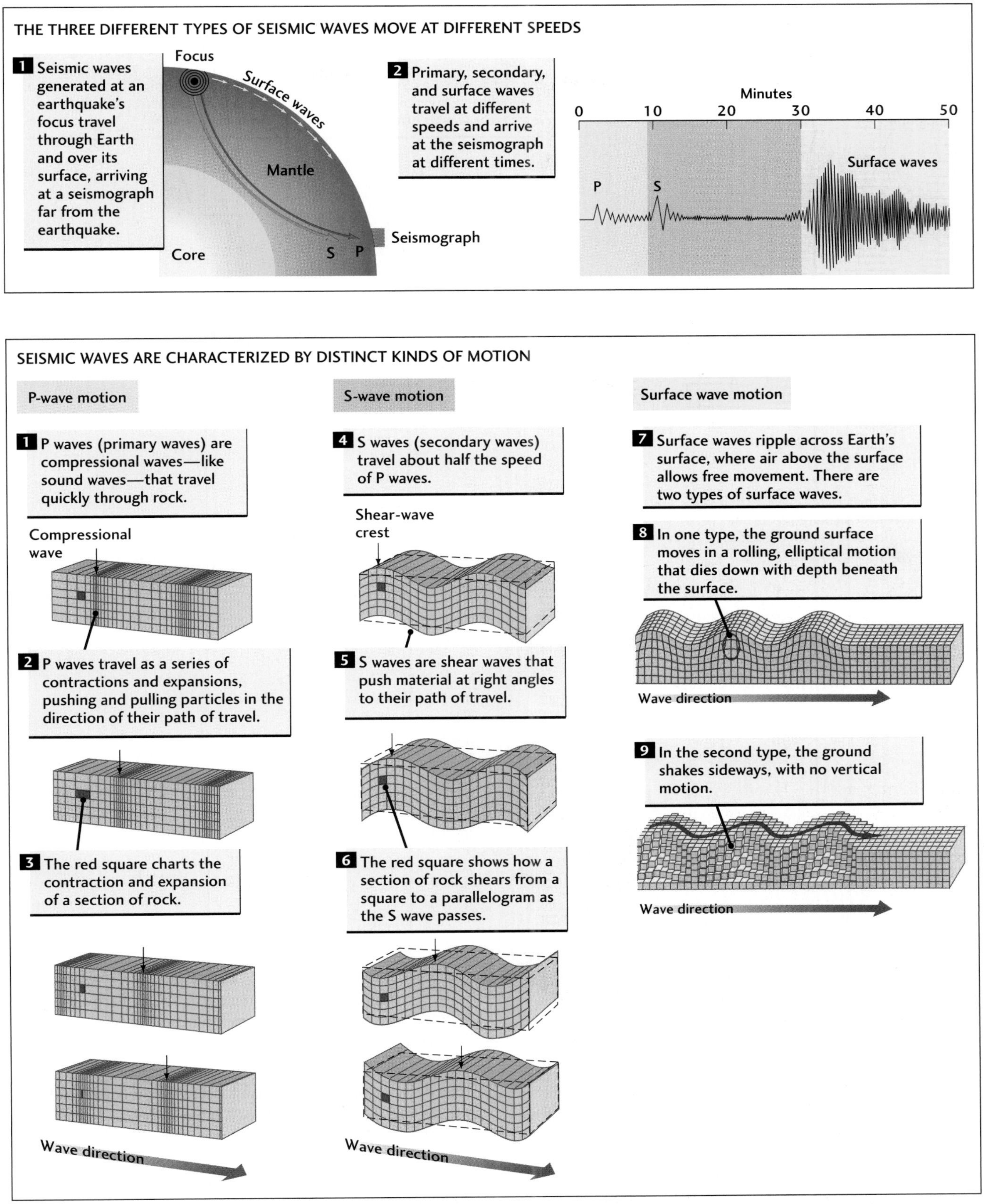

SEISMIC WAVES ARE CHARACTERIZED BY DISTINCT KINDS OF MOTION

P-wave motion

1 P waves (primary waves) are compressional waves—like sound waves—that travel quickly through rock.

Compressional wave

2 P waves travel as a series of contractions and expansions, pushing and pulling particles in the direction of their path of travel.

3 The red square charts the contraction and expansion of a section of rock.

Wave direction

S-wave motion

4 S waves (secondary waves) travel about half the speed of P waves.

Shear-wave crest

5 S waves are shear waves that push material at right angles to their path of travel.

6 The red square shows how a section of rock shears from a square to a parallelogram as the S wave passes.

Wave direction

Surface wave motion

7 Surface waves ripple across Earth's surface, where air above the surface allows free movement. There are two types of surface waves.

8 In one type, the ground surface moves in a rolling, elliptical motion that dies down with depth beneath the surface.

Wave direction

9 In the second type, the ground shakes sideways, with no vertical motion.

Wave direction

Figure Story 19.5 Seismographs detect the three different types of seismic waves.

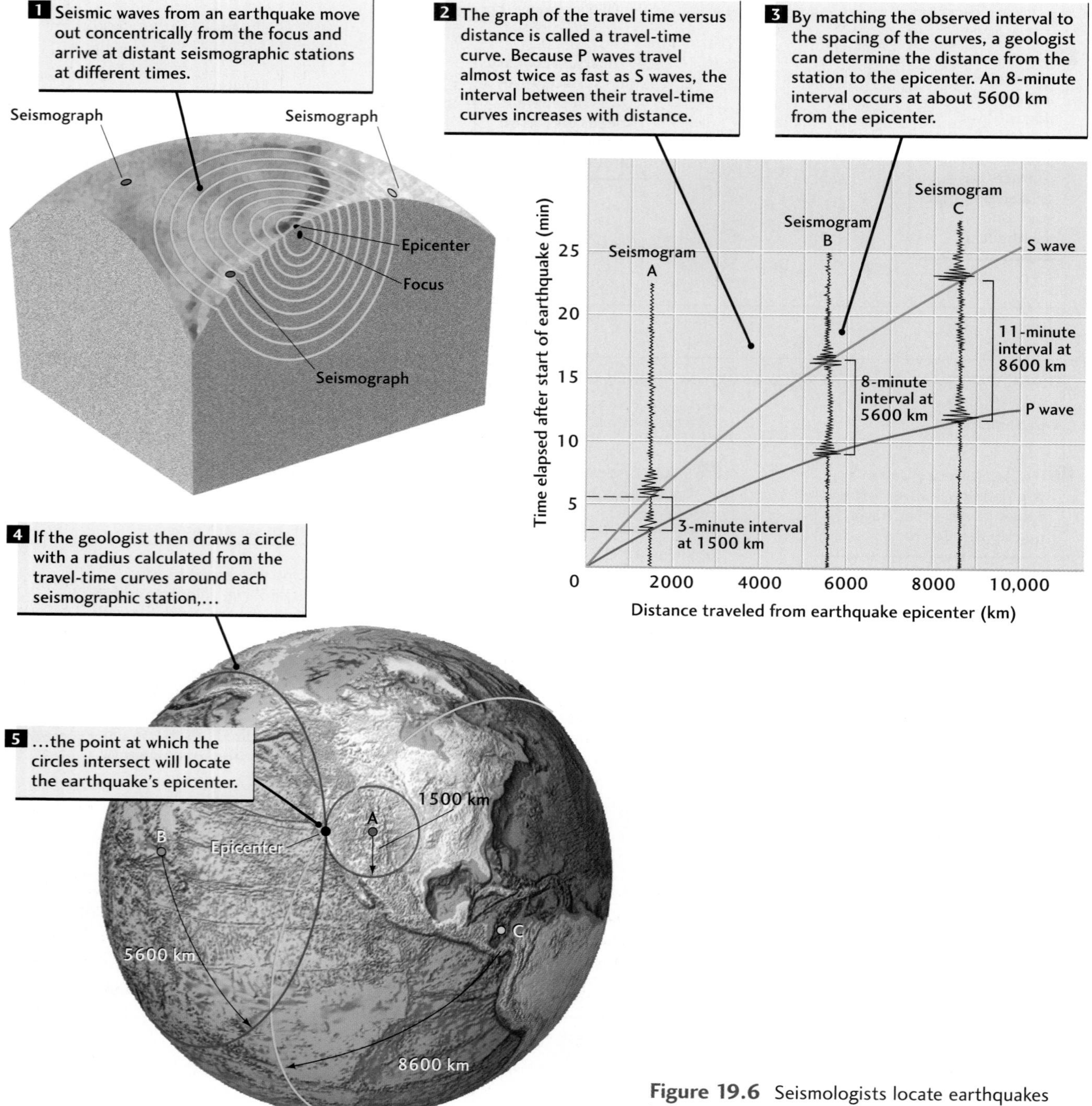

Figure 19.6 Seismologists locate earthquakes by measuring the arrival times of seismic waves.

that is at a known distance from the seismograph. To determine the approximate distance to an epicenter, seismologists read from a seismogram the amount of time that elapsed between the arrival of the first P waves and the later arrival of the S waves. Then they use a table or a graph like the one shown in **Figure 19.6** to determine the distance from the seismograph to the epicenter. If they know the distances from three or more stations, they can locate the epicenter. They can also deduce the time of the shock at the epicenter, because the arrival time of the P waves at each station is known, and it is possible to determine from a graph or table how long the waves took to reach the station. Today, this entire process is carried out repeatedly by a computer until the data from a large number of seismographic stations agree on where the

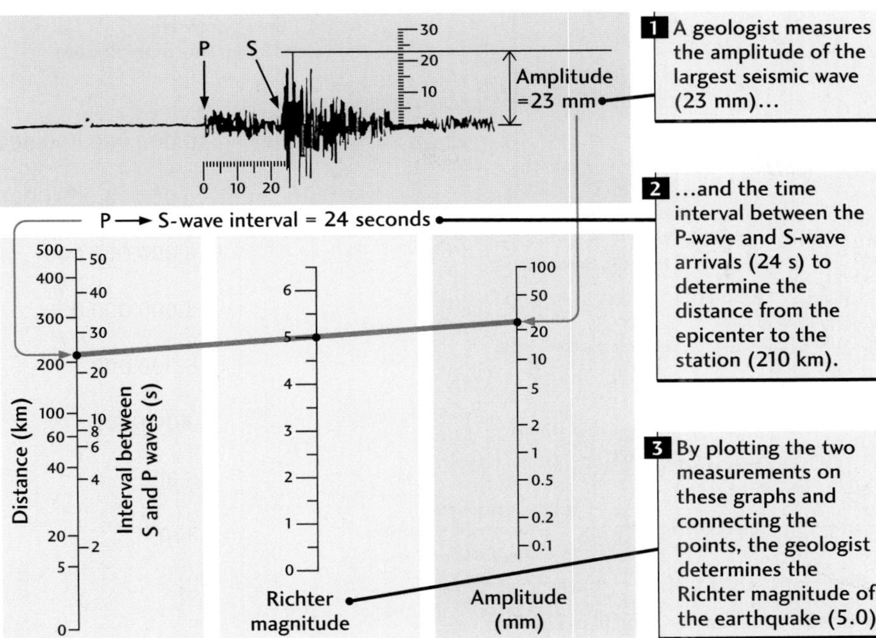

Figure 19.7 The maximum amplitude of the ground shaking and the P–S wave interval, indicated on the seismographic record, are used to assign a Richter magnitude to an earthquake. [California Institute of Technology.]

epicenter is, the time at which the earthquake began, and the depth of the focus below the surface.

Measuring the Size of an Earthquake

Locating earthquakes is only one step on the way to understanding them. The seismologist must also determine their size, or magnitude. Other variables being equal (such as the distance to the focus and the regional geology), an earthquake's magnitude is the main factor that determines the intensity of the seismic waves and thus its potential destructiveness.

Richter Magnitude In 1935, Charles Richter, a California seismologist, devised a simple procedure to determine the size of an earthquake. Richter studied astronomy as a young man and learned that astronomers assign each star a magnitude—a measure of its brightness. Adapting this idea to earthquakes, Richter assigned each earthquake a number, now called the *Richter magnitude.* Just as the brightness of stars varies over a huge range, so do the sizes of earthquakes. To compress his **magnitude scale,** Richter took the logarithm of the largest ground motion registered by a seismograph as his measure of earthquake size. He accounted for the distance between the seismograph and the fault rupture by correcting for the weakening of the seismic waves as they spread away from the focus (**Figure 19.7**).

In Richter's scale, two earthquakes at the same distance from a seismograph that differ in size of ground motion by a factor of 10 differ in magnitude by 1 Richter unit. The ground motion of an earthquake of magnitude 3, therefore, is 10 times that of an earthquake of magnitude 2. Similarly, a magnitude 6 earthquake produces ground motions that are 100 times greater than those of a magnitude 4 earthquake. The energy released as seismic waves increases even more with earthquake magnitude, by a factor of 33 for each Richter unit.

Using Richter's procedure, seismologists anywhere could study their records and in a few minutes come up with nearly the same value for the magnitude of an earthquake, no matter how close or far away their instruments were from the focus. His method came to be used throughout the world.

Moment Magnitude Although the Richter magnitude has become a household term and is an important concept historically, seismologists now prefer a measure of earthquake size more directly related to the physical properties of the faulting that causes the earthquake. This measure, called the *moment magnitude,* is approximately proportional to the logarithm of the area of the fault break. It is also approximately proportional to the logarithm of the seismic energy released during the rupture. Although both Richter's method and the moment method produce roughly the same numerical values, the moment magnitude can be measured more accurately from seismograms, and it can also be deduced directly from field measurements of faulting.

Large earthquakes occur much less frequently than small ones. This observation can be expressed by a very simple relationship between earthquake frequency and magnitude (**Figure 19.8**). Worldwide, approximately 1,000,000 earthquakes with magnitudes greater than 2.0 take place each year, and this number decreases by a factor of 10 for

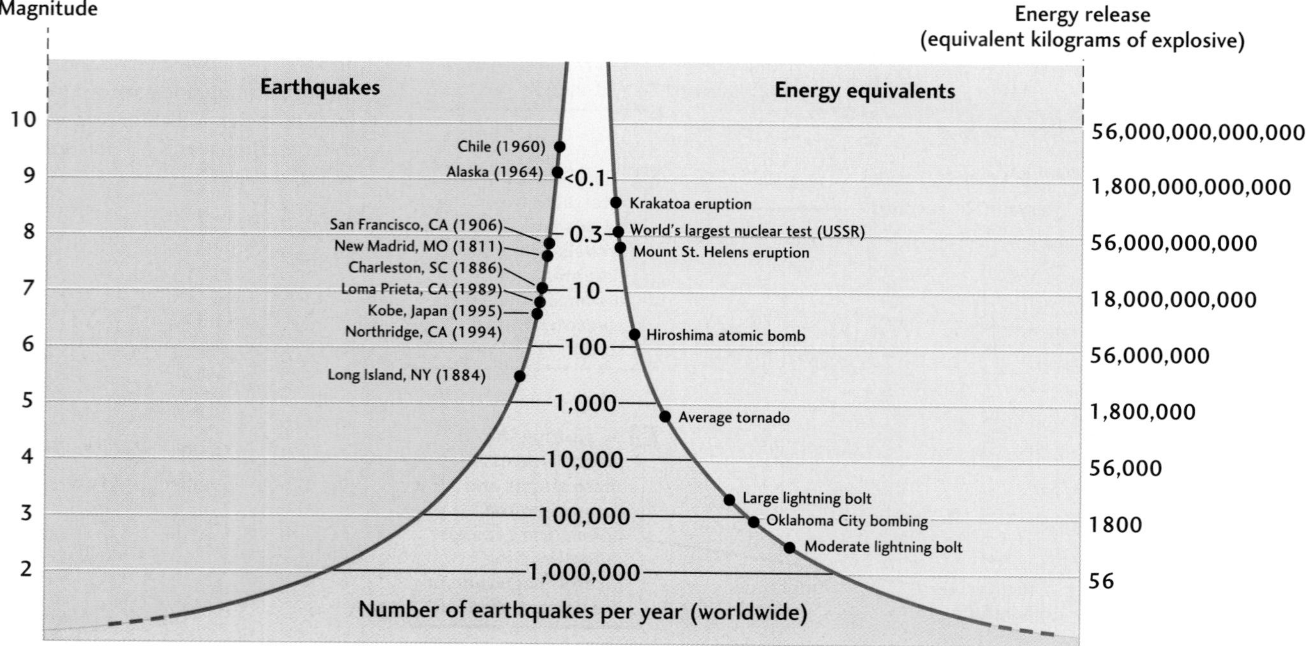

Figure 19.8 Relationship between moment magnitude (scale on left), earthquake energy release (scale on right), number of earthquakes per year worldwide (scale on horizontal lines in colored area), and other large sources of sudden energy release. [IRIS Consortium, http://www.iris.edu.]

each magnitude unit. Hence, there are about 100,000 earthquakes with magnitudes greater than 3, about 1000 with magnitudes greater than 5, and about 10 with magnitudes greater than 7. According to this statistical rule, there should be, on average, about 1 earthquake of magnitude 8 per year. In fact, magnitude 8 earthquakes occur on average only about once every 3 to 5 years, which implies that there is an upper limit to how big fault ruptures can be. Truly huge earthquakes, like the ones that occurred on thrust faults in the subduction zones off Alaska in 1964 (moment magnitude 9.2) and Chile in 1960 (moment magnitude 9.5), appear to be exceptionally rare.

Shaking Intensity Earthquake magnitude does not describe the destructiveness of a particular earthquake. A magnitude 8 earthquake in a remote area far from the nearest city might cause no human or economic losses, whereas a magnitude 6 quake immediately beneath a city is likely to cause serious damage.

In the late nineteenth century, before Richter invented his magnitude scale, seismologists and earthquake engineers developed methods for estimating the intensity of earthquake shaking directly from an event's destructive effects. Table 19.1 shows the scale that remains in most common use today, called the *modified Mercalli intensity* after Giuseppe Mercalli, the Italian scientist who first proposed it (in 1902). This **intensity scale** assigns a value, given as a Roman numeral from I to XII, to the intensity of the shaking at a particular site. For example, a location where an earthquake is just barely felt by a few people is assigned an intensity of II, whereas one where it was felt by nearly everyone is given an intensity of V. Numbers at the upper end of the scale describe increasing amounts of damage. The narrative attached to the highest value, XII, is tersely apocalyptic: "Damage total. Waves seen on ground surfaces. Objects thrown upward into the air." Not a place you would want to be!

By making observations at many sites and interviewing many people who experienced an earthquake, seismologists can make maps showing contours of equal intensity. **Figure 19.9** shows an intensity map for the New Madrid earthquake of December 16, 1811, a magnitude 7.5 event near the southern tip of Missouri. Although the intensities are generally highest near the fault rupture, they also depend on the local geology. For example, for sites at equal distances from the rupture, the shaking tends to be more intense on soft sediments (especially water-saturated sediments near shorelines) than on hard basement rocks. Intensity maps thus provide engineers with crucial data for designing structures to withstand earthquake shaking.

Determining Fault Mechanisms from Earthquake Data

The pattern of ground shaking also depends on the orientation of the fault rupture and the direction of slip, which together specify the **fault mechanism** of an earthquake.

Table 19.1 Modified Mercalli Intensity Scale

Intensity Level	Description
I	Not felt except by a very few under especially favorable conditions.
II	Felt only by a few persons at rest, especially on upper floors of buildings. Delicately suspended objects may swing.
III	Felt quite noticeably by persons indoors, especially on upper floors of buildings. Many people do not recognize it as an earthquake. Standing motor cars may rock slightly. Vibration similar to the passing of a truck.
IV	Felt indoors by many, outdoors by few during the day. At night, some awakened. Dishes, windows, doors disturbed; walls make cracking sound. Sensation like heavy truck striking building. Standing motor cars rocked noticeably.
V	Felt by nearly everyone; many awakened. Some dishes, windows broken. Unstable objects overturned. Pendulum clocks may stop.
VI	Felt by all, many frightened. Some heavy furniture moved; a few instances of fallen plaster. Damage slight.
VII	Damage negligible in buildings of good design and construction; slight to moderate damage in well-built ordinary structures; considerable damage in poorly built or badly designed structures; some chimneys broken.
VIII	Damage slight in specially designed structures; considerable damage in ordinary substantial buildings with partial collapse. Damage great in poorly built structures. Fall of chimneys, factory stacks, columns, monuments, walls. Heavy furniture overturned.
IX	Damage considerable in specially designed structures; well-designed frame structures thrown out of plumb. Damage great in substantial buildings, with partial collapse. Buildings shifted off foundations.
X	Some well-built wooden structures destroyed; most masonry and frame structures destroyed with foundations. Rails bent.
XI	Few, if any (masonry) structures remain standing. Bridges destroyed. Rails bent greatly.
XII	Damage total. Lines of sight and level are distorted. Objects thrown into the air.

The fault mechanism tells us whether the rupture was on a *normal, thrust,* or *strike-slip* fault. If the rupture was on a strike-slip fault, the fault mechanism also tells us whether the sense of motion was *right-lateral* or *left-lateral* (see Chapter 11 for the definition of these terms). We can then use this information to infer the regional pattern of tectonic forces (**Figure 19.10**).

For shallow ruptures that break the surface, we can sometimes deduce the fault mechanism from field observations of the fault scarp. As we have seen, however, most ruptures are too deep to break the surface, so we must infer the fault mechanism from the motions measured on seismographs.

For large earthquakes at any depth, this turns out to be easy to do, because there are enough seismographs around the world to surround the earthquake's focus. Seismologists find that in some directions from the focus, the very first movement of the ground recorded by a seismograph—the P wave—is a *push away* from the focus, causing upward

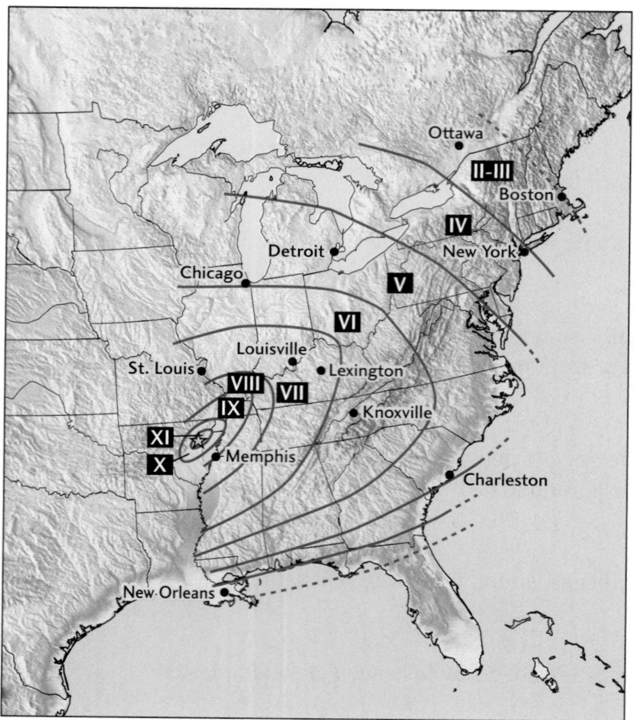

Figure 19.9 Modified Mercalli intensities measured for the New Madrid earthquake of December 16, 1811, a magnitude 7.5 event near the juncture of Missouri, Arkansas, and Tennessee. Regions near the fault rupture show intensities greater than IX, and intensities as high as VI were observed 200 km from the epicenter (see Table 19.1). [Carl W. Stover and Jerry L. Cossman, USGS Professional Paper 1527, 1993.]

motion on a vertical seismograph. In other directions, the initial ground movement is a *pull toward* the focus, causing downward motion on a vertical seismograph. In other words, the slip on a fault looks like a push if you view it from one direction but like a pull if you view it from another. The pushes and pulls can be divided into four sections based on the positions of the seismographic stations, as shown in **Figure 19.11**. One of the two boundaries is the fault orientation; the other is a plane perpendicular to the fault. The slip direction is inferred from the arrangement of pushes and pulls. In this manner, without surface evidence, seismologists can deduce whether the crustal forces that triggered the earthquake were tensional, compressive, or shear.

GPS Measurements and "Silent" Earthquakes

As discussed in Chapter 2, GPS stations (see Figure 2.12) can record the slow movements of plates. These instruments can also measure the strain that builds up from such movements, as well as the sudden slip on a fault when it ruptures in an earthquake. Seismologists are now using GPS observations to study another kind of movement along active faults. It has been known for many years that some sections of the San Andreas fault system creep continuously rather than rupturing suddenly. This creep slowly deforms structures and cracks pavements that cross the fault trace. Recently, new networks of GPS stations have found surface movements at convergent plate boundaries that reflect deeper transient (short-lived) creep events. These slip events may last a few weeks at a time. They have been named *silent earthquakes* because the gradual movements do not trigger destructive seismic waves. Nevertheless, they steadily release large amounts of stored strain energy.

These observations raise many questions that geologists are now trying to answer. Why gradual slip and not sudden breaking? What causes faults to stick and slip catastrophically in some places and creep in others? Will the release of strain energy by slow creep make destructive earthquakes in these regions less frequent or less severe? Can these creep events be used to predict earthquakes? Scientific research is under way to address all these questions.

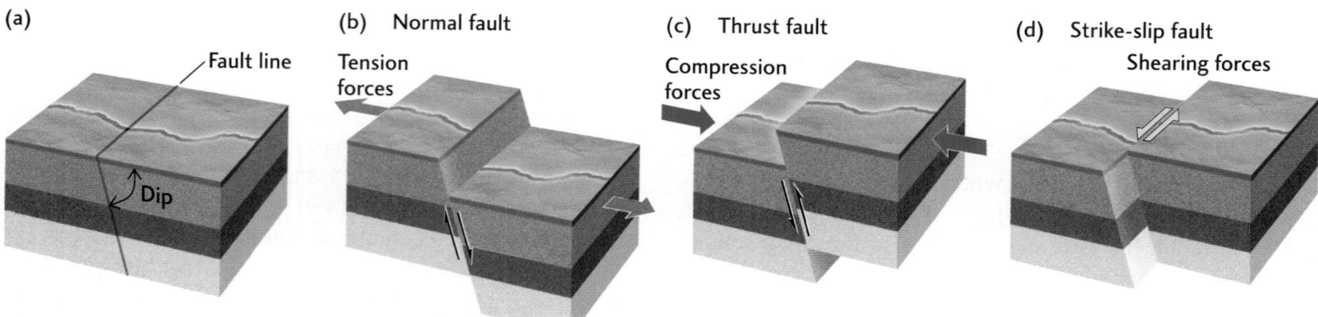

Figure 19.10 The three main types of fault movements that initiate earthquakes and the stresses that cause them. (a) The situation before movement takes place; (b) normal faulting due to tensile stress; (c) thrust (reverse) faulting due to compressive stress; (d) strike-slip (lateral) faulting due to shearing stress.

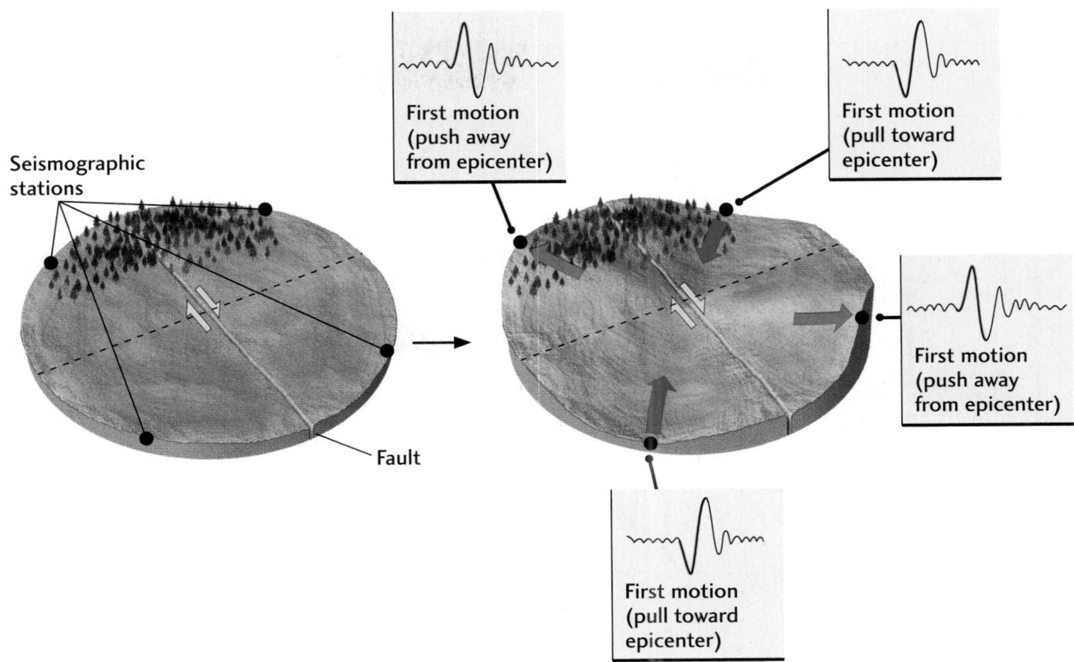

Figure 19.11 The first motion of P waves arriving at seismographic stations is used to determine the orientation of the fault plane and the direction of slip. The case shown here is for the rupture of a right-lateral strike-slip fault. Note that the alternating pattern of pushes and pulls would remain the same if the plane perpendicular to the fault ruptured with left-lateral displacement. Seismologists can usually choose between the two possibilities using additional information, such as field mapping of the fault trace or alignments of aftershocks along the fault plane.

Earthquakes and Patterns of Faulting

Networks of sensitive seismographs have allowed seismologists to locate earthquakes around the globe, measure their magnitudes, and deduce their fault mechanisms. With these tools, they have been able to substantiate the plate tectonics theory and gather new information about tectonic processes on scales much smaller than the plates themselves. In this section, we will summarize the global pattern of earthquake occurrence from the perspective of plate tectonics and show how regional studies of active fault networks are improving our understanding of faulting along plate boundaries and within plate interiors.

The Big Picture: Earthquakes and Plate Tectonics

The **seismicity map** in **Figure 19.12** shows the epicenters of earthquakes recorded around the world in a 27-year period. The most obvious features of this map, known to seismologists for many decades, are the earthquake belts that mark the major plate boundaries.

Divergent Boundaries Seismology gives elegant support to the concept of seafloor spreading. The narrow belts of earthquakes that run through the ocean basins coincide with mid-ocean ridge crests and their offsets on transform faults, as shown in **Figure 19.13**a. The fault mechanisms of the ridge-crest earthquakes, revealed by analysis of the first P-wave motion, correspond to normal faulting. The faults strike parallel to the trend of the ridges and dip toward the mid-ocean rift valley. Normal faulting indicates that tensional forces are at work as the plates are pulled apart during seafloor spreading, explaining why rift valleys develop at the ridge crests. Earthquakes also have normal fault mechanisms in zones where continental crust is being pulled apart, such as in the East African rift valleys and in the Basin and Range province of western North America.

Transform-Fault Boundaries Earthquake activity is even greater along the transform-fault boundaries that offset the ridge segments. These earthquakes show strike-slip fault mechanisms—just as one would expect where plates slide past each other in opposite directions. Moreover, for earthquakes along transform faults between ridge segments, the slip indicated by the fault mechanisms is left-lateral where

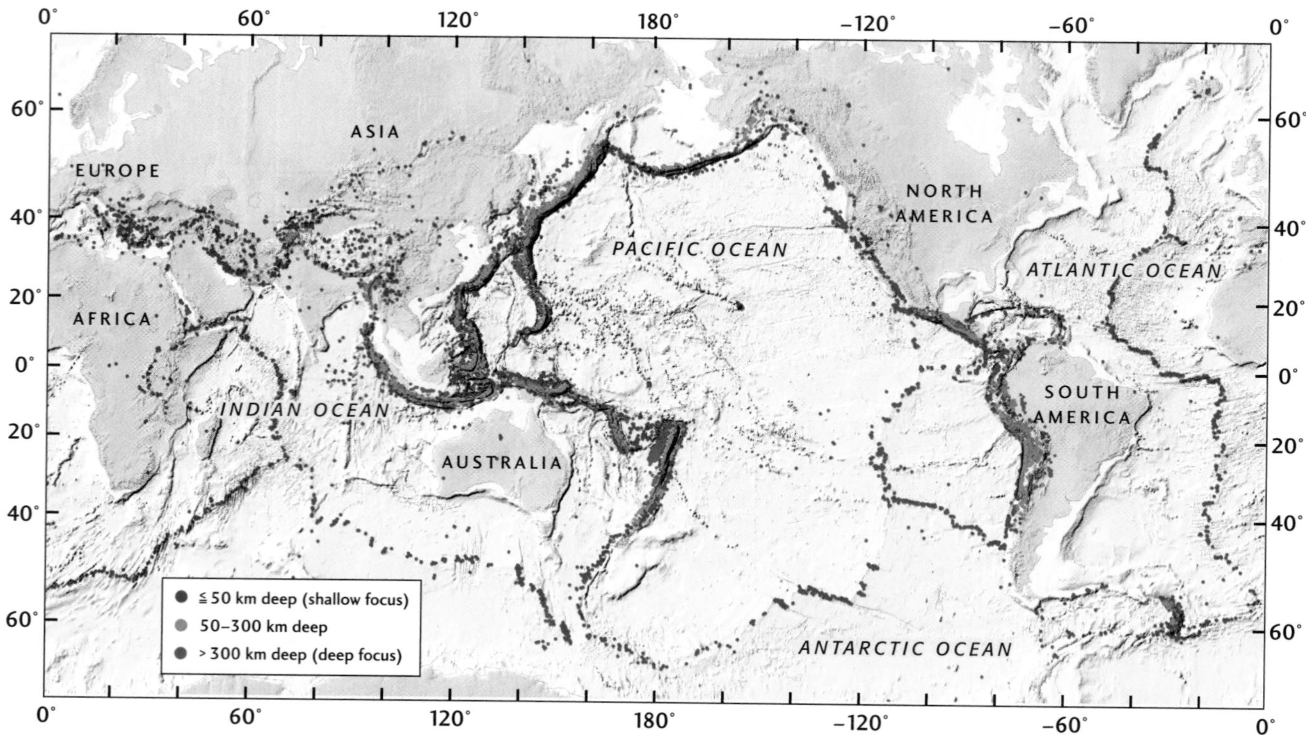

Figure 19.12 World seismicity, from 1976 to 2002. Shallow-focus earthquakes less than 50 km deep are shown in blue, intermediate-focus earthquakes between 50 and 300 km deep are shown in green, and deep-focus earthquakes greater than 300 km deep are shown in red. Intermediate-focus and deep-focus earthquakes mark the major subduction zones. [Data from Harvard CMT catalog; plot by M. Boettcher and T. Jordan.]

the ridge crest steps right and right-lateral where it steps left (see Figure 19.13a). These directions are the opposite of what would be needed to create the offsets of the ridge crest but are consistent with the direction of slip predicted by seafloor spreading. In the mid-1960s, seismologists used this diagnostic property of transform faults to support the hypothesis of seafloor spreading. The motions on transform faults that run through continental crust, such as California's San Andreas fault (right-lateral) and New Zealand's Alpine fault (left-lateral), also agree with the predictions of plate tectonics.

Convergent Boundaries The world's largest earthquakes occur at convergent plate boundaries. Examples include the great earthquakes in Alaska (1964) and Chile (1960). During the latter—the biggest earthquake ever recorded (moment magnitude 9.5)—the crust of the Nazca Plate slipped an average of 10 m beneath the crust of the South American Plate on a fault rupture with an area the size of Kentucky! The fault mechanisms show that these great earthquakes occur by horizontal compression along huge thrust faults, called *megathrusts,* that form the boundaries where one plate subducts beneath the other (Figure 19.13b).

Earth's deepest seismicity also occurs at convergent boundaries. Almost all earthquakes originating below 100 km rupture the descending oceanic plate in subduction zones (see Figure 19.12). The fault mechanisms of these deep earthquakes show a variety of orientations, but they are consistent with the deformation expected within the descending plate as gravity forces it back into the convecting mantle. The deepest earthquakes, located at depths of 600 to 700 km, take place in the oldest—and therefore coldest—descending plates, such as those beneath South America, Japan, and the island arcs of the western Pacific Ocean.

Intraplate Earthquakes Although most earthquakes occur at plate boundaries, the map in Figure 19.12 shows that a small percentage of global seismicity originates within plate interiors. Their foci are relatively shallow, and the majority occur on continents. Among these earthquakes are some of the most destructive in American history: New Madrid, Missouri (1811); Charleston, South Carolina (1886); Boston, Massachusetts (1755). Many of these *intraplate earthquakes* occur on old faults that were once parts of ancient plate boundaries. The faults no longer form plate boundaries but remain zones of crustal weakness.

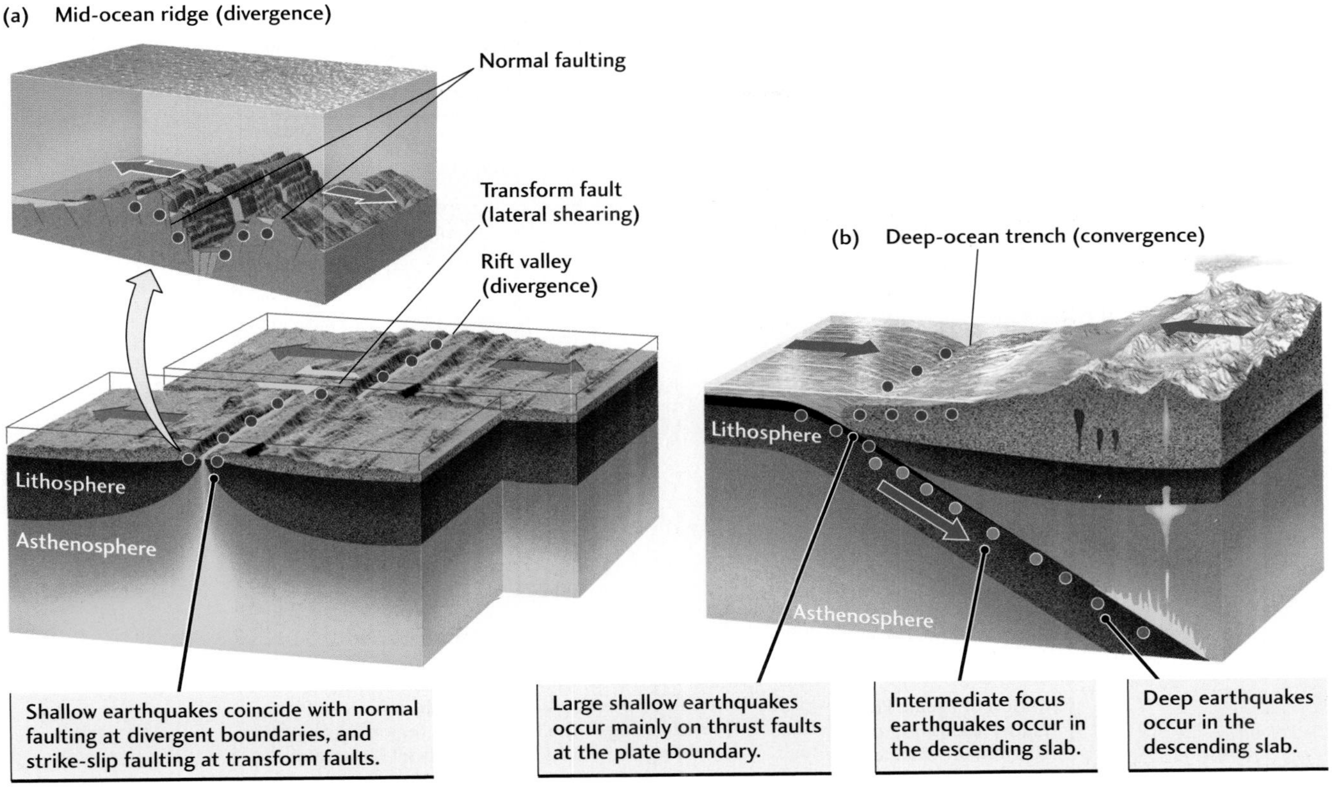

Figure 19.13 Earthquakes at plate boundaries reflect the relative motion between the plates. (a) Shallow-focus normal faulting occurs along diverging mid-ocean ridges, and shallow-focus strike-slip faulting occurs on transform faults that offset the ridge crest. Note that the slip on the transform fault (left-lateral in this case) is opposite to the displacement of the ridge crest by the fault. (b) Reverse faulting occurs along the megathrust where an oceanic plate is subducted. Intermediate-focus and deep-focus earthquakes occur in the interior of the subducted slab as it is pulled by gravity back into the mantle. The deepest earthquakes have focal depths of almost 700 km.

One of the deadliest intraplate earthquakes on record occurred in Gujarat State, western India, in 2001. It is estimated that some 20,000 lives were lost. The epicenter was 1000 km south of the boundary between the Indian Plate and the Eurasian Plate. Indian geologists believe that the local compressive stresses responsible for the Gujarat earthquake originated in the continuing northward collision of India with Eurasia. Apparently, strong crustal forces can still develop and cause faulting within a lithospheric plate far from modern plate boundaries—in this case triggering a previously unknown thrust fault at a depth of about 20 km.

Regional Fault Systems

Although most major earthquakes conform to the types of faulting predicted by plate tectonics, a plate boundary can rarely be described as just one fault, particularly when the boundary involves continental crust. Rather, the zone of deformation between two moving plates usually comprises a network of interacting faults—a *fault system.* The fault system in southern California provides an interesting example (**Figure 19.14**).

The "master fault" of this system is our old nemesis, the San Andreas—a right-lateral strike-slip fault that runs northwestward through California from the Mexican border until it goes offshore in the northern part of the state (see Figure 2.10). There are a number of subsidiary faults on either side of the San Andreas, however, that generate large earthquakes. In fact, most of the damaging earthquakes in southern California during the last century have occurred on these subsidiary faults.

Why is the San Andreas fault system so complex? Part of the explanation has to do with the geometry of the San Andreas itself. Notice in Figure 19.14a that although the strike of the San Andreas is approximately in the direction of the relative motion between the Pacific and North American plates, as expected for a transform-fault boundary, the

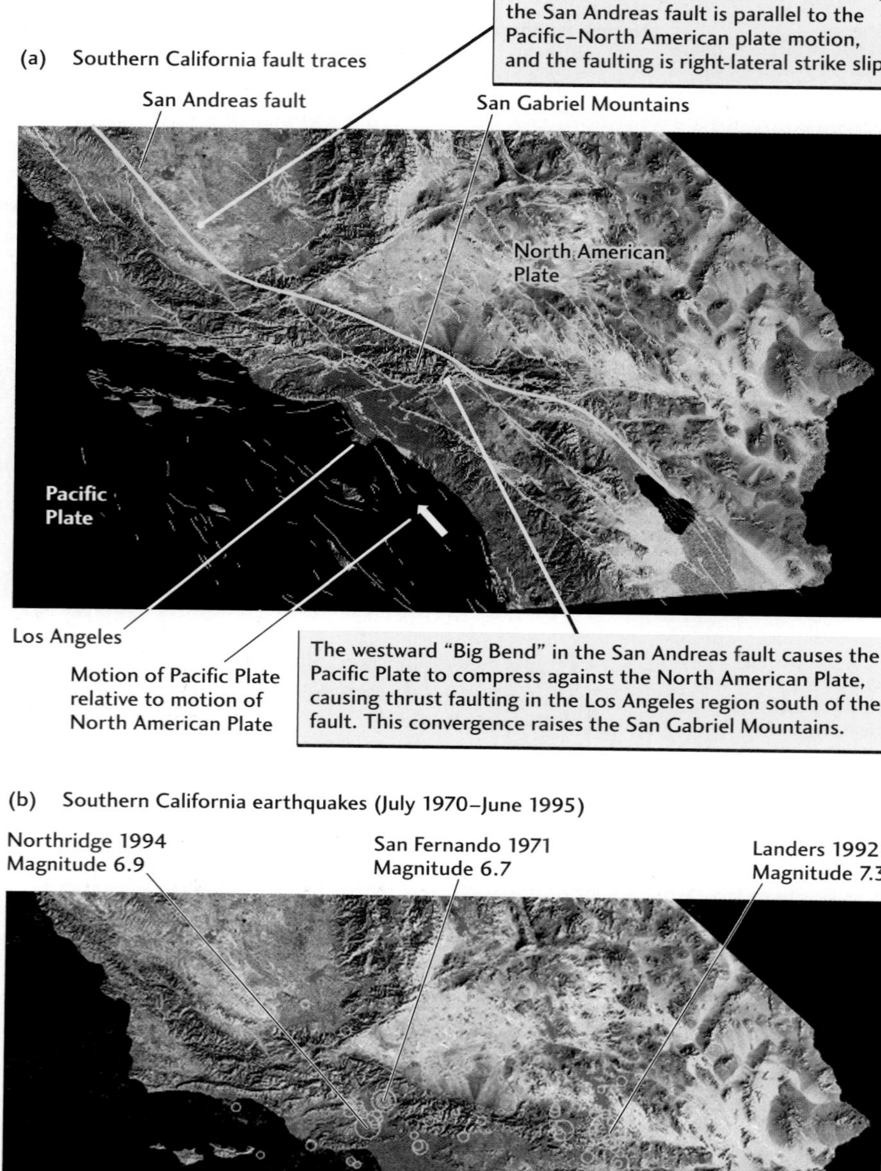

Figure 19.14 (a) Fault system of southern California, showing the San Andreas and other active faults. (b) Locations of damaging earthquakes during the last century. [Southern California Earthquake Center.]

fault is not straight but curves westward just north of Los Angeles. This "Big Bend" causes compression in the region, which is taken up by thrust faulting south of the San Andreas. The thrust faulting was responsible for two recent deadly earthquakes, the San Fernando earthquake of 1971 (magnitude 6.7, 65 people killed) and the Northridge earthquake of 1994 (magnitude 6.9, 58 people killed). Over the past several million years, thrust faulting associated with the Big Bend has raised the San Gabriel Mountains to elevations of 1800 to 3000 m.

Another complication is the extensional deformation that is taking place east of California in the Basin and Range province, which spans the states of Nevada and much of Utah and Arizona (see Chapter 11). This broad zone of divergence connects up with the San Andreas system through a series of faults that run along the eastern side of the Sierra Nevada and through the Mojave Desert. Faults of this system were responsible for the large 1992 Landers earthquake (magnitude 7.3) and the 1999 Hector Mine earthquake (magnitude 7.1), as well as the great 1872 Owens Valley earthquake (magnitude 8.2).

Earthquake Destructiveness

Over the last century, earthquakes worldwide have caused an average of 10,000 deaths per year and hundreds of billions of dollars in economic losses. Table 19.2 describes the geologic effects and destruction caused by recent earthquakes of special interest. Two California earthquakes—

Table 19.2 Recent Earthquakes of Special Interest

Event	Magnitude	Geologic Effects	Destruction
Loma Prieta, California October 1989	7.1	Maximum intensity in parts of Oakland and San Francisco; landslides; soil liquefaction; small tsunami at Monterey	60 killed; 3757 injured; U.S. $7 billion in damage
Landers, California June 1992	7.3	Surface faulting along a 70-km segment with as much as 5.5 m of horizontal displacement and 1.8 m of vertical displacement	1 killed; 400 injured; substantial damage
Northridge, California January 1994	6.9	A maximum uplift of 15 cm occurred in Santa Susana Mountains; many rockslides; ground cracks; soil liquefaction	58 killed; 7000 injured; 20,000 homeless; U.S. $20 million in damage
Northern Bolivia June 1994	8.2	At 637 km depth, the largest deep earthquake on record; first earthquake from this part of South America to have been felt in North America including Canada	Several people killed
Kobe, Japan January 1995	6.9	Surface faulting for 9 km with horizontal displacement of 1.2 to 1.5 m; soil liquefaction	5502 killed; 36,896 injured; 310,000 homeless; severe damage
Northern Iran May–June 1997	7.3	Landslides; rare sequence of large earthquakes	1567 killed; 2300 injured; 50,000 homeless; extensive damage
Papua New Guinea July 1998	7.0	Tsunami as high as 7 m	3000 killed; several villages destroyed
Izmit, Turkey August 1999	7.4	Seventh in a series since 1939; migrating westward along the strike-slip North Anatolian fault; maximum right-lateral displacement of 5 m	15,600 killed; thousands missing
Gujarat, India January 2001	8.0	Intraplate earthquake with no surface rupture	20,000 killed
Denali, Alaska November 2002	7.9	Largest earthquake in continental North America since 1906; multiple event with 400 km surface break; extensive landsliding	Very little in remote wilderness areas; Trans-Alaska pipeline unbroken at specially engineered crossing of Denali fault trace

Figure 19.15 Sixteen people died in the Northridge Meadows apartment building in Los Angeles during the 1994 Northridge earthquake. The victims lived on the first floor and were crushed when the upper levels collapsed. Many more buildings like this one would have collapsed if the newer buildings in the area had not been constructed according to stringent codes for earthquake resistance. [Nick Ut, Files/AP Photo.]

the 1989 Loma Prieta earthquake (magnitude 7.1), which occurred on the San Andreas fault some 80 km south of San Francisco, and the 1994 Northridge earthquake—were among the costliest disasters in the history of the United States. Damage amounted to more than $7 billion in the Loma Prieta earthquake and $20 billion in the Northridge quake because of their proximity to areas of dense population. About 60 people were killed in each event, but the death toll would have been many times higher if stringent earthquake-resistant building codes had not been in place (**Figure 19.15**).

Destructive earthquakes are even more frequent in Japan than in California. The recorded history of destructive earthquakes in Japan, going back 2000 years, has left an indelible impression on the Japanese people. Perhaps that is why Japan is the best prepared of any nation in the world to deal with earthquakes. It has implemented impressive public education campaigns, building codes, warning systems, and other mitigation measures. Despite this preparedness, more than 5000 people were killed in a devastating (magnitude 6.9) earthquake that struck Kobe on January 16, 1995 (**Figure 19.16**). The casualties and the enormous failure of structures (50,000 buildings destroyed) resulted partly from the less stringent building codes that were in effect before 1980, when much of the city was built, and to the location of the earthquake rupture so close to the city.

How Earthquakes Cause Damage

Earthquakes cause destruction in several ways. The *primary hazards* are the breaks in the ground surface that occur when faults rupture, the permanent subsidence and uplift of the ground surface caused by faulting, and the ground shaking caused by seismic waves radiated during the rupture. Ground vibrations can shake structures so hard that they collapse. The ground accelerations near the epicenter of a large earthquake can approach and even exceed the acceleration of gravity, so an object lying on the surface can literally be thrown into the air. Very few structures built by human hands can withstand such severe shaking, and those that do are severely damaged. The collapse of buildings and other structures is the leading cause of economic damage and casualties during earthquakes, including the high death tolls in the Tangshan, China, earthquake of 1976 (more than 240,000 killed); the Spitak, Armenia, earthquake of 1988 (25,000 killed); the Izmit, Turkey, earthquake of 1999 (15,600 killed); and the Gujarat, India, earthquake of 2001 (20,000 killed).

Earthquakes often proceed as chain reactions. The primary effects of faulting and ground shaking generate *secondary hazards* such as slumps, landslides, and other forms of ground failure. When seismic waves intensely shake water-saturated soils, the soils behave like a liquid and can become unstable. The ground simply flows away, taking buildings, bridges, and everything else with it. Soil liquefaction destroyed the residential area of Turnagain Heights, near Anchorage, Alaska, in the 1964 earthquake (see Figure 12.5); the Nimitz Freeway near San Francisco in the 1989 Loma Prieta earthquake; and areas of Kobe in the 1995 earthquake.

In some instances, ground failures can cause more damage than the ground shaking itself. A great earthquake in China's Kansu Province in 1920 triggered an extensive debris flow that covered a region larger than 100 km^2 and resulted in roughly 200,000 deaths. An immense rock and snow landslide (over 80 million cubic meters) triggered by a 1970 earthquake in Peru destroyed the mountain towns of Yungay and Ranrahirca, killing 66,000 (**Figure 19.17**). Many of the 844 people killed in the 2001 El Salvador

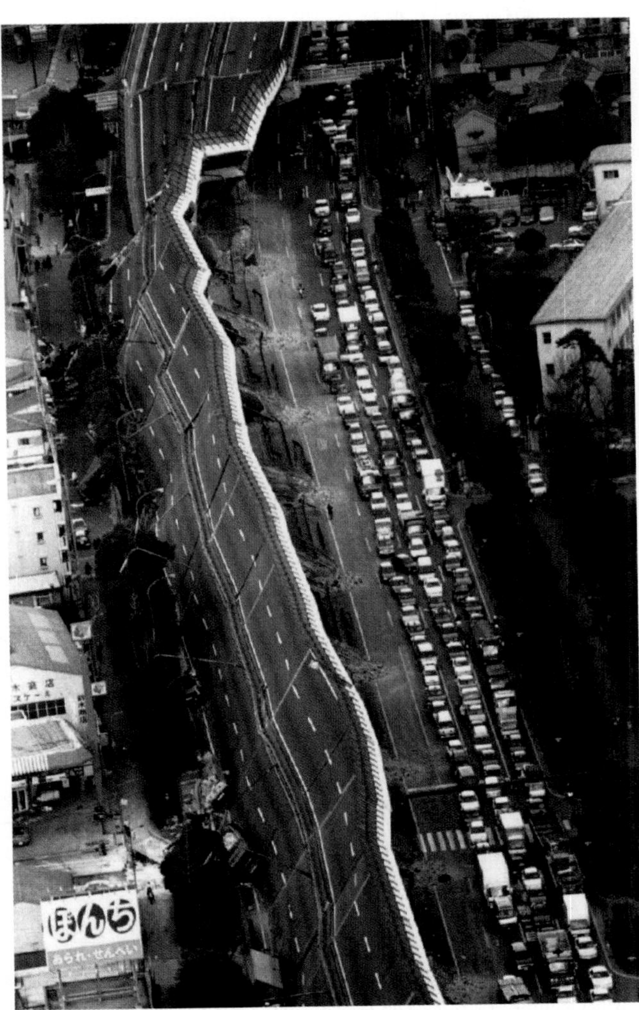

Figure 19.16 This elevated expressway, in Kobe, Japan, was overturned during an earthquake in 1995. [Reuters NewMedia/Corbis.]

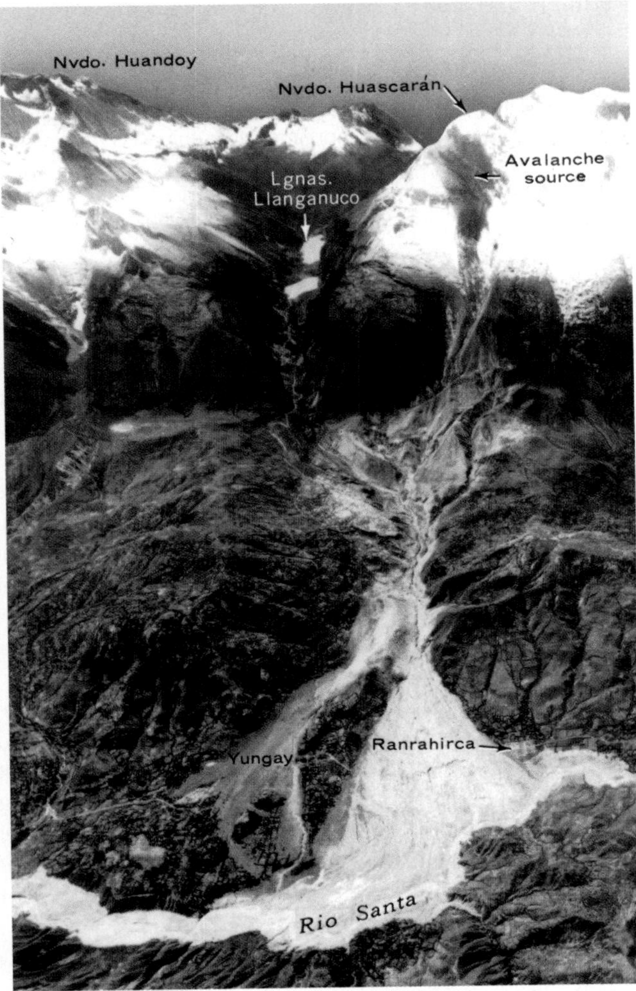

Figure 19.17 The mountain towns of Yungay and Ranrahirca, Peru, were buried when the magnitude 8 earthquake of 1970 triggered a landslide. [Courtesy of Servicio Aerofotografico Nacional de Peru, 13 June 1970/George Plafker/USGS.]

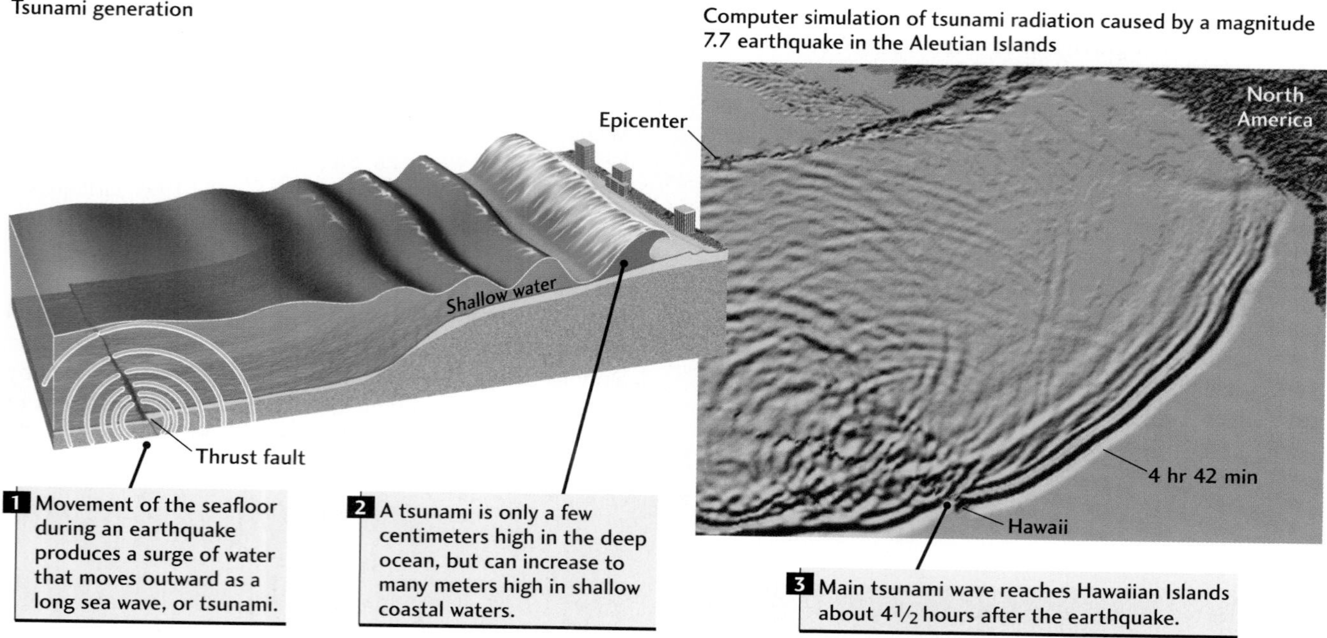

Figure 19.18 A tsunami is a sea wave generated by earthquakes on the seafloor. Destructive tsunamis often occur on shorelines near the epicenter, as illustrated in the drawing, but tsunamis from large earthquakes can also propagate across entire ocean basins. The map shows a computer simulation of the tsunami from a magnitude 7.7 earthquake in the Aleutian Islands that hit Hawaii about 4.5 hours after its origin time. [NOAA, Pacific Marine Environmental Laboratory.]

earthquake were buried by a muddy landslide loosened from a slope in the suburbs of the country's capital, San Salvador. Earthquakes that occur beneath the oceans occasionally generate towering sea waves, commonly called tidal waves but more accurately named **tsunamis** (the Japanese term). Tsunamis can inundate nearby and distant shorelines (**Figure 19.18**). They travel across the ocean at speeds as high as 800 km/hour and can build up walls of water higher than 20 m as they move into shallow waters. Tsunamis cause tremendous damage when they sweep over low-lying coastal areas (see Feature 19.1).

Secondary hazards also include the fires ignited by ruptured gas lines or downed electrical power lines. Damage to water mains in an earthquake can make fire fighting all but impossible—a circumstance that contributed to the burning of San Francisco after the 1906 earthquake. Most of the 140,000 fatalities in the 1923 Kanto earthquake, one of Japan's greatest disasters, resulted from fires in the cities of Tokyo and Yokohama.

Reducing Earthquake Risk

In assessing the possibility of damage from earthquakes, it is important to distinguish between hazard and risk. **Seismic hazard** describes the *intensity* of seismic shaking and ground disruption that can be expected over the long term at some specified location. The hazard depends on the proximity of the site to active faults that might generate earthquakes, and it can be expressed in the form of a seismic hazard map. **Figure 19.19** displays the national seismic hazard map produced by the U.S. Geological Survey.

In contrast, **seismic risk** describes the *damage* that can be expected over the long term for a specified region, such as a county or state, usually measured in terms of average dollar loss per year. The risk depends not only on the seismic hazard but also on two additional factors: the region's *exposure* to seismic damage (its population, number of buildings, and other infrastructure) and its *fragility* (the vulnerability of its built structures to seismic shaking). Because so many geologic and economic variables must be considered, estimating seismic risk is a complex job. The results of the first comprehensive national study, published by the Federal Emergency Management Agency in 2001, are presented in **Figure 19.20**.

The differences between seismic hazard and risk can be appreciated by comparing the two types of national maps. For instance, although the seismic hazard levels in Alaska and California are both high (see Figure 19.19), California's exposure is much greater, yielding a much larger total risk (see Figure 19.20). California leads the nation in seismic

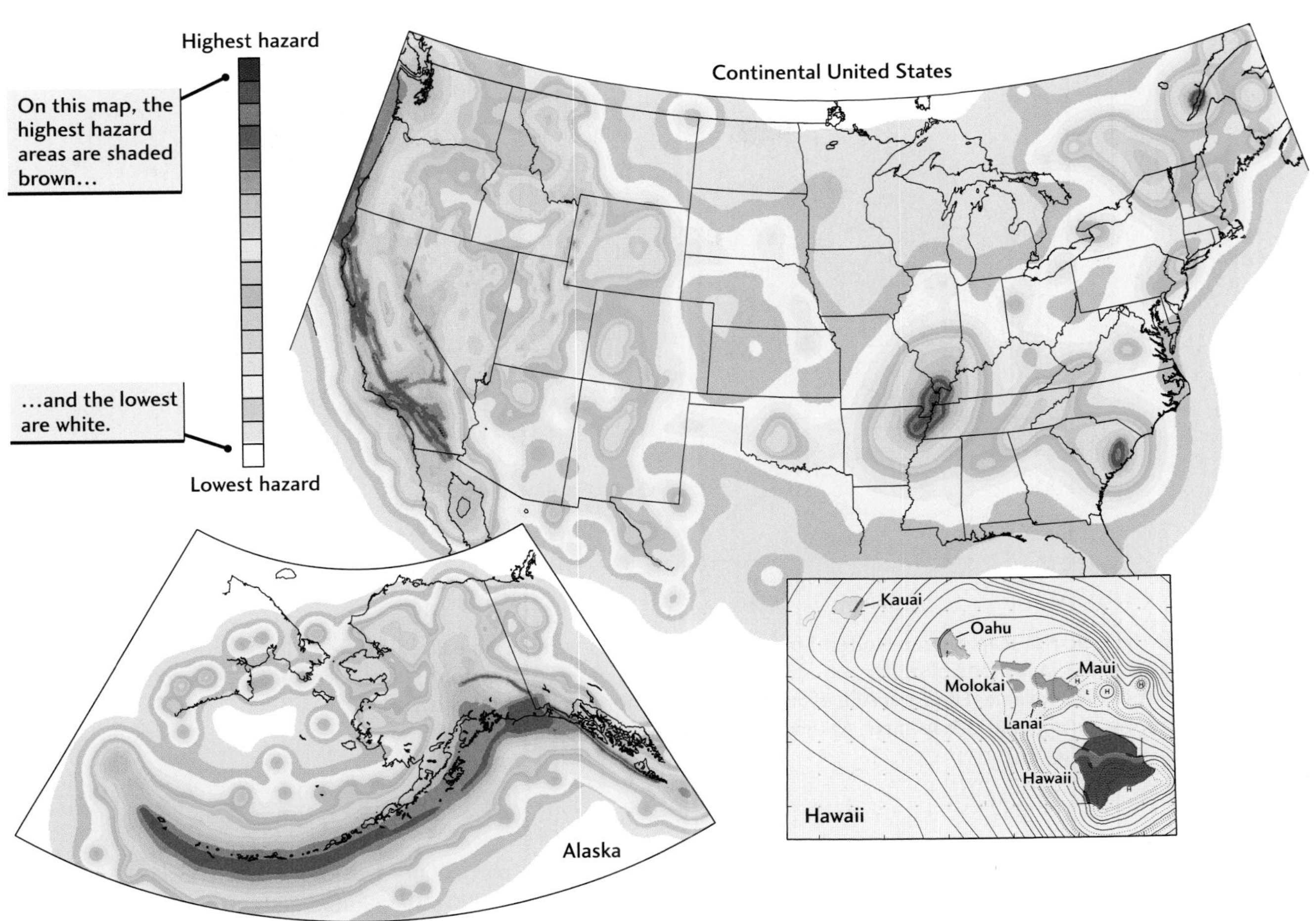

Figure 19.19 Seismic hazard map for the United States. The region of highest hazard lies along the San Andreas fault and the Transverse Ranges in California, with a branch extending into eastern California and western Nevada. High hazards are also found along the coast of the Pacific Northwest and in a zone following the intermountain seismic belt. In the central and eastern United States, the highest hazard areas are New Madrid, Missouri; Charleston, South Carolina; eastern Tennessee; and portions of the Northeast. [U.S. Geological Survey, http://geohazards.cr.usgs.gov/eq/.]

risk, with about 75 percent of the national total; in fact, a single county, Los Angeles, accounts for 25 percent. Nonetheless, the problem is truly national: 46 million people in metropolitan areas outside of California face substantial earthquake risks. Those areas include Hilo, Honolulu, Anchorage, Seattle, Tacoma, Portland, Salt Lake City, Reno, Las Vegas, Albuquerque, Charleston, Memphis, Atlanta, St. Louis, New York, Boston, and Philadelphia.

Not much can be done about seismic hazard because we have no way to prevent or control earthquakes. There are many important steps that society can take, however, to reduce seismic risk.

Hazard Characterization The first step is to follow the old proverb, "Know thine enemy." We still have much to learn about the sizes and frequencies of ruptures on active faults. For example, it is only in the past decade that we have come to appreciate the seismic dangers of the Cascadia subduction zone, which stretches from northern California, through Oregon and Washington, to British Columbia. These dangers became apparent when geologists found evidence of a magnitude 9 earthquake that occurred in 1700, before any written historical accounts of the area existed. This seismic monster caused major ground subsidence along the Cascadia coastline and left a record of flooded, dead coastal forests. A tsunami at least 5 m high hit Japan. Such a large rupture is consistent with plate tectonics. The Juan de Fuca Plate is being subducted under the North American Plate at a rate of about 45 mm/year along the coast of the Pacific Northwest. Geologists have debated whether this motion occurs seismically or

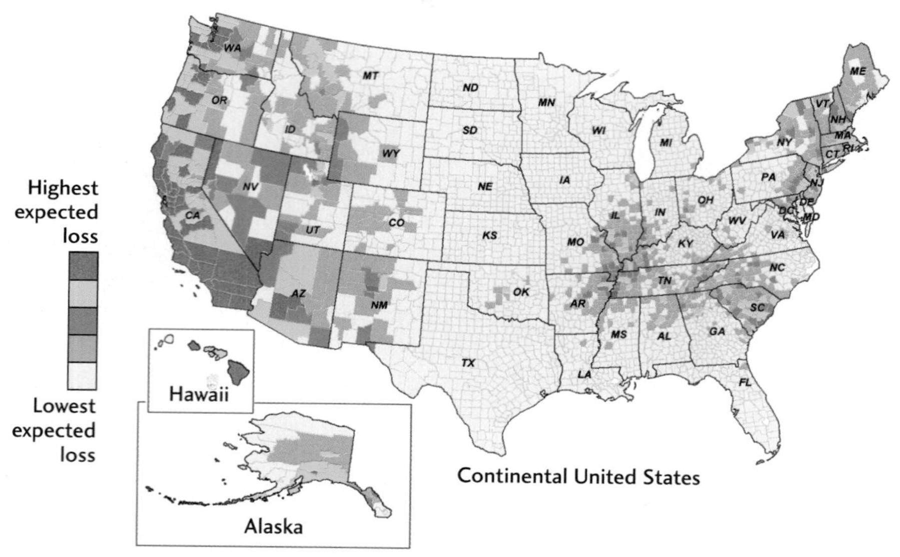

Figure 19.20 Seismic risk map for the United States. The map shows current annualized earthquake losses (AEL) in millions of dollars on a county-by-county basis. Twenty-four states have an AEL greater than $10 million. The total AEL estimated for the entire United States is about $4.4 billion. [Federal Emergency Management Agency, Report 366, Washington, D.C., 2001.]

is taken up by continuous creep or perhaps by silent earthquakes. Current opinion pegs the average time between successive magnitude 9 earthquakes in this subduction zone at 500 to 600 years.

Although we have a good understanding of seismic hazards in some parts of the world—the United States and Japan, in particular—we know much less about others. During the 1990s, the United Nations sponsored an effort to map seismic hazards worldwide as part of the International Decade of Natural Disaster Reduction. This effort resulted in the first global seismic hazard map, shown in **Figure 19.21**. The map is based primarily on his-

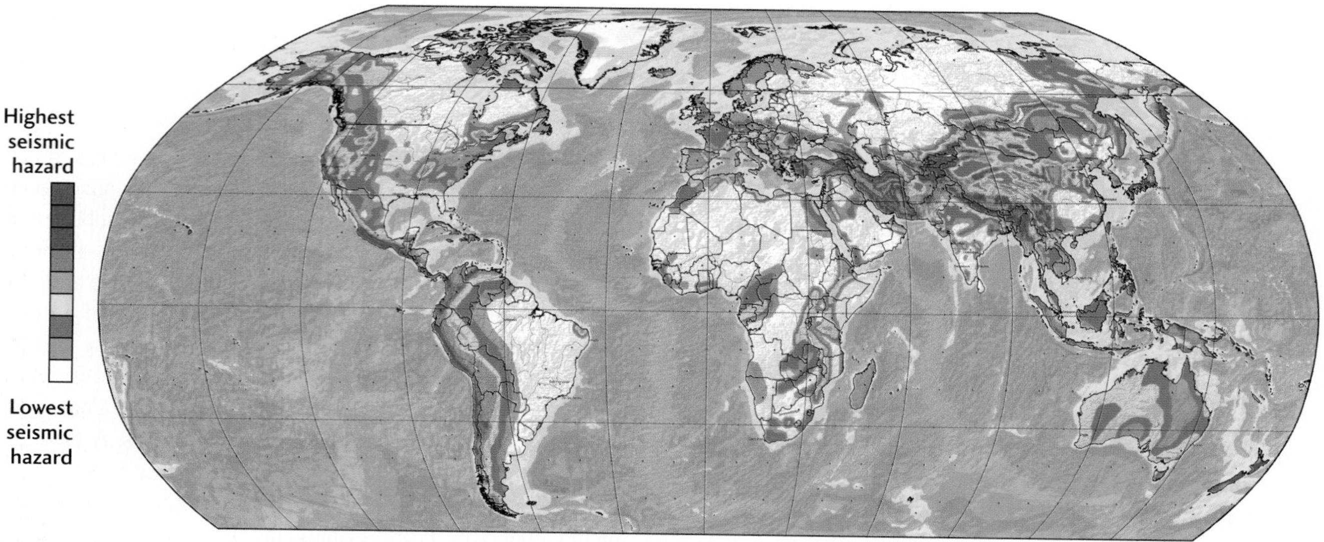

Figure 19.21 World seismic hazard map. [K. M. Shedlock et al., *Seismological Research Letters* 71 (2000): 679–686.]

19.1 Tsunamis

A tsunami is a sea wave that is triggered by an undersea event such as an earthquake or landslide or by the eruption of an oceanic volcano. (A popular name for a tsunami is a "tidal wave"—an unfortunate usage, because tsunamis have nothing to do with tides.) These events push or displace a large mass of the overlying ocean water, and this disturbance is transformed into a wave that travels across the ocean at speeds as high as 800 km/hour. In mid-ocean, where the water is deep, tsunamis are hardly noticeable. When they approach shallow coastal waters, however, the waves steepen and pile up until they become destructively high—perhaps more than 20 m. The tsunami generated by the explosion of the Krakatoa volcano in Indonesia in 1883 reached 40 m in height and drowned 36,000 people on nearby coasts. In 1960, an earthquake off the coast of Chile excited a tsunami that crossed the Pacific and caused property damage and loss of life in Japan. An earthquake off the coast of Alaska in 1964 triggered a tsunami that caused tens of millions of dollars of damage in nearby settlements. It was still destructive when it reached Crescent City, California, where it killed several people.

Disastrous tsunamis are a serious threat to 22 countries surrounding the Pacific Ocean. They have also been known to damage Atlantic coastal countries, though they occur less frequently in the Atlantic than in the Pacific. Because tsunamis can be so destructive, programs have been started (and should be expanded) to reduce their damage. The Pacific Tsunami Warning Center rapidly locates subsea earthquakes, estimates their potential for triggering a tsunami, and quickly warns countries that may be in danger. A warning may be broadcast as much as a few hours before the arrival of the tsunami, allowing time for the evacuation of coastal populations. A more difficult situation arises when a tsunami from an earthquake arrives so quickly that there is no time to warn nearby communities. This was the case in the deadly 1998 Papua New Guinea tsunami, in which as many as 3000 people died. Communities can be protected in such cases by building barrier walls to block the inundation of the sea, but such construction is expensive and is being tried to a significant degree only in Japan.

Here are some dos and don'ts if you live in an area subject to tsunamis:

- If you feel a strong earthquake, move quickly away from the coastal lowlands to higher ground.
- If you hear that an earthquake has occurred under the ocean or in a coastal region, be prepared to move to higher ground.
- Tsunamis sometimes signal their arrival by a precursory rise or fall of coastal water, and this natural warning should be heeded.
- People have lost their lives by going down to the beach to watch for a tsunami. Don't make the same mistake.
- Follow the advice of your local emergency organization.

Tsunami barrier in the town of Taro, Japan. [Courtesy of Taro, Japan.]

torical earthquakes, so it may underestimate the hazard in some regions where the historical record is short. Much more needs to be done to characterize seismic hazards on a global scale.

Land-Use Policies The exposure of buildings and other structures to earthquake risk can be reduced by policies that restrict land use. This approach works well where the hazard is localized, as in the case of fault rupture. Erecting buildings on known active faults, as was done in the residential developments pictured in **Figure 19.22**, is clearly unwise, because few buildings can withstand the deformation to which they might be subjected when the fault ruptures. In the 1971 San Fernando earthquake, a fault ruptured under a densely populated area of Los Angeles, destroying almost 100 structures. The state of California responded in 1972 with a law that restricts the construction of new buildings across an active fault. For existing residences on a fault,

Figure 19.22 Housing tracts constructed within the San Andreas fault zone, San Francisco Peninsula, before the state passed legislation restricting this practice. The white line indicates the approximate fault trace, along which the ground ruptured and slipped about 2 m during the earthquake of 1906. [R. E. Wallace, USGS.]

real estate agents are required to disclose the information to potential buyers. A notable omission is that the act does not cover publicly owned or industrial facilities. Many of the latter facilities house hazardous materials.

Earthquake Engineering Although land-use policies help to reduce seismic risk, they are less effective in addressing seismic hazard, which tends to be distributed across large regions—for example, all of Los Angeles or San Francisco. The hazards resulting from seismic shaking can best be reduced by good engineering and construction. Standards for the design and construction of new buildings are regulated by building codes enacted by state and local governments. A **building code** specifies the level of shaking a structure must be able to withstand, based on the maximum intensity expected from the seismic hazard. In the aftermath of an earthquake, engineers study buildings that were damaged and make recommendations about modifications to the building codes that could reduce future damage from similar earthquakes.

As a result of this procedure, U.S. building codes have been largely successful in preventing loss of life during earthquakes. For example, from 1983 to 2001, 129 people died in eight severe earthquakes in the western United States, whereas more than 160,000 people were killed by earthquakes worldwide. Nevertheless, more can be done to improve earthquake engineering. Damage to structures can be reduced by using specialized construction materials and advanced engineering methods, such as putting entire buildings on rollers and other movable supports to isolate them from the shaking.

For individuals, earthquake preparedness begins at home. Feature 19.2 summarizes some of the steps you can take to protect yourself and your family in an earthquake.

Emergency Response Once an earthquake happens, networks of seismographs can transmit signals automatically to central processing facilities. In a fraction of a minute, these facilities can pinpoint the earthquake focus, measure its magnitude, and determine its fault mechanism. If equipped with strong-motion sensors that accurately record the most violent shaking, these automated systems can also deliver accurate maps in nearly real time of where the ground shaking was strong enough to cause significant damage. Such information can help emergency managers and other officials deploy equipment and personnel as quickly as possible to save people trapped in rubble and to reduce further property losses from fires and other secondary effects. Bulletins about the magnitude and boundaries of the shaking can also be channeled through the mass media, reducing public confusion during disasters and allaying fears aroused by minor tremors.

Real-Time Earthquake Warning With the technology just described, it is possible to issue tens of seconds' warning of the impending arrival of destructive seismic waves. This real-time capability is available because radio waves travel much faster than seismic waves. The most damaging vibrations from a nearby earthquake travel at a velocity of about 3.3 km/s (see Figure 19.6), whereas radio signals travel at the speed of light (300,000 km/s).

Mexico City, where the great earthquake of 1985 killed 10,000 people, provides an example of how the method would work. The earthquakes that induce the strongest ground motion occur some 300 km from the city along the subduction zone where the Cocos Plate is being thrust under the North American Plate. Computers attached to a special network of seismographs in this zone could determine the location and time of an earthquake within seconds, and satellite radio links could transmit the information almost

19.2 Protection in an Earthquake

Before an Earthquake

- Have a battery-powered radio, flashlight, and first-aid kit in your home. Make sure that everyone knows where they are stored. Keep batteries on hand.
- Learn first aid.
- Know the location of your electric circuit breaker box and the gas and water shutoff valves (keep a wrench nearby). Make sure that all responsible members of your family learn how to turn them off.
- Do not keep heavy objects on high shelves.
- Securely fasten heavy appliances to the floor. Anchor heavy furniture, such as cupboards and bookcases, to the wall.
- Devise a plan for reuniting your family after an earthquake in the event that anyone is separated.
- Urge your school board and teachers to discuss earthquake safety in the classroom, have class drills, and secure heavy objects from falling.
- Find out if your office or plant has an emergency plan. Do you have emergency responsibilities? Are there special actions for you to take to make sure that your workplace is safe?

During an Earthquake

- Stay calm. If you are indoors, stay indoors; if outdoors, stay outdoors. Many injuries are sustained as people enter or leave buildings.
- If you are indoors, stand against a wall near the center of the building or get under a sturdy table. Stay away from windows and outside doors.
- If you are outdoors, stay in the open. Keep away from overhead electric wires or anything that might fall (such as chimneys, parapets, and cornices on buildings).
- Do not use candles, matches, or other open flames.
- If you are in a moving car, stop away from overpasses and bridges and remain inside until the shaking is over.
- In a high-rise building, protect yourself under sturdy furniture or stand against a support column. Evacuate if told to do so. Use stairs rather than elevators.
- If at work or school, get under desks or sturdy furniture, facing away from windows.
- If on a playground, stay away from the building.

After an Earthquake

- Check yourself and people nearby for injuries. Provide first aid if needed.
- Check water, gas, and electric lines. If they are damaged, shut off valves.
- Check for leaking gas by odor (*never* use a match). If a leak is detected, open all windows and doors, shut off the gas meter, leave immediately, and report to authorities.
- Turn on the radio for emergency instructions. Do not use the telephone—it will be needed for high-priority messages.
- Do not flush toilets until sewer lines are checked.
- Stay out of damaged buildings.
- Wear boots and gloves to protect against shattered glass and debris.
- Approach chimneys with caution.
- Follow the emergency plan or instructions given by someone in charge.
- Stay away from beaches and waterfront areas where tsunamis could strike, even long after the shaking has stopped.
- Do not go into damaged areas unless authorized. Martial law against looters has been declared after a number of earthquakes.
- Expect aftershocks; they may cause additional damage.

After Bruce A. Bolt, *Earthquakes*, 4th ed. (New York: W. H. Freeman, 1999).

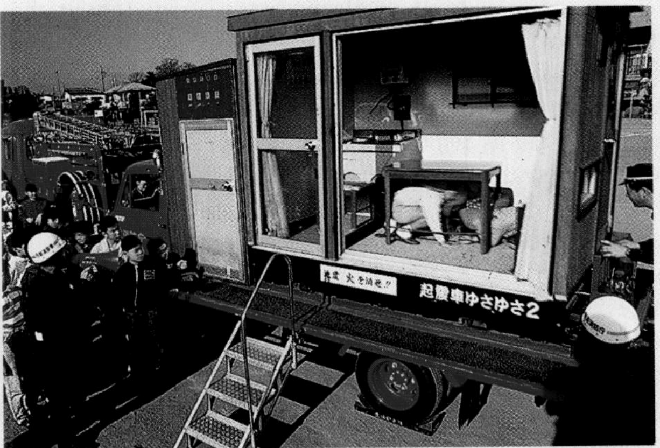

Public demonstration of earthquake safety in Japan. [Yves Gellie/Matrix.]

instantaneously, providing a 50- to 80-s warning before the most destructive waves reached the city.

Even such an extremely short warning would allow schoolchildren to duck under their desks, utilities to shut off the flow of electricity and gas, hospitals to turn on emergency power, nuclear reactors to be prepared to minimize damage, and so forth.

Can Earthquakes Be Predicted?

If we could predict earthquakes accurately, communities could be prepared, people could be evacuated from dangerous locations, and many aspects of an impending disaster might be averted. How well can we predict earthquakes?

To predict an earthquake, one must be able to specify its time, location, and size. By combining the information from plate tectonics and detailed geologic mapping of regional fault systems, geologists can reliably forecast which faults are likely to produce earthquakes over the long term. Specifying precisely when a particular fault will rupture, however, turns out to be very difficult.

Long-Term Forecasting

Ask a seismologist to predict the time of the next great earthquake and the response is likely to be, "The longer the time since the last big shock, the sooner the next one." The **recurrence interval**—the average time between large earthquakes—can be estimated in several ways. According to the elastic rebound model, the recurrence interval is given by the number of years required to accumulate the strain that will be released by fault slip in a future earthquake. The recurrence interval can be calculated from the fault slip rate and the size of the expected slip.

On the San Andreas fault, for example, the slip rate is 35 mm/year, so it takes 170 years to accumulate enough strain to cause a fault displacement of 6 m (about the amount expected for a magnitude 7.5 earthquake). Geologists can also estimate when large earthquakes happened several thousand years in the past by finding soil layers that are offset by fault displacements and then dating these layers. Although these methods usually give similar results, the uncertainty of the prediction turns out to be large—as much as 50 percent of the average recurrence interval. In southern California, the recurrence interval for the San Andreas fault is estimated to be 150 to 300 years. One part of this fault experienced a great earthquake in 1857, whereas another appears to have remained locked since a large earthquake that occurred around 1680. Therefore, an earthquake can be expected at any time—tomorrow, or decades from now.

Because of its large uncertainties (years to decades), this method of earthquake prediction is called *long-term forecasting* to distinguish it from what most people would really want—a *short-term prediction* of a large rupture on a specific fault accurate to within days or even hours of the actual event.

Short-Term Prediction

There have been a few successful short-term earthquake predictions. In 1975, an earthquake was predicted 5 hours before it occurred near Haicheng, in northeastern China. The Chinese seismologists used what they considered to be premonitory events (precursors) to make their predictions: swarms of tiny earthquakes and a rapid deformation of the ground several hours before the main shock. Several million people, prepared in advance by a public education campaign, evacuated their homes and factories in the hours before the shock. Although many towns and villages were destroyed and a few hundred people were killed, there is no doubt that tens of thousands of lives were saved. In the very next year, however, an unpredicted earthquake struck the Chinese city of Tangshan, killing more than 240,000. Obvious precursors like those seen in Haicheng have not been repeated in subsequent large events.

Although many schemes have been proposed, we have not yet found a reliable method of predicting earthquakes within a few days and with few false alarms. In particular, no unambiguous precursors of large earthquakes have been identified, even in well-instrumented areas such as Japan and California. Although we can't say that short-term earthquake prediction is impossible, seismologists are pessimistic that short-term, event-specific prediction will be feasible in the near future.

Intermediate-Term Prediction

Seismologists are more optimistic that the uncertainties in long-term forecasting can be reduced by studying the behavior of regional fault systems. The key is to generalize the elastic rebound model. The simple version depicted in Figure Story 19.1 describes how the tectonic strain steadily building up on an isolated fault segment can be released in a periodic sequence of fault ruptures. However, as we have seen in southern California (see Figure 19.14), faults are rarely isolated from one another. Instead, they are connected together in complex networks. A rupture on one segment changes the stress throughout the surrounding region. Depending on the geometry of the fault network, this stress change can either increase or decrease the likelihood of earthquakes on nearby fault segments. In other words, when and where earthquakes happen in one part of a fault system will influence when and where they happen elsewhere in the system.

Beginning in Chapter 1 and continuing throughout this book, we have discussed how interactions among the components of a geosystem can lead to complex system-wide effects. In the case of regional fault systems, seismologists are just beginning to understand how faults in one part of the system are affected by earthquakes or creep in another part. Such interactions appear to be responsible for some of the variability in earthquake recurrence intervals. Seismologists now believe that a knowledge of how large earthquakes change regional stresses can be used to reduce the uncertainties in earthquake forecasting.

In fact, by monitoring how stress buildup and release changes the regional pattern of earthquake occurrence, scientists might be able to predict earthquakes over time intervals as short as a few years or maybe even a few months. The basic idea is that stress variations can raise or lower the frequency of small events. The events can be recorded on networks of seismographs and thereby provide a regional "stress gauge." Someday you might hear a news report that says, "The National Earthquake Prediction Council estimates that, during the next year, there is a 50 percent probability of a magnitude 7 or larger earthquake on the southern segment of the San Andreas fault." Much current research aims to establish a scientific basis for this type of *intermediate-term prediction.*

The ability to issue such predictions would raise difficult questions. How should society respond to a threat that is neither imminent nor long-term? An intermediate-term prediction would give the probability of an event only on time scales of months to years—not precise enough to evacuate areas that might be damaged by an impending large earthquake. What effect would credible predictions have on property values and other investments in the threatened region? False alarms would be common. How should communities deal with this uncertainty? These are questions more suited to policy makers than to scientists.

SUMMARY

What is an earthquake? An earthquake is a shaking of the ground caused by seismic waves that emanate from a fault that breaks suddenly. When the fault breaks, the strain built up over years of slow deformation by tectonic forces is released in a few minutes as seismic waves.

What determines the depth of an earthquake? Continental earthquakes are rare below 20 kilometers. The crust at those high temperatures and pressures deforms as a ductile material and cannot support brittle fracture. However, in subduction zones, where cold oceanic lithosphere plunges back into the mantle, earthquakes can initiate at depths as great as 690 kilometers.

Where do most earthquakes occur? In keeping with plate tectonics theory, most earthquakes originate in the vicinity of plate boundaries. The small number that occur far from plate boundaries demonstrate the power of tectonic forces to cause faulting within existing plates.

What governs the type of faulting that occurs in an earthquake? In most earthquakes, the fault mechanism is determined by the kind of plate boundary. Normal faulting under tensile stress occurs at boundaries of divergence, thrust faulting under compressive stress occurs at convergent boundaries, and strike slip occurs along transform faults.

What is earthquake magnitude and how is it measured? Earthquake magnitude is a measure of the size of an earthquake. The Richter magnitude is determined from the amplitude of the ground motions measured when seismic waves are recorded on seismographs. Seismologists now prefer to use the *moment magnitude,* because it is more directly related to the physical properties of the faulting that causes the earthquake. It is approximately proportional to the logarithm of the area of the fault break. It is also proportional to the logarithm of the seismic energy released during the rupture.

How frequently do earthquakes occur? About 1 million earthquakes take place each year with magnitudes greater than 2.0, and this number decreases by 10 for each magnitude unit. Hence, there are about 100,000 with magnitudes greater than 3, about 1000 with magnitudes greater than 5, and about 10 with magnitudes greater than 7. However, magnitude 8 earthquakes occur on average only about once every 3 to 5 years. The largest earthquakes, 9 to 9.5 in moment magnitude, are exceptionally rare.

What are the three types of seismic waves? Two types of waves travel through Earth's interior: P (primary) waves, which are transmitted by all forms of matter and move fastest, and S (secondary) waves, which are transmitted only by solids and move at a little more than half the speed of P waves. Surface waves need a free surface to ripple. They move more slowly than S waves.

What causes the destructiveness of earthquakes? Ground vibrations can damage or destroy buildings and other structures and can trigger secondary effects, such as avalanches. Fires are a serious threat after an earthquake. Earthquakes on the seafloor can excite tsunamis, which sometimes cause widespread destruction when they reach shallow coastal waters.

What can be done to mitigate the damage of earthquakes? Construction in earthquake zones can be regulated so that buildings and other structures will be strong enough to withstand destructive vibrations and will not be located on soils that are unstable. People living in earthquake-prone areas should be informed about what to do when an earthquake occurs, and public authorities should plan ahead and be prepared with emergency supplies, rescue teams, evacuation procedures, fire-fighting plans, and other steps to minimize the consequences of a severe earthquake.

Can scientists predict earthquakes? Scientists can characterize the level of hazard in a region, but they cannot consistently predict earthquakes with the degree of accuracy that would be needed to alert a population hours or days in advance. They are searching for premonitory phenomena that may be used to identify the time and place of an earthquake more accurately. They are also increasing their research efforts on more attainable goals, such as improved engineering design of structures, methods of deploying disaster response teams rapidly, and real-time hazard warnings.

Key Terms and Concepts

aftershock (p. 431)
building code (p. 450)
earthquake (p. 428)
elastic rebound theory (p. 428)
epicenter (p. 428)
fault mechanism (p. 436)
focus (p. 428)
foreshock (p. 431)
intensity scale (p. 436)
magnitude scale (p. 435)
P wave (p. 432)
recurrence interval (p. 452)
S wave (p. 432)
seismic hazard (p. 446)
seismic risk (p. 446)
seismic wave (p. 428)
seismicity map (p. 439)
seismograph (p. 431)
slip (p. 428)
surface wave (p. 432)
tsunami (p. 445)

Exercises

MEDIA LINK *This icon indicates that there is an animation available on the Web site that may assist you in answering a question.*

1. Describe two scales for measuring the size of an earthquake. Which is the more appropriate scale for measuring the amount of faulting that caused the earthquake? Which is more appropriate for measuring the amount of shaking experienced by a particular observer?

2. In southern California, a magnitude 5 earthquake occurs about once per year. Approximately how many magnitude 4 earthquakes would you expect each year? Magnitude 3 earthquakes?

3. How does the distribution of earthquake foci correlate with the three types of plate boundaries?

4. What kinds of earthquake faults occur at the three types of plate boundaries?

5. Destructive earthquakes occasionally occur within plates, far from plate boundaries. Why?

6. Seismograph stations report the following S–P time differences for an earthquake: Dallas, S–P = 3 minutes; Los Angeles, S–P = 2 minutes; San Francisco, S–P = 2 minutes. Use a map of the United States and time-travel curves (see Figure 19.6) to obtain a rough epicenter.

7. At a place along a boundary fault between the Nazca Plate and the South American Plate, the relative plate motion is 80 mm/year. The last great earthquake, in 1880, showed a fault slip of 12 m. When should local residents begin to worry about another great earthquake?

Thought Questions

1. Taking into account the possibility of false alarms, reduction of casualties, mass hysteria, economic depression, and other possible consequences of earthquake prediction, do you think the objective of predicting earthquakes is a socially contributing goal?

2. Would you rather live on a planet without earthquakes (implying no plate tectonics)? Why or why not?

Suggested Readings

Bolt, Bruce A. 1999. *Earthquakes,* 4th ed. New York: W. H. Freeman.

Clague, J. J., et al. 2000. Great Cascadia Earthquake Tricentennial. *GSA Today* (November): 14–15.

Earthquakes and Volcanoes. A bimonthly periodical. Washington, D.C.: U.S. Government Printing Office.

Federal Emergency Management Agency. 2001. *HAZUS99 Estimated Annualized Earthquake Losses for the United States.* FEMA Report 366, Washington, D.C.

Hough, S. E. 2002. *Earthquake Science: What We Know (and Don't Know) About Earthquakes.* Princeton, N.J.: Princeton University Press.

Kanamori, H., and E. E. Brodsky. 2001. The physics of earthquakes. *Physics Today* (June): 34–40.

Kanamori, H., E. Hauksson, and T. Heaton. 1997. Real-time seismology and earthquake hazard mitigation. *Nature* 390: 461–464.

National Research Council, Committee on the Science of Earthquakes (T. H. Jordan, chair). 2003. *Living on an Active Earth: Perspectives on Earthquake Science.* Washington, D.C.: U.S. Government Printing Office.

Normile, D. 1995. Quake builds case for strong codes. *Science* 267: 444–446.

Shedlock, K. M., D. Giardini, G. Grünthal, and P. Zhang. 2000. The GSHAP Global Seismic Hazard Map. *Seismological Research Letters* 71: 679–686.

Southern California Earthquake Center. 1995. *Putting Down Roots in Earthquake Country.* http://www.scec.org/ resources/ catalog/roots.html.

Stein, R. S. 2003. Earthquake conversations. *Scientific American* (January): 72–79.

U.S. Geological Survey. 1995. *Reducing Earthquake Losses Throughout the United States.* Fact Sheets 096-95, 097-95, 224-95, 225-95. Available from Earthquake Information Hot Line, USGS, MS 977, Menlo Park, CA 94025, and at http://quake.usgs.gov/prepare/factsheets/index.html.

Shaded relief map of the North American continent. [U.S. Geological Survey.]

CHAPTER

20

Evolution of the Continents

"The only time I have ever seen a man literally foaming at the mouth was when I mentioned continental drift to a distinguished geologist."

SIR EDWARD BULLARD

Nearly two-thirds of Earth's surface—all of the oceanic crust—was created by seafloor spreading during the past 200 million years, an interval that spans a mere 4 percent of Earth's history. To understand how Earth has evolved since its fiery beginnings, we must look to the continents, which contain rocks as old as 4 billion years.

The geologic record in the continental crust is very complex, but our ability to read this record has improved immensely in just the last few years. Geologists have learned to use the concepts of plate tectonics to interpret eroded mountain belts and ancient rock assemblages in terms of closing ocean basins and colliding continents. They have developed new geochemical tools, such as radiometric dating, to decipher the history of continental rocks. They have gathered data from new networks of seismographs and other sensors that allow them to image the structure of the continents far below Earth's surface. This evidence has led them to a startling discovery unanticipated by the theory of plate tectonics: in areas of the oldest continental crust, the plates have deep "keels" that are much thicker than the lithosphere beneath even the oldest ocean basins.

This chapter will examine the 4-billion-year history of the continents and what it tells us about how continental lithosphere formed and how it has evolved over geologic time.

We will see how plate tectonic processes have added new material to the continental crust, how plate convergence has thickened this crust into mountain belts, and how these mountains have been eroded to expose the metamorphic basement rocks found in many older regions of the continents. We also reach back into the earliest period of continental evolution, the Archean eon (4.0 to 2.5 billion years ago), to ponder two of the great puzzles of Earth's history: How did continents and their deep keels form, and how have they been maintained through billions of years of plate tectonics and continental drift?

Continents, like people, are different; individuals show a great variety of features that reflect their parentage and experience over time. Yet, also like people, continents share many similarities in their basic structure and growth. Before considering continents in general, we will discuss some of the major features of one continent we know very well: North America.

The Tectonics of North America

The large-scale geologic features of North America are organized into a pattern that reflects the continent's long-term tectonic evolution (**Figure 20.1**). Parts of the crust built during the most ancient episodes of deformation tend to be found in the northern interior of the continent. This region, which includes most of Canada and the closely connected landmass of Greenland, is tectonically stable. It has remained largely undisturbed in recent episodes of continental rifting, drift, and collision, and it has been eroded flat. On the edges of these older terrains are the younger deformation belts, where most of the present-day mountain chains are found. They form elongated topographic features near the margins of the continent. Examples include the Cordillera, which runs down the western edge of North America, and the Appalachian fold belt, which trends southwest to northeast on its eastern margin.

The Stable Interior

Much of central and eastern Canada is a landscape of very old crystalline basement rocks—a huge tectonic domain (8 million square kilometers) that geologists call the *Canadian Shield* (**Figure 20.2**). It is dominated by granitic and metamorphic rocks, such as gneisses, together with highly deformed and metamorphosed sedimentary and volcanic rocks. This primitive region represents one of the oldest records of Earth's history, much of it of Archean age. It contains major deposits of iron, gold, copper, diamonds, and nickel.

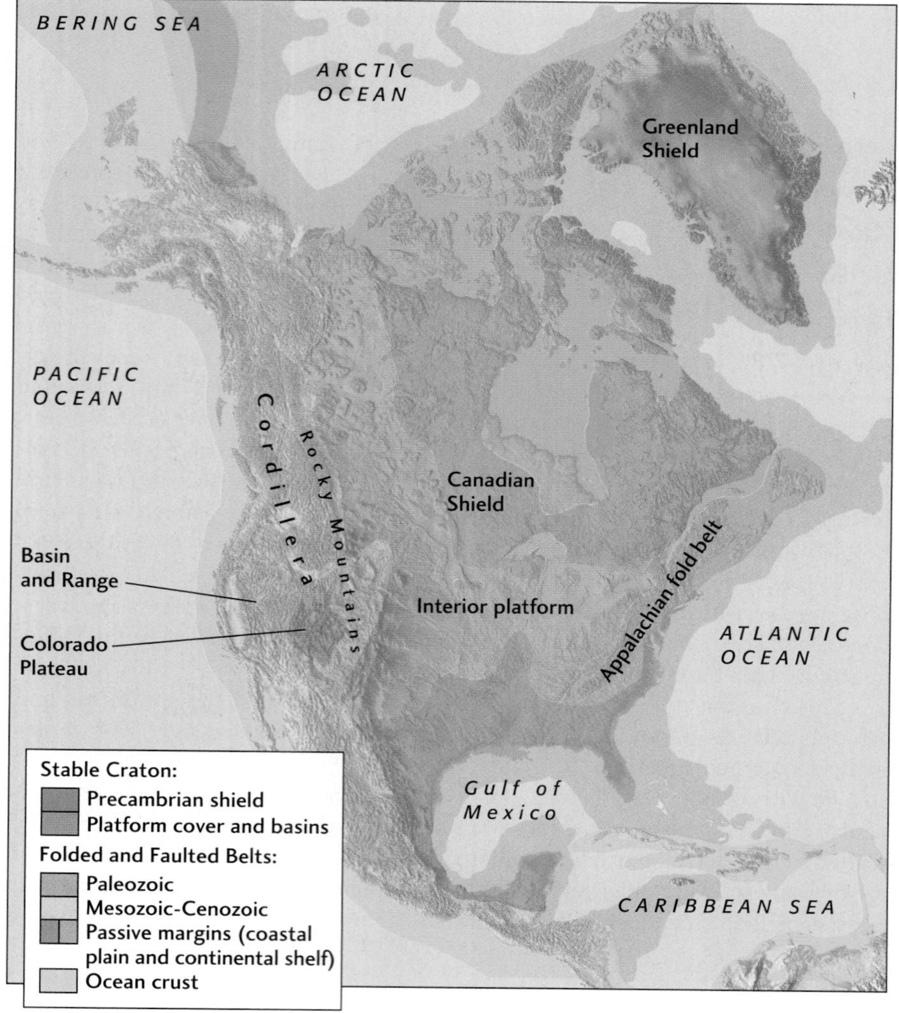

Figure 20.1 Major tectonic features of North America: Canadian Shield; interior platform; Cordilleran orogenic belt, including Basin and Range; Colorado Plateau; Appalachian fold belt; coastal plain and continental shelves of passive margins. [After A. W. Bally, C. R. Scotese, and M. I. Ross, *Geology of North America*, vol. A (Boulder, Colo.: Geological Society of America, 1989).]

Figure 20.2 An aerial view of the ancient, eroded metamorphic rocks exposed on the surface of the Canadian Shield, Nunavut, Canada. [Roy Tanami/Ursus.]

The nineteenth-century Austrian geologist Eduard Suess named these areas continental **shields** because they emerge from the surrounding sediments like a shield partially buried in the dirt of a battlefield. In North America, flat-lying "platform sediments" occur around the periphery of the Canadian Shield and also near its center, beneath Hudson Bay (see Figure 20.1). South of the Canadian Shield is a vast sediment-covered region of low topography called the *interior platform* that comprises the Great Plains of Canada and the United States. The Precambrian basement rocks beneath the interior platform represent a subsurface continuation of the Canadian Shield, although here they are under layers of Paleozoic sedimentary rocks typically less than about 2 km thick.

The North American platform sediments were laid down on the deformed and eroded Precambrian basement under a variety of conditions. The rock assemblages indicate sedimentation in extensive shallow inland seas (marine sandstones, limestones, shales, deltaic deposits, evaporites) as well as deposition on alluvial plains or in lakes or swamps (nonmarine sediments, coal deposits). Many of the continent's deposits of uranium, coal, oil, and gas are found in the sedimentary cover of the interior platform.

Within the platform, broad sedimentary basins are defined by roughly oval depressions where the sediments are thicker than the surrounding platform sediments. The Michigan Basin, a circular area of about 500,000 km^2 that covers much of the Lower Peninsula of Michigan, subsided throughout much of Paleozoic time and received sediments more than 5 km thick in its central, deepest part (**Figure 20.3**). The sandstones and other sedimentary rocks of the basins, laid down under quiet conditions, have remained unmetamorphosed and only slightly deformed to this day. Some geologists believe that the basins subsided in an episode of stretching and thinning of the continental lithosphere.

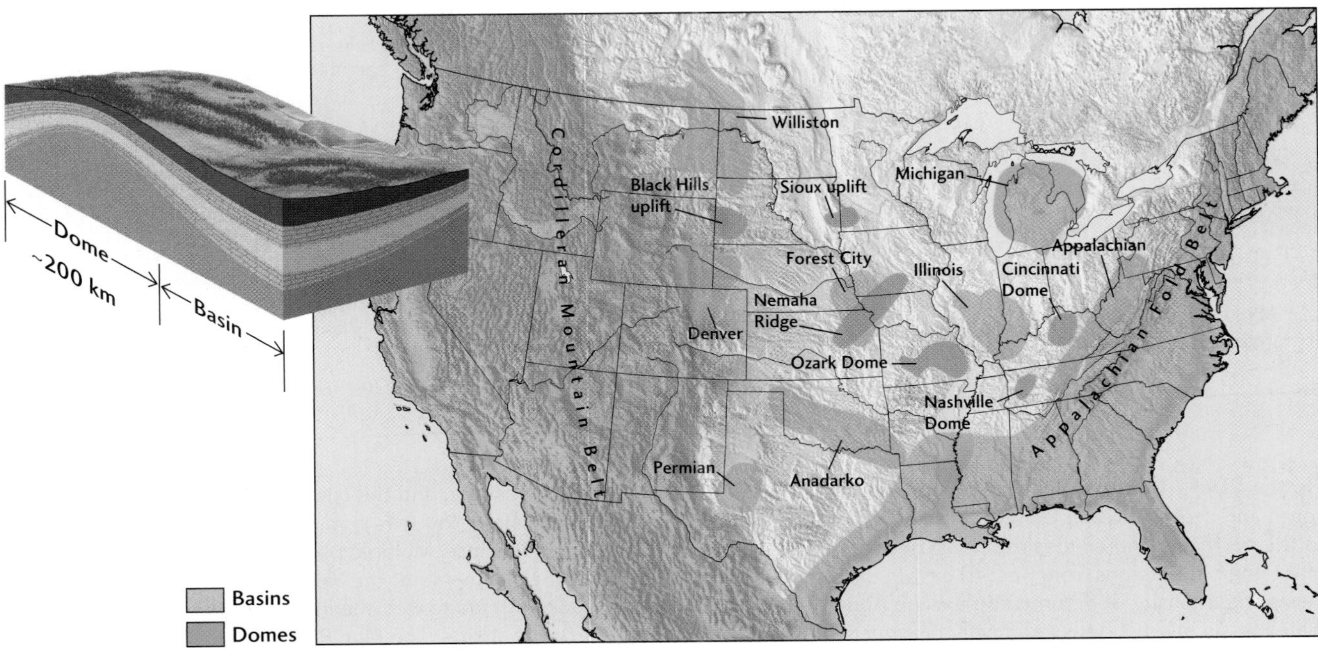

Figure 20.3 A map of the interior platform of North America, showing its basin and dome structure. The basins are nearly circular regions of thick sediments, such as the Michigan and Illinois basins, whereas the domes are regions where the sediments are anomalously thin. Crystalline basement rocks are exposed on the tops of some domes, such as the Black Hills uplift and the Ozark Dome.

The Appalachian Fold Belt

Fringing the eastern side of North America's stable interior are the old, eroded Appalachian Mountains. This classic fold-and-thrust belt, which we first examined in Chapter 11, extends along eastern North America from Newfoundland to Alabama. The rock assemblages and structures found there today resulted from plate divergences and collisions from late Precambrian time to the present. Moving eastward from the mildly deformed plateaus in the west, we encounter regions of increasing deformation (**Figure 20.4**).

- *Valley and Ridge province* Thick Paleozoic sedimentary rocks laid down on an ancient continental shelf were folded

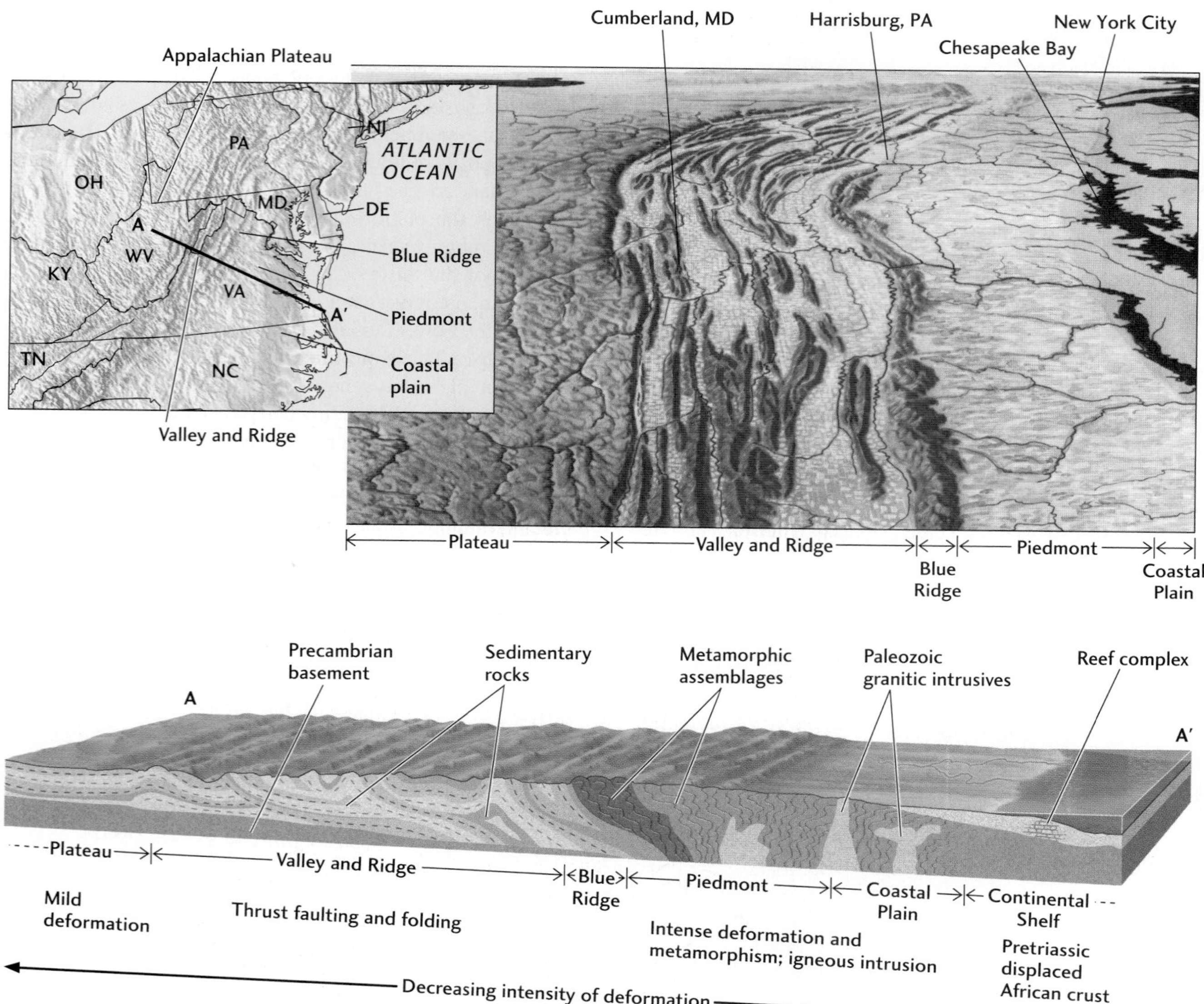

Figure 20.4 The Appalachian Mountain region of the south and central eastern United States, with an aerial view to the northeast and an idealized cross section. The major physiologic subregions from west to east in order of increasing intensity of deformation are the Appalachian plateaus, slightly uplifted, with mildly deformed sediments; the Valley and Ridge province, consisting of long mountain ridges and valleys and erosional features in a fold-and-thrust belt of sedimentary rocks; the Blue Ridge, consisting of mountains of Precambrian crystalline rock; the Piedmont, rough to gentle hilly terrain of metamorphosed sedimentary and volcanic rocks; and the coastal plain and continental shelf, a terrain of low hills grading to flat plains on sediments extending offshore. Multiple plate collisions in the Paleozoic folded the sediments of the Valley and Ridge province and transported them toward the interior of the continent along great thrust sheets. The Blue Ridge and Piedmont sheets were thrust westward over younger sediments. The modern coastal plain and shelf developed after the Triassic-Jurassic splitting of North America from Africa and the opening of the modern North Atlantic Ocean. [After S. M. Stanley, *Earth System History* (New York: W. H. Freeman, 1999).]

and thrust to the northwest by compressive forces from the southeast. The rocks show that deformation occurred in three mountain-building episodes, one beginning in the middle Ordovician, one in the middle to late Devonian, and one in late Carboniferous and early Permian time.

• *Blue Ridge province* These eroded mountains are composed largely of Precambrian and Cambrian crystalline rock, showing much metamorphism. The Blue Ridge rocks were not intruded and metamorphosed in place but were thrust as sheets over the sedimentary rocks of the Valley and Ridge province at the end of the Paleozoic era.

• *Piedmont* This region contains Precambrian and Paleozoic metamorphosed sedimentary and volcanic rocks intruded by granite, all now eroded to low relief. Volcanism began in the late Precambrian and continued into the Cambrian. The Piedmont was thrust over Blue Ridge rocks along a major thrust fault, overriding them to the northwest. At least two episodes of deformation are evident, one in the middle to late Devonian, the other in the Carboniferous.

• *Coastal plain* Relatively undisturbed sediments of Jurassic age and younger are underlain by rocks similar to those of the Piedmont. The continental shelf is the offshore extension of the coastal plain.

The North American Cordillera

The stable interior platform of North America is bounded on the west by a younger complex of mountain ranges and deformation belts (**Figure 20.5**). This region is part of the

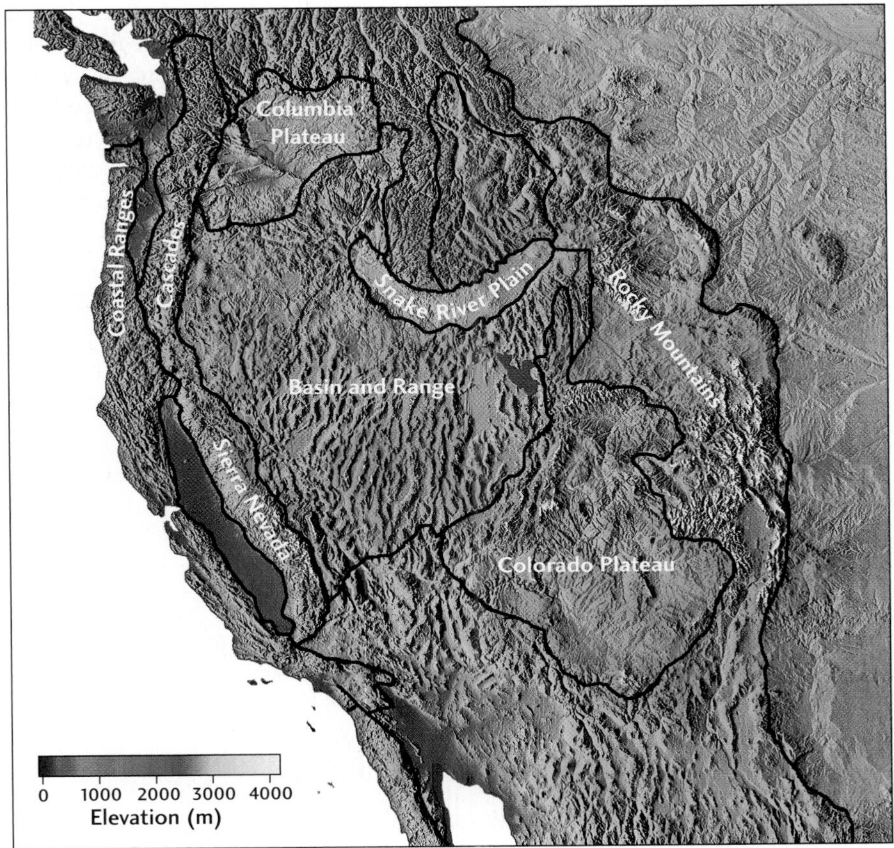

Figure 20.5 Color shaded-relief map of the western United States. Computer manipulation of digitized elevation data produced an image in which the major structural provinces and tectonic history of the area are clearly visible, as if illuminated by a light source low in the west. The long, linear ridges of the Basin and Range contrast with the somewhat smoother Colorado Plateau and the rugged and complex relief of the Rocky Mountains and the Cascade and Coast ranges. The low, flat floor of the Snake River is punctuated by individual volcanic cones. The Columbia Plateau stands out as a relatively flat, low-lying basin between the Cascades and the Rocky Mountains. The smooth, sea-level floor of the Great Valley of California is bounded on the east by the Sierra Nevada and on the west by the conspicuous, linear trend of the San Andreas fault system. [Courtesy of David Simpson, IRIS Consortium.]

North American Cordillera, a mountain belt extending from Alaska to Guatemala, and it contains some of the highest peaks on the continent. Across its middle section, between San Francisco and Denver, the Cordilleran system is about 1600 km wide and includes several contrasting physiographic provinces: the Coast Ranges along the Pacific Ocean; the lofty Sierra Nevada and Cascades; the Basin and Range province; the high tableland of the Colorado Plateau; and the rugged Rocky Mountains, which end abruptly at the edge of the Great Plains on the stable interior.

The history of the Cordillera is a complicated one, with details that vary along its length. It is a story of the interaction of the Pacific, Farallon, and North American plates over the past 200 million years (**Figure 20.6**). Before the breakup of Pangaea, the Farallon Plate occupied most of the eastern Pacific Ocean. As North America moved westward, most of this oceanic lithosphere was subducted eastward under the continent. Today, all that is left of the Farallon Plate are small remnants, which include the Juan de Fuca and Cocos plates (see Figure 2.5). During this period of subduction, the western margin of North America was modified by the accretion of continental and oceanic fragments, thrusting and folding of deep-water and shelf deposits, volcanism and the intrusion of granitic plutons, and metamorphism. Together, these processes led to the juxtaposition of deformed rock assemblages that vary in age and origin and to the reworking of older tectonic features that characterize the Cordillera.

The Cordilleran system is topographically higher and more extensive than the Appalachians because its main phase of mountain building was more recent, having occurred in the last half of the Mesozoic era and in the early Tertiary period (150 to 50 million year ago), and there has been less time for erosion to wear it down. The form and height of the Cordillera that we see today are manifestations of even more recent events in the Tertiary and Quaternary periods, over the past 15 or 20 million years, which resulted in **rejuvenation** of the mountains; that is, the mountains were raised again and brought back to a more youthful stage. At that time, the central and southern Rockies attained much of their present height as a result of a broad regional upwarp. The Rockies were raised 1500 to 2000 m as Precambrian basement rocks and their veneer of later-deformed sediments were pushed above the level of their surroundings. Stream erosion accelerated, the mountain topography sharpened, and the canyons deepened. Other examples of rejuvenation by upwarping include the Adirondacks

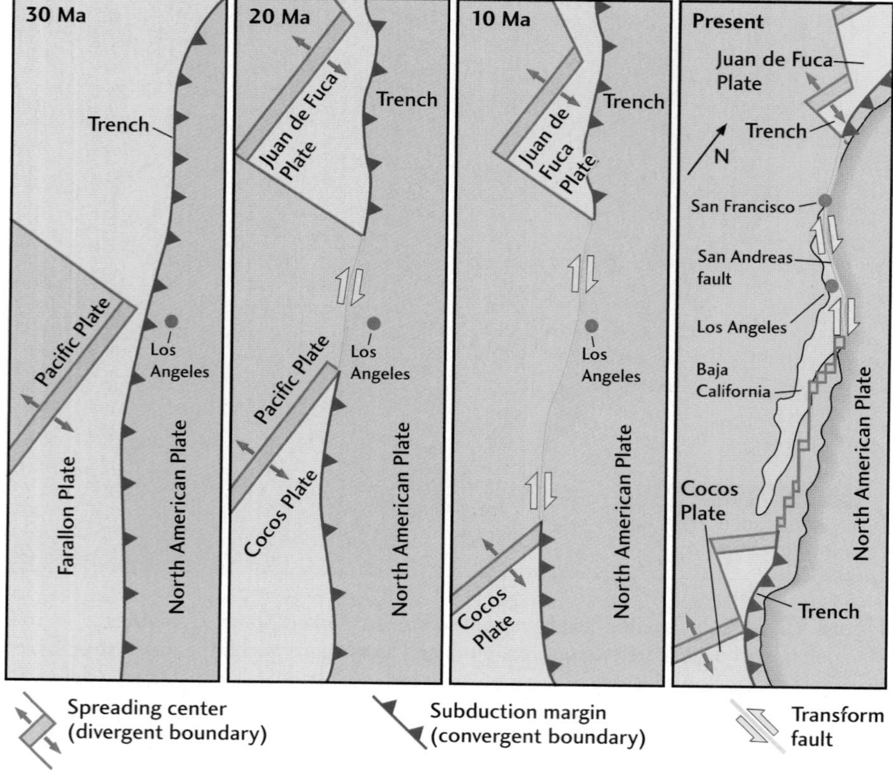

Figure 20.6 Maps illustrating the interaction of the west coast of North America with the shrinking Farallon Plate, as it was progressively consumed beneath the North American Plate, leaving the present-day Juan de Fuca and Cocos plates as small remnants. Large solid arrows show the present-day sense of relative movement between the Pacific and North American plates. [After W. J. Kious and R. I. Trilling, *This Dynamic Earth: The Story of Plate Tectonics* (Washington, D.C.: U.S. Geological Survey, 1996).]

Figure 20.7 Owens Valley and Mount Whitney, on the east side of the Sierra Nevada, illustrate the uplift of the Sierra Nevada along a major normal fault. [Galen Rowell/Corbis.]

of New York and the Labrador Highlands of Canada. As we learned in Chapter 18, there is a growing body of evidence suggesting that rejuvenation is driven not only by tectonic processes but also by an interaction between climate change and tectonics. For example, the increase in relief of some Cordillera mountain chains may have occurred as a result of the onset of the ice ages in North America.

The late structural imprinting responsible for the present-day features of the Basin and Range province and the tilted uplift of the Sierra Nevada of California also occurred in the Tertiary and Quaternary periods. A spectacular episode of faulting occurred in a region extending southeast from southern Oregon to Mexico and encompassing Nevada, western Utah, and parts of eastern California, Arizona, New Mexico, and western Texas (**Figure 20.7**). Thousands of nearly vertical faults sliced the crust into innumerable upheaved and downdropped blocks, forming hundreds of nearly parallel *fault-block mountain ranges* separated by alluvium-filled rift valleys bounded by normal faults. Some geologists think that the faulting can be explained by crustal stretching or extension caused by flow in the mantle below. Other fault-block mountains are the Wasatch Range of Utah and the Teton Range of Wyoming.

The Colorado Plateau seems to be an island of the central stable region, cut off from the interior by the Rocky Mountains. Since the late Precambrian, it has been stable, experiencing no thick basin deposits and no major mountain building. Its rock formations, exposed in the Grand Canyon (see Feature 10.1), resulted mainly from up-and-down movements.

Coastal Plain and Continental Shelf

The Atlantic coastal plain and the continental shelf, its offshore extension (see Figure 20.1), began to develop in the Triassic period with the rifting that preceded the opening of the modern Atlantic Ocean. The rift valleys formed basins that trapped a thick series of nonmarine sediments. As these deposits were accumulating, they were intruded by basaltic sills and dikes. The Connecticut River valley and the Bay of Fundy are such sediment-filled rift valleys.

In the early Cretaceous period, the deeply eroded sloping surface of the Atlantic coastal plain and continental shelf began to subside and to receive sediments from the continent. Cretaceous and Tertiary sediments as much as 5 km thick filled the slowly subsiding trough, and even more material was dumped into the deeper water of the continental rise (see Figure 8.18). This still-active offshore sedimentary basin continues to receive sediments. If the present stage of opening of the Atlantic is reversed some millions of years from now, the sediments in this basin will be folded and faulted in the same kind of process that produced the Appalachians.

The coastal plain and shelf of the Gulf of Mexico are continuous extensions of the Atlantic coastal plain and shelf, interrupted only briefly by the Florida Peninsula, a large carbonate platform. The Mississippi, Rio Grande, and other rivers that drain the interior of the North American continent have delivered sediments to fill a trough some 10 to 15 km deep running parallel to the coast. The Gulf coastal plain and shelf are rich reservoirs of petroleum and natural gas. The Atlantic shelf is currently being explored for these resources.

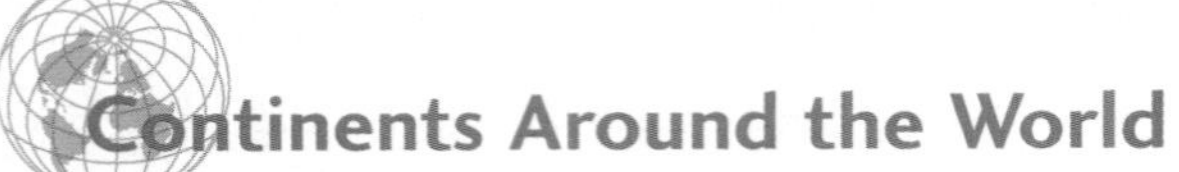

Continents Around the World

We will now expand our view from North America to Earth's other continents (**Figure 20.8**). From a global perspective, the geologic structure of continents exhibits a general, though highly irregular, pattern. The eroded remnants of ancient deformed rocks constitute continental **cratons**—stable nuclei that comprise the continental shields and platforms. Around these cratons are arrayed elongated mountain belts or **orogens** (from the Greek *oros,* meaning "mountain", and *gen,* "be produced") that were formed by later episodes of compressive deformation. The youngest orogenic (mountain-building) systems are found along the active margins of the continents, where plate tectonic motions continue to deform the weak continental crust.

Tectonic Provinces

The general pattern of cratons bounded by orogens can be seen in Figure 20.8a, which summarizes the major tectonic provinces of the continents. The classifications portrayed on this map are closely related to the regions we used to describe the tectonics of North America:

- *Shield* A region of uplifted and exposed crystalline basement rocks of Precambrian age, undeformed in the Phanerozoic eon (543 million years ago to the present). Example: Canadian Shield.
- *Platform* A region where Precambrian basement rocks are overlain by less than a few kilometers of relatively flat-lying sediments. Examples: interior platform of central North America, Hudson Bay.
- *Continental basin* A region of prolonged subsidence where thick sediments have accumulated during the Phanerozoic, with layers dipping away from the margins of the basin. Examples: Michigan and Illinois basins.

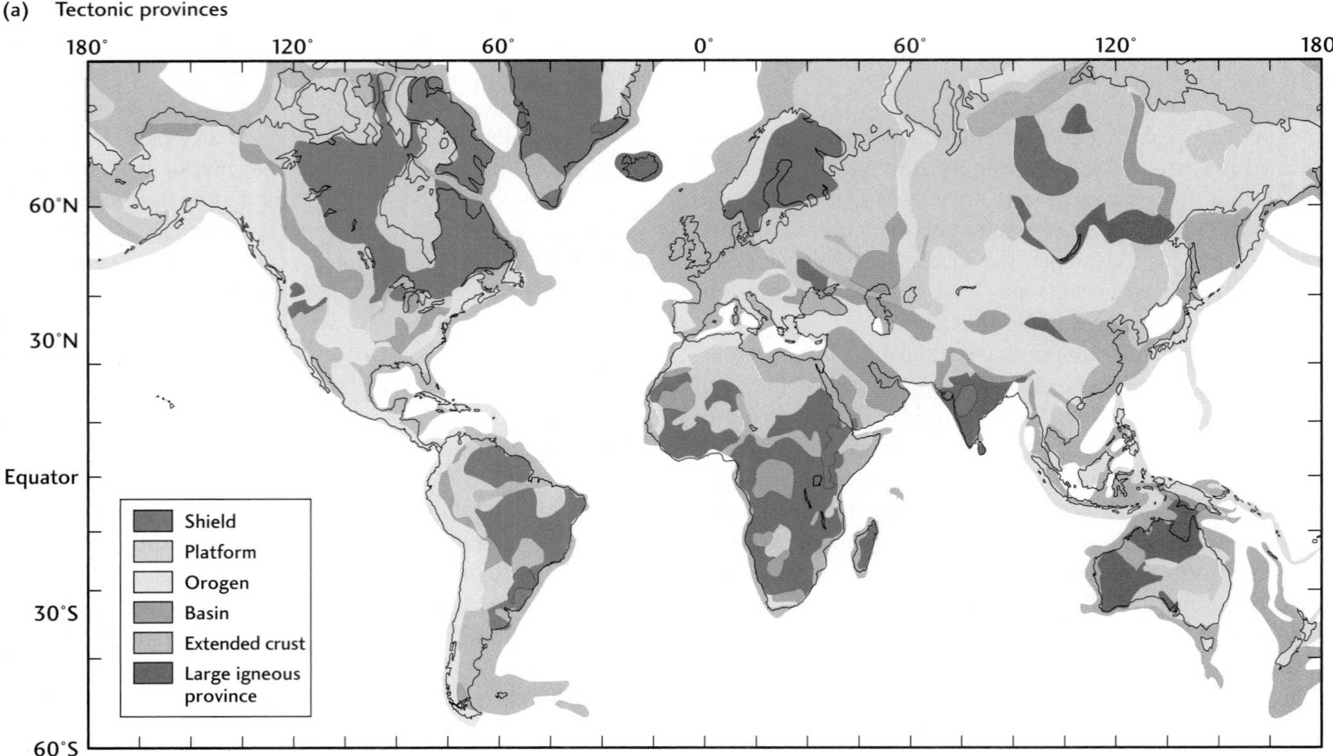

Figure 20.8 Global maps of the continents showing (a) tectonic provinces and (b) tectonic ages. [W. Mooney, U.S. Geological Survey.]

• *Phanerozoic orogen* A region where mountain building has been active during the Phanerozoic. Examples: Appalachian fold belt, Cordillera.

• *Extended crust* A region where the most recent deformation has involved large-scale crustal extension. Examples: Basin and Range province, Atlantic margin.

Tectonic Ages

The tectonic ages shown in Figure 20.8b are grouped according to the geologic eras defined in Chapter 10. What we mean by **tectonic age** deserves some elaboration, however. Most continental basement rocks have survived a long and complex history of repeated deformation, melting, and metamorphism. Geologists can often unravel aspects of this history by using isotopic dating techniques and other age indicators (see Chapter 10), and they can thus assign more than one age to any particular rock. The tectonic ages shown in Figure 20.8b correspond to the last major episode of crustal deformation; that is, to the last time the radiometric "clocks" within the rocks were reset by tectonic activity.

For example, many of the mountain ranges in the southwestern United Statescontain igneous rocks that were originally derived from the melting of crust and mantle 1.9 to 1.6 billion years ago (in the middle Proterozoic) (**Figure 20.9**). However, these rocks were substantially metamorphosed during subsequent periods, including several episodes of compressive deformation in the Mesozoic and rifting in the Cenozoic. Geologists thus assign this region to the youngest age category (Mesozoic-Cenozoic).

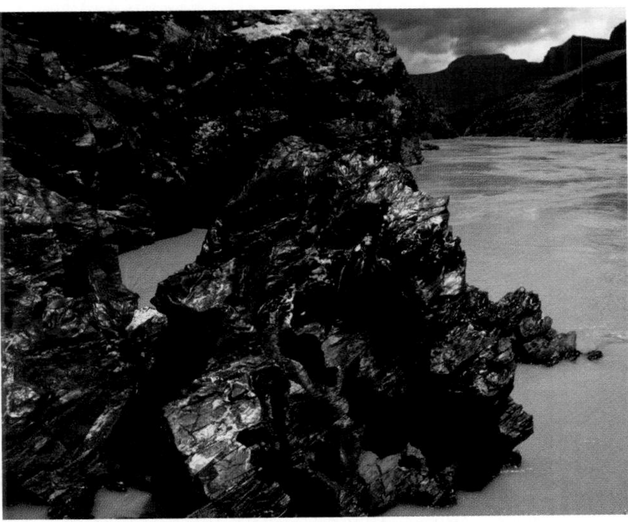

Figure 20.9 The Vishnu schist, part of the middle Proterozoic (1.8 billion years ago) basement found at the base of the Grand Canyon. [Stephen Trimble.]

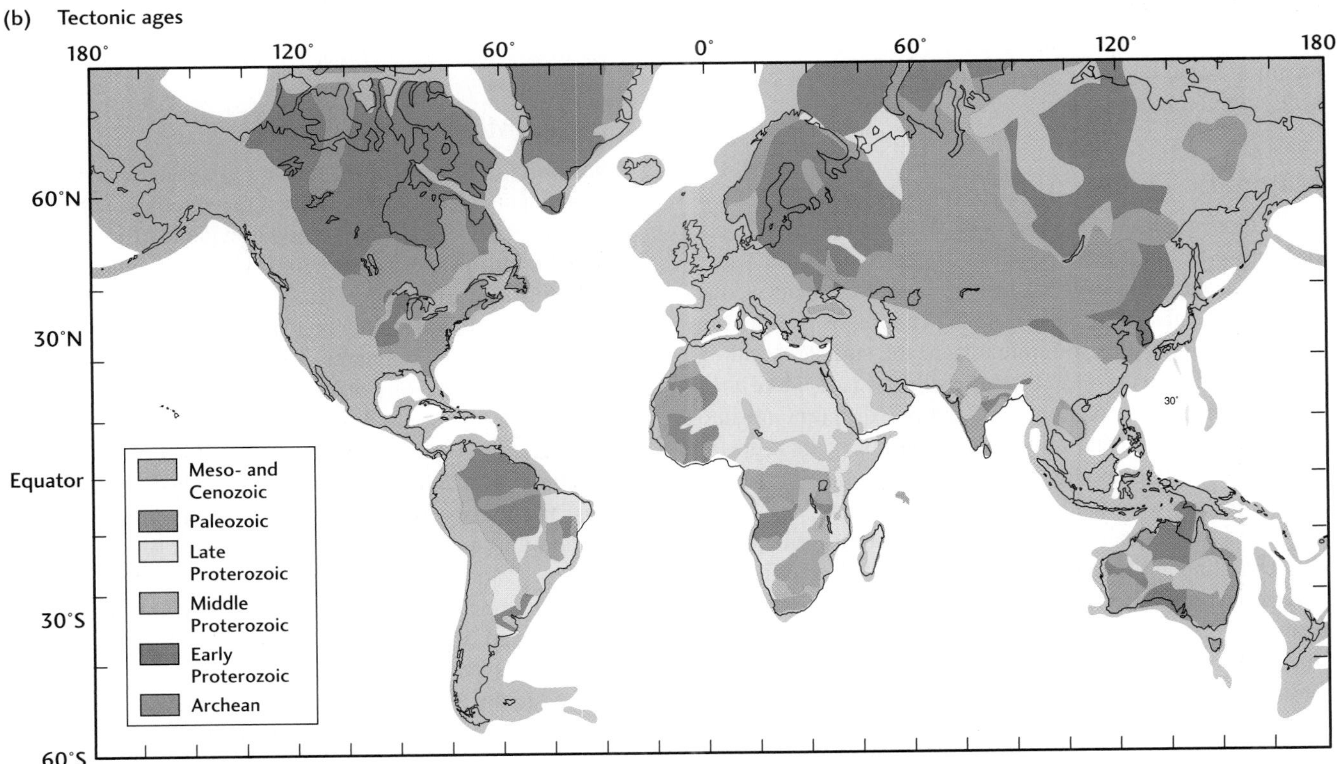

A Global Puzzle

The current distribution of continental provinces and ages represents a giant puzzle whose various pieces have been rearranged by continental rifting, drift, and collision during billions of years of plate tectonics. Only the past 200 million years of these motions are reliably known from existing oceanic crust. Earlier plate motions must be inferred from the indirect evidence found in continental rocks. In Chapter 2, we saw that geologists have made amazing progress in reconstructing earlier configurations of the continents from paleomagnetic and paleoclimate data and from the signatures of deformation exposed in ancient mountain belts.

We will continue the story of the continents back into earlier geologic times. Our focus will be on three key questions of continental evolution: What geologic processes have built the continents we see today? How do these processes fit into the theory of plate tectonics? Can plate tectonics explain the formation of the oldest continents, the Archean cratons? As we will see, these questions have been answered only partially by geologic research.

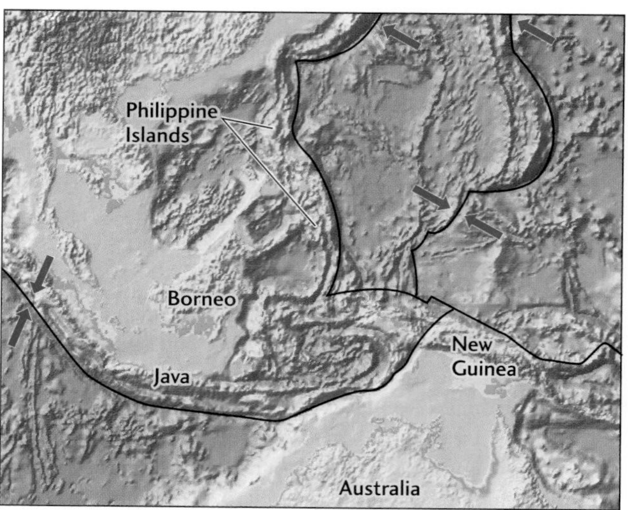

Figure 20.10 The Philippines and other island groups in the southwestern Pacific illustrate how island arcs amalgamate to form protocontinental crust.

How Continents Grow

Over their 4-billion-year history, the continents have grown at an average rate of about 2 km^3 per year. A major debate among geologists concerns whether the growth of continental crust has occurred gradually over geologic time or was concentrated in Earth's early history. In this section, we will examine some of the basic growth processes that are currently operating in the plate tectonic system.

Magmatic Differentiation

In the modern plate tectonic system, the most buoyant, silica-rich crust is born in subduction zones, where the water squeezed out of downgoing slabs melts the mantle wedge above the slabs. The magmas, which are of basaltic to andesitic composition, migrate toward the surface, ponding in magma chambers near the base of the crust. Here they incorporate crustal materials and further differentiate into silica-rich magmas that migrate into the upper crust, forming dioritic and granodioritic plutons capped by andesitic volcanoes (see Chapter 5).

This process can add new crustal material directly to active continental margins. Subduction of the Farallon Plate beneath North America during the Cretaceous period, for example, created the batholiths along the western edge of the continent, including the rocks now exposed in Baja California and the Sierra Nevada. Subduction of the remnant Juan de Fuca Plate continues to add new material to the crust in the volcanically active Cascade Range of the Pacific Northwest, just as subduction of the Nazca Plate is building up the crust in the Andes Mountains of South America.

Buoyant crust is also produced far away from continents, in island arcs where one oceanic plate subducts beneath another. Over time, island arcs can amalgamate into thick sections of "protocontinental" crust, such as those found in the Philippines and other island groups of the southwestern Pacific (**Figure 20.10**). Plate motions eventually attach this crust to active continental margins through accretion.

Continental Accretion

Accretion is a process of continental growth in which buoyant fragments of crust are attached (accreted) to continents during plate motions. The geologic evidence for accretion can be found on the active margins of continents such as North America. In the Pacific Northwest and Alaska, the crust consists of an amalgamation of odd pieces—island arcs, seamounts and remnants of thickened oceanic plateaus, old mountain ranges and other slivers of continental crust—that were plastered onto the leading edge of the continent as it moved across Earth's surface (**Figure 20.11**). The individual, geologically coherent fragments are called **accreted terranes.** Geologists use the term *terrane* (a variation of the more familiar term *terrain,* meaning a tract of land and its topography) to signify a large piece of crust, tens to hundreds of kilometers in geographic extent, with shared characteristics and a common origin, usually transported great distances by plate movements.

The geologic arrangement of accreted terranes can be very chaotic. Adjacent provinces might contrast sharply in the assemblages of rock types, the nature of folding and faulting, and the history of magmatism and metamorphism. Geologists often find fossils indicating that these blocks orig-

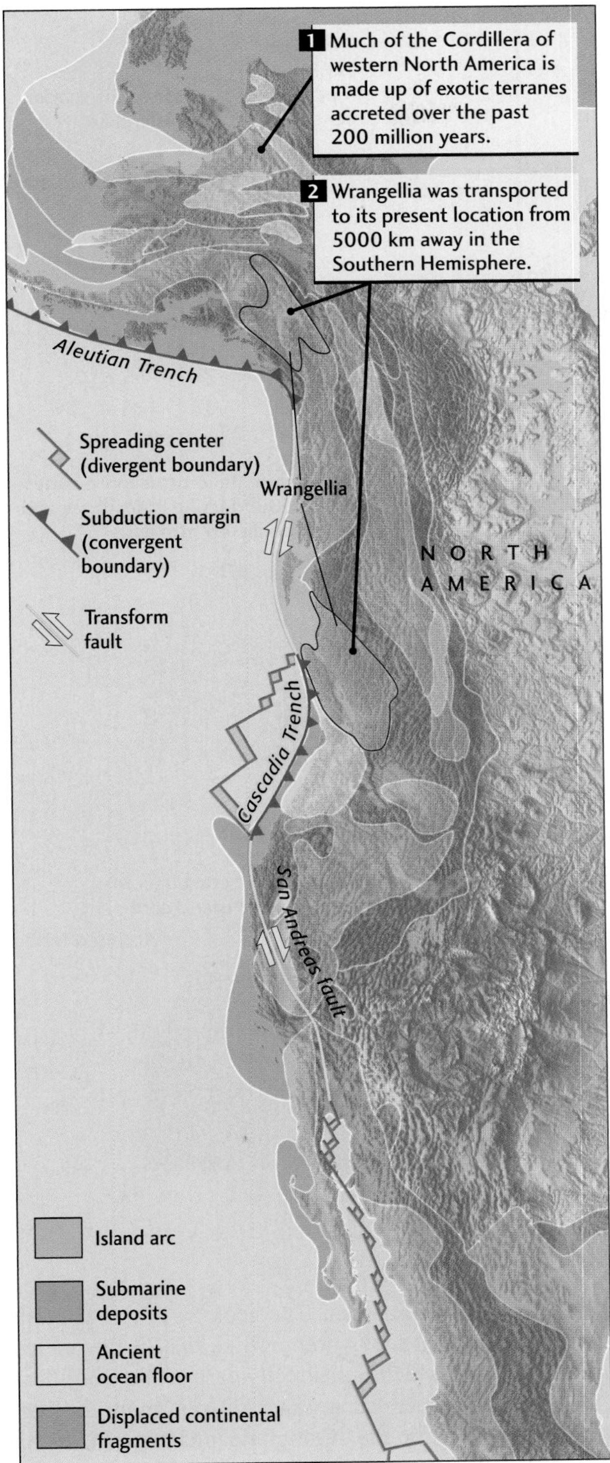

Figure 20.11 Terranes accreted to western North America in the past 200 million years. They are made up of island arcs, ancient seafloor crust (ophiolites, ultramafics), continental fragments, and submarine deposits (e.g., forearc basin sediments), as indicated by different colors in the map. [After D. R. Hutchison, "Continental Margins." *Oceanus* 35 (Winter 1992–1993): 34–44; modified from work of D. G. Howell, G. W. Moore, and T. J. Wiley.]

inated in different environments and at different times from those of the surrounding area. For example, a terrane comprising ophiolites (rocks characteristic of oceanic crust) that contain deep-water fossils might be surrounded by remnants of island arcs and continental fragments containing shallow-water fossils of a completely different age. The boundaries between terranes are almost always major faults that have accommodated substantial slip (although the nature of the faulting is often difficult to discern). Fragments that seem completely out of place are called *exotic terranes*.

Before plate tectonics, exotic terranes were the subject of a fierce debate among geologists, who had difficulty coming up with reasonable explanations of their origins. In recent years, the study of these features, called terrane analysis, has become a specialized field within plate tectonics research. Well over 100 areas of the Cordillera of western North America have been identified as exotic terranes accreted during the last 200 million years (many more than depicted in Figure 20.11). One such terrane, formerly a large oceanic plateau now called Wrangellia, appears to have been transported over 5000 km from the Southern Hemisphere to its current location in Alaska and western Canada (see Figure 20.11). Extensive accreted terranes have also been mapped in Japan, Southeast Asia, China, and Siberia.

In only a few cases have geologists been able to work out precisely where these terranes originated, but they have begun to decipher how plate motions have brought these pieces together. Four mechanisms appear to be important (**Figure 20.12**):

1. The transfer of a crustal fragment from a subducting plate to the overriding plate, which can occur when the fragment is too buoyant to be subducted. The fragments can be small pieces of continent ("microcontinents") or thickened sections of oceanic crust (large seamounts, oceanic plateaus).

2. The closure of a marginal sea that separates an island arc from a continent. A collision with the advancing edge of the continent can attach the thickened island-arc crust to the continent.

3. The transport of terranes laterally along continental margins by strike-slip faulting. Today, the southwestern part of California attached to the Pacific Plate is moving northwestward relative to the North American Plate along the San Andreas transform fault. Strike-slip faulting landward of the trench in oblique subduction zones can also transport terranes for hundreds of kilometers.

4. The suturing of two continental margins by continent-continent collision and the subsequent breakup of this zone by continental rifting.

The fourth mechanism explains some of the accreted terranes found on the passive eastern margin of North America. The Appalachian fold belt, ranging from

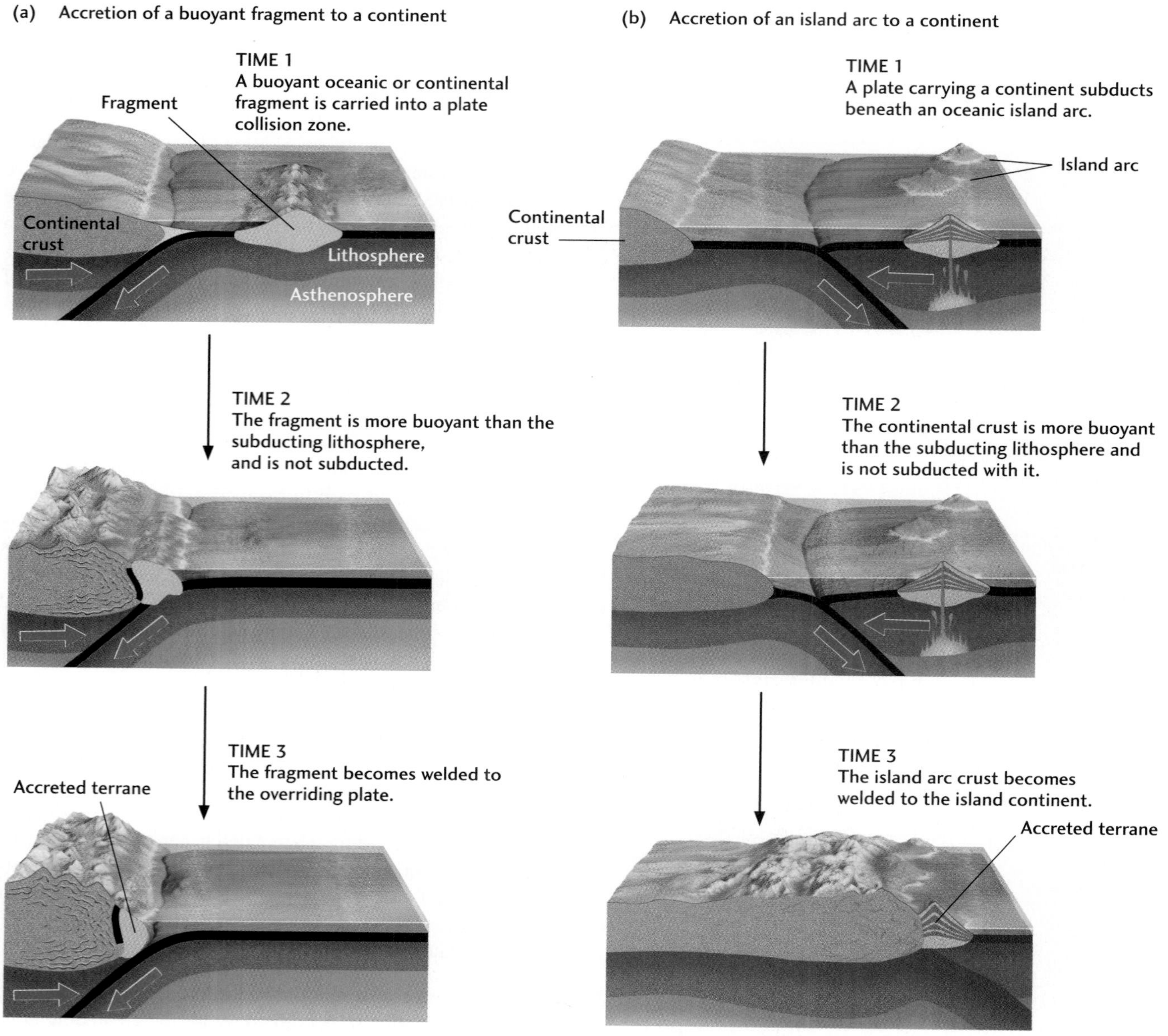

(a) Accretion of a buoyant fragment to a continent

TIME 1
A buoyant oceanic or continental fragment is carried into a plate collision zone.

Fragment

Continental crust

Lithosphere

Asthenosphere

TIME 2
The fragment is more buoyant than the subducting lithosphere, and is not subducted.

TIME 3
The fragment becomes welded to the overriding plate.

Accreted terrane

(b) Accretion of an island arc to a continent

TIME 1
A plate carrying a continent subducts beneath an oceanic island arc.

Island arc

Continental crust

TIME 2
The continental crust is more buoyant than the subducting lithosphere and is not subducted with it.

TIME 3
The island arc crust becomes welded to the island continent.

Accreted terrane

Figure 20.12 Four mechanisms for the accretion of exotic terranes.

Newfoundland to the southeastern United States, contains slices of ancient Europe and Africa, as well as a variety of exotic terranes. Florida's oldest rocks and fossils are more like those in Africa than like those found in the rest of the United States, indicating that most of this peninsula was probably a piece of Africa left behind when North America and Africa split apart about 200 million years ago.

How Continents Are Modified

The Cordilleran region of western Canada comprises many exotic terranes accreted during the drift of North America since the breakup of Pangaea. The geology of this youthful part of the continent looks nothing like that of the ancient Canadian Shield, which lies directly east of the Cordillera. In particular, the accreted terranes do not show the high degree of melting or the high-grade metamorphism that characterizes the Precambrian crust of the shield. Why such a difference? The answer lies in the tectonic processes that have repeatedly modified the older parts of the continent throughout its long history.

Orogeny: Modification by Plate Collision

The continental crust is profoundly altered by **orogeny**—the mountain-building processes of folding, faulting, magma-

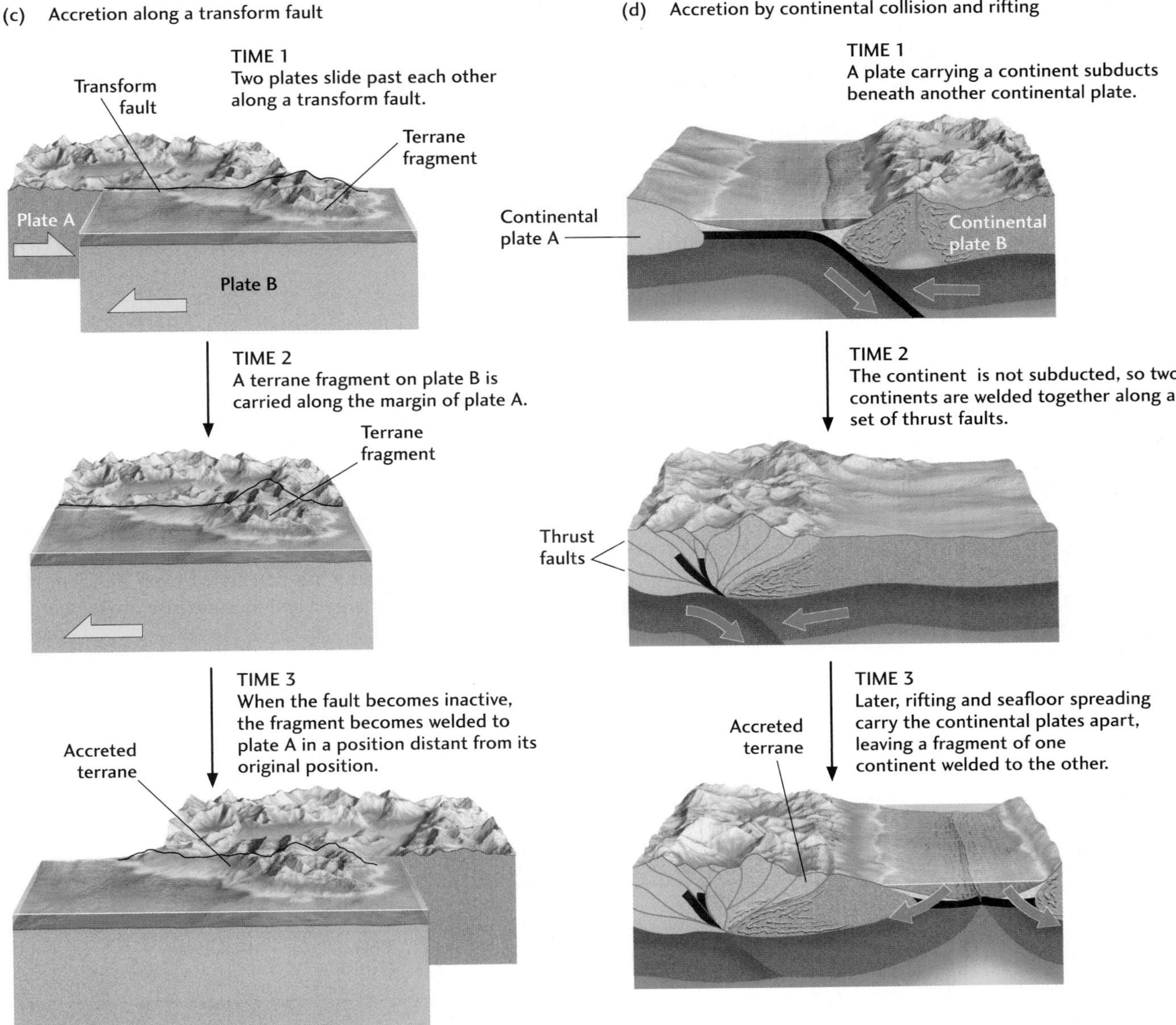

tism, and metamorphism. Most periods of mountain building (orogenies) involve plate convergence. When one or both plates are made of oceanic lithosphere, the convergence is primarily taken up by subduction. Orogenies can result when a continent rides forcefully over a subducting oceanic plate, such as the Andean orogeny now under way in South America, but the most intense orogenies are caused by the convergence of two or more continents. As we observed in Chapter 2, when two continental plates collide, a basic tenet of plate tectonics—the rigidity of plates—must be modified.

Continental crust is much more buoyant than the mantle, so colliding continents resist being subducted with the plates that carry them. Instead, the continental crust deforms and breaks in a combination of intense folding and faulting that can extend hundreds of kilometers from the collision. Low-angle thrust faulting caused by the convergence can stack the upper part of the crust into multiple thrust sheets tens of kilometers thick, deforming and metamorphosing the rocks they contain (**Figure 20.13**). Wedges of continental-shelf sediments can be detached from the basement on which they were deposited and be thrust inland. Compression throughout the crust can double its thickness, causing the rocks in the lower crust to melt. This melting can generate huge amounts of granitic magma, which rises to form extensive batholiths in the upper crust.

The Alpine-Himalayan Orogeny To see orogeny in action, we look to the great chains of high mountains that stretch

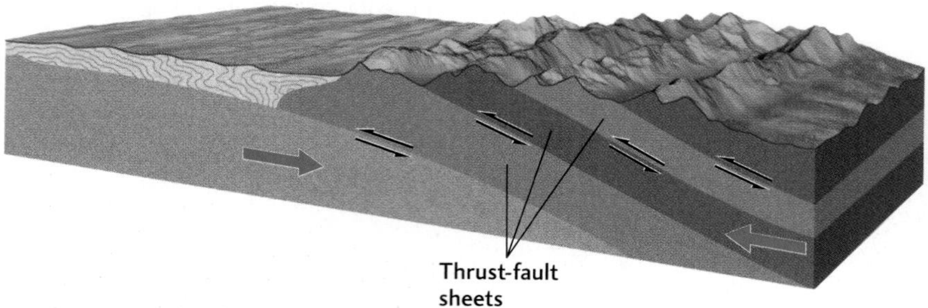

Figure 20.13 When plates bearing continents collide, the continental crust can break into multiple thrust-fault sheets stacked one above the other.

from Europe through the Middle East and across Asia, known collectively as the *Alpine-Himalayan belt* (**Figure 20.14**) The breakup of Pangaea sent the continental crust of Africa, Arabia, and India northward, causing the Tethys Ocean to close as its lithosphere was subducted beneath Eurasia (see Figure 2.15). These former pieces of Gondwanaland collided with Eurasia in a complex sequence, beginning in the western part of Eurasia during the Cretaceous period and continuing eastward through the Tertiary, raising the Alps in central Europe, the Caucasus and Zagros mountains in the Middle East, and the Himalayas and other high mountain chains across central Asia.

The Himalayas, the world's highest mountains, are the most spectacular result of this modern episode of continent-continent collision. About 50 million years ago, the Indian subcontinent, riding on the subducting Indian Plate, first encountered the island arcs and continental volcanic belts that then bounded the Eurasian Plate (**Figure 20.15**). As the landmasses of India and Eurasia merged, the Tethys Ocean disappeared through subduction. Pieces of the oceanic crust were trapped along the suture zone between the converging continents and can be seen today as ophiolites along the Indus and Tsangpo river valleys that separate the high Himalayas from Tibet (see Figure 18.7). The collision slowed India's advance, but the plate continued to drive northward. So far, India has penetrated over 2000 km into Eurasia, causing the largest and most intense orogeny of the Cenozoic era.

The Himalayas were formed from overthrust slices of the old northern portion of India, stacked one atop the other. This process took up some of the compression. Horizontal compression also thickened the crust north of India, causing

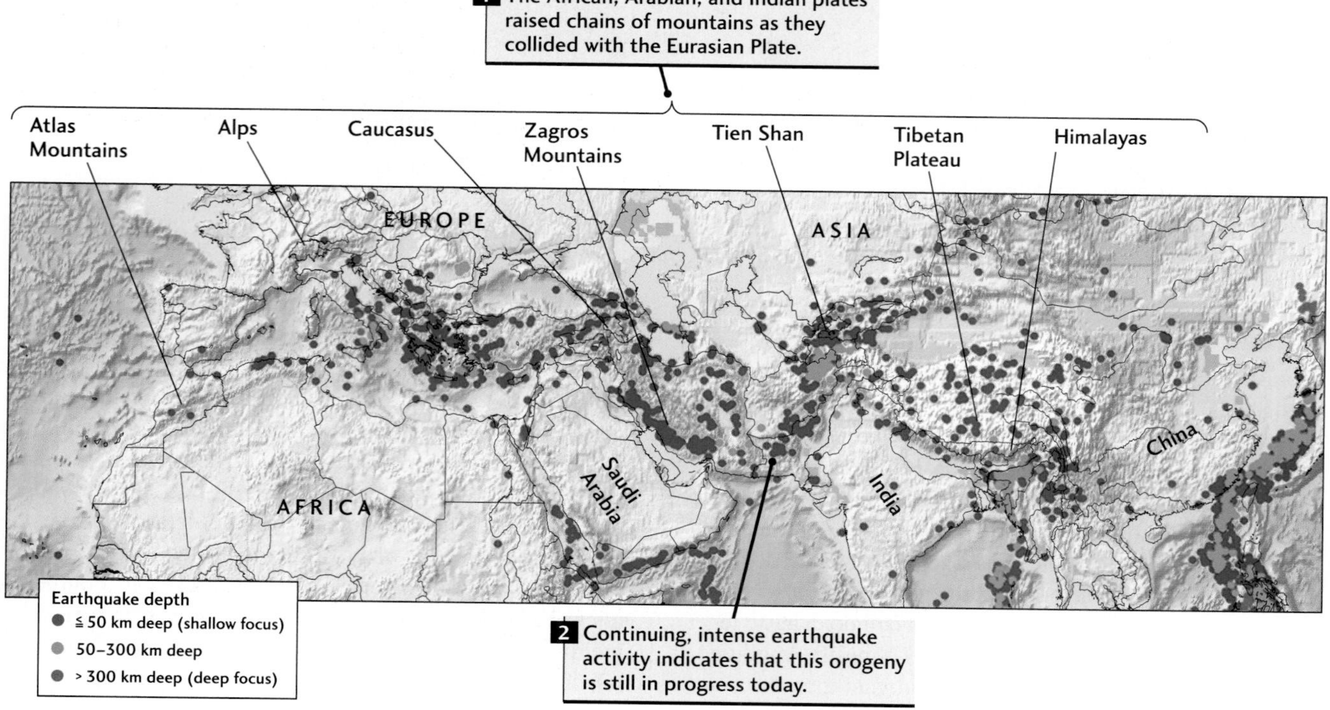

Figure 20.14 The Alpine-Himalayan belt showing the chains of high mountains built by the ongoing collision of the African, Arabian, and Indian plates with the Eurasian Plate.

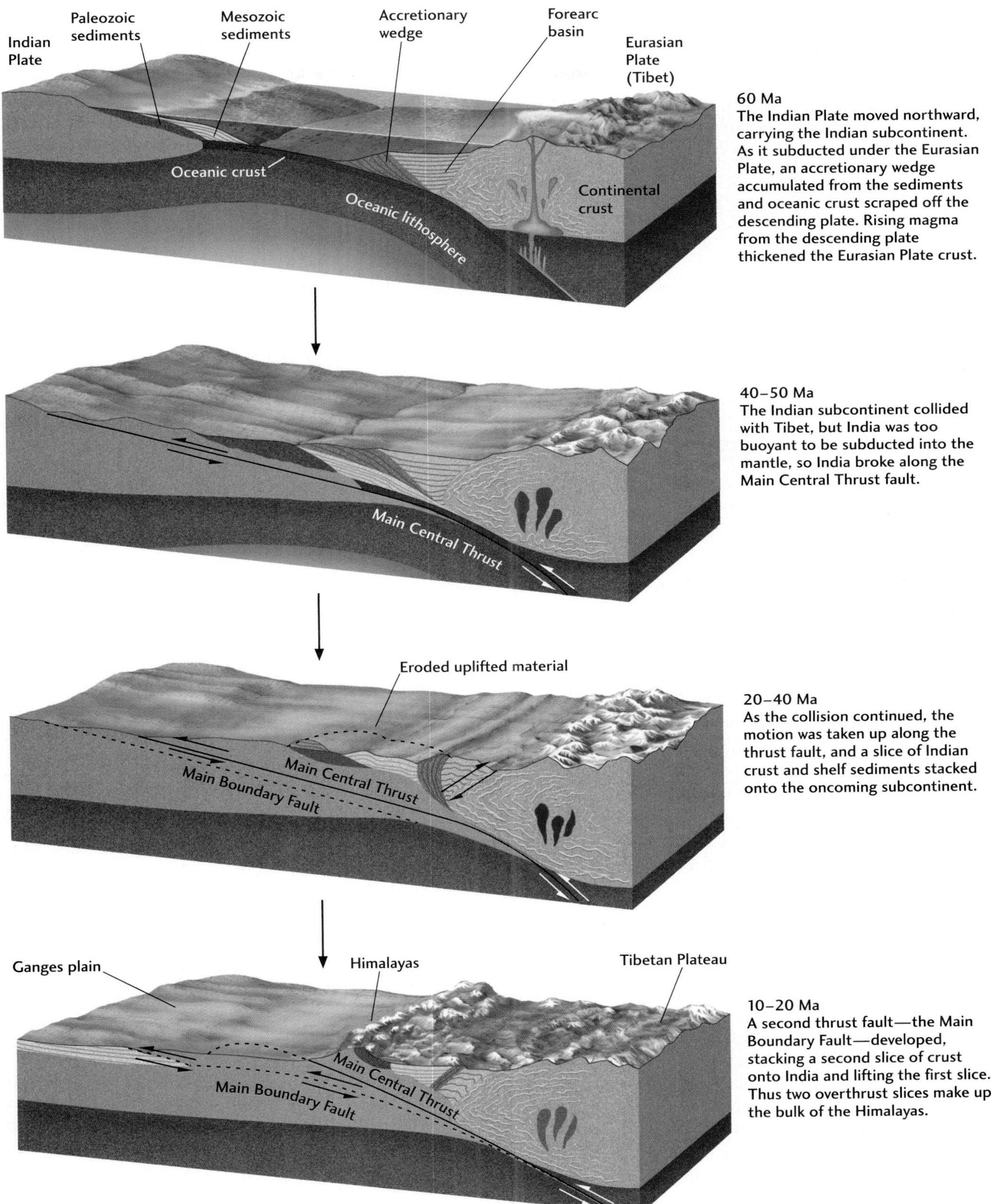

Figure 20.15 Cross sections showing the sequence of events that have caused the Himalayan orogeny, simplified and vertically exaggerated. [After P. Molnar, "The Structure of Mountain Ranges." *Scientific American,* July 1986, p. 70.]

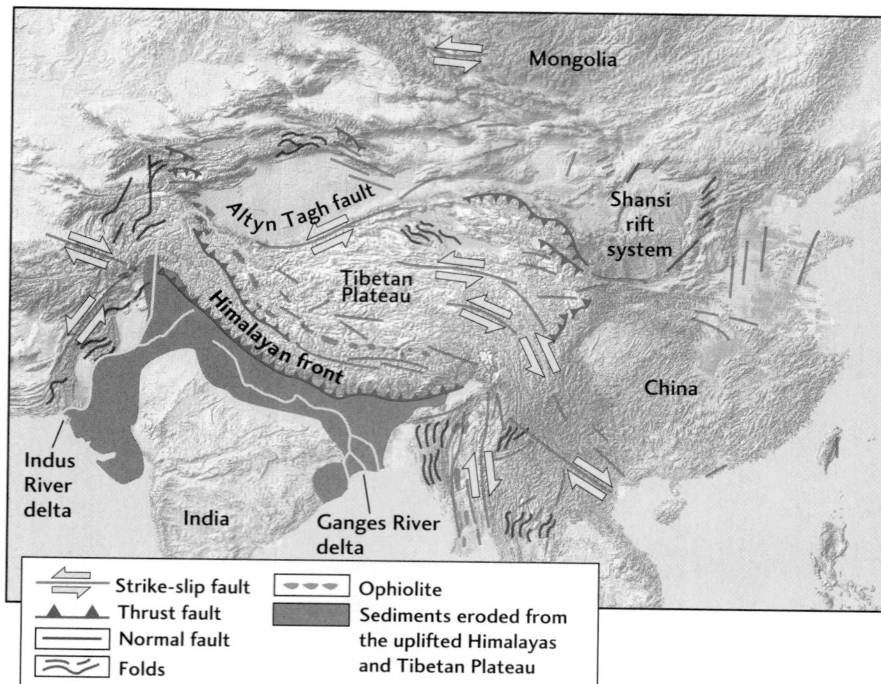

Figure 20.16 Tectonic features associated with the collision between India and Eurasia: large-scale faulting and uplift. [After P. Molnar and P. Tapponier, "The Collision Between India and Eurasia." *Scientific American* (April 1977): 30.]

uplift of the huge Tibetan Plateau, which now has a crustal thickness of 60 to 70 km (almost twice the thickness of normal continental crust) and stands nearly 5 km above sea level. These and other zones of compression account for perhaps half of India's penetration into Eurasia. The other half has been accommodated by pushing China and Mongolia eastward, out of India's way, like toothpaste squeezed from a tube. The movement took place along the Altyn Tagh fault and other major strike-slip faults shown on the map in **Figure 20.16**. The mountains, plateaus, faults, and great earthquakes of Asia, thousands of kilometers from the Indian-Eurasian suture, are thus affected by the Himalayan orogeny, which continues as India ploughs into Asia at a rate of 40 to 50 mm/year.

Paleozoic Orogenies During the Assembly of Pangaea
If we go further back in geologic time, we find abundant evidence of older orogenies caused by earlier episodes of plate convergence. We have already mentioned, for example, that at least three distinct orogenies were responsible for the Paleozoic deformation now exposed in the eroded Appalachian fold belt of the eastern United States. These three periods of mountain building were related to plate tectonic events that led to the assembly of the Pangaean supercontinent near the end of the Paleozoic.

The supercontinent of Rodinia began to break up toward the end of the Proterozoic eon, spawning several paleocontinents (see Figure 2.15). One was the large continent of *Gondwana.* Two of the others were *Laurentia,* which included the North America craton and Greenland, and *Baltica,* comprising the lands around the Baltic Sea (Scandinavia, Finland, and the European part of Russia). In the Cambrian period, Laurentia was rotated almost 90º from its present orientation and straddled the equator; its southern (today, eastern) side was a passive continental margin. To its immediate south was the proto-Atlantic or *Iapetus Ocean* (in Greek mythology, Iapetus was the father of Atlantis), which was being subducted beneath a distant island arc. Baltica lay off to the southeast, and Gondwana was thousands of kilometers to the south. **Figure Story 20.17** shows the sequence of events that ensued as the three continents converged.

The island arc built up by the southward-directed subduction of Iapetus lithosphere collided with Laurentia in the middle to late Ordovician (470 to 440 million years ago), causing the first episode of mountain building—the *Taconic orogeny.* (You can see some of the rocks accreted and deformed during this period when you drive the Taconic Parkway, which runs east of the Hudson River for about 100 miles north of New York City.) The second episode was initiated when Baltica and a connected set of island arcs began to collide in the early Devonian (about 400 million years ago). The convergence of Baltica with Laurentia deformed southeastern Greenland, northwestern Norway, and Scotland in what European geologists refer to as the *Caledonian orogeny.* The deformation continued into present-day North America as the *Acadian orogeny,* when

Figure Story 20.17 Paleogeographic reconstructions of the North Atlantic region, showing the sequence of orogenic events that resulted from the assembly of Pangaea. [Ronald C. Blakey, Northern Arizona University, Flagstaff.]

MULTIPLE OROGENIES RESULTED FROM THE ASSEMBLY OF PANGAEA

Middle Cambrian (510 Ma)
After the breakup of Rodinia, the continent of Laurentia straddled the equator, and its southern side was a passive continental margin, bounded on the south by the Iapetus Ocean.

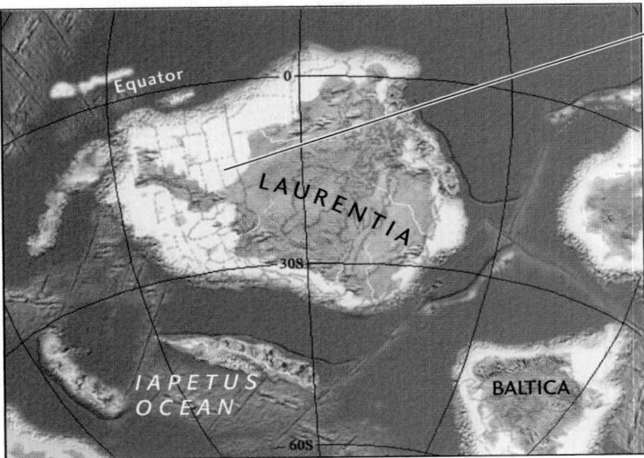

Late Ordovician (450 Ma)
The island arc built up by the southward-directed subduction of Iapetus lithosphere collided with Laurentia in the middle to late Ordovician, causing the Taconic orogeny.

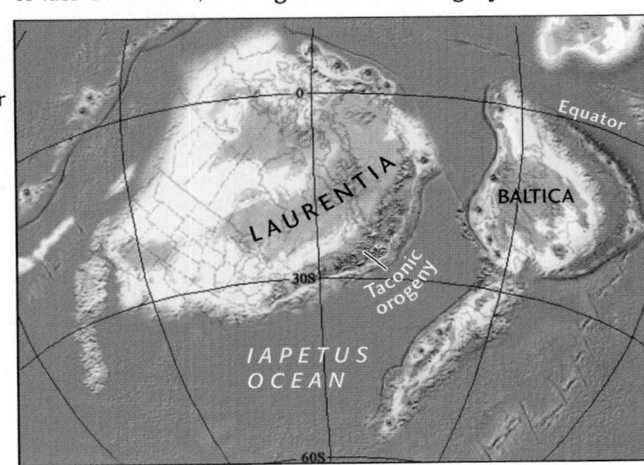

Early Devonian (400 Ma)
The collision of Laurentia with the continent of Baltica caused the Caledonian orogeny and formed Laurussia. The southward continuation of the convergence caused the Acadian orogeny.

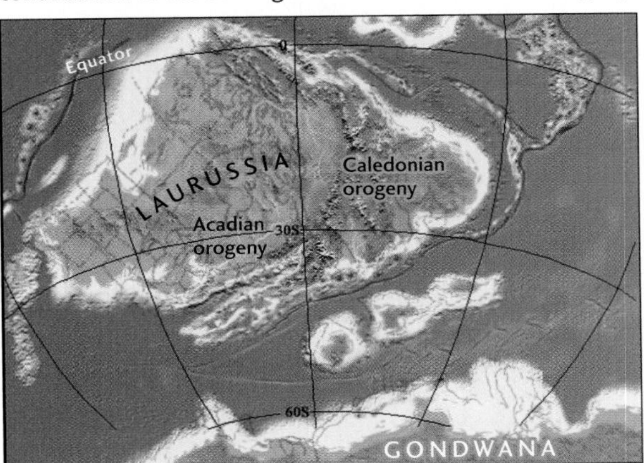

Late Mississippian (340 Ma)
The collision of Gondwana with Laurussia began with the Variscan orogeny in what is now central Europe...

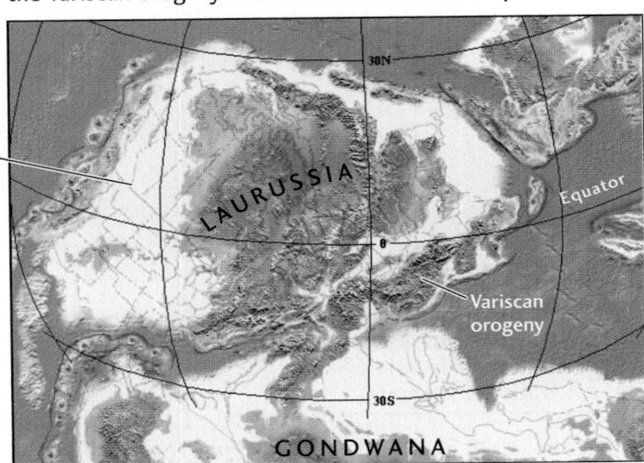

Upper Pennsylvanian (300 Ma)
... and continued along the margin of the North American craton with the Appalachian orogeny. During this terminal cataclysm, Siberia converged with Laurussia in the Ural orogeny to form Laurasia, while the Hercynian orogeny created new mountain belts across Europe and northern Africa.

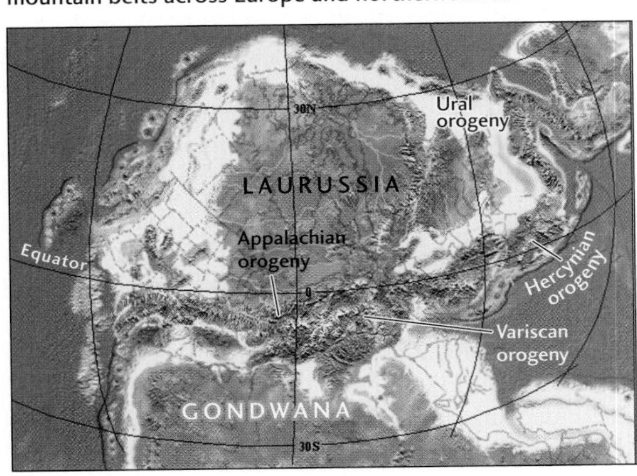

Early Permian (270 Ma)
The end product of these episodes of continental convergence was the supercontinent of Pangaea.

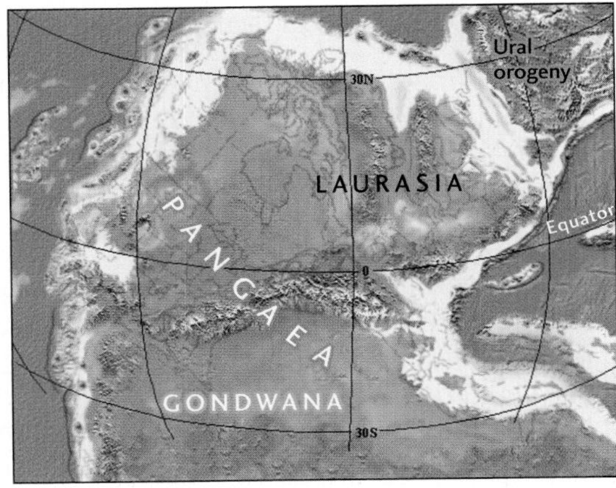

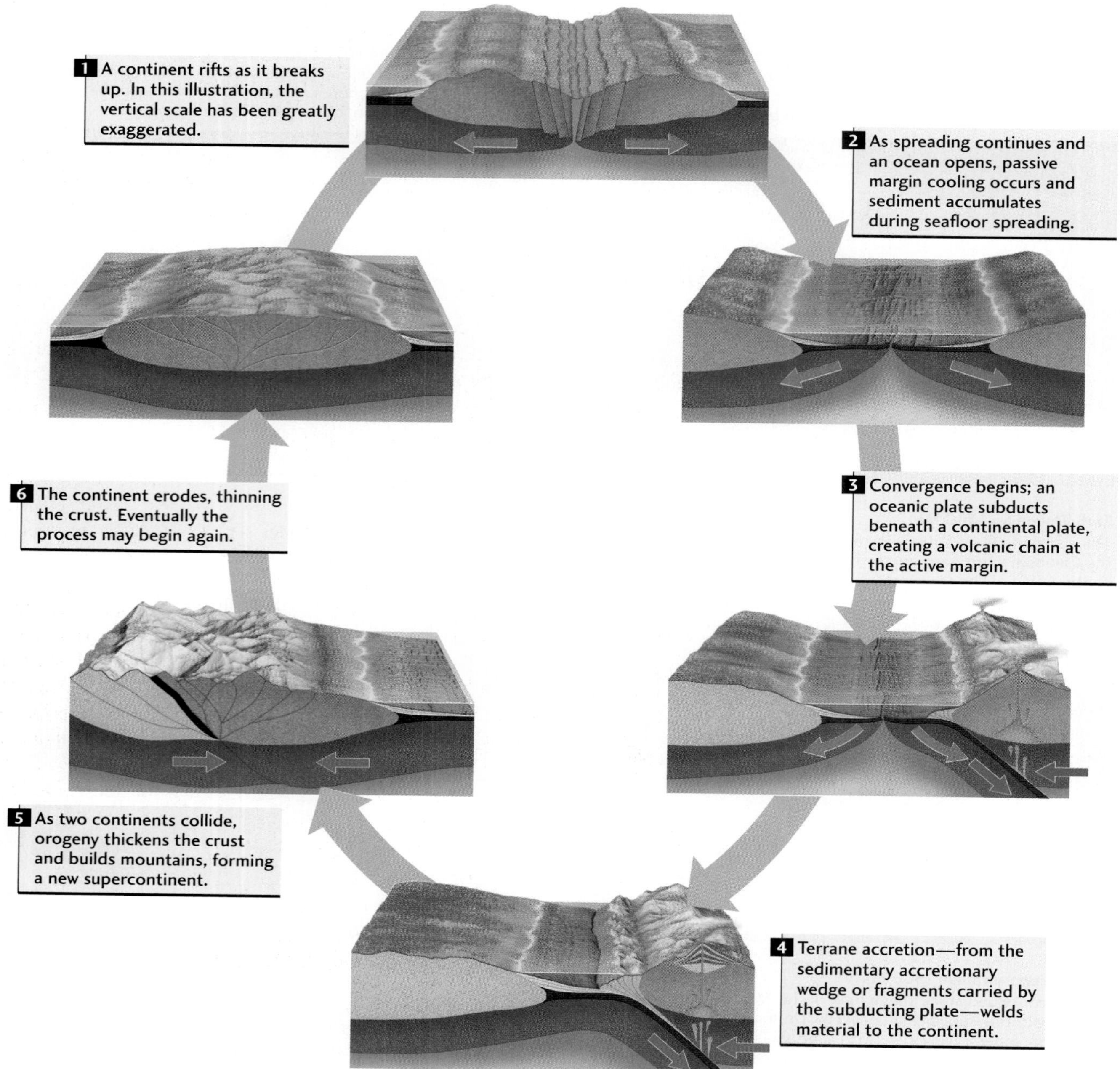

Figure 20.18 Schematic cross sections showing the major phases of the Wilson cycle.

island arcs added the terranes of maritime Canada and New England to Laurentia in the middle to late Devonian (380 to 350 million years ago).

The grand finale in the assembly of Pangaea was the collision of the behemoth landmass of Gondwana with Laurasia and Baltica, by then joined into a continent named *Laurussia* (see Figure Story 20.17). It began around 340 million years ago with the *Variscan orogeny* in what is now central Europe and continued along the margin of the North American craton with the *Alleghenian orogeny* (320 to 270 million years ago). This latter phase of assembly thrust Gondwana crust over Laurentia, lifting the Blue Ridge into a mountain chain perhaps as high as the modern Himalayas and causing much of the deformation now seen in the Appalachian fold belt. Also during this terminal cataclysm, Siberia and other Asian terranes converged with Laurussia, forming the Ural Mountains (and thus *Laurasia*), while extensive deformation created new mountain belts across Europe and northern Africa (the *Hercynian orogeny*).

The crunching together of all these continental masses profoundly altered the structure of the crust. The rigid cratons were little affected, but the younger accreted terranes caught in between were consolidated, thickened, and metamorphosed. The lower parts of this juvenile crust were partially melted, producing granitic magmas that rose to form batholiths in the upper crust and volcanoes at the surface. Uplifted mountains and plateaus were eroded, exposing high-grade metamorphic rocks that were once many kilometers deep and depositing thick sedimentary sequences. Sediments laid down following the first orogeny were deformed and metamorphosed by later mountain-building episodes.

Earlier Orogenies So far, we have investigated two major periods of mountain building: the Paleozoic orogenies associated with the assembly of Pangaea and the Cenozoic continent-continent collisions across the Alpine-Himalayan belt. In Chapter 2, we discussed the late Proterozoic supercontinent of Rodinia. By now, it should not surprise you to learn that major orogenies accompanied the formation of this earlier supercontinent.

Some of the best evidence comes from the eastern and southern margin of the Canadian Shield in a broad belt known as the Grenville province, where new crustal material was added to the continent about 1.1 to 1.0 billion years ago (see Figure 20.8). Geologists believe that these rocks, which are now highly metamorphosed, originally consisted of continental volcanic belts and island-arc terrains that were accreted and compressed by the collision of Laurentia with the western part of Gondwana. They have drawn analogies between what happened during this *Grenville orogeny* and what is happening today in the Himalayan orogeny. A Tibet-like plateau was formed by compressive thickening of the crust through folding and thrust faulting, which metamorphosed the upper crust and partially melted large parts of the lower crust. Once the orogeny ceased, erosion of the plateau thinned the crust and exposed crystalline rocks of high metamorphic grade. Geologists have found orogenic belts of similar age on continents worldwide, and although many of the details remain uncertain, they have reconstructed from this evidence (which includes paleomagnetic data) a general picture of how Rodinia came together between 1.3 and 0.9 billion years ago.

From our brief look at the history of eastern North America, we can infer that the edges of many cratons have experienced multiple episodes in a general plate tectonic cycle that comprises four main phases (**Figure 20.18**):

1. Rifting during the breakup of a supercontinent
2. Passive-margin cooling and sediment accumulation during seafloor spreading and ocean opening
3. Active-margin volcanism and terrane accretion during subduction and ocean closure
4. Orogeny during the continent-continent collision that forms the next supercontinent

Geologists refer to this idealized sequence of events in the opening and closing of ocean basins as the **Wilson cycle,** named after the Canadian pioneer of plate tectonics, J. Tuzo Wilson, who first recognized its importance for the evolution of continents.

The geologic data suggest that the Wilson cycle has operated throughout the Proterozoic and Phanerozoic eons (**Figure 20.19**). Based on the geochronology of ancient rock formations, geologists have postulated the existence of at least two episodes of supercontinent formation prior to Rodinia, one at 1.9 to 1.7 billion years ago (a supercontinent named *Columbia*) and an even earlier assembly at 2.7 to 2.5 billion years ago. The latter date marks the transition from the Archean eon to the Proterozoic eon. Did the Wilson cycle also operate in the Archean eon? We will return to the question of Archean continental evolution shortly.

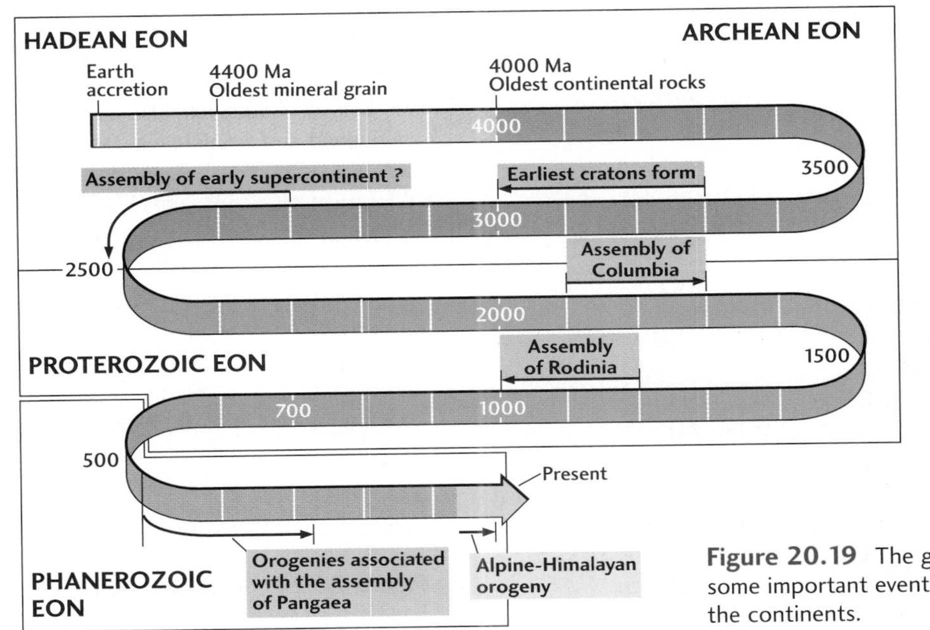

Figure 20.19 The geochronology of some important events in the history of the continents.

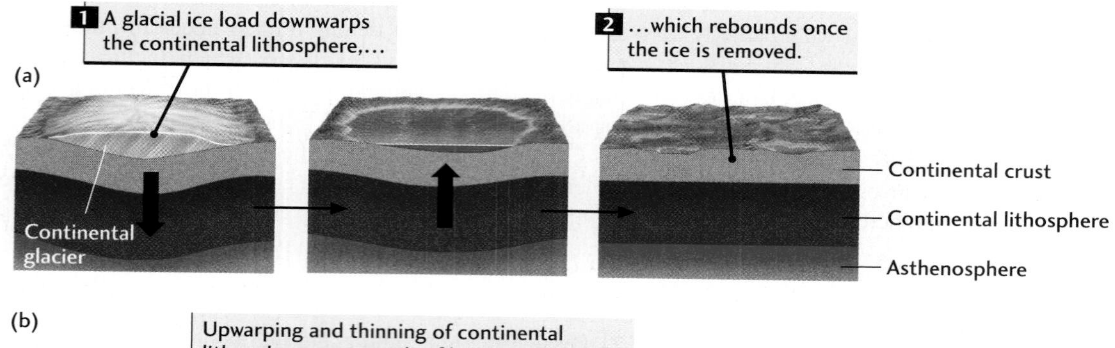

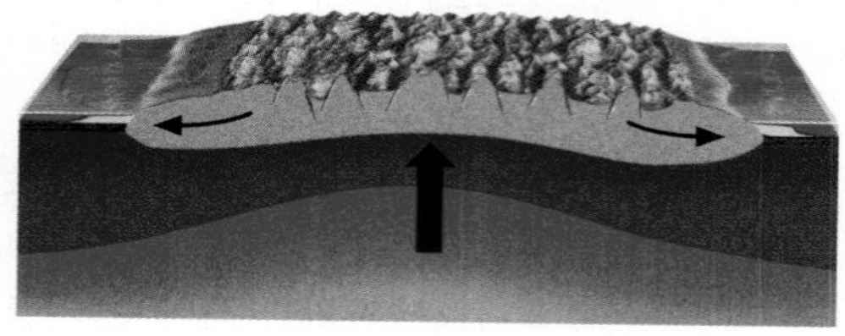

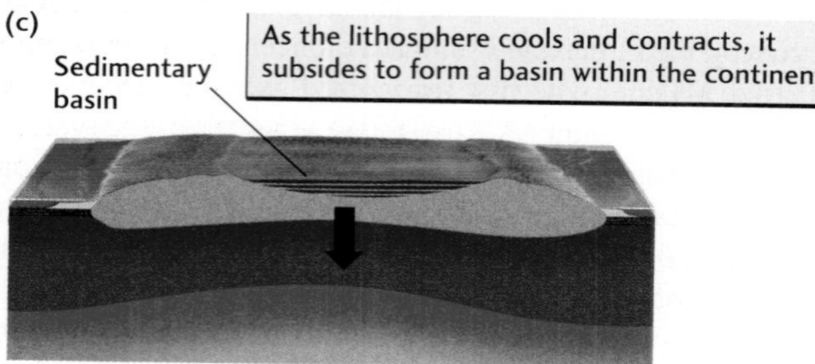

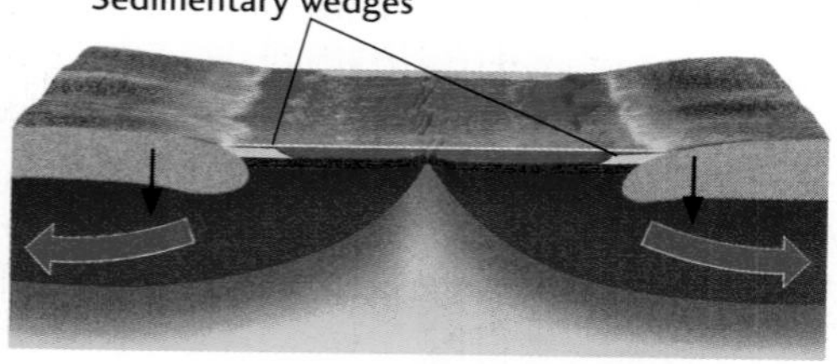

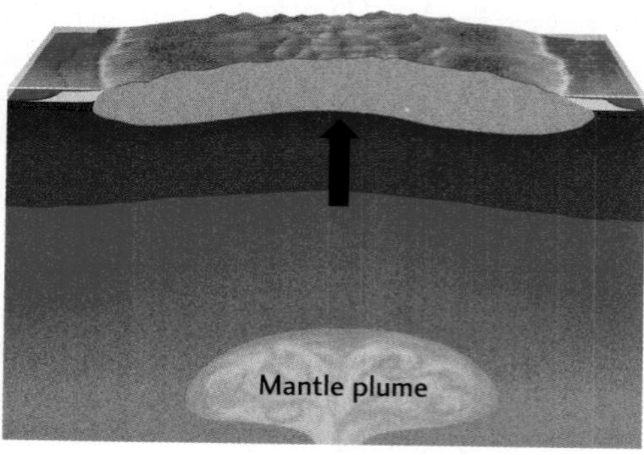

Epeirogeny: Modification by Vertical Motions

So far, our consideration of continental evolution has emphasized accretion and orogeny, processes that involve horizontal plate motions and are usually accompanied by deformation in the form of folding and faulting. Throughout the world, however, sedimentary rock sequences record another kind of motion that has modified the continents: gradual downward and upward movements of broad regions of crust without significant folding or faulting. These vertical motions involve a set of processes called **epeirogeny,** a term coined in 1890 by the American geologist Clarence Dutton (from the Greek *epeiros,* meaning "mainland").

Epeirogenic downward movements usually result in a sequence of relatively flat-lying sediments, such as those found in the stable interior platform of North America. Upward movements cause erosion and gaps in the sedimentary record seen as unconformities. Erosion can lead to the exposure of crystalline basement rocks, such as those found in the Canadian Shield.

Geologists have identified several causes for epeirogenic movements. One example is glacial rebound (**Figure 20.20**a). The weight of large glaciers depresses the continental crust during ice ages. When they melt, the crust rebounds upward for tens of millennia. Glacial rebound explains the uplift of Finland and Scandinavia (see Chapter 21) and the raised beaches of northern Canada (**Figure 20.21**) following the Wisconsin glaciation, which ended about 17,000 years ago. Although rebound seems slow by human standards, it is a rapid process, geologically speaking.

Cooling and heating of the continental lithosphere are important causes of epeirogenic movements on longer time scales. Heating causes rocks to expand, decreasing their density and thus raising the surface (Figure 20.20b). A good example is the Colorado Plateau, which has been uplifted to about 2 km during the last 10 million years or so. Geologists

Figure 20.20 Some proposed mechanisms for vertical crustal movements (epeirogeny) (not to scale).

Figure 20.21 Raised beaches, evidence of upward recovery of the crust after removal of glacial load. Baffin Island, Northwest Territories, Canada. [L. M. Cumming, Geological Survey of Canada.]

think this heating results from an active mantle upwelling (which is also stretching the crust in the Basin and Range province on the western and southern sides of the plateau). The uplift has caused the Colorado River to cut deeply into the plateau, forming the Grand Canyon.

Conversely, the cooling of the lithosphere after heating and rifting increases the density of the lithosphere, making it sink under its own weight to create a sedimentary basin (Figure 20.20c). Cooling of a once-hot area may explain the Michigan and Illinois basins and other deep basins in central North America (see Figure 20.3). When a new episode of seafloor spreading splits a continent apart, the uplifted edges are eroded and eventually subside as they cool, allowing sediments to be deposited and carbonate platforms to accumulate on the passive margins (Figure 20.20d). This process has led to the formation of a thick sedimentary wedge along the east coast of the United States.

One puzzle that intrigues geologists is the South African Plateau, where a craton has been uplifted during the Cenozoic to heights of almost 2 km above sea level—more than twice the elevation of most cratons. The lithosphere in this part of the continent does not appear to be anomalously hot. Some scientists have therefore proposed that forces originating deep within the mantle may be responsible for the uplift of the southern African craton and other broad upwarps. They postulate that a hot and buoyant solid plume rising a few centimeters per year from deep within the mantle (see Chapter 6) applies upward forces at the base of the lithosphere, sufficient to raise the surface by about a kilometer (Figure 20.20e).

None of the proposed mechanisms for epeirogeny explains a central feature of continental cratons: the existence of raised continental shields and subsided platforms. These regions are too vast and have persisted too long for their epeirogenesis to have been caused by the plate tectonic processes we have discussed so far. And there are other mysteries.

The Formation of Cratons

The Wilson cycle of plate tectonics (see Figure 20.18) describes how protocontinental crust can be generated by volcanism on active margins, accreted onto the continents during drift, and transformed through repeated orogenies into the crust of the mature continents. You should note, however, that the Wilson cycle depends on the existence of continental cratons. These stable landmasses comprise the oldest rocks, the **Archean cratonic nuclei,** around which younger parts of the crust have been accreted and between which they have been deformed. Rocks of Archean age (4.0 to 2.5 billion years old) are found on every continental craton (see Figure 20.8). How were these cratonic nuclei created in the first place?

Earth was a hotter planet 4 billion years ago, owing to the heat generated by the decay of radioactive elements, which were more abundant, as well as the energy released by differentiation and during the Heavy Bombardment period (see Chapter 1). Evidence for a hotter mantle comes from a peculiar type of ultramafic volcanic rock found only in Archean crust, called *komatiite* (named after the Komati River in southeastern Africa, where it was first discovered). Komatiites contain a very high percentage (up to 33 percent) of magnesium oxide, and their formation involved a much higher degree of mantle melting than found anywhere on Earth today.

Mantle convection during the Archean was certainly very active, but we don't know to what extent the motions of rocks on Earth's surface resembled modern plate tectonics. The plates may have been smaller and moving more rapidly. Volcanism was widespread, and the crust formed at any spreading centers was probably thicker. Although lithosphere must surely have been recycled back into the mantle, some geologists believe that the plates formed at this time were too thin and light to be subducted in the same way that oceanic plates are consumed in modern subduction zones.

We do know that a silica-rich continental crust existed at this early stage in Earth's history. The Acasta gneiss, in the northwestern part of the Canadian Shield, has been dated at 4.0 billion years, and these rocks look very similar to modern gneisses (**Figure 20.22**). Formations as old as 3.8 billion years have been found on many continents; most are metamorphic rocks evidently derived from even older continental crust. In fact, single grains of the mineral zircon discovered in Australia have been dated at 4.1 to 4.4 billion years. The composition of oxygen isotopes from these grains suggests that liquid water may have been present on Earth's surface.

In the early part of the Archean eon, the continental crust that had differentiated from the mantle was very mobile, perhaps organized into small rafts being rapidly pushed together and torn apart by the intense tectonic activity. The first cratons with long-term stability began to form around 3.3 to 3.0 billion years ago. In North America, the oldest surviving example is the central Slave craton in northwestern Canada, where the Acasta gneiss is found, which

Figure 20.22 The Acasta gneiss in the Archean Slave province of northwestern Canada, dated at 4.0 billion years, is the oldest rock formation so far discovered. It demonstrates that continental crust existed on Earth's surface at the beginning of the Archean eon. [Courtesy of Sam Bowring, Massachusetts Institute of Technology.]

stabilized around 3 billion years ago. The rock assemblages in these Archean cratonic nuclei fall into two major groups (**Figure 20.23**):

1. Granite-greenstone terrains, areas of massive granite intrusions that surround smaller pockets of deformed, metamorphosed volcanic rocks (greenstones), primarily of mafic composition, capped with sediments. The origin of the greenstones is controversial, but many geologists think they were pieces of oceanic crust formed in an extensional setting (perhaps at small spreading centers behind island arcs), incorporated into the continents during compressional episodes, and later engulfed by the granite intrusions.

2. High-grade metamorphic terrains, areas of high-grade (granulite facies) metamorphic rocks derived primarily from the compression, burial, and subsequent erosion of granitic crust.

The high-grade metamorphic terrains found in the Archean cratonic nuclei look similar to the deeply eroded parts of modern orogenic belts, but the geometry of the deformation was different. Modern orogenies typically produce linear mountain belts where the edges of large cratons converge. In the Archean, the style of deformation was less lineated (that is, more circular or sinusoidal in planform), reflecting the fact that the cratons were much smaller with boundaries that were more curved.

By the end of the Archean, 2.5 billion years ago, enough continental crust had been stabilized in the cratonic nuclei to allow the formation of larger and larger continents. Plate tectonics was probably operating much as it does today. It is at about this time that we see the first evidence of major continent-continent collisions and the amalgamation of supercontinents. From this point onward in Earth's history, the evolution of the continents was governed by the plate tectonic processes of the Wilson cycle.

The Deep Structure of Continents

In this chapter, we have surveyed the most important processes that have led to the development of the continental crust. However, we have not yet provided an adequate explanation for one very basic aspect of continental behav-

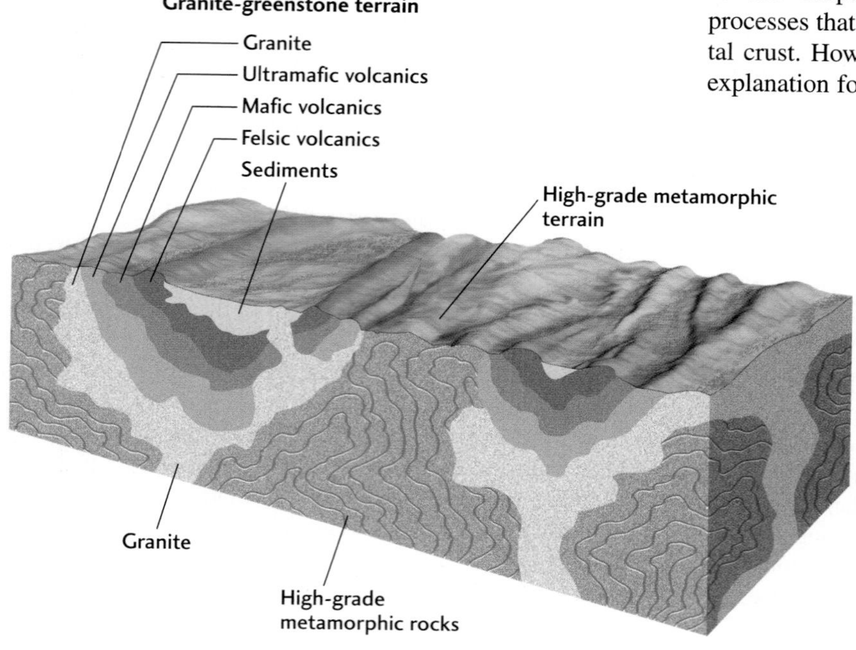

Figure 20.23 Schematic diagram of the two major types of rock formations found in the Archean cratons: granite-greenstone terrains and high-grade metamorphic terrains.

ior: the long-term stability of the cratons. How have the cratons survived being knocked around by plate tectonics for billions of years? The answer to this question lies not in the crust but in the subcontinental mantle.

Cratonic Keels

Using seismological methods that we will describe in the next chapter, geologists have discovered a remarkable fact: the continental cratons are underlain by a thick layer of strong mantle that moves with the plates during continental drift. These thickened sections of lithosphere extend to depths of more than 200 km—over twice the thickness of the oldest oceanic plates.

At 100 to 200 km beneath the oceans (as well as beneath most younger regions of the continents), the mantle rocks are hot and weak. They are part of the convecting asthenosphere that flows relatively easily, allowing the plates to slide across Earth's surface. The lithosphere beneath cratons extends into this region like the hull of a boat into water, so many geologists refer to these mantle structures as **cratonic keels** (**Figure 20.24**). All cratons on every continent appear to have such keels.

The existence of deep continental keels presents many puzzles that scientists are still trying to solve. Less heat is coming from the mantle beneath the cratons than from that beneath the oceanic crust. This has led geologists to infer that the keels are strong because they are several hundred degrees colder than the surrounding asthenosphere. If the rocks in the mantle beneath the cratons are so cold, however, why don't they sink into the mantle under their own weight, just as cold, heavy slabs of oceanic lithosphere sink in subduction zones?

Composition of the Keels

The cratonic keels would indeed sink if their chemical compositions were the same as ordinary mantle peridotites. To get around this problem, geologists have hypothesized that the cratonic keels are made of rocks with different, less dense compositions. The lower density of the rocks that make up the keels counteracts the increased density resulting from colder temperatures. Evidence in support of this hypothesis has come from mantle samples found in kimberlite pipes—the same types of volcanic deposits that produce diamonds.

Kimberlite pipes are the eroded necks (diatremes) of volcanoes that have explosively erupted from tremendous depths (see Chapter 6). Almost all kimberlites that contain diamonds are located within the Archean cratons. A diamond will revert to graphite at depths shallower than 150 km unless it is quickly quenched to low temperatures. Therefore, the presence of diamonds in these pipes demonstrates that the kimberlite magmas come from deeper than 150 km and that they erupted through the keels when magma fractured the lithosphere very rapidly.

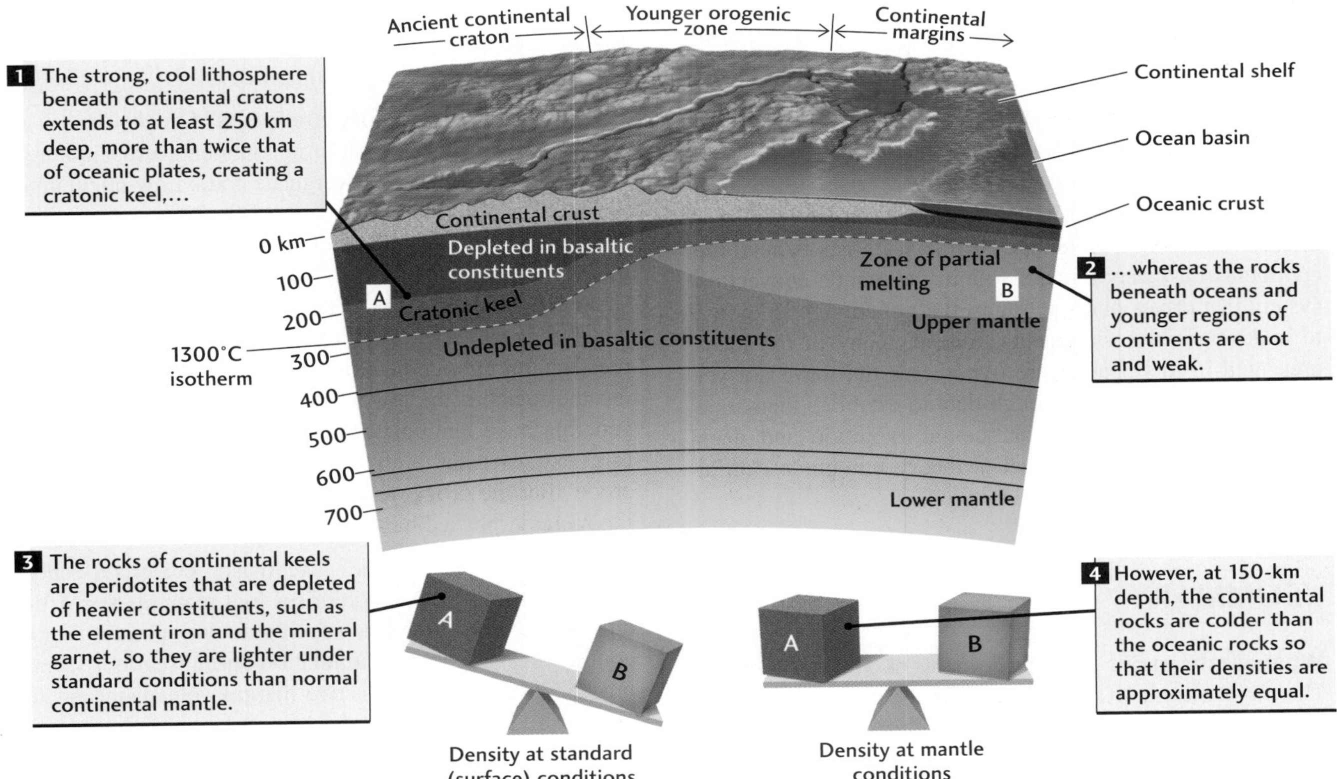

Figure 20.24 Cross section of a continent showing a cratonic keel extending to a depth of 250 km. [After T. H. Jordan, "The Deep Structure of Continents." *Scientific American* (January 1979): 92.]

During a violent kimberlite eruption, fragments of the keel, some containing diamonds, are ripped off and brought to the surface in the magma as mantle *xenoliths* (Greek for "foreign rocks"). The majority of the mantle xenoliths turn out to be peridotites with less iron (a heavy element) and less garnet (a heavy mineral) than ordinary mantle rocks. Such rocks can be produced by extracting a basaltic (or komatiitic) magma from the convecting asthenosphere. In other words, the mantle beneath the cratons is the depleted residue left over from melting sometime earlier in Earth's history. Calculations show that a cratonic keel made of these depleted rocks can still float atop the mantle, despite the fact that the keel is cold (see Figure 20.24).

Age of the Keels

By analyzing the xenoliths from kimberlites and the diamonds they contain, geologists have been able to demonstrate that the cratonic keels are very old. In fact, they are about the same age as the Archean crust above them. (The diamond in your ring or necklace is likely to be several billion years old!) Therefore, the rocks now in the cratonic keels were depleted by the extraction of a basaltlike melt very early in Earth's history, and they were subsequently emplaced beneath the continental crust at the time the Archean cratonic nuclei were stabilized.

In fact, the evidence suggests that the cratons were stabilized by the formation of the cold, depleted, and therefore mechanically strong keel. This theory can explain why at least some of the Archean cratonic nuclei have managed to survive through many continental collisions, including at least four episodes of supercontinent formation, without much internal deformation.

Many aspects of this process are still not understood, however. How did the keels cool down? How did they achieve the balance between temperature and composition illustrated in Figure 20.24? Why are the cratons with the thickest keels of Archean age? Some scientists believe the continents may play a major role in the mantle convection system that drives plate tectonics, but not enough is known to make definitive statements. Indeed, many of the ideas presented in this chapter are hypotheses that have not yet been corroborated with enough data to establish them as a fully accepted theory of continental evolution and deep structure. The search for such a theory remains a central focus of geologic research.

SUMMARY

What are the major geologic features of North America? The most ancient crust is exposed in the Canadian Shield. South of the Canadian Shield is the sediment-covered interior platform, where Precambrian basement rocks are covered by layers of Paleozoic sedimentary rocks. These older regions form the continental craton. Around the edges of the craton are the elongated mountain chains of the younger orogenic belts. The main orogenic belts are the Cordillera, which runs down the western edge of North America, and the Appalachian fold belt, which trends southwest to northeast on its eastern margin. The coastal plains and shelf of the Atlantic Ocean and Gulf of Mexico are parts of a passive continental margin that subsided after rifting during the breakup of Pangaea.

How do continents grow? Buoyant silica-rich rocks are produced by differentiation of magmas, primarily in subduction zones. Plate motions accrete this material to continental margins through four main processes: the transfer of buoyant crustal fragments from a subducting plate to an overriding continental plate; the closure of marginal basins to add thickened island-arc crust to the continent; the transport of crust laterally along continental margins by strike-slip faulting; and the suturing of two continental margins by continent-continent collision and the subsequent breakup of this zone by continental rifting.

What are epeirogeny and orogeny? Forces within the crust can deform large regions of the continents. Some regional movements are simple up-and-down displacements without severe deformation of the rock formations (epeirogeny); examples are the Colorado Plateau and the postglacial uplift of Scandinavia and central Canada. In other cases, horizontal forces arising mainly from plate convergence can produce mountains by extensive and complex folding and faulting (orogeny), as in the Cordillera and Appalachians of North America, the Alps of Europe, and the Himalaya of Asia. The Rockies and the Alps were eroded to low relief, then rejuvenated by recent broad regional uplift.

How does orogeny modify continents? Orogenesis caused by plate convergence can deform continental crust hundreds of kilometers from the convergence zone. Low-angle thrust faulting can stack the upper part of the crust into multiple thrust sheets tens of kilometers thick, deforming and metamorphosing the rocks they contain. Wedges of continental-shelf sediments can be detached from the basement on which they were deposited and then be thrust inland. Compression throughout the crust can double its thickness, causing the rocks in the lower crust to melt. This melting can generate huge amounts of granitic magma, which rises to form extensive batholiths in the upper crust. The mountains erode after the orogeny ends, thinning the crust and exposing metamorphosed basement rocks.

What is the Wilson cycle? The Wilson cycle is a general plate tectonic cycle that comprises four main phases: rifting during the breakup of a supercontinent; passive-margin cooling and sediment accumulation during seafloor spreading and ocean opening; active-margin volcanism and terrane accretion during subduction and ocean closure; and orogeny during the continent-continent collision.

How have the Archean cratons survived billions of years of plate tectonics? The Archean cratons are underlain by a layer of cold, strong mantle more than 200 km thick that moves with the plates during continental drift. These keels appear to be as old as the cratons themselves. They are formed from mantle peridotites that have been depleted by the extraction of mafic magmas, which lowers their density and stabilizes the keel against disruption by mantle convection and plate tectonic processes..

Key Terms and Concepts

accreted terrane (p. 466)
accretion (p. 466)
Archean cratonic nucleus (p. 477)
craton (p. 464)
cratonic keel (p. 478)
epeirogeny (p. 475)
orogen (p. 464)
orogeny (p. 468)
rejuvenation (p. 462)
shield (p. 459)
tectonic age (p. 465)
Wilson cycle (p. 474)

Exercises

1. Draw a rough topographic profile of the contiguous United States from San Francisco to Washington, D.C., and label the major geologic features.

2. Why is the North American Cordillera topographically higher than the Appalachians?

3. Summarize the main features (orogenic belts, shields, platforms, coastal plains, shelves) of the major structural regions of a continent other than North America.

4. Are the interiors of continents usually younger or older than the margins? Why?

5. Summarize the stages of a typical orogenic episode in a series of sketches with legends.

6. Two continents collide, thickening the crust from 35 km to 70 km to form a high, Tibet-like plateau. After hundreds of millions of years, the plateau is eroded down to sea level. (a) What kinds of rocks might be exposed at the surface by this erosion? (b) Estimate the crustal thickness after the erosion has occurred.

7. Evidence of vertical crustal movements is often found in the geologic record. Give some examples of such evidence.

8. How many times have the continents been joined in a supercontinent since the end of the Archean eon? Use this number to give a rough estimate of how fast continents move during plate tectonics.

9. How was orogenesis in the Archean eon different from orogenesis during the Proterozoic and Phanerozoic eons? What factors might explain these differences?

Thought Questions

1. How would you recognize an accreted terrane? How could you tell if it originated far away or nearby?

2. How would you identify a region where active orogeny is taking place today? Give an example.

3. Would you prefer to live on a planet with orogenies or without them? Why?

4. How could the uplift of a mountain range affect climate, the chemistry of the oceans, and the location of oil and mineral deposits?

5. Why are the ocean basins just about the right size to contain all the water on Earth's surface?

6. Where might the mantle material that forms the cratonic keels have come from?

Suggested Readings

Bally, A. W., and A. R. Palmer (eds.). 1989. *The Geology of North America: An Overview.* Boulder, Colo.: Geological Society of America.

Burchfiel, B. C. 1983. The continental crust. *Scientific American* (September): 130.

Cook, F. A., D. S. Albaugh, L. D. Brown, S. Kaufmann, J. E. Oliver, and R. D. Hatcher, Jr. 1979. Thin-skinned tectonics in the crystalline southern Appalachians. *Geology* 7: 563–567.

Geissman, J. W., and A. F. Glazner (eds.). 2000. Focus on the Himalayas. *Bulletin of the Geological Society of America* 112: 323–511.

Jones, D. L., A. Cox, P. Coney, and M. Beck. 1982. The growth of western North America. *Scientific American* (November): 70.

Jordan, T. H. 1979. The deep structure of continents. *Scientific American* (January): 92–107.

Kearey, P., and F. J. Vine. 1990. *Global Tectonics.* London: Blackwell Scientific.

Lithgow-Bertelloni, C., and P. G. Silver. 1998. Dynamic topography, plate driving forces and the African superswell. *Nature* 395: 269–271.

Molnar, P. 1997. The rise of the Tibetan Plateau: From mantle dynamics to the Indian monsoon. *Astronomy and Geophysics* 36(3): 10–15.

Murphy, J. B., and R. D. Nance. 1992. Mountain belts and the supercontinent cycle. *Scientific American* (April): 84.

Murphy, J. B., G. L. Oppliger, G. Brimhall, Jr., and A. J. Hynes. 1999. Mantle plumes and mountains. *American Scientist* 87 (March–April): 146–153.

Taylor, S. R. 1995. The geochemical evolution of the continental crust. *Reviews of Geophysics* 33: 241–265.

Twiss, R. J., and E. M. Moores. 1992. *Structural Geology.* New York: W. H. Freeman.

This open-pit excavation of a kimberlite volcano at Jwaneng, Botswana, is the world's richest diamond mine, producing 12.4 million carats worth more than $1.3 billion in 2001. In addition to diamonds, kimberlite volcanoes bring up rocks from more than 200 km deep, providing critical information about Earth's deep interior. [Peter Essick/Aurora Photos.]

CHAPTER

21

Exploring Earth's Interior

"Below all silken soil-slip, all crinkled earth-crust,
Far deeper than ocean, past rock that against rock grieves,
There at the globe's deepest dark and visceral lust,
Can you hear the groan-swish of magma as it churns and heaves?"

ROBERT PENN WARREN, *"YOUTHFUL TRUTH-SEEKER, HALF-NAKED, AT NIGHT, RUNNING DOWN BEACH SOUTH OF SAN FRANCISCO"*

Ancient thinkers divided the universe into two parts, the Heavens above and Hades below. The sky was transparent and full of light, and they could directly observe its stars and track its wandering planets. The ground was opaque, an inaccessible netherworld of darkness, closed to human view.

This is no longer the case. Geologists can now look downward into Earth's interior, not with rays of starlight but with waves from earthquakes and other seismic sources. In Chapter 19, we saw how the terrible shaking of earthquakes can wreak destruction. Yet this same energy can be harnessed to illuminate Earth's deepest layers, allowing geologists to construct three-dimensional images of oil reservoirs in the crust, rising and falling convection currents in the mantle, and the structure of the inner core.

In this chapter, we will explore Earth's interior down to its center, nearly 6400 km beneath our feet. In addition to the techniques of seismology, we will employ the geologic information from rocks brought up in volcanoes and from rocks magnetized by Earth's magnetic field. We will investigate the high temperatures deep inside the Earth and delve further into the internal workings of its two great heat engines: the geodynamo in the liquid iron outer core, which generates the magnetic field, and mantle convection, which drives plate tectonics.

Exploring the Interior with Seismic Waves

Waves of different types—light, sound, and seismic—have a common characteristic: the velocity at which they travel depends on the material they are passing through. Light waves travel quickest through a vacuum, more slowly through air, and even more slowly through water. Sound waves, on the other hand, travel faster through water than through air and not at all through a vacuum. Why? Sound waves are simply propagating variations in pressure. Without something to compress, such as air or water, they cannot exist. The more force it takes to compress a material, the faster sound will travel through it. The speed of sound in air—Mach 1 in the jargon of jet pilots—is typically 0.3 km/s, or about 670 miles per hour. Water resists compression much more than air, so the speed of sound waves in water is correspondingly higher, about 1.5 km/s. Solid materials are even more resistant to compression, and sound waves travel through them at even higher speeds. In granite rock, sound travels at about 6 km/s—nearly 13,500 miles per hour!

Basic Types of Waves

The push-pull motions of sound waves in a solid are called **compressional waves** to distinguish them from the side-to-side motions of **shear waves** (see Figure Story 19.5). It's harder to compress solids than to shear them, so compressional waves always travel faster than shear waves. This physical principle explains a relationship we discussed in Chapter 19: compressional waves are always the P (primary) arrivals, and shear waves are the S (secondary) arrivals. Another important property of seismic waves is that their shear wave speeds must be zero, because gases and liquids have no resistance to shear. Shear waves cannot propagate through any fluid—air, water, or the liquid iron in Earth's outer core.

From seismograms, geologists can calculate the speed of a P or S wave by dividing the distance traveled by the travel time. The measurements of these wave speeds can then be used to infer which materials the waves encountered along their paths. For example, P and S waves travel about 17 percent more rapidly through rock typical of oceanic crust (gabbro) than through typical upper continental crust (granite), and they travel about 33 percent more rapidly through the upper mantle (peridotite).

The concepts of travel times and wave paths sound simple enough, but complications arise when waves pass through more than one type of material. At the boundary between two different materials, some of the waves bounce off (that is, they are *reflected*) and others are transmitted into the second material—just as light is partly reflected and partly transmitted when it strikes a windowpane. The waves that cross the boundary between two materials are bent, or *refracted,* as their velocity changes from that in the first material to that in the second. **Figure 21.1** shows a laser light beam whose path bends as it goes from air into water, much as a P or an S wave bends as it travels from one material to another. By studying how fast seismic waves travel and how they are refracted and reflected at Earth's internal boundaries, seismologists have been able to measure the layering of Earth's crust, mantle, and core with great accuracy.

Paths of Seismic Waves in the Earth

If Earth were made of a single material with constant properties from the surface to the center, P and S waves would travel from the focus of an earthquake to a distant seismograph along straight lines through the interior. When the first global networks of seismographs were installed about a century ago, however, seismologists discovered that the structure of Earth's interior was much more complicated.

Waves Refracted Through Earth's Interior The first observations of long-distance seismic waves showed that the paths of the P and S waves curved upward through the mantle, as illustrated in **Figure 21.2**. From the travel times and the amount of upward bending, seismologists were able to demonstrate that P waves traveled much faster through rocks at great depths than they did through rocks found on the surface. This was hardly surprising, because rocks sub-

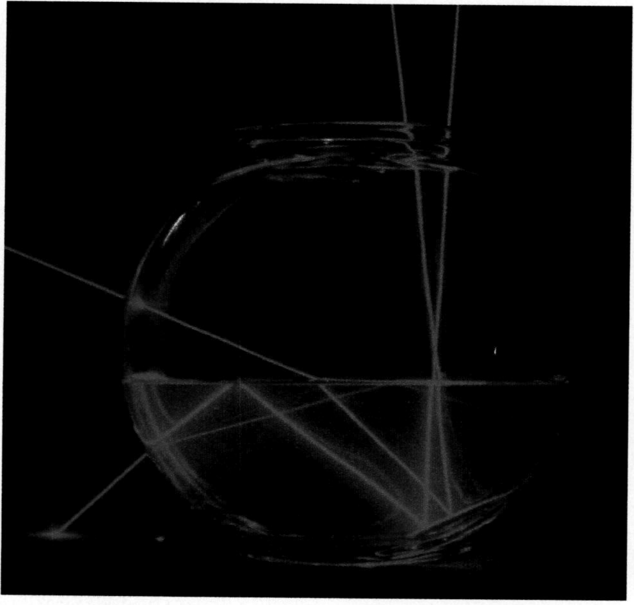

Figure 21.1 In this experiment, two beams of laser light enter a bowl of water from the top. Both beams are reflected from a mirror on the bottom of the bowl. One is then reflected at the water-air interface and passes through the bowl to make a bright spot on the table. Most of the energy from the other beam is bent downward (refracted) as it passes from the water to the air, although a small amount is reflected to form a second spot on the table. You can also trace the paths of other beams reflected by the interfaces. [Susan Schwartzenberg/The Exploratorium.]

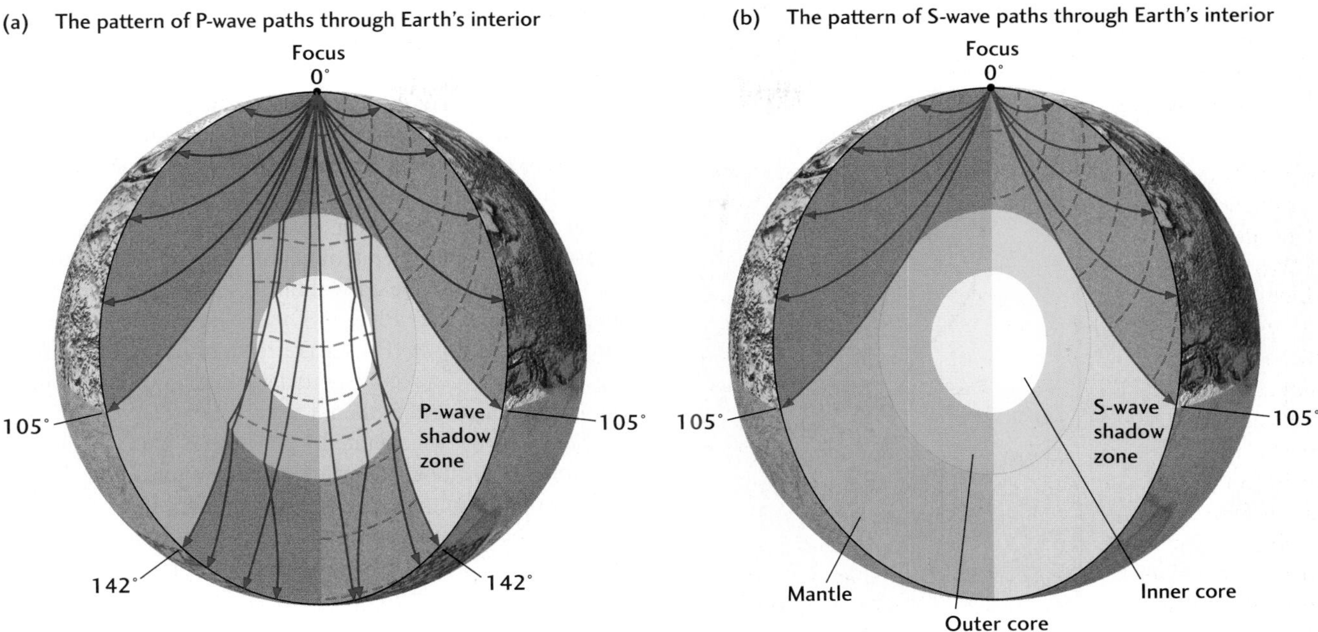

Figure 21.2 (a) The pattern of P-wave paths through Earth's interior. The dashed blue lines show the progress of wave fronts through the interior at 2-minute intervals. Distances are measured in angular distance from the earthquake focus. The P-wave shadow zone extends from 105° to 142°. P waves cannot reach the surface within this zone because of the way they are bent when they enter and leave the core. (b) The larger S-wave shadow zone extends from 105° to 180°. Although S waves strike the core, they cannot travel through its fluid outer region and therefore never emerge beyond 105° from the focus.

jected to great pressures in Earth's interior are squeezed into tighter crystal structures. The atoms in these tighter structures are more resistant to further compression, which causes P waves to travel through them more quickly.

Seismologists were very surprised, however, by what they found at progressively greater distances from an earthquake focus. After the P waves and S waves had traveled beyond about 11,600 km from the earthquake focus, they suddenly disappeared! (Like airplane pilots and ship captains, seismologists prefer to measure distances traveled on Earth's surface in angular degrees, from 0° at the earthquake focus to 180° at a point on the opposite side of Earth. Each degree measures 111 km at the surface, so 11,600 km corresponds to an angular distance of 105°.) When they looked at seismograms recorded beyond this distance, they did not see the distinct P and S arrivals that were so clear on seismograms recorded at shorter distances. Then, beyond about 15,800 km from the focus (142°), the P waves suddenly reappeared as big arrivals, but they were much delayed compared to their expected travel times. The S waves never reappeared.

These observations were first put together in 1906 by the British seismologist R. D. Oldham, and they provided the first evidence that Earth has a liquid outer core. No S waves can travel through the outer core because it is liquid, and there is thus an S-wave **shadow zone** beyond 105° from the earthquake focus (see Figure 21.2b). The propagation of P waves is more complicated (see Figure 21.2a). At 105°, their paths just miss the core, whereas waves that would have traveled to greater distances encounter the core-mantle boundary. At the core-mantle boundary, the P-wave speed drops by almost a factor of two. Therefore, the waves are refracted downward into the core and emerge at greater distances after the delay caused by their detour through the core. This refraction effect forms a P-wave shadow zone at angular distances between 105° and 142°.

Waves Reflected at Earth's Internal Boundaries The core-mantle boundary turned out to be very sharp, so when seismologists looked at records of earthquake waves made at angular distances of less than 105°, they found arrivals corresponding to waves reflected from this boundary. They called the compressional wave that reflects from the top of the outer core PcP and the shear wave ScS. (The lowercase *c* indicates a reflection from the core.) In 1914, a German seismologist, Beno Gutenberg, used the travel times of these core reflections to determine an accurate depth for the core-mantle boundary—just under 2900 km.

Figure 21.3 shows examples of the paths taken by these core-reflected waves, as well as the paths and symbolic names that have been attached to some other prominent arrivals seen on seismograms. For example, a compressional wave reflected once at Earth's surface is called

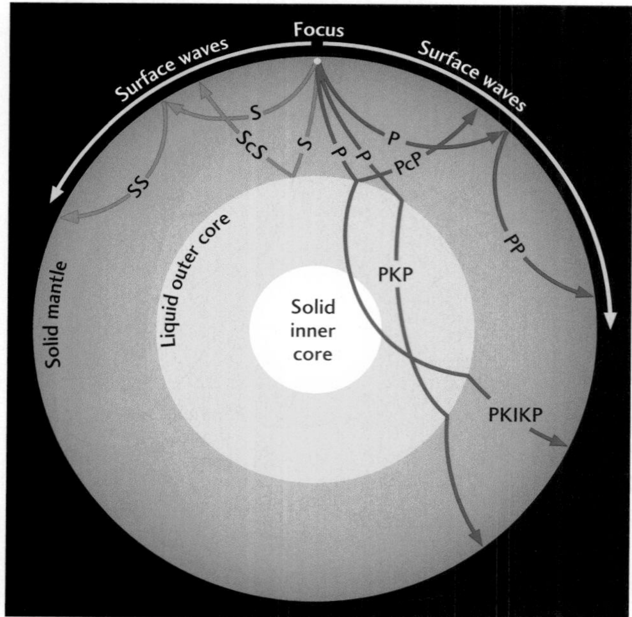

Figure 21.3 P and S waves radiate from an earthquake focus in all directions. This diagram shows the simple labeling scheme seismologists use to describe the various paths the waves take. PcP and ScS are compressional and shear waves that bounce off the core. PP and SS waves are internally reflected from Earth's surface. A PKP wave is transmitted through the liquid outer core, and a PKIKP wave traverses the solid inner core. Surface waves propagate along Earth's outer surface, like waves on the surface of a pond.

PP, and the shear wave with a similar path is called SS. **Figure 21.4** displays several seismograms, recorded at different distances from an earthquake focus, that show these internal reflections.

The path of a compressional wave through the outer core is labeled with a *K* (from the German word for "core"), so PKP describes a compressional wave that propagates from an earthquake through the crust and mantle, into the outer core, and back through the mantle and crust to a surface receiver. In 1936, Danish seismologist Inge Lehmann discovered Earth's inner core by observing the compressional waves refracted from its outer boundary, which she determined to be at a depth of about 5150 km. Paths through the inner core are labeled with an *I*, so the refracted waves she used are called PKIKP. Others have since observed compressional waves (PKiKP) reflected from the top side of the inner core–outer core boundary (the lowercase *i* indicates a reflection rather than a refraction)

Seismic waves generated by artificial sources, such as dynamite explosions, are reflected by geologic structures at shallow depths in the crust. Recording these reflections has proved to be the most successful method for finding deeply buried oil and gas reservoirs. This type of seismic exploration is now a multibillion-dollar industry. Reflected seismic waves are employed in other practical applications, such as measuring the depth of water tables and the thickness of glaciers. At sea, compressional waves can be generated from mechanical sources similar to a loudspeaker, and ships routinely use the underwater sound they produce to measure the depth of the ocean and the thickness of sediments on the seafloor.

Layering and Composition of the Interior

Thousands of sensitive seismographs and highly accurate clocks have enabled seismologists throughout the world to measure the travel times of many types of seismic waves very precisely. Nuclear explosions set off underground at nuclear

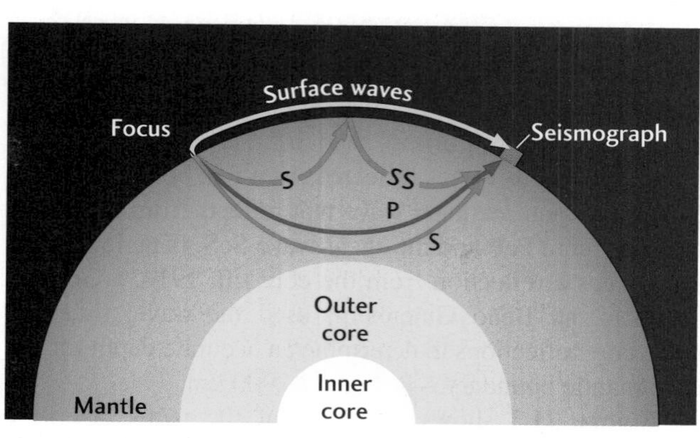

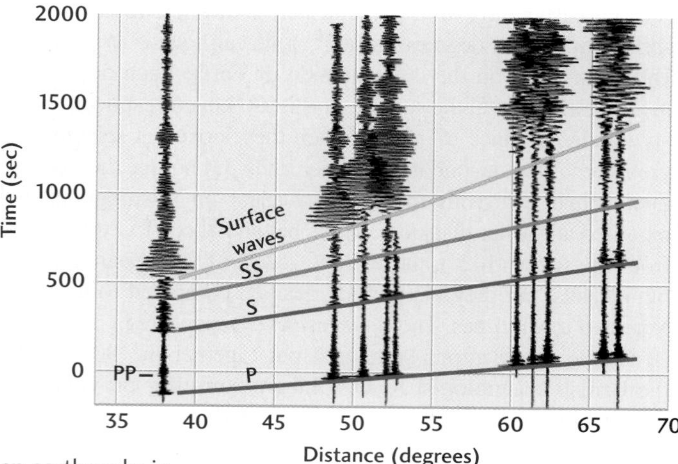

Figure 21.4 Seismograms recorded at various distances from an earthquake in the Aleutian Islands, Alaska, showing P, S, SS, and surface waves.

test sites also excite seismic waves and add valuable data to those derived from earthquakes. From these measurements, seismologists can plot travel-time curves of the kind shown in Figure 19.6 for the various kinds of seismic waves.

The travel times of compressional and shear waves depend on their speeds as they pass through the materials in Earth's interior. The key to making travel times a useful geological tool is to learn how to convert them into a graph or table that shows how the speeds of seismic waves change with depth in the Earth. Solving this problem is something like guessing which of several possible routes a driver took on a trip from Los Angeles to San Francisco and how fast he traveled along the route if you knew that the trip took 6.1 hours and knew the speed limits along the way.

The model that seismologists have devised is displayed in **Figure 21.5**. The illustration shows how the speeds of both compressional and shear waves change with depth and how these changes are related to Earth's major layers. To understand the structure and composition of these layers, scientists must combine the seismological information on this diagram with information from many other types of geologic studies. We will explore what they have found by taking a downward journey through Earth's interior, from its outer crust to its inner core.

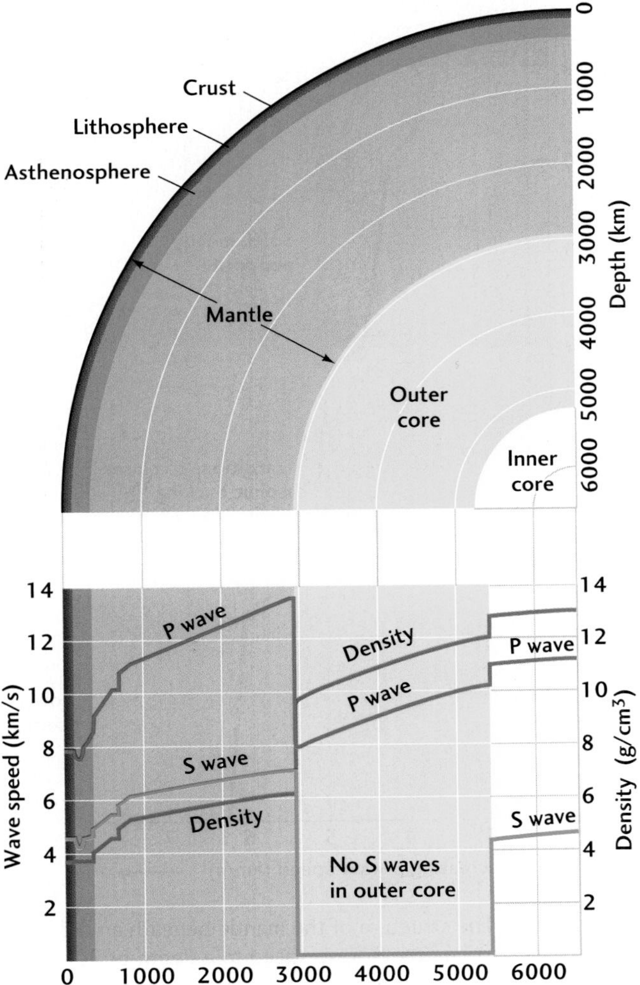

Figure 21.5 Earth's layering revealed by seismology. The lower diagram shows the changes in the compressional and shear speeds and density with depth in the Earth. The upper diagram is a cross section through Earth on the same depth scale, showing how these changes are related to the major layers.

The Crust

Earth's outermost layer, the crust, varies in thickness. It is thin (about 7 km) under oceans, thicker (about 40 km) under continents, and thickest (as much as 70 km) under high mountains. P waves move through crustal rocks at about 6 to 7 km/s. By sampling various materials typical of the crust and mantle and by measuring the velocities of waves passing through these materials in the laboratory, we can compile a library of seismic speeds through all sorts of materials that compose the Earth. Rough values of the compressional wave speeds in igneous rocks, for example, are as follows:

- Felsic rocks typical of the upper continental crust (granite): 6 km/s
- Mafic rocks typical of oceanic crust or the lower continental crust (gabbro): 7 km/s
- Ultramafic rocks typical of the upper mantle (peridotite): 8 km/s

The seismic wave speeds differ because they depend on a rock's density and its resistance to compression and shear, all of which change with composition and crystal structure.

We know from the correlations of wave velocity and rock type that the upper part of the continental crust is made up mostly of granitic rocks and that no granite exists on the floor of the deep ocean. The crust there consists entirely of basalt and gabbro overlain by sedimentation. Below the crust, the velocity of P waves increases abruptly to 8 km/s. This abrupt increase is an indication of a sharp boundary between crustal rocks and underlying mantle, as shown in **Figure 21.6**. The velocity of 8 km/s indicates that the deeper rocks are probably the denser ultramafic rock peridotite. The boundary between the crust and the mantle is called the **Mohorovičić discontinuity (Moho,** for short), after the Yugoslav seismologist who discovered it in 1909. The indications that the crust is less dense than the underlying mantle are consistent with the theory that the crust is made up of lighter materials that floated up from the mantle (see Chapter 1).

The mechanism by which the less dense continents float on the denser mantle, just as an iceberg floats on the ocean, is the *principle of isostasy,* introduced in Chapters 16 and 18. The large volume of less dense continental crust that projects into the denser mantle provides the buoyancy that allows the continents to float, as shown in Figure 21.6. The additional weight of the high mountains is supported by the buoyancy from a deeper root of low-density crustal rocks.

From seismological data, we know that the mantle beneath the crust is solid rock. How can continents float on

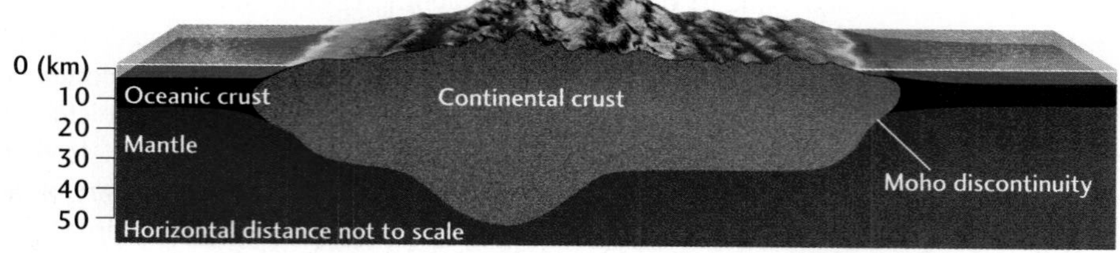

Figure 21.6 Seismic waves reveal the boundary between the crust and underlying mantle and variations in the thickness of the crust. Relatively light continental crust projecting into the denser mantle serves as a buoyant root providing "flotation" for the continent. The root is deeper under mountains, where flotation is required to support the heavier load, in accordance with the principle of isostasy.

solid rock? As we saw in Chapter 1, rocks can be solid and strong over the short term (years) while still being weak over the long term (thousands to millions of years). The principle of isostasy implies that, over long periods, the mantle has little strength and behaves as a viscous solid when it is forced to support the weight of continents and mountains.

Earth's topography shows a bimodal distribution: on average, the elevation of the solid surface tends to be either about 4 to 5 km below sea level or about 0 to 1 km above it. The lower elevation characterizes the seafloor. The higher elevation is typical of continents worn down to near sea level by erosion. This distribution reflects isostasy: the light continental crust "floats" higher than the seafloor (see Feature 21.1).

The Mantle

The primary rock type of the upper mantle, peridotite, is made up mainly of olivine and pyroxene, two silicates that contain magnesium and iron (see Chapters 3 and 5). Laboratory experiments show that changes in pressure and temperature alter the properties and forms of olivine and pyroxene. These minerals begin to melt under the conditions found in the upper mantle. At greater depths, pressure forces the atoms of these minerals closer together into more compact crystal structures. The major changes in mantle peridotites with depth are marked by increases or decreases in S-wave speed (**Figure 21.7**).

In the outermost zone of the mantle, the S-wave speed is typical of the mantle peridotites that constitute the subcrustal parts of the rigid lithosphere. The average thickness of this layer is about 100 km, but it is highly variable geographically, ranging from almost no thickness near the spreading centers where new oceanic lithosphere is forming from hot, rising mantle to over 200 km beneath the cold, stable continental cratons.

Near the base of the lithosphere, the S-wave speed abruptly decreases to form a **low-velocity zone.** The decrease in speed occurs because the temperature of the mantle is increasing with depth. At about 100 km, it approaches

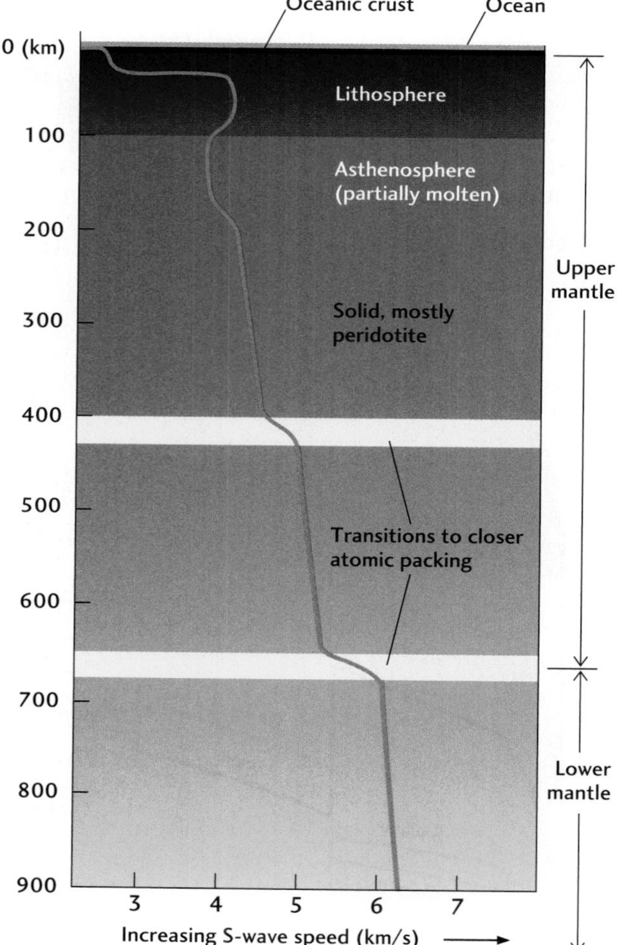

Figure 21.7 The structure of the mantle beneath an old ocean basin, showing the S-wave speed to a depth of 900 km. Changes in speed mark the strong lithosphere, the weak asthenosphere, and two zones in which changes occur because increasing pressure forces a rearrangement of the atoms into denser and more compact crystalline structures. [After D. P. McKenzie, "The Earth's Mantle." *Scientific American* (September 1983): 66.]

21.1 The Uplift of Scandinavia: Nature's Experiment with Isostasy

If you depress a cork floating in water with your finger and then release it, the cork pops up almost instantly. A cork floating in molasses would rise more slowly; the drag of the viscous fluid would slow the process. If we could perform a similar experiment on the Earth, we could learn much about how isostasy works—in particular, about the viscosity of the mantle and how it affects rates of uplift and subsidence. How convenient it would be if we could push the crust down somewhere, remove the depressive force, and then sit back and watch the depressed area rise.

Nature has been good enough to perform this experiment for us in less time than it takes to build and erode a mountain. The force is the weight of a continental glacier—an ice sheet 2 to 3 km thick. Such ice sheets can appear with the onset of an ice age in the geologically short period of a few thousand years. The crust is depressed by the ice load, and a downward bulge develops on its underside to the extent needed to provide buoyant support. At the onset of a warming trend, the glacier melts rapidly. With the removal of the weight, the depressed crust begins to rebound. We can discover the rate of uplift by dating ancient beaches that are now above sea level. Such raised beaches can tell us how long ago a particular stretch of land was at sea level.

Such depression and uplift have occurred in Norway, Sweden, Finland, Canada, and elsewhere in glaciated regions. The ice cap retreated from these regions some 10,000 years ago, and the land has been rising ever since. Figure 20.21 shows a series of raised beaches that have allowed geologists to measure the speed of this glacial rebound and infer the viscosity of the mantle.

TIME 1
A continental glacier starts to form, and continues to thicken over a few thousand years at the start of an ice age.
Ice
Continental crust
Lithosphere

TIME 2
The continental crust bends downward under the ice load to the extent needed to provide buoyant support.

TIME 3
At the end of the ice age, rapid warming melts the glacier. The depressed crust begins to rebound.

TIME 4
Rebound continues long after the glacier has melted, slowly returning to its pre-ice-age elevation.

Isostasy and postglacial uplift.

the melting temperature of the mantle rocks, partially melting some minerals in the peridotite. The small amount of melt (in most places less than 1 percent) decreases the rigidity of the rock, which slows the S waves passing through the rock. Because partial melting also allows the rocks to flow more easily, geologists identify the low-velocity zone with the top part of the asthenosphere—the weak layer across which the rigid lithospheric plates slide. This idea fits nicely with evidence that the asthenosphere is the source of most basaltic magma (see Chapters 5 and 6).

The base of the low-velocity zone occurs at about 200 to 250 km below ocean plates, where the S-wave speed increases to a value consistent with solid peridotite. The low-velocity zone is not as well defined under the stable continental cratons, which comprise colder mantle rocks that can remain unmelted throughout this region.

At depths of about 200 to 400 km, the S-wave speed again increases with depth. Within this layer, pressure continues to increase, but the temperature is not rising as rapidly as it does near the surface, owing to the effects of convection within the asthenosphere. (We will discuss why this is so in the next section.) The combined effects of pressure and temperature cause the amount of melting to decrease with depth and cause the rock rigidity—and thus the S-wave speed—to rise.

About 400 km below the surface, S-wave speed increases by about 10 percent within a narrow zone less than 20 km thick. When this zone of the mantle was discovered in the 1920s, geologists were puzzled. They soon realized, however, that the jump in S-wave speed could be explained by a **phase change** in olivine, the major mineral constituent of the upper mantle, from its ordinary crystal structure to a denser, more closely packed structure. Sure enough, when they subjected olivine to high pressures in the laboratory, they found that the atoms collapsed to a more compact arrangement at the temperatures and pressures corresponding to depths of about 400 km. Moreover, the jump in P- and S-wave speeds measured in the laboratory matched the increase observed for seismic waves at this depth.

In the region from 420 to 650 km below the surface, the mantle properties change little as depth increases. Near 660 km, however, the S-wave speed abruptly increases again, indicating a second major phase change to an even more closely packed crystal structure. Laboratory experiments on olivine confirmed the existence of another major mineralogical phase change at this depth.

Phase changes involve a transition in a rock's mineralogy but not in its chemical composition. Some geologists believe the increase in seismic wave speeds may also signal a change in chemical composition near this depth. The best evidence, however, indicates very little if any chemical change in this region of the mantle.

Below the transition at a depth of 660 km, the seismic wave speeds increase gradually and do not show any unusual features, such as major phase changes, until close to the core-mantle boundary (see Figure 21.5). This relatively homogeneous region, more than 2000 km thick, is called the **lower mantle.**

The Core-Mantle Boundary

At the **core-mantle boundary,** about 2890 km below the surface, we encounter the most extreme change in properties found anywhere in Earth's interior. From the way seismic waves reflect from this boundary, seismologists can tell that it is a very sharp interface. Here the material changes abruptly from a solid silicate rock to a liquid iron alloy. Because of the complete loss of rigidity, the S-wave speed drops from about 7.5 km/s to zero, and the P-wave speed drops from more than 13 km/s to about 8 km/s, causing the core shadow zone. Density, on the other hand, increases by about 4.5 g/cm^3 (see Figure 21.5). This large density jump, which is even greater than the increase in density at Earth's surface, keeps the core-mantle boundary very flat (you could probably skateboard on it!) and prevents any large-scale mixing of the mantle and core.

The core-mantle boundary appears to be a very active place. Heat conducted out of the core should increase the temperatures at the base of the mantle by as much as 1000°C. Indeed, the seismic waves that pass near the base of the mantle show peculiar complications, suggesting a region of exceptional geologic activity. In a thin layer above the core-mantle boundary, seismologists have recently discovered a steep (10 percent or more) decrease in seismic-wave speeds, which may be an indication that the mantle in contact with the core is partially molten, at least in some places. We noted in Chapter 2 that many geologists believe this hot region to be the source of mantle plumes that rise all the way to Earth's surface, creating volcanic hot spots such as Hawaii and Yellowstone.

The lowermost boundary layer of the mantle, a region about 300 km thick, may also be the ultimate graveyard of some lithospheric material subducted at the surface, such as the denser, iron-rich parts of the oceanic crust. Some geologists have speculated that the tectonics in the region above the core-mantle boundary might be an upside-down version of Earth's surface. For example, accumulations of heavy, iron-rich material might form chemically distinct "anticontinents" that are constantly pushed to-and-fro across the core-mantle boundary by convection currents. Seismologists are teaming with other geologists who study mantle and core convection to learn more about whatever geologic processes might be active in this strange land.

The Core

Geologists know quite a bit about the composition of the core, but not from direct observation. Their understanding has been derived from years of research using a combination of astronomical data, laboratory experiments, and seismological data. An iron-nickel composition is consistent with many lines of evidence. These metals are abundant in the cosmos. They are dense enough to explain the mass of the core (about one-third of Earth's total mass) and to be consistent with the theory that the core formed by gravitational differentiation. This hypothesis, first offered in the late nineteenth century, was buttressed by discoveries of meteorites made almost entirely of iron and nickel, which presumably came from the breakup of a planetary body that also had an iron-nickel core.

Laboratory measurements at appropriately high pressures and temperatures have led to a slight revision of this hypothesis. A pure iron-nickel alloy turns out to be about 10 percent too dense to match the data for the outer core. Geologists have therefore proposed that the core includes minor amounts of some lighter element. Oxygen and sulfur are leading candidates, although the precise composition remains the subject of research and debate.

Seismology tells us that the core below the mantle is a fluid, but it is not fluid to the very center of the Earth. As Lehmann first discovered, P waves that penetrate to depths of 5150 km suddenly speed up, indicating the presence of an inner core, a metallic sphere two-thirds the size of the Moon. Seismologists have recently shown that the inner core does transmit shear waves, confirming early speculations that it is solid. In fact, some calculations suggest that the inner core spins at a slightly faster rate than the mantle, acting like a "planet within a planet."

The very center of the planet is not a place you would like to be. The pressures are immense, over 4 million times the atmospheric pressure at Earth's surface. And it's also very hot, as we will see.

Earth's Internal Heat and Temperature

The evidence of Earth's internal heat is everywhere: volcanoes, hot springs, and the elevated temperatures in mines and boreholes. The internal heat fuels convection in the mantle, which drives the plate tectonic system, as well as the geodynamo in the core, which produces Earth's magnetic field.

Heat in the interior comes from several sources, discussed in Chapter 1. During the planet's violent origin, kinetic energy released by infalling chunks of matter heated its outer regions, while gravitational energy released by differentiation of the core heated its deep interior. The disintegration of the radioactive elements uranium, thorium, and potassium continues to generate heat.

After Earth formed, it began to cool, and it is cooling to this day as heat flows from the hot interior to the cool surface. The temperatures of the planet's interior result from a balance between the heat gained and that lost through these processes.

Heat Flow Through Earth's Interior

The Earth cools in two main ways: through the slow transport of heat by conduction and the more rapid transport of heat by convection. Conduction dominates in the lithosphere, whereas convection is more important throughout most of Earth's interior.

Conduction Through the Lithosphere Heat energy exists in a material as the vibration of atoms; the higher the temperature, the more intense the vibrations. The **conduction** of heat occurs when thermally agitated atoms and molecules jostle one another, mechanically transferring the vibrational motion from a hot region to a cool one. Heat is conducted from regions of high temperature to regions of low temperature by this process.

Materials vary in their ability to conduct heat. Metal is a better conductor than plastic (think of how rapidly the metal handle of a frying pan heats up in comparison with one made of plastic). Rock and soil are very poor heat conductors, which is why underground pipes are less susceptible to freezing than those above ground and why wine cellars and other underground vaults have a nearly constant temperature despite large seasonal temperature changes at the surface. Rock conducts heat so poorly that a lava flow 100 m thick takes about 300 years to cool from 1000°C to surface temperatures.

The conduction of heat through the outer surface of the lithosphere causes it to cool slowly over time. As it cools, the thickness of the lithosphere increases, just as the cold crust on a bowl of hot wax thickens over time. Rock, like wax, contracts and becomes denser with decreasing temperature, so the average density of the lithosphere must increase over time, and, by the principle of isostasy, its surface must sink to lower levels. Thus, the mid-ocean ridges stand high because the lithosphere there is young, thin, and hot, whereas the abyssal plains are deep because the lithosphere is old, cold, and thick.

From these considerations, geologists have constructed a simple but precise theory of seafloor topography that can explain the large-scale features of the ocean basins we saw in Figure 17.1. The theory predicts that the depth of the oceans should depend primarily on the age of the seafloor. According to this theory, the ocean depth should increase as the square root of age. In other words, seafloor that is 40 million years old should have subsided twice as much as seafloor that is only 10 million years old (because $\sqrt{40/10} = \sqrt{4} = 2$). This simple mathematical relationship matches seafloor topography near the mid-ocean ridge crests amazingly well, as demonstrated in **Figure 21.8**.

Conductive cooling of the lithosphere accounts for a wide variety of other geologic phenomena, including the subsidence of passive continental margins and the growth of many sedimentary basins. It explains why the heat flowing out of the oceanic lithosphere is high near spreading centers and decreases as the oceanic lithosphere gets older, and it tells us why the average thickness of the oceanic lithosphere is about 100 km. Almost all geologists would agree that the establishment of this theory was one of the great successes of plate tectonics.

Conductive cooling does not explain all aspects of heat flow through Earth's outer surface, however. Marine geologists have found that seafloor older than about 100 million years does not continue to subside as the simple theory would predict. Moreover, simple conductive cooling is far too inefficient to account for the cooling of the Earth over its entire history. It can be shown that if the 4.5-billion-year-old Earth cooled by conduction alone, very little of the heat from depths greater than about 500 km would have reached

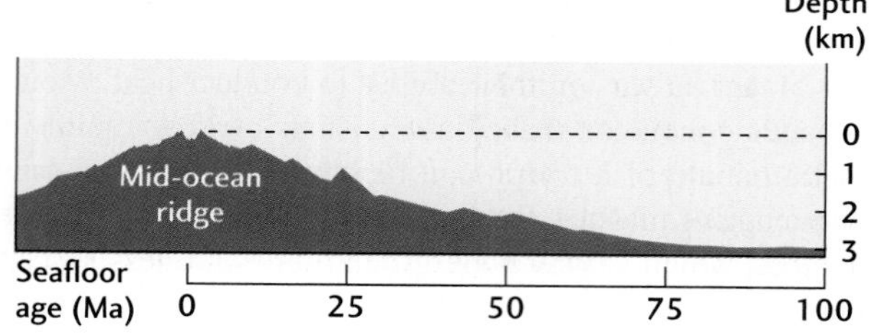

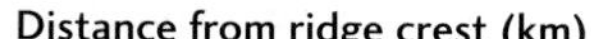

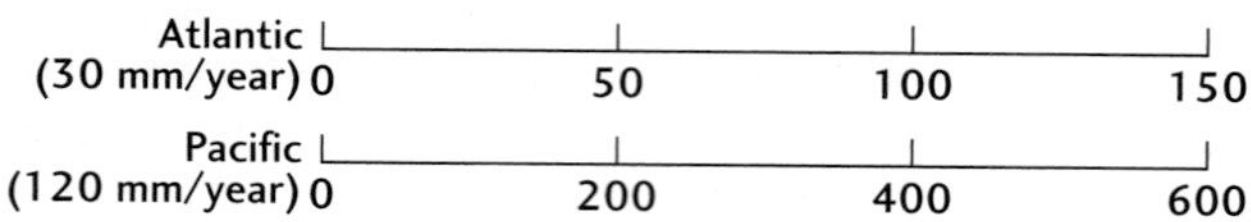

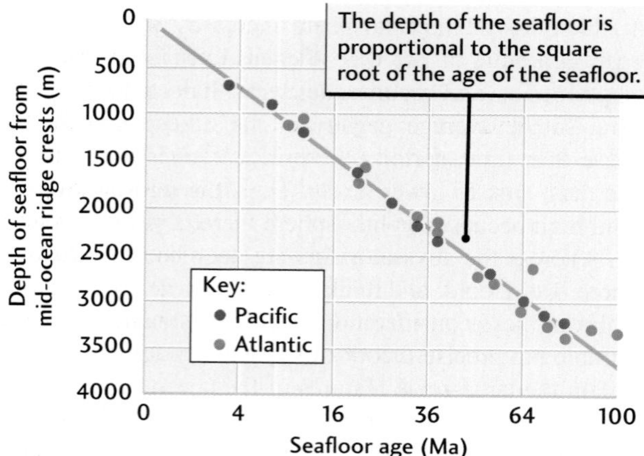

Figure 21.8 Topography of mid-ocean ridges in the Atlantic and Pacific oceans, showing how the water depth increases as the square root of lithosphere age away from the spreading centers. The same theoretical curve, which is derived by assuming the lithosphere cools by conduction, matches the data for both ocean basins, even though seafloor spreading is much faster in the Pacific than in the Atlantic.

the surface. The mantle, which was molten in Earth's early history, would be far hotter than it is now. To understand these facts, we must consider another important, and more efficient, mode of heat transport: **convection.**

Convection in the Mantle and Core Convection occurs when a heated fluid, either liquid or gas, expands and rises because it has become less dense than the surrounding material. Convection moves heat more efficiently than conduction because the heated material itself moves, carrying its heat with it. Colder material flows in to take the place of the hot rising fluid, is itself heated, and then rises to continue the cycle. This is the process by which water is heated in a kettle (see Figure 1.11a). Liquids conduct heat poorly, so a kettle of water would take a long time to heat to the boiling point if convection did not distribute the heat rapidly. Convection moves heat when a chimney draws, when warm tobacco smoke rises, or when clouds form on a hot day.

We have already seen how seismic waves reveal that the outer core is a liquid. Other types of data demonstrate that the iron-rich material in the outer core has a low viscosity (resistance to permanent deformation) and can therefore convect very easily. Convective motions in the outer core move heat through the core very efficiently. There, the heat flows across the core-mantle boundary into the mantle. Convective motions in the outer core also generate Earth's magnetic field, a phenomenon we will examine in more detail later in this chapter.

The existence of convection in the solid mantle is more surprising, but we now know that mantle rocks below the lithosphere are hot enough to flow when subjected to forces over long periods of time. As discussed in Chapters 1 and 2, seafloor spreading and plate tectonics are direct evidence of this solid-state convection at work. The rising hot matter under mid-ocean ridges builds new lithosphere, which cools as it spreads away. In time, it sinks back into the mantle, where it is eventually resorbed and reheated (see Figure 1.11b). This cyclical process is a form of convection; heat is carried from the interior to the surface by the motion of matter.

Geologists are still debating many aspects of mantle convection. Some believe that only the upper mantle is involved in the convection that drives plates, which would imply that the upper and lower mantles do not mix. Many others think that the plate tectonic system of convection extends throughout the mantle, including most of the lower mantle. New means of exploring Earth's interior using seismic waves, as well as evidence from other methods, are beginning to resolve these issues, as we will see below.

Regardless of the specifics, almost all geologists now agree that the convective movement of heat and matter from the interior to the surface is the dominant mechanism by which Earth has cooled over geologic time.

Temperatures in the Earth

Geologists want to know how hot Earth gets deeper in its interior. Temperature and pressure determine whether matter is solid or molten, the degree to which solid matter can creep, and how atoms are packed together in crystals. The higher the temperature at depth, the more rapidly convecting matter will move.

The curve that describes how temperature increases with depth is called the **geotherm.** One possible geotherm is illustrated in Figure 21.9. Near the surface, geologists can directly measure temperature to depths of 4 km in mines and to depths of 8 km in boreholes. They find that the *geothermal gradient* (the change of temperature with depth) is 20° to 30°C per kilometer in normal continental crust. Conditions below the crust can be inferred from lavas and rocks erupted in volcanoes. The data indicate that temperatures near the base of the lithosphere range from 1300° to 1400°C. As Figure 21.9 shows, these temperatures are very near the melting point of mantle rocks, consistent with

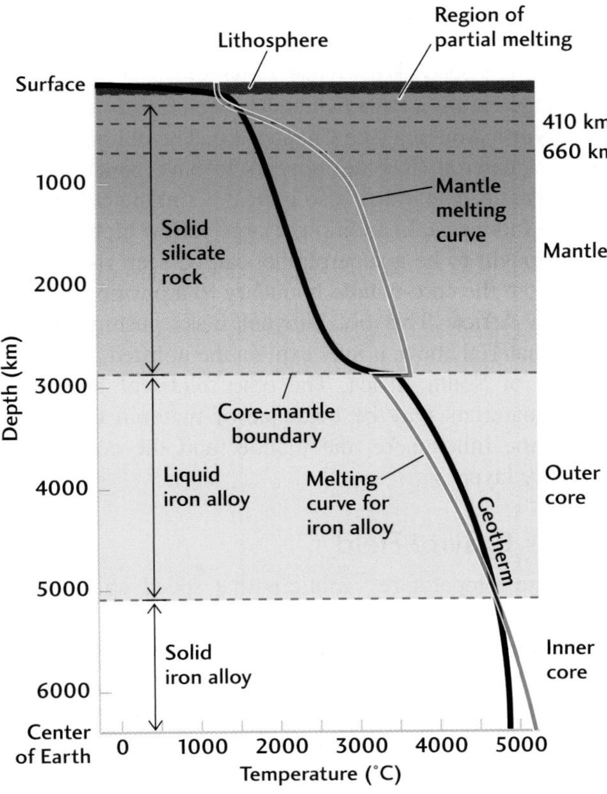

Figure 21.9 One estimate of the geotherm, which describes how temperatures increase with depth in the Earth. The geotherm lies above the temperature at which Earth materials first begin to melt (red line) in the upper mantle, forming the partially molten low-velocity zone, and in the outer core.

widespread observations that basaltic magmas are produced by partial melting in the upper part of the asthenosphere.

The steep geothermal gradient near Earth's surface tells us that heat is transported through the lithosphere by conduction. Below this depth, the temperature does not rise as rapidly. If it did, the temperatures in the deeper parts of the mantle would be so high (tens of thousands of degrees) that the lower mantle would be molten, which is inconsistent with seismological observations. Instead, the change in temperature with depth drops to about 0.5°C per kilometer, which is the geothermal gradient expected in a convecting mantle. This reduction occurs because convection mixes cooler material near the top of the mantle with warmer material at greater depths, averaging out the temperature differences (just as temperatures are evened out when you stir your bathwater).

Geologists can infer the temperature in the mid-mantle from the phase changes observed as steep increases in seismic wave speeds at depths of 410 km and 660 km. The depths (and thus the pressures) at which these phase changes occur can be accurately determined by seismology, so the temperatures at which the phase changes take place can be calibrated using high-pressure laboratory experiments. The values obtained from the laboratory data are consistent with the geotherm shown in Figure 21.9.

Geologists have more limited information about the temperatures at greater depths. Most agree that convection extends throughout the mantle, vertically mixing material and keeping the geothermal gradient low. Near the base of the mantle, however, we expect temperatures to increase more rapidly, because the core-mantle boundary restricts vertical mixing. Motions near the core-mantle boundary, like motions near the surface, are primarily sideways rather than vertical. Close to this boundary, heat is transported from the core into the mantle mainly by conduction, and the geothermal gradients should therefore be high.

Seismology tells us that the outer core is liquid, which implies that its temperature must exceed the melting point of the iron alloy that constitutes it. Laboratory data indicate that this temperature is probably greater than 3000°C, consistent with the high geothermal gradients at the base of the mantle predicted by convection models. The inner core, on the other hand, is solid. If its composition is the same as that of the outer core (by no means a certainty), then the inner core–outer core boundary should correspond to the depth where the geotherm crosses the core's melting curve. According to Figure 21.9, this hypothesis implies that the temperature at Earth's center is slightly less than 5000°C.

Many aspects of this story can be debated, however, especially in regard to the deeper parts of the geotherm. For example, some geologists believe that the temperature at Earth's center may be as high as 6000° to 8000°C. More laboratory experiments and better calculations will be required to reconcile these differences.

The Three-Dimensional Structure of the Mantle

So far, we have investigated how the properties of Earth's materials vary with depth. Such a one-dimensional description would suffice if our planet were a perfect sphere, but of course it is not so symmetric. At the surface, we can see *lateral variations* (geographic differences) in Earth's structure associated with oceans and continents and with the basic features of plate tectonics: spreading centers at mid-ocean ridges, subduction zones at deep-sea trenches, and mountain belts lifted up by continent-continent collisions. Below the crust, we can expect that convection will cause changes in temperature from one part of the mantle to another. Downwelling currents, such as those associated with subducted lithospheric plates, will be relatively cold; whereas upwelling currents, such as those associated with mantle plumes, will be relatively hot.

Computer models tell us that the lateral variations in temperature due to mantle convection should be on the order of several hundred degrees. From laboratory experiments on rocks, we know that a temperature increase of 100°C reduces the speed of an S wave traveling through a mantle peridotite by about 1 percent (or even more if the rock is close to its melting temperature). If the mantle is indeed convecting, the seismic wave speeds should vary by several percentage points from place to place. Geologists can make three-dimensional maps of these small lateral variations in wave speeds—and therefore three-dimensional maps of mantle convection—using the new techniques of seismic tomography.

Seismic Tomography

Seismic tomography is an adaptation of a method now commonly used in medicine to map the human body, called computerized axial tomography (CAT). CAT scanners construct three-dimensional images of organs by measuring small differences in X rays that sweep the body in many directions. Similarly, geologists use the seismic waves from earthquakes recorded on thousands of seismographs all over the world to sweep Earth's interior in many different directions and construct a three-dimensional image of what's inside. The method enables them to find places where seismic waves speed up or slow down. They make the reasonable assumption, consistent with laboratory experiments, that regions where seismic waves speed up are composed of relatively cool, dense rock (for example, subducted ocean plates), whereas regions where seismic waves slow down indicate relatively hot, buoyant matter (for example, rising convection plumes).

Seismic tomography has revealed features in the mantle clearly associated with mantle convection. **Figure Story 21.10** presents a tomographic model of S-wave speed variations in the mantle constructed by researchers at Harvard University. The model is displayed as a series of global maps at depths ranging from just below the crust down to the core-mantle boundary. Near the surface, you can clearly see the structure of plate tectonics. The low S-wave speeds caused by the upwelling of hot asthenosphere along the mid-ocean ridges are shown in warm colors; the high S-wave speeds from cold lithosphere in the old ocean basins and beneath the continental cratons are shown in cool colors. At greater depths, the features become more variable and less coherent with the surface plates, reflecting what geologists infer to be a complex pattern of mantle convection. Some large-scale features stand out, however. For example, you will notice that, just above the core-mantle boundary, there is a red region of relatively low S-wave speeds beneath the central Pacific Ocean, surrounded by a broad blue ring of higher S-wave speeds. Seismologists have speculated that the high speeds represent a "graveyard" of oceanic lithosphere subducted beneath the Pacific's volcanic arcs—the Ring of Fire—during the last 100 million years or so.

Seismic tomography shows that cold lithospheric slabs plunge all the way through the mantle, demonstrating that the plate tectonic system involves convection extending almost as deep as the core-mantle boundary. An example is given in the cross section through the mantle shown in Figure 21.10. The image clearly reveals material from the once-large Farallon Plate, which has been mostly subducted under North America (see Chapter 20). The obliquely sinking slab material (in blue) appears to have penetrated the entire mantle. The image also indicates sinking colder rock beneath Indonesia. In addition, a large yellow blob of hotter rock, thought to be a superplume, can be seen rising at an angle from the core-mantle boundary to a position beneath southern Africa. This hot buoyant mass pushing up the cooler material above it may explain the uplifted, mile-high plateaus of South Africa. The other blobs of hotter and cooler materials may be evidence of material exchanges among the lithosphere, the mantle, and the core-mantle boundary layer.

Earth's Gravity Field

The same temperature variations that speed up and slow down seismic waves also change the densities of mantle rocks. Laboratory experiments show that the expansion of a rock due to a 300°C increase in temperature will reduce its density by about 1 percent. This might seem to be a small effect, but the mass of the mantle is enormous (about 4 billion trillion tons!), so even small changes in the distribution of mass can lead to observable variations in the pull of Earth's gravity.

Geologists can determine the gravity field and features of Earth's mass distribution by observing bulges and dimples in the shape of the planet. Through careful analysis, they have been able to show that the shape measured by Earth-orbiting satellites matches the pattern of mantle convection imaged by seismic tomography (see Feature 21.2). This agreement has allowed geologists to refine their models of the mantle convection system.

Figure Story 21.10 The cross section (upper right) shows the variations in S-wave speeds derived from seismic tomography. Regions with faster S-wave speeds indicate relatively colder, denser rock. The dense rocks sinking obliquely under North America are thought to be cold lithosphere from the subducted Farallon Plate. Cold subducted material is also evident under Indonesia. Regions with slower S-wave speeds (red and yellow) indicate relatively hotter, less dense rock. The yellow blob of hot rock has been identified by geologists as the African "superplume." It is some 1200 km across and rises obliquely 1500 km from the core-mantle boundary. [M. Gurnis, *Scientific American* (March 2001): 40.] The four maps show the variations in S-wave speeds at depths of 70 km, 200 km, 500 km, and 2800 km in the mantle. [S-wave speeds courtesy of G. Ekström and A. Dziewonski, Harvard University; maps courtesy of L. Chen and T. Jordan, University of Southern California.]

SEISMIC TOMOGRAPHY MAPS THE THREE-DIMENSIONAL STRUCTURE OF THE MANTLE

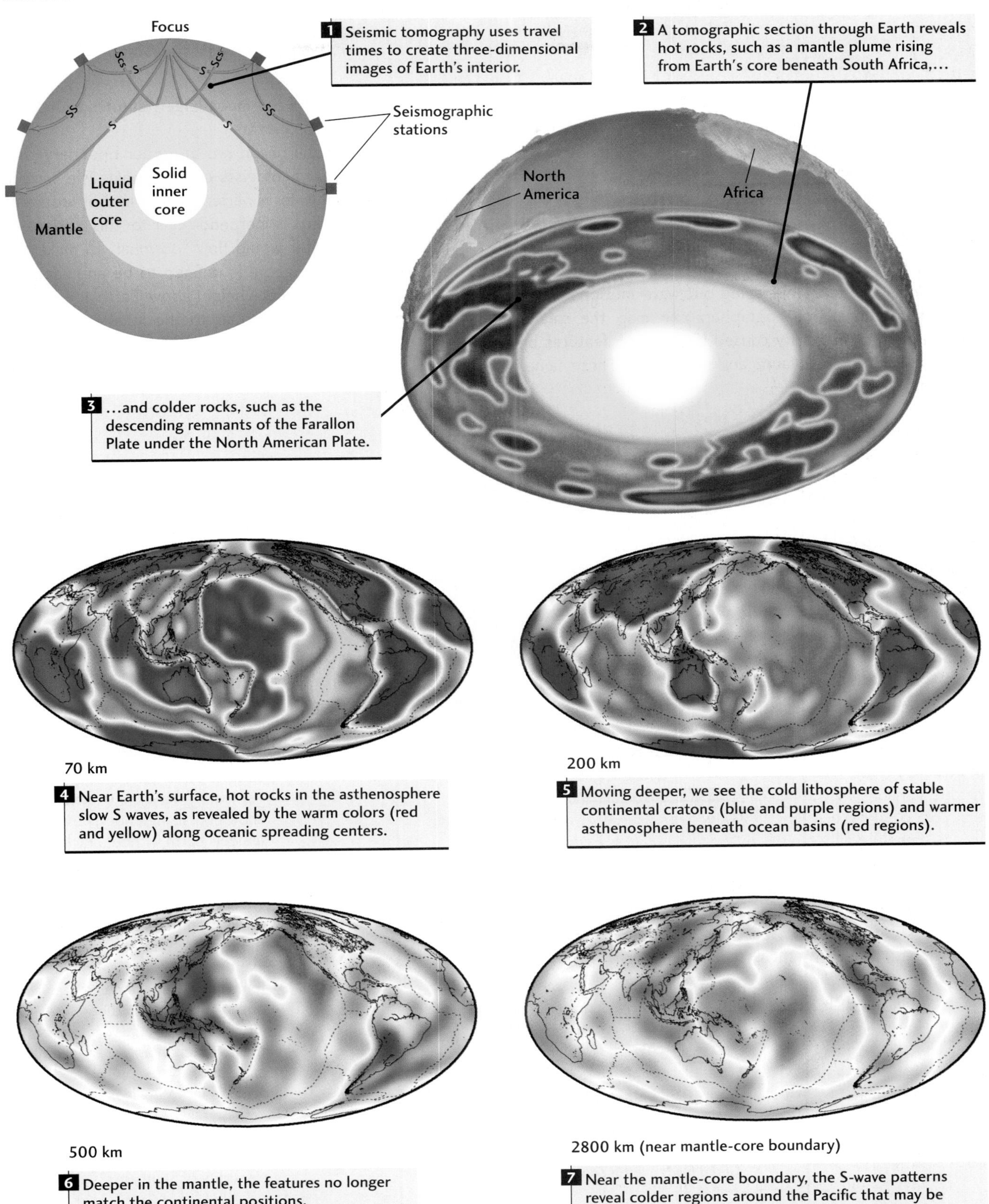

21.2 The Geoid: Shape of Planet Earth

In Chapter 17, we discussed how the gravity field shapes the surface of the ocean. It is warped upward in places where the gravitational pull is stronger and downward where the pull is weaker. The shape of the ocean's surface can be accurately measured by radar altimeters mounted on satellites. By averaging out wave motions and other fluctuations, oceanographers can map the small-scale variations in gravity caused by geologic features on the seafloor, such as faults and seamounts. These same data are also sensitive to the much larger features caused by mantle convection currents.

A perfectly still ocean has an upper boundary that conforms to what geologists call the *geoid*. The surface of a still body of water is perfectly "flat" in the sense that the pull of gravity is perpendicular to this surface—otherwise the water would flow "downhill" to make the surface flatter. The geoid is defined to be an imaginary surface at some reference height above Earth adjusted to be everywhere perpendicular to the local gravitational force. Because the ocean surface approximates the geoid, we usually take the reference height to be sea

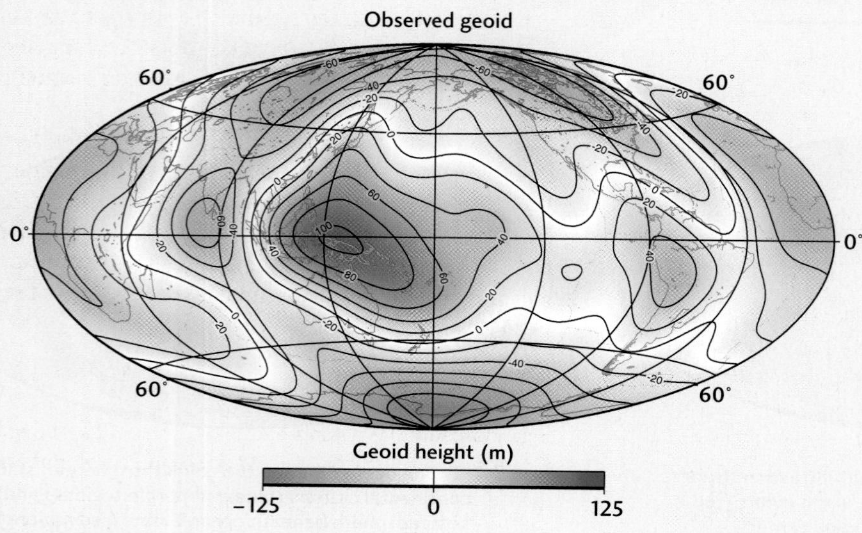

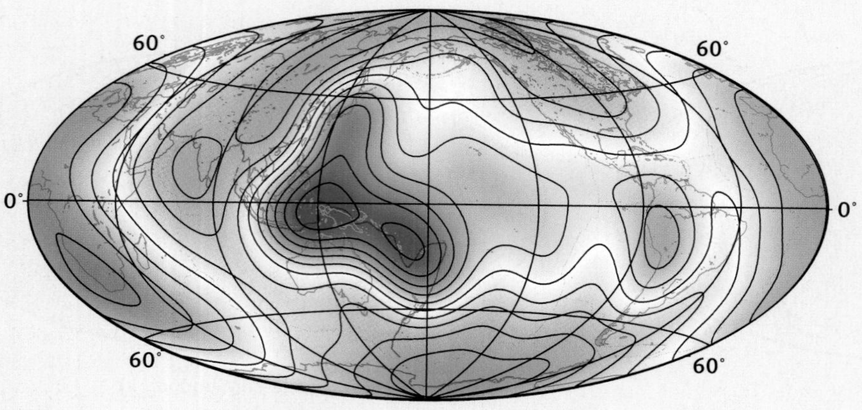

The upper panel of this figure is a smoothed map of the geoid or "shape of the Earth" derived from satellite observations. The contours, given here in meters, show how sea level deviates from an ideal Earth without any lateral variations in rock density. The lower panel is the same map computed from a model of mantle convection that is consistent with the temperature structure of the mantle derived from seismic tomography. By matching the observations with a theoretical model, geologists have improved their understanding of the mantle convection system. [Observed geoid from NASA; geoid from model from B. Hager, Massachusetts Institute of Technology; maps courtesy of L. Chen and T. Jordan, University of Southern California.]

level. When we measure the height of a mountain relative to sea level, we are actually referring to its height above the geoid at that point. In this sense, the geoid is just the "shape of the Earth." Geologists can use the geoid shape to calculate the size and direction of the gravitational force at any point on the planet's surface and infer how the rock density varies in Earth's interior.

Radar altimeters can easily map the geoid over the oceans, but how can we get this information on dry land? It turns out that the geoid can be measured for the entire Earth by tracking orbiting satellites. Three-dimensional mass variations in the mantle exert a small gravitational pull on the satellites, shifting their orbits slightly. By monitoring these shifts for long periods of time, scientists can create a two-dimensional map of the geoid over continents as well as oceans.

A smoothed version of the observed geoid is shown in the first accompanying figure. The highs and lows on this map tell us about the large-scale features in Earth's gravity field. Relative to what sea level would be on an Earth without any lateral variations in mass, the height of the geoid varies from a low of about −110 m at a point near the coast of Antarctica to a high of just over +100 m on the island of New Guinea in the western Pacific.

The geoid shows some similarities to the large-scale features in the deeper parts of the mantle, which you can see by comparing the geoid map with Figure 21.10. The agreement suggests that the three-dimensional variations in the rock density and S-wave speed are both related to temperature differences coming from the large-scale convective flow in the mantle.

This hypothesis can be tested in the following way. Using laboratory data for calibration, geologists first calculate the three-dimensional density differences from the seismic wave speed variations mapped by tomography. They then construct a computer model of convective flow by assuming that the heavier parts of the mantle are sinking while the lighter parts are rising. Finally, they calculate what the geoid shape should be according to this convection model. The second accompanying figure gives the results obtained more than a decade ago by the geophysicists Brad Hager and Mark Richards. You can see that the match to the observed geoid is quite good, especially for the largest features. This agreement has given geologists confidence that temperature variations within the mantle convection system can explain what we see in both the seismic images and the gravity field.

Earth's Magnetic Field

Like the mantle, Earth's outer core transports most of its heat by convection. But the same techniques that have revealed so much about mantle convection—seismic tomography and the study of Earth's gravity field—have provided almost no information about convection in the core. Why not?

The problem has to do with the fluidity of the outer core. The mantle is a viscous solid that flows very slowly. As a result, convection creates regions where the temperatures are significantly higher or lower than the average mantle geotherm. We see these regions as places where the density and seismic wave speeds are lower or higher than they are in the average mantle. The outer core, in contrast, has a very low viscosity—it can flow as easily as water or liquid mercury. Even small density variations caused by convection are quickly smoothed out by the rapid flow of core fluid under the force of gravity. Thus, temperature fluctuations of just a few degrees cannot be sustained in the outer core. Any lateral variations in density and seismic wave speeds caused by convection in the core would be much too small for geologists to see using seismic tomography and would not cause measurable distortions in the shape of the planet.

However, geologists can investigate convection in the outer core through observations of Earth's magnetic field. In Chapter 2, we briefly described the magnetic field, its reversals, and the paleomagnetism of volcanic rocks used to measure seafloor spreading. Here we will investigate further the nature of the magnetic field and its origin in the geodynamo of the outer core.

The Geodynamo System

The British have a long-standing interest in understanding Earth's magnetism, because the magnetic compass has been such an important instrument for that seafaring nation. In 1600, William Gilbert, physician to Queen Elizabeth I, first explained how a magnetic compass works. He offered the proposition that "the whole Earth is itself a great magnet" whose field acts on the small magnet of the compass needle to align it in the north-south direction.

In some ways, Earth's internal magnetic field does indeed behave as if a powerful bar magnet, inclined about 11° from Earth's axis of rotation, were located at Earth's center (**Figure 21.11**). A magnetic field can be visualized as lines of force, such as those revealed by the alignment of iron filings on a piece of paper above a bar magnet. Earth's magnetic lines of force point into the ground at the north magnetic pole and outward at the south magnetic pole. In other words, the lines of force look like a **dipole** (two-pole) magnetic field. A compass needle free to swing under the influence of the magnetic field will rotate into a position parallel to the local line of force, approximately in the north-south direction.

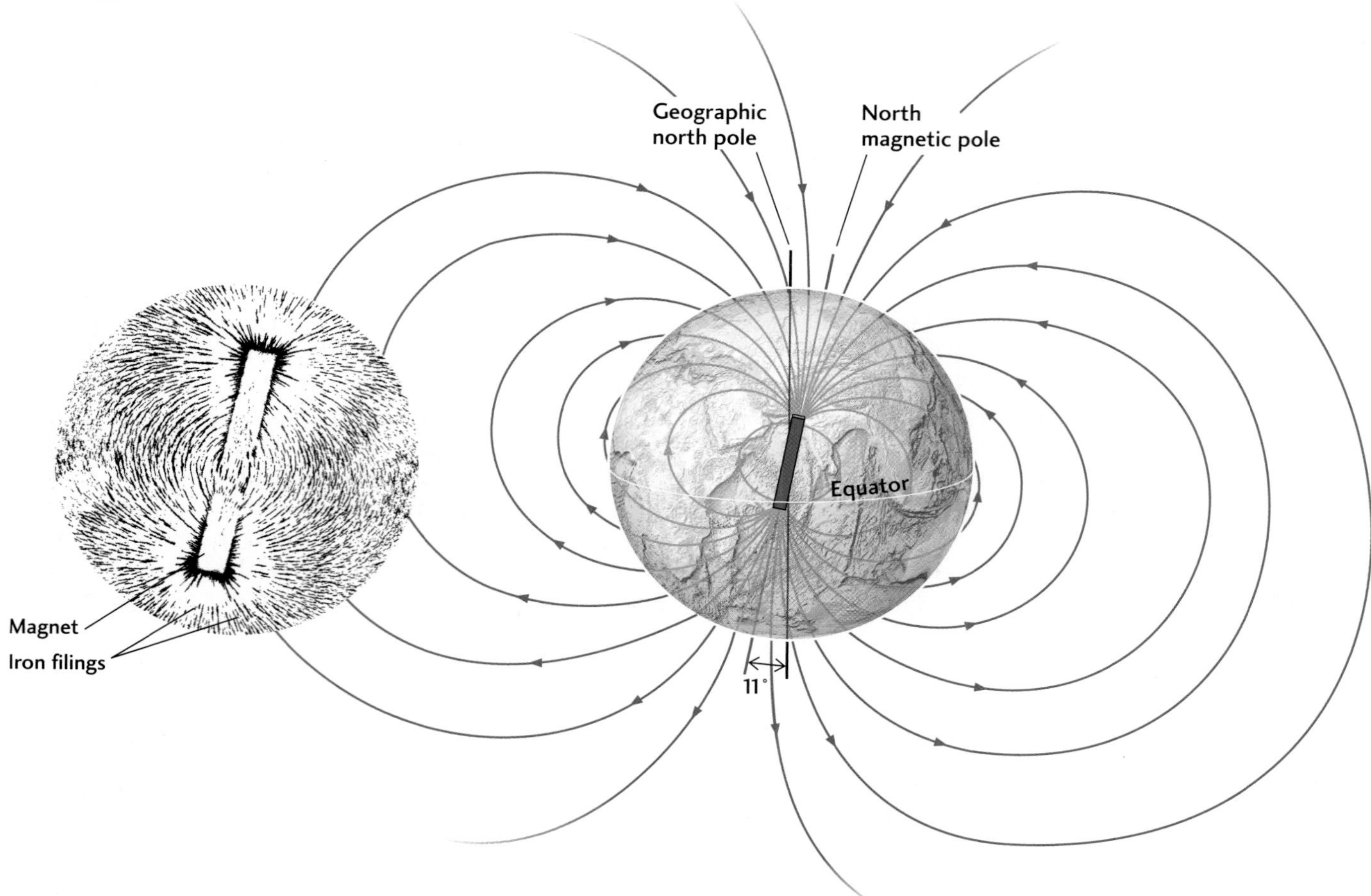

Figure 21.11 (*left*) The magnetic field of a bar magnet is revealed by the alignment of iron filings on paper. [After *PSSC Physics*, 3d ed. (Lexington, Mass.: D. C. Heath, 1971).] (*right*) Earth's magnetic field is much like the field that would be produced if a giant bar magnet were placed at the Earth's center and slightly inclined (11°) from the axis of rotation. Lines of magnetic force produced by such a bar magnet are shown. A compass needle points to the north magnetic pole because it orients in the direction of the local line of force.

Although a permanent magnet at Earth's center can explain the observed magnetic dipole, this model can be easily rejected. Laboratory experiments demonstrate that the field of a permanent magnet is destroyed when the magnet is heated above about 500°C. We have seen that temperatures in Earth's deep interior are much higher than that—thousands of degrees at its center (see Figure 21.9)—so unless the magnetism is constantly regenerated, it cannot be maintained.

Scientists theorize that convection in Earth's outer core generates and maintains the magnetic field. As in the case of mantle convection, the heat flowing out of the system is some combination of the energy that was trapped in the core during Earth's formation and the heat subsequently produced by radioactivity in the core. Why is a magnetic field created by convection in the outer core but not by convection in the mantle? First, the outer core is made primarily of iron, which is a very good electrical conductor, whereas the silicate rocks of the mantle are very poor electrical conductors. Second, the convective motions are a million times more rapid in the liquid, low-viscosity outer core than in the solid, high-viscosity mantle. These rapid motions can stir up electric currents in conducting iron to create a **geodynamo** with a strong magnetic field.

A dynamo is an engine that produces electricity by rotating a coil of conducting wire through a magnetic field. The magnetic field can either come from a permanent magnet or be generated by passing electricity through another coil—an electromagnet. The big dynamos in all commercial power plants use electromagnets (permanent magnets are too weak). In other words, part of the electricity produced by the dynamo is fed back into the system to sustain its magnetic field. The energy needed to keep the magnetic field going, as well as the electricity sent out to customers, comes from the mechanical work required to rotate the coil. In most power plants, this work is done by steam or falling water.

The geodynamo in Earth's outer core operates on the same basic principles, except the work comes from convection powered by the core's internal heat. Similar convective dynamos are thought to generate the strong magnetic fields observed on Jupiter and the Sun.

Behavior of the Magnetic Field

For hundreds of years, explorers and ship captains have measured the direction and strength of Earth's magnetic field. (Not surprisingly, some of the best records come from the British navy.) By the time Gilbert made his famous pronouncement in 1600, careful observers had realized that the magnetic field was not stationary but changed over time. Little did they know that these changes were coming from convective motions deep in Earth's core!

Nondipole Field Measurements at Earth's surface have revealed that only about 90 percent of the magnetic field can be described by the simple dipole illustrated in Figure 21.11. The remaining 10 percent, which geologists refer to as the *nondipole field,* has a more complex structure. This structure can be seen by comparing the strength of the field calculated for a simple dipole (**Figure 21.12**a) with the observed field (Figure 21.12b). If we extrapolate the field lines down to the core-mantle boundary, the size of the nondipole field actually increases relative to the size of the dipole field (Figure 21.12c). The poorly conducting mantle tends to smooth out complexities in the magnetic field, making the dipole field seem bigger than it really is.

Secular Variation Magnetic records for the last 300 years show that both the dipole and nondipole parts of the field are changing over time, but this *secular* (time-related) *variation* is fastest for the nondipole part. The secular variation is very evident in the comparison of today's magnetic field at the core-mantle boundary with maps reconstructed for previous centuries (see Figure 21.12e and f). Change in field strength occur on time scales of decades and indicate that fluid motions within the geodynamo system are on the order of millimeters per second.

Scientists can use the secular variation to help them understand convection in the outer core. With high-performance computers, they have been able to simulate the complex convective motions and electromagnetic interactions in the outer core that might be creating the geodynamo. The magnetic field lines from one such simulation are shown in Figure 21.12d. Away from the core, the field lines can be approximated by a dipole, but they become more complicated near the core-mantle boundary. Within the core itself, they are hopeless entangled by the strong convective motions.

Magnetic Reversals Computer simulations also allow us to understand a remarkable behavior of the geodynamo system: the spontaneous reversals of the magnetic field. As we learned in Chapter 2, the magnetic field reverses its direction at irregular intervals (ranging from tens of thousands to millions of years), exchanging the north and south magnetic poles as if the magnet depicted in Figure 21.11 were flipped 180°. Recent computer simulations of the geodynamo can reproduce these sporadic reversals in the absence of any other external triggers. In other words, Earth's magnetic field spontaneously reverses, purely through internal interactions.

This behavior illustrates a fundamental difference between the geodynamo and the dynamos used in power plants. A steam-powered dynamo is an artificial system engineered by humans to do a particular job. The geodynamo, in contrast, exemplifies a *self-organized natural system*—one whose behavior is not predetermined by external constraints but emerges from internal interactions. The other two global geosystems, plate tectonics and climate, also display a wide variety of self-organized behaviors. Trying to understand how these natural systems organize themselves is the greatest challenge of modern geoscience. We will return to this subject when we discuss the climate system in Chapter 23.

Paleomagnetism

We have seen repeatedly how the geologic record of ancient magnetism, or **paleomagnetism,** has become a crucial source of information for understanding Earth's history. Magnetic stripes mapped on oceanic crust confirmed the existence of seafloor spreading and still provide the best data to explain how plate motions have evolved since the breakup of Pangaea 200 million years ago (see Chapter 2). The paleomagnetism of old continental rocks has been essential for establishing the existence of earlier supercontinents, such as Rodinia (see Chapter 20).

Scientists have also used paleomagnetism to reconstruct the history of Earth's magnetic field. The oldest magnetized rocks found so far formed around 3.5 billion years ago and indicate that Earth had a magnetic field at that time similar to the present one. The presence of magnetism in the most ancient rocks is consistent with the ideas of Earth's differentiation discussed in Chapter 1, which imply that a convecting fluid core must have been established very early in Earth's 4.5-billion-year history.

Let's delve a little more deeply into the rock-forming processes that have allowed geologists to draw these remarkable conclusions.

Thermoremanent Magnetization In the early 1960s, an Australian graduate student found a fireplace in an ancient campsite where the Aborigines had cooked their meals. The stones were magnetized. He carefully removed several stones that had been baked by the fires, first noting their physical orientation. Then he measured the direction of the stones' magnetization and found that it was exactly the reverse of Earth's present magnetic field. He proposed to his disbelieving professor that, as recently as 30,000 years ago, when the campsite was occupied, the magnetic field was the

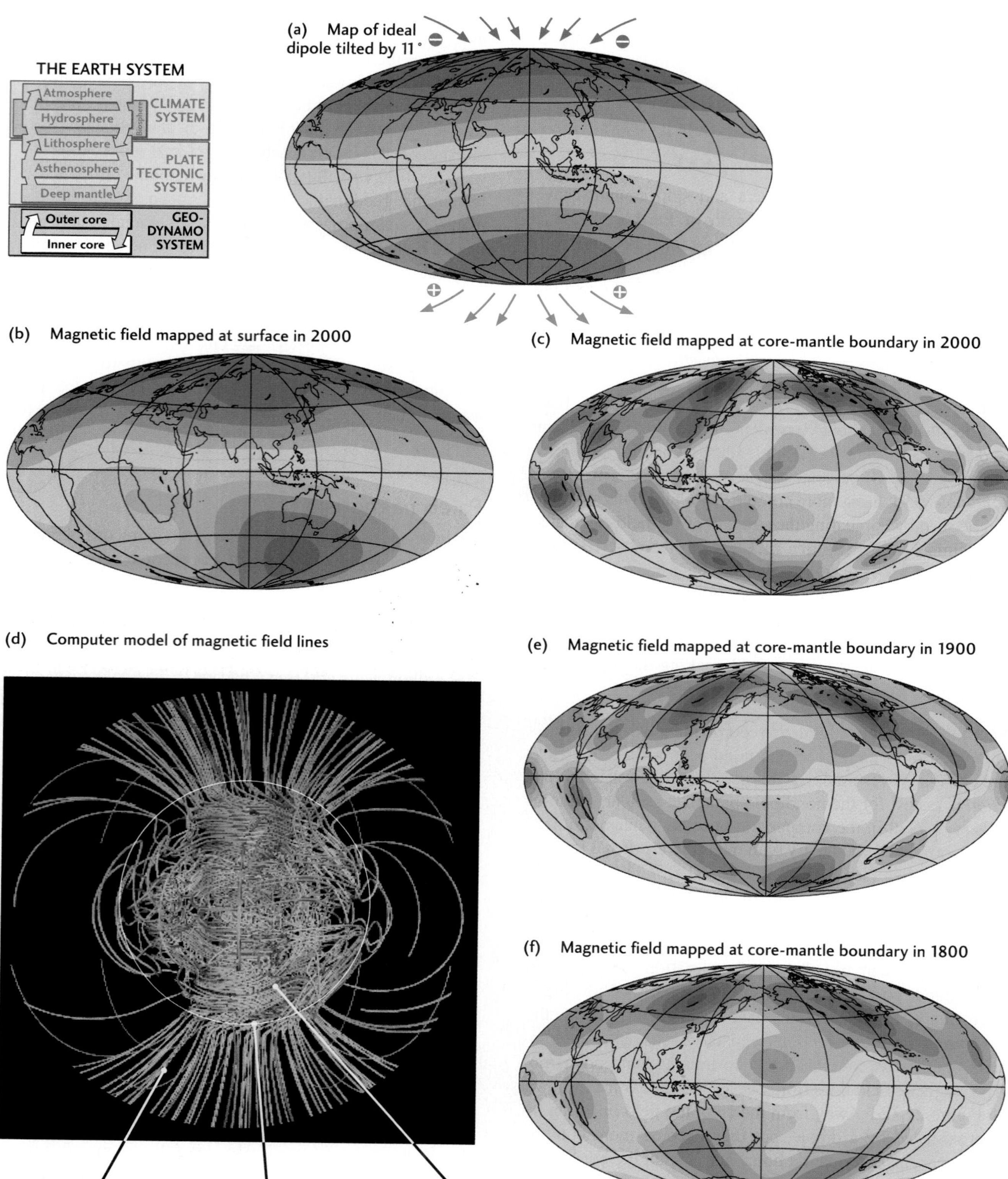
THE EARTH SYSTEM
Atmosphere
Hydrosphere
Biosphere
CLIMATE SYSTEM
Lithosphere
Asthenosphere
Deep mantle
PLATE TECTONIC SYSTEM
Outer core
Inner core
GEO-DYNAMO SYSTEM
(a) Map of ideal dipole tilted by 11°
(b) Magnetic field mapped at surface in 2000
(c) Magnetic field mapped at core-mantle boundary in 2000
(d) Computer model of magnetic field lines
(e) Magnetic field mapped at core-mantle boundary in 1900
(f) Magnetic field mapped at core-mantle boundary in 1800
Magnetic field lines in mantle approximate those of a dipole.
Field mapped at core-mantle boundary reveals complexities in the core.
Field lines in core are entangled owing to convective motions that creates the geodynamics.

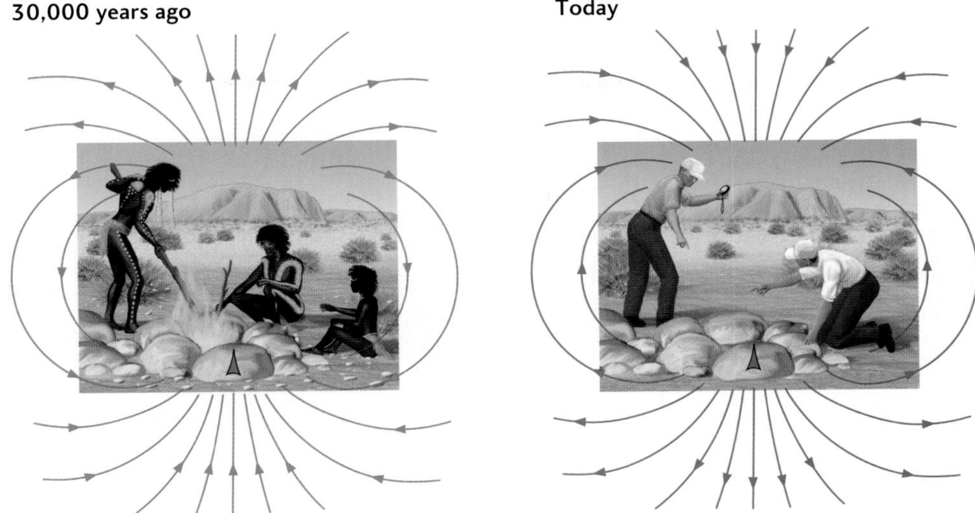

Figure 21.13 Earth's magnetic field 30,000 years ago was the reverse of today's, as evidenced by the discovery of reversely magnetized rocks found in the fireplace of an ancient campsite. The rocks, cooling after the last fire, became magnetized in the direction of the ancient magnetic field, leaving a permanent record of it, just as a fossil leaves a record of ancient life.

reverse of the present one—that is, a compass needle would have pointed south rather than north.

Recall that high temperatures destroy magnetism. An important property of many magnetizable materials is that, as they cool below about 500°C, they become magnetized in the direction of the surrounding magnetic field. This happens because groups of atoms of the material align themselves in the direction of the magnetic field when the material is hot. When the material has cooled, these atoms are locked in place and therefore are always magnetized in the same direction. This process is called **thermoremanent magnetization,** because the magnetization is "remembered" by the rock long after the magnetizing field has disappeared. Thus, the Australian student was able to determine the direction of the field when the stones cooled after the last fire and took on the magnetization of Earth's magnetic field of that time (**Figure 21.13**).

The discovery of magnetic reversals and the means to decipher them was a key ingredient in formulating the theory of plate tectonics. You should review the material in Figure Story 2.11 and its accompanying text, which discusses thermoremanent magnetization and magnetic stratigraphy.

Depositional Remanent Magnetization Some sedimentary rocks can take on a different type of remanent magnetization. Recall that marine sedimentary rocks form when particles of sediment that have settled through the ocean to the seafloor become lithified. Magnetic grains among the particles—chips of the mineral magnetite, for example—become aligned in the direction of Earth's magnetic field as they fall through the water, and this orientation can be incorporated into the rock when the particles becomes lithified. The **depositional remanent magnetization** of a sedimentary rock results from the parallel alignment of all these tiny magnets,

Figure 21.12 Maps comparing the strength of the magnetic field from an ideal dipole (a) with the strength of Earth's magnetic field observed at the surface (b) and calculated at the core-mantle boundary (c). The core-mantle boundary map was constructed by extrapolating the magnetic field lines downward through the mantle, which increases the size of the nondipole part of the field. The nondipole field is caused by complexities in the structure of the core's internal magnetic field. (d) A computer model of what the magnetic field lines might look like in the core and mantle as they are twisted around by convective motions in the fluid outer core. (e, f) Maps comparing the strength of the magnetic field at the core-mantle boundary from observations made in A.D. 1900 and A.D. 1800. By comparing part (c), you can see the secular variation as changes in the nondipole field from century to century. These changes are caused by the complex convective motions in Earth's fluid outer core that generate the geodynamo. [Maps courtesy of J. Bloxham, Harvard University; computer model by G. Glatzmaier, University of California, Santa Cruz.]

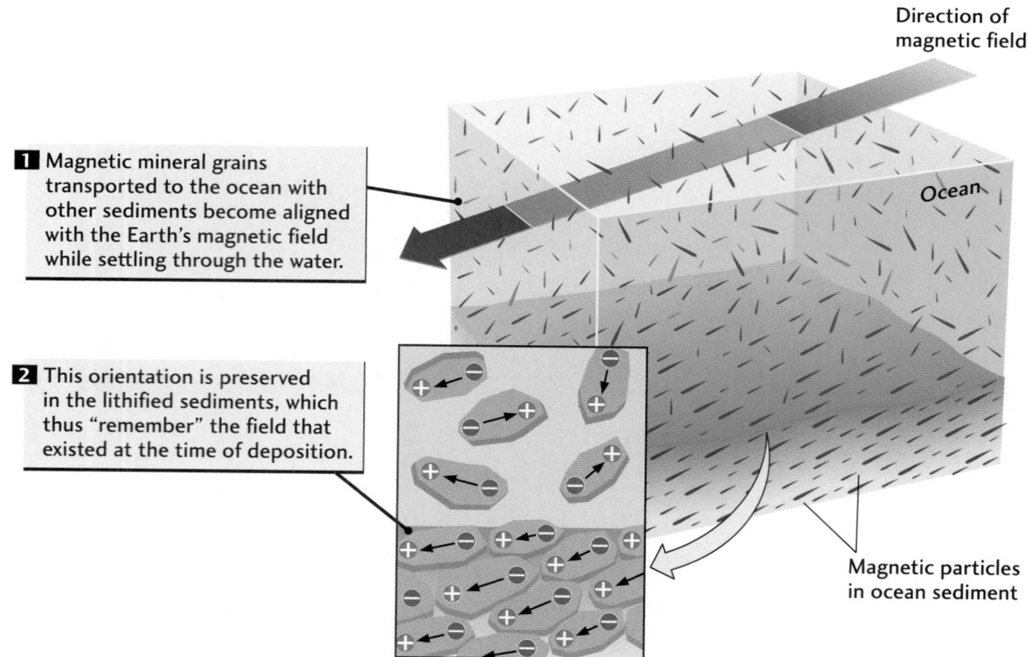

Figure 21.14 Newly formed sedimentary deposits can become magnetized in the same direction as the contemporaneous magnetic field of the Earth.

as if they were compasses pointing in the direction of the field prevailing at the time of deposition (**Figure 21.14**).

Magnetic Stratigraphy Reversals of Earth's magnetic field are clearly indicated in the fossil magnetic record of layered lava flows. Each layer of rocks from the top down represents a progressively earlier period of geologic time, whose age can be determined by radiometric dating methods. The direction of remanent magnetism can be obtained for each layer, and in this way the time sequence of flip-flops of the field—that is, the **magnetic stratigraphy**—can be deduced. Geologists can also get data on Earth's magnetic reversal history by mapping magnetic stripes on the seafloor. From a combination of these data, they have worked out a detailed history of reversals for the last 200 million years (**Figure 21.15**). This information is of use to archeologists and anthropologists as well as geologists. For example, the magnetic stratigraphy of continental sediments has been used to date sediments containing the remains of predecessors of our own species.

About half of all rocks studied are found to be magnetized in a direction opposite that of Earth's present magnetic field. Apparently, then, the field has flipped frequently over geologic time, and normal (same as now) and reversed fields are equally likely. Normal and reversed epochs, now called *magnetic chrons* (from the Greek word for "time"), seem to last about a half-million years. Superimposed on the chrons are transient, short-lived reversals of the field, known as *subchrons,* which may last anywhere from several thousand years to tens of millions of years. The Australian graduate student apparently found a new subchron within the present normal magnetic chron.

SUMMARY

What do seismic waves reveal about the layering of Earth's crust and mantle? Seismic waves were used to discover that beneath Earth's felsic crust lies a denser ultramafic mantle composed mostly of peridotite. The crust and outer part of the mantle make up the rigid lithosphere. Beneath the lithosphere lies the asthenosphere, the weak layer of the mantle across which the lithosphere slides in plate tectonics. At the top of the asthenosphere, the temperature is high enough to partially melt peridotite, forming a zone where S-wave speeds decrease with depth. Below about 200 km, seismic wave speeds increase with depth down to the Earth's core. However, at two places, 410 km and 660 km below the surface, they show jumps caused by phase changes that transform mantle minerals to denser, more closely packed crystal structures stable at higher pressures.

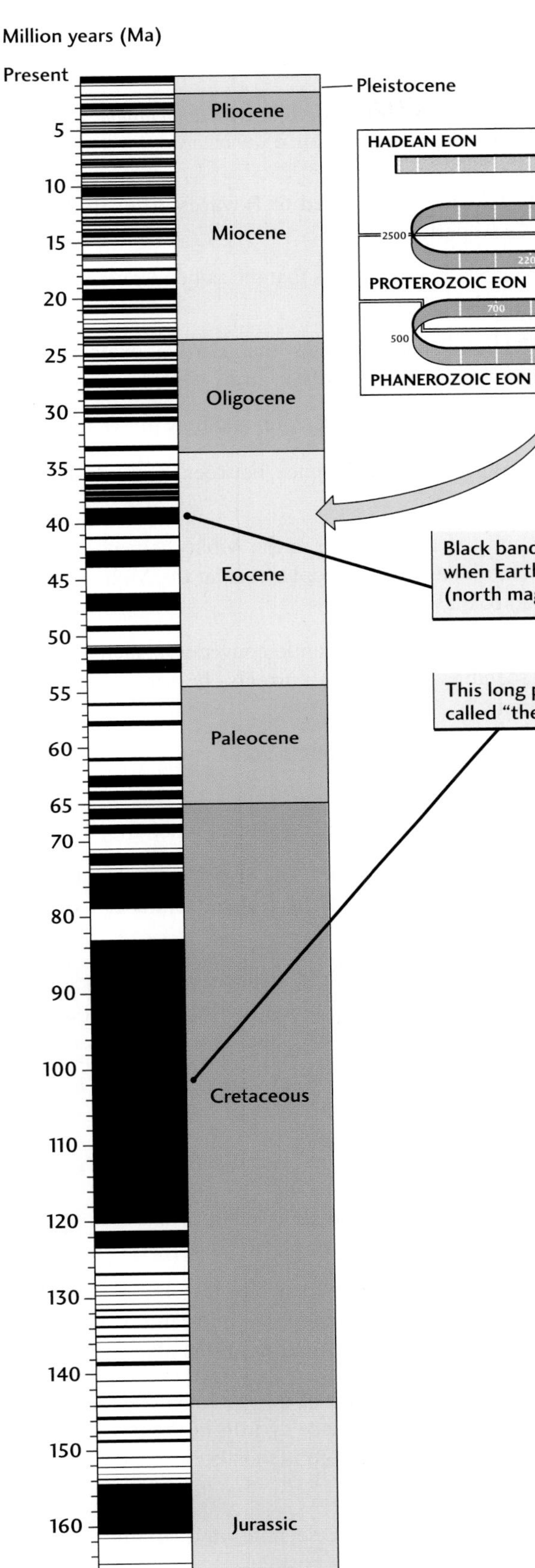

Figure 21.15 The paleomagnetic time scale from 170 million years ago to the present, showing epochs of normal polarity (black bands) and reverse polarity (white bands).

What do seismic waves reveal about the layering of Earth's core? Seismic waves reflected from the core-mantle boundary locate this chemical transition at a depth of 2890 km. The failure of S waves to penetrate below the core-mantle boundary indicates that the outer core is a fluid. Seismic wave speeds indicate that the fluid outer core becomes a solid inner core at a depth of 5150 km. Several lines of evidence show that the core is composed mostly of iron and nickel, with minor amounts of lighter elements such as oxygen and sulfur.

What has seismic tomography revealed about structures in the mantle? Geologists use seismic tomography to make three-dimensional images of Earth's interior. The images show how plates vary from very thin under mid-ocean ridges to very thick under continental cratons. They also reveal many features of mantle convection, such as lithospheric slabs sinking into the lower mantle (some all the way to the core-mantle boundary) and superplumes rising from deep within the mantle.

How hot does it get in Earth's interior? Earth's interior is hot because it still retains much of the heat of its violent formation as well as the heat generated by the decay of radioactive elements. It has cooled over geologic time primarily by convection in the mantle and core and by conduction of heat through the lithosphere. The geotherm describes how temperature increases with depth. Within normal continental crust, it increases at a rate of 20° to 30°C per kilometer. Temperatures near the base of the lithosphere reach 1300° to

1400°C, which is hot enough to begin to melt mantle peridotites. The temperature in the fluid core is probably greater than 3000°C. The temperature at the Earth's center is about 5000°C according to some geologists, or it may be as high as 6000° to 8000°C according to others.

What does Earth's gravity field tell us about the interior? Variations in the pull of gravity over Earth's surface and corresponding distortions in its shape can be measured by satellites. These variations arise primarily from the temperature variations caused by mantle convection, which affect the density of rocks (higher temperatures reduce densities and seismic velocities). The observed gravity field is in agreement with the pattern of mantle convection inferred from seismic tomography.

What does Earth's magnetic field tell us about the fluid outer core? Convective motions in the outer core stir the electrically conducting iron-rich fluid, forming a geodynamo that produces the magnetic field. The magnetic field at the surface is primarily a dipole, but it has a small nondipole part. Maps of the magnetic field derived from compass readings show that it has changed over the last several centuries, which tells us about the type of fluid motions that drive the geodynamo.

What is paleomagnetism and what is its importance? Geologists have discovered that rocks can become magnetized in the direction of Earth's magnetic field at the time they form. This remanent magnetization can be preserved in rocks for millions of years. Paleomagnetism tells us that Earth's magnetic field has reversed (flipped back and forth) over geologic time. The chronology of reversals has been worked out so that the direction of remanent magnetization of a rock formation is often an indicator of stratigraphic age.

Key Terms and Concepts

compressional wave (p. 484)
conduction (p. 491)
convection (p. 492)
core-mantle boundary (p. 490)
depositional remanent magnetization (p. 501)
dipole (p. 497)
geodynamo (p. 498)
geotherm (p. 492)
lower mantle (p. 490)
low-velocity zone (p. 488)
magnetic stratigraphy (p. 502)
Mohorovičić discontinuity (Moho) (p. 487)
paleomagnetism (p. 499)
phase change (p. 490)
seismic tomography (p. 494)
shadow zone (p. 485)
shear wave (p. 484)
thermoremanent magnetization (p. 501)

Exercises

This icon indicates that there is an animation available on the Web site that may assist you in answering a question.

1. How does the speed of P waves differ in granite, gabbro, and peridotite?

2. What evidence suggests that the asthenosphere is probably partially molten?

3. What evidence indicates that Earth's outer core is molten and composed mostly of iron?

4. What is the depth to the core, and how do we know it?

5. What is the difference between heat conduction and convection?

6. Would the temperature at the Moho beneath a continental craton be hotter or cooler than at the Moho beneath an ocean basin?

7. Why can features of mantle convection, such as rising and descending convection currents, be seen by seismic tomography?

8. How can a mountain float on the mantle when both are composed of rock?

9. How do rocks become magnetized when they form?

10. What evidence supports the hypothesis that Earth's magnetic field is generated by a geodynamo in its outer core?

11. Does the magnetic field change by an observable amount over the span of a human lifetime? What does this say about fluid motions in the outer core?

Thought Questions

This icon indicates that there is an animation available on the Web site that may assist you in answering a question.

1. The Moon shows no evidence of tectonic plates or of their motions. What does this observation imply about the state and temperature of the interior of this planetary body?

2. How does the existence of Earth's magnetic field, iron meteorites, and the abundance of iron in the cosmos support the idea that Earth's core is mostly iron and the outer core is liquid?

3. How would you use seismic waves to find a chamber of molten magma in the crust?

4. How would seismic tomography answer the question, "How deep do the subducted slabs go before they are assimilated?"

5. Where might you look in the mantle to find regions of anomalously low S-wave speeds?

Suggested Readings

Bolt, B. A. 1993. *Earthquakes and Geological Discovery.* New York: Scientific American Library.

Gurnis, M. 2001. Sculpting the Earth from inside out. *Scientific American* (March): 40.

Hager, B. H., and M. A. Richards. 1989. Long-wavelength variations in Earth's geoid: physical models and dynamical implications. *Phil. Trans. Roy. Soc. London, Ser. A:* 309–327.

Helffrich, G. R., and B. J. Wood. 2001. The Earth's mantle. *Nature* 412: 501–507.

Kellog, L. H. 1997. Mapping the core-mantle boundary. *Nature* 277: 646–647.

Kerr, R. A. 2001. A lively or stagnant lowermost mantle? *Science* 292: 841.

Lay, T., and T. C. Wallace. 1995. *Modern Global Seismology.* New York: Academic Press.

McKenzie, D. P. 1983. The Earth's mantle. *Scientific American* (September): 66.

Masters, T. G., and P. M. Shearer. 1995. Seismic models of the Earth. In *A Handbook of Physical Constants: Global Earth Physics* (Vol.1), ed. Thomas J. Ahrens, pp. 88–103. Washington, D.C.: American Geophysical Union.

Olson, P., P. G. Silver, and R. W. Carlson. 1990. The large-scale structure of convection of the Earth's mantle. *Nature* 344: 209–215.

Stein, S., and M. Wysession. 2003. *An Introduction to Seismology, Earthquakes, and Earth Structure.* London: Blackwell.

Wysession, M. 1995. The inner workings of the Earth. *American Scientist* 83: 134–146.

Trans-Alaska pipeline crossing coniferous forest, near Gulkana, Alaska. [Danny Lehman/Corbis.]

CHAPTER

22

Energy and Material Resources from the Earth

"Drill for oil? You mean drill into the ground to try and find oil? You're crazy."

DRILLERS WHOM EDWIN I. DRAKE TRIED TO ENLIST FOR HIS OIL-DRILLING PROJECT IN 1859

Earth's internal processes have given us our life-sustaining environment: a breathable atmosphere, oceans, rich soils, a moderate climate. Humans living in the Stone Age survived in this environment at subsistence levels. Humankind progressed to a better quality of life when we learned how to recover and use Earth's minerals. These resources create wealth and comfort by providing the materials and energy needed to grow and process food, build structures, transport things, and manufacture goods of all kinds. Using Earth's finite resources without regard for the fragility of the Earth system can lead to the depletion of those resources and the hazardous accumulation of wastes. It can also trigger climate change, with serious consequences. The challenge to humankind is to use Earth's resources wisely and equitably to ensure a sustainable future.

We have steadily increased the amounts of coal, oil, natural gas, and uranium we take from the Earth to power our complex societies. Minerals are critical to the functioning of a modern nation. Just about everything we use—metals, building stone and cement, sand from which glass and transistors are made—comes from the ground.

In this chapter, we will consider the following questions. How do these resources form? Where are they found? Who does and will control them? How long will the supplies of these critical nonrenewable resources last, and what will we do when they are exhausted? Such questions have aroused concern about the environmental effects of the fuels and minerals we use, about the need to conserve our resources, and about the development of substitutes for them. These concerns result from a new and deeper understanding that we cannot continue to draw wealth from the Earth indefinitely without thinking

about the consequences for our habitat and for the generations to come.

In our increasingly systematic search of the Earth for new sources of the fuel and minerals on which we depend, we use our geological knowledge of how known natural deposits form to determine where we may find more of them. At the same time, we are becoming more sensitive to the finiteness of Earth's resources and the delicacy of its environment. We are paying more attention to extracting and using natural resources more efficiently, without damaging the environment. We are beginning to think about how we can change our use of resources to achieve **sustainable development**—development that will preserve the prospects of future generations.

Resources and Reserves

Two major questions arise in all discussions of materials that we draw from the Earth: How much is left? How long will it last? The amount left consists of more (we hope) than reserves. **Reserves** are deposits that have already been discovered and that can be mined economically and legally at the present time. **Resources,** in contrast, constitute the entire amount of a given material that may become available for use in the future. Resources include reserves, plus discovered but currently unprofitable deposits, plus undiscovered deposits that geologists think they may be able to find eventually (**Figure 22.1**). Often, resources that are too poor in quality or quantity to be worth mining now or that are too difficult to retrieve become profitable when new technology is developed or prices rise. A recent example is the discovery and production of oil and gas from large reservoirs in the continental margin of the Gulf of Mexico, in water depths reaching 3000 m.

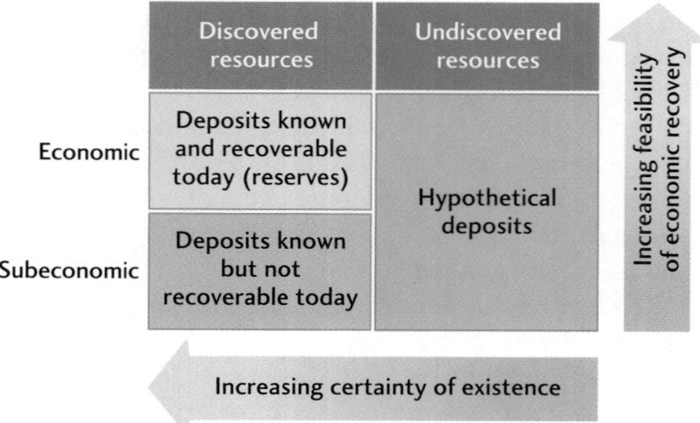

Figure 22.1 Categories that constitute total resources. Discovered resources consist of reserves—known deposits that are economically minable today—and deposits that are known but that are currently subeconomic. Undiscovered resources are hypothetical deposits that we may find.

Reserves are considered a dependable measure of supply as long as economic and technological conditions remain as they are now. As conditions change, some resources become reserves, and vice versa. The conversion of oil resources in the North Sea into productive oil fields, for example, was accelerated by new technology and by price increases resulting from political instability in the oil-exporting nations of the Middle East. The assessment of resources is much less certain than the assessment of reserves. Any figure cited as the resources of a particular material represents only an educated guess of how much will be available in the future.

Most useful geological materials are considered *nonrenewable* because geological processes produce them more slowly than civilization uses them up. Coal and oil, for example, will be exhausted faster than nature can replenish them. This fact makes it increasingly important to develop *renewable* resources such as solar energy, which is essentially infinite in supply, and fuels such as ethanol, which is derived from crops that can be replanted after they are harvested.

Energy Resources

Energy is fundamental to everything. A crisis in the supply of energy can bring a modern society to a halt. Wars have been fought over access to supplies of fuel resources; economic recession and destructive currency inflation have resulted from gyrations in the price of oil. It is not surprising that energy is the biggest business in the world.

Fuel resources are measured in units appropriate to the material, for example, barrels of oil, tons of coal. To make it easier to compare the energy available from different fuel resources, a common unit called the quad is used. A quad is a measure of the energy that can be extracted from a given amount of fuel. The quad is based on a standard measure of energy called the British thermal unit (Btu). One Btu is the amount of energy needed to raise the temperature of 1 pound of water by 1°F. One quad equals 10^{15} Btu. The United States uses about 98 quads of energy a year. **Figure 22.2** shows one estimate of the world's remaining nonrenewable energy resources of all types—about 360,000 quads.

In 1999, the world consumption of energy was 382 quads, of which all but about 35 quads were nonrenewable. (Consumption is estimated to grow to 607 quads by 2020.) Calculations based on these numbers can be deceptive, however. For example, simply dividing total resources by annual consumption might lead us to conclude (mistakenly) that many hundreds of years of resources remain before we have to worry about depletion of the supply. However, some energy sources will give out before others, and the various sources of energy are not readily interchangeable. Also, as we will see later in this chapter and in Chapter 23, each of

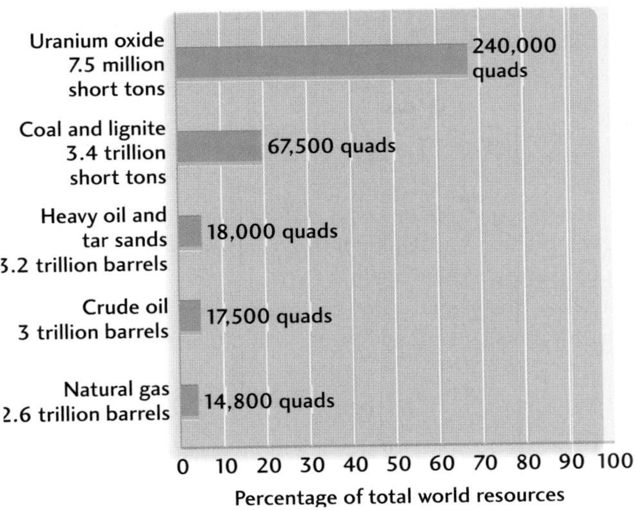

Figure 22.2 A rough estimate of total remaining nonrenewable world energy resources amounts to about 360,000 quads. Amounts are given in conventional units of weight (short tons), volume (barrels), and energy content (quads). A short ton is 2000 pounds, or 907.20 kg; a barrel of oil is 42 gallons. Coal and lignite resources, for example, amount to 3.4 trillion short tons, equivalent to 67,500 quads, or 19 percent of total energy resources. [World Energy Council.]

the nonrenewable sources of energy poses a serious threat to the environment, and continued use of the ones we depend on most—coal, oil, and natural gas—will likely trigger global climate change, with possibly dire consequences.

Energy Use

As the world industrialized, the demand for energy increased and the types of energy used changed. The industrial revolution of the eighteenth and nineteenth centuries was powered by the energy from coal—in Britain, from the coalfields of England and Wales; in continental Europe, from the coal basins of western Germany and bordering countries; and in North America, from the Appalachian coalfields of Pennsylvania and West Virginia. As industrialization expanded, so did the hunger for coal. Geological exploration for this fuel spread over much of the world as coal use climbed.

Half a century after the first oil well in America was drilled in 1859, oil and natural gas were beginning to displace coal as the fuels of choice. Not only did they burn more cleanly, producing no ash, but they could be transported by pipeline as well as by rail and ship.

In the last quarter of the twentieth century, nuclear energy was introduced, with expectations that it would provide a large, new, low-cost, environmentally benign source of energy. These expectations were not realized, however, because safety concerns, the inability to dispose of nuclear wastes, and the escalating costs of stringent safety measures slowed the construction of nuclear power plants.

The advanced nations depend primarily on oil, coal, natural gas, and some nuclear energy. In poorer countries, wood is an important source of fuel. **Figure 22.3** summarizes the history of energy use in the United States. In 1850, coal accounted for less than 5 percent of U.S. energy use. Today, oil, coal, and natural gas supply almost 90 percent of the energy used in the United States. Consumption of this supply is distributed among residential and commercial uses (35.8 percent), industrial use (37.0 percent), and transportation (27.1 percent). Approximately half of the energy produced is lost in distribution and inefficient use.

Fossil Fuels

A century and a half ago, most of the energy used in the United States came from the burning of wood. A wood fire,

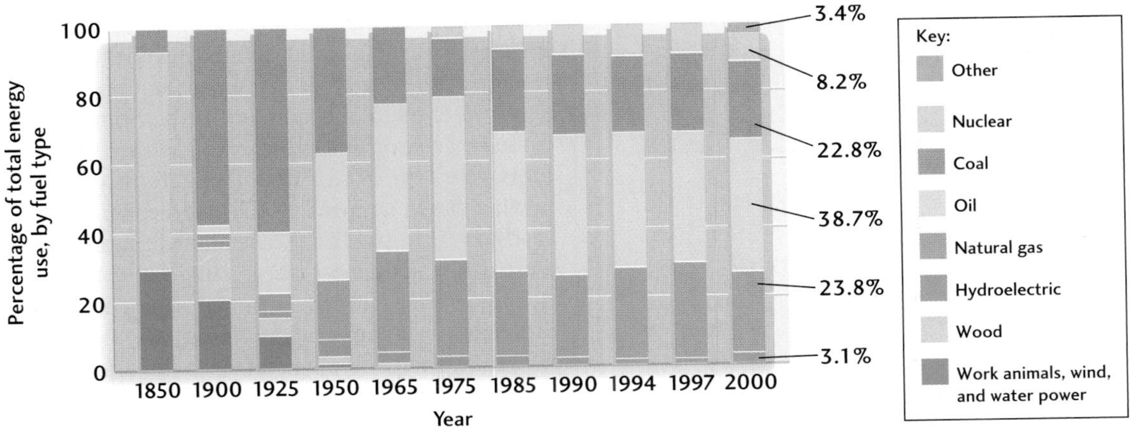

Figure 22.3 Percentages of various types of energy used in the United States from 1850 to 2000. Shown as "Other" for the year 2000 is about 3.4 percent distributed over wood, geothermal, solar, biomass, wind, and other types of energy. [U.S. Energy Information Agency, 2001.]

in chemical terms, is the combustion of organic matter consisting of compounds of carbon and hydrogen. Organic matter is produced by plants and animals. The organic matter in this case is a tree, which obtained its energy for growth by a process called photosynthesis (see Chapter 23). During photosynthesis, plants use the energy supplied by sunlight to convert carbon dioxide and water to carbohydrates. Thus, we can look upon a piece of wood or any piece of plant matter as a photosynthetic product that can be returned by burning or decay to the carbon dioxide and water from which it was made.

If we burn wood that was buried and transformed 300 million years ago into the combustible rock known as coal, we are using energy stored by photosynthesis from late Paleozoic sunlight. We are recovering "fossilized" energy. Crude oil and natural gas were also created by a process of burial and chemical transformation of dead organic matter into a combustible liquid and gas, respectively. We refer to all such resources derived from natural organic materials, from coal to oil and natural gas, as **fossil fuels** (Figure 22.4).

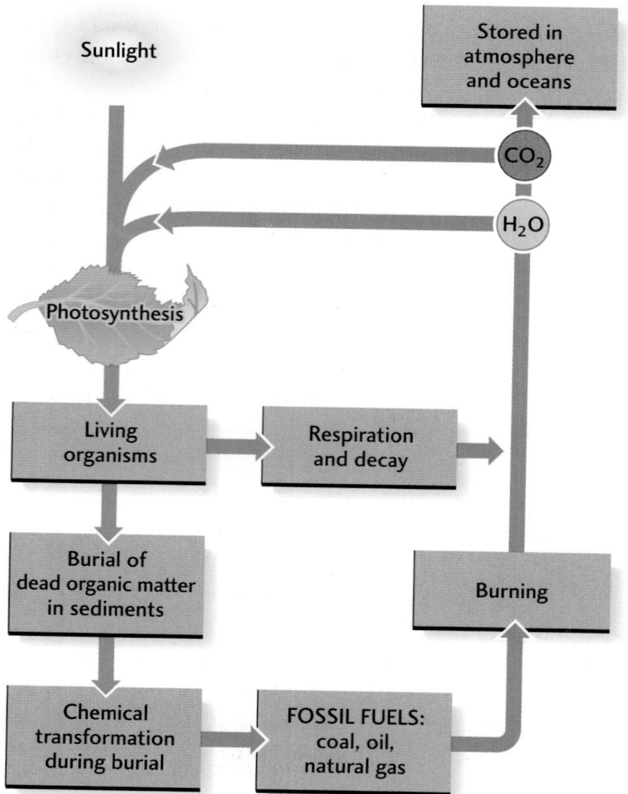

Figure 22.4 Photosynthesis produces organic matter from carbon dioxide (CO_2) and water (H_2O). If dead organic matter is buried and transformed into coal, oil, or natural gas, it becomes a fossilized product of photosynthesis—a fossil fuel. The burning of fossil fuels releases the carbon dioxide and water from which they were made.

More than 85 percent of the world's energy is presently derived from fossil fuels.

Oil and Natural Gas

Crude oil (petroleum) and natural gas in minable form develop under special environmental and geological conditions. Both are the organic debris of former life—plants, bacteria, algae and other microorganisms that have been buried, transformed, and preserved in marine sediments.

How Oil and Gas Form

Oil and gas begin to form when more organic matter is produced than is destroyed by scavengers and decay. This condition exists in environments where the production of organic matter is high—such as in the coastal waters of the sea, where large numbers of organisms thrive—and the supply of oxygen in bottom sediments is inadequate to decompose all the organic matter. Many offshore sedimentary basins on continental shelves satisfy both these conditions. In such environments, and to a lesser degree in some river deltas and inland seas, the rate of sedimentation is high and organic matter is buried and protected from decomposition.

During millions of years of burial, chemical reactions triggered by the elevated temperatures at depth slowly transform some of the organic material into liquid and gaseous compounds of hydrogen and carbon (hydrocarbons). The hydrocarbons are the combustible materials of oil and natural gas. Compaction of muddy organic sediments in their source beds forces the hydrocarbon-containing fluids and gases into adjacent beds of permeable rock (such as sandstones or porous limestones), which we call *oil reservoirs.* The low density of oil and gas causes them to rise to the highest place that they can reach, where they float atop the water that almost always occupies the pores of permeable formations.

Oil Traps Geological conditions that favor the large-scale accumulation of oil and natural gas are combinations of structure and rock types that create an impermeable barrier to upward migration—an **oil trap.** Some oil traps are caused by a structural deformation and are called *structural traps.* One type of structural trap is formed by an anticline in which an impermeable shale overlies a permeable bed of sandstone (Figure 22.5a). The oil and gas accumulate at the crest of the anticline—the gas highest, the oil next—both floating on the groundwater that saturates the sandstone. Similarly, an angular unconformity or displacement at a fault may place a dipping permeable limestone bed opposite an impermeable shale, creating a structural trap for oil (Figure 22.5b). Other oil traps are created by the original pattern of sedimentation, such as when a dipping permeable sandstone bed thins out against an impermeable shale (Figure 22.5c). These are called *stratigraphic traps.* Oil can

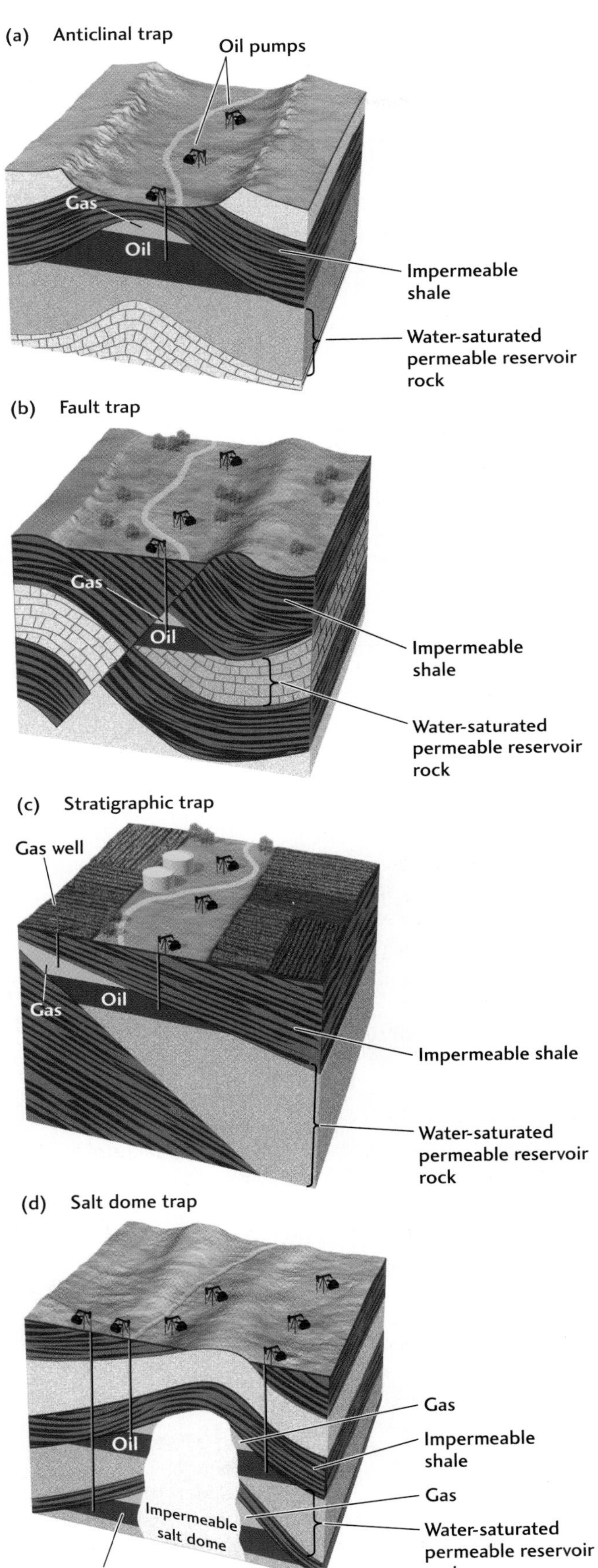

also be trapped against an impermeable mass of salt in a salt dome trap (Figure 22.5d).

Source Beds In their search for oil, geologists have mapped thousands of structural and stratigraphic traps throughout the world. Only a fraction of them have proved to contain any oil or gas, because the traps alone are not enough. A trap will contain oil only if source beds were present, if the necessary chemical reactions took place, and if the oil could migrate into the trap and stay there without being disturbed by subsequent severe heating or deformation. Although oil and gas are not rare, most of the easy-to-find deposits have already been located, and new fields are becoming more difficult to find.

The World Distribution of Oil and Natural Gas

If you were to visit the exploration and research offices of any large oil company, you would find maps and reports of all the regions containing geologic sections of sedimentary rock in which the company has operated, because where there is a thick section of sedimentary rocks, there may be oil. Thirty-one of the 50 states of the United States produce oil for the market, and small, noncommercial occurrences are known in most of the others. Many of Canada's provinces produce oil.

Two of the richest and most important oil-producing regions in the world are the Middle East and the area around the Gulf of Mexico and the Caribbean. The oil fields of the Middle East—including Iran, Kuwait, Saudi Arabia, Iraq, and the Baku region in Azerbaijan—contain about two-thirds of the world's known reserves. The highly productive Gulf Coast–Caribbean area includes the Louisiana-Texas region, Mexico, Colombia, Venezuela, and Trinidad. Saudi Arabia holds the largest reserves. The United States ranks eighth. **Figure 22.6** summarizes oil reserves in various parts of the world.

Oil and the Environment

Pollution is the major problem of offshore drilling. The environment around Santa Barbara, California, suffered great damage in 1969 when oil was accidentally released from an offshore drilling platform. In 1979, a well that was being drilled in the Gulf of Mexico off the Yucatán coast "blew out," spilling as much as 100,000 barrels of oil a day for many weeks before it could be capped. In 1988, an explosion destroyed a drilling platform in the North Sea,

Figure 22.5 Oil traps. (a) Anticlinal trap; (b) fault trap; (c) stratigraphic trap; (d) oil trapped by a salt dome. Natural gas and oil are trapped by an impermeable layer above the permeable oil-producing formation. Oil floats above the waterline.

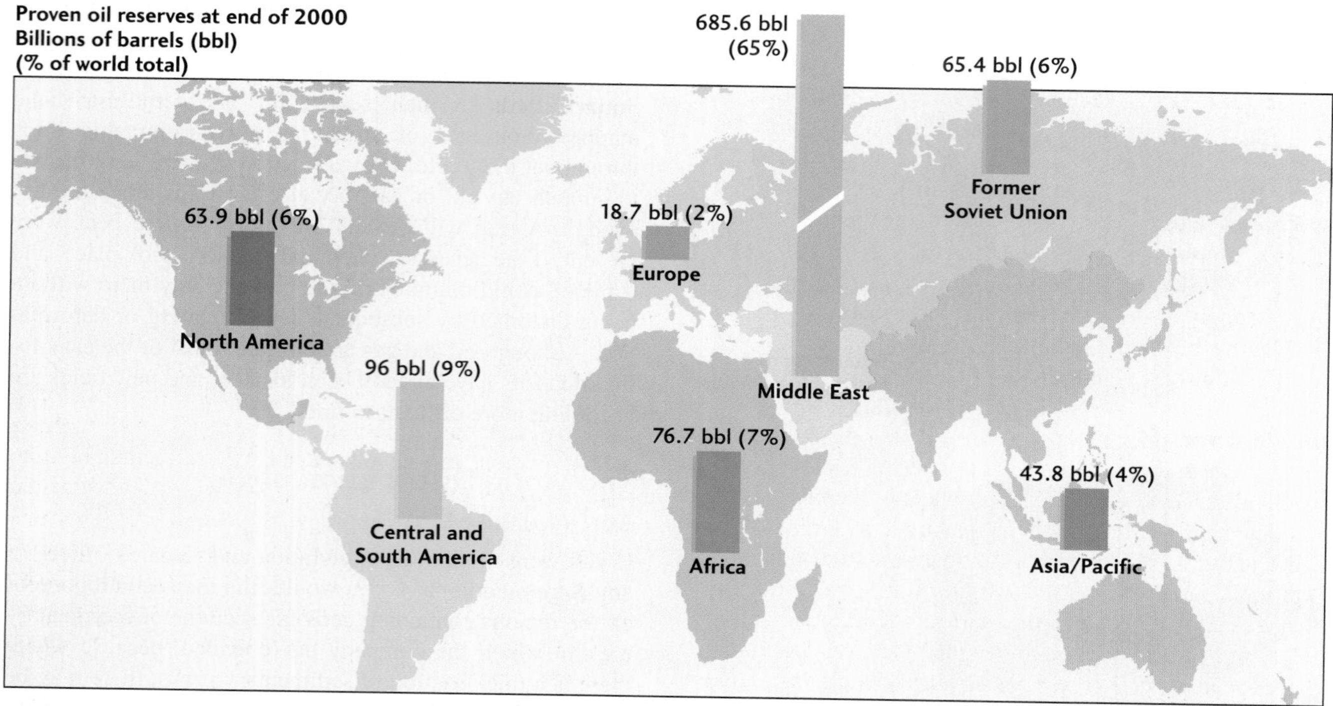

Figure 22.6 Estimated world oil reserves at the end of 2000 by region. [*Oil and Gas Journal*, December 28, 2000; © Pennwell Publications.]

killing many oil workers and marine animals. The grounding of the tanker *Exxon Valdez* off the coast of Alaska in 1989, with the release of 240,000 barrels of crude oil in pristine coastal waters, was covered widely by television and newspapers and heightened public awareness of the severe ecological damage that can result from an oil spill (**Figure 22.7**). Despite such incidents and the difficulty of guaranteeing the safety of a well or a tanker, proponents of oil development believe that careful design of equipment and safety procedures can greatly reduce the chances of a serious accident.

There are large resources of oil and gas under the coastal plain of northern Alaska and under the submerged continental shelves of North America. The deep water of the continental margin in the Gulf of Mexico is one of the best new prospects for oil and gas. Many people argue that eventually these areas will have to be drilled to satisfy the world's growing energy needs. A skeptical public, however, is not convinced that drilling can be done without serious threat to pristine environments.

There is currently a strident political debate in Washington about whether to allow drilling for oil and natural gas in the Arctic National Wildlife Refuge. There is no doubt that these resources would contribute to the national economy. But oil and gas production requires the building of roads, pipelines, and housing in a very delicate ecological environment that is a particularly important breeding area for caribou, musk-oxen, snow geese, and other wildlife.

Figure 22.7 The effect on wildlife of an oil spill from an oil tanker, the *Exxon Valdez*, in Prince William Sound, Alaska. [UPI/Corbis-Bettmann.]

Oil: An Exhaustible Resource

Inconceivable as it may seem to the landowner who sees a gusher spurting oil from a derrick on his property, that oil well will eventually run dry. What the world wants to know is, how soon will *all* the wells run dry? World demand has accelerated rapidly. Twice as much oil was removed from the ground in the past 20 years as in the previous 100 years. It takes millions of years to create oil, and humankind is using it up in centuries. Natural processes cannot replenish the oil supply as quickly as we are using it.

Of the estimated 360,000 quads that make up the world's total energy resources, some 17,500 quads are oil (the equivalent of 3 trillion barrels). How long this supply of oil will last depends on how fast we use oil and on our success in converting resources into reserves. Some experts predict that world oil production will peak within a decade and then begin to decline as the resources are drawn down. Other authorities believe that with new areas available for exploration and new methods of discovering and producing oil, world production will not start to decline until around 2050, and most of the remaining oil will be depleted in about 85 years. The remaining oil supply could last longer with the introduction of new technology to discover new oil fields and produce oil from them more efficiently. The supply could also be extended if we begin to use alternative fuels or if consumption declines, as it did because of conservation and the industrial slowdown that came with the recession of the early 1980s. On the other hand, oil may run out sooner if use picks up, as seems likely. A respected economist and skeptic, Morris Adelman of the Massachusetts Institute of Technology (MIT), challenges pessimistic estimates of remaining supply: "Nobody knows how much hydrocarbon exists or what percentage of that will be recoverable. The tendency to deplete a resource is counteracted by increases in knowledge." He believes that with improved world oil exploration and production technology, oil resources will continue to increase. The U.S. Geological Survey recently boosted its estimates of world oil resources by 20 percent over its estimates of 6 years earlier.

Consumption Exceeds Production in the United States Production of oil in the United States is declining, from 11.3 million barrels per day in 1970 to 8 million barrels per day in 2000. At current use rates, remaining U.S. proved oil reserves (23 billion barrels, up 1 billion from the previous edition of this book) amount to roughly a 10-year supply. Oil imports will fill the gap. At present, the United States consumes about 18 million barrels of oil each day, of which it imports about 10 million barrels. In a decade or two, the United States will become dependent on imports for most of its oil. The cost of imports could amount to many hundreds of billions of dollars annually.

Many people in the United States think that the dependence on foreign sources of oil is potentially destabilizing to the economy. Some oil-exporting nations have not always been reliable suppliers, cutting off exports because of political disagreements and wars. Other industrial nations that depend on oil imports—such as Japan, France, and Germany—have similar concerns. On the other hand, Russia and the countries of the former Soviet Union want to become reliable new suppliers.

Reducing Oil Consumption Oil-importing nations have options to reduce their vulnerability. They can change their patterns of oil use to reduce their need for imports. Automobile and airplane engines can be designed to use fuel much more efficiently. If we could double the average fuel economy of the U.S. automobile fleet, oil imports could be reduced by about 40 percent. Automobiles can run on natural gas; electric batteries; or ethanol, a form of alcohol produced from biomass. (Biomass is material that contains organic carbon, such as plants and wastes that can be used as fuel.) Automobiles can also run on hydrogen. We would need to develop the technology to produce it in sufficient quantities, and prices would need to become competitive. Unfortunately, the production of hydrogen can also release carbon dioxide. The U.S. Department of Energy has initiated a research program to investigate the use of hydrogen in motor vehicles.

These alternative fuels are less polluting than gasoline. Brazil fuels almost all its automobiles with ethanol made from sugarcane and so does not need to import oil for transportation. New technology is needed, however, to improve these alternative fuels, lower their costs, and arrange for efficient large-scale production and distribution. Improvements are also needed in the engines that use these fuels. California is setting the pace for the nation in requiring automobile manufacturers to move to more fuel efficient and less polluting engines.

We will never really "run out" of oil. As the resource diminishes, prices will eventually rise to levels that buyers cannot afford. With perhaps 50 to 100 years' worth of oil remaining beneath the Earth at current rates of use before nature's oil legacy is gone, however, we cannot yet proclaim the end of the age of oil. Yet a century may be barely enough time to plan and put into place an orderly transfer to new systems for transportation, industrial, and residential uses. We will need to make greater use of mass transportation rather than automobiles; design automobiles and airplanes that are more fuel efficient; exploit alternative sources of fuel; and place greater emphasis on energy conservation in industry, commerce, and the home. Without such planning for alternatives and conservation, many nations may face social and economic disruption when transportation systems grind to a halt, production lines stop, and homes grow cold in winter because oil has become so scarce that its price has risen out of reach.

Natural Gas

The resources of natural gas are comparable to those of crude oil (see Figure 22.2) and may exceed them in the decades ahead. Estimates of natural gas resources have been rising in recent years. Exploration for this relatively clean fuel has increased, and geologic traps have been identified in new settings, such as very deep formations, overthrust belts, coal beds, tight (somewhat impermeable) sandstones, and shales. The world's resources of natural gas are less depleted than oil resources because natural gas is a relative newcomer on the energy scene. It has been used on a large scale only in the United States and the former Soviet Union.

The burning of natural gas releases less carbon dioxide per unit of energy than the combustion of coal or oil. Natural gas is mostly methane (CH_4); when it burns, it combines with atmospheric oxygen, releasing energy in the form of heat and producing carbon dioxide and water. Coal and oil contain carbon, hydrogen, oxygen, nitrogen, and sulfur. The combustion of coal and oil produces more carbon dioxide and other pollutants per unit of energy than the combustion of natural gas.

In addition to being less polluting than coal or oil (little ash or precursors of acid rain are released), natural gas is easily transportable. Natural gas from fields in Siberia, for example, is piped to factories and homes in Germany. For these reasons, natural gas is a premium fuel. It accounts for about 24 percent of all fossil fuels consumed in the United States each year, most of it for industry and commerce (55 percent), followed by residential use (24 percent) and the generation of electric power (21 percent). More than half of American homes and a great majority of commercial and industrial buildings are connected to a network of underground pipelines that draw gas from fields in the United States, Canada, and Mexico. Natural gas reserves in the United States should last about 10 years and resources some 35 years at current rates of use—probably longer if we tap accumulations in the new settings mentioned earlier. Imports from Canada and Mexico could also slow depletion.

Coal

The abundant plant fossils found in coal beds are evidence that coal forms from large accumulations of plant materials of the sort that exist in wetlands. As the luxuriant plant growth of a wetland dies, it falls to the waterlogged soil. Rapid burial by falling leaves and immersion in water protect the dead twigs, branches, and leaves from complete decay because the bacteria that decompose vegetative matter are cut off from the oxygen they need. The vegetation

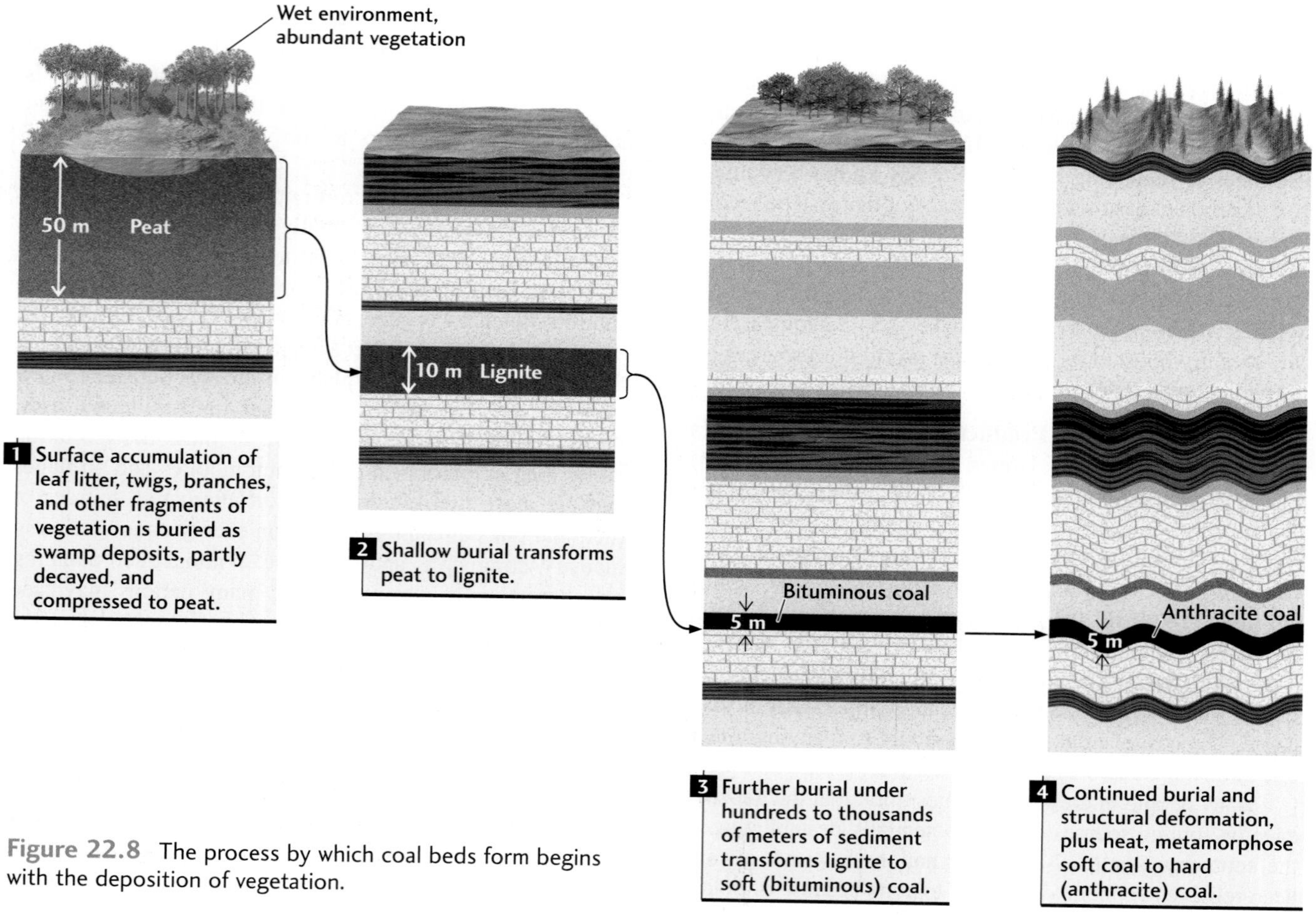

Figure 22.8 The process by which coal beds form begins with the deposition of vegetation.

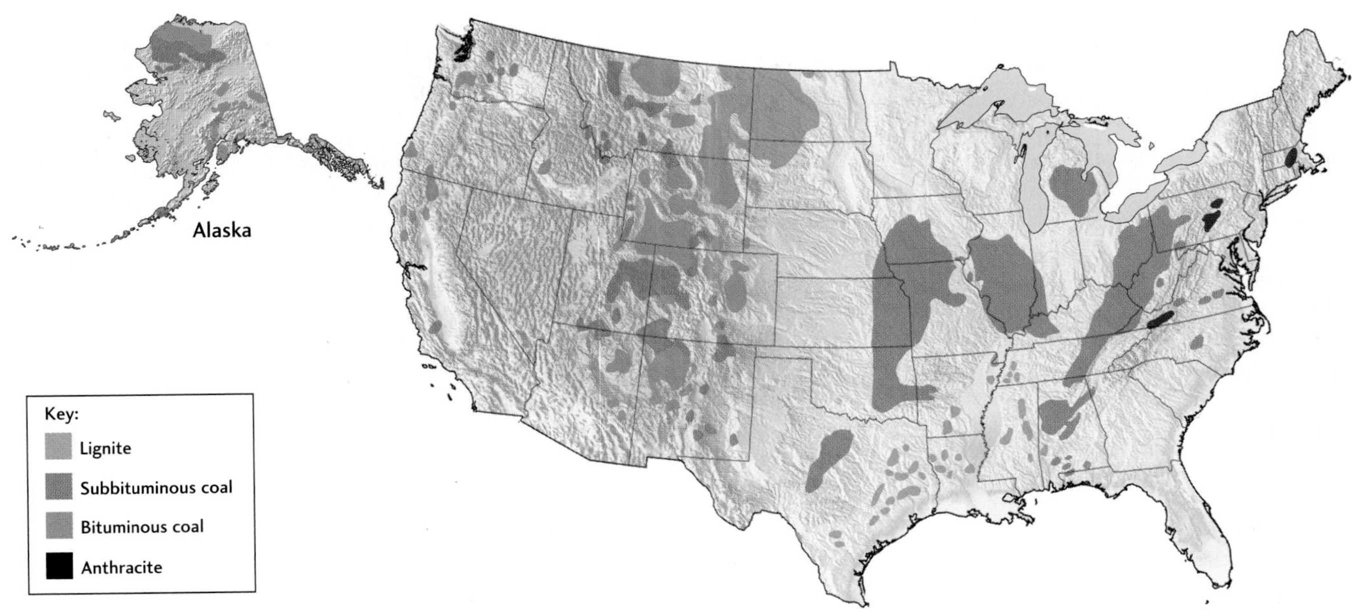

Figure 22.9 Coalfields of the United States. [U.S. Bureau of Mines.]

accumulates and gradually turns into peat, a porous brown mass of organic matter in which twigs, roots, and other plant parts can still be recognized (**Figure 22.8**). The accumulation of peat in an oxygen-poor environment can be seen in modern swamps and peat bogs. When dried, peat burns readily because it is 50 percent carbon.

Over time, with continued burial, the peat is compressed and heated. Chemical transformations increase the peat's already high carbon content, and it becomes *lignite,* a very soft, brownish-black coal-like material containing about 70 percent carbon. The higher temperatures and structural deformation that accompany greater depths of burial may metamorphose the lignite into *subbituminous* and *bituminous coal,* or soft coal, and ultimately into *anthracite,* or hard coal. The greater the metamorphism, the harder and brighter the coal and the higher its carbon content, which increases its heat value. Anthracite is more than 90 percent carbon.

Coal Resources

According to some estimates (see Figure 22.2), the amount of coal remaining in the world is about 3.4 trillion short tons. The leading producers are the United States (**Figure 22.9**), the former Soviet Union, and China, which together hold about 85 percent of the world's coal resources (former Soviet Union, 50 percent; China, 20 percent; United States, 15 percent). Domestic coal resources in the United States would last for a few hundred years at current rates of use—about a billion tons a year. Coal has supplied an increasing proportion of U.S. energy needs since 1975, when the price of oil began to rise, and it currently accounts for about 23 percent of the energy consumed.

Environmental Costs of Coal

There are serious problems with the recovery and use of coal that make it less desirable than oil or gas, whether the coal is burned or converted into synthetic liquid fuel. Much coal contains appreciable amounts of sulfur, which vaporizes during combustion and releases noxious sulfur oxides into the atmosphere. Acid rain, which forms when these gases combine with rainwater, is becoming a severe problem in Canada, Scandinavia, the northeastern United States, and eastern Europe (see Chapter 23). Coal ash is the inorganic residue that remains after coal is burned. It contains metal impurities from the coal, some of which are toxic. Ash can amount to several tons for every 100 tons of coal burned and poses a significant disposal problem. It can escape from smokestacks, posing a health risk to people downwind. Strip mining, the removal of soil and surface sediments to expose coal beds, can ravage the countryside if the land is not restored (**Figure 22.10**). Underground mining accidents take the lives of miners each year, and many more suffer from black lung, a debilitating inflammation of the lungs caused by the inhalation of coal particles.

Government regulations now require that technologies for the "clean" combustion of coal be phased in. The law mandates the restoration of land ravaged by strip mining and the reduction of danger to miners. But these measures are expensive and add to the cost of coal. This drawback, however, is unlikely to prevent the increased use of this fuel,

Figure 22.10 (*left*) Coal strip mine, Buskin, Indiana. (*right*) Coal strip mine after reclamation, Buskin, Indiana. [Photos courtesy of Vigo Coal Company.]

which is so much more abundant than oil. Many countries have no other fuel resources, and some countries will not be able to afford to import oil, which will increase in price as the supply diminishes.

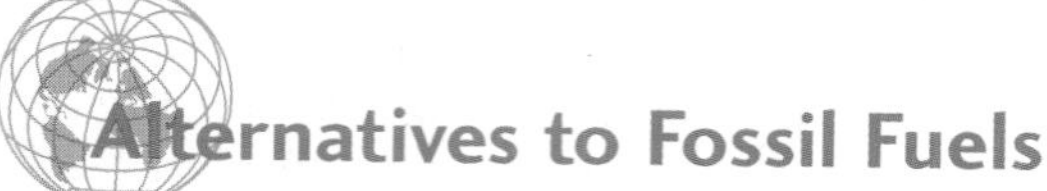

Alternatives to Fossil Fuels

If crude oil and gas continue to be the major resources used to satisfy the world's voracious appetite for energy, the great bulk of the world's supply will be exhausted within a century. Coal will eventually become the predominant fossil fuel in many countries. It may be reassuring to know that at modest annual energy growth rates—say, 3 percent per year—coal and other fossil fuels can meet the world's energy needs for about 100 years or longer. This security, however, may be false. Carbon dioxide released during the combustion of fossil fuels may trigger climate changes that could force a shift away from these traditional fuels before they are depleted (see Chapter 23).

These estimates do not take into account the possibility that we may learn to meet much of our growing need for energy in nontraditional ways: through increased efficiency in the use of fossil fuels and by the development and use of alternative energy sources such as nuclear energy, solar energy, geothermal energy, and energy derived from biomass. To the extent that alternative sources can be used, the pressure on our fossil fuel resources can be reduced and their life extended.

Nuclear Energy Fueled by Uranium

Although the first use of uranium (^{235}U) was in an atomic bomb in 1944, the nuclear physicists who first observed the vast energy released when its nucleus splits spontaneously (called *fission*) foresaw the possibility of peaceful applications of this new energy source. After World War II, their predictions were fulfilled as countries throughout the world built nuclear reactors to produce **nuclear energy:** the fission of ^{235}U releases heat to make steam, which then drives turbines to create electricity. The fission of a piece of ^{235}U releases 3 million times more energy than the combustion of a piece of coal of the same mass. In the United States, some 110 nuclear reactors now provide about 20 percent of the electric energy used. France derives 75 percent of its electric energy from nuclear power. Today, more than 400 nuclear reactors are producing electricity in 25 countries. If its full potential is realized, nuclear energy can meet the world's needs for electric energy for hundreds of years.

Uranium Reserves One aspect of nuclear energy is most definitely in the province of geology: the question of reserves of uranium. Present in very small amounts in Earth's crust, uranium constitutes only 0.00016 percent of the average continental crustal rock. The isotope that fissions and releases energy, ^{235}U, constitutes only 1 of every 139 atoms of uranium mined. In regard to energy content, however, uranium is potentially our largest minable energy resource (see Figure 22.2). It is typically found as small quantities of the uranium oxide mineral uraninite (also called pitchblende) in veins in granites and other felsic igneous rocks. Uranium can also be found in sedimentary rocks. Under near-surface groundwater conditions, uranium in igneous rocks may oxidize and dissolve, be transported in groundwater, and later be reprecipitated as uraninite in sedimentary rocks.

Nuclear Energy Hazards Two nuclear accidents have raised questions about the safety of nuclear energy. The first was at the Three Mile Island reactor in Pennsylvania in 1979. A reactor was destroyed, and radioactive debris was released but confined within the containment building. Although no one was harmed, most experts agree that it was a close call. Much more serious was the destruction of a nuclear reactor in the town of Chernobyl in Ukraine in 1986. The reactor went out of control because of poor design and human error and was destroyed. Radioactive debris spilled into the atmosphere and was carried by winds over Scandinavia and western Europe. Contamination of buildings and soil has made hundreds of square miles of land surrounding Chernobyl uninhabitable. Food supplies in many countries were contaminated by the fallout and had to be destroyed. Excess deaths from cancer caused by exposure to the fallout may be in the thousands over the next 40 years.

The uranium consumed in nuclear reactors leaves behind dangerous radioactive wastes that must be disposed of (see Feature 22.1). A system of safe long-term waste disposal is not yet available, and reactor wastes are being held in temporary storage at reactor sites. In a few years, the limits of space available for temporary storage in the United States will be reached. Although many scientists believe that geological containment—the burial of nuclear wastes in deep, stable, impermeable rock formations—is a workable solution, there is not yet a generally approved plan for storage of the most dangerous wastes for the hundreds of thousands of years required before they cease to be radioactive. France and Sweden have built underground nuclear waste depositories, but the United States is still in the stage of research, development, and testing. It is also embroiled in litigation as the state of Nevada battles the federal government to keep waste repositories from being built within its borders. No one wants to leave to future generations the combined legacy of depleted energy resources and a possibly unmanageable environmental hazard.

In the United States, these unresolved problems have essentially halted the installation of new nuclear power plants, and new installations have been slowed in other countries as well. Concerns about the safety of nuclear reactors, their high cost, and the safe disposal of radioactive waste must be allayed before we can light up the world with nuclear energy.

Solar Energy

An enthusiast for solar energy recently reminded us that "every 20 days Earth receives from sunlight the energy equivalent of the entire planetary reserves of coal, oil, and natural gas!" Because all our fossil energy sources ultimately come from the Sun anyway, why not convert its rays into energy? In principle, the Sun can provide us with all the energy we need, in all the forms we use. Light from the Sun can be converted into heat and electricity. It can even be used to obtain hydrogen, which can be used as a gaseous fuel, from water. **Solar energy** is risk-free and nondepletable—the Sun will continue to shine for at least the next several billion years. Unfortunately, the technology currently available for the large-scale conversion of solar energy into useful forms is inefficient and expensive, although it is improving.

In the near term, the only form of solar energy likely to be available at costs nearly competitive with those of other sources is heat for homes, water, and industrial and agricultural processes. Some homes and factories use solar energy for these purposes, motivated in part by government tax credits and other incentives.

Solar energy can be used to generate electricity in several ways. Solar generating systems are being used for installations where costs are no impediment, such as demonstration projects, or in remote areas where alternatives are not available (**Figure 22.11**). But large-scale solar-generated electricity is not yet an alternative to conventional sources in general use. The efficiency of converting sunlight into electricity is improving but is still too low, and the costs of installing and maintaining the systems are still too high. Moreover, commercial-scale solar electric power plants can present significant environmental problems. A plant with a 100-megawatt electric capacity (about 10 percent of the capacity of a nuclear power plant) located in a desert region

Figure 22.11 Solar cells convert sunlight, a renewable resource, into electric energy at this utility in a remote village in Nepal. [Ned Gillette/Corbis.]

22.1 Subsurface Toxic and Nuclear Waste Contamination

Some decades ago, we knew much less than we know now about the health and environmental effects of toxic wastes. Industrial and military wastes now known to be hazardous were dumped on the ground; disposed of in ponds, lakes, and rivers; or discharged underground. As a result, many nations are left with a legacy of toxic waste sites that can endanger public health or seriously damage regional ecological resources.

In the United States, thanks to the environmental movement, strict laws governing hazardous waste disposal sites were passed in 1980. Collectively known as the Superfund Program, these laws were designed to clean up these sites thoroughly. Retroactive liability requires those responsible for the contamination (if they can be found) to pay the costs. The Environmental Protection Agency (EPA) has identified some 1300 contaminated sites that pose serious threats. Of these, the most hazardous and the most difficult to clean up are federal government facilities that were formerly engaged in the production of nuclear weapons. The cleanup of these facilities constitutes the most expensive environmental remediation project in the world, with estimated costs of $200 billion to $350 billion in the next 70 years. There are more than 100 such sites located in 30 states.

Many geologists, other scientists, and engineers are engaged in a cleanup process that is uniquely difficult because the contaminants are both chemically toxic and radioactive, and the soils, unsaturated zones, and aquifers (see Chapter 13) of many of the sites are polluted. The Hanford site in Washington State is an example. Wastes derive from leaking storage tanks, buried boxes and drums filled with radioactive solids, and contaminated liquids that were discharged into the soil and groundwater in the past. Hanford is located on an alluvial plain composed of sands and gravels, underlain by folded and fractured basalt. In general, waste-bearing fluids flow downward into the unsaturated zone and enter the aquifer, where they flow toward the Columbia River. The flow rates vary according to local geological conditions, such as the permeability of a formation and its mineral and organic content (see the accompanying diagram). Some contaminants become chemically bound to minerals or adhere to the sediments and hardly move; others move readily with the groundwater. Seeps and springs in the area are contaminated. In a 20-year period, some chemically toxic and radioactive wastes have migrated to the Columbia River, prompting the worry that much larger amounts could follow if effective cleanup is not pursued. At Hanford, there are potential health risks to Native Americans who have treaty rights to use the land and the Columbia River and its banks for agriculture, hunting, and fishing. The primary ecological risk comes from the discharge of toxic plumes into the salmon spawning areas of the Columbia River.

Hanford and the other Superfund sites raise a number of challenging economic, social, and philosophic policy questions for the American people. For example, what should be the cleanup goal for these sites? Options include (1) the pristine original "green field" condition; (2) partial cleanup to achieve a state that poses acceptable health risk and does not endanger an important ecological resource such as the salmon spawning grounds; (3) incomplete "brown field" restoration to a state that permits commercial or industrial use but not residential use; and (4) transition to a "stewardship" state in which the land is set aside and the contaminants are contained and monitored to prevent their spread outside the site. At some sites, where options (1) and (2) cannot be achieved with currently known technology, the only recourses are options (3) and (4).

Estimating the effect of a particular clean-up path on health or the environment is called *risk assessment.* The uncertainty of exposure that will result and disagreement about what levels of health and environmental effects are acceptable make risk assessments much debated. Nevertheless, risk assessment is an important tool of decision makers who, by law, must take cleanup action.

Cleanup decisions are made by collective agreement of the U.S. Department of Energy (DOE) and federal and state environmental regulators. In earlier years, nuclear weapons manufacturing activities were kept secret, and the government was insensitive to health risks and environmental effects. DOE now considers socioeconomic, cultural, and ethical concerns in its decisions. It also keeps local governments, affected communities, and other interest groups informed and considers their views.

Many scientists are concerned about cleanup agreements that have been made between DOE and its regulators in which scientists have played little or no role. As a result, some commitments cannot be met by currently available technology, resulting in litigation and a loss of DOE's credibility. In a Catch-22 situation, funding of scientific research to discover new technologies for more effective and less expensive cleanup has been reduced—thus reducing the chance that research will lead to better and cheaper solutions.

Unless the nation recognizes that cleanup of these sites is a national problem and the necessary resources are provided to find and implement new solutions, much money will be spent using existing technology, with disappointingly little to show for it.

Sources: R. E. Gephart, *An Overview of Hanford's Waste Generation History and the Challenges Facing Site Cleanup* (Richland, Wash.: Pacific Northwest National Laboratory, 1998); and National Research Council, *Groundwater and Soil Cleanup: Improving Management of Persistent Contaminants* (1999), http://www. nap.edu/books/0309065496/html/.

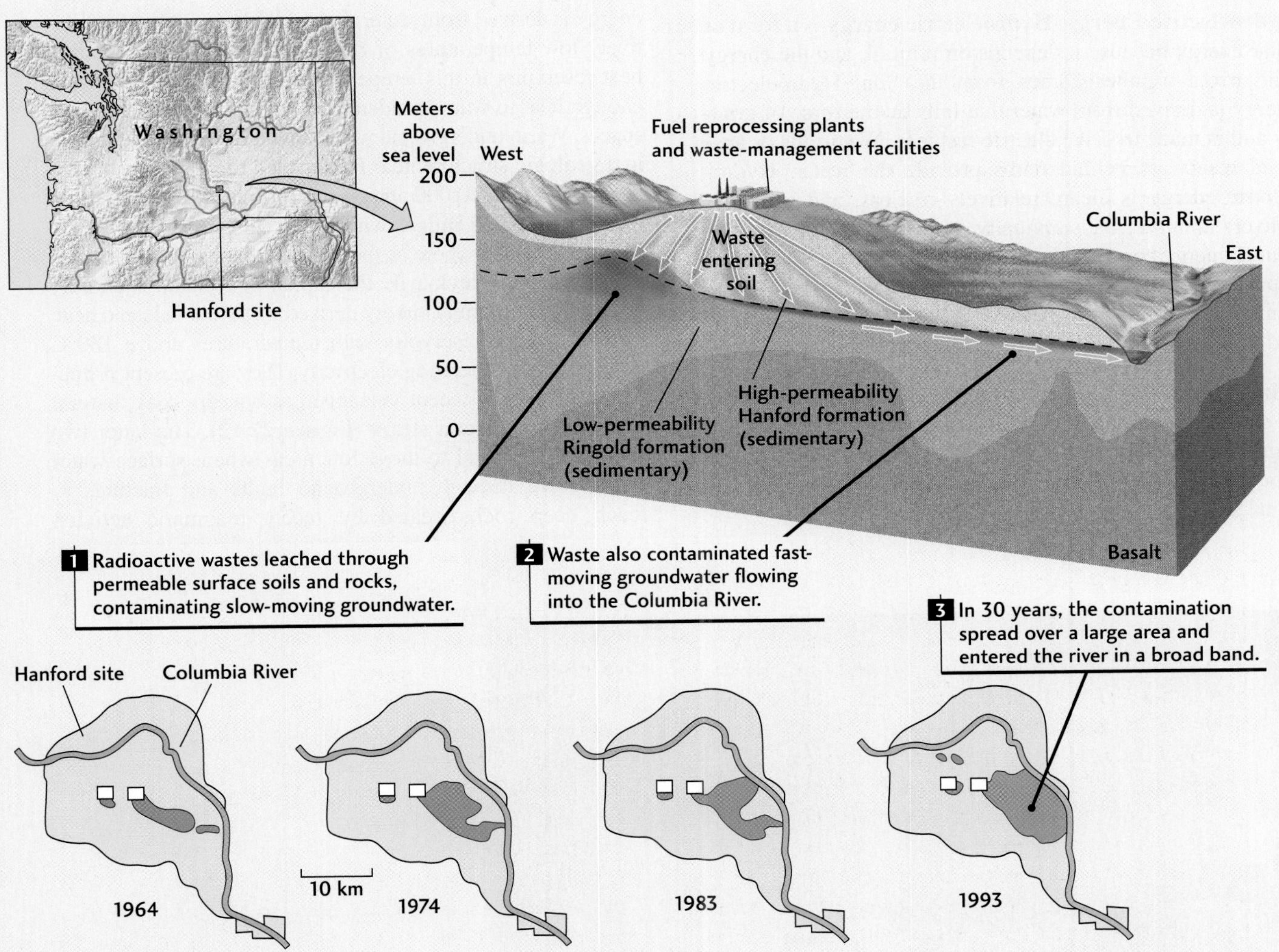

This 35-km cross section of the Hanford site illustrates how liquid waste enters the ground and moves through the unsaturated zone into and through the underlying aquifer. Height in this drawing is exaggerated to show vertical detail. [After R. E. Gephart, *An Overview of Hanford's Waste Generation History and the Challenges Facing Waste Cleanup* (Richland, Wash.: Pacific Northwest National Laboratory, 1998).]

would require at least a square mile of land and might alter the local climate by significantly changing the balance of solar radiation in the area.

Solar energy supplies only a few tenths of 1 percent of consumption. Solar energy enthusiasts believe that some 20 quads per year might be supplied in the United States by the year 2010. This amount is equivalent to about half the oil we now use. Others think it more realistic to figure on less than 10 quads per year. All agree that either scenario would yield important social benefits: conservation of other energy resources, diversification of energy supply so that we are not overly dependent on a single source, and reduction of fuel imports. With adequate research and development, solar energy can probably become economically competitive and a major source of energy in the twenty-first century.

Hydroelectric Energy **Hydroelectric energy** is a form of solar energy because it depends on rainfall, and the energy that drives weather comes from the Sun. Hydroelectric energy is derived from water that falls by the force of gravity and is made to drive electric turbines. Waterfalls or artificial reservoirs behind dams provide the water. Hydroelectric energy is clean, relatively riskless, and cheap. It delivers about 3 quads annually, or about 3 percent of the annual energy consumption in the United States. Significant expansion of the present capacity would be resisted in the United States, however, because it would drown farmlands and wilderness areas under reservoirs behind dams.

Wind Power *Wind power,* or the use of a windmill to drive an electric generator, is also a form of solar energy. Its use is growing in many places as designs improve and costs are brought down to where they are competitive with conventional sources.

Geothermal Energy

In Chapter 2, we learned that Earth's internal heat, fueled by radioactivity, provides the energy for plate tectonics and continental drift, mountain building, and earthquakes. The same internal heat can be harnessed to drive electric generators and heat homes. **Geothermal energy** is produced when underground water is heated as it passes through a subsurface region of hot rocks (a *heat reservoir*) that may be hundreds or thousands of feet deep. The hot water or steam is brought to the surface through boreholes drilled for the purpose. The water is usually naturally occurring groundwater that seeps down along fractures. Less typically, the water is artificially introduced by being pumped down from the surface.

Eighteen countries now use geothermal heat to generate electricity. By far the most abundant form of geothermal energy is derived from water that has been heated to the relatively low temperatures of 80° to 180°C. Water-circulating heat reservoirs in this temperature range are able to extract enough heat to warm residential, commercial, and industrial spaces. Warm underground water drawn from a heat reservoir in a geologic structure near Paris called the Paris Basin now heats more than 20,000 apartments in France. Iceland sits on the Mid-Atlantic Ridge, where upwelling mantle material creates new lithosphere as the North American and Eurasian plates separate. Reykjavík, the capital of Iceland, is entirely heated by geothermal energy derived from this volcanic heat.

Geothermal reservoirs with temperatures above 180°C are useful for generating electricity. They are present primarily in regions of recent volcanism as hot, dry rock; natural hot water; or natural steam (**Figure 22.12**). The latter two sources are limited to those few areas where surface water seeps down through underground faults and fractures to reach deep rocks heated by recent magmatic activity.

Figure 22.12 The Geysers, the world's largest supply of natural steam. The geothermal energy is converted into electricity for San Francisco, 120 km to the south. [Pacific Gas and Electric.]

Naturally occurring water heated above the boiling point and naturally occurring steam are highly prized resources. The world's largest supply of natural steam is at The Geysers, 120 km north of San Francisco. Generators there currently produce more than 600 megawatts of electricity, enough to meet about half the needs of San Francisco. The Geysers is now in its third decade of production and is beginning to show signs of decline, perhaps because of overuse of the steam resource. Some 70 geothermal electricity-generating plants operate in California, Utah, Nevada, and Hawaii, producing 2800 megawatts of power—enough to supply about a million people.

Like most of the other energy sources we have looked at, geothermal energy presents some environmental problems. Regional subsidence can occur if hot groundwater is withdrawn without being replaced. In addition, geothermally heated waters can contain salts and toxic materials dissolved from the hot rock. These waters present a disposal problem if they are not reinjected.

Geothermal energy will probably not make large contributions to the world's energy budget until well into the twenty-first century, if ever.

Conservation

In a real sense, using energy more efficiently is like discovering a new source of fuel. It has been calculated that since the rise in oil prices in 1973, the world has saved more energy than it has gained from all new sources discovered in the same period. Some experts believe that conservation alone could cut energy use in the industrialized nations by half. The savings in the United States could amount to some $200 billion a year, much of it now being spent on imported oil. Savings of this magnitude would reduce the cost of our products and make them more competitive in the world market, reduce our dependence on imported oil, and lower our trade deficit significantly. The kinds of practices that could lead to these savings require the application of mostly familiar technologies: using fluorescent rather than incandescent lighting; better home insulation; more efficient refrigerators, air conditioners, furnaces, and other appliances; more efficient motors, pumps, and other industrial devices; and better-performing automobile engines. Political leadership and public education will be required to induce us to undertake these worthwhile changes today. In the years ahead, dwindling reserves and higher energy prices will force us to do so.

Energy Policy

In view of the diversity and abundance of energy resources, you may well wonder why the world faces an energy crisis. There are several reasons:

- The world demand for energy will increase, driven by the industrialization of China and other developing countries.
- In the next few decades, world oil production will peak and begin to decline.
- The disruption of supply on political grounds by oil-producing nations and the resulting escalation of prices may severely shock the world economic and political order.
- Even without political interruptions in supply, global climate change may restrict our ability to use fossil fuels, precipitating an energy crisis of even greater severity.

The developing nations aspire to grow economically and improve the quality of their people's lives. Their sense of impoverishment and inequitable treatment could be a trigger for conflict. The advanced industrial nations also wish to progress and continue to enjoy their affluence. The challenge for the world is to provide the energy needed to fuel growth in the face of the depletion of oil resources and the possible need to cap or reduce the emission of carbon dioxide by decreasing the use of fossil fuels (**Figure 22.13**). A partial solution, and the cheapest one, is to reduce the waste of energy by using it more efficiently. In the next few decades, however, we may also have to shift to a different mix of energy supply—drawn from natural gas, alternative fuels for transportation, nuclear energy, renewable resources such as solar and biomass energy, and a gradually decreasing dependence on oil and coal. It is to be hoped that nuclear technology will improve in safety and regain public confidence and that advances in the technology for renewable energy sources will lower their costs.

Many experts on resource supply and demand believe that fossil fuels are too cheap in the United States. They are not taxed as much as they are in other advanced nations, so there is too little attention paid to conservation and the introduction of renewable resources. If the full social costs of fossil and nuclear fuels were included in the prices—the cost of cleaning up acid rain, oil spills, and other environmental damage; the cost of storing nuclear wastes; the cost of trade deficits; the cost of global warming; the military cost of defending oil supplies (for example, the defense of Kuwait against the Iraqi invasion)—renewable and geothermal energy resources might compete very well with fossil fuels. Some experts believe that even today the United States can meet 30 percent of its energy demand with renewable resources at competitive prices. Unfortunately, the political will to achieve these results does not yet exist.

Nevertheless, we are in a race against time. We must develop the several options for renewable or indefinitely sustainable energy resources before remaining oil resources are depleted or we are forced to shift away from fossil fuels because of their harmful effects on climate. Whether the transition to an era of energy security is smooth or rough depends on our determination, technological skill, and ability to solve the attendant complex social and political problems. In our view, the technological advances required are

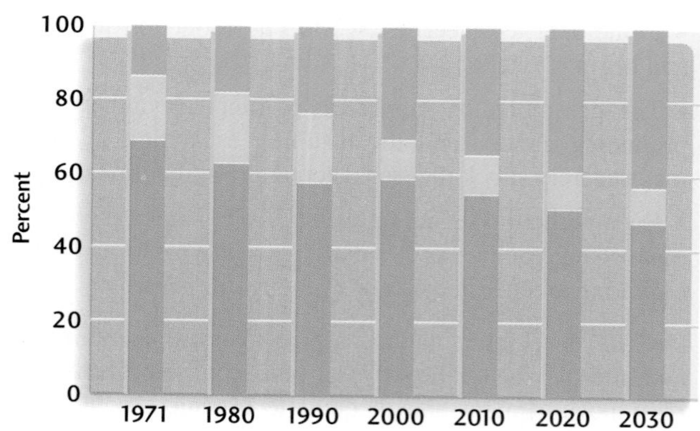

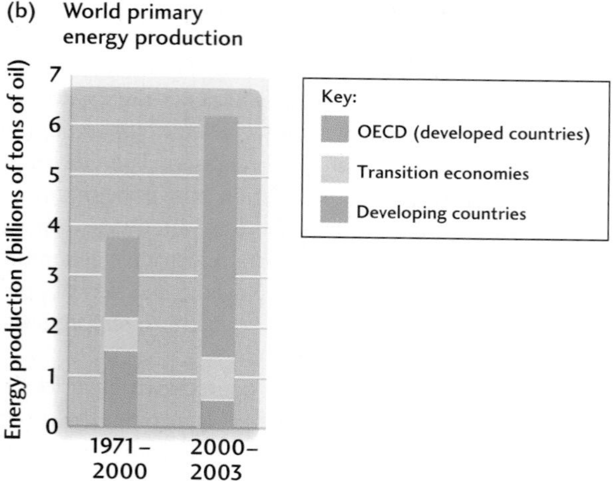

Figure 22.13 World energy demand, 1971–2010. OECD, the Organization for Economic Cooperation and Development, consists primarily of the advanced industrialized nations. Transition economies are the former Soviet Union and Central and Eastern Europe. The energy units are in terms of the energy of a million tons of oil. Data after 1998 are estimates. [Data from *World Energy Outlook,* 1994 and 1998 editions, International Energy Agency, Paris, France.]

achievable if we start to develop them now. How soon the sociopolitical problems will be resolved is less certain.

Mineral Resources

Despite its benefits, mining has been called the original dirty industry. Mining strips more of Earth's surface each year than the natural erosion of rivers. Annually, its waste products exceed those produced by the world's accumulation of municipal garbage. There has been progress in developing clean and environmentally benign mining operations, but mining and processing remain major contributors to environmental degradation and poor health in many countries.

Although mining itself represents only a small part of the national income of countries such as the United States and Canada, these nations depend on materials extracted from the Earth. Without them, there would be no stone for buildings, phosphates for fertilizers, cement for construction, clays for ceramics, sand for silicon transistors and fiber-optic cables, and metals for just about everything. In round numbers, the annual consumption of nonfuel minerals in the United States alone is about 9000 kg per person.

In the following pages, we will survey a broad range of Earth's mineral resources. In addition to their geological settings, we will consider their economic and social contexts, including matters of distribution, depletion, environmental and health costs, conservation, recycling, substitution, and price.

Concentration of Minerals

The chemical elements of Earth's crust are widely distributed in many kinds of minerals, and those minerals are found in a great variety of rocks. In most places, any given element will be found homogenized with other elements in amounts close to its average concentration in the crust. An ordinary granitic rock, for example, may contain a few percentage points of iron, close to the average concentration of iron in Earth's crust.

An element present in higher concentrations has undergone some geologic process that has segregated much larger quantities of the element than normal. High concentrations of elements are found in a limited number of specific geological settings. Some examples are given later in this chapter. These settings are of economic interest, because the higher the concentration of a resource in a given deposit, the lower the cost to recover it.

Ore Minerals Rich deposits of minerals from which valuable metals can be recovered profitably are called **ores;** the minerals containing these metals are *ore minerals.* Ore minerals include sulfides (the main group), oxides, and silicates. Ore minerals in each of these groups are compounds of metallic elements with sulfur, oxygen, and silicon oxide, respectively. The copper ore mineral covelite, for example, is a copper sulfide (CuS). The iron ore mineral hematite (Fe_2O_3) is an iron oxide. The nickel ore mineral garnierite is a nickel silicate, $Ni_3Si_2O_5(OH)_4$. In addition, some metals, such as gold, are found in their native state—that is, uncombined with other elements (**Figure 22.14**).

Concentration Factor The *concentration factor* of an element in an ore body is the ratio of the element's abundance in the deposit to its average abundance in the crust. The *economical* concentration factor varies from element to element, depending on its average abundance (Table 22.1).

Iron, one of the most common elements of the crust, has an average abundance in crustal rocks of 5.8 percent. An

Figure 22.14 Gold occurring in the free state (native gold) on a quartz crystal. [Chip Clark.]

economical iron ore—one that is profitable to mine under current costs of extraction, costs of transportation, and selling prices—must be at least 50 percent iron, about 10 times the average crustal abundance. In other words, an iron ore becomes economical when its concentration factor is about 10. A less abundant metal, such as copper, which has a crustal abundance of 0.0058 percent, has an economical concentration factor of at least 80 to 100. Ore deposits of the rarer elements, such as mercury and gold, require concentration factors in the thousands to hundreds of thousands to be economical.

Supply of Minerals

Elements are so widely distributed in many common rocks that whether a particular deposit should be considered a resource or a reserve depends on the costs of recovery and the selling price. Theoretically, with enough money and energy, we could extract both abundant and rare elements from any rock and never run out of minerals. Our practical concern, however, is the exhaustion of reserves, the identified mineral deposits that are profitable to mine and purify. We can reasonably expect that new discoveries will add to current reserves, but at an uncertain rate. When the highest-grade deposits have been mined out, we will be forced to rely on lower-grade deposits that could be more expensive to recover.

Decreasing Demand for Minerals Current estimates indicate that economically important minerals will be available and affordable for the next century or so. Rates of mineral consumption in the United States and other industrialized nations have slowed. These mature economies are shifting away from construction and manufacturing to service, technology, and other activities that require fewer raw materials. Demand for metals, in particular, has been reduced by *recycling* the metal content of discarded goods and by substituting low-cost ceramics, composites, and plastics.

Table 22.1 Economical Concentration Factors of Some Commercially Important Elements

Element	Crustal Abundance (Percent by Weight)	Economical Concentration Factor[a]
Aluminum	8.00	3–4
Iron	5.8	5–10
Copper	0.0058	80–100
Nickel	0.0072	150
Zinc	0.0082	300
Uranium	0.00016	1,200
Lead	0.00010	2,000
Gold	0.0000002	4,000
Mercury	0.000002	100,000

[a]Concentration factor = abundance in deposit divided by crustal abundance.

Sources: Data from B. J. Skinner, *Earth Resources* (New York: Prentice Hall, 1969); and D. A. Brobst and W. P. Pratt, *Mineral Resources of the U.S.* (USGS Professional Paper 820, 1973).

22.2 Use of Federal Lands in the United States

The land owned and managed by the federal government amounts to about a quarter of the country and is an important part of the political, social, and economic history of the United States. The primary stewards are the Bureau of Land Management and the Forest Service, although the National Park Service and the Department of Defense also manage some of the land.

Through the years, this land was bought at bargain prices, won in wars, or acquired by negotiations with other nations. The Louisiana Purchase, land bought from France in 1803, added a huge area from the mouth of the Mississippi to what is now the state of Montana, at a cost of $15 million. Florida was bought from Spain in 1819 for $7 million. The treaty of 1848 after the war with Mexico added most of the Southwest to the public domain, including parts of Arizona and New Mexico and all of California, Nevada, and Utah. Great Britain gave us the lands that are now the states of Washington, Oregon, and Idaho. The last major acquisition was the purchase of Alaska from Russia for $7 million in 1867.

In addition, the continental shelf beyond a few miles offshore is owned and managed by the federal government. The continental shelf is valuable for its oil and gas resources, fisheries, and mineral resources.

Throughout U.S. history, some 60 percent of lands in the public domain were transferred to states, homesteaders, and various organizations to aid education (the land-grant colleges), to build the privately owned railroads, to reward veterans, and for other purposes.

The federal lands have many uses, some of which conflict. Their commercial value is very large: rents and royalties from private users amounted to $6.2 billion in 1992. About 80 percent of this federal income is from oil and gas producers, mostly offshore, followed by timber (less than 20 percent) and mineral-mining interests (about 1 percent). Cattle ranchers (grazing) and recreational users contribute less than 1 percent each. Recreational users constitute by far the largest group of individual users of the land.

The U.S. mining industry enjoys a huge subsidy from the Mining Act of 1872, an unjustifiable relic of frontier days that still allows mining companies to buy federal land with mineral deposits for about $5 an acre or less. Mining was once given the highest priority for land use. Today, many people would assign greater value to wilderness protection and recreational use.

Federal ownership and management of these holdings is one of the most contentious political issues of the day, as evidenced by congressional debates, election campaigns, lobbying, and lawsuits brought against the federal government by states and private interests. Here are some of the policy issues that must be resolved:

- How much, if any, federal land should be sold off, to whom, and at what price?
- If the government retains ownership, what kind of access should be given to commercial interests, recreational users, and others, and on what terms?
- How much of the land should be set aside for wilderness preservation?

Recycling is a growth industry because it has become less expensive to extract certain elements from trash, such as aluminum cans and automobiles that have been discarded, than from deposits of ore minerals (**Figure 22.15**). About 45 percent of the gold, platinum, and aluminum now consumed is recycled material. Even larger fractions of lead (73 percent), copper (60 percent), and iron and steel (56 percent) are being recycled.

Recycling is a complex issue because it is not always economically or environmentally justified. The only rational way to decide whether to recycle is to compare the costs of mining, smelting, and transporting natural ores; disposing of wastes; and controlling pollution with the corresponding costs of recycling metals.

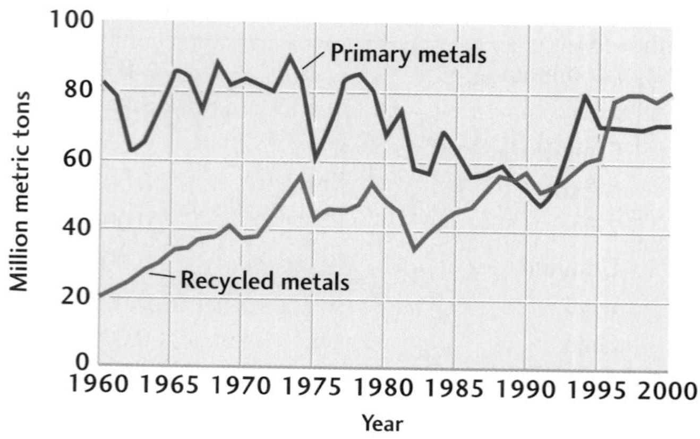

Figure 22.15 The amount of primary metals and recycled metals consumed in the United States, by weight. [G. Matos and L. Wagner, USGS, 1999.]

Imports Although the United States is the world's largest producer of raw minerals, it is also dependent on mineral

• To what degree should state and local governments participate in management of the lands?

A case can be made for federal retention of the lands. The federal government can look after national interests better than state and local governments. The federal government is best suited to organize and monitor the multiple diverse uses of the land. Public agencies can take a longer view of conservation and the needs of future generations than private interests driven by short-term profits.

There are also arguments for selling off much of the land. Private owners would manage the lands more efficiently than federal bureaucrats, who lack management experience and are subject to political pressure. The greatest benefits to society come from the accumulated effect of individual entrepreneurs who work to achieve the greatest personal gain. There is not much evidence that government administration of the land for any purpose is more exemplary than that of the private sector. Jobs created by the private sector are more important than preserving some unimportant subspecies.

The time has come to try to achieve a national consensus on the uses of federal lands.

Source: Much of this feature was drawn from Marion Clawson, *The Federal Lands Revisited* (Washington, D.C.: Resources for the Future, 1983)

Full autumn color surrounds wetlands in the Kanai National Wildlife Refuge, Alaska. [John Hyde/Bruce Coleman/Picture Quest.]

imports. Sometimes imports are cheaper, and sometimes the country lacks reserves of a particular mineral. The United States satisfies about 30 percent of its demand for minerals through imports. Chances are that the graphite in your pencil comes from Mexico or China.

Those who are concerned with U.S. economic and defense security, as well as mining industry representatives worried about sales, are troubled by the country's heavy dependence on imports of several strategically critical metals, such as cobalt, manganese, chromium, titanium, and the platinum group. These are metals without which an entire industry (the aircraft or chemicals industry, for example) could collapse. Since 1939, therefore, the U.S. government has been stockpiling minerals for use in an emergency, such as an economic boycott or a wartime cutoff of supply.

Some people think that the United States should increase domestic supply and decrease dependence on imports by allowing greater access to its federal lands for exploitation of mineral resources. Others think that federal lands should be preserved for public recreational use (see Feature 22.2).

Economists point out that the United States has always depended on imports of many minerals, that alternative markets are always available, and that major interruptions of supply have not occurred. Imports may be advantageous as long as the supply is secure, the price is reasonable, and overall exports and imports are in rough balance. Moreover, even though the United States has consumed more minerals in the past five decades than were consumed in all previous time, most mineral reserves have increased as a result of advances in geologic knowledge and improvements in mining and metal extraction.

World mineral consumption will continue to increase because of the growing demand for raw materials by developing countries. Can this demand be met from existing reserves? Are clean mining and environmental protection—together with recycling, substitution, and other conservation measures—economically feasible for these countries? Developing countries point out that the industrialized nations, now sensitive to the environment, have historically been among the biggest polluters during their own development.

The Geology of Mineral Deposits

Mineral deposits are created by various kinds of geologic processes, most of which we have already discussed in earlier chapters. In general, a mineral deposit forms when three conditions are satisfied:

1. A source of the minerals exists in a place where it is accessible to a natural transport mechanism.

2. A natural transport mechanism is available to move the minerals away from the source.

3. A site exists with a mechanism for the transport agent to deposit the minerals. It is not luck that places ore deposits close to Earth's surface, where humans can reach them. Rocks near the surface contain cracks and open fractures (pressure closes them at greater depths), allowing easier transportation of ore-bearing fluids. In addition, rocks near the surface are cooler, so ore minerals precipitate from the hot fluids that carry them.

Hydrothermal Deposits

Many of the richest known ore deposits crystallized from *hydrothermal solutions.* These hot waters can emanate directly from the magma of an igneous intrusion (the source) and transport the soluble ore constituents of the magma (**Figure 22.16**). Hydrothermal solutions can also form when circulating groundwater contacts heated rock or a hot intrusion, reacts with it, and carries off ore constituents released by the reaction.

Vein Deposits Ore constituents are often deposited in fractured rocks. The hot fluids flow easily through the fractures and joints, cooling rapidly in the process. Quick cooling causes fast precipitation of the ore constituents. The tabular (sheetlike) deposits of precipitated minerals in the fractures and joints are called **vein deposits** or simply **veins.** Some ores are found in veins; others are found in the country rock adjacent to the veins that was altered when the hot solutions heated and infiltrated it. As the solutions react with surrounding rocks, they may precipitate ore minerals together

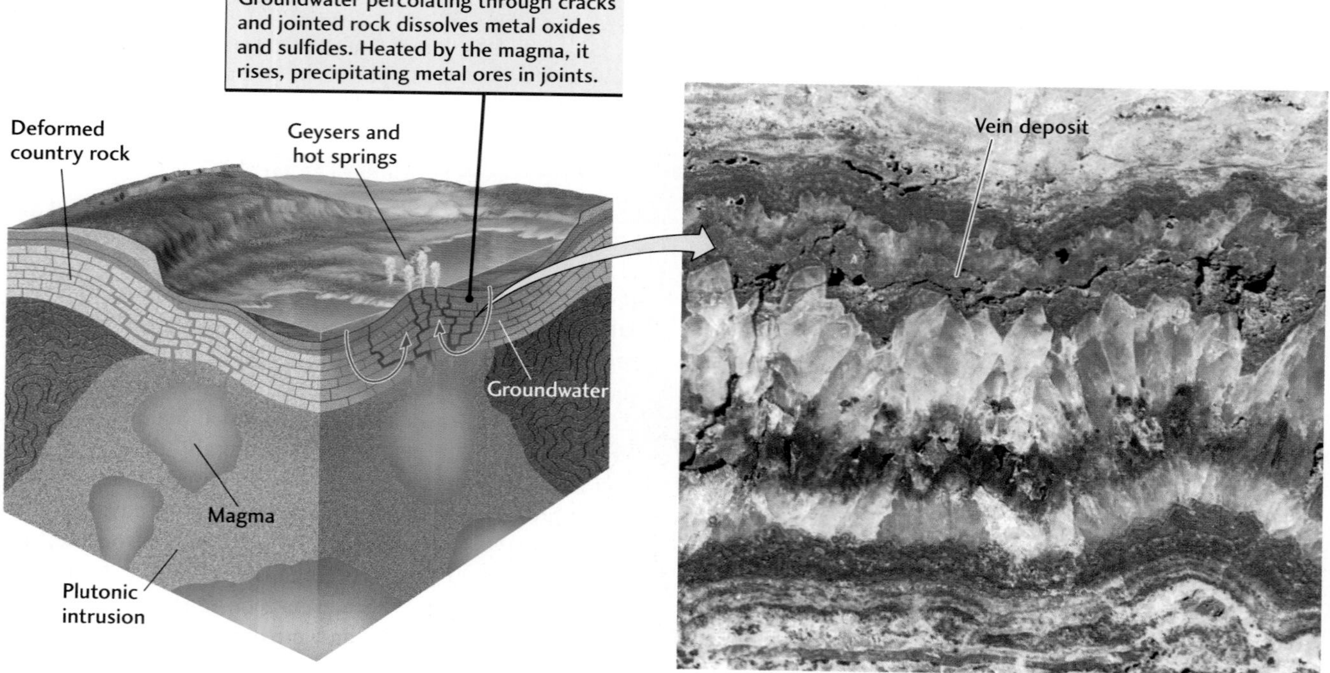

Figure 22.16 Many ore deposits are found in hydrothermal veins formed by hot solutions rising from magmatic intrusions. Quartz vein deposit (about 1 cm thick) containing gold and silver ores. Oatman, Arizona. [Peter Kresan.]

Figure 22.17 Metal sulfide ores. Sulfides are the most common types of metallic ores. [Chip Clark.]

with quartz, calcite, or other common vein-filling minerals. Vein deposits are a major source of gold (Figure 22.16).

Hydrothermal vein deposits are among the most important sources of metal ores. Typically, metallic ores exist as sulfides, such as iron sulfide (pyrite), lead sulfide (galena), zinc sulfide (sphalerite), mercury sulfide (cinnabar)—shown in **Figure 22.17**—and copper sulfide (covelite and chalcocite). Hydrothermal solutions reach the surface as hot springs and geysers, many of which precipitate metallic ores—including ores of lead, zinc, and mercury—as they cool.

Disseminated Deposits Mineral deposits that are scattered through volumes of rock much larger than veins are called *disseminated deposits.* In both igneous and sedimentary rocks, minerals are disseminated along abundant cracks and fractures. Among the economically important disseminated deposits are the porphyry copper deposits of Chile and the southwestern United States. The most common copper mineral in porphyry is chalcopyrite, a copper sulfide (**Figure 22.18**). The copper was deposited when ore-forming minerals were introduced into a great number of tiny fractures in porphyritic felsic intrusives (granitic rocks containing large feldspar or quartz crystals in a finer-grained matrix) and in the country rocks surrounding the higher parts of the plutons. Some unknown process associated with the intrusion or its aftermath broke the rocks into millions of pieces. Hydrothermal solutions penetrated and recemented the rocks by precipitating ore minerals throughout the extensive network of tiny fractures. This widespread dispersal produced a

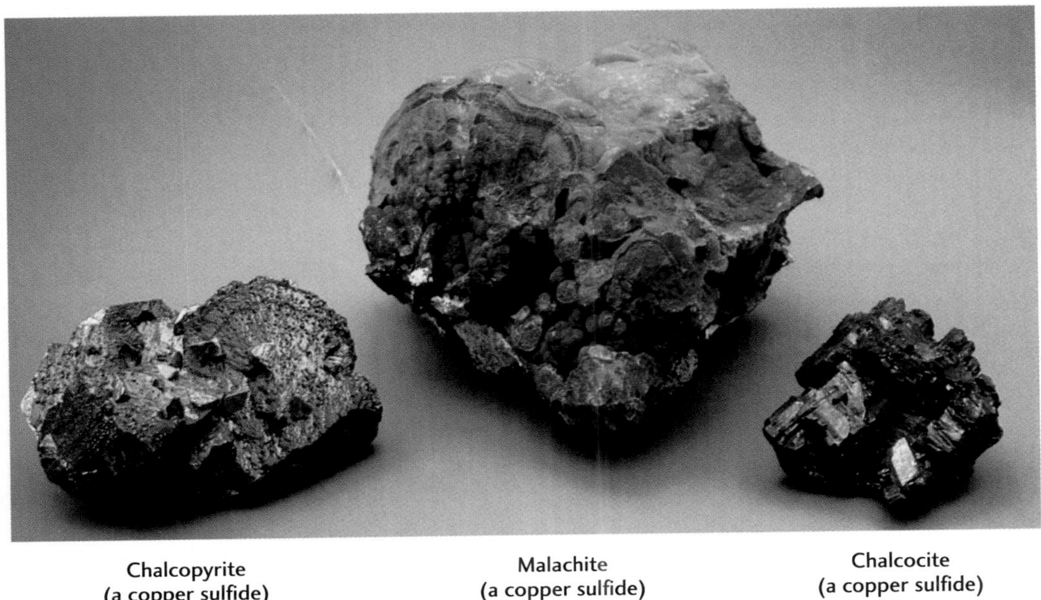

Figure 22.18 Copper ores. Chalcopyrite and chalcocite are copper sulfide ores. Malachite is a carbonate of copper found in association with sulfides of copper. [Chip Clark.]

Figure 22.19 Kennecott Copper Mine, Utah, an open-pit mine. Open-pit mining is typical of the large-scale methods used to exploit widely disseminated ore deposits. [Scott T. Smith/Corbis.]

low-grade but very large resource of many millions of tons of ore, which can be mined economically by large-scale methods (**Figure 22.19**).

Extensive disseminated hydrothermal deposits may also be present in sedimentary rocks. This is the case in the lead-zinc province of the Upper Mississippi Valley, which extends from southwestern Wisconsin to Kansas and Oklahoma. The ores in this province are not associated with a known magmatic intrusion that could have been a source of hydrothermal fluids, so their origin is unknown. Some geologists speculate that the ores were deposited by groundwater. Groundwater may have penetrated hot crustal rocks at great depths and extracted soluble ore constituents, then moved upward into the overlying sedimentary rocks, where it precipitated its mineral load as fillings in cavities. In some cases, it appears that ore fluids infiltrated limestone formations and dissolved some carbonates, then replaced the carbonates with equal volumes of new crystals of sulfide. The major minerals of the hydrothermal deposits in this province are lead sulfide (galena) and zinc sulfide (sphalerite).

Igneous Ore Deposits

The most important *igneous ore deposits*—deposits of ore in igneous rocks—are found as segregations of ore minerals near the bottom of intrusions. The deposits form when minerals crystallize from molten magma, settle, and accumulate on the floor of a magma chamber (see Chapter 5). Most of the chromium and platinum ores in the world, such as the deposits in South Africa and Montana, are found as layered accumulations of minerals that formed in this way (**Figure 22.20**). One of the richest ore bodies ever found, at Sudbury, Ontario, is a large mafic intrusion containing great quantities of layered nickel, copper, and iron sulfides near its base. Geologists believe that these sulfide deposits formed from crystallization of a dense, sulfide-rich liquid that separated from the rest of the cooling magma and sank to the bottom of the chamber before it congealed.

Pegmatites Pegmatites are extremely coarse grained intrusive rocks of granitic composition that are usually found as veins, dikes, or lenses in granitic batholiths. Pegmatites form by fractional crystallization (see Chapter 5) of a granitic magma. As the magma in a large granitic intrusion cools, the last melt to freeze solidifies as pegmatites, in which minerals present only in trace amounts in the parent body are concentrated. Pegmatites may contain rare mineral deposits rich in such elements as boron, lithium, fluorine, niobium, and uranium and in such gem minerals as tourmaline.

Kimberlites One of the most valuable minerals, diamond, exists chiefly in ultramafic igneous rocks called kimberlites, named for Kimberley, South Africa, where they are found in relative abundance. These rocks were forcefully intruded to the surface from deep in the crust and upper mantle in the form of long, narrow pipes. We know that these diamond-bearing kimberlites originate at great depths because diamonds and other minerals found in them can form only under the conditions of extremely high pressure that exist in the upper mantle. Kimberlites erupt to the surface at high

Figure 22.20 Chromite (chrome ore, dark layers) in a layered igneous intrusive. Bushveldt Complex, South Africa. [Spence Titley.]

speed, propelled by pressurized volatiles such as H_2O and CO_2. No one has ever seen a kimberlite eruption. One geologist has likened it to a shotgun blast fired from the mantle through the lithosphere to the surface.

Rich accumulations of diamonds have also been found in alluvial deposits hundreds of kilometers from their kimberlite pipe source, transported there by rivers that picked up fragments eroded from a pipe and carried them downstream. Flowing ice sheets also may transport kimberlite fragments.

Sedimentary Mineral Deposits

Sedimentary mineral deposits include some of the world's most valuable mineral sources. Many economically important minerals segregate by chemical and physical means as an ordinary result of sedimentary processes (see Chapter 8).

Nonmetallic Sedimentary Deposits Limestones—which are chemically precipitated mainly by marine organisms—are used for cement, agricultural lime, and building stone. Pure quartz sands—left behind when waves and currents weather mixed-mineral sands so that all materials other than quartz are removed—are the raw materials for glassmaking and for the fiber-optic cables that are replacing copper wires in communication lines. In many areas of the northern United States and southern Canada, Pleistocene glaciers distributed an abundance of coarse sand and gravel, suitable for construction purposes. These materials are also widely distributed in the channels and former channels of many rivers. Clays of high purity produced by prolonged weathering are used for ceramics in both home and industry. Evaporite deposits of gypsum, separated from seawater by precipitation, are used for plaster. Sodium and potassium salts from evaporites have varied uses, from table salt to fertilizer. Phosphate rocks—marine shales and limestones enriched in phosphate by chemical reaction with deep seawater—are raw materials for the world's fertilizer industry.

Metallic Sedimentary Deposits Sedimentary mineral deposits are also important sources of copper, iron, and other metals. These deposits were chemically precipitated in sedimentary environments to which large quantities of metals were transported in solution. Some of the important sedimentary copper ores, such as those of the Permian Kupferschiefer (German for "copper slate") beds of Germany, may have precipitated from hot brines of hydrothermal origin, rich in metal sulfides, that interacted with sediments on the ocean bottom.

The major iron ores have been found in Precambrian sedimentary rocks. Earth's atmosphere was poor in oxygen early in its history (see Chapter 1), and the low availability of oxygen is now thought to have allowed an abundance of iron in its soluble, lower oxidation (ferrous) form (Fe^{2+}) to be leached in great quantities from the land surface. Groundwater transported the ferrous iron in solution to broad, shallow marine environments where it could be oxidized to its insoluble ferric form (Fe^{3+}) and precipitated. (See Chapter 7 for a discussion of the oxidation states of iron.) In many of these basins, the iron was deposited in thin layers that alternate with layers of siliceous sediments (cherts). Such iron

Figure 22.21 Precambrian banded iron beds. Rust-colored layers are limonite, interbedded with hematite and chert. Hamersley, Australia. [Spence Titley.]

ores are called *banded iron formations* (**Figure 22.21**). The Lake Superior iron deposits, for many years the source of iron for the U.S. steel industry, are of this type.

Placers Possibly the most publicized (and romanticized) type of mineral prospecting is panning for gold. The gold seeker dredges up a flat pan of river sediment and sifts it in the hope of turning up the glint of a nugget. Many rich deposits of gold, diamonds, and other heavy minerals such as magnetite and chromite as found in **placers,** ore deposits that have been concentrated by the mechanical sorting action of river currents.

Because heavy minerals settle out of a current more quickly than lighter minerals such as quartz and feldspar, the heavy minerals tend to accumulate on river bottoms and sandbars, where the current is strong enough to keep the lighter minerals suspended and in transport but is too weak to move the heavier minerals. Similarly, ocean waves preferentially deposit heavy minerals on beaches or shallow offshore bars. The gold panner accomplishes the same thing: the shaking of a water-filled pan allows the lighter minerals to be washed away, leaving the heavier gold in the bottom of the pan. The recent high prices of gold have led to a revival of old-fashioned gold panning

Some placers can be traced upstream to the location of the original mineral deposit, usually of igneous origin, from which the minerals were eroded. Erosion of the Mother Lode, an extensive gold-bearing vein system lying along the western flanks of the Sierra Nevada batholith, produced the placers that were discovered in 1848 and led to the California gold rush. The placers were found before their source was discovered. Placers also led to the discovery of the Kimberley diamond mines of South Africa two decades later.

Ore Deposits and Plate Tectonics

Plate tectonics theory explains the various types of igneous activity in terms of the interactions of plates at boundaries where they separate or collide. Because igneous processes

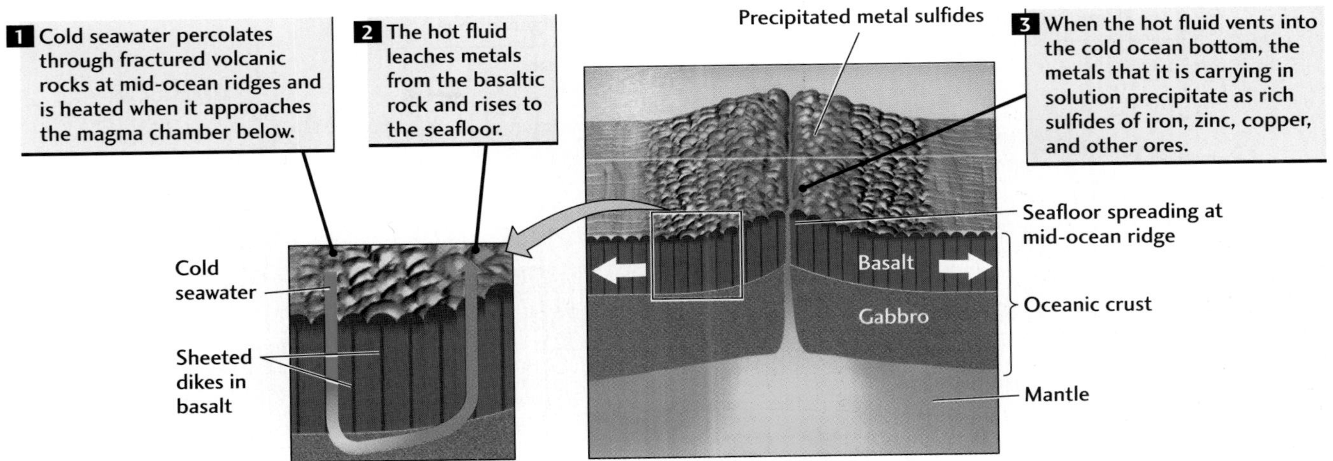

Figure 22.22 Enormous quantities of sulfide ores are found at mid-ocean spreading centers.

bring chemical elements and their mineral compounds from the interior to the surface, the theory of plate tectonics helps us understand the origin of ore deposits and can lead to the discovery of new ones.

In 1979, geologists exploring the seafloor at a plate-separation center (the East Pacific Rise) discovered hot springs, laden with dissolved minerals, venting on the seafloor (see Chapters 6 and 17). These hot springs originate in seawater that circulates through fractures near the rift where the plates separate (**Figure 22.22**). The seawater is heated to temperatures of several hundred degrees Celsius when it comes in contact with magma or hot rocks deep in the crust. The heated seawater leaches minerals from the hot rocks and rises to the seafloor. When the hot waters, now loaded with dissolved minerals, reach the cooler upper crust and near-freezing ocean-bottom waters, the minerals precipitate. This process is the origin of the "black smokers" shown in Figure 17.7. In this manner, enormous quantities of sulfide ores rich in zinc, copper, iron, and other metals are being deposited along mid-ocean spreading centers. High-grade deposits of native gold have been found in one submarine deposit on the Mid-Atlantic Ridge.

When modern spreading centers were recognized as rich sources of mineral deposits, geologists began to look on land for the remains of ancient seafloor, which also might hold valuable resources. Some deposits were found in plate-collision zones, where fragments of ancient oceanic lithosphere are occasionally emplaced on land. These deposits, known as ophiolites, are discussed in Chapter 5. The rich copper, lead, and zinc sulfide deposits in the ophiolites of Cyprus, the Philippines, the Apennines in Italy, and elsewhere probably owe their origin to the process of hydrothermal circulation along ancient mid-ocean rifts. The copper deposits of Cyprus were as important to the economy of ancient Greece as Middle East oil deposits are to the modern economy. Economically important deposits of chromite ores are occasionally found in deeper portions of ophiolites. They may have originated by fractional crystallization within magma chambers that underlie mid-ocean ridges.

Many other types of sulfide ore deposits, of hydrothermal or igneous origin, are found at modern and ancient plate-collision boundaries, including those of the Cordillera of North and South America, the eastern Mediterranean to Pakistan, the Philippine Islands, and Japan. **Figure 22.23** summarizes some of the associations between plate tectonics and mineral deposits. Deposits found in magmatic arcs are thought to result from the igneous activity that typically occurs in collision zones. One hypothesis proposes that some of these collision-boundary deposits represent the second stage in a two-stage ore-forming process. The first stage is the creation of mineral ores by hydrothermal activity at a mid-ocean spreading center. The second stage, separated in time and space from the first, is the subduction and partial melting at a collision zone of oceanic sediments and crust containing these previously concentrated minerals. As the plate descends into increasingly hot regions of the mantle, the metals "boil off"—that is, they melt and rise into the overriding plate along with magma. The iron, copper, molybdenum, lead, zinc, tin, and gold found along convergent plate boundaries could have been created by hydrothermal activity and reborn by igneous processes, all driven by plate tectonic movements.

The seafloor away from plate boundaries, however, may be the first candidate for deep-sea mining because of the widespread occurrence of *manganese nodules,* potato-shaped aggregates of manganese, iron, copper, nickel, cobalt, and other metal oxides (**Figure 22.24**). The nodules vary in size, but most of them are a few centimeters in diameter. They form by the precipitation of metal oxides from seawater, usually on a small nucleus such as a shark's tooth or a chip of rock. They are potentially valuable not only because of the gradual depletion of high-grade deposits of manganese

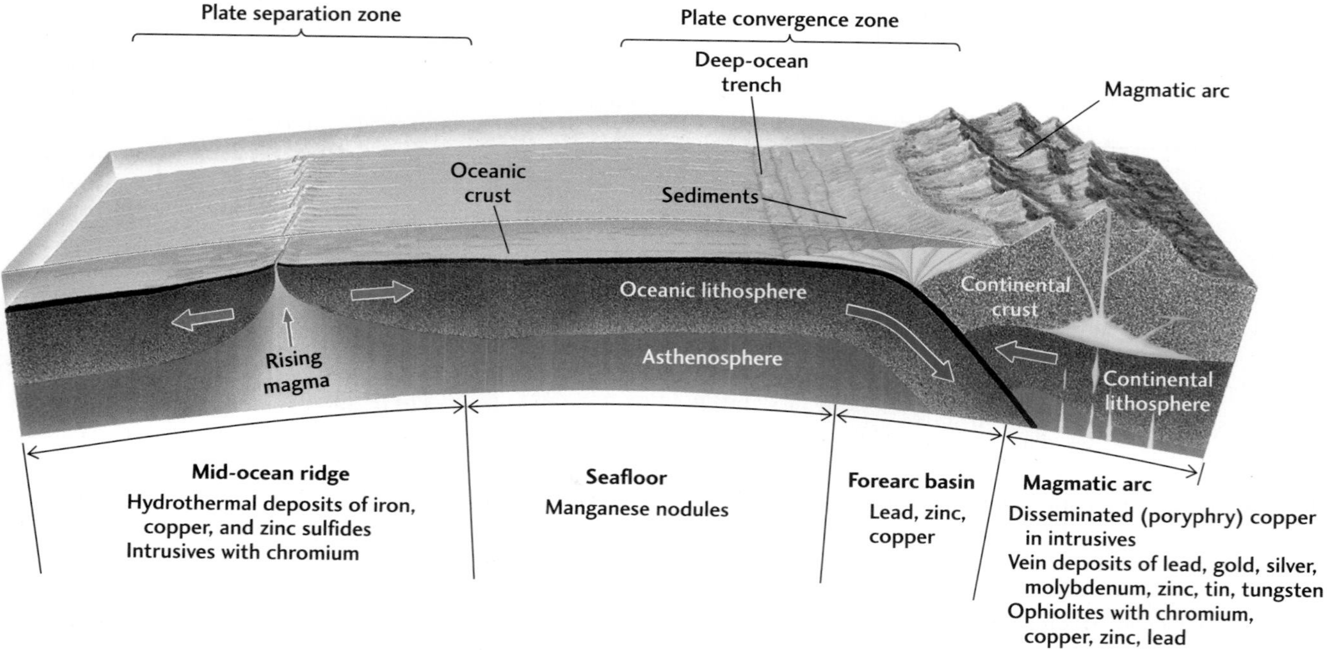

Figure 22.23 The role of plate boundaries in the accumulation of mineral deposits. Ocean sediment and crust are enriched in metals by hydrothermal ore deposition along a mid-ocean ridge. Rising magma in the subduction zone is the source of ores that form the metal-bearing provinces of a magmatic belt such as the Cordillera of North and South America. The melting of subducted sediment and crust may contribute ore constituents. Mineral-bearing oceanic fragments (ophiolites) accrete to the continent in the collision zone.

on land but also because they are rich in other metals. Deposits are estimated to be in the trillions of tons.

This brief summary of the geology of mineral deposits barely touches on the great diversity of geological settings in which various minerals of value are found. Some minerals or ores are found mainly or only in one kind of deposit; others are found in a variety of settings. Although there is probably an abundance of ore bodies on the deep seafloor, most known ore bodies are found on older parts of the continental crust associated with orogenic belts. They either originated on the continent or are remnants of mineralized pieces of ocean crust thrust onto the continent in plate collisions.

Figure 22.24 Manganese nodules are small concretions found on the deep seafloor that are as much as 20 percent manganese and contain smaller amounts of iron, copper, and nickel. (This nodule is about 7.5 cm in diameter.) They have been dredged from the ocean floor to demonstrate the feasibility of mining the seabed. [Chip Clark.]

The Need to Find New Mineral Deposits

The total dollar value of all mineral resources, including fuels, produced in the United States grew some 30-fold over the past four decades. This growth (which includes some inflationary increases resulting from the decrease in the value of the dollar during that time) took place in a highly industrialized society that had already built a huge technological capability and whose population grew by only about 60 percent in the same period. The recovery of many nations after World War II and the growth in their economies led to an enormous increase in consumption and production in the rest of the world, resulting in a relative decline in production and consumption of nonfuel minerals by the United States in comparison with the rest of the world.

Although the growth of mineral demand has now declined in the industrialized nations, unequal sharing of world mineral resources still remains and has far-reaching economic and political repercussions. For instance, North America—with less than 10 percent of the world's population—consumes almost 75 percent of the world's production of aluminum, whereas Asia and Africa—with about two-thirds of the world's population—together use only a little over 5 percent. The same extreme imbalance can be found for other materials. International relations are now deeply affected by struggles over the control of resources. One widespread cause of conflict is the demand by some mineral-rich developing nations for greater control over resources mined on their land by corporations based in North America or Europe. In recent years, several countries in Africa and South America have nationalized foreign-owned mining companies. As the Earth's human population grows and people in less developed nations seek higher standards of living (more food, raw materials, energy, manufactured goods, construction of all types), the demand for mineral resources will be strong.

How important the unexplored deep sea will become is an open question. The answer depends in part on the development of efficient marine technology for deep-sea exploration and mining and the resolution of legal issues regarding ownership of deep-sea deposits. It is generally agreed that a nation has exclusive rights to mineral deposits in the offshore area within 200 nautical miles of its coast—the so-called **exclusive economic zone** (**Figure 22.25**). Still in question is who owns the mineral deposits on the seafloor beyond this zone. Are the deposits owned by all nations in

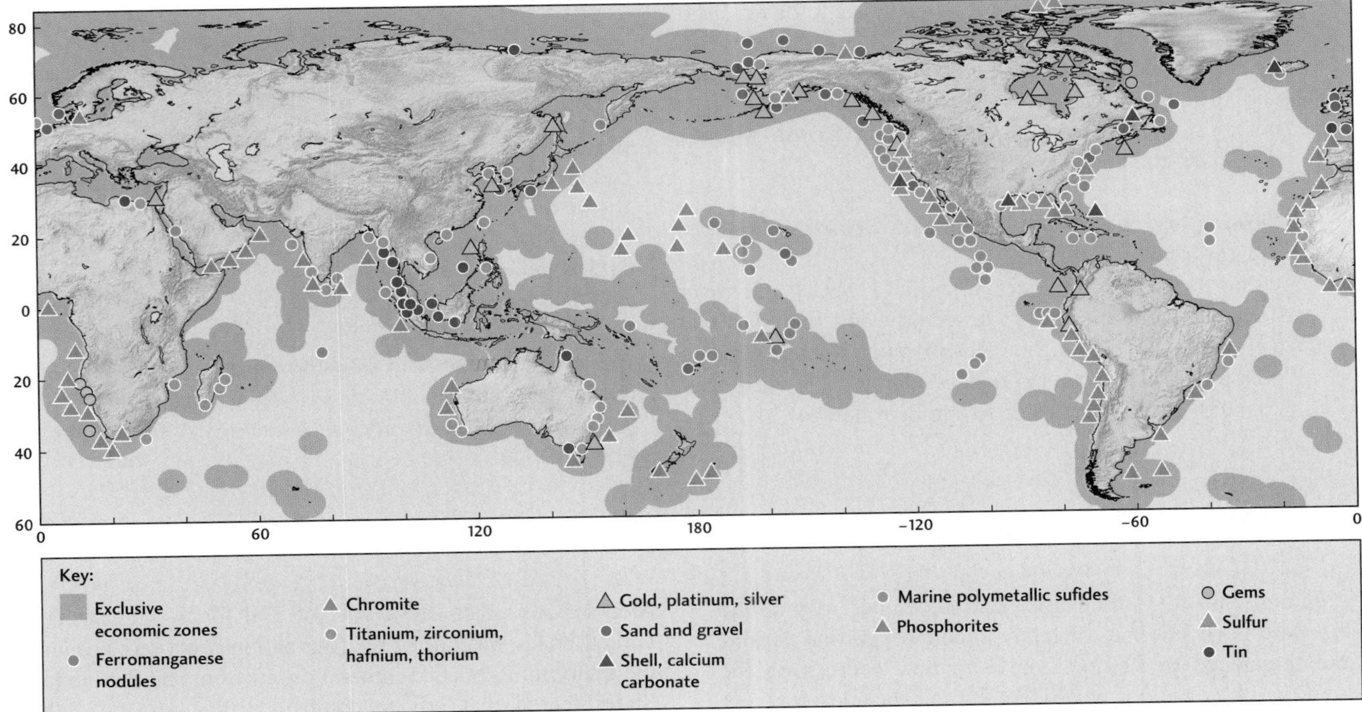

Figure 22.25 Locations of some major nonfuel seabed ore deposits. Note the concentrations at plate boundaries and continental shelves. The gray areas outline exclusive economic zones claimed by individual nations—the sea within 200 nautical miles of their territory. [After J. M. Broadus, "Sea Bed Materials." *Science* 235: 853–859, 1987.]

22.3 Sustainable Development

Scientists who study the biosphere and the oceans, atmosphere, and landmass that are its framework will be important contributors to "sustainable development," a concept that is appearing with increasing frequency in newspapers, public debates, classroom discussions, and scholarly journals. *Sustainable* refers to safeguarding all of life and the support systems on which it depends. These support systems include protecting the environment and its ecosystems and conserving Earth's natural resources from which humankind derives the energy and materials of daily existence. Sustaining Earth as a place of beauty to be appreciated is an important aesthetic, cultural, and recreational consideration. *Development* refers primarily to economic growth but with the stipulation that such growth provide the financial resources to improve the quality of life of people everywhere (for example, housing, health, education, environmental remediation). Debates focus on whether accelerated economic growth can occur without causing irreversible damage to the biosphere and Earth's resources, thus compromising the ability of future generations to meet their needs.

The following sections consist of goals to be achieved in the next 50 years. The outcome will be determined by the interaction of politics, economics, cultural factors, and the ability of scientists to develop innovative technologies that improve productivity and growth without despoiling the environment.

Slowing and Stabilizing Population Growth in Developing Countries

Without such action, scientists and social scientists believe that the demand for energy and materials will increase to the extent that the biosphere will be seriously polluted, global warming will accelerate, and resources will be depleted. According to a joint statement issued by 58 of the world's scientific academies, high fertility rates can be correlated with poverty, high infant mortality, lowered life expectancy, low social status and educational levels of women, low adult literacy, lack of access to reproductive health services, and inadequate availability and acceptance of contraceptives. Cultural influences also play an important role in fertility rates. These influences include the value placed on large families in developing countries, the lack of security for the elderly, and religious and political factors.

Food for an Additional Several Billion People

With a few exceptions—for example, Africa—modern agriculture ("the green revolution" created by scientists) can today provide the world with reliable food production, although hunger still exists because of poverty. However, the demand for food will *triple* by 2050, and yields are not growing rapidly enough to meet the needs. Overfarming and overgrazing are degrading soil quality. In addition, the increased demand for chemicals for irrigation and pest control, water for irrigation, energy, and the expansion of farmlands will present growing environmental problems.

Energy

Energy is crucial to everything—food production, manufacturing, transportation, heating, and cooling. Yet the production and use of energy are arguably the most serious threats to the environment in their ability to pollute the air and trigger global climate change. Providing clean energy for a population growing to some 10 billion by 2050, sufficient to enjoy a reasonable quality of life, will be a major undertaking.

Achieving Sustainable Development

To achieve sustainable development would require a world effort without precedent in history. The commit-

common? By the discoverer and developer? By both? Because the economies of many mineral-exporting nations are threatened by competition from new sources on the seafloor, these nations would like to limit such development and obtain a share of the profits.

These issues of international law have been debated for more than 20 years. In 1982, the United Nations adopted an agreement known as the Law of the Sea Treaty, which provides a legal and regulatory system for the development of deep-sea resources, by a vote of 130 for, 4 against. The United States was one of the four nations opposed, because the treaty limits seabed mineral production, and the United States feared that private companies that developed the technology to mine the seabed and invested in production would be inadequately compensated.

A hard fact of life is that an equal per capita sharing of the world's available reserves would not be enough to bring everyone to a "satisfactory" level of consumption, certainly

ment of political leaders of both the developed and the developing nations is needed to remove obstacles to sustainable development. The developed nations would have to contribute substantially to the necessary investments. Scientists, engineers, and social scientists would play a crucial role because new technologies and new ways of doing things will be required. They could invent more efficient ways to use fossil fuels, reducing CO_2 emissions with the goal of a "decarbonized" energy supply. Perhaps intrinsically safe nuclear energy can be developed. Renewable energy sources such as solar and biomass energy could become widespread if scientists discover how to increase their efficiency and decrease their costs. New biotechnology might create more productive and disease-resistant crop species that need less water and fewer chemical fertilizers and pesticides, thus reducing or eliminating chemical contamination of aquifers and rivers.

A challenge will be to develop an "industrial ecology," one that removes effluents from industrial processes and recycles them. It implies more efficient and less wasteful use of materials, energy, and water. Industrial ecology encourages recycling and substitution, which would reduce the demand for materials and the pollution caused by manufacturing, mining, and smelting operations. An industrial group recently characterized this approach as "cleaner, leaner, lighter, and drier."

To address the correlation of high infant mortality and low life expectancy with increased fertility rates, a global program would include improvements in sanitary engineering; the introduction and use of new vaccines and antibiotics; and the development and dissemination of more effective, simpler, and cheaper contraceptives. Earth scientists would play an important role. Think of natural gas, geothermal energy, sources of clean water, safe deep underground storage of radioactive wastes and other contaminants, minerals and materials, mitigating natural disasters—all discussed in previous chapters. Perhaps their largest contribution will be their special understanding of how Earth's atmosphere, oceans, and landmasses are linked and how its systems and cycles interact and can be perturbed by human activities.

However, science and technology will be to no avail without vision and early action by the world's political leaders in providing moral and material support for goals that stretch well beyond their terms in office. They would have to subscribe to the goal of sustainable development and promote the research required to develop the new technologies required. The authors of this book believe that only by stabilizing population growth, protecting the environment and its ecosystems, and conserving Earth's natural resources can we avoid a calamity as serious as humankind has ever faced.

Thought Question

There are responsible contrary positions that contend that Earth can sustain a larger population. They argue that the only essential resource is the human intellect, and that the aggregate brain power of the larger population will expand the capacity of the human race to create new technologies that will make sustainable development possible. Others believe that the only alternative is for the developed nations to revert to the simpler lifestyles of earlier times, with a slower pace and much lower consumption of resources. In practice, this could mean fewer conveniences of all kinds. Discuss your preference for any of the three courses described in this feature or any other option.

Sources: *Our Common Journey: Towards a Sustainable Transition* (Washington, D.C.: National Research Council, 1999); *Population Summit of the World's Scientific Academies* (Washington, D.C.: National Academy Press, 1993), www.nas.edu.

not to a level anywhere near that of an affluent country in Europe or North America. The problem is compounded by the continued rapid growth of the world's population (see Chapter 23).

Some people argue that searching for more ore bodies to mine is not the solution. The extraction, processing, and waste disposal connected with exploiting materials drawn from the Earth are so damaging that they cannot be sustained. They propose a new "materials economy" based on recycling; greater efficiency in the use of materials; and substitution of advanced, environmentally benign materials for minerals (see Feature 22.3). One example is the use of fiber-optic glass cables made from sand instead of copper wire.

The world clearly faces major readjustments in the decades ahead, and it is certainly not too soon to start working out equitable, humane, and lasting approaches to building a sustainable world. How well we use our planet will depend on how well we understand how it works and the extent to which the people of the world cooperate intelligently in the development and use of its resources—and do so in a manner that protects the environment. We have only one Earth. To continue to live on it, we must learn to use it wisely.

SUMMARY

What is the origin of oil and natural gas? Oil and gas form from organic matter deposited in marine sediments, typically in the coastal waters of the sea. The organic materials are buried as the sedimentary layers grow in thickness. Under heat and high pressure, organic carbon is transformed into liquid and gaseous hydrocarbons, compounds of carbon and hydrogen. Oil and gas accumulate in geologic traps that confine the fluids within impermeable barriers.

What environmental concerns are connected with the production of oil? Pollution in the production and transportation of oil is a major problem. Proponents of oil production believe that careful design of equipment and safety procedures can greatly reduce the chances of oil spills and satisfy the energy needs of our civilization. A significant body of public opinion is unconvinced, however, and exploration for new reserves is therefore restricted in many coastal regions and pristine wilderness areas, such as northern Alaska.

Why is there concern about the world's oil supply? Oil is a finite resource: it will be depleted faster than nature can replenish it. Therefore, as the supply is withdrawn from the oil reservoirs of the world, its availability diminishes, its price rises, and other types of energy sources will have to be found. At current rates of use, the remaining oil available for transportation, heating, and the generation of electricity will be depleted in about 100 years.

What is the origin of coal and how big a resource is it? Coal is formed by the compaction and chemical alteration of wetland vegetation. There are huge resources of coal in sedimentary rocks. We have used only about 2.5 percent of the world's minable coal.

What is the trade-off between risk and benefit in the use of coal? Coal mining and pollution caused by coal burning are risky to human life and to the environment. Coal combustion is a major source of carbon dioxide and the acid emissions that are the precursors of acid rain. Because of its abundance and low cost, however, the use of coal is likely to increase in the next decades to generate electric power and to convert coal into liquid and gaseous fuels.

Is nuclear energy a solution to the world's energy problem? Nuclear power from the fission of ^{235}U can be a major energy source but only if its costs do not keep escalating and the public can be assured of its safety. It has the advantage that it does not release carbon dioxide and the disadvantage that safe repositories must be found to store radioactive wastes for hundreds of thousands of years. Known high-grade reserves of ^{235}U ore can support the projected use of conventional nuclear power plants for a few decades, longer if advanced reactors are installed early in this century. The use of nuclear energy could extend our resources of fossil fuels.

What are the prospects for alternative energy sources? Alternative sources include hydroelectric energy, wind power, solar energy, biomass energy, and geothermal energy, none of which has any immediate prospect of being an adequate response to world energy needs. With advances in technology and reductions in cost, however, solar and biomass energy could become major sources in this century.

What should be the goal of energy policy? Energy policy should guide the nations of the world through the transition from hydrocarbon fuels to nonpolluting, renewable energy sources. Carbon dioxide emissions must be reduced to decrease the impact of global climate change. The more efficient use of energy, the greater use of natural gas, and the introduction of a safe nuclear energy technology would facilitate this changeover. We have the resources and technology to achieve this goal peacefully and without economic dislocation if the sociopolitical problems can be worked out.

What is hydrothermal mineral deposition? Hydrothermal deposits, which are some of the most important ore deposits, are formed by hot water that emanates from igneous intrusions or by heated circulating groundwater or seawater. The heated water leaches soluble minerals in its path and transports them to cooler rocks, where they are precipitated in fractures, joints, and voids. These ores may be found in veins or in disseminated deposits, such as the porphyry copper type.

What are some processes that lead to the formation of sedimentary mineral deposits? Ordinary chemical and mechanical sedimentary processes segregate such economically important minerals as limestone, sand and gravel, and evaporite salt deposits. Sedimentary ores of copper and iron have formed as precipitates in special sedimentary environments, the iron ores chiefly in Precambrian times. Placers are ore deposits, rich in gold or other heavy minerals, that were laid down by currents.

How do igneous ore deposits form? Igneous ore deposits typically form when minerals crystallize from molten magma, settle, and accumulate on the floor of a magma chamber. They are often found as layered accumulations of minerals. The rich ore body at Sudbury, Ontario, for example, is a mafic intrusion that contains great quantities of layered nickel, copper, and iron sulfides near its base.

What are some mineral policy issues modern society faces? We must discover new mineral deposits to support an increasingly industrialized world civilization. Prospects for finding new resources are good. The sea is a largely untapped resource. Although technological advances can postpone the day of reckoning, stocks will become scarcer as the world population increases. Therefore, conservation, recycling, and substitution of alternative materials will become increasingly important in the years ahead.

Key Terms and Concepts

exclusive economic zone (p. 533)
fossil fuel (p. 510)
geothermal energy (p. 520)
hydroelectric energy (p. 520)
nuclear energy (p. 516)
oil trap (p. 510)
ore (p. 522)
placer (p. 530)
reserve (p. 508)
resource (p. 508)
solar energy (p. 517)
sustainable development (p. 508)
vein deposit (vein) (p. 526)

Exercises

This icon indicates that there is an animation available on the Web site that may assist you in answering a question.

1. Do you think the continental slope and rise might be good places to drill for oil if we can invent the necessary technology? Why or why not?

2. Which of the following factors are most important in estimating the future supply of oil and gas? (a) Rate of oil accumulation, (b) rate of natural seepage of oil, (c) rate of pumping of oil from known reserves, (d) rate of discovery of new reserves, (e) total amount of oil now present in the Earth.

3. Taking benefits and risks into consideration, rank all the forms of fossil fuels by relative importance and explain how their rankings might differ at the end of the twenty-first century.

4. Name four regions of the world that are major sources of oil.

5. Which three countries have the largest coal reserves?

6. Contrast the risks and benefits of nuclear fission and coal combustion as energy sources.

7. How would you use knowledge of the distribution of plate boundaries to make a map showing the areas of Earth most likely to be sources of geothermal power?

8. What social costs should be included in evaluating the true costs of energy derived from fossil fuels, from nuclear fission, and from geothermal heat?

9. What do you think will be the major sources of energy in the year 2050? In the year 3000?

10. What are the characteristics of an economical ore deposit?

11. Describe the creation of an ore body by hydrothermal activity.

12. List several examples of minerals in igneous ore bodies.

13. Contrast the process of ore formation in the sediment and crust of the deep sea with that in a convergence zone.

14. Why are the conservation and recycling of useful materials drawn from the Earth and the creation of substitutes important to the future of humankind?

Thought Questions

This icon indicates that there is an animation available on the Web site that may assist you in answering a question.

1. A tax increase of $1 per gallon on gasoline would generate $100 billion of income for the federal government that could serve many useful purposes. It would also motivate drivers to switch to more fuel efficient automobiles, thereby reducing imports of oil. Even with this increase, the price of gasoline would still be lower in the United States than in many European countries. Despite all these advantages, such a tax is unlikely to be enacted. What are the arguments against it?

2. Should we stockpile oil against future shortages caused by political disruptions? What are the costs and benefits of doing so?

3. Taking into account the costs of oil imports to the economy, the risks of relying on foreign exporters, and the environmental consequences, do you think we should remove restrictions on the production of oil from the continental shelves and from wildlife preserves? Explain.

4. Explain the statement that increased conservation is the cheapest new energy source.

5. You are the U.S. representative to the United Nations. You have just made a speech extolling the importance of sustainable development. How would you respond to the representative of a developing country that has to increase energy production to grow economically, who claims that her country does not have the capital to invest in renewable energy and criticizes the United States for being profligate in the use of resources?

6. Should the United States spend large sums to stockpile strategically important minerals (those essential for economic or defense needs) for use in an emergency in the event that foreign supply is cut off?

7. In 1992, many nations with interests in Antarctica signed a treaty agreeing not to mine for minerals on that continent. Why did they do so? Do you agree with their decision?

8. Will plate tectonics theory contribute to the search for ore bodies? How?

9. Do you think people in developed countries will be willing to reduce consumption or shift to a new "materials economy" to make materials more readily available to people in less developed countries?

Short-Term Team Projects

Alternatives to Oil

Regions of the world that produce oil often suffer conflict over access to and use of this nonrenewable resource. The Nigerian activist Ken Saro-Wiwa was executed by his own government, despite worldwide protest, for agitating against oil companies accused by knowledgeable people of polluting the Niger delta. Choose a classmate as a partner for this project. Considering the turmoil that often surrounds the use of oil, investigate and report on an alternative energy source that could be widely used because of global availability.

Suggested Readings

Baldwin, S. F. 2002. Renewable energy: Progress and prospects. *Physics Today* (April): 62–68.

Barnes, H. L., and A. W. Rose. 1998. Origins of hydrothermal ores. *Science* 279: 2064–2065.

Briskey, J. A., et al. 2001. It's time to know the planet's mineral resources. *Geotimes* (March): 14–22.

Broadus, J. M. 1987. Sea bed materials. *Science* 235: 853–859.

Carr, D. D., and N. Herz (eds.). 1988. *Concise Encyclopedia of Mineral Resources.* Cambridge, Mass.: MIT Press.

Clark, J. P., and F. R. Field III. 1985. How critical are critical materials? *Technology Review* (August/September).

Dorr, A. 1987. *Minerals: Foundations of Society.* Alexandria, Va.: American Geological Institute.

The Economist. 2001. A survey of energy. February 10.

Edwards, J. D. 1998. Crude oil and alternate energy production forecasts for the twenty-first century: The end of the hydrocarbon era. *AAPG Bulletin* 81: 1292–1305.

Energy Information Administration. 2002. *Annual Energy Review.* Washington, D.C.: U.S. Department of Energy.

Glasby, G. P. 2000. Lessons learned from deep sea mining. *Science* 289:551–554.

Goldenberg, J. 1995. Energy needs in developing countries and sustainability. *Science* 269: 1058–1059.

Guilbert, J. M., and C. F. Park, Jr. 1986. *The Geology of Ore Deposits.* New York: W. H. Freeman.

Heaton, G., R. Repetto, and R. Sobin. 1991. *Transforming Technology: An Agenda for Environmentally Sustainable Growth in the 21st Century.* Washington, D.C.: World Resources Institute.

Hodges, C. A. 1995. Mineral resources, environmental issues, and land use. *Science* 268: 1305–1312.

Keary, P., and F. J. Vine. 1990. Plate tectonics and economic geology. Chap. 11 in *Global Tectonics.* London: Blackwell Scientific.

Monastersky, R. 1998. Can methane hydrates fuel the 21st century? *Science News* 154: 312–314.

Moniz, E. J., and Melanie A. Kenderdine. 2002. Meeting energy challenges: Technology and policy. *Physics Today* (April): 40–46.

National Research Council. 1990. *Competitiveness of the U.S. Minerals and Metals Industry.* Washington, D.C.: National Academy Press.

National Research Council. 1991. *Policy Implications of Greenhouse Warming.* Washington, D.C.: National Academy Press.

National Research Council. 1992. *Radioactive Waste Repository Licensing.* Washington, D.C.: National Academy Press.

National Research Council. 1993. *Solid-Earth Sciences and Society.* Washington, D.C.: National Academy Press.

Parsons, E. A., and D. W. Keith. 1998. Fossil fuels without CO_2 emissions. *Science* 282: 1053–1054.

Physics Today. 2002. Special Issue: The Energy Challenge. April.

Resource Reserve Definitions. 1980. Circular 831. Washington, D.C.: U.S. Geological Survey.

Sawkins, F. J. 1984. *Metal Deposits in Relation to Plate Tectonics.* New York: Springer-Verlag.

Science Summit on World Population. 1993. Report of conference of national academies of sciences of 80 countries. New Delhi, October 24–27.

Smith, J. R. (ed.). 1999. *Colloquium on Geology, Mineralogy, and Human Welfare.* Washington, D.C.: National Academy Press.

Stone, J. L. 1993. Photovoltaics: Unlimited electrical energy from the Sun. *Physics Today* (September): 22–29.

U.S. Bureau of Mines. 1994. *Minerals Yearbook.* Washington, D.C.: U.S. Government Printing Office.

Coal-fired power plants such as this one in San Juan County, New Mexico, are converting fossil carbon previously buried in Earth's crust into carbon dioxide, a greenhouse gas. Fossil-fuel burning also produces gases that cause smog as well as acid rain. [Larry Lee Photography/Corbis.]

CHAPTER

23

Earth's Environment, Global Change, and Human Impacts

"Although for decades civilization has pillaged and polluted the Earth, in the past those attacks were relatively localized; now, with the global changes caused by greenhouse gases and ozone erosion, man has altered the most elemental process of life everywhere, and the outdoors, nature itself, has been turned into the equivalent of an enormous heated room."

BILL MCKIBBEN, *THE END OF NATURE*

The human habitat is a thin interface between Earth and sky, where great forces interact to shape the face of our planet. Tectonic forces within the lithosphere, driven by heat from the deep interior, generate earthquakes, erupt volcanoes, and lift up mountains. Meteorological forces within the atmosphere and hydrosphere, driven by heat from the Sun, produce storms, floods, glaciers, and other agents of erosion. Interactions between the global geosystems of plate tectonics and climate maintain an equable environment at Earth's surface in which human society can prosper and grow.

Indeed, our numbers and activities are multiplying at phenomenal rates. From 1930 to 2000, the global population rose by 300 percent, from 2 billion to 6 billion. This total is projected to exceed 8 billion within the next 30 years. Moreover, total energy usage has increased by 1000 percent over the last 70 years and is now rising twice as fast as the population.

Humans have modified the environment by deforestation, agriculture, and other types of land usage throughout recorded history, but the effects in earlier times were usually restricted to a local or regional habitat. Today's society affects the environment on an entirely new scale: our activities can have global consequences. The magnitude of current human activities relative to the plate tectonic and climate systems that govern Earth's surface can be illustrated by some simple statistics:

- Reservoirs built by humans now trap about 30 percent of the sediment transported down the world's rivers.

• In most developed countries, civil engineers move more tons of soil and rock each year than do all natural processes of erosion combined.

• Within 50 years of the invention of the coolant *freon,* enough man-made chlorofluorocarbons had leaked out of refrigerators and air conditioners and floated up into the stratosphere to damage Earth's protective ozone layer.

• Since the industrial revolution began in the early nineteenth century, deforestation and the burning of fossil fuels have increased the amount of carbon dioxide in the atmosphere by over 30 percent. Atmospheric carbon dioxide content is rising at the unprecedented rate of 4 percent per decade and will likely cause significant global warming during the lifetimes of most people now in college.

In this final chapter, we examine how human activities might change the global environment. We consider several environmental issues, but we focus on the problem many scientists judge to be the most pressing of all: the potential for global climate change arising from the increasing amounts of carbon dioxide and other greenhouse gases humans emit into the atmosphere.

To appreciate the potential for this type of human-induced global change, we need to learn more about how the climate system works and how it varies as a result of natural processes.

The Earth System Revisited

Throughout this book, we have described Earth as a complex system of many interacting components (**Figure 23.1**). We have shown how geologists learn about the Earth system through observations of terrestrial processes that are active today or have left their imprints on the geologic record. We start this closing chapter by summarizing in general terms what we have learned about the use of the "geosystems approach" to interpret geologic observations.

Geosystems and the Scientific Method

The first step in the analysis of a geologic process is to identify the appropriate geosystem: the proper subset of components and interactions within the Earth system needed to explain an interesting type of observed behavior. Consider, for instance, the eruptions produced by an active volcano or the earthquakes generated by an active fault. Geologists represent a volcanic system in terms of the rock that partially melts to produce magma, the transport of magma upward through the lithosphere, its eruption and solidification at the surface, and the construction and deformation of the resulting volcanic edifice (see Chapter 6). Similarly, they represent an active fault system in terms of faults in a brittle rock

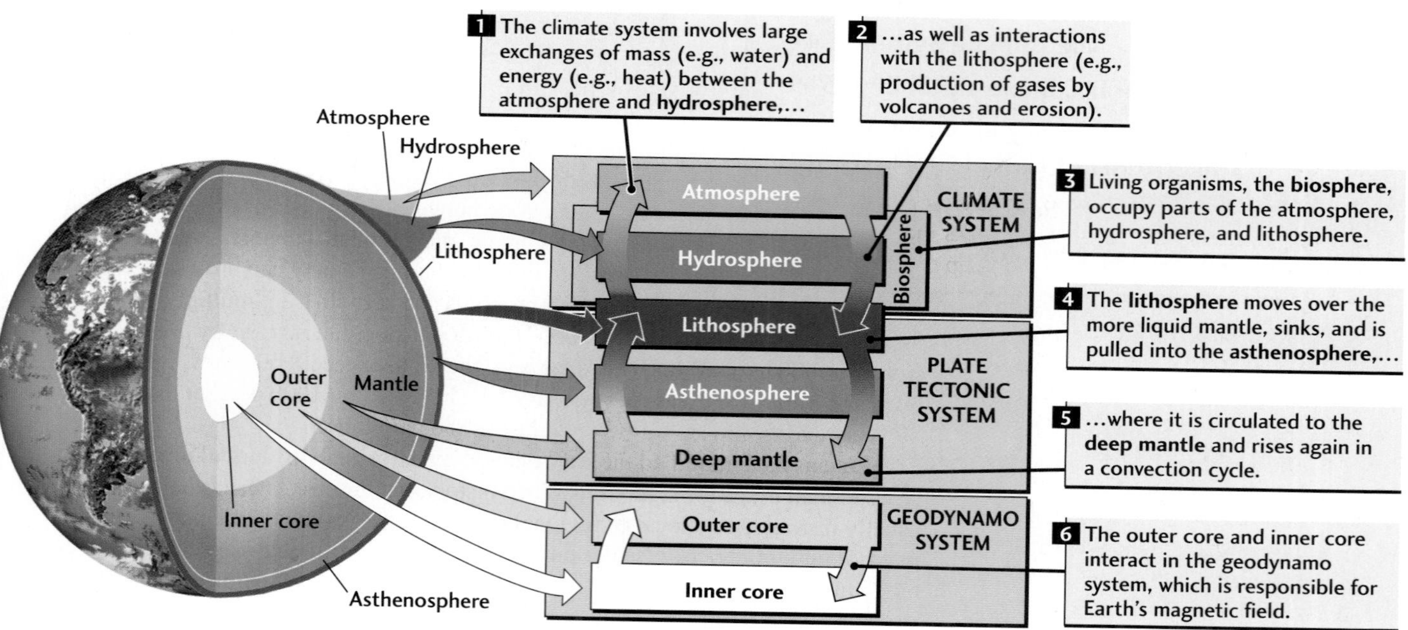

Figure 23.1 Interactions among components of the Earth system that form the global geosystems of climate, plate tectonics, and the geodynamo.

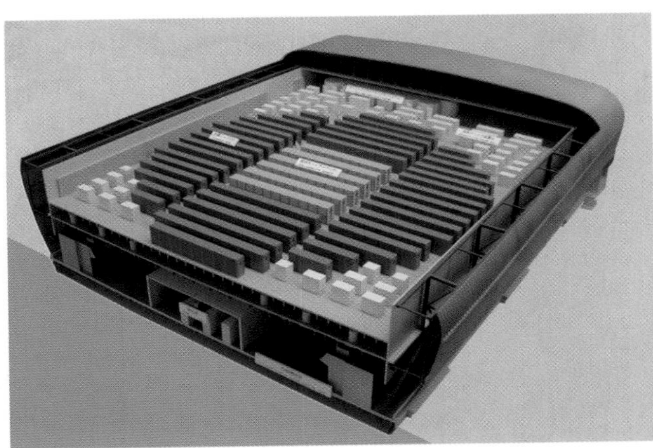

Figure 23.2 The world's largest computer is the Earth Simulator at the Yokohama Institute for Earth Sciences. This computer is specifically designed to simulate geosystems such as Earth's climate, mantle convection, and the geodynamo. The computer has over 5000 vector processors, occupies an area that covers four tennis courts, and has achieved a speed of 35.6 trillion floating point operations per second—making it five times faster than its nearest competitor. [Courtesy of Earth Simulator Center.]

layer that overlies more ductile layers being deformed by plate motion (see Chapter 11), and they describe an earthquake in terms of the sudden slippage that occurs on a fault once the stresses caused by the plate motion exceed the fault's frictional strength (see Chapter 19).

Based on such concepts, geologists use both established theories (for example, about how rocks melt or break) and more speculative hypotheses (for example, about the geometry of a particular magma body or fault surface) to construct a quantitative representation—a model—of the geosystem. The geometry and interactions needed to describe a real geosystem are so complex that millions of numbers and manipulation by large computers may be required (**Figure 23.2**). Geologists then adjust the model parameters—such as the rock's melting temperature or breaking strength—to simulate the observed behavior—such as the volumes of magmatic eruptions or the sizes of earthquakes—most accurately. They gain confidence in a model if just a few parameters with reasonable values can account for many data, and even more confidence if the model accurately predicts new data not used in its original construction. If the model doesn't fit the data no matter how the parameters are adjusted, they go back to the drawing boards, revising their hypotheses and maybe even their theories to set up a better model.

Geologists thus understand geosystems through successive attempts to match models to an ever-widening array of data. Models that successfully and repeatedly predict new types of observations are taken as evidence that the basic workings of a geosystem are properly described by the underlying theories and hypotheses. As we have seen throughout this book, plate tectonics is a good example of a successful theory. All models, however, remain subject to constant challenge from new data. No model, no matter how successful, proves that the theories and hypotheses on which it is based are true. In thinking about this last statement, you might want to review the discussion of the scientific method outlined in Figure 1.1.

Global Geosystems

Volcanic eruptions in Hawaii and earthquake faulting in southern California can be represented by regional geosystems with outer dimensions of only a few hundred kilometers or less. Interactions between a regional geosystem and other parts of the Earth system can usually be handled by considering the energy and mass fluxes across the outer boundaries of the geosystem—in other words, by considering it to be an open system.

Nothing in this type of analysis restricts the scale of the system, however. The same methods can be used to study natural systems on any scale, including the global geosystems introduced in Chapter 1. Numerical models of mantle convection, the engine that drives the plate tectonic system, have been used to investigate a wide range of observations related to seafloor spreading, the subduction of oceanic lithosphere, the rise of mantle plumes, and the forces acting on the plates. Much of what we said about these phenomena at the end of Chapter 2 was based on such models. Numerical models of the geodynamo, the convective engine in the outer core that produces Earth's magnetic field, have also provided many new insights into the phenomena described in Chapter 21: magnetic reversals, nondipole components, and the secular variation. The geodynamo and mantle convection are clearly open systems. For example, heat flows out of the geodynamo into mantle convection across the core-mantle boundary.

The geosystems approach provides us with an appropriate framework to discuss the global geosystem whose observed behavior is the most complex of all—Earth's climate system.

The Climate System

In Chapter 1, we distinguished climate from weather based on the time period of observation. Climate is the long-term average of weather (as well as some measure of its variability) in a specified place and at a specified time of day and time of year. The **climate system** includes all parts of the Earth system and all the interactions among these components needed to describe how climate behaves in space and time (**Figure 23.3**).

Components of the Climate System

The main components of the climate system are the atmosphere, hydrosphere, lithosphere, and biosphere. Each component plays a different role in the climate system, depending on its ability to store and transport energy.

The Atmosphere Earth's atmosphere is the most mobile and rapidly changing part of the climate system. The wind belts pictured in Figure 15.1 transport a typical parcel of air eastward around the globe in about one month (which is why it takes a few days for storms to blow across the country).

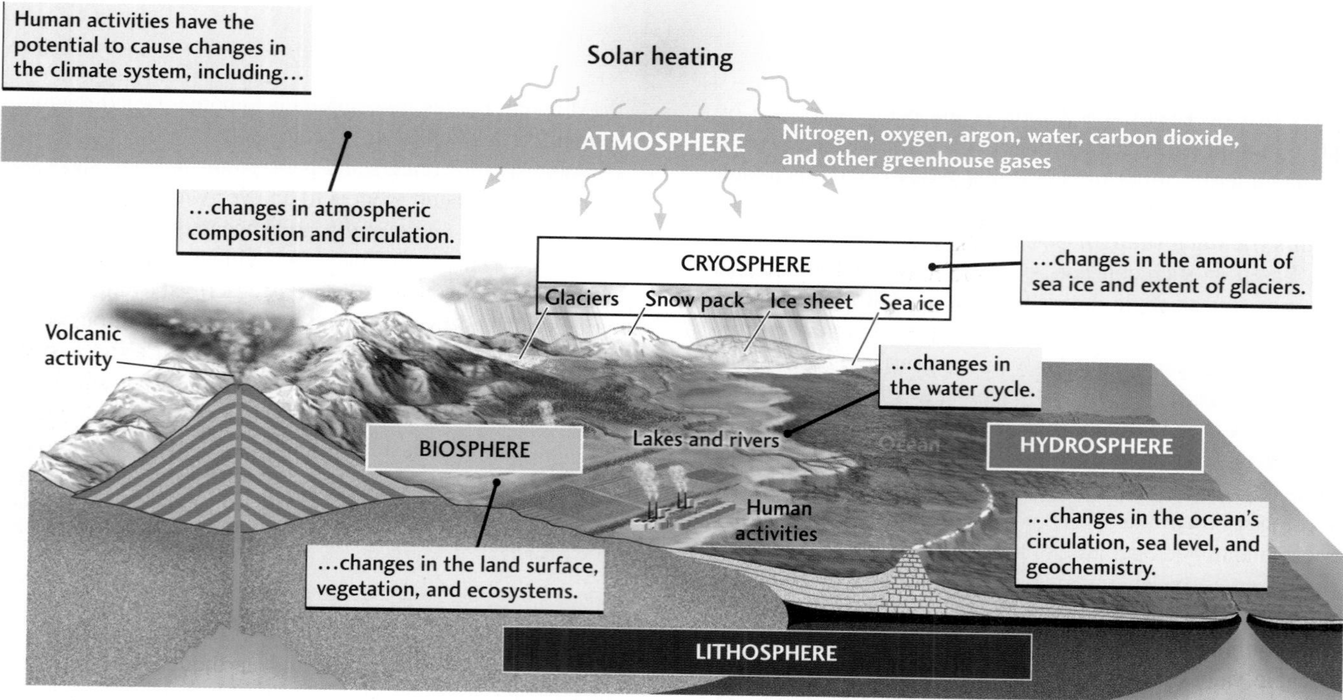

Figure 23.3 Components of Earth's climate system and some changes that might be caused by human activities.

Moreover, the circulation of air in a north-south direction significantly affects the climate system because this movement across latitudes transports energy from the hotter equatorial regions to the colder polar regions.

The atmosphere is a mixture of gases, mostly molecular nitrogen (78 percent by volume in dry air) and oxygen (21 percent by volume in dry air). The remaining 1 percent consists of argon (0.93 percent), carbon dioxide (0.035 percent), and other minor gases (0.035 percent). Among the minor gases is water vapor, which is concentrated near Earth's surface in highly variable amounts (up to 3 percent, but typically about 1 percent). Water vapor (H_2O) and carbon dioxide (CO_2) are the principal greenhouse gases. If there were no greenhouse gases, heat generated by solar radiation would pass out easily through the atmosphere, and Earth's climate would be much colder. Greenhouse gases absorb solar radiation at certain wavelengths, raising the surface temperature. We will discuss the greenhouse effect in more detail shortly.

Another minor constituent of the atmosphere is ozone (O_3^+), a highly reactive gas produced primarily by the ionization of molecular oxygen by ultraviolet (UV) radiation from the Sun. In the lower part of the atmosphere, ozone exists in only tiny amounts, although it is a strong enough greenhouse gas to play a significant role in the atmospheric heat budget. Most of the ozone is concentrated in the stratosphere. Absorption of ultraviolet energy by the stratospheric "ozone shield" filters out this potentially damaging part of the solar radiation, protecting the biosphere at Earth's surface.

The Hydrosphere The hydrosphere comprises all the water on Earth's surface, including oceans, lakes, rivers, and groundwater. In setting up climate models, scientists have found it important to differentiate the solid components (ice and snow) from the liquid components (ocean and surface waters). Almost all the liquid water is in the global ocean (1350 million cubic kilometers); lakes, rivers, and groundwater constitute a mere 1 percent (15 million cubic kilometers). However, these continental components play a vital role in the climate system as reservoirs for land moisture, as well as providing the transportation system for returning precipitation back into the oceans and supplying the oceans with salt and other minerals.

Although water moves more slowly in the oceans than air moves in the atmosphere, water can store a much larger amount of heat. For this reason, ocean currents transport energy very effectively. Winds blowing across the ocean generate surface currents and large-scale circulation patterns within the ocean basins. As in the atmosphere, the most important currents for regulating climate are those that transport heat from equatorial to polar regions. These currents involve vertical convection as well as horizontal motion. An example is the Gulf Stream, which flows along the western Atlantic margin, bringing waters from the Caribbean Sea that warm the climate of the North Atlantic and Europe. The water cooled by this exchange of energy sinks below the surface in the North Atlantic, eventually moving southward in a convection system known as the *thermohaline circulation* (see Figure 16.28). Scientists think that changes in the volume of water cycled from the equator to the poles by thermohaline circulation might strongly influence the global climate.

The ice component of the climate system (33 million cubic kilometers), called the **cryosphere,** comprises the polar ice caps, glaciers, and other surface ice (**Figure 23.4**). The role of the cryosphere in the climate system differs from the role of the liquid hydrosphere, because ice is relatively immobile and white, reflecting almost all of the solar energy that falls on it. Large masses of water are exchanged between the cryosphere and the liquid hydrosphere during glacial cycles. At the last glacial maximum 18,000 years ago, sea level was about 130 m lower than it is today, and the volume of the cryosphere was three times larger (see Chapter 16).

Figure 23.4 Sea ice is an important part of the cryosphere. This satellite image shows sea ice flowing through the Bering Strait in May 2002. According to a recent study by the U.S. Navy, global warming since the 1950s has shrunk the polar ice cap by as much as 40 percent and reduced the ice extent by 20 percent, disrupting whale migrations and the habitats of seals and polar bears that live at the ice edge. [NASA MODIS satellite.]

The Lithosphere The part of the lithosphere most important to the climate system is the land surface, which makes up 30 percent of Earth's total area. The composition of the land surface affects how solar energy is absorbed and returned to the atmosphere. As the temperature rises, the land radiates more energy as infrared waves back into the atmosphere, and more water evaporates from the land surface. Evaporation requires considerable energy, so soil moisture and other factors that influence evaporation—such as vegetation and the subsurface flow of water—are very important in controlling surface temperatures.

Topography has a direct effect on climate through its influence on atmospheric winds and circulation (see Chapter 18). Air masses that flow over large mountain ranges dump rain on the windward side, creating a rain shadow in the lee of the mountains (see Figure 13.3). In fact, the overall asymmetry of continents, a direct consequence of plate tectonics, induces hemispheric asymmetries in the global climate system.

Geologists have documented many long-term changes in the climate system associated with the rearrangement of land areas by plate tectonics. Changes in the shape of the seafloor due to seafloor spreading induce sea-level changes, and the drift of continents over the poles leads to the growth of continental glaciers. Tectonic movements can also block ocean currents or open gateways through which currents can flow. The emergence of the Isthmus of Panama, for example, closed a passage connecting the Atlantic and Pacific oceans. If future geologic activity were to close the narrow channel between the Bahamas and Florida through which the Gulf Stream flows, temperatures in western Europe would drop drastically.

Volcanism, which occurs in the lithosphere, affects climate by changing the composition of the atmosphere. Major volcanic eruptions can inject aerosols and other particulates into the stratosphere, blocking solar radiation from reaching the surface and temporarily lowering atmospheric temperature. After the massive April 1815 eruption of Mount Tambora in Indonesia, New England suffered through a "year without a summer" in 1816. According to a diarist in Vermont, "no month passed without a frost, nor one without a snow," and crop failures were common. Careful studies have shown that more recent large volcanic eruptions, including Krakatoa (1883), El Chichón (1982), and Mount Pinatubo (1991), produced an average dip of 0.3°C in global temperatures about 14 months after the eruption (local temperature variations can be much larger, of course). Temperatures returned to normal in about 4 years. Although the effects of most eruptions on global climate are short-lived, they provide an important way to calibrate numerical climate models.

The Biosphere The amount of energy contained and transported by living organisms is small, but the biosphere is strongly coupled to the climate system through its interactions with the atmosphere, hydrosphere, and lithosphere. We have mentioned the effects of land vegetation on surface moisture and temperature. Life also regulates the composition of the atmosphere, including greenhouse gases such as carbon dioxide and methane. Through photosynthesis, marine and terrestrial plants extract carbon dioxide from the atmosphere. Some of this carbon dioxide is precipitated as shells made of calcium carbonate or buried as organic matter in sediments. The biosphere thus plays a central role in the carbon cycle.

Of course, we ourselves—perhaps the most active agents for global climate change—are part of the biosphere.

The Greenhouse Effect

The Sun is a yellow star that puts out about half its radiation in the visible part of the electromagnetic spectrum. The other half is split between infrared waves, which have longer wavelengths and lower energies, and ultraviolet waves, which have shorter wavelengths and higher energies. (The fact that the Sun's radiation is most intense in the visible band is no coincidence, by the way; our eyes evolved to optimize their sensitivity to yellow sunlight.) At Earth's position in the solar system, the average amount of radiation its surface receives throughout the year is 342 W/m^2 (1 watt = 1 joule per second; a joule is a unit of energy or heat). In comparison, the average amount of heat flowing out of Earth's deep interior from mantle convection is minuscule, only 0.06 W/m^2. Essentially all of the energy driving the climate system ultimately comes from the Sun (**Figure 23.5**).

For the global temperature averaged over daily and seasonal cycles to remain the same, Earth's surface must radiate energy back into space at a rate of precisely 342 W/m^2. Any less would cause the surface to heat up; any more would cause it to cool down. How is this *radiation balance*—the equality between incoming and outgoing electromagnetic energy—achieved?

A Planet Without Greenhouse Gases Suppose Earth had no atmosphere at all but was a rocky sphere like the Moon. Some of the sunlight falling on the surface would be reflected back into space and some would be absorbed by the rocks, depending on the color of the surface. A perfectly white body would reflect all the solar energy, whereas a perfectly black body would absorb it all. The fraction of the energy reflected is called the planet's **albedo** (from the Latin word *albus,* meaning "white"). Although a full Moon looks bright to us, the rocks on its surface are mainly dark basalts, so its albedo is only about 7 percent. In other words, the Moon is very nearly black.

The energy radiated by a black body goes up rapidly with increasing temperature, by an amount that can be precisely calculated from a basic equation of physics. A cold bar of iron is black and gives off little heat. If you heat the bar to 100°C, it gives off warmth in the form of infrared waves (like a steam radiator). If you heat the bar to 1000°C, it becomes bright orange, radiating heat at visible wave-

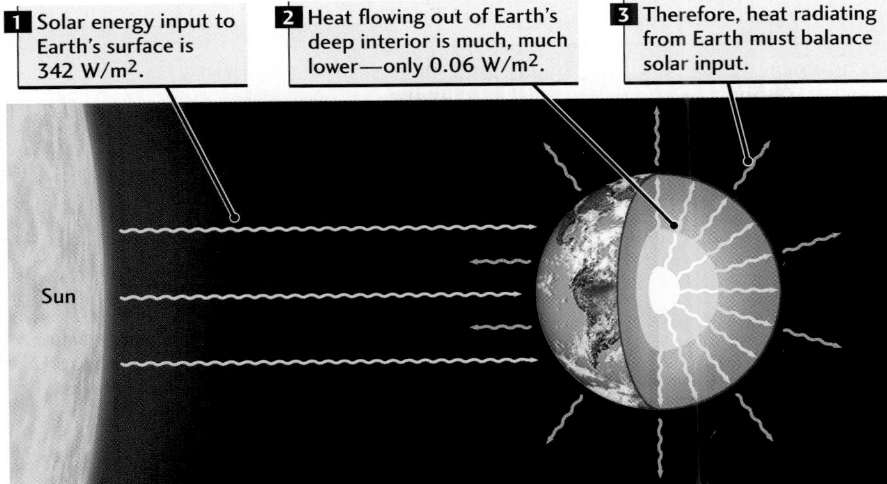

Figure 23.5 Earth's energy balance is achieved by radiation of the incoming solar energy back into space. Heat from Earth's interior is negligible in comparison to solar energy.

lengths (like the burner on an electric stove). Raising the temperature decreases the wavelength of black-body radiation, thus increasing the radiated energy.

A black body exposed to the Sun heats up until its temperature is just the right value for it to radiate the incoming solar energy back into space. The same principle applies to a "gray body" like the Moon, except you must exclude the reflected energy from this balance. The Moon's daytime temperatures rise to 130°C and its nighttime temperatures drop to –170°C. Not a pleasant environment!

Earth rotates much faster than the Moon (once per day rather than once per month), which evens out the day-and-night extremes of temperature. Earth's albedo, about 31 percent, is much higher than the Moon's, because Earth's blue oceans and white clouds are more reflective than dark lunar basalts. If Earth's atmosphere did not contain greenhouse gases, the average surface temperature required to balance the nonreflected solar radiation would be about –19°C (–2°F), cold enough to freeze all the water on the planet. Instead, Earth's average surface temperature remains a balmy 14°C (57°F). The difference of 33°C is a result of the greenhouse effect.

Earth's Greenhouse Atmosphere At a given temperature, a planet with an atmosphere containing greenhouse gases radiates solar energy back into space less efficiently than a planet with no atmosphere. Therefore, its surface temperature must rise to achieve the radiation balance. In this way, the atmosphere acts like the glass in a greenhouse, allowing solar radiant energy to pass through but trapping heat. This is known as the **greenhouse effect.**

Greenhouse gases such as water vapor and carbon dioxide absorb infrared energy selectively at certain wavelengths and radiate infrared energy at other wavelengths in all directions, including downward to Earth's surface. The net effect is to trap heat within the atmosphere by increasing the temperature of the surface relative to the temperature at higher levels of the atmosphere.

How Earth's atmosphere balances incoming and outgoing radiation is illustrated in **Figure 23.6**. The incoming solar radiation not directly reflected is absorbed by the atmosphere and the surface. To achieve radiation balance, Earth radiates this same amount of energy back into space as infrared waves. Because of the heat trapped by the greenhouse gases, the amount of energy transported away from Earth's surface, both by radiation and by the flow of hot air and moisture from the surface, is significantly larger than the amount Earth receives as direct solar radiation. The excess is exactly the energy received as Earthward infrared radiation from the greenhouse gases. It is this "back radiation" that causes Earth's surface to be 33°C warmer than it would be if the atmosphere contained no greenhouse gases.

Balancing the System Through Feedbacks

How does the climate system actually achieve the radiation balance described in Figure 23.6? Why does the greenhouse effect yield an overall warming of 33°C and not some other temperature, larger or smaller? The answers to these questions are not simple, because they depend on interactions among the many components of the climate system. The most important interactions involve feedbacks.

We first explored this topic in Chapter 18, where we discussed the feedbacks between erosion and uplift and between climate and tectonics. We saw that there are two basic types: *positive feedback,* in which a change in one component is *enhanced* by the changes it induces in other components, and *negative feedback,* in which a change in one component is *reduced* by the changes it induces in other components. Positive feedbacks tend to amplify changes in the system, whereas negative feedbacks tend to stabilize the system against change.

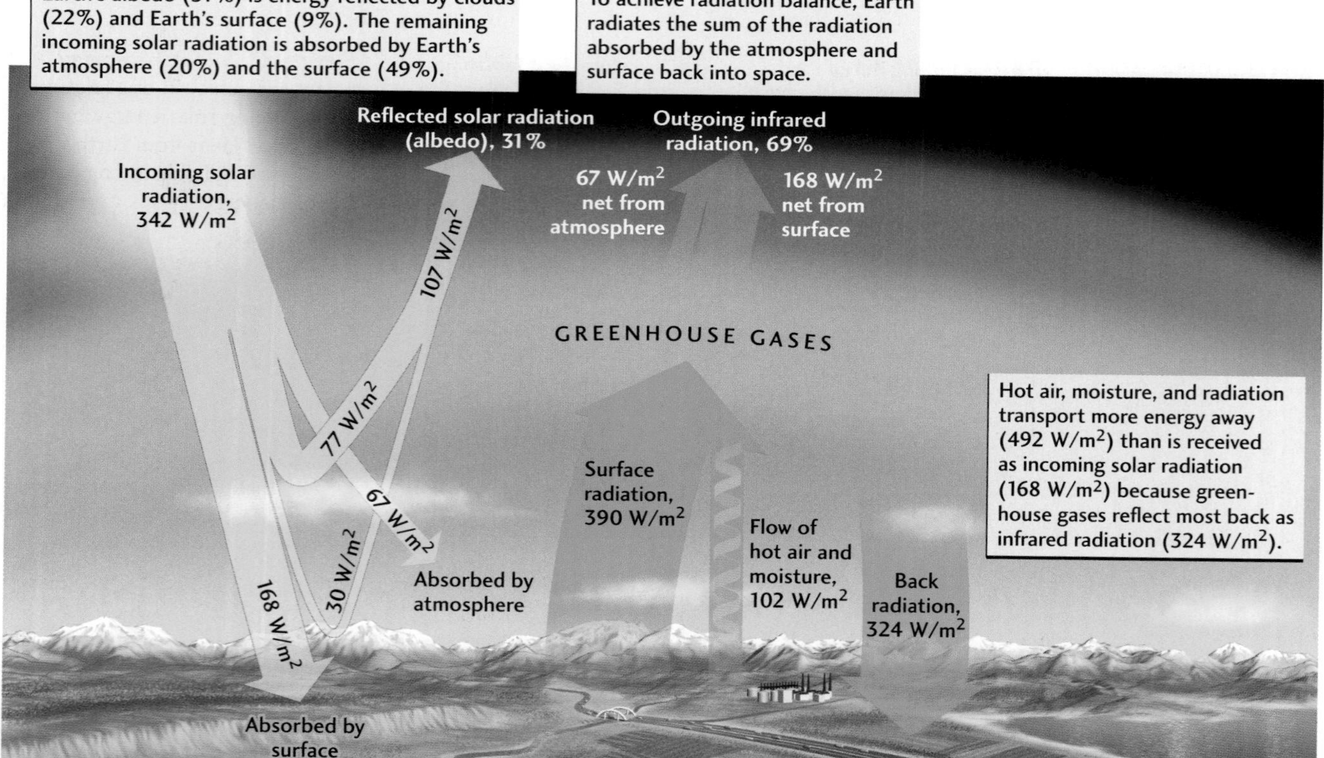

Figure 23.6 A diagram showing the balance of incoming and outgoing radiation in Earth's atmosphere. Of the total incoming solar radiation, 22 percent is directly reflected by clouds and other components of the atmosphere, and 9 percent is directly reflected from the surface; the sum of these two fractions gives Earth's albedo (31 percent). The remainder is absorbed by the atmosphere (67 W/m^2 or 20 percent) and the surface (168 W/m^2 or 49 percent). Earth radiates the sum of the latter two amounts (235 W/m^2 or 69 percent) back into space as infrared waves. Because of the heat trapped by the greenhouse gases, the energy transported away from Earth's surface as radiation (390 W/m^2) and by the flow of hot air and moisture from the surface (102 W/m^2) is significantly larger than the amount it receives as direct solar radiation (168 W/m^2). The excess (324 W/m^2) is just the energy received as downward infrared radiation from the greenhouse gases. It is this "back radiation" that causes Earth's surface to be 33°C warmer than it would be if the atmosphere contained no greenhouse gases. [IPCC, *Climate Change 2001: The Scientific Basis.*]

Here are some of the feedbacks within the climate system that can significantly affect the temperatures achieved by radiation balance.

- *Water vapor feedback* A rise in temperature increases the amount of water vapor in the atmosphere. Water vapor is a greenhouse gas, so this increase enhances the greenhouse effect and thus the temperature rise—a positive feedback.

- *Albedo feedback* Another example of positive feedback involves the interaction between atmospheric temperature and planetary albedo. A rise in temperature reduces the accumulation of ice and snow in the cryosphere, which decreases Earth's albedo and increases the energy its surface absorbs. The increased warming of the atmosphere enhances the temperature rise.

- *Radiation feedback* A primary negative feedback is "radiative damping": a rise in atmospheric temperature strongly increases the amount of infrared energy radiated back into space, which reduces the temperature rise. Radiative damping stabilizes Earth's climate against major changes, keeping the oceans from freezing up or boiling off and thus maintaining an equable habitat for water-loving life.

- *Plant growth feedback* A secondary negative feedback comes from the stimulation of plant growth by carbon dioxide. An increase in atmospheric CO_2 causes temperatures to rise by enhancing the greenhouse effect. However, this increase also stimulates plant growth, which removes CO_2 from the atmosphere by converting it to carbon-rich organic matter, thus reducing the greenhouse temperature rise.

Feedbacks can involve much more complex interactions among components of the climate system. For example, an increase in atmospheric water vapor produces more clouds. Because clouds reflect solar energy, they increase the planetary albedo, which sets up a negative feedback between water vapor and temperature. On the other hand, clouds absorb infrared radiation efficiently, so increasing the cloud cover enhances the greenhouse effect, thus providing a positive feedback between water vapor and temperature. Does the net effect of clouds produce a positive or negative feedback?

Scientists have found it surprisingly difficult to answer such questions by direct observation. Earth has only one climate system. The components of this system are joined through an amazingly complex web of interactions on a scale far beyond experimental control. Consequently, it is often impossible to gather data that isolate one type of feedback from all the others. Scientists must therefore turn to computer models to understand the inner workings of the climate system.

Climate Models and Their Limitations

Generally speaking, a **climate model** is any representation of the climate system that can reproduce one or more aspects of climate behavior. Some models are designed for studying local or regional climate processes, such as the relationships between water vapor and clouds, but the most interesting representations are global models used to predict how the climate might change in the future as a result of human activities, such as the massive burning of fossil fuels.

At the heart of such global models are schemes for computing the motions within the atmosphere and oceans based on the fundamental laws of physics—the conservation of mass, energy, and momentum. These *general circulation models* represent the currents of air and water driven by solar energy on scales ranging from small disturbances (storms in the atmosphere, eddies in the oceans) to global circulations (wind belts in the atmosphere, thermohaline convection in the oceans). Scientists represent the basic physical variables (temperature, pressure, density, velocity, and so forth) on three-dimensional grids comprising millions of geographic points. They use supercomputers to solve numerically the mathematical equations that describe how the variables change with time at each of these points.

You see the results of this type of calculation whenever you tune in the weather report on your favorite TV station. These days, most weather predictions are made by setting up a model that accurately describes the current conditions observed at thousands of weather stations and running it forward in time using an atmospheric general circulation model. Numerical weather prediction thus employs the same basic computer programs that are used for climate modeling. Climate modeling is intrinsically more difficult, however. In predicting weather a few days from now, scientists can ignore such slow processes as changes in atmospheric greenhouse gases or ocean circulation. Climate predictions, on the other hand, require that these slow processes, including all important feedbacks, be properly modeled, in addition to modeling the fast motions of air masses. Moreover, the simulation must be extended for years or decades into the future. Such enormous calculations require weeks of time on the world's largest supercomputers.

Owing to their complexity, current models of the climate system must be regarded with a fair measure of skepticism. Although they incorporate the information from decades of climate studies by hundreds of scientists, many questions about how the climate system works—for instance, the role of clouds in regulating atmospheric temperatures—have not been fully answered. The predictions of the models are therefore subject to errors that can be difficult to assess. These uncertainties have engendered much debate among experts and governmental authorities who must deal with the regulation and consequences of human-induced climate change.

Natural Climate Variability

Much of the controversy surrounding global climate change comes from skepticism about our ability to separate natural variations of climate from human-induced effects. Here the geologic record of past climates can provide key information.

Natural variations cause the climate to change on a wide range of scales in both time and space. Some variations occur in the external forcing of climate; examples include fluctuations in the amount of solar radiation received at Earth's surface and changes in the distribution of land surface caused by continental drift. Others result from internal variations within the climate system itself, such as oscillations of mass and energy among its components. Both types of variations, external and internal, can be amplified or suppressed by feedbacks.

Long-Term Variations

The geologic record indicates that long periods of global warmth have been punctuated by much colder epochs characterized by extensive continental glaciations. In Chapter 16, we discussed how rearrangements of the continents by plate tectonics, coupled with feedbacks in the climate system, may have led to long-term changes in Earth's climate. We saw, for example, how a runaway albedo feedback might explain why Earth's surface may have been completely covered with ice in the late Precambrian—the controversial Snowball Earth hypothesis—when the supercontinent of Rodinia formed at high latitudes. We also learned that the oscillatory pattern of ice ages during the last 1.8 million years (in the Pleistocene epoch) appears to have been synchronized by subtle variations in solar radiation caused by periodicities in Earth's orbit around the Sun, the so-called *Milankovitch cycles.*

The best records of climate variations during the last half-million years come from ice cores drilled in the Antarctic and Greenland ice sheets. **Figure 23.7** displays temperatures and the concentrations of two important greenhouse gases, carbon dioxide and methane (CH_4), recovered from the famous Vostok ice core in East Antarctica (see Feature 16.1).

The temperature record shows much variation, but the largest swings correspond to the sawtooth pattern of the glacial cycles—a gradual decline of about 6° to 8°C from a warm interglacial period into a cold ice age, followed by a

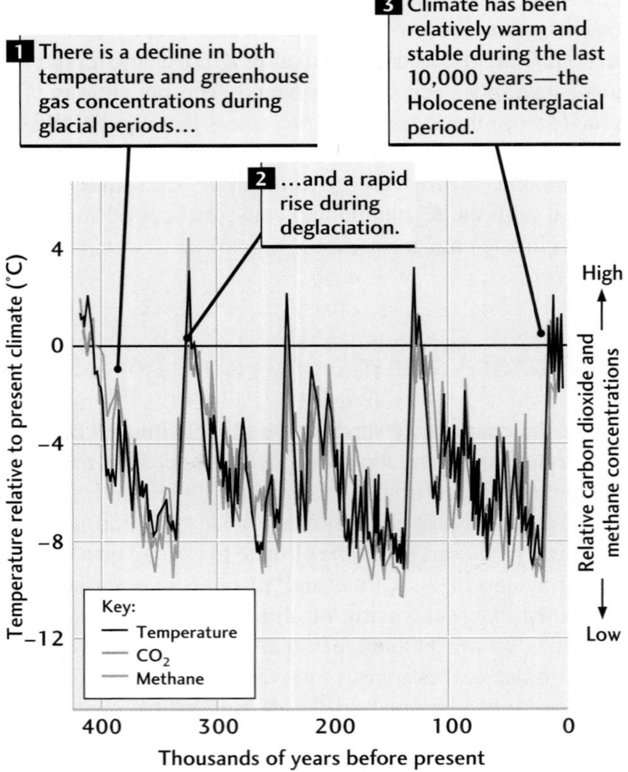

Figure 23.7 A graph showing three types of data recovered from the Vostok ice core in East Antarctica: temperatures inferred from oxygen isotopes (black line) and the concentrations of two important greenhouse gases, carbon dioxide (blue line) and methane (red line). The gas concentrations come from measurements of air samples trapped as tiny bubbles within the Antarctic ice. [IPCC, *Climate Change 2001: The Scientific Basis.*]

rapid rise during a short interval of deglaciation. Temperatures began to drop in the last glacial cycle about 120,000 years ago, reaching their lowest values 25,000 to 18,000 years ago (the Wisconsin glacial maximum). Temperatures then rebounded to the warm interglacial climate of the last 10,000 years, which geologists call the **Holocene epoch.** Three other ice ages are visible in this record; they have minimum temperatures at about 140,000, 260,000, and 350,000 years ago.

The 100,000-year spacing of ice ages matches the times of high orbital eccentricity, when Earth received less radiation on average from the Sun (see Figure 16.27). This correlation has led scientists to conclude that glacial cycles are externally forced by variations in the amount of solar radiation. The small changes in solar radiation during the Milankovitch cycles, however, cannot completely explain the large changes observed in Earth's surface temperature. Some strong positive feedback within the climate system must be operating to amplify the external forcing. The data in Figure 23.7 clearly indicate that this feedback involves the greenhouse gases CO_2 and CH_4, because their concentrations precisely track the temperature variations throughout the glacial cycles—warm interglacial intervals are marked by high concentrations, whereas cold glacial intervals have low concentrations. Exactly how this feedback works has not yet been fully explained, but it demonstrates the importance of the greenhouse effect in long-term climate variations.

Short-Term Variations

The temperatures shown in Figure 23.7 do not vary smoothly over time. Superposed on the 100,000-year glacial cycles are climate fluctuations of shorter duration, some nearly as large as the changes from glacial to interglacial periods. Geologists have combined the information from cores in polar ice sheets, mountain glaciers, lake sediments, and deep-sea sediments to reconstruct a decade-by-decade—and in some cases, a year-by-year—history of short-term climate variations during the last glacial cycle. Important aspects of this remarkable chronicle were discussed in Chapter 16. Here we summarize some of the basic features:

- During the last ice age, Earth's climate was highly variable, with shorter (1000-year) oscillations between warm and cold superposed on longer (10,000-year) cycles. The most extreme variations appear to have been in the North Atlantic region, where mean local temperatures rose and fell by as much as 15°C. Each 10,000-year cycle comprised a set of progressively cooler 1000-year oscillations and ended with an abrupt warming. The massive discharges of icebergs and fresh water resulting from the warming altered the thermohaline "conveyor belt" circulation in the oceans and dumped large amounts of glacial drift into deep-sea sediments.

- The main phase of warming from the last ice age occurred in the interval between 14,500 and 10,000 years ago. It was not a smooth transition but involved two main stages, with a pause in deglaciation and a return to cold conditions during the Younger Dryas event 13,000 to 11,500 years ago. The extremely abrupt increases in temperature at 14,500 and 11,500 years ago are perhaps the most astonishing aspect of this jerky transition. Broad regions of Earth experienced almost synchronous changes from glacial to interglacial temperatures during intervals as short as 10 to 30 years. Evidently, atmospheric circulation can reorganize very rapidly, flipping the entire climate system from one state (glacial cold) to another (interglacial warmth) in less than a human lifetime! This raises the possibility that human-induced global climate change could involve abrupt shifts to a new (and unknown) climate state, rather than just a gradual warming.

- The warm interglacial climate of the Holocene, since about 10,000 years ago, has been much more stable than the previous parts of the glacial cycle. The warmest temperatures occurred during the beginning of this epoch. Geologists have documented regional variations of about 5°C on time scales of 1000 years or so, but the global changes during this period are much smaller in amplitude, with a total range of only 2°C. It is interesting that the Holocene appears to be the longest warm "stable" period over the last 400,000 years. Equable Holocene conditions no doubt promoted the rapid rise of agriculture and civilization that followed the end of the last ice age.

Regional Patterns of Variability

Local and regional climates are much more variable than global climate, because averaging over large surface areas, like averaging over time, tends to smooth out small-scale fluctuations. Over periods of years to decades, the predominant regional variations result from interactions between atmospheric circulation and the sea and land surfaces. They generally occur in distinctive geographical patterns, although their timing and amplitudes can be highly irregular.

The most famous example is an anomalous warming of the eastern Pacific Ocean that occurs every 3 to 7 years and lasts for a year or so. Peruvian fishermen call such an event **El Niño** ("the Christ Child" in Spanish) because the warming typically reaches the surface waters off the coast of South America about Christmastime. El Niño events can be calamitous for local ecosystems, which depend on upwellings of cold water for their nutrient supply. Besides disrupting fishing in the eastern Pacific, they can trigger changes in wind and rain patterns over much of the globe, precipitating disastrous floods, prolonged droughts, forest fires, landslides, and the spread of waterborne diseases. Scientists have shown that El Niño and the complementary cooling events, known as La Niña, are part of a natural variation in the exchange of heat between the atmosphere and the tropical Pacific Ocean. This variation is known as ENSO, or El Niño–Southern Oscillation (see Feature 23.1).

Climate scientists have identified similar patterns of weather and climate variability in other regions. One example is the North Atlantic Oscillation, a highly irregular fluctuation in the barometric pressure between Iceland and the Azores Islands that has a strong influence on the movement of storms across the North Atlantic and thus affects weather conditions throughout Europe and parts of Asia. Understanding these patterns is improving long-range weather forecasting and may provide important information about the regional effects of human-induced climate change.

Twentieth-Century Warming: Human Fingerprints of Global Change?

Systematic measurements of weather and climate using thermometers and other scientific instruments began only a few hundred years ago. By the mid-nineteenth century, temperatures around the world were being reported by enough meteorological stations on land and on ships at sea to allow accurate estimation of Earth's average annual surface temperature. Between the end of the nineteenth century and the beginning of the twenty-first, the mean surface temperature rose by about 0.6°C, a climate trend called the **twentieth-century warming** (**Figure 23.8**).

Fossil-fuel burning and deforestation have substantially increased the amount of atmospheric CO_2 during the same interval, and human activities have also generated significant rises in other greenhouse gases, such as methane. We know that humans are responsible for the CO_2 increase because the carbon isotopes of fossil fuels have a distinctive signature that precisely matches the changing isotopic composition of

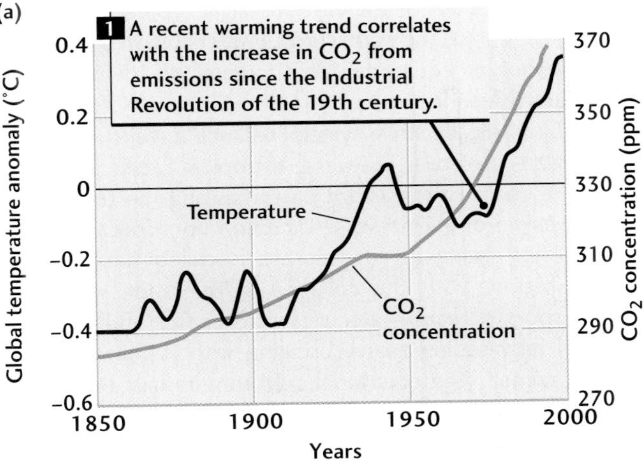

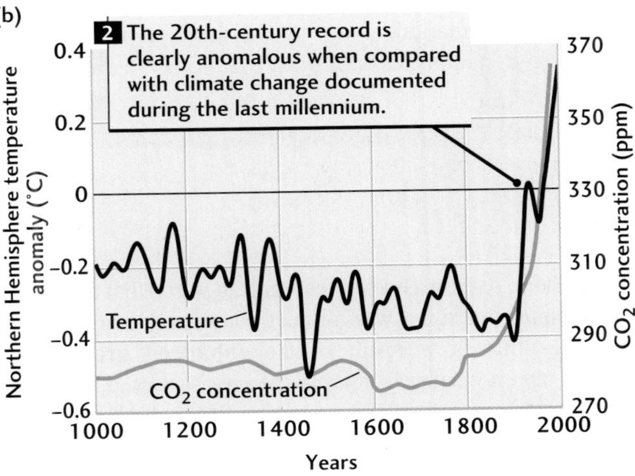

Figure 23.8 Earth heats up. A comparison of Earth's average annual surface temperature with CO_2 concentrations in the atmosphere, showing that the recent warming trend correlates with the increase in CO_2 caused by human activities since the industrial revolution. [IPCC, *Climate Change 2001: The Scientific Basis*.]

23.1 El Niño: The Wayward Child

El Niño is caused by an upset in the delicate balance between winds and currents of the tropical Pacific. The tropical Pacific absorbs an enormous amount of solar heat—more than any other ocean. Normally, the warmer waters of the western tropical Pacific off Indonesia cause lower pressure, towering thunderstorms, and heavy rainfall in that region. In contrast, air pressure is higher and rainfall is less over the cooler waters in the eastern tropical Pacific. These prevailing pressure patterns drive the trade winds that blow from east to west, pushing the heated tropical Pacific waters westward, where they pile up and maintain a large warm pool. In the eastern Pacific, colder deep water rises to replace the warm surface waters blown to the west, producing an equatorial "cold tongue" off the west coast of South America.

Sporadically, this system balance breaks down. Air pressure rises over the western tropical Pacific and drops in the central and eastern parts, causing the trade winds to weaken or even reverse direction upon occasion. This recurring swing in air pressure is called the Southern Oscillation. With the collapse of the trade winds, the transport of warm water westward ceases and the warm pool migrates eastward, bringing with it thunderstorms and rainfall. The equatorial cold tongue fails to develop, and temperatures across the tropical Pacific that normally are warmer in the west and cooler in the east become equalized. These regional changes can trigger anomalous weather patterns worldwide. The jet stream, the atmospheric circulation, and the transport of moisture are changed in large parts of both the Northern Hemisphere and Southern Hemisphere. For example, cool wet winters become more frequent in the southeastern United States, and milder winters tend to prevail in the Pacific Northwest and New England.

The 1997–1998 El Niño was the strongest on record, and its particularly severe climatic effects were felt throughout the world. Some examples of unusual weather attributed to this event were drought in Australia; record drought and forest fires in Indonesia; severe hurricanes and typhoons in the Pacific and milder hurricanes in the Caribbean; severe storms in California that caused landslides and floods; and heavy rains and flash floods in Peru, Ecuador, and Kenya. Crops failed and the fish catch was decimated in many areas. A single blizzard paralyzed Denver, and warmer than usual temperatures stole Christmas in New York City. In regions with increased rainfall, such as parts of South America and Africa, waterborne diseases (cholera, dysentery, typhoid, hepatitis, rift valley fever) increased. According to one estimate, the global disruption in weather patterns and ecosystems may have cost 23,000 lives and caused $33 billion in damage.

El Niño has a flip side, called La Niña ("the girl child"), that sometimes (but not always) follows El Niño by a year or so. La Niña is characterized by stronger trade winds, colder sea-surface temperatures in the eastern tropical Pacific, and warmer temperatures in the western tropical Pacific than are the norm. Global weather anomalies are generally the opposite of those that occur during El Niño. Some scientists attribute the severe droughts on the East Coast of the United States in the summer of 1999 to a severe La Niña event in the same year.

atmospheric carbon. But how certain can we be that the twentieth-century warming was a direct consequence of the CO_2 increase—that is, a result of the **enhanced greenhouse effect**—and not some fortuitous change associated with natural climate variability? This question lies at the heart of the global warming controversy.

Most, though not all, experts on Earth's climate are now convinced that the twentieth-century warming was in part human-induced and that the warming will continue into the twenty-first century as the levels of atmospheric greenhouse gases continue to rise. They base this judgment on two principal lines of reasoning: the climate-change record and their understanding of how the climate system works.

The twentieth-century warming lies within the range of temperature variations that have been inferred for the Holocene. In fact, the average temperatures in many regions of the world were probably warmer 10,000 to 8000 years ago than they are today. The twentieth-century record is clearly anomalous, however, when compared with the pattern and rate of climate change documented during the last millennium. Based on data from tree rings, corals, ice cores, and other climate indicators, scientists have drawn two major conclusions about climate during the 1000 years that preceded the twentieth century: (1) there was an irregular but steady global cooling of about 0.2°C over this entire interval, and (2) the maximum fluctuation in mean

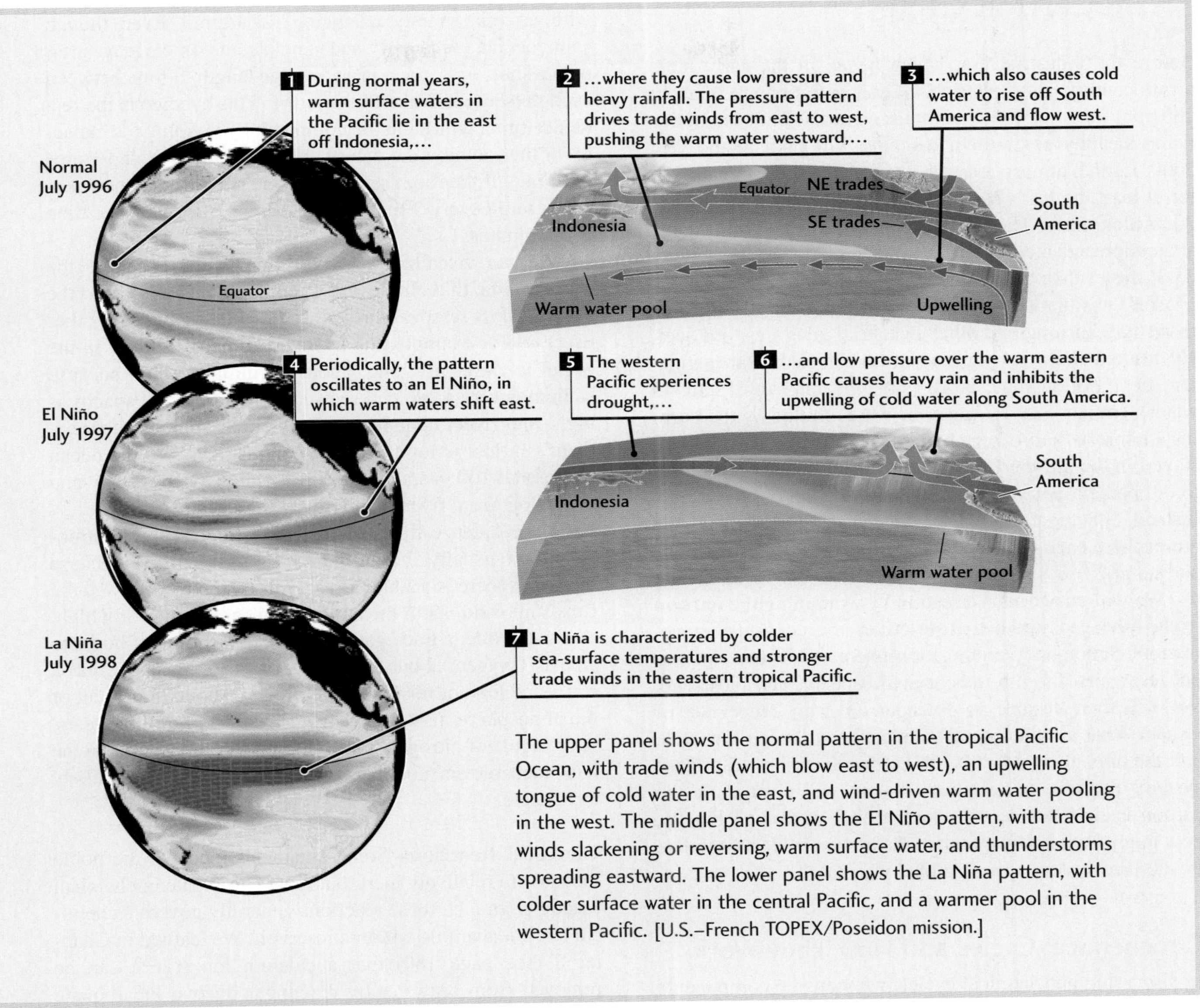

The upper panel shows the normal pattern in the tropical Pacific Ocean, with trade winds (which blow east to west), an upwelling tongue of cold water in the east, and wind-driven warm water pooling in the west. The middle panel shows the El Niño pattern, with trade winds slackening or reversing, warm surface water, and thunderstorms spreading eastward. The lower panel shows the La Niña pattern, with colder surface water in the central Pacific, and a warmer pool in the western Pacific. [U.S.–French TOPEX/Poseidon mission.]

temperatures during any one of the nine previous centuries was probably less than 0.3°C. Against this background, the twentieth-century warming appears to be very abnormal (see Figure 23.8b).

The second argument, and to many scientists a more compelling one, comes from the agreement between the observed pattern of warming and the pattern predicted by the best climate-system models of the enhanced greenhouse effect. Models that include changes in atmospheric greenhouse gases not only satisfy the global temperature rise but also reproduce the pattern of temperature change both geographically and with altitude in the atmosphere—what some scientists have called the "fingerprints" of the enhanced greenhouse effect. For example, models of the enhanced greenhouse effect predict that, as global warming occurs, nighttime low temperatures at the surface should increase more rapidly than daytime high temperatures, thus reducing the daily variation of temperature. Climate data for the last century confirm this prediction.

As we emphasized earlier in this chapter, poorly understood aspects of the climate system may introduce substantial errors in climate model predictions. Nevertheless, the consistency of the measured trends with the basic physics of the enhanced greenhouse effect lends powerful support to the hypothesis that we ourselves are the agents responsible for the recent global warming.

The Carbon Cycle

Before the industrial era, which began in the early nineteenth century, atmospheric CO_2 concentrations were about 280 ppm for at least several thousand years. They have been rising steadily since then, reaching 370 ppm in the year 2000. Earth's atmosphere has not contained this much CO_2 for at least the last 400,000 years and probably for the last 20 million years. The CO_2 concentration is now increasing at an unprecedented rate of 0.4 percent per year (see Figure 23.8), faster than at any time in recent geologic history.

Yet the situation could be worse. In the 1980s, through fossil-fuel burning and other industrial activities, our society emitted about 5.4 gigatons of carbon each year (a gigaton, or 1 billion tons, is 10^{12} kg, the mass of 1 km^3 of water), and another 1.7 gigatons were emitted by the burning of forests and other changes in land use. If it had all stayed in the air, the CO_2 increase would have been closer to 0.9 percent per year, more than twice the observed rate. Instead, 3.8 gigatons of carbon were removed from the atmosphere each year by natural processes. Where did this carbon go?

We will answer this question by examining the **carbon cycle**—the geosystem that describes the continual movement of carbon between the atmosphere and its other principal reservoirs, the lithosphere, hydrosphere, and biosphere. We will then be able to consider an even deeper set of issues: What are the interactions between the carbon cycle and the climate system? Is the net feedback between these systems positive or negative—that is, will changes in the carbon cycle enhance or reduce global warming? Finally, how might humans intervene in the carbon cycle to stabilize the climate system against global warming?

Geochemical Cycles and How They Work

In discussing geochemical cycles, we view the components of the Earth system—atmosphere, hydrosphere, lithosphere, and biosphere—as **geochemical reservoirs** for holding terrestrial chemicals, linked by processes that transport chemicals among them. Geochemical cycles trace the flow, or *flux*, of chemicals from one reservoir to another. By quantifying the amounts of the chemicals that are stored in and moved among the various reservoirs, we can gain new insights into the workings of the Earth system.

Residence Time When the inflow of an element into a reservoir equals the outflow, the average time that an atom of the element spends in the reservoir before leaving is called the element's **residence time.** Think of a crowded bar where many more people want to get in than are allowed by the fire code. After the room fills up or reaches its *capacity*, the bouncer begins holding people at the door. During the most active hours, when people are waiting to get in, the bar is filled to capacity, or *saturated*, and is in a **steady state,** with arrivals exactly balancing departures. Even though some people come early and stay late and others leave after only a short time, there is an average length of time between each person's arrival and departure. This average is the residence time, which can be computed by dividing the capacity of the room by the rate of arrivals (*inflow*) or departures (*outflow*). If the room's capacity is 30 people and a new person is let in every 2 minutes on average, the residence time is 60 minutes.

We can visualize residence time in the ocean as the average time that elapses between an atom's entry into the ocean and its removal through sedimentation or some other process. For example, the residence time of sodium in the ocean is extremely long, about 48 million years, because sodium is very soluble in seawater (the reservoir capacity is high) and rivers contain relatively small amounts (the element's inflow is low). Iron, in contrast, stays in the ocean only about 100 years, because its solubility is very low and the inflow from rivers is relatively high.

The residence times of trace constituents in the atmosphere are usually shorter than those in the ocean, because the atmosphere is a smaller reservoir than the ocean and the fluxes into and out of the atmosphere can be relatively high. Sulfur dioxide, a trace gas, has a residence time of hours to weeks. Oxygen, about 21 percent of the atmosphere, has a residence time of 6000 years. Nitrogen, about 78 percent of the atmosphere, has a residence time of 400 million years. A molecule of nitrogen that went into the atmosphere in the late Paleozoic era, about 300 million years ago, is still likely to be there.

Chemical Reactions Some chemicals (such as the noble metals) are relatively inert, but most participate in chemical reactions. In fact, these reactions generally govern a chemical's residence time within a reservoir. We learned in Chapter 8 (see page 180) that a calcium ion (Ca^{2+}) can be removed from seawater by reacting with two bicarbonate ions ($2HCO_3^-$) to form carbonic acid (H_2CO_3) and calcite ($CaCO_3$). Carbonic acid can dissociate into water (H_2O) and carbon dioxide (CO_2), both of which can escape to the atmosphere, and calcite can precipitate out of the ocean in limestone sediments. How much calcium can be dissolved in seawater thus depends on the availability of bicarbonate ions, which in turn depends on the influx of carbon dioxide into the seawater. As this example illustrates, chemical reactions can couple the geochemical cycle of one element (calcium) to that of another (carbon).

This coupling implies that geochemical cycles can rarely be considered in isolation but must be treated as interacting geosystems. Moreover, the chemical reactions depend on the physical conditions of the reservoirs, such as the local pressure and temperature. This dependence can couple geochemical cycles to the climate system itself.

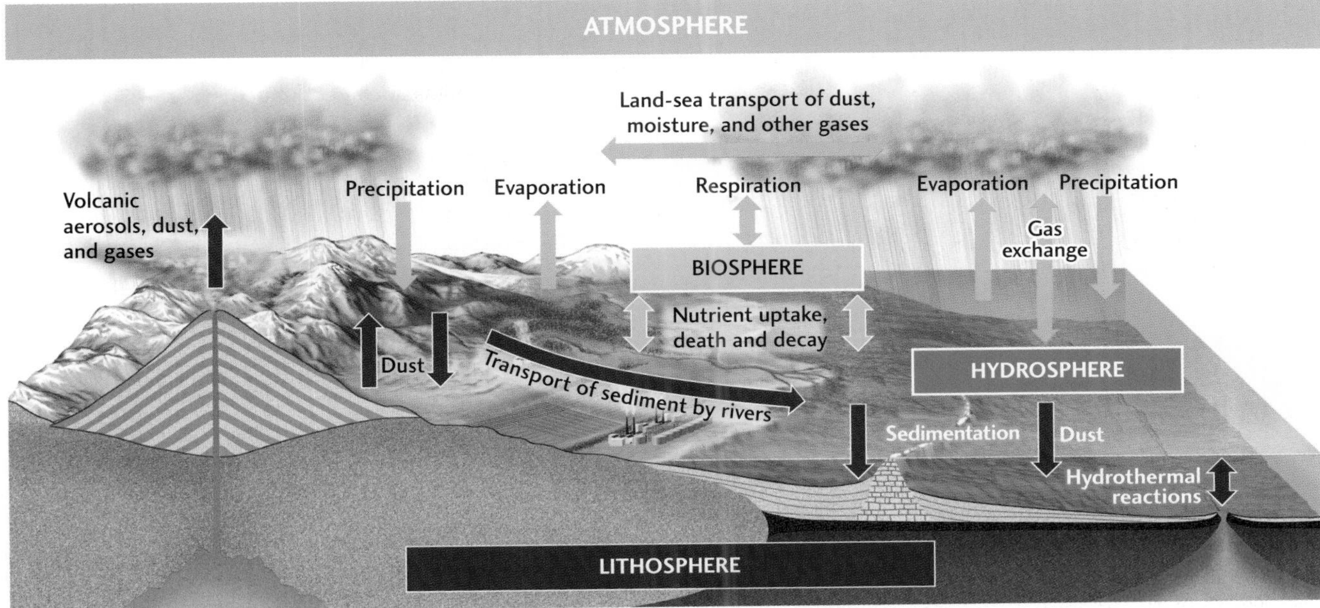

Figure 23.9 Transport processes between components of the climate system.

Transport Across Interfaces Fluxes between reservoirs are governed by processes that transport chemicals across their interfaces. The atmosphere interacts with the oceans and the land along the thin layer of air immediately overlying Earth's surface (**Figure 23.9**). Volcanic processes transport gases, aerosols, and dust from the lithosphere into the atmosphere (see Chapter 6). Wind also lifts dust into the atmosphere, whereas gravity pulls it back to the surface (see Chapter 15). Windborne dust is an important mechanism for transferring minerals from the lithosphere to the hydrosphere, although by far the largest flux comes from the dissolved or suspended minerals in the outflow from rivers (see Chapter 14).

Evaporation and precipitation transport huge amounts of water between the atmosphere and the surfaces of both land and ocean. At the sea surface, gas molecules escape from their dissolved state in the water and enter the atmosphere. Their escape is promoted by the evaporation of sea spray, which releases dissolved gases as well as dissolved salt in the form of tiny crystals. The gas exchange is balanced by the dissolution of atmospheric constituents into sea spray and rain falling on the ocean or by the dissolution of gases directly across the ocean surface.

Sedimentation is the great flux that keeps the ocean in a steady state, primarily by counterbalancing the influx of river water. As sediments are buried, they become part of the oceanic crust. There they stay until they move into the mantle through subduction or become part of the continental crust through accretion. The uplift of continental regions by plate convergence and other mountain-building processes exposes crustal rocks to weathering and erosion, maintaining the balance of fluxes among all the reservoirs.

The biosphere is a unique reservoir, because each individual organism maintains an active interface with its environment, which often includes all three of the other main surface reservoirs—atmosphere, hydrosphere, and lithosphere. The most important transport processes are the inflow and outflow of atmospheric gases by respiration, the inflow of nutrients from the lithosphere and hydrosphere, and the outflow of these nutrients through death and decay of the organisms. When the biosphere plays a key role in the flux of a chemical, geologists refer to the geosystem that describes the flux as a *biogeochemical cycle.* The carbon cycle, which depends critically on the organic pumping of carbon into and out of the atmosphere, is clearly a biogeochemical cycle.

Example: The Calcium Cycle The **calcium cycle** provides a fairly simple illustration of the concepts involved in geochemical cycles (**Figure 23.10**). The ocean contains about 560,000 Gt of calcium, dissolved in a total ocean mass of about 1.4×10^9 Gt. Calcium steadily enters this reservoir in large quantities through the rivers of the world, which transport dissolved and suspended calcium derived from the weathering of such minerals as calcite, gypsum, calcium feldspars, and other calcium silicates. A much smaller amount enters the ocean via transport by windblown dust.

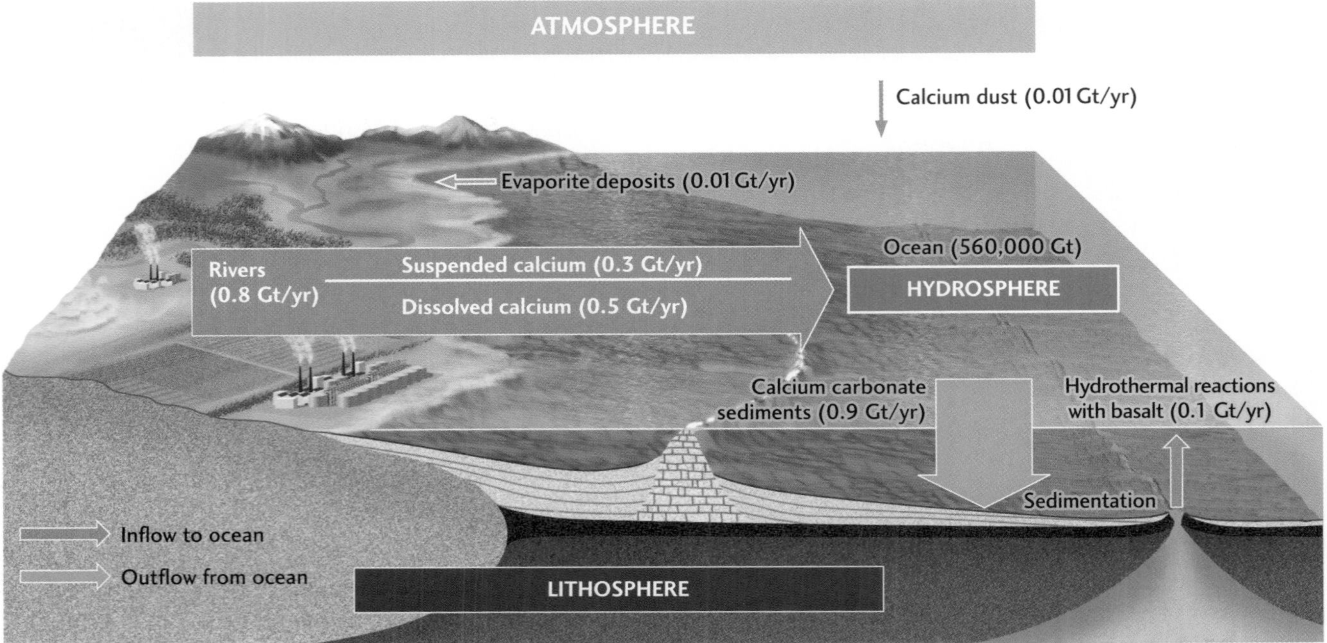

Figure 23.10 The calcium cycle, showing fluxes into and out of the ocean. Fluxes are in units of gigatons (10^{12} kg) per year. The inflow of calcium to the ocean is approximately equal to the outflow, resulting in a steady state.

If the ocean continually received calcium liberated by weathering, without there being a way to remove excess calcium, it would quickly become supersaturated with this element. The flux that removes the majority of excess calcium from the ocean is the *sedimentation of calcium carbonate,* which we described in Chapter 8. A smaller amount of calcium is precipitated as gypsum in evaporite deposits.

The system diagram in Figure 23.10 shows this steady-state situation. The ocean's calcium capacity (the amount of calcium in the ocean) is much larger than the inflow and outflow of calcium, so calcium has a long residence time in the ocean. By dividing the total annual influx (0.9 gigaton/year) by the ocean's calcium capacity (560,000 gigatons), we obtain a residence time of about 600,000 years.

The Carbon Budget

The carbon cycle, depicted in **Figure 23.11**, involves four main reservoirs: the atmosphere; the global ocean, including marine organisms; the land surface, including terrestrial plants and soils; and the lithosphere. We can describe the flux of carbon among these reservoirs in term of four basic subcycles. During times when Earth's climate is stable, each subcycle is in a steady state and can therefore be characterized by a constant flux.

Air-Sea Gas Exchange The exchange of CO_2 across the air-sea interface amounts to a carbon flux of about 90 gigatons/year. The process depends on many factors, including air and sea temperatures and the composition of the seawater, but it is particularly sensitive to the wind velocity, which increases gas transfer by stirring up the surface water and generating spray.

Photosynthesis and Respiration in the Terrestrial Biosphere The greatest carbon flux, 120 gigatons/year, comes from the cycling of atmospheric CO_2 by land plants and animals through photosynthesis, respiration, and decay. Plants take in this entire amount during photosynthesis and respire about half of it back into the atmosphere. The other half is incorporated into plant tissues—leaves, wood, and roots. Animals eat the plants, and microorganisms promote their decay; both processes oxidize plant tissues and respire CO_2. A small fraction (about 3 percent) is directly oxidized through forest fires and other combustion.

Dissolved Organic Carbon A small fraction of the CO_2 incorporated into plant tissues (0.4 gigaton/year) is dissolved into surface waters and transported by rivers to the ocean, where it is respired back into the atmosphere.

Carbonate Weathering and Precipitation The weathering of carbonate rocks removes about 0.2 gigaton of carbon per year from the lithosphere and an equal amount from the atmosphere. This inorganic carbon is dissolved into surface waters, primarily as bicarbonate ions, and is transported by

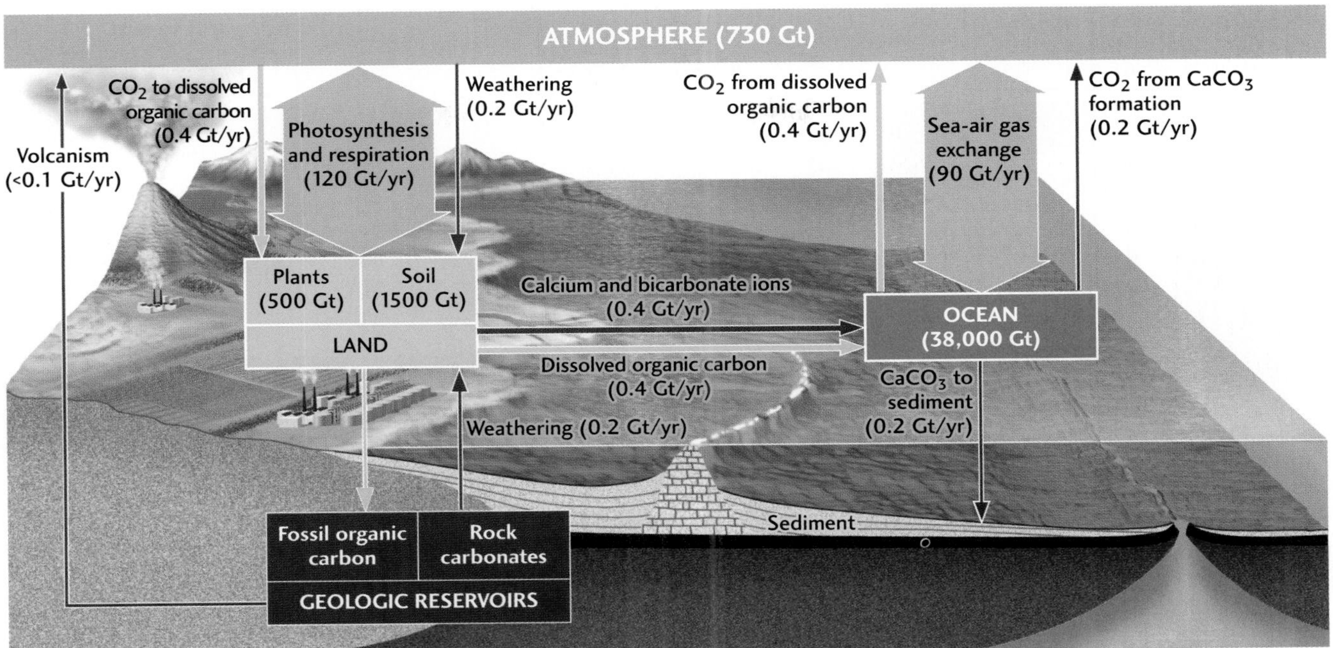

Figure 23.11 The carbon cycle, showing reservoirs and fluxes for the four balanced subcycles described in the text: air-sea gas exchange (right blue arrow), photosynthesis and respiration (left blue arrow), dissolved organic carbon (green arrows), and carbonate weathering and precipitation (brown arrows). The inflow of carbon dioxide into each reservoir is balanced by the outflow from that reservoir. [IPCC, *Climate Change 2001: The Scientific Basis.*]

rivers to the ocean. Here shell-forming marine organisms reverse the weathering reaction, precipitating calcium carbonate and releasing an equal amount of carbon back into the atmosphere as CO_2 (see Figure 7.6).

Other geological processes contribute to the carbon budget. Volcanism releases minor amounts of CO_2 into the atmosphere, and the weathering of silicate rocks and the burial of organic carbon by sedimentation consume CO_2. The flux of carbon by these processes is relatively small (<0.1 gigaton/year), so they are usually neglected in considerations of short-term climate change. Over the long term, however, their effects can be substantial. For example, regional uplift might cause changes in the carbon cycle by accelerating physical erosion and chemical weathering of the newly exposed silicate rocks in the mountains. In simplified form, the weathering reaction that removes atmospheric carbon dioxide is

silicate rock weathering

$$CaSiO_3 + CO_2 \rightarrow CaCO_3 + SiO_2$$

According to one controversial hypothesis, the uplift of the Himalaya and the Tibetan Plateau, which began about 40 million years ago, increased weathering rates enough to draw down CO_2 in the atmosphere. By weakening the greenhouse effect, this may have contributed to the subsequent climate cooling and growth of ice sheets.

Human Perturbations to the Carbon Cycle

With this background, we now return to the fate of human carbon emissions. Careful studies have provided good estimates of how human activities perturbed the carbon cycle during the 1980s. Over this decade, the average mass of carbon that human activities emitted into the atmosphere was 7.1 gigatons/year. **Figure 23.12** shows what happened to this carbon. Only 46 percent of the total (3.3 gigatons/year) stayed in the atmosphere as CO_2. The remainder was absorbed in nearly equal amounts by the oceans and the land surface. On the land surface, the uptake was primarily by plant growth at temperate latitudes in the Northern Hemisphere, caused in part by the "fertilization effect" of increasing CO_2 (plants love the stuff!) and in part by the rapid growth of new forests on land previously used for agriculture.

How will these numbers change as the atmospheric CO_2 concentration continues to rise? Scientists believe that the percentage of human carbon emissions taken up by the ocean and the land surface will decrease as the reservoirs become more saturated. Moreover, they also believe that the enhanced greenhouse warming expected from the rise in atmospheric CO_2 will further decrease the capacity of the oceans to absorb human carbon emissions. The magnitude of these effects is highly uncertain, but it does appear that changes in the carbon cycle may eventually enhance the greenhouse warming by human activities.

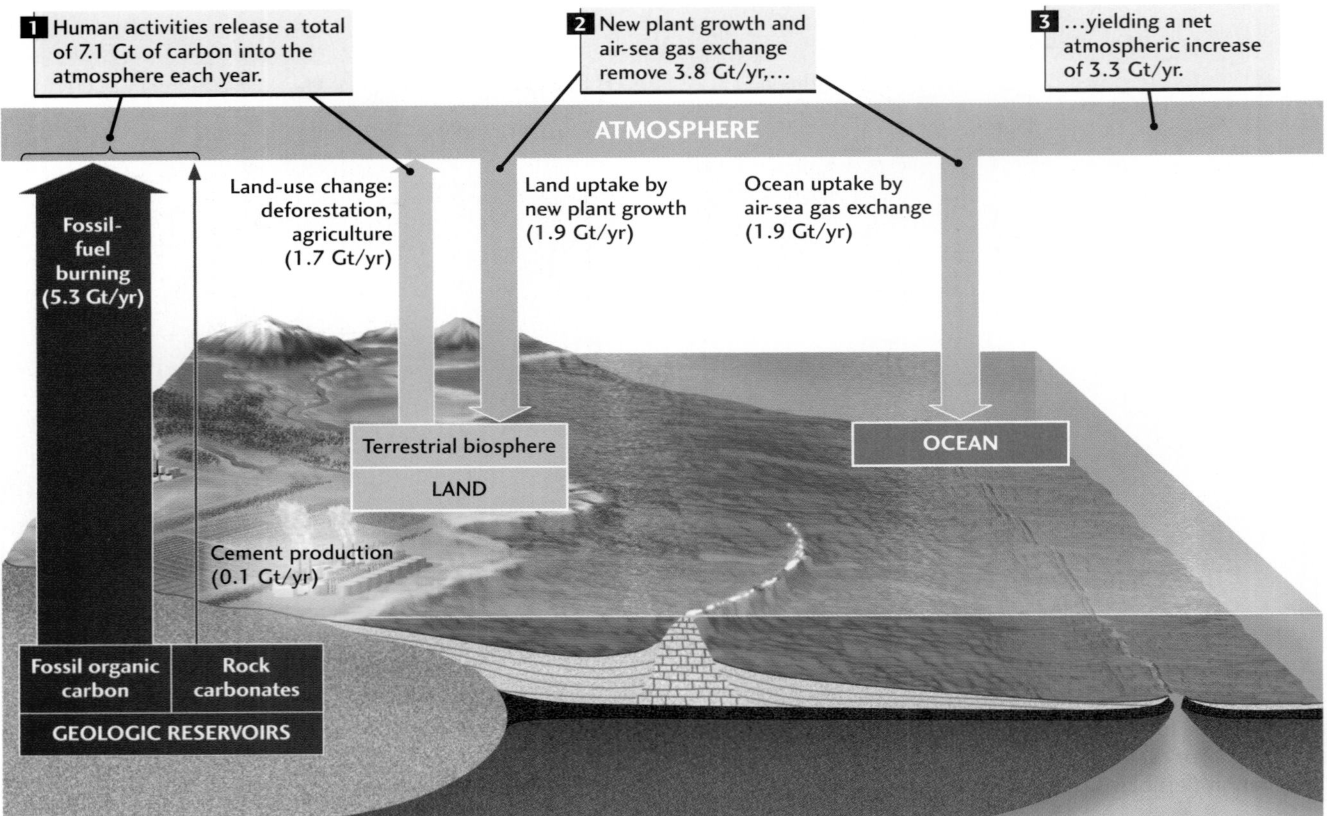

Figure 23.12 Humans add CO_2 to the atmosphere by burning fossil fuels, producing cement, and releasing carbon by deforestation, agriculture, and other land-use practices. The climate system responds by absorbing some of this carbon into the oceans and increasing plant production on land. The remainder stays in the atmosphere, increasing the concentration of CO_2. [IPCC, *Climate Change 2001: The Scientific Basis*.]

Human Activity and Global Change

The expression **global change** entered the world's vocabulary as evidence mounted that emissions from human activities could alter the chemistry of the atmosphere, with disastrous worldwide consequences that might include

- Mass die-offs due to acid rain
- Increased exposure to ultraviolet rays due to stratospheric ozone depletion
- Global warming due to an enhanced greenhouse effect

The recognition of these potentially dire consequences has motivated nations to work together to reduce the adverse effects of global environmental change. Governments have enacted new regulations to address regional environmental problems, and new treaties have been formulated to reduce human impacts on global geosystems.

The three issues just listed are rather different in scope and magnitude. Acid rain is primarily a regional problem, although if left unabated, it could have global consequences. Stratospheric ozone depletion is clearly a global problem, but its cause is mainly chlorofluorocarbons of industrial origin, which are now severely regulated through an international treaty—a major environmental success. In contrast, there appears to be little agreement on what we should do about global warming and related aspects of human-induced climate change.

Acid Rain

In many industrialized areas of the world, the air is greatly polluted with sulfur-containing gases such as sulfur dioxide (SO_2). These gases are emitted from the smokestacks of power plants that burn coal containing large amounts of the mineral pyrite (iron sulfide, FeS_2), from smelters of sulfide ores, and from some factories. Coals mined in the eastern and midwestern regions of the United States contain more of these pollutants than do coals mined in the western states. Although volcanoes and coastal marshes also add sulfur gases to the atmosphere, more than 90 percent of the sulfur emissions in eastern North America are of human origin.

Figure 23.13 A monument before and after deterioration caused by acid rain. [Westfälisches Amt für Denkmalpflege.]

Sulfur gases in the atmosphere react with oxygen and rainwater to form sulfuric acid, which is far stronger than the carbonic acid formed by carbon dioxide and rainwater. Some nitric acid is formed in the same way from nitrogen oxide gases (NO_x) emitted from smokestacks and automobile exhausts. Small amounts of sulfuric and nitric acids turn harmless rainwater into **acid rain.** Although it is much too weak to sting human skin, acid rain causes widespread damage to water and air, delicate organisms, and solid rock.

By acidifying sensitive lakes and streams, particularly those underlain by soils with limited ability to neutralize acidic compounds, acid rain has caused massive kills of fish in many lakes in Canada, the northeastern United States, and Scandinavia. A survey of more than 1000 lakes and thousands of miles of streams in the United States showed that acid rain has affected 75 percent of the lakes and 50 percent of the streams. In some acidified lakes and streams, fish species such as the brook trout have been completely eradicated. The salmon habitat of eastern Canada has been seriously affected. Acid rain also damages mountain forests, particularly those at higher elevations. Acid moisture in the air reduces visibility.

Acid rain causes noticeable damage to fabrics, paints, metals, and rocks, and it rapidly weathers stone monuments and outdoor sculptures (**Figure 23.13**). In Canada alone, acid rain causes about $1 billion in damage every year to buildings and monuments.

Coal Burning The relationship between acid rain and the burning of coal has been firmly established. Agencies of the U.S. and Canadian governments, as well as independent scientific panels, have tracked sulfur gases emitted by smokestacks to downwind locations where they are precipitated as acid rain. Careful tracing of pollutants from their sources to the sites of acid rain is necessary to demonstrate that coal-burning power plants are responsible for much of the problem.

Most of the sulfurous emissions in North America come from coal-fired electric utility plants in the midwestern and eastern United States, and most of the fallout occurs in these regions (**Figure 23.14**) and in eastern Canada. About 50 percent of the sulfate deposits in Canada originate in the United States—a source of friction between the two countries. The problem is equally critical in Europe and Asia, where scientists have noted damage to lakes and forests caused by acid rain. The damage is particularly heavy in China, in eastern European countries, and in Russia.

The Clean Air Act Concerned scientists have recommended the restriction of sulfur emissions from power plants and smelters. For a long time, however, political resistance prevented effective control. Finally, in 1990, after many years of wrangling, the U.S. Congress passed major amendments to the 1970 Clean Air Act. The legislation required coal-burning power plants to reduce their annual SO_2 emissions by 10 million tons and their annual NO_x emissions by 2 million tons by the year 2000. Deregulation of U.S. railroads has lowered transportation costs, making it less costly to ship cleaner western coal to midwestern and eastern power plants and thus making it easier to meet the new emissions standards. The current federal administration's proposal to loosen the 1990 standards, however, has reopened this contentious debate.

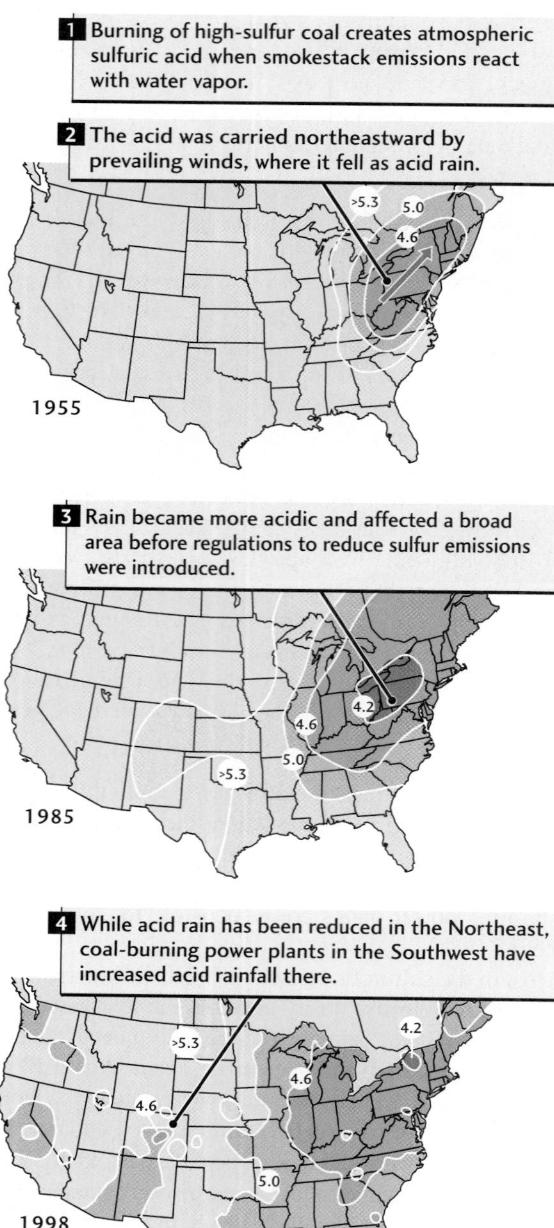

Figure 23.14 Acidity of precipitation across the contiguous United States and Canada. Units are in pH; the lower the value, the higher the acidity. The most acidic precipitation occurs in the East, primarily as a result of burning high-sulfur coal. [National Atmospheric Deposition Program/National Trends Network, 2003.]

Stratospheric Ozone Depletion

Near Earth's surface, ozone gas (O_3^+) is a major constituent of smog. It undermines health, damages crops, and corrodes materials. Low-lying ozone forms when sunlight interacts with nitrogen oxides and other chemical wastes from industrial processes and automobile exhausts.

Ozone in Earth's stratosphere, 10 to 50 km above the surface, is another matter. There, solar radiation ionizes oxygen gas (O_2) into ozone, forming a protective layer that shields Earth by absorbing cell-damaging UV radiation. Skin cancer, cataracts, impaired immune systems, and reduced crop yields are attributable to excessive UV exposure.

In 1995, the Nobel Prize in chemistry was awarded to Paul Crutzen, Mario Molina, and Sherwood Rowland for the hypotheses they had advanced more than 20 years earlier about how the protective ozone layer can be depleted by reactions involving human-made compounds. One class of compounds called chlorofluorocarbons, or CFCs, which are used as refrigerants, spray-can propellants, and cleaning solvents, are stable and harmless except when they migrate to the stratosphere. High above Earth, the intense sunlight breaks down CFCs, releasing their chlorine. Molina and Rowland proposed that chlorine reacts with the ozone molecules in the stratosphere and thins the protective ozone layer. Molina and Rowland's hypothesis was confirmed when a large hole in the ozone layer was discovered over Antarctica in 1985 (**Figure 23.15**). Subsequently, **stratospheric ozone depletion** was found to be a global phenomenon. Elevated surface levels of UV radiation were observed to correspond to low ozone levels in 1992 and 1993.

In the 1980s, when scientists were trying to convince government and industry officials that the ozone layer was possibly being depleted due to CFCs, a senior government official remarked that the solution was for people to wear hats, sunscreen, and dark glasses. Fortunately, political wisdom prevailed. In 1989, a group of nations entered into a global treaty to protect the ozone layer. This treaty, called the Montreal Protocol, now includes 140 countries. The Montreal Protocol phased out CFC production in 1996 and set up a fund, paid for by developed nations, to help developing nations switch to ozone-safe chemicals. The phase-out of CFCs has been successful; safer alternatives to CFCs are being manufactured, and the destruction of the ozone layer has diminished (**Figure 23.16**).

The Montreal Protocol has become a model for how scientists, industrial leaders, and government officials can work together to head off an environmental disaster.

Global Warming

Fossil-fuel burning and the changes in land use since the beginning of the industrial revolution have caused significant CO_2 pollution of the atmosphere and the rise of other greenhouse gases. Recognizing the potential problems that these trends pose for global climate change, the United Nations established an Intergovernmental Panel on Climate Change (IPCC) in 1988 to assess the risk of human-induced climate change, its potential impacts, and options for adaptation and mitigation. The IPCC provides a continuing forum for hundreds of scientists, economists, and policy experts to work together to understand these issues.

In its latest major assessment report, published in 2001, the IPCC drew the following conclusions:

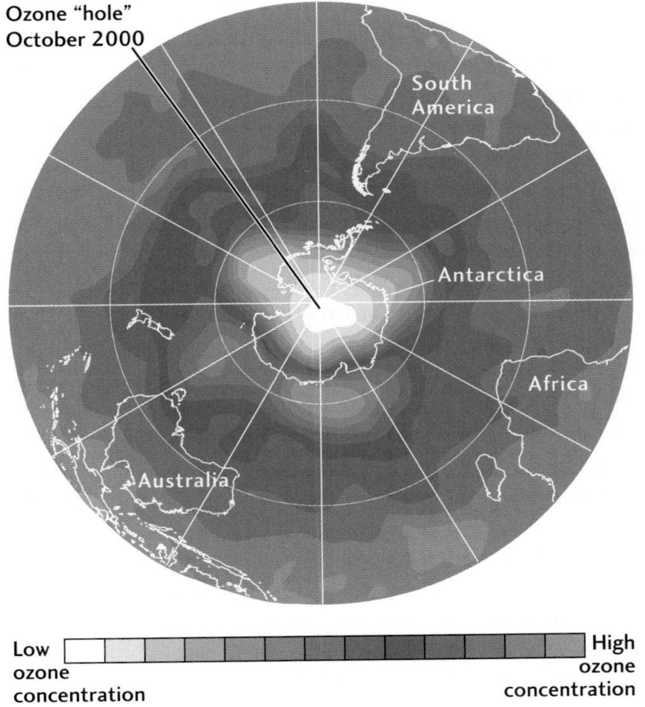

Figure 23.15 The ozone hole is a well-defined, large-scale destruction of the ozone layer over Antarctica that occurs each Antarctic spring. The word *hole* is a misnomer; the hole is really a significant thinning or reduction of ozone concentrations, which results in the destruction of up to 70 percent of the ozone normally found over Antarctica. The whiter areas represent thinning in the ozone layer that appeared over Antarctica in October 2000. [NOAA TOVS satellite.]

- Human-induced increases in greenhouse gases are probably responsible for much of the twentieth-century warming (see Figure 23.8).
- Greenhouse gases will continue to rise throughout the twenty-first century primarily because of human activities, although the increase depends on a number of socioeconomic factors that will govern the rate of greenhouse gas emissions, including active steps by society to limit these emissions. Examples of some scenarios are illustrated in **Figure 23.17**.
- The rise in greenhouse gases will cause significant global warming during this century. Projections of the amount of warming are highly uncertain because of doubt about the amount of future greenhouse gas emissions and an incomplete knowledge of how the climate system works. The range of temperature increases accepted by most experts is 1.4° to 5.8°C (see Figure 23.17). Note that the lower end of this range is more than twice the value of the twentieth-century warming.

The IPCC has also concluded that global warming during this century will likely be accompanied by a variety of changes in regional climates. The warming will probably be greater over land than over the ocean, and regional climates will probably become more variable. Other possible effects include increasing rainfall in the tropics, decreasing rainfall at midlatitudes, continued thinning of polar sea ice, and more frequent and intense El Niño events.

Consequences of Global Warming The most recent annual climate report by the World Meteorological Organization provides a preview of what might lie in store. The year 2002 was the second warmest year on record. (The warmest year was 1998, and the 10 warmest years have all occurred since 1987.) The extent of sea ice cover in the Arctic Ocean in September 2002 was the lowest for that month since satellite records began in 1978, continuing a downward trend, and the surface melt of the Greenland ice sheet was the largest on record. Regional climate events included a prolonged heat wave in India that cost hundreds of lives; record floods in central Europe, which claimed more than 100 lives; and drought in Africa that threatens millions with starvation. Some regional events, such as drought and record temperatures in Australia, were associated with El Niño conditions in the eastern Pacific, which affected global climate in the second half of 2002 and contributed to the year's ranking as the second warmest.

Such phenomena, though not necessarily caused directly by the enhanced greenhouse effect, have raised public consciousness of the seriousness of global climate change in the decades ahead. Because human-induced warming will be more rapid than any that occurs naturally,

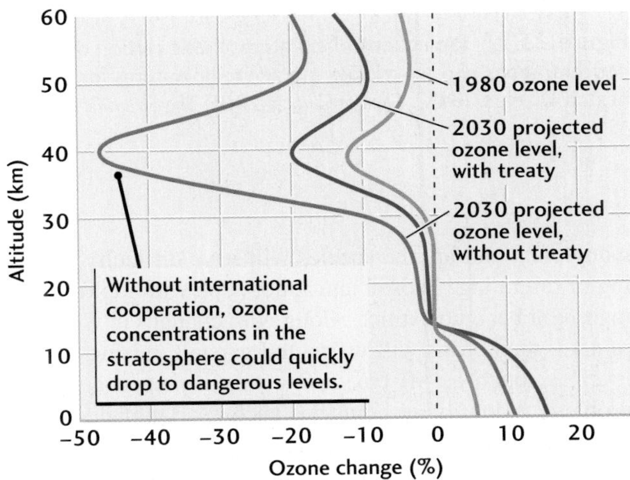

Figure 23.16 Comparison of ozone levels in the atmosphere in 1980 with projected ozone levels for 2030 with and without the Montreal Protocol. The 140 nations that signed this international treaty have been successful in eliminating emissions of CFCs, which attack the ozone layer. Without such international cooperation, ozone concentrations in the stratosphere would have quickly dropped to dangerous levels. [After T. E. Graedel and P. J. Crutzen, *Atmospheric Change* (New York: W. H. Freeman, 1993).]

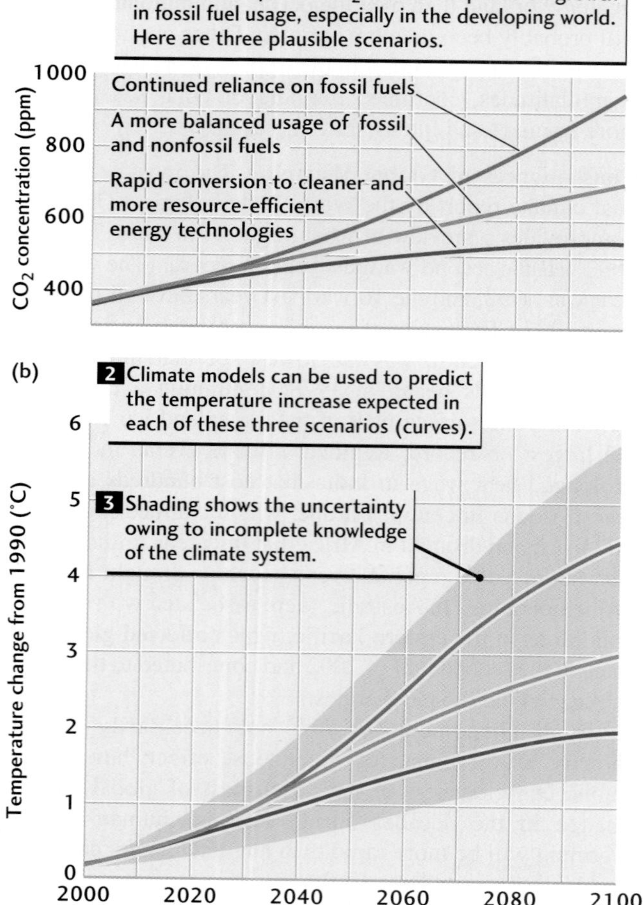

Figure 23.17 Projections of (a) atmospheric carbon dioxide concentrations and (b) average surface temperatures for the next 100 years. [IPCC, *Climate Change 2001: The Scientific Basis.*]

many plant and animal species will have difficulty adjusting or migrating. Those that cannot cope with rapid warming might become extinct. Along with changes in wind and rainfall, the oceans will warm and expand, raising the sea level as much as 90 cm during this century—a serious problem in low-lying countries such as Bangladesh. The north polar ice cap will continue to shrink rapidly, and much of the Arctic Ocean is expected to become ice-free within a few decades (**Figure 23.18**). If the continental ice sheets start to melt, the sea level could rise even higher. Table 23.1 summarizes the effects of climate change on various systems.

Reducing Greenhouse Gas Emissions No one can say at this time how much global warming will ultimately occur or to what effect, but we must consider the possibility that a climatic crisis could prevent us from using our remaining resources of coal and other fossil fuels fully. Such a crisis could last a long time; even if all human activities that generate carbon dioxide were to stop, it could take as long as 200 years for atmospheric carbon dioxide to return to its preindustrial level.

This uncertainty presents a problem for policy makers. How much money should we spend to curb human-generated carbon dioxide emissions, and will the benefits justify the costs? On the one hand, too much spending could depress the economy and cause job losses. On the other hand, prevention might be less costly than coping with a disaster after it happens.

More than 2000 of America's leading economists, including 6 Nobel laureates, declared in a statement (published in *Global Change,* electronic edition, February 1997): "As economists we believe that global climate change carries with it significant environmental, economic, social, and geopolitical risks and that preventive steps are justified. Economic studies have found that there are many potential policies to reduce greenhouse gas emissions in which total benefits outweigh total costs." What they had in mind was a "no regrets" policy—one in which actions taken now

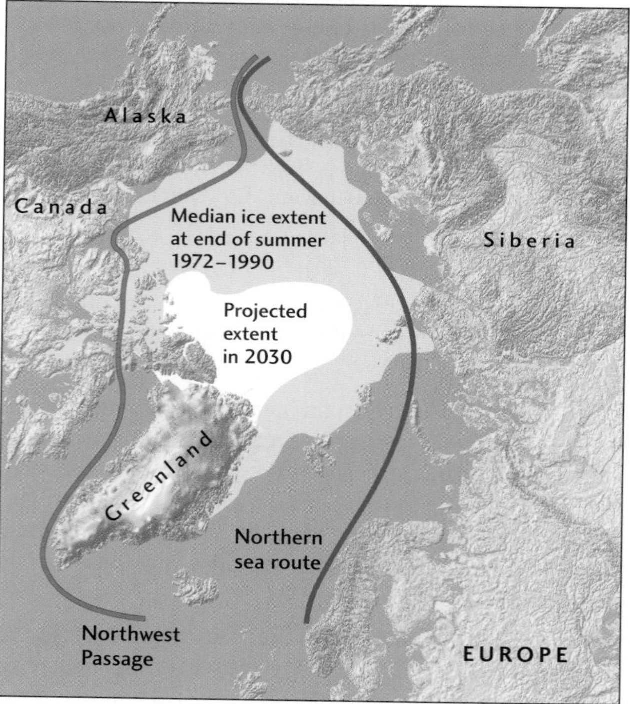

Figure 23.18 Global warming is melting the north polar ice cap. This map of the Arctic compares the extent of the polar ice at the end of the summer for the period 1972–1990 with the projected extent in 2030. The change is expected to disrupt Arctic ecosystems. One benefit for humans may be the opening of the Northwest Passage and other shorter sea routes between the Atlantic and Pacific oceans within the twenty-first century. [U.S. Navy.]

Table 23.1 Potential Climate-Change Effects on Various Systems

Systems	Potential Effects
Forests and terrestrial vegetation	Migration of vegetation Reduction in inhabited range Altered ecosystem composition
Species diversity	Loss of diversity Migration of species Invasion of new species
Coastal wetlands	Inundation of wetlands Migration of wetlands
Aquatic ecosystems	Loss of habitat Migration to new habitats Invasion of new species
Coastal resources	Inundation of coastal development Increased risk of flooding
Water resources	Changes in supplies Changes in drought and floods Changes in water quality and hydropower production
Agriculture	Changes in crop yields Shifts in relative productivity and production
Human health	Shifts in range of infectious diseases Changes in heat-stress and cold-weather afflictions
Energy	Increase in cooling demand Decrease in heating demand Changes in hydropower output
Transportation	Fewer disruptions of winter transportation Increased risk for summer inland navigation Risks to coastal roads

Source: Office of Technology Assessment, U.S. Congress.

make good sense economically and socially even if global warming does not occur. They view this policy as low-cost "insurance"—something done cheaply now to avoid high costs later.

Many actions could be taken: raising energy taxes to encourage conservation; using energy more efficiently; further developing and using renewable resources (wind, water, and solar power); accelerating scientific research to develop new carbon-free energy sources; and finding other ways to adapt to global warming. Shifting from coal and oil to natural gas would reduce the emission of CO_2 per unit of energy used. New, intrinsically safe nuclear reactors could provide a carbon-free alternative within a decade or two. The nations of the world all agree to such ideas in principle, but many—including the United States—emphasize the costs of moving too quickly (see Feature 23.2).

Policy makers must also face the issue of fairness in international politics. The United States, Canada, the European Union, and Japan—with only 25 percent of the world's population—are responsible for about 75 percent of the global increase in CO_2 and other greenhouse gases. These rich, industrially advanced nations will be able to adjust to

23.2 The Kyoto Accords and the Politics of Global Warming

At long last, scientists have convinced most world political leaders that using the atmosphere as a dump for carbon dioxide and other greenhouse gases could lead to global warming in the next 50 to 100 years, with serious consequences. The readiness of governments to take joint action culminated in a historic conference that took place on December 10, 1997, in Kyoto, Japan, where 160 nations agreed to limit emissions of greenhouse gases. Despite the fact that curtailing the use of fossil fuels—the major source of greenhouse gas emissions—could slow economic growth, all agreed to the Kyoto Protocol. Upon ratification, the protocol could become a legally binding treaty.

The agreement requires 39 industrial countries currently responsible for three-fourths of fossil-fuel burning to limit their annual CO_2 emissions between 2008 and 2012 to a percentage of their 1990 levels. For the European Union, the reduction is 8 percent below 1990 releases; for Japan, 6 percent; for the United States, 7 percent; and for developing countries, no reduction. (The United States is the largest emitter of CO_2 from fossil-fuel usage.) Without concessions to individual countries on the amount of reduction required, agreement would not have been achieved. The treaty comes into force when it is ratified by at least 55 countries, representing 55 percent of the 1990 developed world CO_2 emissions. By 2003, 104 countries (including the European Union, China, and Japan), which together include more than two-thirds of the global population, had ratified the protocol. However, the aggregate CO_2 emissions of these nations is only 43.9 percent, so the treaty is not yet in effect. Russia could put it over the top and is leaning in that direction, but it is holding out for some additional concessions.

President Clinton's administration signed the protocol in 1997, but the succeeding administration of President Bush rejected it in 2002 and refused to submit it to the Senate, which is the ratifying body. The Bush government and large segments of business and labor oppose the treaty, claiming that deep reductions in emissions would reduce energy usage by 25 to 35 percent below that expected for the year 2010. They argue that the dire predictions of adverse climate change are unproved and that the economic costs would be too high for uncertain benefits to the environment. They state that the economic losses could be substantial, as much as 2.4 percent of the gross domestic product (GDP), about \$240 billion each year. Instead of adherence to the Kyoto Protocol, they propose a voluntary program that would provide incentives to industry to reduce greenhouse gas emissions. Funds to develop nonpolluting energy technologies would be increased. The U.S. Senate, in a 95-to-0 vote, signaled that it would not ratify the treaty if it did not include emissions limits for the developing countries—which, with their growing populations and increasing use of energy, could become the major polluters of the future.

Unfortunately, even if implemented, the Kyoto Protocol will only slow the increase of CO_2 and, at most, delay warming for about 10 years. However, it does put governments, businesses, and the public on notice that they should be concerned about energy use and other practices that increase CO_2 in the atmosphere.

standards for reducing emissions more easily than the developing countries. China, for example, depends on its huge coal deposits for its economic growth. Developing nations argue for financial and technological support to help them cope with the demand to reduce emissions.

The authors of this book believe that the consequences of climate change could be so serious that it is not too soon for our leaders to start planning for reduced reliance on fossil fuels. The United States could reduce emissions of greenhouse gases by as much as 60 percent from 1990 levels at little cost by improving the efficiency of energy use—for example, by insulating buildings; replacing incandescent lights with fluorescent lights; increasing the fuel efficiency of motor vehicles a few miles per gallon; and making greater use of natural gas, which emits less carbon dioxide when burned than coal does. These modest steps constitute a low-cost insurance policy against global warming that also confers substantial benefits, including reductions in expensive oil imports, lowered manufacturing costs, and increasingly healthful air quality.

Intervention in the Climate System We are inadvertently changing the climate system by emitting greenhouse gases, thereby perturbing the carbon cycle and enhancing the greenhouse effect. What about the possibility of deliberately manipulating the carbon cycle to reduce the accumulation of greenhouse gases in the atmosphere? We have seen that deforestation feeds CO_2 into the atmosphere, whereas reforestation withdraws it, in surprisingly large amounts. Land-use policies that encourage reforestation and other biomass production, discourage deforestation, suppress fires, and reduce the loss of carbon from soils might help to mitigate the global warming problem.

Another possibility, much more controversial, is fertilization of the marine biosphere. We know that marine organisms pump CO_2 out of the atmosphere by fixing carbon in organic tissues and calcium carbonate shells. In most regions of the ocean, the limitations to marine productivity come from a lack of nutrients, such as iron. Preliminary experiments suggest that the growth of plankton can be stimulated by dumping modest amounts of iron into the ocean. Plankton blooms are a source of food for carbon-fixing marine animals, so fertilizing the ocean in this manner could increase the power of its biological pump. Any large-scale use of this technique to remove CO_2 from the atmosphere would require a much better understanding of marine ecosystems to avoid altering the ocean in unforeseen ways. This type of "planetary bioengineering," however, is not so far-fetched as it once might have seemed.

SUMMARY

What is the climate system? The climate system includes all parts of the Earth system and all the interactions among these components needed to describe how climate behaves in space and time. The main components of the climate system are the atmosphere, hydrosphere, lithosphere, and biosphere. Each component plays a different role in the climate system, depending on its ability to store and transport energy.

What is the greenhouse effect? For the global temperature averaged over daily and seasonal cycles to remain the same, Earth's surface must radiate as much energy back into space at it receives from the Sun. Much of the radiant energy from the Sun passes through the atmosphere and is absorbed by Earth's surface. The warmed surface radiates heat back to the atmosphere as infrared rays. Carbon dioxide and other trace gases absorb infrared rays selectively at certain wavelengths and radiate infrared energy at other wavelengths in all directions, including downward to Earth's surface. The net effect is to trap heat within the atmosphere by increasing the temperature of the surface relative to the temperature at higher levels in the atmosphere. In this way, the atmosphere acts like the glass in a greenhouse, allowing solar radiant energy to pass through but trapping heat.

How has Earth's climate changed over time? Natural variations of climate occur on a wide range of scales in both time and space. Some occur in the external forcing of climate, such as fluctuations in the amount of solar radiation received at Earth's surface or changes in the distribution of land surface by continental drift. Others result from internal variations within the climate system itself, such as oscillations of mass and energy between its components. The largest changes are the 100,000-year glacial cycles, with surface temperature changes of 6° to 8°C. Superposed on the glacial cycles are climate fluctuations of shorter duration, some as large as the changes from glacial to interglacial periods and occurring during intervals as short as 10 to 30 years. The warm interglacial climate of the Holocene, since about 10,000 years ago, has been much more stable than the previous parts of the glacial cycle.

Was the twentieth-century warming caused by human activities? Global warming of about 0.6°C during the twentieth century correlates with the significant rise in atmospheric CO_2 and other greenhouse gases caused by fossil-fuel burning, deforestation, and other human activities. Most, though not all, experts on Earth's climate are now convinced that the twentieth-century warming was in part human-induced and that the warming will continue into the twenty-first century as the levels of atmospheric greenhouse gases continue to rise.

What are geochemical cycles? Geochemical cycles trace the flux of Earth's elements from one reservoir to another, making it possible to quantify the amounts of elements that are stored in and moved among the oceans, atmosphere, land surface, crust, and mantle. The calcium cycle shows how rivers transport calcium derived from rock weathering to the oceans and how sedimentation transports calcium back into the lithosphere. If the cycle is in steady state, inflow balances outflow, and the residence time can be calculated as the total amount of the element in the reservoir divided by the inflow.

What is the carbon cycle? The carbon cycle is the geosystem that describes the continual movement of carbon between the atmosphere and its other principal reservoirs—the lithosphere, hydrosphere, and biosphere. Major processes include gas exchange between the atmosphere and the ocean; photosynthesis, respiration, and the burning of organic materials; and the weathering and precipitation of calcium carbonate. Outflows of carbon dioxide from the atmosphere also include the weathering of silicate rocks and the burial of carbon in the crust, which releases oxygen to the atmosphere. Inflows of carbon into the atmosphere also include volcanism.

How can internal geologic processes cause climatic change? Over the short term of a few years, sulfuric acid aerosols emanating from large volcanic eruptions can absorb solar radiation before it reaches the lower atmosphere and thus decrease global temperatures. Over the long term of millions of years, plate tectonic movements can cause continents to drift over the poles, stabilizing polar ice caps; block or open gateways to ocean currents; and cause uplift, which alters weather systems and rates of chemical erosion that draw on atmospheric CO_2.

What is acid rain, and what is being done about it? Acid rain is rain that contains sulfuric acid or nitric acid or both, formed when sulfur and nitrogen gases react with rainwater.

The sulfur gases are waste gases from power plants that burn coal containing pyrite and from smelters of sulfide ores. Nitric acid is a product of coal combustion and automobile exhausts. Acid rain can destroy fish stocks in lakes and rivers, damage mountain forests, and damage buildings and monuments. In the United States, the Clean Air Act was amended in 1990 to reduce acid emissions.

What is ozone depletion, and why should it concern us? CFCs are chemicals used as refrigerants and in industrial processes. When they reach the upper atmosphere, their breakdown products react with ozone (O_3^+) and convert it into oxygen (O_2). Ozone absorbs harmful ultraviolet radiation streaming in from the Sun. Without the ozone shield in the upper atmosphere, many life-forms would die off. An international agreement has been reached to stop the production of chemicals that destroy the ozone shield.

How much global warming will there be in the twenty-first century, and what will be the consequences? Greenhouse gases will continue to rise throughout the twenty-first century primarily because of human activities, although the increase depends on a number of socioeconomic factors that will govern the rate of greenhouse gas emissions, including active steps by society to limit these emissions. Projections of the amount of warming are highly uncertain because of the uncertainty in greenhouse gas emissions as well as an incomplete knowledge of how the climate system works, but the range accepted by most experts is from 1.4° to 5.8°C. This warming will disrupt plant and animal habitats, changing ecosystems and increasing the rate of species extinctions. The oceans will warm and expand, raising the sea level as much as 90 cm. The north polar ice cap will continue to shrink rapidly, and much of the Arctic Ocean is expected to become ice-free.

Key Terms and Concepts

acid rain (p. 559)
albedo (p. 546)
carbon cycle (p. 554)
calcium cycle (p. 555)
climate model (p. 549)
climate system (p. 544)
cryosphere (p. 545)
El Niño (p. 551)
enhanced greenhouse effect (p. 552)
geochemical reservoir (p. 554)
global change (p. 558)
greenhouse effect (p. 547)
Holocene epoch (p. 550)
residence time (p. 554)
steady state (p. 554)
stratospheric ozone depletion (p. 560)
twentieth-century warming (p. 551)

Exercises

1. What is a greenhouse gas, and how does it affect Earth's climate?

2. Why is it wrong to assert that greenhouse gases prevent heat energy from escaping to outer space?

3. Give an example of a positive and a negative feedback in the climate system not discussed in this chapter.

4. From the information given in Figure 23.11, estimate the residence time of carbon dioxide in the ocean.

5. List two inflows of chemical components into the ocean and two outflows of the same chemical components from the ocean.

6. List three causes of climate change that result from the interaction of Earth's internal and external systems.

7. Will the reduction of sea ice cause sea level to rise?

8. Make a list of human activities that have negative effects on the atmosphere and hydrosphere.

Thought Questions

This icon indicates that there is an animation available on the Web site that may assist you in answering a question.

1. How would the calcium geochemical cycle be affected by a global increase in chemical weathering?

2. Devise a simple geochemical cycle for the element sodium, which is found in marine evaporites (halite) and clay minerals.

3. What would the carbon cycle have looked like after life had originated but before photosynthesis had evolved?

4. Assume that we keep pumping carbon dioxide into the atmosphere at a steadily increasing rate and Earth warms significantly in the next 100 years. How might this affect the global carbon cycle?

5. Can scientists offer credible advice about such politically and socially charged issues as population growth, global climate change, and relations between the developed and developing nations? What should a government leader do when 8 of 10 scientists say that human-induced global warming will occur unless something is done to head it off, whereas 2 of 10 disagree?

6. An economist once wrote: "The predicted change in global temperature due to human activity is less than the difference in winter temperature between New York and Florida, so why worry?" Do you think the economist was right?

7. A high-level government official once suggested that ozone depletion is not worrisome, because people could just use sunglasses, sunscreen lotion, and wide-brimmed hats to minimize the danger from UV rays. Do you think the official was right?

8. Scientists are working hard to learn how to predict reliably El Niño events. If they succeed, how could this knowledge be used to reduce the loss of life and property and to mitigate economic disruption?

9. Do you think we should act now to reduce carbon emissions or delay until the science of climate change is better understood?

10. Is the United States justified in insisting that developing countries, which now use the least amount of fossil fuels, agree to limit their future carbon emissions?

11. Do you think that future scientists and engineers will be able to intervene and prevent catastrophic changes in the Earth system?

Short-Term Team Project

Investigate the climate at your locality by compiling data on the maximum and minimum annual temperatures and average rainfall during the last 40 years. Good places to start are the Web sites of the National Climate Diagnostics Center (http://www.cdc.noaa.gov/USclimate/) and the National Climatic Data Center (http://www.ncdc.noaa.gov/oa/ncdc.html). Can you see evidence of global warming in your local climate record?

Suggested Readings

American Geophysical Union. 1999. Climate change and greenhouse gases. *EOS* 80: 453–458.

Berner, R. A., and A. C. Lasaga. 1989. Modeling the geochemical carbon cycle. *Scientific American* (March): 74–81.

Bonnet, S., and A. Crave. 2003. Landscape response to climate change: Insights from experimental modeling and implications for tectonic versus climactic uplift of topography. *Geology* 31(2): 123–126.

Cook, E. 1996. *Marking a Milestone in Ozone Protection: Learning from the CFC Phase-Out.* Washington, D.C.: World Resources Institute.

Costanza, R., et al. 1997. The value of the world's ecosystem services and natural capital. *Nature* 387: 253–260.

Graedel, T. E., and P. J. Crutzen. 1993. *Atmospheric Change: An Earth System Perspective.* New York: W. H. Freeman.

Intergovernmental Panel on Climate Change. 2001. *Climate Change 2001: The Scientific Basis.* Cambridge: Cambridge University Press.

National Academy of Sciences. 1992. *Policy Implications of Greenhouse Warming.* Washington, D.C.: National Academy Press.

National Research Council. 1993. *Solid Earth Sciences and Society.* Washington, D.C.: National Academy Press.

National Research Council. 1999. *From Monsoons to Microbes: Understanding the Ocean's Role in Human Health.* Washington, D.C.: National Academy Press.

National Research Council. 2001. *Climate Change Science: An Analysis of Some Key Questions.* Washington, D.C.: National Academy Press.

Ruddiman, W. F. 2001. *Earth's Climate: Past and Future.* New York: W. H. Freeman.

Webster, P. J., and J. A. Curry. 1998. The oceans and weather. *Scientific American Presents: The Oceans: The Ultimate Voyage Through Our Watery Home* (Fall): 38–43.

APPENDIX 1

Conversion Factors

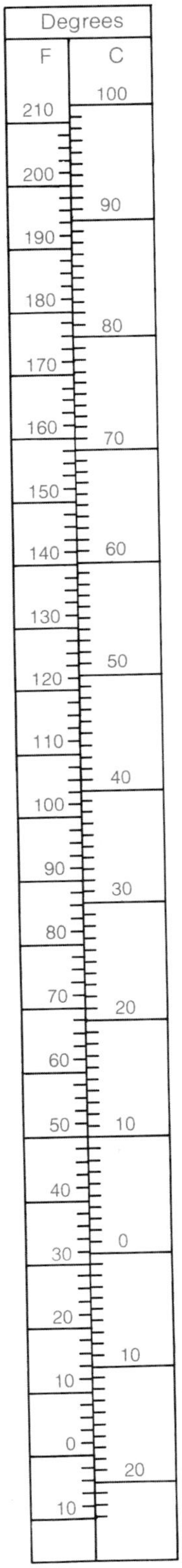

LENGTH

1 centimeter	0.3937 inch
1 inch	2.5400 centimeters
1 meter	3.2808 feet; 1.0936 yards
1 foot	0.3048 meter
1 yard	0.9144 meter
1 kilometer	0.6214 mile (statute); 3281 feet
1 mile (statute)	1.6093 kilometers
1 mile (nautical)	1.8531 kilometers
1 fathom	6 feet; 1.8288 meters
1 angstrom	10^{-8} centimeter
1 micrometer	0.0001 centimeter

VELOCITY

1 kilometer/hour	27.78 centimeters/second
1 mile/hour	17.60 inches/second

AREA

1 square centimeter	0.1550 square inch
1 square inch	6.452 square centimeters
1 square meter	10.764 square feet; 1.1960 square yards
1 square foot	0.0929 square meter
1 square kilometer	0.3861 square mile
1 square mile	2.590 square kilometers
1 acre (U.S.)	4840 square yards

VOLUME

1 cubic centimeter	0.0610 cubic inch
1 cubic inch	16.3872 cubic centimeters
1 cubic meter	35.314 cubic feet
1 cubic foot	0.02832 cubic meter
1 cubic meter	1.3079 cubic yards
1 cubic yard	0.7646 cubic meter
1 liter	1000 cubic centimeters; 1.0567 quarts (U.S. liquid)
1 gallon (U.S. liquid)	3.7853 liters

MASS

1 gram	0.03527 ounce
1 ounce	28.3495 grams
1 kilogram	2.20462 pounds
1 pound	0.45359 kilogram

PRESSURE

1 kilogram/square centimeter	0.96784 atmosphere; 0.98067 bar; 14.2233 pounds/square inch
1 bar	0.98692 atmosphere; 10^5 pascals

ENERGY

1 joule	0.239 calorie; 9.48451×10^{-11} Btu
1 quad	10^{15} Btu

POWER

1 watt	0.001341 horsepower (U.S.); 0.05688 Btu/minute

Degrees

F C

APPENDIX 2

Numerical Data Pertaining to Earth

Equatorial radius	6378 kilometers
Polar radius	6357 kilometers
Radius of sphere with Earth's volume	6371 kilometers
Volume	1.083×10^{27} cubic centimeters
Surface area	5.1×10^{18} square centimeters
Percent surface area of oceans	71
Percent surface area of land	29
Average elevation of land	623 meters
Average depth of oceans	3.8 kilometers
Mass	5.976×10^{27} grams
Density	5.517 grams/cubic centimeter
Gravity at equator	978.032 centimeters/second/second
Mass of atmosphere	5.1×10^{21} grams
Mass of ice	$25–30 \times 10^{21}$ grams
Mass of oceans	1.4×10^{24} grams
Mass of crust	2.5×10^{25} grams
Mass of mantle	4.05×10^{27} grams
Mass of core	1.90×10^{27} grams
Mean distance to Sun	1.496×10^{8} kilometers
Mean distance to Moon	3.844×10^{5} kilometers
Ratio: Mass of Sun/mass of Earth	3.329×10^{5}
Ratio: Mass of Earth/mass of Moon	81.303
Total geothermal energy reaching Earth's surface each year	2.39×10^{-20} calorie; 80 quads
Earth's daily receipt of solar energy	1.49×10^{22} joules
U.S. energy consumption, 2000	97.7 quads

APPENDIX 3

Important Events That Led to the Theory of Plate Tectonics

"In science . . . the work of the individual is so bound up with that of his scientific predecessors and contemporaries that it appears almost as an impersonal product of his generation."

—ALBERT EINSTEIN

1915

Alfred Wegener, of Germany, advances first theory of continental drift based on geological data on different continents.

1929

Arthur Holmes (Durham University) proposes model of continental drift driven by subcrustal convection currents. His theory includes elements now known as plate divergence and convergence and mantle convection.

1946 – 50

Scientists begin to use echo sounders, hydrophones, magnetometers, navigational aids, and other instruments developed in World War II to explore ocean basins. They show that ocean crust is basaltic and 5 km thick, in contrast to 40-km-thick granitic continental crust.

1954

Hugo Benioff (Caltech) describes pattern of deep-focus earthquakes that lie along a dipping plane beneath ocean trenches and active island arcs. This would later become associated with subducting plates.

1956

Cambridge University physicists Patrick M. S. Blackett, Edward Irving, and S. Keith Runcorn argue that their paleomagnetic measurements of rocks of different continents support continental movements according to Wegener's theory.

1959

Columbia University scientists Bruce Heezen and Marie Tharp develop first detailed topographic maps of the seafloor showing global mid-ocean ridge–rift system.

1962

Princeton geologist Harry Hess, using newest results from seafloor exploration, buttresses the case that continents drift by riding on top of seafloor that spreads laterally from mid-ocean ridges. Seafloor is consumed by sinking back into the mantle at deep-sea trenches. Proposes mantle convection as the driving force.

1963

Allan Cox of Stanford and U.S. Geological Survey scientists Richard Doell and Brent Dalrymple use reversals in magnetization of continental layered basalt flows and the radiometric (absolute) ages of these flows to establish a new geomagnetic time scale to date rocks.

1963

In a major breakthrough, Frederick Vine and Drummond Matthews of Cambridge University relate the bands of alternately reversed magnetization of seafloor crust to lateral spreading of seafloor from mid-ocean ridges. Canadian scientist Lawrence Morley independently advances this concept.

(*Continued*)

1965

Canadian geologist J. Tuzo Wilson proposes transform faults as the explanation of ocean-ridge offsets.

1966

Columbia University scientists Neil Opdyke, Walter Pitman, and J. R. Heirtzler link the geomagnetic and radiometric time scale of continental basaltic flows, the geomagnetic and paleontological time scales of deep-sea sediment cores, and the time scales of seafloor magnetic bands. The ages of the seafloor and spreading rates for all of the oceans can now be determined.

1968

Drilling ship *Glomar Challenger* begins to drill and recovers cores of sediment and basaltic crust of seafloor that corroborate and fill in details of seafloor spreading.

Late 1960s

Tuzo Wilson suggests pattern of rigid plates; Dan McKenzie (Cambridge), Robert Parker (Scripps), W. Jason Morgan (Princeton), and French scientist Xavier LePichon work out the shapes of plates and the geometry and history of their motion on a sphere.

1968

Final chapter. Columbia seismologists Jack Oliver, Lynn Sykes, and Bryan Isaacs show that precise locations, depths, and mechanisms of earthquakes conform to plate-tectonics predictions for mid-ocean ridge spreading centers, transform faults, and subduction zones.

APPENDIX 4

Chemical Reactions

Electron Shells and Ion Stability

Electrons surround the nucleus of an atom in a unique set of concentric spheres called electron shells. Each shell can hold a certain maximum number of electrons. In the chemical reactions of most elements, only the electrons in the outermost shells interact. In the reaction between sodium (Na) and chlorine (Cl) that forms sodium chloride (NaCl), the sodium atom loses an electron from its outer shell of electrons and the chlorine atom gains an electron in its outer shell (see Figure 3.4).

Before reacting with chlorine, the sodium atom has one electron in its outer shell. When it loses that electron, its outer shell is eliminated and the next shell inward, which has eight electrons (the maximum that this shell can hold), becomes the outer shell. The original chlorine atom had seven electrons in its outer shell, with room for a total of eight. By gaining an electron, its outer shell is filled. Many elements have a strong tendency to acquire a full outer electron shell, some by gaining electrons and some by losing them in the course of a chemical reaction. The stability of ions with fully occupied outer shells is related to the interactions of electrons in various orbitals around the nucleus.

Many chemical reactions entail gains and losses of several electrons as two or more elements combine. The element calcium (Ca), for example, becomes a doubly charged cation, Ca^{2+}, as it reacts with two chlorine atoms to form calcium chloride. (In the chemical formula for calcium chloride, $CaCl_2$, the presence of two chloride ions is symbolized by the subscript 2. Chemical formulas thus show the relative proportion of atoms or ions in a compound. Common practice is to omit the subscript 1 next to single ions in a formula.

The periodic table (see Figure 3.3) organizes the elements (from left to right along a row) in order of atomic number (number of protons), which also means increasing numbers of electrons in the outer shell. The third row from the top, for example, starts at the left with sodium (atomic number 11), which has one electron in its outer shell. The next is magnesium (atomic number 12), which has two electrons in its outer shell, followed by aluminum (atomic number 13), with three, and silicon (atomic number 14), with four. Then come phosphorus (atomic number 15), with five; sulfur (atomic number 16), with six; and chlorine (atomic number 17), with seven. The last element in this row is argon (atomic number 18), with eight electrons, the maximum possible, in its outer shell. Each column in the table forms a vertical grouping of elements with similar electron-shell patterns.

Elements That Tend to Lose Electrons

The elements in the leftmost column of the table all have a single electron in their outer shells and have a strong tendency to lose that electron in chemical reactions. Of this group, hydrogen (H), sodium (Na), and potassium (K) are found in major abundance at Earth's surface and in its crust.

The second column from the left includes two more elements of major abundance, magnesium (Mg) and calcium (Ca). Elements in this column have two electrons in their outer shells and a strong tendency to lose both of them in chemical reactions.

Elements That Tend to Gain Electrons

Toward the right side of the table, the two columns headed by oxygen (O), the most abundant element in the Earth, and fluorine (F), a highly reactive toxic gas, group the elements that tend to gain electrons in their outer shells. The elements in the column headed by oxygen have six of the possible eight electrons in their outer shells and tend to gain two electrons. Those in the column headed by fluorine have seven electrons in their outer shells and tend to gain one.

Other Elements

The columns between the two on the left and the two headed by oxygen and fluorine have varying tendencies to gain, lose, or share electrons. The column toward the right side of the table headed by carbon (C) includes silicon (Si), of major abundance in the Earth. Both silicon and carbon tend to share electrons.

The elements in the last column on the right, headed by helium (He), have full outer shells and thus no tendency either to gain or to lose electrons. As a result, these elements, in contrast with those in other columns, do not react chemically with other elements, except under very special conditions.

Properties of the Most Common Minerals of Earth's Crust

	Mineral or Group Name	Structure or Composition	Varieties and Chemical Composition	Form, Diagnostic Characteristics	Cleavage, Fracture	Color	Hardness
LIGHT-COLORED MINERALS, VERY ABUNDANT IN EARTH'S CRUST IN ALL MAJOR ROCK TYPES	FELDSPAR	FRAMEWORK SILICATES	*POTASSIUM FELDSPARS* $KAlSi_3O_8$ Sanidine, Orthoclase, Microcline	Cleavable coarsely crystalline or finely granular masses; isolated crystals or grains in rocks, most commonly not showing crystal faces	Two at right angles, one perfect and one good; pearly luster on perfect cleavage	White to gray, frequently pink or yellowish; some green	6
			PLAGIOCLASE FELDSPARS $NaAlSi_3O_8$ Albite, $CaAl_2Si_2O_8$ Anorthite		Two at nearly right angles, one perfect and one good; fine parallel striations on perfect cleavage	White to gray, less commonly greenish or yellowish	
	QUARTZ		SiO_2	Single crystals or masses of 6-sided prismatic crystals; also formless crystals and grains or finely granular or massive	Very poor or nondetectable; conchoidal fracture	Colorless, usually transparent; also slightly colored smoky gray, pink, yellow	7
	MICA	SHEET SILICATES	*MUSCOVITE* $KAl_3Si_3O_{10}(OH)_2$	Thin, disc-shaped crystals, some with hexagonal outlines; dispersed or aggregates	One perfect; splittable into very thin, flexible, transparent sheets	Colorless; slight gray or green to brown in thick pieces	$2\text{–}2\frac{1}{2}$
DARK-COLORED MINERALS, ABUNDANT IN MANY KINDS OF IGNEOUS AND METAMORPHIC ROCKS			*BIOTITE* $K(Mg, Fe)_3AlSi_3O_{10}(OH)_2$	Irregular, foliated masses; scaly aggregates	One perfect; splittable into thin, flexible sheets	Black to dark brown; translucent to opaque	$2\frac{1}{2}\text{–}3$
			CHLORITE $(Mg,Fe)_5(Al,Fe)_2Si_3O_{10}(OH)_8$	Foliated masses or aggregates of small scales	One perfect; thin sheets flexible but not elastic	Various shades of green	$2\text{–}2\frac{1}{2}$
	AMPHIBOLE	DOUBLE CHAINS	*TREMOLITE-ACTINOLITE* $Ca_2(Mg,Fe)_5Si_8O_{22}(OH)_2$	Long, prismatic crystals, usually 6-sided; commonly in fibrous masses or irregular aggregates	Two perfect cleavage directions at 56° and 124° angles	Pale to deep green; Pure tremolite white	5–6
			HORNBLENDE Complex Ca, Na, Mg, Fe, Al silicate				
	PYROXENE	SINGLE CHAINS	*ENSTATITE-HYPERSTHENE* $(Mg,Fe)_2Si_2O_6$	Prismatic crystals, either 4- or 8-sided; granular masses and scattered grains	Two good cleavage directions at about 90°	Green and brown to grayish or greenish white	5–6
			DIOPSIDE $(Ca,Mg)_2Si_2O_6$			Light to dark green	
			AUGITE Complex Ca, Na, Mg, Fe, Al silicate			Very dark green to black	

	Mineral or Group Name	Structure or Composition	Varieties and Chemical Composition	Form, Diagnostic Characteristics	Cleavage, Fracture	Color	Hardness
	OLIVINE	ISOLATED TETRAHEDRA	$(Mg,Fe)_2SiO_4$	Granular masses and disseminated small grains	Conchoidal fracture	Olive to grayish green and brown	$6\frac{1}{2}$–7
	GARNET		Ca, Mg, Fe, Al silicate	Isometric crystals, well formed or rounded; high specific gravity, 3.5–4.3	Conchoidal and irregular fracture	Red and brown, less commonly pale colors	$6\frac{1}{2}$–7
LIGHT-COLORED MINERALS, TYPICALLY AS ABUNDANT CONSTITUENTS OF SEDIMENTS AND SEDIMENTARY ROCKS	CALCITE	CARBONATES	$CaCO_3$	Coarsely to finely crystalline in beds, veins, and other aggregates; cleavage faces may show in coarser masses; calcite effervesces rapidly, dolomite slowly, only in powders	Three perfect cleavages, at oblique angles; splits to rhombohedral cleavage pieces	Colorless, transparent to translucent; variously colored by impurities	3
	DOLOMITE		$CaMg(CO_3)_2$				$3\frac{1}{2}$–4
	CLAY MINERALS	HYDROUS ALUMINO-SILICATES	*KAOLINITE* $Al_2Si_2O_5(OH)_4$	Earthy masses in soils; bedded; in association with other clays, iron oxides, or carbonates; plastic when wet; montmorillonite swells when wet	Earthy, irregular	White to light gray and buff; also gray to dark gray, greenish gray, and brownish depending on impurities and associated minerals	$1\frac{1}{2}$–$2\frac{1}{2}$
			ILLITE Similar to muscovite +Mg,Fe				
			SMECTITE Complex Ca, Na, Mg, Fe, Al silicate + H_2O				
	GYPSUM	SULFATES	$CaSO_4 \cdot 2H_2O$	Granular, earthy, or finely crystalline masses; tabular crystals	One perfect, splitting to fairly thin slabs or sheets; two other good cleavages	Colorless to white; transparent to translucent	2
	ANHYDRITE		$CaSO_4$	Massive or crystalline aggregates in beds and veins	One perfect, one nearly perfect, one good; at right angles	Colorless, some tinged with blue	3–$3\frac{1}{2}$
	HALITE	HALIDES	NaCl	Granular masses in beds; some cubic crystals; salty taste	Three perfect cleavages at right angles	Colorless, transparent to translucent	$2\frac{1}{2}$
	OPAL-CHALCEDONY	SILICA	SiO_2 [Opal is an amorphous variety; chalcedony is a formless microcrystalline quartz.]	Beds in siliceous sediments and chert; in veins or banded aggregates	Conchoidal fracture	Colorless or white when pure, but tinged with various colors by impurities in bands, especially in agates	5–$6\frac{1}{2}$
DARK-COLORED MINERALS, COMMON IN MANY ROCK TYPES	MAGNETITE	IRON OXIDES	Fe_3O_4	Magnetic; disseminated grains, granular masses; occasional octahedral isometric crystals; high specific gravity, 5.2	Conchoidal or irregular fracture	Black, metallic luster	6

(Continued)

	Mineral or Group Name	Structure or Composition	Varieties and Chemical Composition	Form, Diagnostic Characteristics	Cleavage, Fracture	Color	Hardness
	HEMATITE	IRON OXIDES	Fe_2O_3	Earthy to dense masses, some with rounded forms, some granular or foliated; high specific gravity, 4.9–5.3	None; uneven, sometimes splintery fracture	Reddish brown to black	5–6
	"LIMONITE"		*GOETHITE* [the major mineral of the mixture called "limonite," a field term] $HFeO_2$	Earthy masses, massive bodies or encrustations, irregular layers; high specific gravity, 3.3–4.3	One excellent in the rare crystals; usually an early fracture	Yellowish brown to dark brown and black	$5–5\frac{1}{2}$
LIGHT-COLORED MINERALS, MAINLY IN IGNEOUS AND METAMORPHIC ROCKS AS COMMON OR MINOR CONSTITUENTS	KYANITE	ALUMINO-SILICATES	Al_2SiO_5	Long, bladed or tabular crystals or aggregates	One perfect and one poor, parallel to length of crystals	White to light-colored or pale blue	5 parallel to crystal length 7 across crystals
	SILLIMANITE		Al_2SiO_5	Long, slender crystals or fibrous, felted masses	One perfect parallel to length, not usually seen	Colorless, gray to white	6–7
	ANDALUSITE		Al_2SiO_5	Coarse, nearly square prismatic crystals, some with symmetrically arranged impurities	One distinct; irregular fracture	Red, reddish brown, olive-green	$7\frac{1}{2}$
	FELDSPATHOIDS		*NEPHELINE* $(Na,K)AlSiO_4$	Compact masses or as embedded grains, rarely as small prismatic crystals	One distinct; irregular fracture	Colorless, white, light gray; gray-greenish in masses, with greasy luster	$5\frac{1}{2}–6$
			LEUCITE $KAlSi_2O_6$	Trapezohedral crystals embedded in volcanic rocks	One very imperfect	White to gray	$5\frac{1}{2}–6$
	SERPENTINE		$Mg_6Si_4O_{10}(OH)_8$	Fibrous (asbestos) or platy masses	Splintery fracture	Green; some yellowish brownish or gray; waxy or greasy luster in massive habit; silky luster in fibrous habit	4–6
	TALC		$Mg_3Si_4O_{10}(OH)_2$ masses or aggregates	Foliated or compact masses or aggregates	One perfect, making thin flakes or scales; soapy feel	White to pale green; pearly or greasy luster	1
	CORUNDUM		Al_2O_3	Some rounded, barrel-shaped crystals; most often as disseminated grains or granular masses (emery)	Irregular fracture	Usually brown, pink, or blue; emery black Gemstone varieties: ruby, sapphire	9

	Mineral or Group Name	Structure or Composition	Varieties and Chemical Composition	Form, Diagnostic Characteristics	Cleavage, Fracture	Color	Hardness
DARK-COLORED MINERALS, COMMON IN METAMORPHIC ROCKS	EPIDOTE	SILICATES	$Ca_2(Al,Fe)Al_2Si_3O_{12}(OH)$	Aggregates of long prismatic crystals, granular or compact masses, embedded grains	One good, one poor at greater than right angles; conchoidal and irregular fracture	Green, yellow-green, gray, some varieties dark brown to black	6–7
	STAUROLITE		$Fe_2Al_9Si_4O_{22}(O,OH)_2$	Short prismatic crystals, some cross-shaped, usually coarser than matrix of rock	One poor	Brown, reddish, or dark brown to black	$7–7\frac{1}{2}$
METALLIC LUSTER, COMMON IN MANY ROCK TYPES, ABUNDANT IN VEINS	PYRITE	SULFIDES	FeS_2	Granular masses or well-formed cubic crystals in veins and beds or disseminated; high specific gravity, 4.9–5.2	Uneven fracture	Pale brass-yellow	$6–6\frac{1}{2}$
	GALENA		PbS	Granular masses in veins and disseminated; some cubic crystals; very high specific gravity, 7.3–7.6	Three perfect cleavages at mutual right angles, giving cubic cleavage fragments	Silver-gray	$2\frac{1}{2}$
	SPHALERITE		ZnS	Granular masses or compact crystalline aggregates; high specific gravity, 3.9–4.1	Six perfect cleavages at 60° to one another	White to green, brown, and black; resinous to submetallic luster	$3\frac{1}{2}–4$
	CHALCOPYRITE		$CuFeS_2$	Granular or compact masses; disseminated crystals; high specific gravity, 4.1–4.3	Uneven fracture	Brassy to golden-yellow	$3\frac{1}{2}–4$
	CHALCOCITE		Cu_2S	Fine-grained masses; high specific gravity, 5.5–5.8	Conchoidal fracture	Lead-gray to black; may tarnish green or blue	$2\frac{1}{2}–3$
MINERALS FOUND IN MINOR AMOUNTS IN A VARIETY OF ROCK TYPES AND IN VEINS OR PLACERS	RUTILE	TITANIUM OXIDES	TiO_2	Slender to prismatic crystals; granular masses; high specific gravity, 4.25	One distinct, one less distinct; conchoidal fracture	Reddish brown, some yellowish, violet, or black	$6–6\frac{1}{2}$
	ILMENITE		$FeTiO_3$	Compact masses, embedded grains, detrital grains in sand; high specific gravity, 4.79	Conchoidal fracture	Iron-black; metallic to submetallic luster	5–6
	ZEOLITES	SILICATES	Complex hydrous silicates; many varieties of minerals, including analcime, natrolite, phillipsite, heulandite, and chabazite	Well-formed radiating crystals in cavities in volcanics, veins, and hot springs; also as fine-grained and earthy bedded deposits	One perfect for most	Colorless, white, some pinkish	4–5

APPENDIX 6

Topographic and Geologic Maps

A map is a quantitative representation of the spatial distribution of some attribute or property of the Earth. It is a kind of graph in which the axes are lines of latitude and longitude and the positions of points on the surface (or beneath it) are plotted in relation to those axes or some other established reference. Geologists also want maps that describe the geological materials at or near the surface so that they can construct a three-dimensional mental picture of the geology from this two-dimensional graph. The practiced map reader can become proficient at deducing much of the geologic structure and history of an area.

The use of topographic maps (see Chapter 18) and geologic maps (see Chapter 11) has spread widely throughout our culture. To the more traditional users of such maps—such as geologists and surveyors—have been added city planners, industrial zoning commissions, and many members of the public seeking recreational areas for hiking, camping, fishing, and other activities. In 1993 the U.S. Geological Survey distributed nearly 7 million copies of its 70,000 published topographic maps and 18,000 copies of geologic and hydrologic maps from its open file reports. Maps are a necessity for all kinds of geological and mineral resource studies, as well as for studies of groundwater, flood control, soil management, and such environmental concerns as land-use planning, which involves the locations of highways, industrial areas, oil and gas pipelines, and recreational areas.

Scale

Because the size of the area covered, and thus the amount of detail that can be shown, is always important, maps are always drawn to an appropriate *scale*—that is, the relationship of a distance (or area) on the map to the true distance on the Earth. A map's scale is stated as a ratio, such as 1:24,000, which indicates that a distance of one unit on the map represents a distance of 24,000 such units on the Earth. It does not matter what the units are: a map of scale 1:24,000 is the same whether we use the metric or the English system. The scale can be thought of in any convenient units desired: 1 in. = 2000 ft, or 1 m = 24 km, or 10 cm = 2.4 km. For convenience, maps have a graphic scale, usually at the bottom margin, in which a distance such as 1 km or 1 mile, usually with subdivisions, is shown as it would appear on the map. A common scale for detailed topographic and geologic maps is 1:24,000, used by the U.S. Geological Survey for most modern maps. The scale used for regional maps covering much larger areas is 1:250,000. Scales of 1:1,000,000 are used for aeronautical charts. In 1976 the U.S. Geological Survey introduced the first of a new series of 1:100,000 all-metric topographic maps on which graphic scales show both kilometers and miles.

Topographic Maps

A topographic map shows a region's landforms and elevations. On most maps, natural and constructed features of the surface are represented by conventional symbols. Those used by the U.S. Geological Survey are typical: rivers, lakes, and oceans are shown in blue; topography is shown in brown; constructed features are shown in black, with main highways and urban areas in red; green shaded areas show wooded land. Some special symbols may be shown in the explanation, or *legend,* of the map, which is usually displayed along the bottom margin. The most complex representations are the topographic elevations of the surface, usually shown on maps by contours (see Chapter 18). Special maps are sometimes prepared to show environmental variables, such as the distribution of slopes of various steepness.

Geologic Maps

Geologic maps are a representation of the distribution of rocks and other geologic materials of different lithologies and ages over the Earth's surface or below it. The geologist perceives the Earth not only in its surface expression of topography and patterns of land and water but also in terms of its pattern of subsurface structures, stratigraphic sequences, igneous intrusions, unconformities, and other geometric relationships of rocks. Just as an anatomist can visualize the muscles and bones beneath the skin, so can a geologist visualize details of the Earth's subsurface.

Detailed geologic maps are normally constructed on a topographic map base (Figure A6.1). This makes it easy to locate geologic structures with respect to surface features of the Earth. It is also important because topography is so often related to the nature of the underlying rocks and their structures. Because it contains so much more information than a topographic map alone, a geologic map is the most valuable for many of the purposes for which maps are used.

Traditional geologic maps are made by a geologist who roams over the area and notes the kinds of rocks, sediments, and soils and their structural and stratigraphic relationships. Geologists cover the ground on foot to see most if not all of the outcrops; they are helped enormously by the automobile, sometimes by a helicopter, and in some places even by a horse or donkey.

Figure A6.1 Topographic map (*above*) and geologic map with cross sections (*facing page*) of folded sedimentary rocks in the Valley and Ridge province of the Appalachian Mountains. Contours show the pronounced trends of valleys and ridges that reflect the parallel folds. The ridges have developed along the formations that are resistant to erosion, some at the crests of anticlines, such as Jack Mountain north of Crab Run, and others along the flanks of folds. The valleys are in the easily eroded formations, some in synclines, such as Jackson River, some on anticlines,

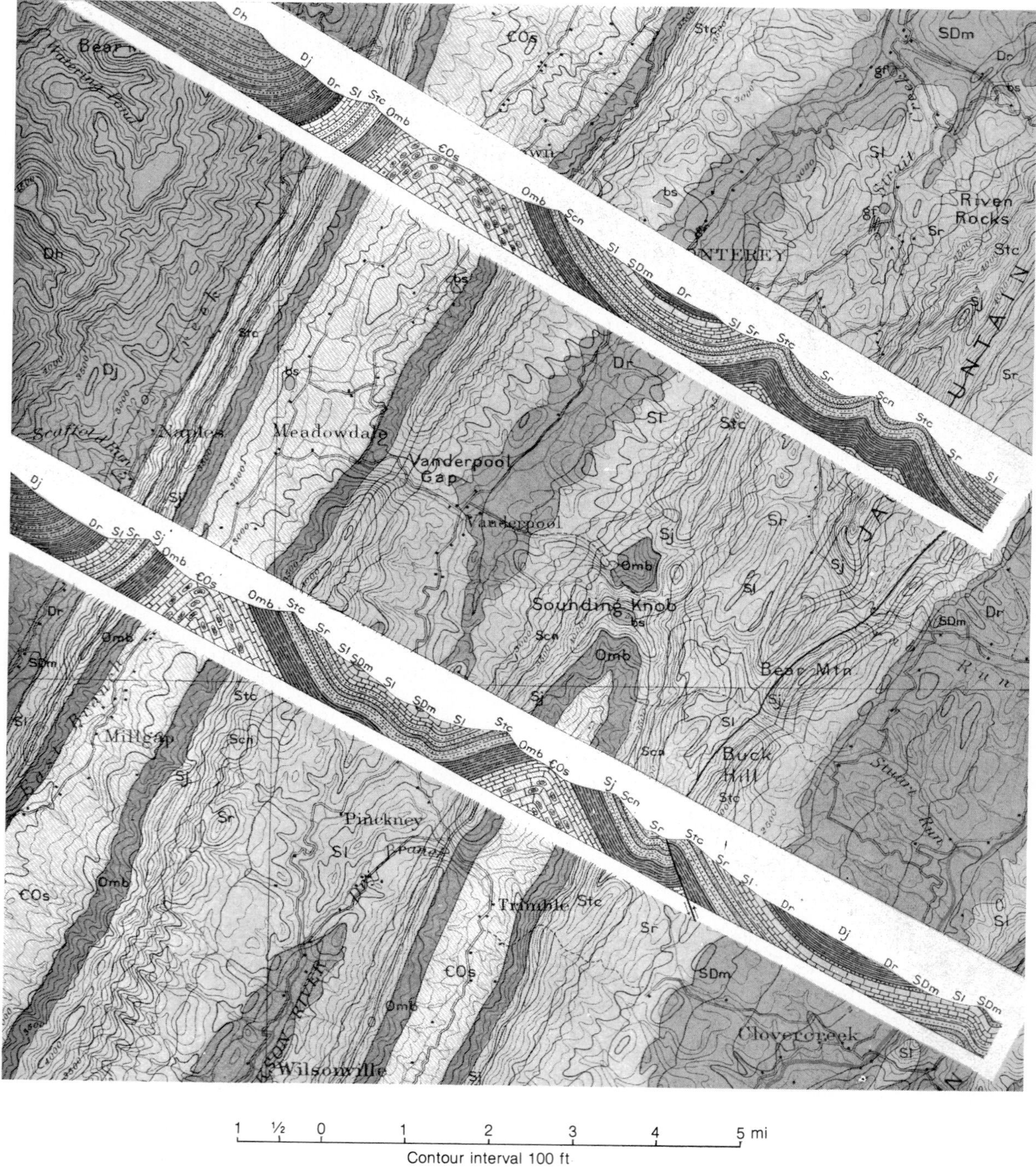

(*Continued from page AP-11*)

such as East Branch, and some in the flanks of folds, such as Back Creek. On the geologic map the pattern of anticlines and synclines can be read from the positions of formations of different age, such as at East Branch, where the oldest rocks, Cambrian and Ordovician formations (COs), are at the surface bordered on both sides by younger formations of Ordovician and Silurian age (Omb, Stc, Sj, and others). The cross sections make these relationships clearer and add some detail. [U.S. Geological Survey.]

Figure A6.2 Satellite photo of stream drainage in the folded mountain belt of Al Ghaydah, Yemen. This image shows clear examples of trellis drainage (*center*) and dendritic drainage (*above*).

Geologic mapping proceeds by the following steps:

1. *Principal observations.* Description of outcrop location, lithology, age, fossil content, and structural attitude as measured by dip and strike, direction of fault movement, fold axes, and so forth (see Chapter 10). Plotting observations on work map.
2. *First integration.* Conceptualizing the spatial relationship of one outcrop to another by stratigraphic correlation of rocks of the same age, facies, degree of metamorphism and deformation. Grouping of mappable rock units into formations. Drawing of lines on the primitive geologic map of inferred connections where formations are hidden. Compilation of the complete or composite stratigraphic sequences and ages of deformational or igneous intrusive events.
3. *Synthesizing the map.* Visualizing the larger pattern of geologic relationships and constructing the map, together with geologic cross sections made both to help geologists in their thinking and to illustrate more detail and inference from the map.

Analyses of rock composition, absolute age, and seismic, gravity, and magnetic data may be incorporated into the map. The geologist further draws on the geologic literature or personal experience of the geology of nearby and similar kinds of regions. The finished geologic map is a codified mass of information from which anyone familiar with geology can quickly read the nature of the Earth's crust in the area and a good deal of its geologic history.

In the last few decades much of traditional geologic mapping has been supplemented or even entirely supplanted by remote sensing by aerial and satellite photography or geophysical instruments. Inaccessible areas, such as those in some polar or desert regions, may be mapped almost entirely by this method, with the geologist ground-checking in scattered places. The mapping of the Moon is an extreme example of this approach. Mars is being mapped with no ground check at all except for the area immediately surrounding the *Viking* landing site.

Aerial images are now obtained by aircraft aerial photography, manned and unmanned satellites, and the space shuttle. These images may be in ordinary color, or they may be taken in infrared or other parts of the light spectrum, or they may be false-color radar images. Digital processing systems allow the information gathered by remote sensing devices to be processed into specialized maps that show geologic structures, erosional patterns, bodies of water, and the distribution and types of vegetation and urbanization. Figure A6.2, for example, is a satellite image that provides a map of stream drainage in a mountainous area of Yemen.

There are numerous environmental applications of remote mapping techniques. Aerial and satellite images can track changes in coastal zones, including alteration of sandy beaches, which can be a guide to shoreline engineering. Satellite mapping of the sea surface can be used for imaging the topography of the seafloor, and details of seafloor topography can be mapped by side-scanning radar (see Chapter 17). Satellites can also be used to monitor landslides, floods, volcanic eruptions, and strain changes that could signal an earthquake (see Chapter 18) and to assess the damages from these natural hazards.

There are many kinds of geologic maps. The most common shows the bedrock geology and gives a picture of what the land would look like if all soil were stripped away. Surficial geologic maps, on the other hand, emphasize the nature of soils, unconsolidated river sediment, sand dunes, and whatever other materials, including outcrops, appear at the surface. A special kind of surficial geologic map is used for environmental hazards. One kind shows areas of high-angle or unsupported slopes that are likely to slump or slide (see Chapter 12). Whatever the geologic purpose, there is a map that can be made to show the relevant data. There is no question that the map is both the best device for geological research into the origin of the distribution of important geologic characteristics over the Earth and the best way to illustrate the patterns discovered from such research.

Glossary

Words in *italic* have separate entries in the Glossary. Specific minerals are defined in Appendix 3.

ablation The total amount of ice that a *glacier* loses each year.

abrasion The erosive action that occurs when suspended and saltating *sediment* particles move along the bottom and sides of a *stream channel.*

abyssal hill A hill on the slope of a mid-ocean ridge, typically 100 m or so high and lineated parallel to the ridge crest, formed primarily by normal *faulting* of the basaltic oceanic *crust* as it moves out of the *rift valley.*

abyssal plain A wide, flat plain that covers large areas of the ocean floor at depths of about 4000 to 6000 m.

accreted terrane An individual, geologically coherent fragment of an amalgamation of odd pieces of *crust—island arcs, seamounts;* remnants of thickened oceanic *plateaus;* old mountain ranges; and other slivers of continental crust—that were plastered onto the leading edge of a continent as it moved across Earth's surface.

accretion A process of continental growth in which buoyant fragments of *crust* are attached (accreted) to continents during plate motions.

acid rain Acidic precipitation caused by the pollution of air with sulfur gases emitted by the smokestacks of power plants that burn *coal* containing large amounts of the *mineral* pyrite, from smelters of sulfide *ores,* and from some factories; and with nitrogen dioxide gases emitted by smokestacks and automobile exhausts. These gases react with water to form sulfuric acid and nitric acid.

active margin A *continental margin* characterized by volcanic activity and frequent *earthquakes* and associated with *subduction* or *transform faulting.*

aftershock An *earthquake* that occurs as a consequence of a previous earthquake of larger magnitude.

A-horizon The uppermost layer of a *soil,* containing organic matter and *clay* and insoluble *minerals* such as quartz.

albedo The fraction of *solar energy* reflected by the surface of a planet or moon.

alluvial fan A cone- or fan-shaped accumulation of *sediment* deposited where a *stream* widens abruptly as it leaves a mountain front for an open *valley.*

amphibolite A usually nonfoliated *metamorphic rock* made up mainly of amphibole and plagioclase feldspar. Foliated amphibolites can be produced by deformation.

andesite A *volcanic rock* type intermediate in composition between *rhyolite* and *basalt.* The extrusive equivalent of *diorite.*

angle of repose The maximum angle at which a slope of loose material will lie without cascading down.

angular unconformity An *unconformity* in which the upper layers overlie lower beds that have been folded by tectonic processes and then eroded to a more or less even plane.

anion A negatively charged *ion.*

antecedent stream A *stream* that existed before the present *topography* was created and so maintained its original course despite changes in the structure of the underlying rocks and in topography. (Compare *superposed stream.*)

anticline A large upfold or arch of layered rocks. (Compare *syncline.*)

aquiclude A relatively impermeable bed that bounds an *aquifer* above or below and acts as a barrier to the flow of *groundwater.*

aquifer A bed that stores and transmits *groundwater* in sufficient quantity to supply wells.

Archean cratonic nucleus A stable landmass comprising the oldest rocks, around which younger parts of the *crust* have been accreted and between which they have been deformed.

arête The sharp, jagged crest along the *divide* between *cirques,* resulting from the headward *erosion* of the walls of adjoining cirques.

artesian flow A water flow produced by the greater pressure of *groundwater* in a *confined aquifer* than in an *unconfined aquifer*.

artesian well A well in a *confined aquifer* drilled at a point where the *elevation* of the ground surface is lower than that of the *groundwater table* in the *recharge* area, which causes the water to flow out of the well spontaneously.

asthenosphere The weak layer of soft but solid rock comprising the lower part of the upper *mantle* (below the *lithosphere*). Movement in the asthenosphere occurs by plastic deformation. (From the Greek *asthenes,* meaning "weak.")

atom The smallest unit of an element that retains the element's physical and chemical properties.

atomic mass The sum of the masses of an element's *protons* and *neutrons.*

atomic number The number of *protons* in the *nucleus* of an *atom.*

axial plane An imaginary surface that divides a fold as symmetrically as possible, with one limb on either side of the plane.

badland A deeply gullied *topography* resulting from the fast *erosion* of easily erodible *shales* and *clays.*

barchan A crescent-shaped *eolian dune* produced by limited *sand* supply and unidirectional winds. The horns of the crescent point downwind.

barrier island A long offshore sandbar that builds up to form a barricade between open ocean waves and the main shoreline.

basal slip The sliding of a *glacier* along its base. (Compare *plastic flow.*)

basalt A fine-grained, dark, *mafic igneous rock* composed largely of plagioclase feldspar and pyroxene. The extrusive equivalent of *gabbro.*

base level The *elevation* at which a *stream* ends by entering a large standing body of water, such as a lake or ocean.

basin A bowl-shaped depression of rock layers in which the beds *dip* radially toward a central point.

batholith A great irregular mass of coarse-grained *igneous rock* that covers at least 100 km^2; the largest *pluton.*

bauxite An *ore* composed of aluminum hydroxide that forms when *clay minerals* derived from weathered silicates continue to weather until they have lost all their silica and *ions* other than aluminum. The major source of aluminum metal.

bed load The material a *stream* carries along its bed by sliding and rolling. (Compare *suspended load.*)

bedding The formation of parallel layers of *sediment* as particles settle to the bottom of the sea, a *river*, or a land surface.

bedding sequence A pattern of interbedded and vertically stacked layers of *sandstone, shale,* and other *sedimentary rock* types.

bedrock The solid underlying rock beneath loose surface materials, such as *soil.*

B-horizon The intermediate layer in a *soil,* below the *A-horizon* and above the *C-horizon,* consisting of solid *minerals* and iron oxides and containing little organic matter.

biochemical sediment The undissolved *mineral* remains of organisms as well as minerals *precipitated* by biological processes.

bioclastic sediment A shallow-water *sediment* consisting predominantly of two calcium carbonate *minerals*—calcite and aragonite—in variable proportions.

biosphere The sum total of all living organisms and the organic matter they produce.

bioturbation The process by which organisms rework existing *sediments* by burrowing through *muds* and *sands.*

blowout dune A crescent-shaped *eolian dune,* typically one blown back from a beach, that has its convex *slip face* oriented downwind.

blueschist A *metamorphic rock* formed under conditions of high pressure and moderate temperature, often containing glaucophane, a blue amphibole.

bolide A large meteoroid that impacts Earth.

bottomset bed A thin, horizontal bed of *mud* deposited seaward of a *delta* and then buried by continued delta growth.

braided stream A *stream* whose *channel* divides into an interlacing network of channels, which then rejoin in a pattern resembling braids of hair.

brittle material A material that undergoes little change under increasing force, until it breaks suddenly. (Compare *ductile material.*)

building code Standards for the design and construction of new buildings that specify the level of shaking a structure must be able to withstand in an *earthquake,* based on the maximum intensity expected from the *seismic hazard.*

calcium cycle A geochemical cycle that includes all fluxes of calcium into and out of *reservoirs.*

caldera A large, steep-walled, *basin*-shaped depression formed after a violent eruption in which large volumes of *magma* are discharged, when the overlying volcanic structure collapses catastrophically through the roof of the emptied *magma chamber.*

capacity The total *sediment* load carried by a flow. (Compare *competence.*)

carbon cycle The *geosystem* that describes the continual movement of carbon between the atmosphere and its other principal *reservoirs*—the *lithosphere, hydrosphere,* and *biosphere.*

carbonate environment A marine setting where calcium carbonate, principally of biochemical origin, is the main *sediment.*

carbonate platform A shallow, extensive flat area where both biological and nonbiological carbonates are deposited.

carbonate rock A *sedimentary rock* formed from the accumulation of carbonate *minerals precipitated* organically or inorganically.

carbonate sediment A *sediment* formed from the accumulation of carbonate *minerals precipitated* organically or inorganically.

cation A positively charged *ion.*

cementation A major chemical diagenetic change in which *minerals* are *precipitated* in the pores of *sediments,* forming cements that bind *clastic sediments* and rocks.

channel The trough through which the water in a *stream* flows.

chemical and biochemical sediments New chemical substances that form by precipitation when some of a rock's components dissolve during *weathering* and are carried in *river* waters to the sea.

chemical reaction The interaction of the *atoms* of two or more chemical elements in certain fixed proportions that produces a new chemical substance.

chemical sediment The dissolved products of *weathering precipitated* from water (usually seawater) by *chemical reactions* and formed at or near their place of deposition.

chemical stability A measure of a substance's tendency to remain in a given chemical form rather than to react spontaneously to become a different chemical substance.

chemical weathering The *weathering* that occurs when the *minerals* in a rock are chemically altered or dissolved.

chert A *sedimentary rock* made up of chemically or biochemically *precipitated* silica.

C-horizon The lowest layer of a *soil,* consisting of slightly altered *bedrock,* broken and decayed, mixed with *clay* from *chemical weathering.*

cirque An amphitheater-like hollow, usually shaped like half of an inverted cone, formed at the head of a *glacial valley* by the plucking and tearing action of ice.

clastic particle A physically transported rock fragment produced by the *weathering* of preexisting rocks.

clastic sediment An accumulation of *clastic particles* laid down by running water, wind, or ice and forming layers of *sand, silt,* or *gravel.*

clay The most abundant component of fine-grained *sediments* and *sedimentary rocks,* consisting largely of clay *minerals.* Clay-sized particles are less than 0.0039 mm in diameter.

claystone A rock made up exclusively of *clay*-sized particles.

cleavage (1) The tendency of a *crystal* to break along flat planar surfaces. (2) The geometric pattern produced by such breakage.

climate model Any representation of the *climate system* that can reproduce one or more aspects of climate behavior.

climate system A *geosystem* that includes all parts of the *Earth system* and all the interactions among these components needed to describe how climate behaves in space and time.

coal A biochemically produced *sedimentary rock* composed almost entirely of organic carbon formed by the *diagenesis* of swamp vegetation.

cohesion An attractive force between particles of a solid material that are close together.

color A property of a *mineral* imparted by light—either transmitted through or reflected by *crystals,* irregular masses, or a *streak.*

compaction A decrease in the volume and *porosity* of a *sediment* that occurs when the grains are squeezed closer together by the weight of overlying sediment.

competence The ability of a flow to carry material of a given size. (Compare *capacity.*)

compressional wave A wave that propagates by a push-pull motion. (Compare *shear wave.*)

compressive force A force that squeezes together or shorten a body.

concordant intrusion An *intrusive igneous rock* whose boundaries lie parallel to layers of preexisting bedded rock. (Compare *discordant intrusion.*)

conduction The mechanical transfer of the vibrational energy of thermally agitated *atom*s and molecules, which constitutes heat energy, by the mechanism of atomic or molecular impact. (Compare *convection.*)

confined aquifer An *aquifer* that is bounded both above and below by *aquicludes.* (Compare *unconfined aquifer.*)

conglomerate A *sedimentary rock* composed of pebbles, cobbles, and boulders. The lithified equivalent of *gravel.*

consolidated material *Sediment* that is compacted and bound together by *mineral* cements. (Compare *unconsolidated material.*)

contact metamorphism Changes in the *mineralogy* and *texture* of rock resulting from the heat and pressure in a small area, such as the rocks near and in contact with an igneous intrusion.

continental drift The large-scale movement of continents over the globe. Evidence for continental drift comes from the jigsaw-puzzle fit of the coasts—and from the similarity of rocks, geologic structures, and *fossils*—on opposite sides of the Atlantic.

continental glacier An extremely slow moving, thick sheet of ice that covers a large part of a continent. (Compare *valley glacier.*)

continental margin The shoreline, shelf, and slope of a continent.

continental rise An apron of muddy and sandy *sediment* extending from the *continental slope* into the main ocean *basin.*

continental shelf A broad, flat, *sand-* and *mud-*covered platform that is a slightly submerged part of a continent and extends to the edge of the *continental slope.*

continental shelf deposit *Sediment* supplied from *erosion* of the land adjacent to a *thermal sag basin.*

continental slope A steep, *mud-*covered slope between the *continental shelf* and *continental rise.*

contour A line that connects points of equal *elevation* on a topographic map.

convection A mechanism of heat transfer in which a heated fluid expands and rises because it has become less dense than the surrounding material. Convection moves heat more efficiently than *conduction* because the heated material itself moves, carrying its heat with it. Colder material flows in to take the place of the hot rising fluid, is itself heated, and then rises to continue the cycle. (Compare *conduction.*)

core The central part of the Earth below a depth of 2900 km, comprising a liquid outer core and a solid inner core. The core is composed of iron and nickel, with minor amounts of some lighter element, such as oxygen or sulfur.

core-mantle boundary The boundary between the *crust* and underlying *mantle,* about 2890 km below the surface, where we encounter the most extreme change in properties found anywhere in Earth's interior.

country rock The rock surrounding an *intrusive igneous rock.*

covalent bond A bond between *atoms* in which the outer *electrons* are shared. (Compare *ionic bond.*)

craton A stable nucleus composed of the eroded remnants of ancient deformed rocks that comprises the continental *shields* and platforms.

cratonic keel A part of the *lithosphere* that extends into the convecting *asthenosphere* at 100 to 200 km beneath the oceans (as well as beneath most younger regions of the continents) like the hull of a boat into water.

creep The slow downhill *mass movement* of *soil* or other debris at a rate ranging from about 1 to 10 mm/year.

crevasse A large vertical crack in the surface of a *glacier* caused by the movement of brittle surface ice as it is dragged along by the *plastic flow* of the ice below.

cross-bedding A *sedimentary structure* comprising sets of bedded material deposited by currents of wind or water in the direction in which the bed slopes downward and is inclined at angles as large as 35° from the horizontal.

crust The thin (from about 7 to 70 km thick) outer layer of the Earth, consisting of relatively light materials that melt at low temperatures. Continental crust consists largely of *granite* and *granodiorite.* Oceanic crust is mostly *basalt.*

cryosphere The ice component of the *climate system,* comprising the polar ice caps, *glaciers,* and other surface ice.

crystal An ordered three-dimensional array of *atoms* in which the basic arrangement is repeated in all directions.

crystal habit The shape in which a *mineral's* individual *crystals* or aggregates of crystals grow.

crystallization The growth of a solid from a gas or liquid whose constituent *atoms* come together in the proper chemical proportions and crystalline arrangement.

cuesta An asymmetrical ridge in a tilted and eroded series of beds with alternating weak and strong resistance to *erosion.* One side has a long, gentle slope; the other is a steep cliff formed at the edge of the resistant bed where it is undercut by erosion of a weaker bed beneath.

dacite A light-colored, fine-grained *extrusive igneous rock* with the same general composition as *andesite.* The extrusive equivalent of *granodiorite.*

Darcy's law A summary of the relationships among the volume of water flowing through an *aquifer* in a certain time, the vertical drop of the flow, the flow distance, and the *permeability* of the aquifer.

debris avalanche A fast downhill *mass movement* of *soil* and rock that usually occurs in humid mountainous regions. The speed results from the combination of high water content and steep slopes.

debris flow A fluid *mass movement* of rock fragments supported by a muddy matrix. Debris flows differ from *earthflows* in that they generally contain coarser material and move faster than earthflows.

debris slide A *mass movement* of rock material and *soil* largely as one or more units along planes of weakness at the base of or within the rock material.

decompression melting Melting that occurs when *mantle* material rises to an area of lower pressure at a mid-ocean ridge. As the mantle material rises and the pressure decreases below a critical point, solid rocks melt spontaneously, without the introduction of any additional heat.

deflation The removal of dust, *silt,* and *sand* from dry *soil* by strong winds that gradually scoop out shallow depressions in the ground.

delta A depositional platform built of *sediments* deposited in an ocean or lake at the mouth of a *stream.*

dendritic drainage An irregular *stream drainage network* that resembles the limbs of a branching tree.

density The mass per unit volume of a substance, commonly expressed in grams per cubic centimeter (g/cm^3). (Compare *specific gravity.*)

depositional remanent magnetization A weak magnetization created in *sedimentary rocks* by the parallel alignment of magnetic particles in the direction of Earth's magnetic field as they settle.

desert pavement A remanent ground surface of *gravel* too large for the wind to transport, left when continued *deflation* removes the finer-grained particles from a mixture of *gravel, sand,* and *silt* in *sediments* and *soils.*

desert varnish A distinctive dark brown, sometimes shiny, coating found on many rock surfaces in the desert that is a mixture of *clay minerals* with smaller amounts of manganese and iron oxides. Desert varnish is hypothesized to form very slowly from a combination of dew, *chemical weathering,* and the adhesion of windblown dust to exposed rock surfaces.

desertification Changes in a region's climate that transform semiarid lands into deserts.

diagenesis The physical and chemical changes—including pressure, heat, and *chemical reactions*—by which buried *sediments* are lithified into *sedimentary rocks.*

diatreme A volcanic vent formed by the explosive escape of gases and often filled with breccia.

differentiation The transformation of random chunks of primordial matter into a body whose interior is divided into concentric layers that differ both physically and chemically.

dike A tabular igneous intrusion that cuts across layers of *bedding* in *country rock.*

diorite A coarse-grained *intrusive igneous rock* with composition intermediate between *granite* and *gabbro.* The intrusive equivalent of *andesite.*

dip The amount of tilting of a rock layer; the angle at which the bed inclines from the horizontal.

dipole Two oppositely polarized magnetic poles.

discharge (1) The volume of *groundwater* leaving an *aquifer* in a given time. (Compare *recharge.*) (2) The volume of water that passes a given point in a given time as it flows through a *channel* of a certain width and depth.

disconformity An *unconformity* in which the upper set of layers overlies an erosional surface developed on an undeformed, still-horizontal lower set of beds.

discordant intrusion An *intrusive igneous rock* that cuts across the layers of the *country rock* that it intrudes. (Compare *concordant intrusion.*)

distributary A smaller *stream* that receives water and *sediment* from the main *channel,* branches off downstream, and thus distributes the water and sediment into many channels. (Compare *tributary.*)

divide A ridge of high ground along which all rainfall runs off down one side of the rise or the other.

dolostone An abundant *carbonate rock* composed primarily of dolomite and formed by the *diagenesis* of *carbonate sediments* and *limestones.*

dome A broad circular or oval upward bulge of rock layers.

drainage basin An area of land, bounded by *divides,* that funnels all its water into the network of *streams* draining the area.

drainage network The pattern of connections of tributaries, large and small, of a *stream* system.

drift All material of glacial origin found anywhere on land or at sea.

drumlin A large streamlined hill of *till* and *bedrock* that parallels the direction of ice movement in a *continental glacier* terrain.

dry wash A desert *valley* that carries water only briefly after a rain. Called a *wadi* in the Middle East.

ductile material A material that undergoes smooth and continuous plastic deformation under increasing force and does not spring back to its original shape when the deforming force is released. (Compare *brittle material.*)

dune An elongated ridge of *sand* formed by wind or water.

Earth system All the parts of our planet and all their interactions, taken together.

earthflow A fluid *mass movement* of relatively fine grained materials, such as *soils,* weathered *shales,* and *clay,* at speeds of up to a few kilometers per hour.

earthquake The violent motion of the ground that occurs when rocks being stressed suddenly break along a new or preexisting *fault.*

eclogite A *metamorphic rock* formed under very high pressure and moderate to high temperature, typically containing *minerals* such as coesite (a very dense, high-pressure form of quartz).

effluent stream A *stream* or portion of a stream that receives some water from *groundwater discharge* because the stream's *elevation* is below the *groundwater table.* (Compare *influent stream.*)

El Niño An anomalous warming of the eastern Pacific Ocean that occurs every 3 to 7 years and lasts for a year or so. El Niño events can be calamitous for local ecosystems, which depend on upwellings of cold water for their nutrient supply.

elastic rebound theory A theory of *fault* movement and *earthquake* generation holding that faults remain locked while strain energy accumulates in the rock *formations* on both sides, temporarily deforming them until a sudden *slip* along the fault releases the energy.

electron A negatively charged particle that moves around the *nucleus* of an *atom.*

electron sharing The mechanism by which a *covalent bond* is formed between the elements in a *chemical reaction.*

electron transfer The mechanism by which an *ionic bond* is formed between the elements in a *chemical reaction.*

elevation The altitude, or vertical distance, above or below sea level.

enhanced greenhouse effect Global warming during the twentieth century caused by the human-induced increase of atmospheric carbon dioxide (CO_2).

eolian Pertaining to geological processes powered by the wind.

eon The largest division of *geologic time,* embracing several *eras;* for example, the Phanerozoic eon, from 543 million years ago to the present.

epeirogeny The gradual downward and upward movements of broad regions of *crust* without significant *folding* or *faulting.*

epicenter The geographic point on Earth's surface directly above the *focus* of an *earthquake.*

epoch A subdivision of a geologic *period,* often chosen to correspond to a stratigraphic sequence. Also used for a division of time corresponding to a paleomagnetic interval.

era A division of *geologic time* including several *periods,* but smaller than an *eon.* Commonly recognized eras are the Paleozoic, Mesozoic, and Cenozoic.

erosion The set of processes that loosen *soil* and rock and move them downhill or downstream, where they are deposited as layers of *sediment.*

erratic A large boulder, often found in *till,* that was carried by a *glacier* away from the *outcrop* from which it was derived, often into an area underlain by a rock type different from that of the boulder.

esker A long, narrow, winding ridge of *sand* and *gravel* found in the middle of a ground *moraine.*

evaporite rock A *sedimentary rock* formed from *evaporite sediments.*

evaporite sediment An accumulation of materials *precipitated* inorganically from evaporating seawater and from water in arid-region lakes that have no *river* outlets.

exclusive economic zone The zone from a country's coast to 200 nautical miles offshore in which the country has exclusive rights to *mineral* deposits.

exfoliation A *physical weathering* process in which large flat or curved sheets of rock fracture and are detached from an *outcrop.*

exhumation A process in which subducted *metamorphic rocks* rise to the surface because of buoyancy and circulation in the *subduction* zone.

extrusive igneous rock A fine-grained or glassy *igneous rock* formed from a rapidly cooled *magma* that erupts at the surface through a *volcano.*

fault A fracture with relative movement of the rocks on both sides of it, parallel to the fracture.

fault mechanism The type of *fault* rupture (normal, thrust, or strike-slip) that produced an *earthquake;* it is determined by the orientation of the fault rupture and the direction of *slip.*

faulting The processes by which crustal forces cause a rock *formation* to break and *slip* along a *fault.*

felsic rock A light-colored *igneous rock* that is poor in iron and magnesium and rich in high-silica *minerals* such as quartz, orthoclase feldspar, and plagioclase feldspar.

ferric iron (Fe^{3+}) Iron in which the *atoms* have lost three of the *electrons* they have in metallic form. All iron oxides formed at Earth's surface are ferric.

ferrous iron (Fe^{2+}) Iron in which the *atoms* have lost two of the *electrons* they have in metallic form. Found in silicate *minerals* such as pyroxene.

firn Old, dense, compacted granular snow.

fissure eruption A volcanic eruption emanating from an elongate fissure rather than a central vent.

fjord A former *glacial valley* with steep walls and a U-shaped profile, now occupied by the sea.

flexural basin A *basin* that develops within zones of tectonic convergence, where one lithospheric plate pushes up over the other and the weight of the overriding plate causes the overridden plate to bend or flex downward.

flood basalt An immense basaltic lava *plateau* extending many kilometers in flat, layered flows originating from *fissure eruptions.*

floodplain A flat area about level with the top of a *channel* that lies on either side of the channel. It is the part of the *valley* that is flooded when a *stream* spills over its banks, carrying with it *silt* and *sand* from the main channel.

fluid-induced melting Melting that takes place when water-laden *sediments* on a subducting oceanic plate are carried downward into the *subduction* zone. The increase in pressure squeezes water out of the *minerals* in the outer layers of the descending slab. The water rises buoyantly into the *mantle* wedge above the slab. Because water lowers the melting temperature of rock, it induces melting in the mantle wedge.

focus The point along a *fault* at which *slip* initiates in an *earthquake.*

folding The processes by which crustal forces deform an area of *crust* so that layers of rock are pushed into folds.

foliated rock A *metamorphic rock* that displays *foliation.* Foliated rocks include *slate, phyllite, schist,* and *gneiss.*

foliation A set of flat or wavy parallel planes produced by deformation.

foraminifera A group of tiny single-celled organisms that live in surface waters and whose secretions and calcite shells account for most of the ocean's *carbonate sediments.*

foraminiferal ooze A sandy and silty *sediment* composed of the shells of dead foraminifers.

foreset bed A gently inclined deposit, resembling a large-scale *cross-bed,* of fine-grained *sand* and *silt* on the outer front of a *delta.*

foreshock A small *earthquake* that occurs in the vicinity of, but before, a main shock.

formation A series of rock layers in a region that has about the same physical properties and may contain the same assemblage of *fossils.*

fossil Trace of an organism of past geologic ages that has been preserved in the *crust.*

fossil fuel An energy *resource* derived from natural organic materials, from *coal* to *oil* and natural *gas.*

fractional crystallization The process by which the *crystals* formed in a cooling *magma* are segregated from the remaining liquid at progressively lower temperatures.

fracture The tendency of a *crystal* to break along irregular surfaces other than *cleavage* planes.

frost wedging A *physical weathering* process in which the expansion of freezing water in a crack breaks a rock.

gabbro A dark gray, coarse-grained *intrusive igneous rock* with an abundance of *mafic minerals,* particularly pyroxene. The intrusive equivalent of *basalt.*

gas A fluid *organic sediment* formed by the *diagenesis* of organic material in the pores of *sedimentary rocks,* mainly *sandstones* and *limestones.*

geochemical reservoir One of the components of the *Earth system*—atmosphere, *hydrosphere, lithosphere,* and *biosphere*—that is a *reservoir* for holding terrestrial chemicals.

geodynamo The mechanism that sustains Earth's magnetic field, essentially driven by *convection* in the outer *core.*

geologic time (1) A timeline based on a *stratigraphic succession* that provides a chronological record of the history of a region. (2) The entire span of time since the Earth formed.

geomorphology The study of landscapes and their evolution.

geosystem A specialized subsystem of the Earth that encompasses interesting types of terrestrial behavior. Geosystems include the *plate tectonic system,* the *climate system,* the *geodynamo,* and other smaller subsystems. The *Earth system* can be thought of as the collection of all these open, interacting (and often overlapping) geosystems.

geotherm The curve that describes how temperature increases with depth in the Earth.

geothermal energy Energy produced when underground water is heated as it passes through a subsurface region of hot rocks.

glacier A large mass of ice on land that shows evidence of being in motion or of once having moved. (See also *continental glacier; valley glacier.*)

global change A change in climate that has worldwide effects on the *biosphere,* atmosphere, and other components of the *Earth system.*

gneiss A light-colored, poorly *foliated rock* with coarse bands of segregated light and dark *minerals* throughout, produced by high-pressure, high-temperature metamorphism.

graded bedding A bed that formed horizontal or nearly horizontal layers at the time of deposition, in which the coarsest particles are concentrated at the bottom and grade gradually upward into fine *silt.*

graded stream A *stream* in which the slope, velocity, and *discharge* combine to transport its *sediment* load, with neither sedimentation nor *erosion.*

granite A *felsic,* coarse-grained, *intrusive igneous rock* composed of quartz, orthoclase feldspar, sodium-rich plagioclase feldspar, and micas. The intrusive equivalent of *rhyolite.*

granoblastic rock A nonfoliated *metamorphic rock* composed mainly of *crystals* that grow in equant shapes, such as cubes and spheres, rather than in platy or elongate shapes. Granoblastic rocks include *hornfels, quartzite, marble, greenstone, amphibolite,* and *granulite.*

granodiorite A light-colored, coarse-grained *intrusive igneous rock* similar to *granite* in having abundant quartz, but its predominant feldspar is plagioclase, not orthoclase. The intrusive equivalent of *dacite.*

granulite A medium- to coarse-grained *granoblastic rock* formed under conditions of relatively high pressure and temperature.

gravel The coarsest *clastic sediment,* consisting of particles larger than 2 mm in diameter and including pebbles, cobbles, and boulders.

greenhouse effect A global warming effect in which a planet with an atmosphere containing greenhouse gases radiates *solar energy* back into space less efficiently than a planet with no atmosphere. The net effect is to trap heat within the atmosphere by increasing the temperature of the surface relative to the temperature at higher levels of the atmosphere.

greenschist A *schist* containing chlorite and epidote (which are green) and formed by low-pressure, low-temperature metamorphism of *mafic volcanic rocks.*

greenstone A *granoblastic rock* produced by the *low-grade metamorphism* of *mafic volcanic rock.* Chlorite accounts for the greenish cast.

groundwater The mass of water stored beneath Earth's surface.

groundwater table The boundary between the *unsaturated zone* and the *saturated zone.*

guyot A large, flat-topped *seamount* resulting from *erosion* of an island *volcano* when it was above sea level.

half-life The time required for one-half of the original number of radioactive *atoms* in an element to decay.

hanging valley A *valley* left by a melted *tributary glacier* that enters a larger *glacial valley* above its base, high up on the valley wall.

hardness A measure of the ease with which the surface of a *mineral* can be scratched.

hematite The principal iron *ore;* the most abundant iron oxide at Earth's surface.

hogback A *formation* similar to a *cuesta* with steep, narrow, more or less symmetrical ridges.

Holocene epoch A *geologic time* interval consisting of the past 10,000 years.

hornfels A *granoblastic rock* of uniform grain size that has undergone little or no deformation. Usually formed by *contact metamorphism* at high temperatures.

hot spot A volcanic center found at the beginning of progressively older aseismic ridges or within a continent far from a plate boundary. Hypothesized to be the surface expression of a *mantle plume.*

humus The remains and waste products of the many plants, animals, and bacteria living in a *soil.*

hydration The absorption of water by a *mineral,* usually in *weathering.*

hydraulic gradient The ratio between the difference in *elevation* of the *groundwater table* at two points and the flow distance that the water travels between those two points.

hydroelectric energy A form of *solar energy* derived from water that falls by the force of gravity and is made to drive electric turbines.

hydrologic cycle The cyclical movement of water from the ocean to the atmosphere by evaporation, to the surface through rain, to *streams* through *runoff* and *groundwater,* and back to the ocean.

hydrology The science that studies the movements and characteristics of water on and under Earth's surface.

hydrosphere Earth's surface waters, including all oceans, lakes, *rivers,* and *groundwaters.*

hydrothermal activity The circulation of water through hot *volcanic rocks* and *magmas,* producing hot springs and geysers on the surface.

hydrothermal vein A *vein* filled with *minerals* that contain large amounts of chemically bound water and are known to crystallize from hot-water solutions.

hydrothermal water Hot water deep in the *crust.*

ice stream A current of ice in a *continental glacier* that flows faster than the surrounding ice.

igneous rock A rock formed by the solidification of a *magma,* before or after it reaches the surface.

infiltration The movement of *groundwater* or *hydrothermal water* into rock or *soil* through pores and *joints.*

influent stream A *stream* or portion of a stream that *recharges groundwater* through the stream bottom because its *elevation* is above the *groundwater table.* (Compare *effluent stream.*)

intensity scale A scale for estimating the intensity of *earthquake* shaking directly from an event's destructive effects.

intermediate igneous rock An *igneous rock* midway in composition between mafic and felsic, neither as rich in silica as the *felsic rocks* nor as poor in it as the *mafic rocks.*

intrusive igneous rock A coarse-grained *igneous rock* that crystallized slowly when *magma* intruded into *country rock* deep in Earth's *crust.*

ion An *atom* or group of atoms that has gained or lost *electrons* and so has a net positive or negative electrical charge.

ionic bond A bond between *atoms* formed by electrical attraction between *ions* of the opposite charge. (Compare *covalent bond.*)

iron formation A *sedimentary rock* that usually contains more than 15 percent iron in the form of iron oxides and some iron silicates and iron carbonates.

island arc A linear or arc-shaped chain of volcanic islands formed on the seafloor at a convergent plate boundary. The islands are formed in the overriding plate from rising melt derived from the subducted plate and from the *asthenosphere* above that plate.

isochron A *contour* that connects points of equal age on the world's ocean floors as determined by magnetic reversal data and *fossils* from deep-sea drilling.

isotopes *Atoms* of the same *atomic number* that have differing *atomic masses.*

joint A crack in a rock along which there has been no appreciable movement.

kame A small hill of *sand* and *gravel* dumped at or near the edge of a retreating *glacier* by meltwater *streams.*

kaolinite The white to cream-colored *clay* produced by the *weathering* of feldspar.

karst topography An irregular hilly terrain characterized by *sinkholes,* caverns, and a lack of surface *streams;* formed in regions of high rainfall with extensively jointed *limestone formations* and an appreciable *hydraulic gradient.*

kettle A hollow or undrained depression that often has steep sides and may be occupied by ponds or lakes; formed in glacial deposits when *outwash* was deposited around a residual block of ice that later melted.

lahar A torrential *mudflow* of wet volcanic debris produced when pyroclastic or lava deposits mix with rain or the water of a lake, *river,* or melting *glacier.*

laminar flow A flow in which straight or gently curved streamlines run parallel to one another without mixing or crossing between layers. (Compare *turbulent flow.*)

landform A characteristic landscape feature on Earth's surface that attained its shape through the processes of *erosion* and sedimentation; for example, a hill or a *valley.*

large igneous province (LIP) A voluminous emplacement of predominantly *mafic extrusive* and *intrusive igneous rock* whose origins lie in processes other than "normal" *seafloor spreading.* LIPs include continental *flood basalts,* ocean *basin* flood basalts, and aseismic ridges.

laterite A deep-red *soil* produced by fast, intense *weathering* in warm, humid climates, in which feldspar and other silicates have been completely altered, leaving behind mostly aluminum and iron oxides and hydroxides.

limestone A biochemical *sedimentary rock* lithified from *carbonate sediments* and composed mainly of calcium carbonate in the form of the *mineral* calcite.

linear dune A long *eolian dune* whose orientation is parallel to the direction of the prevailing wind. Most areas covered by linear dunes have a moderate *sand* supply, a rough pavement, and winds that may shift but are always in the same general direction.

liquefaction The temporary transformation of solid material to a fluid state caused by the pressure of water in the pores of the material.

lithification The process that converts *sediments* into solid rock by *compaction* or *cementation.*

lithosphere The strong, rigid outer shell of the Earth that encases the *asthenosphere* and contains the *crust,* the uppermost part of the *mantle* down to an average depth of about 100 km, the continents, and the plates.

loess A blanket of unstratified, wind-deposited, fine-grained *sediment* rich in *clay minerals.*

longitudinal profile The smooth, concave-upward curve that represents a cross-sectional view of a *stream,* from notably steep near its head to almost level, near its mouth.

longshore current A shallow-water current that is parallel to the shore.

lower mantle A relatively homogeneous region, beginning nearly 660 km below the surface and more than 2000 km thick, where *S-wave* speed abruptly increases, indicating a major *phase change* to a more closely packed *crystal* structure.

low-grade (burial) metamorphism Metamorphism in which buried *sedimentary rocks* are altered by the progressive increase in pressure exerted by overlying *sediments* and sedimentary rocks and by the increase in heat associated with increased depth of burial in the Earth.

low-velocity zone An area near the base of the *lithosphere* where the *S-wave* speed abruptly decreases.

luster The way in which the surface of a *mineral* reflects light to produce the shine of its surface, described by such subjective terms as dull, glassy, or metallic. (See Table 3.3.)

mafic rock A dark-colored *igneous rock* containing *minerals* (such as pyroxenes and olivines) rich in iron and magnesium and relatively poor in silica.

magma A mass of melted rock that originates deep in the *crust* or upper *mantle.*

magma chamber A *magma*-filled cavity in the *lithosphere* that forms as ascending drops of melted rock push aside surrounding solid rock.

magmatic differentiation A process by which rocks of varying composition can arise from a uniform parent *magma.* Various *minerals* crystallize at different temperatures, and the composition of the magma changes as it is depleted of the chemical elements withdrawn to make the crystallized minerals.

magnetic anomaly A pattern of long, narrow bands of high and low magnetic fields on the seafloor that are parallel to and almost perfectly symmetrical with respect to the crest of a mid-ocean ridge. The detection of such patterns was one of the great discoveries that confirmed *seafloor spreading* and led to *plate tectonics* theory.

magnetic stratigraphy The time sequence of reversals of Earth's magnetic field as indicated in the *fossil* magnetic record of layered lava flows.

magnetic time scale The detailed history of Earth's magnetic-field reversals going back into *geologic time,* as determined by measuring the *thermoremanent magnetization* of rock samples.

magnitude scale A scale for estimating the size of an *earthquake* using the logarithm of the largest ground motion registered by a *seismograph* (Richter magnitude) or the logarithm of the area of the *fault* rupture (moment magnitude).

mantle The region that forms the main bulk of the solid Earth, between the *crust* and the *core,* ranging from depths of about 40 km to 2900 km. It is composed of rocks of intermediate *density,* mostly compounds of oxygen with magnesium, iron, and silicon.

mantle plume A narrow, cylindrical jet of hot, solid material rising from deep within the *mantle* and thought to be responsible for intraplate *volcanism.*

marble A *granoblastic rock* produced by the metamorphism of *limestone* or *dolostone.*

mass movement A downhill movement of masses of *soil* or rock under the force of gravity.

mass wasting All the processes by which masses of rock and *soil* move downhill under the influence of gravity.

meander A curve or bend in a *stream* that develops as the stream erodes the outer bank of a bend and deposits *sediment* against the inner bank. Meanders are usual in streams flowing on low slopes in plains or lowlands, where *channels* typically cut through *unconsolidated* sediments or easily eroded *bedrock.*

mélange A heterogeneous mixture of rock materials produced by high-pressure, low-temperature metamorphism, found where a plate carrying a continent on its leading edge converges with a subducting oceanic plate.

mesa A small, flat elevation with steep slopes on all sides created by differential *weathering* of *bedrock* of varying hardness.

metallic bond A *covalent bond* in which freely mobile *electrons* are shared and dispersed among *ions* of metallic *elements,* which have the tendency to lose electrons and pack together as *cations.*

metamorphic facies Groupings of *metamorphic rocks* of various *mineral* compositions formed under different grades of metamorphism from different parent rocks.

metamorphic P–T path The history of the changing conditions of pressure and temperature that occurred during metamorphism. The P–T path can be a sensitive recorder of many important factors that influence metamorphism and thus may provide insight into the plate tectonic settings responsible for metamorphism.

metamorphic rock A rock formed by the transformation of preexisting solid rocks under the influence of high pressure and temperature.

metasomatism A change in a rock's bulk chemical composition by fluid transport of chemical components into or out of the rock.

meteoric water Rain, snow, or other forms of water derived from the atmosphere.

migmatite A mixture of *igneous* and *metamorphic rock* produced by incomplete melting. Migmatites are badly deformed and contorted, and they are penetrated by many *veins,* small pods, and lenses of melted rock.

mineral A naturally occurring, solid crystalline substance, generally inorganic, with a specific chemical composition.

mineralogy (1) The branch of geology that studies the composition, structure, appearance, stability, occurrence, and associations of *minerals.* (2) The relative proportions of a rock's constituent minerals.

Mohorovičić discontinuity (Moho) The boundary between the *crust* and the *mantle,* at a depth of 5 to 45 km, marked by an abrupt increase in *P-wave* velocity to more than 8 km/s.

Mohs scale of hardness An empirical, ascending scale of mineral *hardness* based on the ability of one *mineral* to scratch another. (See Table 3.2.)

moraine An accumulation of rocky, sandy, and clayey material carried by glacial ice or deposited as *till.*

mud A *clastic sediment,* mixed with water, in which most of the particles are less than 0.062 mm in diameter.

mudflow A fluid *mass movement* of material mostly finer than *sand,* along with some rock debris, containing large amounts of water. The water tends to make mudflows move faster than *earthflows* or *debris flows.*

mudstone A blocky, poorly bedded, fine-grained *sedimentary rock* produced by the *lithification* of *mud.*

natural levee A ridge of coarse material that confines a *stream* within its banks between floods, even when water levels are high.

nebular hypothesis The idea that the solar system originated from a diffuse, slowly rotating cloud of gas and fine dust (a "nebula") that contracted under the force of gravity and eventually evolved into the Sun and planets.

negative feedback A process in which one action produces an effect (the feedback) that tends to slow the original action and stabilize the process at a lower rate. (Compare *positive feedback.*)

neutron An electrically neutral elementary particle in the *nucleus* of an *atom,* having an *atomic mass* of 1.

nonconformity An *unconformity* in which the upper beds overlie *metamorphic* or *igneous rock.*

nuclear energy Energy produced by the fission of uranium (^{235}U), which releases heat to make steam, which then drives turbines to create electricity.

nucleus The center of an *atom,* comprising *protons* and *neutrons* and containing virtually all the mass of the atom.

obsidian A dense, dark, glassy *volcanic rock,* usually of *felsic* composition.

oil A fluid *organic sediment* formed by the *diagenesis* of organic material in the pores of *sedimentary rocks,* mainly *sandstones* and *limestones.*

oil trap An impermeable barrier to upward migration of *oil* or *gas* that allows it to collect beneath the barrier.

ophiolite suite An unusual assemblage of rocks, characteristic of the seafloor but found on land, consisting of deep-sea *sediments,* submarine basaltic lavas, and *mafic* igneous intrusions. The assemblage comprises fragments of oceanic *crust* that were transported by *seafloor spreading* and then raised above sea level and thrust onto a continent in a later episode of plate collision.

ore A *mineral* deposit from which valuable metals can be recovered profitably.

organic sedimentary rock A *sedimentary rock* that consists entirely or partly of organic carbon–rich deposits formed by the decay of once-living material that has been buried.

original horizontality, principle of The principle that *sediments* are deposited as essentially horizontal beds.

orogen An elongated mountain belt arrayed around a continental *craton* and formed by a later episode of compressive deformation.

orogeny Mountain building, particularly by the *folding* and thrusting of rock layers, often with accompanying magmatic activity.

outcrop A segment of *bedrock* exposed to the atmosphere.

outwash *Drift* that has been caught up and modified, sorted, and distributed by meltwater *streams.*

overturned fold A fold in which a limb has tilted beyond the vertical so that the older strata are uppermost.

oxbow lake A crescent-shaped, water-filled loop created in the former path of a *stream* when it abandons a *meander* and takes a new course.

oxidation The *chemical reaction* of an element with oxygen that occurs during *chemical weathering.*

P wave The first *seismic wave* to arrive from the *focus* of an *earthquake.*

paleomagnetism The remanent magnetization recorded in ancient rocks; allows the reconstruction of Earth's ancient magnetic field and the positions of continents and supercontinents.

paleontology The study of the history of ancient life from the *fossil* record.

Pangaea A supercontinent that coalesced in the latest Paleozoic *era* and comprised all present continents. The breakup of Pangaea began in Mesozoic time, as inferred from paleomagnetic and other data.

partial melting The incomplete melting of a rock that occurs because the *minerals* that compose it melt at different temperatures.

passive margin A *continental margin* far from a plate margin, with no *volcanoes* and few *earthquakes.*

peat A rich organic material that contains more than 50 percent carbon.

pedalfer A common *soil* type of temperate regions; the upper and middle layers contain abundant insoluble *minerals* such as quartz, *clay* minerals, and iron alteration products.

pediment A broad, gently sloping platform of *bedrock* left behind as a mountain front erodes and retreats from its *valley.*

pedocal A common *soil* type of arid regions; rich in calcium from calcium carbonate and other soluble *minerals* and low in organic matter.

pegmatite A *vein* of extremely coarse grained *granite,* crystallized from a water-rich *magma* in the late stages of solidification, that cuts across a much finer grained *country rock.* Pegmatites provide *ores* of many rare elements, such as lithium and beryllium.

pelagic sediment An open-ocean *sediment* composed of fine-grained terrigenous and biochemically *precipitated* particles that slowly settle from the surface to the bottom.

perched water table The *groundwater table* in a shallow upper surface of an *aquifer* that is perched above and separated from the main body of *groundwater* by an *aquiclude.*

peridotite A coarse-grained *ultramafic intrusive igneous rock* composed of olivine with small amounts of pyroxene and amphibole. The dominant rock in Earth's *mantle* and the source of basaltic rocks formed at mid-ocean ridges.

period A division of *geologic time* representing one subdivision of an *era.*

permafrost A permanently frozen aggregate of ice and *soil* occurring in very cold regions. Any rock or soil remaining at or below 0°C for 2 or more years.

permeability The ability of a solid to allow fluids to pass through it.

phase change A transition in a rock's *mineralogy* (but probably not its composition) to a denser, more closely packed *crystal* structure, signaled by a change in *seismic-wave* velocity.

phosphorite A chemical or biochemical *sedimentary rock* composed of calcium phosphate *precipitated* from phosphate-rich seawater and formed diagenetically by the interaction between muddy or *carbonate sediments* and the phosphate-rich water.

phyllite A *foliated rock* that is intermediate in grade between *slate* and *schist.* Small *crystals* of mica and chlorite give it a more or less glossy sheen.

physical weathering *Weathering* in which solid rock is fragmented by mechanical processes that do not change its chemical composition.

pillow lava A pile of ellipsoidal, pillowlike blocks of *basalt* about a meter wide formed in an underwater eruption when tongues of molten basaltic lava develop a tough, plastic skin on contact with the cold ocean water.

placer An *ore* deposit that has been concentrated by the mechanical *sorting* action of a *river* current.

plastic flow The total of all the small movements of the ice *crystals* that make up a *glacier,* resulting in a large movement of the whole mass of ice. (Compare *basal slip.*)

plate tectonic system A *geosystem* that includes all parts of the *Earth system* and all the interactions among these components needed to describe how *plate tectonics* works in space and time.

plate tectonics The theory proposing that the *lithosphere* is broken into about a dozen large plates that move over Earth's surface. Each plate acts as a distinct rigid unit that rides on the *asthenosphere,* which also is in motion. The theory attempts to explain seismicity, *volcanism,* mountain building, and evidence of *paleomagnetism* in terms of these plate motions.

plateau A large, broad, flat area of appreciable *elevation* above the neighboring terrain.

playa A flat bed of *clay*, sometimes encrusted with *precipitated* salts, that forms by the complete evaporation of a *playa lake.*

playa lake A permanent or temporary lake that occurs in arid mountain *valleys* or *basins.* As the lake water evaporates, dissolved *weathering* products are concentrated and gradually *precipitated.*

plunging fold A fold whose axis is not horizontal but *dips.*

pluton A large igneous intrusion ranging in size from a cubic kilometer to hundreds of cubic kilometers, formed at depth in the *crust.*

plutonic rock An *intrusive igneous rock.*

point bar A curved sandbar deposited along the inside bank of a *stream,* where the current is slower.

polymorph One of two or more alternative possible *crystal* structures for a single chemical compound; for example, the *minerals* quartz and cristobalite are polymorphs of silica (SiO_2).

porosity The percentage of a rock's volume consisting of open pores between grains.

porphyroblast A large *crystal* surrounded by a much finer grained matrix of other *minerals* in a *metamorphic rock.*

porphyry A lava of mixed *texture* in which large *crystals* (phenocrysts) "float" in a predominantly fine crystalline matrix.

positive feedback A process in which one action produces an effect (the feedback) that tends to speed up the original action and stabilize the process at a faster rate. (Compare *negative feedback.*)

potable water Water that tastes agreeable and is not dangerous to health.

pothole A hemispherical hole in the *bedrock* of a streambed, formed by *abrasion* of small pebbles and cobbles in a strong current.

precipitate To drop out of a saturated solution as *crystals.* The crystals that drop out of a saturated solution.

pressure-temperature-time path The changing regimes of pressure and temperature recorded by changes in the assemblages and chemical compositions of the *minerals* in regionally metamorphosed rocks.

principle of original horizontality See *original horizontality, principle of.*

principle of superposition See *superposition, principle of.*

principle of uniformitarianism See *uniformitarianism, principle of.*

proton An elementary particle in the *nucleus* of an *atom,* having an *atomic mass* of 1 and a positive electrical charge of +1.

pumice A frothy mass of volcanic glass with a great number of holes (vesicles) that remain after trapped gas has escaped from the solidifying melt.

pyroclast A *volcanic rock* fragment ejected into the air during an eruption.

pyroxene granulite A coarse-grained *metamorphic rock* containing pyroxene and calcium plagioclase. The highest grade of metamorphosed *mafic volcanic rock.*

quartzite A very hard, white *granoblastic rock* derived from quartz-rich *sandstones.*

radiometric age The actual number of years that have passed since a rock formed. (Compare *relative age.*)

radiometric dating The use of naturally occurring radioactive elements to determine the ages of rocks.

rain shadow An area of low rainfall on the leeward slope of a mountain range.

recharge The *infiltration* of water into any subsurface *formation,* often by rain or snow meltwater from the surface.

recurrence interval The average time interval between the occurrence of two geological events of a given magnitude.

reef A moundlike or ridgelike organic structure constructed of the carbonate skeletons of millions of organisms.

regional metamorphism Metamorphism caused by high pressures and temperatures that extend over large regions, as happens where plates collide.

regolith The layer of loose, heterogeneous material overlying *bedrock,* including *soil,* particles of weathered and unweathered parent rock, *clay minerals,* iron and other metal oxides, and other products of *weathering.*

rejuvenation Renewed uplift in a mountain chain on the site of earlier uplifts, returning the area to a more youthful stage.

relative age The age of a rock in relation to the ages of other rocks. (Compare *radiometric age.*)

relative humidity The amount of water vapor in the air, expressed as a percentage of the total amount of water the air could hold at that temperature if saturated.

relative plate velocity The velocity at which one plate moves relative to another plate.

relief The difference between the highest and lowest *elevations* in a particular area.

reserve A deposit that has already been discovered and that can be mined economically and legally at the present time. (Compare *resource.*)

reservoir A place in which water is stored, including the oceans; *glaciers* and polar ice; *groundwater;* lakes and *rivers;* the atmosphere; and the *biosphere.* A source or place of residence for elements in a chemical cycle or *hydrologic cycle.*

residence time The average time that an *atom* of a particular element spends in a *reservoir* before leaving.

resource The entire amount of a given material that may become available for use in the future, including *reserves,* plus discovered but currently unprofitable deposits, plus undiscovered deposits that geologists think they may be able to find eventually. (Compare *reserve.*)

rhyolite A light-brown to gray, fine-grained *extrusive igneous rock* with a *felsic* composition. The extrusive equivalent of *granite.*

rift basin A deep, narrow, elongate *basin* with thick successions of *sedimentary rocks* and also *extrusive* and *intrusive igneous rocks.* Current examples include the *rift valleys* of East Africa, the rifted valley of the Rio Grande, and the Jordan Valley in the Middle East.

rift valley A long, narrow trough bounded on each side by one or more parallel normal *faults.* Created by *tensional forces.*

ripple A *sedimentary structure* consisting of a very small *dune* of *sand* or *silt* whose long dimension is at right angles to the current.

river A major branch of a *stream* system.

rock avalanche The rapid *mass movement* of broken rock material, typically triggered by an *earthquake.* Rock avalanches move at velocities of tens to hundreds of kilometers per hour.

rock cycle The set of geologic processes that convert each type of rock into the other two types.

rock flour A glacial *sediment* of extremely fine (*silt*- and *clay*-sized) ground rock formed by *abrasion* of rocks at the base and sides of a *glacier.*

rockfall A very rapid *mass movement* in which newly detached individual blocks of rock plummet suddenly in free fall from a cliff or steep mountainside.

rockslide The rapid *mass movement* of large blocks of detached *bedrock* sliding more or less as a unit.

runoff The sum of all rainwater that flows over the surface, including the fraction that may temporarily infiltrate near-surface *formations* and then flow back to the surface.

S wave The second *seismic wave* to arrive from the *focus* of an *earthquake.* S waves do not exist in liquids or gases.

salinity The total amount of dissolved substances in a given volume of seawater.

saltation An intermittent jumping motion of *sand* or fine *sediment* along a streambed, in which grains are sucked up into the flow by turbulent eddies, move with the current for a short distance, and then fall back to the bottom.

sand A *clastic sediment* consisting of medium-sized particles, ranging from 0.062 to 2 mm in diameter.

sand budget The inputs and outputs caused by *erosion* and sedimentation of a beach.

sandblasting The *erosion* of rock *outcrops,* boulders, and pebbles by *abrasion* caused by the impact of high-speed *sand* grains carried by the wind.

sandstone A clastic rock composed of grains of quartz, feldspar, and rock fragments, bound together by a cement of quartz, carbonate, or other *minerals,* or by a matrix of *clay* minerals. The lithified equivalent of *sand.*

saturated zone The level in the *groundwater table* in which the pores of the *soil* or rock are completely filled with water. (Compare *unsaturated zone.*)

schist A *metamorphic rock* characterized by the pervasive coarse, wavy *foliation* known as schistosity.

scientific method A general research plan, based on methodical observations and experiments, by which scientists propose and test hypotheses that explain some aspect of how the universe works.

seafloor metamorphism Metamorphism associated with mid-ocean ridges, in which changes in a rock's bulk chemical composition are produced by fluid transport of chemical components into or out of the rock.

seafloor spreading The mechanism by which new seafloor is created along the rift at the crest of a mid-ocean ridge as adjacent plates move apart. The *crust* separates along the rift, and new seafloor forms as hot new crust upwells into these cracks. The new seafloor spreads laterally away from the rift and is replaced by even newer crust in a continuing process of plate creation.

seamount A submerged *volcano,* usually extinct, found on the *abyssal plain.*

sediment A material deposited at Earth's surface by physical agents (wind, water, and ice), chemical agents (precipitation from oceans, lakes, and *rivers*), or biological agents (organisms, living and dead).

sedimentary basin A region of considerable extent (at least 10,000 km^2) where the combination of deposition and *subsidence* has formed thick accumulations of *sediment* and *sedimentary rock.*

sedimentary environment A geographic location characterized by a particular combination of geologic processes and environmental conditions.

sedimentary rock A rock formed as the burial product of layers of *sediments* (such as *sand, mud,* and calcium carbonate shells), whether they were laid down on the land or under the sea.

sedimentary structure Any kind of *bedding* or other surface (such as *cross-bedding, graded bedding,* and *ripples*) formed at the time of deposition.

seismic hazard The intensity of shaking and ground disruption in an *earthquake* that can be expected over the long term at some specified location, expressed in the form of a seismic hazard map.

seismic risk The *earthquake* damage that can be expected over the long term for a specified region, such as a county or state, usually measured in terms of average dollar loss per year.

seismic tomography A method that uses the *seismic waves* from *earthquakes* recorded on thousands of *seismographs* all over the world to sweep Earth's interior in many different directions and construct a three-dimensional image of what's inside.

seismic wave A ground vibration produced by *earthquakes* or explosions. (See also *P wave; S wave; surface wave.*)

seismograph An instrument that magnifies and records the motions of Earth's surface caused by *seismic waves.*

sequence stratigraphy The description, correlation, and classification of *unconformity*-bounded sequences of sedimentary strata. The unconformities that define sequences represent fluctuations in sea level that allowed land *erosion* to take place. The beds that make up the sequences show internal patterns that are diagnostic of changes in sedimentation.

settling velocity The speed with which particles of various weights suspended in a *stream* settle to the bottom.

shadow zone (1) A zone beyond 105° from the *focus* of an *earthquake* where *S waves* are absent because the waves are not transmitted through the liquid outer *core.* (2) A zone at an angular distance of 105° to 142° from the focus of an earthquake where *P waves* are absent because they are refracted downward into the core and emerge at greater distances after the delay caused by their detour through the core.

shale A fine-grained *clastic rock* composed of *silt* plus a significant component of *clay,* which causes it to break readily along *bedding* planes.

shear wave A wave that propagates by a side-to-side motion. Shear waves cannot propagate through any fluid—air, water, or the liquid iron in Earth's outer *core.* (Compare *compressional wave.*)

shearing force A force that pushes two sides of a body in opposite directions.

shield A large region of stable, ancient crystalline basement rocks within a continent.

shield volcano A broad, shield-shaped *volcano* many tens of kilometers in circumference and more than 2 km high built by successive flows of fluid basaltic lava from a central vent.

shock metamorphism Metamorphism that occurs when *minerals* are subjected to the high pressures and temperatures of shock waves generated when a meteorite collides with Earth.

silica ooze A biochemically *precipitated pelagic sediment* consisting of the remains of the silica shells of diatoms and radiolarians.

siliceous environment A special deep-sea *sedimentary environment* in which silica-secreting organisms grow in surface waters where nutrients are abundant. Their shells settle to the ocean floor and accumulate as layers of siliceous *sediment.*

sill A tabular, sheetlike igneous intrusion formed by the injection of *magma* between parallel layers of preexisting bedded rock.

silt A *clastic sediment* in which most of the grains are between 0.0039 and 0.062 mm in diameter.

siltstone A clastic rock that contains mostly *silt* and looks similar to *mudstone* or very fine grained *sandstone.* The lithified equivalent of silt.

sinkhole A small, steep depression in the land surface above a cavernous *limestone formation.*

slate The lowest grade of *foliated rock,* easily split into thin sheets; produced primarily by the metamorphism of *shale.*

slip The distance that one face of a *fault* is displaced relative to the other.

slip face The steep lee slope of a *dune* on which *sand* is deposited in *cross-beds* at the *angle of repose.*

slump A slow *mass movement* of *unconsolidated material* that travels as a unit.

soil The topmost layer of *regolith,* composed of fragments of *bedrock, clay minerals,* and organic matter.

solar energy Energy derived from the Sun, including energy from solar generating systems, hydroelectric systems, and wind power.

sorting The tendency for variations in current velocity to segregate *sediments* according to size.

specific gravity The weight of a *mineral* in air divided by the weight of an equal volume of pure water at 4°C. (Compare *density.*)

spheroidal weathering A *physical* and *chemical weathering* process in which curved layers split off from a rounded boulder, leaving a spherical inner core. Similar to *exfoliation* but on a smaller scale.

spreading center The area surrounding a mid-ocean ridge, where new *crust* is being formed by *seafloor spreading.*

stalactite A long, narrow spike of carbonate suspended from the ceiling of a cave. It is deposited by evaporation and precipitation from solutions seeping through *limestone.*

stalagmite An inverted icicle-shaped deposit that builds up on a cave floor beneath a *stalactite* and is formed by the same process as a stalactite.

steady state A state in which the inflow of an element into a *reservoir* equals the outflow.

stock An irregular mass of coarse-grained *igneous rock* less than 100 km^2 in area.

stratification The characteristic layering or *bedding* of *sedimentary rocks.*

stratigraphic succession A vertical set of strata, providing a chronological record of the geologic history of a region.

stratigraphy The description, correlation, and classification of strata in *sedimentary rocks.*

stratospheric ozone depletion The thinning of the layer of ozone in the stratosphere caused by reactions involving human-made compounds, such as chlorofluorocarbons (CFCs). The stratospheric ozone forms a protective layer that shields Earth by absorbing cell-damaging ultraviolet radiation.

stratovolcano A concave-shaped *volcano* containing alternating layers of lava flows and beds of *pyroclasts.*

streak The *color* of the fine deposit of mineral dust left on an abrasive surface, such as a tile of unglazed porcelain, when a *mineral* is scraped across it.

stream Any flowing body of water, large or small. (Compare *river.*)

stream power The rate at which a *stream* can erode *bedrock* and transport its load; the product of stream slope and *discharge.*

stress The force per unit area, acting on any surface within a solid body. Consists of confining pressure and directed pressure.

striation A scratch or groove left on *bedrock* and boulders by overriding ice, showing the direction of glacial motion.

strike The compass direction of a rock layer as it intersects with a horizontal surface.

subduction The sinking of an oceanic plate beneath an overriding plate at a convergent plate boundary. The overriding plate may be oceanic or continental.

submarine canyon A deep *valley* eroded into the *continental slope* and the *continental shelf* behind it.

subsidence A depression or sinking of the *crust* induced partly by the additional weight of *sediment*s on the crust but driven mostly by tectonic mechanisms, such as regional downfaulting.

superposed stream A *stream* that erodes a gorge in resistant *formations* because its course was established at a higher level on uniform rocks before downcutting began. A superposed stream tends to continue the pattern it developed earlier rather than adjusting to its new conditions. (Compare *antecedent stream.*)

superposition, principle of The principle that each layer of *sedimentary rock* in a tectonically undisturbed sequence is younger than the one beneath it and older than the one above it.

surface tension The attractive force between molecules at a surface.

surface wave The last *seismic wave* to arrive from the *focus* of an *earthquake;* a surface wave travels around Earth's surface only rather than through the interior.

surge A sudden period of fast movement of a *valley glacier,* sometimes occurring after a long period of little movement and lasting several years.

suspended load All the material temporarily or permanently suspended in the flow of a *stream.* (Compare *bed load.*)

sustainable development Growth in economies and in living standards that can last indefinitely without harming the environment.

syncline A large downfold whose limbs are higher than its center. (Compare *anticline.*)

talus An accumulation of blocks of fallen rock at the foot of a steep *bedrock* cliff.

tectonic age The age of a rock that corresponds to the last major episode of crustal deformation; that is, to the last time the radiometric "clock" within the rock was reset by tectonic activity.

tensional force A force that stretches a body and tends to pull it apart.

terrace A flat, steplike surface that lines a *stream* above the *floodplain,* often paired one on each side of the stream, marking a former floodplain that existed at a higher level before regional uplift or an increase in *discharge* caused the stream to erode into the former floodplain.

terrigenous sediment *Sediment* eroded from the land surface.

texture The sizes and shapes of a rock's *mineral crystal*s and the way they are put together.

thermal sag basin A *basin* produced in the later stages of rifting, when newly formed continental plates are drifting away from each other. The *lithosphere* that was thinned and heated during the earlier rifting stage cools, leading to an increase in *density,* which in turn leads to *subsidence* and the development of offshore thermal sag basins.

thermoremanent magnetization A permanent magnetization acquired by *minerals* in *igneous rock*s when groups of *atoms* of the mineral align themselves in the direction of the magnetic field when the material is hot. When the material has cooled, these atoms are locked in place and therefore are always magnetized in the same direction.

tidal flat A muddy or sandy area that lies above low *tide* but is flooded at high tide.

tide The twice-daily rise and fall of the sea caused by the gravitational attraction of the Moon and, to a lesser degree, the Sun.

till An unstratified and poorly sorted *sediment* containing all sizes of fragments from *clay* to boulders and deposited by glacial action.

tillite Ancient lithified *till.*

topography The general configuration of varying heights that give shape to Earth's surface.

topset bed A horizontal bed of *sediments* formed at the top of a *delta* and overlying the *foreset beds.*

transform fault A plate margin at which the plates slide past each other and *lithosphere* is neither created nor destroyed. Relative displacement occurs along the *fault* as horizontal *slip* between the adjacent plates.

transpiration The release of water vapor by plants into the atmosphere.

transverse dune An *eolian dune* whose long axis is at right angles to the prevailing wind direction. Transverse dunes form in arid regions where abundant *sand* dominates the landscape and vegetation is absent.

tributary A *stream* that *discharges* water into a larger stream.

tsunami A fast-moving, towering sea wave generated by an *earthquake* that occurs beneath the ocean.

tuff Any *volcanic rock* lithified from *pyroclasts.*

turbidite A *graded bed* of *sand, silt,* and *mud* deposited by a *turbidity current* on the *abyssal plain.*

turbidity current A flow of turbid, muddy water down a slope. The *suspended load* of *mud* makes the turbid water denser than the overlying clear water, so that the turbidity current flows beneath the clear water.

turbulent flow A flow in which streamlines mix, cross, and form swirls and eddies. (Compare *laminar flow.*)

twentieth-century warming A rise in Earth's mean surface temperature of about 0.6°C between the end of the nineteenth century and the beginning of the twenty-first.

ultramafic rock An *igneous rock* consisting primarily of mafic *minerals* and containing less than 10 percent feldspar.

unconfined aquifer An *aquifer* that is not overlain by an *aquiclude,* so that the water travels through beds of more or less uniform *permeability* that extend to the surface in both *discharge* and *recharge* areas. (Compare *confined aquifer.*)

unconformity A surface between two layers that were not laid down in an unbroken sequence.

unconsolidated material *Sediment* that is loose and uncemented. (Compare *consolidated material.*)

uniformitarianism, principle of A principle of geology that can be summarized as "the present is the key to the past." It holds that the geologic processes we see in action today have worked in much the same way throughout *geologic time.*

unsaturated zone The level in the *groundwater table* in which the pores of the *soil* or rock are not completely filled with water. (Compare *saturated zone.*)

U-shaped valley A deep *valley* with steep upper walls that grade into a flat floor; the typical shape of a valley eroded by a *glacier.*

valley The entire area between the tops of the slopes on both sides of a *stream.*

valley glacier A river of ice that forms in the cold heights of mountain ranges where snow accumulates, usually in preexisting *valleys,* and flows down the *bedrock* valleys. (Compare *continental glacier.*)

varve A pair of alternating coarse and fine layers formed in one year by the seasonal freezing of a lake surface.

vein A sheetlike deposit of *minerals precipitated* in fractures or *joints* that are foreign to the host rock.

ventifact A wind-faceted pebble having several curved or almost flat surfaces that meet at sharp ridges. Each surface or facet is made by *sandblasting* of the pebble's windward side.

viscosity A measure of a fluid's resistance to flow.

volcanic ash Extremely small fragments, usually of glass, that form when escaping gases force a fine spray of *magma* from a *volcano.*

volcanic geosystem The total system of rocks, *magmas,* and interactions needed to describe the entire sequence of events from melting to eruption.

volcanic rock An *extrusive igneous rock.*

volcanism The process by which *magma* rises from the Earth's interior through the *crust,* emerges onto the surface as lava, and cools into hard *volcanic rock.*

volcano A hill or mountain constructed from the accumulation of lava and other erupted materials.

wadi A desert *valley* that carries water only briefly after a rain. Called a *dry wash* in the western United States.

wave-cut terrace A level surface formed by wave *erosion* of coastal *bedrock* beneath the surf zone. May be visible at low *tide.*

weathering The general process that breaks up rocks into fragments of various sizes by a combination of physical fracturing and chemical decomposition.

Wilson cycle A general plate tectonic cycle that comprises (1) rifting during the breakup of a supercontinent, (2) *passive-margin* cooling and *sediment* accumulation during *seafloor spreading* and ocean opening, (3) *active-margin volcanism* and terrane *accretion* during *subduction* and ocean closure, and (4) *orogeny* during the continent-continent collision that forms the next supercontinent.

zeolite A class of silicate *minerals* containing water in cavities within the *crystal* structure and formed by metamorphism at very low temperatures and pressures.

Index

Italics refers to figures **Boldface** refers to glossary terms *t* refers to tables

Earth's plates today, showing the plate boundaries and the relative plate velocities (mm/year)

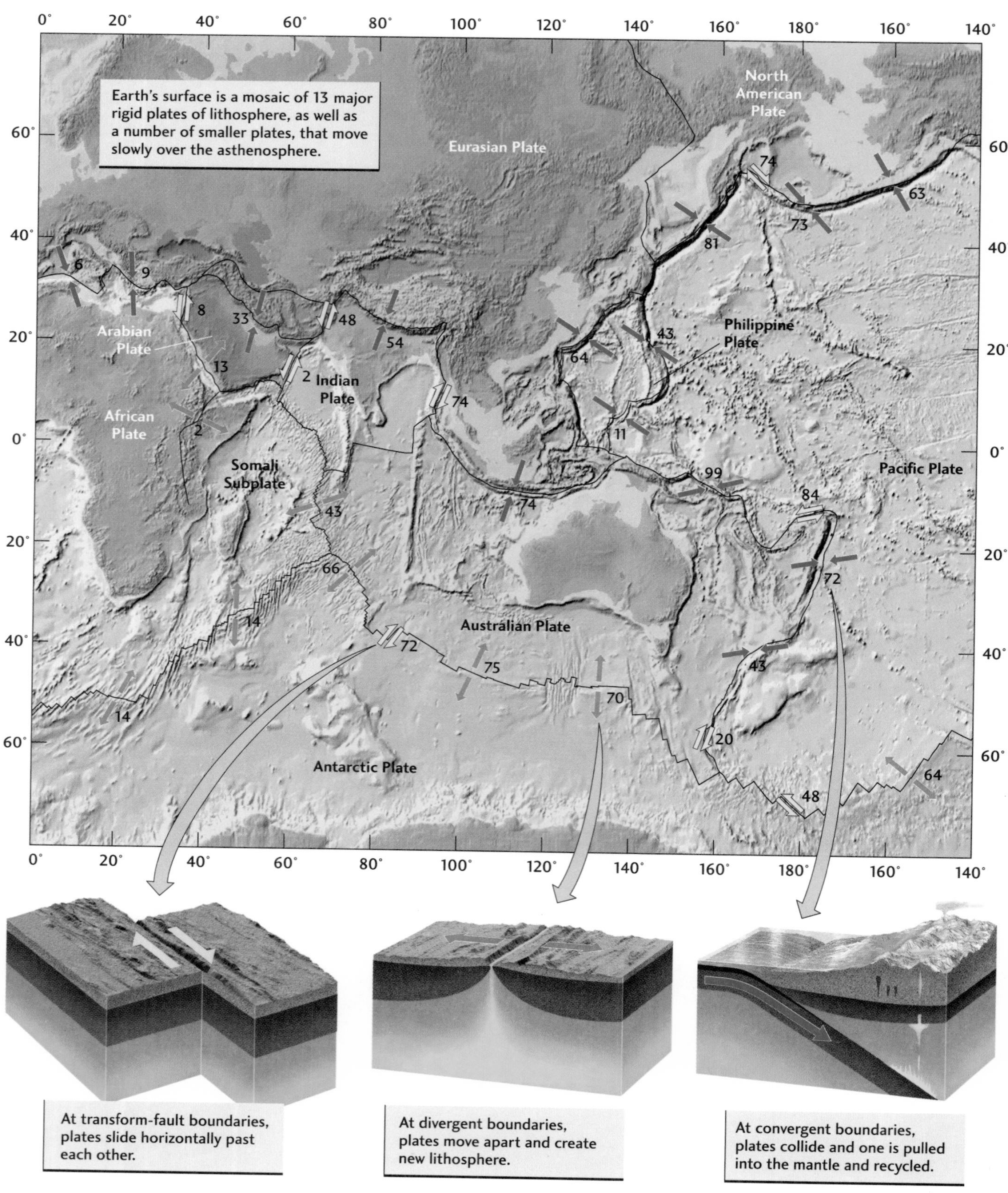

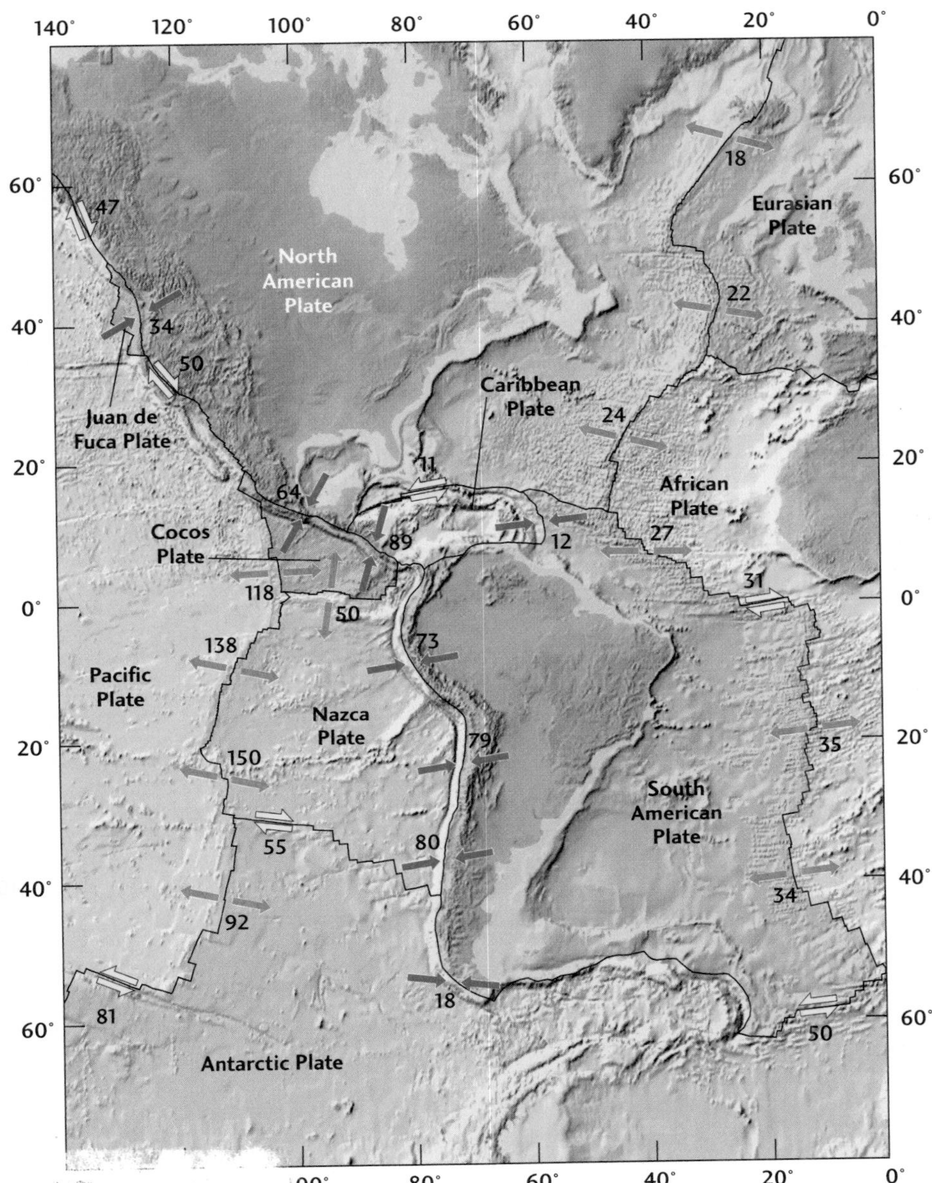

140°
120°
100°
80°
60°
40°
20°
0°
60°
40°
20°
0°
20°
40°
60°
North American Plate
Eurasian Plate
Caribbean Plate
African Plate
Juan de Fuca Plate
Cocos Plate
Pacific Plate
Nazca Plate
South American Plate
Antarctic Plate
18
47
22
34
50
24
11
64
89
12
27
118
31
50
138
73
79
35
150
55
80
34
92
18
81
50